MW00844672

OIL AND GAS PIPELINES

# INTERNATIONAL ADVISORY BOARD

# OIL AND GAS PIPELINES
# Integrity and Safety Handbook

Edited by

R. WINSTON REVIE

Published by John Wiley & Sons, Inc., Hoboken, New Jersey
Published simultaneously in Canada



***Library of Congress Cataloging-in-Publication Data applied for.***

Printed in the United States of America

10 9 8 7 6 5 4 3 2 1

# CONTENTS

## PART II MANUFACTURE, FABRICATION, AND CONSTRUCTION

## PART III THREATS TO INTEGRITY AND SAFETY

## PART V INSPECTION AND MONITORING

## PART VII RISK MANAGEMENT

# PREFACE

The objective of preparing this Handbook was to make available, in one volume, essential scientific, engineering, and technical knowledge on integrity and safety of oil and gas pipelines. The aim was to help ensure the safe operation of pipeline networks that we depend on for reliability in service, deliverability of energy, protection of the environment, and safety of the public.

This book, containing 52 chapters, is divided into the following eight parts:

Part I: Design
Part II: Manufacture, Fabrication, and Construction
Part III: Threats to Integrity and Safety
Part IV: Protection
Part V: Inspection and Monitoring
Part VI: Maintenance, Repair, Replacement, and Abandonment
Part VII: Risk Management
Part VIII: Case Histories

The breadth of coverage will help make the book useful to engineers, scientists, technicians, and students from a wide range of disciplines. The depth of coverage is intended to meet the needs of those who require a high level of detail. In addition to the case histories presented in Part VIII, there are practical examples with solutions to problems and analytical discussion throughout the book. References to other books, journals, and standards that readers can consult for further information are listed.

The 86 authors who have contributed the 52 chapters come from different sectors, industry, academia, and government, and from 14 countries, bringing diverse perspectives to pipeline integrity and a balance of practice and theory.

I would like to convey my thanks to the team of authors for contributing chapters in their areas of specialty. I would also like to highlight the contributions of the reviewers who, in anonymity, reviewed the draft chapters and provided their comments, suggestions, and recommendations for the authors, so as to enhance the quality of the chapters, documenting the current state of knowledge. In addition, the International Advisory Board provided helpful advice and guidance in the preparation of the Handbook.

I would like to thank Bob Esposito, Michael Leventhal, and the staff at John Wiley & Sons, Inc. for their encouragement. The concept for this book was developed in discussion with Bob Esposito about the Banff Workshops on managing pipeline integrity, which I organized over a period of nearly 20 years, beginning with the 1993 Workshop. These Workshops continue to be held biennially.

If this Handbook helps in the quest to achieve the objective that we all share of zero incidents—no incidents, in the world of engineering and business where "zero" means "zero" and "no" means "no"—then the efforts of the entire team who worked tirelessly in preparing it will have been rewarded.

R. Winston Revie

*Ottawa, Ontario, Canada*

# CONTRIBUTORS

**Gusai H. Al-Aithan**, Saudi Aramco, Dhahran, Saudi Arabia

**Jon Allin**, Permasense Ltd., Horsham, UK

**Faisal M. Al-Mutahar**, Saudi Aramco, Dhahran, Saudi Arabia

**Ted L. Anderson**, Quest Integrity Group, LLC, Boulder, CO, USA

**John A. Beavers**, Det Norske Veritas (U.S.A.), Inc., Dublin, OH, USA

**Lynsay A. Bensman**, Det Norske Veritas (U.S.A.), Inc., Dublin, OH, USA

**David H. Boteler**, Earth Sciences Sector, Natural Resources Canada, Ottawa, Ontario, Canada

**Holger Brauer**, Salzgitter Mannesmann Line Pipe GmbH, Hamm, Germany

**Stephan Brockhaus**, ROSEN Technology & Research Center, Lingen, Germany

**William A. Bruce**, Det Norske Veritas (U.S.A.), Inc., Dublin, OH, USA

**Dean Carnes**, Canadian Natural Resources Ltd., Calgary, Alberta, Canada

**Homero Castaneda**, Chemical and Biomolecular Engineering Department, National Center for Research and Education in Corrosion and Materials Performance, The University of Akron, Akron, OH, USA

**Frederic Cegla**, Mechanical Engineering Department, Imperial College London, South Kensington, UK

**Ljiljana Djapic-Oosterkamp**, Statoil, Kårstø, Norway

**David Dorling**, Starco Engineering, Calgary, Alberta, Canada

**Ayman Eltaher**, MCS Kenny, Houston, TX, USA

**Merv Fingas**, Spill Science, Edmonton, Alberta, Canada

**Doug Fisher**, Willowglen Systems Inc., Edmonton, Alberta, Canada

**Ming Gao**, Blade Energy Partners, Houston, TX, USA

**James Gianetto,** CanmetMATERIALS, Natural Resources Canada, Hamilton, Ontario, Canada

**Ray Goodfellow**, IRISNDT – Engineering, Calgary, Alberta, Canada

**J. Malcolm Gray**, Microalloyed Steel Institute Inc., Houston, TX, USA

**Lorna Harron**, Enbridge Pipelines Inc., Edmonton, Alberta, Canada

**Shokrollah Hassani**, BP America Inc., Houston, TX, USA

**Holger Hennerkes**, ROSEN Headquarters, Stans, Switzerland

**Douglas Hornbach**, Lambda Technologies, Cincinnati, OH, USA

**Nobuyuki Ishikawa**, JFE Steel Corporation, Hiroshima, Japan

**John J. Jonas**, Department of Mining and Materials Engineering, McGill University, Montreal, Canada

**Katherine Jonsson**, IRISNDT – Engineering, Calgary, Alberta, Canada

**Paul Jukes**, Wood Group, Houston, TX, USA

**Lynne C. Kaley**, Trinity Bridge, LLC, Houston, TX, USA

**Christoph Kalwa**, EUROPIPE GmbH, Mülheim an der Ruhr, Germany

**Russell D. Kane**, iCorrosion LLC, Houston, TX, USA

**Shawn Kenny**, Memorial University, St. John's, Newfoundland, Canada; Carleton University, Ottawa, Ontario, Canada

**Leo A.I. Kestens**, Department of Materials Science and Engineering, Ghent University, Ghent, Belgium; Department of Materials Science and Engineering, Delft University of Technology, Delft, The Netherlands

**John Kiefner**, Kiefner and Associates, Inc., Worthington, OH, USA

**Gerhard Knauf**, Salzgitter Mannesmann Forschung GmbH, Duisburg, Germany

**Franz Martin Knoop**, Salzgitter Mannesmann Großrohr GmbH, Salzgitter, Germany

**Angel R. Kowalski**, Det Norske Veritas (U.S.A.), Inc., Dublin, OH, USA

**Klaus Kraemer**, Vallourec & Mannesmann Deutschland GmbH, Düsseldorf, Germany

**Ravi Krishnamurthy**, Blade Energy Partners, Houston, TX, USA

**Axel Kulgemeyer**, Salzgitter Mannesmann Forschung GmbH, Duisburg, Germany

**Rolf Kümmerling**, Vallourec & Mannesmann Deutschland GmbH, Düsseldorf, Germany

**Jason S. Lee**, Naval Research Laboratory, Stennis Space Center, Mississippi, USA

**John Leeds**, DC Voltage Gradient Technology & Supply Ltd., Wigan, UK

**Sarah Leeds**, DC Voltage Gradient Technology & Supply Ltd., Wigan, UK

**Keith G. Leewis**, Dynamic Risk Assessment Systems, Inc., Calgary, Alberta, Canada

**Brian N. Leis**, B N Leis Consultant, Inc., Worthington, OH, USA

**Arnold L. Lewis II**, Research and Development Center, Saudi Aramco, Dhahran, Saudi Arabia

**Hubert Lindner**, ROSEN Technology & Research Center, Lingen, Germany

**Brenda J. Little**, Naval Research Laboratory, Stennis Space Center, Mississippi, USA

**Hendrik Löbbe**, Salzgitter Mannesmann Line Pipe GmbH, Hamm, Germany

**Jim Mason**, Mason Materials Development, LLC, Birdsboro, PA, USA

**Brenton S. McLaury**, Erosion/Corrosion Research Center, The University of Tulsa, Tulsa, OK, USA

**Robert E. Melchers**, Centre for Infrastructure Performance and Reliability, The University of Newcastle, Newcastle, NSW, Australia

**Rumi Mohammad**, Willowglen Systems Inc., Edmonton, Alberta, Canada

**Lars Vendelbo Nielsen**, MetriCorr ApS, Glostrup, Denmark

**Mavis Sika Okyere**, Bluecrest College, Kumasi, Ghana

**Robert O'Grady**, Wood Group Kenny, Galway, Ireland

**Alan Pentney**, National Energy Board of Canada, Calgary, Alberta, Canada

**Roumen H. Petrov**, Department of Materials Science and Engineering, Ghent University, Ghent, Belgium; Department of Materials Science and Engineering, Delft University of Technology, Delft, The Netherlands

**Ben Poblete**, ATKINS, Houston, TX, USA

**Daniel E. Powell**, Williams, Tulsa, OK, USA

**Buddy Powers**, Clock Spring Company L.P., Houston, TX, USA

**Kathleen O. Powers**, Trinity Bridge, LLC, Houston, TX, USA

**Gail Powley**, Willowglen Systems Inc., Edmonton, Alberta, Canada

**Paul Prevéy**, Lambda Technologies, Cincinnati, OH, USA

**Konrad Reber**, Innospection Germany GmbH, Stutensee, Germany

**R. Winston Revie**, Consultant, Ottawa, Ontario, Canada

**Hernan E. Rincon**, ConocoPhillips Company, Houston, TX, USA

**Kenneth P. Roberts**, Erosion/Corrosion Research Center, The University of Tulsa, Tulsa, OK, USA

**Randy L. Roberts**, N-SPEC Pipeline Services, Business Unit of Coastal Chemical Co., L.L.C./A Brenntag Company, Broussard, LA, USA

**Omar Rosas**, Chemical and Biomolecular Engineering Department, National Center for Research and Education in Corrosion and Materials Performance, The University of Akron, Akron, OH, USA

**Edmund F. Rybicki**, Erosion/Corrosion Research Center, The University of Tulsa, Tulsa, OK, USA

**John R. Shadley**, Erosion/Corrosion Research Center, The University of Tulsa, Tulsa, OK, USA

**Abdelmounam M. Sherik**, Research and Development Center, Saudi Aramco, Dhahran, Saudi Arabia

**Siamack A. Shirazi**, Erosion/Corrosion Research Center, The University of Tulsa, Tulsa, OK, USA

**Binder Singh**, Genesis-Technip USA Inc., Houston, TX, USA

**Robert Smyth**, Petroline Construction Group, Nisku, Alberta, Canada

**Tom Steinvoorte**, ROSEN Europe, Oldenzaal, The Netherlands

**Larisa Trichtchenko**, Earth Sciences Sector, Natural Resources Canada, Ottawa, Ontario, Canada

**William Tyson**, CanmetMATERIALS, Natural Resources Canada, Hamilton, Ontario, Canada

**Neb I. Uzelac**, Neb Uzelac Consulting Inc., Toronto, Ontario, Canada

**Michael VanderZee**, Willowglen Systems Inc., Edmonton, Alberta, Canada

**Doug Waslen**, Sherwood Park, Alberta, Canada

**Nader Yoosef-Ghodsi**, C-FER Technologies, Edmonton, Alberta, Canada

**Greg Zinter**, Applus RTD, Edmonton, Alberta, Canada

# PART I

# DESIGN

# 1

# PIPELINE INTEGRITY MANAGEMENT SYSTEMS (PIMS)

RAY GOODFELLOW AND KATHERINE JONSSON
*IRISNDT – Engineering, Calgary, Alberta, Canada*

## 1.1 INTRODUCTION

Effective management of pipeline system integrity is essential for safe and reliable pipeline operation. Pipeline integrity management systems (PIMS) provide the overarching, integrated framework for effective pipeline asset management.

Significant failures in both gas and liquid pipelines have made global headlines. Although pipelines are statistically very safe and reliable, pipeline failures have resulted in fatalities, environmental damage, and an erosion of public confidence in the pipeline industry. Some examples of catastrophic pipeline failures that resulted in fatalities include the sweet gas line rupture in Carlsbad, New Mexico, in 2000, the gasoline pipeline failure in Bellingham, Washington, in 1999, and the gas line rupture in San Bruno, California, in 2010. Failures in oil pipelines such as the Kalamazoo River oil spill in 2010, the Red Deer River spill in 2012, and the Mayflower, Arkansas, spill in 2013 also generated significant public concern regarding environmental impacts. Failure investigations have identified that the significant contributing factors to the cause and the size of pipeline releases are directly related to flaws in the company's management systems. As such, an effective PIMS is critical to prevent failures. Additional information on pipeline failures and causes is publicly available on websites such as Pipeline and Hazardous Materials Administration (PHMSA), the U.S. National Transportation Safety Board (NTSB), and Alberta Energy Regulator (AER).

Pipeline integrity management requirements and expectations have been continuously evolving and will continue to change in the future. There is no single correct "formula" for developing an integrity management system; however, this chapter outlines the fundamental basics of an effective management system that have been successfully integrated in companies across the world. Industry groups such as International Association of Oil and Gas Producers [1] and the American Petroleum Institute (API) [2] have developed guidance documents that can be used as additional references for developing management systems. This chapter covers downstream, midstream, and upstream oil and gas pipelines. Pipelines have different operating practices and different consequences of failure; however, the fundamental principles of an effective integrity management system apply to all pipeline operations.

Although this chapter focuses on PIMS, it is important to note that the pipeline industry is shifting toward safety management systems (SMS), which are more comprehensive than traditional PIMS. For example, SMS emphasize safety culture and process safety management more than conventional PIMS. Significant changes are expected within CSA Z662 Section 3.0 Safety Loss Management to reflect the transition to SMS principles. API Recommended Practice 1173—Pipeline Safety Management System Requirements [2] (Draft Version 11.2 issued in June 2014) is an example of an industry SMS document. The reader is encouraged to review and understand the latest industry and regulatory documents pertaining to both SMS and PIMS. SMS are outside the scope of this chapter; however, PIMS and SMS are closely linked and the principles governing PIMS development can be extended to SMS.

*Oil and Gas Pipelines: Integrity and Safety Handbook,* First Edition. Edited by R. Winston Revie.

## 1.2 LESSONS LEARNED AND THE EVOLUTION OF PIPELINE INTEGRITY

In the early days of the oil and gas pipeline industry, integrity-related activities such as coating application, cathodic protection, corrosion inhibition, and weld inspection were implemented in response to pipeline failures and incidents. As the pipelines have aged and the size and complexity of pipeline networks have grown, the impact of pipeline failures has increased. More specifically, there is more public awareness and concern regarding pipeline failures. In response, the industry improved prevention, mitigation, monitoring, and inspection technologies for pipeline hazards. For example, coatings have improved, in-line inspection (ILI) was developed (and has since evolved dramatically), corrosion inhibitors have become much more effective, and alternative materials to carbon steels, such as spoolable composites, were developed.

In addition to improving technology, companies started utilizing risk assessments to focus and prioritize integrity management-related activities. Risk assessments provide a more structured and analytical approach to identify and address pipeline integrity issues.

As a result of the combined efforts of technology improvements and risk assessment application, the number of failures in the North American pipeline industry declined significantly. For example, the upstream pipeline failure rate in Alberta declined from 5.0 failures per 1000 km of pipelines in 1990 to 1.5 failures per 1000 km by 2012 [3].

The regulators and the oil and gas industry agree that technological improvements and risk management are positive changes that are reducing the number of failures; however, they also understand that these advancements alone are not sufficient to address all pipeline hazards. Integrity management systems provide a more encompassing and integrated approach to addressing pipeline risks. At the June 2013 National Energy Board (NEB) Pipeline Safety Forum, the NEB paper "Emerging issues in oil and gas industry safety management" listed three key aspects that must be in place for an effective management system. This paper states that the management system must be

- Consistently applied
  "The system elements are applied consistently across operational programs (worker safety, asset integrity, damage prevention, environmental protection, and emergency management), facilities and geographic regions."
- Highly integrated
  "There are multiple interdependencies between management system elements and so the management system is designed to share information and intelligence to promote better decisions."
- Assign accountability
  "All officers and employees have a role to play in meeting the safety, security and environmental protection goals of the organization. These responsibilities must be clearly assigned and communicated. Performance must be measured and improvement required."

Pipeline failure incident investigations provide important insight into why an effective management system is necessary. The examples below are summarized from pipeline investigations published on regulatory or industry websites describing contributing factors to major pipeline failures:

- Lack of training and competency: examples include operator error that led to overpressurization of a system, inadequate inspector training that led to construction-related failures, and incorrect ILI analysis that misinterpreted a defect signal.
- Lack of effective change management: examples include production changes that altered corrosion rates, staffing changes that resulted in critical tasks incomplete or communicated improperly, and material substitution that resulted in an unsuitable material being used incorrectly in the wrong environment.
- Lack of records: improper or missing/lost documentation.
- Lack of understanding hazards: a lack of understanding of potential hazards resulting in an unaccounted for corrosion mechanism that resulted in pipeline failure, and an incorrect assumption of pipeline condition (the pipeline was assumed to be "dry" but rather the pipeline catastrophically ruptured due to internal corrosion).

As explained above, pipeline failures commonly link to management system failures. Developing and implementing an effective PIMS requires significant, albeit critical, work to prevent pipeline failures due to ineffective management systems. The success of PIMS requires the integration of a company vision, a well-developed structure, and a company safety culture supported by process management. The following sections will define and outline the core structure of an effective PIMS.

## 1.3 WHAT IS A PIMS?

The principles of asset integrity management apply to a wide range of industries, including power generation, aircraft, nuclear, pharmaceutical, transportation, and defense. Common asset management system principles apply to any system with physical assets that are core to achieving the company's business objectives.

There are many industry documents, standards, and recommended practices that describe asset integrity management systems. One standard that is frequently used as a guideline document for asset management is the Publicly Available Specification (PAS) 55 [4], which is administered by the British Standards Institution (BSI). This standard was developed by the Institute of Asset Management and has been adopted by utility, transportation, mining, process, and manufacturing industries worldwide. PAS 55 has been incorporated by the International Organization for Standardization (ISO) and has been updated as the ISO 55000 Asset Management series of standards [5].

In PAS 55-1:2008 [4], asset management is described as

> Systematic and coordinated activities and practices through which an organization optimally and sustainably manages its assets and asset systems, their associated performance, risks and expenditures over their life cycles for the purpose of achieving its organizational strategic plan.

An organizational strategic plan is described as

> Overall long-term plan for the organization that is derived from and embodies its vision, mission, values, business polices, stakeholder requirements, objectives and the management of its risks.

Figure 1.1 demonstrates the interdependencies of company vision/strategic plan, the management system structure, and the company culture. All three elements are required for success.

In the oil and gas industry, assets can include pipelines, pressure equipment, piping, and tanks. Integrity management systems can be developed for the conventional oil and gas assets mentioned, and can also extend to rotating equipment, electrical, and structural systems.

The purpose of pipeline integrity management is the effective execution, documentation, and communication of the technical work throughout the pipeline life cycle, and PIMS provides the framework (structured and integrated system elements) to execute effective pipeline integrity management. The technical programs can include, but are not limited to, quality control and inspection, cathodic protection, ILI, chemical inhibition, coating selection, excavation (dig) programs, corrosion monitoring, and river block valve maintenance. These are the activities that require plans, programs, processes, and procedures to be managed, scheduled, executed, tracked, documented, communicated, and reported.

There are multiple interdependencies between the many functions within an organization. Effective integrity management requires the cultural aspects of understanding, communication, and collaboration between operations, engineering, management, finance, and safety—in fact, almost every function within an organization through business unit integration. A great technical program may fail for one of the following reasons: people are not trained and competent, change is not effectively managed, key information, records, and documents cannot be found, and roles and responsibilities are not clearly defined.

## 1.4 REGULATORY REQUIREMENTS

The codes, standards, and regulations that govern the pipeline industry continue to change in response to lessons learned from industry failures. These regulatory documents

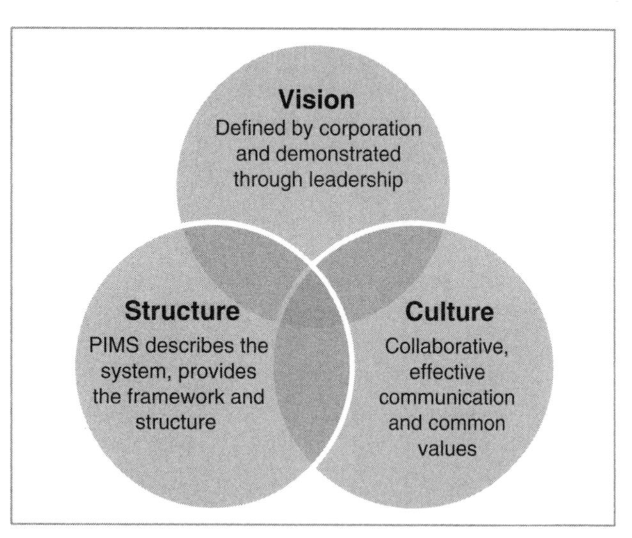
The vision for purpose
The structure to execute
The culture to succeed
Vision
Defined by corporation and demonstrated through leadership
Structure
PIMS describes the system, provides the framework and structure
Culture
Collaborative, effective communication and common values

**FIGURE 1.1** PIMS: vision, structure, and culture.

provide a defined basis that is used by industry to determine pipeline integrity requirements. It is important to realize that regulatory changes take time and typically lag public expectations and industry practices of progressive companies.

Companies often fall into the trap of building their integrity management systems to meet, but not exceed, regulatory requirements. In the 2013 NEB Safety Forum, Jeff Weise, Associate Deputy Director of PHMSA, commented to the audience that "regulatory requirements are 'the floor' on which you stand and meeting regulatory requirements is nothing to be proud of." He emphasized that companies need to be proud of their safety and performance achievements, not of meeting minimum requirements. If you are only meeting regulations, then you are already well behind the industry leaders and good integrity management practices.

Regulatory requirements fall into two very different categories. The specific and prescriptive "shall and must" aspects of regulations provide clearly defined minimum requirements. For example, the AER Pipeline Rule that "the licensee of a pipeline that crosses water or unstable ground shall at least once annually inspect the pipeline right of way" is easy to understand and apply. The minimum compliance is therefore one right of way (ROW) inspection a year; however, an effective and comprehensive approach to risk management suggests that companies conduct ROW inspections at an appropriate frequency for their operating conditions. The "appropriate frequency" can be quarterly or monthly, and in some cases, operators perform daily aerial surveillance to mitigate risk.

The second category of regulatory requirements is more general and requires interpretation. Regulations regarding management systems require each company to interpret the intent and incorporation of the stated requirement. For example, the management systems requirement for management of change (MOC) is only prescriptive in the sense that MOC is required. Each company must determine what specifically is required to ensure MOC exists in their organization, including documenting, communicating, and archiving MOC documentation such as the MOC process or completed MOC forms. Logically, MOC for a small upstream company may be simpler than MOC for a major transmission pipeline company.

Regulations for management systems, on the other hand, are more general and require each company to interpret how to meet the intent of the stated requirement. For example, a regulatory audit investigates a company's document and records management system. The audit may determine that "yes, there is a document and records management system and it meets the audit requirements." However, this conclusion does not mean that the company has an effective system that ensures all critical records, documents, and information are verified and are readily accessible for use in supporting risk assessment, fitness for service, MOC, and other important processes. From a process maturity perspective, an adequate system is a long way from a more mature "competent" or "excellent" system.

A maturity grid approach can be used to determine a system's relative state. A maturity grid is a description of the characteristics/criteria of the company operating at various levels of maturity; for example, maturity levels may include Innocence, Awareness, Understanding, Competence, and Excellence. This is a qualitative assessment technique that can provide an effective way of communicating and illustrating the current state of the management system as well as the future state that the management system aims to achieve. This approach has been used in many industries and a good example can be found in the book "Uptime: Strategies for Excellence in Maintenance Management" [6]. There can be a staged approach to achieving the desired level. For example, a company may wish to first achieve Understanding and then implement a plan to progress to Competence or Excellence.

It is important for each company to evaluate major incidents and review their own programs for weaknesses and deficiencies. Canadian and U.S. regulatory and industry bodies (such as the NTSB, AER, NEB, and PHMSA) will include interim directives, current incident investigation reports, and other communications on industry lessons learned. It may also be beneficial to consider regulations and industry practices that are not mandatory in the regulatory regime in which your company operates. Other pipeline or industry regulations and practices may be in a more advanced state or may contain useful approaches to more comprehensive integrity management systems. As previously mentioned, the API Recommended Practice 1173 addresses process safety management and safety culture in more detail than this chapter provides, and the reader is encouraged to review this document for additional information.

## 1.5 CORE STRUCTURE AND PIMS ELEMENTS

All pipeline companies have some degree of a PIMS in place; however, the PIMS elements typically are not fully integrated between business units and/or not effectively executed. So, what does a fully integrated and effectively executed PIMS look like? As mentioned in Section 1.1, there is no single "right" PIMS; however, the purpose of this chapter is to introduce a method to effectively structure and develop a PIMS.

A complex, interdependent system such as a PIMS can be difficult to describe and illustrate. One common way to represent a dynamic PIMS is using a Plan–Do–Check–Act (PDCA) cycle. The PDCA cycle can be drawn and interpreted in many ways and the reader is encouraged to customize the approach used here to meet their own needs. The PDCA cycle shown in Figure 1.2 has been adapted and

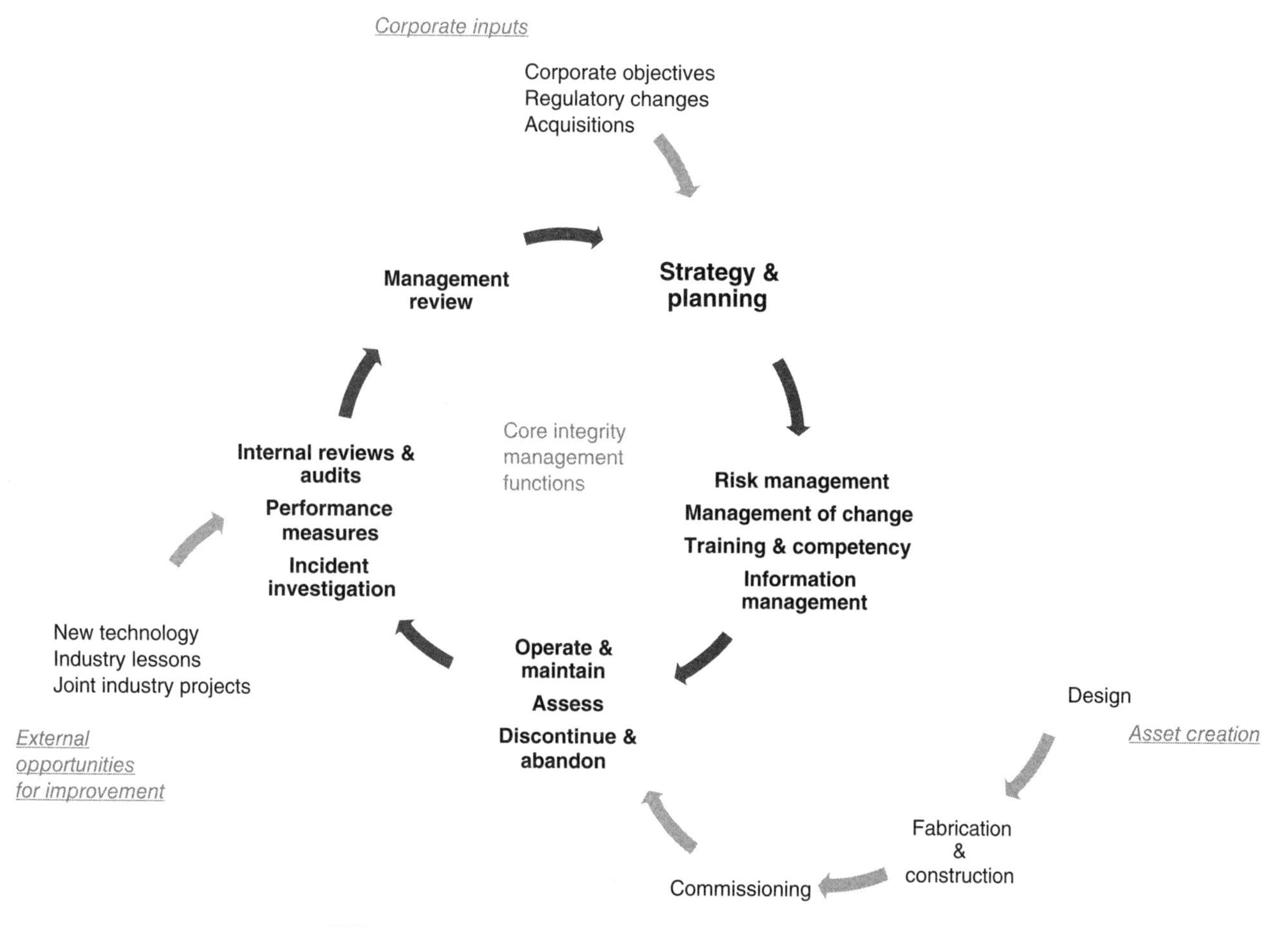


**FIGURE 1.2** Integrity management Plan–Do–Check–Act cycle.

modified from PAS 55-1:2008 [4]. PAS 55 describes the importance of strategic planning (Plan), program enable and execution (Do), assurance and verification (Check), and management reviews (Act). In this chapter, we will examine each of these sections in terms of what they mean and present ideas for development.

The center circle in Figure 1.2 represents the ongoing PDCA cycle that must be in place to keep the integrity management system functioning to mitigate both asset and business risks. For clarification purposes, the primary inputs into the ongoing integrity management cycle are illustrated as arrows feeding into the circle. The PDCA cycle inputs include the following:

- New asset design–fabrication–construction–commissioning that tie in the asset life cycle.
- Improvement opportunities from outside the company to illustrate adopting industry lessons, new technology, and new concepts.
- Inputs that can impact asset integrity planning such as changes in company objectives, significant regulatory changes, new acquisitions, and so on.

| | |
|---|---|
| **Plan**: strategic and operational plans | Planning sets risk reduction direction, incorporates new ideas and technologies, supports process improvement, and ensures appropriate resources are available. |
| **Do**: execute | This includes both processes that<br>• are common corporate processes such as MOC and training that provide the structure and skills to support core life cycle activities;<br>• support core life cycle technical activities that address pipeline hazards in design, operations, and maintenance through to abandonment. |
| **Check**: assurance and verification | Processes to verify the effectiveness of the integrity management system. |
| **Act**: management review | Providing input for planning and strategic decisions. |

## 1.6 PIMS FUNCTION MAP

A function map, such as that shown in Figure 1.3, is another method that can illustrate the PIMS core elements. The intent of Figure 1.3 is to provide an overview of common elements; as such, it must be customized for each company. The same four elements of Figure 1.2 (Plan–Do–Check–Act) are incorporated into Figure 1.3. Function maps are particularly useful for structuring and organizing the systems a company uses to execute asset management activities.

## 1.7 PLAN: STRATEGIC AND OPERATIONAL

Figure 1.4 shows an expanded function map for the Plan element. Operational planning for integrity programs includes both annual activities and longer term strategic plans. In addition, operational planning often coincides with the annual budget cycle. An annual budget is created for ILI and excavations (digs), cathodic protection, risk assessments, scheduled training, facilities programs, and line replacements.

Strategic planning is a more comprehensive exercise and must incorporate lessons learned from management reviews as well as changes in corporate objectives and plans for new major projects or acquisitions. Strategic planning includes considering new concepts and ideas, addressing long-term risk reduction, and ensuring new major initiatives are properly planned and resourced. Strategic planning is about stepping back and ensuring there is a long-term focus on sustaining and improving integrity programs.

Strategic planning should also include the assessment of new technology, participating in joint industry projects, and funding research and development projects. In addition, technologies, ideas, and concepts from outside the oil and gas industry may provide a long-term benefit to the organization.

A strategic plan is more comprehensive than a 1-year annual budget cycle. Strategic plans often forecast multiple years in advance to plan the progressive advancement of key initiatives to improve pipeline integrity programs. For example, if a company's goal is to move from Innocence to Excellence (using a process maturity perspective) for information management, strategic planning would outline a multiyear approach and achievable targets to progressively

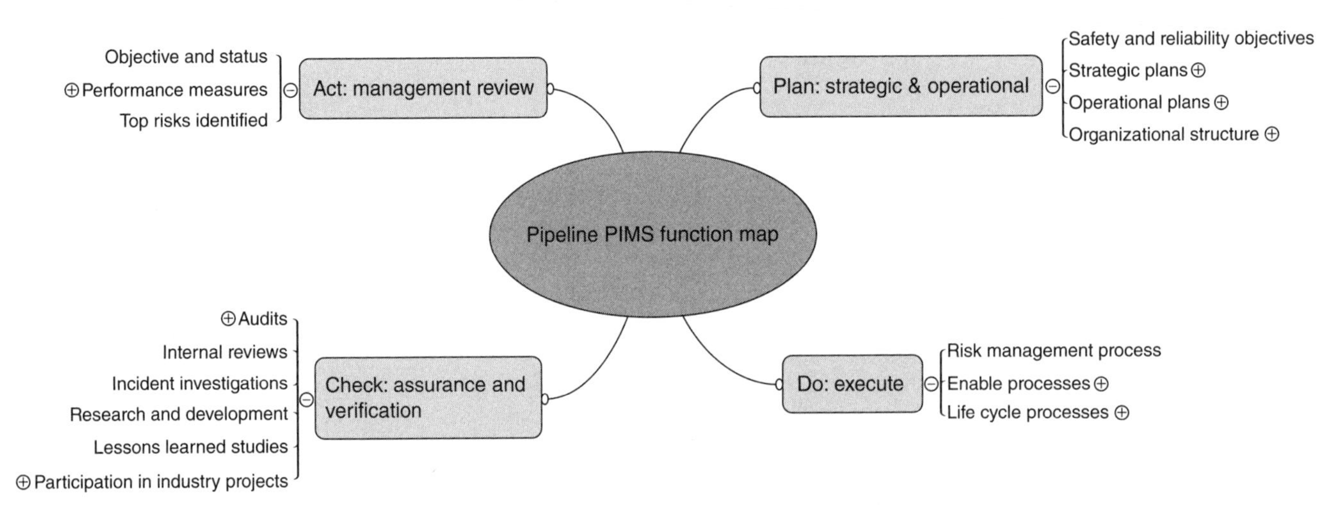


**FIGURE 1.3** Pipeline integrity management system (PIMS) function map.

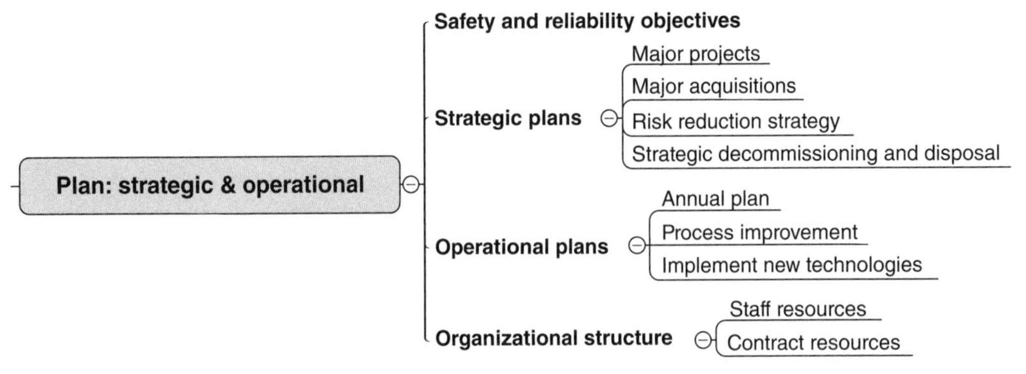


**FIGURE 1.4** Plan function map.

advance the information management system at a manageable pace. Another example of strategic planning is the divestiture of low-value but high-risk pipeline assets.

## 1.8 DO: EXECUTE

The Do cycle shown in Figure 1.3 has three main components, which are expanded in Figure 1.5. The Do section includes the core life cycle technical activities that address pipeline hazards, documentation, and communication.

*Risk management* is a key element of integrity management and determines how pipeline risks are assessed and controlled by the organization. Risk management is a comprehensive framework of principles, processes, and procedures to ensure the safety and integrity of pipelines and facilities.

The ISO 31000:2009 [7] standard for risk management provides principles, framework, and a process for managing risk that can be applied to pipeline integrity. Risk management is not a stand-alone activity that is separate from the main activities and processes of the organization; rather, risk management is a part of management's responsibilities and an integral part of all organizational processes, including strategic planning and all project and change management processes.

Risk management includes identifying threats, hazards, and/or degradation mechanisms that could lead to failure, characterizing risks, prioritizing actions, and executing risk control from design and construction through mitigation, monitoring, and inspection programs.

An effective risk management process will include the following:

- Process management to ensure effective execution and integration of the processes and supporting activities throughout the organization.
- A common understanding and effective communication between operations, engineering, maintenance, asset integrity, and other groups regarding their respective roles, responsibilities, and contributions to achieving safe and reliable operations.
- Defined physical and operational threats that could result in product release that could potentially have serious safety, environmental, production, or regulatory impacts.
- Processes to assess risks and guarantee that the appropriate prioritized controls are in place to reduce the likelihood of a release and to limit the consequence if a release occurs.
- Defined requirements to assess and continuously improve the risk management process through internal reviews, incident investigations, and the use of key performance indicators (KPIs). This may include developing metrics and defining categories (bounds) to indicate performance levels.

*Enable processes* are those that provide the support structure for pipeline integrity work. They are the common management processes that are used to guide and support the work activities, and they are applicable to all aspects of the company's business. Examples include MOC, training and competency programs, and information management.

The effective execution of these processes is essential for a successful PIMS. These processes need to be in place to ensure that the knowledge, skills, and tools are available to manage risk and support the life cycle elements. As previously mentioned, failures still occur in companies that have corrosion controls and inspection programs in place; however, important hazards are still missed or are not properly addressed, so their risk management programs are ineffective for all relevant hazards. Examples of significant contributing factors to pipeline failures include a lack of training and

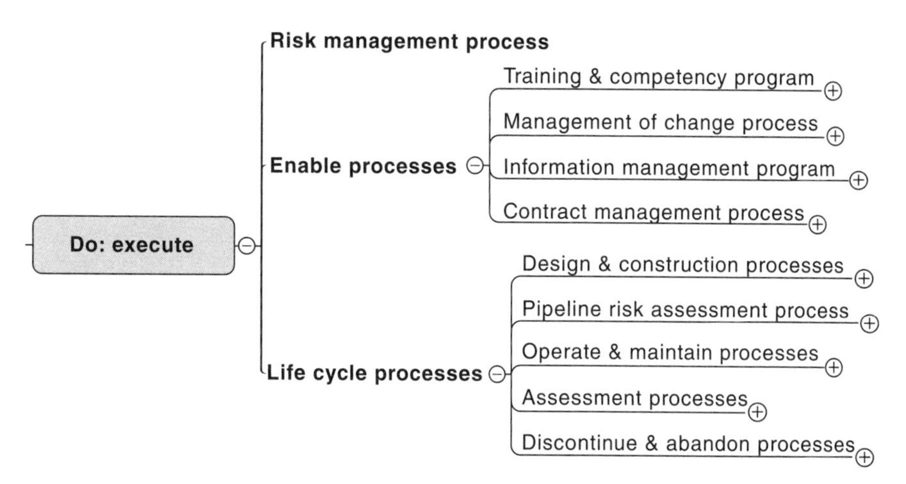


**FIGURE 1.5** Do (execute) function map: risk management, enable, and life cycle.

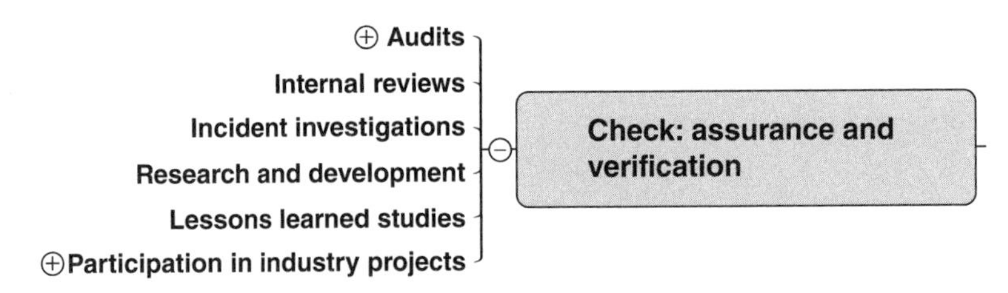


**FIGURE 1.6** Check: assurance and verification.

knowledge, inadequate documentation and records management, and poorly applied MOC.

*Life cycle processes* of a management system include the technical processes and activities that are required to effectively execute integrity programs and activities. Life cycle processes are the technical core of pipeline integrity activities, and include the technologies and activities that are directly applied to the pipeline to prevent failures. For pipeline integrity, the activities and tasks can be broadly grouped into design and construction, maintenance (mitigation, monitoring, and inspection activities), assessment, and abandonment.

Design and construction includes materials selection, life cycle design, fabrication inspection, construction quality control, and commissioning that must be completed before a pipeline (new and repaired) is transferred to operations. Integrity issues that are "built" into a pipeline (such as weld seam defects, poorly applied coatings, construction dents, and incapability with ILI tools) either create additional hazards to manage during operation or limit the ability of operations to manage hazards. A well-thought-out life cycle design and the elimination/reduction of fabrication and construction defects can significantly improve pipeline safety and reliability.

Maintenance life cycle processes include mitigation (cathodic protection and chemical inhibition), monitoring (corrosion coupons and slope stability), and inspection (ILI, direct assessment, and ROW surveillance). These are the primary operation activities to address and control potential pipeline hazards such as external and internal corrosion, geotechnical hazards, and cracks.

Assessment processes are in place to ensure all hazards are understood and addressed, fitness for service methodology is applied to pipeline defects, and sound engineering judgment is used to make good decisions based on operation, monitoring, and inspection data.

Discontinue and abandonment processes are required to ensure end-of-life programs are in place and any residual hazards due to a pipeline are addressed.

## 1.9 CHECK: ASSURANCE AND VERIFICATION

The next element in the PDCA cycle is the Check (assurance and verification) step. This element includes external audits, internal reviews or assessments, incident investigation, and research and development. This also includes external inputs such as new technologies, industry lessons learned, involvement in joint industry projects (JIPs), and participating in industry task forces. Figure 1.6 outlines the expanded function map for the Check element.

The Check part of the PDCA cycle is critical, but often not well executed. During a review of major incidents, the NEB determined that capturing lessons learned and implementing learnings is an area often overlooked. The quote below is from NEB Safety Forum 2013 paper titled "Emerging issues in oil and gas industry safety management":

> The assessment indicted that while most organizations involved in the accidents had management systems or programs developed, they were not effectively implemented or reviewed on a regular basis to ensure adequacy and effectiveness.

A well-thought-out internal review process can be a useful way to assess the internal process and program effectiveness. The internal review process should include both the traditional audit check ("are we doing what we say we are doing") and a way of measuring or quantifying effectiveness. The highest value obtained from an internal assessment will be a measure of program effectiveness and recommendations for improvements. A maturity grid or other assessment models can be used to support the assessment of effectiveness.

## 1.10 ACT: MANAGEMENT REVIEW

The "Act" portion of the Plan–Do–Check–Act cycle is shown in Figure 1.7. The management review is an assessment of the outcomes of pipeline integrity work, including

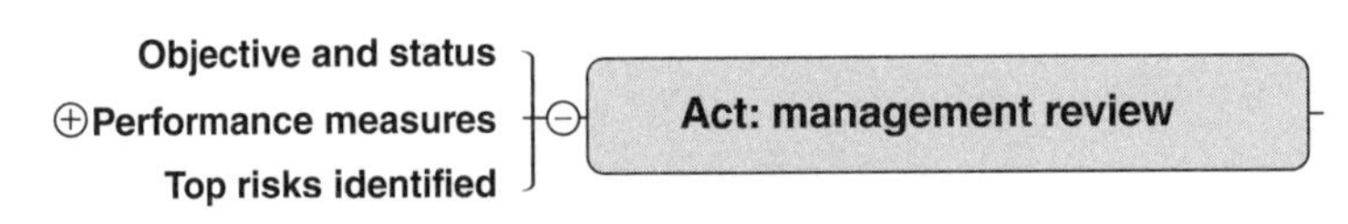


**FIGURE 1.7** Act: management review.

performance measures, audit and review results, reviews of major projects, outcomes from participation in industry task groups, and evaluation of important company and industry lessons learned. Management reviews can be conducted annually or as a more frequent, less formal, assessment.

Many companies have a dashboard summary report for senior management. The objective of a dashboard is to succinctly communicate a few key metrics that represent the ongoing performance of the pipeline integrity program and represent a "status or progress gauge." The metrics or performance measures incorporated in a dashboard are often tied to a company's specific safety, reliability, and production targets.

Management reviews often include high-level assessments such as a summary of the "top 5 risks." This allows the most significant risks to the company to be understood and addressed at senior level in the organization. Addressing these risks can become part of the company's strategic plan.

Management assessments provide direction for future activities. This direction can be expectations for process changes, audit improvements, and training requirements. The outcomes of the management reviews can be used to determine changes required or additional resources to support integrity programs or as inputs into longer term strategic plans.

## 1.11 CULTURE

The processes, procedures, skills, knowledge, and technologies required for pipeline integrity management are only one part of an effective PIMS; the other critical aspect is a process safety culture. Safety can be considered in two ways: occupational safety that addresses personal safety issues such as "slips, trips, and falls" and process safety that addresses preventing releases of any substance (ruptures and leaks) that could lead to injury/fatality and/or environmental damage. Process safety started in the chemical process industry after a series of serious equipment-related failures that resulted in numerous fatalities such as the Bhopal, India, disaster. Process safety includes both the mechanical integrity of the equipment and the safety culture required to ensure that the structure and company culture are aligned.

The importance of a safety culture is emphasized in CSA Z662 Safety Loss Management [8] and in API Recommended Practice 1173 [2]. As detailed in Section 1.1, the pipeline industry has had far too many major failures resulting in public fatalities (Carlsbad, Sand Bruno) and environmental damage (Kalamazoo, Red Deer River) [9]. Pipeline operators have been adopting process safety and cultural change into their overall approach to pipeline integrity management, with the goal to minimize, and ultimately eliminate, pipeline failures.

This overview of PIMS provides the framework and ties together the supporting processes and procedures to provide the overall structure to ensure pipelines are designed, built, operated, and maintained in a way that achieves pipeline safety and reliability. The functions described in the PIMS include the roles, responsibilities, and accountabilities that, when fully understood and implemented, can drive culture change.

## 1.12 SUMMARY

To be effective, a PIMS must be fully integrated within a company's core business functions. The multiple interdependencies within an organization regarding integrity management need to be defined and understood. Operations, engineering, asset integrity, and many other groups have essential roles in achieving safe and reliable operations. The responsibilities and accountabilities must be clearly assigned and communicated.

In addition to role clarity between departments, the PIMS elements must be applied consistently across all operational departments, facilities, and geographic regions. The risks associated with lower value assets need to be understood and managed with the same diligence as high-value assets. The elements of integrity management should have the same focus and attention as the company's commitment to personal safety and production targets.

The PIMS is a comprehensive, continually improving and evolving system that is essential to achieve safe and reliable operations. The PIMS does not need to be overly complex or heavily dependent on processes—it is fundamentally a simple system intended to effectively manage risks and assets. If companies create too much structure and implement complex controls, the system can become bogged down and rendered ineffective. PIMS can be successful if the interdependencies and relationships of the elements shown in Figure 1.2 are understood, the processes are in place to support these elements, and the roles and responsibilities are defined. The system requires regular reviews, assessments, and improvements to ensure it will continue to be effective and, as a result, both the company and the pipeline industry will change for the better.

## REFERENCES

1. International Association of Oil and Gas Producers (2008) Asset integrity: the key to managing major accident risks. Report No. 415. Available at http://www.ogp.org.uk/pubs/415.pdf.
2. API (2014) API Recommended Practice 1173: Pipeline Safety Management System Requirements.
3. Alberta Energy Regulator (2013) Pipeline Performance in Alberta, 1990–2012. Report 2013-B. Available at http://www.aer.ca/documents/reports/R2013-B.pdf.

4. British Standards Institution (2008) PAS 55 Specifications for Asset Management. PAS 55-1:2008—Asset Management Part 1: Specification for the optimized management of physical assets. PAS 55-2:2008—Asset Management Part 2: Guidelines for the application of PAS 55-1.
5. ISO (2014) ISO 55000 Standards for Asset Management. ISO 55000 Asset management—Overview, principles and terminology. ISO 55001 Asset management—Management systems—Requirements. ISO 55002 Asset management—Management systems—Guidelines for the application of ISO 55001.
6. Campbell, J.D. and Reyes-Picknell, J.V. (2006) *Uptime: Strategies for Excellence in Maintenance Management*, 2nd ed., Productivity Press (Figure 1-8: Maintenance maturity grid).
7. ISO (2009) ISO Standards for Risk Management. ISO 31000:2009, Risk management—Principles and guidelines. ISO Guide 73:2009, Risk management—Vocabulary. ISO/IEC 31010:2009, Risk management—Risk assessment techniques.
8. Canadian Standards Association (2011) CSA Z662-11: Safety Loss Management.
9. Hayes, Jan and Hopkins, Andrew (2014) *Nightmare Pipeline Failures: Fantasy Planning, Black Swans and Integrity Management*, CCH Australia Limited, Sydney, Australia.

# 2

# SCADA: SUPERVISORY CONTROL AND DATA ACQUISITION

Michael VanderZee, Doug Fisher, Gail Powley, and Rumi Mohammad
*Willowglen Systems Inc., Edmonton, Alberta, Canada*

## 2.1 INTRODUCTION

The daily operation of pipelines has long since been computerized by "SCADA" (supervisory control and data acquisition) systems. In its simplest definition, SCADA is a computer network that gathers real-time pipeline operational data and brings them into a control room for humans to view and make control actions on [1]. In other words, SCADA can be thought of as the eyes and ears of the people who control the pipeline. SCADA has brought measurable and tangible benefits to the owners and operators of pipeline assets, most notably in terms of pipeline integrity and safety due to the real-time nature of the SCADA control systems. SCADA can and should be deeply integrated into a pipeline company's safety protocols and incident avoidance and mitigation.

Due to the proven benefits that SCADA provides to pipeline owners as well as to the environment and public at large, SCADA systems are considered mission-critical and vital equipment to pipelines. In fact, most jurisdictions around the world have legislated that SCADA systems be installed as a necessary part of all petroleum pipeline systems.

The SCADA system consists of several parts, including sensors, RTUs/PLCs, computer servers, computer workstations, communications infrastructure, computer networking equipment, and various peripherals (such as a GPS accurate time reference source). Some equipment is, by necessity, located out at field site locations, whereas the focal point of the SCADA system is located in one or more pipeline control rooms. Common terms used in SCADA are defined in Table 2.1.

Unfortunately, the industry use of the term SCADA is somewhat ambiguous; some use the term to refer only to the software displaying operational data in the pipeline control room (such as pressures, temperatures, densities, and flow rates), whereas others using the term also include the sensors and the RTUs. In this chapter, the term SCADA will be used in reference to both the pipeline control software and the RTUs that sit out in the field and interface to the sensors.

SCADA systems can generally be classified as either enterprise-level or simpler HMI systems. HMI SCADA systems are less capable and historically were limited to real-time data gathering and presentation, whereas full-fledged enterprise SCADA goes further by including many additional features. Examples of features distinguishing HMI and enterprise-level SCADA would be the following:

- Retention of past data values in a historical database.
- Interfacing to other corporate software (e.g., pipeline scheduling, leak detection, accounting, and ERP).
- Ability for additional "emergency backup" control rooms.
- The quantity of I/O points the SCADA system can manage.
- Hardened system to protect against hackers and cyber threats.
- A "human factors" interface to avoid human operator error and enhance efficiency.

*Oil and Gas Pipelines: Integrity and Safety Handbook*, First Edition. Edited by R. Winston Revie.

**TABLE 2.1 SCADA Terminology and Definitions**

| SCADA Term | Definition |
|---|---|
| SCADA | Supervisory control and data acquisition: a computerized control system |
| RTU | Remote terminal unit: manages data from sensors wired into it |
| LAN | Local area network: an Ethernet computer network |
| Real time | The prompt collection and processing of data such that operational decisions can be made before undesired consequences occur |
| Controller | A human who uses SCADA to perform pipeline operations |
| PLC | Programmable logic controller: similar to an RTU; it manages data from sensors |
| DCS | Distributed control system: similar to SCADA but designed for processing plants and other "campus-sized" assets |
| HMI (MMI) | Human–machine interface (man–machine interface): a simpler SCADA system focused on data presentation with less (or no) emphasis on higher level functions |
| I/O | Input/output: one I/O point represents a single piece of data at the RTU |
| GUI | Graphical user interface: a software interface providing a rich graphical display to human users |
| Mbps | Mega bits per second |
| Pop-up | A small SCADA display window that shows additional information about a single data point |
| RAID | Redundant array of independent disks: an arrangement of hard drives that provides high-availability service |
| Telco | Telecommunications company |

- Ability to replay historical events captured by the SCADA system for incident review or training.
- Operator training scenarios.
- Advanced alarm management.
- Operator shift handover report.
- Rapid polling of screens being viewed by operators.
- Interfacing to related computer systems such as leak detection, batch tracking, energy management system, equipment maintenance systems, and pipeline hydraulic simulation.

Sometimes, there can be confusion with those not familiar with the industry about a similar product called DCS (distributed control system). Although very similar in many regards in that both SCADA and DCS ultimately collect and display data from sensors in a real-time fashion, DCS was designed to be used in processing plants where the physical distances between sensor and computer control system are relatively small and often used to manage a process (such as refining crude oil into products such as jet fuel and diesel). SCADA, on the other hand, can span almost any geographical distance. Large pipelines often exceed 5000 km in length with sensors spread across the entire length, and this is the domain of SCADA systems, whereas DCS would not be able to do this.

## 2.2 SCADA COMPUTER SERVERS

The backbone of the SCADA equipment in the control rooms is the SCADA server computers. These computers communicate with and gather data from the RTUs and PLCs out in the field, maintain the SCADA real-time database, maintain a historical database of past values, interface to other computer systems, and generally do everything except providing the graphical interface for the human SCADA operators [2].

Best practices for maximizing pipeline integrity require that each SCADA server contain internal redundancy in the form of dual power supplies, RAID hard drives, dual Ethernet interfaces (connected to dual SCADA LANs), and possibly dual CPUs.

To further the concept of redundancy, SCADA servers are delivered in pairs so as to provide a main unit and a hot standby. A good SCADA system will improve upon this by load sharing between the two servers under normal circumstances, and reverting to one or the other server if either should fail.

SCADA servers may be Windows based, but UNIX servers are often better suited to industrial applications such as SCADA due to the Windows operating system not having been designed for "real-time" applications.

In all but the smallest SCADA systems, the servers are usually delivered without video display, mouse, or keyboard. Rather, they are normally rack-mounted computer hardware providing the background work of the SCADA system. The SCADA system's user interface is provided through the SCADA workstations.

## 2.3 SCADA COMPUTER WORKSTATIONS

The SCADA workstations exist to present the GUI to the human operators and this is the method through which pressures, temperatures, flow rates, and other parameters are displayed. SCADA workstations are often outfitted with multiple LCD panels with the specific arrangement designed with the physical control room layout in mind. Common arrangements are $2 \times 3$, $1 \times 4$, and $2 \times 2$.

The quantity of SCADA workstations in each control room should be a function of the pipeline operations and is proportional to the length of the pipeline(s) being controlled (larger companies usually have a network consisting of numerous pipelines all being controlled from within one control room). Obviously, the minimum is one SCADA workstation, whereas the largest pipeline control rooms contain upward of 30 SCADA workstations.

Due to the client/server software model, SCADA workstation hardware need not be as hardened as the SCADA servers. Often high-end off-the-shelf PCs are the best choice for SCADA workstations. Usually the most important parameter of the SCADA workstation is its graphics ability, which is a function of the video card in the computer.

Although there is no need for high fidelity, most SCADA systems produce alarm sounds when a process variable exceeds its nominal range. As such, SCADA workstations should have volume-adjustable audio capability.

At present, the human operator interfaces with the SCADA workstation mostly by use of the mouse and, to a lesser extent, the keyboard. However, emerging trends are increasingly augmented with haptic touchscreens [3].

## 2.4 HIERARCHY

Enterprise-level SCADA systems should be able to be implemented in a hierarchical fashion. In a large distributed system, it may be desirable, for example, to have completely independent SCADA systems for each separate region, but a higher level master SCADA system that receives summary data from each of the lower SCADA systems, thus forming a hierarchy. In fact, this idea of a hierarchy of SCADA systems can consist of more than just two levels; conceptually, it should support three or more levels, for example, SCADA at a municipal level, then provincial level, and finally national level. Or alternatively a pipeline company could have separate SCADA systems for its gas pipelines versus crude pipelines versus liquid product pipelines, and an overall SCADA system on top whose database is the sum of the three lower level SCADA systems.

Depending on the pipeline operator's needs, it is not always necessary to push the entire database up to the next level of hierarchy. Instead, it should be possible to identify the specific data in the lower level database that are needed in the higher level database. Thus, by picking and choosing data points, a customized summary of data can be pushed up to the next level of hierarchy.

From a pipeline integrity point of view, one benefit this approach can provide is additional personnel to observe and review operations in real time.

## 2.5 RUNTIME AND CONFIGURATION DATABASES

Although not universal in all SCADA systems, best practices in SCADA technology require that there be separate databases for the live runtime data and the system configuration. The main reason for this is to allow for system changes to be made, and possibly tested out on a development server before being committed to the runtime system. Any SCADA package that does not provide this database separation will have system development being done on the live runtime (production) database. Obviously, the consequences of making an error when editing a database are more severe if those changes are made on the live runtime database.

The purpose of the configuration database is to hold the SCADA system parameters. This would be information such as how many RTUs and PLCs there are, how many I/O data points each contains, the computer address of each one, the log-in user IDs for each human SCADA operator, and generally all system information.

The purpose of the live runtime database is to use the configuration database as a framework, but to populate it with live data.

These two databases are linked through a process called a commit.

The best SCADA systems will offer a feedback mechanism between the live runtime database and the configuration database. The meaning of "feedback" in this context is a software mechanism to allow the human SCADA operators to fine-tune certain system parameters on the live runtime database and to have those parameters automatically updated in the offline configuration database. This is a unique and uncommon action as the data flow is normally from the offline configuration database into the live runtime database. Feedback is normally automated for only a small subset of the overall system parameters.

## 2.6 FAULT TOLERANCE

In any SCADA system that includes interfacing to more than a single RTU or PLC, it is an important design criterion that the SCADA host should be fault tolerant. The general design question to ask is: "Is there a single component (software or hardware) that if it fails will prevent SCADA from monitoring and controlling the critical field devices?" Note that a failure of a single RTU will prevent monitoring and control of field devices connected to that RTU, but should not affect the other RTUs, and so usually failure to communicate with a single RTU is not considered critical. However, if there is a single RTU that is considered critical, then that RTU itself should have some fault tolerance built into it, such as a dual communication line, a dual processor, or the entire RTU could have a hot standby RTU.

The simplest level of SCADA fault tolerance is called "cold standby" and is achieved by providing two identical pieces of hardware installed with identical software. If any component fails, then the failed device is manually swapped out with the spare from the cold standby. As this requires a skilled technician on call, the main drawback of this method is the time it takes to fix a problem, and the second drawback is that it requires a regular maintenance step to ensure that the spare devices are working properly and available to be swapped in. The second level of fault tolerance is called "warm standby." It is similar to the cold standby setup, but the hardware is physically installed and running. If any component fails, it should be set up so that it can be easily swapped out as a whole unit by moving a switch or moving some cables. This significantly shortens the time to fix a problem, and lessens the requirements for the technician who initiates the swap. The warm standby still should be checked regularly to make sure it is functional. In this setup, there are often procedures that must be followed after a switchover as the most current configuration might not be installed on the warm standby computer and recent information such as current process variable set points would have to be reentered. The third level of fault tolerance is called "hot standby." It is similar to the warm standby setup except that databases on the two hosts are kept up to date and it is possible to easily switch control between the two computers. How up to date the hot standby server is varies by SCADA vendor and can be as little as a fraction of a second to as high as several hours old. For a hot standby setup, the communication links to the RTUs must be designed to be easily switched from one SCADA server to the other. Ethernet connections are fairly easy to switch over; serial connections will require a changeover switch. In this setup, the SCADA host software should be able to automatically initiate a swap if it detects a failure of the active host, or the operator should be able to initiate a swap. The fourth level of fault tolerance is called "function-level fault tolerance." It is similar to hot standby except that if a single function fails, for example, a serial port used to talk to an RTU, it is no longer necessary to fail over the entire host computer. In this mode, individual SCADA features can be swapped over from one server to another. This makes an overall SCADA system much more tolerant of small failures and makes it much easier to test standby hardware. This also opens up possibility for load sharing when a standby would otherwise be unused. In any fault-tolerant system, it is very important that the standby hardware be tested on a regular basis. The most reliable means of testing this is by initiating a swap on a regular basis such as once a month. Depending on the hardware, it might be possible to test backup lines automatically. For example, to test backup communications to an RTU, the RTU must be able to support talking with two hosts at the same time. Remember that a backup system that is never tested is very likely to not be working when it is really needed and thus would not provide any fault tolerance.

## 2.7 REDUNDANCY

Beyond the above-mentioned software concept of fault tolerance, physical redundancy is one of the most effective tools a pipeline operator can employ to improve pipeline safety and integrity. The basic notion is to provide two pieces of equipment to do a single job: a main and a backup. This concept of main and backup can be simultaneously employed at three levels. The lowest level of physical redundancy exists within the computer servers themselves. The SCADA servers should be designed with dual power supplies, dual Ethernet interfaces, RAID hard drives, and dual CPUs.

Once each SCADA server is built with inherent redundancy, the second step is to provide not just one such server but two such SCADA servers in the control room. In this way, even if one SCADA server is rendered ineffective, a backup SCADA server remains available to take over. This is important because even though many components of each server are provided with a backup, not all components are. For example, the motherboard itself is not redundant. Although it is unlikely that all those components would fail, if they did fail, then the entire SCADA server would be out of commission and so it makes sense to have a backup.

Beyond these is the third and final level of redundancy, which is redundancy of entire control rooms. This is a growing trend in SCADA systems and has legitimate value. History has shown that natural forces can be unpredictable and intense. Threats of tsunami, earthquake, tornado, fire, hurricane, and so on make an entire control room vulnerable. Even if a control room contains two SCADA servers, each of which has redundant components, a tsunami will knock out both servers if it knocks out one, hence the logic for a second control room situated some distance away, often 50–100 km distance between control rooms.

Although the concepts of fault tolerance (described above) and redundancy overlap, there do exist some differences.

Clearly, the best approach is to use both redundancy and fault tolerance.

## 2.8 ALARM RATIONALIZATION, MANAGEMENT, AND ANALYSIS

Alarms notify human operators about abnormal situations, so they can respond in a timely and effective manner to resolve problems before they escalate. Studies have shown that the ability of operators to react to alarms depends largely on how well alarms are rationalized and managed.

In the early days of SCADA, alarms had to be hardwired to physical constructs such as flashing lights and sirens. These physical limitations forced designers to be very selective about which alarms to annunciate. In comparison, modern software-based alarms are often considered free. This creates a common mindset where anything that can be alarmed upon is configured to do so. The result is a SCADA control room with a constant overload of alarms, often unnecessary or unimportant, which operators are expected to filter through when searching for "real" alarms with serious consequences.

Efficient alarm management begins with proper alarm rationalization. Every alarm configured in the system must be considered carefully by a team of experts with operational domain knowledge. Consequences with regard to safety, finance, and the environment must be factored in. The goal is to create as few alarms as possible, while accounting for all serious consequences. The approved alarms and procedures for responding to each alarm should be captured in a master alarm database (MADB), which is then periodically reviewed by the team. The MADB serves as the foundation for configuring alarms within the SCADA system [4].

Alarms must capture an operator's attention, and depending on the severity of the alarm, this is done using a combination of graphic effects such as flashing colors and repeated sounds. It should be easy for the operator to distinguish the process variable in alarm from other process variables that are not in alarm. For example, a red flashing border stands out against an otherwise gray screen. Operators should then examine and acknowledge the alarm. Information that can help the operator react to the alarm should be easily accessible from the context of the alarm.

A summary list of recent alarms helps operators identify when an alarm occurred, the nature of the alarm, its severity and scope, whether it is still in alarm, and whether an operator has acknowledged it. A searchable list of historical alarms helps both operators and analysts to track alarms over long periods of time to detect important patterns. These lists can be filtered and sorted.

Sometimes, alarms need to be suppressed. Operators can suppress alarms for a short period, for example, when a sensor is temporarily malfunctioning. Alarms can be configured to be suppressed for long periods of time, for example, when a point is still being commissioned. Dynamic logic can also suppress alarms, for example, suppressing a low-pressure alarm for a pipe when the corresponding valve is closed.

Long-term operational use reveals badly configured alarms. There is an arsenal of techniques to correct these configuration issues, based on analysis of recurring trends. Analog alarms that constantly chatter between low and high values can apply deadbands to mute the noise. Digital alarms that quickly switch states multiple times can apply on-delays and off-delays to stabilize the state transitions. Incidental alarms for a device power outage can be presented as a single group of alarms. Analytical tools can hone in the most frequent offenders, which when dealt with can reduce unnecessary alarms by large percentages. As established earlier, fewer alarms make alarms more manageable for operators in real-time environments.

## 2.9 INCIDENT REVIEW AND REPLAY

In any SCADA system, situations will arise that management will want to have reviewed to find out what exactly happened and whether the operator had sufficient information available to properly handle the situation. A SCADA system should provide tools to make it possible to review these situations. Some SCADA systems include audio recorders on all phone lines and radio channels for this kind of review. It is also possible to record the operator screens to verify whether the operator went to the correct screens to diagnose a problem. The disk storage requirements for this are quite high, so the number of hours of storage will have to be limited. There are also potential personnel issues with this level of monitoring, which should be carefully reviewed. A SCADA system should also provide tools to analyze the precise order that things occurred in, preferably to the millisecond. The activities that need to be monitored are as follows: when field devices such as valves or circuit breakers changed status, when operator commands were sent out, and when operators acknowledged alarms. Additional information such as updates for analog values from the field can also be very useful in analyzing an incident, to check whether an analog value changed before a command or in response to a command. A simple tool might look something like a text log of each change, but that is difficult and time consuming to review, but does contain very precise information. A better system will replay the incident using the same screens that are available to the operator showing the current state of the system. The best system will include both the text and GUI interface and allow for slow replay and pauses. In summary, it is important that an incident that occurred can be recorded and made available for review in an offline system.

## 2.10 DATA QUALITY

One of the many aspects critical to the proper use of a SCADA system is the integrity and quality of the data displayed to the human operator in the control room. There are several conditions that may adversely affect the data being displayed and it is important that the operator be aware of what these conditions might be and when they are occurring. Because SCADA systems are computerized and are thus aware of the context in which the data were gathered, they can also infer the quality of the data and advise the operator if the data are questionable. For example, if the communication link between the SCADA server computers and an RTU is broken for whatever reason, most SCADA systems can inform the human operator that the data being displayed are stale. It is important for the operator to know that the data currently displayed for RTU in question have lost their "real-time" nature.

**TABLE 2.2 Data Quality Terms, Symbols, and Meanings**

| Term | Symbol | Meaning |
|---|---|---|
| Communications failed | F | The communication link is no longer passing data, so the last received data are still being displayed but these are now stale |
| Manual mode | M | A human operator has performed a data value override and has manually entered a value into the runtime database |
| Out of range | R | Used only for analog values (pressure, temperature, etc.): the data value has exceeded the sensor's calibrated range and the displayed value is no longer accurate and should be considered as approximate |
| Blocked alarm | B | Alarms have been suppressed for this data point |

Although the list of possible data quality conditions is quite long, Table 2.2 shows some typical data quality conditions.

SCADA systems use a visual flag to indicate when any of the data quality conditions are present. In order to not dominate screen space, the flag is usually represented as a single letter associated with the data point in question. The single letters shown in Table 2.2 are examples of those that could be shown in a SCADA display screen. Once a pop-up is rendered, the SCADA system should provide a more descriptive indication of the data quality condition if such is present as shown in Table 2.2.

More sophisticated SCADA systems maintain a hierarchy of data quality conditions and can arbitrate which to display if simultaneous conditions exist. Also, better SCADA systems will be able to propagate data quality conditions present with calculation inputs through to the output results of those calculations. Data quality should also be maintained in the historical database so that these metadata are preserved when historical values are recalled.

## 2.11 OPERATOR LOGBOOK AND SHIFT HANDOVER

The 49 CFR Parts 192 and 195 [5] regarding control room management (CRM) provide a clear definition on the needs for shift handover. "Pipeline operators must also establish a means for recording shift changes and other situations in which responsibility for pipeline operations is handed over from one controller to another. The procedures must include the content of information to be exchanged during the turnover." The intent is to capture important information that has been provided, or has occurred during the shift—and that the next or future shifts should be aware of in order to responsibly operate the pipeline.

Operators have long kept physical handwritten logbooks; however, electronic logbooks and shift handover software tools are now the norm. These electronic tools can range from being an integrated part of the SCADA system to separate stand-alone software. There are advantages to both approaches. With the advance of these software tools, control room operators now have much better access to fuller (and now searchable) information that has happened in previous shifts. Information access can also be role-based at the operator, shift supervisor, manager, and other levels. Systems that allow operators to select entries based on drop-down selections (thus require less typing) tend to obtain fuller information, and the information is more standardized and sortable as well.

Key information captured in operator logbook and shift handover tools can include the following:

- Operations/maintenance orders/schedules for the day provided by management/supervisors.
  - With operator checklist ability that the orders have been read, are underway, completed, and/or if delayed understanding on why.
- Ability to search for past "similar experiences" to aid in troubleshooting unusual situations.
- Lessons learned.
- Personnel planning: easy access regarding past/future personnel on shift for planning purposes.
  - With now a more trackable understanding of their abilities.
- Multifaceted information capture: more sophisticated integrated systems allow for the "drag and drop" capture of rich information regarding process graphics and trends—that allow a full "snapshot" of operations for full information transfer.
- Integrated reporting.
- Integrated procedure management.
- And more.

With this information now more easily available, today's operators are better equipped to avoid information transfer loss during shift handover, which has in the past been a significant contributor to pipeline incidents and/or operational inefficiencies.

## 2.12 TRAINING

There are many aspects to consider when it comes to SCADA systems and how they relate to pipeline integrity. One often forgotten aspect is the human element, and the need for training. All too often when a SCADA system is commissioned, the pipeline operating staff receive a set of training courses on the use of the SCADA system, but what about eventual staff turnover? When new SCADA staff are hired, who will teach them the system, and with what credentials? Once a new SCADA system has been delivered, it rarely happens that the pipeline owner hires the SCADA vendor to return to the facility to teach new staff members.

Indeed, training via experienced workers offers several benefits, but can sometimes also have some disadvantages, as experienced workers are not necessarily skilled teachers. A full-featured SCADA system can compensate, in part, by providing training tools.

The ultimate in training aids is a full hydraulic model built onto a SCADA simulation workstation that is identical to a real SCADA workstation. Such a facility would not only look, but also feel like the actual SCADA system, with accurate hydraulic calculated results for the SCADA operator's actions. For many companies, however, such a hydraulic model is prohibitively expensive, as these packages often cost more than the entire SCADA system itself.

A reasonable compromise is a training tool that simulates the look of the SCADA system, but is not connected to a hydraulic model (i.e., a tool that looks like, but does not feel like the real system). Lacking the hydraulic model means the trainee cannot take actions and observe the results of his/her decisions, but such a training tool can provide a different training benefit: there is at least one SCADA system available that has a SCADA training simulator that functions by replaying previously recorded time periods of SCADA activity. It allows a trainer to insert pauses and questions for the trainee to respond to, but is limited in that whatever decision a human trainee makes, when he/she continues the replay, there is only one decision tree that can be followed. Regardless, such a tool would give the trainee hands-on time at the controls of an identical SCADA system and could be given prerecorded scenarios to follow. The scenarios could be both best practices in terms of how to do regular duties and the possibility to replay accidents or incidents.

With training tools, new employees can be trained more quickly and more effectively by running through the steps to operate valves, compressors, pumps, and so on. Training also affords the opportunity to have each human SCADA operator certified for a variety of standard daily procedures as well as emergency response. SCADA training can also be used to recertify experienced operators from time to time. All these ideas will improve the daily operation of the pipeline, avoid costly accidents, and thus contribute toward greater pipeline integrity.

## 2.13 SCADA USER PERMISSIONS AND AORs

Every user in the SCADA system is assigned a set of permissions and areas of responsibility (AORs). Permissions define what actions a SCADA operator is allowed to do or not to do. AORs define the parts of the system where the SCADA operator can perform his/her list of allowable actions.

Permissions can be used to grant or restrict actions from simply being able to browse information to restoring databases to earlier versions or setting the system time. Any combination of permissions can be granted to a user. Typically, permissions are grouped by user roles. A visitor or observer of the system is limited to browse permissions only, with the ability to read information from the system, but has no ability to change any information or send any commands. An operator typically has permission to send commands to field devices and acknowledge alarms. A configuration engineer has permission to edit the database. At the highest level, a system administrator has permission to restore database and manage user accounts, and do everything.

A user can choose to temporarily disable some permissions. An operator who normally has the ability to acknowledge all alarms at once may choose to disable that ability under normal circumstances. He/she can choose to enable it again if needed.

AORs divide users into different areas of the system. Areas that do not belong to a SCADA user are filtered out. One operator can still inspect other areas of the system, but is prevented from taking any action in areas he/she is not authorized for. A single user may have multiple AORs. A single AOR may have multiple users assigned to it. A single database point typically exists in a single AOR. A point that is not assigned to an AOR is considered global, and therefore accessible for all users.

A user can choose to temporarily disable some AORs. A system administrator who is authorized to access all AORs may choose to disable AORs that are currently manned by other operators. If one of the other operators leaves early, the first operator can choose to re-enable that AOR.

## 2.14 WEB CONNECTION

Ever since the 1990s it has been commonplace for pipeline operators to publish certain SCADA data outside of the SCADA control network. Other users within the organization often want certain SCADA data (e.g., accounting department) as well as the customers of the pipeline company, and this can be done by granting access to the SCADA network and SCADA servers to all these other users. Doing so does, however, present a weakness in SCADA security, and that in turn can compromise pipeline integrity.

To mitigate this risk, a properly designed SCADA system will offer a protected way of publishing its real-time data.

This is best achieved by adding a SCADA web server computer, connected through a firewall and a tightly configured "managed Ethernet switch," to the production SCADA servers. At the software application layer, the web server and production SCADA servers will guarantee a one-way data flow meaning that "outside" web users can view certain published SCADA data, but they will not have any way to send SCADA commands or otherwise corrupt the production servers. Such an arrangement will allow for approved SCADA data to be sent to the web server computer in real time, meaning that interested parties within the pipeline company or its customers can continuously get accurate and timely updates on the operational parameters of interest to them.

Granting such data access, in a protected way, will not compromise pipeline integrity, and even has the potential to actually improve it: with more trained eyes watching more real-time system data, it is more likely that any odd and potentially harmful conditions will be noticed sooner and with corrective action by the SCADA operators pipeline integrity can be enhanced.

## 2.15 SCADA SECURITY

Since SCADA is fundamental to pipeline operations, it follows that SCADA security is fundamental to pipeline security and integrity. SCADA, as a mission-critical element to pipeline operations, is vulnerable to several threats, and thus needs protection.

The first type of SCADA protection to discuss is physical security. The SCADA infrastructure (servers, workstations, communication equipment, RTUs) needs to be physically protected by access control systems. It should not be possible for the general public to gain access to these assets. Access control for an RTU is often achieved by putting the RTU in a weather-protected enclosure and then locking the enclosure. Sometimes, the enclosure also has a perimeter fence around it to add an additional barrier. This can be augmented with a prosecution-quality CCTV system.

As for the SCADA servers and workstations, this equipment is usually located in an office building; thus, it is quite easy to have this equipment behind locked doors. To increase access control to the SCADA control room, it is also possible to add fingerprint scanners. Such facilities usually have a separate environmentally controlled server room to host the SCADA and other computer servers, and additional physical security should be added to this room.

Beyond physical access control, the SCADA equipment needs to be protected from the environment to continue to operate properly: SCADA servers should be air-conditioned, RTUs should be protected from rain, and the entire control room should be located out of harm's way and away from natural disaster zones.

In addition to physical security in the form of access control and environmental protection, a SCADA system needs to be hardened against cyber threats. This means that the SCADA servers and workstations should prevent unauthorized programmatic access by locking down all I/O ports (USB ports, CD drives, Ethernet ports, serial ports, etc.). Connections between SCADA computers should also be made via secure channels such as SSL.

When possible, the link between the SCADA servers and RTUs should be secured. Applying encryption to the data link is one way to do this, but this is always not practical when different vendors have supplied the SCADA system and the RTU, but when a single vendor has supplied the equipment at both ends of the communication line, it becomes possible to add an encryption algorithm to the communication channel. This will prevent eavesdroppers from decoding the data and commands taking place on the link, especially when wireless communications are used (as is the case with microwave communications). Taking this idea further, when one vendor supplies both RTU and SCADA, a software authentication method can be added that will prevent unauthorized users from injecting commands to control and possibly damage pipeline assets.

Because the data contained in a SCADA system are useful to other departments of a pipeline company, it is very common for there to be a connection between the SCADA system and other business systems. For maximum protection this should be avoided, but doing so is usually not practical. The use of a firewall between the SCADA and business networks will mitigate the possibility of a virus or other cyber threats from reaching the mission-critical SCADA network.

## 2.16 HUMAN FACTORS DESIGN IN SCADA SYSTEMS

The pipeline industry in recent years has formally acknowledged the value of human factors design in SCADA systems—shown through recent U.S. regulations such as the amendments made to 49 CFR Parts 192 and 195 regarding CRM, and the Pipeline Inspection, Protection, Enforcement, and Safety (PIPES) Act requiring a human factors plan to CRM. The value of assuring that central control systems are designed to provide early detection of abnormal situations, in a manner that does not provide a stressful environment for the human operator while also recognizing the need of fatigue management, has been recognized in the refining and process industries for many years. There is a wealth of knowledge from the ISA, NPRA, and other organizations that can be applied and adapted to the pipeline industry [6–8].

The key elements of an effective human factors plan for control room management (see references [3,9–13]) require the design of the SCADA system to equip the operator with the information needed to properly operate the pipeline safely and reliably. The manner appropriate for delivering this information needs to consider the ways humans process information, at all times of the day, after many hours on shift, as well as how to capture and transfer this information through to the next shift.

After the National Transportation Safety Board intensively investigated the root cause of past pipeline incidents, they identified five areas for improvement: graphic displays, alarm management, operator training, fatigue management, and leak detection. Graphic displays and fatigue management have common elements, but in addition the physical control room design has a significant role to play in fatigue management via elements such as room lighting, monitor placement, and separation of areas of responsibility.

As computer graphic displays are the primary method for control room operators to visualize the big picture of the state of their entire pipeline operations, this is often considered the top opportunity for improvement when first implementing a "human factors approach." The recommended practice for Pipeline SCADA Displays API RP 1165 [14] is expected to be followed, and expert practitioners have applied the research from industrial psychologists to achieve significant improvements in operator comprehension and speed of response. Some of the main considerations when developing graphics optimized for human factors are to "selectively use reserved colors" against a gray-scale background versus the traditional "black background utilizing the full-color spectrum." Also providing a confirmation identification element that can help color-insensitive personnel confirm their understanding of the situation is key to allowing a clear diagnosis of the issues by the general control room population. These advancements in graphic display designs have reliably shown improvements to comprehension and response time of over 50%, and in addition have shown that the learning curve is significantly reduced for inexperienced personnel who tend to be the most vulnerable during times of emergency. Despite the recognized value of using a "human factors" gray-scale approach, it is often difficult for existing operators to be enthusiastic about losing their black background graphics. As with any change, there is an adjustment time, additional work, and retraining. However, operators who move on to a new area of responsibility do appreciate the human factors gray-scale approach, and are able to master the new area earlier than ever before.

Of course, taking a human factors approach to pipeline design, operations, and maintenance has applications and benefits beyond the control room—as is shown in Chapter 9.

## 2.17 SCADA STANDARDS

There are several standards for SCADA that exist as of this writing. Most standards focus on a specific area of SCADA, such as alarm management, security, or graphic display. Some standards apply to process industries in general, whereas others are for individual industries such as oil and gas, pipeline, electrical, and transportation. Some are created by formal regulatory bodies that require organizations to adhere strictly to requirements, whereas others are from established consortiums that recommend guidelines that may form the basis of future regulatory requirements.

Standards often borrow from each other, building upon legacy documents to add variations or new requirements. Core themes from various standards include the following:

- Ensuring only authorized users have access to sensitive information and actions.
- Providing greater awareness for operators using SCADA displays [11].
- Reducing overall stress levels for operators to improve responsiveness.
- Providing alternate views of data in various contexts for various purposes.
- Maintaining data integrity for audits and analyses.
- Using simple symbols to convey common objects.
- Scoping data at various levels, allowing users to drill down from overviews.
- Interactivity between different modules for seamless workflows.

Essential standards for the pipeline industry include the following:

- Safety: PHMSA 49 CFR 192/195 [5].
- Alarms: ISA 18.2 [15], API 1167 [16], EEMUA 191 [4].
- Control rooms: API 1168 [17], PHMSA CRM.
- SCADA: API 1165 [14], ISA 5.5.
- Security: API 1164 [18], ISA 99, ISO 27001.

While some standards or requirements are very specific and detailed, others tend to leave room for interpretation. Thus, different SCADA providers may provide different solutions to meet selected standards. In general, standards compliance by organizations is achieved by best practices that a SCADA system can help achieve. Understanding of human factors and industry workflows is essential to interpreting and implanting standards.

## 2.18 PIPELINE INDUSTRY APPLICATIONS

Although at its most basic definition a SCADA system simply presents operational data to humans in a control room for them to take action, the reality is that after years of R&D and continued refinement, SCADA systems now offer much more than this basic functionality. One way that SCADA systems offer more is by adding optional software application modules. Although there may be many such applications to choose from, three important ones for pipeline integrity are leak detection, batch tracking, and dynamic line coloring.

### 2.18.1 Leak Detection

Obviously, it is necessary for pipeline operators to know whether they have a leak in their system. Several reputable companies offer software modules that can do this, and some SCADA systems have this ability built-in. Either way, these software application modules must interface to SCADA in order to use the real-time data to constantly monitor for leaks. By using live operational data from the SCADA system, these modules run algorithms to compute whether there is any product missing from their pipeline system. If a leak is detected, the leak detection module can push data back into the SCADA system to alert the human operator.

There remains some debate as to which is the better approach: to build the leak detection module directly into the SCADA system, or have it as a separate add-on module supplied by a third-party vendor. The advantage of having it built in is a tightly integrated solution that should be seamless between SCADA and leak detection. However, the advantage of a third-party leak detection package is technical ability. It does stand to reason that the core strength of a SCADA company is its SCADA software and to a lesser degree the ability of its leak detection add-on, whereas a dedicated leak detection company will focus its efforts primarily on its leak detection software. As long as the SCADA system and third-party leak detection package can successfully exchange data, something more easily accomplished these days by OPC, it should be a powerful combination.

### 2.18.2 Batch Tracking

Batch tracking is an optional add-on module that is used for liquid product pipelines (as opposed to gas or crude oil pipelines). In total, there are maybe 100 different refined products that can be output from a refinery (e.g., gasoline, kerosene, jet A, jet B, and diesel). These different products follow one another in the same pipeline and thus it is necessary for the pipeline operator to know which batch is at which location in the pipeline, both for injecting additional products and for extracting products at delivery points.

Like the leak detection module example above, batch tracking can be built into the SCADA system or provided by a third-party vendor.

### 2.18.3 Dynamic Pipeline Highlight

Proper operation of the pipeline is, of course, vital to pipeline integrity, and any tool that can assist the proper operation or avoid improper operation is important to operational integrity. One feature that quality SCADA systems exhibit is "dynamic pipeline highlight." This is a feature that can visually indicate to the human SCADA operator whether any given section of pipeline is flowing or not flowing. In other words, the graphics on the screen that show the pipeline can use color or other highlighting to indicate whether any given section of pipe is safe or not, and it can do this in real time based on upstream valves being open or closed. Surprisingly, not all SCADA systems have this ability, yet it can be easily understood that this information is valuable for the daily operation of the pipeline and thus for ensuring its integrity.

## 2.19 COMMUNICATION MEDIA

While the RTUs gather data from a variety of instrumentation, these data must be communicated up to the SCADA server computers, and a variety of communication media can be employed to effect this communication. The selection of communication medium is influenced by factors such as purchase and installation price, operating price, reliability, availability, bandwidth, and, of course, geographical considerations. Ultimately, however, the greater the bandwidth, the more rapidly the SCADA database can be updated with real-time data. Any communication medium that can pass data more quickly will enhance safety and integrity. Another aspect is suitability for data transmission via serial or Ethernet methods. The most common communication media are Cat5 data cable, leased line, microwave, dial-up line, optical fiber, and satellite [19].

### 2.19.1 Cat5 Data Cable

This medium is for use only with Ethernet communications. As such, it is appropriate mostly in urban settings where the necessary infrastructure is in place. Cable lengths are limited to 100 m, but that is not to suggest that the RTUs must be within 100 m of the SCADA servers; rather, repeaters or other active equipment may be used if necessary to bring the data into a telco for transmission over arbitrary distances, and at rates of up to 1000 Mbps. Cat5 cable and the equipment needed to support it are relatively inexpensive but there are usually monthly fees associated with using them. In some areas of the world, these rates can be prohibitively expensive. In addition to very high bandwidth, Cat5 cable medium usually has a very low bit error rate and thus very high communications integrity.

### 2.19.2 Leased Line

A leased line is a pair of copper wires rented from a telco to carry serial data. Modems must be installed at both ends of the leased line to use it. Leased line is usually less expensive than Ethernet and offers lower bandwidth, but is still fit for transmitting RTU data. Communications integrity is usually very good with leased lines.

### 2.19.3 Microwave

Microwave infrastructure is a form of wireless radio data communications. Many factors come into play regarding the efficacy of microwave data transmission such as distance, repeaters, and geography. Even weather, which changes hour by hour, can affect wireless communications bit error rate, meaning microwave data rates are less consistent than other media. Depending on the amount of existing infrastructure, microwave implementations can vary dramatically regarding their installation price. For example, erecting a 50 m antenna mast is very expensive, whereas renting bandwidth on existing infrastructure can be very affordable. One advantage of microwave, of course, is that it remains a tangible option for sending data long distances over undeveloped terrain without existing infrastructure. Microwave offers an additional benefit of being flexible enough to transmit serial or Ethernet data.

### 2.19.4 Dial-Up Line

Very similar to leased line medium is serial communications over dial-up lines. Dial-up lines provide serial data transmission only at lower bandwidth, but the data rates are still suitable for most SCADA applications. Two of the greatest advantages of dial-up communications are the low installation price and low ongoing monthly fees. Dial-up communications also enjoy comparatively low bit error rates, making dial-up line a fairly robust choice to transmit SCADA data.

### 2.19.5 Optical Fiber

Fiber-optic communications offer the highest data rates of any existing communication medium, rates that are far greater than needed for passing field data up to the SCADA control room. Optical fiber is also an extremely reliable medium for data transmission. The reason that fiber-optic communication is often not used is the very high price of installing the fiber. Long-distance communications—especially through undeveloped areas—are prohibitively expensive, but if there is existing fiber that can be used then this option may be a good choice. Fiber is also a good choice for electric power companies, as it avoids the liability associated with copper spanning distances where a high-voltage ground fault can vaporize buried cable. Since optical fiber is made of glass and thus not electrically conductive, it is an excellent choice in areas that can experience ground faults.

### 2.19.6 Satellite

Sending data by satellite is a viable option in almost any area of the world, making it an attractive choice for remote locations. However, a satellite is among the least reliable choices for consistency of data transmission. In real-world use cases, satellite is not as reliable as other media. Depending on the volume of data being sent, satellite also has the potential to be expensive.

## 2.20 COMMUNICATIONS INFRASTRUCTURE

Depending on the communication medium employed for any given SCADA system, a collection of appropriate infrastructure equipment is required.

For those SCADA systems using Ethernet-based media between the SCADA servers and RTUs, Ethernet switches are needed with at least one port for each RTU, which could number up into the high hundreds for large systems. Some users design for two (or more) communication paths to each RTU, meaning even more Ethernet ports.

The communications infrastructure needed for serial-based communications can be somewhat more involved. For example, dial-up lines will need one serial port for each RTU (or additional ports for each RTU if redundant communication paths are desired). But computer servers do not have so many physical serial ports—usually just one or two; therefore, serial port expanders can be used. Such devices usually present a single Ethernet connection to the server on one side, and offer several serial ports on the other side (from 1 to 8, 16, 32, or more).

Once the necessary number of serial ports have been added to a serial communications design, the next consideration is a changeover switch, which at its most basic conceptual idea is a three-ended device with two serial port inputs and one serial port output. The idea is that the changeover switch will connect either the A input port or the B input port to the output serial port. The purpose for this switch is to route the serial data line from an RTU to either the primary SCADA server (A) or the backup SCADA server (B). This idea is then repeated for each serial communication line such that all RTU serial lines are connected through the A or B port, ultimately reaching either the primary or the backup SCADA server. In this way, if one SCADA server should fail, the RTU communication lines can be swung over to the other SCADA server to continue the essential service of retrieving RTU data. Due to the potentially large quantity of serial ports, a changeover switch should ideally be rack mounted with the ability to add "slide-in" modules allowing the changeover switch to be grown to the appropriate size with the ability to increase in size to account for future expansion.

Beyond the changeover switch are the modems. Each serial communication line will require a modem at the RTU end and at the SCADA control room end. Although it is quite common for these modems to be individual units, some companies provide modem solutions that are able to reduce the overall modem infrastructure. For example, a "6-pack modem" uses just one serial port from the SCADA servers to give access to six separate serial modem lines. While this type of solution offers great savings in terms of communications hardware and provides simplicity, it does imply a custom implementation where the modem vendor needs to make application-specific versions of the communications protocol software to drive the 6-pack modem.

Although the above paragraphs describe some common infrastructure for a dial-up serial implementation, similar hardware elements would be needed for a leased line implementation, while a fiber-optic or satellite implementation would deviate further.

Some communications hardware may be optional and not necessarily required for communications. For example, hardware encryptors can be added that can provide a layer of cyber security to the communication path. Or some radio communication networks can benefit from data repeaters to amplify the wireless data signal.

In contrast to the medium-dependent infrastructure, all SCADA implementations will have an Ethernet LAN for the SCADA servers, SCADA workstations, printers, and so on. In fact, the majority of SCADA control rooms use dual LANs for network redundancy. Single or dual, an Ethernet switch is the focal point of each LAN. In addition to these switches, most SCADA systems need to pass data up to business systems and to isolate the SCADA control room network from these business systems a firewall can be used. But use caution while putting too much trust in a firewall; indeed, it adds a measure of security to the mission-critical SCADA LAN, but additional measures should also be employed.

Quite often elements of the communications infrastructure hardware maintain statistics that can provide insight into the degree of success or indications of problems in transferring data from the RTU up into the SCADA servers. Since RTU communications are vital to a SCADA system's operation, and since SCADA plays a role with pipeline integrity, there should be someone assigned to periodically review these communication statistics.

## 2.21 COMMUNICATIONS INTEGRITY

One aspect of overall pipeline integrity is communications integrity, meaning that the operational data contained in the RTU/PLC database be accurately transmitted to the SCADA servers and vice versa. The inability to do this will surely compromise pipeline operations. To protect against this, several strategies are simultaneously employed.

One of the ideas used to improve communications integrity is that of cyclic redundancy checks (CRCs) or checksums on data packets. These are mathematically computed codes added by the sender of each message. These codes depend on the message contents. When each message is received, its code is recomputed and compared with the transmitted code. If they are identical, the message has been received intact; otherwise, it is known to be corrupted. The ability to use these codes depends very much on the computer protocol used between sender and receiver. CRCs and checksums are mostly used to protect against noise on a communication channel.

A layer of communications integrity that can be used in addition to CRCs or checksums is encryption. This is a method whereby the transmitting computer uses a nontrivial number sequence (a "key") to obfuscate the entire data message before it is sent. The receiving computer must have the same encryption key to recompose the message after it is received. Encryption enhances pipeline integrity by hiding details of the SCADA messages to prevent them from being reverse engineered when the communication protocol is known, something that is quite easy to do especially when using wireless data communications. Encryption must be supported by the computer protocol used between the SCADA servers and RTUs/PLCs.

Authentication is an additional method to protect pipeline integrity by using certificates between the two ends of any communication channel. These certificates ensure that messages are being exchanged between authorized devices. For example, in a non-authenticated scenario, it could be that an intruder is injecting protocol-compliant messages into the SCADA hosts or PLCs/RTUs and in doing so takes control of the infrastructure. While this threat is diminished by adding encryption, a capable hacker can also add matching encryption to his/her malicious data packets. But adding authentication will invalidate the data packets injected by a hacker. However, much like the strategies mentioned above, authentication depends very much on the computer protocol being used. Authentication is not compatible with many popular protocols (such as Modbus).

In closing, it should be understood that SCADA plays a vital role in pipeline integrity and SCADA is, at its root, all about sending data over communication lines. It follows, therefore, that pipeline integrity depends to some degree upon SCADA communications integrity.

## 2.22 RTUs AND PLCs

RTUs and PLCs are basically functionally equivalent. Both RTUs and PLCs connect to a collection of sensors and actuators, collect the raw signals, and convert those signals into engineering values. These engineering values are stored in a database, and when interrogated, these database values are transmitted to the SCADA host. Most SCADA systems utilize anywhere from one to several hundred RTUs or PLCs.

To enhance pipeline integrity and safety, some RTUs and PLCs can be supplied with redundant CPUs, or power supplies. RTUs are usually a better technical choice over PLCs, as they are designed for rugged outdoor environments and can withstand harsh conditions better than PLCs. Obviously, a failed PLC would compromise pipeline operations, so RTUs remain the best choice.

The choice of RTU also relates to the communication protocol used. A protocol that can support report by exception (RBE) is inherently more efficient as the RTU will transmit only new or modified data up to the SCADA host when interrogated. Doing so as compared with always sending all the RTU data, even if unchanged, will save bandwidth, which in turn will allow more real-time data to be brought to the human operators in the SCADA control room. Bandwidth saved is pipeline safety earned.

## 2.23 DATABASE

RTUs and PLCs maintain an internal database that represents the current value of all the process variables connected to them. These data are ultimately communicated up to the SCADA hosts and as these represents vital pipeline operational data, their integrity must be maintained by the RTU or PLC, and there are several methods to achieve this.

First, the CPU card of an RTU should have some method of keeping the RAM database alive temporarily if the power to the RTU is lost for any reason. Older RTUs and PLCs achieve this via a backup battery, but a better approach is a supercapacitor, the reason being that battery performance decays as the battery ages. A supercapacitor should be able to keep the RTU's RAM data alive for several days, ample time to repair or swap out a dead power supply or otherwise fix the power problem.

In addition to the live data maintained in RAM, the database configuration should be maintained in nonvolatile memory such as a flash ROM on the CPU card. Proper RTU maintenance practices would dictate that copies of the configuration database also be held in an offline repository such as on a laptop and removable media.

The better RTUs will also maintain a historical data buffer. This is used to store RTU data values during periods of SCADA communication outages. In such an event, the RTU will buffer new data values from the live field process variables and once communications to the SCADA hosts are repaired, it will relay the buffered data up to the SCADA hosts. Although this feature does not solve the problem of failed communications, it does allow the SCADA host to insert old data values into its permanent historical database. Although this important feature is available in better RTUs, the best ones will relay this old buffered data up to the SCADA servers as a lower priority task once communications are restored. By doing so, the RTU will first relay the critical live data values, and will report the older values as bandwidth allows at a lower priority.

However, keeping copies of the live runtime data and configuration data does not go far enough to ensure database integrity. The databases themselves should have a checksum or other validation code to ensure they are not corrupted.

## 2.24 USER-DEFINED PROGRAMS

Modern RTUs have the capability to add user-defined programs. This is a feature that allows the pipeline owner/operator to build arbitrary functions into the RTU based on any input trigger condition or schedule. Although proprietary methods do exist for some RTU vendors, there is an internationally recognized standard for doing this, which is IEC-61131. These user-defined programs can be anything from a simple sum of multiple flows to very complex behavior that can optimize operations in real time, or increase safety of pipeline operations, including pipeline integrity.

Some RTUs come preloaded from the factory with some user-defined programs to put the RTU into a "safe mode" if it should lose communications to the SCADA servers. The specific definition of what a "safe mode" is depends on different pipeline operators, but generally it means adjusting set points and other parameters away from maximum operational limits and toward moderate values without shutting down the pipeline. By moving the set points away from maximum operational limits during communication outages, it means that for periods of time that the human pipeline operators are unable to observe or affect the pipeline operations, the pipeline will automatically revert to a more conservative operating capacity, the specifics of which should be specified by each different pipeline owner/operator, and possibly customized further for each separate RTU. By doing so, pipeline integrity is automatically preserved even while live operational data are not available.

## 2.25 RTU/PLC INTEGRITY

Reliable data are vital to pipeline integrity and as such the equipment that generates and transmits operational data has a dramatic effect on pipeline integrity. RTUs and PLCs are the frontline equipment that provides these operational data and as such pipeline integrity is linked to the integrity of these devices.

RTUs and PLCs should be robust and capable of performing properly even in the event of physical upsets such as noncritical component failure. In other words, RTUs and PLCs should be "high-availability" devices. This is best achieved by making the critical components redundant. Some vendors offer redundant power supplies for their RTUs, or dual CPU cards, but the ultimate in high availability is entirely redundant RTUs, meaning that for each RTU needed, two are actually delivered. Doing so provides dual CPUs, dual power supplies, and dual I/O cards. Although very rare, there are vendors who can offer such RTUs and in doing so assist pipeline owners/operators achieve greater pipeline operational integrity.

Another feature of RTUs and PLCs that can contribute toward greater integrity is the ability to "hot swap" its components. This means, for example, that a PLC I/O card can be removed and replaced with another without the need to power down the entire PLC. Similarly for RTUs or PLCs that have dual CPU cards or dual power supplies, a failed component can be removed without shutting down the device while a replacement component is added.

## REFERENCES


1. Boyer, S.A. (2010) *SCADA Supervisory Control and Data Acquisition*, 4th ed., International Society of Automation (ISA), Research Triangle Park, NC.
2. Engineering Equipment Materials Users' Association (EEMUA) (2002) Process Plant Control Desks Utilising Human–Computer Interfaces: A Guide to Design, Operational and Human Interface Issues, Publication No. 201, EEMUA, London.
3. Bullemer, P.T., Reising, D.V., Errington, J., and Hadjukiewicz, J. (2004) Interaction requirements methods for effective operator interfaces. ASM® Consortium Technical Book, version 1.0.
4. Engineering Equipment Materials Users' Association (EEMUA) (2013) Alarm Systems: A Guide to Design, Management and Procurement, 3rd ed., Publication No. 191, EEMUA, London.
5. Department of Transportation (2009) 49 CFR Parts 192 and 195, Amended Regulations, Pipeline and Hazardous Materials Safety Administration (PHMSA), Department of Transportation, Washington, DC.
6. Endsley, M.R. (2012) Building situation awareness in oil & gas operations. Presentation at API Pipeline Conference and Cybernetics Symposium, April 2012.
7. Endsley, M.R. (1988) Situation awareness global assessment technique (SAGAT). Proceedings of the National Aerospace and Electronics Conference (NAECON), IEEE, New York, pp. 789–795.
8. Willowglen Systems (1990–2013) SCADACOM Reference Manuals, Willowglen Systems, Edmonton, Alberta.
9. ASM (2000) ASM Consortium Guidelines: Effective Operator Display Design, version 2.01.
10. Endsley, M. and Jones, D. (2012) *Designing for Situational Awareness: An Approach to User-Centered Design*, 2nd ed., CRC Press, Boca Raton, FL.
11. Errington, J. (2013) Human factors planning for control room management. Banff/2013 Pipeline Workshop, April 2013, Banff, Alberta.
12. Few, S. (2006) *Information Dashboard Design: The Effective Visual Communication of Data*, O'Reilly Media, Sebastopol, CA.
13. Mohammad, R. (2013) Control centre human factors graphics and interface design. Banff/2013 Pipeline Workshop, April 2013, Banff, Alberta.
14. American Petroleum Institute (API) (2007) Recommended Practice for Pipeline SCADA Displays, RP 1165 (R2012), 1st ed., API, Washington, DC.
15. ISA (2009) ANSI/ISA-18.2-2009: Management of Alarm Systems for the Process Industries.
16. American Petroleum Institute (API) (2010) Pipeline SCADA Alarm Management, RP 1167, 1st ed., API, Washington, DC.
17. American Petroleum Institute (API) (2008) Pipeline Control Room Management, RP 1168, 1st ed., API, Washington, DC.
18. American Petroleum Institute (API) (2009) Pipeline SCADA Security, STD 1164, API, Washington, DC.
19. Boyes, W. (2011) *Instrumentation Reference Book*, 4th ed., Butterworth-Heinemann.

# 3

# MATERIAL SELECTION FOR FRACTURE CONTROL*

William Tyson
*CanmetMATERIALS, Natural Resources Canada, Hamilton, Ontario, Canada*

## 3.1 OVERVIEW OF FRACTURE CONTROL

"Fracture control" has as its objective the prevention of leaks and ruptures caused by crack growth. There are many conditions that can lead to loss of containment, including various mechanisms for flaw initiation and growth such as fatigue and stress corrosion cracking (SCC), and material loss by corrosion and subsequent rupture. Failure by growth of a dominant crack is controlled by the applied crack driving force and the material resistance. Final failure may be by unstable growth of the crack or by overload tensile failure such as occurs to terminate a tensile test.

The important subject of prevention of failure by material loss through general corrosion warrants separate treatment, beyond the scope of this chapter. Much effort has been expended in deriving equations to predict the failure pressure of pipe as a function of the geometry of corroded areas. Much of this work is based on finite element analysis (FEA) parameterized in the form of equations for practical use, taking into account the geometry of the pipe and flaw and the tensile properties of the pipe steel. Popular references including practical procedures to characterize the shape of the corroded area and determine the maximum allowable operating pressure (although based on semiempirical relationships rather than FEA) are ASME B31G [1] and RSTRENG [2]; Zhu and Leis [3] have recently reviewed and evaluated the commonly used and three new proposed criteria. Similarly, SCC is a separate subject and will not be dealt with here. Practical approaches to SCC including prevention, investigation, management, and mitigation may be found in a document by CEPA (Canadian Energy Pipeline Association) [4]. Failure originating from dents and gouges, also a special topic, will also not be discussed; an extensive treatment of this subject can be found in Ref. [5] and an example of recent work in Ref. [6].

The scope of this chapter will be limited to fracture prevention by material selection and design, focusing on the general principles involved. Details of the steps to be followed to ensure fracture control can be found in the relevant sections of applicable pipeline standards, for example, CSA Z662, "Oil and Gas Pipeline Systems" [7].

Crack growth typically occurs through several stages. A crack may develop from an initial flaw in a weld (e.g., lack of fusion or hydrogen crack), by fatigue from a site of stress concentration (e.g., at the toe of a weld), or by environmental attack (e.g., stress corrosion cracking). When stress is applied, the flaw initially blunts by plastic flow at its tip. The region of intense plastic flow and associated material damage at the crack tip is called the "process zone." In the process zone, voids may grow in response to hydrostatic (triaxial) stresses and microcracks may form at hard spots. Eventually, the blunted crack tip sharpens by linking with damaged regions in the process zone and growth may occur by either ductile micromechanisms (microvoid coalescence) or brittle fracture (cleavage). Brittle fracture characteristically occurs suddenly, sometimes at relatively low applied stress



*Oil and Gas Pipelines: Integrity and Safety Handbook*, First Edition. Edited by R. Winston Revie.

when the bulk of the structure remains elastic, and can lead to a catastrophe. Ductile fracture occurs more gradually, generally requiring an increase in applied crack driving force to extend the crack.[1] However, rapid ductile propagation can occur if a point of instability is passed; this occurs when the rate of increase of material resistance with crack growth (the slope of the *R*-curve) is lower than the rate of increase of crack driving force.

Fracture control has traditionally been categorized as "initiation control" and "propagation control." The distinction is becoming somewhat blurred for modern steels that are designed for high strength and toughness. These materials generally show at least some "*R*-curve behavior" in which initiation is a gradual continuous process, and the early stages of growth could be termed "propagation," although this generally occurs in a controlled fashion. In this chapter, "initiation control" refers to design and material selection to limit the extension of a flaw to a small amount, such as 0.2 or 0.5 mm beyond the initial flaw size, and "propagation control" refers to prevention of fast-running fractures that can extend over many pipe sections. This distinction is related to the concept of "leak before break" or "leak versus rupture." The idea contained in these terms is that it is much better to design for containment of a failure within a short distance,[2] thereby limiting the rate of release of fluid, than for the possibility of a long-running crack that could be catastrophic. This is possible if the fluid pressure to cause loss of containment by a flaw of a given length breaking through the wall (producing a leak) is less than the pressure required to extend the flaw along the pipe (producing a break). This is a plausible approach, and early work was focused on achieving "leak before break" conditions. Much of this work was done at the Battelle Memorial Institute in Columbus, OH, where Maxey and his coworkers pioneered the application of fracture mechanics to pipeline fracture control. The seminal document in the field was presented at the Fifth Symposium on Line Pipe Research in 1974 [9].

## 3.2 TOUGHNESS REQUIREMENTS: INITIATION

The equation in Figure 3.1 is the classic "ln sec" relation relating failure stress to toughness and flaw size for through-thickness cracks. It was developed from the "strip yield" model of Dugdale–Bilby–Cottrell–Swinden described in textbooks on fracture mechanics (see, for example, [8]). It has the property of describing brittle fracture in the limit of low toughness (in which failure stress depends on toughness) and "flow-stress-dependent" fracture in the limit of high toughness (in which failure stress depends on material strength). The equation was modified so that it could be used for surface flaws in both toughness-dependent and flow-stress-dependent limits. To evaluate the material toughness, Maxey proposed that $K_c$ be related to the Charpy absorbed energy $C_v$ (the "Charpy" test is discussed in Section 3.4.1) through Equation 3.1:

$$\frac{K_c^2 \pi}{8c\bar{\sigma}^2} = \ln \sec \frac{\pi}{2}\left[\frac{M_T \sigma_T}{\bar{\sigma}}\right]$$

where

$\sigma_T$ = the hoop stress level at failure (namely, $PR/t$)
$P$ = the internal pressure level at failure
$\bar{\sigma}$ = the flow stress of the material (yield strength + 10,000 psi)
$M_T$ = the "Folias" correction, a function of $\frac{2c}{\sqrt{Rt}}$ as shown in Figure J-3
$R$ = the radius of the pipe
$t$ = the wall thickness
$2c$ = the length of the through-wall flaw
$K_c^2$ = a parameter related to the material's resistance to fracture

**FIGURE 3.1** The "ln sec" equation of Maxey and coworkers for a through-wall flaw. The "Folias" correction is $M_T \approx (1 + 1.255c^2/(Rt) - 0.0135c^4/(R^2t^2))^{1/2}$. (Adapted from Ref. [9]; Figure J-3 is in Ref. [9]. © Pipeline Research Council International, Inc.)

$$\frac{12C_v}{A_c} = \frac{K_c^2}{E} \tag{3.1}$$

where $C_v$ is the Charpy shelf energy in ft lb, $A_c$ is the area of fracture surface of a Charpy V-notch specimen (in.$^2$), $K_c$ is the toughness (psi in.$^{1/2}$), and $E$ is Young's modulus (psi). (The factor 12 is introduced simply to convert ft lb into in. lb for consistency of the units.)

Equation 3.1 was an important step forward, because it gave a simple method to estimate $K_c$ using a parameter (Charpy shelf energy) that is widely specified and used in the pipeline industry. In spite of the fact that although the test uses a relatively blunt notch and is performed in impact and is then applied to predict the behavior of sharp flaws under slow loading, the simple "ln sec" equation with Equation 3.1 for the toughness parameter has proven to be remarkably robust, at least for the relatively low-strength

[1] A plot of the driving force, characterized by the *J*-integral or CTOD, as a function of the crack growth is called a resistance curve, or *R*-curve. Discussions of the definition and significance of *J* and CTOD may be found in Ref. [8].

[2] This is particularly important for compressible fluids (e.g., gas or $CO_2$). A full-bore rupture may initially form, but may be contained within a short "propagation distance" (e.g., within the initiation joint, or traversing a few joints, determined by the distribution of propagation resistance within the pipes in the pipeline).

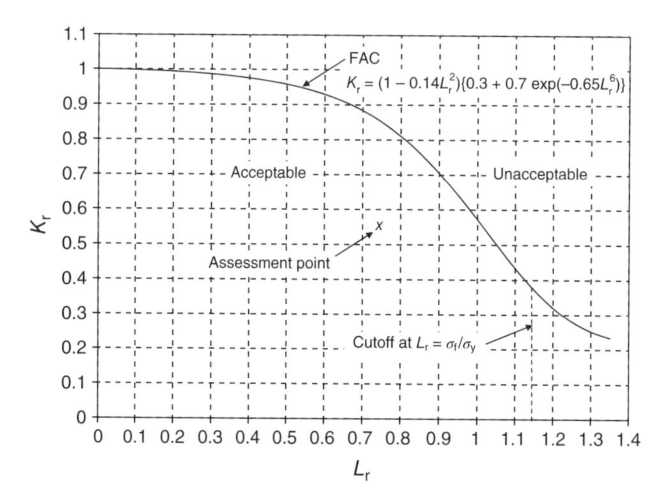


**FIGURE 3.2** Failure assessment diagram for stress-based design.[3]

and low-toughness steels available at the time Maxey carried out his work.

The "ln sec" equation demonstrates an early form of two-parameter fracture mechanics in which failure by crack growth is controlled by a toughness parameter $K_c$ and plastic collapse is controlled by a strength parameter $\bar{\sigma}$. In the toughness-dependent limit, the "ln sec" equation in Figure 3.1 reduces to $\sigma_T = K_c/(M_T\sqrt{(\pi c)})$. In fracture mechanics terms, this states that the crack driving force (stress intensity factor $K = M_T\sigma_T\sqrt{(\pi c)}$) is equal to the material toughness $K_c$. Similarly, in the flow-stress-dependent limit $M_T\sigma_T = \bar{\sigma}$, which states that the concentrated hoop stress $M_T\sigma_T$ is equal to the material flow stress $\bar{\sigma}$. Note that the "Folias factor" $M_T$, which increases with the variable $2c/\sqrt{(Rt)}$, acts as a stress concentration factor that amplifies the hoop stress by the action of the fluid pressure to cause bulging at the crack tips.

Similarly, current methodologies for "stress-based design" are based on two-parameter fracture mechanics. In stress-based design, structural integrity is ensured for applied hoop stresses that are designed to remain below the yield stress. This is achieved by ensuring that the driving force on the largest flaw in the pipe remains smaller than the material toughness, and that the stress on the net section containing the flaw remains smaller than the plastic collapse stress. The "failure assessment curve" (FAC) in the "failure assessment diagram" (FAD) shown[4] in Figure 3.2 is used in several structural integrity standards, notably the UK-based fitness-for-service standard BS 7910 [10] and API 579 [11]. Structural integrity is assured by confirming that the assessment point plotted at $K_r = K_e/K_c$ (the ratio of the applied elastic stress intensity factor to the toughness) and $L_r = \sigma/\sigma_{pc}$ (the ratio of the applied stress to the collapse stress) falls within the FAC. $K_r$ is often calculated as $(\delta_e/\delta)^{1/2}$, where $\delta$ is the crack tip opening displacement (CTOD), and $L_r$ is often calculated as $\sigma_{ref}/\sigma_y$, where $\sigma_{ref}$ is a reference stress and $\sigma_y$ is the yield stress. Note that the shape of the FAC reflects the fact that the elastic–plastic crack driving force increases with increasing stress more rapidly than the elastic crack driving force $K_e$ (i.e., $K_r$ falls below unity as $L_r$ increases), and that there is some load-bearing capacity above the cutoff at $L_r = \sigma_f/\sigma_y$, where $\sigma_f$ is the flow stress (the average of the yield and ultimate strengths) so that the FAC does not drop to zero above the cutoff. In practice, the FAD is often used to generate curves of allowable flaw size for a given design stress, displayed as length versus height, which can be used to assess detected flaws. It must be remembered in this case, however, that adequate safety factors must be applied.

It is sometimes necessary to construct pipelines through unstable terrain where there is risk of substantial deformation of the pipe because of soil movement. This could subject the pipe to stresses above the yield strength. To deal with such situations, "strain-based design" (SBD) is being actively developed. Obviously, SBD requires exceptional toughness to withstand crack growth in the plastic strain field at large stresses. The fundamental principle, as in stress-based design, is to ensure that any crack growth is within acceptable limits and that the load-bearing capacity of the net section is not exceeded. To achieve this objective, extensive FEA calculations have been performed to enable estimation of the crack driving force as a function of flaw geometry, material properties, and applied strain. Examples of the approach may be found in Annex C of CSA Z662 [7] and in DNV RP-F108 [12]. The latest edition of the R6 fitness-for-purpose code [13] also contains an SBD procedure, formulated in a similar fashion to the FAD. It is also necessary to measure or estimate the material resistance (toughness) under the constraints appropriate to field conditions. Conventional toughness tests subject the material to the highest possible constraint to generate a conservative result, but such conservatism may make SBD uneconomic. In response, test methods to generate more appropriate toughness measures are being developed. This will be discussed in Section 3.4.2.

## 3.3 TOUGHNESS REQUIREMENTS: PROPAGATION

If an axial crack begins to run in a pipeline, it could travel a considerable distance if the toughness is not sufficient to arrest it. This is of greater concern for pipelines carrying gas

[3] CMAT, Proposed revision to CSA Z662 Annexes J and K to enable ECA by FAD, unpublished work, CanmetMATERIALS, Hamilton, June 2011.

[4] The FAC in Figure 3.2 is used in the so-called level 2 and is normally the default level because it is not material specific. At the cost of requiring more details on material properties, a "level 3" approach is available that enables a more accurate assessment.

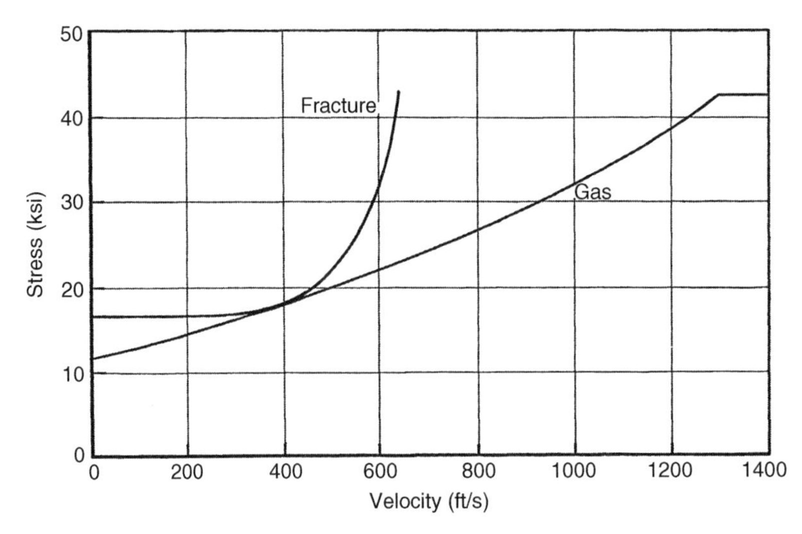


**FIGURE 3.3** Stress required to propagate a crack through pipe steel ("fracture") and that provided by the gas pressure ("gas") as a function of velocity of the crack and of the escaping gas. (Adapted from Ref. [9]. © Pipeline Research Council International, Inc.)

rather than oil because the latter has low compressibility and pressure is relieved rapidly at a breach. Brittle fracture is the most dangerous mode of this type, and pipeline standards contain provisions to guard against brittle fracture in the form of requirements that the fracture appearance of full-scale specimens of pipe fractured in impact (drop weight tear test (DWTT), see Section 3.4.1) should be ductile. However, even if the fracture is 100% ductile, it is still possible for fast-running fractures to occur. Estimation of the toughness required to arrest a running crack is an important consideration in pipeline design.

The pioneering work of the Battelle Columbus labs in this area has stood the test of time and attests to the insightful engineering intuition of Maxey and his coworkers. Figure 3.3, adapted from the NG18 report mentioned earlier [9], shows the main features of what is now referred to as the "Battelle two-curve model" (BTCM, or TCM). The velocity of the running crack is determined by a balance between the resistance (the "fracture" curve) and the driving force (the "gas" curve). The "fracture" curve is a property of the material, and moves up or down on the stress axis for steels of higher or lower toughness. This curve shows that the stress for fracture rises as the crack velocity increases. The curve rises sharply when the inertial forces needed to accelerate the material ahead of the crack in the process zone from rest to the displacement required for fracture become large. This typically occurs, for pipe steel, at a velocity of several hundred m/s (recall that 1 ft = 0.305 m and 1 ksi = 6.895 MPa). (In comparison, brittle fracture by cleavage, requiring only small displacements in the process zone, can occur at several thousand m/s.) As the crack propagates, the stress driving the crack drops from the operating hoop stress level before fracture to a value dependent on the speed of the gas escaping from the open rupture. As understood intuitively, the velocity of the escaping gas decreases as the pressure decreases. The velocity of a decompression wave depends on the gas density, which in turn depends on the gas pressure (that can be converted directly into the hoop stress on the pipe), and the BTCM models the rupture process as a "race" between the fracture front and the decompression wave. If the toughness is low, the "fracture" curve may fall below the "gas" curve and consequently intersect it at two points. The lower stress point is unstable because for an increase in velocity the gas pressure rises faster than the resistance. Following the same reasoning, the higher stress intersection is stable, and hence the crack can run freely. However, if the toughness is high enough, the resistance lies above the driving force curve for all velocities and the crack cannot propagate. Estimation of the toughness required for arrest has been the objective of a number of programs throughout the world [14].

Maxey, based on his model for initiation (Figure 3.1) and the relation between Charpy absorbed energy $C_v$ and toughness, proposed that the arrest toughness would correspond to a certain value of $C_v$. This value would depend on the gas characteristics and pipeline design parameters (pressure, diameter, etc.). The shape of the so-called *J*-curve (the "fracture" curve in Figure 3.3, not to be confused with the *J*-integral of Section 3.4.2) was semiempirically determined to rise in proportion to the sixth power of the velocity. The "gas" curve was developed using a model for isentropic (adiabatic) expansion of an ideal gas. Based on the knowledge then available of these characteristics, computer models were developed and run to generate toughness values required for arrest in pipe carrying "lean gas" (essentially methane) using pipe steel current at the time (X52 to X65, with CVN absorbed energy $C_v$ of 100 J or less). The critical arrest toughness for buried pipelines could be fitted to the following equation:

$$C_v = 0.0873\sigma_H^2(Rt)^{1/3}A_c \quad (3.2)$$

where the units are $C_v$ (ft lb), $\sigma_H$ (ksi), $R$ and $t$ (in.), and $A_c$ (in.$^2$). This formula has demonstrated its utility for materials within the limits for which it was calibrated. Other groups around the world have proposed semiempirical formulas of the same general form. For example, in CSA ([7], Clause 5.2.2.3) an equation that can be used for buried pipelines and lean gas is given by

$$C_V = 0.00036S^{1.5}D^{0.5} \quad (3.3)$$

where $C_v$ is the full-size CVN absorbed energy (J), $S$ is the operating stress or gas pressure test hoop stress (MPa), and $D$ is the pipe diameter (mm). This relation is the "AISI" (American Iron and Steel Institute) formula, one of several similar equations developed from numerous burst tests carried out in the 1960s and 1970s. The broad similarity between Equations 3.2 and 3.3 is evident. It should be noted, however, that these equations have limited scope and should not be used for materials and fluids outside of the range for which the equations were calibrated. For example, in many current projects, the fluid being transported is rich gas with characteristics more complicated than the ideal gas depicted in Figure 3.3, and the "gas" curve is therefore different and is often estimated using GASDECOM (see [15]).

Full-scale burst testing is still necessary today to demonstrate inherent crack arrestability. New steels with improved strength and toughness are continually being developed and used at higher stress (pressure) levels, and it has been found that the conventional formulas such as Equations 3.2 and 3.3 are progressively nonconservative as pipe grades and gas pressures increase. "Fudge factors" to adjust (increase) the required Charpy energies have been used successfully for grades up to X80, for lack of a better validated approach to toughness measurement. The Charpy test is convenient and familiar, has a well-earned role as a means to demonstrate fracture resistance, and will continue to be used in specifications. However, it is limited in several respects. The thickness of a Charpy specimen is frequently less than the pipe wall thickness and so does not develop the full constraint of a crack in a pipe. Also, for high-toughness steels much of the Charpy energy is absorbed in crack initiation rather than propagation. Indeed, for some new steels the specimen bends and does not break completely during the test, raising serious doubts about whether the test characterizes crack propagation meaningfully. The short crack path in the Charpy specimen does not allow the crack to develop fully into the form it would have in a full-scale pipe, implying that even if the propagation energy is extracted from the Charpy test, it does not characterize the same morphology as a full-scale running fracture. For these reasons, alternatives to the Charpy test have been intensively developed over the past few decades. In particular, the DWTT [16] has been studied because the DWTT specimen is of full pipe wall thickness and has a crack path long enough to enable development of the same morphology as in a full-scale pipe. Also, to focus on crack propagation rather than initiation, the crack tip opening angle (CTOA) is being studied as a parameter that is much more appropriate for characterizing the propagation toughness compared with the Charpy absorbed energy. However, CTOA methods have not yet been developed sufficiently to confidently predict the arrest toughness for the full range of modern pipelines, hence the need for more full-scale burst tests. There are other complicating factors as well that must be taken into account, such as the type of backfill used to bury a pipeline. The evolution of ductile fracture control methods since the introduction of the BTCM has been well described by Zhu and Leis [17] and Zhu [18].

## 3.4 TOUGHNESS MEASUREMENT

"Toughness" may be broadly defined as the ability of a material to absorb energy during fracture. Good toughness is an essential component of fracture control to ensure that should failure occur it will be progressive and not catastrophic.

"Notch toughness" is the most familiar characterization, and refers to the energy-absorbing capacity of a specimen with a notch. It is commonly measured using the Charpy test (see Section 3.4.1) as the absorbed energy $C_v$. Its major use is to identify materials that are susceptible to brittle (low-energy-absorbing) fracture through characterization of a brittle-to-ductile transition temperature. Early fracture control plans, a typical example being the methodology developed by Pellini and his coworkers at the Naval Research Labs in the United States, were designed to avoid brittle fracture by ensuring that the critical parts of the structure of interest operate above the transition temperature. Although this approach has been largely superseded by quantitative fracture mechanics, the reality of the ductile-to-brittle transition phenomenon in structural steels and the practical utility of the notch toughness test will ensure the continued use of the Charpy test. For example, standards for line pipe (e.g., CSA Z245.1 [19]) routinely specify notch toughness as a requirement. Another important characteristic of Charpy data is the "upper shelf" absorbed energy, used to characterize the resistance to ductile fracture, although caution is required in interpreting this value for high-toughness steels for which the Charpy specimens may not actually break in two.

"Fracture toughness" characterizes the resistance to propagation of a sharp crack. It can be quantified in a number of ways. The earliest to be used was the "plane strain fracture toughness" $K_{Ic}$ (see test standard ASTM E399 [20]), defined as the stress intensity factor at fracture initiation and primarily elastic conditions, that is, where the plastic zone at the crack tip is much smaller than the specimen size. However, if fracture does occur under such "small-scale yielding" conditions, the material would be completely unsuitable for a pipeline, and so E399 is irrelevant for pipe steel. Rather, "elastic–plastic fracture mechanics" has evolved to address the need for toughness parameters and tests involving substantial plasticity. The associated parameters are the *J*-integral and the CTOD, defined and discussed at length in standard textbooks (see, for example, [8]). The CTOA is another elastic–plastic parameter to characterize propagating cracks, and has been standardized for sheet material by ASTM [21]. Practical methods to measure CTOA for pipe steels are under active development.

### 3.4.1 Toughness Measurement: Impact Tests

The Charpy test (ASTM E23 [22]) was standardized in the early 1900s. The full-size Charpy specimen is a small bar, dimensions $55 \times 10 \times 10\ mm^3$, with a 2 mm deep notch in the center. The specimen is fractured in impact, usually in a pendulum machine, and the absorbed energy $C_v$ is measured. Line pipe specifications normally require a certain absorbed energy at a specified temperature related to the pipeline design temperature. The test is often used to identify a ductile-to-brittle transition temperature (DBTT), defined as the temperature at which the value of $C_v$ is equal to some predetermined value in the transition range or the temperature at which the cleavage portion of the fracture surface reaches a specified fraction (often 50%, but in the pipeline industry 85% is required for DWTT tests). Typical CVN transition curves are shown in Figure 3.4 for a high-strength pipe steel and its weldment.

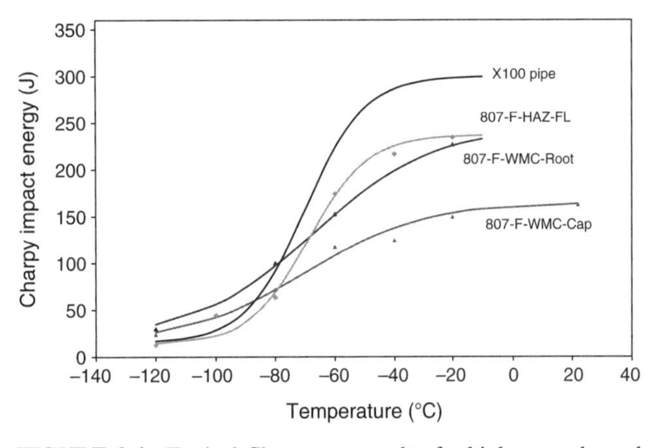


**FIGURE 3.4** Typical Charpy test results, for high-strength steel X100 and weldment 807-F, showing results for pipe (base metal), HAZ-FL (heat-affected zone, notch centered on the fusion line), and weld metal WMC (weld metal centerline: specimen extracted in either root region or cap region). (Adapted from Ref. [23].)

The DWTT (API RP 5L3 [24] or ASTM E436 [16]) uses a significantly larger specimen of full pipe wall thickness and in-plane dimensions of $3 \times 12\ in.^2$ ($76 \times 305\ mm^2$). The specimen is placed in either a drop tower or pendulum machine and impacted at a velocity no less than 4.88 m/s. The shear area fraction is measured on the fracture surface, and should be at least 85% to ensure that a fracture running in the pipe will propagate primarily by a ductile mode. With the increasing sophistication of mill test labs, there is growing interest in instrumenting the test to enable measurement of absorbed energy and its components (initiation and propagation) and the steady-state CTOA, using DWTT specimens. This should give a much more realistic quantification of the full-scale propagation resistance than can be obtained using Charpy specimens.

### 3.4.2 Toughness Measurement: *J*, CTOD, and CTOA

Fracture mechanics has provided quantitative tools to describe crack phenomena. The practical importance of this for the design of pipelines is that the significance of flaws, notably imperfections in girth welds, can be quantitatively assessed by engineering critical assessment (ECA) using methods such as those described in Section 3.2. Also, steady advances are being made in the description of fast ductile fracture characterized by the CTOA.

Flaws can grow by micromechanisms that may be either brittle (cleavage or intergranular fracture) or ductile (microvoid coalescence). The latter is clearly preferable, as brittle fracture can be catastrophic. However, even ductile fracture can lead to unstable propagation, and it is important to characterize the resistance to fracture by either the *J*-integral or CTOD. Either of these parameters is sufficient to quantify the driving force. The CTOD was championed by researchers in the United Kingdom, and the British standard [25] remains in popular use in the pipeline industry. ASTM was the first to standardize *J*-integral testing, and ASTM E1820 [26] remains preeminent in many industries in North America. Increasing globalization has led to pressures to consolidate ("harmonize") standards, and the International Organization for Standardization now has toughness test standards for homogeneous materials [27] and for welds [28].

The basic methodology in elastic–plastic fracture mechanics testing involves loading a precracked standard specimen, the most popular in the pipeline industry being a three-point bend bar, and measuring the crack size as a function of the applied force. In the most commonly used procedure, the specimen is instrumented to measure the crack mouth opening displacement (CMOD), which enables calculation of the *J*-integral and CTOD using equations developed with the use of FEA. The specimen is periodically subjected to unload–load cycles to measure the CMOD compliance and thereby calculate the crack size from established relations between the compliance and the crack size. In earlier standards, the CTOD was estimated using a model in which the specimen was idealized as rotating about a plastic hinge in the ligament, but the current trend is to evaluate CTOD directly from the $J$-integral using correlations developed using FEA. Unfortunately, the two methods do not agree in all cases, and this raises the issue of the use of CTOD-from-$J$ in ECA procedures that were validated using CTOD toughness measured with the older standards. If the older methods gave lower values of CTOD, the ECA procedure could be non-conservative with the use of current measures of CTOD. This issue remains to be resolved.

A three-point bend bar generates significant triaxial stresses (i.e., high constraint) at the crack tip, and intentionally provides lower bound estimates of the toughness. With current trends to use higher strength steels designed to withstand plastic strains in service, the lower bound estimates may be excessively conservative. It is important to provide toughness data that are relevant to the service conditions. Pipe steel is relatively thin-walled compared with other engineering structures. It is difficult to generate high triaxiality in thin structures, especially if plastic yielding occurs, and so the constraint is significantly lower (and the toughness higher) for cracks in pipe than for cracks in highly constrained test specimens. The low-constraint toughness can be higher than the high-constraint toughness by a factor of the order of 2 for shallow cracks. Consequently, there has been considerable interest in developing a low-constraint test that simulates service conditions more closely. Det Norske Veritas (DNV) was the first to standardize such a test [12] using a single-edge-notched tensile (SE(T) or SENT) specimen loaded in tension. The DNV test requires at least six specimens to develop an $R$-curve, and to simplify the procedure efforts are under way to standardize a single-specimen test. Some typical $J$–$R$ curves measured using a single-specimen test procedure are shown in Figure 3.5.

For resistance to fast fracture, the first objective is to ensure that the fracture occurs by a ductile mechanism. Current line pipe standards specify that the fracture mode should be primarily shear in full-wall-thickness specimens tested in impact using the DWTT [16,24]. The second objective is to ensure an adequate level of toughness. As related in Section 3.3, current standards characterize the energy-absorbing capacity of a pipe steel using the Charpy test. This is well known to overestimate the resistance for high-strength steels, and as noted in Section 3.4.1, current efforts are under way to measure the CTOA as a suitable alternative parameter. ASTM E2472 [21] provides a standardized procedure for quasistatically loaded sheet material; however, for pipe applications, considerably thicker material must be tested in impact. A round robin managed by CanmetMATERIALS is now under way using a proposed procedure based on a simplified single-specimen method [30] to assess its practicality with the intent of developing a standard for CTOA testing of pipe steel.

## 3.5 CURRENT STATUS

Fracture control for stress-based design (with operating stresses below the yield stress) is well understood. Pipeline standards include procedures to guard against rupture at locations where the pipe wall has thinned owing to general corrosion, and methods are available including inspection and remediation to combat stress corrosion cracking. Fracture mechanics methods have been developed to assess fitness for service in the presence of flaws, enabling evaluation of weld imperfections. Suitable test standards to measure material toughness are in general usage.

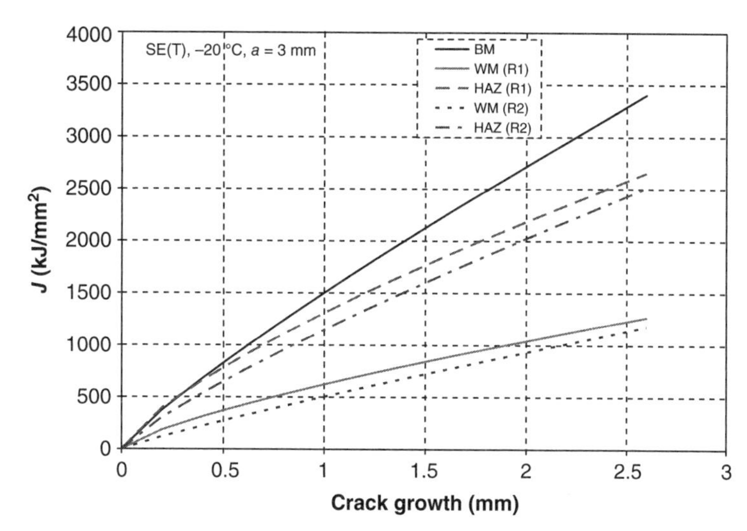


**FIGURE 3.5** Typical low-constraint SE(T) $J$–$R$ curves for a high-strength steel weldment: BM (base metal), WM (weld metal), and HAZ (heat-affected zone), measured using shallow-cracked ($a/W = 0.17$) single-edge tension specimens at −20 °C: R1 and R2 denote two rounds of nominally identical X100 steels and welds. (Adapted from Ref. [29].)

The primary objective incorporated in existing standards is the avoidance of brittle fracture. Although modern steels are being produced with increasing strength and design stresses are increasing accordingly, these steels can also be made with excellent toughness and so brittle fracture in the pipe body is normally not a concern. However, during welding the carefully designed fine-grained microstructure is changed and some regions of the HAZ can have a coarse-grained structure of low toughness. Care must be taken to ensure that the toughness properties of the weld are well characterized and that flaw acceptance standards recognize the possible presence of low-toughness regions in welds.

Prevention of fracture propagation in the form of fast-running ductile rupture can be achieved automatically for oil lines (because of the low compressibility of oil) and for gas lines using the two-curve method for "normal" pipeline designs using steels of modest strength and toughness and lean gas. However, for demanding situations of highly stressed high-strength pipe and possibly rich gas, traditional methods have been shown to be inadequate and special measures must be taken to guard against running ductile fracture. Methods of design and material characterization to achieve this are under development.

The state of the art in practical approaches to fracture control has been well summarized by Widenmaier and Rothwell [31].

## REFERENCES


1. American Society of Mechanical Engineers (ASME) (2012) ASME B31G: Manual for Determining the Remaining Strength of Corroded Pipelines, American Society of Mechanical Engineers, New York.
2. Kiefner & Associates, Inc. (1989) RSTRENG: A Modified Criterion for Evaluating the Remaining Strength of Corroded Pipe, Kiefner & Associates, Inc., Report to PRCI, Catalog Number L51688B.
3. Zhu, X.-K. and Leis, B.N. (2012) Assessment criteria and burst pressure prediction for pipelines with long blunt defects. Proceedings of the 9th International Pipeline Conference (IPC 2012), September 24–28, 2012, Calgary, Alberta, Canada, Paper IPC2012-90625.
4. Canadian Energy Pipeline Association (CEPA) (2007) Stress Corrosion Cracking Recommended Practices, 2nd Edition, Canadian Energy Pipeline Association, Calgary, Alberta, Canada.
5. Mechanical Damage, Final report by Michael Baker Jr., Inc. in association with Kiefner & Associates, CC Technologies, C-FER Technologies, and Macaw Engineering to U.S. Department of Transportation, Pipeline and Hazardous Materials Safety Administration, Office of Pipeline Safety, Integrity Management Program, Delivery Order DTRS56-02-D-70036, April 2009. Available at http://primis.phmsa.dot.gov/gasimp/docs/Mechanical_Damage_Final_Report.pdf.
6. Zarea, M., Batisse, R., Leis, B., Cardin, P., and Vignal, G. (2012) Full scale experimental database of dent and gouge defects to improve burst and fatigue strength models of pipelines. Proceedings of the 9th International Pipeline Conference (IPC 2012), September 2012, Calgary, Alberta, Canada, Paper 90620.
7. Canadian Standards Association (2011) CSA Z662: Oil and Gas Pipeline Systems, Canadian Standards Association, Mississauga, Ontario, Canada.
8. Anderson, T.L. (2005) *Fracture Mechanics: Fundamentals and Applications*, 3rd ed., CRC Press, Boca Raton, FL.
9. Maxey, W.A. (1974) Fracture initiation, propagation and arrest. Fifth Symposium on Line Pipe Research, NG18, November 1974, Houston, TX, PRCI Catalogue No. L30174e, reprinted in IPC2012. The full copy of the report can be found at www.prci.org/. PRCI bears no responsibility for the interpretation, utilization, or implementation of any information contained in the report.
10. British Standards Institution (2013) BS 7910: Guide on Methods for Assessing the Acceptability of Flaws in Metallic Structures, British Standards Institution, London, UK.
11. American Petroleum Institute (2007) API RP 579-1/ASME FFS-1: Fitness for Service, 2nd ed., American Petroleum Institute, Washington, DC.
12. Det Norske Veritas (DNV) (2006) Fracture Control for Pipeline Installation Methods Introducing Cyclic Plastic Strain, Recommended Practice RP-F108, DNV, Høvik, Norway.
13. EDF Energy Nuclear Generation Ltd. (2013) R6: Assessment of the Integrity of Structures Containing Defects, EDF Energy Nuclear Generation Ltd., Gloucester, UK.
14. Widenmaier, K.A. and Rothwell, A.B. (2008) The practical application of fracture control to natural gas pipelines. Proceedings of the 7th International Pipeline Conference, September–October 2008, Calgary, Alberta, Canada, Paper IPC2008-64557.
15. Eiber, R.J., Bubenik, T.A., and Maxey, W.A. (1993) Fracture Control Technology for Natural Gas Pipelines, PRCI Catalog No. L51691.
16. ASTM International (2014) ASTM E436: Standard Test Method for Drop-Weight Tear Tests of Ferritic Steels, ASTM International, West Conshohocken, PA.
17. Zhu, X.-K. and Leis, B.N. (2012) CVN and DWTT energy methods for determining fracture arrest toughness of high strength pipeline steels. Proceedings of the 9th International Pipeline Conference (IPC 2012), September 2012, Calgary, Alberta, Canada, Paper IPC2012-90624.
18. Zhu, X.-K. (2013) Existing methods in ductile fracture propagation control for high-strength gas transmission pipelines. Proceedings of ASME Pressure Vessels & Piping Division Conference (PVP 2013), July 2013, Paris, France, Paper PVP2013-98052.
19. Canadian Standards Association (2014) CSA Z245.1: Steel Pipe, Canadian Standards Association, Mississauga, Ontario, Canada.
20. ASTM International (2012) ASTM E399: Standard Test Method for Linear-Elastic Plane-Strain Fracture Toughness $K_{Ic}$ of Metallic Materials, ASTM International, West Conshohocken, PA.

21. ASTM International (2012) ASTM E2472: Standard Test Method for Determination of Resistance to Stable Crack Extension Under Low-Constraint Conditions, ASTM International, West Conshohocken, PA.
22. ASTM International (2012) ASTM E23: Standard Test Methods for Notched Bar Impact Testing of Metallic Materials, ASTM International, West Conshohocken, PA.
23. Gianetto, J.A., Tyson, W.R., Park, D.Y., Shen, G., Lucon, E., Weeks, T.S., Quintana, M.A., Rajan, V.B., and Wang, Y.-Y. Small-Scale Tensile, Charpy V-Notch, and Fracture Toughness Tests, CANMET Report No. 2011-07(TR).
24. American Petroleum Institute (2008) API RP 5L3: Recommended Practice for Conducting Drop-Weight Tear Tests on Line Pipe, 3rd ed., American Petroleum Institute, Washington, DC.
25. British Standards Institution (1991 to 2005) BS 7448: Fracture Mechanics Toughness Tests, (published in four parts) British Standards Institution, London, UK.
26. ASTM International (2013) ASTM E1820: Standard Test Method for Measurement of Fracture Toughness, ASTM International, West Conshohocken, PA.
27. International Organization for Standardization (2002) ISO 12135: Metallic materials—Unified method of test for the determination of quasistatic fracture toughness, International Organization for Standardization, Geneva, Switzerland.
28. International Organization for Standardization (2010) ISO 15653: Metallic materials—Method of test for the determination of quasistatic fracture toughness of welds, International Organization for Standardization, Geneva, Switzerland.
29. Park, D.-Y., Tyson, W.R., Gianetto, J.A., Shen, G., and Eagleson, R.S. (2012) Fracture toughness of X100 pipe girth welds using SE(T) and SE(B) tests. Proceedings of the 9th International Pipeline Conference (IPC 2012), September 2012, Calgary, Alberta, Canada, Paper IPC2012-90289.
30. Xu, S., Bouchard, R., and Tyson, W.R. (2007) Simplified single-specimen method for evaluating CTOA. *Engineering Fracture Mechanics*, 74, 2459–2464.
31. Widenmaier, K.A. and Rothwell, A.B. (2008) The practical application of fracture control to natural gas pipelines. Proceedings of the 7th International Pipeline Conference (IPC2008), Calgary, Alberta, Canada, paper IPC2008-64557.

# 4

# STRAIN-BASED DESIGN OF PIPELINES

Nader Yoosef-Ghodsi
*C-FER Technologies, Edmonton, Alberta, Canada*

## 4.1 INTRODUCTION AND BASIC CONCEPTS

### 4.1.1 Overview of Strain-Based Design

Historically, stress-based methods have been used for designing pipelines. In a traditional stress-based pipeline design, the applied stress is kept below a prescribed limiting stress, typically the specified minimum yield stress (SMYS), by an amount dictated by the adopted safety factor. Since the SMYS for line pipe is typically defined as the stress measured at 0.5% total strain, stress-based design generally limits the longitudinal strain to a value less than 0.5%. In contrast, strain-based design of pipelines aims at maintaining pipeline integrity and serviceability under larger longitudinal strain levels (typically greater than 0.5%), by recognizing and taking advantage of steel's ability to deform plastically without loss of integrity. Large longitudinal strain levels can result from installation practices or from deformations induced by ground movement. For onshore pipelines, high strain levels generally arise when the line is in service due to seismic activity (including fault movement and slope liquefaction[1]), slope instability, frost heave,[2] thaw settlement,[3] or mine subsidence.[4] For offshore pipelines, high strain levels more commonly arise during pipe laying, but in-service events such as ice scour[5] can also produce high strain levels.

Strain-based design involves the consideration of both strain demand (the strain applied to the pipe by its surrounding environment) and strain capacity (the strain limit beyond which the pipe would experience a drop in load-carrying ability and/or material failure). The two basic limiting conditions for strain-based design are tensile rupture and compressive local buckling. In this context, a strain-based approach requires that both tensile and compressive strain demands be kept below the tensile and compressive strain capacities, respectively.

Strain-based methods may be applied in two different situations: first, at the design stage where displacement-controlled loading events are anticipated; and second, when an existing pipeline is subjected to (or anticipated to be subjected to) displacement-controlled loading not expected or accounted for at the design stage. Strain-based assessment of an existing pipeline can serve to demonstrate continued safety of the pipeline in operation or inform decisions on integrity maintenance measures required to meet safety criteria.[6]

Several pipeline codes including DNV-OS-F101, CSA Z662, ASME B31.8, ABS, API RP 1111, BS 8010 Parts 1

[1] Liquefaction is the loss of soil strength and stiffness under an applied stress, which is mainly caused by seismic activity.
[2] Frost heave is an upward ground movement due to the growth of ice lenses around the pipeline. This freezing process is caused by sufficiently low temperatures of the transported oil or gas.
[3] Thaw settlement is a downward ground movement due to the melting of the ice-rich soil near the pipeline. This melting process is caused by sufficiently high temperatures of the transported oil or gas.
[4] Mining subsidence is a ground movement caused by collapse of underlying mines.
[5] Ice scour is the impact of a pipeline by the passage of keels of floating ice.
[6] The term "strain-based analysis" has been suggested by Law [1] as an alternative to "strain-based design and assessment."

*Oil and Gas Pipelines: Integrity and Safety Handbook,* First Edition. Edited by R. Winston Revie.

and 3, ISO 16708:2006, and the Dutch standard NEN 3650 recognize strain-based design as an alternative to stress-based design. While some of these codes such as the DNV and CSA codes set out detailed requirements and provisions for the use of strain-based methods, other codes such as B31.8 do not provide explicit provisions related to strain-based design.

### 4.1.2 Deterministic versus Probabilistic Design Methods

The primary goal of structural design and assessment procedures is to achieve an economical design with an adequate level of safety. Modern structural design and assessment methods can be categorized as either deterministic or probabilistic. Deterministic methods generally employ design checks containing one or more explicit safety factors and a number of implicit measures to achieve the intended level of safety.[7] The implicit measures typically include the use of conservative estimates of material strength parameters, and analysis methods that give lower bound estimates of failure loads. The probabilistic (or reliability-based[8]) methods treat the load effects and structural resistances as uncertain quantities, which are explicitly characterized by probability distributions. Probabilistic methods generally yield an estimate of the probability of component failure, and the design and/or assessment approach involves ensuring that the calculated probability levels do not exceed a prescribed threshold.[9]

By explicitly considering analysis uncertainties on a case-by-case basis, probabilistic methods have the potential to achieve greater consistency in operating safety levels. It is noted, however, that probabilistic analysis methods can and have been used to assess and calibrate the safety factors for deterministic methods in some design codes. In this way, some of the advantages of a probabilistic approach are reflected in some deterministic methods.

### 4.1.3 Limit States

A limit state can be defined as a state beyond which a structure no longer satisfies a particular design requirement. There are two basic limit state categories that are typically addressed in civil engineering codes:

1. Ultimate limit states (ULS) pertain to loss of the primary structural function. They often refer to loss of strength or stability and may have adverse safety and environmental consequences. Burst and tensile rupture are prime examples of ultimate limit states for pipelines.
2. Serviceability limit states (SLS) pertain to the ability of the system to meet its functional requirements. They usually refer to excessive deformations that affect functionality without jeopardizing the structural integrity, and thus do not lead to safety or environmental impacts. Ovalization and local buckling are two examples of serviceability limit states for pipelines.

Some pipeline codes define other limit state categories that overlap the ULS category. For instance, DNV [3] and ISO [4] define fatigue limit states (FLS) and accidental limit states (ALS) as separate categories. Fatigue limit states relate to failures resulting from cyclic loading, and accidental limit states address severe, rare accidental loading events such as fires or falling objects. Furthermore, Annex O of CSA Z662 [2] defines leakage limit states (LLS) as small leaks that do not lead to significant safety consequences, and reserve the ULS category for rupture failures only.

Local buckling is typically considered to be a serviceability limit state. However, if local buckling progresses to the extent that it results in almost full blockage of the flow or excessive axial or hoop tensile strains, or if the buckle is subject to significant cyclic axial loading (i.e., ratcheting), it should be reclassified as an ultimate limit state.

A comprehensive design and/or assessment approach involves the consideration of all applicable limit states. Using a deterministic approach, separate checks are required for each limit state. Using a probabilistic approach, the probability of failure due to each applicable limit state within each limit state category is evaluated, and the total combined probability of failure is compared with the prescribed threshold value for that category.

### 4.1.4 Displacement Control versus Load Control

Load-controlled loading is through direct application of loads that are not relieved (or affected in any way) as the pipe deforms. In contrast, displacement-controlled loading can be relieved with the deformation of the pipe. Pipeline loadings are often a combination of displacement- and load-controlled components. Pressure (be it internal and external) is load-controlled, whereas ground movement is usually displacement-controlled.[10] Thermally induced loads and loads induced by Poisson's effects are displacement-controlled in most situations. Pipe laying tension is typically

[7] Working stress design (WSD), also known as allowable stress design (ASD), involves the use of design checks with a single safety factor intended to account for all uncertainties associated with the applied loads and the component resistance capacity. Load and resistance factor design (LRFD) involves the use of design checks with separate partial factors for the applied load and component resistance.

[8] In this context, reliability is defined as one minus the probability of failure.

[9] An example of probabilistic design included the reliability-based design and assessment (RBDA) method as set out in Annex O of CSA Z662 [2].

[10] Some ground movements that can cause severe integrity concerns are not purely displacement-controlled. For example, the sudden failure of a slope or mine subsidence may not be purely displacement-controlled.

caused by a combination of displacement- and load-controlled loadings.

A wide variety of intermediate conditions are possible between displacement-controlled and load-controlled situations. Strains resulting from load-controlled loading can often be readily calculated. When they occur under displacement-controlled loading, or are intermediate in type, they may require nonlinear elastic–plastic analysis.

The resistance of a pipeline to load- and displacement-controlled loads is governed by the strength and deformation capacity of the pipe, respectively. Therefore, stress-based and strain-based criteria are typically used for load-controlled and displacement-controlled limit states, respectively. However, strain-based design can also be applied in predominantly load-controlled conditions if the occurrence of plastic strains is to be accommodated.

### 4.1.5 Strain-Based Design Applications

The development of plastic strains during the installation of offshore pipelines has been a reality for many years since reeling[11] of small diameter steel pipes was first practiced in the 1940s [5]. Cold bending of pipes before installation[12] has also been successfully practiced for many years. These installation techniques have over time been extended to higher strength and larger diameter pipes. To adequately deal with high-strain installation processes and displacement-controlled loadings during operation, especially ground movement, many pipeline projects have utilized strain-based design in recent years. Table 4.1 presents a sample of recent worldwide projects that have used strain-based design. Cases of in-service plastic strain have also been observed in pipelines as a result of ground movement (e.g., slope movement, mining subsidence, and seismic loadings). The resistance of steel pipelines to these loadings and better understanding of pipe behavior have led pipeline designers to employ strain-based methods to address in-service plastic strains.

## 4.2 STRAIN DEMAND

### 4.2.1 Overview

A variety of installation and operating conditions can produce loads sufficient to induce large, potentially inelastic strain levels in a pipeline. The magnitude of these strains is defined as the *strain demand*. The following outlines some of the conditions leading to high strain levels and provides some guidance for determining the strain demand for different loading conditions.

### 4.2.2 Challenging Environments and Strain Demand

Increasingly challenging environments such as Arctic and offshore regions must be accommodated to facilitate the development of new oil and gas fields. For pipelines in Arctic and sub-Arctic regions, designers are required to deal with ice scour, permafrost thaw, frost heave, and installation techniques. Also, an increasing number of onshore pipelines traverse mountainous regions and earthquake-prone areas, which expose the pipelines to geohazards such as landslides, progressive slope movements, and soil liquefaction. Offshore pipelines can experience significant longitudinal strains if they are reeled prior to laying, are laid in deep water, or operate at high temperatures and high pressures.

Global buckling of offshore pipelines can be caused or assisted by high temperature and pressure, wave and current loading, and trawling interference. Onshore pipeline could also experience upheaval buckling due to significant temperature differentials and reduced support from the surrounding soil. Global buckling can lead to failure modes such as local buckling, fracture, and fatigue.[13] Complete failures of offshore pipelines have occurred due to cyclic thermal fatigue of global buckles within relatively short periods after start-up. The cyclic strain range of unplanned, upheaval or lateral thermal buckles can exceed 1%, which may lead to pipeline section collapse or high strain–low cycle fatigue and failure by either leakage or rupture [6].

### 4.2.3 Strain Levels and Analysis Considerations

***4.2.3.1 Installation Strains*** Field cold bends typically have a curvature of 1–3° per diameter,[14] which translates into the maximum longitudinal strains of approximately 1.5–3.2% as the pipe is bent with a mandrel inside to prevent significant wrinkling and ovalization. Minor wrinkling (rippling) of the pipe wall may occur on the compression side to varying sizes in different cold bending instances of pipes from the same heat.[15] This problem is generally more severe

[11] Reeling is a part of a pipeline (typically offshore) installation process where the pipe is fabricated into a long segment, wrapped around a circular reel, transported to the installation site, and then unwound from the reel.

[12] Cold bends (or field bends) serve to adapt the pipeline profile to that of terrain or accommodate the deviations required in the pipeline route plan.

[13] Note that according to DNV-OS-F101 [3], global buckling may be allowed provided that (1) the global buckling is not entirely load-controlled; (2) pipeline integrity is maintained in post-buckling configurations (against failure modes such as local buckling, fracture, and fatigue); and (3) displacement of the pipeline is acceptable.

[14] Note that while B31.8 [7] and API RP 1111 [6] set a variable cold bend curvature of approximately 1.9°/$D$ to 3.2°/$D$ as a function of the pipe outer diameter ($D$), CSA Z662-11 [2] prescribes a cold bend curvature limit of 1.5°/$D$ unless the effects of further bending on the mechanical properties of the pipe are determined to be within acceptable limits.

[15] The reader may refer to the work by Bilston and Murray [8] for a method to determine the critical wrinkling (which they term buckling) stress.

**TABLE 4.1 Examples of Pipelines Utilizing Strain-Based Design**

| Pipeline | Application |
|---|---|
| Northstar for BP | Shallow subsea in Alaskan Arctic |
| Haltenpipe for Statoil | Design strain limits near 0.5%, mostly for spanning on uneven seabed |
| Norman Wells for Enbridge | Onshore pipeline across permafrost, strain-based acceptance of on-slope design |
| Badami for BP | River crossings in Alaskan Arctic |
| Nova Gas Transmission Line in Alberta | Strain-based acceptance for discontinuous permafrost |
| TAPS fuel gas pipeline | Strain-based acceptance of upheaval buckling in permafrost |
| Ekofisk II pipelines for ConocoPhillips | Limit state design over subsiding seabed |
| Malampaya for Shell | Limit state design for seismic events and seabed movement |
| Erskine replacement line for Texaco | Limit state design for HP/HT (high-pressure/high-temperature) pipe-in-pipe replacement |
| Elgin/Franklin flowlines and gas export line | Limit state design for pipeline bundles |
| Mallard in North Sea | Limit state design for pipe-in-pipe |
| Sakhalin Island for ExxonMobil | Onshore pipelines in seismic area |
| Liberty in offshore Alaska for BP | Shallow water Arctic pipeline |
| Thunder Horse for BP | Limit state design for HP/HT flow lines |
| Mackenzie Gas Project for Imperial Oil, ConocoPhillips Canada, Shell Canada, and ExxonMobil Canada | Strain-based design for frost heave and thaw settlement in discontinuous permafrost |
| Alaska Pipeline Project for TransCanada and ExxonMobil | Strain-based design for frost heave and thaw settlement in discontinuous permafrost |
| China's Second West–East Pipeline | Extensive use of strain-based design |
| Burma–China Oil and Gas Pipelines | Extensive use of strain-based design |

Most entries adapted with permission from Ref. [5].

for higher grade, high-*Y*/*T* (yield to tensile strength ratio) materials [5]. Furthermore, cold bends in older pipelines are more likely to contain wrinkles than newer pipelines mainly due to lack of or less effective wrinkling prevention measures in the older cold bending processes. While cold bend ripples have little effect on burst pressure, they may grow fatigue cracks through cycling, or in extreme cases, the wrinkles may impede pigging (thus exceeding a serviceability limit). The presence of cold bends can be accounted for in strain demand analysis by incorporating the residual stresses and strains due to the cold bending process.

Strains during installation of offshore pipelines can typically be put in one of the three categories: strains before the pipe is released (i.e., reeling strains), strains as the pipe is released (i.e., overbend in S-lay), and strains in the area of laying (i.e., sagbend in J-lay and S-lay). A close examination of these situations indicates that the loading condition at various stages of offshore pipeline installations is typically a combination of load- and displacement-controlled conditions [5]. While longitudinal strains of up to 2% can be generated by S-lay and J-lay processes [9], strains on the order of 2–4% can develop in coiled pipe on a reel [10]. Pipe sections that include items other than standard pipe such as buckle arrestors or cathodic protection anodes may need a special treatment in the assessment of the laying process as strain concentrations may occur at the transitions.

***4.2.3.2 Strains due to Ground Movement*** Pipelines can be subjected to high longitudinal strains due to movement of the surrounding or underlying soil. Ground movement can result from slope movement (triggered by rainstorms, snowmelt, earthquakes, and various human activities), fault displacement and soil liquefaction (mainly due to earthquakes), ground subsidence (caused by mining or natural sinkholes due to solution of bedrock), thaw settlement, and frost heave. Slope movement, which is typically the most common geohazard for onshore pipelines outside Arctic regions, can be either progressive (i.e., creep) or sudden (i.e., slip). Permafrost thaw settlement and frost heave are usually the most common geohazards for onshore pipelines in Arctic regions. Ground movements can lead to pipeline failure due to excessive tensile and/or compressive longitudinal strains depending on the orientation of the pipeline with respect to the ground movement.

Seismic events can impact a pipeline through transient or permanent ground movement effects. Transient ground movements include near-surface ground deformations caused by ground waves that propagate from a seismic source. Shell and beam buckling are typical failure modes for buried pipelines when exposed to strong ground shaking as a result of excessive compressive axial load [11]. Permanent ground movements associated with earthquakes include landslides (i.e., failing slopes during or shortly after an earthquake due to additional inertial forces or increased pore water pressure), liquefaction (due to the weakening effect of increased pore water pressure in a saturated cohesionless soil), and fault displacement. Since estimates of ground motion in any particular geological setting may be subject to large uncertainties, standards are normally used to define the maximum expected values for general design. Japanese standards have addressed both temporary ground deformation such as seismic wave motion during an earthquake and permanent ground motion, including soil liquefaction [12,13]. The pipe axial strains due to temporary ground deformation have been found to be limited to ±0.41%. The Japanese standards provide two levels of ground motion for permanent ground deformation: level 1 soil motion occurs once or twice during the pipeline lifetime with pipe axial strains within ±1% and level 2 constitutes very strong seismic motion with a low probability of occurrence, resulting in pipe axial strains within ±3%, which may also apply to liquefaction cases.[16]

There are both analytical and numerical (finite element analysis (FEA)) methods for modeling a pipeline's response to ground movement. Analytical models are typically based on solutions to differential equations for a "beam on elastic foundation" as developed for specific pipe layouts (usually straight) and ground motions relative to the pipeline. Newmark and Hall [14], Kennedy et al. [15], and Wang and Yeh [16] developed approximate methods for predicting pipeline response to large fault displacements.[17] Miyajima et al. [19] and O'Rourke [20] developed simple analytical models for pipeline response to spatially distributed transverse ground movement, where a finite length of soil is assumed to move perpendicular to the pipe direction.[18] Rajani et al. [21] developed an analytical model for pipelines subject to longitudinal soil movements,[19] which was later improved by Yoosef-Ghodsi et al. [22]. The use of analytical methods for pipe–soil interaction is best suited for preliminary design or the screening of an existing pipeline and determination of the need for more detailed calculations.

Finite element analysis is the preferred method to assess pipeline response to imposed ground movements as it allows for realistic modeling of the pipe and ground geometry and can account for nonlinearities in both the behavior of the pipe and soil material. The expanded use of FEA for analyzing pipelines in recent years is due to the widespread availability of high-speed, desktop computers and relatively inexpensive, user-friendly FEA software.

The common approach in pipe–soil interaction analysis is to model the pipeline with beam elements and to represent the soil loading on the pipeline using discrete spring or pipe–soil interaction elements. The beam elements should be capable of accounting for the biaxial state of stress due to the presence of internal pressure in addition to axial loading, as the hoop stresses from internal pressure can reduce the axial stress at yield. The finite element formulation should also be able to accommodate large displacements and strains, nonlinear soil springs, and nonlinear stress–strain curves for the pipe material. Beam elements can adequately represent the pipe response up to the point of peak moment prior to wrinkling of the pipe. Thus, in order to determine the valid range for an FEA using beam elements, the analysis should be performed in conjunction with evaluating compressive strain capacity, which is typically defined at peak moment (see Section 4.3.2). The soil springs are defined such that the spring forces act in the axial, horizontal, and vertical directions relative to the pipeline. Comprehensive guidelines and recommendations for pipe–soil interaction analysis can be found in Refs [23,24].

Three-dimensional (3D) continuum models, where pipe and soil are, respectively, represented by 3D shell and continuum elements, have also been developed for computing pipeline response to large ground displacements. While these models can explicitly address many of the limitations inherent in the simplified pipe–soil interaction using beam elements and soil springs, several significant obstacles remain to be overcome before continuum models can be considered superior to beam element and soil spring representations for routine engineering applications [24]. Some of these obstacles are (1) large size of the continuum models, which makes them computationally intensive (typically requiring significant effort to create as well); (2) inability of the continuum models to capture noncontinuum behavior including flow and fracture behavior with slip planes developing within the soil mass; (3) the continuum modeling requirement for more detailed characterization of soil properties than is required for the simple soil spring analogy and is typically available in the industry; and (4) limited validation of the continuum models.

## 4.3 STRAIN CAPACITY

### 4.3.1 Overview

This section discusses the pipe strain capacity with respect to the main failure modes or limit states that are typically associated with strain-based design. These are local buckling on the compression side and tensile fracture on the tension side of a pipeline, which are serviceability and ultimate limit

[16] More guidance on seismic loading and analysis of pipelines subjected to seismic events can be found in PRCI seismic guidelines [10] and Japanese seismic codes [12,13].

[17] Detailed discussions of these models can be found in the Guideline for the Seismic Design of Oil and Gas Pipelines [17,18].

[18] Detailed discussions of these models can be found in Ref. [18].

[19] The authors also developed an analytical model for transverse soil movements in the same reference; however, the model is based on elastic pipe behavior and was primarily used for evaluating pipe stresses, not strains.

states, respectively. Load-controlled failure modes such as burst, collapse, and fatigue are more commonly addressed by stress-based criteria. Similarly, dent and ovalization failure modes (mainly serviceability limit states) are addressed by pipeline codes through limiting the cross-sectional deformation to maintain piggability and unhindered fluid flow as well as eliminating the potential for fatigue failure rather than through strain capacity analysis.

### 4.3.2 Compressive Strain Capacity

During installation or in service, pipelines may experience strains high enough to precipitate local buckling (or wrinkling), in which the pipe wall buckles under axial compressive stresses. The formation of a local buckle can lead to severe cross-sectional deformation and ovalization that could obstruct the passage of pigs through the pipeline (i.e., a serviceability limit state). Continued bending of the pipe following the initiation of local buckling can lead to excessive tensile strains on the tension side of the pipe or at the wrinkle on the compression side resulting in pipe rupture and loss of containment (i.e., an ultimate limit state). Cyclic ratcheting of the wrinkle and a shortened fatigue life due to the wrinkle may also pose integrity threats.

It is current practice to define compressive strain capacity as the average compressive strain over a gauge length of $1D$ to $2D$ (more commonly $1D$, where $D$ is the pipe outer diameter) at peak bending moment at the wrinkle location. Even though strain localization on the compression side starts typically before the peak bending moment is attained, visible wrinkle and major ovalization generally do not occur until after the peak moment has been reached. Compressive strain capacity is primarily influenced by the pipe $D/t$ ratio (where $t$ is the pipe wall thickness), with higher $D/t$ ratios resulting in lower strain capacities. Compressive strain capacity is generally increased by the presence of internal pressure and decreased by the following: higher misalignment values at the girth weld, larger pipe body imperfections, higher yield strengths, and lower strain hardening slopes (including the presence of a Lüder's yield plateau in the pipe body stress–strain curve).

The preferred approach to the determination of compressive strain capacity for a given pipeline is to use a combination of full-scale testing and supporting FEA. However, it is recognized that this can be impractical (e.g., due to lack of representative pipe for test), prohibitively expensive, or simply not warranted if compressive strain capacity is not likely to be the governing limit state. The alternative is the use of semiempirical models.

Early versions of compressive strain capacity models were developed based on elastic shell buckling theory, small-scale tests, and a limited number of full-scale tests. Since the early 1990s, various research organizations have developed compressive strain capacity equations based on significant number of full-scale tests in combination with supporting nonlinear finite element analyses. Selected compressive strain capacity models are presented below, followed by a general discussion of these models.[20]

***4.3.2.1 CSA Z662-11*** CSA Z662-11 [2] incorporates a modified version of the compressive strain capacity equations developed by Gresnigt [28].[21] The CSA equations differ from those of Gresnigt by leaving out the ovalization effects present in Gresnigt's equations and introducing a cutoff on the strain limit for internal pressure values greater than 40% of the yield pressure.[22] Moreover, while Gresnigt's equations are based on the average (i.e., mid-wall) pipe diameter, the CSA equations are based on outside diameter of the pipe. The CSA and Gresnigt's equations produce very close strain capacity predictions for internal pressure values up to 40% of the yield pressure, especially at higher $D/t$ values.

The CSA strain capacity equations show an explicit dependence on the inverse of the $D/t$ ratio and a correction term for pressure. However, the CSA equations do not explicitly account for yield strength or the strain hardening characteristics of the pipe material and do not distinguish between plain and girth-welded pipes. Furthermore, the CSA equations do not directly account for geometric imperfections (i.e., pipe body imperfection and misalignment at girth weld), and they do not explicitly account for the effect of axial force. Finally, no particular gauge length is assigned to the predicted strain limits.

***4.3.2.2 DNV-OS-F101*** The offshore pipelines standard DNV-OS-F101 [3] contains a compressive strain capacity model for pipelines subjected to bending moment, axial force, and internal overpressure (i.e., internal pressure > external pressure) in a displacement-controlled loading condition that is valid for $D/t \leq 45$. Based on work by Yoosef-Ghodsi et al. [29], the DNV standard acknowledges the significant impact of girth welds on compressive strain capacity by applying a reduction factor for $D/t > 20$. The basic form of the equation has similarities to Gresnigt's equation, where there is a dependence of strain capacity on the inverse of the $D/t$ ratio and a correction term for pressure. There is a second correction term to account for the shape of the stress–strain curve, as represented by the $Y/T$ ratio. The DNV model also makes a distinction between plain

[20] The reader can also refer to the compressive strain capacity model recently developed by CRES [25–27], which is not further described in this chapter.

[21] Gresnigt developed a semianalytical model based on an elastic–plastic pipe material to represent the behavior of pipes subjected to internal pressure, axial force, and bending. The compressive strain capacity was defined at peak bending moment and incorporated the ovalization of the pipe cross section due to bending.

[22] The yield pressure is an internal pressure corresponding to a nominal hoop stress equal to the pipe SMYS.

and girth-welded pipes, even though it does not directly account for the magnitude of geometric imperfections. Furthermore, the effect of axial force is not explicitly accounted for in the DNV equation. Finally, no particular gauge length is associated with the predicted strain limits.

*4.3.2.3 Dorey's Model* Dorey [30] developed a compressive strain capacity model based on parametric FEA that was validated by a series of full-scale bend tests performed under a variety of internal pressure and axial force levels based on practical operating conditions. Dorey defined the critical buckling strain over a 1*D* gauge length at the initiation of compressive strain localization. Dorey's model distinguishes between plain and girth-welded pipes also between pipe materials represented by a rounded stress–strain curve at yield or a distinct yield plateau. Dorey's compressive strain capacity model was the first to account for the magnitude of geometric imperfections. While Dorey's model does not account for the degree of strain hardening,[23] it recognizes the reduction in capacity due to the presence of a Lüder's yield plateau. Finally, although the effect of axial force is not explicitly accounted for in Dorey's equations, the parametric FEA used to derive the model incorporated a temperature increase of 45 °C and the Poisson's effect due to pressurization (always resulting in a net compressive axial force).

*4.3.2.4 Discussion on Compressive Strain Capacity Models* As discussed earlier, most existing models do not explicitly account for influential variables other than $D/t$ and internal pressure, those being gauge length, geometric imperfection levels, axial forces, and post-yield (i.e., strain hardening) stiffness. The existing models generally produce conservative predictions of compressive strain capacity. However, it should be noted that these models can produce nonconservative predictions for high-pressure, high-strength pipes, and the DNV equation can lead to nonconservative predictions for unpressurized pipes.

Furthermore, since the available compressive strain capacity models do not explicitly account for cold bends, caution is advised when applying these models to cold bend pipe segments. Cold bends can negatively impact the compressive strain capacity, particularly in the closing mode where the cold bend curvature is subject to increase. Cold bending processes result in both altered material properties and amplification of pipe body imperfections. Although some studies have been carried out on this topic, more work needs to be done to incorporate the effect of cold bends in strain limit equations.

Finally, the bending of a pipeline under relatively high external pressures can lead to a collapse (i.e., flattening of the cross section) failure mode [31]. The above compressive strain capacity equations are only valid for the local buckling failure mode and not the collapse failure mode, which may be a concern for offshore pipelines.

### 4.3.3 Tensile Strain Capacity

Defect-free steel pipe, constructed using arc welded butt joints (i.e., girth welds), is normally very ductile and can carry large axial tensile strains typically exceeding 4% [18]. The Guideline for the Seismic Design of Oil and Gas Pipelines [17] suggests that maximum tensile strain limits on the order of 2–5% may be reasonable for a well-designed and constructed pipeline.[24] In contrast, older steel pipelines constructed using gas-welded joints often cannot carry large tensile strains before rupture, and welded slip joints also do not perform as well as butt-welded joints. In addition, it is noted that the tensile strain capacity can be significantly reduced due to the presence of weld flaws.

In conventional stress-based design, girth-weld flaws are implicitly addressed through workmanship criteria. For a flaw outside the workmanship limits, welding standards may allow the acceptance of the rejected flaw when an engineering critical assessment (ECA) is performed based on fitness-for-service acceptance criteria. The acceptance criteria used in an ECA are typically based on stress-based fracture mechanics considerations and a plastic collapse criterion.[25] Note, however, that such ECA methods generally produce conservative estimates provided relevant material property data are available. This conservatism is even more pronounced for flaws in overmatched[26] welds [32]. Furthermore, since stress-based ECA methods do not provide any information on the degree of conservatism, such assessments do not necessarily lead to the most economical solutions.

Tensile strain capacity is primarily influenced by the dimensions of the flaw that is present in the girth weld or HAZ (heat-affected zone) relative to the pipe wall thickness and the weld/HAZ material toughness. In addition, tensile strain capacity is negatively impacted by lower levels of weld strength overmatch [33], higher internal pressure values,[27] higher misalignment values at the girth weld [34], and lower

[23] The parameters $n$ (i.e., the strain hardening exponent in Ramberg–Osgood equation) and $Y/T$ used by some models can be considered as indicators of the strain hardening slope.

[24] Note that these statements apply to high strain hardening pipes made of quenched and tempered materials. Some modern microalloyed TMCP steels can have uniform elongation strains less than 4–5%.

[25] The combination of ductile fracture and plastic collapse criteria is usually represented by a failure assessment diagram (FAD), which divides the plane of interaction between fracture and plastic collapse resistances into safe and unsafe zones.

[26] Overmatched welds are those with an overmatch ratio of greater than 1. Weld overmatch ratio is the ratio of the weld tensile strength to that of the parent pipe material.

[27] Tensile strain capacity decreases by about 50% as internal pressure increases from zero up to a certain (high) level of internal pressure, where the strain capacity has its minimum. Higher pressure levels typically result in a small increase of the strain capacity [35].

strain hardening slopes (typically represented by higher *Y/T* ratios) [33].

A number of research programs have been carried out in recent years to characterize the tensile strain capacity of pipeline girth welds and to develop strain-based ECA methods. A selection of these methods is described below,[28] followed by a discussion on the role of full-scale and curved wide plate testing.

***4.3.3.1 CSA Z662-11*** A strain-based ECA method was first introduced in Annex C of the 2007 edition of the Canadian pipeline code CSA Z662 [38], which is also included in the latest edition of the code [2]. The CSA standard offers two different equations to calculate the tensile strain capacity for surface breaking and embedded flaws. The input parameters for these equations include the pipe diameter and wall thickness, the flaw dimensions (flaw depth and length, also the embedment depth in the case of embedded flaws), the weld/HAZ apparent toughness,[29] and the *Y/T* ratio for the parent pipe material. The CSA approach has the advantage of not requiring the so-called crack resistance curve[30] (i.e., *J* or CTOD *R*-curve) as an input, which is not readily available for most in-service pipelines. The CSA method does not allow for weld strength undermatching (i.e., where the weld strength is less than that of the parent pipe), but it does not take advantage of potential gains in tensile strain capacity due to weld strength overmatching.

Similar to stress-based ECA methods, the CSA strain-based ECA method does not explicitly account for the effects of pressure-induced biaxial loading in a pipeline. Several studies in recent years have shown that although the material resistance to fracture appears to be similar under uniaxial and biaxial loading, the tensile strain capacity of girth welds can be significantly reduced under biaxial loading due to an increase in the crack driving force.[31] This implies that the CSA strain-based ECA method results in different levels of conservatism for different pressure values (i.e., less conservative tensile strain capacities at higher pressure values).

[28] The reader can also refer to the tensile strain capacity models developed by SINTEF [36] and University of Ghent [9,37], which are not further described in this chapter.

[29] The term "apparent" implies that the traditional single-parameter-based fracture mechanics does not strictly apply to large crack tip plasticity. The apparent toughness represents the combined effects of the material toughness and structural response. For large-scale test specimens, the CTOD values associated with final failure are generally well above 0.5 mm, sometimes much larger at 1.0–1.5 mm. These values are much greater than the single-parameter "valid" CTOD value (0.1 mm or less) for a tension CTOD specimen with a through-wall crack [39].

[30] Crack resistance curve, or *R*-curve, is a plot of crack driving force (*J* or CTOD) versus crack extension.

[31] Crack driving force is defined as either the applied *J*-integral or crack tip opening displacement (CTOD).

***4.3.3.2 DNV-OS-F101*** The DNV-OS-F101 [3] standard provides two strain-based ECA procedures[32] as presented below. The common input parameters for these procedures include the pipe, weld, and flaw dimensions (including weld misalignment) and the stress–strain curve for the parent pipe material.

*ECA Static—Full* Assessments for strains exceeding 0.4% are carried out according to BS 7910 [40] at assessment level 3B with amendments and adjustments as described in DNV-OS-F101. For strains less than 0.4%, however, it is acceptable to carry out a level 2B assessment in accordance with BS 7910, again with specified amendments and adjustments. Both level 2B and 3B assessment procedures require that material-specific stress–strain curves be established. While a crack resistance curve (*J* or CTOD *R*-curve) for the weld/HAZ is required for a level 3B assessment, only a single parameter fracture toughness is required for level 2B (CTOD or *J*). This method does not account for weld strength overmatch, but it requires evidence that the weld metal stress–strain curve overmatches the parent pipe stress–strain curve. Furthermore, this procedure is not applicable where there is internal overpressure (i.e., internal pressure > external pressure) unless axial strain is below 0.4%. Segment specimen testing[33] or full-scale testing is required where more than one strain increment is applied (e.g., the strain cycles during the installation of an offshore pipeline).

*Finite Element Fracture Mechanics Analysis* It is recommended to use general solid 3D FE fracture mechanics analyses, but utilizing other dedicated software programs is also acceptable if the geometry and flaw sizes assessed are benchmarked against dedicated solid 3D FE fracture mechanics analyses that fulfill the requirements specified in the standard. This assessment requires a crack resistance curve (*J* or CTOD *R*-curve) for the weld/HAZ and the stress–strain curve of the weld metal in addition to that of the parent pipe. The capacity assessment can be based on either of the following two analysis options:

a. Multiple stationary flaw analyses, where a minimum of three FE analyses of stationary flaws with different heights but equal length are performed, and the resulting crack driving force is plotted versus the flaw height. The instability point may be determined as the tangency point between the crack driving force curve based on the FE predictions and the crack resistance curve established by testing of the relevant material condition.

[32] The DNV standard also includes a stress-based ECA procedure for strains up to 2.25%.

[33] Segment specimen testing involves uniaxial tension straining of material cut from the welded pipe wall.

b. One FE analysis capable of simulating crack growth, for instance, using the Gurson–Tvergaard–Needleman formulation crack growth model. However, the crack growth model must be calibrated and validated against relevant experimental results.

***4.3.3.3 ExxonMobil ECA Framework*** ExxonMobil recently developed a three-tier approach to calculate the tensile strain capacity of pipeline girth welds containing flaws, which explicitly accounts for the pressure-induced biaxial loading [41,42]. The capacity assessment is done using either simplified closed-form parametric equations or detailed FEA based on the limiting condition defined as the point where the crack driving force curve becomes tangential to the *R*-curve. The common input parameters for all assessment levels include the pipe diameter and wall thickness, flaw dimensions, weld overmatch ratio, and internal pressure.

An increase in the ECA level (from 1 to 3) implies an increase in complexity and accuracy of the assessment procedure and a decrease in conservatism. Level 1 and 2 procedures have some limitations in terms of the range of input parameters. For example, both procedures cover only pipe grades from X60 to X80, and are based on an assumed *Y/T* ratio of 0.9 for level 1 and 0.92 for level 2.[34] Furthermore, in the interest of simplicity and conservatism, the level 1 procedure has these additional fixed inputs: weld misalignment = 3 mm, uniform elongation strain (UEL) = 6%, and a lower bound CTOD *R*-curve, which is the lowest of the three specified *R*-curves available for the level 2 procedure. In addition to detailed FEA, the level 3 procedure requires project-specific testing. Level 3 appears to be most suitable for cases involving high strain demand, higher strength pipe (X80+), or other step out factors. Levels 1 and 2 would be suitable for evaluating preliminary designs and conducting strain-based ECA for cases involving relatively routine strain capacities and conventional pipe grades.[35]

***4.3.3.4 CRES ECA Framework*** Center for Reliable Energy Systems (CRES) has developed a four-level ECA approach that can be adapted to the scale of the project and its design and maintenance requirements [26,27,44,45]. The common input parameters for all assessment levels include the pipe diameter and wall thickness, flaw dimensions, weld misalignment, yield and tensile strength and uniform elongation of the pipe material, tensile strength of the weld material, and internal pressure.

The CRES approach allows the use of a wide variety of material toughness test options and explicitly accounts for the pressure-induced biaxial loading. The level 1 procedure provides estimated tensile strain capacity in a tabulated form for quick initial assessment, where the apparent toughness is estimated from upper shelf Charpy impact energy. The level 2 procedure includes parametric equations, where the apparent toughness is obtained from either upper shelf Charpy energy or upper shelf toughness of standard CTOD test specimens. The level 3 procedure uses the same equations as in level 2 with the distinction that the toughness values are obtained from low-constraint tests (e.g., single-edge-notched tension or SENT test). In the level 3 procedure, two limit state options are available. Level 3a is based on initiation control limit state, which implies the use of apparent toughness as the toughness parameter, and level 3b is based on ductile instability limit state (or tangency method) using resistance curve as the toughness parameter. The level 4 procedure is based on direct FEA to develop crack driving force relations using the same limit state options as those in level 3. The level 4 procedures are recommended to be used by experts only and in special circumstances where lower level procedures are judged inappropriate or inadequate.

## 4.4 ROLE OF FULL-SCALE AND CURVED WIDE PLATE TESTING

Full-scale tension (FST) testing of pipes containing flaws in their girth welds is the preferred experimental method to determine the tensile strain capacity, as FST testing realistically accounts for the biaxial effects of pressure loading. Thus, major projects often use FST testing to validate/calibrate project-specific tensile strain capacity models. In fact, full-scale verification of any tensile strain capacity model should be considered for any given project to ensure its applicability.

Curved wide plate[36] (CWP) testing can more economically address variability through repeated testing at the expense of requiring adjustments for pressure effects. Denys et al. [32] suggest that for welds having a strength comparable to that of the parent pipe (i.e., even matched), the strain capacity as determined from CWP tests needs to be reduced by a factor of 2 to account for the adverse biaxial stress effects that would exist in a pipeline operating at pressure.[37]

[34] These choices for the *Y/T* ratio are on the conservative side.

[35] The reader can also refer to the recent tensile strain capacity model developed by ExxonMobil [43], which is not further described in this chapter.

[36] The CWP test involves a curved (i.e., unflattened) girth-welded pipe segment containing a surface-breaking circumferentially oriented crack-like notch (or possibly a real flaw) in the weld or HAZ at mid-length of the specimen, which is axially loaded in tension to failure.

[37] Denys et al. suggest that a lower reduction factor would be appropriate for overmatched welds [32]. Other pressure correction factors have recently been developed in independent studies with similar outcomes [41,45,46].

## 4.5 SUMMARY

The following summarizes the key aspects of the strain-based design and assessment of pipelines:

- Strain-based design is the preferred way to design new or assess existing pipelines subject to very high (inelastic) longitudinal strains resulting from extreme service and/or installation environments, particularly where the loads are displacement-controlled.
- The primary limit states (or failure conditions) addressed by strain-based design are tensile rupture and compressive local buckling.
- Assessment involves comparing the strain demand resulting from applied loads with the strain capacity as affected by pipe geometry, material properties, and the presence of defects.
- Two design and assessment philosophies exist, deterministic and probabilistic, which differ in how the analysis uncertainties are handled (i.e., implicitly through safety factors or explicitly through calculation of probability of failure and comparison with probability limits). Probabilistic methods, while more difficult to implement, have the potential of achieving more consistent safety levels by more accurately accounting for analysis uncertainties.
- The compressive strain capacity is better understood and better addressed by existing models than tensile strain capacity, and tensile strain capacity models are limited in their range of applicability due to reliance on validation by tests.

## REFERENCES


1. Law, M. (2007) Review of Strain Based Analysis for Pipelines. ANSTO Report R06M132, Australian Nuclear Science and Technology Organisation (ANSTO).
2. Canadian Standards Association (CSA) (2011) CSA Standard Z662-11: Oil and Gas Pipeline Systems, Canadian Standards Association, Mississauga, Ontario, Canada.
3. DNV (2013) Offshore Standard OS-F101: Submarine Pipeline Systems, Det Norske Veritas Classification A/S.
4. ISO (2006) ISO Standard 16708: Petroleum and natural gas industries—Pipeline Transportation systems—Reliability based limit state methods, 1st ed.
5. Mohr, W. (2003) Strain-Based Design of Pipelines. Technical Report 45892GTH, Edison Welding Institute, 68 pp.
6. API (2011) API Recommended Practice 1111: Design, Construction, Operation, and Maintenance of Offshore Hydrocarbon Pipelines (Limit State Design)—Downstream Segment, API, Washington, DC.
7. ASME (2012) ASME B31: Gas Transmission and Distribution Piping Systems—ASME Code for Pressure Piping.
8. Bilston, P. and Murray, N. (1993) The role of cold field bending in pipeline construction. 8th Symposium on Line Pipe Research, American Gas Association, No. 27, pp. 1–19.
9. Hertelé, S. (2012) Coupled experimental–numerical framework for the assessment of strain capacity of flawed girth welds in pipelines. Ph.D. thesis, Ghent University.
10. Honegger, D.G. and Nyman, D.J. (2004) Seismic Design and Assessment of Natural Gas and Liquid Hydrocarbon Pipelines, Pipeline Research Council International Inc., No. L51927.
11. Mitsuya, M., Sakanoue, S., and Motohashi, H. (2013) Beam-mode buckling of buried pipeline subjected to seismic ground motion. *Journal of Pressure Vessel Technology*, 135, 021801-1–021801-10.
12. JGA (2004) Seismic Design Codes for High-Pressure Gas Pipelines, JGA-206-03, Japan Gas Association.
13. JGA (2001) Seismic Design Codes for High-Pressure Gas Pipelines: Considering Liquefaction-Induced Permanent Ground Deformation, JGA-207-01, Japan Gas Association.
14. Newmark, N.M. and Hall, W.J. (1975) Pipeline design to resist large fault displacement. Proceedings of the U.S. National Conference on Earthquake Engineering, EERI, Ann Arbor, MI.
15. Kennedy, R.P., Chow, A.W., and Williamson, R.A. (1977) Fault movement effects on buried oil pipeline. *Journal of the Transportation Engineering Division, ASCE*, 103 (TE5), 617–633.
16. Wang, L.R.-L. and Yeh, Y.-H. (1985) A refined seismic analysis and design of buried pipeline for fault movement. *Earthquake Engineering & Structural Dynamics*, 13 (1), 75–96.
17. Committee on Gas and Liquid Fuel Lifelines (1984) The Guideline for the Seismic Design of Oil and Gas Pipelines, ASCE.
18. O'Rourke, M.J. and Liu, X. (1999) Response of Buried Pipelines Subject to Earthquake Effects. Multidisciplinary Center for Earthquake Engineering Research, University at Buffalo.
19. Miyajima, M., Kitawa, M., and Nomura, Y. (1990) Study on response of buried pipelines subjected to liquefaction-induced permanent ground displacement. *International Journal of Rock Mechanics and Mining Sciences and Geomechanics Abstracts*, 27 (2), A128.
20. O'Rourke, T.D. (1989) Seismic design considerations for buried pipelines. *Annals of the New York Academy of Sciences*, 558 (1), 324–346.
21. Rajani, B.B., Robertson, P.K., and Morgenstern, N.R. (1995) Simplified design methods for pipelines subject to transverse and longitudinal soil movements. *Canadian Geotechnical Journal*, 32 (2), 309–323.
22. Yoosef-Ghodsi, N., Zhou, J., and Murray, D.W. (2008) A simplified model for evaluating strain demand in a pipeline subjected to longitudinal ground movement. International Pipeline Conference, September 28–October 3, ASME, Calgary, Alberta..
23. American Lifelines Alliance (2005) Guidelines for the design of buried steel pipe, ASCE.

24. C-CORE, D.G Honegger Consulting, and SSD, Inc. (2009) Guidelines for Constructing Natural Gas and Liquid Hydrocarbon Pipelines Through Areas Prone to Landslide and Subsidence Hazards. Final report, Pipeline Research Council International Inc.
25. Liu, M., Wang, Y.-Y., Zhang, F., Wu, X., and Nanney, S. (2013) Refined compressive strain capacity models. Proceedings of the 6th International Pipeline Technology Conference, October 6–9, Ostend, Belgium.
26. Wang, Y.-Y. and Liu, M. (2013) Real world considerations for strain-based design and assessment. Proceedings of the 6th International Pipeline Technology Conference, October 6–9, Ostend, Belgium.
27. Liu, M., Wang, Y.-Y., Zhang, F., and Kotian, K. (2013) Realistic Strain Capacity Models for Pipeline Construction and Maintenance. Contract No. DTPH56-10-T-000016, Final Report to U.S. Department of Transportation Pipeline and Hazardous Materials Safety Administration (U.S. DOT PHMSA), Office of Pipeline Safety.
28. Gresnigt, A.M. (1986) Plastic design of buried steel pipelines in settlement areas. *Heron*, 31 (4), 1–113.
29. Yoosef-Ghodsi, N., Kulak, G.L., and Murray, D.W. (1994) Behaviour of Girth Welded Line-pipe. Structural Engineering Report No. 23, Department of Civil Engineering, University of Alberta.
30. Dorey, A. (2001) Critical buckling strains in energy pipelines. Ph.D. thesis, Department of Civil and Environmental Engineering, University of Alberta, Edmonton, Alberta, Canada.
31. Gresnigt, A.M. and van Foeken, R.J. (1996) Experiences with strain based limit state design in the Netherlands. Advances in Subsea Pipeline Engineering and Technology (Aspect '96), November 27–28, Aberdeen, UK, Paper No. ASPECT-96-111.
32. Denys, R.M., Hertelé, S., and Lefevre, A.A. (2013) Use of curved-wide-plate (CWP) data for the prediction of girth-weld integrity. *Journal of Pipeline Engineering*, 12 (3), 245.
33. Denys, R.M., Lefevre, A.A., and De Baets, P. (2002) A rational approach to weld and pipe material requirements for a strain based pipeline design. Proceedings of the International Conference on Application and Evaluation of High-Grade Linepipes in Hostile Environments, Pacifico Yokohama, Japan, pp. 121–158.
34. Kibey, S., Minnaar, A.K., Issa, J.A., and Gioielli, P.C. (2008) Effect of misalignment on the tensile strain capacity of welded pipelines. Proceedings of the 18th International Offshore and Polar Engineering Conference, Vancouver, British Columbia, Canada, pp. 90–95.
35. Gordon, J.R., Zettlemoyer, N., and Mohr, W.C. (2007) Crack driving force in pipelines subjected to large strain and biaxial stress conditions. Proceedings of the 17th International Offshore and Polar Engineering Conference, Lisbon, Portugal, pp. 3130–3140.
36. Østby, E. (2005) Fracture control—offshore pipelines: new strain-based fracture mechanics equations including the effects of biaxial loading, mismatch and misalignment. Proceedings of the International Conference on Ocean, Offshore and Arctic Engineering, Halkidiki, Greece, Paper No. OMAE2005-67518.
37. Denys, R., Hertelé, S., Verstraete, M., and De Waele, W. (2011) Strain capacity prediction for strain-based pipeline designs. Proceedings of the Internal Workshop on Welding of High Strength Pipeline Steels, CBMM, Araxá, Brazil.
38. Canadian Standards Association (CSA) (2007) CSA Standard Z662-07: Oil and Gas Pipeline Systems, Canadian Standards Association, Mississauga, Ontario, Canada.
39. Wang, Y.-Y., Liu, M., and Song, Y. (2011) Second Generation Models for Strain Based Design, Contract PR-ABD-1, Project 2, Final Report to Pipeline Research Council International, Prepared for U.S. Department of Transportation Pipeline and Hazardous Materials Safety Administration (PHMSA).
40. BSI (2005) BS 7910:2005: Guide to Methods for Assessing the Acceptability of Flaws in Metallic Structures. British Standards Institution (BSI), London, UK.
41. Fairchild, D.P., Kibey, S.A., Tang, H.S., Krishnan, V.R., Wang, X., Macia, M.L., and Cheng, W. (2012) Continued advancements regarding capacity prediction of strain-based pipelines. Proceedings of the 9th ASME International Pipeline Conference, Calgary, Alberta, Canada, Paper IPC2012-90471.
42. Fairchild, D.P., Macia, M.L., Kibey, S., Wang, X., Krishnan, V.R., Bardi, F., Tang, H., and Cheng, W. (2011) A multi-tiered procedure for engineering critical assessment of strain-based pipelines. Proceedings of the 21st International Offshore and Polar Engineering Conference (ISOPE), June 19–24, Maui, HI.
43. Tang, H., Fairchild, D., Panico, M., Crapps, J., and Cheng, W. (2014) Strain Capacity Prediction of Strain-Based Pipelines. Proceedings of the 10th ASME International Pipeline Conference, Calgary, Alberta, Canada, Paper IPC2014-33749.
44. Wang, Y.-Y. and Liu, M. (2013) Status and applications of tensile strain capacity models. Proceedings of the 6th International Pipeline Technology Conference, October 6–9, Ostend, Belgium.
45. Liu, M., Wang, Y.Y., Song, Y., Horsley, D., and Nanney, S. (2012) Multi-tier tensile strain models for strain-based design. Part 2. Development and formulation of tensile strain capacity models. Proceedings of the 9th ASME International Pipeline Conference, Calgary, Alberta, Canada, Paper IPC2012-90659.
46. Verstraete, M. (2013) Experimental–numerical evaluation of ductile tearing resistance and tensile strain capacity of biaxially loaded pipelines. Ph.D. thesis, Ghent University.

# 5

# STRESS-BASED DESIGN OF PIPELINES

Mavis Sika Okyere
*Bluecrest College, Kumasi, Ghana*

## 5.1 INTRODUCTION

In stress-based design, the pipeline is designed so that the stress on the pipeline is maintained below a prescribed limiting value, the specified minimum yield stress (SMYS), by a safety factor. In this chapter, reference is made to standards that are used in different jurisdictions, and some comparisons are made among British, Canadian, European, international, and U.S. standards. For more detailed information, readers should refer to the current editions of the relevant standards.

As discussed in CSA Z662-11, Section 4.3.5 [1], the design formula for straight pipe is

$$P = \frac{2St}{D} \times F \times L \times J \times T \quad (5.1)$$

where $P$ is the design pressure (MPa), $S$ is the specified minimum yield strength (MPa), $t$ is the design wall thickness (m), $D$ is the outside diameter of pipe (m), $F$ is the design factor, $L$ is the location factor, $J$ is the joint factor, and $T$ is the temperature factor.

## 5.2 DESIGN PRESSURE

The design pressure is the pressure that is used in all equations and stress calculations and is the maximum internal pressure of the pipeline during its design life. The maximum allowable operating pressure (MAOP), maximum operating pressure (MOP), and surge pressure should be considered with the design pressure. Unduly high design pressures require the use of excessively thick pipe.

For a gas pipeline, the pressure in the pipeline does not vary greatly due to elevation, but where the elevation is extreme (hundreds of meters) below the inlet point, the highest pressure in the pipeline and the location of the highest pressure need to be determined. For liquid pipelines, there is always a need to consider elevation, especially where the pipeline is below the inlet point.

Normally it is better to keep a fixed design pressure for the whole pipeline being designed. However, for pipelines with large positive elevation change (i.e., hilly and mountainous zones), the design pressure can be reduced as the elevation increases. This will result in different pipes for different sections, complicating the line pipe order and construction, and it makes any future uprating by adding additional pumping stations limited or impossible.

Normally for gas pipelines, the maximum operating pressure does not exceed the design pressure.

### 5.2.1 Maximum Allowable Operating Pressure

As per Clause 805.214 of ASME B31.8, the maximum allowable operating pressure is defined as the maximum pressure at which a gas pipeline system is allowed to operate. The value of the MAOP varies and depends on the location and prescribed test pressure [2].

Due to lack of surge pressure in a gas pipeline, the MAOP should not be higher than the design pressure but often is set slightly (5% or less) below the design pressure. For liquid pipelines, the MAOP can be 10% or more below the design

*Oil and Gas Pipelines: Integrity and Safety Handbook,* First Edition. Edited by R. Winston Revie.

**TABLE 5.1 Percentage Overpressure Permitted [1–9]**

| | ASME B31.8 | ASME B31.4 | PD 8010 | IGE/ TD/1 | ISO 13623 | EN 1594 | CSA Z662 |
|---|---|---|---|---|---|---|---|
| Relating to | N/A | DP | DP | MOP | MAOP | MOP | MOP |
| % | 0 | 10 | 10 | 10 | 10 | 15 | 10 |

**TABLE 5.2 Test Pressures [2]**

| | Test Pressure (TP) | Pressure Factor (PF) |
|---|---|---|
| Installed pipeline system | TP = MAOP × 1.25 | 1.25 |
| Offshore platform piping | TP = MAOP × 1.4 | 1.4 |
| Offshore pipeline risers | TP = MAOP × 1.4 | 1.4 |

pressure due to surge problems. The difference between the MAOP and the design pressure permits shutdown alarms and other protective devices to be set to ensure that the pipeline does not exceed the design pressure [2–9].

$$\text{MAOP} = \frac{\text{test pressure } (TP)}{\text{pressure factor (PF)}} \tag{5.2}$$

### 5.2.2 Maximum Operating Pressure

The maximum operating pressure is equal to the maximum pressure to which the piping system will be subjected in operational conditions. This includes static pressure and the pressure to overcome friction [2].

Above MOP, a warning alarm might be set, but this will not shut down the pipeline until the pressure goes above the MAOP. To raise the MOP above the MAOP, there is the need to retest the pipeline to a higher pressure in order to raise the MAOP.

***5.2.2.1 Overpressure*** The amount of overpressure that is allowed varies between design codes. Table 5.1 outlines the available percentages.

### 5.2.3 Surge Pressure

Surge pressures in a liquid pipeline are produced by a change in the velocity of the moving stream that results from shutting down of a pump station or pumping unit, closing of a valve, or blockage of the moving stream. Surge pressure decreases in intensity as it moves away from its point of origin [2].

The surge pressure depends on the density of fluid, velocity of fluid, pipe length, speed of closure or shutdown, fluid pressure, and sonic velocity of the fluid. The design codes have limiting values for surge pressure to be added to the MAOP. Surge calculations should be made, and adequate controls and protective equipment should be provided.

Gas pipelines do not suffer excessively from surge due to compressibility within the fluid. However, this can become a major issue for liquid pipelines that have a relatively high fluid velocity (>2.5 m/s) or are subject to sudden closure of valves or pumps.

### 5.2.4 Test Pressure

Test pressure is set by the design codes to verify that the pipeline is fit for purpose and free from material or construction defects. The setting of the pressure test levels needs to follow the methodology of the design code (e.g., Clause 847.2 of ASME B31.8), but care also needs to be taken not to overstress the pipeline either at its lowest point or when the design calculations have used empty weight of the pipe, as in gas pipes, and not the temporary weight of the hydrotest water [2].

Pipelines in mountainous or hilly regions are often sectioned into different lengths for testing.

$$\text{Test pressure} = \text{MOP} \times \text{PF}, \quad 1.1 < \text{PF} < 1.4 \tag{5.3}$$

The test pressures used in different situations are listed in Table 5.2.

## 5.3 DESIGN FACTOR

Design factors (Table 5.3) have been used since pipeline design codes were first established, to provide a defined level of safety and mechanical strength. The design factor to be used is sometimes limited by legislation in a particular country [2–9].

The design codes state where in this range a specific design factor should be used, but generally allow a higher factor to be used provided that this action is supported by a safety evaluation or subject to scrutiny by the regulatory safety authority, which in the United Kingdom would be the Health and Safety Executive (HSE).

It is appropriate to specify the design factor at a level that takes into consideration the boundary between defect arrest and propagation, also known as leak/break boundary.

**TABLE 5.3 Design Factor [2–9]**

| | ASME B31.8 | ASME B31.4 | PD 8010 | IGE/TD/1 | ISO 13623 | EN 1594 |
|---|---|---|---|---|---|---|
| Design factor (liquid) | – | 0.72 | 0.72 | – | 0.77 | – |
| Design factor (gas) | 0.40–0.80 | – | 0.30–0.72 | 0.30–0.80 | 0.45–0.83 | 0.72 |

The two important design factors are

- 0.30 for pipelines operating in locations where a line break is not acceptable;
- 0.80 for pipelines operating in open country where a line break is unlikely but could be tolerated.

## 5.4 DETERMINATION OF COMPONENTS OF STRESS

### 5.4.1 Hoop and Radial Stresses

Hoop stress, $\sigma_H$, is the stress in a pipe of wall thickness $t$ acting circumferentially in a plane perpendicular to the longitudinal axis of the pipe, produced by the pressure $P$ of the fluid in a pipe of diameter $D$ and is determined by Barlow's formula [2]:

$$\sigma_H = \frac{PD}{2t} \tag{5.4}$$

where $\sigma_H$ is the hoop stress (MPa), $P$ is the internal design pressure (gauge) (MPa), $t$ is the pipe wall thickness (m), and $D$ is the pipe diameter (m).

***5.4.1.1 Thick Cylinders*** The full Lamé equations are simplified for the design of thick wall pipelines. A pipeline with $D/t < 20$ is known as a thick wall pipeline. Taking into consideration a thick wall pipe, subjected to an internal pressure, $P_i$, with zero external pressure [10],

$$\text{hoop stress, } \sigma_H = A + \frac{B}{r^2} \tag{5.5}$$

$$\text{radial stress, } \sigma_r = A - \frac{B}{r^2} \tag{5.6}$$

The two well-known conditions of stress that allow the Lamé constants $A$ and $B$ to be determined are

$$\text{at } r = R_1 \quad \sigma_r = P_i$$
$$\text{at } r = R_2 \quad \sigma_r = 0$$

$$\text{Lamé constant, } A = -P_i\left(\frac{R_1^2}{R_1^2 - R_2^2}\right) \tag{5.7}$$

$$\text{Lamé constant, } B = P_i\left(\frac{R_1^2 R_2^2}{R_1^2 - R_2^2}\right) \tag{5.8}$$

Therefore, from Equations 5.5–5.8, the radial and hoop stresses are calculated as

$$\sigma_r = -P_i\left(\frac{R_1^2}{R_2^2 - R_1^2}\right)\left(1 - \frac{R_2^2}{r^2}\right) \tag{5.9}$$

$$\sigma_H = -P_i\left(\frac{R_1^2}{R_2^2 - R_1^2}\right)\left(1 + \frac{R_2^2}{r^2}\right) \tag{5.10}$$

The maximum radial and hoop (circumferential) stresses occur at $r = R_1$ when $\sigma_r = P_i$. The negative sign indicates tension.

$$\therefore \sigma_H = -P_i\left(\frac{R_1^2 + R_2^2}{R_2^2 - R_1^2}\right) \tag{5.11}$$

$$\sigma_r = P_i\left(\frac{R_1^2}{R_2^2 - R_1^2}\right) \tag{5.12}$$

where $\sigma_H$ is the hoop stress (MPa), $\sigma$ is the radial stress (MPa), $R_1$ is the internal radius (m), $R_2$ is the external radius (m), $r$ is the radius at point of interest (measured from pipeline center), $P_e$ is the external pressure (gauge) (MPa), and $P_i$ is the internal pressure (gauge) (MPa).

***5.4.1.2 Thin Wall Pipeline*** A pipeline with $D/t > 20$ is known as a thin wall pipeline. A basic approach is known as thin wall hoop stress theory. Since the maximum hoop stress is normally the limiting factor, it is this stress that will be considered [10–12].

It is predictably accurate for $D/t > 20$. The hoop stress is then calculated as follows:

$$\sigma_H = (P_i - P_e)\frac{D_2}{2t}, \quad D/t > 20 \tag{5.13}$$

Hoop stress developed in the pipe wall at the internal design pressure is given by

$$\sigma_H = (P_i)\frac{D_2}{2t}, \quad D/t > 20, \quad P_e = 0 \tag{5.14}$$

where $\sigma_H$ is the hoop stress (MPa), $P_i$ is the internal design pressure (gauge) (MPa), $P_e$ is the external pressure (gauge) (MPa), $t$ is the design thickness (m), $D_1$ is the inside pipe diameter (m), and $D_2$ is the outside pipe diameter (m).

### 5.4.2 Longitudinal Stress

The estimation of the longitudinal stress in a section of pipeline requires the individual stress components to be identified knowing external restraining conditions.

The axial (longitudinal) stress in a pipeline depends wholly on the limiting conditions (imposed boundary condition) experienced by the pipeline, that is, whether the pipeline is unrestrained, restrained, or partially restrained. The boundary conditions can include the effects of soil reaction loads, anchor restraints, line pipe bend resistance, and residual pipe-lay tension forces [10–12]. The longitudinal stress in a thin cylindrical shell is calculated as half the hoop stress.

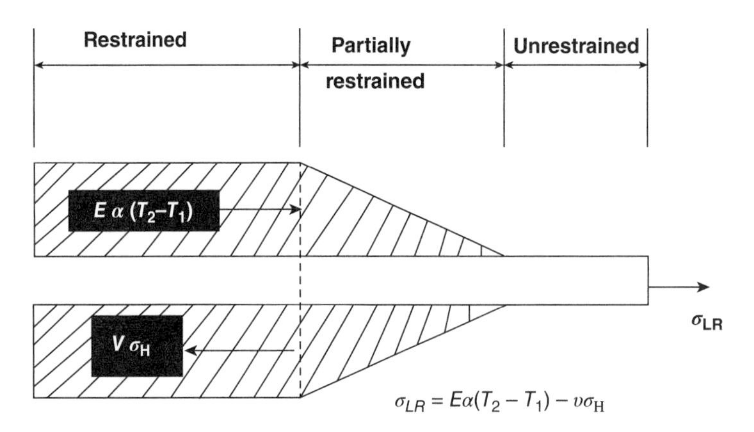


**FIGURE 5.1** Longitudinal tensile stresses in a fully restrained pipe section.

The total longitudinal stress should be the sum of the stresses arising from the following (see Sections 5.4.2.1–5.4.2.7):

- pressure;
- bending;
- temperature;
- weight;
- other sustained loads;
- occasional loadings.

A pipeline should be considered totally restrained when axial movement and bending resulting from temperature or pressure change are totally prevented.

***5.4.2.1 Fully Restrained Pipeline*** A fully end constrained boundary condition can occur at an anchor block or pig trap, and pipeline end manifold (PLEM) or pipeline end termination (PLET) sled. For a fully end constrained pipeline (Figure 5.1), the longitudinal strain ($\varepsilon_l = 0$) and deflection ($\Delta = 0$) components are zero and the longitudinal stress response can be determined from Equation 5.15, assuming a constant uniform temperature field [10–12].

Piping in which soil or supports prevent axial displacement of flexure at bends is restrained. Restrained piping may include the following [2,6]:

- straight sections of buried piping;
- bends and adjacent piping buried in stiff or consolidate soil;
- sections of above-ground piping on rigid supports.

The net longitudinal compressive stress in a restrained pipe is calculated based on Clause 419.6.4 of ASME B31.4 as

$$\sigma_{LR} = E\alpha(T_2 - T_1) - \upsilon\sigma_H \quad (5.15)$$

where $\sigma_{LR}$ is the restrained longitudinal stress (MPa), $\alpha$ is the thermal linear coefficient of expansion (mm/mm/°C), $T_2$ is the operating temperature (maximum or minimum metal temperature) (°C), $T_1$ is the installation temperature (°C), $\upsilon$ is the Poisson's ratio ($\upsilon = 0.30$ for steel), $\sigma_H$ is the hoop stress (MPa), and $E$ is the modulus of elasticity (GPa).

Piping and equipment should be supported, so as to prevent or reduce excessive vibration, and should be anchored sufficiently to prevent undue strains on connected equipment. Supports, hangers, and anchors should be so installed as not to interfere with the free expansion and contraction of the piping between anchors. Suitable spring hangers, sway bracing, and so on should be provided where necessary [2].

Anchor blocks are used to stop axial movement of a pipeline. Anchors are normally required when the pipeline comes above ground, prior to pig hatches, other branches, or manifolds. Connection of the pipeline to the anchor block is normally done by the addition of slip-on flanges to the pipeline, which are fillet welded in place. A large concrete block is then constructed around the pipeline to resist the expansion forces.

Restrained portions are always prevented from moving by installing anchors and guides, but in a buried line a large portion is fully restrained by soil friction only [12].

The axial compressive force required to restrain a pipeline can be calculated as follows:

$$\text{Thin wall}: \quad F = A(E\alpha(T_2 - T_1) + 0.5S_h - \upsilon S_h) \quad (5.16)$$

$$\text{Thick wall}: \quad F = A\left(E\alpha(T_2 - T_1) + \frac{S_h}{k^2 + 1} - \upsilon\left(S_h - \frac{P}{10}\right)\right) \quad (5.17)$$

where $F$ is the axial force (N), $E$ is the modulus of elasticity (GPa), $T_1$ is the installation temperature (°C), $S_h$ is the hoop stress (MPa), $\upsilon$ is the Poisson's ratio (0.3 for steel), $A$ is the cross-sectional area of pipe wall (m$^2$), $\alpha$ is the coefficient of thermal expansion (°C$^{-1}$), $T_2$ is the maximum or minimum metal temperature (°C), $k$ is the ratio of outside diameter to inside diameter, and $P$ is the internal design pressure (gauge) (MPa).

*5.4.2.2* ***Unrestrained Pipeline*** An end free boundary condition can occur at locations where no physical longitudinal restraint exists, for example, at a riser bend extending from the seabed to the production platform. Since the pipeline is not restrained axially, the Poisson effect and thermal expansion component do not produce stress in the line pipe. The pipeline longitudinal stress is only due to the end cap effect [2–5].

Piping that is freed to displace axially or flex at bends is unrestrained. Unrestrained piping may include the following:

- aboveground piping that is configured to accommodate thermal expansion or anchor movements through flexibility;
- bends and adjacent piping buried in soft or unconsolidated soil;
- an unbackfilled section of otherwise buried pipeline that is sufficiently flexible to displace laterally or that contains a bend;
- pipe subject to an end cap pressure force.

For unrestrained sections of a pipeline, the longitudinal tensile stress should be calculated as follows (refer to BS 8010, Part 2, Section 2.8, and Clause 2.9.3.2):

$$\sigma_{LU} = \frac{\sigma_H}{k^2+1} + \frac{1000M_b \times i}{Z} \tag{5.18}$$

For a thin wall pipeline, use $k=1$.

When the internal pressure is greater than the external pressure, the stress will be positive, tensile, as expected.

The unrestrained longitudinal stress can generally be expressed as

$$\sigma_{LU} = \sigma_E = D_1^2 \frac{(P_i - P_e)}{\left(D_2^2 - D_1^2\right)} \tag{5.19}$$

Expanding Equation 5.19, an approximate thin wall expression is given as

$$\sigma_{LU} = \frac{D_1^2(P_i - P_e)}{4t(D_2 - t)} \propto \frac{D_2(P_i - P_e)}{4t}, \quad D/t \geq 20 \tag{5.20}$$

where $\sigma_{LU}$ is the unrestrained longitudinal stress (MPa), $M_b$ is the bending moment (N m), $D_1$ is the inside diameter (m), $D_2$ is the outside diameter (m), $Z$ is the pipe section modulus ($m^3$), $\sigma_E$ is the end cap stress (MPa), $i$ is the stress intensification factor (see Section 5.6.1), and $k$ is the ratio $D_2/D_1$.

*5.4.2.3* ***Partially Restrained Pipeline*** At the free end of a pipeline, the total friction force acting on the pipeline increases as one progresses from the free end. This friction is passive and acts against the forces that try to displace the line [2,4,5].

For a constant frictional force per unit length of line, the rate of change of longitudinal stress is constant. The expression for the longitudinal stress experienced by a pipeline that is partially restrained is given as

$$\sigma_{LP} = \sigma_E - \mu \frac{W_s X}{A_s}, \quad 0 \geq X \geq L_A \tag{5.21}$$

$$\sigma_E = D_1^2 \frac{(P_i - P_e)}{\left(D_2^2 - D_1^2\right)} \tag{5.22}$$

Expanding the above, an approximate thin wall expression is given as [5]

$$\sigma_{LP} = \frac{D_2(P_i - P_e)}{4t} - \mu \frac{W_s X}{A_s}, \quad D/t \geq 20 \tag{5.23}$$

where $\mu$ is the seabed coefficient of friction, $W_s$ is the submerged weight of pipeline, $X$ is the distance from free end of pipeline (m), $A_s$ is the area of steel ($A_s = \pi D t$) ($m^2$), and $L_x$ is the length between anchors (m).

*Note*

| | |
|---|---|
| $X=0$ | when $\sigma_{LP}=\sigma_E=\sigma_{LU}$ |
| $X=L_x$ | when $\sigma_{LP}=\sigma_{LR}$ |

*5.4.2.4* ***Bending Stress*** A pipe must sustain installation loads and operational loads. In addition, external loads such as those induced by waves, current, uneven seabed, trawlboard impact, pullover, and expansion due to temperature changes need to be considered. A pipe subjected to increasing bending may fail due to local buckling/collapse or fracture, but it is the local buckling/collapse limit state that usually dictates the design [13].

The pipe (beam) bending is analyzed using the engineer's theory of bending (ETB) and simple beam theory. The theory associates the bending stress at a point to the moment imposed on the section or the curvature experienced. It is used to calculate the bending stress at any point in a pipe.

*Assumptions*

- Transverse planes before bending remain transverse after bending, no warping.
- Pipe material is homogenous and isotropic and obeys Hooke's law with $E$ the same in tension or compression.
- The pipeline (beam) is straight and has constant or slightly tapered cross section.
- Load does not cause twisting or buckling. This is satisfied if the loading plane coincides with the section's symmetry axis.

- The pipeline is subject to pure bending. This means that the shear force is zero, and that no torsional or axial loads are present.

The equation is valid for elastic pure bending and does not take into account the geometrical deformations that occur, particularly in hollow cylinders (ovality) under bending. The equation is expressed as

$$\frac{\sigma_x}{y} = \frac{M}{I} = \frac{E}{R} \tag{5.24}$$

This is called the engineer's theory of bending. The standard form of writing this equation is

$$\sigma_x = \frac{My}{I} = \frac{Ey}{R}, \quad 0 \leq y \leq \text{OD}/2 \tag{5.25}$$

$$(\sigma_x)_{\max} = \frac{MD_0}{2I} = \frac{ED_0}{2R}, \quad y = \text{OD}/2 \tag{5.26}$$

where $\sigma_x$ is the bending stress ($0 \leq \sigma_x \leq \sigma_y$) (MPa), $M$ is the bending moment (N m), $y$ is the distance from the neutral axis ($0 \leq y \leq \text{OD}/2$) (m), $I$ is the moment of inertia (kg m$^3$), $E$ is the Young's modulus of elasticity (GPa), and $R$ is the radius of curvature (m).

Therefore, knowing the applied bending moment and the location of the centroid, the second moment of area and the stresses along the depth of the pipe section can be calculated.

In the simple theory of bending of beams, it is assumed that no appreciable distortion of the cross section takes place so that there is no displacement of material either toward or away from the neutral axis. For a thin-walled pipe subjected to bending, movement of the fibers toward the neutral axis does occur [14].

The maximum bending moment and hence the bending stress at any point in the cross section can be determined by assuming a beam configuration.

*Beam Configurations* A fixed–fixed beam under a uniformly distributed load gives a maximum bending moment ($M_{\text{ff}}$) at the fixed ends of the beam.

$$M_{\text{ff}} = \frac{WL^2}{12} \tag{5.27}$$

A beam with pinned supports under a uniformly distributed load gives a maximum bending moment ($M_{\text{pp}}$) at midspan.

$$M_{\text{pp}} = \frac{WL^2}{8} \tag{5.28}$$

It is commonly accepted that a real beam configuration for an unrestrained pipeline resting on the seabed is somewhere between the fixed–fixed and the pinned–pinned cases, and is given a value as follows:

$$M_{fp} = \frac{WL^2}{10} \tag{5.29}$$

where $M_{\text{ff}}$ is the fixed–fixed bending moment, $M_{\text{pp}}$ is the pinned–pinned bending moment, $W$ is the uniformly distributed load (per meter), $L$ is the length of pipeline span, and $M_{\text{fp}}$ is the bending moment halfway between the fixed–fixed and the pinned–pinned cases.

The stages of longitudinal stresses connected with bending are as follows:

- Elastic bending up to yield at outer fiber. At this point, the bending moment is equal to the first yield moment.
- Elastoplastic bending up to the development of a full plastic hinge. When plastic hinge is developed, it is known as the full plastic moment.

On the other hand, this can be solved by drawing the bending moment diagram, that is, if the value of the bending moment, $M$, is determined at various points of the beam and plotted against the distance $x$ measured from one end of the beam. It is further facilitated if a shear diagram is drawn at the same time by plotting the shear, $V$, against $x$.

This approach facilitates the determination of the largest absolute value of the bending moment in the beam.

#### 5.4.2.5 *Stress due to Sustained Loads*

Sustained loads are the sum of dead weight loads, axial loads caused by internal pressure, and other applied axial loads that are not caused from temperature and accelerations [3].

In accordance with ASME Boiler and Pressure Vessel Code, Section III, Subsections NC and ND, the calculated stresses due to pressure, weight, and other sustained mechanical loads must meet the allowable $1.5S_{\text{h}}$, that is,

$$\frac{B_1 P D_0}{2t} + \frac{B_2 M_{\text{A}}}{Z} \leq 1.5 S_{\text{h}} \tag{5.30}$$

where $T$ is the temperature derating factor, $P$ is the internal design pressure (MPa), $D_0$ is the outside diameter of pipe (m), $Z$ is the section modulus of pipe (m$^3$), $M_{\text{A}}$ is the resultant moment loading on cross section due to weight and other sustained loads (N m), $S_{\text{h}} = 0.33 S_u T$, at the maximum installed or operating temperature (MPa), and $S_{\text{u}}$ is the specified minimum ultimate tensile strength (N/m$^2$).

#### 5.4.2.6 *Stress due to Occasional Loads*

Occasional loads are loads such as wind, earthquake, breaking waves or green sea impact loads, and dynamic loads such as pressure relief, fluid hammer, or surge loads. In accordance with ASME Boiler and Pressure Vessel Code, Section III, Subsections

NC and ND, the calculated stress due to pressure, weight, other sustained loads, and occasional loads must meet the allowable stress as follows [3]:

$$\frac{B_1 P_{max} D_0}{2t} + \frac{B_2(M_A + M_B)}{Z} \leq kS_h \quad (5.31)$$

where $P_{max}$ is the peak pressure (MPa) and $M_B$ is the resultant moment loading on cross section due to occasional loads, such as thrusts from relief and safety valves, loads from pressure and flow transients, and earthquake, if required. For earthquake, use only one-half of the range. Effects of anchor displacement due to earthquake may be excluded if they are included under thermal expansion. $kS_h = 1.85S_h$ for upset condition but not greater than $1.5S_y$, $2.25S_h$ for emergency condition but not greater than $1.85S_y$, and $3.0S_h$ for faulted condition but not greater than $2.0S_y$. $S_h = 0.33S_uT$, at the maximum installed or operating temperature (MPa), $S_u$ is the specified minimum ultimate tensile strength (N/m$^2$), and $S_y$ is the material yield strength at temperature consistent with loading under consideration.

***5.4.2.7 Stress due to Thermal Expansion*** Thermal expansion may be detrimental for the pipe itself, flanges and bolts, branch connections, pipe supports, and connected equipment such as pumps and compressors. Sufficient pipe flexibility is necessary to prevent such detrimental loads [2].

Stresses due to expansion for those portions of the piping without substantial axial restraint shall be combined in accordance with the following equation (refer to Clause 833.8 of ASME B31.8):

$$\frac{M_E}{Z} = S_E \quad (5.32)$$

$$M_E = \left[(i_i M_i)^2 + (i_o M_o)^2 + M_t^2\right]^{1/2} \quad (\text{N m}) \quad (5.33)$$

The cyclic stress range $S_E \leq S_A$, where

$$S_A = f[(1.25S_c + 0.25S_h) - S_L] \quad (5.34)$$

If Equation 5.32 is not met, the piping may be qualified by meeting the following equation:

$$\frac{PD_0}{4t} + \frac{0.75iM_A}{Z} + \frac{iM_E}{Z} \leq S_h + S_A$$

$$(0.75i \text{ should not be less than } 1.0) \quad (5.35)$$

where $S_E$ is the stress due to expansion (MPa), $S_A$ is the allowable stress range for expansion stress, $f$ is the stress range reduction factor, $M_E$ is the range of resultant moment due to thermal expansion (N m), also includes moment effects of anchor displacements due to earthquake if anchor displacement effects were omitted from occasional loadings, $S_c = 0.33S_uT$, at the minimum installed or operating temperature (MPa), $S_h = 0.33S_uT$, at the maximum installed or operating temperature (MPa), $S_u$ is the specified minimum ultimate tensile strength (N/m$^2$), $S_L$ is the unrestrained longitudinal stress (N/m$^2$), $T$ is the temperature derating factor, $i$ is the stress intensification factor (see Section 6 or Appendix E of ASME B31.8), $M_i$ is the in-plane bending moment (N m), $M_t$ is the torsional moment (N m), $M_o$ is the out-of-plane bending moment (N m), $i_o$ is the out-of-plane stress intensification factor (refer to Appendix E of ASME B31.8), and $i_i$ is the in-plane stress intensification factor (refer to Appendix E of ASME B31.8).

### 5.4.3 Shear Stress

Shear stress in a pipeline should be minimized. The shear stress should be calculated from the torque and shear force applied to the pipeline using the following equation:

$$\tau = \frac{1000T}{2Z} + \frac{2F_s}{A} \quad (5.36)$$

where $\tau$ is the shear stress (N/m$^2$), $T$ is the torque applied to the pipeline (N m), $F_s$ is the shear force applied to the pipeline (N), $A$ is the cross-sectional area of the pipe (m$^2$), and $Z$ is the section modulus of the pipe (m$^3$).

Sections 5.4.3.1 and 5.4.3.2 show how to determine shear stress due to torsion and spanning, which makes up the total shear stress.

***5.4.3.1 Shear Stress due to Torsion*** Torsion is the twisting of a straight bar when it is loaded by twisting moments or torques that tend to produce rotation about the longitudinal axes of the bar. When subjected to torsion, every cross section of a circular shaft remains plane and undistorted, and the bar is said to be under pure torsion [15].

The shear stress on a uniform cylindrical shaft that is under a uniform torsion is given by

$$\tau = \frac{Tr_x}{J} = \frac{Tr_x}{2I_x} = \frac{T}{2Z} \quad (5.37)$$

For a thin cylindrical shaft (or thin-walled tube) with $t < R/10$,

$$J = R^2(2\pi rt) = 2\pi R^3 t = \frac{\pi}{4}D^3 t \quad (5.38)$$

$$\tau_{max} = \frac{TD}{2\left((\pi/4)D^3 t\right)} = \frac{2T}{\pi D^2 t} \quad (5.39)$$

The maximum shear stress due to torsion can be calculated as follows:

$$\tau_{max} = \frac{TD_o}{2J} = \frac{TD_o}{4I_x} = \frac{16TD_o}{\pi\left(D_o^4 - D_i^4\right)} \quad (5.40)$$

where $\tau$ is the shear stress (N/m$^2$), $T$ is the torque or twisting moment (N m), $R$ is the radial distance from the longitudinal axis (m), $I_x$ and $I_y$ are the moments of inertia about the $x$- and $y$-axis, respectively (kg m$^3$), $Z$ is the section modulus (m$^3$), and $J$ is the polar moment of inertia.

*5.4.3.2 Shear Stress due to Spanning* The shear stress due to spanning is composed of vertical shear and a longitudinal shear due to the bending. The maximum shear force acting on a simple span is equal to the maximum support reaction. This is in turn equal to the change in shear force at the reaction. The maximum vertical shear stress is defined as the force per unit area. The maximum vertical shear stress is calculated as follows [15]:

$$\tau_{max} = \frac{R_{max}}{A_s} = \frac{2F_s}{A_s} \quad (5.41)$$

where $F_s$ is the shear force applied to the pipeline (N), $A_s$ is the cross-sectional area of the pipe (m$^2$), and $R_{max}$ is the maximum vertical reaction on the pipe (N).

### 5.4.4 Equivalent Stress

Pressure and temperature as well as other operating conditions such as bending can create expansion and flexibility problems and therefore stress criteria are specified in all codes limiting the level of combined stresses allowed in a pipeline.

The design factor relates only to hoop stress; if other stresses are significant, then these could contribute to the pipeline steel exceeding its yield stress. Design codes vary in the way they calculate the combined or equivalent stress, but the following equation is typical [2–9].

The equivalent stress corresponds to the total stress in the pipeline resulting from a combination of all the stresses.

The equivalent stress can be calculated using the following equation (refer to DNV 2012, Clause 103):

$$\sigma_e = \sqrt{\sigma_H^2 + \sigma_L^2 - \sigma_H\sigma_L + 3\tau^2} \quad (5.42)$$

where $\sigma_e$ is the equivalent stress (N/m$^2$), $\sigma_H$ is the hoop stress (N/m$^2$), $\sigma_L$ is the longitudinal stress (N/m$^2$), and $\tau$ is the shear stress (N/m$^2$).

In accordance with Clause 833.4 of ASME B31.4, the maximum allowable equivalent stress is 90% of the SMYS.

### 5.4.5 Limits of Calculated Stress

Pipe structures should be designed by considering the limit states at which they would be unfit for their intended use by applying appropriate factors.

With reference to BS 8010, Part 3, Clause 4.2.5.4, stress in the pipeline system should satisfy the following inequality:

$$\text{allowable stress} = \text{design factor} \times \text{specified minimum yield strength} \quad (5.43)$$

The allowable stress depends on the pipe material used, the location of the pipe, the operating conditions, and other limitations imposed by the designer in conformance with the code used. The allowable stresses for various grades and types of material are tabulated in Table 402.3.1(a) of ANSI B31.4, 1992 edition.

In accordance with ASME B31.4, for an unrestrained pipeline the allowable effective stress is $0.72S_y$, and for a restrained line the allowable stress is $0.9S_y$.

With reference to Clause 833.3 of ASME B31.8, for a restrained pipe the allowable longitudinal stress is $0.9S_yT$, where $S_y$ is the specified minimum yield strength (MPa) and $T$ is the temperature derating factor.

Based on Clause 833.6 of ASME B31.8, for an unrestrained pipe the allowable longitudinal stress is $\sigma_{AL} \leq 0.75S_yT$, where $S_y$ is the specified minimum yield strength (MPa) and $T$ is the temperature derating factor [2].

*5.4.5.1 Allowable Hoop Stress* The allowable hoop stress may be calculated using the following equation (refer to Clause 805.234 of ASME B31.8 and Clause 201.4.1 of API RP 1111):

$$\sigma_{aH} = f \times e \times T \times S_y \quad (5.44)$$

where $\sigma_{aH}$ is the allowable hoop stress, $S_y$ is the specified minimum yield strength, $f$ is the design factor, $e$ is the weld joint factor, and $T$ is the temperature derating factor.

The allowable hoop stress for cold worked pipe is 75% of the above value (Clause 201.4.4 of API RP 1111). The design factor is 0.72 for pipelines and liquid risers, 0.60 for gas risers, and 0.50 for gas platform piping [2].

*5.4.5.2 Allowable Equivalent Stress*

$$\sigma_{ae} \leq FS_y \quad (5.45)$$

In accordance with Clause 833.4 of ASME B31.4, the maximum allowable equivalent stress is 90% of the SMYS.

$$\sigma_{ae} = 0.9S_y \quad (5.46)$$

where $\sigma_{ae}$ is the allowable equivalent stress and $S_y$ is the specified minimum yield strength.

*5.4.5.3 Limits of Calculated Stress due to Sustained Loads* The sum of the longitudinal stresses due to pressure, weight, and other sustained external loads shall not exceed $0.72S_A$, where $S_A = 0.75S_y$ ($S_y$ is the specified minimum yield strength) [6].

***5.4.5.4 Limits of Calculated Stress due to Occasional Loads*** The sum of the longitudinal stresses produced by pressure, live and dead loads, and other sustained loadings and of the stresses produced by occasional loads, such as wind or earthquake, may be as much as $1.33S_h$ [6].

***5.4.5.5 Limits of Calculated Stress due to Expansion Loads*** The computed displacement stress range (expansion stress range) $S_E$ in a pipeline should not exceed the allowable displacement stress range $S_A$ [6].

$$S_E < S_A = f(1.25S_c + 0.25S_h) \tag{5.47}$$

When $S_h$ is greater than $S_L$, the difference between them may be added to the term $0.25S_h$; in that case, the allowable stress range is calculated as

$$S_A = f[(1.25(S_c + S_h) - S_L] \tag{5.48}$$

where $S_E$ is the expansion stress range $= \left(S_h^2 + 4S_t^2\right)^{1/2}$ (MPa), $M_E$ is the resultant bending stress $= \left[(i_i M_i)^2 + (i_o M_o)^2\right]^{1/2}/Z$ (MPa), $S_t$ is the torsional stress $= M_t/2Z$ (MPa), $M_i$ is the in-plane bending moment (N m), $M_o$ is the out-of-plane bending moment (N m), $M_t$ is the torsional moment (N m), $i_i$ is the in-plane stress intensification factor, $i_o$ is the out-of-plane stress intensification factor, $Z$ is the section modulus of pipe (m$^3$), $S_c = 0.33S_uT$, at the minimum installed or operating temperature (MPa), $S_h = 0.33S_uT$, at the maximum installed or operating temperature (MPa), $S_u$ is the specified minimum ultimate tensile strength (MPa), and $f$ is the stress range reduction factor (obtained from Table 302.3.5 of ASME B31.3) or calculated as follows:

$$f = 6.0(N)^{-0.2} \leq 1.0 \tag{5.49}$$

where $N$ is the equivalent number of full displacement cycles during the expected service life of the piping system.

## 5.5 FATIGUE

Fatigue is a structural damage that occurs when a pipe material is subjected to cycles of stress or strain [16]. Such stresses are normally concentrated locally by structural discontinuities, geometric notches, surface irregularities, damage defects, and so on.

ASME B31.4 shows how to design the pipeline against fatigue failure.

Pipelines can vary in pressure over hourly, daily, or yearly cycles [16]. Pressure cycling can cause small weld defects to grow in time to a critical size and can be a major factor in determining the fatigue life of welded steel gas pipelines, particularly those pipelines designed for use as line-pack storage. (Gas can be stored temporarily in the pipeline system through a process called line packing.)

### 5.5.1 Fatigue Life

ASTM defines fatigue life as the number of stress cycles of a specified character that a specimen sustains before failure of a specified nature occurs. Fatigue life may be affected by cyclic stress state, geometry, surface quality, material type, residual stresses, size and distribution of internal defects, air or vacuum, direction of loading, grain size, environment, temperature, and crack closure [17].

To avoid fatigue failure of a pipeline, the following should be adhered to [8]:

- One stress cycle per day of 125 N/mm$^2$ could lead to failure after 40 years (15,000 cycles). This is very important for gas pipelines operating under line-pack conditions.
- Determine the number of stress cycles and the stress range expected during the design life of a pipeline. If the value of stress cycles comes within reach of 15,000, then the pipe material should be changed or the frequency of cycling or the stress range of cycling should be reduced over the pipe life.
- Revalidate the pipeline by hydrostatic testing, if the number of stress cycles reaches 15,000.
- A crack detection tool can be used to determine the condition of the pipeline.
- Keep the log of pressures and cycles throughout the design life of the pipeline.

### 5.5.2 Fatigue Limit

The significance of the fatigue limit is that if the material is loaded below this stress, then it will not fail, regardless of the number of times it is loaded [16]. In accordance with IGE/TD/1, 15,000 cycles at 125 N/mm$^2$ has been set as the maximum permissible fatigue life.

### 5.5.3 *S–N* Curve

A very useful way to visualize time to failure for a specific material is with the *S–N* curve. This is a graph of the magnitude of a cyclic stress ($S$) against the logarithmic scale of cycles to failure ($N$).

A fatigue life test should be conducted for the pipe material and an *S–N* curve drawn [16]. For instance, a specimen of the pipe material is placed in a fatigue testing machine and loaded repeatedly to a certain stress, $\sigma_1$. The

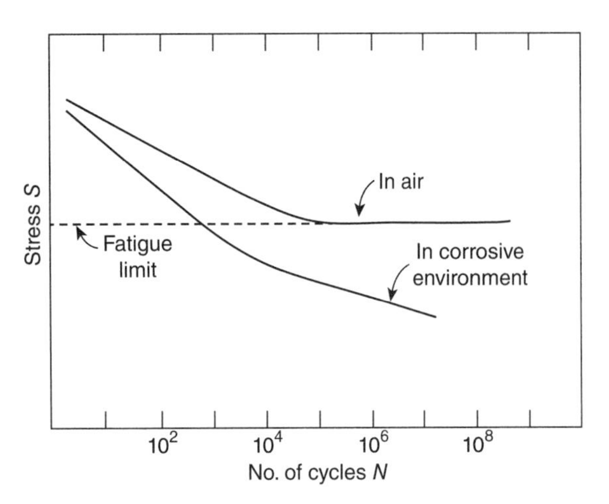


**FIGURE 5.2** Typical generic *S–N* curves for steel.

loading cycles are continued until failure occurs and the number, $n$, of loading cycles to failure is noted. Then the test is repeated for a different stress, say $\sigma_2$; if $\sigma_2$ is larger than $\sigma_1$, the number of cycles to failure will be less. If it is smaller, the number of cycles to failure will be more [16].

Eventually, enough data are accumulated to plot an *S–N* curve. Such curves have the general shape shown in Figure 5.2, where the vertical axis is usually a linear scale and the horizontal axis is a log scale.

From Figure 5.2, the smaller the stress, the larger is the number of cycles to produce failure. Fatigue strength curves (*S–N* curves) for particular material or structural weldment formation can be found in design standards such as ASME Boiler and Pressure Vessel Code, Section VIII, Division 2, Appendix 5, and BS 7910.

## 5.6 EXPANSION AND FLEXIBILITY

Avoid expansion bends and design the entire pipeline to take care of its own expansion. Maximum flexibility is obtained by placing supports and anchors so that they will not interfere with the natural movement of the pipe. Allow for stress intensification factors in components [10].

Expansion joints may be used to avoid pipeline bending (flexure) stress due to the movement of supports or the tendency of the pipe to expand under temperature change. The following should be considered for the use of expansion joints [2]:

- Select expansion joint carefully for maximum temperature range (and deflection) expected so as to prevent damage to expansion fitting.
- Provide guides to limit movement at expansion joint to direction permitted by joint.
- Provide adequate anchors at one end of each straight section or along their middle length, forcing movement to occur at expansion joint yet providing adequate support for pipeline.
- Mount expansion joints adjacent to an anchor point to prevent sagging of the pipeline under its own weight and do not depend upon the expansion joint for stiffness (it is intended to be flexible).
- Give consideration to effects of corrosion, since corrugated character of expansion joints makes cleaning difficult.

Formal flexibility analysis for an unrestrained piping system may not be required.

### 5.6.1 Flexibility and Stress Intensification Factors

The stress intensification factor (SIF) is defined as the ratio of the maximum stress state (stress intensity) to the nominal stress, calculated by the ordinary formulas of mechanics [2].

In piping design, this factor is applied to welds, fittings, branch connections, and other piping components where stress concentrations and possible fatigue failure might occur. Usually, experimental methods are used to determine these factors.

The structural analysis of risers, expansion loops, or tee assemblies entails the use of flexibility stress intensification factors applicable to accurately model the structural behavior of bends and tees within the system.

It is recognized that some of the SIFs for the same components are different for different codes. In some cases, different editions of the same code provide different SIFs for a given component. The way that the SIFs are applied to moment loadings is also different for different codes. The B31.1 and ASME Section III codes require that the same SIF be applied to all the three-directional moments, while the B31.3, B31.4, B31.5, and B31.8 codes require that different SIFs be applied to the in-plane and out-of-plane moments, with no SIF required for torsion.

Therefore, the stress analyst has to ensure that the appropriate SIFs from the applicable code are used [2].

Flexibility is added in a pipe system by changes in the run direction (offsets, bends, and loops) or by use of expansion joints or flexible couplings of the slip joint, ball joint, or bellow type. In addition, more or less flexibility can be added by changing the spacing of pipe supports and their function (e.g., removal of a guide close to a bend to add flexibility).

Another way to increase the flexibility is to change the existing piping material to a material with a higher yield or tensile strength, or to a material quality that does not need additional corrosion and erosion allowance, and thereby obtains a reduction in the wall thickness that again gives more flexibility, since the moment of inertia is reduced with a reduction of the pipe wall thickness.

The flexibility factors and stress intensification factors that can be used are listed in standards, for example, ASME B31.8, Table E1 and CSA Z662-11, Table 4.8.

## 5.7 CORROSION ALLOWANCE

Additional wall thickness is added to account for corrosion when water is present in a fluid along with contaminants such as oxygen, hydrogen sulfide ($H_2S$), and carbon dioxide ($CO_2$). This is one of the several methods available to mitigate the effects of corrosion, and often the least recommended [18].

Corrosion allowance is made to account for corrosion loss during service, damage during fabrication, transportation, and storage. A value of 1/16 in. may be appropriate. A thorough assessment of the internal corrosion mechanism and rate is necessary before any corrosion allowance is taken. Refer to BS 8010, ASME codes, ISO, API 5L, and other governing codes for the use of corrosion allowance in the design of pipelines [19].

### 5.7.1 Internal Corrosion Allowance

There is no need for internal corrosion allowance if the substance being transmitted is noncorrosive, for example, dry natural gas [18]. An internal corrosion allowance should be added to the wall thickness calculation when corrosive substances such as wet gas, hydrocarbon liquid, or two-phase flow are being transported through the pipeline [19].

### 5.7.2 External Corrosion Allowance

A wall thickness allowance for corrosion is not required if pipe and components are protected against corrosion in accordance with the requirements and procedures prescribed in ASME B31.4, ASME B31.8, and other governing codes (i.e., coated and cathodically protected) [2–9].

### 5.7.3 Formulas

This section shows how to add corrosion allowance to the calculation of the nominal wall thickness of a pipeline, using any of the engineering codes listed below [2–9].

1. Using BS 8010:

$$t = \frac{P \times D}{20 S_y \times E \times F} \tag{5.50}$$

$$t_{nom} = (t + t_{corr}) \times \text{manufacturing tolerance} \tag{5.51}$$

2. Using ASME B31.8:

$$t = \frac{P \times D}{20 S_y \times E \times F} + t_{corr} \tag{5.52}$$

3. Using ISO 13623:

$$t = \frac{P \times D}{(20 S_y \times E \times F) + P} \tag{5.53}$$

$$t_{nom} = (t + t_{corr}) \times \text{manufacturing tolerance} \tag{5.54}$$

where $t_{corr}$ is the corrosion allowance (m), $t_{nom}$ is the nominal wall thickness (m), $P$ is the design pressure (MPa), $S_y$ is the specified minimum yield strength (MPa), $t$ is the design wall thickness (m), $D$ is the outside diameter of pipe (m), $F$ is the design factor, and $E$ is the joint factor.

## 5.8 PIPELINE STIFFNESS

With reference to DNV 2012, Clause 205, possible strengthening effect of weight coating on a steel pipe is not normally taken into account in the design against yielding. Coating that adds significant stiffness to the pipe may increase the stress in the pipe at discontinuities in the coating. When appropriate, this effect should be taken into account.

For buried pipe, resistance to external loading is a function of pipe stiffness and passive soil resistance under and adjacent to the pipe.

The overall stiffness of a long section of pipeline should be calculated from the value for the moment of inertia and should be used to calculate the overall deflections and induced bending moments for concrete-coated pipe [20].

### 5.8.1 Calculation of Pipeline Stiffness

Under load, the individual components of the pipe wall (steel, mortar lining, and, when applicable, mortar coating) act together as laminated rings. The combined action of these elements increases the overall moment of inertia of the pipe, over that of the steel pipe alone [20].

The pipe wall stiffness is the sum of the stiffness of the bare pipe, lining, and coating.

The pipe wall stiffness ($EI$) is the sum of the stiffness of the bare pipe, lining, and coating.

$$(EI) = E_s I_s + E_L I_L + E_c I_c \tag{5.55}$$

$$I = \frac{t^3}{12} \tag{5.56}$$

where $t$ is the wall thickness of pipe, lining, or coating, $(EI)$ is the pipe wall stiffness per inch of pipe length (in./lb), $E_L I_L$ is the stiffness of lining, $E_c I_c$ is the stiffness of coating (e.g., concrete), $E_s I_s$ is the stiffness of steel pipe wall, $E$ is the modulus of elasticity (207 GPa for steel and 27.6 GPa for cement mortar), and $I$ is the transverse moment of inertia per unit length of pipe wall (in.$^3$ or mm$^3$).

The stiffness of each of the laminar rings (i.e., steel pipe, cement mortar lining, and cement mortar coating) is calculated using the modulus of elasticity of the component in GPa and the moment of inertia as a per unit length value, defined as $t^3/12$.

*5.8.1.1* ***Deflection*** Pipe stiffness and passive soil resistance of backfill play a significant role in predicting deflection. M.G. Spangler [21] of Iowa State University published the Iowa formula in 1941.

$$\text{Pipe deflection} = \frac{\text{load on pipe}}{\text{pipe stiffness} + \text{soil stiffness}} \quad (5.57)$$

Deflection of a pipeline is calculated using the modified Iowa deflection formula as follows:

$$\Delta x = \frac{D_l K W R^3}{(EI) + 0.061 E' R^3} \quad (5.58)$$

where $\Delta x$ is the horizontal deflection of the pipe (m), $D_l$ is the deflection lag factor (1.0–1.5), $K$ is the bedding constant (0.1), $R$ is the pipe radius (m), $(EI)$ is the pipe wall stiffness per meter of pipe length (in./lb, m/kg), $E'$ is the modulus of soil reaction (GPa), and $W$ is the external load per unit length of pipe (dead load (earth load) + live load).

Some data on loading are presented in Table 5.4.

Spangler hypothesized that if the lateral movement of various points on the pipe ring were known, the distribution of lateral pressures could be determined by multiplying the movement of any point by the modulus of passive resistance, $E'$. For mathematical convenience, this lateral pressure was assumed to be a simple parabolic curve embracing only the middle 100° arc of the pipe (see Figure 5.3).

He also assumed that the total vertical load was uniformly distributed across the width of the pipe, and the bottom vertical load was distributed uniformly over the width of the pipe bedding [20].

The following terms can be introduced to describe the three separate factors that affect the pipe deflection:

**TABLE 5.4 Standard HS-20 Highway and E-80 Railroad Loading [20]**

| Highway HS-20 Loading | | Railroad E-80 Loading | |
|---|---|---|---|
| Height of Cover (ft) | Load (psi) | Height of Cover (ft) | Load (psi) |
| 1 | 12.5 | 2 | 26.4 |
| 2 | 5.6 | 5 | 16.7 |
| 3 | 4.2 | 8 | 11.1 |
| 4 | 2.8 | 10 | 7.6 |
| 5 | 1.7 | 12 | 5.6 |
| 6 | 1.4 | 15 | 4.2 |
| 7 | 1.2 | 20 | 2.1 |
| 8 | 0.7 | 30 | 0.7 |

- load factor $(D_l K W)$;
- ring stiffness factor $(EI/R^3)$;
- soil stiffness factor $(0.061E')$. Values of $E'$ are listed in Table 5.5.

The modified Iowa formula can be represented as

$$\Delta X = \frac{\text{load factor}}{\text{ring stiffness factor} + \text{soil stiffness factor}} \quad (5.59)$$

Information on the load factor, ring stiffness factor, soil stiffness factor, bedding constant, and deflection lag factor can be obtained from standards of American Concrete Pipe Association [20–22].

The steel pipe is designed as a flexible conduit; considerable deflection can occur without damaging the pipeline. Deflection limitations are a function of the rigidity of the specific lining and coating being used.

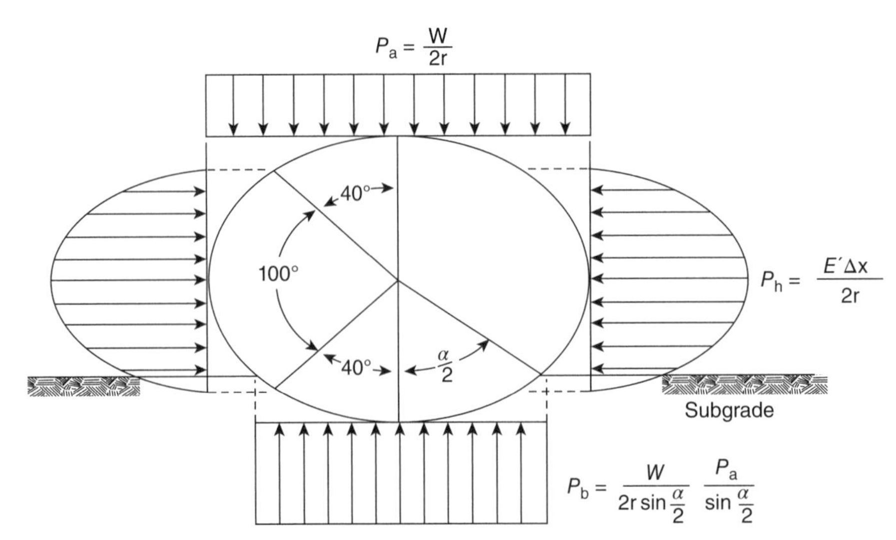


**FIGURE 5.3** Spangler assumptions for pressure distribution [21]. ( Published with permission of American Concrete Pipe Association.)

**TABLE 5.5 $E'$ Modulus of Soil Reaction [20]**

| Type of Soil | Depth of Cover (ft) | $E'$ Modulus of Soil Reaction (psi) | | | |
|---|---|---|---|---|---|
| | | 85%[a] | 90%[a] | 95%[a] | 100%[a] |
| Fine-grained soils with less than 25% sand content (CL, ML, and CL—ML) | 0–5 | 500 | 700 | 1000 | 1500 |
| | 5–10 | 600 | 1000 | 1400 | 2000 |
| | 10–15 | 700 | 1200 | 1600 | 2300 |
| | 15–20 | 800 | 1300 | 1800 | 2600 |
| Coarse-grained soil with fines (SM and SC) | 0–5 | 600 | 1000 | 1200 | 1900 |
| | 5–10 | 900 | 1400 | 1800 | 2700 |
| | 10–15 | 1100 | 1700 | 2300 | 3300 |
| | 15–20 | 1300 | 2000 | 2700 | 3800 |
| Coarse-grained soil with little or no fines (SP, SW, GP, and GW) | 0–5 | 700 | 1000 | 1600 | 2500 |
| | 5–10 | 1000 | 1500 | 2200 | 3300 |
| | 10–15 | 1050 | 1600 | 2400 | 3600 |
| | 15–20 | 1100 | 1700 | 2500 | 3800 |
| Crushed stone | N/A | 3000 | 3000 | 3000 | 3000 |

[a]Standard AASHTO relative compaction.

The calculated deflection is limited to 5%, although larger deflections may not affect pipe performance. Limits for pipe deflection for various forms of coating or lining should be obtained from governing standards and codes [20].

### 5.8.2 Calculation of Induced Bending Moment

Coating that adds significant stiffness to the pipe may increase the stress in the pipe at discontinuities in the coating. When appropriate, this effect should be taken into account.

Maximum stresses will develop along the pipeline where there is no concrete coating (i.e., at the field joint) and may be calculated for a known curvature from Equation 5.60.

$$\sigma_B = \frac{E \times C}{R} \times \frac{I_o}{I_s} \tag{5.60}$$

where $\sigma_B$ is the bending stress, $C$ is the outside radius of steel pipe, $R$ is the imposed bending radius, $I_o$ is the overall moment of inertia, $I_s$ is the moment of inertia of bare pipe, and $E$ is the modulus of elasticity.

## 5.9 PIPELINE OVALITY

Ovality is generally associated with bending of pipes and affects the integrity of pipe bends. The control of ovality in pipe bending is discussed in this section [13].

The ovality of a pipe section depends on the dimensional tolerances imposed during the pipe bending while installing the pipe to fit the terrain [14].

Based on DNV 2012, Clause 202, ovality of a pipeline is defined in Equation 5.61. This will affect the structural capacity of the pipeline and shall be taken as the maximum ovality prior to loading. Advantage of ovality less than 0.5% is not allowed. Ovality in excess of 3% shall be assessed in line with DNV 2012, Clause D900. Ovalization caused during construction stage should be included in the total ovality to be used for design [7]. Pipe ovality is calculated as

$$f_0 = \left(\frac{D_{max} - D_{min}}{D}\right) \leq 0.03 \tag{5.61}$$

The out of roundness is calculated as

$$O = D_{max} - D_{min} \tag{5.62}$$

where $D_{max}$ is the maximum measured inside or outside diameter, $D_{min}$ is the minimum measured inside or outside diameter, $D$ is the nominal outside diameter, $f_0$ is the ovality, and $O$ is the out of roundness.

With reference to DNV 2012, Clause D1100, a maximum allowable ovalization of 3% applies for the pipeline as installed condition. Under DNV 2012, Clause D400, a minimum ovalization of 0.5% is to be accounted for in the system collapse check and the combined loading [7].

In accordance with DNV 2012, Clause D1100, the ovality of a pipeline exposed to bending strain may be calculated using Equation 5.63.

$$f'_0 = \frac{f_0 + \left[0.03(1 + (D/120t))(2\varepsilon_c(D/t))^2\right]}{1 - (P_e/P_c)} \tag{5.63}$$

where $f'_0$ is the total ovalization due to unidirectional bending and external pressure, $\varepsilon_c$ is the characteristic bending strain resistance, $f_0$ is the initial ovalization, $P_e$ is the external pressure (MPa), $P_c$ is the characteristic collapse pressure (MPa), and $t$ is the nominal wall thickness of pipe (not corroded) (m).

### 5.9.1 Brazier Effect

The ovalization mechanism results in loss of stiffness in the form of limit point instability, referred to as "ovalization instability" or Brazier effect. Brazier effects describe the influence of ovalization on the buckling of thin shells. Ovalization can be enhanced by applying bending moments and internal or external pressure [23].

Brazier projected that the ovality was related to longitudinal bending strain by a correlation as follows [5,24]:

$$F_b \propto \left(\frac{\varepsilon_b D}{t}\right)^2 \tag{5.64}$$

Based on the work of Calladine, the correlation to be used in resolving the flattening induced by bending is calculated as [5,23,24]

$$F_b = \frac{3}{8}\left(\frac{\varepsilon_b D_o}{t}\right)^2 \tag{5.65}$$

where $\varepsilon$ is the percentage bending strain, $D$ is the nominal diameter (m), and $t$ is the nominal wall thickness (m).

This correlation may be used for diameter to wall thickness ratio less than 35.

### 5.9.2 Ovality of a Buried Pipeline

A buried pipe will ovalize under the effects of earth and live loads. The modified Iowa deflection formula may be used to calculate the pipe ovality under earth and live loads [20,22].

$$\frac{\Delta y}{\text{OD}} = \frac{D_l KP}{(EI)_{eq}/R^3 + 0.061E'} \tag{5.66}$$

where OD is the pipe outside diameter (m), $\Delta y$ is the vertical deflection of pipe (m), $D$ is the deflection lag factor (1.0–1.5), $K$ is the bedding constant (0.1), $P$ is the pressure on pipe due to soil load $P_v$ plus live load $P_p$ (MPa), $R$ is the pipe radius (m), $(EI)_{eq}$ is the equivalent pipe wall stiffness per inch of pipe length (in./lb, m/kg), and $E'$ is the modulus of soil reaction (MPa).

The pipe wall stiffness, $(EI)_{eq}$, is the sum of the stiffness of the bare pipe, lining, and coating [22].

$$(EI)_{eq} = EI + E_L I_L + E_c I_c \tag{5.67}$$

$$I = \frac{t^3}{12} \tag{5.68}$$

where $E_L I_L$ is the stiffness of lining, $E_c I_c$ is the stiffness of coating, and $t$ is the wall thickness of pipe, lining, or coating.

## 5.10 MINIMUM PIPE BEND RADIUS

In engineering, the minimum bend radius of a material is a measure of how tightly a piece can be bent before it breaks. Materials that can withstand a high degree of curvature are preferred in construction because they are more versatile, lending themselves to wider array of industrial applications [11].

Many factors affect the pipe minimum bend radius and knowing these factors is crucial to building sound structures. In pipe fitting and sheet metal construction, the minimum bending radii of materials depend on their thickness, composition, and skill of the fabricator [25].

### 5.10.1 Minimum Pipe Bend Radius Calculation Based on Concrete

$$R = \frac{E \times C}{\sigma_B} \tag{5.69}$$

where $R$ is the bending radius (m), $C$ is the pipe radius + enamel thickness + concrete thickness (m), $\sigma_B$ is the bending stress (MPa), and $E$ is the modulus of elasticity for concrete ($10^3$ MPa).

### 5.10.2 Minimum Pipe Bend Radius Calculation Based on Steel

$$\sigma_L + \sigma_B \leq (F)(S_y) \tag{5.70}$$

$$\sigma_B = (F)(S_y) - \sigma_L \tag{5.71}$$

$$\sigma_L = \frac{PD}{4t}$$

$$R = \frac{E \times C}{\sigma_B}$$

$$\therefore R = \frac{E \times C}{((FS_y) - (PD/4t))} \tag{5.72}$$

where $\sigma_L$ is the longitudinal stress (MPa), $\sigma_B$ is the bending stress (MPa), $S_y$ is the pipe specified minimum yield strength (MPa), $P$ is the design pressure (MPa), $D$ is the pipe outer diameter (m), $t$ is the pipe wall thickness (m), $R$ is the bending radius (m), $E$ is the modulus of elasticity ($10^3$ MPa), $C$ is the pipe radius (m), and $F$ is the stress factor.

Generally, the minimum pipe bend radius is required during the installation load case and the operational (in-service) load case [11].

### 5.10.3 Installation Condition

It is usually mandatory to specify the minimum radius of curvature permitted during the installation stage of a pipeline, whether it is S-lay, J-lay, or Tow-out, reeling of a pipeline [11].

The minimum bend radius for installation condition should be calculated as

$$R = \frac{E \times C}{\sigma_B}$$

$$\sigma_L + \sigma_B \leq \sigma_{ae}$$

From Equation 5.45,

$$\sigma_{ae} = FS_y$$

$$\sigma_L + \sigma_B \leq FS_y$$

$$\sigma_B = (F)(S_y) - \sigma_L$$

$$(\sigma_B)_{perm} = (FS_y) - \sigma_L$$

$$\sigma_L = \frac{PD}{4t}$$

$$\therefore (\sigma_B)_{perm} = (FS_y) - \frac{PD}{4t}$$

$$R_{perm} = \frac{E \times C}{(\sigma_B)_{perm}} \quad (5.73)$$

where $(\sigma_B)_{perm}$ is the permissible bending stress (MPa).

All other parameters are as defined in Section 5.10.2.

### 5.10.4 In-Service Condition

$$R_a = \frac{E \times C}{E_{perm}} \quad (5.74)$$

where $R_a$ is the minimum as-laid radius and $E_{perm}$ is the maximum permissible permanent elastic strain.

***5.10.4.1 Pipeline Located Offshore*** On the other hand, for a pipeline located offshore, the minimum bend (curvature) in order to prevent slippage is calculated by the following equation:

$$R_{sl} = \frac{T_o}{W_s F} \quad (5.75)$$

where $R_{sl}$ is the minimum bend to prevent slippage (limiting slippage curvature), $T_o$ is the normal on bottom tension, $W_s$ is the submerged unit weight, and $F$ is the lateral friction coefficient or the on-bottom friction coefficient.

$$T_o = T_i - WH$$

where $T_0$ is the maximum tension, $H$ is the water depth, and $T_i$ is the tension at inflexion point.

## 5.11 PIPELINE DESIGN FOR EXTERNAL PRESSURE

### 5.11.1 Buried Installation

***5.11.1.1 Check for Buckling*** Buried pipelines supported by a well-compacted, granular backfill will not buckle due to vacuum (i.e., when the gauge pressure is below atmospheric pressure) in the pipeline system. To confirm the stability of a pipeline, an analysis of the external loads relative to the pipe stiffness can be performed.

If the soil and surface loads are excessive, the pipe cross section could buckle. The ring buckling depends on limiting the total vertical pressure load on pipe [20,22].

$$\sigma_{rb} = \frac{1}{FS}\sqrt{32R_w B' E' \frac{(EI)_{eq}}{D^3}} \quad (5.76)$$

where $\sigma_{rb}$ is the allowable buckling pressure (MPa), FS is the factor of safety, $C$ is the depth of soil cover above pipe (m), $D$ is the outside diameter of pipe (m), $R_w$ is the water buoyancy factor $= 1 - 0.33(h_w/C)$, $0 < h_w < C$, $h_w$ is the height of water surface above top of pipe, $(EI)_{eq}$ is the equivalent pipe stiffness, $E'$ is the modulus of soil reaction (GPa), and $B'$ is the empirical coefficient of elastic support.

$$B' = \frac{1}{1 + 4e^{(-0.065C/D)}} \quad (5.77)$$

In steel pipelines, buckling occurs when the ovality reaches about 20%.

The sum of the external loads should be less than or equal to the pipe's allowable buckling pressure, $\sigma_{rb}$.

Confirming the resistance of a pipeline to bucking involves an analysis of the external loads relative to the pipe stiffness [20]. The total external load acting on a pipe must be less than or equal to the allowable buckling pressure of the pipe, and can be calculated using the following equation:

$$\gamma_W h_W + R_w \frac{W_c}{D} + P_v \leq \sigma_{rb} \quad (5.78)$$

where $\sigma_{rb}$ is the allowable buckling pressure (MPa), $\gamma_W$ is the specific weight of water (0.0361 lb/in.$^3$), $h_w$ is the height of water above pipe (in., m), $R_w$ is the water buoyancy factor, $W_c$ is the vertical soil load on pipe per unit length (lb/in., kg/m), $D$ is the outside diameter (m), and $P_v$ is the internal vacuum pressure (MPa).

The total live load acting on a pipe must be less than or equal to the allowable buckling pressure of the pipe. When analyzing the possibility of potential buckling of a pipeline, there is the need to determine the total live loads acting on a pipeline, using the following equation:

$$\gamma_W h_W + R_w \frac{W_c}{D} + \frac{W_L}{D} \le \sigma_{rb} \quad (5.79)$$

where $W_L$ is the live load on pipe per unit length (lb/in., kg/m).

When the allowable buckling pressure is not sufficient to resist the buckling loads, the soil envelope should first be investigated to increase the allowable $E'$ [20].

### 5.11.2 Above-Ground or Unburied Installation

It is recommended to use propagation criteria for pipeline diameters under 16 in. and collapse criterion for pipeline diameters above or equal to 16 in.

The propagation criterion is out of date and should be used where optimization of the wall thickness is not required or for pipeline installation methods not compatible with the use of buckle arrestors such as reel and tow methods [20].

It is generally economical to design for propagation pressure for diameters less than 16 in. For greater diameters, the wall thickness penalty is too high. When a pipeline is designed based on the collapse criteria, buckle arrestors are recommended.

***5.11.2.1 Collapse Criterion*** When a pipeline installed above ground is subjected to vacuum, the wall thickness must be designed to resist collapse due to the vacuum. Analysis should be based on the pipe functioning in the open atmosphere, absent of support from any backfill material [20]. Collapse pressure, $p_c$, is the pressure required to buckle a pipeline.

The collapse pressure should be calculated using Timoshenko's theory for collapse of a round steel pipe as follows [20]:

$$P_c = \frac{2E_s(t_s/d_n)^3}{(1-v_s{}^2)} + \frac{2E_I(t_I/d_n)^3}{(1-v_I{}^2)} + \frac{2E_c(t_c/d_n)^3}{(1-v_c{}^2)} \quad (5.80)$$

where $P_c$ is the collapsing pressure (MPa), $t_s$ is the steel cylinder wall thickness (m), $t_I$ is the cement coating thickness (m), $d_n$ is the diameter to neutral axis of shell (m), $E_s$ is the modulus of elasticity for steel ($30 \times 10^6$ psi, $207 \times 10^3$ MPa), $E_I$ and $E_c$ are the moduli of elasticity for cement mortar ($4 \times 10^6$ psi, $27.6 \times 10^3$ MPa), $v_s$ is the Poisson's ratio for steel (0.30), and $v_I$ and $v_c$ are the Poisson's ratios for cement mortar (0.25).

The mode of collapse is a function of $D/t$ ratio, pipeline imperfections, and load conditions. Safety factor of 1.3 is recommended. When a pipeline is designed using the collapse criterion, a good knowledge of the loading conditions is required.

***5.11.2.2 Propagation Criterion*** Propagating pressure, $P_p$, is the pressure required to continue a propagating buckle. A propagating buckle will stop when the pressure is less than the propagating pressure (refer to DNV 2012, Clause 501). The recommended formula for calculating the propagation criterion is the latest given by AGA [18].

$$P_p = 33 S_y \left(\frac{t_{nom}}{D}\right)^{2.4} \quad (5.81)$$

The nominal wall thickness ($t_{nom}$) should be determined such that

$$P_p \ge 1.3 P_e \quad (5.82)$$

where $P_p$ is the propagation pressure (MPa) and $P_e$ is the external pressure (MPa).

The recommended safety factor of 1.3 is to account for uncertainty in the envelope of data points used to derive Equation 5.82.

## 5.12 CHECK FOR HYDROTEST CONDITIONS

In order to check that the pipeline is fit for the purpose for which it was designed, a pressure test of a harmless fluid prior to commissioning is required by most design codes and safety legislation. Most pressure testing of subsea pipelines is done with water, but on some occasions, nitrogen or air has been used.

The minimum hydrotest pressure for gas pipelines is equal to 1.25 times the design pressure for pipelines [18]. Codes do not require that the pipeline be designed for hydrotest conditions, but sometimes give a tensile hoop stress limit 90% of the SMYS.

The pressure test level should be based on design codes and is aimed at

- showing the integrity of the pipeline;
- removing defects;
- locating the presence of small leaks and pinholes.

## 5.13 SUMMARY

This chapter on stress-based design of pipelines has provided a summary of the following:

- The information needed for a stress-based design of pipelines and design consideration.
- What to include in the pipeline design in order to ensure pipeline integrity.
- Standardized methods to achieve safe and reliable pipeline design.
- Pipeline integrity management approach that would meet industry needs.

Pipeline design, materials, and construction techniques are documented in design codes, client standards, handbooks, research papers, and so on.

Based on research, there is regular improvement in pipeline design, construction methods, and materials, and so it is important to access the most recent information on currently accepted and proven technologies.

## REFERENCES


1. Canadian Standards Association (2011) CSA Z662-2011: Oil and Gas Pipeline Systems, Canadian Standards Association, Mississauga, Ontario, Canada.
2. American Society of Mechanical Engineers (2003) ASME B31.8-2003: Gas Transmission and Distribution Piping Systems. American Society of Mechanical Engineers, New York.
3. American Society of Mechanical Engineers (2000) ASME B31.3-2000: Code for Pressure Piping, American Society of Mechanical Engineers, New York.
4. British Standard Institute (1992) BS 8010-1992: Code of Practice for Pipelines, British Standard Institute, UK.
5. JP Kenny Group (1994) Engineering Design Guideline, JP Kenny Group, UK.
6. American Society of Mechanical Engineers (2006) ASME B31.4-2006: Pipeline Transportation Systems for Liquid Hydrocarbons and Other Liquids, American Society of Mechanical Engineers, New York.
7. Det Norske Veritas (2012) DNV-OS-F101-2012: Submarine Pipeline Systems—Offshore Standard. Det Norske Veritas.
8. Institution of Gas Engineers (2001) IGE/TD/1-2001: Steel Pipelines for High Pressure Gas Transmission, 4th ed., Institution of Gas Engineers, London, UK.
9. Det Norske Veritas (2008) DNV-RP-D101-2008: Structural Analysis of Piping Systems, Det Norske Veritas, Norway.
10. Symonds, J., Vidosic, J.P., Hawkins, H.V., and Dodge, D.D. (1996) Strength of materials. In: *Marks' Standard Handbook for Mechanical Engineers*, 10th ed., McGraw-Hill, New York, Section 5.
11. Stewart, M. (1997) Pipeline Engineering 2, International Training and Development, Jakarta.
12. Chung, P.L. (1978) Stress analysis methods for underground pipe lines. OT Pipe Line Industry, April/May, Parts 1 and 2.
13. Hauch, S. and Bai, Y. (1999) Bending moment capacity of pipes. Proceedings of Offshore Mechanics and Arctic Engineering (OMAE), Houston, TX, Paper No. PL-99-5033.
14. Alexander, C. (2012) Evaluating the effects of ovality on the integrity of pipe bends. Proceeding of the 9th International Pipeline Conference, American Society of Mechanical Engineers, Calgary, Alberta, Canada, Paper No. IPC2012-90582.
15. Roylance, D. (2000) Shear and Torsion, Massachusetts Institute of Technology, Cambridge.
16. Timoshenko, S.P. and Gere, J.M. (1998) Fatigue. In: *Mechanics of Materials*, 3rd ed., Stanley Thornes Publishers Ltd, UK, p. 116.
17. Chernilevsky, D.V., Lavrova, E.V., and Romanov, V.A. (1984) Fatigue. In: *Mechanics for Engineers*, MIR Publishers, Moscow, Russia, pp. 347–381.
18. Guo, B. and Ghalambor, A. (2005) *Natural Gas Engineering Handbook*, 2nd ed., Gulf Publishing, Houston, TX, Chapter 11.
19. Hopkins, P. (2003) *The Structural Integrity of Oil and Gas Transmission Pipelines: Comprehensive Structural Integrity*, Elsevier, Vol. 1.
20. American Spiral Weld Pipe (2013) ASWP Manual—2013. Steel Pipe Design, American Spiral Weld Pipe, North Carolina, USA.
21. American Concrete Pipe Association (2001) ACPA-2001. Buried Facts: Structural Design Consideration, American Concrete Pipe Association, USA.
22. American Lifeline Alliance (2001) ALA-2001: Guidelines for the Design of Buried Steel Pipe, American Lifeline Alliance, USA (with addenda through February 2005). Available at http://www.americanlifelinesalliance.com/pdf/Update061305.pdf.
23. Calladine, C.R. (1989) *Theory of Shell Structures: The Brazier Effect in the Buckling of Bent Tubes*, Cambridge University Press, Cambridge, UK, p. 595.
24. Ellinas, C.P., Raven, P.W.J., Walker, A.C., and Davies, P. (1987) Limit state philosophy in pipeline design. Proceedings of Offshore Mechanics and Artic Engineering (OMAE).
25. Emerson, E. (2010) Factors that affect the bending radius of pipe cable. Available at www.had2know.com.

# 6

# SPIRAL WELDED PIPES FOR SHALLOW OFFSHORE APPLICATIONS

Ayman Eltaher
*MCS Kenny, Houston, TX, USA*

## 6.1 INTRODUCTION

The oil and gas (O&G) offshore industry has been interested in spiral welded pipe (SAWH) for some time due to the potential economic benefits and simpler technology involved. The latter implies that a wider range of nations would be able to be involved in their own O&G projects, which has been a target that more nations started to work toward. However, a concern of the industry has been that the performance of the SAWH pipes may be substandard compared with the more established and better accepted types such as seam welded (SAWL), high-frequency induction (HFI) welded, and electric resistance welded (ERW) line pipes and that it does not meet the requirements of the industry standards. To alleviate the concern, a comprehensive program to qualify the type of pipes would be required; such a program would need to include testing and analysis that cover known aspects of mechanical behavior of that pipe type and potential deviation in its response from that of the other better recognized pipe technologies. This chapter presents main areas of reassurance and concern, with the aim being to help the different parties (operators, designers, etc.) interested in the technology.

As discussed in Ref. [1], the use of SAWH pipes has indeed been popular in onshore and very shallow water low-pressure O&G applications, water pipelines, and ship-borne piping. Further interest in the technology has increased, as new SAWH pipe mills started to employ better fabrication technology, with the resulting chemical compositions, mechanical properties, and dimensional tolerances perceived to be comparable to those of the SAWL pipe. Other advantages of the SAWH line pipes are that they can be manufactured and coated in 80 ft lengths, which makes it faster/more economical to install. However, the use of SAWH in high-pressure hydrocarbon transportation has lagged due to concerns related to uncertainties (as opposed to actual negative observations) in the mechanical response of the pipe type as well as issues related to the variability in produced pipe quality. The former is thought to be able to overcome with further research and the latter with improvements in the fabrication process. Indeed, for spiral welded pipes, DNV-OS-F101 [2] recommends further investigation on a number of areas, including fracture arrest, resistance of external pressure, and designing for displacement-controlled conditions.

Too few initiatives have been made by the offshore O&G industry to investigate the validity of the above technology gaps and possible approaches to close them, with the majority having been proprietary limited scope projects. As one of the bigger published efforts to date, Det Norske Veritas (USA), Inc. (DNV) and MCS Kenny led a joint industry project (JIP) that investigated and published information on the suitability of SAWH pipe for shallow offshore applications [1,3]. Since minimal work is otherwise available in the public domain, this paper depends mainly on the work carried out in that JIP (hereafter referred to as "the JIP") and released to the public domain in Refs [1,3].

The first phase of the JIP [1] carried out a state-of-the-art review of the use of SAWH line pipe for offshore applications and identified knowledge gaps where further work has a

*Oil and Gas Pipelines: Integrity and Safety Handbook,* First Edition. Edited by R. Winston Revie.

potential to clarify how and where this type of line pipe can be used in subsea pipelines. In order to assess the readiness of SAWH pipe for use offshore and to identify which critical issues need to be considered in the JIP, a technology assessment per DNV-RP-A203 [4] was performed, with the criticality of an issue identified as a function of both its severity and expected frequency. A review of a number of design and line pipe standards (namely, DNV-OS-F101 [2], API RP 1111 [5], ASME B31.8 [6], and API 5L/ISO 3183 [7,8]) was also carried out as part of the JIP in order to see how they address SAWH pipe, with the conclusion having been that none of the reviewed standards fully cover the use of SAWH pipe for offshore applications, which supported the need for further work.

The second phase of the JIP [3] focused on qualifying commonly manufactured spiral welded pipe sizes for shallow offshore applications, using finite element analysis (FEA). Limiting the scope to shallow water also puts an emphasis on the S-lay pipelay procedure and conditions. In particular, the FEA qualification process aimed at comparing the limit states of SAWH line pipe under different loading conditions (e.g., axial tension and bending) with those of the more commonly used seamless and UOE (SAWL) pipes.

## 6.2 LIMITATIONS OF THE TECHNOLOGY FEASIBILITY

As part of the JIP investigation, relevant installation and operation conditions (e.g., water depths, environmental loading, and typical installation vessel characteristics) were studied together with readily available pipe sizes. The conclusion of the study was that SAWH pipes are not a relevant option beyond the fairly shallow water depth of about 240 m. Spiral welded pipelines in water depths shallower than 240 m were then studied in more detail, with S-lay considered as the relevant method of installation, and refined operational, environmental, and installation vessel configuration (e.g., fixed or floating stinger) and motion data were addressed. In the study, data were compiled from actual fields where SAWH pipe might be an option. The lines in these fields were found to have diameters that were in the lower range of what is generally considered feasible for SAWH pipe. Oil, gas, water injection, and water disposal service were identified as areas where SAWH pipe could be an option.

## 6.3 CHALLENGES OF OFFSHORE APPLICATIONS

### 6.3.1 Design Challenges

The challenges below were considered by the JIP [1] as key to possible deviation in the response of SAWH pipes compared with other types more commonly used by the industry. As stated in DNV-OS-F101, these issues may not be critical but are highlighted due to lack of knowledge.

- Collapse: The following were identified as characteristics that could potentially affect the collapse capacity of the SAWH pipe compared with SAWL and seamless pipes, and therefore addressed by the JIP: spiral weld, overmatch or undermatch of the spiral weld, residual stresses and flaws induced by manufacturing, propagating buckle behavior, and geometric imperfections characteristic to spiral weld.
- Displacement-controlled loading: Codes usually allow for strain-based design of components subject to displacement-controlled loading, which implies allowing for global yield of the material. Global yielding at weld locations could lead to localized yielding of the weld material if it is undermatching. Consequences of undermatching seam and girth welds are well understood but need further investigation for the case of spiral weld.
- Fracture arrest: In a bursting gas pipeline, the decompression speed (speed of sound) in the gas is lower than the speed of the propagating crack, which may result in the fracture propagating for long distances, possibly over multiple pipe joints. The common mitigation is to ensure adequate toughness of the base material to arrest the propagating crack. A complicating factor for SAWH pipes is that the arrest length in base material may only be on the order of a diameter before the crack crosses the spiral weld. If the properties of this weld are poor, the running fracture may run along this weld and thus no arrest is achieved, as shown in Figure 6.1.

### 6.3.2 Stress Analysis Challenges

It is generally believed that certainty and reliability of stress analysis of SAWH pipes are rather lower than those for other types commonly used by the O&G industry, due to the challenges discussed below, and the thought is that allowable levels of stresses and strains should be accordingly reduced (again, relative to those established for other types of pipes).

As discussed in Ref. [1], dimensional quality of produced SAWH pipe welds greatly depends on the dimensional quality of the coil. In particular, pipe out of roundness, weld peaking, and weld bead height and shape are of importance to offshore applications, especially for deeper water. Although one could argue that these parameters (for common currently fabricated SAWH pipes) are not consistently up to the offshore standards, they could be improved with careful control of production processes. On the other hand, the following concern is attributed to the characteristic shape of the spiral weld and not apt to be solved through

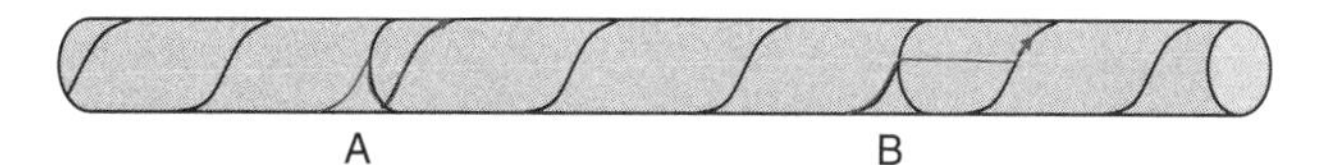


FIGURE 6.1 Illustration of possible running fracture paths. (Adapted with permission from Ref. [1].)

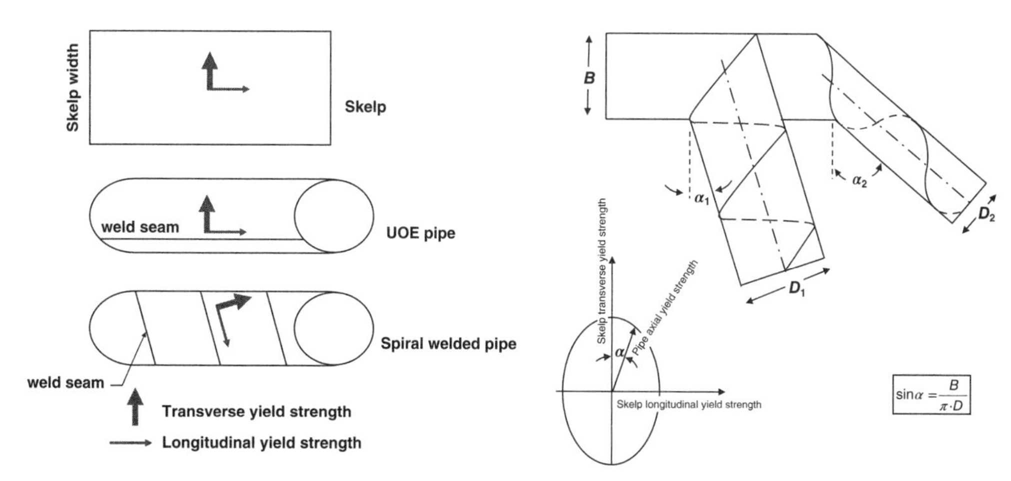


FIGURE 6.2 Material anisotropic and principal directions of spiral welded pipe. (Adapted with permission from Ref. [1].)

improvement of the production processes: namely, the fact that principal directions of the steel plate material properties are different from the principal directions of the pipe geometry. This is due to the hot rolling process of manufacturing of the steel plates, as shown in Figure 6.2. Depending on the specific pipe helical angle, axial and hoop strengths and stiffnesses could be higher or lower than anticipated for homogeneous materials.

Until more experience is accumulated in this area, it is currently believed that spiral welded pipe needs to be modeled and analyzed to the highest level of detail possible. As reported in Refs [1,3], detailed fully nonlinear three-dimensional (3D) finite element models have been used in the industry and by the JIP. Details of the welds including misalignment were modeled in full (solid-element) 3D, while most of the *plain* tubular section was modeled with shell elements, as shown in Figure 6.3. Pipe ovality was incorporated when deemed important to the expected response (e.g., collapse), with the anisotropic properties included in most cases since the extent of their influence was not certain. Away from the area of special concern, the SAWH pipe could be modeled using pipe elements, which can be kinematically coupled to the part of the pipe modeled with 3D solid and shell elements. Submodeling was also used to study local effects in further detail.

### 6.3.3 Materials and Manufacturing Challenges

Material and manufacturing parameters of SAWH pipes are important input to design; these include achievable wall thickness and diameter and quality and consistency of material and dimensions. Following are the more prominent challenges related to material properties and failure modes of SAWH pipe [1]:

- For offshore pipelines, weldability of girth welds is important, due to the industry requirements regarding

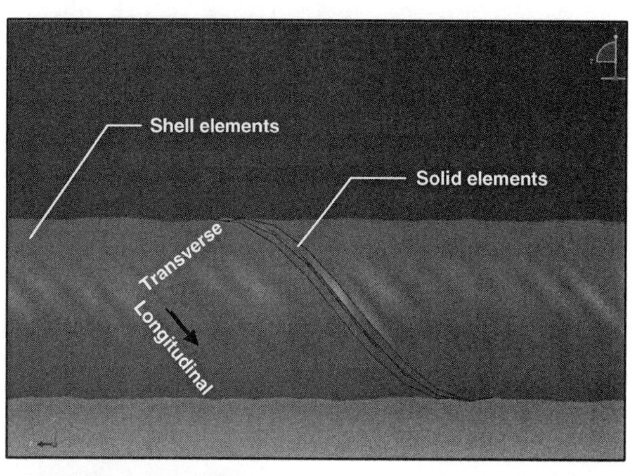


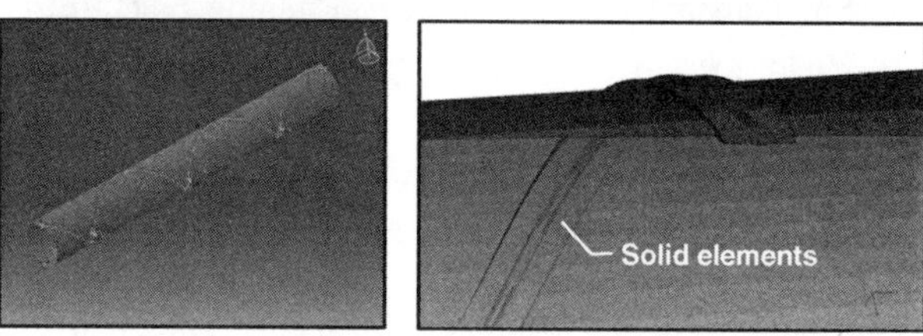


FIGURE 6.3 Finite element modeling of spiral welded pipe. (Adapted with permission from Ref. [3].)

efficiency and reliability of welds and NDE, which are rather more stringent than those of onshore industries. Furthermore, for offshore projects, line pipe manufacturers typically are required to provide extensive, detailed, and specific weldability data that spiral pipe manufacturers do not typically collect.

- For SAW pipe, local brittle zones (LBZs) in the heat-affected zone (HAZ) of the seam weld have been observed. Before having confidence in the adequacy of SAWH pipes for high-pressure offshore application, the extent and impact of those LBZs along the spiral weld would need to be assessed. In particular, LBZs may have an impact on the resistance of SAWH pipe to running fracture.
- Performance of SAWH pipe in sour service conditions has been a concern to the offshore industry due to the level of residual stresses in the pipe after manufacturing. Residual stress level and distribution are believed to be the main differences between UOE pipe and SAWH pipe, particularly as (in most cases) SAWH pipe is not expanded or heat treated after welding.
- Not currently adequately standardized, the variability in the steelmaking process can have a significant impact on the expected mechanical response of line pipe. Therefore, a good understanding is required of the processes used in the manufacturing of a specific SAWH pipe, especially that of the coil thermomechanical control process (TMCP). The TMCP uses controlled rolling and accelerated cooling to achieve required mechanical properties (e.g., grain refining improvement through accelerated cooling [9]) and therefore has a significant impact on the mechanical response of the pipe.

## 6.4 TYPICAL PIPE PROPERTIES

As discussed in Ref. [1], test and inspection data of spiral pipe actually supplied for actual projects were compiled as part of the JIP study. It was observed that, in general, the data met the requirements of DNV-OS-F101; however, the following were also noted:

- Charpy values sometimes being below requirements per Supplementary Requirement F (Running Fracture).
- Yield to tensile ratio being at the acceptable limit according to DNV-OS-F101.
- Unclear methodology of implementation of NDE.
- Missing significant information.

Further testing and gathering of test data, particularly on the following, was recommended to fill in the above gaps [1]: dimensional tolerances, ductile–brittle transition curves, weldability, effects of thermal aging, weld residual stress distribution, and NDT procedures and inspection data.

## 6.5 TECHNOLOGY QUALIFICATION

Reference [1] described the technology qualification process adopted by the JIP, which followed the approach described in DNV-RP-A203 [4]. Per the RP, the technology under investigation (SAWH pipe) was broken into manageable components that could be assessed. Then, the outcome of the technology assessment was used as input to the failure mode effects and criticality analysis, with the identified critical components and their defined functions evaluated to identify failure modes. Threat assessment was also performed as part of the procedure, aiming to find potential failure modes and mechanisms that could occur in the components identified as critical during the technology assessment. The impact and consequences of those failures were analyzed and ranked based on their risk.

The second phase of the JIP focused on implementing an important step of the pipe qualification plan, namely using FEA [3]. The FEA qualification process aimed at identifying the response of SAWH line pipe to different loading conditions (e.g., axial tension and bending) as well as identifying its limit states and comparing these limit states with those of the more commonly used seamless and UOE pipes. However, Ref. [3] emphasized that for the comparison between the analytical responses of the SAWH and seamless/UOE pipes to be valid and meaningful, the spiral pipe material and fabrication quality need to be comparable to that of the seamless/UOE pipe; in other words, the spiral pipe should conform to line pipe requirements of the industry, stated in its codes and standards, such as DNV-OS-F101 [2] and API Spec 5L [7]. Fracture arrest was not discussed in Ref. [3].

Limit states relevant to SAWH line pipes were identified by carrying out FE simulations, using the nonlinear general-purpose FE package, Abaqus [10]. The analysis parameters of these simulations were selected based on guidance from known limit states of seamless and UOE pipes, particularly as described in DNV-OS-F101 [2], with the following limit states considered: burst, collapse, yielding (due to axial tension), local buckling (due to bending and combined loading), and fatigue. Results of the finite element analysis were compared with a number of relevant benchmark solutions:

- FEA of UOE pipe (with same material properties and weld misalignment).
- DNV limit: based on Equation 5.19a of DNV-OS-F101 [2], with the $\alpha$ and $\gamma$ factors set to 1.0, to calculate the actual/characteristic limit state, with no margin of safety, and discount the effect of expansion.

- API limit: for the burst and collapse limit states, this is based on API RP 1111 Equations 2(a) or (b) and 6(a), respectively [5]. For bending conditions, this is the allowable moment from Equation 3.2.3-1c of API RP 2A [11] multiplied by 1.67 to convert the calculated "allowable" to "ultimate capacity."
- Elastic limit: this is the analytical/closed-form elastic solution.
- Plastic limit: this is the analytical/closed-form plastic solution, with no strain hardening.

Limiting the scope to shallow water puts an emphasis on the S-lay pipelay procedure and loads and conditions associated therewith, and as the industry has limited experience with SAWH line pipes subjected to large strains, only pipe with overmatching welds and subject to load-controlled situations was studied.

Limited discussion was presented in Ref. [3] as to the conclusion on the performance of the spiral welded pipe compared with seamless and UOE counterparts. However, the example data presented for combined (tension and bending) loading show that failure combinations of the spiral welded pipe fell on or outside the envelope predicted by DNV-OS-F101, which indicates that that specific pipe configuration is likely to behave at least as good as seamless and UOE pipes, under addressed conditions.

Fatigue was also deemed a loading condition of interest, as it is of concern during installation conditions using the S-lay procedure and in free span situations [3]. To address this limit state, the pipe models were subjected to axial and bending loading conditions, and stress concentrations at the hotspots were investigated. Special focus was placed on the spiral weld geometric characteristics and tolerances. Girth welds and their interaction with spiral welds (which could potentially generate atypical stress concentrations) were not addressed, and no conclusion regarding the criticality of the fatigue issue was given in Ref. [3].

## 6.6 ADDITIONAL RESOURCES

Due to the limited amount of information in the subject of SAWH pipes, and specifically for offshore applications, we reference in this section work from other industries concerning spirally welded tubular structures. In particular, Ref. [12] is part of the European research project Combitube and presents data from full-scale four-point bending tests on SAWH structural tubes. Another European project that has also started recently is SBD-SPIPE, which investigates the use of SAWH tubes for demanding applications both onshore and offshore that require good performance under application of large strains.

Comparison between data from the investigations of the two industries will surely be useful when the investigations reach reasonable maturity levels. However, caution should be exercised before utilizing such data in the O&G offshore industry, due to the difference in the intended purposes, relevant ranges of *D*/*t* ratio, and referenced codes.

## 6.7 SUMMARY

A state-of-the-art review regarding offshore application of spiral welded pipe is presented in this chapter. However, due to the limited pool of technical references on the topic, majority of the presented material was derived from information published out of the spiral welded pipe JIP [1,3]. Following is a summary of the ideas and conclusions discussed:

- Offshore use of SAWH pipe is not common practice, although it has been applied in some cases. However, for offshore applications, SAWH pipe should be considered primarily for shallow water pipelines.
- No design code fully covers the use of SAWH pipe for offshore applications.
- Material properties and weld properties obtained from actual mill production seem to meet recognized offshore standards; it is thus reasonable to assume that these can be met for an actual project.
- As further analysis and testing is being conducted, a main objective of the industry and projects needs to be addressing identified failure modes and defining efficient safeguards in the form of improved design practices, project-specific testing or qualification, and/or efficient inspection and quality control procedures.
- The spiral welded pipe JIP started out a qualification process (including an FE study) of the SAWH pipe for shallow offshore applications (<240 m water depth) and to identify and quantify its relevant limit states. For this, the SAWH pipe was assumed to conform to the industry-relevant specifications. Detailed conclusion of the qualification process is presented in Refs [1,3]. Results from the qualification study have generally increased the confidence in SAWH pipe as no clear showstoppers were identified.
- Making use of information from similar efforts of other industries will surely be useful, while making sure of the validity of such results to oil and gas offshore applications.

## REFERENCES

1. Heiberg, G., Eltaher, A., Sharma, P., Jukes, P., and Viteri, M. (2011) Spiral wound linepipe for offshore applications. Proceedings of the Offshore Technology Conference, Houston, TX, Paper OTC 21795.

2. Det Norske Veritas (2012) DNV-OS-F101: Submarine Pipeline Systems, Det Norske Veritas.
3. Eltaher, A., Jafri, S., Panikkar, A., Jukes, P., and Heiberg, G. (2012) Advanced Finite Element Analysis for Qualification of Spiral Welded Pipe for Offshore Application, ISOPE, Rhodes, Greece.
4. Det Norske Veritas (2001) DNV-RP-A203: Qualification Procedures for New Technology, Det Norske Veritas.
5. American Petroleum Institute (1999) API RP 1111: Design, Construction, Operation and Maintenance of Offshore Hydrocarbon Pipelines (Limit State Design), C 3rd ed., American Petroleum Institute, Washington, DC.
6. ASME (2007) ASME B31.8-2007: Gas Transmission and Distribution Piping Systems, ASME, New York.
7. ANSI/API Spec 5L (2008) Specification for Line Pipe, 44th ed., American Petroleum Institute, Washington, DC.
8. ISO 3183 (2007) Petroleum and Natural Gas Industries: Steel Pipe for Pipeline Transportation Systems, International Standard, 2nd ed., ISO, Geneva, Switzerland.
9. Gorni1, A.A. and da Silveira, J.H.D. (2008) Accelerated cooling of steel plates: the time has come. *Journal of ASTM International*, 5 (8), Paper ID JAI101777.
10. SIMULIA (2011) Abaqus 6.11: User Manual, Dassault Systèmes Simulia Corp., Providence, RI.
11. API RP 2A-WSD (2005) Planning, Designing and Constructing Fixed Offshore Platforms: Working Stress Design, American Petroleum Institute, Washington, DC.
12. Van Es, S.H.J., Gresnigt, A.M., Kolstein, M.H., and Bijlaard, F.S.K. (2013) Local Buckling of Spirally Welded Tubes: Analysis of Imperfections and Physical Testing, ISOPE, Anchorage, Alaska.

# 7

# RESIDUAL STRESS IN PIPELINES

Paul Prevéy and Douglas Hornbach
*Lambda Technologies, Cincinnati, OH, USA*

## 7.1 INTRODUCTION

This chapter is intended to provide an overview of residual stresses for engineers engaged in the design, installation, and maintenance of pipelines. Residual stresses introduced into pipelines by welding and forming during assembly, installation, or repair and their effect on service life are often overlooked. The goal of this chapter is to provide a general presentation of residual stresses, how they relate to applied stresses, and their influence in service performance of pipelines and related applications. Avoiding creating detrimental tension when possible and dealing with unavoidable tension by thermal stress relief or the introduction of beneficial compression for improved performance are covered. The effects of residual stresses on stress corrosion cracking (SCC) and fatigue failure mechanisms are briefly discussed, but are covered in detail elsewhere in this volume. Practical methods of measuring residual stresses that may be of use to a pipeline engineer are included, with emphasis on their relative advantages and limitations. In addition to the macroscopic residual stresses of primary interest to the engineer, a brief discussion of cold working and "microstress" and their effect upon corrosion behavior is included.

### 7.1.1 The Nature of Residual Stresses

Residual stresses, sometimes referred to as "self-stresses" or "internal stresses," are those stresses that are present in a body when it is free of any externally applied forces. Residual stresses are as real as applied stresses, and will generally be retained in a component, such as a section of welded pipe, assembled coupling, or flange joint for the entire service life. Residual stresses are additive to the applied stresses with which engineers are primarily concerned in the design of a pipeline, and the summation of stress governs performance in service. Tensile residual stresses are generally detrimental, but compressive residual stress can markedly improve performance. The magnitude of residual stresses can often exceed the applied stresses, and can therefore have a major impact on the performance in service.

Residual stresses can be considered to be of two types—macroscopic and microscopic, which are distinguishable by X-ray or neutron diffraction. Macroscopic residual stresses extend over distances that are large relative to the size of the crystals or grains of the material on the order of millimeters. Macroscopic residual stresses are directly additive to the applied stresses, which engineers must consider, due to mechanical loads and pressures on a pipeline. Microstresses are the result of plastic deformation, or cold work, of the metallic material and are very local stresses that extend over distances between the dislocations present within the individual crystals in the metal, on the order of fractions of a micrometer. Microstresses can alter material properties, such as yield strength and corrosion behavior, influencing how the sum of applied and macroscopic residual stresses affect performance.

Although the term "cold working" is sometimes used to imply introducing macroscopic residual compression, as by shot peening (SP) or roller burnishing, it is important to realize that there is no definite relationship between macro- and microstresses, or residual stress and cold work. Stress is a

*Oil and Gas Pipelines: Integrity and Safety Handbook*, First Edition. Edited by R. Winston Revie.

tensor property, whereas cold work is a scalar property describing the degree of plastic deformation the material has experienced. A body can be uniformly highly cold worked, but nearly stress free. Alternatively, simple bending can produce residual stresses equal to the yield strength with very little cold working. The term residual stress, used here, refers to macroscopic residual stress. Cold work refers to the combination of dislocation density and microstrain causing work hardening.

Residual stresses are caused by nonuniform plastic deformation of either thermal or mechanical origin, or by phase transformations, such as case hardening of steels. Although residual stresses are caused by plastic deformation, regardless of the complex thermal–mechanical history, the residual stresses remaining once the deformation occurred and the body is in equilibrium are necessarily entirely elastic. No matter how complex is the thermal mechanical history of smelting, rolling, forming, grinding, welding, and so on that may have occurred to produce a state of stress, the residual stresses remaining after that complex processing will be entirely elastic. The residual stresses are necessarily limited to less than the yield strength of the material in its then cold-worked or heat-treated condition. The strains due to the elastic residual stresses can then be measured by mechanical dissection or diffraction techniques, and used to calculate the residual stresses present.

### 7.1.2 Sources of Residual Stresses

Residual stresses encountered in pipelines may arise from a variety of sources, such as the original fabrication of the pipe sections, welding, forming, machining, grinding, handling, and even assembling of the pipeline. The final state of residual stress in any section of pipe will arise from the differences in the amount of plastic and elastic strains created by the combined thermal–mechanical history.

Pipe sections fabricated as rolled, longitudinally welded sections can contain residual stresses from both the forming into the cylindrical shape and the welding of the pipe along the seam [1]. Helical welded pipe will have a more complex stress distribution, again due to the forming and welding. Extruded pipe, although seamless, may have a through-wall stress distribution created in the piercing and drawing operations [2].

Any bending of pipe to form curved sections will necessarily produce tensile residual stresses on the inside of the bend (intradose), which is driven into compression during bending, and compressive residual stresses on the outside (extradose) of the bend, which is stretched in tension. The highest tensile residual stresses are often encountered at the point of tangency just entering into the bend on the inside surface, because the highest plastic strain gradient during bending is at the beginning of the bend. Flaring or other sizing of pipe or tubing to alter diameter will produce residual stresses that can be in high tension at the point of transition into the increased diameter. Any process that nonuniformly deforms a metal component will result in residual stresses, potentially up to the yield strength of the material. Tensile stresses will be produced in the areas that are plastically deformed in compression during the forming operation; compressive stresses will occur in the regions deformed in tension.

The residual stress distribution in any body must be in equilibrium after external forces and tractions are removed. Equilibrium requires that the sum of the moments and forces acting on any plane entirely through the body must be zero. Equilibrium does not imply that there should be equal and opposite residual tension and compression in adjacent locations in a component. For example, a layer of equal magnitude tension does not occur immediately beneath the compressive layer produced by shot peening a section of pipe. Low-magnitude tension extending through the thick wall will balance the thin compressive layer. A region of residual tension may exist through the entire thickness of a component, such as a pipe wall that has been heated locally so that it is deformed in compression by thermal expansion entirely through the wall, and then cooled. The zone of through-thickness tension is supported in equilibrium by surrounding material in lower magnitude compression.

Welding to join pipe sections is a primary source of residual stress [3]. Welding creates regions that are heated to a liquid or highly plastic state, fused, and then allowed to cool locally. The contraction of the metal upon cooling stretches the fusion zone into high tension, typically up to the yield strength. A zone of residual tension then extends out into the heat-affected zone (HAZ) on either side of the weld [4]. The complex distribution of residual stress, with regions of high tension in a 304 stainless steel T-weld, is shown as a contour plot in Figure 7.1. Multipass welding deposits layers of filler metal that are alternately reheated by the next pass. The result can be complex residual stress distributions created by the partial stress relaxation in the first layer as it is reheated, combined with new tension in the newly deposited layer. Preheating the work before welding is a common means of reducing the thermal strain gradient during weld fusion, and thus the tension created when the weld cools.

Postweld thermal stress relief is commonly used to reduce the tensile stresses created by welding, but complete thermal stress relief can be difficult to achieve. To avoid introducing additional residual stresses from thermal strains, the temperature throughout the component must be uniform during the heating cycle. This is especially important in complex welded assemblies where large thermal stress gradients can develop during heating or cooling. Figure 7.2 shows the residual stress depth profiles for a low carbon steel weld before and after stress relief and shot peening.

Because fatigue and SCC initiate at the surface, the stresses produced by finish machining or grinding can be

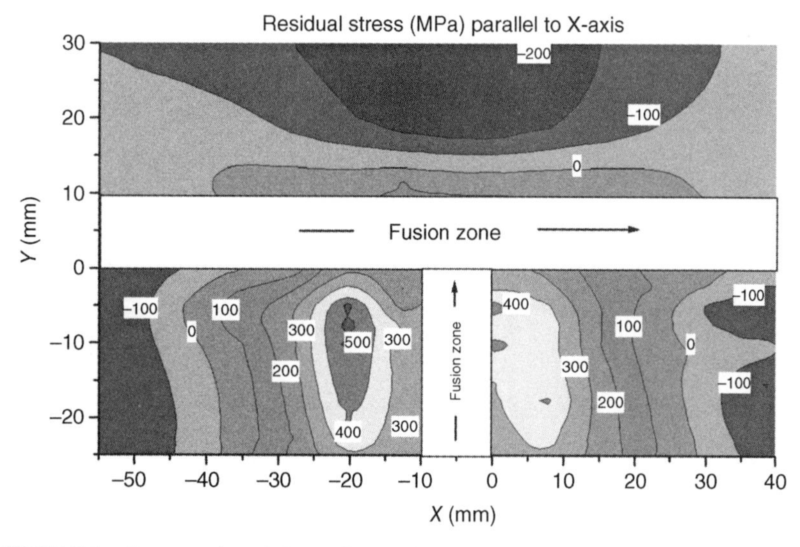


**FIGURE 7.1** Contour plot of the surface residual stress distribution in a 304 SS T-Weld.

critically important for subsequent performance. Surface integrity is the discipline dealing with how the condition of the surface, including the residual stress and cold work produced by manufacturing, influences performance in service. Machining and grinding produce residual stress distributions that are shallow, but can be of very high magnitude [5]. The residual stress layer produced is generally less than 0.25 mm deep, but stresses can range from yield strength tension to compression.

Rapid local surface heating is a primary source of highly detrimental residual tension. Machining or grinding practices that quickly heat the surface can produce surface tension and grinder "burn." The small patch of material at the point of contact with the grinding wheel or cutting tool is quickly heated to incandescence. Local temperatures can be sufficient to cause phase transformations in steels, leaving brittle, cracked, untempered martensite on the surface, a common fatigue initiation mechanism [5]. In any alloy, the locally

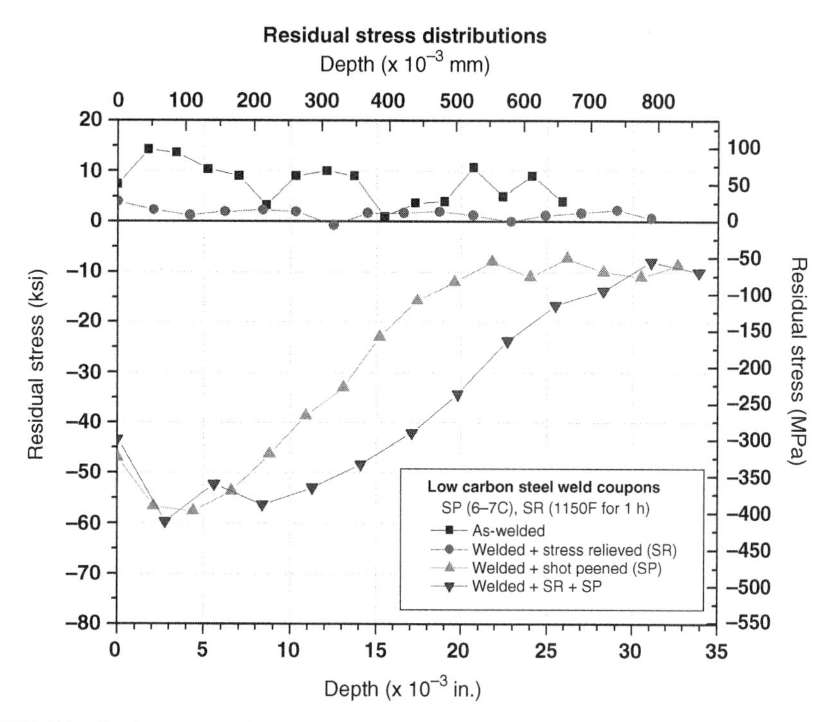


**FIGURE 7.2** Residual stress depth distributions for a low carbon steel-butt weld, after thermal stress relief, and shot peening.

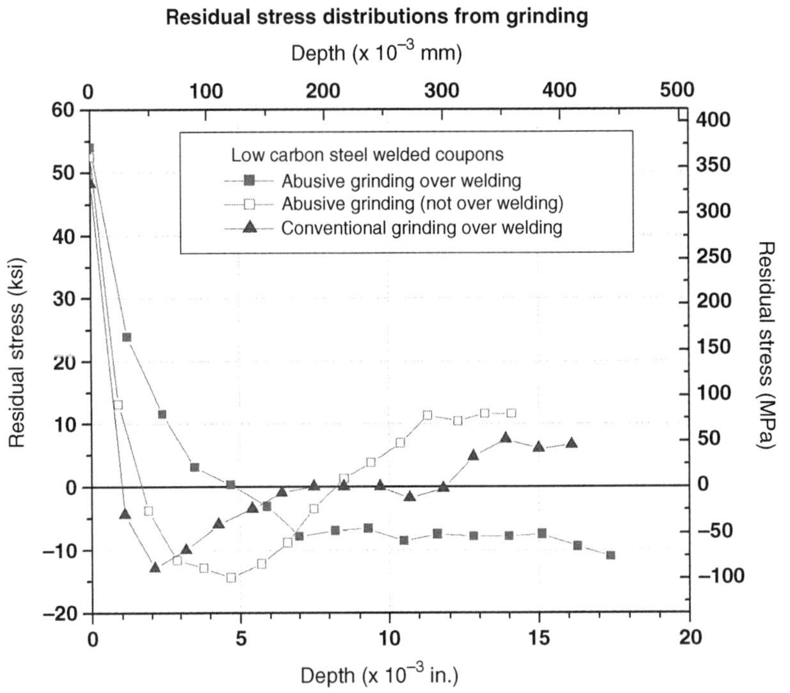

**FIGURE 7.3** Subsurface residual stress distributions in ground low carbon steel welds.

heated surface zone expands, but is constrained by the cool material below, briefly creating a compressive thermal stress sufficient to yield the hot zone in compression. When the tool or grinding wheel moves on, the heated zone rapidly cools by self-quenching into the cool substrate material. After yielding in compression when hot, the heated zone is then stretched into tension upon cooling. The residual tension may be so high as to cause surface cracking, forming fatigue and SCC initiation sites.

Cold machining or abrasive operations typically produce shallow, but often high residual compression. Polishing and grinding are chip forming operations on a fine scale that deform the surface in tension, and will produce a shallow layer of surface compression if performed with a liquid coolant so that the surface is not significantly heated. The surface yields in tension as chips are formed, leaving it in residual compression. "Gentle" milling, turning, and similar chip forming operations performed with coolant, sharp tools, shallow depths of cut, and low feeds that minimize heating also leave a compressive surface layer. Wire brushing also produces compression if heat is avoided. Figure 7.3 shows the residual stresses left behind by grinding in low carbon steel weld coupons. Note the high surface tension.

Grinding or machining for weld preparation deserves special mention with regard to pipelines. Weld preparation by hand grinding or machining can highly cold work the surface layer, increasing the yield strength of the material adjacent to the weld. Contraction of the weld upon cooling then stretches the cold-worked material in the HAZ into high residual tension, leaving the HAZ subject to fatigue or SCC failure in service. This is a primary cause of SCC failures of 304 stainless steel pipe welds made in boiling water nuclear reactor piping.

Installation or "fit-up" stresses are sometimes considered residual stresses in the assembled pipeline or component assembly. Stresses from suspending a pipeline with hangers, loss of support caused by erosion, or flexing of pipe sections in order to make connections before welding or flange joining can be considered to be residual stresses in the completed assembly.

Surface damage in transport, handling, and assembly can produce dents, gouges, cold-worked zones, local areas of high residual stress, and stress concentrations that serve as fatigue and SCC initiation sites [6].

## 7.2 THE INFLUENCE OF RESIDUAL STRESSES ON PERFORMANCE

The macroscopic residual stresses of concern to engineers are inherently elastic and are additive to applied stresses. Residual stresses remain in equilibrium in the body indefinitely after creation by thermal–mechanical processes that cause nonuniform plastic deformation. In the case of pipelines, stresses from pressurization, suspended loads, vibration during operation from pumping cycles, earthquakes, soil settling, and so on would be directly summed with existing residual stresses in the pipe. Pipelines are subject to failure in service primarily by fatigue, SCC, or corrosion fatigue (CF), all of which require a net tensile stress at the surface for initiation. Tensile residual stresses from any origin render the pipeline more likely to fail, whereas compressive stresses are generally beneficial. It is useful to consider how residual stresses influence each of these failure mechanisms separately.

### 7.2.1 Fatigue

Fatigue failure is caused by cyclic loading to stress levels above the endurance limit of the material. Cracks initiate at the surface or from some internal flaw, such as an inclusion in the metal. Stress concentrations such as corrosion pits or scratches due to handling are common initiation sites. Fatigue is classified as low cycle fatigue (LCF) or high cycle fatigue (HCF), depending upon the stress level and number of cycles to failure. In LCF, the maximum alternating stress exceeds the proportional limit causing plastic deformation. Cracks initiate immediately, and life is determined by crack propagation, usually in tens of thousands of cycles or less. Cyclic plastic deformation in LCF reduces the effect of any residual stresses, and even applied mean stresses. HCF occurs more commonly in pipelines, under essentially elastic cyclic loading, and failures occur only after $10^5$ or more cycles. Residual stresses add to the mean applied stress during fatigue, and can strongly influence HCF performance. Welding residual stresses have a dominant influence on the mean stress and fatigue performance in pipelines [7]. Once a pipeline is assembled and in operation, the vibratory stresses contributing to fatigue are generally so small that only HCF needs to be considered.

All fatigue failures initiate in shear, and will invariably initiate at the surface of a body under cyclic load because the shear stresses are maximum at the free surface. When the maximum stress is raised to the fatigue endurance limit, those few crystals on the surface of a pipeline that happen to be favorably oriented to exceed the critically resolved shear stress (i.e., having slip planes tipped 45° into the direction of loading) will begin to ratchet on slip planes with each loading cycle. Dislocations are pumped along the slip planes, alternately forming notches and extruding metal slightly from the surface. After the majority of the fatigue life in HCF, the ratcheting process eventually creates a notch on the order of a full crystalline grain. Fatigue cracks then grow from the notch, and propagate by the rules of fracture mechanics to failure in the final stages of the fatigue process.

The contribution of mean stresses and residual stresses to fatigue performance can be understood in terms of the Haigh diagram. The alternating stress allowed for a given fatigue life is plotted on the vertical axis as a function of the mean stress, plotted on the horizontal axis, which ranges from tension to compression. The sum of the alternating and mean

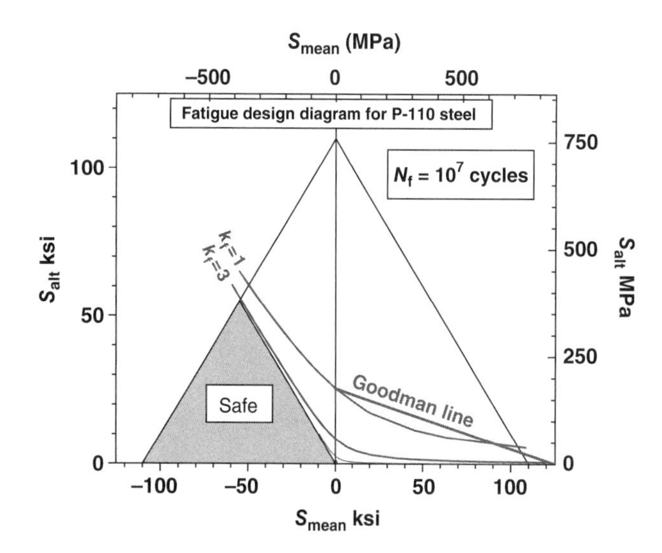


**FIGURE 7.4** Fatigue design diagram for P110 steel.

stress is limited to the material yield strength at the triangular boundary. The Goodman line, an estimate for only tensile mean stresses, is plotted from the fully reversed (zero mean stress) fatigue strength to the ultimate strength. An example for P110 steel oil field tubulars for a fatigue life of $10^7$ cycles is shown in Figure 7.4. The Smith–Watson–Topper curve plotted for no damage, $k_f = 1$, gives the upper bound for the allowed alternating stress with mean stresses extended into compression. Adding Neuber's rule to account for damage provides a fatigue design diagram (FDD) useful for determining the amount of compression needed to mitigate damage [8]. The mean stress is the sum of the applied mean stress and the residual stresses at the surface of the pipe. In pipeline applications, the alternating stress could be the magnitude of the transient stress from vibrations or pressure pulsations from a pumping station.

The FDD shows that under applied mean or residual compression, the allowed alternating stress increases significantly, a great benefit in fatigue. Damage increases $k_f$, reducing the allowed alternating stress with positive mean stress. The allowed alternating stress must be less than the $k_f = 3$ curve for a $10^7$ cycle life when damage reduces the fatigue strength to one-third of the original. Even with severe damage, such as an existing fatigue crack, if the sum of residual and applied mean stresses moves the operating condition into the gray triangle labeled "SAFE," the material is always in compression, and fatigue cracks cannot grow. Any surface, even with fatigue-initiating damage, operating within this safe triangle cannot fail in fatigue. This is the basis for introducing beneficial residual compression using surface enhancement treatments. The maximum possible alternating stress is achieved at the center of the "SAFE" triangle with the net mean stress, residual plus applied, equal to half of the yield strength.

### 7.2.2 Stress Corrosion Cracking

Stress corrosion cracking (SCC) refers to the propagation of cracks under a static tensile load in the presence of a chemically active environment to which the material is susceptible. SCC is also termed "environmentally assisted" cracking. Ferritic pipeline steels can be subject to hydrogen sulfide ($H_2S$) SCC or "sulfide stress cracking" (SSC) in sour well environments [9,10]. Carbonate ($CO_3$) cracking is also possible and chloride SCC is common to many alloys, notably stainless steels, and is exacerbated by tensile residual stresses [11]. Pipelines transporting ethanol, even as a minor fraction mixed with petroleum products, are subject to internal SCC due to the affinity of ethanol for water and the ability of ethanol to support corrosion [12]. The combination of chemical and mechanical elements makes SCC more complex, and less well understood, than fatigue. The mechanisms and occurrence of SCC in various alloys and environments are discussed in detail elsewhere in this volume.

SCC can only occur if three conditions are met, as shown schematically in the Venn diagram in Figure 7.5. First, the material must be susceptible to cracking in the service environment. Second, the material must be exposed to the chemically active environment with the ions present that will cause cracking to occur. And third, there must be a tensile stress, residual plus applied, present at the surface of the material that exceeds some threshold level.

A possible, but often prohibitively expensive, way to mitigate SCC is to replace the alloy with one that is not susceptible in the environment. In the case of ferritic steels, the use of lower strength grades with higher fracture toughness can reduce the propensity to fracture, but with the performance limitations of the lower strength. A protective coating can be used to isolate the surface from the chemical environment, but the coating may be breached, allowing failure.

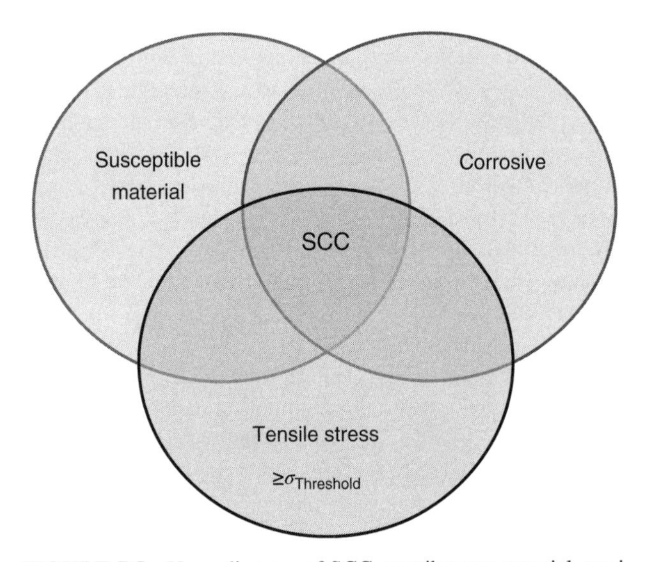


**FIGURE 7.5** Venn diagram of SCC contributors: material, environment, and tension.

As in the case of fatigue, residual stresses influence SCC by being additive to the mean stresses applied to the pipeline. If a tensile residual stress, for example, from welding, added to the applied stress exceeds the threshold for SCC for that material and exposure, cracking can be expected to occur. Welding procedures and parameters can be modified to control the magnitude and distribution of residual stresses developed by pipeline welding [13]. If the surface that is in contact with the environment is kept in compression by a layer of residual stress, SCC can be eliminated. It may be possible to control welding methods to produce compressive residual stresses in the pipe [14]. In addition to inter- and transgranular SCC cracking, corrosion pitting has been reported to occur preferentially in areas with the highest tensile residual stress [15].

### 7.2.3 Corrosion Fatigue

Corrosion fatigue is the combination of mechanical fatigue-generated cracks that are further propagated in a chemically aggressive environment by SCC. The fatigue strength of many alloys is reduced by exposure to chloride solutions, generally salt water. In fatigue with the absence of corrosion, many alloys, including pipeline steels, will exhibit an endurance limit. Below this cyclic stress level, fatigue cracks will not initiate, and the fatigue life is essentially infinite. A prominent effect of corrosion fatigue, illustrated in the case studies section, is the elimination of any endurance limit, so that fatigue cracks propagate even at very low alternating stress levels. In high-strength steels, the loss of fatigue strength in the presence of a corrosive environment can be spectacular. High-strength 300M steel exposed to 3.5% NaCl solution has only 20% of the fatigue strength of baseline material at $10^7$ cycles, with no evident endurance limit. As in simple fatigue or static SCC, the presence of residual stresses adds to the mean stress, and can either exacerbate or mitigate SCC. The corrosion debit during fatigue in pipeline steels requires tension exceeding a threshold level, and is mitigated by reducing tensile or introducing compressive residual stress [15].

### 7.2.4 Effects of Cold Working and Microscopic Residual Stresses

Any polycrystalline metal, such as pipeline steel, when mechanically processed by grinding, machining, bending, flaring, weld shrinkage, or other mechanical means will necessarily be plastically deformed to some degree. Even a delicate metallographic polishing process is a chip forming grinding operation on a microscopic scale, plastically deforming the surface to a shallow depth. Plastic deformation of metals forms vast numbers of dislocations along slip planes in the individual grains or crystals of the metal. The increase in dislocation density changes the properties of the metal in ways that can be as important as the macroscopic residual stresses. The degree of plastic deformation can be quantified during X-ray diffraction (XRD) residual stress measurement as an increase in X-ray peak broadening, while the macroscopic residual stress causes a shift in the peak position.

Cold working influences the corrosion behavior of many metals, leaving the deformed surface more chemically active. For example, cold working of P110 steel casing reduces the open-circuit potential (OCP) in salt water, effectively making the metal less noble than chemically identical P110 material that is not so severely cold worked. The cold-worked areas corrode preferentially and faster than the undisturbed material. The corrosion rates of pipeline steels have been shown to increase with cold working in sour gas ($H_2S$) environments [16,17].

Cold working influences the stability of residual stresses at elevated temperatures. Once the dislocation density of the material becomes high enough, residual stresses can relax extremely rapidly by dislocation annihilation. This effect can result in the loss of beneficial compression introduced by shot peening or other mechanical working processes that introduce high levels of cold work [18].

Cold working and the resulting dislocation density produce an increase in the yield strength of many materials. A surface cold worked by shot peening or other means will have a variation in yield strength with depth through the deformed layers, from the highly cold-worked surface to the unworked interior. Austenitic alloys, like stainless steels, can have a 400% increase in yield strength from surface work hardening. If a surface layer of cold-worked metal is then deformed, as may occur in the weld HAZ during cooling and contraction, the metal will yield different amounts at each depth, depending upon the state of residual stress and the yield strength in each layer. A compressive surface layer produced by shot peening or low stress grinding can be left in residual tension after subsequent deformation.

Because cold working can alter mechanical behavior, unnecessary cold working should be avoided. Austenitic alloys with face-centered cubic structures are particularly susceptible to work hardening. Minimizing cold work ensures that the finished pipeline or component will behave in a manner to be expected for the base material properties.

## 7.3 RESIDUAL STRESS MEASUREMENT

This section is intended to introduce the pipeline engineer to the practical methods of residual stress determination that should be considered in the event that measurement is necessary, and to support understanding of residual stress studies presented in the literature. The advantages, disadvantages, and limitations of the established methods are described. Methods that have been investigated but remain largely in the laboratory and are of limited use to a pipeline engineer are not included. The reader is referred to handbooks with comprehensive chapters covering the various methods for detailed discussion of the practical techniques and experimental methods [19,20].

Although commonly used, the phrase "residual stress measurement" is actually a misnomer. Stress, like density, is an extrinsic property that cannot be directly measured, but can only be calculated from measurable properties. Residual stress measurement or determination methods can be considered to be in two categories: those that measure strain and calculate the stress by linear elasticity theory, and those that utilize some other property influenced by residual stress, such as magnetic or acoustical behavior.

The linear elastic methods that measure strain, or changes in strain, and calculate the stress using elasticity theory include the mechanical and diffraction methods. Residual stresses are necessarily entirely elastic once the body is in equilibrium, so the residual strain is proportional to the residual stress.

The first group of linear elastic mechanical methods includes all of the dissection methods of slitting, sectioning, center hole drilling, and ring core or trepanning that use electrical resistance or mechanical strain gauges to measure the change in strain as residual stresses are relaxed. The residual stresses present before sectioning are calculated from these strain measurements. Most of the layer removal and slitting methods rely upon simplifying assumptions of component geometry and stress field to facilitate solution. Only the portion of the residual stress that is relaxed can be calculated, and care must be taken not to introduce residual stresses during sectioning.

The second group of linear elastic methods includes the X-ray diffraction (including synchrotron) and neutron diffraction techniques. In the diffraction methods, the strain is measured between planes of atoms in the crystals of the metal, in this case the pipeline steel. Tension will increase and compression will decrease the lattice spacing in the direction of the stress. The residual stress is calculated from the strain measured in the crystal lattice. The diffraction techniques are nondestructive, but require sampling of a large number of small crystals to provide an accurate statistical average of the macroscopic residual stress. Measurement in coarse-grained material, such as castings and weld fusion zones, can be difficult. Conventional X-ray diffraction penetration is very shallow, providing only surface values without removing material. Neutron and synchrotron penetration can be up to centimeter and millimeter, respectively, but require highly specialized facilities and independent knowledge of the unstressed lattice spacing to calculate the stress present.

Residual stress measurement by mechanical methods involves removing or sectioning material to relieve a portion of the residual stress, and measuring the resulting change in strain or material displacement, from which the stress is calculated. These methods can be categorized as destructive or semidestructive depending upon the relative amount of material removed. Displacements are typically measured with electrical resistance strain gauges. Air gauge and optical systems can also be used to measure material displacements. The residual stress that was relaxed by sectioning is then calculated using equations developed usually assuming some symmetry to the residual stress field to allow algebraic solution. Both sectioning and hole drilling methods are commonly applied to welded pipelines [21]. For the general application, semidestructive mechanical measurement methods such as center hole drilling and ring core are preferred because symmetry of the residual stress field is not required.

The nonlinear elastic techniques form the other major class of residual stress determination methods. These include magnetic and ultrasonic techniques, where pressure- or stress-sensitive changes in properties other than strain are measured. All of these require measurement of some nonlinear, often high-order effect with high accuracy. All are strongly influenced by other material properties, such as hardness, dislocation density, and crystal orientation, and interpretation of the residual stress contribution is very difficult in practice. The Barkhausen noise magnetic method is the only nonlinear elastic method that will be further considered here [22].

### 7.3.1 Center Hole Drilling Method

The center hole drilling method [20,23,24] is a mechanical technique that allows the determination of bulk or incremental residual stress versus depth. Hole drilling involves introducing a small hole at the center of a specially designed three- or six-element strain gauge and measuring the resulting strain relaxation, as shown schematically in Figure 7.6. Depths of up to 2 mm (0.08 in.) can typically be measured with this method. The hole drilling method is often referred to as semidestructive because the hole that is introduced does not significantly affect the usefulness of the work piece for some applications, or may be repaired after measurement. It has been used extensively for both laboratory and field studies of residual stresses, including welding stresses in pipe [21].

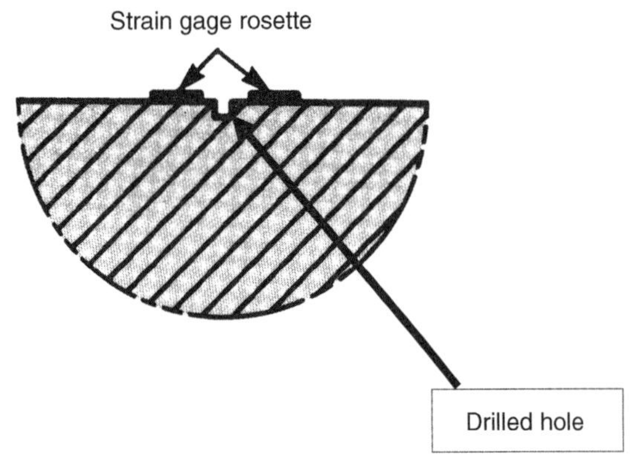


FIGURE 7.6 Schematic of center hole drilling test.

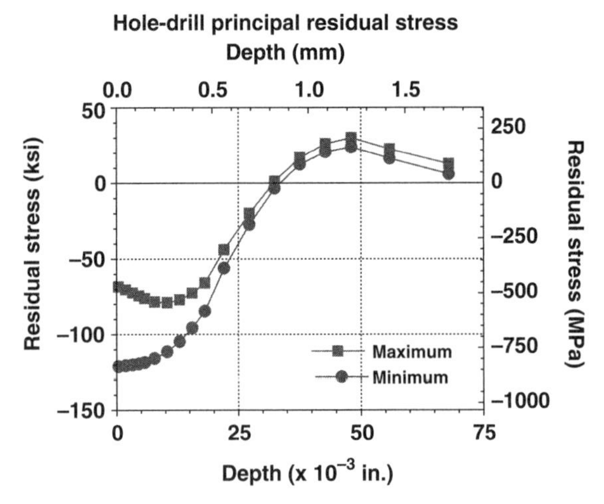


FIGURE 7.7 Example of hole drilling data set.

Principal residual stresses and their angular orientation are determined from the relaxed strain and calibration coefficients for each specific strain gauge geometry and hole size. An example of the subsurface principal residual stress distributions produced bylow plasticity burnishing® (LPB) on a test coupon and measured by the incremental center hole drilling method is shown in Figure 7.7.

The hole drilling procedure has been standardized in ASTM Standard Test Method E837 under the auspices of Subcommittee E28.13 on residual stress measurement [23]. This method applies in cases where material behavior is linear elastic. In theory, it is possible for local yielding to occur due to the stress concentration around the drilled hole when the residual stress exceeds nominally half of the yield strength, and stresses calculated to be significantly higher may be subject to higher experimental error. The theoretical basis, underlying assumptions and limitations, are most thoroughly developed in Micro-Measurements Tech Note TN-503, which is strongly recommended as a training guide [24].

Hole-drilling equipment and supplies are commercially available and a qualified stress technician can successfully perform the test by adhering to the ASTM standard. Tests can be performed either in a laboratory or in the field on components with a variety of sizes and shapes.

### 7.3.2 Ring Core Method

The ring core or trepanning [25–27] is another mechanical residual stress measurement method that allows for deeper

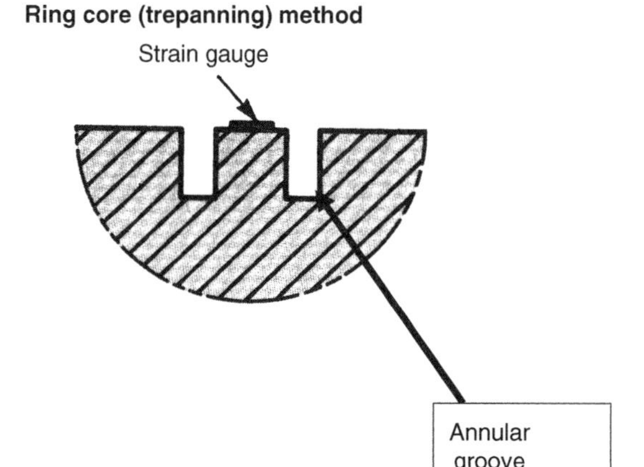


**FIGURE 7.8** Schematic of ring core test.

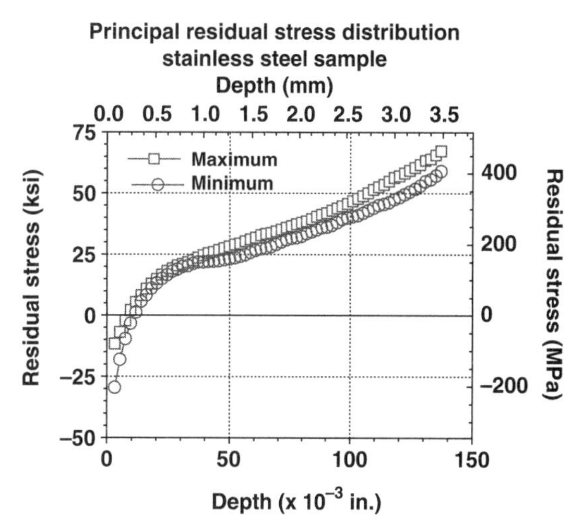


**FIGURE 7.9** Example of ring core data set.

residual stress measurements to be made compared with the hole drilling method. The ring core method involves introducing an annular groove around a three-element strain gauge. Because the cylindrical core is completely separated from the rest of the component, more of the residual strain is relaxed, and the ring core technique can provide nominally an order of magnitude greater measurement sensitivity than hole drilling. The cylindrical core containing the strain gauge is relaxed and the resulting strain relaxations are measured. Because there is no stress concentration at the gauge, residual stresses up to the yield strength can be measured accurately. Principal residual stresses and their angular orientation at each depth are determined from the relaxed strain using relaxation coefficients determined by calibration or finite element calculation. Figure 7.8 is a schematic of the basic ring core setup.

Depths of up to 5 mm (~0.2 in.) can typically be measured in a single ring core. Measurement depth can be increased through the use of larger diameter coring tools; however, resolution is reduced as a trade-off. Although the strain response attenuates with greater depths, there is a limit to which residual stresses can be measured. In order to collect deeper residual stresses, the cylindrical post can be removed and a new strain gauge applied to continue further in depth. An example of the principal residual stress distributions measured by the incremental ring core method in a welded 304 SS sample is shown in Figure 7.9.

### 7.3.3 Diffraction Methods

In X-ray diffraction residual stress measurement, the strain in the crystal lattice is measured directly and the residual stress producing the strain is calculated, assuming a linear elastic distortion of the crystal lattice. The basic "sine-squared psi" method was well developed by the SAE Fatigue Design and Evaluation Committee in the 1960s, and the reader is referred to SAE HS784 for details of the theoretical basis and practical experimental techniques [28]. XRD is the only nondestructive method that determines residual stress directly from measured strain that can be used in the field or laboratory with commercially available apparatus. The pipeline engineer interested in determining residual stress will benefit by understanding the similarity of XRD to mechanical elasticity-based methods. Therefore, a brief overview with derivation from linear elasticity theory is presented.

X-ray diffraction residual stress measurement is applicable to materials that are crystalline and relatively fine grained and that produce diffraction for any orientation of the sample surface. To determine the stress, the strain in the crystal lattice must be measured for at least two orientations relative to the sample surface. Samples may be metallic or ceramic, provided a diffraction peak of suitable intensity and free of interference from neighboring peaks can be produced in the high back-reflection region with the radiations available. X-ray diffraction residual stress measurement is unique in that macroscopic and microscopic residual stresses can be determined nondestructively and separately from the diffraction peak position and breadth, respectively.

***7.3.3.1 Principles of X-Ray Diffraction Stress Measurement*** Figure 7.10 shows the diffraction of a monochromatic beam of X-rays at a high diffraction angle ($2\theta$) from the surface of a stressed sample for two orientations of the sample relative to the X-ray beam. The angle $\psi$, defining the orientation of the sample surface, is the angle between the normal of the surface and the incident and diffracted beam bisector, which is also the angle between the normal to the diffracting lattice planes and the sample surface.

Diffraction occurs at an angle $2\theta$, defined by Bragg's law: $n\lambda = 2d \sin\theta$, where $n$ is an integer denoting the order of diffraction, $\lambda$ is the X-ray wavelength, $d$ is the lattice spacing of crystal planes, and $\theta$ is the diffraction angle. For the

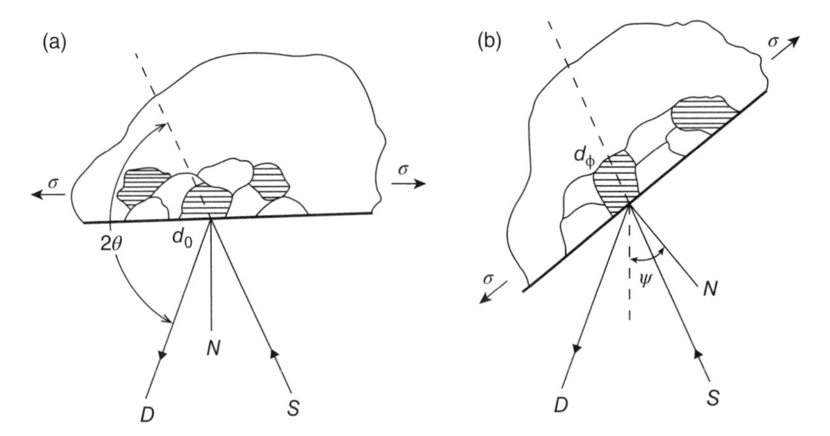


**FIGURE 7.10** Principles of X-ray diffraction stress measurement. (a) $\psi = 0$. (b) $\psi = \psi$ (sample rotated through some known angle $\psi$). D: X-ray detector; S: X-ray source; N: normal to the surface.

monochromatic X-rays produced by the metallic target of an X-ray tube, the wavelength is known to be 1 part in $10^5$. Any change in the lattice spacing $d$ results in a corresponding shift in the diffraction angle $2\theta$.

Figure 7.10a shows the sample in the $\psi = 0$ orientation. The presence of a tensile stress in the sample results in a Poisson's ratio contraction, reducing the lattice spacing and slightly increasing the diffraction angle $2\theta$. If the sample, or apparatus, is then rotated through some known angle $\psi$ (Figure 7.10b), the tensile stress present in the surface increases the lattice spacing over the stress-free state and decreases $2\theta$. Measuring the change in the angular position of the diffraction peak for at least two orientations of the sample defined by the angle $\psi$ enables calculation of the stress present in the sample surface lying in the plane of diffraction, which contains the incident and diffracted X-ray beams. To measure the stress in different directions at the same point, the sample is rotated about its surface normal to coincide the direction of interest with the diffraction plane.

The residual stress determined using X-ray diffraction is the arithmetic average stress in a volume of material defined by the irradiated area, which may vary from square centimeters to square millimeters, and the depth of penetration of the X-ray beam. The linear absorption coefficient of the material for the radiation used governs the depth of penetration, which can vary considerably. In steels, 50% of the radiation is diffracted from a layer approximately 0.005 mm (0.0002 in.) deep for the Cr $K_\alpha$ radiation generally used for stress measurement. Electropolishing is used to remove thin layers to expose new surfaces for subsurface measurement.

*7.3.3.2 Plane Stress Elastic Model* X-ray diffraction stress measurement is confined to the surface of the sample in a layer so thin that a condition of plane stress is assumed to exist. That is, a stress distribution described by principal stresses $\sigma_1$ and $\sigma_2$ exists in the plane of the surface, and no stress is assumed perpendicular to the surface, $\sigma_3 = 0$. However, a strain component perpendicular to the surface, $\varepsilon_3$, exists as a result of the Poisson's ratio contractions caused by the two principal stresses, shown in a three-dimensional view in Figure 7.11.

The strain, $\varepsilon_{\phi\psi}$ in the direction defined by the angles $\phi$ and $\psi$ is

$$\varepsilon_{\phi\psi} = \left[\frac{1+v}{E}\sigma_\phi \sin^2\psi\right] - \left[\left(\frac{v}{E}\right)(\sigma_1 + \sigma_2)\right] \tag{7.1}$$

where $E$ is the modulus of elasticity, $v$ is the Poisson's ratio, and $\sigma_\phi$ is the stress of interest in the surface.

If $d_{\phi\psi}$ is the spacing between the lattice planes measured in the direction defined by $\phi$ and $\psi$, the strain can be expressed in terms of changes in the linear dimensions of the crystal lattice:

$$\varepsilon_{\phi\psi} = \frac{\Delta d}{d_0} = \frac{d_{\phi\psi} - d_0}{d_0} \tag{7.2}$$

where $d_0$ is the stress-free lattice spacing.

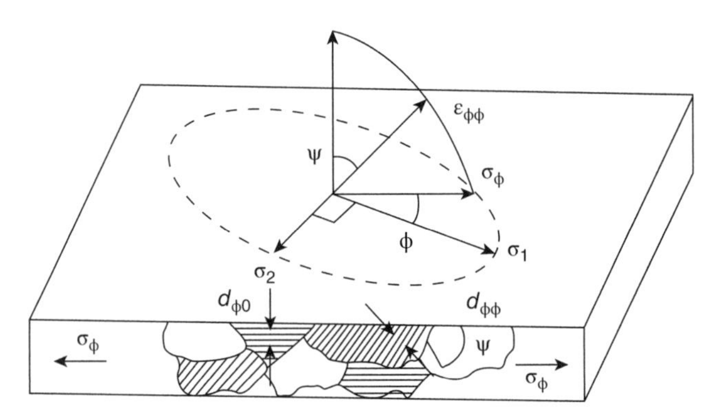


**FIGURE 7.11** Plane stress elastic model.

Substituting for strain in the crystal lattice, the lattice spacing for any orientation, then, is

$$d_{\phi\psi} = \left[\left(\frac{1+\nu}{E}\right)_{(hkl)} \sigma_\phi d_0 \sin^2\psi\right] - \left[\left(\frac{\nu}{E}\right)_{(hkl)} d_0(\sigma_1+\sigma_2)+d_0\right] \tag{7.3}$$

This is the fundamental relationship between lattice spacing and the biaxial stresses in the surface of the sample. The elastic constants $(1+\nu/E)_{(hkl)}$ and $(\nu/E)_{(hkl)}$ are not the bulk values, but the values for the crystallographic direction normal to the lattice planes in which the strain is measured, as specified by the Miller indices $(hkl)$. Because of elastic anisotropy, the elastic constants in the $(hkl)$ direction commonly vary significantly from the bulk mechanical values.

The lattice spacing $d_{\phi\psi}$ is then a linear function of $\sin^2\psi$. If the lattice spacing is measured at two or more angles $\psi$ (two-angle or sine- squared psi methods), then the slope of the plot is

$$\frac{\partial d_{\phi\psi}}{\partial \sin^2\psi} = \left(\frac{1+\nu}{E}\right)_{(hkl)} \sigma_\phi d_0 \tag{7.4}$$

Solving for the stress of interest, $\sigma_\phi$ is

$$\sigma_\phi = \left(\frac{E}{1+\nu}\right)_{(hkl)} \frac{1}{d_0}\left(\frac{\partial d_{\phi\psi}}{\partial \sin^2\psi}\right) \tag{7.5}$$

The X-ray elastic constants can be determined empirically, but the unstressed lattice spacing $d_0$ varies with alloy composition and cold working, and is generally unknown. However, because for metals, $E \gg (\sigma_1+\sigma_2)$, the value of $d_{\phi=0}$ differs from $d_0$ by not more than $\pm 1\%$ and $\sigma_\phi$ may be approximated to that accuracy using

$$\sigma_\phi = \left(\frac{E}{1+\nu}\right)_{(hkl)} \frac{1}{d_{\phi 0}}\left(\frac{\partial d_{\phi\psi}}{\partial \sin^2\psi}\right) \tag{7.6}$$

The method then becomes a differential technique and no stress-free reference standards are required to determine $d_0$ for the biaxial stress case. This is a significant advantage over synchrotron and neutron diffraction methods that must assume a three-dimensional stress state, and therefore require independent knowledge of the unstressed crystal lattice spacing in order to calculate the stress tensor from the measured strains.

Because the depth of material sampled is very shallow, any secondary abrasive treatment, such as wire brushing or sand blasting, radically alters the surface residual stresses, generally producing a shallow, highly compressive layer completely altering the original residual stress distribution.

To measure inaccessible locations, such as the inside surface of pipes and tubing, the sample must be sectioned to provide clearance for the diffraction apparatus and the incident and diffracted X-ray beams. Unless prior experience with the sample under investigation indicates that no significant stress relaxation occurs upon sectioning, electrical resistance strain gauge rosettes should be applied to the measurement area to record the strain relaxation that occurs during sectioning. Following X-ray diffraction residual stress measurements, the total stress before sectioning can be calculated by subtracting the sectioning stress relaxation from the X-ray diffraction results.

***7.3.3.3 Sources of Error*** Geometric precision is required for diffraction techniques. Because the diffraction angles must be determined to accuracies of nominally $\pm 0.01°$, the sample must be positioned in the X-ray beam at the true center of rotation of the $\psi$ and $2\theta$ axes and the angle $\psi$ must be known and constant throughout the irradiated area. Precise positioning of the sample to accuracies of approximately 0.025 mm (0.001 in.) is critical. The size of the irradiated area must be limited to an essentially flat region on the sample surface.

Instrument alignment requires coincidence of the $\theta$ and $\psi$ axes of rotation and positioning of the sample such that the diffracting volume is centered on these coinciding axes. Instrument alignment can be checked readily using a stress-free powder sample [29]. If the diffraction apparatus is properly aligned, a loosely compacted stress-free powder sample should indicate not more than $\pm$14 MPa ($\pm$2 ksi) apparent stress.

Excessive sample surface roughness or pitting, curvature of the surface within the irradiated area, or interference of the sample geometry with the diffracted X-ray beam can result in systematic error. Residual stress generally cannot be measured reliably using X-ray diffraction in samples with coarser-grain sizes, common in castings and weld beads.

A major source of potential systematic proportional error arises in determination of the X-ray elastic constants $(E/1+\nu)_{(hkl)}$. The residual stress calculated is proportional to the value of the X-ray elastic constants, which may differ by as much as 40% from the bulk value due to elastic anisotropy. The X-ray elastic constant can be determined empirically by loading a sample of the material to known stress levels and measuring the change in the lattice spacing as a function of applied stress and $\psi$ tilt [30,31]. The X-ray elastic constant can then be calculated from the slope of a line fitted by least-squares regression through the plot of the change in lattice spacing for the $\psi$ tilt used function of applied stress. X-ray elastic constants for many steels have been published.

X-ray diffraction measures the entire strain present before any sectioning or removal of material, but the shallow penetration requires removal of layers to expose the subsurface material for strain measurement. There are two important corrections that must be applied to subsurface measurements. First, in the presence of a steep stress gradient, as is common on machined or ground surfaces, the exponential attenuation of X-rays penetrating the layers of varying stress gives a weighted average stress that can be unfolded if measurements are made in sufficiently fine layers. Second, the removal of stressed material alters the residual stress in the exposed layers, causing an error that accumulates with depth, but can be corrected with an integral solution analogous to the layer removal mechanical residual stress measurement methods. Appropriate corrections for both sources of error in subsurface measurement are well established and described in SAE HS784 [28].

### 7.3.4 Synchrotron X-Ray and Neutron Diffraction: Full Stress Tensor Determination

Conventional X-ray diffraction stress determination uses low-energy, shallow-penetrating radiation allowing stress determination in the biaxial case, eliminating the need for a stress-free reference sample to calculate the residual stress. This is a major advantage, allowing reliable measurement in materials, such as welds, with local variation in composition. In principle, the full tensor can be determined nondestructively using either high-energy synchrotron X-radiation or thermal neutrons from a suitable reactor source. Both synchrotron X-ray and neutron diffraction methods have been developed for determining the full triaxial stress tensor [20]. Penetration depths can range from millimeter for synchrotron X-rays to centimeter for thermal neutrons. Both require expensive stationary facilities generally available only at major government laboratories. In all other respects, both are diffraction techniques measuring lattice strain, and are subject to the same limitations and error sources as the conventional, portable, plain stress biaxial X-ray method.

Unlike the plane stress X-ray method, determination of the triaxial stress tensor requires independent knowledge of the unstressed lattice spacing $d_0$ at the accuracy required for strain measurement (1 part in $10^5$) to calculate the residual stress from the measured strains. It is not possible to solve for three-dimensional stresses from the measured strains without separate knowledge of the unstressed lattice spacing as a zero-stress reference. Unfortunately, the unstressed lattice spacing varies with cold working (due to dislocation density) and local variation in chemical composition. In many cases, including surfaces plastically deformed by machining, weld fusion and heat-affected zones, or in case-hardened carburized steels, the lattice spacing varies as a result of deformation, composition, or heat treating, precluding independent determination of the unstressed lattice spacing with sufficient precision to make reliable stress tensor calculations.

The equipment requirements, lack of portability, extensive data collection required, and dependence on absolute knowledge of $d_0$ limit the full tensor method primarily to research applications. Mechanical methods, primarily center hole drilling and ring core, or conventional plane stress biaxial X-ray diffraction methods, are recommended for general use for determining the residual stress distributions in pipelines.

### 7.3.5 Magnetic Barkhausen Noise Method

The Barkhausen Noise method [20,32] is a nondestructive, high-speed technique applicable to ferritic materials and, therefore, well suited for steel pipe inspections [33]. The Barkhausen technique relies upon analyzing the "noise" observed on the magnetic hysteresis loops when an alternating current is used to alternately magnetize the metal in North–South and South–North orientations. A pickup coil then detects the resultant magnetic field. The individual magnetic domains within the crystalline grains that make up the material are typically oriented in alternating N–S and S–N orientations. Under stress, the domains can flip their orientation, resulting in changes in the magnitude of the noise imposed upon the magnetic hysteresis loop, shown schematically in Figure 7.12. This noise must be analyzed by electronically separating and filtering a certain frequency range that is sensitive to stress. As noted below, the noise level is affected by a variety of other metallurgical properties, such as hardness, grain size, and texture (preferred orientation), the effects of which must be separated from that of residual stress.

The Barkhausen noise level is expressed as a "magnetoelastic parameter" (MP), defined as the root mean square (RMS) level of the Barkhausen noise after amplification and filtering, relative to the unstressed state of the material. The depth of the material sampled is attenuated exponentially with depth, and diminishes with increasing frequency. The depth of penetration is governed by the "skin" effect.

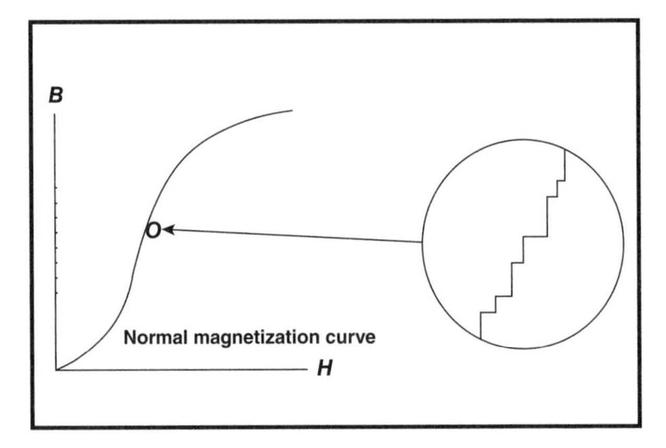


**FIGURE 7.12** Barkhausen noise caused by magnetic domain reversals on the magnetization hysteresis curve for steel [19]. ( Reproduced with kind permission from Society for Experimental Mechanics, Inc.)

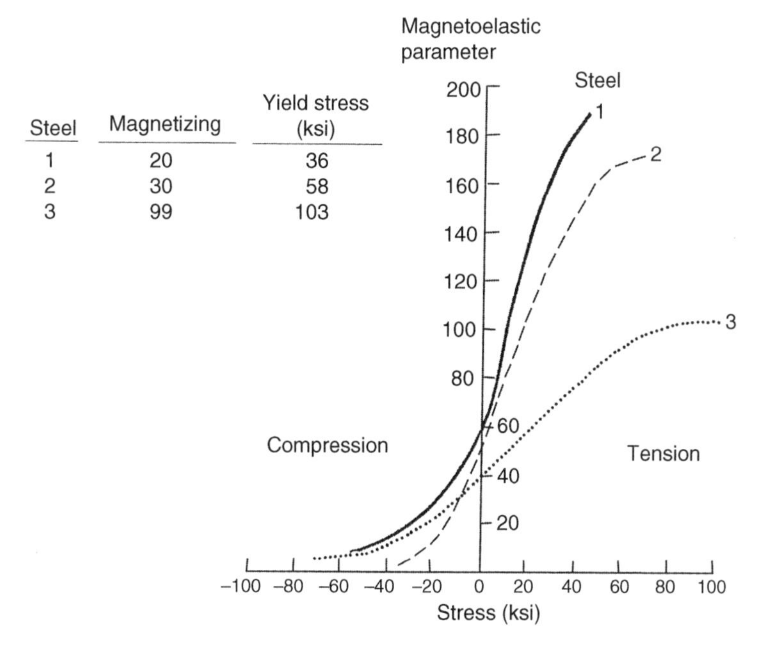


**FIGURE 7.13** Barkhausen MP calibration curves for steels of different hardness and yield strength showing nonlinearity, higher sensitivity to tensile stress, and reduced sensitivity at higher strengths.

Excitation frequencies from 3 to 15 kHz provide a depth of penetration of nominally 0.20 mm. Higher frequencies of 70–200 kHz sample shallower depths on the order of 0.02 mm. Barkhausen provides a sampling depth between that of the shallower X-ray diffraction and the deeper hole drilling or ring core mechanical techniques. The maximum range of penetration reportedly ranges from 0.10 to 3.0 mm for the practical frequencies and instrumentation.

The Barkhausen noise response to stress is asymmetric, being larger for tensile stresses and less sensitive to compression. The response curve is a general sigmoid "S" shape, shown in Figure 7.13. The magnetoelastic parameter (MP) is higher for softer, more easily magnetized low-strength steels, and lower for harder, higher strength steels. Calibration is required for the material, frequency, noise analysis (filtration) method, and instrumentation used. Individual transducers may not be interchangeable, so calibration using strain gauge beams or pipe sections loaded to known stress levels is recommended for each instrument and application.

The Barkhausen technique is sensitive to a range of material properties other than strain, as is true of virtually all of the nonlinear-elastic measurement methods. Barkhausen is sensitive to the grain size of the material. For fine-grained material, the MP initially increases rapidly with grain size, and then less rapidly for larger grains. Texture, or the crystallographic orientation, strongly influences the Barkhausen technique, which samples the switching of the direction of magnetization of magnetic domains along the specific crystallographic directions. Hardness and yield strength of the material have a strong influence, as seen in Figure 7.13. The hardness dependence is so pronounced that the same Barkhausen technology used for stress measurement is used to distinguish hardness and detect grinder burn in steels. The MP level for a given tensile stress can range from nearly a factor of 4, from a high output of 200 for soft 35 HRC steel, to as low as 50 for fully hardened 60 HRC material. The dependence on material properties other than stress makes use of the Barkhausen technique highly material and component geometry specific.

Calibration of the Barkhausen probe is necessary to account for the effect that hardness, texture, grain size, and other material properties have in the presence of a varying residual stress field. In the case of pipe, a calibration tool can be made using a section of the pipe instrumented with electrical resistance strain gauges to provide a means of measuring the MP as a function of applied stress. In the case of welds, where there is a wide range of grain size, hardness, and even composition within the weld HAZ, none of the nonstress properties can be assumed to be constant, and interpretation of the Barkhausen data must be used with caution. Rapid detection of potential zones of tension may be then confirmed by other residual stress measurement techniques such as XRD or hole drilling.

The Barkhausen magnetic method can be summarized as (i) being suitable for only ferromagnetic materials, (ii) limited to measuring surface layers of the material, (iii) limited in measurement range by saturation of the Barkhausen noise response when testing compressive residual stresses, particularly in hard steels, and (iv) sensitive to a variety of microstructural variations (grain size, hardness, and texture), making separation of those contributions from stresses difficult [32].

## 7.4 CONTROL AND ALTERATION OF RESIDUAL STRESSES

The engineer has a variety of tools for modifying the residual stresses that are encountered to reduce or eliminate the adverse effects of tensile residual stress. Machining and grinding practices can be modified to avoid surface tension. Weld shrinkage-induced tension can be reduced by well-established practices of preheating or postweld thermal stress relief. However, it is also possible to introduce beneficial residual compression using surface enhancement methods to further improve performance beyond that achievable with the available materials, even in a stress-free state. When material substitution is not an option, and applied stresses cannot be reduced, surface enhancement can ensure reliable performance, especially in highly stressed critical locations. Several of the available surface enhancement methods are briefly described.

### 7.4.1 Shot Peening

Shot peening is the most widely used surface enhancement method. Metallic or ceramic shot, ranging in size from nominally 0.25 mm (0.010 in.) to as large as 3.18 mm (0.125 in.), impacts the surface of the component, producing spherical indentations. The shot peening process forms a compressive layer through a combination of subsurface compression developed by Hertzian loading combined with lateral displacement of the surface material around each of the dimples. As the dimples overlap with random impacts, the entire surface is effectively elongated, driving the surface layer of deformed material into compression as its expansion is resisted and supported by the equilibrating tension in the material below.

Metallic shot is often either cast steel or cut wire blasted against a carbide plate to form a nearly spherical shape. Cut wire shot can be manufactured from virtually any alloy to avoid elemental contamination. Ceramic shot is typically zirconium oxide or glass bead. The shot will wear and fracture during use; therefore, the shot peening media should be constantly screened to remove broken shot and dust.

The shot peening process is defined by parameters that include the size and type of shot used, the Almen intensity achieved, and the coverage. Almen intensity is a measure of the deflection of the Almen A, C, or N strip that occurs during the peening cycle, and is a measure of the elastic energy stored in the 1070 steel Almen strip by the formation of a layer of residual compression. Almen intensity is only a process control tool. The Almen strip arc height is related to the area under the residual stress depth curve in the Almen strip, and it does not uniquely define a depth or magnitude of compression in either the strip or the component being shot peened. Coverage is the percentage of the surface impacted by shot, and is a function of time under the shot stream. In order to ensure uniform treatment of the surface, coverage is often specified at 100% or higher, implying that each point on the surface was impacted at least one or more times.

The shot is accelerated to sufficient velocity to deform the surface on impact, either by air blast or wheel machines. In wheel shot peening machines, the shot is thrown from a rapidly spinning radially bladed wheel. Air blast machines propel the shot in a stream of compressed air through a hardened carbide nozzle. Flapper peening is a form of controlled shot peening without free flying shot. The flapper tool has captive shot embedded, like rivets, in rubberized fabric flaps that are attached radially to rotate around a shaft. The shaft is rotated so that the flaps successively impact the surface peening of a local area. Flapper peening is used for repair or limited access work or when free shot cannot be tolerated, as in field-welded joints in pipelines [34].

By its very nature, shot peening relies upon random impacts of shot. In order to achieve the coverage requirements, some regions will receive numerous impacts before adjacent areas are impacted at all. The result is a nonuniform and often very highly cold-worked surface, especially for high coverage peening. Cold work levels range from 40% to over 100% during creation of the layer of surface compression. In the work hardening materials, such as austenitic alloys, peening-induced cold working can exhaust the ductility of the material, leaving a brittle surface layer. Shot peening damage in the form of "laps and folds" creates stress concentrations that reduce fatigue performance.

Shot peening is a very practical surface enhancement method, provided the components are not exposed to elevated temperatures or mechanical overload. Research has shown that the highly cold-worked shot-peened surface will relax more completely and much faster than a low cold-worked surface at the same state of compressive stress [18].

### 7.4.2 Roller or Ball Burnishing and Low Plasticity Burnishing

In roller or ball burnishing, a wheel-like axle-mounted roller or a hydrostatically supported ball is pressed against the surface. The tool is then moved over the surface, repeatedly deforming the surface layer to create compression up to nominally 1 mm, or more, deep. Like shot peening, the compressive residual stress creating mechanism is a combination of subsurface yielding by Hertzian loading and lateral displacement of the surface material. Conventional roller burnishing deliberately creates a highly cold-worked layer that is also in high residual compression. Roller burnishing is generally performed in lathe operations, advancing the tool with each revolution at a constant feed rate.

LPB differs from conventional roller and ball burnishing primarily by imparting the minimal amount of cold working needed to create the depth and magnitude of residual compression required to achieve the desired damage tolerance and fatigue or stress corrosion performance. Low cold work provides both thermal and mechanical stability of the beneficial compression. LPB uses a constant volume hydrostatic tool design to "float" the burnishing ball continuously during operation, regardless of the force applied, to avoid damaging the surface. The LPB burnishing force can be optimized for each application and varied under closed-loop control to impart the depth and magnitude of compression required in just the critical failure-prone locations. By controlling the burnishing force, LPB produces a depth of compression ranging from a few thousandths of an inch to over a full centimeter under closed-loop control.

### 7.4.3 Laser Shock Peening

The laser shock peening process introduces compression with low cold working using shock waves to yield the material. Laser peening utilizes high-speed, high-powered lasers to focus a short-duration energy pulse on a coating, usually black tape, on the surface of the work piece to absorb the energy of the laser beam. A transparent layer of flowing water covers the tape to direct the shock wave energy into the surface of the material. When the laser is fired periodically, the laser beam passes through the water, explodes the tape, and creates a shock wave sufficient to deform the material to depths typically on the order of a millimeter. The shocking process is repeated in a computer-controlled pattern across the surface, creating a series of slight indentations and regions of residual compression. Several cycles or "layers" of peening are often required. Although limited by cost, quality control, and logistical issues, laser peening has been studied as a means of introducing residual compression in pipelines and the residual stress distributions documented [35].

Figure 7.14 shows subsurface stress profiles for each surface treatment process and the cold work induced in titanium.

### 7.4.4 Thermal Stress Relief

Thermal stress relief is widely used to reduce residual stresses. Postweld heat treatment of pipeline joints reduces the detrimental tensile stresses generated by welding. The reader is referred to the API bulletin 939E for a detailed discussion of procedures for postweld heat treatment of pipe sections.

The primary process at work in thermal stress relaxation is a creep mechanism. Residual stresses relax exponentially with both time and temperature, reducing the overall strain energy. Both tensile and compressive residual stresses are reduced as the temperature of the material is increased. No stresses higher than the yield strength at the maximum exposure temperature can be retained in the component. The conventional rule of thumb for thermal stress relief of steels based upon this time–temperature relaxation mechanism is exposure to 611 °C (1150 F) for a period of 1 h for each 25 mm (1 in.) of thickness, with the minimum of 1 h for any thickness being stress relieved.

Another mechanism more recently recognized is the rapid relaxation of residual stress in a highly cold-worked material [36]. At elevated temperatures, the deformed surface of machined, ground, shot-peened, and other surfaces that have high dislocation density can relax virtually immediately by a dislocation annihilation mechanism, which is then followed by the slow exponential relaxation. This effect can cause a loss of beneficial compression from shot peening with elevated temperature exposure, particularly in austenitic alloys.

Residual stresses do not relax uniformly, that is, not in proportion to the original levels present at each point in the body, because the amount and rate of stress relaxation depend upon the initial stress magnitude. Furthermore, the residual stress distribution will re-equilibrate as the regions of higher tension or compression relax more rapidly, and by a greater amount, than the lower stress regions. Stresses may actually increase in some areas following thermal stress relief as more highly stressed regions relax and the body re-equilibrates. Because the rate of relaxation depends upon the stress level, complete thermal stress relief is nearly impossible to achieve. Once the stresses are reduced, the driving force for further relaxation diminishes, and some level of stress is invariably retained.

Successful thermal stress relief requires uniform temperature throughout the component during the process. If there are large thermal strains created during either the heating or the cooling stages of the thermal cycle, the plastic strains introduced at the temperature can result in even higher residual stresses after cooling than were present before heating. Quenching from elevated temperatures is an extreme case of high thermal strain, and the source of residual "quench stresses." Complex geometries such as heat exchangers require very slow heating and cooling. It may be impossible to thermally stress relieve components composed of dissimilar materials with different coefficients of thermal expansion. The key to a successful thermal stress relief is to keep the entire body at a nominally uniform temperature, which is slowly raised to the critical temperature, held for the required exposure time, and then cooled back to ambient temperature without creating thermal strains that could cause plastic deformation.

## 7.5 CASE STUDIES OF THE EFFECT OF RESIDUAL STRESS AND COLD WORK

Thermal stress relaxation can only reduce the magnitude of undesirable residual stresses, but cannot introduce beneficial

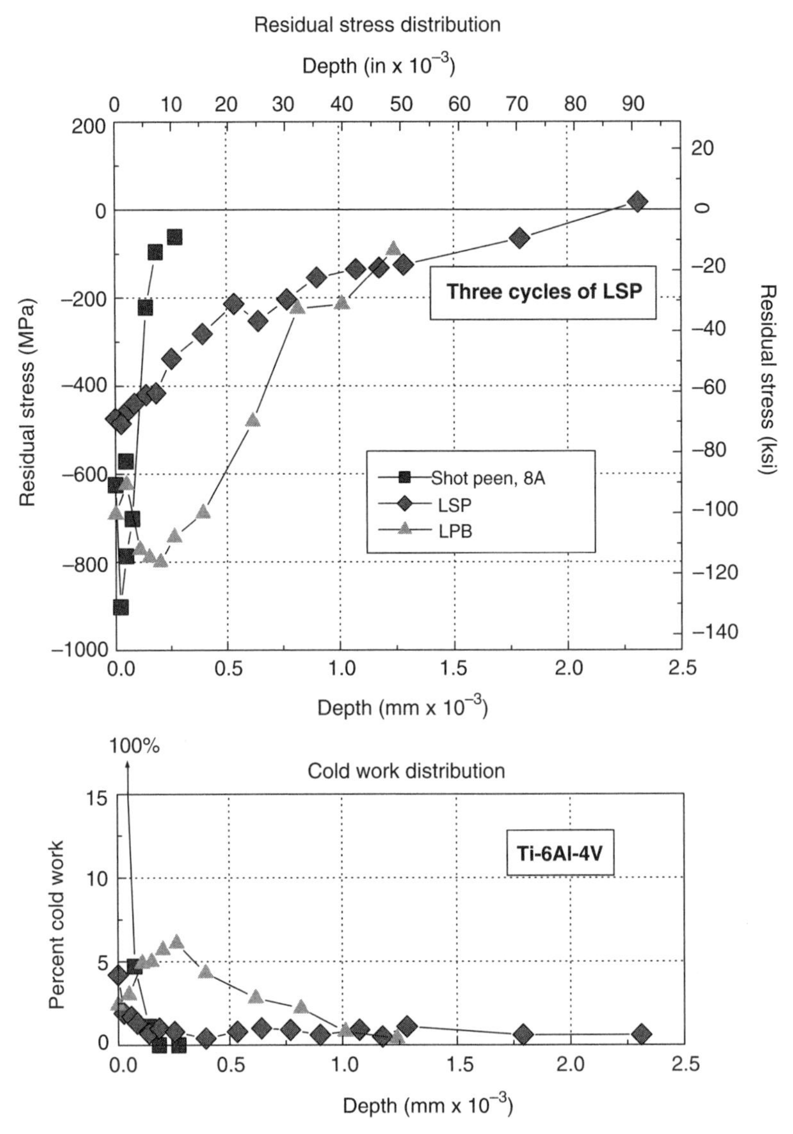


**FIGURE 7.14** Stress profiles and cold work from SP, LPB, and LSP.

compression to advantage. It must also generally be applied to all or large portions of a component to avoid creating large thermal strains. Mechanical surface enhancement treatments introduce beneficial compressive residual stress to critical locations to eliminate detrimental tension or improve performance at the high-stress critical locations. The following case studies provide examples of use of surface enhancement to mitigate SCC and improve fatigue performance in pipeline-related applications.

### 7.5.1 Case Study 1: Restoration of the Fatigue Performance of Corrosion and Fretting Damaged 4340 Steel

Corrosion pits commonly form at the site of inclusions on the material surface. Pits can be rounded or branched as the pit grows into the surface, and form fatigue crack initiating stress concentrations. Corrosion pitting damage generally increases with time. The fatigue debit from pitting is typically on the order of half of the endurance limit of the material. The fatigue strength can be restored by introducing a layer of compressive residual stress that is deeper than the corrosion pits to prevent fatigue crack growth from the bottom of the pits.

The effect of introducing compressive residual stress on fatigue performance was studied for 4340 steel with corrosion pitting in a salt fog chamber [37]. The benefits of surface enhancement were investigated with fatigue tests conducted in four-point bending at a stress ratio $S_{min}/S_{max} = 0.1$ after exposure to salt fog for 100 and 500 h. Samples were tested with and without wire brushing to remove the corrosion product, as might be performed during maintenance.

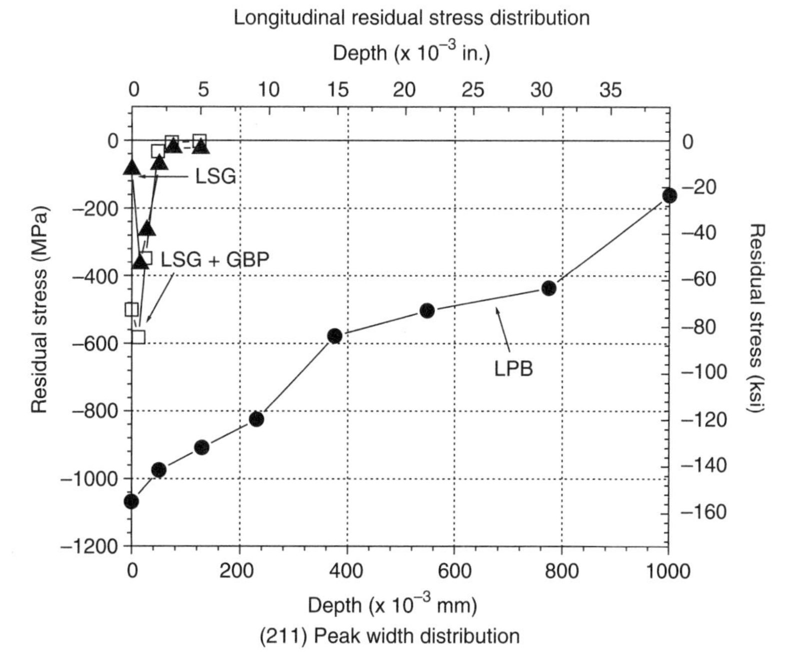


**FIGURE 7.15** Subsurface residual stress distributions of LSG, GBP, and LPB 4340 steel.

Residual stress distributions measured by X-ray diffraction are shown in Figure 7.15. Surfaces were low-stress ground (LSG), glass bead peened (GBP), and LPB. LSG is the common baseline condition for fatigue sample preparation, producing shallow compression typical of cold abrasion processes. GBP produced an equally shallow (0.05 mm) layer of higher −600 MPa magnitude. LPB produced a layer of compression over 1 mm deep starting at over −1000 MPa on the surface.

The fatigue results in Figure 7.16 show that pitting from 100 and 500 h salt fog exposures reduced the $10^7$ cycle fatigue strength by nominally 25 and 50%, respectively. Fatigue cracks initiated in corrosion pits in all cases. LPB applied to the corroded surfaces after superficial cleaning to

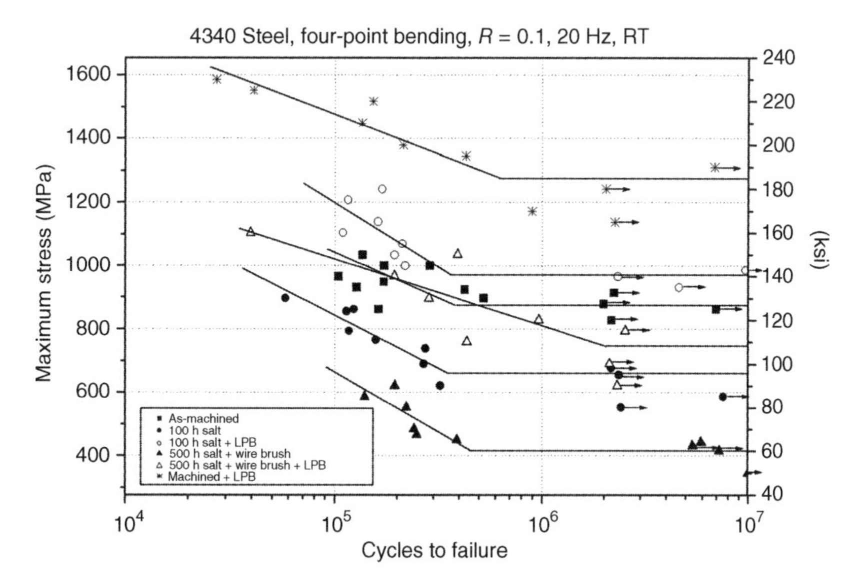


**FIGURE 7.16** High cycle fatigue S/N data.

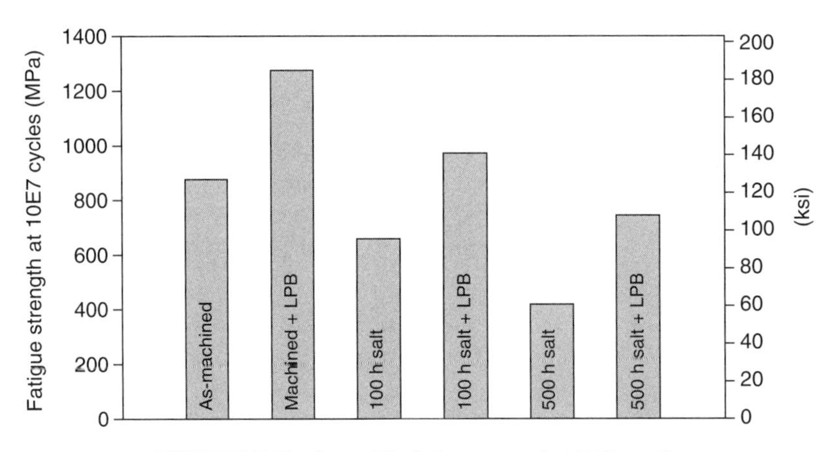


**FIGURE 7.17** Long-life fatigue strength (4340 steel).

remove loose corrosion product restored the fatigue strength to 110% of the original strength after 100 h exposure and 85% after 500 h. The relative fatigue strength at $10^7$ cycles, summarized in Figure 7.17, is attributed to the increased depth of pitting with exposure time.

Fretting damage causes fatigue failures in components that are clamped together and vibrate, such as bolted flange joints in pipelines. The combination of clamping stress pressing the surfaces together and alternating normal stresses in the plane of the surface produces high shear stresses at the edges of contact that initiate shear cracks. The shear stresses are high only at the surface, and the shear cracks produced are generally limited to depths of less than 0.08 mm (0.003 in.). The shallow shear cracks ultimately propagate in fatigue to failure driven by the alternating normal stresses. The fatigue debit from fretting is typically on the order of half of the endurance limit of the material, but can be mitigated by arresting fatigue crack propagation with a layer of residual compressive stress deeper than the maximum depth of the shear cracks.

A "bridge" fretting fatigue apparatus [38] was used with flat samples loaded in four-point bending to study the effect of surface treatments on 4340 steel, 50 HRC, before, after, and with repeated fretting damage. The multiple fretting exposures were intended to simulate combinations of damage in service, overhaul with treatment, and return to service. Figure 7.18 shows the results of a study of GBP and LPB, with fretting damage either before or after surface enhancement.

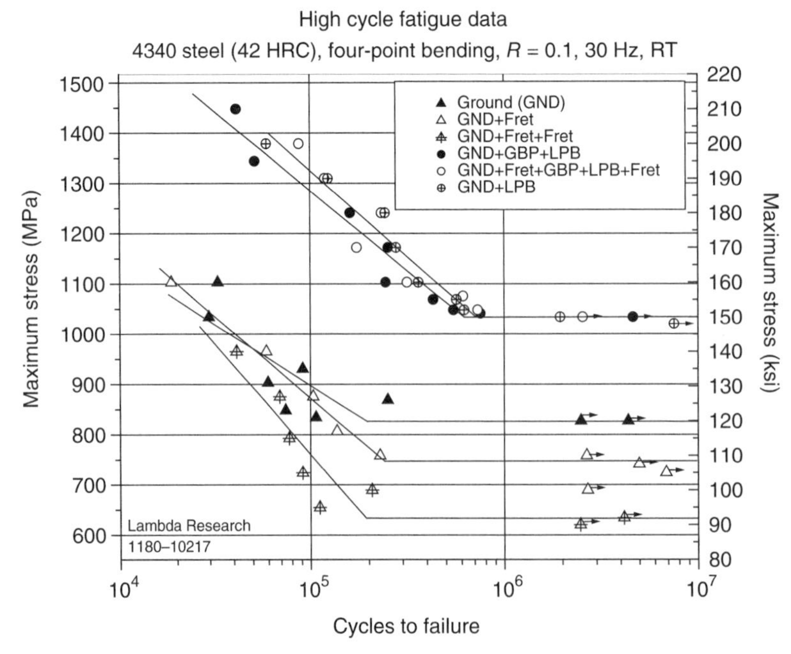


**FIGURE 7.18** 4340 Steel fretting fatigue results, as ground and with prior LPB treatment.

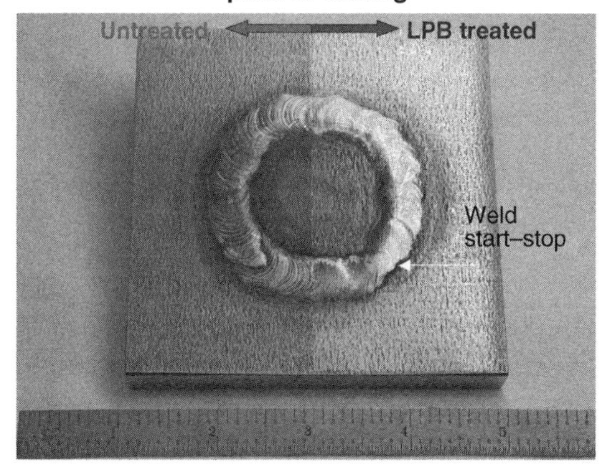


FIGURE 7.19 304L SS butt-welded plate specimen, half LPB processed.

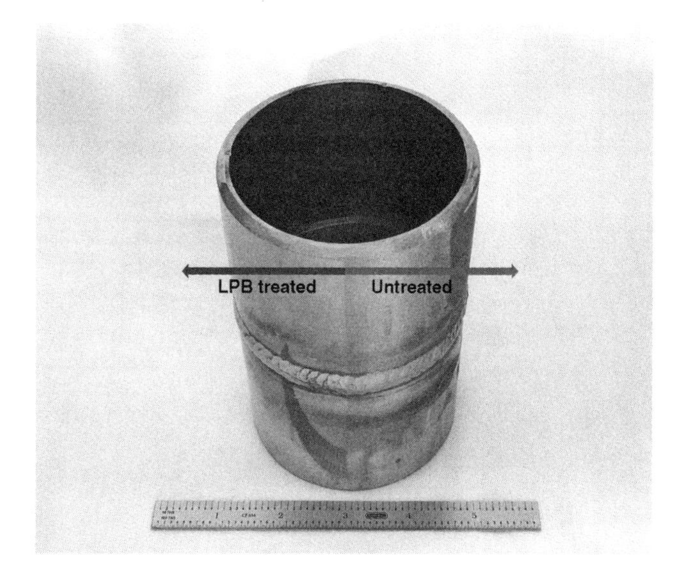


FIGURE 7.20 316L SS circular welded plate specimen, half LPB processed.

The fretting fatigue debit was entirely eliminated with a layer of compression deeper than the fretting damage, and surface enhancement increased the HCF strength 25% over the baseline ground condition.

### 7.5.2 Case Study 2: Mitigating SCC in Stainless Steel Weldments [39]

SCC can occur only in a susceptible alloy where the net tensile stress (residual and applied) exceeds a threshold level in a corrosive environment. Austenitic stainless steels, including 304L and 316L, are SCC-susceptible alloys, especially in the "sensitized" condition. Sensitization occurs in the HAZ of welds that have reached a temperature at which Cr can precipitate out of solution forming carbide inclusions along the grain boundaries. Because the sensitization temperature is nominally the same as that required for thermal stress relaxation, thermal stress relief of these welds generally is not possible. If the SCC environment cannot be avoided, and thermal stress relief is not practical, the net stress level can be reduced to less than the SCC threshold by introducing a layer of residual compression into the surface.

Two different symmetrical weld geometries were prepared and processed using LPB to produce a 1 mm depth of compressive residual stress on half of each specimen. The samples were a circular weld deposit and a butt-welded pipe, as shown in Figures 7.19 and 7.20.

Residual stresses produced by welding were first measured by X-ray diffraction as functions of distance and depth on the untreated and treated sides of the 316L circular welded plate. Figure 7.21 shows compressive residual stresses within the treated zone and tensile stresses from weld contraction within the untreated zone, extending to beyond the measured maximum depth of 0.50 mm (0.020 in.).

The effectiveness of SCC mitigation was aggressively tested in heated $MgCl_2$ solution, followed by dye penetrant inspection. The untreated welds suffered severe SCC damage due to the combination of residual tension from the welding operation and sensitization during the welding heat cycle. The fluorescent dye penetrant examination photos in Figure 7.22 reveal complete mitigation of SCC by the deep compressive residual stress layer. Propagating cracks are arrested at the boundary of the treated half.

The sectional view in Figure 7.23 reveals that the cracks penetrated entirely through the 12.5 mm (0.5 in.) thick welded plate. Cracks originating in the untreated half of the circular weld samples and propagating entirely through the thickness were arrested without penetrating under the compressive layer of the treated material.

The 304L butt-welded pipe sections showed the same behavior as the circular welds. Florescent dye penetrant inspection shown in Figure 7.24 revealed that no SCC occurred on the compressive surface of the treated side.

Introduction of beneficial compression to eliminate residual tension in critical locations, such as sensitized austenitic stainless steel welds, can dramatically improve resistance to SCC.

### 7.5.3 Case Study 3: Mitigation of Sulfide Stress Cracking in P110 Oil Field Couplings [40]

Another form of SCC, sulfide stress cracking and hydrogen embrittlement (HE), limits the use of high-strength carbon steels in $H_2S$-containing "sour" service environments. High-magnitude tensile stresses are generated in threaded connections during power makeup of downhole tubular components. These coupling stresses can be considered to be

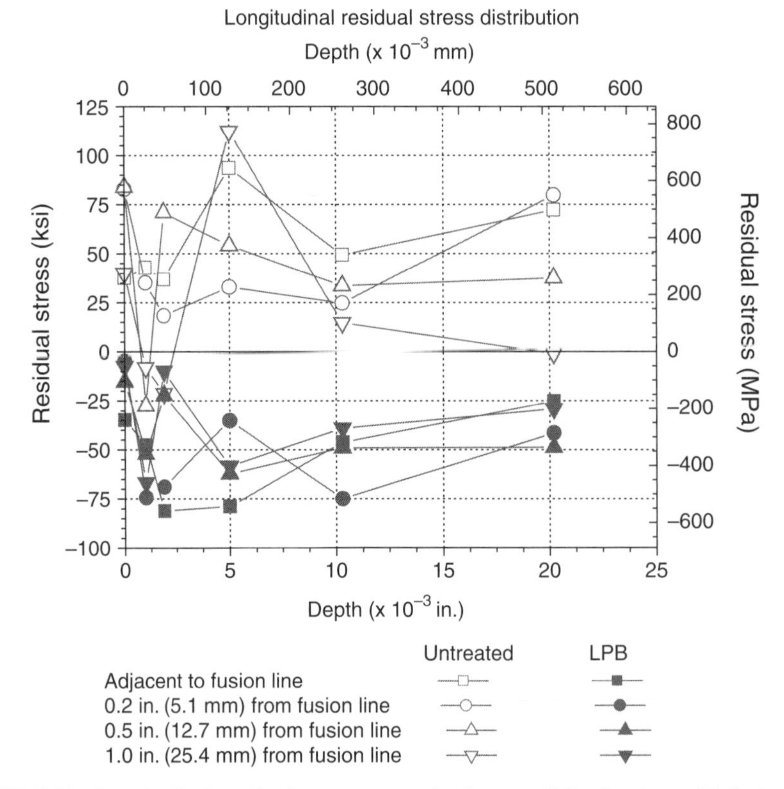


**FIGURE 7.21** Longitudinal residual stress versus depth on a 316L circular welded plate. Compression extends over 25 mm (1.0 in.) through the fusion line and HAZ.

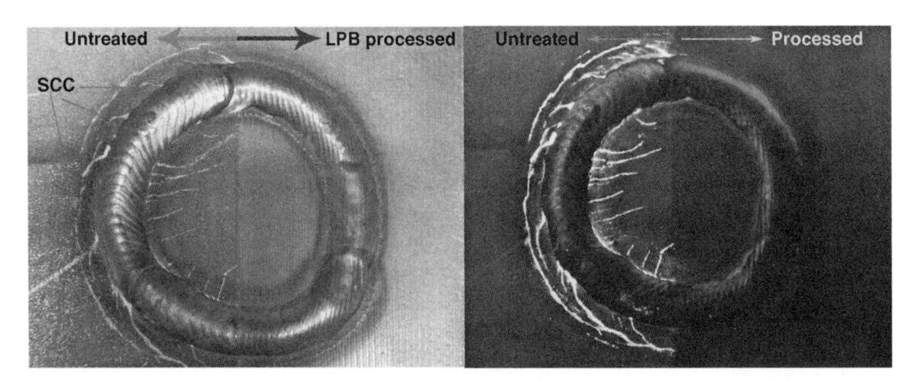


**FIGURE 7.22** Fluorescent dye penetrant reveals severe SCC on 304L circular weld specimen (ambient and UV lighting).

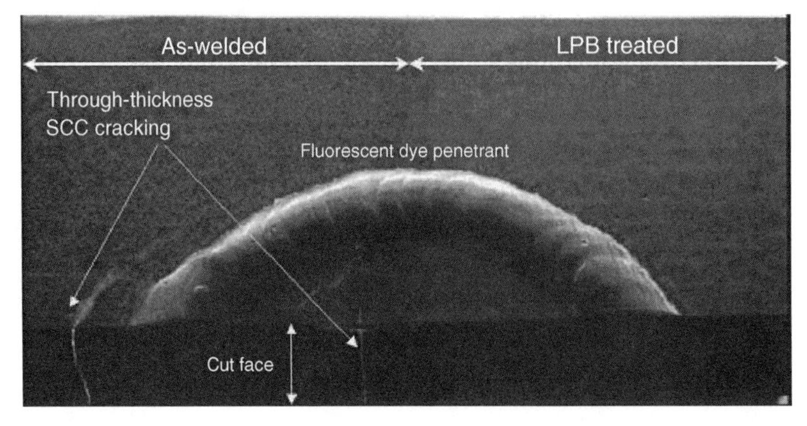


**FIGURE 7.23** Fluorescent dye penetrant photo showing through-thickness SCC cracking on untreated side of 304L SS circular weld specimen.

**FIGURE 7.24** SCC on untreated side of 304L SS pipe weld specimen (left). Fluorescent dye shows no cracking on the LPB-processed regions.

"residual" fit-up stresses once the connections are made and placed in service. Service loads can then increase the total stress to exceed the threshold tensile stress for SSC initiation. High-strength steel grades are generally preferred, but have inherently lower $K_{ISCC}$, and are at higher risk of SSC in sour service than the low-strength grades. The conventional options are to use more expensive SSC-resistant alloys or to accept the strength limitations of lower strength grades to reduce the chance of SSC failure.

Introduction of stable, high-magnitude compressive residual stresses into less-expensive carbon steel alloys alleviates the tensile stresses and mitigates SSC, while also improving fatigue strength. Less-expensive alloys could then be used in sour environments. The benefits of introducing a 1 mm deep layer of high compression on SSC performance were evaluated in C-ring specimens and on full-size pressurized quench and tempered API P110 steel coupling blanks with a yield strength of 910 MPa (132 ksi). The C-ring and the full-size 4.5 in. coupling blanks specimens are shown in Figure 7.25.

Both specimen types were statically loaded to fixed fractions of the specified minimum yield stress (SMYS) and exposed to 100% NACE TM0177 $H_2S$ solution A. No SSC failures occurred in LPB-treated test specimens stressed up to 85% SMYS hoop tension, exceeding the NACE 720 h exposure requirement.

Figure 7.26 shows that untreated coupling blanks failed in 37.5 h at 45% SMYS, but treated couplings run out the 720 h test duration even at 95% SMYS. The C-ring sample results, shown in Figure 7.27, were comparable. Untreated, baseline samples failed in 10 h at 45% SMYS. Treated samples exceeded 840 h exposures stressed at 90% SMYS.

The SSC results confirm that the compressive residual stress layer at 1 mm deep isolated the material from the corrosive $H_2S$ environment and eliminated sour well cracking. As observed for the welded stainless steel in chloride SCC, the total stress in the compressive surface layer remains below the threshold for SSC even under applied service stresses up to the yield strength of the alloy. Surface enhancement mitigates $H_2S$ cracking allowing high-strength steel to be used in an otherwise prohibitive environment.

### 7.5.4 Case Study 4: Improving Corrosion Fatigue Performance and Damage Tolerance of 410 Stainless Steel [41]

CF and SCC occur in ferritic and martensitic stainless steel components. Shot peening is widely used to improve fatigue

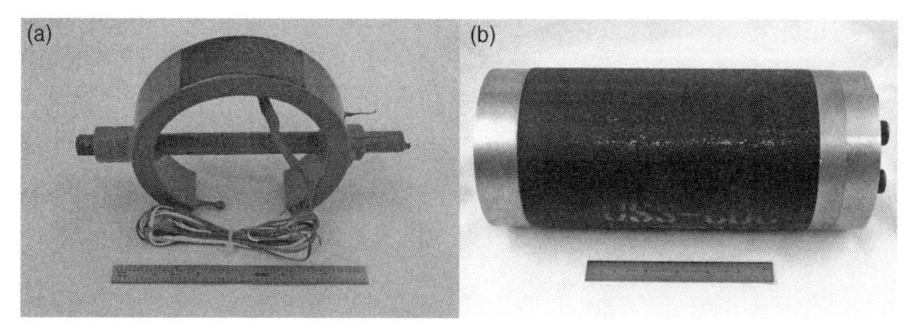


**FIGURE 7.25** (a) C-ring specimen and (b) full-sized coupling blank in test fixture.

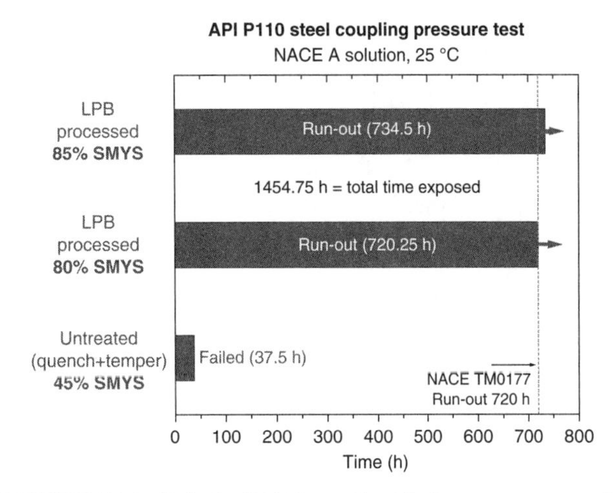


**FIGURE 7.26** Full-sized P110 coupling blank pressure test results (in NACE A solution).

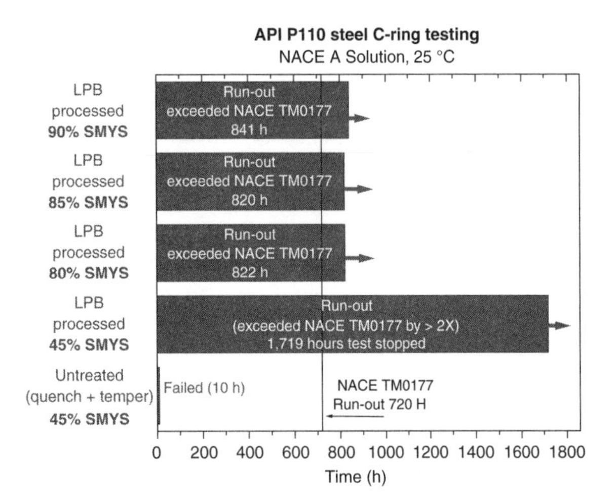


**FIGURE 7.27** P110 C-ring test results (in NACE A solution).

performance; however, the repeated random impacts from SP can produce a highly cold-worked surface. Corrosion pits or mechanical damage can penetrate the relatively shallow SP compressive layer, providing initiation sites for SCC and CF. Low plasticity burnishing produces deeper compression with lower cold working of the surface. Residual stress distributions obtained using X-ray diffraction are plotted as a function of depth in Figure 7.28 for samples processed using SP, LPB, and LSG used for fatigue sample preparation.

High cycle fatigue tests were conducted on type 410 stainless steel to determine the benefits of LPB and SP in mitigating surface damage by introducing notches to simulate foreign object damage, pitting damage, or erosion prior to testing. The effects of 0.25 and 0.50 mm (0.01 and 0.02 in.) mechanical damage on fatigue life are shown in Figure 7.29. Because the residual compression from LPB was deeper than the damage in the samples, LPB provided nominally a 100× improvement in fatigue life compared with the shallow compression from SP.

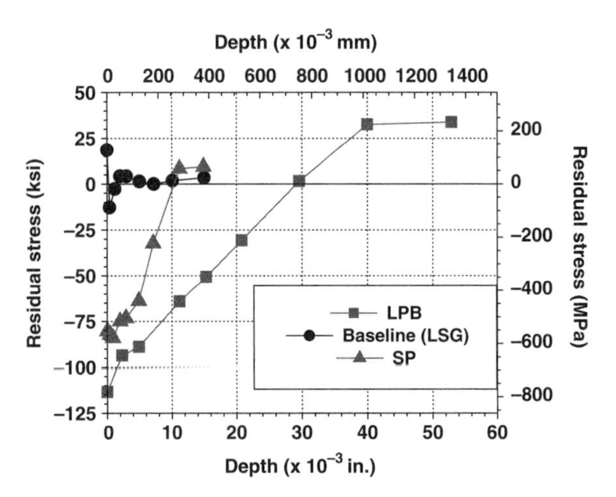


**FIGURE 7.28** Residual stress profiles for low plasticity burnishing, low stress grinding, and shot peening.

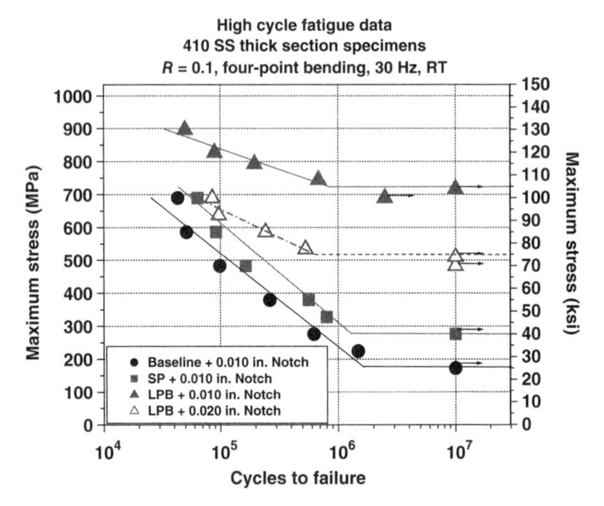


**FIGURE 7.29** High cycle fatigue results for 410 SS with mechanical damage.

High cycle corrosion fatigue tests were conducted in an active corrosion (AC) medium of 3.5% weight NaCl solution on type 410 stainless steel to determine the effect of LPB and SP on corrosion fatigue performance. The S–N curves for samples subject to prior SCC damage and fatigue tested in active corrosion are shown in Figure 7.30. Baseline and SP samples have similar fatigue performance. Prefatigue SCC damage with active corrosion during fatigue introduces a fatigue debit of 50% of the endurance limit. The debit from corrosion was nearly as high as that resulting from the 0.25 mm (0.01 in.) deep notch. LPB-treated samples have nominally twice the fatigue strength of the baseline and SP samples and an increase in life nominally 50× that of the SP or baseline conditions.

The cold work associated with the surface treatments can adversely affect the corrosion behavior. Polarization curves for LPB- and SP-treated samples are shown in Figure 7.31.

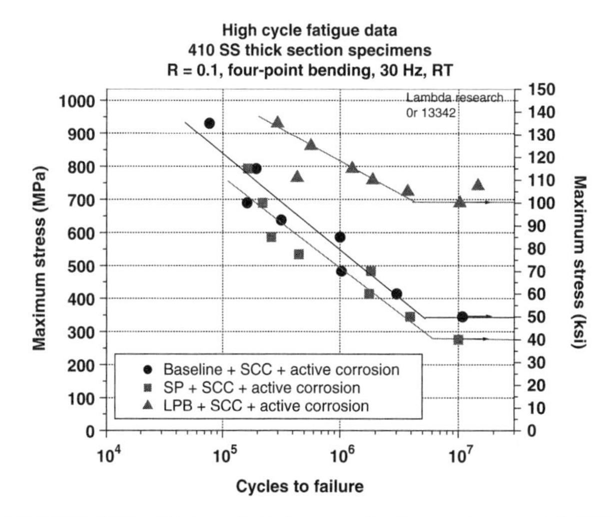


FIGURE 7.30 High cycle fatigue results for specimens with SCC and active corrosion.

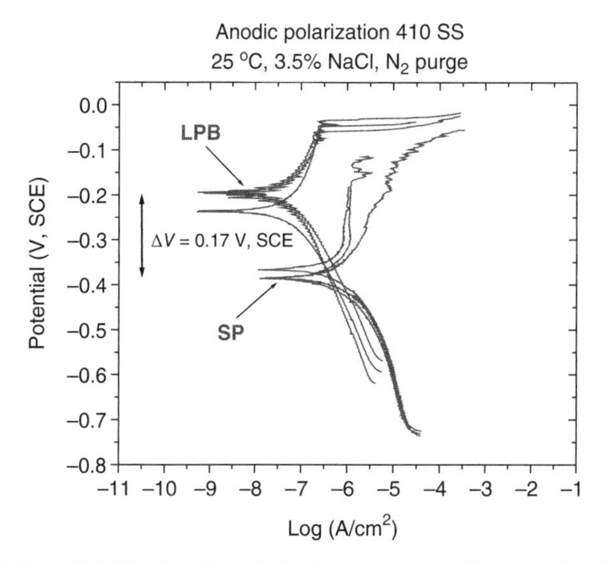


FIGURE 7.31 Anodic polarization curves for shot-peened and LPB-treated 410 SS.

The shift in the OCP indicates greater electrochemical activity and susceptibility to corrosion at the surface of the more highly cold-worked SP surface. Tafel slope extrapolations of the curves yield a corrosion rate of 14.7 mm per year (0.58 mils per year) for the SP surface, and 0.8 mm per year (0.03 mils per year) for LPB. The SP surface is corroding at 20× the corrosion rate of the lower cold-worked LPB surface.

### 7.5.5 Case Study 5: Improving the Fatigue Performance of Downhole Tubular Components [42]

The P110 couplings used in horizontal drilling are subjected to high bending stresses and high internal pressures, especially in fracking operations. Fatigue failure exacerbated by

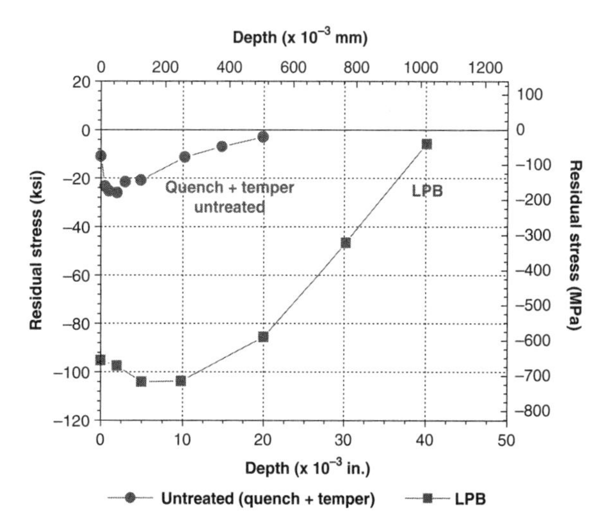


FIGURE 7.32 Residual stress distribution for as-received and LPB-processed API-P110 couplings.

mechanical damage and corrosive environments, such as $H_2S$ and NaCl, is a major problem in these downhole tubular components. Surface enhancement introduces a layer of beneficial compressive residual stress to alleviate applied tensile stresses and mitigate damage and SSC to improve fatigue performance.

Quench and tempered API P110 grade steel couplings were LPB treated on the OD for surface enhancement. Residual stress measurement by X-ray diffraction shown in Figure 7.32 revealed compression of −700 MPa (−100 ksi) extending 1 mm into the surface after treatment.

Untreated and LPB-treated P110 samples were tested with mechanical damage or corrosive damage at ambient temperature in four-point bending at 30 Hz and stress ratio $S_{min}/S_{max} = 0.1$. Active corrosion fatigue testing was performed in a neutral 3.5 wt.% NaCl solution. Several fatigue samples were given prior to SCC exposure in NaCl solution to simulate extended service exposure.

Figure 7.33 shows fatigue results as stress versus life (S–N) curves. The $10^7$ cycle fatigue strength of as-machined, baseline, P110 with 0.5 mm (0.020 in.) deep mechanical damage is only 170 MPa (25 ksi), compared with 345 MPa (50 ksi) with LPB. Surface enhancement provided nominally 2× the fatigue strength and a 50× increase in life, even with fatigue initiating surface damage present. Corrosion damage reduced the fatigue strength of untreated P110 to less than 138 MPa (20 ksi). P110 pipe after LPB in active corrosion fatigue has 1.75× greater fatigue strength of 241 MPa (35 ksi), and nominally an order of magnitude increase in life.

Polarization testing conducted on low cold-worked (electropolished) and high cold-worked (shot-peened) P110 steel test samples is shown in Figure 7.34. The shift in OCP indicates greater electrochemical activity and susceptibility to corrosion at the surface of the highly cold-worked shot-peened surface.

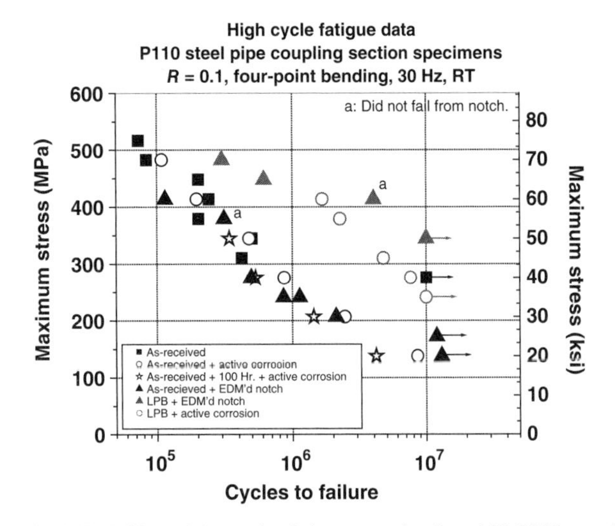


**FIGURE 7.33** High cycle fatigue results for API-P110 steel specimens with damage and corrosion.

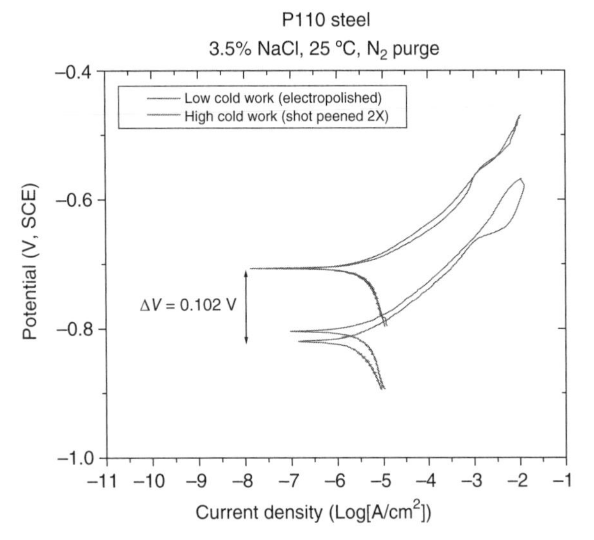


**FIGURE 7.34** Anodic polarization curves low cold-work and shot-peened P110 steel.

## REFERENCES


1. Walker, T.R. and Pick, R.J. (1990) Approximation of the axial strains developed during the roll forming of ERW pipe. *Journal of Materials Processing Technology*, 22(1), 29–44.
2. Weng, C.C. and Pekoz, T. (1990) Residual stresses in cold-formed steel members. *Journal of Structural Engineering*, 116 (6), 1611–1625.
3. Scaramangas, A. and Goff, P.R. (1985) Residual stresses in cylinder girth butt welds. Offshore Technology Conference.
4. Bate, S.K., Green, D., and Buttle, D. (1997) *A Review of Residual Stress Distributions in Welded Joints for the Defect Assessment of Offshore Structures*, HSE Books.
5. Brinksmeier, E., Cammett, J.T., Konig, W., Leskovar, P., Peters, J., and Tonshoff, H.K. (1982) Residual stresses: measurements and causes in machining processes. *CIRP Annals: Manufacturing Technology*, 31(2), 491–510.
6. Antaki, G.A. (2003) *Piping and Pipeline Engineering: Design Construction, Maintenance, Integrity and Repair*, CRC Press, Vol. 159, pp. 411–412.
7. Shi, Y.W., Chen, B.Y., and Zhang, J.X. (1990) Effects of welding residual stresses on fatigue crack growth behaviour in butt welds of a pipeline steel. *Engineering Fracture Mechanics*, 36(6), 893–902.
8. Jayaraman, N. and Prevey, P. (2007) A design methodology to take credit for residual stresses in fatigue limited designs. *Residual Stress Effects on Fatigue and Fracture Testing*, STP1497, American Society for Testing and Materials, Lancaster, PA. pp. 69–84.
9. Snape, E. (1967) Sulfide stress corrosion of some medium and low alloy steels. *Corrosion*, 23(6), 154–172.
10. Makarenko, V.D. and Khalin, A.N. (2005) Effects of hydrogen on the corrosion failure of welded pipelines. *Chemical and Petroleum Engineering*, 41(7–8), 448–450.
11. Ghosh, S., Rana, V.P.S., Kain, V., Mittal, V., and Baveja, S.K. (2011) Role of residual stresses induced by industrial fabrication on stress corrosion cracking susceptibility of austenitic stainless steel. *Materials and Design*, 32(7), 3823–3831.
12. Beavers, J. (2010) Recent progress in understanding and mitigating SCC of ethanol pipelines. Corrosion 2010, NACE International, Houston, TX.
13. Chandra, U. (1985) Determination of residual stresses due to girth-butt welds in pipes. *Journal of Pressure Vessel Technology*, 107(2), 178–184.
14. Brust, F.W. and Kanninen, M.F. (1981) Analysis of residual stresses in girth welded type 304 stainless steel pipes. *Journal of Materials for Energy Systems*, 3(3), 56–62.
15. Chen, W., Van Boven, G., and Rogge, R. (2007) The role of residual stress in neutral pH stress corrosion cracking of pipeline steels—Part II: crack dormancy. *Acta Materialia*, 55(1), 43–53.
16. Huang, H. and Shaw, W.J.D. (1992) Electrochemical aspects of cold work effect on corrosion of mild steel in sour gas environments. *Corrosion*, 48(11), 931–939.
17. Huang, H. and Shaw, W.J.D. (1993) Cold work effects on sulfide stress cracking of pipeline steel exposed to sour environments. *Corrosion Science*, 34(1), 61–78.
18. Prevey, P.S. (2000) The effect of cold work on the thermal stability of residual compression in surface enhanced IN718. 20th ASM Materials Solutions Conference & Exposition, ASM, St. Louis, MO, pp. 426–434.
19. Lu, J. (editor) and Society for Experimental Mechanics, Inc. (1996) *Handbook of Measurement of Residual Stresses*, Fairmont Press, Inc., Lilburn, GA.
20. Schajer, G. (editor) (2013) *Practical Residual Stress Measurement Methods*, John Wiley & Sons. Inc., New York
21. Zeinoddini, M. (2013) Repair welding influence on offshore pipeline residual stress fields: an experimental study. *Journal of Constructional Steel Research*, 86, 31–41.

22. Lu, J. (editor) and Society for Experimental Mechanics, Inc. (1996) Magnetic methods. *Handbook of Measurement of Residual Stresses*, Fairmont Press, Inc., Lilburn, GA.
23. ASTM (2008) ASTM E837: Standard Test Method for Determining Residual Stresses by the Hole-Drilling Strain-Gage Method. American Society for Testing and Materials.
24. Vishay Precision Group (2010) Measurement of residual stresses by the hole-drilling strain gage method. Micro-Measurements Tech Note T-503-6.
25. Keil, S. (1992) Experimental determination of the residual stresses with the ring-core method and an on-line measuring system. *Experimental Techniques*, 16(5), 17–24.
26. Lu, J. (editor) and Society for Experimental Mechanics, Inc. (1996) *Handbook of Measurement of Residual Stresses*, The Fairmont Press, Inc., Lilburn, GA.
27. Hornbach, D. (2009) Incremental ring-core determination of the principal residual stress in a duplex stainless steel centrifuge. Lambda Research, Diffraction Notes 35.
28. SAE (2003) Residual Stress Measurement by X-ray Diffraction. SAE HS784, Society of Automotive Engineers.
29. ASTM (1984) *Standard Method for Verifying the Alignment of X-Ray Diffraction Instrumentation for Residual Stress Measurement*. ASTM E915-10, Vol. 03.01, American Society for Testing and Materials, Philadelphia, PA, pp. 809–812.
30. ASTM (2009) Standard Test Method for Determining the Effective Elastic Parameter for X-Ray Diffraction Measurements of Residual Stress. ASTM E1426-98, American Society for Testing and Materials, Philadelphia, PA.
31. Prevey, P.S. (1976) *Advances in X-Ray Analysis*, 19, 709.
32. Lu, J. (editor) (1996) Magnetic methods. *Handbook of Measurement of Residual Stresses*. Fairmont Press, Inc., Lilburn, GA.
33. Bate, S.K., Green, D., and Buttle, D. (1997) *A Review of Residual Stress Distributions in Welded Joints for the Defect Assessment of Offshore Structures*, HSE Books.
34. Fonseca, M.C., Teodosio, J.R., Rebello, J.M., and da Cruz, A.C. (2001) Residual stress state behaviour under fatigue loading in pipeline welded joints. *The Journal of Strain Analysis for Engineering Design*, 36(5), 465–472.
35. Kong, D.J., Zhou, C.Z., and Hu, A.P. (2011) Effect of laser shock on the mechanical properties of weld joint of X70 steel pipeline. *Journal of Jilin University (Engineering and Technology Edition)*, 41(5), 1507–1512.
36. Cammett, J.T., Prevey, P.S., and Jayaraman, N. (2005) The effect of shot peening coverage on residual stress, cold work and fatigue in a nickel-base superalloy. Proceedings of ICSP 9, Paris, France.
37. Cammett, J.T. and Prevey, P.S. (2001) Fatigue strength restoration in corrosion pitted 4340 alloy steel via low plasticity burnishing. Available at http://www.lambdatechs.com/documents/228.pdf. (last accessed December 2013).
38. Frost, N.E., Marsh, K.J., and Pook, L.P. (1974) *Metal Fatigue*, Clarendon Press, Oxford, p. 366.
39. Hornbach, D.J., Jayaraman, N., and Scheel, J.E. (2011) Engineered residual stress to mitigate stress corrosion cracking of stainless steel weldments. Available at http://www.lambdatechs.com/documents/232.pdf.
40. Chelette, D., Moore, P., Hornbach, D., Prevey, P., and Scheel, J. (2011) Mitigation of sulfide stress cracking in down hole P110 components via low plasticity burnishing. Advances in Materials for Oil & Gas Production, STG, Houston, TX.
41. Hornbach, D. and Scheel, J. (2012) Improving corrosion fatigue performance and damage tolerance of 410 stainless steel via LPB. Available at http://www.lambdatechs.com/documents/285.pdf. (last accessed 2013).
42. Hornbach, D.J. and Scheel, J.E. (2012) The effect of surface enhancement on improving the fatigue and sour service performance of downhole tubular components. ASME 2012 International Mechanical Engineering Congress and Exposition, Houston, TX.

# 8

# PIPELINE/SOIL INTERACTION MODELING IN SUPPORT OF PIPELINE ENGINEERING DESIGN AND INTEGRITY

Shawn Kenny[1,2] and Paul Jukes[3]
[1]Memorial University, St. John's, Newfoundland, Canada
[2]Carleton University, Ottawa, Ontario, Canada
[3]Wood Group, Houston, TX, USA

## 8.1 INTRODUCTION

Hydrocarbon products, such as oil, natural gas, and natural gas liquids, are transported from the production field to hubs, batteries, tank farms, and facilities via gathering lines or flow lines. Feeder lines or laterals transport the hydrocarbon products to long-distance transmission pipelines that may cross national or international boundaries and extend over tens to hundreds of kilometers in length. In Canada, there is over 100,000 km of transmission line ranging in diameter from 101.6 to 1212 mm [1], whereas in the United States the natural gas transmission pipeline network increases to more than 477,000 km onshore and 7175 km offshore [2].

These transmission pipeline systems navigate through regions with varied terrain units, topography, physical environment, geology, and, for onshore pipelines, hydrological characteristics. Energy pipelines are typically buried, particularly within urban onshore areas and shallow water offshore regions, to mitigate risk from potential damage due to external force arising from geohazards, natural processes, and anthropogenic activities that may impact mechanical integrity. The external forces are transmitted through the soil and impose geotechnical loads on the buried pipeline. The soil forces and relative soil displacement may be developed through environmental conditions (e.g., soil self-weight and groundwater), operational loading conditions (e.g., pipe self-weight and thermal expansion), anthropogenic activities (e.g., subsidence due to subsurface mining, mechanical damage, and blasting), and natural geohazards (e.g., slope instability, seismic fault movement, and offshore ice gouging). As shown in Figure 8.1, a complex relationship exists between the coupled pipeline/soil system with respect to the demand (i.e., loads and hazards), response (i.e., pipeline/soil interaction, load transfer, and load effects), and capacity (i.e., pipeline mechanical resistance and integrity). As an illustrative example, in the context of ice gouging hazards for offshore pipelines, Kenny et al. [3] present a more detailed discussion on each element within this relationship.

For buried pipelines, the system demand (i.e., loads and hazards) is primarily related to anthropogenic activities and natural events. The processes and mechanisms governing system demand are not addressed in this chapter where the technical issues involve complex physics within multidisciplinary fields of expertise. The treatment of external force and geohazards is examined in greater detail within other chapters of this book (e.g., mechanical damage) and other publications [4–19]. For pipelines in frontier regions, such as deep water and Arctic locations, there are unique geohazards that include strudel scour, permafrost-related thaw settlement or frost heave, and offshore ice gouging [20–42].

For small deformation loading events, the system response is primarily governed by equations of equilibrium (i.e., primary loading conditions) with the soil and pipeline system capacity (i.e., mechanical response) dominated by elastic behavior. Current guidelines, codes, and standards provide significant advice on engineering practice to assess

*Oil and Gas Pipelines: Integrity and Safety Handbook*, First Edition. Edited by R. Winston Revie.

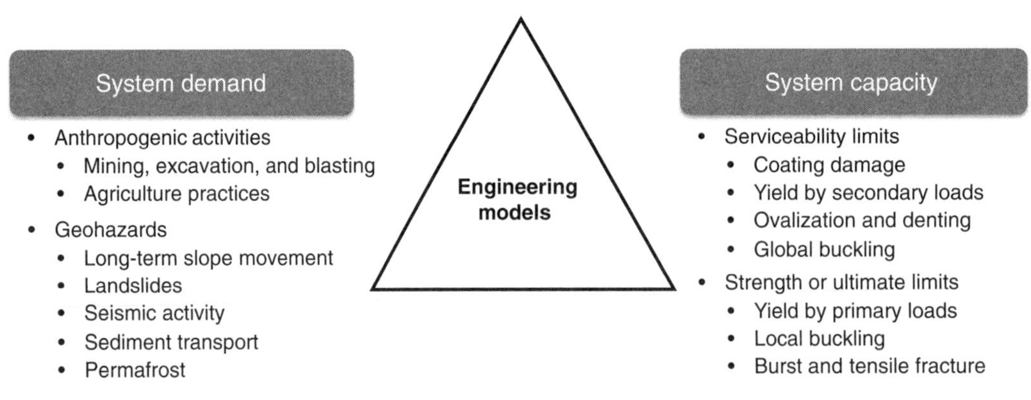


**FIGURE 8.1** Relationship among system demand, response, and capacity for a buried pipeline system.

the system demand and capacity for stress-based design of pipeline systems [43–49]. Natural geohazards, such as long-term slope movement, landslides, seismic fault movement, and offshore ice gouging, may result in pipelines experiencing large deformations with local plastic strain response that may be associated with mechanisms such as buckling, fracture, and plastic collapse. Laws of compatibility govern the system response and capacity, which is associated with secondary loading events, where limited guidance is provided in current engineering practice [46–49]. These large deformation geohazards drive technology development programs to assess system response and capacity. The development of serviceability and ultimate limit states or other mechanical performance acceptance criteria is not addressed in this chapter with reference to other publications for guidance [3,38,41,50–72]. The consequences of pipeline failure have varying significance that is dependent upon the product type (e.g., oil, natural gas, and gas liquids), pipeline diameter and wall thickness, operating pressure, and location [73–78]. The system demand and system response are not dependent on the product type or consequences and risk level.

The response of surficial or partially buried pipelines is an important design element for the offshore environment that has received significant attention over the last decade due to increasing operational pressure and temperature requirements for pipeline transportation systems. The design challenges with pipe/soil interaction and pipe mechanical response have been the subject of research programs and joint industry projects [48,79–85]. A recent handbook provides a comprehensive overview of the technical issues, reference to supporting publications, and guidance on engineering practice for offshore surface-laid risers and flow lines [33]. Although the key parameters influencing the response of surficial and partially embedded pipeline/soil interaction events have been established, the topic remains an area of current interest and research. These reference publications provide a comprehensive discussion and technical guidance on the subject matter that is not addressed in this chapter.

This chapter is specifically focused on the system response elements, as shown in Figure 8.1, for fully buried pipelines. The discussion addresses how engineering tools may be used to estimate the effects of geotechnical loads on buried pipelines (i.e., pipeline/soil interaction, load transfer, and soil failure mechanisms), for long-term ground movement events, and predict the pipeline mechanical response (i.e., geotechnical load effects). Other important factors influencing pipeline/soil interaction events (e.g., effects of loading rate, pore water pressure, and interface behavior) are also highlighted in this discussion.

As a prelude to this primary discussion expanded upon later in this chapter, aspects of geotechnical engineering that support pipeline/soil interaction analysis are highlighted in Section 8.2. The use of physical modeling techniques at full scale, large scale, and reduced scale (i.e., centrifuge), which is primarily focused on the soil mechanical response and failure mechanism, is reviewed in Section 8.3.2. A basic understanding of physical modeling techniques is required as this underpins the development and advancement of engineering tools to assess pipeline/soil interaction events and system response.

Engineering practice using analytical tools (Section 8.3.3.1) and structural numerical modeling procedures (Section 8.3.3.2) to assess geotechnical load effects on buried pipelines, with reference to current practice documents [43–46], is examined. Further guidance on how to improve the current state of practice (Section 8.3.3.2) through incremental refinement is presented in Section 8.3.4. An overview of emerging trends in advanced numerical simulation of pipeline/soil interaction events, addressing system response, is provided in Section 8.3.5. Linked with these refinements and advancements, in the computational simulation of pipeline/soil interaction events, is the incorporation of more robust but technically challenging constitutive models that imposes greater requirements on the input data, field programs, laboratory testing, and skill sets for the analysis and interpretation of numerical pipeline/soil interaction simulations. In reference to recent studies, guidance on the application and refinement of common soil constitutive models within this numerical simulation framework is presented in Section 8.3.6. Finally, a high-level outlook on the integration of advancements in the state of art to become an improved state of practice is discussed in Section 8.3.7.

From this perspective, a key goal of this chapter is to establish a progressive narrative, where as an outcome the reader develops an appreciation and understanding of how the current state of practice (Section 8.3.3) is in a state of change that is evolving from an idealized structural modeling basis (e.g., Winkler-type foundation model) toward a more robust and comprehensive technology framework (Sections 8.3.4–8.3.7). The primary motivation was the need to develop advanced engineering solutions that address the conservatism and technical uncertainty in current practice, while satisfying governing constraints (e.g., logistics and economics) for energy pipeline transportation systems in harsh frontier environments, such as deep water and Arctic regions. The integration of conventional geotechnical engineering practices (i.e., field, laboratory, and analysis studies) with recent advances in computational software and hardware platforms has provided this technology framework to improve the state of practice for assessing geotechnical load effects on buried pipelines.

Although current engineering practice continues to principally serve the needs of industry, in this chapter it is argued that refinements to this technical approach are needed. Advancements in the state of practice will occur through a technology framework that incorporates laboratory testing, physical modeling, and advanced computational modeling procedures, which are explored in this chapter. The need for advanced user skill sets to pursue this technology path is recognized; however, it is expected that these outcomes from the emerging applied research and development activities will ultimately be formulated within a practical design framework. For example, the approach used to develop the load resistance factored design (LRFD) philosophy used for offshore pipelines and geotechnical structures could be adopted [49,86–89]. This technology framework and advancement in the current state of practice will provide industry with a sound technical basis to develop practical engineering solutions with reduced uncertainty and improved confidence in predictable, reliable outcomes.

## 8.2 SITE CHARACTERIZATION AND GEOTECHNICAL ENGINEERING IN RELATION TO PIPELINE SYSTEM RESPONSE ANALYSIS

### 8.2.1 Overview

In order to conduct the pipeline system response analysis, the geotechnical conditions along the pipeline route must be established and assessed through site characterization that includes desktop studies, remote sensing, reconnaissance, and field investigations. The current state of practice integrates data and knowledge from the disciplines of remote sensing, terrain analysis, pipeline routing, geotechnical engineering, and geohazard management to achieve successful engineering design solutions [8,15–20,90–96]. A detailed discussion is presented in these cited references, where the primary objectives are to identify the preferred pipeline route that evaluates and balances other factors and constraints that include cost, crossing of water bodies and infrastructure, potential environmental impact, and land use. The pipeline route assessment and selection involves multidisciplinary fields of study that must assess the biological, physical, geophysical, geotechnical, hydrological, and socioeconomic environments.

The site characterization and geotechnical engineering studies must be linked with the geophysical and geotechnical surveys in order to establish the engineering parameters for use in analysis and design. The scope and level of detail required is dependent on the pipeline engineering design requirements related to the pipeline location (i.e., onshore, offshore, water or shore crossing, and facility approaches), route length, expected variability in soil conditions, geohazard characteristics (i.e., type, magnitude, and impact), and project-level design philosophy (i.e., stress-based or strain-based design) [96–101]. These activities may also be influenced by other factors related to project execution (e.g., lack of data, uncertainty, and long lead times) and operational strategies (e.g., monitoring, assessing, and mitigating geohazards) [8,10,53,93,94]. Furthermore, the planning, logistics, work scope, level of detail, resource requirements, and cost increase with the pipeline engineering project gates or phases associated with concept selection, front-end engineering design (FEED), and detailed design activities.

In the following subsections, a brief overview of the key factors that influence pipeline routing, site characterization, and geotechnical investigations in support of pipeline system response analysis is reviewed. This discussion highlights the major considerations for conventional or standard pipeline engineering projects and identifies key elements that are required to address specific geohazards and special design considerations (e.g., deep water and northern harsh environments). Reference to other authoritative studies and engineering practice is also provided.

### 8.2.2 Pipeline Routing

Similar to other linear infrastructure in the civil built environment (e.g., highways and transmission lines), geotechnical site characterization and geohazard assessment are integral components of the route planning and selection process and pipeline engineering design activities [8,10,15–17,90,91,93, 94,96–98,102]. The reader is directed to these comprehensive references for a more detailed discussion on the importance of pipeline route selection for offshore and onshore pipelines and the important connection with geotechnical investigations and geohazard assessment in support of pipeline engineering analysis, design, and construction activities. One of the primary outcomes is to support effective decision making in pipeline engineering design with respect to safety, economics, and pipeline security and integrity. As discussed by Palmer and King [94], significant technical, logistical, and economic issues can be realized through uninformed or misinformed engineering decision making based on poor quality or lack of data.

In the route planning and selection process, geotechnical site characterization is integrated with knowledge from other scientific disciplines including meteorology, hydrology, surficial geology, seismology, physiography, biology, and zoology, and other factors related to public safety, land use, and sustainability. One of the key issues when selecting an optimal pipeline route is to establish an alignment that mitigates terrain effects and impact of geohazards on pipeline design, construction, and operations within the permitted right of way. Furthermore, the development of construction, operational, and integrity management plans may also be influenced by these geotechnical considerations. For example, the pipeline route may pass through regions with sensitive clays, ground subsidence or acid rock drainage due to mining activities, active slope failures, seismic fault zones and soils with liquefaction potential, limited seasonal access (e.g., muskeg), or permafrost for northern pipelines. Although there are common requirements, onshore and offshore pipeline systems present unique challenges and constraints during surveying and route planning activities that must account for the pipeline life cycle [8,10,15–17,19, 90,91,93,94,96,102].

### 8.2.3 Geotechnical Investigations

The vast majority of onshore and offshore pipeline systems are resting on the ground (seabed surface) or fully buried with a soil cover above the pipeline crown. The pipeline may be embedded or buried within the soil to address requirements for flow assurance (e.g., thermal resistance), operational loads (e.g., global buckling), and external forces due to wave and current loads (e.g., on-bottom stability), to afford protection from natural hazards (e.g., slope movement, ice gouge events, frost heave, and thaw settlement), outside force (e.g., fishing gear and anchor impact), and third-party damage (e.g., blasting, surface loads, excavators, and farm equipment) [10–42,44,46,79–85,96,102]. This section is focused on the geotechnical investigations and data needed to support engineering studies on pipeline system response analysis that are used in pipeline engineering design and integrity assessment.

A detailed discussion on the requirements of geotechnical surveys and investigations to support engineering analysis and design is provided in several comprehensive references and recommended practice documents [91,93,94,96,99–102]. The requirements for planning and conducting the work scope for obtaining the soil parameters used in engineering analysis and design with respect to field investigations, data acquisition and sampling, *in situ* and laboratory testing, and reporting are examined in detail.

The primary geotechnical parameters include the physical and strength parameters for clay or cohesive soil (i.e., particle size distribution, Atterberg limits, unit weight, and undrained shear strength) and sand or noncohesive soil (i.e., particle size distribution, unit weight, relative density, and internal friction angle). For conventional stress-based pipeline design, there is typically limited scope on special considerations in the pipeline routing, site characterization, and geotechnical investigation activities. Conventional engineering practices can be used with successful outcomes realized in the engineering design and pipeline operations [12,19,33, 45–47,49,99–101].

However, for large deformation, displacement-controlled pipeline response mechanisms (e.g., global buckling and ratcheting) and geohazards (e.g., long-term slope movement, frost heave, thaw settlement, seismic fault movement, liquefaction, and offshore ice gouging), where strain-based design methods are typically used, a more comprehensive geotechnical investigation framework is required [10,15,29,42,

45,46,48,96–102]. For these large deformation geohazards, the type, spatial variability and intensity, and influence on pipeline mechanical response have a significant impact on the need and requirements for site characterization studies and geotechnical investigations in support of pipeline/soil system response analysis. The characterization of geotechnical conditions and assessment of geohazards should not be confined to a focused corridor, such as the pipeline right of way, but also assess the surrounding region and terrain units that may affect the pipeline mechanical integrity. A number of factors require thoughtful and practical assessment that will enhance decision making and provide a sound basis for pipeline design, project execution, and operational pipeline integrity management. These factors include an evaluation of the data type (e.g., index properties and laboratory tests), level of detail (e.g., quantity, model or data uncertainty, and spatial variability of soils), and significance of any geohazard (e.g., frequency, system demand, or load intensity) that drive the requirements for geotechnical site characterization and field investigations in support of pipeline engineering design activities.

The importance of geotechnical investigations for defining the geotechnical and geophysical attributes for use in conducting pipeline analysis and design, such as soil type, mechanical properties (e.g., cohesion and friction angle), groundwater conditions (e.g., spatial and seasonal variation), and unique features (e.g., bedrock, aggressive chemicals, and geohazards), cannot be understated. The ultimate goal is to provide baseline data for conducting physical tests or numerical simulation to assess the geotechnical load effects on buried pipelines and predict the pipeline mechanical response.

## 8.3 PIPELINE/SOIL INTERACTION ANALYSIS AND DESIGN

### 8.3.1 Overview

Over the past 50 years, the technical engineering approaches to assess pipeline/soil interaction events have evolved from empirical observations and idealized closed-form solutions through to a suite of advanced numerical simulation techniques. The primary goal is to estimate load effects on pipeline mechanical performance in support of engineering design and operational integrity. The advanced computational engineering tools provide a technically robust and cost-effective framework to conduct detailed investigations on pipeline mechanical response to pipeline/soil interaction events across a range of practical design parameters. These engineering tools, however, are fundamentally dependent on input data from engineering surveys to define the pipeline alignment and profile, field programs to collect route specific geotechnical conditions, laboratory tests to define parameters used in constitutive models of the engineering computational tools, and physical models to provide confidence in these numerical simulation tools. In this chapter, components of this integrated technology framework in support of pipeline engineering design are addressed in terms of an overview of the current state of practice, evaluation of the technical requirements, advantages and constraints for each technology approach, and an assessment of the current state of art.

### 8.3.2 Physical Modeling

A physical model is typically a smaller physical copy of an object. The geometry of the model and the object it represents are often similar in the sense that one is a rescaling of the other. Similarity of pipe/soil interaction models is also required in terms of mechanical response. Mechanical and geotechnical similarity can be considered using dimensionless analysis to ensure that physical models are representative [103–105].

In other subsections within this chapter, the contributions of physical modeling studies, investigating the soil mechanical response and pipe/soil interaction mechanisms, with respect to the development of current knowledge base, formulation of existing industry practice and guidance documents, and foundation for areas of emerging research are referenced and discussed. Examples of simple elemental lateral, vertical, or axial pipe/soil interaction events, with a rigid pipe segment, are examined. The importance for integrating physical modeling studies to develop practical engineering tools in support of pipeline design, and to establish confidence in more advanced computational modeling procedures, addressing state-of-the-art pipe/soil interaction studies, is presented. For example, the historical literature indicates that the drained lateral interaction load factors for noncohesive soils, based on physical model tests, may differ significantly. Recent studies have shown that these differences can be explained by consideration of the pipe size and contribution of soil self-weight to the soil mechanical resistance. The importance of effective stress level in soil response is also an essential consideration. These issues are discussed in further detail later in this chapter. Furthermore, physical models of larger pipeline/soil interaction systems examining large-scale ground movement events, such as ice gouging, seismic fault movement, and frost heave mitigation, have provided important insight and are referenced in the discussion.

For pipeline/soil interaction events when soil weight effects are important, physical modeling tests are best conducted under large, near full-scale, field-scale conditions or in reduced-scale models using a geotechnical centrifuge. There are at least three large-scale geotechnical physical modeling facilities in North America focused on pipeline testing at Queen's University, Cornell University, and University of British Columbia [106–111]. There are a larger

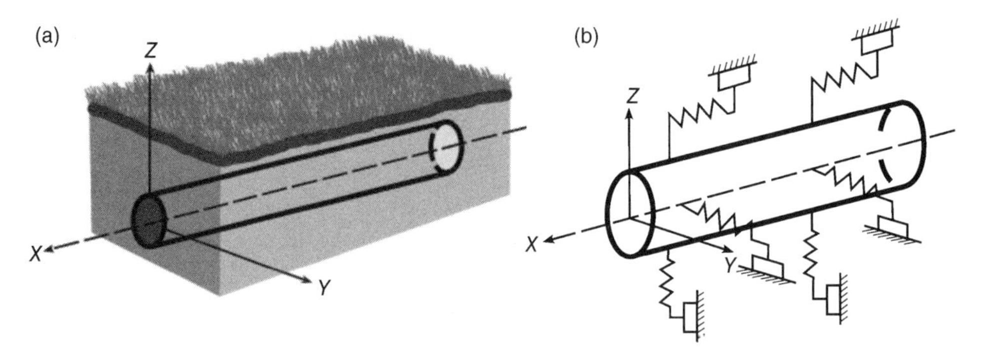


**FIGURE 8.2** Schematic illustration of the (a) continuum pipe/soil interaction problem and (b) mechanical idealization using beam and spring structural elements.

number of geotechnical centrifuge modeling centers with more than 100 facilities operating worldwide [33,112,113] with the principles of geotechnical centrifuge modeling effectively summarized in several publications [33,114].

Physical modeling is most beneficial when complemented by other engineering studies within an integrated framework that includes field investigations, laboratory testing, and computational simulation. These concepts are reinforced in the following subsections where excellent comparisons between physical modeling and numerical simulation of pipe/soil interaction events are discussed and referenced.

### 8.3.3 Computational Engineering Tools

***8.3.3.1 Analytical Solutions*** Analytical solutions characterizing the mechanical response of surface-laid and buried pipelines have generally evolved from the concept of a subgrade reaction [115]. The Winkler model was initially developed to examine rail/ground surface interaction problems through a linear solution relating the foundation pressure with the foundation modulus and deflection. The foundation was modeled as a series of independent or decoupled, discrete springs with specified stiffness. The Winkler model was extended through multiparameter models that accounted for the beam and soil behavior [116–121]. As discussed by Konuk et al. [120], the parameters for multiparameter models are established through knowledge of the corresponding solution to the applicable continuum problem. These structural-type models introduce simplifying assumptions, idealizations, and constraints with respect to the spatial distribution of the displacement and stress fields, which may not necessarily produce conservative results [121–124].

There exists a range of simple empirical and closed-form solutions to estimate geotechnical loads on buried pipeline [44,46,125–132]. These methods are constrained by the governing idealizations but provide a relatively simple engineering tool for use in preliminary engineering design iterations and as a confidence check for more rigorous and complex analysis. The engineering tools and key design parameters derived were generally associated with an assessment of the limit load as defined by the ultimate soil load capacity and yield displacement.

***8.3.3.2 Structural Pipe/Spring Models for Buried Pipelines***

*Overview* Using numerical techniques such as finite difference and finite element methods, the Winkler model was extended to more complex, nonlinear problems with large deformations, such as the analysis of buried pipeline/soil interaction events. Engineering guidelines for the analysis of buried pipeline/soil interaction events were developed in the 1980s and have been refined through advances in research [12,43,45,46]. These engineering guidelines model the pipe/soil system as a series of connected structural beam and spring elements (Figure 8.2). The analysis techniques have evolved and engineering tools have been applied to the analysis of fault movement due to seismic events, ground settlement, upheaval and lateral buckling instability, ice gouging, frost heave and thaw settlement, and riser touchdown [3,26,120,133–138].

This subsection provides an overview of pipe/soil interaction models for buried pipeline systems for conventional pipeline engineering design considerations in the onshore and offshore domains. The examination of special topics, such as Arctic pipelines and surficial pipeline systems, is presented in Section 8.4.

Euler–Bernoulli or Timoshenko beam theory defines the beam element behavior that may also have additional variables to account for the effects of internal pressure, thermal expansion, section ovalization and warping, and stiffness or flexibility characteristics for curved sections or elbows. The pipeline constitutive model is generally defined as isotropic, elastoplastic behavior with piecewise representation of the stress–strain relationship. The pipe plastic material response is generally defined by the von Mises yield surface with isotropic hardening rule [139,140]. The plasticity models may also need to account for kinematic hardening, which is important for cyclic loading problems such as fatigue, ratcheting, and offshore reel-lay installation [141–143].

The soil continuum response is idealized through a series of discrete springs, which represent the generalized mechanical response of a segmented soil slice, connected to the pipe. The spring elements represent the soil force–displacement response per unit length of pipe that act on three mutually orthogonal axes to the pipeline centerline defined along the longitudinal, transverse horizontal, and transverse vertical directions. The soil spring force–displacement relationships are defined by engineering guidelines [43,45,46] with bilinear (i.e., elastic-perfectly plastic), piecewise multilinear, or hyperbolic functions to represent the nonlinear response characteristics.

The soil spring load–displacement relationships define the ultimate load and yield displacement through geometric properties (i.e., pipe diameter and pipe springline depth), soil physical properties (e.g., adhesion, soil unit weight, and pipe/soil interface friction angle), soil strength properties (i.e., cohesion and internal friction angle), and interaction factor (i.e., bearing capacity factors). The soil spring parameters, such as yield load, yield displacement, and empirical coefficients for bearing interaction, have been primarily established through scaled physical modeling studies in uniform soil conditions [144–149]. Although it is recognized that the post-peak residual force may be less than the ultimate soil load that may be associated with strain softening mechanisms, the guidelines generally consider a constant post-peak soil force. The loading and unloading response of the soil springs traces along the same force–displacement relationship.

In the following sections, the basis for soil spring formulation in the longitudinal, transverse lateral, and transverse vertical directions is examined and critiqued. Guidance on the state of practice and state of art in support of pipeline engineering design is discussed in Section 8.3.4. The development of emerging engineering tools for pipeline/soil interaction analysis, through research and development studies, that may advance current practice is explored in Section 8.3.5.

In the following subsections, a mixed formulation is used to characterize the soil ultimate force ($t_u$, $p_u$, $q_{uu}$, and $q_{ud}$) as a function of strength parameters, as recommended by the industry guidance documents [43,45,46] for simple axial, lateral, or vertical pipe/soil interaction load events. The mixed formulation includes total stress ($\alpha$-method) and effective stress ($\beta$-method) approaches. From the author's viewpoint, the use of mixed formulation to define the soil mechanical response *is not recommended.* The use of effective stress formulation to represent soil mechanical response and characterize pipe/soil interaction events *is recommended* and considered to be more representative of soil behavior during pipeline/soil interaction events [14,108–111,148–157]. The $\beta$-method can account for undrained and partially drained loading conditions when the excess pore pressure is known or estimated. For example, in saturated cohesive soils, the friction angle is typically small (i.e., <5°); however, as the water content is decreased, the importance of the friction angle on strength mobilization will increase for the same adhesion resistance [8].

*Longitudinal Axial Soil Response* Estimates on soil axial forces were established using pile theory where the limit skin friction per unit area ($f_s$) at the pipeline/soil interface can be expressed by the $\alpha$-method (i.e., total stress response) as [150]

$$f_s^\alpha = \alpha s_u \tag{8.1}$$

and by the $\beta$-method (i.e., effective stress response) as

$$f_s^\beta = H_s \gamma' \left(\frac{1+K_0}{2}\right) \tan\delta = \beta\sigma'_v \tag{8.2}$$

The engineering guidelines [45,46] provide a mixed formulation (i.e., combining total stress and effective stress concepts) defining the ultimate or maximum longitudinal soil force as

$$t_u = \pi D \alpha s_u + \pi D H_s \gamma' \left(\frac{1+K_0}{2}\right) \tan\delta \tag{8.3}$$

The ultimate soil displacement is defined as 8–10 mm for stiff to soft clay and 3–5 mm for dense to loose sand [45,46]. The axial soil load–displacement relationship is generally defined as bilinear, elastic-perfectly plastic response.

The adhesion factor ($\alpha$) can be estimated through back calculation from pile load tests for undrained loading conditions. In pipeline engineering design, however, drained conditions are typically assumed for axial loading events [145,150–152]. Except for fast loading conditions (e.g., seismic fault movement and liquefaction), it is expected during axial pipe/soil interaction events that the excess pore pressures, generated by shear stress, will rapidly dissipate through a thin soil layer circumscribing the outside pipe diameter. This would result in a drained loading condition that requires effective stress analysis. For the axial response, using one-dimensional consolidation theory, the effects of load duration on drained versus undrained behavior can be assessed using the degree of consolidation at specified depth and time [151]. Although methods exist to correlate the adhesion factor and undrained shear strength ($s_u$) with effective stress state [152], other factors influence the undrained shear strength soil for pipe/soil interaction events that include soil sensitivity, strain rate effects, and reduction in the soil water content with increased pore water suction, consolidation, soil desiccation, and cementation.

These factors (i.e., displacement rate effects, water content, and consolidation) may only partly account for the uncertainty and significant scatter in the adhesion values as proposed by the ALA guidelines [46]. Experimental studies and well-controlled full-scale field tests indicate significantly lower values for the adhesion factor [150,153,154,158]. In addition, recent studies have indicated that slight axial misalignment (i.e., oblique pipe/soil interaction) or pipe asperity (e.g., larger diameter transition and in-line valve) can increase the longitudinal soil resistance [113,159,160]. However, displacement rate effects may also play a role in the differences between the adhesion factors as discussed by industry guidelines [43,45,46]. The validity of adopting the $\alpha$-method from pile theory for application to buried pipeline systems for both cohesive and noncohesive (frictional) soils, however, remains an open question [150,154].

For pipeline designs incorporating the $\alpha$-method to estimate longitudinal soil resistance, higher adhesion factors should be selected when the maximum longitudinal soil forces are important for pipe stress (e.g., von Mises stress) and stability (e.g., global buckling) analysis, whereas lower adhesion factor should be selected to determine soil strength (e.g., virtual anchor lengths for thermal expansion analysis).

The effective stress parameter ($\beta$-method) is influenced by the effective normal stress state and pipe/soil interface friction angle ($\delta$), which can vary from $0.5\phi'$ to $1.0\phi'$, and is dependent on the characteristics of the external pipe coating and surrounding soil [3,45,145,161]. These friction angle estimates, however, do not account for the effects of damage, degradation, or long-term creep response of the external pipe coating material [154]. Based on direct shear box tests and comparison with field trials, axial pipe frictional load magnitude was governed by characteristics of the external pipe coating roughness and compliance with limited effect from the soil backfill conditions [150]. Pipeline coatings with a smooth and hard surface will minimize axial loads for both cohesive and noncohesive (frictional) soils. For offshore pipelines in noncohesive soil, the interface friction angle ($\delta$) can be estimated as [152]

$$\delta = \phi' - 5 \tag{8.4}$$

For cohesive soils with drained loading conditions, the peak interface friction angle can be estimated as

$$\tan \delta_{\mathrm{p}} = f \tan \phi'_{\mathrm{p}} \tag{8.5}$$

The axial soil load is also significantly influenced by the lateral earth pressure coefficient at rest ($K_0$). The ALA and PRCI guidelines [45,46] do not provide any recommendation on the value, but a value of 0.5 is typical with the general expression $K_0 = 1 - \sin\phi'$ recommended. Further discussion on the determination of the lateral earth pressure coefficient at rest is presented in Section 8.3.6.2.

The lateral earth pressure coefficient at rest accounts for the general stress state at the pipe springline and does not account for the actual stress state or local deformations during the pipe/soil interaction event. For loose sand conditions, axial pullout tests are consistent with industry guidelines [43,45,46] as expressed in Equation 3.3. In dense sand, however, the measured axial loads were significantly greater than the recommended guidelines where increased resistance was due to constrained dilation, in response to shear deformation, within a thin annular zone circumscribing the pipeline [108–111,150,154,156–158,162]. For larger diameter pipeline, the NEN 3650 [43] guideline states that the pipe self-weight and contents should also be considered when determining frictional forces during axial pipe/soil interaction events. The use of hard and smooth coatings was recommended to mitigate pipe/soil interface friction and axial loads irrespective of the soil conditions [157].

*Transverse Lateral Soil Response* Based on bilinear model, the maximum transverse lateral soil force per unit length of pipe segment is expressed as [45,46]

$$p_{\mathrm{u}} = N_{\mathrm{ch}} s_{\mathrm{u}} D + N_{\mathrm{qh}} \gamma' H_{\mathrm{s}} D \tag{8.6}$$

Estimates of the maximum transverse lateral soil force have been established through analytical solutions (e.g., passive earth pressure theory and strip footings), physical models (e.g., anchor plates, piles, and pipe), and numerical simulation (e.g., finite element analysis of anchors and pipe) [108,111,145,146,148,149,156,160,163–189].

For slow loading rates in cohesive material, effective soil strength parameters for noncohesive soil (e.g., sand) can be used to predict the lateral load–displacement response in cohesive soil (e.g., clay) [148,158,190].

The ultimate soil displacement at the peak transverse lateral soil load is expressed as a function of the pipe burial depth [45,46]:

$$\Delta_{\mathrm{p}} = 0.04\left(H_{\mathrm{s}} + \frac{D}{2}\right) \leq 0.10D \text{ to } 0.15D \tag{8.7}$$

The force–deflection relationship for lateral pipe/soil interaction events exhibits nonlinear response characteristics with strain hardening or strain softening behavior. The soil spring force–deflection relationship can be represented using bilinear, elastic-perfectly plastic behavior or hyperbolic functions through piecewise approximation [148,149,165, 171,191–194].

Project-specific data including the results from physical tests or numerical simulations may also be used to establish the soil load–displacement relationship. Interpretation of the soil load–deflection response and establishing the maximum transverse lateral load and corresponding soil displacement at the peak load on a consistent basis are required. This can be achieved through definition of rational selection criteria. These criteria are established in reference to the characteristic shape of the soil load–deflection response that include (1) peak load associated with a unique maximum (e.g., classical strain softening response) or asymptote (e.g., classical strain hardening), (2) intersection of a tangent line to the strain hardening response with the ordinate, (3) intersection of tangent lines to the elastic modulus and strain hardening response, (4) secant line with a slope equal to one-quarter of the elastic modulus ($k_4$ method), and (5) secant line with an upper load limit fraction (0.70–0.95) of the maximum peak load [167–169,171,191–196].

Transverse lateral pipe/soil load events involve complex interactions with dependence on parameters that include soil strength characteristics (e.g., undrained shear strength, consolidation ratio, and internal friction angle), burial depth, loading or displacement rate (i.e., drained or undrained conditions), large deformation soil behavior (e.g., strain hardening, strain softening, and dilation), interaction and contact processes (e.g., trailing edge separation or breakaway, negative or suction pressure, and infilling), trench effects (e.g., backfill properties and trench geometry), and failure mechanisms (i.e., plastic wedge formation with uplift at shallow burial depth and local punching failure with flow at deeper burial depth). These parameters will influence the maximum load, displacement at peak load, and interaction factors.

The interaction factor ($N_{ch}$) for cohesive soil, as defined by industry guidelines [43,45,46], is generally based on the work of Hansen [165], assuming undrained conditions (i.e., total stress analysis). Several studies have demonstrated the importance of loading rate, associated with time-dependent consolidation of cohesive soils, on the lateral interaction loads [113,148,190]. As shown in Figure 8.3, for rapid loading events, where undrained conditions prevail, the interaction factors were consistent with the immediate separation (breakaway) results of Rowe and Davis [171] but less than the estimates based on the approach by Hansen [165]. These observations are consistent with the findings of Popescu et al. [179] for physical and numerical models of pipe/soil interaction in soft and stiff clay. For slower loading rates, where drained conditions and effective stress state become more prevalent, the current state of practice, based on total stress analysis, underestimates the load transfer to the pipe [148,190].

Recent studies have also indicated the importance of burial depth, soil weight, and failure mechanism on the lateral bearing interaction factor [113,190,195–197]. Through continuum finite element simulations, the predicted transverse lateral bearing interaction factor ($N_{ch}$) was consistent with the study by Hansen [165]. The key finding from

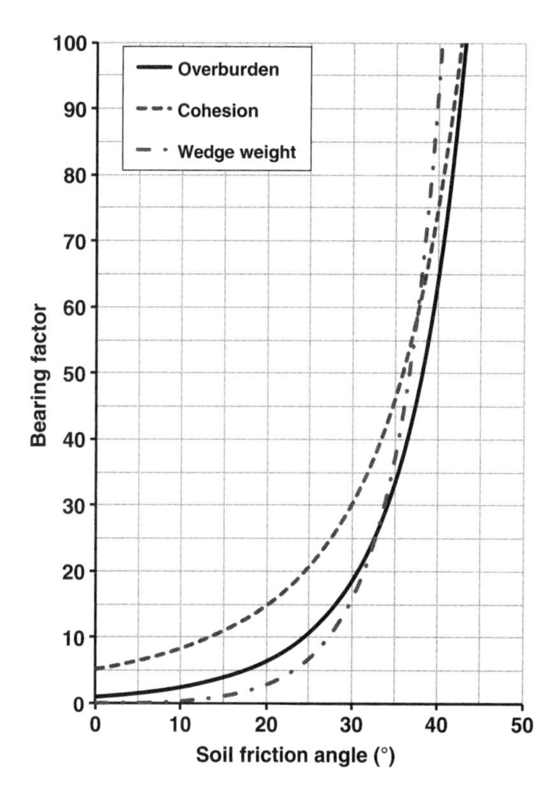


**FIGURE 8.3** Vertical bearing downward interaction factors.

these studies, however, was the contribution of the soil weight to the interaction factor for shallow buried pipelines associated with a passive wedge failure mechanism. Based on the work of Rowe and Davis [171], Phillips et al. [113] proposed a modified interaction relationship that accounted for the effects of soil weight within the passive wedge:

$$N_{\mathrm{ch}} = \min\left(N_{\mathrm{ch}}^{*} + \beta\frac{\gamma H_{\mathrm{s}}}{s_{\mathrm{u}}}, N_{\mathrm{ch}}^{\max}\right) \tag{8.8}$$

A total stress analysis was conducted with undrained conditions assumed and effective stress parameters defined at the pipe/soil interface. The coefficient $\beta$ relates the effect of the soil weight relative to the vertical stress at the pipe springline and normalized with the soil undrained shear strength. The numerical studies [113,195–197] indicate that the significance of the coefficient $\beta$ diminishes at pipe springline burial depth to pipe diameter ratios ($H_s/D$) greater than 5, which is related to the transition in the governing failure mechanism. The coefficient $\beta$ was defined as a constant value of 0.85 [113], but recent numerical modeling studies [195–197] suggest that the coefficient may be a function of the pipeline burial depth and intensity of equivalent plastic strain within the failure zone. The upper limit for the bearing factor $(N_{ch}^{max})$ is approximately 11, which is consistent with the lateral loading of piles in cohesive soil [113]. A decrease in the upward movement of the pipeline was also observed with increasing burial depth ratio [195–197].

For noncohesive soil, the interaction factor ($N_{qh}$), as defined by industry guidelines [43,45,46], is based on the work of Hansen [165] that increases with increasing burial depth ratio ($H_s/D$) and soil internal friction angle across the range of 20–45°. The studies by Neely et al. [167], Audibert and Nyman [168], and Popescu et al. [179] were consistent with industry guideline [43,45,46] estimates of the lateral bearing interaction factor for noncohesive soil. Other studies predict a reduction in the lateral bearing interaction factor by a multiplier of 0.5–0.67 [148,172,198,199]. Recent studies have reconciled this issue through accounting for the effects of pipe size and soil weight [113,138,160,183,197,200–202]. Further studies are required to provide reliable estimates of the lateral bearing factor in partially saturated noncohesive soil conditions due to the complex relationship with soil shear strength, dilation, deformation response, and failure mechanisms.

*Transverse Vertical Uplift Soil Response* In contrast to the axial and lateral soil load–displacement behavior, the vertical downward bearing and uplift response mechanisms can be characterized as asymmetric for most pipeline design burial depths. The soil load–displacement relationships account for the differences in stiffness, yield displacement, and yield load for the respective directions.

The maximum transverse vertical uplift soil force per unit length of pipe segment is expressed as [45,46]

$$q_{uu} = s_u N_{cv} D + \gamma' H N_{qv} D \tag{8.9}$$

The vertical uplift soil load–displacement relationship is based on theoretical solutions, physical tests, and finite element analysis [146,171,172,203]. Over the past decade, there has been significant number of studies examining the uplift behavior of partially or fully buried pipelines with respect to global instability mechanisms due to upheaval buckling [149]. The following discussion is focused on static pipe/soil interaction events that are stable from the kinematic perspective.

For cohesive soil, the vertical uplift bearing factor ($N_{cv}$) was based on the theoretical solution for buried cylinders by Vesic [203], adopted by Audibert et al. [204], and is considered valid for springline depth ratios ($H/D$) less than 10. The vertical uplift bearing factor is bounded by finite element analysis of horizontal plate anchor with assumed immediate breakout (slip) and fully bonded conditions [45,46]:

$$N_{cv} = 2\left(\frac{H}{D}\right) \leq 10 \tag{8.10}$$

In noncohesive soil, the vertical uplift bearing factor ($N_{qv}$) was based on physical modeling of buried pipelines and numerical simulation of horizontal plates [146,172]:

$$N_{qv} = \frac{\phi H}{44D} \leq N_q \tag{8.11}$$

where the overburden bearing capacity factor ($N_q$) was based on Meyerhof [205] bearing capacity factor defined as

$$N_q = K_p\, e^{\pi \tan\phi} = \tan^2\left(45 + \frac{\phi}{2}\right) e^{\pi \tan\phi} \tag{8.12}$$

The ultimate yield displacement is defined as

$$\Delta_{qu} = 0.1H \text{ to } 0.2H < 0.1D \tag{8.13}$$

for stiff to soft cohesive material and

$$\Delta_{qu} = 0.01H \text{ to } 0.02H < 0.1D \tag{8.14}$$

for dense to loose noncohesive soil.

*Transverse Vertical Downward Soil Response* The vertical downward soil load–displacement relationship can be established using bearing capacity theory based on Meyerhof [205]. The relationship between soil friction angle and the overburden ($N_q$), cohesion ($N_c$), and failure wedge bearing interaction factor ($N_\gamma$) is illustrated in Figure 8.3. The maximum transverse vertical uplift soil force per unit length of pipe segment is expressed as [45,46]

$$q_{ud} = s_u N_{cv} D + \gamma' H N_{qv} D + \frac{\gamma}{2} N_\gamma D^2 \tag{8.15}$$

The soil cohesion bearing interaction factor ($N_c$) is defined as

$$N_c = \left(N_q - 1\right)\cot\phi \tag{8.16}$$

where $N_c = 5.14$ in the limit with an internal soil friction angle of 0°. The failure wedge bearing interaction factor ($N_\gamma$) is defined as

$$N_\gamma = \left(N_q - 1\right)\tan(1.4\phi) \tag{8.17}$$

The ultimate yield displacement ($\Delta_{qd}$) is defined as 0.2$D$ for cohesive soil and 0.1$D$ for noncohesive soil.

### 8.3.4 Guidance on Best Practice to Enhance Computational Pipe/Soil Interaction Analysis

***8.3.4.1 Overview*** Modeling is an inherent characteristic and fundamental component for engineering analysis and problem solving. Over the past 30 years, there has been significant advancement in computational hardware and software technology that has provided a technically robust, efficient, and cost-effective tool for the numerical simulation of geotechnical engineering problems including pipeline/soil interaction.

The technology framework relies on laboratory tests to support the development and refinement of constitutive models and physical modeling to calibrate and verify the numerical modeling procedures [8,33,92,95,105,206–211]. Field investigations are conducted to provide focus for these specialized studies in support of engineering design activities in order to develop practical engineering solutions.

In Section 8.3.3, the state of practice for pipeline/soil interaction analysis using structure-based modeling techniques was presented. In the following sections, engineering guidance on the use of structure- and continuum-based finite element modeling procedures for pipeline/soil interaction analysis is presented. The motivation for this discussion is to provide an overview of the inherent limitations, constraints, advantages, and future trends for each technical approach. A discussion on physical modeling studies, for research and in support of engineering design, with respect to pipeline/soil interaction events was presented in Section 8.3.2.

#### 8.3.4.2 Structure-Based Models

*Overview* The current state of practice for pipeline/soil interaction analysis using structure-based numerical tools, through beam and spring elements to idealize the continuum response (Section 8.3.3.2), has served industry needs for the past three decades. Pipeline/soil interaction problems, such as fault movement due to seismic events, ground settlement, upheaval and lateral buckling instability, ice gouging, frost heave and thaw settlement, and riser touchdown, have been examined using these structural pipe/soil interaction simulation tools [14,26,120,127,128,134–137,212–217]. In comparison with other numerical techniques, such as continuum-based finite element methods (discussed in Section 8.3.7), structure-based tools require a relatively lower degree of technical proficiency in numerical modeling, limited computational hardware resources, and less computational time for equivalent pipeline systems being analyzed. However, there exists uncertainty on the adequacy and reliability of these structure-based pipeline/soil interaction tools particularly with respect to large deformation ground movement events that may have non-monotonic loading path or complex boundary conditions. Key factors that support this statement are highlighted in the following subsections.

A primary criticism of the structural beam/spring models is the idealization of a continuum with discrete elements that represent the generalized or composite soil mechanical response. This structure-based idealization cannot provide an adequate characterization of more realistic soil mechanical response including load coupling between mutually orthogonal axes, strength behavior with respect to path dependence, pore pressure effects and strain softening and hardening, and shear-induced dilation and compaction. Some of these issues are further explored in the following subsections.

*Soil Spring Functional Relationship* The soil load–displacement relationships may be defined by bilinear, multilinear, or hyperbolic mathematical formulations with guidance provided by engineering practice documents [43,45,46]. The peak soil loads should be defined in terms of total stress or effective stress formulation. The criterion specifying how the peak load and corresponding yield displacement magnitude should be determined will influence the soil stiffness response [167–169,193–197]. In general, the multilinear and hyperbolic formulations will have greater stiffness (i.e., less compliance) for sub-yield loading events than the bilinear counterpart. Furthermore, the hyperbolic formulation may be more representative of the constitutive behavior for soft clay and loose sand but may not provide an adequate characterization of the constitutive behavior for strain softening materials such as dense sand or stiff clay.

*Influence of Trench Backfill and Native Soil* In general, the spring formulations assume that homogenous soil conditions prevail and do not account for the trench boundary conditions and potential interaction between the pipeline within the trench backfill and the surrounding native soil. Several studies have demonstrated that the pipe diameter, pipe burial depth, trench geometry, and relative soil strength differences (i.e., between the native and backfill soils) will influence the development of peak loads, yield displacement, and failure mechanisms [111,113,130,156,218–220]. Structural beam/spring models can be tailored to simulate soil backfill and trench wall effects but only within a generalized framework or case-specific application. These idealized numerical procedures, however, cannot account for complex interactions and failure mechanisms that may occur at multimaterial interfaces such as the pipe/soil, trench boundary, or other physical barrier (e.g., geotextile). The use of physical modeling and numerical simulation techniques is needed to address these issues. The pipeline engineer will need to assess all possible design options to mitigate load effects on buried pipelines that may include engineered backfill with optimal trench configurations, preferential route alignment, low friction coatings, and geotextiles. Consideration and optimization of issues such as logistics, project execution risk, and cost will be required.

*Interaction Factors* There is general agreement on the general characteristics and parameters influencing the interaction factor, which include pipe diameter, pipe burial depth, soil strength, and pipe/soil interface friction factor. As

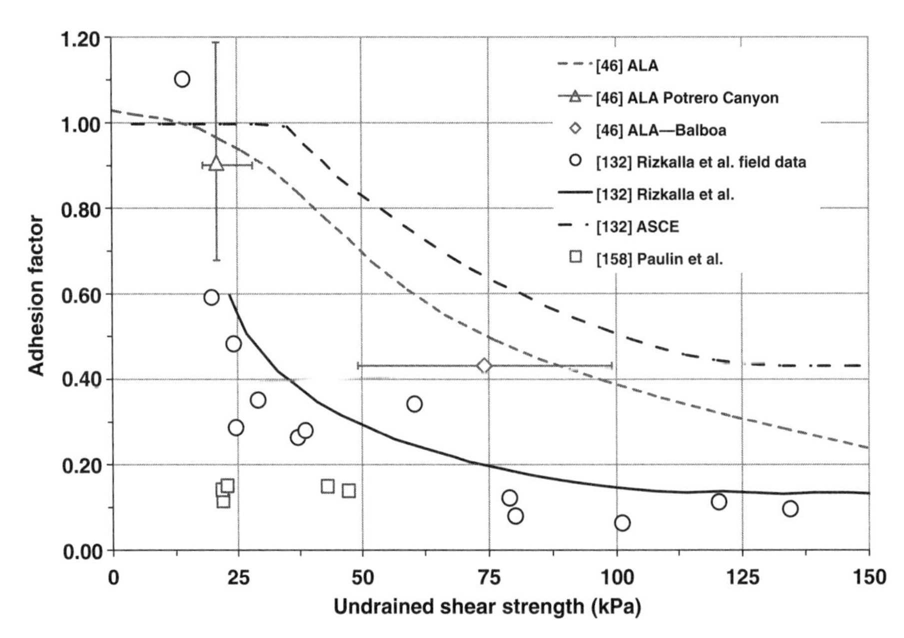


**FIGURE 8.4** Variation of adhesion factor with undrained shear strength.

discussed in Section 8.3.3.2, the interaction (i.e., bearing) factors have been derived through analytical solutions, physical models, and numerical simulations for analog problems (e.g., plate anchor) and buried rigid pipes. However, recent studies have raised questions on the basis for these interaction factors as defined by current practice. Some of the key factors include the basis for the interaction factors, soil stress state (i.e., total versus effective stress conditions), pipe size and soil stress effects, local pipe/soil interface effects, and soil failure mechanisms. An overview of these factors is presented in the following paragraphs.

Axial pipe/soil interaction is dominated by the interaction, load transfer, and failure mechanism developed at the local annular pipe interface with the soil. The interface behavior can be characterized by nonlinear deformation, shear-induced volume change, and excess pore pressure for undrained conditions [108–111,132,156,157,160,200,221]. For cohesive soil, assuming total stress analysis with undrained conditions, these effects are averaged or smeared through the use of an adhesion factor (Section 8.3.3.2). However, there is significant variation in the adhesion factor, as shown in Figure 8.4 [45,46,222], that may be attributed to test conditions, misalignment of the pipe axis from pure axial motion, and soil desiccation and cementation [132,148,149, 153,158,222]. For the pipeline engineer, consideration of the soil type, groundwater conditions, load obliquity, and loading rates is needed when interpreting the adhesion factor. In dense noncohesive soil, recent studies have shown that the axial resistance may significantly increase due to constrained dilation that is underestimated by current pipeline/soil interaction guidelines [108–111,150,154,158]. The use of an effective stress formulation with frictional properties defining the pipe/soil interface may be more representative of longitudinal interface behavior and load transfer mechanisms [149–152,154,156,157,222–224].

Recent studies have identified areas of uncertainty with the determination of lateral bearing factor ($N_{qh}$) [148,160, 183,196,197]. For example, the test results reported by Trautman and O'Rourke [145,147] are consistent with the predictions by Ovesen [198,199] but may differ by a factor of 2 when compared with the tests by Audibert and Nyman [144], which are consistent with the model by Hansen [165]. This variability is illustrated in Figure 8.5 and may be attributed to the soil conditions (i.e., gradation, density, and friction angle), experimental procedures (e.g., data acquisition and reporting and loading rate), and boundary conditions (i.e., plate anchor versus pipe). Recent studies have attributed the differences to the effects of pipe size, soil weight, and failure mechanism [113,138,160,183,197, 200–202]. For practical range of diameters for energy pipelines, however, the scale effect, as discussed by Guo and Stolle [183], on lateral soil bearing resistance is negligible for diameters greater than 273 mm. Furthermore, as discussed in Section 8.3.3.2, for shallow burial depths, the soil weight being lifted through the lateral pipe/soil interaction has been shown to be a significant parameter on soil resistance that is associated with a passive wedge failure mechanism [113,195–197]. For deeper pipe burial depths, the failure mechanisms can be characterized by plastic flow around the pipe circumference.

Uplift resistance models have generally been developed to establish peak load and mobilization distance parameters for monotonic loading conditions. A significant volume of work has been conducted on vertical uplift resistance for offshore

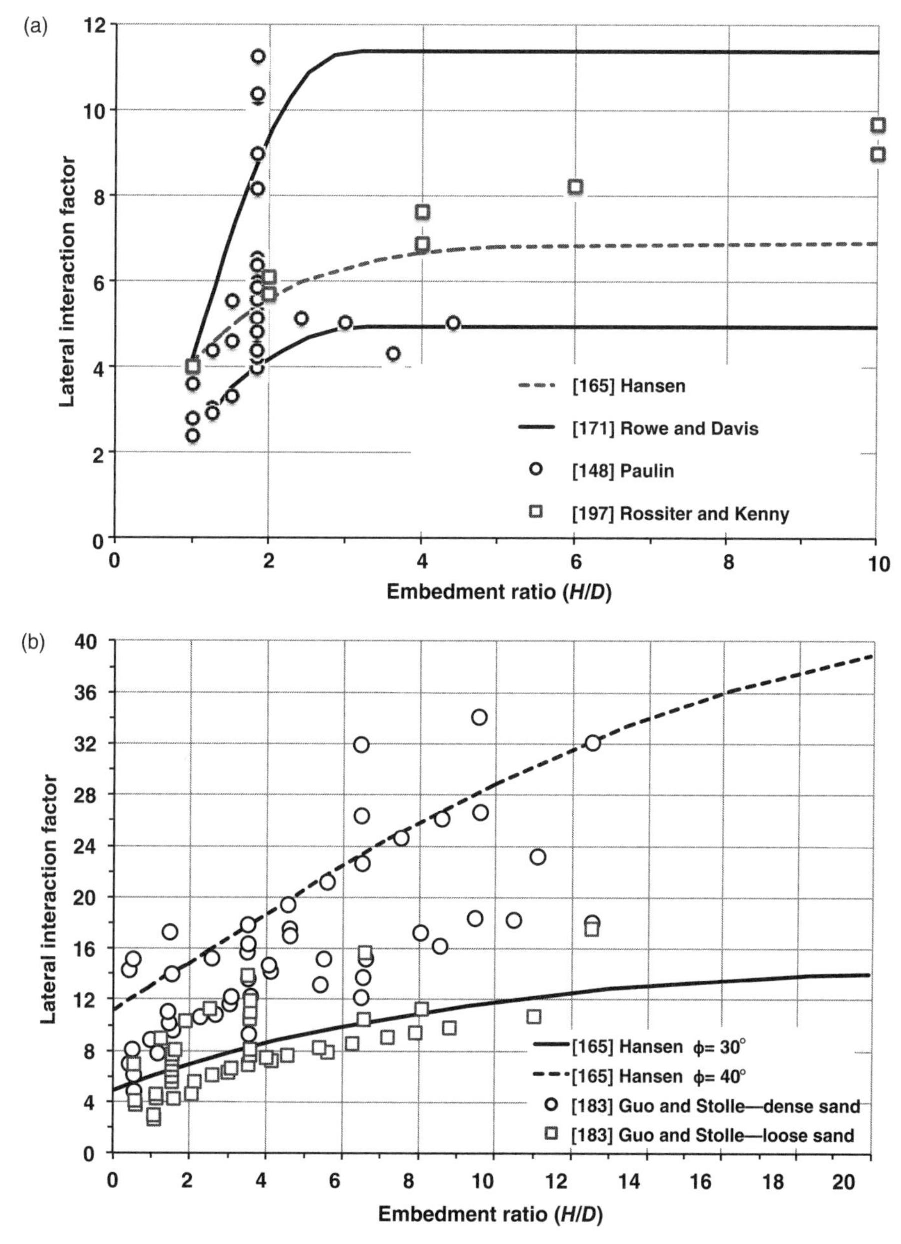


**FIGURE 8.5** Interaction factors for transverse lateral pipe/soil loading events in (a) cohesive soil and (b) noncohesive soil.

pipelines to mitigate the effects of upheaval buckling mechanisms [48,83,104,213,223–235]. A complex relationship exists between key design parameters, including pipe (e.g., diameter, weight, burial depth, and interface friction) and soil (e.g., type, gradation, soil state, strength, and dilation) characteristics, with the interaction factor and mobilization distance. These parameters also influence the failure mechanism developed during the pipe/soil interaction event that may be characterized by wedge failure (vertical or inclined) with slip lines or plastic flow model. As the pipeline lifts upward, the soil may fill the void located at the pipe invert and haunches.

For pipeline systems with cyclic operational loading conditions (e.g., start-up, shutdown, and restart) or variable operational parameters (e.g., transient pressure), this may result in the continued vertical upward movement of the pipe through a geotechnical ratcheting mechanism [48,83,149,231–235]. In response to the operational loads and out-of-straightness imperfections, the pipe lifts vertically and, in weak cohesive or loose noncohesive soil conditions, the soil may fill the void beneath the pipe. This action prevents the pipe from returning to the initial position and results in the systematic upward creep or ratcheting of the pipe vertically through the soil column. This may lead to a sufficient loss of soil cover that results in pipe global buckling response. These recent studies have provided guidance on upheaval buckling with respect to predicting uplift resistance and mobilization distance, and the significance of geotechnical ratcheting mechanisms.

### 8.3.5 Emerging Research

***8.3.5.1 Oblique Loading*** As discussed in Section 8.3.3.2, conventional engineering practice for the analysis of pipeline/soil interaction events is generally performed using decoupled, structural beam/spring models. However, for pipelines subjected to multiaxial loads [236,237] and foundations subjected to oblique or inclined loads [238–245], a coupled mechanical response along principal directions has been observed. A schematic illustration of oblique loading scenarios in three orthogonal planes is shown in Figure 8.6. The integration of shear coupling between soil springs has also been examined in past studies [123–125,137]. As previously discussed, these structure-based parameter models (i.e., Winkler-type foundation) cannot account for the complex soil mechanical response with respect to strength evolution (e.g., peak stress, residual stress, and dilation), load and stress path dependence, and deformation mechanisms and strain localization effects.

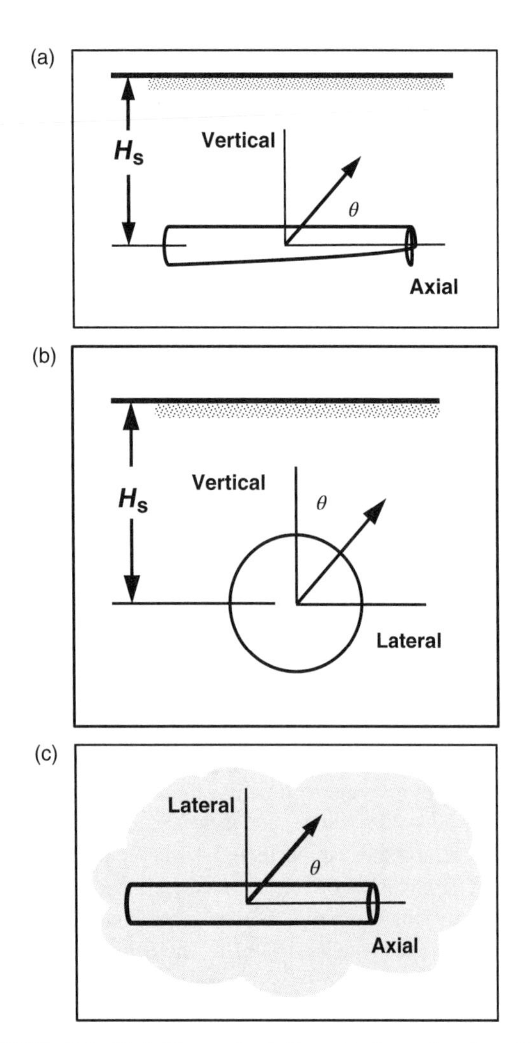


**FIGURE 8.6** Definition of relative attack angles for oblique loading in (a) vertical–axial plane, (b) vertical–lateral plane, and (c) lateral–axial plane.

One of the primary goals for investigating load coupling effects during oblique pipeline/soil interaction events is to improve the current basis of the structural beam/spring models. As discussed in Section 8.3.3.2, structural models provide a relatively simple and cost-effective tool that is well suited for pipeline/soil interaction studies involving a large parameter matrix or probabilistic framework [3,26,30,246]. For these problems, the use of more robust computational tools, such as continuum finite element methods, would be limited to a select number of design cases that must be identified through screening analysis with specified assessment criteria. There exists uncertainty, however, on the use of this idealized structural model to simulate complex stress paths, load transfer processes, and failure mechanisms for problems with severe nonlinear behavior involving contact and strain localization [3,120,133,217,247].

Enhanced structural models can be developed using macroelements or refined spring elements that account for the effects of mechanical load and displacement coupling [138,248,249]. These coupled formulations define the soil yield envelope, in terms of the peak load and yield displacement within the three orthogonal planes (i.e., vertical–axial, lateral–axial, and vertical–lateral). Load coupling effects are incorporated within the soil spring formulation through knowledge on the shape of the peak load and yield displacement failure envelopes for specific design conditions, such as pipe diameter, pipe burial depth, soil type and strength properties, and attack angle. For macroelements, parameters defining the shape of the yield surface and plastic potential function or flow rule must also be established. Physical models, which may be integrated with numerical simulations to extend the parameter database, provide the basis to define the nonlinear soil response due to soil deformations, material behavior, and load coupling.

Another goal for investigating soil load coupling effects is to gain confidence in advanced computational tools, such as continuum finite element methods, to simulate demanding and multifaceted problems, such as ice gouging that involves complex load paths, pipe trajectories, failure mechanisms, clearing processes, contact mechanics, and interface behavior [196,217,221,247,250–253]. Confidence in these advanced simulation tools can be established through the calibration and verification of problems with reduced complexity, such as 2D oblique pipeline/soil interaction events, through laboratory testing and physical modeling [113,138,159,160,196,197,211].

Early studies on the effects of oblique loading on soil mechanical response for pipeline/soil interaction events were based on analytical solutions (e.g., limit load analysis) [254] that were complemented by analog events (e.g., inclined plate anchor) [167,239]. More recently, an increasing number of studies have investigated the load coupling effects during oblique pipeline/soil interaction events. These studies have focused on pipeline-specific configurations using physical models and numerical simulation tools aimed at the advancement of current practice. Key technical aspects and outcomes from these investigations with respect to oblique load effects on the soil mechanical response are highlighted and referenced in this section.

Buried pipelines may experience oblique loading in the vertical–axial and vertical–lateral planes due to physical processes such as upheaval buckling, soil liquefaction, frost heave, and slope movement. The effect of load coupling in the vertical–axial plane has not been systematically examined, particularly with respect to the axial interface behavior, and requires further study [138,248,249]. The primary requirement is the advancement of appropriate soil constitutive models and physical testing for the verification of numerical simulation tools.

*Vertical–Axial Interaction* A recent study on oblique vertical–axial pipe/soil interaction in cohesive soil was conducted using continuum finite element modeling procedures [138,200]. Through examination of the peak loads developed, a two-phase failure envelope, related to the loading angle or attack angle, was observed (Figure 8.7). The linear segment (dashed lines) was associated with low attack angles and failure within an annular layer at the pipe/soil interface that was influenced by the increased normal stress with oblique loading (Section 8.3.3.2). For low attack angles, less than 10°, the axial soil strength and interface properties governed the soil failure mechanism with an increase in the axial resistance by a factor of 1.25. The nonlinear response was related to bearing failure and work done in lifting the soil mass that resulted in a decreased vertical resistance by a factor of 0.8 at low attack angles. The general characteristics of the vertical–axial interaction curve are consistent with recent studies by Cocchetti et al. [248,249]. For pure vertical bearing interaction, the peak resistance was consistent with previous studies on horizontal anchor plates and dependent on soil strength parameters (i.e., friction angle and dilation angle) [235,255,256].

The numerical study, conducted by Daiyan [138,200], observed the oblique vertical–axial failure envelope expanded with increasing soil friction angle ($\phi = 35°$, 40°, and 45°), pipe/soil interface friction factor ($f = 0.5$ and 0.8), and burial depth ratio ($H/D = 2$, 4, and 7). Across the range of parameters investigated, an expression defining the nonlinear response of the oblique vertical–axial failure envelope was defined:

$$N_{qv\theta}^2 + 0.3N_{t\theta}^2 = N_{qv90}^2 \tag{8.18}$$

where $N_{qv90}$ is the interaction factor for pure vertical uplift condition. Results from this numerical parameter study require physical tests for calibration and verification of the simulation procedures.

*Vertical–Lateral Interaction* A greater volume of literature exists for vertical–lateral pipeline/soil interaction in cohesive soil, where Nyman [254] first proposed a mathematical expression defining the effects of load coupling. A limit load analysis was conducted, based on the analog problem

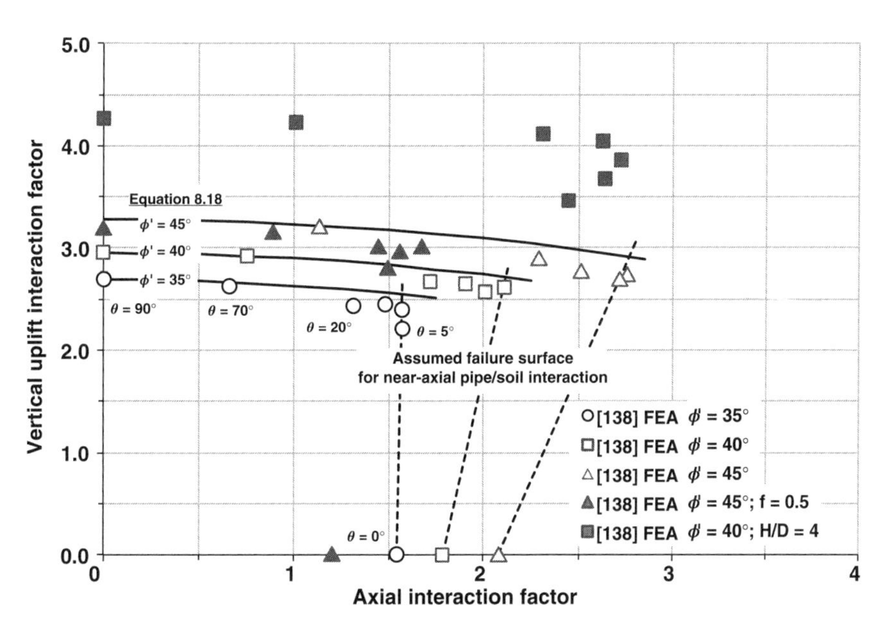


**FIGURE 8.7** Failure envelope for oblique vertical–axial pipe/soil interaction in noncohesive soil for $H/D = 2$ and $f = 0.5$ unless otherwise noted.

for inclined anchor plates [238], where an expression was developed for the peak soil resistance ($p_{u\theta}$)

$$p_{u\theta} = (1 - \beta)q_u + \beta p_u \tag{8.19}$$

in the oblique loading direction ($\theta$) measured from the vertical axis with the interaction factor ($\beta$) defined as

$$\beta = \frac{0.25\theta}{90° - 0.75\theta} \tag{8.20}$$

The peak lateral soil restraint ($p_u$) was based on Hansen [165] bearing capacity factor, and the peak vertical load ($q_u$) was based on Vesic [203] uplift breakout factor. The solution accounts for passive wedge mechanism that is applicable for shallow pipe burial depths. The ultimate oblique displacement was defined as a constant percentage of the springline burial depth with the pipe motion assumed collinear with the oblique load vector. A recent study suggests that this assumption may be best suited for deep pipe burial conditions [138].

As shown in Figure 8.8, a number of studies using continuum finite element methods have confirmed the hypothesis proposed by Nyman [254] and extended the knowledge base for buried pipelines subject to oblique vertical–lateral interaction scenarios in cohesive soil [138,195–197,257]. The normalized interaction factor relates the interaction factor determined for each attack angle examined divided by the normalizing parameter as defined by pure vertical uplift ($N_{cv}$, $N_{qv}$) or pure lateral ($N_{ch}$, $N_{qh}$) pipe/soil interaction.

Discrete points within the failure envelope, illustrated in Figure 8.8, were established based on the peak or yield lateral and vertical uplift soil force for each attack angle ($\theta$), as shown in Figure 8.9. The results suggest that nonlinear coupling effects are more prominent at attack angles greater than 15° [195–197,257]. This observation compares well with numerical simulations of oblique pipeline/soil interaction events in loose sand where a critical attack angle of 30° was observed [229]. However, further work is required to better delineate the critical attack angle that is associated with a change in the soil failure mechanism.

Alternate expressions defining the upper and lower bound yield surfaces (Figure 8.8) for oblique vertical–lateral interaction in cohesive soil have also been developed, which can be characterized by nominal elliptical and circular forms [257]. The transition from an elliptical to circular failure surface was found to be dependent on the pipe diameter, pipe burial depth, and soil shear strength profile with depth, particularly at shallow embedment depths [178,195–197,257]. For deeper pipe burial depths, the influence of soil shear strength profile on the shape of the failure surface was only observed for oblique attack angles greater than 45° [195–197].

Overburden coefficients ($\beta_h$, $\beta_v$) were also established that relate the nonlinear interaction factor behavior with pipe embedment ($H/D$) and overburden pressure to strength ($\gamma H/s_u$) ratio [195–197]. These effects were attributed to the shape of the failure surface and size or intensity of the plastic strain field as a function of the burial depth (i.e., $H/D$

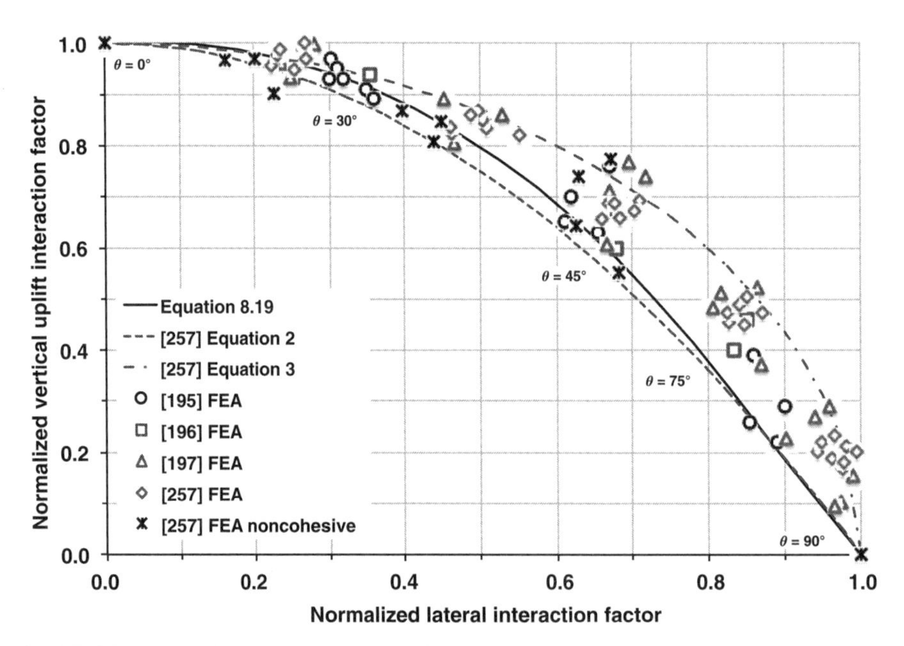


**FIGURE 8.8** Normalized vertical–lateral oblique loading failure envelope in cohesive soil unless otherwise noted.

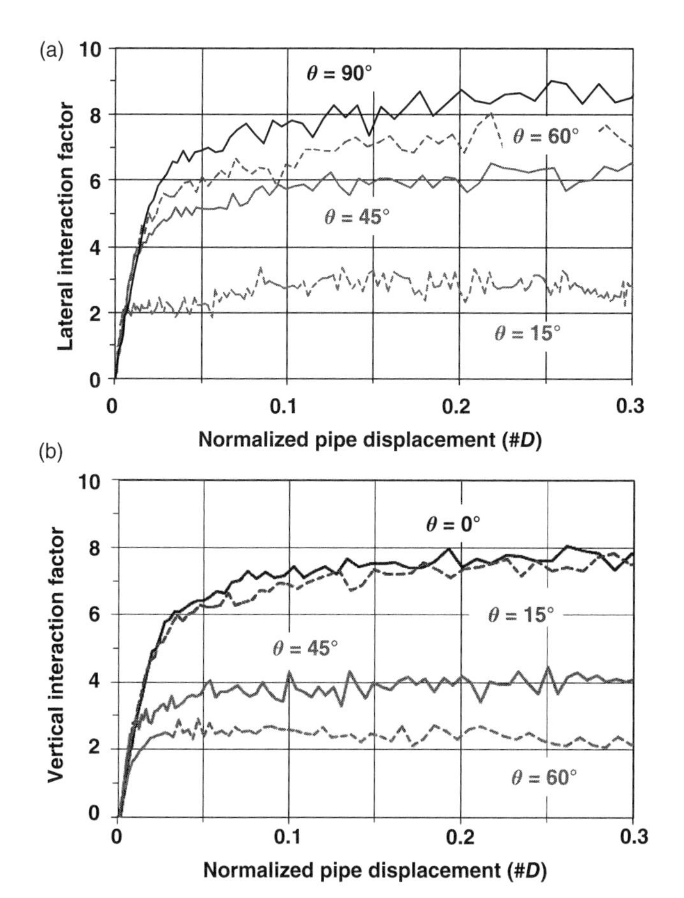


**FIGURE 8.9** Normalized load–displacement relationship for (a) lateral and (b) vertical bearing as a function of the attack angle ($\theta$) in cohesive soil.

ratio) and attack angle ($\theta$) with potential influence of the soil undrained shear strength.

Based on the mobilized uplift and lateral force response (Figure 8.9), the upper bound lateral ($N_{ch}$) and vertical ($N_{cv}$) bearing factors are consistent with current engineering practice (Section 8.3.3.2). The observed variability on the mobilized peak soil force of up to 120%, as shown in Figure 8.8, may be attributed to those measures used to evaluate the yield condition as discussed in Section 8.3.4.2. In general, the peak load occurs between displacement having 10 and 15% of the pipe diameter, which is consistent with current engineering guidelines [45,46]. Correspondence between the interaction factors (i.e., lateral ($N_{ch}$) and vertical ($N_{cv}$)) and failure mechanisms (i.e., shallow and deep) with respect to the pipe embedment ($H/D$) ratio, overburden pressure to strength ($\gamma H/s_u$) ratio, and pipe loading angle of attack ($\theta$) was also established [195–197]. Similar to the observations for pure lateral loading, as discussed in Section 8.3.3.2, the failure mechanism for oblique loading at shallow burial depth was characterized by a passive wedge with surface heave and work done through vertical lifting of the soil wedge weight. At deeper burial depths, the failure mechanism was associated with soil flow around the pipe circumference.

Studies have also examined oblique vertical–lateral pipeline/soil interaction in noncohesive soil with research outcomes and failure envelopes similar to the previous discussion on cohesive soil [229,258–261]. In these studies, the ultimate soil resistance increased with oblique attack angles within the range of 30–45° (Figure 8.10). These observations suggest that, for practical values of soil imperfection profiles, upheaval buckling mechanisms are

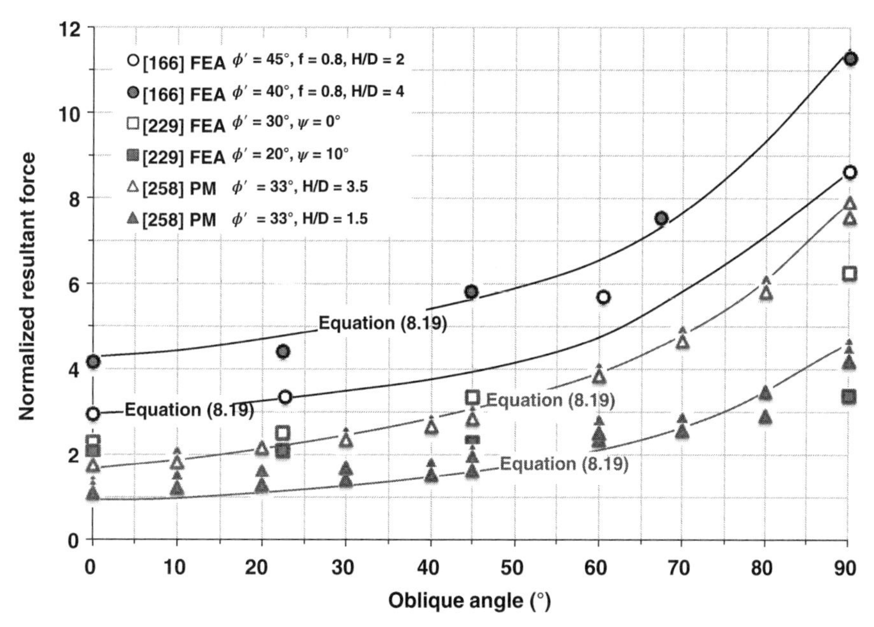


**FIGURE 8.10** Variation of normalized resultant force during vertical–lateral oblique loading in noncohesive soil.

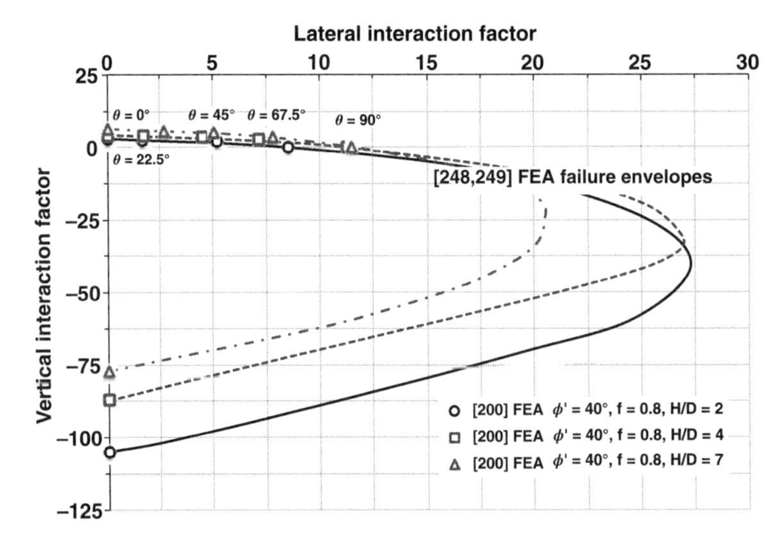


**FIGURE 8.11** Interaction factors for vertical–lateral oblique loading in noncohesive soil.

dominated by the vertical soil uplift resistance [149,229]. For limited numerical modeling studies, there is general consistency in the failure envelope response (Figure 8.11). A comprehensive physical test program is required to calibrate and verify the numerical simulation tools. In developing the yield envelopes for vertical–axial interaction plane, consideration of scale effects in the physical model, with respect to determination of the peak soil resistance and mobilization distance to yield, is required [227,228,262].

*Lateral–Axial Interaction* In reference to the previous discussion in this section on oblique pipeline/soil interaction events and two-phase failure envelopes, similar observations and conclusions may be formulated for oblique lateral–axial loading events in cohesive [113,138,160,221] and noncohesive [138,201,202,259,260] soils. An earlier study by Kennedy et al. [128] reported that the effective axial resistance increased with lateral soil pressure that was accounted for through the pipe/soil interface friction factor.

Based on reduced-scale centrifuge tests and numerical simulations, an expression defining the lateral–axial yield envelope was developed (Figure 8.12) [113]:

$$N_{qh\theta}^2 + 3N_{t\theta}^2 = N_{qh90}^2 \tag{8.21}$$

where $N_{qh90}$ is the interaction factor for pure lateral loading. The finite element analysis assumed undrained conditions

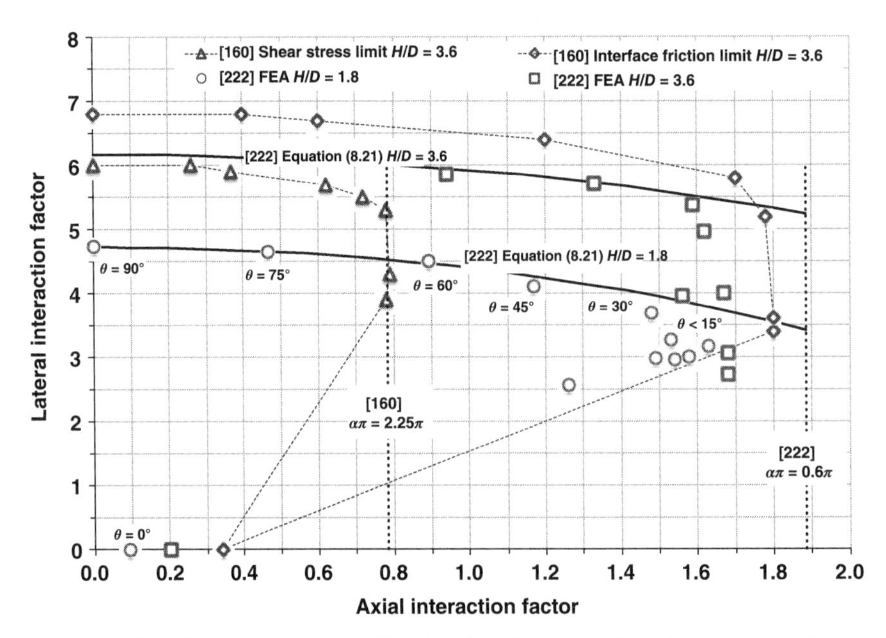


**FIGURE 8.12** Lateral–axial oblique loading failure envelope for cohesive soil.

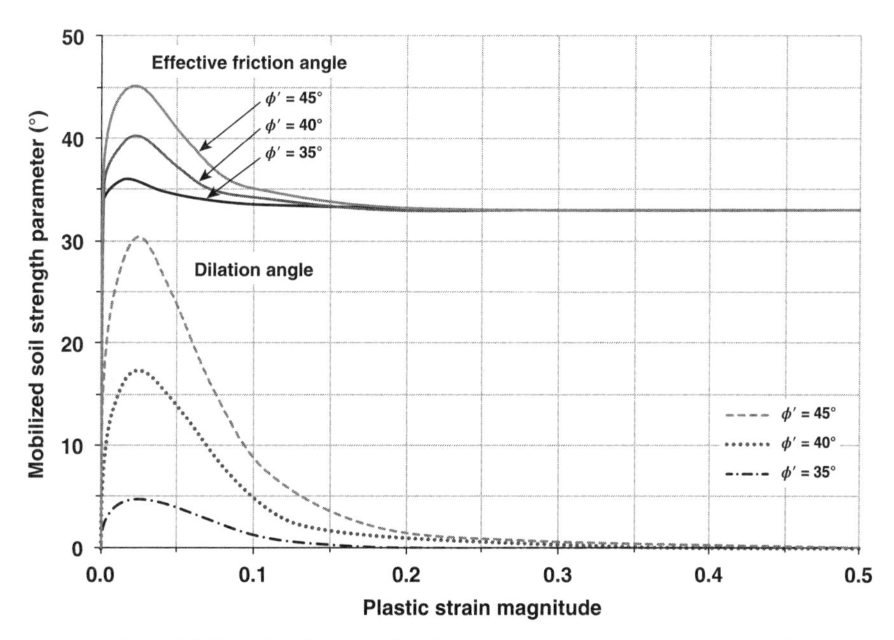


**FIGURE 8.13** Mobilization of soil strength parameters with plastic strain.

with an effective interface friction angle of 20°. The yield envelope for lateral–axial interaction (Equation 8.21) has a similar form to the failure surface for vertical–axial interaction (Equation 8.18).

The interaction diagram (Figure 8.12) illustrates that large axial resistance may be developed at low attack angles (<10°) where the failure mechanism is controlled by shear and pipe/soil interface properties. As discussed by Phillips et al. [113], confidence in the observed failure envelope was developed through comparison with direct physical evidence and correspondence with similar interaction diagrams for suction caissons. For larger attack angles, generally greater than 30°, the failure mechanism was controlled by lateral bearing interaction, which may be shear through the soil mass or flow around the pipe circumference (see Section 8.3.3.2).

Lateral–axial failure envelopes have also been developed for noncohesive soil based on reduced-scale centrifuge tests and numerical parameter studies using continuum finite element methods [138]. The numerical parameter study examined the influence of soil friction angle ($\phi = 35°$, 40°, and 45°), interface friction factor ($f = 0.5$ and 0.8), and pipe embedment ratio ($H/D = 2$ and 4). A user subroutine was developed to account for the mobilization of the soil strength parameters (Figure 8.13), including the soil friction angle and dilation angle, in terms of the soil plastic strain response and potential function (i.e., flow rule) that can account for strain softening response [138,195,207,211,263,264]. The mathematical expression developed by Phillips et al. [113] (Equation 8.21) defining the nonlinear response of the lateral–axial failure envelope for cohesive soil was also found to be representative for noncohesive soil as shown in Figure 8.14.

Recent studies have highlighted the need to reexamine the effects of pipe/soil interface properties and contact formulation on the coupled lateral–axial soil response, particularly for low attack angles [160,221]. As shown in Figure 8.12, for attack angles less than 70°, the normalized axial reaction force exhibits sensitivity with the defined soil shear stress limit at the pipe/soil interface. The normalized axial load was observed to double when the interface shear stress varied from a fraction of the undrained shear strength (i.e., $\tau_{max} = 0.5 s_u$) to a condition where the interface shear stress was equal to the static interface friction coefficient multiplied by the applied normal stress (i.e., $\tau_{max} = \mu\sigma_n$). The studies conducted by Phillips et al. [113] and Seo et al. [159] are consistent with the interface defined by a static interface coefficient (i.e., $\tau_{max} = \mu\sigma_n$). The lower bound failure envelope (Figure 8.12) was established using a clay sensitivity, $S_t = 2$, that corresponds to a normalized axial load, $N_{y\,max} = \pi/4$, with an adhesion factor of 0.25, consistent with current practice [46].

In these numerical investigations [160,221], a void on the trailing (leeward) face of the pipe due to the relative lateral displacement of the pipe through the soil mass was observed. The soil constitutive model was defined by elastic–plastic behavior with von Mises yield criterion that did not account for the effects of tension cracking, slumping, consolidation, and repeated desiccation/saturation cycles during long-term interaction events. These factors may result in greater circumferential contact between the pipe and soil that would tend to facilitate load transfer processes and increase pipe axial loads. In addition, there exists some uncertainty on the mobilized normal and traction forces generated during oblique pipeline/soil interaction events due to element formation,

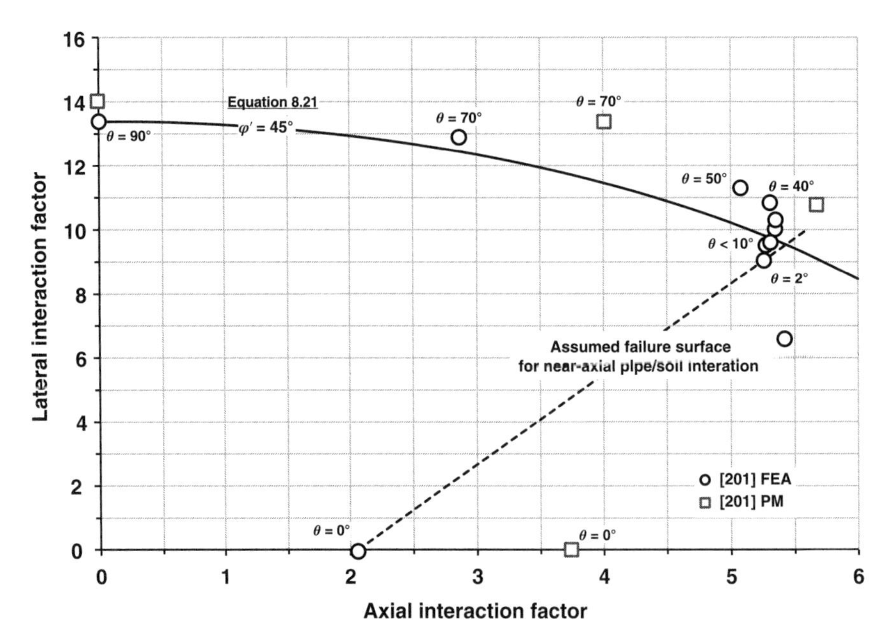


FIGURE 8.14 Lateral–axial oblique loading failure envelope for noncohesive soil with heavy pipe configuration at an embedment ratio $H/D=2$ and interface friction factor $f=0.5$.

interface properties, and contact mechanics implemented within the numerical algorithms [160,221]. Based on these observations and the two failure envelopes illustrated in Figure 8.12, a family of interaction curves may exist for practical oblique loading scenarios that may be influenced by other factors including burial depth, soil type, soil strength, pipe coatings, and trench configuration (see Figures 8.14 and 8.15).

As highlighted in this section, significant insight into the soil mechanical coupling response during oblique load events has been acquired through physical modeling and numerical simulation investigations. Key parameters have

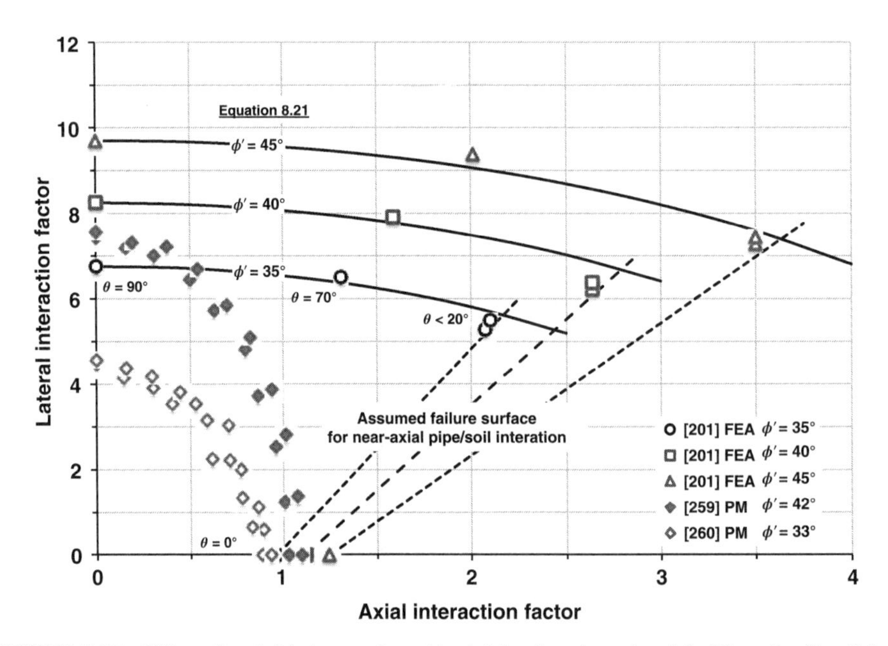


FIGURE 8.15 Effect of peak friction angle and burial depth on lateral–axial oblique loading failure envelope for noncohesive soil with normal pipe configuration at an embedment ratio $H/D=2$ and interface friction factor $f=0.5$.

been identified that include characteristics of the design conditions (e.g., pipe diameter, burial depth, and pipe/soil interface properties), soil parameters (e.g., type, strength properties, and stress history), load event (e.g., angle of attack, relative pipe trajectory, and small or large deformation), and modeling basis (e.g., element formulation, contact mechanics, and pipe/soil interface behavior). However, further investigations should be conducted to provide completeness across a wider parameter range, address other factors not yet examined (e.g., effective stress and pore pressure, load path, loading rate), and reduce uncertainty on the significance of these parameters with respect to the effects of load coupling on soil behavior, load transfer response, and failure mechanisms. The largest data gap exists in the vertical–axial oblique loading plane for cohesive and noncohesive soils, whereas the greatest discrepancy between research studies exists for lateral–axial oblique loading in noncohesive soils.

Furthermore, additional studies are needed to establish those design conditions and parameters where the use of conventional structure-based pipeline/soil interaction models may be effective, and where the application of more complex simulation tools (e.g., continuum finite element modeling procedures) is required. From the perspective of pipeline engineering projects, resolving these questions within academia or applied research institutions is a much needed first step where the outcomes can then be augmented by industry through joint industry projects. The use of advanced numerical modeling procedures will impose additional requirements in terms of resource commitment, planning, and logistics to support input from subject matter experts, field programs, laboratory testing, and physical modeling investigations. From this perspective, there will be a need for a comprehensive integrated program of research that includes laboratory tests to refine constitutive models and physical tests to validate numerical simulation tools.

***8.3.5.2 Loading Rates*** The idealized soil spring formulation does not account for the effects of loading rate or strain rate on soil behavior, such as the influence on stiffness, strength, consolidation, dilation, and excess pore pressure [113,148,158,265]. Laboratory tests have demonstrated that the soil stiffness and undrained strength typically increase with strain rate or loading rate, and can be influenced by other factors such as soil physical properties (e.g., gradation, plasticity index, and mass density), soil type and state (e.g., NC clay, dense sand, saturated or dry conditions, and frozen or unfrozen soil), and loading conditions (e.g., confining pressure, effective stress path, and pore pressure) [266–271]. These effects of increased forces with pulling speed have been observed through investigations of offshore submarine ploughs used to construct pipeline trenches [272–274]. However, strain rate effects on soil mechanical behavior are a complex process where interpretation of laboratory tests for application in full-scale problems may not yield conservative results and may not be representative of *in situ* soil behavior with respect to strength evolution, failure mechanisms, interface behavior, and effective stress path [148,158,267–269,275].

For example, physical modeling studies on the axial pullout indicated that slow loading rates (0.5 mm/h) resulted in a 25% higher load relative to faster loading rate (10 mm/h) [158]. A series of reduced-scale centrifuge tests extended these observations to lateral loading in cohesive soil where the maximum loads at slow loading rates (0.0095 m/day) could be 2.5 times greater than the measured peak loads at faster loading rates (0.74 m/day) [148]. The behavior was attributed to the relationship between loading rate and dissipation of excess pore pressure [113,148,190,276,277]. Through coupled finite element analysis for cohesive soil, Phillips et al. [113] extended these observations and related the effect of loading rate and transition from undrained to drained behavior to the pipe diameter, relative movement rate, and coefficient of consolidation (Figure 8.16).

Similar observations on increasing soil load with strain rate were observed in reduced-scale centrifuge tests of lateral loading in saturated dense sand [275]. The dilative behavior caused an increase in the soil volume that resulted in negative pore pressure with larger effective stress, which in turn results in greater applied loads on the pipeline. These effects were also observed through axial pullout tests on dense sand due to constrained dilation [110,111,184].

***8.3.5.3 Large Deformations: Load Coupling and Superposition*** Recent studies have illustrated deficiencies in the structural beam/spring modeling approach for pipeline/soil interaction problems that are primarily associated with relatively large soil deformations, multiaxial loading events, and complex loading paths. In comparison with physical models and continuum finite element analysis, the uncoupled soil spring formulation fails to account for realistic soil behavior that may result in conservative and nonconservative estimates of soil loads and pipe deformations [3,120,138,159,160,217,222]. These studies have shown that for multiaxial load events, such as pipe/soil interaction with combined axial, lateral, and vertical soil deformations, a complex interaction develops where the failure envelope reduced soil capacity due to the load coupling (see Section 8.3.5.1). For ice gouge problems, arguments have been formulated that a superposition error primarily accounts for the observed discrepancy between the structure-based and continuum-based numerical simulation tools [133]. Other studies suggest that the discrepancy can be attributed to errors in load coupling [3,138,253]. Furthermore, the soil spring model does not account for the effects of stress history and stress path on soil type (e.g., normally consolidated versus overconsolidated cohesive soil) and stress state (e.g., plane strain versus triaxial stress state).

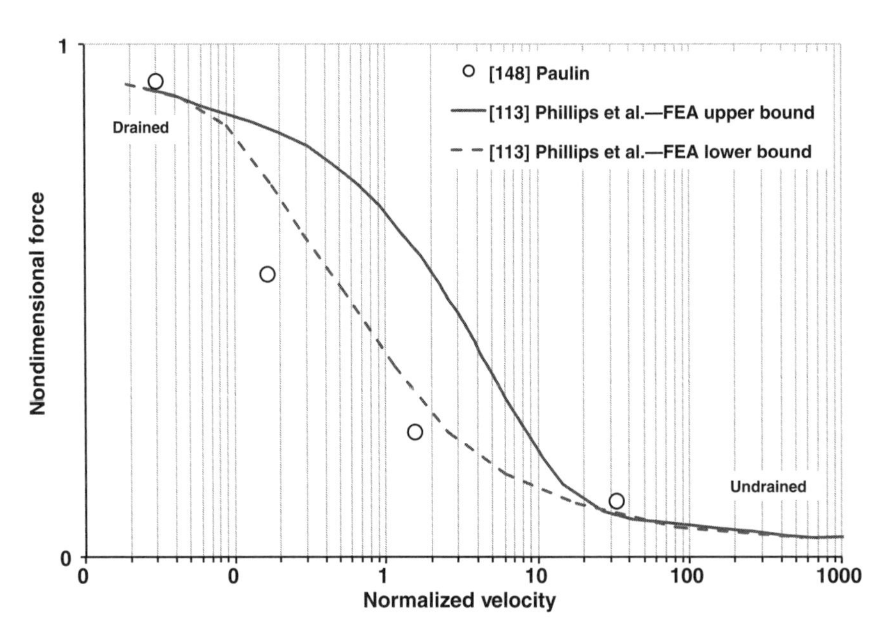


**FIGURE 8.16** Effects of loading rate on the peak lateral force in cohesive soil.

***8.3.5.4 Pipe Diameter and Model Scale Effects*** The discussion presented in the following paragraphs is intended to provide guidance on the issues of pipe diameter (i.e., size) and scale (i.e., laws of similitude) effects. While the emphasis is focused on pipeline/soil interaction events, these issues affect a range of engineering problems in materials science, concrete design, geomechanics, and ice mechanics [278–286]. Pipe size effect refers to physical models with varying pipe diameter conducted at $1g$, while other parameters (e.g., burial depth, soil density, and friction angle) are held constant [144,147,183,187,258–260]. Model scale effects arise when there are differences between the reduced dimensional scale physical model, conducted in the geotechnical centrifuge at an artificial gravity using laws of similitude, and the prototype [113,190,201,262,287].

These factors of size and scale may affect strength evolution and limit forces (i.e., equilibrium and interaction factors), kinematics (i.e., mobilization displacement and failure mechanism), and strain localization mechanisms (e.g., shear band). If these factors are not properly addressed, the model behavior may not adequately represent prototype conditions, with incompatible results as an outcome. These concepts are discussed in the following paragraphs in order to provide a basis for interpreting and evaluating results from physical and numerical modeling investigations.

In physical modeling of soil/structure interaction problems, there are potential scale effects that result from the relationship among soil grain size relative to the minimum structure dimension (e.g., pipe diameter), thickness (width) of the soil rupture zone or shear band, and mobilization distance that is associated with changing rates of dilation through the evolution from peak strength to critical state conditions [210,288].

Consequently, the stress (i.e., forces) developed across the localization (e.g., shear band) and the associated relative displacement and mobilization (i.e., kinematics) for this mechanism may not be adequately represented in the physical model, with respect to size or scale, for the corresponding prototype conditions. Examination of all factors, including strength evolution, mobilization distance, kinematics, propagation of instabilities, and strain localization mechanisms, is required to demonstrate consistency between the model and prototype [190,262,286,287].

In pipe/soil interaction problems, the key issue for consideration is the proportional relationship between the soil grading curve, characterized by the mean particle size ($d_{50}$), and the pipe diameter [286,289]. To address model scale effects, when comparing the results from full-scale to reduced-scale physical models, the ratio of mean grain size ($d_{50}$) to minimum structural dimension should be at least 10, as discussed in Section 8.3.2. Randolph and House [288] state that where discrete rupture surfaces are formed, with dilation followed by strain softening, the required ratio of minimum structural dimension to shear bandwidth to achieve an asymptotic response applicable to full-scale structures is approximately 20. Several studies suggest that the shear band thickness ranges from $8d_{50}$ to $20d_{50}$ [188,189,210,290]. In the context of model scale issues with respect to load transfer effects and evolution of failure mechanisms, to adequately assess the shear band evolution, assuming a composite $d_{50}$ of 0.6 mm, the minimum structural dimensions should range from 96 to 240 mm.

A recent study by Guo and Stolle [183] observed that the lateral bearing interaction force decreased with increasing pipe diameter for a constant burial depth ($H/D$). A parameter study was conducted, using finite element methods, to examine the effects of pipe diameter and model scale, which were termed size effect and model effect, respectively. The lateral bearing interaction factor decreased across the range of 19.7, 10.8, and 9.9 for pipe diameters of 33, 330, and 3300 mm, respectively. Studies conducted by Audibert and Nyman [144], Hsu et al. [258–260], and Trautman and O'Rourke [147], however, did not observe any scale effect for pipeline diameters across the range of 150–610 mm. The minimum grain size used in these studies satisfied the criteria as defined by Randolph and House [288]. Further insight is provided in Section 8.3.4.2, discussion by Peek [289], and closure by Guo and Stolle [183] on the concepts of size versus model scale effects in reference to pipe dimension, soil stress state, and laws of similitude.

Conversely, through numerical modeling investigations, Badv and Daryani [187] observed similar effects on lateral bearing interaction factor due to a variation in the pipe diameter as discussed by Guo and Stolle [183]. For an increase in the pipeline diameter, from 0.3 to 2.0 m, the lateral bearing interaction factor decreased by a multiplier of 0.7, based on Ref. [187]. In comparison, for the same relative change in pipe diameter, then Guo and Stolle [183] would estimate the interaction factor to be reduced by a multiplier of 0.92. Based on the study by Guo and Stolle [183], there is a significant increase in the lateral bearing interaction factor for smaller pipe diameter (e.g., <50 mm) where the interaction factor may be two times the corresponding factor for a 330 mm diameter pipe. However, for practical ranges of energy pipeline diameters, say 150–1200 mm, the expected variation in the lateral bearing factor, estimated by Guo and Stolle [183], would be from 12 to 10. This is a relatively minor variation in the interaction factor, due to pipe size effects, that is more consistent with the historical physical evidence.

The relationships established by Badv and Daryani [187] and Guo and Stolle [183] are based on numerical simulations. Finite element simulations based on conventional, local constitutive models cannot account for scale effects. Furthermore, mesh-based numerical simulations are influenced by size effects associated with the mesh topology (i.e., element type and mesh density) that may influence limit loads, kinematics, failure mechanisms, strain localization, and propagation of instabilities [210,211,290]. To address these issues in numerical simulations, several techniques have been developed that include simplified mathematical approaches (e.g., local perturbations or imperfections), tailored constitutive models (e.g., strength evolution with shear strain), and localization limiters based on enriched continuum models such as nonlocal models (e.g., averaging or smearing effects and fracture) and gradients of internal variables (e.g., strain gradient theory) [207,208,263,278–282,290–303]. For pipe/soil interaction events, there remains considerable uncertainty where further investigations are required to evaluate the effects of pipe size and scale for physical models and numerical simulation tools.

### 8.3.6 Soil Constitutive Models

*8.3.6.1* ***Overview*** Over the last decade, more complex and robust computational mechanics tools (e.g., finite difference, finite element, and mesh-free methods) have been used to solve problems in geomechanics and pipeline engineering. These numerical modeling procedures can provide a robust and cost-effective tool for conducting parameter studies to expand the knowledge base. However, calibration and verification of the simulation tools with a physical basis is paramount in order to establish confidence. A central requirement for these computational tools is the constitutive model defining the stress–strain behavior of the pipeline and soil.

In this section, the technical basis and fundamental characteristics of major constitutive models, used in numerical simulation of pipeline/soil interaction events, are examined. The focus here is to inform the reader and provide sufficient information for the selection of an appropriate soil constitutive model for application in engineering practice, when using these more advanced computational tools, with respect to estimating geotechnical loads and load effects on buried pipelines. Reference to classical studies, which formed the technical basis, and more recent investigations, which have advanced the state of practice, is presented. To support these goals, guidance on the engineering requirements for the determination of mechanical properties and constitutive model parameters based on soil index tests and laboratory investigations is provided.

In addition to solving laws of equilibrium (i.e., forces) and kinematics (i.e., compatibility), numerical simulations of soil mechanical behavior require constitutive models to quantify the relationship between stress and strain. A constitutive model defines the physical behavior of a continuous medium, in mathematical terms at a defined scale (i.e., microscopic or macroscopic), consistent with empirical data and physical laws. Identification of governing constitutive parameters forms a basis for mathematical characterization of these concepts defining material behavior. The goal is to implement realistic, robust yet efficient constitutive models within a numerical framework to evaluate load effects, deformation response, and governing mechanisms during pipe/soil interaction events.

For the practicing engineer, a balance is required among factors that include complexity of the soil constitutive model (i.e., number of independent parameters defining relationship between stress and strain), quantity and type of field or laboratory test required to establish these constitutive parameters, relative accuracy and computational efficiency of the implemented constitutive models to simulate realistic soil behavior, and expected sources of information characterizing soil conditions and properties along the pipeline route through field investigations during the design process.

A range of supporting fundamental knowledge in the field of soil mechanics and technical experience on pipe/soil interaction events are not addressed in this section as they are presented in comprehensive detail within this chapter and other authoritative references. In addition, the treatment and discussion of materials science, mechanics, closed-form solution, approximate methods (e.g., equilibrium stress field and limit load analysis), and soil behavior to load and deformation are examined in detail within other resources, including textbooks, conferences, journals, guidelines, standards, and engineering handbooks [8,44–46,48,92,95, 105,206,304–318].

*8.3.6.2 Elastic Parameters* Classical solutions for determining the stress and deformation response in soils have applied the theory of elasticity (e.g., Hooke's law) to solve engineering design problems for structural elements such as footings, retaining walls, foundations, and excavations [8,105,206]. The general form of the elastic constitutive relationship relating stress with strain is

$$\{\sigma_{ij}\} = [D]\{\varepsilon_{kl}\} \tag{8.22}$$

where $[D]$ is the total stress stiffness (constitutive) matrix. For an isotropic material, only two independent elastic constants are needed to define the constitutive matrix $[D]$, which becomes symmetric. In geotechnical engineering, the elastic shear modulus ($G$), which relates the change in shear stress with shear strain,

$$G = \frac{E}{2(1+\nu)} \tag{8.23}$$

and bulk modulus ($K$), which relates changes in mean stress ($\bar{p}$) with volumetric strain ($\varepsilon_V$),

$$K = \frac{d\bar{p}}{d\varepsilon_V} = \frac{E}{3(1-2\nu)} \tag{8.24}$$

are often used to define the soil elastic behavior.

The Poisson's ratio ($\nu$) can also be determined using

$$\nu = \frac{K_0}{1+K_0} \tag{8.25}$$

Typical values of Poisson's ratio include 0.5 for saturated soil with undrained loading conditions for total stress analysis, range of 0.2–0.4 for cohesive soil in drained loading conditions, 0.3–0.4 for dense cohesionless soil in drained loading conditions, and 0.1–0.3 for loose cohesionless soil in drained loading conditions [8]. There exists some evidence to consider the Poisson's ratio as isotropic and constant for a defined void ratio [319–321].

For loose sand and normally consolidated soils, the lateral coefficient of earth pressure at rest ($K_0$) can be defined as [8,322]

$$K_0 = \frac{\sigma'_h}{\sigma'_v} = 1 - \sin\phi' \tag{8.26}$$

where the effective stress friction angle, $\phi'$, is obtained from a triaxial test with the confining pressure equal to the *in situ* horizontal ground stress.

The coefficient of earth pressure at rest for dense sand and overconsolidated cohesive soils, the lateral coefficient of earth pressure at rest ($K_0$), can be defined as [8,323,324]

$$K_0 = (1 - \sin\phi')\mathrm{OCR}^{\sin\phi'} \tag{8.27}$$

where OCR is the overconsolidation ratio. Based on experimental results, an alternative relationship for the lateral earth pressure coefficient at rest ($K_0$) can be expressed as [162]

$$K_0 = 1 - \sin\phi' + 5.5\left(\frac{\gamma_d}{\min(\gamma_d)} - 1\right) \tag{8.28}$$

Unlike the constitutive behavior of other materials (e.g., metals), the linear isotropic model provides a poor characterization of realistic elastic soil constitutive behavior. The mechanical behavior of soils is dependent on a number of physical properties, including the soil type, composition, particle size, and relative distribution of particles. The material constants and correlation between stress and strain exhibit nonlinear behavior that has a complex relationship with density (void ratio), stress history, confining pressure, shear (deviator) stress, load path, volume change, and pore pressure [8,105,206]. For example, several studies have demonstrated the nonlinear response of geomaterials at low strain levels [8,325–327]. Consequently, the use of linear, elastic material constants, within elastic, isotropic constitutive models, does not provide adequate representation of realistic soil behavior.

The Cauchy and hypoelastic models can be used to represent the observed nonlinear elastic response of soils. From the pipeline engineering perspective, the deformation theory of plasticity (e.g., Ramberg–Osgood relationship) used to represent the deformation response of metals is a familiar example. These models, however, do not account for deformation history to evaluate the current stress state, rate of loading, and shear-induced dilation (i.e., volume change). Although a number of hyperelastic models have been proposed, the major limitation has been the difficulty in determining the constitutive model parameters [304].

Although geomaterials may exhibit anisotropic behavior on loading, there is evidence to support the use of homogenous, isotropic constitutive models for the nonlinear elastic response [321,328,329]. For hypoelastic models, the primary advantages of these model formulations include the mathematical simplicity, as related to numerical coding and limited number of constitutive parameters, and practicality of determining model parameters using standard geotechnical tests. In a hypoelastic model proposed by Janbu [330], the constrained soil elastic modulus can be defined as

$$E = K^0 p_a \left(\frac{\sigma_3}{p_a}\right)^n \quad (8.29)$$

that accounts for the effects of confining pressure. The constrained modulus coefficient, $K^0$, and exponent, $n$, can be derived through triaxial tests examining the logarithmic relationship between the normalized elastic modulus ($E/p_a$) and normalized confining pressure ($\sigma_3/p_a$). The slope represents the exponent, $n$, and the constrained modulus is determined as the elastic modulus where $\sigma_3/p_a = 1$. In the absence of triaxial test data, the parameters can be established through direct shear box tests and consolidation tests. The parameters may also be estimated based on soil type and *in situ* density with reference to available published data.

The constrained modulus should be determined from the loading–unloading curve rather than the initial loading response that may be influenced by nonlinear behavior [8,193]. Approximate value for the constrained modulus, $K^0$, is 2000–3000 for dense sand, 1000–2000 for medium sand, and 500–1000 for loose sand [8]. The exponent, $n$, can vary from 0.4 to 0.7 for noncohesive soil, with a typical value of 0.5, whereas an exponent of 1.0 is used for normally consolidated cohesive soil [8].

Although widely used, the Janbu model [330], as presented in Equation 8.29, is technically restricted to static load events with axisymmetric triaxial stress state for drained conditions. In addition, the conservation of energy principle is not satisfied for cyclic loading events and closed stress loops [331]. This latter issue has more significance for the cyclic loading of soils. A number of other hypoelastic models have also been proposed for geomaterials [321,332,333].

Lade and Nelson [321] proposed an isotropic hypoelastic relationship to predict the elastic modulus of noncohesive material that satisfies the principle of conservation of energy for closed stress loops or strain paths. The elastic modulus, $E$, was a function of the mean normal stress ($\bar{p}$) and deviatoric stress ($\sigma_{ij}$):

$$E = p_a M \left[\left(\frac{I_1}{p_a}\right)^2 + R\frac{J_2}{p_a^2}\right]^{\lambda} \quad (8.30)$$

The Poisson's ratio for this model is assumed to be constant. Examining the logarithmic relationship between the normalized elastic modulus ($E/p_a$) and stress invariants $[(I_1/p_a)^2 + R(J_2/p_a^2)]$, the modulus parameter, $M$, and exponent, $\lambda$, can be derived. The slope represents the exponent, $\lambda$, and the stress invariant modulus, $M$, can be determined as the value of $(E/p_a)$ at $[(I_1/p_a)^2 + R(J_2/p_a^2)] = 1$. Typical values for the stress invariant modulus, $M$, range from 400 to 1200 and exponent, $\lambda$, from 0.2 to 0.4 [334].

For cohesive soils, knowing the undrained shear strength ($s_u$), overconsolidation ratio (OCR), and plasticity index (PI), the shear modulus, $G_{50}$, can be approximated using the rigidity index ($I_R$) [251,252,335]

$$I_R = \frac{e^{(137-\mathrm{PI})/23}}{\left[1 + \ln(1 + (\mathrm{OCR} - 1)/26)^{3.2}\right]^{0.8}} \quad (8.31)$$

The shear modulus at 50% of maximum shear strength, $G_{50}$, can be estimated as

$$G_{50} = I_R s_u \quad (8.32)$$

The undrained rigidity index ($I_R$) represents the ratio of the shear modulus to shear strength ratio that may be interpreted from triaxial stress–strain curve, pressure meter tests, and empirical relationships.

The undrained elastic modulus for cohesive soils may also be defined in terms of the plasticity index (PI), overconsolidation ratio (OCR), and undrained shear strength ($s_u$), as shown in Figure 8.17 [8,305].

$$K_c = \frac{E_u}{s_u} \quad (8.33)$$

where the modulus coefficient ($K_c$) provides an empirical correlation between the undrained stiffness and undrained strength for cohesive soils [305].

As discussed in this section, the initial elastic, nonlinear response is correlated with the confining pressure and the constitutive behavior may be described by a nonlinear, hypoelastic relationship. The stress and deformations are assumed to be fully recoverable. Guidance on methods to establish the constitutive parameters has been provided. Increasing load or deformation results, however, in irrecoverable behavior where the peak strength is proportional to confining pressure. Under these loading conditions, the stress state within the geomaterial may exceed a specified failure criterion (i.e., yielding) where the constitutive behavior can be characterized using plasticity models.

#### 8.3.6.3 *Plasticity Models*

*Overview* The primary elements of plasticity models include the definition of a yield function, plastic potential function or flow rule, and hardening or softening rules. Other primary considerations for classical plasticity theory include the yield stress being independent of hydrostatic pressure (i.e.,

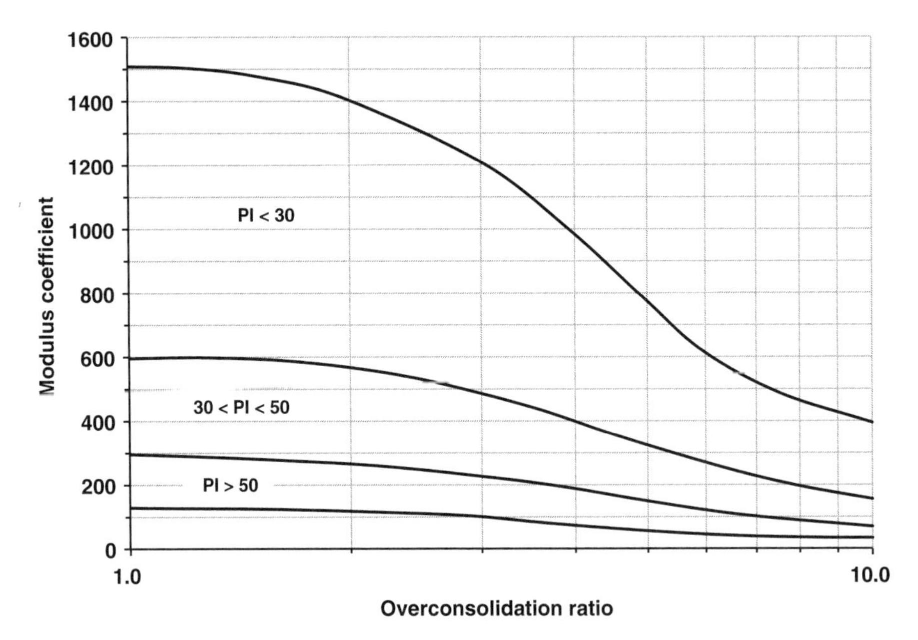


**FIGURE 8.17** Correlation factor to estimate the undrained elastic modulus for cohesive soils.

mean stress), incompressible behavior (i.e., no volume change), material response does not exhibit different yield strengths as a function of loading sense (i.e., Bauschinger effect), and plastic response is not influenced by the loading rate. These concepts have been integrated within constitutive models (e.g., von Mises and Tresca) that have been primarily developed for ductile metals where significant experience and literature are available [306,307].

If the material behavior is assumed to be isotropic, then the yield function can be described as

$$F(\sigma_1, \sigma_2, \sigma_3) = F(I_1, J_1, J_2) = k \qquad (8.34)$$

with respect to the principal stress ($\sigma_1$, $\sigma_2$, and $\sigma_3$) or stress invariants ($I_1$, $J_2$, and $J_3$). The general characteristics of the classical von Mises and Tresca plasticity models, including the yield function, flow rule, and hardening rules, are illustrated in Figure 8.18 and discussed in the following paragraphs and subsections.

The yield function, defined in terms of stress invariants or components, characterizes the stress condition and state parameters (e.g., perfect plasticity, strain, or work hardening response) for plastic material behavior that is associated with the intersection of the stress path with the yield surface (Figure 8.18a).

The flow rule is a function defining the post-yield incremental plastic strain direction vector for a defined stress state. For most problems in metal plasticity, the plastic potential function is assumed to be the same as the yield function (i.e., the yield and plastic potential function surfaces coincide). This is known as an associated flow rule where the plastic strain increment vector is normal to the yield surface and the plastic strain is along the same direction as yielding, which is also known as the normality condition (Figure 8.18b). The plastic potential function governs dilatancy where most geomaterials use a nonassociated flow rule to control excessive plastic, shear-induced volume change [8,33,206]. For nonassociated flow, the plastic strain increment is normal to the potential surface that is no longer normal to the yield surface and does not coincide with direction of yielding.

The hardening or softening rule characterizes the evolution of the yield surface by relating the state parameters, as defined in the yield function, with the plastic strain increment and quantifying the scalar multiplier in the plastic potential function, which defines the magnitude of the plastic strain components. If the yield surface remains fixed in space (i.e., perfect plasticity), the state parameters are constant. The hardening or softening rules may be related to accumulated plastic strain or work done. Uniform expansion of the yield surface is defined by isotropic rules, and translation of the yield surface is defined by kinematic rules (Figure 8.18b). Mixed hardening rules integrate both characteristics.

For geomaterials, early studies focused on the extension of these elastic, perfectly plastic models, which were developed for ductile metals, to solve nonlinear plastic deformation problems. A comprehensive discussion on the stress–strain behavior of geomaterials, for drained and undrained loading conditions, and the estimation of soil strength properties through laboratory tests and empirical correlations with field measurements and site characterization is provided in textbooks and handbooks [8,33,92,95,206]. The mechanical behavior of soils (e.g., peak strength values, strain hardening and strain softening, and dilation) and the relationship with

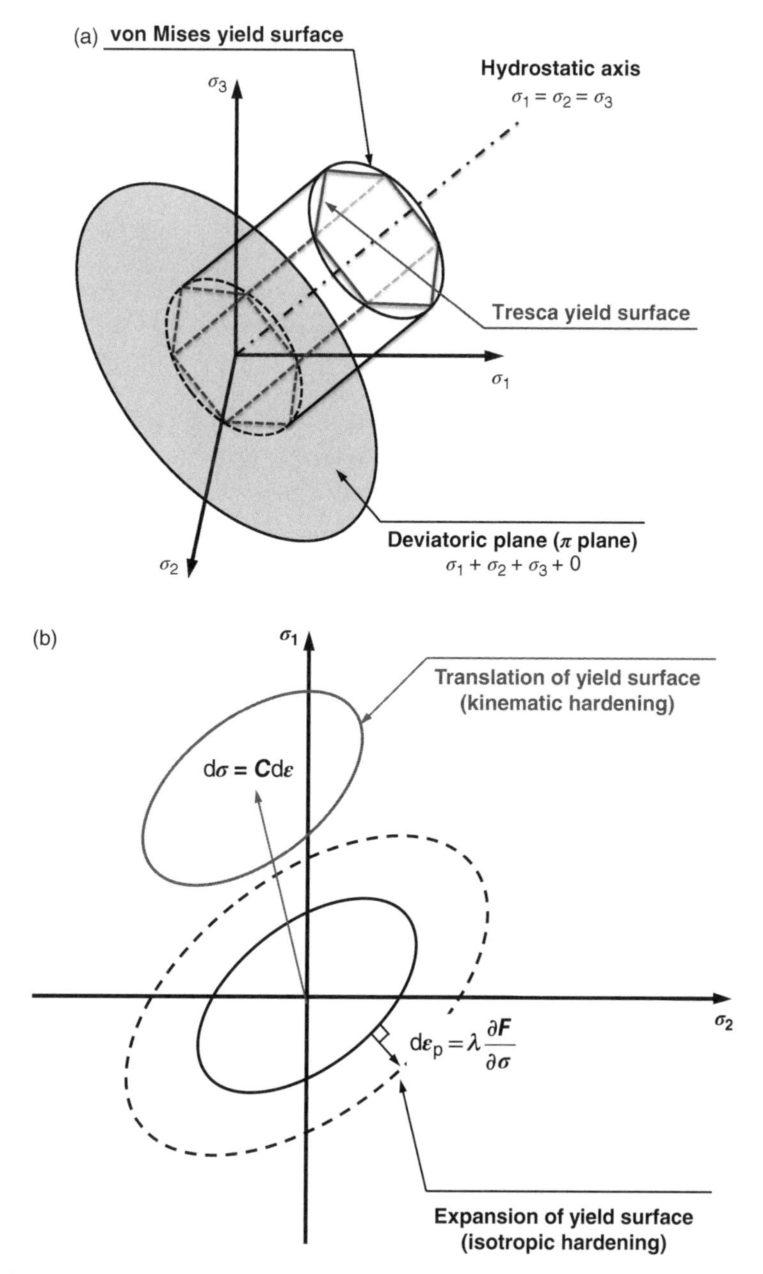


**FIGURE 8.18** (a) Yield surfaces for von Mises and Tresca plasticity models in three-dimensional principal stress space and (b) von Mises yield surface in two-dimensional principal stress space.

other parameters (e.g., mean stress, deviatoric stress, void ratio, pore pressure, and soil state) are also presented.

Based on this review, it can be realized that the relatively simple elastic, perfectly plastic constitutive models developed for ductile metals do not adequately address many elements of realistic soil behavior [8,33,105,206,308–310]. Some of the key characteristics of soil mechanics not addressed by these classical plasticity models include soil strength that is dependent on pressure, soil yield accompanied by volume change (i.e., dilation and compaction) that is pressure dependent, shear induced dilative behavior that may have associated or nonassociated flow response, strain hardening and softening behavior dependent on initial soil state and pressure, different strengths in compression and tension (i.e., Bauschinger-type effect), and effects of pore pressure (i.e., multiphase behavior).

The classical plasticity (i.e., strength) models can be used to solve a class of problems in soil mechanics, provided the underlying idealizations and limitations are realized. These models can be improved and enhanced to address these shortcomings to provide a more representative characterization of realistic soil behavior. The development of critical state plasticity models has provided significant advancements for the application of plasticity theory in geomechanics using numerical methods. The critical state concept accounts for post-yield soil behavior to exhibit shear-induced deformation at constant effective stress state and void ratio [8,33,105,206,310]. These issues will be further explored in the following subsections with guidance on requirements to establish the constitutive model parameters, general implementation with numerical simulation procedures, and application for practical pipeline engineering problems.

*Classical Plasticity Models* The classical strength models include constitutive relationships based on the von Mises, Drucker–Prager, Mohr–Coulomb, and Tresca yield criteria. Although these plasticity models consider a linear, isotropic elastic response with irrecoverable plastic deformation, the nonlinear elastic soil models presented in Section 8.3.6.2 can be incorporated. These plasticity models are developed for monotonic loading conditions without rate dependence and isotropic hardening. Early application of these elastic, perfectly plastic response models was to predict limit loads and strength capacity. In the early development of numerical procedures, however, there was preference for the von Mises and Drucker–Prager yield functions due to the smoothness of the yield surface that addressed problems with uniqueness, associated with the partial differential of the yield and plastic potential functions, at corners of the Tresca and Mohr–Coulomb yield surfaces [33,105,206]. Algorithms have since been developed to address this issue of uniqueness and problems with numerical convergence.

In soil mechanics, there is general preference for the Tresca and Mohr–Coulomb models due to assumptions that are more consistent with geotechnical engineering practice [8,33,105,206]. For example, the Tresca formulation defines a critical shear stress related to principal stress, whereas the von Mises yield criterion is based on critical distortional energy. Through consideration of different strength parameters for compressive and tensile loading conditions, the Drucker–Prager and Mohr–Coulomb yield functions are extensions of the von Mises and Tresca yield criteria, respectively. For cohesive soils, the Mohr–Coulomb criterion reduces to the Tresca yield function where the irregular hexagonal cone failure surface evolves into a regular hexagonal failure surface in principal stress space. The Tresca and von Mises models are within a total stress analysis framework, whereas the Mohr–Coulomb and Drucker–Prager models use effective stress parameters.

In plasticity theory, constitutive models are generally defined by stress and strain invariants, which are quantities of the stress and tensor independent (invariant) of the coordinate system. For geomaterials, common set of stress invariants includes the mean stress ($p$), deviatoric stress ($q$), and Lode angle or deviatoric polar angle ($\theta$), which are defined in the following expressions [33,206]:

$$p = \frac{1}{3}\sigma_{ii} = \frac{1}{3}(\sigma_{11} + \sigma_{22} + \sigma_{33}) = \frac{1}{3}(\sigma_1 + \sigma_2 + \sigma_3) = \frac{1}{3}I_1 \quad (8.35)$$

$$q = \sqrt{\frac{3}{2}J_2} = \sqrt{\frac{3}{2}s_{ij}s_{ij}} \quad (8.36)$$

$$\cos(3\theta) = \frac{3\sqrt{3}}{2}\frac{J_3}{J_2^{3/2}} \quad (8.37)$$

where $I_1$ is the first principal invariant of stress. The deviatoric stress tensor can also be defined as

$$s_{ij} = \sigma_{ij} - p\delta_{ij} \quad (8.38)$$

where $\delta_{ij}$ is the Kronecker delta. The deviatoric stress invariants are defined as

$$J_2 = \frac{1}{2}s_{ij}s_{ij} = \frac{1}{6}\left[(\sigma_1 - \sigma_2)^2 + (\sigma_2 - \sigma_3)^2 + (\sigma_3 - \sigma_1)^2\right] \quad (8.39)$$

$$J_3 = \frac{1}{3}s_{ij}s_{jk}s_{ki} \quad (8.40)$$

The Lode angle may vary between $\pm\pi/6$ ($\pm 30°$) and can be used in the plasticity model rather than the third invariant of deviatoric stress ($J_3$). The principal stresses can be calculated from the stress invariants:

$$\begin{Bmatrix} \sigma_1 \\ \sigma_2 \\ \sigma_3 \end{Bmatrix} = \begin{Bmatrix} p \\ p \\ p \end{Bmatrix} + \frac{2\sqrt{J_2}}{\sqrt{3}} \begin{Bmatrix} \cos(\theta) \\ \cos\left(\theta - \frac{2\pi}{3}\right) \\ \cos\left(\theta + \frac{2\pi}{3}\right) \end{Bmatrix} \quad (8.41)$$

The Tresca and von Mises constitutive models are best suited for the analysis of undrained loading conditions within total stress analysis. In the deviatoric plane ($\pi$-plane), a circle with a radius $\rho = \sqrt{2J_2}$ defines the von Mises yield surface, whereas the Tresca yield surface is represented by a regular hexagon (Figure 8.19).

For the Tresca model, the yield function is defined as

$$F(\sigma, k) = \sigma_1 - \sigma_3 - 2s_u = 0 \quad (8.42)$$

where $s_u$ can be determined through a conventional unconfined compression test for cohesive soils. The Tresca yield function can be redefined in terms of stress invariants:

$$F(\sigma, k) = J_2 \cos\theta - s_u \quad (8.43)$$

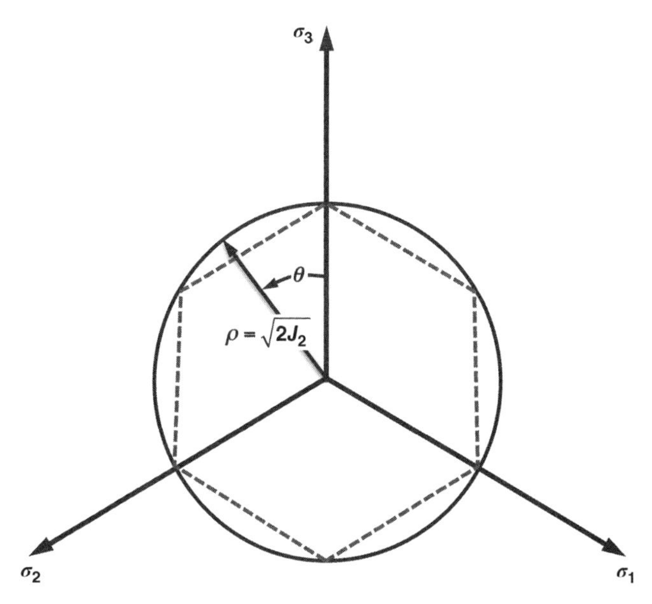


**FIGURE 8.19** Yield surfaces for the von Mises and Tresca plasticity models in the deviatoric plane ($\pi$-plane).

The von Mises yield criterion is

$$F(\sigma, k) = J_2 - \alpha = 0 \tag{8.44}$$

where $\alpha$ is material parameter representative of the soil shear strength defined as

$$\alpha = \frac{s_u}{\cos\theta} \tag{8.45}$$

A rationale must be established to relate the material parameter, $\alpha$, used in the von Mises yield criterion with the soil undrained shear strength ($s_u$). In the deviatoric stress plane, the von Mises circular failure surface will circumscribe the Tresca hexagonal failure surface when the Lode angle ($\theta$) is $\pm 30°$, which is $\alpha = 1.155 s_u$, as shown in Figure 8.19. The material parameter is $\alpha = 1.0 s_u$ when the von Mises circular failure surface inscribes the Tresca hexagonal failure surface (i.e., $\theta = 0°$).

The Mohr–Coulomb failure criterion defines a maximum shear stress as

$$\tau = c' + \sigma'_n \tan\phi' \tag{8.46}$$

that is a function of the soil material parameters for cohesion ($c'$) and shearing resistance due to internal angle of friction ($\phi'$). The Mohr–Coulomb model becomes equivalent to the Tresca model for frictionless materials where the representation in the deviatoric plane is shown in Figure 8.20. The yield function can be defined in terms of effective stress parameters [33,206]:

$$F(\sigma', k) = \sigma'_1 \frac{1 - \sin\phi'}{2c' \cos\phi'} - \sigma'_3 \frac{1 + \sin\phi'}{2c' \cos\phi'} - 1 = 0 \tag{8.47}$$

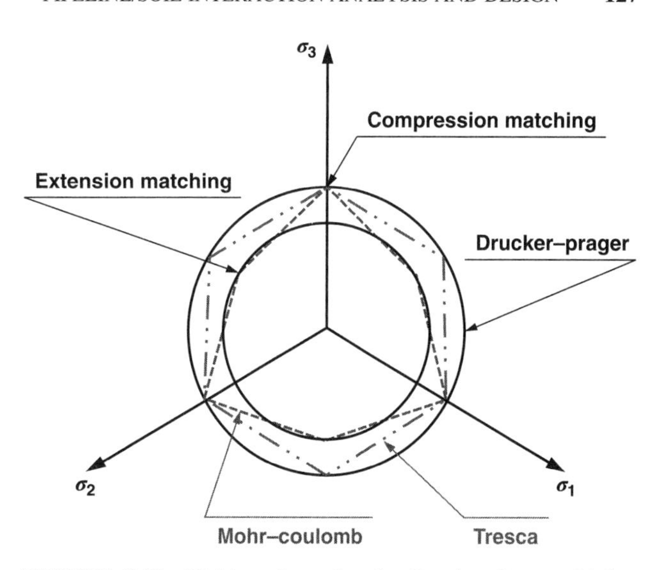


**FIGURE 8.20** Yield surfaces for the Drucker–Prager, Mohr–Coulomb, and Tresca plasticity models in the deviatoric plane ($\pi$-plane).

and can be redefined using stress invariants as

$$F(\sigma', k) = J_2 - \left(\frac{c'}{\tan\phi'} + p\right) g(\theta) = 0 \tag{8.48}$$

where the material parameter, $g(\theta)$, is related to the shearing resistance ($\phi'$):

$$g(\theta) = \frac{\sin\phi'}{\cos\theta + (\sin\theta \sin\phi')/\sqrt{3}} \tag{8.49}$$

for associated flow rule. The soil state parameters $k = f(c', \phi')$ are assumed constant and independent of plastic strain or plastic work. For a nonassociated flow rule, the material parameter, $g(\theta)$, is

$$g(\theta) = \frac{\sin\psi'}{\cos\theta + (\sin\theta \sin\psi')/\sqrt{3}} \tag{8.50}$$

Thus, the Mohr–Coulomb model has five state parameters that include the elastic modulus ($E$), Poisson's ratio ($\nu$), effective cohesion ($c'$), effective friction angle ($\phi'$), and effective dilation angle ($\psi'$). These soil state parameters and hardening response can be estimated from the results of triaxial cell tests, with multiple confining pressures of interest subject to compressive and tensile loading, and analyzed in the meridional ($q$–$p$) plane. Further detailed discussion on the determination of soil strength parameters as state variables in plasticity constitutive models is presented in Refs [8,105,206,310].

For the Tresca and Mohr–Coulomb yield surfaces shown in Figure 8.20, the flow direction can change significantly when the stress state has two equal principal stress values, which occurs at the failure surface apex. The classical Mohr–Coulomb model with an associated flow rule results in the predicted volumetric strains exceeding realistic soil behavior with no control or limit on soil dilation [206]. Although soils will dilate at low mean pressures, with increasing deformation the behavior will reach a critical state with constant volume conditions. As discussed in the next section, strategies to overcome these deficiencies include using nonassociated flow rule and defining the plastic material parameters ($c', \phi', \psi'$) to be dependent on the deviatoric plastic strain.

The Drucker–Prager model can be viewed as an extended version of the von Mises yield function where the yield function is [206]

$$F(\sigma', k) = J_2 - \left(\frac{c'}{\tan\phi'} + p\right) g(\theta) = 0 \qquad (8.51)$$

where the material parameter, $g(\theta)$, is related to the shearing resistance ($\phi'$):

$$g(\theta) = \frac{\sin\phi'}{\cos\theta + (\sin\theta \sin\phi')/\sqrt{3}} \qquad (8.52)$$

In the deviatoric stress plane, the Drucker–Prager circular failure surface will circumscribe the Mohr–Coulomb failure surface when the Lode angle ($\theta$) is 30° for triaxial extension and −30° for triaxial compression. If the Drucker–Prager circular failure surface inscribes the Mohr–Coulomb failure surface, then the Lode angle ($\theta$) is replaced by the inscribed Lode angle ($\theta_{ins}$) [206]:

$$\theta_{ins} = \tan^{-1}\left(\frac{\sin\phi'}{\sqrt{3}}\right) \qquad (8.53)$$

For nondilatant soil behavior with nonassociated flow rule, the Mohr–Coulomb parameters can be related to the Drucker–Prager material parameters for internal friction using the following relationship for plane strain conditions:

$$\tan\beta = \sqrt{3}\sin\phi' \qquad (8.54)$$

where $\beta$ is the Drucker–Prager friction angle and the Drucker–Prager cohesion, $d'$, is defined as

$$d' = c'\sqrt{3}\cos\phi' \qquad (8.55)$$

The Tresca, with total stress parameters, and Mohr–Coulomb, with total stress parameters, plasticity models assume that the yield function and strength behavior are independent of the intermediate principal stress ($\sigma_2$). In these models, the soil strength is related to differences in the major and minor principal stress states. The von Mises and Drucker–Prager models, however, consider the yield function and strength behavior to be dependent on the intermediate principal stress ($\sigma_2$).

*Advancement of Classical Plasticity Models for Application in Soil Mechanics* Over the past few decades, the classical plasticity models have been modified and enhanced to improve the capabilities in numerical simulation tools. The major technical areas include refinement of the yield function surface to improve numerical convergence, and the correlation of strength parameters with plastic strain magnitudes to characterize for strain hardening and softening behavior [8,33,105,206,310–312].

For example, to improve numerical convergence the tension cutoff apex of the Mohr–Coulomb failure surface (Figure 8.20) can be characterized by an eccentricity parameter to create a radius of curvature for the flow potential in the meridional plane ($p$–$q$ space) [313]. However, the accuracy and confidence in simulations may be questioned particularly when flow localization may be important. Other strategies may include defining an artificial cohesion to mitigate convergence issues due to tension failure.

The use of field variables and user subroutines in numerical simulation of large deformation problems in soil mechanics allows for the characterization of state variables (e.g., soil strength parameters) to vary (i.e., softening or hardening rules) as a function of the plastic strain or work done [195,207,208,211,263,264,336]. The procedures can be established and calibrated using triaxial cell and direct shear box tests with verification against physical models. A series of numerical simulations are performed to correlate the material state parameters (e.g., $c', \phi', \psi'$) with engineering variables characterizing plastic behavior (e.g., plastic strain and plastic work done).

For large deformation problems in soil mechanics, the formation of shear bands or strain localizations has been investigated in theoretical studies and numerical simulations and observed in experimental investigations [207,209–211, 263,314]. The occurrence of discontinuous deformation mode within a continuum mechanics framework presents technical challenges for the numerical analyst. The strain localization phenomenon (e.g., width of the shear band) is associated with a length scale that is dependent on soil grain size, confining pressure, and numerical solution discretization scheme (i.e., grid or mesh dimensions). Recent studies have presented a methodology to account for the mesh size effect on the propagation of bifurcation modes and shear bands by scaling the residual, or critical, plastic shear strain with element size [209–211].

***8.3.6.4 Critical State Models*** The enhancement of classical plasticity models to limit volumetric plastic strain and control shear-induced dilation was through the definition of a cap hardening surface that accounts for hydrostatic compression [8,33,206,310,315]. These improvements were identified through the development of a critical state soil mechanics (CSSM) conceptual framework [316–318]. The conceptual framework describes that the behavior of geomaterials subject to deformation will tend toward a well-defined critical state where shear deformations occur without any further change in the mean effective stress, deviator stress, and void ratio. CSSM addresses the importance of changes in the soil volume and effective stress state in the understanding of soil behavior subject to load and deformation [310,318]. CSSM provides the rational framework for characterizing the effects of volume change and effective stress history, and linkage with the development of numerical modeling procedures simulating realistic soil mechanical behavior. The CSSM framework explains why a soil under drained shear when normally consolidated will compress and exhibit strain hardening behavior; whereas the same overconsolidated soil will be stiffer, dilate, and exhibit strain softening behavior. The original critical state models were based on isotropic elastic response with plastic behavior that accounted for hardening and softening flow rules, soil consolidation, and shear deformation. The classical critical state models (e.g., Cam clay and modified Cam clay) were formulated for triaxial space and require additional treatment for use in general numerical analysis [206]. The critical state models define five main state variables that include elastic, strength, and consolidation parameters. Guidance on the determination of these parameters and calibration of the constitutive model is presented in several studies [8,33,206,310].

### 8.3.7 Advancing the State of Art into Engineering Practice through an Integrated Technology Framework

The current state of practice for estimating geotechnical load effects on buried pipelines was founded on an idealized approach to define the soil mechanical behavior as a series of discrete, independent soil springs [43,45,46]. The soil spring behavior (i.e., load–displacement response) is defined using simple empirical expressions that are based on a mixed stress state (i.e., total stress and effective stress) formulation. This structural modeling approach has served industry needs and has provided a practical and generally effective analysis tool for specific pipeline engineering design applications (e.g., stress based or small displacement and small strain loading events).

There exists technical uncertainty in this idealized approach that may not be sufficient to address pipeline/soil interaction events, load transfer processes, soil failure mechanisms, and localization effects (e.g., shear banding, interface behavior, and contact) for a class of engineering problems. Some of the major issues have been discussed in previous sections. This discussion on uncertainty in the Winkler-type soil model, however, is not complete. For example, the importance of multiphase soil behavior (i.e., soil and water), pore pressure, stratified soil layers, cyclic and transient load events (e.g., pipe start-up and shut-in cycles, environmental wetting and desiccation cycles, seismic events, and liquefaction response) in soil mechanical behavior and the required modeling techniques to best represent this soil response have not been thoroughly addressed in this chapter [4–9,12–14,33].

There is a need to refine these structure-based engineering tools and advance current state of art into an improved, more robust yet practical state of practice. Advancing knowledge, reducing uncertainty, and establishing reliability in predictable outcomes will provide a solid basis for the development of practical, safe, and cost-effective design solutions for buried energy pipeline systems. Successful outcomes have been achieved through an integrated technology framework that includes elements of laboratory testing, physical modeling, and numerical simulation, which have been discussed at various levels of detail throughout this chapter [3,14,33,38,41,79,82,83,113,138, 154–158,160,179,211,222,223,231,232,248,249,253,264, 336–342]. Laboratory tests are fundamental to establish basic soil properties and state variables that characterize the evolution of strength mobilization with deformation. This knowledge can be used to develop numerical algorithms to enhance existing or build new constitutive models for use in advanced numerical simulation tools (Section 8.3.6). Thoughtful integration of physical modeling studies (Section 8.3.2) and full-scale observations will provide the objective evidence to support the calibration and verification of the computational models, which is paramount to establish confidence in the simulation tools. This approach will provide the technical basis to reaffirm and delineate the performance envelope for existing state of practice, establish and integrate new knowledge and technology in support of engineering design, and promote the advancement of engineering practice through verification, refinement, and incorporation of the state of art.

## NOMENCLATURE

| | |
|---|---|
| $D$ | outside pipe diameter (m) |
| $E$ | elastic modulus (MPa) |
| $f$ | interface friction factor accounting for the pipe coating material at the pipe/soil interface |
| $f_s^\alpha$ | skin friction per unit area based on total stress (kPa) |
| $f_s^\beta$ | skin friction per unit area based on effective stress (kPa) |
| $G$ | shear modulus (MPa) |
| $H_s$ | burial depth to pipe springline (m) |
| $I_1$ | first stress invariant or effective stress tensor (N/m$^2$) |
| $J_2$ | second invariant of the deviatoric stress tensor (N/m$^2$) |

$K$ bulk modulus (MPa)
$K_0$ lateral earth pressure coefficient at rest
$K_p$ passive earth pressure coefficient
$M$ dimensionless modulus in the Lade–Nelson hypoelastic model
$N_{ch}$ transverse lateral bearing capacity factor for cohesive soil
$N_{qh}$ transverse lateral bearing capacity factor for noncohesive soil
$P_a$ atmospheric pressure (kPa)
$p_u$ ultimate transverse lateral soil load per unit pipe segment length (kN/m)
$q_{ud}$ ultimate transverse vertical downward soil load per unit pipe segment length (kN/m)
$q_{uu}$ ultimate transverse vertical upward soil load per unit pipe segment length (kN/m)
$R$ material constant, $6(1+\nu)/(1-2\nu) = 9K/G$
$S_u$ undrained soil shear strength (kPa)
$t_u$ ultimate axial soil load per unit pipe segment length (kN/m)
$x$ axial soil displacement (m)
$y$ lateral soil displacement (m)
$z$ vertical upward and downward soil displacement (m)
$\alpha$ adhesion factor
$\delta$ pipe/soil interface friction angle, $\delta = f\phi'$ (°)
$\phi'$ effective internal friction angle of the soil (°)
$\gamma_d$ dry unit weight of the soil (kN/m$^3$)
$\gamma'$ effective unit weight of the soil (kN/m$^3$)
$\lambda$ dimensionless exponent in the Lade–Nelson hypoelastic model
$\nu$ Poisson's ratio

## ACKNOWLEDGMENTS

The authors would like to recognize the contributions from several colleagues, including Lynden Penner (JD Mollard and Associates), Ryan Phillips (C-CORE), Kenton Pike (Ph.D. candidate at Memorial University of Newfoundland), and Moness Rizkalla (Via Plus), on their insightful discussions and thoughtful comments.

## REFERENCES

1. Canadian Energy Pipeline Association (CEPA) (2014) Available at http://www.cepa.com.
2. Pipeline and Hazardous Materials Safety Administration (PHMSA) (2014) Available at http://phmsa.dot.gov.
3. Kenny, S., Barrett, J., Phillips, R., and Popescu, R. (2007) Integrating geohazard demand and structural capacity modelling within a probabilistic design framework for offshore arctic pipelines. Proceedings of the International Offshore and Polar Engineering Conference, ISOPE2007-SBD-03, 9 pp.
4. Grivas, D.A., Schultz, B.C., O'Neil, G.D., and Simmonds, G.R. (1996) Phenomenological models to predict rainfall-induced movements. Proceedings of the International Conference on Offshore Mechanics and Arctic Engineering (OMAE), Vol. 5, pp. 355–362.
5. Bughi, S., Aleotti, P., Bruschi, R., Andrei, G., Milani, G., Scarpelli, G., and Sakellariadi, E. (1996) Slow movements of slopes interfacing with pipelines: modelling and monitoring. Proceedings of the International Conference on Offshore Mechanics and Arctic Engineering (OMAE), Vol. 5, pp. 363–372.
6. Vulliet, L. (1999) Modelling creeping slopes. *Rivista Italiana di Geotecnica*, 23(1), 71–76.
7. Leroueil, S. (2001) Natural slopes and cuts: movement and failure mechanisms. *Géotechnique*, 51(3), 197–243.
8. Rowe, K. (editor) (2001) *Geotechnical and Geoenvironmental Engineering Handbook*, Kluwer Academic Publishers, 1088 pp.
9. Calvello, M. and Cascini, L. (2006) Predicting rainfall-induced movements of slides in stiff clays. Proceedings of Geohazards, 8 pp.
10. Rizkalla, M. (editor) (2008) *Pipeline Geo-Environmental Design and Geohazard Management*, American Society of Mechanical Engineers, 352 pp.
11. Baum, R.L., Galloway, D.L., and Harp, E.L. (2008) Landslide and Land Subsidence Hazards to Pipelines. U.S. Geological Survey Open-File Report 2008-1164, 192 pp.
12. Honegger, D.G., Hart, J.D., Phillips, R., Popelar, C., and Gailing, R.W. (2010) Recent PRCI guidelines for pipelines exposed to landslide and ground subsidence hazards. Proceedings of the International Pipeline Conference, IPC-31311, 10 pp.
13. Cascini, L., Calvello, M., and Grimaldi, G.M. (2010) Groundwater modeling for the analysis of active slow-moving landslides. *Journal of Geotechnical and Geoenvironmental Engineering*, 136(9), 1220–1230.
14. Bruschi, R., Bughi, S., and Spinazzè, M. (2010) The role of FEM in the operation of pipelines on unstable slopes. *Journal of Pipeline Engineering*, 9(3), 167–177.
15. Mitchell, T., Hitchcock, C., and Amine, D. (2010) Surface, sub-surface mapping, geohazard identification and associated risk mitigation for pipelines. Proceedings of the International Pipeline Conference, IPC-31338, 7 pp.
16. McElroy, R.A. (2010) Using geospatial solutions to meet distribution integrity management requirements. Proceedings of the International Pipeline Conference, IPC-31147, 6 pp.
17. Ripley, N., Simpson, N., and Leir, M. (2010) Lessons learned from supporting a geohazard management program. Proceedings of the International Pipeline Conference, IPC-31339.
18. EPFL (2011) SafeLand—Living with Landslide Risk in Europe: Assessment, Effects of Global Change, and Risk Management Strategies. 7th Framework Programme, Cooperation Theme 6 Environment (including climate change), Sub-Activity 6.1.3 Natural Hazards, EC FP7, Project No. 226479, 73 pp.

19. Rizkalla, M. (2012) A global perspective on pipeline geohazards. *Pipelines International*, 14, 42.
20. William, B. (1972) Geotechnical investigations for arctic gas pipelines. Proceedings of the Society of Petroleum Engineers, SPE-4008, 6 pp.
21. Mathews, A.C. (1977) Natural gas pipeline design and construction in permafrost and discontinuous permafrost. Proceedings of the Society of Petroleum Engineers, SPE-6873, 8 pp.
22. Nixon, J., Stuchly, J., and Pick, A. (1984) Design of Norman Wells pipeline for frost heave and thaw settlement. Proceedings of the International Conference on Offshore Mechanics and Arctic Engineering (OMAE), pp. 69–76.
23. Williams, P.J. (1989) *Pipelines and Permafrost*, Carleton University Press, 129 pp.
24. Nixon, J.F. and Burgess, M. (1999) Norman Wells pipeline settlement and uplift movements. *Canadian Geotechnical Journal*, 36(1), 119–135.
25. Oswell, J.M. and Hanna, A.J. (1997) Aspects of geotechnical engineering in permafrost regions. In: Tan, C.K. (editor), *Innovative Design and Construction*, Geotechnical Special Publication 73, American Society of Civil Engineering, pp. 32–50.
26. Kenny, S., Phillips, R., McKenna, R.F., and Clark, J.I. (2000) Response of buried arctic marine pipelines to ice gouge events. Proceedings of the International Conference on Offshore Mechanics and Arctic Engineering (OMAE), OMAE00-5001, 8 pp.
27. Oswell, J.M. (2002) Geotechnical aspects of northern pipeline design and construction. Proceedings of the International Pipeline Conference, IPC02-27327.
28. Oswell, J.M., O'Hashi, K., and Hirata, M. (2003) Some engineering challenges to design, construction and operation of export oil and gas from Russia to the Far East. Proceedings of the World Gas Congress, Working Group 9.
29. Doblanko, R.M., Oswell, J.M., and Hanna, A.J. (2002) Right-of-way and pipeline monitoring in permafrost: the Norman Wells Pipeline experience. Proceedings of the International Pipeline Conference, IPC-27357, 10 pp.
30. Kenny, S., Phillips, R., Clark, J., and Nobahar, A. (2005) PRISE numerical studies on subgouge deformations and pipeline/soil interaction analysis. Proceedings of the Conference on Ports and Ocean Engineering Under Arctic Conditions (POAC), 10 pp.
31. Phillips, R., Clark, J., and Kenny, S. (2005) PRISE studies on gouge forces and subgouge deformations. Proceedings of the Conference on Ports and Ocean Engineering Under Arctic Conditions (POAC), 10 pp.
32. Lanan, G., Cowin, T.G., and Johnston, D.K. (2011) Alaskan Beaufort Sea pipeline design, installation and operation. Proceedings of the Offshore Technology Conference, OTC-22110, 9 pp.
33. McCarron, W.O. (editor) (2011) *Deepwater Foundations and Pipeline Geomechanics*, J. Ross Publishing, 336 pp.
34. Barrette, P. (2011) Offshore pipeline protection against seabed gouging by ice: an overview. *Cold Regions Science and Technology*, 69(1), 3–20.
35. Barrette, P. and Sudom, D. (2012) Physical simulations of seabed scouring by ice: review and database. Proceedings of the International Offshore and Polar Engineering Conference (ISOPE), pp. 381–388.
36. Palmer, A. and Croasdale, K. (2012) *Arctic Offshore Engineering*, World Scientific, 357 pp.
37. Kenny, S. (2012) Working through permafrost: pipelines in arctic terrain. *Pipelines International*, 14, 44–45.
38. Panico, M., Lele, S.P. Hamilton, J.M., Arslan, H., and Cheng, W. (2012) Advanced ice gouging continuum models: comparison with centrifuge test results. Proceedings of the International Offshore and Polar Engineering Conference (ISOPE), 7 pp.
39. Paulin, M. (2013) Arctic offshore pipeline design and installation challenges. Proceedings of the Society of Petroleum Engineers, SPE-166881, 12 pp.
40. C-CORE (2013) 3rd Ice Scour and Arctic Marine Pipelines (ISAMP) Workshop.
41. Lele, S.P., Hamilton, J.M., Panico, M., and Arslan, H. (2013) Advanced continuum modeling to determine pipeline strain demand due to ice-gouging. *Journal of International Society of Offshore and Polar Engineers*, 23(1), 22–28.
42. Kenny, S. (2014) Polar and arctic technology. In: Charlton, J. Jukes, P., and Sang, C.Y. (editors), *Encyclopedia of Marine and Offshore Engineering*, World Press, 6 Vols.
43. NEN (1991) Eisen voor stalen transportleidingsystemen (Requirements for steel pipeline transportation systems). Publikatie Uitsluitend ter Kritiek, Nederlandse Norm NEN 3650.
44. O'Rourke, M.J. and Liu, X. (1999) Response of Buried Pipelines Subject to Earthquake Effects, Multidisciplinary Center for Earthquake Engineering Research, Buffalo, NY, 249 pp.
45. PRCI (2004) Guidelines for the Seismic Design and Assessment of Natural Gas and Liquid Hydrocarbon Pipelines, Pipeline Research Council International, 196 pp.
46. ALA (2005) Guidelines for the Design of Buried Steel Pipe, American Lifelines Alliance.
47. Canadian Standards Association (2010) CSA Z662: Oil and Gas Pipeline Systems, Canadian Standards Association, 566 pp.
48. DNV (2007) DNV RP F110: Global Buckling of Submarine Pipelines—Structural Design Due to High Temperature/High Pressure, C Det Norske Veritas, 64 pp.
49. DNV (2013) DNV OS-F101: Submarine Pipeline Systems, Det Norske Veritas, 372 pp.
50. Zimmerman, T.J.E., Stephens, M.J., DeGeer, D.D., and Chen, Q. (1995) Compressive strain limits for buried pipelines. Proceedings of the International Conference on Offshore Mechanics and Arctic Engineering (OMAE), Vol. 5, pp. 365–378.
51. Bruschi, R., Monti, P., Bolzoni, G., and Tagliaferri, R. (1995) Finite element method as numerical laboratory for analysing pipeline response under internal pressure, axial load and bending moment. Proceedings of the International Conference on Offshore Mechanics and Arctic Engineering (OMAE), Vol. 5, pp. 389–401.

52. Dorey, A.B. (2001) Critical buckling strains in energy pipelines. Ph.D. thesis, University of Alberta, 346 pp.
53. Glover, A., Zhou, J., and Blair, B. (2002) Technology approaches for northern pipeline developments. Proceedings of the International Pipeline Conference, IPC2002-27293, 13 pp.
54. Zimmerman, T., Timms, C., Xie, J., and Asante, J. (2004) Buckling resistance of large diameter spiral welded linepipe. Proceedings of the International Pipeline Conference, IPC04-0364, 9 pp.
55. Kyriakides, S. and Corona, E. (2007) *Mechanics of Offshore Pipelines: Buckling and Collapse*, Elsevier, Vol. 1, 401 pp.
56. Kibey, S.A., Minnaar, K., Issa, J.A., and Gioielle, P.C. (2008) Effect of misalignment on the tensile strain capacity of welded pipelines. Proceedings of the International Offshore and Polar Engineering Conference (ISOPE), pp. 90–95.
57. Cravero, S., Bravo, R.E., and Ernst, H.A. (2009) Fracture mechanics evaluation of pipes subjected to combined load conditions. Proceedings of the International Conference on Offshore Mechanics and Arctic Engineering, OMAE2009-80025, pp. 219–225.
58. Hamilton, J.M. and Jones, J.E. (2009) Sakhalin–1: technology development for frontier arctic projects. Proceedings of the Offshore Technology Conference, OTC-20208, 7 pp.
59. Kibey, S.A., Minnaar, K., Cheng, W., and Wang, X. (2009) Development of a physics-based approach for the prediction of strain capacity of welded pipelines. Proceedings of the International Offshore and Polar Engineering Conference (ISOPE), pp. 132–137.
60. Kibey, S., Wang, X., Minnaar, K., Macia, M.L., Fairchild, D.P., Kan, W.C., Ford, S.J., and Newbury, B. (2010) Tensile strain capacity equations for strain-based design of welded pipelines. Proceedings of the International Pipeline Conference, IPC2010-31661, 9 pp.
61. Suzuki, N., Tajika, H., Igi, S., Okatsu, M., Kondo, J., and Arakawa, T. (2010) Local buckling behavior of 48", X80 high-strain line pipes. Proceedings of the International Pipeline Conference, IPC201-31367, 8 pp.
62. Ahmed, A.U., Roger Cheng, J.J., and Zhou, J. (2010) Development of failure criteria for predicting tearing of wrinkled pipe. Proceedings of the International Pipeline Conference, IPC-31117, 9 pp.
63. Igi, S., Sakimoto, T., Suzuki, N., Muraoka, R., and Arakawa, T. (2010) Tensile strain capacity of X80 pipeline under tensile loading with internal pressure. Proceedings of the International Pipeline Conference, IPC-31281, 10 pp.
64. Zhang, J., Cheng, J.J., and Zhou, J. (2010) Experimental study on low cycle fatigue behaviour of wrinkled pipes. Proceedings of the International Pipeline Conference, IPC-31286, 10 pp.
65. Bian, Y., Collins, L., Mackenzie, R., and Penniston, C. (2010) Evaluation of UOE and spiral-welding line pipe for strain-based designs. Proceedings of the International Pipeline Conference, IPC-31315, 10 pp.
66. Stephens, M.J., Petersen, R.T., Wang, Y.-Y., Gordon, R., and Horsley, D. (2010) Large scale experimental data for improved strain-based design models. Proceedings of the International Pipeline Conference, IPC-31396, 9 pp.
67. Al-Showaiter, A., Taheri, F., and Kenny, S. (2011) Effect of misalignment and weld induced residual stresses on the local buckling response of pipelines. *Journal of Pressure Vessel Technology*, 133, doi: 10.1115/1.4002858.
68. Fatemi, A. and Kenny, S. (2012) Ovality of high-strength linepipes subject to combined loads. Proceedings of the Offshore Technology Conference, OTC-23783, 11 pp.
69. Fatemi, A. and Kenny, S. (2012) Characterization of initial geometric imperfections for pipelines and influence on compressive strain capacity. Proceedings of the International Offshore and Polar Engineering Conference, ISOPE-12-TPC-0304, 7 pp.
70. Vitali, L., Bruschi, R., Bartolini, L., Torselletti, E., and Spinazzè, M. (2013) Strain based design: crossing of local features in arctic environment. Proceedings of the International Conference on Offshore Mechanics and Arctic Engineering, OMAE-10328, 13 pp.
71. Xiaotong, H., Kenny, S., and Martens, M. (2014) Mechanical integrity evaluation of unequal wall thickness transition joints in transmission pipelines. Proceedings of the International Pipeline Conference, IPC-33141, 9 pp.
72. Bakhtyar, F. and Kenny, S. (2014) Development of a fatigue life assessment tool for pipelines with local wrinkling through physical testing and numerical modeling. Proceedings of the International Conference on Offshore Mechanics and Arctic Engineering, OMAE-24082, 10 pp.
73. Zimmerman, T.J.E., Hopkins, P., and Sanderson, N. (1998) Can limit states design be used to design a pipeline above 80% SMYS? Proceedings of the International Conference on Offshore Mechanics and Arctic Engineering (OMAE), 9 pp.
74. Zhou, J., Nessim, M., Zhou, W., and Rothwell, B. (2009) Reliability-based design and assessment standards for onshore natural gas transmission pipelines. *Journal of Pressure Vessel Technology*, 131(3), 6.
75. Nessim, M.A., Yue, H.K., and Zhou, J. (2010) Application of reliability based design and assessment to maintenance and protection decisions for natural gas pipelines. Proceedings of the International Pipeline Conference, IPC-31555, 10 pp.
76. Nessim, M.A., Yoosef-Ghodsi, N., Honegger, D., Zhou, J., and Wang, S. (2010) Application of reliability based design and assessment to seismic evaluations. Proceedings of the International Pipeline Conference, IPC-31557, 11 pp.
77. Nessim, M., Stephens, M., and Adianto, R. (2012) Safety levels associated with the ultimate limit state reliability targets in Annex O of CSA Z662. Proceedings of the International Pipeline Conference, IPC-90450, 7 pp.
78. Nessim, M. (2012) Limit states design and assessment of onshore pipelines. Proceedings of the International Pipeline Conference, IPC-90449, 9 pp.
79. Bruton, D.A.S., White, D.J., Carr, M., and Cheuk, J.C.Y. (2008) Pipe–soil interaction during lateral buckling and pipeline walking—the SAFEBUCK JIP. Proceedings of the Offshore Technology Conference, OTC-19589, 20 pp.

80. Sun, J. and Jukes, P. (2009) From installation to operation: a full-scale finite element modelling of deep-water pipe-in-pipe system. Proceedings of the International Conference on Offshore Mechanics and Arctic Engineering, OMAE-79519, 8 pp.
81. Jukes, P., Eltaher, A., Sun, J., and Harrison, G. (2009) Extra high-pressure high temperature (XHPHT) flowlines: design consideration and challenges. Proceedings of the International Conference on Offshore Mechanics and Arctic Engineering, OMAE-79537, 10 pp.
82. Bruton, D.A.S., Sinclair, F., and Carr, M. (2010) Lesson learned from observed walking of pipelines with lateral buckles, including new driving mechanisms and updated analysis models. Proceedings of the Offshore Technology Conference, OTC-20750, 12 pp.
83. Collberg, L., Carr, M., and Levold, E. (2011) Safebuck design guideline and DNV RP F110. Proceedings of the Offshore Technology Conference, OTC-21575, 7 pp.
84. Haq, M. and Kenny, S. (2013) Lateral buckling response of subsea HTHP pipelines using finite element methods. Proceedings of the International Conference on Offshore Mechanics and Arctic Engineering, OMAE2013-10585, 8 pp.
85. Haq, M.M. and Kenny, S. (2014) Assessment of parameters influencing lateral buckling of deep subsea pipe-in-pipe system using finite element modeling. Proceedings of the International Conference on Offshore Mechanics and Arctic Engineering, OMAE-23731, 12 pp.
86. Phoon, K.K., Honjo, Y., and Gilbert, R.B. (2003) *Limit State Design in Geotechnical Engineering Practice*, World Scientific.
87. European Committee for Standardization (2004) EN 1997-1: Eurocode 7: Geotechnical Design—Part 1: General Rules. CEN/TC 250/SC7, European Committee for Standardization, 168 pp.
88. Phoon, K.K. (2008) *Reliability-Based Design in Geotechnical Engineering: Computations and Applications*, CRC Press, 544 pp.
89. Fenton, G.A. and Griffiths, D.V. (2010) Reliability-based geotechnical engineering. GeoFlorida Conference, Keynote lecture, American Society of Civil Engineers, 40 pp.
90. Singhroy, V.H. (1988) Case studies on the application of remote sensing data to geotechnical investigations in Ontario, Canada. In: Johnson, A.I. and Pettersson, C.B. (editors), *Geotechnical Applications of Remote Sensing and Remote Data Transmission: A Symposium*, ASTM, STP 967, pp. 9–45.
91. American Society of Civil Engineers (1998) Pipeline Route Selection for Rural and Cross-Country Pipelines. ASCE Manuals and Reports on Engineering Practice No. 46, 97 pp.
92. McCarthy, D.F. (2002) *Essentials of Soil Mechanics and Foundations: Basic Geotechnics*, Prentice-Hall, 864 pp.
93. Mohitpour, M., Golshan, H., and Murray, A. (2003) *Pipeline Design and Construction: A Practical Approach*, ASME Press, 656 pp.
94. Palmer, A.C. and King, R.A. (2004) *Subsea Pipeline Engineering*, PennWell Corporation, 570 pp.
95. Das, B.M. (2006) *Principles of Geotechnical Engineering*, Cengage Learning, 688 pp.
96. Cosford, J.I., van Zeyl, D., and Penner, L.A. (2014) Terrain analysis for pipeline design, construction and operation. *Journal of Pipeline Engineering*, 13(3), 149–165.
97. Imperial Oil Resources Ventures Limited (2004) Application for Approval of the Mackenzie Valley Pipeline—Volume 1: Pipeline Overview. Report IPRCC.PR.2004.05, 156 pp.
98. Imperial Oil Resources Ventures Limited (2004) Application for Approval of the Mackenzie Valley Pipeline—Volume 3: Engineering Design. Report IPRCC.PR.2004.05, 164 pp.
99. Norwegian Petroleum Industry (2004) Norsok G-001: Marine Soil Investigations, Norwegian Petroleum Industry, 66 pp.
100. OSIG (2004) Guidance Notes on Geotechnical Investigations for Marine Pipelines, Pipeline Working Group of the Offshore Soil Investigation Forum, 47 pp.
101. Danson, E. (editor) (2005) *Geotechnical and Geophysical Investigations for Offshore and Nearshore Developments*, International Society for Soil Mechanics and Geotechnical Engineering, 94 pp.
102. Sweeney, M. (2005) Terrain and geohazard challenges facing onshore oil and gas pipelines. Institution of Civil Engineers International Conference, 733 pp.
103. Palmer, A.C. (2008) *Dimensional Analysis and Intelligent Experimentation*, World Scientific, 154 pp.
104. Wang, J., Haigh, S., Forrest, G., and Thusyanthan, N. (2012) Mobilization distance for upheaval buckling of shallowly buried pipelines. *Journal of Pipeline Systems Engineering and Practice*, 3(4), 106–114.
105. Wood, D.M. (2004) *Geotechnical Modelling*, Spon Press, 488 pp.
106. Munro, S.M., Moore, I.D., and Brachman, R.W.I. (2009) Laboratory testing to examine deformations and moments in fiber-reinforced cement pipe. *Journal of Geotechnical and Geoenvironmental Engineering*, 135(11), 1722–1731.
107. O'Rourke, T.D., Jezerski, J.M., Olson, N.A., Bonneau, A.L., Palmer, M.C., Stewart, H.E., O'Rourke, M.J., and Abdoun, T. (2008) Geotechnics of pipeline system response to earthquakes. Proceedings of the Geotechnical Earthquake Engineering and Soil Dynamics IV, Keynote address, 38 pp.
108. Wijewickreme, D. and Weerasekara, L. (2010) Full-scale testing of buried extensible pipes subject to axial soil loading. Proceedings of the International Conference on Physical Modelling in Geotechnics (ICPMG), pp. 657–662.
109. Wijewickreme, D., Weerasekara, L., and Johnson, G. (2008) Soil load mobilization in axially loaded buried polyethylene pipes. Proceedings of the IACMAG Conference, pp. 561–569.
110. Wijewickreme, D., Karimian, H., and Honegger, D. (2009) Response of buried steel pipelines subjected to relative axial soil movement. *Canadian Geotechnical Journal*, 46(7), 735–752.
111. Karimian, H., Wijewickreme, D., and Honegger, D. (2006) Buried pipelines subjected to transverse ground movement: comparison between full-scale testing and numerical

modeling. Proceedings of the International Conference on Offshore Mechanics and Arctic Engineering, OMAE-92125, 7 pp.

112. Ha, D., Abdoun, T., O'Rourke, M.J., Symans, M.D., O'Rourke, T.D., Palmer, M.C., and Stewart, H.E. (2010) Earthquake faulting effects on buried pipelines: case history and centrifuge study. *Journal of Earthquake Engineering*, 14(5), 646–669.
113. Phillips, R., Nobahar, A., and Zhou, J. (2004) Trench effects on pipe–soil interaction. Proceedings of the International Pipeline Conference, IPC04-0141, pp. 321–327.
114. Taylor, R.N. (editor) (1995) *Geotechnical Centrifuge Technology*, Blackie Academic and Professional, 296 pp.
115. Winkler, E. (1867) *Die leher von der elastizitat und festigkeit*, Dominicus, Prague.
116. Reissner, E. (1958) Deflection of plates on viscoelastic foundation. *Journal of Applied Mechanics*, 80, 144–145.
117. Kerr, A.D. (1964) Elastic and visco-elastic foundation models. *Journal of Applied Mechanics*, 31, 491–498.
118. Faeli, Z., Fakher, A., and Sadatieh, S.R.M. (1985) Allowable differential settlement of oil pipelines. *International Journal of Engineering*, 4(4), 308–320.
119. Rajani, B. and Morgenstern, N. (1992) Behaviour of a semi-infinite beam in a creeping medium. *Canadian Geotechnical Journal*, 29(5), 779–788.
120. Konuk, I., Yu, S., and Fredj, A. (2006) Do Winkler models work: a case study for ice scour problems. Proceedings of the International Conference on Offshore Mechanics and Arctic Engineering, OMAE-92335, 7 pp.
121. Ieronymaki, E.S. (2011) Response of continuous pipelines to tunnel induced ground deformations. M.Sc. thesis, Massachusetts Institute of Technology, 208 pp.
122. Vlazov, V.Z. and Leontiev, U.N. (1966) Beams, Plates and Shells on Elastic Foundations, Israel Program for Scientific Translations, Jerusalem.
123. Selvadurai, A.P.S. (1979) *Elastic Analysis of Soil–Foundation Interaction. Developments in Geotechnical Engineering*, Elsevier, Amsterdam, The Netherlands, Vol. 17.
124. Pasternak, P.L. (1954) On a new method of analysis of an elastic foundation by means of two foundation constants. Gosudarstvennoe Izdatelstro Liberaturi po Stroitelstvui Arkhitekture, Moscow (in Russian).
125. Kettle, R.J. (1984) Soil–pipeline interaction: a review of the problem. Proceedings of Pipelines and Frost Heave, Caen, France, pp. 35–37.
126. Vortser, T.E.B., Klar, A., Soga, K., and Mair, R.J. (2005) Estimating the effects of tunnelling on existing pipelines. *Journal of Geotechnical and Geoenvironmental Engineering*, 131(11), 1399–1410.
127. Newmark, N.M. and Hall, W.J. (1975) Pipeline design to resist large fault displacement. U.S. National Conference on Earthquake Engineering, Earthquake Engineering Research Institute, pp. 416–425.
128. Kennedy, R.P., Chow, A.W., and Williamson, R.A. (1977) Fault movement effects on buried oil pipeline. *Journal of Transportation Engineering Division, ASCE*, 103(5), 617–633.
129. O'Rourke, M.J. (1989) Approximate Analysis Procedures for Permanent Ground Deformation Effects on Buried Pipelines. Technical Report NCEER-89-0032, Multidisciplinary Center for Earthquake Engineering Research, pp. 336–347.
130. Rizkalla, M., Poorooshasb, F., and Clark, J.I. (1992) Centrifuge modelling of lateral pipeline/soil interaction. Proceedings of the International Conference on Offshore Mechanics and Arctic Engineering (OMAE),13 pp.
131. Trigg, A. and Rizkalla, M. (1994) Development and application of a closed form technique for the preliminary assessment of pipeline integrity in unstable slopes. Proceedings of the International Conference on Offshore Mechanics and Arctic Engineering (OMAE), Vol. 4, pp. 127–139.
132. Rizkalla, M., Trigg, A., and Simmonds, G. (1996) Recent advances in modelling of longitudinal pipeline/soil interaction. Proceedings of the International Conference on Offshore Mechanics and Arctic Engineering (OMAE), Vol. 5, pp. 325–332.
133. Peek, R. and Nobahar, A. (2012) Ice gouging over a buried pipeline: superposition error of simple beam-and-spring models. *International Journal of Geomechanics*, 12(4), 508–516.
134. Xie, X., Symans, M.D., O'Rourke, M.J. Abdoun, T.H., O'Rourke, T.D., Palmer, M.C., and Stewart, H.E. (2013) Numerical modeling of buried HDPE pipelines subjected to normal faulting: a case study. *Earthquake Spectra*, 29(2), 609–632.
135. Einsfeld, R., Murray, D.W., and Yoosef-Ghodsi, N. (2003) Buckling analysis of high-temperature pressurized pipelines with soil–structure interaction. *Journal of Brazilian Society of Mechanical Sciences and Engineering*, 25(2), 164–169.
136. Nixon, J.F., Palmer, A., and Phillips, R. (1996) Simulations for buried pipeline deformations beneath ice scour. Proceedings of the International Conference on Offshore Mechanics and Arctic Engineering (OMAE), Vol. 5, pp. 383–392.
137. C-FER (1995) Development of Pipe–Soil Interaction Models for Frost Heave Analysis. Report Submitted to National Energy Board, C-FER, Edmonton, Alberta.
138. Daiyan, N. (2013) Investigating soil/pipeline interaction during oblique relative movements. Ph.D. thesis, Memorial University of Newfoundland, 204 pp.
139. Hill, R. (1950) *The Mathematical Theory of Plasticity*, Oxford University Press, Oxford, 355 pp.
140. Haslach, H.W. and Armstrong, R.W. (2004) *Deformable Bodies and Their Material Behaviour*, John Wiley & Sons, Inc., Hoboken, NJ, 532 pp.
141. Kaiser, T.M.V., Yung, V.Y.B., and Bacon, R.M. (2005) Cyclic mechanical and fatigue properties for OCTG materials. International Thermal Operations and Heavy Oil Symposium, SPE/PS-CIM/CHOA-97775, 8 pp.
142. Jukes, P., Wang, S., and Wang, J. (2008) The sequential reeling and lateral buckling simulation of pipe-in-pipe flowlines using finite element analysis for deepwater applications. Proceedings of the International Offshore and Polar Engineering Conference, ISOPE08-209, 8 pp.

143. Tkaczyk, T., Pépin, A., and Denniel, S. (2012) Fatigue and fracture performance of reeled mechanically lined pipes. Proceedings of the International Offshore and Polar Engineering Conference, ISOPE12-602, 10 pp.
144. Audibert, J.M.E. and Nyman, K.J. (1977) Soil restraint against horizontal motion of pipes. *Journal of the Geotechnical Engineering Division, ASCE*, 103(NGT10), 1119–1142.
145. Trautmann, C.H., and O'Rourke, T.D. (1983) Behavior of Pipe in Dry Sand Under Lateral and Uplift Loading. Geotechnical Engineering Report 83-7, Cornell University.
146. Trautmann, C.H., O'Rourke, T.D., and Kulhawy, F.H. (1985) Uplift force–displacement response of buried pipe. *Journal of Geotechnical Engineering*, 111(9), 1061–1076.
147. Trautmann, C.H. and O'Rourke, T.D. (1985) Lateral force–displacement response of buried pipe. *Journal of Geotechnical Engineering*, 111(9), 1077–1092.
148. Paulin, M. (1998) An investigation into pipelines subjected to lateral soil loading. Doctoral thesis, Memorial University of Newfoundland, 600 pp.
149. Cathie, D.N., Jaeck, C., Ballard, J.-C., and Wintgens, J.-F. (2005) Pipeline geotechnics: state of art. Proceedings of the International Symposium on Frontiers in Offshore Geotechnics (ISFOG), pp. 95–114.
150. Scarpelli, G., Sakellariadi, E., and Furlani, G. (2003), Evaluation of soil–pipeline longitudinal forces. *Rivista Italiana di Geotecnica*, 4(3), 24–41.
151. Oliphant, J. and Maconochie, A. (2007) The axial resistance of buried and unburied pipelines. Proceedings of the International Offshore Site Investigation and Geotechnics Conference, pp. 125–132.
152. Finch, M., Fisher, R., Palmer, A., and Baumgard, A. (2000) An integrated approach to pipeline burial in the 21st century. Deep Offshore Technology International Conference and Exhibition.
153. Sladen, J.A. (1992) The adhesion factor: applications and limitations. *Canadian Geotechnical Journal*, 29(2), 322–326.
154. Cappelletto, A., Tagliaferri, R., Giurlani, G., Andrei, G., Furlani, and Scarpelli, G. (1998) Field full scale tests on longitudinal pipeline–soil interaction. Proceedings of the International Pipeline Conference (IPC), Vol. 2, pp. 771–778.
155. Bruschi, R., Glavina, S., Spinazze, M., Tomassini, D., Bonanni, S., and Cuscuna, S. (1996) Pipelines subject to slow landslide movements: structural modelling vs field measurement. Proceedings of the International Conference on Offshore Mechanics and Arctic Engineering (OMAE), Vol. 5, pp. 343–353.
156. Karimian, S.A. (1995) Response of buried steel pipelines subjected to longitudinal and transverse ground movement. Ph.D. thesis, UBC, 352 pp.
157. Scarpelli, G., Sakellariadi, E., and Furlani, G. (2003) Longitudinal pipeline–soil interaction: results from field full scale and laboratory testing. *Rivista Italiana di Geotecnica*, 4, 24–40.
158. Paulin, M.J., Phillips, R., Clark, J.I., Trigg, A., and Konuk, I. (1998) A Full-scale investigation into pipeline/soil interaction. Proceedings of the International Pipeline Conference (IPC), Vol. 2, pp. 779–787.
159. Seo, D., Kenny, S., and Hawlader, B. (2011) Yield envelopes for oblique pipeline/soil interaction in cohesive soil using ALE procedure. Proceedings of the Pan-American CGS Geotechnical Conference, 7 pp.
160. Pike, K.P. and Kenny, S.P. (2012) Lateral–axial pipe/soil interaction events: numerical modeling trends and technical issues. *Proceedings of the International Pipeline Conference*, IPC-90055, 6 pp.
161. Kulhawy, F.H.C., Trautmann, C.H., Beech, J.F., O'Rourke, T.D., and McGuire, W. (1983) Transmission Line Structure Foundations for Uplift-Compression Loading. Report No. EL-2870, Electric Power Research Institute.
162. Sherif, M.A., Fang, Y.S., and Sherif, R.I. (1984) $K_a$ and $K_0$ behind rotating and non-yielding walls. *ASCE Journal of Geotechnical Engineering*, 110(1), 41–56.
163. Meyerhof, G.G. (1953) The bearing capacity of foundations under eccentric and inclined loads. Proceedings of the International Conference on Soil Mechanics and Foundation Engineering, Vol. 1, pp. 440–445.
164. Mackenzie, T.R. (1955) Strength of deadman anchors in clay: pilot tests. M.Sc. thesis, Princeton University.
165. Hansen, J.B. (1961) The ultimate resistance of rigid piles against transversal forces. Bulletin 12, Danish Geotechnical Institute, pp. 5–9.
166. Terzaghi, K. and Peck, R.B. (1967) *Soil Mechanics in Engineering Practice*, John Wiley & Sons, Inc., New York, 729 pp.
167. Neely, W.J., Stuart, J.G., and Graham, J. (1973) Failure loads of vertical anchor plates in sand. *Journal of Soil Mechanics and Foundations Division, ASCE*, 99, 669–685.
168. Audibert, J.M.E. and Nyman, K.J. (1975) Coefficients of subgrade reaction for the design of buried piping. Proceedings of the American Society of Civil Engineers: Structural Design of Nuclear Plant Facilities, pp. 109–141.
169. Wantland, G.M., O'Neill, M.W., Reese, L.C., and Kalajian, E.H. (1979) Lateral stability of pipelines in clay. Proceedings of the Offshore Technology Conference, OTC-3477, 10 pp.
170. Ranjan, G. and Aurora, V.B. (1980) Model studies on anchors under horizontal pull in clay. Proceedings of the Australia–New Zealand Conference on Geomechanics, Vol. 1, pp. 65–70.
171. Rowe, R.K. and Davis, E.H. (1982) The behaviour of anchor plates in clay. *Géotechnique*, 32(1), 9–23.
172. Rowe, R.K. and Davis, E.H. (1982) The behaviour of anchor plates in sand. *Géotechnique*, 32(1), 25–41.
173. Randolph, M.F. and Houlsby, G.T. (1984) The limiting pressure on a circular pile loaded laterally in cohesive soil. *Géotechnique*, 34(4), 613–623.
174. Rajani, B.B., Robertson, P.K., and Morgenstern, N.R. (1993) A simplified method for pipelines subject to transverse soil movements. Proceedings of the International Conference on Offshore Mechanics and Arctic Engineering (OMAE), Vol. 5, pp. 157–165.

175. Poorooshasb, F., Paulin, M.J., Rizkalla, M., and Clark, J.I. (1994) Centrifuge modeling of laterally loaded pipelines. Transportation Research Record No. 1431, pp. 33–40.
176. Ng, P.C.F. (1994) Behaviour of buried pipelines subjected to external loading. Ph.D. thesis, University of Sheffield, 339 pp.
177. Rajani, B.B., Robertson, P.K., and Morgenstern, N.R. (1995) Simplified design methods for pipelines subject to transverse and longitudinal soil movements. *Canadian Geotechnical Journal*, 32, 309–323.
178. Merifield, R.S., Sloan, S.W., and Yu, H.S. (2001) Stability of plate anchors in undrained clay. *Géotechnique*, 51(2), 141–153.
179. Popescu, R., Phillips, R., Konuk, I., Guo, P., and Nobahar, A. (2002) Pipe–soil interaction: large-scale tests and numerical modeling. Proceedings of the International Conference on Physical Modelling in Geotechnics (ICPMG), pp. 917–922.
180. Turner, J.E. (2004) Lateral force–displacement behavior of pipes in partially saturated sand. M.Sc. thesis, Cornell University, 365 pp.
181. Yimsiri, S., Soga, K., Yoshizaki, K., Dasari, G.R., and O'Rourke, T.D. (2004) Lateral and upward soil–pipeline interactions in sand for deep embedment conditions. *Journal of Geotechnical and Geoenvironmental Engineering*, 130(8), 830–842.
182. Yoshizaki, K. and Sakanoue, T. (2004) Analytical study on soil–pipe interaction due to large ground deformation. Proceedings of the World Conference on Earthquake Engineering, WCEE-1402, 13 pp.
183. Guo, P. and Stolle, D. (2005) Lateral pipe–soil interaction in sand with reference to scale effect. *Journal of Geotechnical and Geoenvironmental Engineering*, 131(3), 338–349.
184. Karimian, S.A. (2006) Response of buried steel pipelines subjected to longitudinal and transverse ground movement. Ph.D. thesis, University of British Columbia, 352 pp.
185. Olson, N.A. (2009) Soil performance for large-scale soil–pipeline tests. Ph.D. thesis, Cornell University.
186. Oliveira, J.R.M.S., Almeida, M.S.S., Almeida, M.C.F., and Borges, R.G. (2010) Physical modeling of lateral clay–pipe interaction. *Journal of Geotechnical and Geoenvironmental Engineering*, 136(7), 950–956.
187. Badv, K. and Daryani, K.E. (2010) An investigation into the upward and lateral soil–pipeline interaction in sand using finite difference method. *Iranian Journal of Science and Technology*, 34(B), 433–445.
188. Jung, J.K. (2011) Soil–pipe interaction under plane strain conditions. Ph.D. thesis, Cornell University, 303 pp.
189. Jung, J.K. and Zhang, K. (2011) Finite element analyses of soil–pipe behavior in dry sand under lateral loading. ASCE Pipelines 2011 Conference, pp. 312–324.
190. Paulin, M.J., Phillips, R., and Boivin, R. (1996) An experimental investigation into lateral pipeline interaction. *Proceedings of the International Conference on Offshore Mechanics and Arctic Engineering (OMAE)*, Vol. 5, pp. 313–323.
191. Kondner, R.L. (1963) Hyperbolic stress–strain response: cohesive soils. *Journal of the Soil Mechanics and Foundations Division, ASCE*, 89, 115–143.
192. Chin, F.K. (1970) Estimation of the ultimate load of piles not carried to failure. Proceedings of the Southeast Asia Conference on Soil Engineering, pp. 81–90.
193. Duncan, J.M. and Chang, C. (1970) Nonlinear analysis of stress and strain in soils. *Journal of the Soil Mechanics and Foundations Division, ASCE*, 96(5), 1629–1653.
194. Decourt, L. (1999) Behavior of foundations under working load conditions. Proceedings of the CGS Soil Mechanics and Geotechnical Engineering, Vol. 4, pp. 453–488.
195. Daiyan, N., Kenny, S., Phillips, R., and Popescu, R. (2009) Parametric study on vertical–lateral pipeline/soil interaction events in clay. Proceedings of CSCE, IEMM-005, 10 pp.
196. Pike, K. and Kenny, S. (2011) Advancement of CEL procedures to analyze large deformation pipeline/soil interaction events. Proceedings of the Offshore Technology Conference, OTC-22004-PP, 10 pp.
197. Rossiter, C. and Kenny, S. (2012) Evaluation of vertical–lateral pipe/soil interaction in clay. Proceedings of the Offshore Technology Conference, OTC-23735, 13 pp.
198. Ovesen, N.K. (1964) Anchor slab calculation methods and model tests. Bulletin 16, Danish Geotechnical Institute, Copenhagen, Denmark, 40 pp.
199. Ovesen, N.K. and Stromann, H. (1972) Design methods for vertical anchor slabs in sand. Proceedings of the Speciality Conference on Performance of Earth and Earth-Supported Structures, American Society of Civil Engineers, Vol. 1, pp. 1481–1500.
200. Daiyan, N., Kenny, S., Phillips, R., and Popescu, R. (2011) Numerical investigation of axial–vertical and lateral–vertical pipeline/soil interaction in sand. Proceedings of the Pan-American CGS Geotechnical Conference, CGS-233, 9 pp.
201. Daiyan, N., Kenny, S., Phillips, R., and Popescu, R. (2011) Investigating pipeline/soil interaction under axial/lateral relative movements in sand. *Canadian Geotechnical Journal*, 48 (11), 1683–1695.
202. Daiyan, N., Kenny, S., Phillips, R., and Popescu, R. (2010) Numerical investigation of oblique pipeline/soil interaction in sand. Proceedings of the International Pipeline Conference, IPC2010-31644, 6 pp.
203. Vesic, A.S. (1971) Breakout resistance of objects embedded in ocean bottom. *Journal of Soil Mechanics and Foundations Division, ASCE*, 97(SM9), 1183–1205.
204. Audibert, J.M.E., Lai, N.W., and Bea, R.G. (1979) Design of pipelines: Sea bottom loads and restraints. Proceedings of the ASCE: Pipelines in Adverse Environments: A State of the Art, Vol. 1, pp. 187–203.
205. Meyerhof, G.G. (1955) Influence of roughness of base and ground-water conditions on the ultimate bearing capacity of foundations. *Géotechnique*, 5, 227–241.
206. Potts, D.M. and Zdravković, L. (1999) *Finite Element Analysis in Geotechnical Engineering: Theory*, Thomas Telford Publishing, 440 pp.
207. Nobahar, A., Popescu, R., and Konuk, I. (2001) Parameter calibration of strain hardening/softening of sand from direct shear tests. Proceedings of the International Conference on

Computer Methods and Advances in Geomechanics, pp. 971–976.

208. Yap, T.Y. and Hicks, M.A. (2001) An investigation of element-size dependency in strain localization. Proceedings of the International Conference on Computer Methods and Advances in Geomechanics, pp., 607–610.
209. Anastasopoulos, I., Gazetas, G., Bransby, M., Davies, M., and El Nahas, A. (2007) Fault rupture propagation through sand: finite-element analysis and validation through centrifuge experiments. *Journal of Geotechnical and Geoenvironmental Engineering*, 133(8), 943–958.
210. Anastasopoulos, I. (2009) Fault rupture propagation through sand: finite-element analysis and validation through centrifuge experiments. *Journal of Geotechnical and Geoenvironmental Engineering*, 135(6), 846–850.
211. Pike, K., Kenny, S., and Hawlader, B. (2013) Advanced analysis of pipe/soil interaction accounting for strain localization. Proceedings of the CGS, 6 pp.
212. Ju, G.T. and Kyriakides, S. (1988) Thermal buckling of offshore pipelines. *Journal of Offshore Mechanics and Arctic Engineering*, 110(4), 355–364.
213. Klever, F.J., van Helvoirt, L.C., and Sluyterman, A.C. (1990) A dedicated finite-element model for analyzing upheaval buckling response of submarine pipelines. Proceedings of the Offshore Technology Conference, OTC-6333, pp. 529–538.
214. Svanö, G., Jostad, H.P., and Bjaerum, R.O. (1992) Finite element response analysis of a lay-away pipeline expansion curve on soft clay. Proceedings of the Offshore Technology Conference, OTC-6878, pp. 45–52.
215. Bruschi, R., Monti, P., Bolzoni, G., and Tagliaferri, R. (1995) Finite element method as numerical laboratory for analysing pipeline response under internal pressure, axial load, bending moment. Proceedings of the International Conference on Offshore Mechanics and Arctic Engineering (OMAE), Vol. 5, pp. 389–401.
216. Kim, H.S., Kim, W.S., Bang, I.W., and Oh, K.H. (1998) Analysis of stresses on buried natural gas pipeline subjected to ground subsidence. *Proceedings of the International Pipeline Conference (IPC)*, Vol. 2, pp. 749–756.
217. Nobahar, A., Kenny, S., and Phillips, R. (2007) Buried pipelines subject to subgouge deformations. *International Journal of Geomechanics*, 7(3), 206–216.
218. Guo, P. and Popescu, R. (2002) Trench effects on pipe/soil interaction. Proceedings of the Canadian Specialty Conference on Computer Applications in Geotechnique, pp. 261–269.
219. Kouretzis, G.P., Sheng, D., and Sloan, S.W. (2013) Sand–pipeline–trench lateral interaction effects for shallow buried pipelines. *Computers and Geotechnics*, 54, 53–59.
220. Monroy, M., Wijewickreme, D., and Honegger, D. (2012) Effectiveness of geotextile-lined pipeline trenches subjected to relative lateral seismic fault ground displacements. Proceedings of the WCEE, WCEE-2534, 10 pp.
221. Pike, K. and Kenny, S. (2012) Numerical pipe/soil interaction modelling: sensitivity study and extension to ice gouging. Proceedings of the Offshore Technology Conference, OTC-23731, 8 pp.
222. Phillips, R., Nobahar, A., and Zhou, J. (2004) Combined axial and lateral pipe–soil interaction relationships. Proceedings of the International Pipeline Conference, IPC-0144, 5 pp.
223. Schaminée, P., Zorn, N., and Schotman, G. (1990) Soil response for pipeline upheaval buckling analyses: full-scale laboratory tests and modelling. Proceedings of the Offshore Technology Conference, OTC-6486, pp. 563–572.
224. Finch, M. (1999) Upheaval buckling and floatation of rigid pipelines: the influence of recent geotechnical research on the current state of the art. Proceedings of the Offshore Technology Conference (OTC), Vol. 1, pp. 27–43.
225. Croll, J.G.A. (1997) A simplified analysis of imperfect thermally buckled subsea pipeline. Proceedings of the International Offshore and Polar Engineering Conference (ISOPE), Vol. 2, pp. 666–676.
226. Guijt, J. (1990) Upheaval buckling of offshore pipelines: overview and introduction. Proceedings of the Offshore Technology Conference, OTC-6487, pp. 573–580.
227. Palmer, A.C., Ellinas, C.P., Richards, D.M., and Guijt, J. (1990) Design of submarine pipelines against upheaval buckling. Proceedings of the Offshore Technology Conference, OTC-6335, Vol. 2, pp. 551–560.
228. Palmer, A.C., Carr, M., Maltby, T., McShane, B., and Ingram, J. (1994) Upheaval buckling: what do we know, and what don't we know? Proceedings of the Offshore Pipeline Technology Seminar, IBC Technical Services Ltd.
229. Vanden Berghe, J.-F., Cathie, D., and Ballard, J.-C. (2005) Pipeline uplift mechanisms using finite element analysis. Proceedings of the International Conference of Soil Mechanics and Foundation Engineering, Osaka, Japan.
230. Richards, D.M. (1990) The effect of imperfection shape on upheaval buckling behaviour. In: Ellinas, C.P. (editor), *Advances in Subsea Pipeline Engineering and Technology*, Springer, pp. 51–66.
231. Thusyanthan, N.I., Ganesan, S.A., Bolton, M.D., and Allan, P. (2008) Upheaval buckling resistance of pipelines buried in clayey backfill. Proceedings of the International Offshore and Polar Engineering Conference, ISOPE-TPC-499, 7 pp.
232. Thusyanthan, N.I., Mesmar, S., Wang, J., and Haigh, S.K. (2008) Uplift resistance of buried pipelines and DNV-RP-F110 guideline. Proceedings of the OPT, 20 pp.
233. Nielsen, N.J.R., Petersen, P.T., Grundy, A.K., and Lyngberg, B. (1988) New design criteria for upheaval creep of buried sub-sea pipelines. Proceedings of the International Conference on Offshore Mechanics and Arctic Engineering (OMAE), Vol. 5, pp. 243–250.
234. Nielsen, N.-J.R., Lyngberg, B., and Petersen, P.T. (1990) Upheaval buckling failures of insulated buried pipelines: a case story. Proceedings of the Offshore Technology Conference, OTC-6488, pp. 581–591.
235. Cheuk, C.Y., White, D.J., and Bolton, M.D. (2008) Uplift mechanisms of pipes buried in sand. *Journal of Geotechnical and Geoenvironmental Engineering*, 134(2), 154–163.

236. Hauch, S. and Bai, Y. (1999) Bending moment capacity of pipes. Proceedings of the International Conference on Offshore Mechanics and Arctic Engineering, OMAE-PL-99-5033, 12 pp.

237. Huo, X., Kenny, S., and Martens, M. (2014) Mechanical integrity evaluation of unequal wall thickness transition joints in transmission pipeline. Proceedings of the International Pipeline Conference, IPC-33141, 9 pp.

238. Meyerhof, G.G. (1973) Uplift resistance of inclined anchors and piles. Proceedings of the International Conference on Soil Mechanics and Foundation Engineering, Vol. 2, pp. 167–172.

239. Meyerhof, G.G. and Hanna, A.M. (1978) Ultimate bearing capacity of foundation on layered soil under inclined load. *Canadian Geotechnical Journal*, 15(4), 565–572.

240. Das, B.M. (1985) Resistance of shallow inclined anchors in clay. In: Clemence, S.P. (editor), *Uplift Behavior of Anchor Foundations in Soil*, American Society of Civil Engineers, pp. 86–101.

241. Das, B.M. and Puri, V.K. (1989) Holding capacity of inclined square plate anchors in clay. *Soils and Foundations*, 29(3), 138–144.

242. Bransby, M.F. and Randolph, M.F. (1998) Combined loading of skirted foundations. *Géotechnique*, 48(5), 637–655.

243. Taiebat, H.A. and Carter, G.T. (2000) Numerical studies of the bearing capacity of shallow foundations on cohesive soil subjected to combined loading. *Géotechnique*, 50(4), 409–418.

244. Martin, C.M. and Houlsby, G.T. (2001) Combined loading of spudcan foundations on clay: numerical modelling. *Géotechnique*, 51(8), 687–699.

245. Aubeny, C.P., Han, S.W., and Murff, J.D. (2003) Inclined load capacity of suction caissons. *International Journal of Numerical Analytical Methods in Geomechanics*, 27, 1235–1254.

246. Nobahar, A., Kenny, S., King, T. McKenna, R., and Phillips, R. (2007) Analysis and design of buried pipelines for ice gouging hazard: a probabilistic approach. *Journal of Offshore Mechanics and Arctic Engineering*, 129, 219–228.

247. Pike, K., Kenny, S., Kavanagh, K., and Jukes, P. (2012) Pipeline engineering solutions for harsh arctic environments: technology challenges and constraints for advanced numerical simulations. Proceedings of the Offshore Technology Conference, OTC-23731, 8 pp.

248. Cocchetti, G., Prisco, C., Galli, A., and Nova, R. (2009) Soil–pipeline interaction along unstable slopes: a coupled three-dimensional approach. Part 1. Theoretical formulation. *Canadian Geotechnical Journal*, 46(11), 1289–1304.

249. Cocchetti, G., Prisco, C., and Galli, A. (2009) Soil–pipeline interaction along unstable slopes: a coupled three-dimensional approach. Part 2. Numerical analysis. *Canadian Geotechnical Journal*, 46(11), 1305–1321.

250. Abdalla, B., Pike, K., Eltaher, A., Jukes, P., and Duron, B. (2009) Development and validation of a coupled Eulerian Lagrangian finite element ice scour model. Proceedings of the International Conference on Offshore Mechanics and Arctic Engineering (OMAE), pp. 87–95.

251. Pike, K. and Kenny, S. (2012) Advanced continuum modeling of the ice gouge process: assessment of keel shape effect and geotechnical data. Proceedings of the International Offshore and Polar Engineering Conference, ISOPE-12-TPC-0464, 7 pp.

252. Rossiter, C. and Kenny, S. (2012) Assessment of ice/soil interactions: continuum modeling in clays. Proceedings of the International Offshore and Polar Engineering Conference, ISOPE-12-TPC-0304, 8 pp.

253. Phillips, R., Barrett, J., and Al-Showaiter, A. (2010) Ice keel-seabed interaction: numerical modeling validation. Proceedings of the Offshore Technology Conference, OTC2010-20696, 13 pp.

254. Nyman, K. (1984) Soil response against oblique motion of pipes. *Journal of Transportation Engineering*, 110(2), 190–202.

255. Merifield, R.S. and Sloan, S.W. (2006) The ultimate pullout capacity of anchors in frictional soils. *Canadian Geotechnical Journal*, 43, 852–868.

256. Rowe, R.K. (1978) Soil structure interaction analysis and its application to the prediction of anchor behaviour. Ph.D. thesis, University of Sydney.

257. Guo, P. (2005) Numerical modeling of pipe–soil interaction under oblique loading. *Journal of Geotechnical and Geoenvironmental Engineering*, 131(2), 260–268.

258. Hsu, T.W. (1996) Soil restraint against oblique motion of pipelines in sand. *Canadian Geotechnical Journal*, 33(1), 180–188.

259. Hsu, T., Chen, Y., and Hung, W. (2006) Soil restraint to oblique movement of buried pipes in dense sand. *Journal of Transportation Engineering*, 132(2), 175–181.

260. Hsu, T., Chen, Y., and Wu, C. (2001) Soil friction restraint of oblique pipelines in loose sand. *Journal of Transportation Engineering*, 127(1), 82–87.

261. Ghaly, A.M. (1997) Soil restraint against oblique motion of pipelines in sand. Discussion. *Canadian Geotechnical Journal*, 34, 156–157.

262. Palmer, A.C., White, D.J., Baumgard, A.J., Bolton, M., Barefoot, M.D., Finch, M., Powell, T., Faranski, A.S., and Baldry, A.S. (2003) Uplift resistance of buried submarine pipelines: comparison between centrifuge modelling and full-scale tests. *Géotechnique*, 53(10), 77–883.

263. Nobahar, A., Popescu, R., and Konuk, I. (2001) Estimating progressive mobilization of soil strength. Proceedings of the Canadian Geotechnical Conference, Vol. 2, pp. 1311–1318.

264. Mahdavi, H., Kenny, S., Phillips, R., and Popescu, R. (2008) Influence of geotechnical loads on local buckling behavior of buried pipelines. Proceedings of the International Pipeline Conference, IPC2008-64054, 9 pp.

265. House, A.R., Oliveira, J.R.M., and Randolph, M.F. (2001) Evaluating the coefficient of consolidation using penetrometer tests. *International Journal of Physical Modelling in Geotechnics*, 1(3), 17.

266. Ishihara, K. (1996) *Soil Behaviour in Earthquake Geotechnics*, Oxford Engineering Science Series 46, Clarendon Press, Oxford, 353 pp.

267. Lefebvre, G. and LeBoeuf, D. (1987) Rate effects and cyclic loading of sensitive clays. *Journal of Geotechnical Engineering*, 113(5), 476–489.

268. Rodriguez, J., Alvarez, C., and Velandia, E. (2008) Load rate effects on high strain tests in high plasticity soils. In: dos Santos, Jaime Alberto (editor), *The Application of Stress Wave Theory to Piles: Science, Technology and Practice, Proceedings of StressWave 2008*, IOS Press, pp. 131–134.

269. Berre, T. and Bjerrum, L. (1973) Shear strength of normally consolidated clays. Proceedings of the Soil Mechanics and Foundation Engineering, Vol. 1, pp. 39–49.

270. Zhu, J. and Yin, J. (2000) Strain-rate-dependent stress–strain behaviour of overconsolidated Hong Kong marine clay. *Canadian Geotechnical Journal*, 37(6), 1272–1282.

271. Díaz-Rodríguez, J.A., Martínez-Vasquez, J.J., and Santamarina, J.C. (2009) Strain-rate effects in Mexico City soil. *Journal of Geotechnical and Geoenvironmental Engineering*, 135(2), 300–305.

272. Reece, A.R. and Grinsted, T.W. (1986) Soil mechanics of submarine ploughs. Proceedings of the Offshore Technology Conference, OTC-5341, pp. 453–461.

273. Palmer, A.C. (1999) Speed effects in cutting and ploughing. *Géotechnique*, 49(3), 285–294.

274. Cathie, D.N. and Wintgens, J.F. (2001) Pipeline trenching using plows: performance and geotechnical hazards. Proceedings of the Offshore Technology Conference, OTC-13145, 14 pp.

275. Krstelj, I. (1996) Behavior of laterally loaded pipes in dry and saturated sand (centrifuge testing). Ph.D. thesis, Department of Civil Engineering and Operations Research, Princeton University.

276. House, A.R., Oliveira, J.R.M., and Randolph, M.F. (2001) Evaluating the coefficient of consolidation using penetrometer tests. *Journal of Physical Modelling in Geotechnics*, 1(3), 17.

277. Oliphant, J. and Maconochie, A. (2006) Axial pipeline–soil interaction. Proceedings of the International Offshore and Polar Engineering Conference, ISOPE-06-172, 8 pp.

278. Bazant, Z.P. (1993) Scaling laws in mechanics of failure. *Journal of Engineering Mechanics*, 119(9), 1828–1844.

279. Bazant, P.Z. (1997) Scaling of structural failure. *Applied Mechanics Reviews*, 50(10), 593–620.

280. Bazant, P.Z. (1999) Size effect on structural strength: a review. *Applied Mechanics*, 69, 703–725.

281. Bazant, Z.P. (2000) Size effect. *International Journal of Solids and Structures*, 37, 69–80.

282. Carpinteri, A. (editor) (2008) *Size-Scale Effects in the Failure Mechanics of Materials and Structures*, Taylor & Francis, 594 pp.

283. Gui, M.W. and Bolton, M.D. (1998) Geometry and scale effects in CPT and pile design. In: Robertson, P.K. and Mayne, P.W. (editors), Proceedings of the Geotechnical Site Characterization, pp. 1063–1068.

284. Zhu, F., Clark, J.I., and Phillips, R. (2001) Scale effect of strip and circular footings resting on dense sand. *Journal of Geotechnical and Environmental Engineering*, 127(7), 613–621.

285. Cerato, A.B. and Lutenegger, A.J. (2007), Scale effects of shallow foundation bearing capacity on granular material. *Journal of Geotechnical and Environmental Engineering*, 133(10), 1192–1202.

286. Stone, K.J.L., Newson, T.A., and Bransby, M.F. (2005) Discussion: Uplift resistance of buried submarine pipelines: comparison between centrifuge modelling and full-scale tests. *Géotechnique*, 55(4), 338–340.

287. Stone, K.J.L. and Newson, T.A. (2006) Uplift resistance of buried pipelines: an investigation of scale effects in model tests. Proceedings of the International Conference on Physical Modelling in Geotechnics (ICPMG), pp. 741–746.

288. Randolph, M.F. and House, A.R. (2001) The complementary roles of physical and computational modeling. *International Journal of Physical Modelling in Geotechnics*, 1(1), 1–8.

289. Peek, R. (2006) Discussion of "Lateral pipe–soil interaction in sand with reference to scale effect". *Journal of Geotechnical and Geoenvironmental Engineering*, 132(10), 1371–1372.

290. Anastasopoulos, I., Callerio, A., Bransby, M.F., Davies, M.C.R., El Nahas, A., Faccioli, E., and Rossignol, E. (2008) Numerical analyses of fault–foundation interaction. *Bulletin of Earthquake Engineering*, 6(4), 645–675.

291. Aifantis, E.C. (1992) On the role of gradients in the localization of deformation and fracture. *International Journal of Engineering Science*, 30, 1279–1299.

292. Aifantis, E.C. (1984) On the microstructural origin of certain inelastic models. *Transactions of ASME: Journal of Engineering Materials and Technology*, 106(4), 326–330.

293. Fleck, N.A. and Hutchinson, J.W. (1997) Strain gradient plasticity. In: Hutchinson, J.W. and Wu, T.Y. (editors), *Advances in Applied Mechanics*, Academic Press, New York, Vol. 33, pp. 295–361.

294. Fleck, N.A. and Hutchinson, J.W. (2001) A reformulation of strain gradient plasticity. *Journal of the Mechanics and Physics of Solids*, 49, 2245–2271.

295. Sluys, L. and Estrin, Y. (2000) The analysis of shear banding with a dislocation based gradient plasticity model. *International Journal of Solids and Structures*, 37, 7127–7142.

296. Vardoulakis, I. and Aifantis, E. (1991) A gradient flow theory of plasticity for granular materials. *Acta Mechanica*, 87, 197–217.

297. Bazant, Z.P. and Chang, T.P. (1984) Instability of nonlocal continuum and strain averaging. *Journal of Engineering Mechanics*, 110, 1666–1692.

298. Bazant, Z.P. and Pijaudier-Cabot, G. (1988) Nonlocal continuum damage, localization instability and convergence. *Journal of Applied Mechanics*, 55, 287–293.

299. Bazant, Z.P. and Guo, Z. (2002) Size effect and asymptotic matching approximations in strain-gradient theories of microscale plasticity. *International Journal of Solids and Structures*, 39, 5633–5657.

300. Kulkarni, M. and Belytschko, T. (1988) On the effect of imperfections and spatial gradient regularization in strain softening viscoplasticity. *Mechanics Research Communications*, 18, 335–343.

301. Tomita, Y. (1994) Simulations of plastic instabilities in solid mechanics. *Applied Mechanics Reviews*, 47, 171–205.
302. Tejchman, J. (2004) Effect of heterogeneity on shear zone formation during plane strain compression. *Archives Hydro-Engineering and Environmental Mechanics*, 51(2), 149–181.
303. Jirásek, M. (2002) Objective modeling of strain localization. *Revue Française de Génie Civil*, 6(6), 1119–1132.
304. Chen, W.F. and Saleeb, A.F. (1982) *Constitutive Equations for Engineering Materials*, John Wiley & Sons, Inc.
305. USACE (1990) Settlement Analysis. U.S. Army Corps of Engineers, EM-1110-1-1904, 205 pp.
306. Hill, R. (1950) *The Mathematical Theory of Plasticity*, Oxford University Press, 355 pp.
307. Chakrabarty, J. (2006) *Theory of Plasticity*, Elsevier, 895 pp.
308. Chen, W.F. (2008) *Limit Analysis and Soil Plasticity*, J. Ross Publishing, 638 pp.
309. Desai, C.S. (2001) *Mechanics of Materials and Interfaces: Disturbed State Concept*, CRC Press, 698 pp.
310. Wood, D.M. (1990) *Soil Behaviour and Critical State Soil Mechanics*, Cambridge University Press, 462 pp.
311. Kolymbas, D. and Herle, I. (1997) Hypoplasticity: a framework to model granular materials. In: Cambou, B. (editor), *Behaviour of Granular Materials*, Springer, pp. 239–268.
312. Mroz, Z. (1980) On hypoelasticity and plasticity approaches to constitutive modelling of inelastic behavior of soils. *International Journal for Numerical and Analytical Methods in Geomechanics*, 4(1), 45–55.
313. Menétrey, Ph. and Willam, K.J. (1995) Triaxial failure criterion for concrete and its generalization. *ACI Structural Journal*, 92, 311–318.
314. Vardoukalis, I. and Sulem, J. (1995) *Bifurcation Analysis in Geomechanics*, Blakie Academic and Professional, 447 pp.
315. Drucker, D.C., Gibson, R.E., and Henkel, D.J. (1957) Soil mechanics and work hardening theories of plasticity. *Transactions of the American Society of Civil Engineers*, 122, 338–346.
316. Roscoe, K.H., Schofield, A.N., and Wroth, C.P. (1958) On the yielding of soils. *Géotechnique*, 8, 22–53.
317. Roscoe, K.H. and Burland, J.B. (1968) On the generalized stress-strain behaviour of wet clay. In: Heyman, J. and Leckie, F.A. (editors), *Engineering Plasticity*, Cambridge University Press, pp. 535–609.
318. Schofield, A.N. and Wroth, C.P. (1968) *Critical State Soil Mechanics*, McGraw-Hill, London, UK, 218 pp.
319. Rowe, P.W. (1971) Theoretical meaning and observed values of deformation parameters for soil. In: Parry, R.H.G. (editor), Proceedings of the Roscoe Memorial Symposium on Stress–Strain Behavior of Soils, pp. 143–194.
320. Daramola, O. (1980) On estimating $k_0$ for overconsolidated granular soils. *Géotechnique*, 30(3), 310–313.
321. Lade, P.V. and Nelson, R.B. (1987) Modelling the elastic behaviour of granular materials. *International Journal for Numerical and Analytical Methods in Geomechanics*, 11, 521–542.
322. Jaky, J. (1948) Pressure in silos. Proceedings of the Soil Mechanics and Foundations Engineering, Vol. 1, pp. 103–107.
323. Schmidt, B. (1966) Discussion paper. Earth pressure at rest related to stress history. *Canadian Geotechnical Journal*, 3(4), 239–242.
324. Mayne, P.W. and Kulhawy, F.H. (1982) $K_o$–OCR relationships in soil. *Journal of the Geotechnical Engineering Division*, 108(6), 851–872.
325. Tatsuoka, F. and Kohata, Y. (1994) Stiffness of hard soils and soft rock in engineering applications. In: Shibuya, S., Mitachi, T., and Miura, S. (editors), Proceedings of the Conference on Pre-Failure Deformation Characteristics of Geomaterials, Vol. 2, pp. 947–1063.
326. Jardine, R.J. (1992) Some observations on the kinematic nature of soil stiffness. *Soils and Foundations*, 32(2), 111–124.
327. Jamiolkowski, M., Lancelotta, R., and Lo Presti, D.C.F. (994) Remarks on the stiffness at small strains of six Italian clays. In: Shibuya, S., Mitachi, T., and Miura, S. (editors), Proceedings of the Conference on Pre-Failure Deformation Characteristics of Geomaterials, Vol. 2, pp. 816–836.
328. Krizek, R.J. (1977) Fabric effects on strength and deformation of kaolin clay. Proceedings of the International Conference on Soil Mechanics and Foundation Engineering, pp. 169–176.
329. Wong, P.K.K. and Mitchell, R.J. (1975) Yielding and plastic flow of sensitive cemented clays. *Géotechnique*, 25(4), 763–782.
330. Janbu, N. (1963) Soil compressibility as determined by oedometer and triaxial tests. Proceedings of the European Conference on Soil Mechanics and Foundation Engineering, Vol. 1, pp. 19–25.
331. Zytynski, M., Randolph, M.F., Nova, R., and Wroth, C.P. (1978) On modelling the loading and reloading behaviour of soils. *International Journal for Numerical and Analytical Methods in Geomechanics*, 2, 78–94.
332. Coon, M.D. and Evans, R.J. (1972) Recoverable deformation of cohesionless soils. *Journal of the Soil Mechanics and Foundations Division, ASCE*, 97(SM2), 375–391.
333. Corotis, R.B. Farzin, M.H., and Krizeck, R.J. (1974) Nonlinear stress–strain formulations for soil. *Journal of the Geotechnical Engineering Division*, 100(GT9), 993–1008.
334. Lade, P.V. (2005) Single hardening model for soils: parameter determination and typical values. In: Yamamuro, J.A. and Kaliakin, V.N. (editors), *Soils Constitutive Models. Evaluation, Selection, and Calibration*, American Society of Civil Engineers, pp. 290–309.
335. Keaveny, J. and Mitchell, J.K. (1986) Strength of fine-grained soils using the piezocone. Use of In-Situ Tests in Geotechnical Engineering, GSP 6, American Society of Civil Engineers, pp. 668–685.
336. Mahdavi, H. (2011) Influence of soil confinement on local buckling behavior of buried pipelines. Ph.D. thesis, Memorial University, St. John's, Newfoundland, Canada, 588 pp.
337. Been, K., Sancio, R.B., Ahrabian, D., Van Kesteren, W., Croasdale, K., and Palmer, A. (2008) Subscour displacement

in clays from physical model tests. Proceedings of the International Pipeline Conference, IPC-64186, pp. 239–245.

338. Piercey, G., Volkov, N., Phillips, R., and Zakeri, A. (2011) Assessment of frost heave modelling of cold gas pipelines. Proceedings of the Pan-American CGS Geotechnical Conference, 8 pp.

339. Morgan, V., Hawlader, B., and Zhou, J. (2006) Mitigation of frost heave of chilled gas pipelines using temperature cycling. Proceedings of the International Pipeline Conference, IPC-2006-10169, 5 pp.

340. Gaudin, C., Cluckey, E.C., Garnier, J., and Phillips, R. (2010) New frontiers for centrifuge modelling in offshore geotechnics. Proceedings of the International Symposium on Frontiers in Offshore Geotechnics (ISFOG), pp. 155–188.

341. Mahdavi, H., Kenny, S., Phillips, R., and Popescu, R. (2013) Significance of geotechnical loads on local buckling response of buried pipelines with respect to conventional practice. *Canadian Geotechnical Journal*, 50, 68–80.

342. Almahakeri, M., Moore, I., and Fam, A. (2012) The flexural behaviour of buried steel and composite pipes pulled relative to dense sand: experimental and numerical investigation. Proceedings of the International Pipeline Conference, IPC2012-90158, 9 pp.

# 9

# HUMAN FACTORS

LORNA HARRON
*Enbridge Pipelines Inc., Edmonton, Alberta, Canada*

## 9.1 INTRODUCTION

As important as the technical components of a design, construction, operation, and maintenance program is the human component of the activities being performed. According to the International Labor Organization (ILO) Encyclopedia of Occupational Health and Safety [1], a study in the 1980s in Australia identified that at least 90% of all work-related fatalities over a 3-year period involved a behavioral factor. When an incident investigation is performed, human factors are often lumped into the category of "human error" as either the immediate cause or a contributory cause. While this acknowledges the human as a contributor to an undesirable event, it does not provide sufficient details to facilitate reduction of the potential for recurrence of the human error. When up to 90% of incidents are related to a human factor, it is important to understand human factors and how to reduce the potential for human error in our business. This chapter is designed to provide the reader with a better understanding of human factors and how to manage the potential for human error in his/her organization.

## 9.2 WHAT IS "HUMAN FACTORS"?

"Human factors" is the study of factors and tools that facilitate enhanced performance, increased safety, and/or increased user satisfaction [2]. These three end goals are attained through use of equipment design changes, task design, environmental design, and training. "Human factors" is therefore goal oriented, focused on achieving a result rather than following a specific process. When studying human factors, the end product focuses on system design, accounting for those factors, psychological and physical, that are properties of the human element. "Human factors" centers the design process on the user to find a system design that supports the user's needs rather than making the user fit the system.

Other technical fields interact with or are related to human factors, such as cognitive engineering and ergonomics. Cognitive engineering focuses on cognition (e.g., attention, memory, and decision making). Cognitive engineering impacts how a human or machine thinks or gains knowledge about a system. Ergonomics is related to human factors, focusing on the interaction of the human body with a workspace or task.

It is well understood in industry that improvements at the design stage of a project cost significantly less than improvements post-construction, so management of human factors at the design stage of a project will provide greatest value (Table 9.1).

## 9.3 LIFE CYCLE APPROACH TO HUMAN FACTORS

In the pipeline industry, human factors can create the potential for a human error at many points along the life cycle of a pipeline. At the Banff/2013 Pipeline Workshop, a tutorial on human factors was conducted using a life cycle analysis to identify where there is a potential for human error. Figure 9.1 illustrates the results of this discussion.

*Oil and Gas Pipelines: Integrity and Safety Handbook,* First Edition. Edited by R. Winston Revie.

**TABLE 9.1 Benefits of Incorporating Human Factors at the Design Stage**

1. Increased sales
2. Decreased development costs
3. Decreased maintenance costs
4. Decreased training cost
5. Increased productivity
6. Fewer errors by human, machine, or human–machine interface (HMI)
7. Increased employee satisfaction
8. Decrease in number of incidents
9. Decrease in medical expenses and number of sick days taken by employees

Using a life cycle approach to manage human factors provides an organization with the capability to integrate human factors into programs, standards, procedures, and processes using a disciplined approach. In order to use the human factors life cycle, organizations may wish to assess themselves against the areas identified for the presence of programs, standards, procedures, and processes to manage human error potential.

An examination of the potential for human error using a life cycle approach also highlights the interrelationship between human factors management and management systems. Common management system elements, such as change management and documentation, are reflected in each stage of the human factors life cycle.

Opportunities exist to reduce the potential for human error in design, construction, operation, maintenance, and decommissioning through use of robust quality control processes, such as self-assessments. Incorporation of human factors assessment techniques such as job hazard analysis (JHA), job task analysis (JTA), human factors failure modes and effect analysis (FMEA), and workload analysis can help operators understand their current state and potential gaps from a desired state. Specialists in human factors can be engaged to assess and mitigate any/all of the areas identified in the life cycle provided in Figure 9.1.

To illustrate this, an example case study is provided below.

### 9.3.1 Example Case Study

XYZ Pipeline Company has a Pipeline Integrity Program that includes threat mitigation for corrosion, cracking, deformation and strain, third-party damage, and incorrect operations. XYZ Pipeline Company has decided to perform a gap analysis for integration of human factors using the life cycle approach. In the "maintain" category of the life cycle, the company is confident that they use the best available technology to monitor and assess the integrity condition of their pipelines. They use Vendor A, who is a leader in inspection technology in the industry, to perform pipeline inspections and to provide detailed analyses on the results of these inspections. Using the life cycle approach, XYZ Pipeline Company evaluates how they have incorporated human factors into the identified areas. First, they ask whether there is a potential for human error in the determination of inspection frequency. The process used to determine inspection frequency involves a review process by at least three senior engineering professionals, and it is determined that the potential for an error at this point is minimized through the existing process. The next question asked is whether there is potential for human error in the selection of the inspection tool used to evaluate the threat. The existing process relies on the expertise and experience of integrity management engineers at the company. While today these individuals have over 20 years of industry experience, some are preparing to retire in the next 2–5 years. This risk is highlighted as a potential gap and action planning identifies the need to create a guideline for tool selection. During the discussions on this topic, it is also identified that the technology is changing rapidly and that understanding tool capability requires constant monitoring of the inspection market. Another action is then identified to create a work task for a senior engineer to monitor the inspection market at least annually to ensure the guideline is current. Next, they question the potential for an error in the inspection report that they base integrity decisions on. This discussion highlights several potential gaps, including a lack of quality control on the vendor reports received by XYZ Pipeline Company, a lack of understanding of the experience and expertise of the people performing the defect analysis within the tool inspection company, the lack of expertise in XYZ Pipeline Company in reading and understanding defect signals for all the inspection tools, and the lack of processes between the time the tool inspection is performed and the mitigation plan is prepared. Action planning for these gaps required creation of a small task force to focus on mitigating actions to manage these gaps. At this point, XYZ Pipeline Company decided that this was as far as they could progress in a single year, so a multiyear assessment protocol was developed to review the Integrity Management Program over a 5-year term to assure integration and management of human factors through the life cycle of their pipelines.

## 9.4 HUMAN FACTORS AND DECISION MAKING

Human factors impact a pipeline system when a nonoptimal decision has been made. There are many factors that impact how information is processed and subsequent decision making.

Human factors impact how we process information for decision making. The steps involved in decision making

| Human factors life cycle | | | | |
|---|---|---|---|---|
| Areas/topics with potential for human error | | | | |
| **Design** | **Build** | **Operate** | **Maintain** | **Decommission** |
| Design parameters | Site safety practices/guidelines | Fatigue | Inspection scheduling (ILI frequency) | Safety of practices for decommissioning |
| Design change (MOC) | Construction practices | Incorrect operation | Design of inspection program (tool choice) | Records management/documentation |
| Assumptions used | Construction schedule | Training effectiveness | Process management | Accuracy of drawings (MOC) |
| Interpretation (scope) | Quality control process | Experience of operators | Records management/documentation | ID of repurposing for pipeline |
| ID of design limitations | Experience of people/contractors | Competency of operators | Management of integrity threats | Classification of line |
| ID of design constraints | Quality of mill inspection | Response to changes (MOC) | Motivation of personnel/organization | Decision to use line for other product |
| Experience level of engineer | Interpretation of build scope | Not following procedures | Budget process/constraints | Decommission experience |
| Understanding of environment | Accuracy of as-built drawings | Poor written procedures | Prioritization of dig sites | Interpretation of decommission scope |
| Useful life assumed | Field level modification (MOC) | Quality of communications | Prioritization of PL segments to inspect | Field level modification (MOC) |
| Budget process/constraints | Records management/documentation | Interactions with externals (e.g., landowners) | Balancing capacity needs and integrity | |
| Interaction of design elements | Constructability of the chosen design | Documentation/records management | Interpretation of organizational culture | |
| Interpretation of regulations | Communication methods used | Understanding/managing risks | Incorrect interpretation of data (vendors) | |
| Determining current needs | Internal/external communications | Decision making under stressful situations | Incorrect interpretation of data (company) | |
| Determining future needs | Contractor management | Vendor management | Quality of data received from vendor | |
| Prediction of future service | | | Decision making in short time frame | |

| **Common elements** |
|---|
| Change management |
| Records management/documentation |
| Interpretation |
| Communication |

FIGURE 9.1 Human factors life cycle, Banff/2013 Pipeline Workshop.

include receipt of information, processing of information, and deciding on a course of action (the decision). Receipt of information focuses on information cues, while processing of information focuses on memory and cognition [2]. Deciding on a course of action involves heuristics (simplifying decision-making tools) and biases.

### 9.4.1 Information Receipt

Information is received into the body as cues. These cues may be visual such as printed words in a procedures document, or auditory such as speech or alarms. The quality of the cues provided will increase the probability of receipt of the information into awareness and eventually action. Enhancement of cues can be provided through targeting, which involves making a cue conspicuous and highly visible while maintaining the norms for a cue. For example, if a cue to an operator to perform an action is a light turning on, then from a human factors perspective a few simple steps can ensure that the appropriate action is clear to the operator. First, an evaluation of what action the cue is supposed to elicit is required. If the light indicates an upset condition and the resulting action is to turn off a device, then the normal color anticipated by the operator would be red. If the light indicates a desired state and the action is to start up a piece of equipment, then the normal color anticipated by the operator would be green. If the colors used were not consistent with the expectations of the operator, then the likelihood of a human error such as an incorrect reaction to the cue is heightened.

An example of expectancy is what we anticipate we will see at a traffic light. If we see the traffic light on the left-hand side of Figure 9.2, this meets our expectation of what a traffic light should look like, considered as meeting our norms. The traffic light on the right-hand side of the figure, however, does not meet our expectations and norms and would therefore be likely to contribute to a human error and subsequent traffic accident.

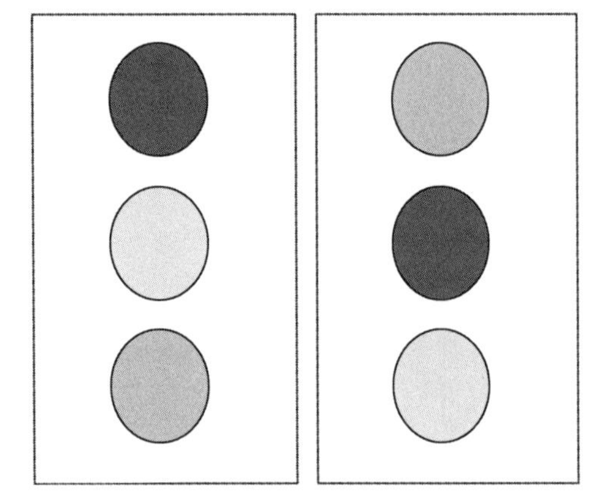

**FIGURE 9.2** Expectancy.

### 9.4.2 Information Processing

The manner in which we process information is impacted through our memory systems. Prior knowledge is stored in long-term memory and information is manipulated in working memory.

Working memory is like a shelf of products that are available for a short period of time but that require a lot of effort to maintain. Information processing depends on cognitive resources, which is a pool of attention or mental effort available for a short period of time that can be allocated to processes such as rehearsing, planning, understanding, visualizing, decision making, and problem solving. How much effort we put into working memory will be based on the value we place on the information. If we put effort into only a few areas, the result is "selective attention." Effort may be influenced by fatigue, where the effort to perform a task is perceived to be greater than the anticipated value. There are some limitations to working memory based on its transitory nature and limited amount of cognitive resources. Working memory is limited to $7 \pm 2$ chunks of information [3], where a chunk of information may be grouped information such as a phone number area code or PIN for your computer. Working memory is also limited by time, decaying as time progresses. Another limitation of working memory is confusion of similar pieces or chunks of information, causing erroneous information to be processed. When translated to an operating environment, the amount of activity in the operating environment can impact the ability to process information in working memory effectively. Some tips for managing working memory are listed in Table 9.2.

Long-term memory is a way to store information for retrieval at a later time. When we learn something, we are processing stored information into long-term memory. Training is specific instruction or procedures designed to facilitate learning. Long-term memory is distinguished by whether it involves memory for general knowledge (semantic memory) or memory for specific knowledge (event memory) [4].

Material in long-term memory has two important features that determine the ease of later retrieval: its strength and its

**TABLE 9.2 Tips to Manage Working Memory**

1. Minimize working memory load by using written information rather than verbal
2. Provide visual echoes/redundancy
3. Provide placeholders for sequential tasks
4. Exploit chunking:
   - physical chunk size 3–4 letters or numbers per chunk
   - meaningful sequences
   - letters better than numbers
   - separating numbers from letters
5. Minimize confusability/use of negation (do not)
6. Consider working memory limits in instructions

**TABLE 9.3 Tips to Manage Long-Term Memory**

1. Encourage regular use of information to increase frequency and recency
2. Encourage active verbalization or reproduction of information that is to be recalled
3. Standardize
4. Use memory aids
5. Carefully design information to be remembered
6. Design to support development of correct mental models

associations. Strength is determined by the frequency and recency of its use. For example, a PIN or password used frequently will have strength, and if the PIN or password is associated with something relevant to the individual, it will also have an association and therefore be easier to recall.

Associations occur when items retrieved in long-term memory are linked or associated with other items—if associations are not repeated, the strength of association weakens over time. Tips for managing long-term memory are listed in Table 9.3.

Metaknowledge or metacognition refers to people's knowledge or understanding about their own knowledge and abilities. Metacognition impacts people's choices, in the actions they choose to perform, the information or knowledge they choose to communicate, and choices of whether additional information is required to make a decision. Knowledge can be based on perceptual information and/or memory. The accuracy of information may be greater for perceptual information, but it may require greater effort than use of information stored in memory. According to resource theory [5], we make decisions under limited time with limited information. Our scarce mental resources are shared by various tasks, and more difficult tasks require greater resources. Some concurrent tasks may therefore see performance decline (see Table 9.4).

***9.4.2.1 Stressors*** Information processing is influenced by stressors that are outside of the information itself. These stressors may be physical, environmental, or psychological such as life stressors and work or time stressors [2]. Examples of common stressors include noise, vibration, heat, lighting, anxiety, and fatigue.

The effect of stressors may be physical and/or psychological. Some stressors produce a physiological reaction, such as raised heart rate, nausea, or headache. Some stressors may produce a psychological response such as mood swings. The duration and intensity of a stressor may lead to long-term negative effects on the health of the individual. In most cases, the stressors affect information receipt and processing.

*Work Overload* Work overload occurs when there are more decisions to be made than cognitive resources available to make optimal decisions. When an overload is encountered, the decision maker becomes more selective in the receipt of cues. Limiting the inputs received through a selective attention filtering process allows the decision maker more cognitive resources for the task at hand—making a decision! Another result of work overload may be weighing information that is received to limit the amount of information that is processed. This will result in better use of available cognitive resources, if the screening is performed accurately. The filtering and screening processes will result in choosing the "easiest" decision. The results of this could be less accuracy in the decision-making process and potentially cognitive tunneling or fixation on a specific course of action [6]. Work overload can manifest in fatigue or in sleep disruption.

Stress can occur because we have too many things to do in too short a time. Management of work overload as a stressor requires an evaluation of current workload [7]. To determine workload, consider which tasks need to be accomplished, when these tasks need to be performed, and how long the typical tasks take to accomplish. When considering the workload, consideration of time to perform a task needs to include cognitive tasks such as planning, evaluating alternatives, and monitoring success (Table 9.5). It is difficult to quantify the impact of workload. Will each individual perform a task exactly the same way? Will individuals process information and make decisions at the same speed? Are there

**TABLE 9.4 Ways to Address a Multitasking Environment**

| Method | Results/Impact |
|---|---|
| 1. Task redesign | Reduce working memory limitations |
| 2. Interface redesign | Direct selective attention (e.g., use voice for tasks when eyes are required for a different task) |
| 3. Training | Develop automaticity through repeated and consistent practice; improve resource allocation through training in attention management |
| 4. Automation | Limit resource demands through number of tasks; direct selective attention |

**TABLE 9.5 Tips to Manage Workload Overload**

1. Critical task analysis to determine task priorities and time requirements
2. Task redesign (redistribute tasks to others; automate some tasks; reduce potential for overlapping or interfering tasks)
3. Develop a display design so that the most important tasks are obvious to the user (example is red alarm for a control panel)
4. Training on tasks to increase speed
5. Training on time management skills

**TABLE 9.6 Tips to Manage Fatigue**

1. Ensure adequate hours of sleep
2. Napping when feeling the impact of sleep disruption (at least 15 min in duration)
3. Training on fatigue management for shift workers
4. Avoid stimulants, such as caffeine

**TABLE 9.7 Tips to Manage Vigilance**

1. Limit vigil duration (length of time between rest breaks)
2. Enhance cues/signals to make them more obvious
3. Use of music, noise, and conversation
4. Ensure operators are not sleep deprived
5. Use automation carefully, as this may create vigilance tasks

other distractions or overlapping information that would impact the task and the speed of accomplishing a task?

*Fatigue* Fatigue is the state between being alert and being asleep. Fatigue may be mental, physical, or both in nature. In the pipeline industry, it is commonly understood that long shifts, in particular night shifts, may result in fatigue [8]. When fatigue occurs, performance level will be impacted. Management of fatigue includes scheduling rest breaks and enforcement of maximum hours to work in one day (Table 9.6).

Sleep disruption is a major contributor to fatigue. Sleep disruption is influenced by the amount of time you sleep each night. Sleep deprivation occurs when an individual receives less than 7–9 h of sleep per night. For shift workers, sleep disruption occurs when tasks are performed at the low point of natural circadian rhythms, which equates to the early hours of the morning. Circadian rhythms are the natural cycles of the body regulated by light/darkness that impact physiological functions such as body temperature, sleeping, and waking [9]. For those who travel extensively, disruption of circadian rhythms may occur from jet lag. Understanding the impact of sleep disruption is important when evaluating the potential for a human error to occur. Tasks requiring input from visual cues are impacted by sleep disruption. Vigilance tasks, common for control center operations, may result in inattention and poor decision making as a result of sleep disruption.

*Vigilance* Vigilance is defined as the detection of rare, near-threshold signals in noise. In a control room, the rare, near-threshold signal may be a small fluctuation in a process variable that precedes a larger change that is picked up by the alarm system. Flying an airplane is considered to be a vigilance task, with takeoff and landing as the two main tasks accomplished while in flight. Monitoring of the equipment gauges requires some mental effort, but action is only required when a nonstandard condition arises. Similar to this aviation example, in the pipeline industry a vigilance task may be monitoring a control panel in a control center. A concern with vigilance tasks is the increased number of misses that may occur in relation to the length of the vigilance task, called vigilance decrement. For a control center operator, management of alarms or other cues on a regular basis rather than having long periods of time between cues would reduce this concern. This and other tips for managing vigilance are listed in Table 9.7.

*Lighting and Noise* Lighting and noise are two environmental stressors that can be remedied in a work environment (Table 9.8). Performance can be degraded when the amount of light and/or noise is too high or too low [2]. If continued high exposure (e.g., noise) or low exposure (e.g., light) occurs, then health issues may also result. These effects may not be obvious until some point in the future, so correlation with this stressor may be difficult.

Readers who wish to apply the above tips can refer to the following guidance documents:

- Criteria for a Recommended Standard: Occupational Exposure to Noise, U.S. Department of Health, Education and Welfare, National Institute for Occupational Safety and Health, 1972.
- Attwood, D., Deeb, J., and Danz-Reece, M. *Ergonomic Solutions for the Process Industries*, Gulf Publishing (Elsevier), Burlington, MA, 2004 (Chapter 4).

*Vibration* Vibration is an environmental stressor that can lead to both performance issues and health issues (e.g., repetitive motion disorders and motion sickness) [10]. High-frequency vibration can impact either a specific portion of the body or the whole body. Types of tasks that affect whole-body vibration include off-road driving and working around large compressors or pumps. The impact of vibration that affects the whole body

**TABLE 9.8 Tips to Manage Lighting and Noise**

1. Reduce glare on visual surfaces such as computer screens
2. Ensure sufficiently high level of light for operating tasks
3. Promote use of sunglasses for appropriate conditions and tasks (e.g., driving)
4. Use of noise-muffling devices for loud operating environments (>85 dB)
5. Consider background noise levels when designing sounds that require an action (e.g., alarm or voice command)
6. Follow alarm guidelines in appropriate regulations (e.g., control room management guidelines)
7. Minimize potential for false alarms that could create complacency

**TABLE 9.9 Tips to Manage Vibration**

1. Ensure that text fonts are larger than the minimum specified for stable environments
2. Insulate the user/interface from vibration, for example, hand-tool vibration coverings, improved seating for off-road vehicles

includes stress, risk of injury [10], difficulty using devices such as touchscreens or handheld devices, difficulty with eye–hand coordination, and blurred vision. At lower vibration frequencies, the impact of vibration can be motion sickness that in turn impacts performance through inattention. Tips for managing vibration are listed in Table 9.9.

*Temperature* Temperature impacts performance and health when it is either too low or too high [11]. Both excessive heat and excessive cold can produce performance degradation and health problems. Health issues can include a physical reaction to excessively high or low temperatures, such as heatstroke and frostbite. The performance impact can be cognitive, focusing on how information is processed by an individual. For managing problems related to temperature, some tips are listed in Table 9.10.

*Air Quality* Air quality issues can negatively impact performance and health of an individual. Air quality is not limited to internal spaces, such as an office, but also external spaces such as a facility. Poor air quality can result from inadequate ventilation, pollution, smog, or the presence of carbon monoxide. Poor air quality can impact cue recognition, task performance, and decision-making capability [12].

*Psychological Stress* Psychological stress can be triggered by a perception of potential harm or potential loss of something of value to the individual. It is extremely difficult to design for this type of stressor, since it is related to perception, and perception varies with individuals. The natural question is "What impacts the perception of harm or loss to an individual?" The way a person can interpret a situation will impact the stress level they feel. Cue recognition and information processing are key elements of this interpretation. Psychological stress is a real phenomenon that is hard to express and manage (Table 9.11). Physiological reactions may occur due to the psychological stressors, such as elevated heart rate. At an optimal amount of psychological stress, we can see performance improvement. When this optimal level is exceeded, however, the result can be elevated feelings of anxiety or danger, or "overarousal." Information processing when in a state of overarousal can be altered, impacting memory and decision-making capability.

**TABLE 9.10 Tips to Manage Temperature**

1. Ensure air movement in hot work environments through use of fans or natural air currents/breeze
2. Reduce physical work requirements in extreme hot or cold temperatures
3. Reduce amount of time exposed to extreme hot or cold temperatures
4. Ensure there is lots of water/fluids to replenish hydration
5. Choose clothing appropriate for the environment

**TABLE 9.11 Tips to Manage Psychological Stress**

1. Simplification of displays, controls, and procedures to impact cue reception and information processing
2. Training on stressors to build individual awareness of psychological stress triggers

**TABLE 9.12 Tips to Manage Life Stress**

1. Implement stress management programs
2. Provide counseling services
3. Train leaders in the identification and management of stress in employees

*Life Stress* Life itself may be stressful outside the work environment. Life events, such as a health concern with a loved one, can impact work performance and potentially cause an incident. When an incident occurs related to human factors, the classification may be "inattention." The cause of inattention is likely the diversion of attention from work tasks and toward the source of life stress. Some tips for managing life stress are listed in Table 9.12.

## 9.5 APPLICATION OF HUMAN FACTORS GUIDANCE

Guidelines for management of working memory are available in other industries, including aviation, medicine, and road transportation. The following are some examples of references for the reader to explore and apply.

1. HARDIE Design Guidelines Handbook: Human Factors Guidelines for Information Presentation by ATT Systems, 1996.
2. Development of Human Factors Guidelines for Advanced Traveler Information Systems and Commercial Vehicle Operations: Driver Memory for In-Vehicle Visual and Auditory Messages. Publication No. FHWA-RD-96-148, 1996.
3. NCHRP Report 600: Human Factors Guidelines for Road Systems, 2nd ed., 2012.

4. Do It by Design: An Introduction to Human Factors in Medical Devices, U.S. Department of Health and Human Services.
5. Human Factors Guidelines for Aircraft Maintenance Manual. ICAO Document 9824 AN/450, 2003.
6. Health and Safety Executive—Managing Shift Work: Health and Safety Guidance, 2006.
7. Applying Human Factors and Usability Engineering to Optimize Medical Device Design—Draft, 2011.

## 9.6 HEURISTICS AND BIASES IN DECISION MAKING

Decision making is a task involving selection of one option from a set of alternatives, with some knowledge and some uncertainty about the options. When we make decisions, we plan and select a course of action (one of the alternatives) based on the costs and values of different expected outcomes.

There are various decision-making models in the literature, including rational or normative models, such as utility theory and expected value theory, and descriptive models [13,14]. If there is good information and lots of time available to make a decision, a careful analysis using rational or normative models is possible. If the information is complex and there is limited time to make a decision, simplifying heuristics are often employed. The focus of this section is on descriptive models such as heuristics and biases.

When decisions are made under limited time and information, we rely on less complex means of selecting among choices [15,16]. These shortcuts or rules of thumb are called heuristics. The choices made using heuristics are often deemed to be "good enough." Because of the simplification of information receipt and processing employed using heuristics, they can lead to flaws or errors in the decision-making process. Despite the potential for errors, use of heuristics leads to decisions that are accurate most of the time. This is largely due to experience and the resources people have developed over time.

Cognitive biases result when inference impacts a judgment. Biases can result from the use of heuristics in decision making.

There are a number of specific heuristics and biases that will be explored in this section, including

- satisficing heuristic;
- cue primacy and anchoring;
- selective attention;
- availability heuristic;
- representativeness heuristic;
- cognitive tunneling;
- confirmation bias;
- framing bias.

### 9.6.1 Satisficing Heuristic

As may be surmised from the name, the satisficing heuristic is related to a decision that satisfies the needs of the user. The person making the decision will stop the inquiry process when an acceptable choice has been found, rather than continuing the inquiry process until an optimized choice has been found [17]. This heuristic can be very effective, as it requires less time. As it does not provide an optimal solution, however, it can lead to biases and poor decisions. This heuristic can be advantageous when there are several options for achieving a goal and a single optimized solution is not required, such as a decision on providing an example of a data set for display in a presentation. This heuristic could lead to less than optimal result when the decision requires an extensive review to determine a single solution, such as a decision to choose a method for defect growth prediction.

### 9.6.2 Cue Primacy and Anchoring

This heuristic is illustrated when the first few cues that have been received receive a greater weight or importance versus information received later [15]. This leads the decision maker to support a hypothesis that is supported by information received early. This heuristic could lead to decision errors if extensive analysis or increased depth of analysis occurs as part of the inquiry process. For example, use of cue primacy and anchoring would not be appropriate for decision making during a root cause investigation.

### 9.6.3 Selective Attention

Selective attention occurs when the decision maker focuses attention on specific cues when making a decision [2]. A decision maker has limited resources to process information, so we naturally "filter out" information and focus attention on certain cues. If an operator has many alarms occurring at once, then focusing on a few selective cues/alarms influences the action that is taken. Selective attention is impacted by four factors—salience (prominence or conspicuous), effort, expectancy, and value. Cues that are more prominent, such as cues at the top of a display screen, will receive attention when there are limited resources and become the information used to make a decision. Selective attention is an advantage in a well-designed environment where many potential cues are present. However, it can lead to poor decision making if the important cues are not the most salient in the environment.

### 9.6.4 Availability Heuristic

The meaning of the availability heuristic is represented by its name. Decision makers generally retrieve information or decision-making hypotheses that have been considered recently or frequently [18]. This information comes foremost to the mind of the decision maker and decisions can be made based on the ability to recall this information easily and quickly. For example, if a pipeline designer has recently completed a design for a pipe in rocky conditions and then asked what considerations should be included in the design of a different pipeline, the designer will readily recall soil conditions and comment that this should be considered for the new project. This heuristic is of great benefit when brainstorming various options to consider, but can be limiting if a full analysis of all options is required.

### 9.6.5 Representativeness Heuristic

Once again, the meaning of the representativeness heuristic is as its name implies. Decision makers will focus on cues that are similar to something that is already known [16]. The propensity is to make the same decision that was successful in a previous situation, since the two cues "look the same." For example, when an analyst is evaluating ILI data, he/she may see two signals that look very similar and due to the representativeness heuristic use the same characterization for both features. When the two images are identical, this heuristic provides a fast and effective means of decision making, or in this case characterizing features. If, however, the signals are similar but not identical, then the representativeness heuristic may lead to a decision-making error.

### 9.6.6 Cognitive Tunneling

Cognitive tunneling may be colloquially known as "tunnel vision." In this instance, once a hypothesis has been generated or a decision chosen to execute, subsequent information is ignored [19]. In decision making, cognitive tunneling can easily lead to a decision error. In order to avoid cognitive tunneling, the decision maker should actively search for contradictory information prior to making a decision. For example, asking for a "devil's advocate" opinion is one means of testing a hypothesis prior to finalizing a decision.

### 9.6.7 Confirmation Bias

In confirmation bias, the decision maker seeks out information that confirms a hypothesis or decision [20]. Similar to cognitive tunneling, focusing only on confirming evidence, and ignoring other evidence that could lead to an optimal decision, can lead to a decision error. When relying on memory, a tendency to fail to remember contradictory information, or to underweigh this evidence in favor of confirming evidence, impacts our ability to effectively diagnose a situation, prepare a hypothesis, and subsequently make an optimal decision. The main difference between cognitive tunneling and confirmation bias is that cognitive tunneling ignores all information once a hypothesis or decision has been chosen. With confirmation bias, however, information is filtered for only that information that confirms the chosen hypothesis or decision.

**TABLE 9.13 Factors that Influence Decision Making**

1. Inadequate cue integration
2. Inadequate or poor-quality knowledge the person holds in long-term memory
3. Tendency to adopt a single course of action and fail to consider the problem broadly
4. Working memory capacity; limits to attention
5. Poor awareness to changing situation
6. Poor feedback

### 9.6.8 Framing Bias

While many of the biases that occur are on a subconscious level, it is possible to use framing bias deliberately when trying to influence others. Framing bias involves influencing the way material is received to impact the decision that is made by another person [21]. For example, when writing a procedure about a requirement to perform an activity under specific circumstances, there is a framing bias imposed if the wording indicates the activity is not done unless certain criteria are met instead of that the activity is done when certain criteria are met. The decision maker is influenced to decide that the activity is not performed, rather than performed, based on the selected wording.

### 9.6.9 Management of Decision-Making Challenges

Some factors that influence decision making are listed in Table 9.13. The most effective means of correcting poor decisions is through feedback that is clearly understood with some diagnostics included. In many organizations, the largest contributor to incidents involves motor vehicles. When driving, it is difficult to learn from previous behavior and make better decisions due to the lack of feedback received while performing the task. When operating a pipeline, the operator receives feedback through alarms to indicate if a decision was effective. The alarm feedback may not lead to an optimal decision, however, so use of simulators to provide feedback enhances optimal decision making.

***9.6.9.1 Methods to Improve Decision Making*** There are many ways, as we have described, to negatively impact decision making and subsequently increase the potential for human error to occur. The process of making a decision is influenced by design, training, and the presence of decision aids [2]. We often jump to the conclusion that poor performance in decision making means that we must do something "to the person" to make him or her a better decision maker. The category of "human error," when used during an incident investigation, tends to result in activities related to the person who was performing the task at the time the incident occurred. A more holistic approach would be to consider the human being as well as opportunities to improve the way this human being could make a decision in a similar situation. Such an approach involves, potentially, task redesign, the use of decision support systems, and training.

A decision support system is an interactive system designed to improve decision making by extending cognitive decision-making capabilities. These support systems can focus on the individual, using rules or algorithms that require a consistent decision to be made by an individual. For tasks that result in a pass/fail decision with little deviation in the information received this approach works well, since decision consistency is more important than optimizing decisions made in unusual situations [22]. Focus on the individual can fail when there is a potential for unusual/nonroutine information receipt or when data entry mistakes can occur.

Alternatively, decision support systems can focus on tools to support decision making. For tasks that require the ability to make decisions in the presence of unusual/nonroutine information, this type of support system can complement the human in the decision-making process.

Examples of tools to support decision making include decision matrices for a risk management approach to decision making, spreadsheets to reduce cognitive loading, simulations to perform "what–if analyses," expert systems such as computer programs to reduce error for routine tasks, and display systems to manage visual cues. Some of these will utilize automation, so it is important to keep in the mind the potential to create vigilance tasks. Balancing the amount of automation for optimized decision making is an art rather than a science.

#### *9.6.9.2 Case Study: USS Vincennes*

*Overview* On July 3, 1988, the USS Vincennes accidentally shot down an Iranian airbus (Flight 655) resulting in the death of 290 people [23]. This occurred in the Strait of Hormuz in the Persian Gulf. Aboard the ship, the cruiser was equipped with sophisticated radar and electronic battle gear called the AEGIS system. The crew tracked the oncoming plane electronically, warned it to keep away, and, when it did not, fired two standard surface-to-air missiles. The Vincennes' combat teams believed the airliner to be an Iranian F14 jet fighter. No visual contact was made with the aircraft until it was struck and blew up about six miles from the Vincennes. Although the United States and Iran were not at war, there were some short but fierce sea battles in the Gulf between the United States and Iran. The year before, on May 17, there was an incident involving the near-sinking of the USS Stark by an Iraqi fighter-bomber. The Iraqi fighter launched a missile attack that killed 37 U.S. sailors. Information recently received from U.S. intelligence predicted a high-profile attack for July 4th, the day after this incident occurred. The USS Vincennes and another U.S. vessel were in a gunfight with an Iranian gunboat when the "fighter" appeared on the radar screen. The airstrip from which the airbus left was the same airstrip used by military aircraft. Vincennes tried to determine "friend or foe" using electronic boxes that are in military planes. These warnings were sent out on civilian and military channels with no response from the airbus. The airbus flight path was over the Vincennes and during the flight path the crew of the Vincennes interpreted the airbus to be descending. The resulting decision was for the Vincennes to launch two missiles resulting in the destruction of the airbus and all passengers.

*Cue Analysis* In order to understand the human factors at play, it is important to understand the cues that were perceived and how these cues impacted the decision-making process. It was noted that the radio was using the "friend or foe" on the military channel because the operator had earlier been challenging an Iranian military aircraft on the ground at Bandar Abbas, and had overlooked to change the range setting on his equipment. Flight 655 was actually transmitting IFF Mode III, which is the code for a civilian flight. Although the warning was broadcast from the USS Vincennes and the cue received by the airbus crew, the civilian frequency warning was ignored. The airbus was interacting with air traffic control and likely did not suspect that the military channel warning was intended for them.

Noise and activity level impacted the auditory cues as some officers thought the airbus was commercial, while others thought it was a fighter jet, but only the first information (fighter jet) was "heard."

*Heuristics Analysis* Several heuristics impacted the decision that was made to launch the two missiles.

1. Availability heuristic: The surprise attack on USS Stark the previous year may have impacted the decisions made. The Stark incident became the best available exemplar; every approaching radar trace tended to indicate a potential surprise attack.
2. Representativeness heuristic: The attack was similar to the USS Stark as it had not responded to warnings and flew at the boat.
3. Expectation: Captain Rogers was aware of the rumors that the Iranians were planning some sort of token Independence Day attack. Here, too, every approaching radar trace became the target of intense suspicion.

4. Cognitive framing: On repeated occasions, Flight 655 was reported as a descending warplane, when Vincennes' own data tapes indicate that it was transmitting the correct Mode III signal continuously, steered a straight course for Abu Dhabi (its destination on the opposite shores of the Persian Gulf), and was climbing steadily throughout the proceedings.

*Case Study Summary* In understanding the events that led up to this event, it is easy to determine that they "should have known better" or use hindsight bias. The presence of information that indicated a potential attack, under stressful environmental and psychological conditions coupled with limited time for decision making, resulted in the use of heuristics and subsequent decision errors.

## 9.7 HUMAN FACTORS CONTRIBUTION TO INCIDENTS IN THE PIPELINE INDUSTRY

Could human factors impact the pipeline industry? It has in the past. From incident reports on the NTSB website (http://www.ntsb.gov/investigations), the following examples were found:

1. Natural gas transmission pipeline rupture and fire in Cleburne, Texas, on June 7, 2010.

   The National Transportation Safety Board (NTSB) determined that the probable cause of the rupture and fire was a contractor's puncturing the unmarked, underground natural gas pipeline with a power auger. Contributing factors were the lack of permanent markers along the pipeline and failure of the pipeline locator to locate and mark the pipeline before the installation of a utility pole in the pipeline right-of-way.

   From a human factors perspective, the presence of visual cues to make decisions is critical. Unmarked pipelines represent a lack of visual cues, which contributed to an incident.

2. Natural gas transmission pipeline rupture and fire in Palm City, Florida, on May 4, 2009.

   The National Transportation Safety Board determined that the probable cause of the accident was environmentally assisted cracking under a disbonded polyethylene coating that remained undetected by the integrity management program. Contributing to the accident was the failure to include the pipe section that ruptured in their integrity management program. Contributing to the prolonged gas release was the pipeline controller's inability to detect the rupture because of SCADA system limitations and the configuration of the pipeline.

   From a human factors perspective, the ability of the operator to make a decision about the presence of a leak was impacted by the visual cues provided by the SCADA system.

3. Hazardous liquid pipeline rupture in Marshall, Michigan, on July 25, 2010.

   The National Transportation Safety Board determined that the probable cause of the pipeline rupture was corrosion fatigue cracks that grew and coalesced from crack and corrosion defects under disbonded polyethylene tape coating, producing a substantial crude oil release that went undetected by the control center for over 17 h. The rupture and prolonged release were made possible by pervasive organizational failures that included the following:

   - Deficient integrity management procedures, which allowed well-documented crack defects in corroded areas to propagate until the pipeline failed.
   - Inadequate training of control center personnel, which allowed the rupture to remain undetected for 17 h and through two start-ups of the pipeline.
   - Insufficient public awareness and education, which allowed the release to continue for nearly 14 h after the first notification of an odor to local emergency response agencies.

   From a human factors perspective, following a human factors life cycle approach can minimize the potential for human error and decision-making flaws. Additionally, the ability of an operator to make decision based on visual cues is highlighted. This incident also illustrates the need to consider human factor interfaces external to an organization, including the public and emergency responders.

4. Natural gas transmission line rupture in San Bruno, California, on September 9, 2010.

   The National Transportation Safety Board determined that the probable cause of the accident was inadequate quality assurance and quality control in 1956 during a line relocation project, which resulted in the installation of a substandard, poorly welded pipe section with a visible seam weld flaw. Over time, the flaw grew to a critical size, causing the pipeline to rupture during a pressure increase stemming from poorly planned electrical work; an inadequate pipeline integrity management program failed to detect and repair the defective pipe section.

   Contributing to the severity of the accident were the lack of either automatic shutoff valves or remote control valves on the line and flawed emergency response procedures and delay in isolating the rupture to stop the flow of gas.

   Similar to the previous example, following a human factors life cycle approach can minimize the potential for human error and decision-making flaws. In addition to integrity management, the ability to make decisions under emergency conditions, and the decision-making tools required to make optimized decisions, is highlighted.

## 9.8 HUMAN FACTORS LIFE CYCLE REVISITED

Application of human factors in an organization using the life cycle approach may occur in various ways based on the size and complexity of an organization. Some common elements to consider in development of a human factors management plan/program include the following:

1. Design standards that consider the human element. Ensure that standards consider how a human will technically design a system as well as how a human will use the design in operation and maintenance. For example, a change management process that requires all changes to be reviewed by operations and maintenance stakeholders could reduce the potential for operability issues with an asset.
2. Construction practices that consider where a human error can occur with specific controls to manage this potential. For example, implementing a maximum workday of 12 h could reduce the potential for fatigue-related human errors during excavation.
3. Running a high-quality in-line inspection (ILI) tool prior to operation of a pipeline. A baseline ILI run could identify construction incidents requiring an investigation.
4. Procedures designed to minimize human error. For example, physically attaching a procedure that is infrequently used to a physical asset can reduce the potential of an operator error.
5. Implementation of an observation program for procedures frequently accomplished as well as those infrequently accomplished. Observation programs can reduce the potential for operator error and create a supportive work culture.
6. Use of technology to automate repetitive processes that are prone to human error, such as data entry. Automation of some processes may reduce the potential for copy/paste human errors, for example.
7. Use of expert systems to reduce judgment in rule-based analyses. For example, use of an expert system to evaluate ILI data could reduce the potential for judgment errors caused by heuristics.
8. Training personnel on heuristics and biases. Training could reduce human error through awareness of heuristics and biases.
9. Investigation of all incidents with incorrect operations or human error root causes to a greater depth of understanding. For example, investigation into why a human error occurred (e.g., state of busy, too many tasks to accomplish in required time, complacency, focusing on the wrong hazard) could result in the reduction of human error through analysis of workload, environmental factors, current design, and presence of stressors.
10. Training program designed for decision-making complexity. For example, providing advanced knowledge training for nonroutine activities reduces the potential for a judgment error. Automation opportunities exist for routine activities that are less complex in nature.
11. Evaluation of the workforce for skills, expertise, and experience. Understanding the current capabilities of resources, both internal and external to the organization, can ensure that known knowledge and skill levels match requirements. For example, development of a certification program with assessment protocols can assure that a workforce has required competency levels and reduces the potential for human error due to lack of training or inadequate qualifications.
12. Experience and knowledge transfer program, such as pairing a junior employee with an experienced mentor. This type of program can reduce the potential of a human error during nonroutine tasks.
13. Incorporate current standards such as control room management.
14. Learn from other industries, such as Aviation, Nuclear, and Medicine. For example, consider the relevance of the Aviation Industry "Dirty Dozen" to an organization (https://www.faasafety.gov/files/gslac/library/documents/2012/Nov/71574/DirtyDozenWeb3.pdf)

## 9.9 SUMMARY

In the oil and gas pipeline industry, everything we do requires some level of human intervention, from designing an asset, to turning a valve, to operating the pipeline, to analyzing data related to pipe condition. Understanding as an organization where these human interventions occur and ensuring some level of evaluation of the potential for human error in decision making is critical to the reduction in the frequency and/or magnitude of errors associated with human factors. An understanding of human factors, including what people see, what they hear, what they interpret, and subsequently how they respond can help an engineer to design systems and processes that reduce the potential for human error. The human factors life cycle approach illustrated in this chapter is an important tool on the path to the holistic design, construction, operation, maintenance, and decommissioning of reliable systems.

## REFERENCES


1. Feyer, A. and Williamson, A. (2011) Human factors in accident modelling. In: Stellman, J.M. (editor), *Encyclopedia of Occupational Health and Safety*, International Labor Organization, Geneva.
2. Wickens, C., Lee, J., Liu, Y., and Becker, S. (2004) *An Introduction to Human Factors Engineering*, 2nd ed., Pearson Education Inc., New Jersey.
3. Miller, G. (1956) The magical number seven, plus or minus two: some limits on our capacity for processing information. *Psychological Review*, 63 (2), 81–97.
4. Tulving, E. (1972) *Episodic and Semantic Memory*, Academic Press, New York.
5. Fiedler, F.E. and Garcia, J.E. (1987) *New Approaches to Leadership: Cognitive Resources and Organizational Performance*, John Wiley & Sons, Inc., New York.
6. Woods, D. and Cooke, R. (1999) Perspectives on human error: hindsight biases and local rationality. In: Durso, F. (editor), *Handbook of Applied Cognition*, John Wiley & Sons, Inc., New York.
7. Svenson, O. and Maule, A.J. (editors), (1993) *Time Pressure and Stress in Human Judgment and Decision Making*, Plenum Press, New York.
8. Pipeline and Hazardous Materials Safety Administration (2009) 49 CFR Parts 192 and 195. Pipeline Safety: Control Room Management/Human Factors, Office of the Federal Register.
9. Refinetti, R. (2010) *Circadian Physiology*, 2nd ed., CRC Press, Boca Raton, FL.
10. British Standards Institution (1987) BS 6841: Measurement and Evaluation of Human Exposure to Whole-Body Mechanical Vibration, British Standards Institution, London.
11. Pilcher, J., Nadler, E., and Busch, C. (2002) Effects of hot and cold temperature exposure on performance: a meta-analytic review. *Ergonomics*, 54 (10), 682–698.
12. Wyon, D. (2004) The effects of indoor air quality on performance and productivity. *Indoor Air*, 14 (7), 92–101.
13. Fishhoff, B. (1982) Debiasing. In: Kahneman, D., Slovic, P., and Tversky, A. (editors), *Judgment Under Uncertainty: Heuristics and Biases*, Cambridge University Press, New York.
14. Simon, H.A. (1957) *Models of Man*, John Wiley & Sons, Inc., New York.
15. Tversky, A. and Kahneman, D. (1974) Judgment under uncertainty: heuristics and biases. *Science*, 185 (4157), 1124–1131.
16. Kahneman, D., Slovic, P., and Tversky, A. (1982) *Judgment Under Uncertainty: Heuristics and Biases*, Cambridge University Press, New York.
17. Simon, H.A. (1956) Rational choice and the structure of the environment. *Psychological Review*, 63 (2), 129–138.
18. Anderson, J. (1990) *Cognitive Psychology and Its Implications*, 3rd ed., W.H. Freeman, New York.
19. Cook, R. and Woods, D. (1994) Operating at the sharp end: the complexity of human error. In: Bogner, M.S. (editor), *Human Error in Medicine*, Erlbaum, Hillsdale, NJ.
20. Einhorn, H. and Hogarth, R. (1978) Confidence in judgment: persistence of the illusion of validity. *Psychological Review*, 85, 395–416.
21. Kahneman, D. and Tversky, A. (1984) Choices, values and frames. *American Psychologist*, 39, 341–350.
22. Zachary, W. (1988) Decision support systems: designing to extend the cognitive limits. In: Helander, M. (editor), *Handbook of Human–Computer Interaction*, North-Holland, Amsterdam.
23. Fogarty, W. (1988) Formal Investigation into the Circumstances Surrounding the Downing of a Commercial Airliner by the USS Vincennes, USN, 28.


## BIBLIOGRAPHY

Human factors information is available in a number of journals, including *Human Factors*, *Human Factors and Ergonomics in Manufacturing*, and *Human–Computer Interaction*. Some articles that may be of interest to readers include the following:


Chapanis, A. and Moulden, J.V. (1990) Short-term memory for numbers. *Human Factors*, 32, 123–137.

Craik, F. and Lockhart, R. (1972) Levels of processing: a framework for memory research. *Journal of Verbal Learning and Verbal Behavior*, 11, 671–684.

CUPE (2002) Enough Overwork: Taking Action on Workload, Canadian Union of Public Employees, Ottawa, Ontario.

Enander, A. (1989) Effects of thermal stress on human performance. *Scandinavian Journal of Work and Environmental Health*, 15 (Suppl. 1), 27–33.

Haselton, M., Nettle, D., and Andrews, P.W. (2005) The evolution of cognitive bias. In: Buss, D.M. (editor), *The Handbook of Evolutionary Psychology*, John Wiley & Sons, Inc., Hoboken, NJ.

ISO (1997) ISO 2631-1: Mechanical vibration and shock—Evaluation of human exposure to whole-body vibration. International Organization for Standardization.

OR-OSHA 103: Conducting a Job Hazard Analysis (JHA), Oregon Occupational Safety and Health Administration.

OSHA (2002) Job Hazard Analysis, U.S. Occupational Safety and Health Administration 3071.

Peterson, L. and Peterson, M. (1959) Short-term retention of individual verbal items. *Journal of Experimental Psychology*, 58, 193–198.


There are a number of books in addition to those identified in the reference section that may be useful in furthering your understanding of human factors, including the following:


Attwood, D., Deeb, J., and Danz-Reece, M. (2004) *Ergonomic Solutions for the Process Industries*, Gulf Publishing (Elsevier), Burlington, MA.

Broadbent, D. (1958) *Perception and Communications*, Pergamon Press, London.

CCPS (2007) *Human Factors Methods for Improving Performance in the Process Industries*, Center for Chemical Process Safety, John Wiley & Sons, Inc., Hoboken, NJ.

CCPS (2010) *A Practical Approach to Hazard Identification for Operations and Maintenance Workers*, Center for Chemical Process Safety, John Wiley & Sons, Inc., Hoboken, NJ.

Davies, D. and Jones, D. (1982) Hearing and noise. In: Singleton, W. (editor), *The Body at Work*, Cambridge University Press, New York.

Eastman-Kodak (1983) *Kodak's Ergonomic Design for People at Work*, 2nd ed., Van Nostrand Reinhold, New York.

Grandjean, E. (1988) *Fitting the Task to the Man: A Textbook of Occupational Ergonomics*, 4th ed., Taylor & Francis.

LaBerge, D. (1995) *Attentional Processing: The Brain's Art of Mindfulness*, Harvard University Press, Cambridge, MA.

Proctor, R.W. and Van Zandt, T. (2008) *Human Factors in Simple and Complex Systems*, 2nd ed., CRC Press, Taylor & Francis Group, New York.

Roughton, J. and Crutchfield, N. (2008) *Job Hazard Analysis: A Guide for Compliance and Beyond*, Elsevier, New York.

Sanders, M.S. and McCormick, E.J. (1993) *Human Factors in Engineering and Design*, 7th ed., McGraw-Hill, New York.

Stanton, N., Baber, C., and Young, M. (2004) Observation. In: Stanton, N., Hodge, A., Brookhuis, K., Salas, E., and Hendrick, H. (editors), *Handbook of Human Factors and Ergonomic Methods*, CRC Press, New York.

Vicente, K. (2004) *The Human Factor: Revolutionizing the Way We Live with Technology*, Vintage Canada, Toronto.

Wickens, C. and McCarley, J. (2008) *Applied Attention Theory*, CRC Press, New York.

Woodson, W., Tillman, B., and Tillman, P. (1992) *Human Factors Design Handbook*, 2nd ed., McGraw-Hill, New York.

# PART II

# MANUFACTURE, FABRICATION, AND CONSTRUCTION

# 10

# MICROSTRUCTURE AND TEXTURE DEVELOPMENT IN PIPELINE STEELS

ROUMEN H. PETROV,[1,2] JOHN J. JONAS,[3] LEO A.I. KESTENS,[1,2] AND J. MALCOLM GRAY[4]

[1]*Department of Materials Science and Engineering, Ghent University, Ghent, Belgium*
[2]*Department of Materials Science and Engineering, Delft University of Technology, Delft, The Netherlands*
[3]*Department of Mining and Materials Engineering, McGill University, Montreal, Canada*
[4]*Microalloyed Steel Institute Inc., Houston, TX, USA*

## 10.1 INTRODUCTION

The predictions of the U.S. Energy Information Administration's World Energy Outlook are that fossil fuels will remain the primary source of energy until 2035 and that primary gas consumption will almost double between 2006 and 2035, whereas oil demand will rise by 1.6% per year for the same period (Figure 10.1) [1]. Transportation of such amounts of gas and crude oil requires well-developed systems that operate under severe safety restrictions and varied environmental conditions. According to Ref. [2], the total length of high-pressure transmission pipelines around the world in 2006 has been estimated at 3,500,000 km, split as follows: ~64% carry natural gas, ~19% carry petroleum products, and ~17% carry crude oil. It is apparent that such transportation systems require special material properties and for many years the development of pipeline steel grades for pipelines has been a subject of exceptional interest for steel producers.

The main challenges in the field are well known: pipeline grades should simultaneously fulfill a quite complex set of requirements. These include excellent weldability, very high toughness and strength over a wide temperature range, and exceptional crack arrestability. The specific sites of the largest gas and petroleum sources require the mechanical, chemical, and technological properties to be stable over a very large range. For example, pipelines should be able to operate at temperatures between +50 and −50 °C (and even lower). Increasing needs for the transportation of larger volumes also require pipelines to operate at higher pressures. This requirement is often solved by using pipes with thicker walls. However, the requirements for pipeline grades are more complex than the simplified description above, which refers mainly to weldability, strength, and toughness. In fact, in addition to excellent strength, toughness, and weldability, pipeline grades should have high corrosion resistance against sour hydrocarbons containing $H_2S$ and $CO_2$, resistance to stress corrosion cracking, high fatigue strength, and resistance to collapse. Another important characteristic of pipeline steels is the anisotropy of the mechanical properties—strength and toughness. And last but not least, the above-mentioned combination of properties should be achieved at an affordable price.

The already mentioned set of mechanical (strength, toughness, and crack arrestability) and technological requirements (weldability) resulted in the development of the contemporary pipeline steel grades. The only way to control the mechanical properties of these grades is by the control of microstructure and crystallographic texture of the steel plates. This is possible (i) by the design of an appropriate chemical composition of the steel, and (ii) by an appropriate thermomechanically controlled processing, that is, by applying the appropriate rolling reduction at the appropriate temperature and employing the right cooling rate. Such an approach is known as thermomechanically controlled processing (TMCP) and the exact parameters are dependent on steel chemistry, initial (before rolling) and final plate thicknesses, and of course the capacity of the rolling equipment. By this approach, metallurgical engineers are able

*Oil and Gas Pipelines: Integrity and Safety Handbook,* First Edition. Edited by R. Winston Revie.

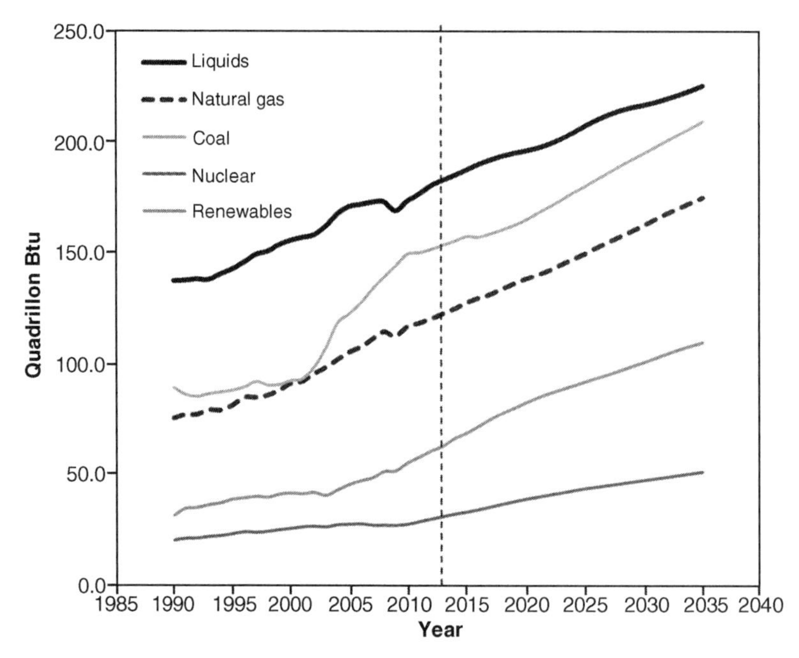


**FIGURE 10.1** World energy consumption by fuel, 1990–2035 [1].

to use effectively the various mechanisms contributing to the strength and toughness of the steel under the restrictions imposed by the thermo-physical and geometrical parameters of the material. These include the thermal conductivity, quenchability, and plate thickness.

The requirements for excellent weldability of the pipeline steels impose severe restrictions on their carbon content, which should be kept as low as possible. The low carbon content results in general in a decrease in the strengthening potential of the steel and in a decrease in the brittle-to-ductile transition temperature, which contributes to better low-temperature toughness. Under such circumstances, the most effective way of improving both strength and toughness is grain refinement. Grain refinement is in turn brought about via control of the recrystallization and transformation of the austenite to body-centered transformation product behavior. One of the prerequisites for such control is the microalloying strategy applied to pipeline steel grades. Microalloying of the pipeline steel has a multipurpose goal. On the one hand, microalloying is used to control the austenite grain size during hot rolling via pinning of the austenite grain boundaries and via the control (delay or even suppression) of austenite recrystallization, which results in finer transformation products at room temperature. On the other hand, the microalloying elements precipitate in the BCC phase and therefore contribute by precipitation hardening to the strength of the steel.

## 10.2 SHORT HISTORY OF PIPELINE STEEL DEVELOPMENT

A short historical overview of microalloyed steel grades shows that they were first introduced in the late 1930s [3] and initially the alloying elements V, Ti, and Nb were added separately in amounts between 0.005 and 0.01% (mass). Later on, with the advance in metallurgical understanding, they were added in combination with the intention of precipitating at different temperatures during rolling. Such a strategy extended the rolling mill function from merely being a tool for shape changing to a powerful tool for shape and property control. This change in approach has given rise to the development of the current generation of high-strength low-alloy (HSLA) steels.

The first report of a hot rolled HSLA steel for a pipeline application came from Europe (Mannesmann) [3] in around 1952. This material was a normalized API Grade X52 microalloyed with vanadium. According to Refs [4,5], it was later applied to API Grades X56 and X60 in 1953 and 1962, respectively. In North America, some hot rolled steels utilizing the microalloying concept were reported in 1959 [5–8] and these replaced normalized steels by 1972 [3–5]. By contrast, normalized medium carbon (0.17% C) steel grades were in use in the Soviet Union until the mid-1990s. Table 10.1 displays the chemical compositions of the niobium microalloyed steels proposed by Clarence L. Altenburger and Frederic A. Bourke [9].

**TABLE 10.1 Chemical Composition Range of Niobium Microalloyed Steels**

| Element | Range (%) | Preferred Range (%) |
|---|---|---|
| Carbon | 0.05–0.30 | 0.15–0.20 |
| Manganese | 0.30–1.15 | 0.60–0.90 |
| Silicon, maximum | 0.10 | 0.10 |
| Niobium (columbium) | 0.005–0.2 | 0.005–0.2 |

*Source:* Adapted from Ref. [9].

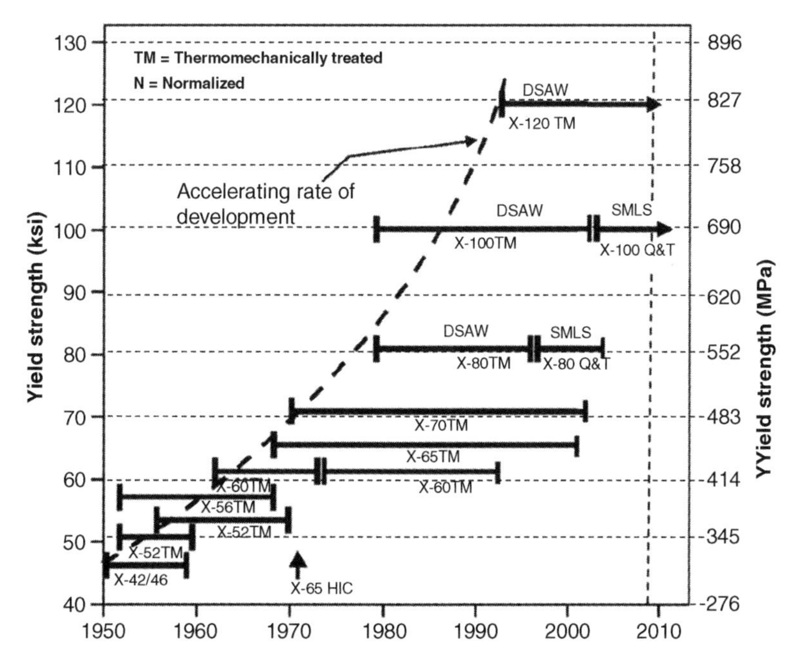


**FIGURE 10.2** Timescale of the development of different HSLA steel grades and the corresponding level of yield strength [3]. DSAW: double submerged arc welded; Q&T: quench and tempered; SMLS: seamless pipe. (Adapted from Ref. [3].)

The evolution of pipeline grades over the period 1949 to the present day and their associated yield strengths are discussed in detail in Ref. [3]. These developments are shown schematically in Figure 10.2. However, due to a number of reasons, the X70 and X80 grades are still the reference materials of today. The most important reason for restricting the application of the X100 and X120 pipeline grades despite their very high yield strengths is that these steels exhibit limited uniform elongations, that is, they cannot fulfill the requirements of strain-based design. This prevents their employment in earthquake regions. Nevertheless, X100 steel has already been used for demonstration and test purposes by the TransCanada Corporation in a pipeline [3].[1] Currently, researchers are concentrating on enhancing the ductilities of these steels so that they can be considered for the construction of the next generation of pipelines [10]. In recent studies, Zhang et al. [11], Zhou et al. [12], and Nafisi et al. [13] have demonstrated that by appropriate combinations of alloying, TMCP, and heat treatment, significant improvements in the ductility and anisotropy of X120 can be achieved.

The approximate chemical compositions of the most widely used steel grades are shown in Table 10.2. The combination of microalloying additions with TMCP allows the formation of a large range of microstructures varying from ferrite–pearlite to different types of bainite and bainite with islands of martensite and retained austenite (MA constituents). These approaches combine several simultaneously operating strengthening mechanisms, among which are solid solution strengthening, grain refinement, precipitation hardening, and work hardening (dislocation strengthening), all of which can be effectively controlled via the TMCP parameters.

Finally, the specific steel composition and combination of microalloying additions to be used depends not only on the engineering requirements but also on the pipe wall thickness, alloy costs, and available processing equipment.

An example of the progress in microstructure development, corresponding yield strength, and toughness, as well as the contributing strengthening mechanisms through the years, has been summarized by Vervynckt et al. [14]. This review surveys the general tendencies in microstructure formation as a tool to control the strength and toughness of pipeline steel grades. These developments can be expressed in general as follows:

1. Decrease the carbon content from 0.2 to 0.05%.
2. Increase the number of microalloying elements.
3. Decrease the pearlite fraction and change from "ferrite–pearlite"-type microstructures to "ferrite–bainite" or even "bainite–martensite"-type microstructures.
4. Change the general processing strategy from hot rolling and normalization to thermomechanically controlled processing in combination with accelerated cooling and/or direct quenching.

[1] Application of TransCanada Corporation ("TransCanada") and Alaskan Northwest Natural Gas Transportation Company ("ANNGTC")—submitted to the Alaska Department of Revenue, Pursuant to AS 43.82.120 for approvals under the Alaska Stranded Gas Development Act, June 2004.

**TABLE 10.2 Typical Compositions of Most Frequently Used Pipeline Steel Grades [3,11,12]**

| Steel Grade | Chemical Composition in Mass% | | | | | | | | |
|---|---|---|---|---|---|---|---|---|---|
| X60 | C: 0.07 | Mn: 0.95 | Si: 0.20 | S: 0.003 | P: 0.015 | Al: 0.03 | Nb: 0.04 | Ti: 0.012 | N: 0.005 | Other: nil |
| X70 | C: 0.12 | Mn: 1.5 | Si: 0.25 | S: 0.003 | P: 0.015 | Al: 0.025 | Nb: 0.03 | V: 0.08 | N: 0.007 | Other: nil |
| X80 | | | | | | | | | | |
| Nb-V | C: 0.08 | Mn: 1.63 | Nb: 0.04 | S: <0.003 | V: 0.08 | Mo: 0.10 | Cr: nil | Ti: 0.013 | N: <0.008 | Other[a]: Cu, Ni |
| Nb-Mo | C: 0.04 | Mn: 1.68 | Nb: 0.08 | S: <0.003 | V: nil | Mo: 0.30 | Cr: nil | Ti: 0.015 | N: nil | Other[a]: 0.3 Cu, 0.2 Ni |
| Nb-Cr | C: 0.03 | Mn: 1.70 | Nb: 0.095 | S: <0.003 | V: nil | Mo: 0.10 | Cr: 0.028 | Ti: 0.013 | N: nil | Other[a]: Ni |
| X100 | C: 0.056 | Mn: 1.97 | Nb + Ti + V: <0.13 | S: 0.002 | B: nil | Mo: 0.41 | Ni: nil | Cr: nil | P: 0.004 | Other[a] |
| X120 | C: 0.05 | Mn: 1.9 | Nb: 0.048 | S: 0.002 | B: 0.0013 | Mo: 0.30% | Cr: 0.22 | Ti: 0.015 | P: 0.004 | Other[a] |
| X120 | C: 0.05 | Mn: 1.99 | Nb + Ti: 0.08 | S: 0.002 | B: 0.0017 | Mo: 0.27 | Ni: 0.62 | Cr + Cu: 0.53 | P: 0.008 | Other[a] |

[a] Al: 0.02–0.05.

A summary of the influence of the individual alloying elements on the microstructure and properties of contemporary pipeline steels according to Ref. [14] is given in Table 10.3 and more details can be found in Refs [4,14], whereas an extensive review of the evolution of the strength of microalloyed pipeline steels can be found in Refs [3,4,15,16].

### 10.2.1 Thermomechanically Controlled Processing of Pipeline Steels

The specific chemical composition of a steel is designed to ensure that the required combination of the functional and technological properties is satisfied. As already mentioned, the pipeline steels are in general low carbon grades alloyed with Mn and with well-balanced microalloying additions. The chemical composition aims to balance two conflicting criteria: (i) good weldability and (ii) sufficient toughness, strain hardening capacity, and ductility. If the limiting criterion is only the strength, this can be readily achieved by quenching and tempering steels with sufficient carbon content to provide good hardenability. However, high carbon contents are in direct conflict with the requirements for good weldability, toughness, and ductility.

To fulfill these complex and even conflicting criteria, the carbon content in the pipeline grades is kept as low as possible and a specific combination of the processing steps known as TMCP is utilized. TMCP was developed in the 1960s [15] as an advanced variant of HSLA steels. The general approach in TMCP is to use a relatively low carbon content (to provide good weldability) in combination with sufficient Mn to decrease the austenite-to-ferrite transformation temperature. In combination with a critical quench or cooling rate and minor additions of alloying elements such as Nb, V, Ti, Mo, and so on (Table 10.2), sufficient steel strengthening is produced via various mechanisms such as grain refinement, solid solution strengthening, precipitation hardening, the introduction of hard second phases, and increased dislocation density (Figure 10.3). Because of their extreme harmfulness with regard to embrittlement, it should be explicitly mentioned that extreme cleanliness of these steel grades is required with respect to sulfur (S) and phosphorus (P), which must be reduced to the part per million level [3].

TMCP additionally involves the controlled distribution of the rolling reductions in the regions above and below the austenite no-recrystallization temperature ($T_{nr}$) in order to create the required very fine-grained microstructure. More precisely, the final hot rolling passes must be applied below the $T_{nr}$ temperature so that the heavily deformed austenite does not recrystallize. The microstructure of austenite that has been heavily deformed before the $\gamma$–$\alpha$ transformation consists of a large number of internal defects—dislocation tangles, slip bands, shear bands, high- and low-angle grain

**TABLE 10.3 Influence of Alloying Elements on Properties of Pipeline Steels**

| Element | Amount in Mass% | Influence |
|---|---|---|
| C | | Strengthener |
| Mn | 0.5–2 | Austenite stabilizer. Delays austenite decomposition during accelerated controlled cooling (ACC). Decreases ductile-to-brittle transition temperature (DBTT). Mild solid solution strengthener |
| Si | 0.1–0.5 | Deoxidizer. Solid solution strengthener. Suppresses carbide formation |
| Al | >0.02 | Deoxidizer. Restricts austenite grain growth (AlN) |
| Nb | 0.02–0.1 | Strong ferrite strengthener. [Nb(CN)]. Suppresses austenite recrystallization. Grain size control. Delays $\gamma$–$\alpha$ transformation |
| Ti | 0–0.06 | Grain size control (TiN formation). Strong ferrite strengthener |
| V | 0–0.1 | Strong ferrite strengthener; (VN) |
| N | <0.012 | Strengthener. Forms TiN, VN, and AlN |
| Mo | 0–0.3 | Promotes bainite formation; ferrite strengthener |
| Ni | 0–0.5 | Austenite stabilizer. Increases fracture toughness |
| Cu | 0–0.55 | Improves corrosion resistance; ferrite strengthener |
| Cr | 0–1.25 | Improves atmospheric corrosion resistance. Decreases critical quench rate |

*Source:* Adapted from Ref. [14].

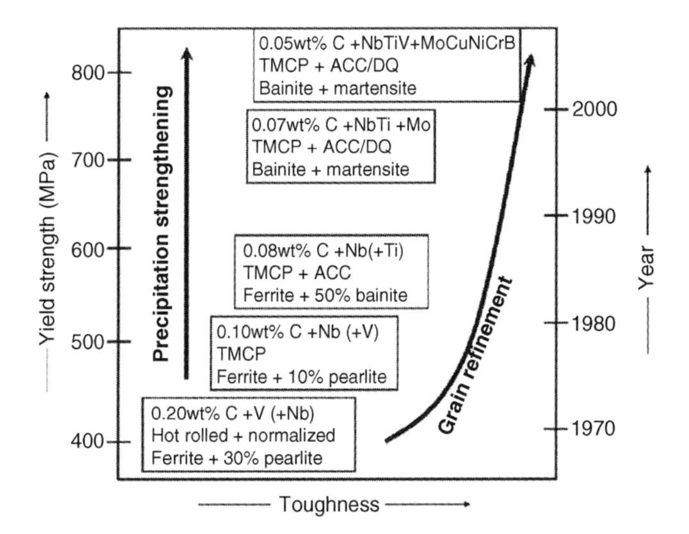


**FIGURE 10.3** Development of pipeline steels as example of HSLA steel research. TMCP: thermomechanically controlled processing; ACC: accelerated cooling; DQ: direct quench [14].

boundaries (HAGB and LAGB), which provide abundant potential nucleation sites for the BCC and BCT constituents (ferrite, bainite, or martensite). As a consequence, a very fine-grained microstructure is developed that contributes to both the high strength and high toughness of the TMCP grades.

A general scheme of microstructure formation during thermomechanically controlled processing is shown in Figure 10.4 [14].

In an integral processing schedule after slab casting, there are several important stages where microstructure formation can be intentionally controlled: (i) slab reheating, (ii) rough rolling, (iii) finish rolling, and (iv) controlled cooling.

The "classical" organization of a (sheet) strip rolling mill and its main ancillary equipment for the rolling of pipeline steel grades are shown in Figure 10.5.

Only some of the pipeline steels are produced in the form of sheet, as a significant part is rolled into plate. The difference between plate and sheet involves the final thickness of the product. According to Dieter [17], plates generally have thicknesses greater than 6 mm, whereas sheet and strip are thinner than 6 mm. However, in pipeline steel production, the above-mentioned thickness limits are not rigid and the term "sheet" is employed for final products that are coiled, whereas plates are flat products that are not coiled. Nowadays, coilers are so powerful that they can coil flat products with thicknesses up to 24 mm and so the boundary between plate and sheet involves the coiling or absence of coiling in the final processing stage. The final lengths of plates can extend to 24 m and widths to 4.5 m, but in general the dimensions selected depend on the target pipe dimensions and can vary significantly. Note that coils are used not only for spiral welded pipes but also (after flattening) for cut-to-length sheets for LSAW (longitudinal submerged arc welding) as well as for relatively small-diameter ERW/HFI welded pipes. An example of a typical plate rolling mill and its ancillary equipment is provided in Figure 10.6.

***10.2.1.1 Slab Reheating*** The slab reheating stage is the first stage in the production line for conventional controlled rolling (CCR) and TMCP. Here the as-cast slabs are transferred to the reheat furnace next to the rolling line. The as-cast slab microstructure is formed under conditions of slow cooling with a very large temperature gradient between the surface and the central zone. It is characterized by the presence of large dendrites as well as of the segregation of the alloying

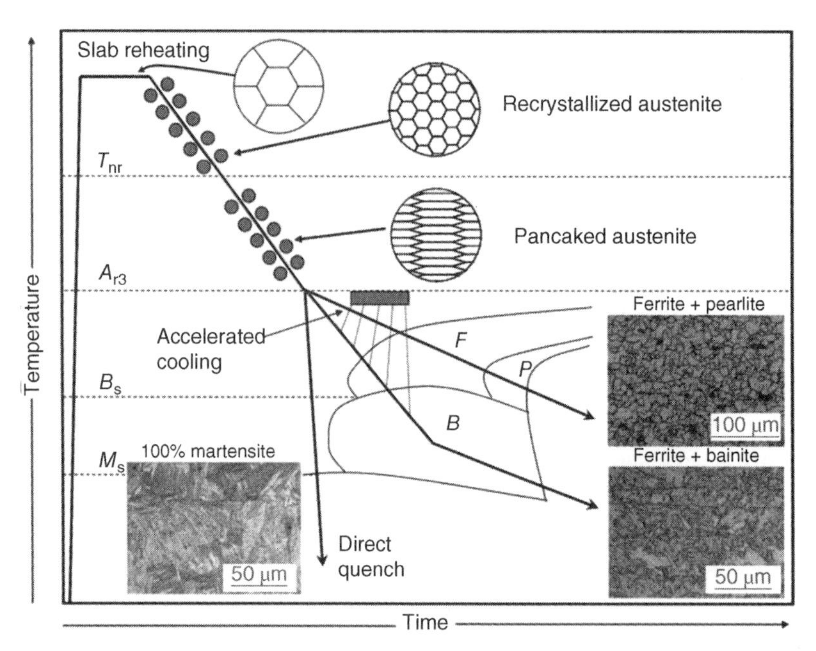


**FIGURE 10.4** Schematic diagram of thermomechanically controlled processing (TMCP) and the microstructures that result from this process [14].

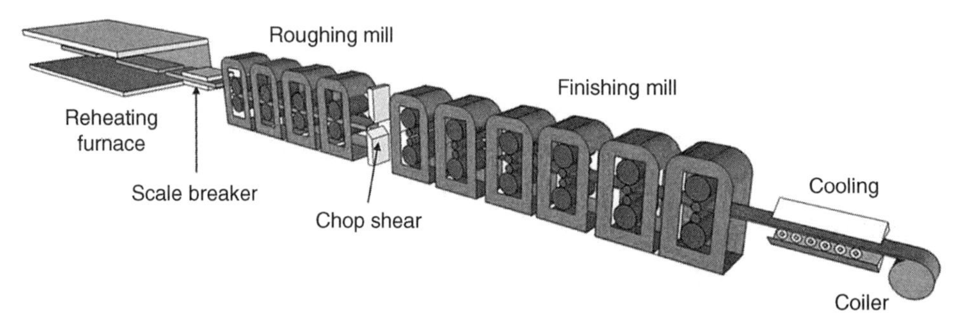


**FIGURE 10.5** A "classical" arrangement of a strip rolling mill for the hot rolling of steel slabs with ancillary equipment.

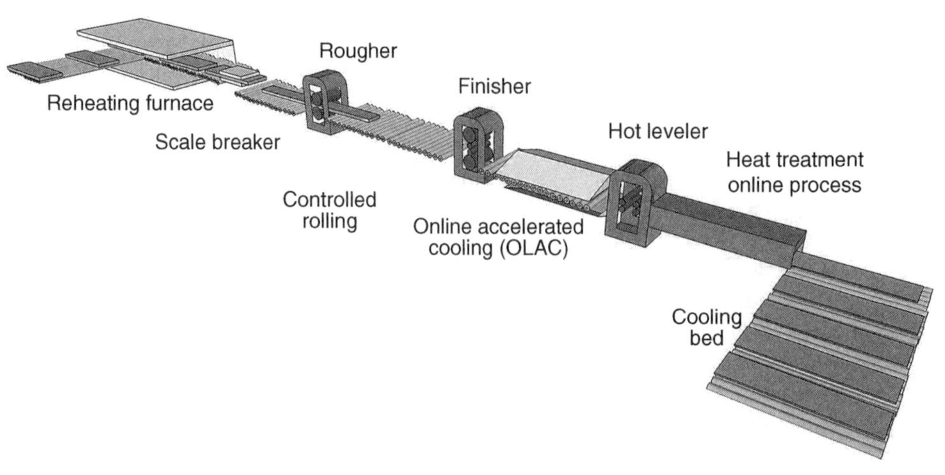


**FIGURE 10.6** An example of the arrangement of rolling mills for plate rolling with ancillary equipment.

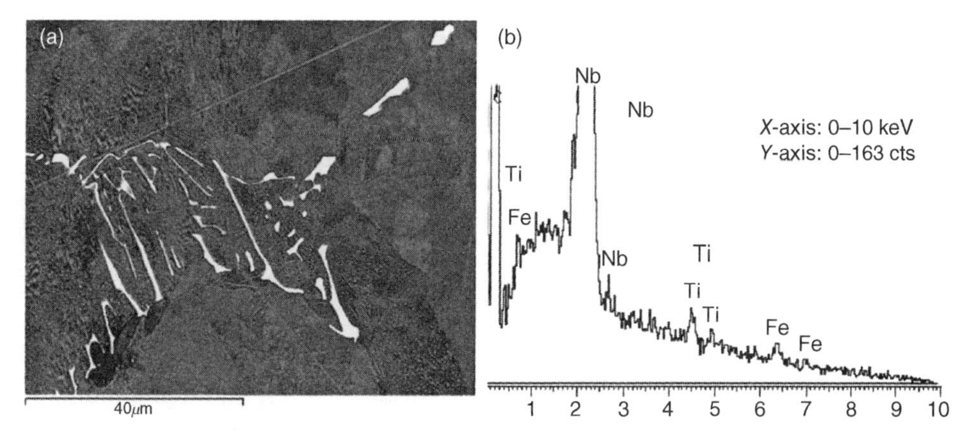


**FIGURE 10.7** (a) Large eutectic Nb(C, N) particles observed at the centerline position of a slab. (b) EDS spectrum from one of these particles. (Reprinted from Ref. [18] with kind permission from Maney Publishing, permission conveyed through Copyright Clearance Center, Inc.)

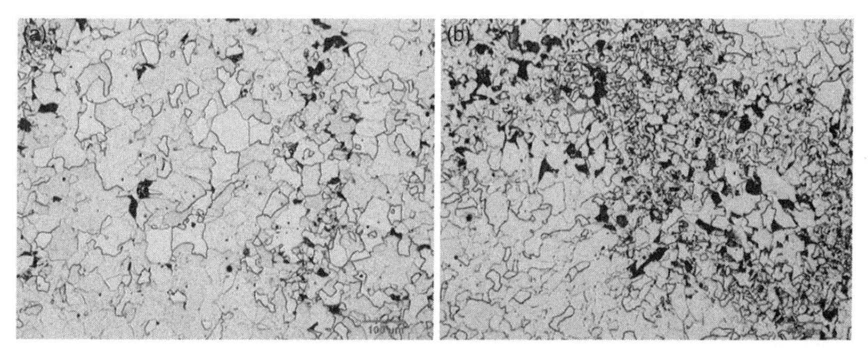


**FIGURE 10.8** Optical micrographs showing near-surface (a) and centerline (b) segregation in a pipeline slab of composition: 0.03% C, 0.25% Si, 1.70% Mn, 0.008% P, 0.005% S, 0.052% Al, 0.082% V, and 0.063% Nb. (Reprinted from Ref. [18] with kind permission from Maney Publishing, permission conveyed through Copyright Clearance Center, Inc.)

elements in the interdendritic spaces (Figure 10.7a and b). This is generally accompanied by typical centerline segregation, which is a large-scale segregation phenomenon (Figure 10.8). Macrosegregation is an important microstructural parameter because it changes the through-thickness composition of the plate, which in turn creates different microstructures and properties through the plate thickness. Davis and Strangwood [18] provided a detailed analysis of macro- and microsegregation phenomena, but it is also well documented in a large number of studies [18–20].

Currently, continuous casting is generally used in pipeline steel production. The solidification rate during continuous casting is higher than in the ingot casting method, which minimizes the secondary dendrite arm spacing (SDAS). The latter are preferential sites for local microscale segregation. However, macrosegregation is still present, which is manifested as a difference in composition between the core and the case. Macrosegregation in slabs appears not only across the slab thickness but also along the slab width. In summary, the local average composition depends on the macrosegregation behavior, and the spatial separation of solute-rich and solute-poor regions (i.e., the SDAS) depends on the casting characteristics (particularly heat and fluid flow), which are different for various casting methods and casting speeds. Such segregation effects can have an undesirable influence on the mechanical properties of the plate because they create prerequisites for variations in the grain size of the plates (i.e., they lead to bimodal grain size distributions) and even for the formation of different microstructures along the centerline of the plate [18,20]. It is shown in the same work [18] that both centerline macrosegregation and microsegregation can in most cases be successfully predicted by the methods of equilibrium thermodynamics. Nevertheless, the equilibrium thermodynamics approach to centerline segregation is not successful when more than one solid phase is formed during final solidification.

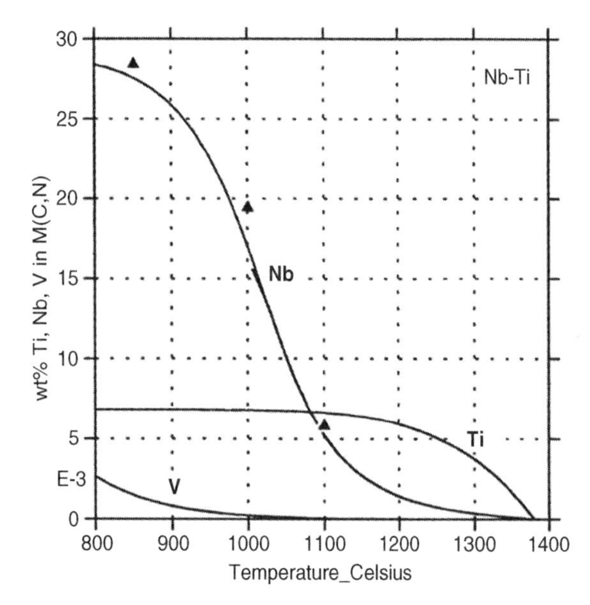


**FIGURE 10.9** Calculated curves for the precipitation of TiN, Nb (C, N), and V(C, N) and experimental data for Nb(C, N) in a Nb-Ti microalloyed steel with 0.013% Nb and 0.007% Ti. (Reproduced from Ref. [19], Figure 16, with kind permission from Springer Science+Business Media.)

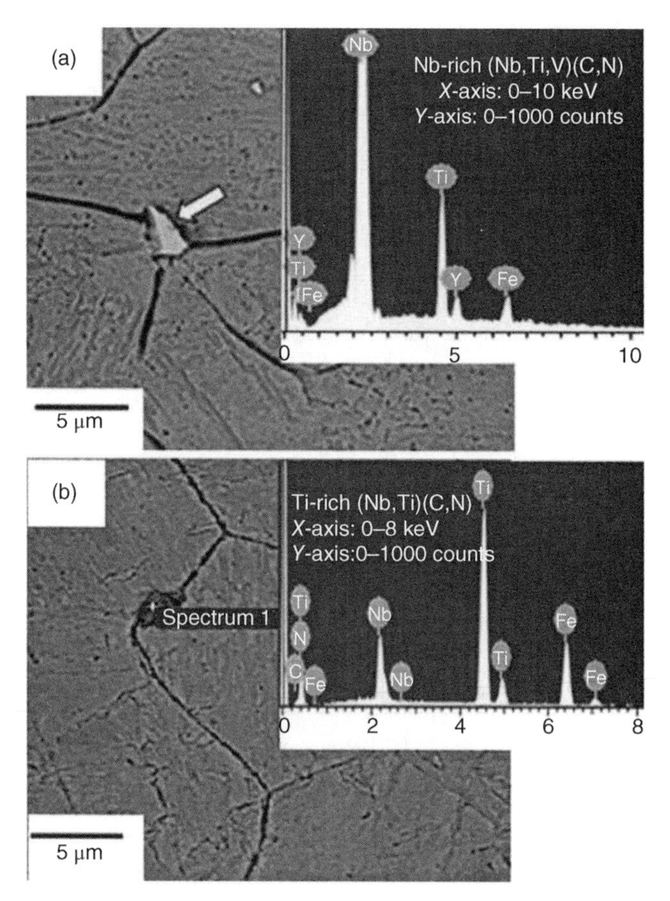


**FIGURE 10.10** Change in the composition of the microalloy precipitates with increase in reheat temperature from (a) 1348 K (1075 °C) (Nb-rich (Nb, Ti, V) (C, N)) to (b) 1473 K (1200 °C) (Ti-rich (Nb, Ti)(C, N)), as indicated by EDS analysis. (Reproduced from Ref. [21], Figure 10, with kind permission from Springer Science+Business Media.)

Slab reheating cannot completely remove segregation, but it creates conditions suitable for the further control of the microstructure through determination of the austenite grain size and the level of dissolution of the alloying elements in the austenite. In fact, the austenite grain size and the level of dissolution of the elements are connected because the incomplete dissolution of the alloying elements suppresses austenite grain growth. Therefore, selection of the slab reheating temperature is made considering two important restrictions:

1. It has to be high enough to dissolve a large fraction of the microalloying additions.
2. It has to be low enough to prevent unwanted austenite grain growth.

In their work, Zajac and Jansson [19] studied the thermodynamics of the Fe–Nb–C–N system and the solubility of niobium carbonitrides in austenite. They showed by thermodynamic calculations, well supported by experimental results, that the dissolution temperatures of the microalloying elements depend on the actual combination of alloying elements and presented a method for the accurate prediction of the solubility of niobium in the presence of titanium, aluminum, boron, and vanadium. It is also important to note that all the alloying elements are completely dissolved in the austenite only at temperatures above 1380 °C. (Figure 10.9). However, such high reheat temperatures are not used in rolling practice and reheat temperatures usually do not exceed 1250 °C. In this way, small amounts of Ti (CN) and Nb (CN) remain undissolved and prevent unwanted austenite grain growth (Figure 10.10) [21].

In summary, the reheating parameters generally take into account the required dissolution temperatures of the various carbonitride types in order to promote appropriate control of the austenite grain size before rough rolling. Segregation of the elements through the slab thickness is an important microstructural parameter that is difficult to control but must be considered.

***10.2.1.2 Microstructure Formation During Rough Rolling*** After leaving the reheat furnace, the slabs enter the scale breaker where the thick oxide layer formed on the slab surface is removed from the slab surface by powerful water jets. Immediately after the scale breaker, the slab is moved

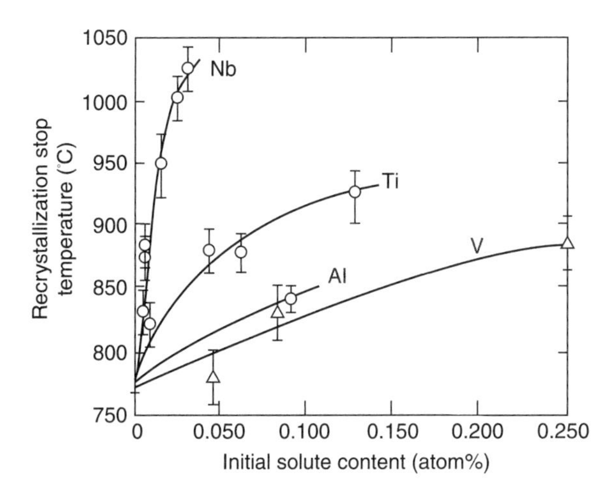


**FIGURE 10.11** Effect of different elements on the austenite recrystallization stop temperature for a 0.07C–1.4 Mn–0.25 Si steel. (Reproduced from Ref. [22] with kind permission from the Minerals, Metals & Materials Society.)

to the first roughing mill where rough rolling begins. This is the stage of rolling applied above the austenite nonrecrystallization temperature ($T_{nr}$), also known as the recrystallization stop temperature. The latter is the temperature above which austenite recrystallizes completely between rolling passes. This temperature is not a physical constant of the material but an important technological parameter of the process. The austenite nonrecrystallization temperature for a specific steel grade in a given rolling process depends on the chemical composition of the steel (alloying with elements such as Mo, Nb, Ti, V, and Al) and on the rolling parameters (strain per pass, interpass time, and cooling rate). The nonrecrystallization temperature (Figure 10.11) increases with the concentration of Nb, Ti, Al, and V and, in particular, the level of Nb addition has the strongest effect [21–23]. TMCP steels with niobium additions above 0.10% are also referred to as HTP (high-temperature processing) steels.

An empirical correlation between the $T_{nr}$ and the chemical composition was proposed by Boratto et al. [24] and $T_{nr}$ values for microalloyed steels have frequently been reported in the literature [24–26]. In most cases, $T_{nr}$ values are determined by measuring the mean flow stress (MFS) of the steel at different temperatures in the rolling mill or in laboratory tests (most often torsion tests). After plotting the MFS as a function of $1000/T$, the $T_{nr}$ is defined as the point at which there is a slope change in the experimental data [14,24–26]. Recently, Gautam et al. [27] proposed a more accurate way of analyzing the temperature dependence by detecting changes in the rate of strain hardening at different temperatures instead of the MFS.

An accepted practice in the rolling of pipeline grades is that rolling reductions accumulating to approximately 60% are given above the austenite nonrecrystallization temperature ($T_{nr}$) and rolling reductions accumulating to at least 70% are applied below the $T_{nr}$ [28]. The main goal of rolling above the $T_{nr}$ is to refine the austenite grain size by complete recrystallization of the austenite between passes. The high rolling temperatures and complete recrystallization allow high rolling reductions to be applied in a relatively low number of passes. However, there are no universal recipes for the reduction percentages. These depend on the capacities of the rolling mills, crop shear capabilities, desired transfer bar dimensions, final product thickness, and so on.

Depending on the slab thickness and the requirements of the final product, there are two main possibilities: (i) the final product is the plate for longitudinally welded pipes; and (ii) the final product is the plate for spiral welded pipes. In the case of longitudinally welded pipes with a diameter of 1.43 m, the width of the plate must be ~4.5 m, which requires the first passes of rough rolling to produce the required width dimension. Therefore, the slab is first cross-rolled to a width of ~4.5 m and then is rotated 90° and conventionally rolled above and below the $T_{nr}$ to the final thickness. The change in rolling direction (RD) in the broadside passes not only leads to the exact width of the plate but also minimizes the segregation in the transverse direction (TD) of the slab. The exact lateral dimensions are obtained by rolling with vertical rolls in the first rolling passes. In cases where the slab is intended for spiral welded pipes, the width of the plate is close to the initial width of the slab and the rolling direction is maintained constant during rough rolling. Here again, vertical rolls are used for lateral dimension control (TD). These two approaches impose specific restrictions with regard to the interpass times between the rough rolling passes or between rough rolling and finish rolling, which can influence the austenite microstructure by causing grain growth after rough rolling.

In Figure 10.12 the evolution of the austenite grain size after leaving the reheating furnace and cooling to the start temperature of finish rolling is illustrated for a steel containing 0.06% C, 1.6% Mn, 0.055% Nb, and (Mo + Ni + Cu + Cr) < 0.1%. The initial austenite grain size of 58.1 μm after reheating at 1250 °C for 1 h (Figure 10.12a) is reduced to an average grain diameter of 26.1 μm after three rough rolling passes at 1250, 1220, and 1180 °C subsequently (Figure 10.12b). However, during 1 min of cooling at the temperature interval from 1180 to 1060 °C, grain growth of the recrystallized austenite takes place and the average austenite grain diameter increases to 34.5 μm (Figure 10.12c) [29]. The austenite grain size during rough rolling is controlled by the chemical composition of the steel [30] and the distribution of the rolling reductions between the passes. Detailed studies of the recrystallization and strain accumulation behavior especially with regard to high Nb steels can be found in Refs [23,28,31,32].

Deformation, precipitation, solute drag, recovery, and recrystallization are the main metallurgical phenomena that control microstructure formation over the entire

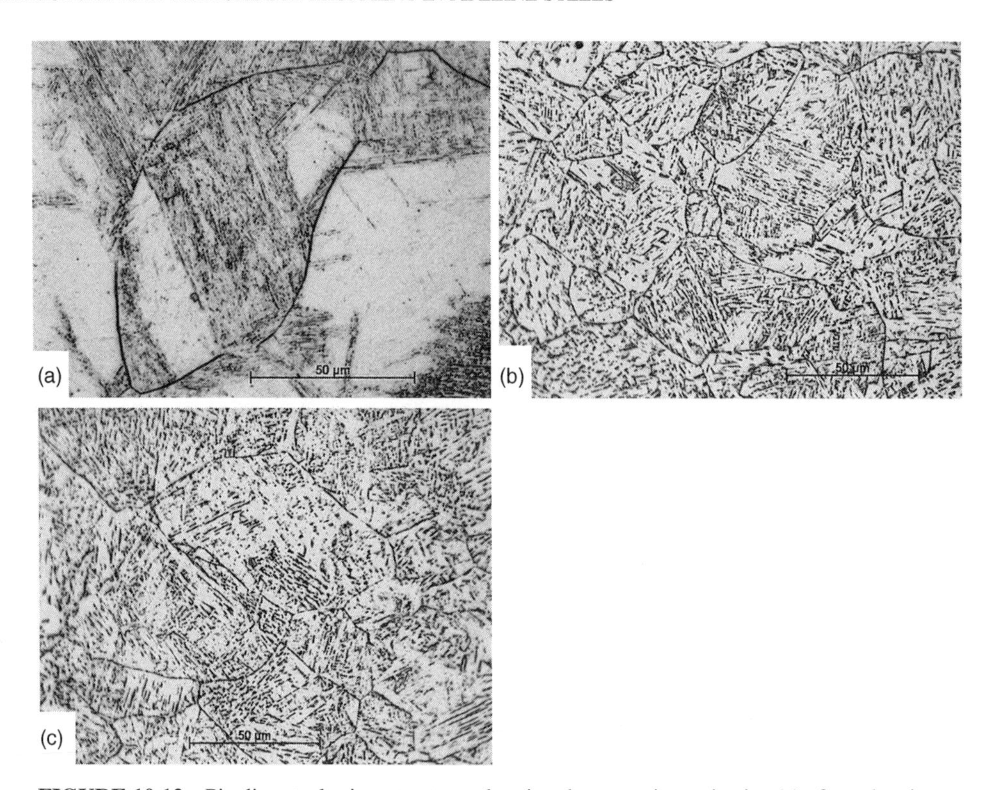


**FIGURE 10.12** Pipeline steel microstructures showing the austenite grain size (a) after reheating at 1250 °C, 1 h, $d_{av} = 58.6\,\mu m$; (b) after three rough rolling passes at 1250, 1220, and 1180 °C, $d_{av} = 26.1\,\mu m$; and (c) at a temperature of 1060 °C, which was reached ~1 min after the last roughing pass at 1180 °C, $d_{av} = 34.5\,\mu m$ [29]. All the samples were water quenched from the above temperatures and etched using the Bechet–Beaujard technique [33]. Scale bar is 50 μm. (Reproduced with permission from Trans Tech Publications.)

thermomechanically controlled processing route. Their interactions are reviewed in detail in Ref. [23]. They can be successfully controlled by adjusting the chemical composition and the rolling parameters, including the reheat temperature, the temperatures at which the rolling passes are applied, the interpass time, the rolling reduction per pass, the strain rate, and so on, as already described.

***10.2.1.3 Microstructure Formation During Finish Rolling and Cooling*** After rough rolling, the slabs are transferred to the finish rolling mill where the final plate thickness and length are established. The number of finish rolling passes depends on the final plate thickness and the available rolling equipment. As can be seen from Figure 10.13, the finish rolling passes are usually executed with decreasing reductions per pass accompanied by decreases in the rolling temperature. In most cases the overall rolling reduction applied during finishing (i.e., below the $T_{nr}$) is ~70%. Because the individual rolling reduction is not additive (the plate finishing reductions in Figure 10.13 add up to more than 100%), it is common to add the reductions in the form of true strains. According to this measure, total von

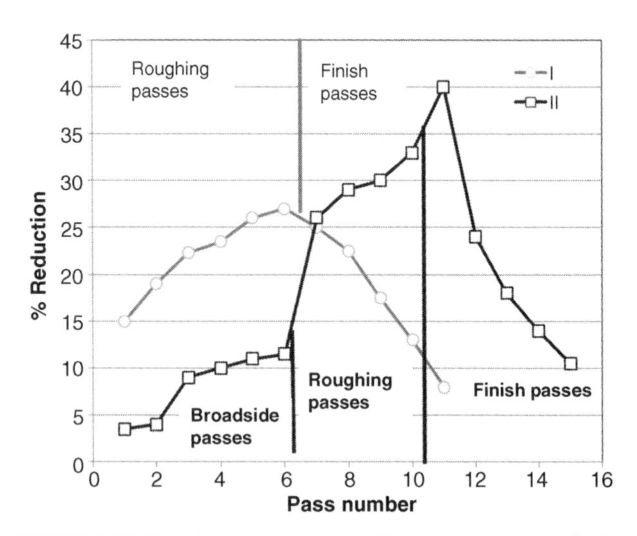


**FIGURE 10.13** Examples of the rolling reductions applied in different rolling schedules. I: Rolling of plates for spiral pipes (gray) without broadside passes. II: Rolling of plates for longitudinally welded pipes where the first six roughing passes are broadside passes. (Adapted from Ref. [28].)

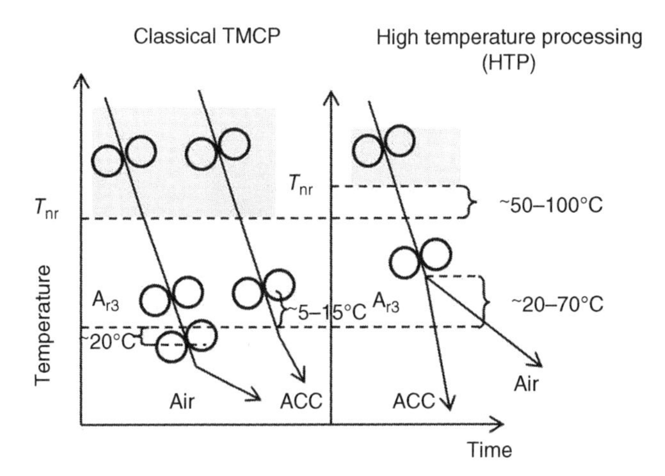


**FIGURE 10.14** Schematic of classical TMCP and HTP. (Adapted from Ref. [28].)

Mises strains of at least 1.4 should be applied below the $T_{nr}$ (or at least 1.2 when specified in plane strain units).

The temperature distribution between rolling passes, the reductions per pass, and finish rolling temperature play important roles in the microstructure formation of the plate. Selection of the rolling parameters depends on the specific steel grade; Figure 10.14 shows schematically an example of the distribution of the finish rolling passes in the case of classical TMCP and the so-called HTP method, which is applied to high Nb content (~0.1% Nb) pipeline grades [28]. As can be seen, the HTP rolling schedules are usually terminated at higher finish rolling temperatures of ~20–70 °C above the $Ar_3$, whereas in the case of "classical" TMCP the finish rolling temperature is much closer to the $Ar_3$ or in some cases the final rolling pass can even be slightly below the $Ar_3$, that is, in the intercritical temperature region where some volume fraction of ferrite is formed before or during the last rolling pass. However, finish rolling in the intercritical region is not a common practice because, although it contributes to increasing the yield strength of the plates, it often decreases the impact toughness and can lead to delamination effects associated with the development of a bimodal grain structure and strengthening of the undesirable $\{001\}\langle 110\rangle$ texture components [34,35].

Microstructure formation during finish rolling is controlled by the pretransformation state of the deformed austenite—grain size, grain morphology, and distribution of in-grain defects (in-grain shear bands, deformation twins, dislocation in-grain substructures, etc.). All of the above-mentioned defects have different effects, which depend on their efficiency as potential BCC nucleation sites during the $\gamma$–$\alpha$ phase transformation that takes place during the final cooling stage. In addition, precipitation of the microalloying elements (Nb, V, and Ti) continues during finish rolling making the description of microstructure formation even more complex. Nevertheless, all the internal defects present in the deformed austenite before transformation contribute to grain refinement of the product BCC structure. The latter is the important microstructural parameter that determines the strength and toughness of the plate.

Depending on the chemical composition of the steel, the start and finish temperatures of finish rolling, the distribution of the rolling reduction between passes, and the cooling rate, different types of microstructures can be obtained in the rolled plates. It is important to mention that often the final microstructures vary in the plate thickness due to local variations in chemistry and cooling rate and the through-thickness strain distribution. However, the design of an appropriate technological process primarily involves minimization or if possible the complete avoidance of heterogeneities in the through-thickness microstructure and properties. As pipeline steels are produced under conditions of continuous cooling, the microstructure is generally a mixture of constituents such as ferrite, bainite (upper, lower, and granular), degenerated pearlite (Figure 10.15a and b), martensite, martensite–austenite islands, and carbides (Figure 10.15). It is thus difficult to characterize the microstructural constituents exactly because sometimes they have similar morphologies, such as bainite and acicular ferrite, or lower bainite and martensite. Nevertheless, electron backscatter diffraction (EBSD) analysis provides the possibility of distinguishing more clearly between some of the constituents on the basis of crystallography or lattice misorientation. These approaches are discussed by Zajac et al. [36] and are illustrated in Figure 10.16.

By analyzing the misorientation angles in different pipeline microstructures, Zajac et al. [36] concluded that lower bainite displays a misorientation spectrum with a well-pronounced maximum at ~60° (Figure 10.16c), whereas the spectrum of upper bainite displays a maximum at around 10° (Figure 10.16b). The misorientation spectrum of granular bainite is more blurred and contains peaks at around 10° as well as in the range between 45° and 60° (Figure 10.16a). However, the overlaps in the misorientation profiles of lower bainite and martensite, as well as of upper bainite and ferrite [38], prevent the application of this technique to mixed microstructures containing combinations of these constituents.

A very important parameter of the finish rolling sequence is the cooling schedule that, together with the parent austenite microstructure, determines the final microstructure of the plate. As already mentioned (Figure 10.4), depending on the composition and processing parameters, various pipeline steel grades can form mixed microstructures of ferrite and pearlite, often denominated as "degenerate" pearlite (Figure 10.15a and b) [37], or of acicular ferrite and bainite in combination with small amounts of martensite and retained austenite (Figure 10.15c), or of bainite and martensite, as described in Refs [3,14]. In all cases, the morphology of the BCC

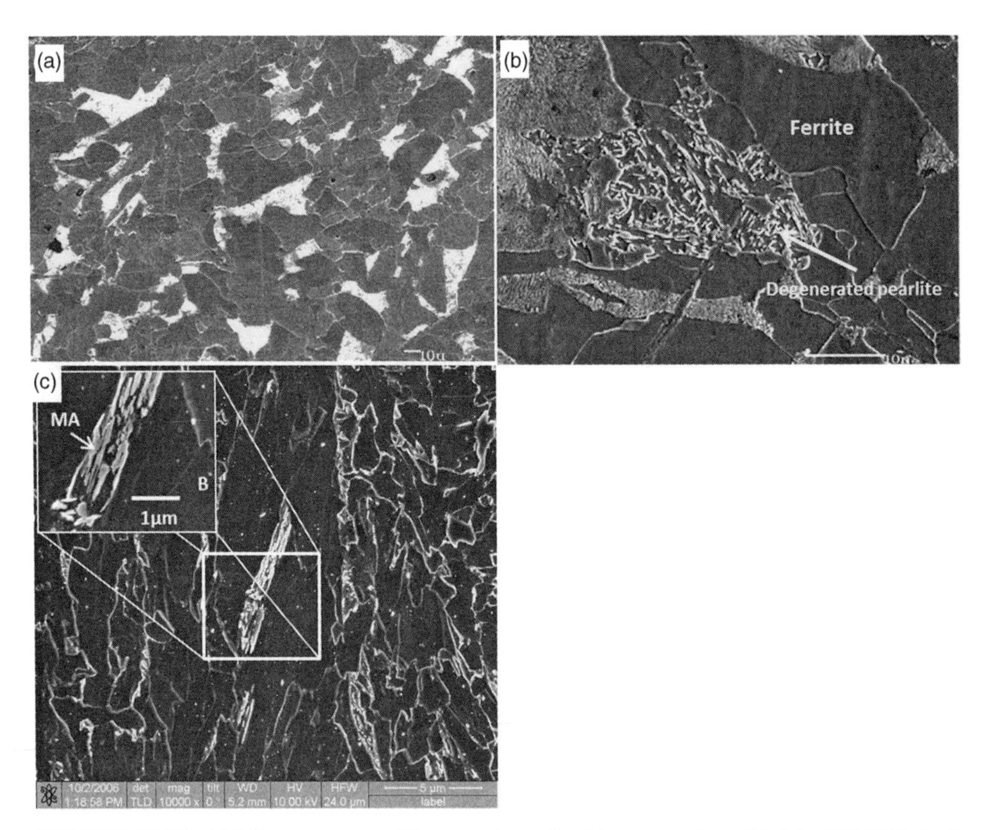


**FIGURE 10.15** (a) Microstructure of ferrite and pearlite (degenerated pearlite) in API X52 steel. (Reprinted from Ref. [37] with permission from Elsevier.) (b) The microstructure of degenerated pearlite. (Reprinted from Ref. [37] with permission from Elsevier.) (c) Microstructure of X80 pipeline grade with upper bainite (B) and zones with lower bainite and martensite–austenite islands (MA) (original research by Petrov). The white spots in the image were identified as carbides situated on the interlath spaces of the bainite.

transformation products inherits the elongated shape of the hot rolled and unrecrystallized austenite, creating a morphological (structural) shape anisotropy. In the undeformed austenite, the nucleation of the body-centered phases (or constituents) starts below the $Ar_3$ temperature; however, in the strained austenite, the $Ar_3$ temperature shifts to higher values and is often denominated as an $Ar_3^d$ temperature [39].

Nucleation of the body-centered phases in this case is controlled by the defects in the austenite and the preferential nucleation sites are high-angle grain boundaries and in-grain shear bands (Figure 10.17a and b) [39]. In the latter case, it appears that the HAGBs are the preferential nucleation sites, whereas the in-grain defects are less effective.

The body-centered transformation products have specific morphologies that are characterized by their pancaked shapes and orientations [40,41]. These features are illustrated in Figure 10.18a and b and are characterized in more detail in Refs [40,41].

In most cases, the product grains can be approximated by ellipsoids with long, short, and mid-length axes (Figure 10.18b). Figure 10.19 displays equal-area pole figure plots of the longest and shortest grain axes with respect to the plate coordinate system—rolling direction (RD), normal direction (ND), and transverse direction (TD). Here more than 200 individual grains (upper or lower bainite) have been sampled. The grain shape data were obtained by 3D EBSD analysis of the mid-thickness of an 18 mm steel plate of API X80 steel.

The data in Figure 10.19 indicate that rolling followed by accelerated cooling contribute to the formation of a microstructure with an increased density of grains with their longest axes oriented along or close (~30–45°) to the plate RD and TD (Fig.19 a). This observation can be associated with the elongated shapes of the parent austenite grains and the subsequent nucleation and growth of the BCC phase along the strain-generated structure [39]. More quantitative characterizations of the influence of temperature, cooling schedule, and the distribution of rolling strain between finishing passes can be found in Ref. [42].

By the systematic study of grain size evolution as a function of the finish rolling parameters, namely, rolling reduction per pass, start finish rolling temperature, and the cooling rate in plates rolled to a final thickness of 12 mm,

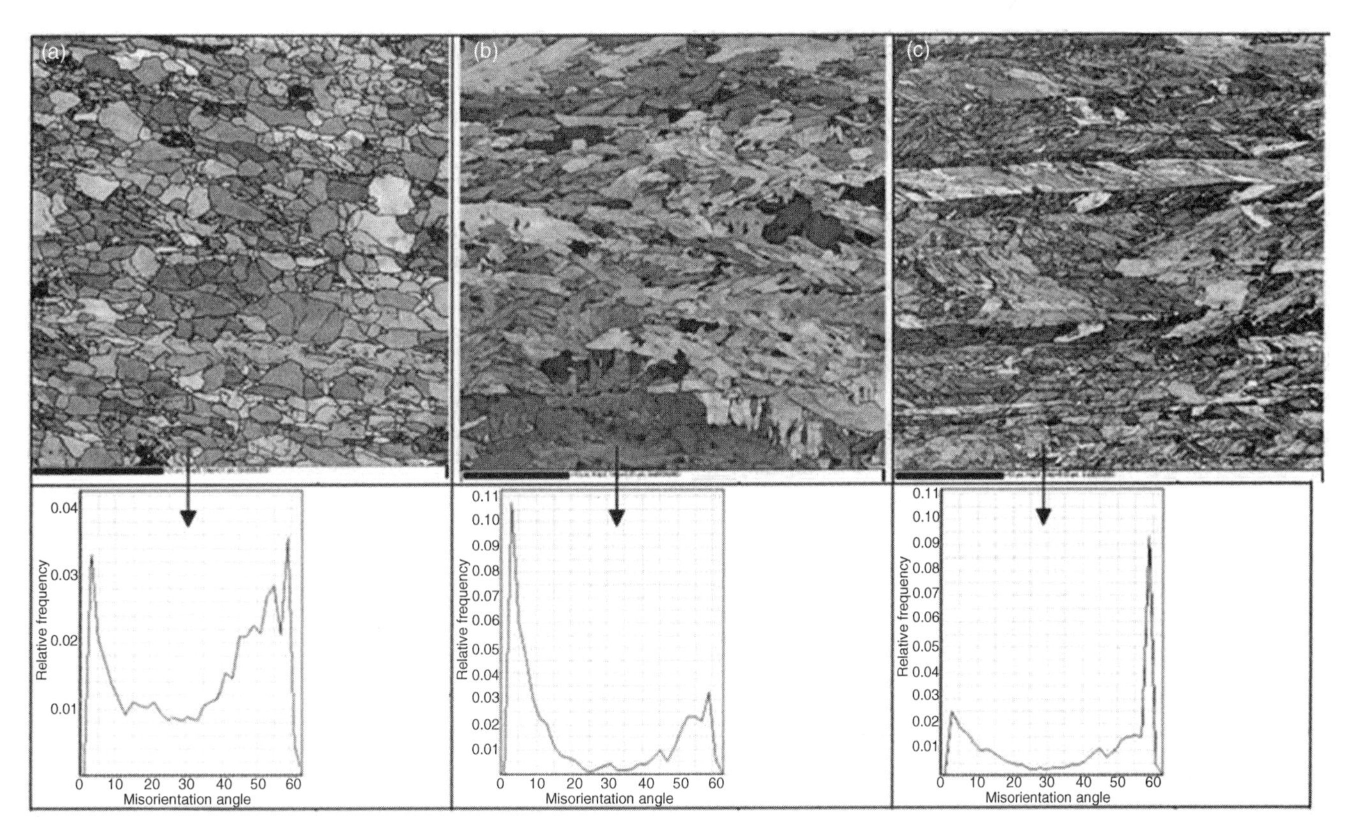


**FIGURE 10.16** EBSD-orientation maps and distributions of the misorientation angles of the ferritic grains in (a) granular bainite (GB), (b) upper bainite (UB), and (c) lower bainite (LB). (Reproduced with permission from Ref. [36].)

Carretero Olalla et al. [42] confirmed that accelerated cooling makes the most significant contribution to grain refinement. A summary of the results of this study is provided in Figures 10.20a and b and 10.21, where the microstructure of a pipeline grade steel is represented by EBSD maps displaying only the high-angle (>15°) and low-angle (2–15°) grain boundaries as a function of rolling reduction per pass, start finish rolling temperature, and cooling rate

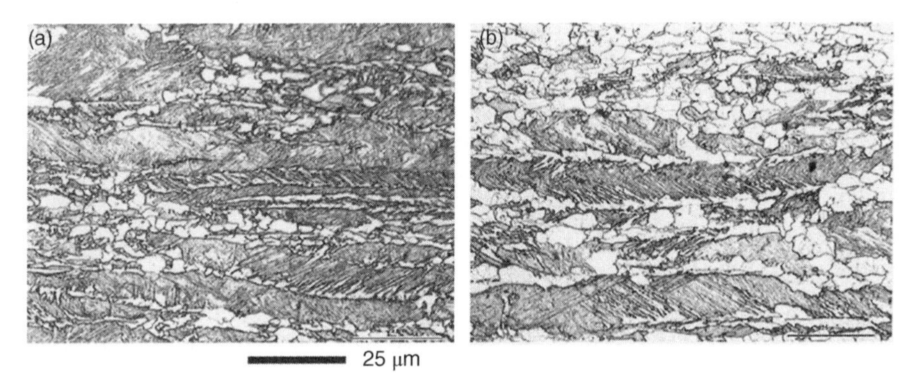


**FIGURE 10.17** Microstructure of steel with 0.082% C, 1.54% Mn, 0.35% Si, 0.055% Nb, and 0.078% V finish rolled at 850 °C and subsequently quenched at (a) 800 °C and at (b) 720 °C, respectively, to reveal the evolution of the BCC and BCT phases as a function of temperature. $Ar_3 = 790$ °C; $Ar_3^d = 840$ °C. White zones are ferrite and gray zones are martensite. (Reprinted from Ref. [39] with kind permission from Elsevier.)

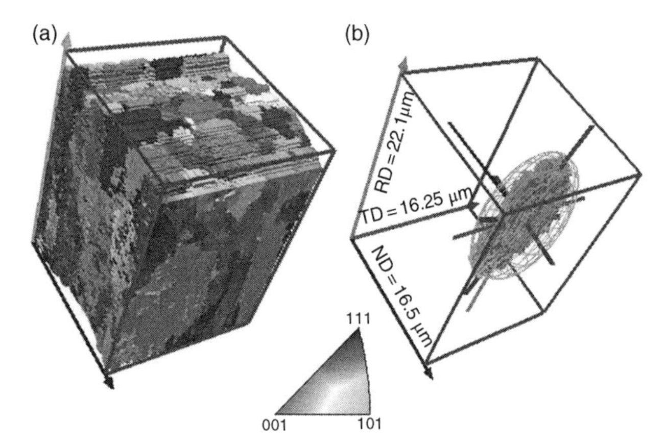


**FIGURE 10.18** (a) Three-dimensional EBSD reconstruction of a volume of 16.5 × 16.25 × 22.1 μm$^3$ from the central part of an API X80 pipeline plate 18 mm thick. (b) Selected representative grains contained in a volume of 108 μm$^3$ from the interior. The gray shades represent the crystallographic orientations of the grains with respect to the rolling plane (see the unit triangle).

after the final rolling pass—accelerated cooling (ACC) (Figure 10.20) or air cooling (Figure 10.21). The rolling schedules were carried out under exactly the same rough rolling conditions but various finish rolling parameters, namely, high start finish rolling temperature and increasing reduction per pass (schedule I, Figures 10.20a and 10.21a), low finish rolling temperature and increasing reduction per pass (schedule II, Figures 10.20c and 10.21c), high start finish rolling temperature and decreasing reduction per pass (schedule III, Figures 10.20b and 10.21b), low finish rolling temperature and decreasing reduction per pass (schedule IV, Figures 10.20d and 10.21d), and accelerated cooling (ACC) (Figure 10.20a–d) or air cooling (air) (Figure 10.21a–d). It should be pointed out that the rolling schedule of the plate with the microstructure shown in Figure 10.21c began and finished at temperatures 20 °C higher than the rest of the low-temperature rolling experiments. This is the reason for the larger grain size in Figure 10.21c.

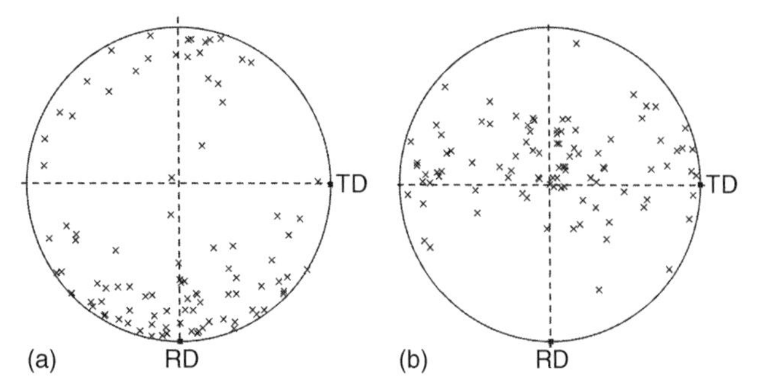


**FIGURE 10.19** Equal area plots of the longest (a) and shortest (b) axes of ~200 individual BCC grains with respect to the sample coordinate system (ND, RD, TD) of an 18mm thick TMCP plate of API X80 steel. (Adapted from Ref. [41].)

Another important finding is that significant grain refinement can be successfully achieved by starting finish rolling at low temperatures in both the accelerated and air-cooled conditions. Figure 10.22 shows that a combination of a low start finish rolling temperature and air cooling also results in significant grain refinement (Figures 10.21d and 10.22). Note that the finish rolling temperatures were higher for low FRT (air)–increasing reduction/pass than for low FRT (ACC)–increasing reduction/pass.

An interesting alternative for producing excellent combinations of strength and toughness was proposed recently in Refs [43,44] and involves quenching and partitioning [45]. The quenching and partitioning heat treatment is applied after the final rolling pass and consists of interrupted quenching to a temperature below Ms but above Mf of the particular steel grade. This step is followed by subsequent reheating to temperatures slightly above Ms in order to stabilize the untransformed austenite by carbon diffusion. The resulting microstructure is a low carbon martensite with excellent strength in which fine retained austenite islands are spread homogeneously. These islands are expected to make significant contributions to the toughness by the so-called transformation-induced plasticity (TRIP) effect.

In summary, the final microstructure in pipeline plates is characterized by a complex mixture of microstructural constituents produced under highly nonequilibrium conditions. The required grain refinement responsible for the high strength and toughness can be effectively produced in two stages of processing: (i) rough rolling for initial refinement of the austenite grains and (ii) finish rolling for the creation of a high density of nucleation sites for the formation of the necessary BCC transformation products. Accelerated cooling controls the type of the transformation product formed, which depends in turn on the alloy composition. These products include mixtures of various morphologies, such as ferrite, pearlite, and a wide range of bainitic and martensitic structures.

## 10.3 TEXTURE CONTROL IN PIPELINE STEELS

Crystallographic texture is one of the important microstructural parameters that control the mechanical behavior of pipeline steels. The two principal properties affected by the texture are (i) the overall fracture toughness and especially the tendency to form delaminations or splits and (ii) the in-plane anisotropy of the fracture toughness. The texture of pipeline steel grades can be controlled successfully via TMCP and the factors affecting the formation of desirable and undesirable texture components during processing are reviewed herein.

Crystallographic texture is a statistical characteristic of a polycrystalline material, pertaining to the presence of

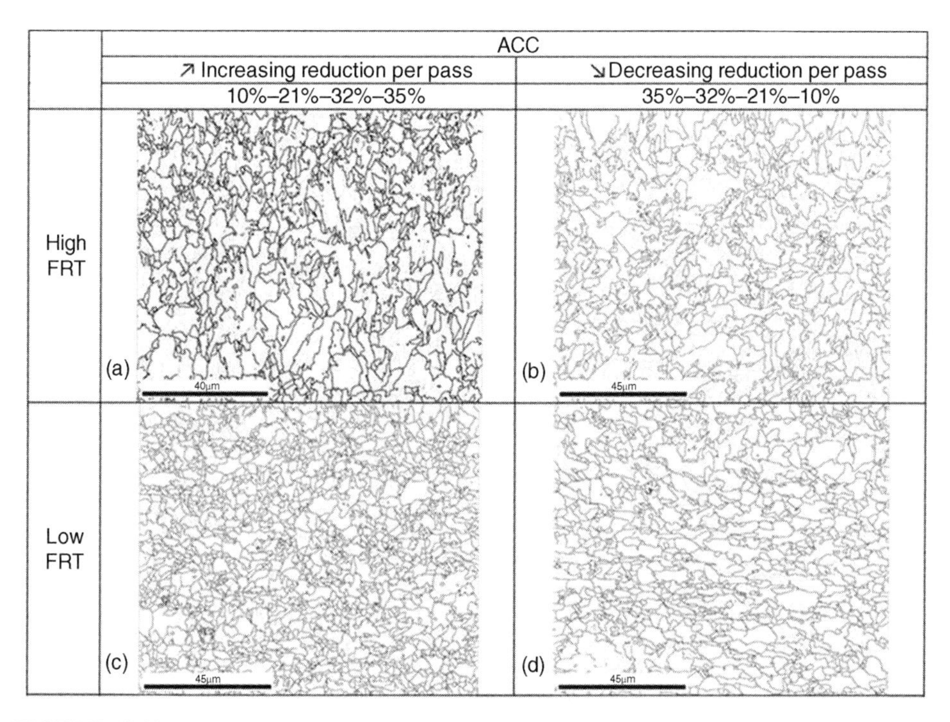


**FIGURE 10.20** Grain boundary maps of (a) schedule I, (b) schedule III, (c) schedule (II), and (d) schedule IV showing HAGB >15° misorientations (black) and LAGB >2–15° (gray). Accelerated cooling.

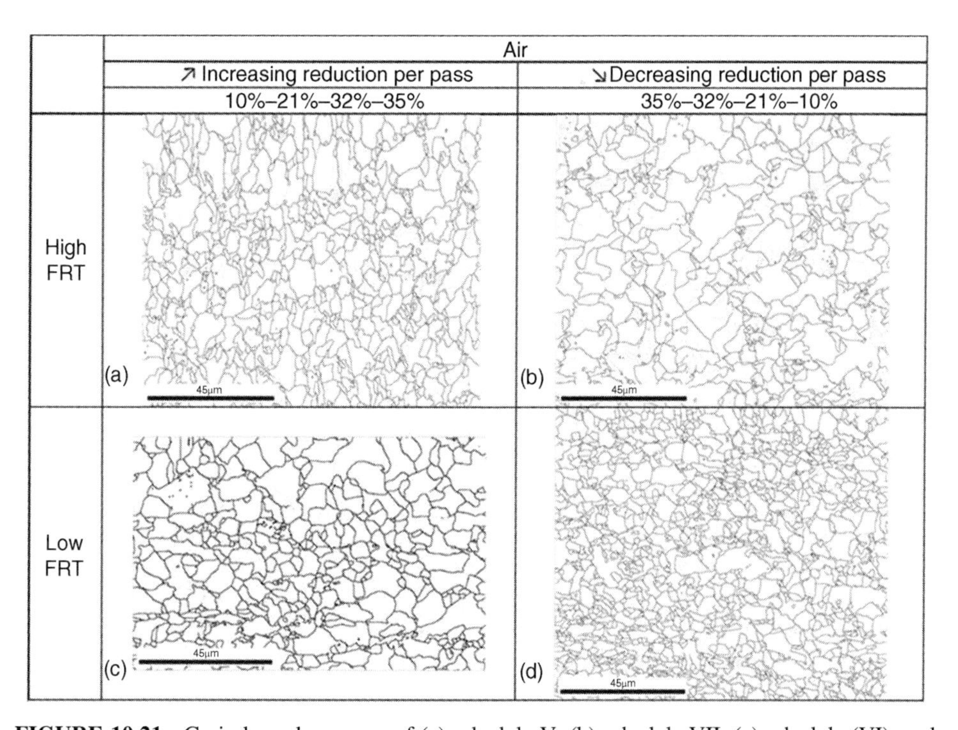


**FIGURE 10.21** Grain boundary maps of (a) schedule V, (b) schedule VII, (c) schedule (VI), and (d) schedule (VIII) showing HAGB >15° misorientations (black) and LAGB >2–15° (gray). Air cooling.

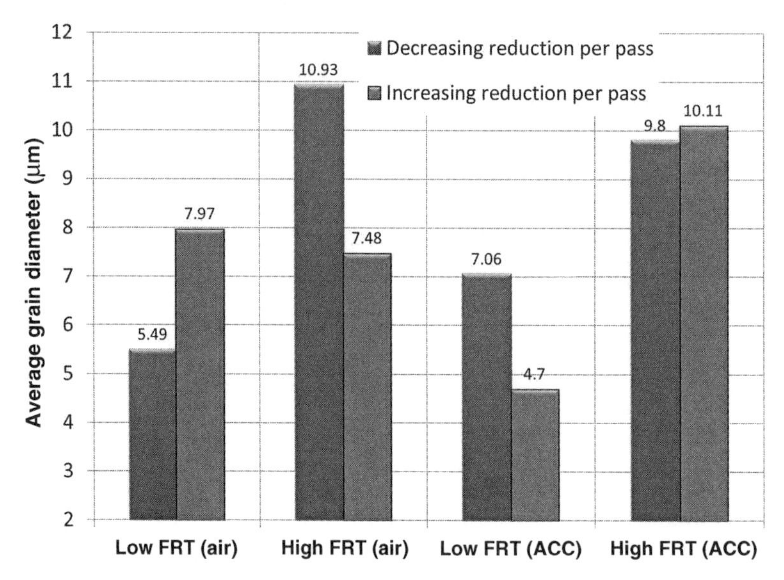


**FIGURE 10.22** Dependence of the average grain diameter on the TMCP parameters: start finish rolling temperature; accelerated cooling (ACC), air cooling (air), and increasing or decreasing reduction per pass.

preferred crystallographic orientations. The orientations of the individual crystals in a polycrystalline aggregate are represented commonly by specifying the crystallographic planes and directions with respect to the sample coordinate system. In the case of rolling, the sample coordinate system is associated with the rolling direction, the transverse direction, and normal direction (Figure 10.23). To uniquely characterize the orientation of an individual crystal, one needs to indicate the crystallographic plane, with Miller indices (*hkl*), that is parallel to the rolling plane and the crystallographic direction, with indices [*uvw*], that is parallel to the rolling direction. For example, the orientation of the crystal depicted in Figure 10.23 is (110) [001].

Alternatively, the orientation of a specific crystal can be specified in terms of Euler angles. The Euler angles represent three consecutive rotations, which bring the sample coordinate system into coincidence with the crystal coordinate system. There are several ways of defining such rotations, but the one most frequently used is that proposed by Bunge [46]. Euler angles can be represented in a 3D space, which is commonly referred to as Euler space. A set of three Euler angles ($\varphi_1$, $\Phi$, $\varphi_2$) identifies the exact position of each orientation in Euler space. As Euler angles are of a cyclic nature, the boundaries of Euler space are determined by $\varphi_1 = 0$–$360°$, $\Phi = 0$–$180°$, and $\varphi_2 = 0$–$360°$, but these boundaries are reduced to the limits $\varphi_1 = 0$–$90°$, $\Phi = 0$–$90°$, and $\varphi_2 = 0$–$90°$ for cubic crystal lattices and the orthorhombic symmetry of rolled samples.

The distribution of *individual crystallographic orientations* in a polycrystalline material can be plotted in a pole figure or in an inverse pole figure. The pole figure represents the distribution of a specific crystal pole $\langle uvw \rangle$ with respect to the sample coordinate system (Figure 10.24a), whereas the inverse pole figure displays the distribution of a sample axis (e.g., the ND direction) with respect to the crystal coordinate system (Figure 10.24b).

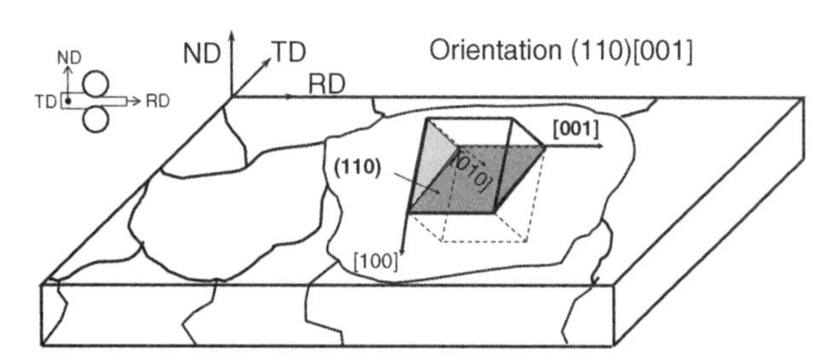


**FIGURE 10.23** Schematic representation of a crystal of orientation (110)[001].

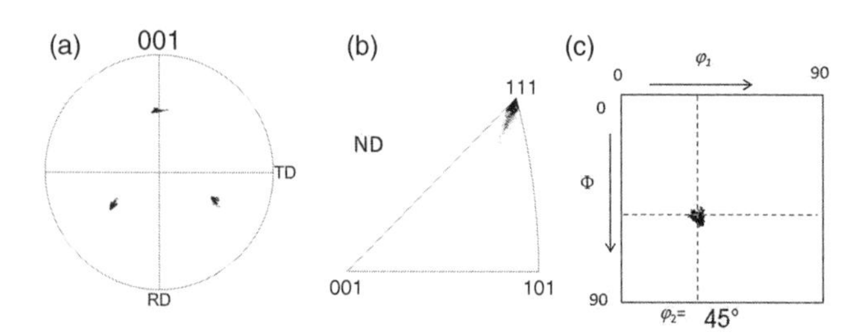


**FIGURE 10.24** Representation of the orientation of a particular BCC grain with (a) a pole figure, (b) an ND-inverse pole figure, and (c) in a $\varphi_2 = 45°$ of the Euler space as a discrete plot. The orientations are slightly scattered because there is an orientation gradient in the grain.

Figure 10.24a shows the (100) pole figure of a particular BCC grain with Miller indices ~(111) [11−2], which corresponds to Euler angles $\varphi_1 = 34–35°$, $\Phi = 51.5–52.7°$, and $\varphi_2 = 44.6–45.3°$; Figure 10.24b displays the orientation of the same grain in the ND inverse pole figure, whereas Figure 10.23c shows the position of this orientation in the $\varphi_2 = 45°$ section of Euler space. The above methods of representation are semiqualitative and do not give a complete quantitative description of the texture. The orientation distribution function (ODF) $f(g)$ is an orientation density function such that $f(g)dg$ represents the volume fraction of crystallites with an orientation within an infinitesimal environment $g \pm dg$ around orientation $g$. Figure 10.25a shows a discrete plot of orientations in the $\varphi_2 = 45°$ section of Euler space and Figure 10.25b displays the orientation distribution function calculated from the discrete orientations.

The isointensity lines of Figure 10.25b exhibit the intensity of each texture component with respect to the random texture. For BCC metals, the most important texture components of rolled sheet (either in the deformed or annealed condition) can be found in the $\varphi_2 = 45°$ section of Euler space, whereas for FCC metals the dominant texture components of rolled sheet appear in the sections $\varphi_2 = 45°$, $\varphi_2 = 65°$, and $\varphi_2 = 90°$(or 0°). Figure 10.26, which displays the ideal position of the most common BCC crystallographic orientations in the $\varphi_2 = 45°$ section, can be employed as a key to read the experimental ODFs discussed later in the text.

### 10.3.1 Fracture of Pipeline Steels

The initiation of pipeline fractures can usually be attributed to corrosion, construction damage, or earth settlement and earthquakes and is therefore largely beyond the control of the process metallurgist. The extent of propagation of such failures, on the other hand, is clearly under the influence of the metallurgist, as it depends on the toughness and microstructure of the steel. This is particularly an issue in high-pressure gas pipelines, which operate at pressures in excess of 100 atm and in which, unlike in the case of oil pipelines, a rupture does not lead to the immediate release of the pressure. The texture and microstructure, in turn, are sensitively affected by the cycles of recrystallization and transformation that take place during steel processing. In what follows, the stages of controlled rolling during which recrystallization is to be promoted, or in turn avoided, will be examined closely, with particular focus on its effects on the texture and consequently on the toughness. Control of the microstructure [47] has already been discussed.

The toughness is critical because it determines the speed of propagation of a crack, once initiated. When a steel is sufficiently tough, the rate of propagation is low and, in particular, well below the speed at which the pressure drop due to the escaping gas travels down the pipe. As a result, the driving force for propagation decreases below that necessary for crack growth and, ideally, the fracture does not extend beyond a single length of pipe. Conversely, in a low toughness steel, cracks can propagate much more quickly than the drop in gas pressure and the fracture can extend for many kilometers before the pressure decreases sufficiently to arrest the crack [48,49].

Microalloyed steels that exhibit low fracture toughness are also subject to the phenomenon of "splitting" or separation. Although splits are commonly observed in low-toughness test specimens [50], these separations do not in fact play a direct role in fracture. They are rather an indication of the presence of the {100}⟨011⟩ texture component, which is in turn responsible, at least in part, for the poor fracture properties [47,51]. Before considering how to reduce the intensity of this undesirable component, it will be useful to first consider why its presence should be avoided.

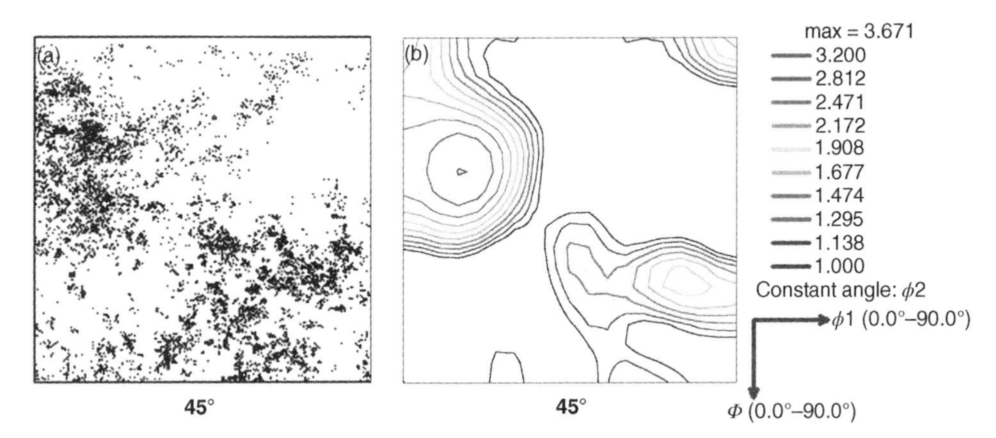


FIGURE 10.25 (a) Discrete plot of orientations of a pipeline steel in the $\varphi_2 = 45°$ section of Euler space. (b) ODF calculated from the discrete plot. The lines show the texture intensities in multiples of random distribution (MRD).

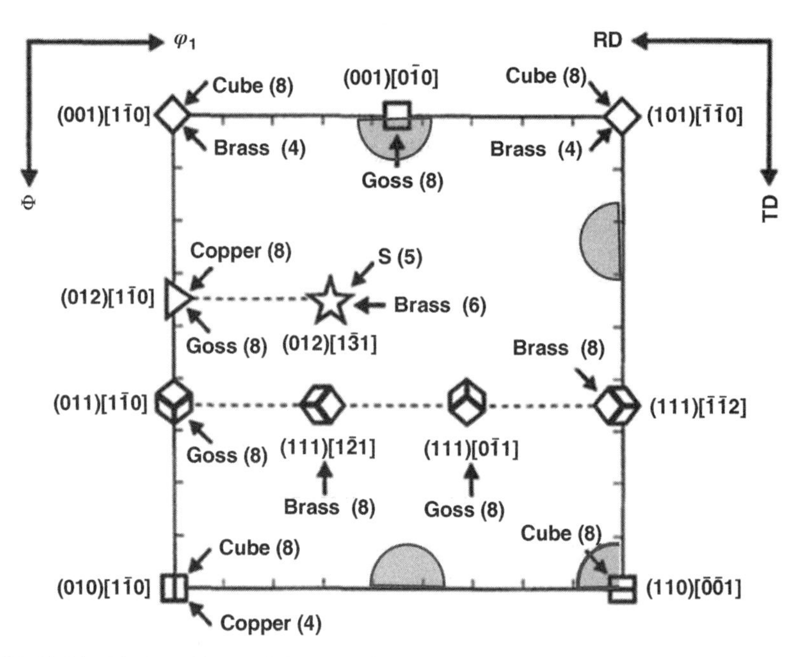


**FIGURE 10.26** Ideal positions of the most important BCC texture components in the $\varphi_2=45°$ section of Euler space. The positions of the main texture components of the austenite, cube, copper, brass, and Goss are represented as shaded semicircles. The arrows identify the parent FCC orientations from which specific BCC orientations originate.

***10.3.1.1 Properties of the {100}⟨011⟩ Component*** A schematic representation of a {100}⟨011⟩ crystal in a steel plate is provided in the central sketch of Figure 10.27. Here the grain is of cubic shape and its faces consist of {100} surfaces, which have low strengths in cleavage. These faces are inclined at 45° to the hoop stress of the pipe in the case of longitudinally welded products (LHS) and are approximately perpendicular to the hoop stress in the case of spiral welded pipe (RHS). Thus, the presence of grains in this orientation will promote brittle behavior during crack initiation and propagation and lower the fracture toughness in this way.

Because of the cubic symmetry, the "top" and "bottom" faces of both the "longitudinal" and "spiral" {100} grains are parallel to the outer surfaces of the pipe and are therefore responsible for splitting when it occurs. However, during operation, stresses perpendicular to the surface cannot be generated and so there is no driving force for cleavage on these planes. It is instead the occurrence of necking and plate thinning during crack propagation that is responsible for the generation of the through-thickness triaxial stresses, which in turn produce the splits (Grey, J.M. and Fonzo, A. (2003) Private communication). Thus, the separations are the *result*

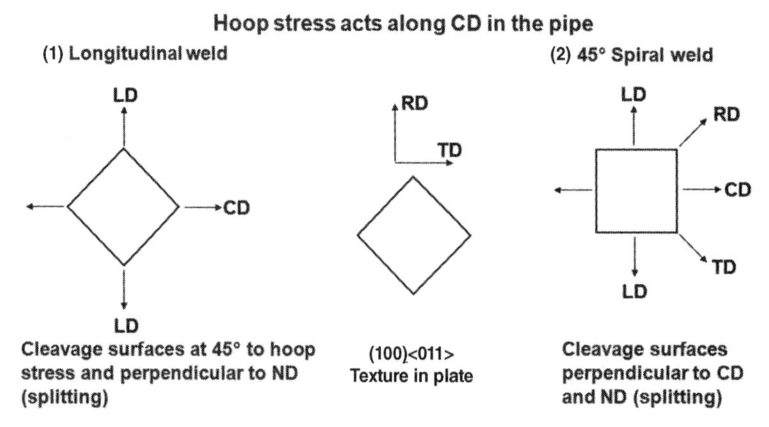


**FIGURE 10.27** Schematic diagrams of a {100}⟨011⟩-oriented grain in a rolled plate (center), a longitudinally welded pipe (LHS), and a spiral welded pipe (RHS). RD: rolling direction; LD: longitudinal direction; CD: circumferential (hoop) direction.

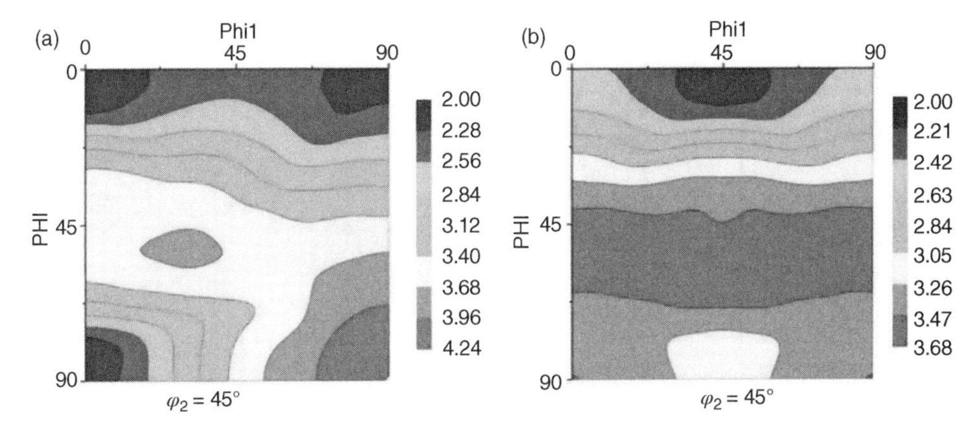


**FIGURE 10.28** Taylor factor map for plane strain tension along the circumferential direction calculated using the "full constraint" theory of crystal plasticity and plotted in two 2D cross sections of Euler space. (a) Longitudinally welded pipe. (b) ±45° spiral welded pipe.

of fracture propagation and do not in fact play a direct role in advancing the crack.

Another characteristic associated with the $\{100\}\langle 011\rangle$ component is its contribution to the local yield strength of the steel, which is illustrated in Figure 10.28. Here it can be seen that the cube and rotated cube components have the lowest yield strengths of all crystal orientations. At 2.2–2.4, their Taylor factors are almost half that of the Goss component in Figure 10.28a and only about 60% that of the $\{111\}$ fiber in Figure 10.28b. (The yield strength of a given grain is obtained by multiplying the critical resolved shear stress for dislocation motion by the Taylor factor.) Thus, the presence of this component has a local weakening effect on the plastic properties of the plate and therefore on the fracture toughness. Two other undesirable features associated with the cube fiber (not illustrated here) are (i) that it has the lowest rate of work hardening (proportional to the Taylor factor) of all the texture components, an additional source of weakening, and (ii) that it also has the lowest mean R-value of all texture components, which means that it provides the least resistance to thinning of the plate during tensile deformation and therefore during crack propagation.

### 10.3.2 Effect of Phase Transformation on the Texture Components

Steel processing generally involves the operation of five texture change mechanisms [52]. These are the ones that are brought about by (i) austenite deformation, (ii) austenite recrystallization, (iii) the $\gamma$–$\alpha$ phase transformation, (iv) ferrite deformation, and (v) ferrite recrystallization. In what follows, the conversion of the deformation and recrystallization austenite texture components into their ferrite or bainite counterparts is taken to occur largely following the Kurdjumov–Sachs (KS) and Nishiyama–Wassermann (NW) correspondence relations [53,54]. In the first instance, this involves rotations of +/−90° about the 12 $\langle 112\rangle$ axes, leading to 24 possible variants. In the latter case, the rotations do not have simple descriptions, but lead to 12 variants, each of which is an average of two KS variants. The austenite texture components introduced by rolling will be described below, together with their transformation products. In a similar manner, the principal recrystallization component, the cube texture, will also be described, together with its main transformation products.

### 10.3.3 Effect of Austenite Recrystallization on Plate Texture

When austenite is rolled above the nonrecrystallization temperature $T_{nr}$, the $\gamma$ deformation textures introduced are converted, by recrystallization, into the cube or $\{100\}\langle 001\rangle$. The presence of this component is illustrated in a $\varphi_2 = 45°$ Euler space cross section in Figure 10.29a [55], whereas the transformation texture products predicted to appear (by KS and NW) after cooling and phase transformation are identified in Figure 10.29b [55]. An example of the transformation products appearing in a 0.11% C TRIP steel after cooling is presented in Figure 10.29c, where it can be seen that the FCC cube component has been transformed into the rotated cube $\{100\}\langle 011\rangle$, the Goss $\{110\}\langle 001\rangle$, and the rotated Goss $\{110\}\langle 110\rangle$, with the rotated cube predominating at an intensity of 8+ [56]. (The TRIP steel has been chosen here to demonstrate the transformation behavior of microalloyed steels because the microstructure contains retained austenite, the presence of which enables determination of the recrystallization texture in the FCC phase. In contrast, in conventional microalloyed steels, there is no austenite present at room temperature from which to determine the FCC texture.)

An illustration of the grain orientations produced by transformation is provided in Figure 10.30, which shows

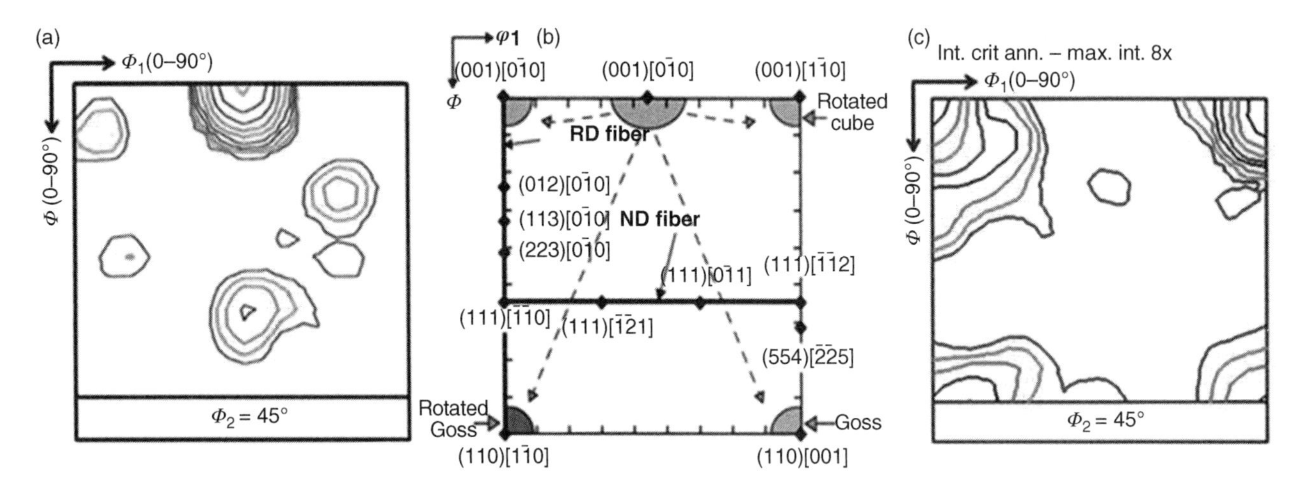


**FIGURE 10.29** Transformation of the FCC cube texture into its BCC counterparts. (a) Hot rolled and recrystallized Fe–28% Ni alloy (X-ray diffraction). (b) The rotated cube, Goss, and rotated Goss components are formed from the FCC cube $\varphi_2 = 45°$ section. (c) Transformation texture in a laboratory hot rolled (FRT: 950 °C) and intercritically annealed TRIP steel (0.11% C–1.53% Mn–1.26% Si).

an austenite grain of cube orientation as well as that of its twin in Figure 10.30a and those of the bainite and ferrite transformation products in Figure 10.30b [57]. Thus, it is clear that when the cube component predominates in hot rolling, something that is brought about by austenite recrystallization, the final texture after cooling and transformation will contain high-volume fractions of the undesirable {100}⟨011⟩ component. For this reason, the occurrence of austenite recrystallization during the final stages of steel processing is very much to be avoided. This is accomplished by ensuring that the final stages of rolling are carried out well below the $T_{nr}$ and that, in particular, the interiors of heavy plates are below this temperature.

### 10.3.4 Effect of Austenite Pancaking on the Rolling Texture

When austenite is rolled below the $T_{nr}$, the deformation texture components are not removed by recrystallization, but are retained during cooling down to the transformation temperature. These components are the copper {112}⟨111⟩, brass {110}⟨112⟩, S {123}⟨634⟩, and Goss {110}⟨001⟩, which, together with their intermediate neighbors, form the rolling tube or β fiber. The locations of the austenite rolling fiber components in 3D Euler space are illustrated in Figure 10.31a and in the $\varphi_2 = 45°$ 2D cross section of Euler space in Figure 10.31b [52,57]. (The S component does not appear in this cross section.)

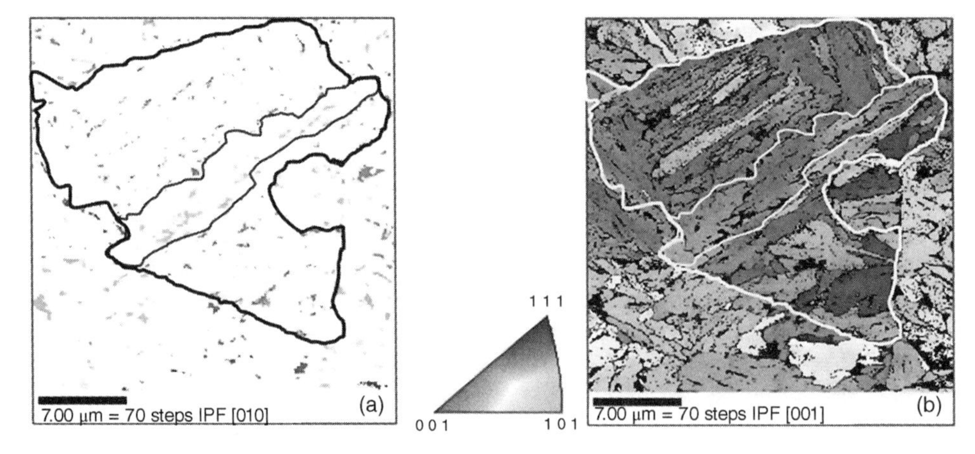


**FIGURE 10.30** (a) A recrystallized cube-oriented austenite grain (upper and lower extremities of the outlined grain) containing a twin in the central part. (b) The bainite decomposition products of rotated cube, rotated Goss, and Goss orientation. EBSD inverse pole figures: as-hot rolled TRIP steel; composition: 0.22% C–1.50% Mn–1.56% Si–0.045% Nb.

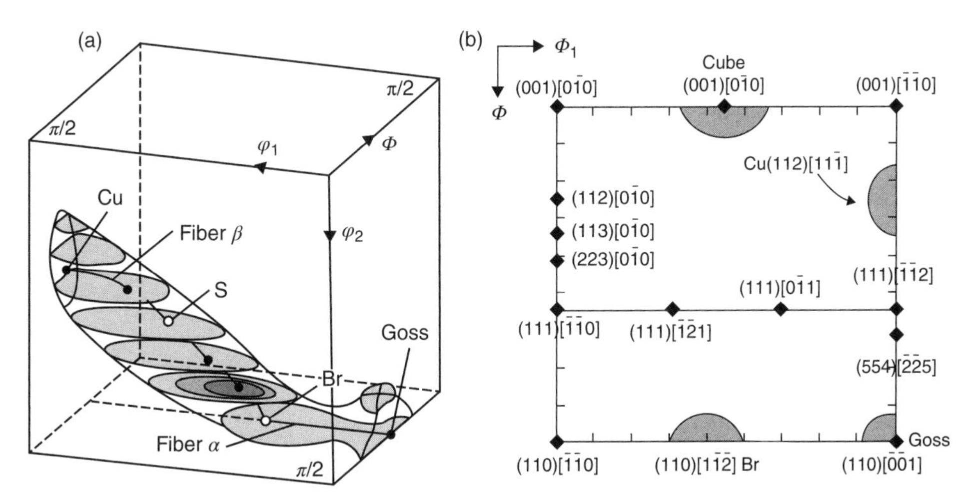


**FIGURE 10.31** (a) The austenite rolling fiber represented in 3D Euler space showing the orientations of the Cu, S, Br, and Goss components. (b) The locations of the Cu, Br, and Goss components on a $\varphi_2 = 45°$ cross section of Euler space.

The effect of phase transformation on these components is depicted in Figure 10.32 [52,57]. The important points to note are the following: (i) The copper component transforms primarily into the $\{112\}\langle 110\rangle$, also known as the "transformed Cu," with a much smaller fraction appearing in the vicinity of the rotated Goss. The remaining KS components do not appear in the $\varphi_2 = 45°$ cross section and are not relevant here. (ii) The S component transforms in part into orientations near the $\{112\}\langle 131\rangle$; the rest are dispersed elsewhere in the 3D Euler space and are not implicated in the present discussion. (iii) The Goss component (a minority component during rolling) transforms partly into the rotated Cu, contributing to the intensity of this component, and partly into the $\{111\}\langle 110\rangle$, with the balance located outside the $\varphi_2 = 45°$ cross section. (iv) Finally, the brass or Br component transforms primarily into the "transformed Br" located in the vicinity of $\{111\}\langle 112\rangle$ and $\{554\}\langle 225\rangle$, with minor amounts rotating to $\{112\}\langle 131\rangle$ and $\{100\}\langle 011\rangle$.

The appearance of a deformed austenite grain of Br orientation in a TRIP steel is depicted in Figure 10.33, together with the ferrite/bainite orientations appearing after transformation [57]. Because of the retained austenite present in TRIP steels after transformation, it is possible to measure both the austenite and ferrite textures in the same sample. The FCC and BCC textures determined in this way are shown in

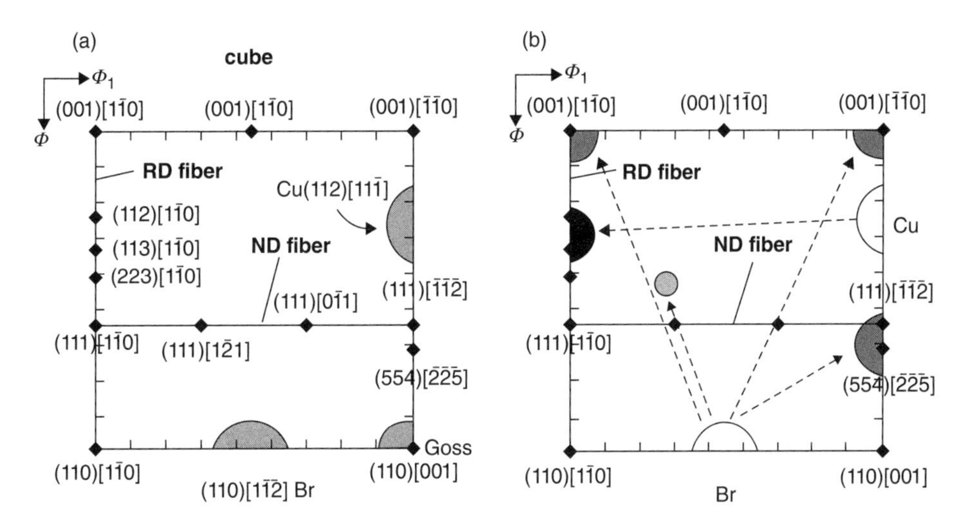


**FIGURE 10.32** Transformation of the FCC rolling fiber into its BCC counterparts. (a) Locations of the Cu, Br, and Goss components; the RD fiber, containing the $\{100\}\langle 011\rangle$, is produced by rolling a BCC material ($\varphi_2 = 45°$ section). (b) The BCC transformation products of the Cu and Br FCC orientations ($\varphi_2 = 45°$ section).

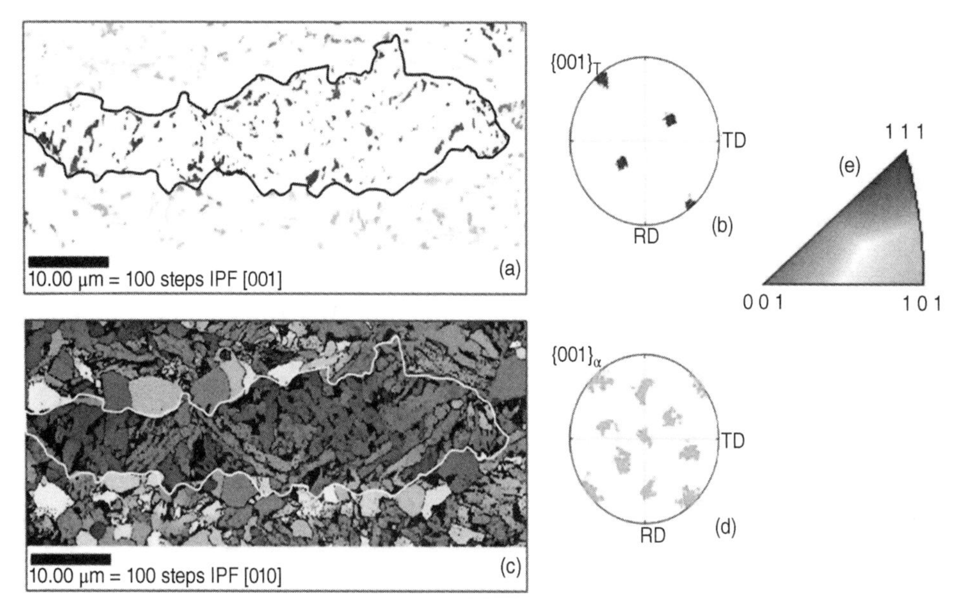


**FIGURE 10.33** A pancaked austenite grain of Br orientation (a) (upper micrograph) with its orientation identified in the {001} pole figure (b); the bainite transformation products (c) consisting of components centered around {554}⟨225⟩, {112}⟨131⟩, and {100}⟨011⟩ represented in {001} pole figure (d). Key for orientations in the inverse pole figures (e).

Figure 10.34, where the presence of the Cu, Br, and Goss components in the austenite is clearly evident and the typical "transformation texture" that results from controlled rolling is also shown. In this example, the intensity of the rotated cube component (at 3+) is well below that of any of the components of the transformation fiber and the volume fraction of this component is appropriately low.

With regard to the deleterious rotated cube orientation, the first point to note is that, whether rolling is completed above or below the $T_{nr}$, in both cases the rotated cube component will be present to some degree after transformation. However, it is of particular importance here that, when recrystallized material is transformed, the rotated cube is the most intense, that is, the highest volume fraction component. Conversely, when a "pancaked" plate is transformed, the {100}⟨011⟩ orientation is only a minor component, as it is a low-intensity by-product of just one of the four principal preferred orientations of the rolling fiber. Thus, transformation of a pancaked plate will generally lead to volume fractions of the rotated cube that are approximately an order of magnitude lower than that resulting from the transformation of a recrystallized plate.

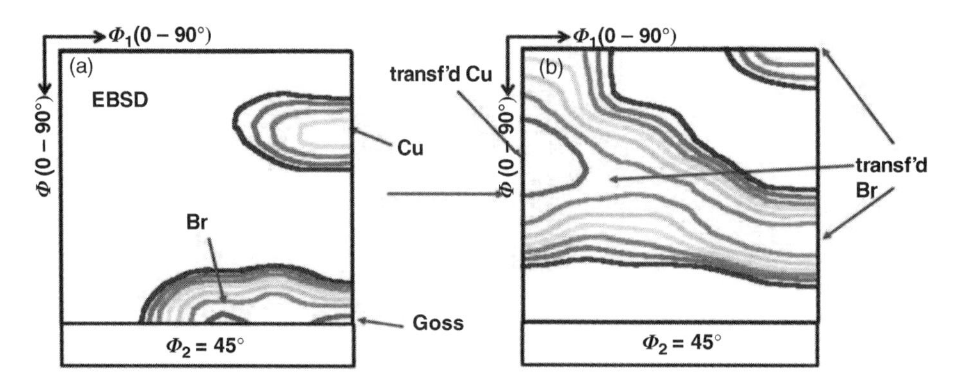


**FIGURE 10.34** FCC and BCC textures determined by EBSD on the as-hot rolled, pancaked TRIP steel of Figure 10.26. (a) Retained austenite texture illustrating the presence of the Cu, Br, and Goss rolling components ($\varphi_2 = 45°$ section). (b) Bainite texture running from the transformed Cu (LH axis) transversely along the "transformation fiber" to the "transformed Br" ($\varphi_2 = 45°$ section). The much lower intensity of the rotated cube (upper left and upper right corners) can be seen.

### 10.3.5 Effect of Finish Rolling in the Intercritical Region

It remains to consider the effect of deformation and recrystallization of the ferrite phase on the final texture. This is because it is sometimes desirable to finish rolling in the intercritical region so as to increase the flow stress of the ferrite phase present under these conditions and also to increase the yield strength of the final product. Here it must be pointed out that the {100}⟨011⟩ component is part of the ferrite rolling fiber and so a certain amount will be introduced by rolling in that temperature domain (Figure 10.32a). Furthermore, when such rolled ferrite is able to recrystallize, the principal orientation produced is the rotated cube [58]. The presence of the rotated cube component in combination with a bimodal grain size distribution, as discussed above and reported in Refs [35,47], is particularly undesirable as it leads to poor crack arrestability. For these reasons, when the fracture propagation properties are of importance, it is preferable to avoid finish rolling in the intercritical temperature range.

In summary, the presence of the rotated cube component, {100}⟨011⟩, in pipeline steels is associated with low toughness values, which are in turn responsible for rapid crack propagation. This is why it is essential to minimize the volume fraction of this component in the plate. It should be added that the presence of the {100}⟨011⟩ texture component appears to be particularly injurious to the fracture properties when elevated volume fractions of this orientation are present in the outer layers of the plate [58].

The complete recrystallization of austenite during hot rolling leads to the formation of the cube texture. On transformation to ferrite and bainite, the cube is converted into the rotated cube, the Goss, and the rotated Goss, with the rotated cube predominating. For this reason, it is important to avoid austenite recrystallization during the finishing stages of the rolling of pipeline steels and to do all the final reductions below the $T_{nr}$.

The pancaking of austenite during rolling below the $T_{nr}$ leads to the presence of the Cu, S, Br, and Goss components. On transformation, these lead to the appearance of the well-known transformation (mostly {111}) fiber. Although this texture includes minor quantities of the {100}⟨011⟩, it is the preferred texture with respect to ensuring high toughness values in the steel. Rolling in the intercritical temperature range is to be avoided, as it leads to the introduction of additional quantities of the undesirable rotated cube component.

The above principles are of particular importance during the rolling of plate for heavy-gauge pipelines. Due to the large thickness, appreciable temperature gradients can develop between the surface and the interior of the plate. As a result of the temperature difference, the external layers of the plate or bar may be below the $T_{nr}$ during finishing, while the core is above the $T_{nr}$. An example of the textures formed in a plate rolled under these conditions is provided in Figure 10.35 [58]. The surface texture of such a plate is illustrated in Figure 10.35a, while the core texture is shown in Figure 10.35b. The layers close to the surface were finished below the $T_{nr}$, so that the resulting ferrite transformation texture originated from pancaked austenite. In contrast, the central layers of the bar were finished above the $T_{nr}$, so that the austenite was able to recrystallize prior to transformation. As a consequence, the main BCC texture component at this location was the {100}⟨011⟩, which forms from recrystallized austenite. These problems can be avoided by ensuring that sufficient cooling takes place before finishing. The $T_{nr}$ temperature can also be raised, providing a wider pancaking temperature range, by employing higher concentrations of Nb (Figure 10.14).

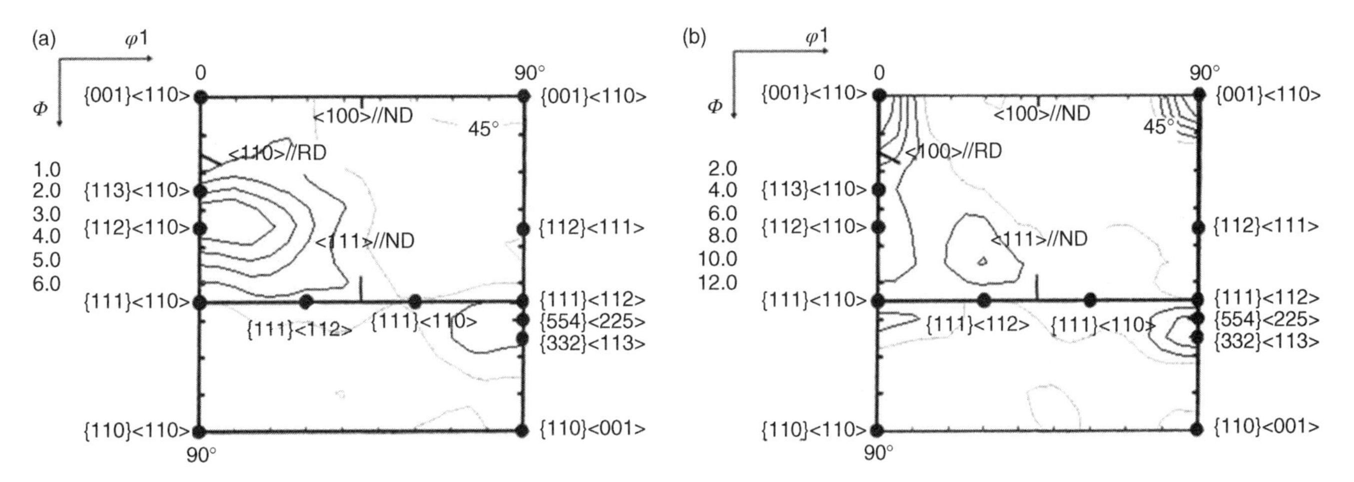


**FIGURE 10.35** $\varphi_2 = 45°$ ODFs of heavy gauge plate. (a) Close to the surface. (b) Close to the mid-thickness of the plate. The core texture contains high intensities of the {001}⟨110⟩ BCC transformation component that originates from recrystallized austenite [58].

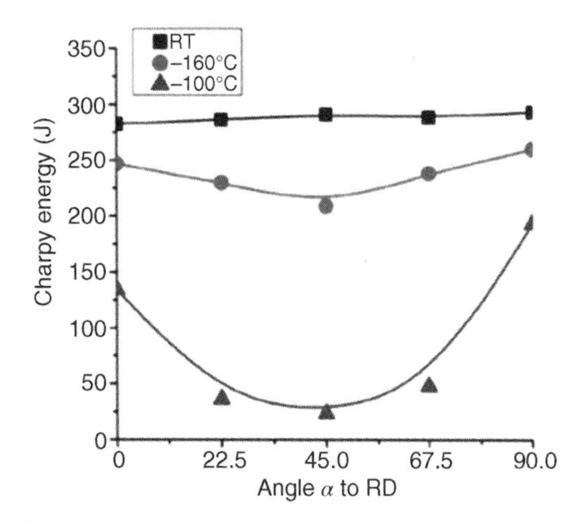


**FIGURE 10.36** Planar anisotropy in Charpy V-notch energy.

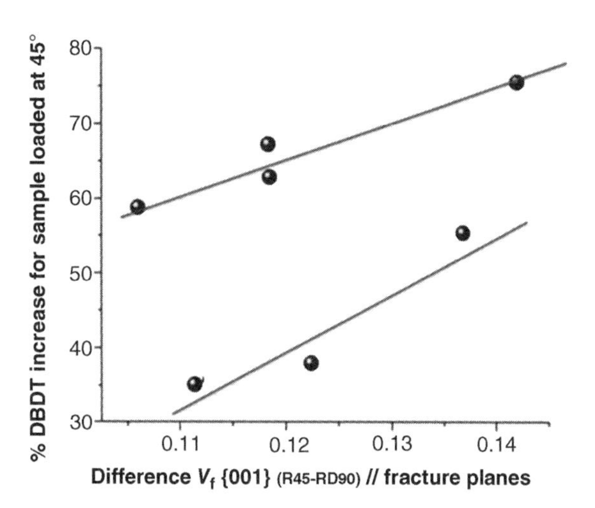


**FIGURE 10.38** Percentage increase in DBTT as a function of the difference in {001} volume fraction parallel to the fracture plane between samples loaded at 45° and 90° [38].

## 10.4 EFFECT OF TEXTURE ON IN-PLANE ANISOTROPY

The anisotropy in Charpy impact energy of an X80 grade pipeline steel (Figure 10.36) was studied in detail in Ref. [38]. The Charpy tests were carried out at various temperatures, with the long axis of the sample enclosing different angles of inclination $\alpha$ (between 0° and 90°) with respect to the RD and the notch perpendicular to the plane of the sheet. It can be seen that at room temperature there is virtually no anisotropy, whereas at −60 and −100 °C a distinct reduction in toughness is observed at $\alpha = 45°$, when the long axis of the Charpy sample is parallel to the diagonal axis of the sheet.

The texture of this steel (Figure 10.37) is a typical hot band transformation texture originating from a pancaked austenite rolling texture, with characteristic components {112}⟨110⟩, {112}⟨131⟩, and {554}⟨225⟩, which are the transformation products of the austenite rolling components Cu ({112}⟨111⟩), S ({123}⟨634⟩), and Br ({110}⟨112⟩), respectively.

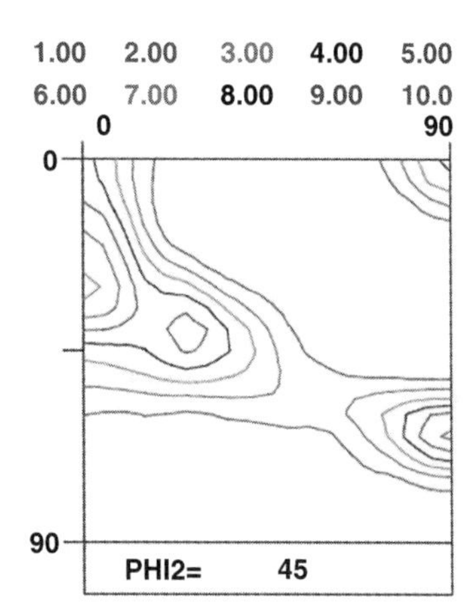


**FIGURE 10.37** Texture ($\varphi_2 = 45°$ section) of an X80 grade pipeline steel exhibiting the toughness profile shown in Figure 10.36 [38].

For any given texture, the volume fraction of grains for which the {001} crystal plane is parallel to the macroscopic fracture plane of the Charpy V-notch sample can be readily calculated. This volume fraction depends on the angle $\alpha$ along which the Charpy sample is loaded. As the {001} planes have the lowest strength in cleavage, variations in the {001} volume fraction can explain the observed toughness anisotropy. The dependence of the latter on crystallography is confirmed by the data shown in Figure 10.38, from which it can be seen that the rise in the ductile-to-brittle transition temperature (DBTT) at 45° (i.e., the vertical axis of the graph) increases with the difference in {001} volume fraction between the samples loaded along 45° and 90° (i.e., the horizontal axis of the graph).

## 10.5 SUMMARY

Descriptions are given of the microstructure and texture in the most commonly used pipeline steel grades. The composition ranges of these grades are summarized, together with a brief discussion of segregation effects. After a short historical review, the general principles of microstructure and texture formation during the thermomechanically controlled processing of plate and sheet are discussed. These are illustrated with examples taken from the recent literature. Special attention is paid to texture formation, methods of texture control, and the influence of texture on the mechanical properties, anisotropy, and fracture behavior of these microalloyed steels.

## ACKNOWLEDGMENTS

The authors thank Eng. Tuan Nguen Minh, Eng. Victor Carretero Olalla from Ghent University, Belgium, and Dr. Alexey Gervasyev from Rosniti, Russian Federation for interesting and fruitful discussions and help in the preparation of the figures. L.K. is grateful for financial support from the FWO-Odysseus Program under the project "Engineering of 3D microstructures in metals: bridging ten length scales of functionality." J.J.J. acknowledges with gratitude funding received from the Natural Sciences and Engineering Research Council of Canada.

## REFERENCES

1. EIA (2011) International Energy Outlook (2011). Report No. DOE/EIA-0484(2011). Available at http://www.eia.gov/forecasts/ieo/world.
2. Hopkins, P. Oil and gas pipelines: yesterday and today. C Pipeline Systems Division (PSD), International Petroleum Technology Institute, American Society of Mechanical Engineers (ASME). Available at http://www.engr.mun.ca/~spkenny/Courses/Undergraduate/ENGI8673/Reading_List/2007_Hopkins.pdf.
3. Gray, J.M. and Siciliano, F. (2009) High strength microalloyed linepipe: half a century of evolution. Proceedings of the 5th International Conference on Pipeline Technology, Ostend, Belgium.
4. De Ardo, A.J., Hua, M.J., Cho, K.G., and Garcia, C.I. (2009) On strength of microalloyed steels: an interpretive review. *Materials Science and Technology*, 25(9), 1074–1082.
5. Peters, P.A. and Gray, J.M. (1992) Genesis and development of specifications and performance requirements for modern linepipe: strength, toughness, corrosion resistance and weldability. International Convention, Australian Pipeline Industry Association, Inc., Hobart-Tasmania, Australia, October 24–29, 1992.
6. Gray, J.M. and Peter, P.A. (2002) Technical demands and specifications for linepipe during the past decades. CBMM/TSNIICHERMET Seminar: 25 Years of Cooperation, Moscow, Russia, September 5–6, 2002.
7. Altenburger, C.L. (1960) Columbium (niobium) treated, low carbon semi-killed steel. AISI Regional Technical Meeting No. 59, Buffalo, New York, NY.
8. Barkow, A.G. (1960) Columbium steel in high pressure line pipe service. AISI Regional Technical Meeting No. 59, Buffalo, New York, NY.
9. Altenburger, C.L. and Bourke, A. (1961) Columbium containing steels, process for their manufacture and articles prepared therefrom. U.S. Patent 3,010,822, November 28, 1961.
10. Demofonti, G., Mannucci, G., Spinelly, C.M., Barsanti, L., and Hillenbrand, H.G. (2003) Large diameter X100 gas linepipes: fracture propagation evaluation by full-scale burst test, Europipe. Proceedings of the 14th Joint Technical Meeting on Pipeline Research, Berlin, Germany, May 19–23, 2003, pp. 1–17.
11. Zhang, J.-M., Sun, W.-H., and Sun, H. (2010) Mechanical properties and microstructure of X120 grade high strength pipeline steel. *Journal of Iron and Steel Research, International*, 17(10), 63–67.
12. Zhou, M., Du, L.-X., and Liu, X.-H. (2011) Relationship among microstructure and properties and heat treatment process of ultra-high strength X120 pipeline steel. *Journal of Iron and Steel Research, International*, 18(3), 59–64.
13. Nafisi, S., Arafin, M.A., Collins, L., and Szpunar, J. (2012) Texture and mechanical properties of API X100 steel manufactured under various thermomechanical cycles. *Materials Science and Engineering A*, 531, 2–11.
14. Vervynckt, S., Verbeken, K., Lopez, B., and Jonas, J.J. (2012) Modern HSLA steels and role of non-recrystallisation temperature. International Materials Reviews, 57(4), 187–207.
15. Pickering, F.B. (1975) High strength, low-alloy steels: a decade of progress. Proceedings of the MicroAlloying '75 Conference, Washington, DC.
16. DeArdo, A.J. (2003) Niobium in modern steels. *International Metals Reviews*, 48, 371–402.
17. George, E. (1988) *Dieter Mechanical Metallurgy*, McGraw-Hill Book Company, p. 587.
18. Davis, C.L. and Strangwood, M. (2009) Segregation behaviour in Nb microalloyed steels. *Materials Science and Technology*, 25(9), 1126–1133.
19. Zajac, S. and Jansson, B. (1998) Thermodynamics of the Fe–Nb–C–N system and the solubility of niobium carbonitrides in austenite. *Metallurgical and Materials Transactions B*, 29B, 163–176.
20. Mendoza, R., Alanis, M., Perez, R., Alvarez, O., Gonzalez, C., and Juarez-Islas, J.A. (2002) On the processing of Fe/C/Mn/Nb steels to produce plates for pipelines with sour gas resistance. *Materials Science and Engineering*, A337, 115–120.
21. Roy, S., Chakrabarti, D., and Dey, G.K. (2013) Austenite grain structures in Ti- and Nb-containing high-strength low-alloy steel during slab reheating. *Metallurgical and Materials Transactions A*, 44A, 717–728.
22. Cuddy, L.J. (1982) In: De Ardo, A.J. et al. (editors), *Themomechanical Processing of Microalloyed Austenite*, The Metallurgical Society of AIME, Warrendale, PA, 129–140.
23. Vervynckt, S. (2010) Control of the non-recrystallization temperature in high strength low alloy (HSLA) steels, Ph.D. thesis, Ghent University.
24. Boratto, F., Barbosa, R., Yue, S., and Jonas, J.J. (1988) In: Tamura, I. (editor), *THERMEC-88*, Iron and Steel Institute of Japan, Tokyo, Japan, Vol. 1, pp. 383–390.
25. Mousavi Anijdan, S.H. and Yue, S. (2011) The necessity of dynamic precipitation for the occurrence of no-recrystallization temperature in Nb-microalloyed steel. *Materials Science and Engineering A*, 528, 803.
26. Bai, D.Q., Yue, S., Sun, W.P., and Jonas, J.J. (1993) Effect of deformation parameters on the no-recrystallization temperature in Nb-bearing steels. *Metallurgical Transactions A*, 24A, 2151.

27. Gautam, J., Miroux, A., Moerman, J., Barbatti, C., van Liempt, P., and Kestens, L.A.I. (2012) Determination of the non-recrystallisation temperature ($T_{nr}$) of austenite in high strength C–Mn steels. *Materials Science Forum*, 706–709, 2722–2727.
28. Stalheim, D. (2010) The evolution, review and optimization next steps related to the high temperature processing (HTP) alloy design for pipeline steels in China. Presentation on International Pipeline Technology Conference, Beijing, China.
29. Carretero Olalla, V., Petrov, R.H., Thibaux, P., Liebeherr, M., Gurla, P., and Kestens, L. A.I. (2011) Influence of rolling temperature and cooling rate on microstructure and properties of pipeline steel grades. *Materials Science Forum*, 706–709, 2710–2715.
30. Irvine, K.J., Pickering, F.B., and Gladman, T. (1967) Grain refined C–Mn steels. *Journal of Iron Steel Institute*, 205, 161–182.
31. Graaf, M., Schröder, J., Schwinn, V., and Hulka, K. (2002) Production of large diameter grade X70 pipes with high toughness using acicular ferrite microstructures. Proceedings of the International Conference on Application and Evaluation of High-Grade Linepipes in Hostile Environments, PacifiCorp, Yokohama, Japan, 2002, pp. 323–338.
32. Miao, S.L., Shang, S.J., Zhang, G.D., and Subramanian, S.V. (2010) Recrystallization and strain accumulation behaviors of high Nb-bearing line pipe steel in plate and strip rolling. *Materials Science and Engineering A*, 527, 4985–4992.
33. Bechet, S. and Beaujard, L. (1955) New reagent for the micrographical demonstration of the austenite grain of hardened or hardened-tempered steels. *Reviews in Metals*, 52, 830–836.
34. Baczynski, G.J., Jonas, J.J., and Collins, L.E. (1999) The influence of rolling practice in notch toughness and texture development in high-strength linepipe. *Metallurgical and Materials Transactions A*, 30A, 3045.
35. Vanderschueren, D., Kestens, L., Van Houte, P., Aernoudt, E., Dilewijns, J., and Meers, U. (1990) Influence of transformation induced recrystallization on hot rolling textures of low carbon steel sheet. *Materials Science and Technology*, 6, 1247–1259.
36. Zajac, S., Schwinn, V., and Tacke, K.-H. (2005) Characterisation and quantification of complex bainitic microstructures in high and ultra-high strength linepipe steels. *Materials Science Forum*, 500–501, 387–394.
37. Shanmugam, S., Misra, R.D.K., Mannering, T., Panda, D., and Jansto, S.G. (2006) Impact toughness and microstructure relationship in niobium- and vanadium-microalloyed steels processed with varied cooling rates to similar yield strength. *Materials Science and Engineering A*, 437(2), 436–445.
38. Mouriño, N.S. (2010) Crystallographically controlled mechanical anisotropy of pipeline steel. Ph.D. thesis, Ghent University.
39. Petrov, R., Kestens, L., and Houbaert, Y. (2004) Characterization of the microstructure and transformation behavior of strained and not-strained austenite in Nb–V alloyed C–Mn steel. *Materials Characterization*, 53(1), 51–61.
40. Petrov, R., Leòn García, O., Mulders, J.J.L., Reis, A.C.C., Kestens, L., Bae, J.H., and Houbaert, Y. (2007) Three dimensional microstructure–microtexture characterisation of pipeline steel. *Materials Science Forum*, 550, 625–630.
41. Petrov, R., León García, O., Sánchez Mouriño, N., Kestens, L., Bae, J.H., and Kang, K.B. (2007) Microstructure: texture related toughness anisotropy of API-X80 pipeline steel characterized by means of 3D-EBSD technique. *Materials Science Forum*, 558–559, 1429–1434.
42. Carretero Olalla, V., Sanchez Mouriño, N., Thibaux, P., Kestens, L.A.I., and Petrov, R.H. (2013) Physical simulation of hot rolling steel plate and coil production for pipeline applications. Proceedings of the 7th International Conference on Physical and Numerical Simulation of Materials Processing, ICPNS'13, Oulu, Finland, June 16–19, 2013.
43. Lejiang, Xu (2010) Innovative steels for low carbon economy. In: Weng, Y. et al. (editors), *Advanced Steels*, Springer, Berlin, pp. 9–14.
44. Nobuyuki, I., Nobuo, S., and Joe, K. (2008) Development of ultra-high strength linepipes with dual-phase microstructure for high strain application. JFE Technical Report, No. 12, October 2008.
45. Speer, J., Matlock, D.K., De Cooman, B.C., and Schroth, J.G. (2003) Carbon partitioning into austenite from martensite transformation. *Acta Materialia*, 51, 2611–2622.
46. Bunge, H.J. (1982) *Texture Analysis in Materials Science*, Butterworths, London.
47. Pyshmintsev, I., Gervasyev, A., Petrov, R.H., Olalla, V.C., and Kestens, L. (2012) Crystallographic texture as a factor enabling ductile fracture arrest in high strength pipeline steel. *Materials Science Forum*, 702–703, 770–773.
48. Buzzichelli, G. and Scopesi, L. (1999) Fracture propagation control in very high strength gas pipelines. ATS International Steelmaking Conference, Paris, December 8–9, 1999, pp. 1409–1416.
49. Baldi, G. and Buzzichelli, G. (1978) Critical stress for delamination fracture in HSLA steels. *Metal Science*, 12, 459–472.
50. Yang, L., Guanfa, L., Lixia, Z., Yaorong, F., Chunyong, H., Xiaodong, H., and Xinli, H. (2008) Effect of micro-texture on fracture separation in an X80 line pipe steel. Proceedings of the X80 and HGLPS 2008, Xi'an, China.
51. Jonas, J.J. He, Y., and Godet, S. (2006) Transformation textures in as-hot rolled TRIP steels. *Steel Research International*, 77, 650–653.
52. Ray, R.K. and Jonas, J.J. (1990) Transformation textures in steels. *International Materials Reviews*, 35, 1–36.
53. Ray, R.K., Butrón-Guillén, M.P., Jonas, J.J., and Ruddle, G.E. (1992) Effect of controlled rolling on texture development in a plain carbon and a Nb microalloyed steel. *ISIJ International*, 32, 203–212.
54. Jonas, J.J., Petrov, R., and Kestens, L. (2006) Transformation behaviors of intercritically annealed and as-hot rolled TRIP steels. In: Qihua, X. and Lejiang, X. (editors), *Proceedings of the 2nd Baosteel Biennial Academic Conference*, Baosteel Research Institute, Shanghai, China, Vol. 1, pp. 319–325.
55. Kestens, L. and Jonas, J.J. (2005) Transformation and recrystallization textures associated with steel processing.

*ASM Handbook Volume 14A: Metalworking: Bulk Forming*, Standardsmedia, pp. 685–700.

56. Ray, R.K., Jonas, J.J., and Hook, R.E. (1994) Cold rolling and annealing textures in low carbon and extra low carbon steels. *International Materials Reviews*, 39, 129–172.
57. He, Y., Godet, S., Jacques, P.J., and Jonas, J.J. (2006) Crystallographic features of the $\gamma$-to-$\alpha$ transformation in a Nb-added transformation-induced plasticity steel. *Metallurgical and Materials Transactions A*, 37, 2641–2653.
58. Pyshmintsev, I. Yu., Arabey, A.B., Gervasyev, A.M., and Boryakova, A.N. (2009) Effects of microstructure and texture on shear fracture in X80 linepipes designed for 11.8 MPa gas pressure. Pipeline Technology Conference, Ostend, October 12–14, 2009, Technical. Paper No. Ostend 2009-028.

# 11

# PIPE MANUFACTURE—INTRODUCTION

GERHARD KNAUF AND AXEL KULGEMEYER
*Salzgitter Mannesmann Forschung GmbH, Duisburg, Germany*

## 11.1 PIPE MANUFACTURING BACKGROUND

Most gas, crude oil, and petroleum products are transported via pipelines. Compared with transport via road and rail, pipelines are the safest means for fossil fuel transportation [1,2], especially over long distances. To ensure a high safety margin, the main focus for development and production of line pipes is on high-quality properties, especially with regard to geometry, strength, ductility, toughness, and corrosion resistance.

The first modern-day steel pipes were produced in the early 1800s in England and were made from steel strip by forge welding. In 1885, Max and Reinhard Mannesmann invented the manufacturing of seamless pipe by a pierce-rolling process [3,4]. Today, line pipes are produced by various production routes.

Seamless pipes are manufactured by cross-roll piercing, pilger rolling, plug rolling, mandrel rolling, and other processes. Welded line pipes are manufactured by cold forming—either from steel strip or using steel plates. Welding is carried out with filler material, for example, in the case of UOE and spiral welded pipes, or without filler material, in the case of electric resistance welded pipe.

For more than 100 years, pipelines have been used to transport hydrocarbons over long distances. One of the first long-distance natural gas pipelines was built in 1891 in the United States from Indiana to Chicago with a length of about 190 km (120 miles) [5]. In 1906, one of the first long-distance kerosene pipelines was put into operation, between Baku and Batumi, using Mannesmann seamless pipes, connecting the Black Sea and the Caspian Sea. At 835 km (519 miles) in length, it was for a long time the longest kerosene pipeline worldwide [6,7]. However, major progress in metallurgy, pipe manufacturing, and welding techniques was needed to establish the start of the construction of pipeline networks in the United States and in Europe in the 1950s and 1960s. Over the past decades, pipeline networks in the industrialized countries were extended continuously in order to keep pace with the growing demand for fossil fuels and to guarantee the stability of supply, accompanied by the demand for pipe with improved properties and quality.

## 11.2 CURRENT TRENDS IN LINE PIPE MANUFACTURING

There is a strong interest worldwide to transport large gas volumes from remote areas and increasingly hostile environments to the market. The challenges that have to be overcome include low operating temperatures, below −40 °C, for example, in arctic areas, and the ability to withstand plastic deformation caused by seismic activity or frost heave. At the same time, operators aim to increase the level of operating pressure for economic reasons [8,9].

For several decades, the increase of the strength level of line pipe steels was the focus of materials development. This was achieved by optimization of the alloy design and of the thermomechanically controlled processing (TMCP) for heavy plate and strip production. While in the 1960s and 1970s high-strength pipelines were built from API grade X60 to X70, after 2000 high-strength steels in grade X80 became state of the art, and line pipe grades up to X120 have been

*Oil and Gas Pipelines: Integrity and Safety Handbook,* First Edition. Edited by R. Winston Revie.

developed on the industrial scale [10–12]. With increasing strength level, other material properties were affected. The capacity of the material to withstand plastic deformation, crack arrest behavior, and field weldability are attributed to the higher alloy content that is required for high strength levels. These factors have limited up to now the large-scale applications of high-strength grades above the X80 strength level.

Continuous research on pipe materials as well as on pipe manufacturing processes such as forming and welding is aiming to shift the current limitations. Current and future challenges for pipe development and production are pipes and pipelines for areas with ground movement, for arctic and deep sea application, sour gas transportation, and transportation of hydrogen, $CO_2$, and other fluids.

Together with the improvement of the pipe properties and the adjustment to the increasing demands, the reduction of project cost plays an increasing role for pipe manufacturing. Especially, the increase of strength resulted in lowered project cost because of the reduced quantity of steel required, lower pipe transportation cost, and lower pipe laying and construction cost [13].

In the following four chapters, the main manufacturing routes for line pipe are described, including longitudinal submerged arc welded (Chapter 12), spiral submerged arc welded (Chapter 13), high-frequency electric resistance welded (Chapter 14), and seamless pipes (Chapter 15). Quality control, ranges of grades, dimensions, and typical fields of applicability are discussed. The major standards for line pipe manufacturing and testing are reviewed in Chapter 16.

## REFERENCES


1. Furchtgott-Roth, D. (2013) Pipelines are safest for transportation of oil and gas, Issue Brief No. 23, June 2013, Manhattan Institute for Policy Research.
2. Hopkins, P. (2005) High design factor pipelines: integrity issues. *The Journal of Pipeline Integrity*, 2, Quarter 69–97.
3. Pipeline Equities—Pipeline Appraisal & Recovery (August 2010). A Brief History of Steel Pipe, Available at http://www.pipelineequities.com/A-Brief-History-of-Steel-Pipe.php.
4. Mannesmann Aktiengesellschaft (1965) Rohre gab es immer schon, Publisher Mannesmann AG, Düsseldorf, Germany.
5. Overview of Natural Gas—History (2013) Available at http://www.naturalgas.org/overview/history.asp.
6. Beagel Baku-Batumi pipeline (July 2008) Available at http://en.wikipedia.org/wiki/Baku%E2%80%93Batumi_pipeline.
7. Wessel, Horst A. (2003) Kontinuität im Wandel, Düsseldorf 1990, S. 130; Ost-West Contact 49 (2003), Heft 11, S. 90.
8. Pishmintsev, I.Y., Lubanova, T.P., Arabey, A.B., Sozonov, P.M., and Struin, A.O. (2009) Crack arrestability and mechanical properties of 1420 mm X80 grade pipes designed for 11.8 MPa operation pressure. Proceedings of the Pipeline Technology Conference, 2009, Ostend, Belgium, Paper No. 2009-078.
9. Glover, A., Zhou, Joe, and Horsley, David (2003) Design, application and installation of an X100 pipeline. Proceedings of the 22nd International Conference on Offshore Mechanics and Arctic Engineering (OMAE 2003), June 8–13, 2003, Cancun, Mexico.
10. Petersen, C.W., et al. (2004) Improving long distance gas transmission economics: X120 development overview 3. Proceedings of Pipeline Technology Conference, May 9–13, 2004, Ostend, Belgium.
11. Asahi H., et al. (2004) Metallurgical design of high strength steels and development of X120 UOE linepipe. Proceedings of Pipeline Technology Conference, May 9–13, 2004, Ostend, Belgium.
12. Hillenbrand, H.-G., et al. Development of high strength and pipe production technology for grade X120 line pipe. Proceedings of International Pipeline Conference IPC, October 4–8, 2004, Calgary, Canada.
13. Hillenbrand, H.-G., et al. (2005) Development and production of linepipe steels in grade X100 and X120. Proceedings of X120 Grade High Performance Pipe Steels, Technical Conference, July 28–29, 2005, Beijing.

# 12

# PIPE MANUFACTURE—LONGITUDINAL SUBMERGED ARC WELDED LARGE DIAMETER PIPE

Christoph Kalwa
*EUROPIPE GmbH, Mülheim an der Ruhr, Germany*

## 12.1 INTRODUCTION

For transporting large quantities of hydrocarbons over long distances, longitudinal submerged arc welded (LSAW) pipes have been used for decades. When operational conditions require high wall thickness due to high internal or external pressures, LSAW pipes are commonly the most economical solution. The experience gained over many years of production and operation of this type of pipe has resulted in a thorough understanding of the pipe material with detailed knowledge so that the material can be modified to meet specific and demanding requirements. This is appreciated very much in the pipeline industry, especially when harsh conditions with respect to corrosion, deformation capability, or low-temperature toughness are involved.

## 12.2 MANUFACTURING PROCESS

In LSAW pipe mills, processing is carried out by cold forming heavy carbon steel plates, the pipe pre-material on which many of the pipe properties depend. The state of the art is to roll slabs by thermomechanically controlled processing (TMCP), where each rolling step is defined in a narrow range of temperature and of deformation. Subsequent accelerated cooling enhances the strength of the plate. Due to lean chemistry, good weldability is achieved for the pipe.

The pipe manufacturing designation describes the forming process as U-forming–O-forming Expansion (UOE), as shown in Figure 12.1. Further pipe manufacturing processes are three-roll bending (3RB) and J-forming–C-forming–O-forming (JCO) with a forming press. The processes that are used have a strong influence on the limitations of the pipe mill in terms of the dimensions of the pipe and the productivity of the mill. The UOE-process provides very good productivities, up to 360 m/h, but is limited, by the available press forces, to about 50 mm pipe wall thickness. Depending on the equipment, up to 18 m long pipe can be manufactured. The 3RB process is usually limited to 25 mm wall thickness and 12 m pipe length. The JCO process opens the wall thickness ranging above 50 mm, but the productivity stays below 100 m/h. Some mills with the JCO process provide pipes with length up to 18 m. In all cases, the pipe diameter is limited by the forming tools and, more importantly, by the available plate width.

Figure 12.2 shows an example of the production scheme of UOE pipes. It is evident in this scheme that there are as many production steps as quality control steps.

Most important for all processes is the submerged arc welding process for the longitudinal weld (Figure 12.3). In most cases it is performed as a two-layer technique with one layer welded from the inside and followed by a second layer from the outside. For high productivity, a sufficient deposition rate is mandatory, which is achieved by multi-wire technique with up to five wires. Especially for high wall thicknesses, the heat input is high and affects the zone around the weld significantly, which may result in low toughness in localized areas. For joining the plate edges after forming, a continuous tack weld is applied, which also serves as

*Oil and Gas Pipelines: Integrity and Safety Handbook*, First Edition. Edited by R. Winston Revie.

**FIGURE 12.1** Forming stages of UOE process (reproduced with permission of EUROPIPE GmbH).

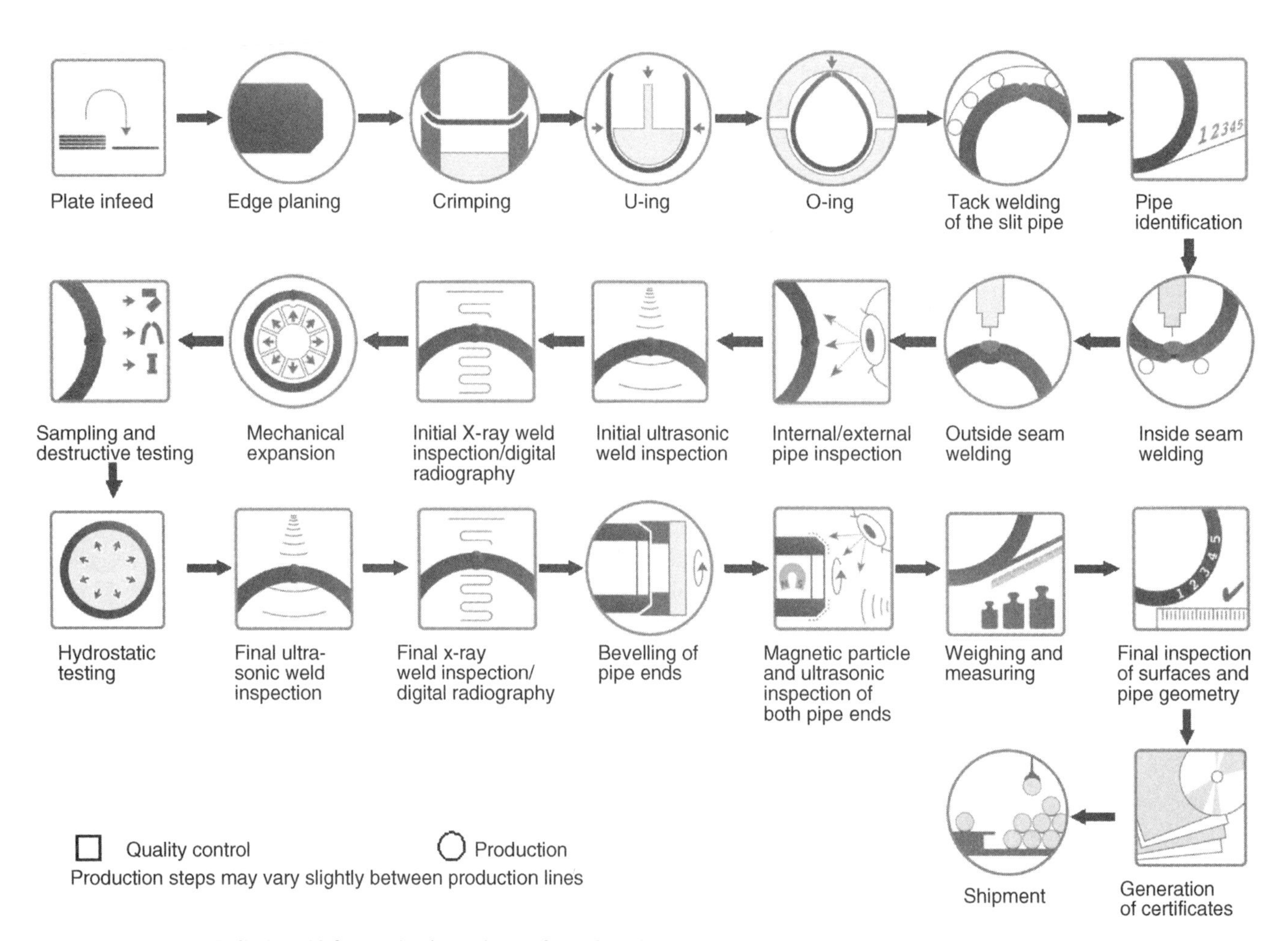


**FIGURE 12.2** Production scheme for UOE pipes (reproduced with permission of EUROPIPE GmbH).

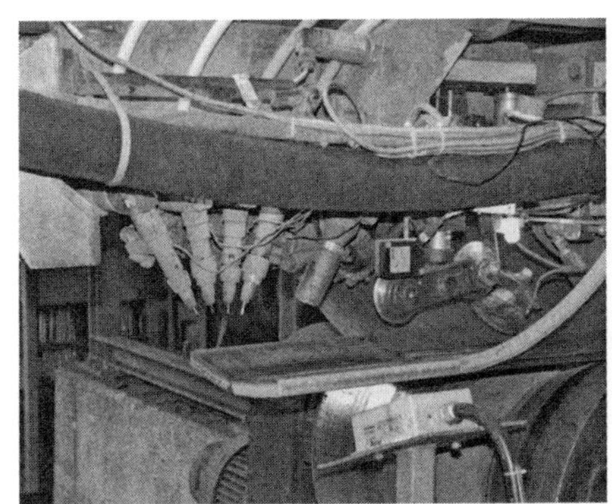

**FIGURE 12.3** Inside and outside welding process (reproduced with permission of EUROPIPE GmbH).

backing for the inside weld. To avoid any influences from starting and stopping, run-on– run-off tabs are assembled on the pipe ends to ensure the stability of the welding process when the electric arc is entering the pipe body. After the submerged arc welding, the tabs can be cut and scrapped.

In many cases, a subsequent mechanical expansion completes the forming process. The main purpose for expansion is to shape the pipe perfectly round for a better girth welding performance. Besides this issue, there are beneficial influences with respect to homogenisation and decrease of residual stresses. Because the mechanical expansion stresses the pipes above yield strength, the pipe is stressed more than that in the hydrostatic test, giving an additional assurance of pipe integrity.

In the final pipe production steps, the pipe ends are prepared for the girth welding and ultrasonic testing in the field. To assemble equipment for automatic narrow gap welding and automatic nondestructive testing devices, the inside and outside weld reinforcements are removed. The pipe ends are machined plane or are given a welding phase to prepare for the girth welding in the field.

## 12.3 QUALITY CONTROL PROCEDURES

The entire process is accompanied by various quality assurance and quality control steps. Besides the systematic approaches and the quality control equipment, an understanding of the interaction between all the processes is essential to ensure product quality.

For the incoming pre-material, a definite identifier has to be chosen that is applied to the plate to be able to fully trace the pipe in each step of production and quality control. Modern pipe mills use electronic support for tracking recording time and parameters of each working step. The availability of electronic basic data of the plates is helpful to avoid material exchange at the interface plate infeed and guarantees traceability of the material back to the steel mill. The result of groups of production steps has to be proven with technical measures and well educated inspecting personnel.

The most important step for the integrity of pipes in pipelines is a reliable longitudinal weld. The precondition of a sound weld is a proper preparation that allows a stable welding process with stable welding parameters and a low-indication rate. The cleanliness of the weld groove and a high quality continuous tack weld has to be assured before welding. In most of the worldwide rules, weld repair is forbidden after mechanical expansion, which is an important issue for pipeline integrity. Therefore the quality of welding should be proven before expanding the pipe mechanically, considering the requirements of the specifications. The proof should be done on the entire length with two techniques, first, visually with respect to visible surface flaws as undercuts and then ultrasonically with respect to flaws in the weld deposit volume and lack of fusion or lack of penetration.

With the expansion, the main production steps are finalized, giving the right situation for comprehensive quality control. Rings have to be cut and sampled for mechanical testing, representing a certain lot of similar pipes. Experience has shown that one test pipe adequately represents all pipes from one heat. A hydrostatic test stresses the pipe at a pressure that applies a stress to a certain percentage, in most cases specified at 90%, of the specified minimum yield strength. Expanded pipes are stressed at a higher level due to the fact that expansion stresses the pipe to a controlled plastic deformation above yield strength.

The entire weld is tested with an automatic ultrasonic system (Figure 12.4) with probes fixed in such a way that they indicate longitudinal and transverse flaws in the pipe volume and the fusion line. Regularly a NDT reference pipe with artificial defects, bore holes and notches, has to be fed in to the ultrasound equipment to prove the selectivity of the system. Indications should be marked explicitly with paint or electronic measures and examined with radiography and manual ultrasonics to exclude false echoes or acceptable indications.

**FIGURE 12.4** Automatic ultrasonic testing equipment (reproduced with permission of EUROPIPE GmbH).

A special focus is put on the pipe ends with respect to good preconditions for girth welding and nondestructive testing on site. The pipe ends are X-rayed usually over the extreme 200 mm at both pipe ends. For repeatable results and good performance, digital radiography is available (Figure 12.5). Ultrasonic testing on laminations of the extreme 50 mm as well as surface testing of the welding phase by magnetic particle inspection or eddy current testing at both pipe ends, support good girth weld quality.

Besides nondestructive testing of the weld, the pipe quality is defined by its geometry. Low tolerances to the perfect round shape at the pipe ends are beneficial for the performance of the pipe-laying process. This issue becomes more important when operational reasons, such as high strain conditions in the longitudinal direction caused by ground movement, require low offset for the girth welds. For deep-sea applications, ovality is an issue for promoting buckling under high external pressures. For this application, low values are required for the entire pipe length. Modern pipe mills are using automatic laser based measuring systems that allow accurate, personnel-independent evaluation of geometric properties. Nevertheless manual measuring methods are in place, which give less restrictive results.

**FIGURE 12.5** Digital radiography (reproduced with permission of EUROPIPE GmbH).

## 12.4 RANGE OF GRADES AND DIMENSIONS

LSAW pipes are used in sizes from 16" (406 mm) to 60" (1520 mm) with wall thicknesses from 7 to 60 mm. Pipe size and wall thickness depend mainly on the available plates, but the processes that are used have limitations with respect to forming forces and these also limit pipe size and wall thickness.

Also in terms of grades the main trigger is given by the pre-material. Usually grades with strength classes up to L 485 (X70) are available, but also grade L 555 (X80), are frequently used for pipelines worldwide as standard grades. Even higher grades, L 690 (X100) and L 830 (X120) have been developed, but the pipeline market is reluctant to use these very high-strength pipe grades at this time. Higher grades allow reduction of the pipe weight with smaller wall thickness at the same diameter and operational pressure, which is beneficial for pipe transport and laying activities.

## 12.5 TYPICAL FIELDS OF APPLICATION

Pipelines are the arteries of industrial development. The areas where the gas and oil are explored rely on safe operation of the pipeline grid to secure continuous prosperity from sales of gas and oil. The areas where the users are located depend on a safe supply for the continuous operation of their industry. Environmental and population safety aspects trigger high-quality requirements and appeal to the responsibility of all involved bodies to put all efforts into safe pipeline operation. In many cases pipelines are running onshore with no special requirements. For this use, international standards describing the required properties of the pipes are sufficient. The different areas where the resources are explored, the different composition of media to be transported, and the different areas pipelines are running through all demand different requirements of the pipe material over and above the standard specifications.

Offshore pipelines link platforms at subsea wells with the processing devices onshore. Offshore pipelines may link the national pipeline grid of supply countries with the grid of user countries, crossing in some cases deep-sea areas. Those pipelines could be very long, for example, the export pipeline for the Ichthys project (890 km), the Nord Stream pipeline (two lines, 1220 km each) or the South Stream pipeline (up to four lines, 930 km each). In many cases, the gas is compressed at the infeed side to a high pressure and is transported to its destination without any intermediate compression. The high pressure requires heavy walls for the pipes, which can be reduced with decreasing pressure along the pipeline. The requirements of the pipes are also controlled by the pipe-laying process. Because the pipes hang from the pipe-laying vessel, the material has to withstand the weight forces in the longitudinal direction. For an efficient pipe-laying process with respect to welding, a close to perfect round shape of the pipe ends is necessary. A low ovality of the entire pipe and a high wall thickness is needed for deep-sea laying to avoid pipe collapse due to the high external pressure in deep water.

More and more pipelines are being installed in polar environments in Canada, Russia, and Norway, where they are exposed to very low temperatures. Carbon steels have a temperature dependent transition from high to low toughness. For arctic use, pipe material has been developed with the transition below −50 °C.

If the medium contains considerable amounts of hydrogen sulfide, the pipeline operator has to consider corrosion with respect to sulfide stress cracking (SSC) and hydrogen-induced cracking (HIC). These subjects are discussed in more detail in Chapters 24 and 17, respectively. Both phenomena are based on affecting the carbon steel structure by hydrogen embrittling the material in the first case and damaging it in the second case. To avoid these phenomena and let the hydrogen pass through the steel without affecting it, the material must exhibit a high steel purity with well controlled segregation and precipitates. Segregating elements that enhance the strength of the steel, such as carbon, manganese, niobium, and so on, must be limited. In the past, there were sour grades available up to the strength level of X 65. Recent developments have shown that X 70 and even X 80 can be produced as sour resistant grades.

When pipelines are laid through areas with ground movement by frost heave and thaw settlement or landslides or earthquake, the pipe must be deformable to the extent required by the ground movement, in order to ensure safe pipeline operation. In that case, strain-based design criteria are used that require adequate deformation properties for longitudinal straining. Strain-based design is discussed in Chapter 4. Close cooperation between those responsible for pipeline design and for pipe manufacturing is necessary to ensure that the requirements of the design are consistent with the properties of the material.

Finally a pipe manufacturer has to cope with the requirements of a project in tailoring the entire set of properties to the many requirements coming from the conditions in the area the pipeline will traverse.

# 13

# PIPE MANUFACTURE—SPIRAL PIPE

Franz Martin Knoop
*Salzgitter Mannesmann Großrohr GmbH, Salzgitter, Germany*

## 13.1 MANUFACTURING PROCESS

Spiral welded pipe was first introduced on an industrial basis in the United States in 1880 [1], but the technology used for pipe production today traces back to 1960s when submerged arc welding(SAW) became applicable and new pipe-forming machines were made available. The terms spiral welded pipe and helical seam pipe are considered interchangeable and generally used for spiral formed line pipe with a single helical weld seam.

Nowadays spiral welded pipe is used in gas and oil pipeline projects all over the world. The progress in steel and coil processing combined with the technological development in pipe production and quality control, have pushed the use of this product for the transport of gas and oil. Despite this the predominant portion of large-diameter pipes produced all around the world for gas and oil pipelines >20 in. is produced as longitudinally welded pipe. Compared with UOE mills, the capital costs of a spiral pipe plant are very low. Also the versatility of the process to produce a wide range of pipe sizes with a small range of steel coil width has pushed the construction of mills all around the world. The politically driven desire for local pipe manufacturing facilities in combination with the growing availability of high-quality line pipe steels is another reason for the growing acceptance of this type of pipe. As a consequence the major part of the existing plants is characterized by very simple techniques for manufacturing and testing not being suitable for the production of high-pressure pipelines. Consequently, the pipeline operators have many reservations against this type of pipe or they even refuse completely to use this type for pipeline projects. Therefore, it is a commonly held perception that spiral line pipe is an inferior product. The opinions are mostly based on anecdotal evidence, poor experience from more than 30 years ago or poor experience with inferior pipe produced by mills with inadequate equipment and manufacturing procedures [2].

The production of spiral welded pipe follows one of the two process routes commonly referred to as "one-step" and "two-step" [3]. In both routes, the steel coil, known as "skelp," is unwound and then formed into spiral-seamed pipe. Heavy plate material has also been used in the past due to the limited availability of heavy gauge coils, but is not an option anymore. The forming process of helically welded large-diameter pipes generally consists of three main forming steps:

- Uncoiling of the hot-rolled wide strip
- Leveling in a multi-roller leveling unit
- Forming of the pipe in a three-roll bending system with an outside roller cage and/or inside roller tool.

The produced pipe diameter ($D$) depends on the helix entry angle ($\alpha$) at which the material enters the forming unit and the width ($B$) of the hot-rolled coil (Figure 13.1).

Theoretically an angle $\alpha$ between 15° and 73° can be used on the basis of actual standards with theoretical coil width between 0.8D and 3.0D [4]. In practice the working range in most pipe mills is less than this. Coil width produced by modern hot rolling mills today range between 1200 and 1800 mm. With increasing coil width the productivity of the process is increased and the weld seam length is reduced.

*Oil and Gas Pipelines: Integrity and Safety Handbook*, First Edition. Edited by R. Winston Revie.

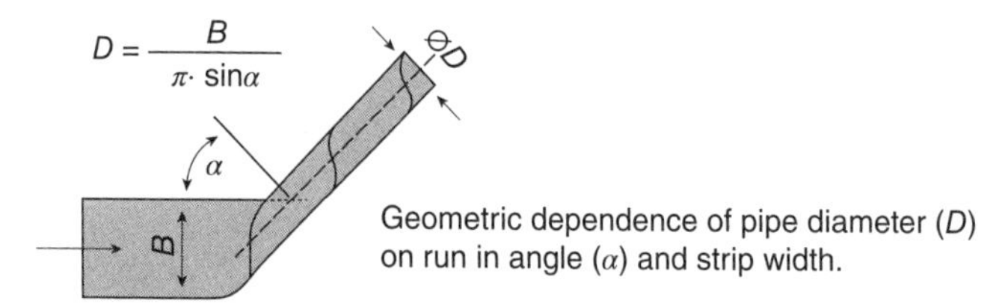

$D = \frac{B}{\pi \cdot \sin\alpha}$
Geometric dependence of pipe diameter ($D$) on run in angle ($\alpha$) and strip width.


**FIGURE 13.1** Principle scheme of spiral pipe forming. (Reproduced with permission of Salzgitter Mannesmann Großrohr GmbH.)

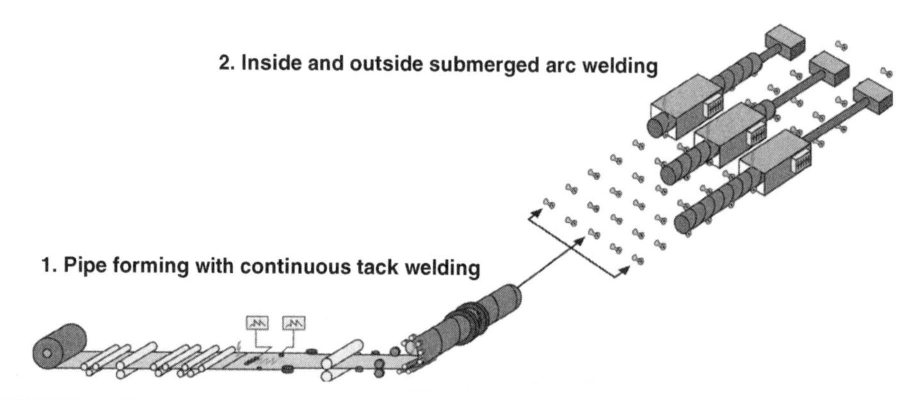


**FIGURE 13.2** Two-step spiral welded pipe manufacturing process [6]. (Reproduced with permission of Salzgitter Mannesmann Großrohr GmbH.)

It is important to mention that during formation, the skelp can be over or under bend with the circumference too tight or to loose and finally resulting in residual stresses in the pipe. These stresses can result in opening or closing of a formed pipe ring when cut in longitudinal direction after forming. Depending on the adjustment of the three-roll bending system different forming strategies can be realized. It is possible and necessary to reduce and minimize such residual by optimizing the initial forming setup. Low residual stresses present in the final welded pipe are essential for an optimized pipeline construction and when the pipe is used for sour service. Also, the pre-bending of the skelp edges needs careful control in the forming process to avoid peaking of the weld and the adjacent areas on the pipe body. Peaking of the weld seam and the coil edges can dramatically reduce the fatigue resistance and the number of pressure cycles in operation [5].

In the one-step process, the coil is leveled and uncoiled by a straightening device. The leading end of the incoming coil is welded to the trailing end of the foregoing coil. This can be done by SAW or other welding methods, depending if the skelp end weld keeps an integral part of the final pipe or if the skelp weld will be cut off. First the coil edges are trimmed or milled and then, after pre-bending of the edges, the steel passes through a roller bending machinery to form a cylindrical pipe. The formed pipe is joined by SAW internally, at the point where the skelp edges first come together, and then after half a revolution, externally, also by SAW.

The principle of the helical two-step (HTS) manufacturing process is shown in Figure 13.2. For this method, production is split into pipe forming combined with continuous tack welding and internal and external SAW on separate welding stations.

In the first step the hot-rolled wide strip is formed into a pipe in the pipe-forming machine and continuously tack welded. This forming unit consists of a three-roll bending system with an outside roller cage as illustrated in Figure 13.3. The function of the roller cage is to set the pipe axis and guarantee the roundness of the pipe.

In the forming unit, the converging strip edges of the pipe are joined using shielded gas metal arc welding. As the continuous tack welded pipe then leaves the forming machine, a plasma cutter—moving with the tube—cuts the individual lengths required by the customer. The tack welding is done automatically and typically done by means of a laser-guided weld head.

To optimize the pipe and weld gap geometry the run-out angle is also permanently controlled and adjusted by an automatic gap control system. Any change in the coil width because of the variations in the coil dimensions before or after milling does not affect the final pipe geometry.

The formed and continuously tack welded pipes are then subsequently fed into separate off-line welding stands for final computer-controlled internal and external SAW with multiple wire systems. Two to four wires are used internally, with up to three wires externally.

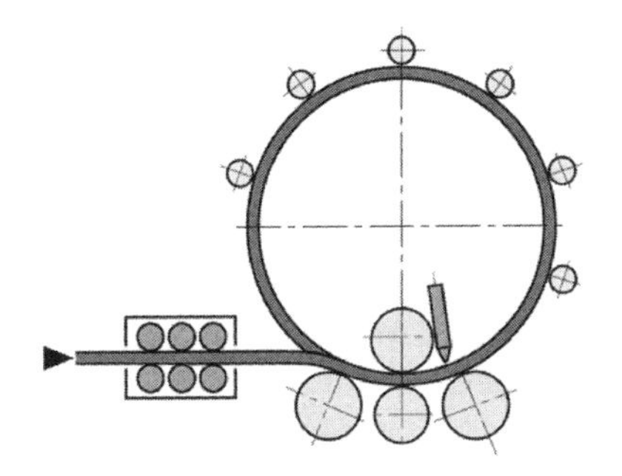

**FIGURE 13.3** Pipe forming and tack welding [6]. (Reproduced with permission of Salzgitter Mannesmann Großrohr GmbH.)

Using special roller tables, each pipe rotates with a precise screw-like motion while the SAW is carried out—first internally, then externally (Figure 13.4)—using the multiwire technique. A laser-controlled seam tracking system guarantees exact positioning and optimum overlapping of the weld seam. The tack weld carried out during the pipe-forming stage serves as a backing for the weld and is fully remolten. The use of laser controls combined with precise gearboxes and electric motors for the pipe movement ensure low tolerances on the weld geometry (e.g., good penetration and minimized misalignment). It is also state of the art in two-step mills that all welding operations are computer controlled and continuously recorded.

The two-step process offers qualitative and economical advantages. Due to splitting of pipe forming and SAW the forming speed can be increased up to 15 m/min. Exact geometrical tolerances are achieved by concentrating on the pipe geometry within the pipe-forming machine—and this without being influenced by the SAW process. Thus, the process-automated technique for carrying out the subsequent SAW of the pipe—which is already tacked to be dimensionally stable—is used with a consistently high quality and without being influenced by pipe forming. The advantages of the HTS process compared with the conventional spiral process are manifold:

**Pipe geometry**

- Better and more precise root gap control.
- Fine, automatic control of forming combined with a better control of strip camber and thickness.
- Improved pipe geometry, reduced peaking and plate offset, low ovality, and narrow diameter tolerances over the full pipe length.

Arrangement of submerged arc
welding heads internally (2–4 heads)
and externally (2–3 heads).

**FIGURE 13.4** Submerged arc welding of the spiral weld [6]. (Reproduced with permission of Salzgitter Mannesmann Großrohr GmbH.)

**Weld seam**

- Improved welding stability (especially at higher wall thicknesses).
- No weld repairs due to start and stop of the pipe forming.
- No risk of weld cracks due to relative motion of edges especially at high wall thicknesses.
- The internal welding position can be adjusted separately to improve the geometry of the inside weld.
- The physical stability of welding stands is higher than in the one-step process.

**Productivity and costs**

- Process efficiency and production rates are significantly increased.

The same welding methods, welding consumables, and even the same software tools and quality controls used in modern UOE plants are also used in the two-step process. In terms of process control the forming process and the pipe geometry can be controlled and documented *in situ*.

Another particularity of spiral pipes is the so-called skelp end weld. They may or may not be accepted by the purchasers, and some pipe mills use the skelp end weld sections for testing [7]. In contrast to manual welded skelp welds, high-quality skelp end welds to be used as integral part of the pipes are completed as double submerged arc welds with the first welding step performed before pipe forming and the second welding step on the outside of the individual pipe after other welding and cutting operations. In particular for pipes with high single weights, the customer's acceptability of the skelp end weld can be essential for the commercial efficiency and yield of this process. The same welding consumables used for the helical seam shall be also used for the skelp end welding. Also, computer-assisted welding controls and recording of weld parameters are applied from the beginning. Thanks to the similarity of the welding process, the structure and the quality assessment of this skelp end weld is equivalent to the spiral weld. The same usage factor can be applied for pipes with helical seam only or combined with a skelp end weld. Fatigue tests performed on spiral welded pipes exposed to dynamically pulsating stress demonstrate full equivalency for spiral and longitudinal submerged arc welded pipe, if forming and welding controls and limits are carefully defined and guaranteed during pipe production [5,8].

## 13.2 QUALITY CONTROL PROCEDURES

Steel quality and composition make a major contribution to the quality of spiral welded pipe as they do for UOE pipe. However, steelmaking, from the liquid product, through casting and rolling to final coiling, is a complex operation with many variables, which is within the control of the steel production rather than the pipe production. Nevertheless, the pipe manufacturer and his clients need to be confident that the material supplied will give an appropriate pipe quality.

Strict test requirements necessitate that each pipe is tested with regard to its usability by means of extensive individual tests (e.g., hydrostatic, ultrasonic, X-ray testing, visual, and geometrical inspections). Material properties, such as strength, toughness, hardness, and ductility are tested in specified intervals according to standards and/or project specifications in the course of the manufacturing process. For all welding operations throughout the production of spiral welded pipe, the welding parameters need to be monitored and adjusted to ensure defect-free welds with adequate penetration optimized bead shape and height.

State of the art mills prepare documents about testing facilities and quality systems in which all process, production, and testing steps are detailed for each individual pipe length. Typical process flows for a conventional spiral pipe mill and for a two-step (offline) mill are shown in Figure 13.5 and Figure 13.6, respectively. In case of the two-step process, all welding and testing steps for pipes with skelp end welds are included. Although variations of these idealized process flows can be found in existing mills around, a comprehensive traceability up to the steel making has to be assured with full data acquisition and storage (electronic pipe log) along the complete manufacturing and testing route.

Pipes shall be identifiable through all stages of manufacture and testing up to final inspection by the pipe production number on the pipe. With the pipe production number, the commission number, the coil number, and the heat number(s), which are recorded on the pipe record card or in the electronic data processing system, full traceability is maintained from the end product through all stages of fabrication to the melting of the base material (steelmaking, casting, hot rolling).

On completion of the final inspection each pipe may additionally receive a shop inspection number, which is also recorded in the electronic data processing system and in the final certificate.

## 13.3 RANGE OF GRADES AND DIMENSIONS

The dimensions of spiral weld pipes are continuously adjustable, so that any diameter, as desired by the customer, can be produced from a base material of the same width. Unlike the longitudinal-weld pipe, it is not necessary to use a large number of forming tools for the various separate sizes. Spiral welded pipes are manufactured in steel grades from API 5L Grade B to X80, and to various other national and international standards. The size capability of each mill will depend upon the forming and welding capability, and the grades of

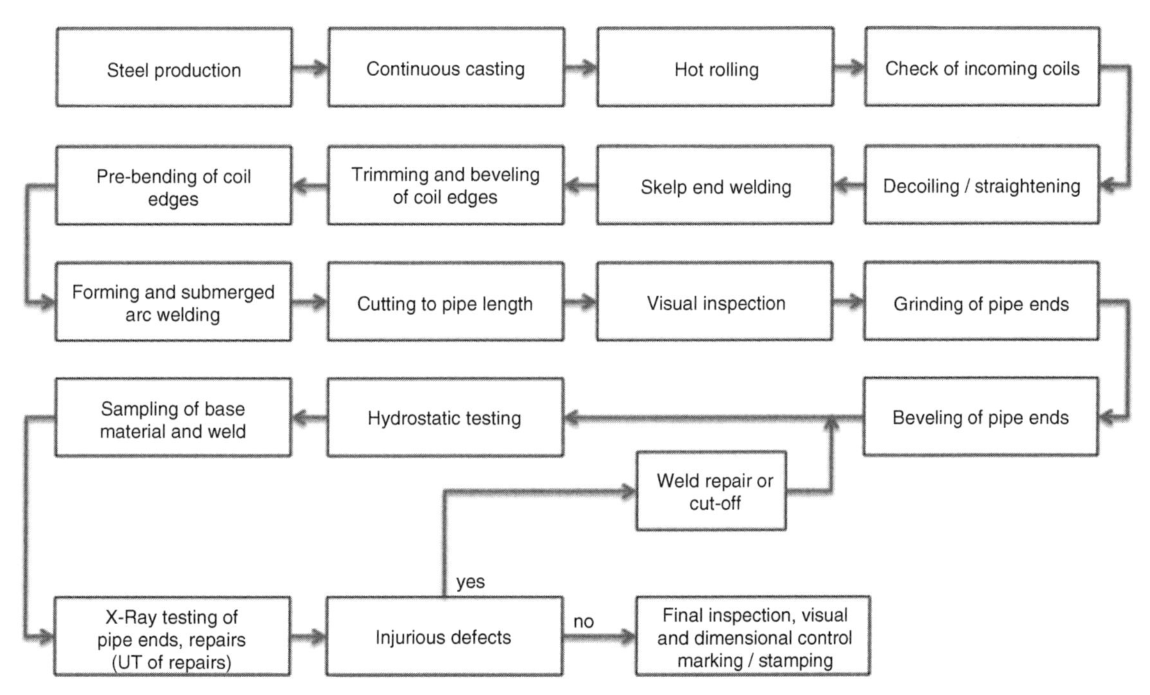


**FIGURE 13.5** Production and test flow in a conventional spiral pipe mill. (Reproduced with permission of Salzgitter Mannesmann Großrohr GmbH.)

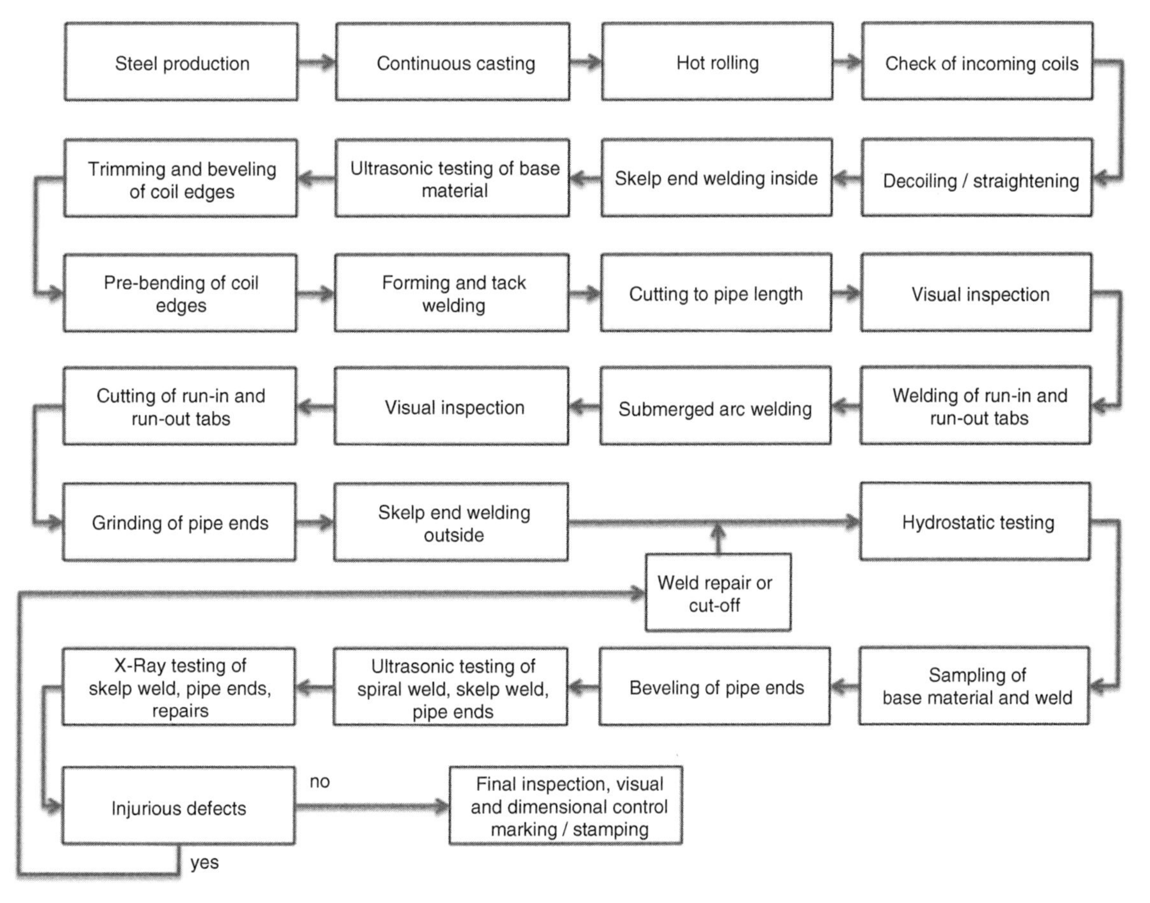


**FIGURE 13.6** Production and test flow in a two-step spiral pipe mill. (Reproduced with permission of Salzgitter Mannesmann Großrohr GmbH.)

steel used. Spiral welded pipe can be produced in diameters up to 120 in. and in wall thicknesses up to 25.4 mm, depending on the client request. Some pipe manufacturers using the one-step process can produce pipe for piling or water service in sizes up to or in excess of 2540 mm (100 in.). The limiting dimensions arise from forming restrictions and the strip width. For oil and gas service the typical diameter range is 20–56 in. with 7–20 mm wall thickness.

## 13.4 TYPICAL FIELDS OF APPLICABILITY

Spiral welded pipe of suitable quality for both sweet and sour oil and gas service is currently available. It is, however, necessary to draw a distinction between pipe mills with facilities, quality controls and experience needed to produce such pipe, and those that are only capable of producing low-grade pipe for water supply and construction applications.

Spiral welded pipe has been used for gas transmission pipes in Europe and North America for decades, but recent projects in the USA and China have clearly demonstrated the competitiveness of this type of pipe. The Cheyenne pipeline project and the REX-pipeline in the USA or the 2nd West–East Gas Pipeline in China are excellent examples of this development. Also the willingness to use X80 spiral pipes for high-pressure gas pipelines has generally changed from sporadically used portions for short pipeline lengths to be a real alternative for pipeline projects of greater lengths [9–11].

Even though most offshore pipelines constructed today are designed for high water depth with heavy gauge plate materials, spiral welded pipes become increasingly attractive as more shallow water offshore projects are planned and the technical and cost advantages are recognized [12–14].

The use of spiral welded pipe in sour service applications has been very limited due to negative experiences with inadequate sour service resistance in operation. Failures have been reported by Canadian pipeline operators but also in Mexico and Saudi Arabia [15,16]. Provided that stable process conditions during continuous casting, hot rolling, pipe forming, and welding are guaranteed and assumed that modern steel chemistry and cleanness is used, severe sour service requirements can be fulfilled even for high-strength steel grades [17,18].

With regard to the time and cost optimized field installation of the pipes, the requirements for a good forming capability and weldability are irrevocable basic requirements. Modern thermomechanically treated pipe steels are characterized by low carbon contents and carbon equivalents as characteristic for a good weldability.

Like longitudinally welded pipes as well, spiral welded pipes can be cold-bent directly on site or hot-bent within the pipe bending facility without any restrictions. Thanks to the forming process, stringent geometrical requirements and narrow tolerances on diameter, ovality and straightness can be guaranteed for pipe ends and the entire pipe body. As opposed to longitudinally welded pipes, no cold expansion is necessary for spiral welded pipes and no difference in tolerances between pipe body and pipe ends is needed. Basically, spiral welded pipes show a preferred orientation of the weld seam and of the direction of the hot-rolled wide strip toward the main direction of the tensions acting on the pipe. The safety of spiral welded pipes is out of the question. Many comparative full scale fracture propagation tests carried out by independent institutions show that the fracture behavior and crack propagation resistance of spiral welded pipes are at least equivalent to those of longitudinally welded pipes. More than 120 full-scale burst tests on longitudinal and spiral welded pipe carried out worldwide have been collected and compared on behalf of the European Pipeline Research Group [19].

Cold-bent pipes are commonly used in pipeline construction in order to conform the pipeline to the geometry of the trench (vertical bend) or to change direction as required following the route of the pipeline (horizontal bend). Cold field bending is used for bends having a large radius and a small bending angle in gas and oil pipelines. Spiral welded pipe can exhibit anisotropic mechanical properties with higher yield in longitudinal than in hoop direction. Also, placing the weld seam in the neutral zone as done for longitudinal pipe is not possible. Although most international engineering standards do not differentiate between spiral and longitudinal welded pipe some recommendations based on practical experience can be given. For high $D/t$ ratio, small bending radius and for pipe with skelp end weld a mandrel shall be used. At least one pipe diameter, with a minimum of 0.5 m, shall be left straight at either end of a bent pipe. The bending radius should be larger than a specified minimum value, increasing with the pipe diameter, and the bending angle at each bending step should be limited to a specified value that decreases when the pipe diameter increases [20,21].

In situations where pipeline integrity may be a function of geography and additional external load action may be due to ground movement, strain-based design concepts may be considered. Depending on the area of application, significant strain demand due to different types of ground movement can take place, reflecting in specific requirements for line pipe material in terms of strain resistance capability. Quite recently detailed investigations and full-scale testing on spiral welded and longitudinal welded large diameter pipe have been carried out. Even though materials were not specifically developed for strain-based applications, all pipes exhibited good behavior in terms of strain capacity and critical curvature at buckling onset and demonstrated to be suitable for SBD applications [22,23].

## REFERENCES


1. The Iron Age, March1, 1888, 359-360, Spiral Weld Tube Machine.
2. Jones, B.L. (1999) Large diameter line pipe for high pressure oil and gas transmission; Petromin, August 1999, 44–49.
3. Knoop, F.M. and Sommer, B. (2004) Manufacturing and use of spiral welded pipes for high-pressure service—state of the art. Proceedings of IPC 2004, October 4–8 2004, Calgary, Alberta, Canada, pp. 1761–1770.
4. API (2012) *API 5L: Specification for Line Pipe*, 45th ed., American Petroleum Institute.
5. DIN 2413; Steel pipes: calculation of wall thickness subjected to internal pressure; Deutsche Normen, English translation; June 1972.
6. Spiral Welded Large Diameter Pipes; Brochure and Product Information Salzgitter Mannesmann Grossrohr GmbH, 2008.
7. Wang, Y.-Y., Liu, M., Rapp, S., and Collins, L. (2012) Recommended ITP for the quality assurance of skelp-end welds in spiral pipes. Proceedings of IPC 2012, September 24–28, 2012, Calgary, Alberta, Canada, Paper IPC2012-90663.
8. Knoop, F.M., Marewski, M., Mannucci, G., Steiner, M., Zarea, M., Wolvert, G., and Owen, R. Effect of mean stress on the fatigue behaviour of gas transmission pipes, special edition 1/2006. *3R International*, 43–48.
9. Collins, L.E. (2006) Production of high strength line pipe steel by Steckel mill rolling and spiral pipe forming. CBMM-TMS International Symposium Microalloyed Steels for the Oil & Gas Industry, 2006, Araxa/Brazil. pp. 221–238.
10. Knoop, F.M., Boppert, C., and Schmidt, W. (2008) Perfecting the Two-Step, World Pipelines, March 2008, pp. 29–38.
11. Zhiling, T., et al. (2008) The high performance X80 pipe line steel in the second West-East Gas Pipeline Project. Proceedings of the Taiwan 2008 International Steel Technology Symposium, November 3–5, China Steel Corporation, Kaohsiung.
12. Knoop, F.M., Marewski, Groß-Weege, J., and Zimmermann, S. (2007) HTS pipes (helical seam two step) optimised for offshore applications special edition 2/2007. *3R International*, 19–27.
13. Tolkemit, H.J., Mally, G., Koch, F.O., and Dormagen, D. (1988) SAW spiral pipes for offshore pipelines. Seventh International Conference on Offshore Mechanics and Arctic Engineering, February 7–12 1988, Volume V, Houston, TX. pp. 159–163 (88-1).
14. Heiberg, G., Eltaher, A., Sharma, P., Jukes, J., and Viteri, M. Spiral Wound Linepipe for Offshore Applications, OTC 21795.
15. Lawson, V.B., Duncan, C., and Treseder, R.S. (1987) Pipeline failures in the Grizzly Valley sour gas pipeline. CORROSION 1987, Houston, TX, Paper No. 52.
16. Roberts and Roberts (1998) Risk Engineering and Management, 1998 SOHIC incident in Saudi Arabia. www.roberts-roberts.com
17. Mirkovic, D., Flaxa, V., and Knoop, F.M. (2012) Development and production of helical two-step (HTS) pipes: grades up to API X70 for sour service application. Proceedings of the 9th International Pipeline Conference, September 24–28, 2012, Calgary, Alberta, Canada, Paper IPC2012-90438.
18. Kurashi, T. and Yumang, N. (1996) Production technology and evaluation of line pipe with helical seam for sour service. Proceedings of the 7th Middle East Corrosion Conference, February 26–28, 1996, Vol. 1, Manama. pp. 429–452.
19. Pistone, V. and Mannucci, G. Fracture arrest criteria for spiral welded pipes. Pipeline Technology Conference, 2000, Brugge. pp. 455–469.
20. Peeck, A. (1985) Cold bending behaviour of large pipe of high strength. *3R International*, 10, 4–8.
21. Koch, F.O. and Hofmann (1967) Herstellung von Rohrbogen aus Spiralrohren durch Kaltbiegen, Bänder, Bleche, Rohre, 12. pp. 822–831.
22. Spinelli, C., Demofonti, G., Fonzo, A., Lucci, A., Ferino, J., Di Biagio, M., Flaxa, V., Zimmermann, S., Kalwa, C., C and Knoop, F.M. (2011) Full scale investigation on strain capacity of high grade large diameter pipes 2011, special edition 1/2011. *3R International*, 14–26.
23. Zimmermann, S., Karbasian, H., and Knoop, F.M. (2013) Helical submerged arc welded line pipe engineered for strain based design. The 23rd International Ocean and Polar Engineering Conference, June 30–July 5, 2013, Anchorage Convention Center, Alaska, USA.

# 14

# PIPE MANUFACTURE—ERW PIPE

HOLGER BRAUER AND HENDRIK LÖBBE
*Salzgitter Mannesmann Line Pipe GmbH, Hamm, Germany*

## 14.1 INTRODUCTION

Nowadays, more than a third of the global annual pipe production of low alloyed steels is high-frequency electric resistance welded (HFW). The longitudinal seam welding of the tube is performed by a continuous resistance welding by heating the strip edges mostly with the use of high-frequency conductive or inductive heating and pressing them together.

## 14.2 MANUFACTURING PROCESS

The high-frequency welding technology for manufacturing longitudinally welded pipes was introduced in the early 1960s and optimized in the next decades. This welding technology for production of pipes without the need for additional filler material is an established and recognized method and can be relied on as an optimal solution both technically and economically.

Figure 14.1 shows a schematic overview of the HFW steel pipe production process [2]. In a first step the strip edges are milled in order to ensure that the strip material possesses the smooth and dimensionally accurate edges ideal for longitudinal HFW welding (Figure 14.1, step 1). Forming of the thus prepared strip can be performed with different forming processes. The most common ones are the central tool adjustment (CTA) process, cage roll/fin pass forming, and the Nakata orbital die forming (ODF) process. For the CTA process, universal forming rolls are mounted in the roll cage that is centrally and infinitely adjusted to allow the production of a huge tube diameter range. This is especially advantageous for an economic production of smaller lot sizes and intermediate sizes. The two-step process, cage roll forming line for "U-forming" (Figure 14.1, step 2) followed by a fin pass line for "O-forming" (Figure 14.1, step 3), is shown in Figure 14.2. Nakata developed a new pipe forming method named as "orbital die forming" method, where a multiplicity of die blocks moving in the circumferential direction on an endless track are connected together to provide a tool surface with a very large curvature radius and work just like a huge roll.

The high efficiency of the HFW process is caused by the concentrated heat input via conduction (contact) or induction (contactless) (Figure 14.3) and associated with high welding speeds. Today, the high-frequency induction (HFI) process is state of the art due to its advantage with regard to heat input, surface quality, weld quality, and welding speed. As an example, for an outer pipe diameter of 610.0 mm (24 in.), welding speeds of up to 24 m/min, depending on wall thickness and intended application, are achieved for the HFI welding process at Salzgitter Mannesmann Line Pipe GmbH (MLP, Germany) (Figure 14.1, step 4) [3]. Welding speeds up to or above 100 m/min are quite reachable for pipes with lower diameter and wall thickness. Figure 14.4 shows a schematic and a real view of this process. The open pipe to be welded is inserted into the welding bench and gripped by upset rolls, by means of which the machined edges entering in a wedge configuration at an angle are first forced together. The high-frequency current fed in by the welding generator generates around the annular inductor an electromagnetic field that induces in the "open pipe" an AC voltage that is equivalent to a current in the circumferential direction

*Oil and Gas Pipelines: Integrity and Safety Handbook,* First Edition. Edited by R. Winston Revie.

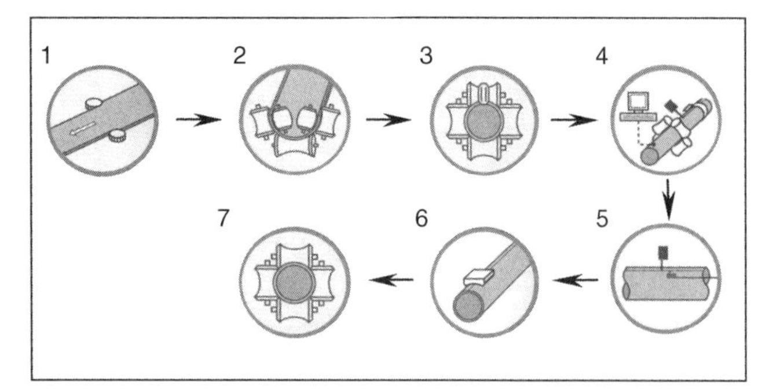


**FIGURE 14.1** Schematic overview of the HFW steel pipe production process steps. 1—strip edge milling; 2—U-forming; 3—O-forming; 4—process-controlled induction welding; 5—internal and external flash removing; 6—inductive weld annealing, single or double; 7—sizing [1]. (Reproduced with permission of Salzgitter Mannesmann Line Pipe GmbH.)

**FIGURE 14.2** Cage roll forming device. (Reproduced with permission of Salzgitter Mannesmann Line Pipe GmbH.)

(Figure 14.5). The circuit is deflected at the open slit edges and runs along the right edge via the welding point and along the left edge back to the inductor circumferential level, to close on the back of the pipe (Figure 14.6) The heated edges (Figure 14.7) are forced together by the upset rolls and welded (Figure 14.8a). The internal and external flash in this process is scraped off from the finished weld (-Figure 14.1, step 5). Figure 14.8b shows the HFI weld in its condition following the welding process. Because of the high welding speeds and the localized heat input, high cooling rates are present in the weld metal and the heat-affected zone. Martensitic and bainitic microstructures are formed with corresponding mechanical properties. Therefore, localized reheating of the weld and the heat-affected zone is necessary in order to improve the mechanical properties. The weld and its surrounding area are inductively heat treated (Figures 14.9 and 14.1, step 7). Different heat

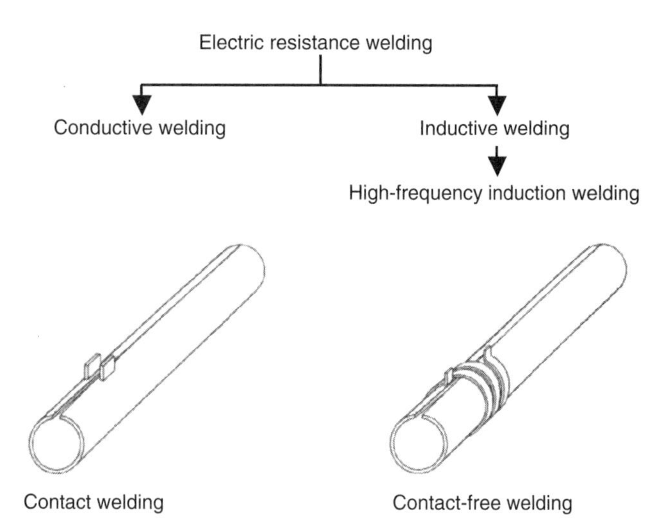


**FIGURE 14.3** Conduction and induction welding [1]. (Reproduced with permission of Salzgitter Mannesmann Line Pipe GmbH.)

treatments can be performed, for example, normalizing, double normalizing, tempering, quenching and tempering, and so on. As an alternative or addition, a full-body heat treatment for reduction of residual stresses, normalization, or quenching and tempering can be carried out. The final step in the welding line is calibration on a straightening machine, by means of compression rollers (Figure 14.1, step 7). A reduction of approximately 0.5% in pipe diameter occurs at this stage. This step, sometimes followed with an additional shaping in a rotary straightening stand, and the process computer-controlled production cycle ensure for HFI pipes high geometrical accuracy of diameter and uniform straightness.

## 14.3 QUALITY CONTROL PROCEDURES

In HFI/electric resistance welding (ERW)-pipe mills today, quality management systems are state of the art and well established in the entire value-added chain. With regard to the pipe production itself as a technical production process, various quality control steps are performed. The inspection techniques are partly normative given as, for example, a tensile test or an ultrasonic test of the weld and are also driven by the know-how of the pipe producers. Of course, required tests for the sensitivity are dependent on the product, customer specification, and intended use. In the following, some aspects are given, from the point of view of both the production technology and the testing technology itself.

Traceability is one of the key aspects in pipe production today. That means that every single production step can be traced backward from the clearly marked pipe together with the corresponding certificate to the coil and heat. For that purpose, a huge variety of data handling, paper work, and

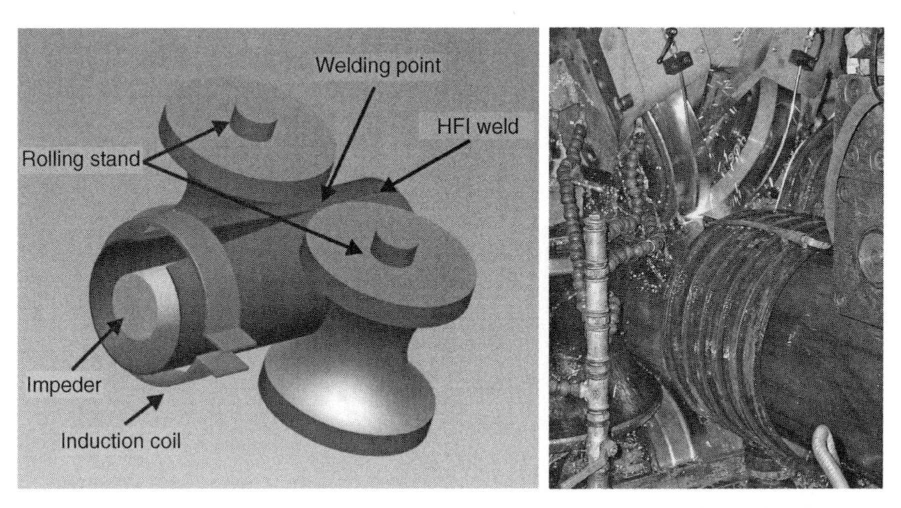


**FIGURE 14.4** High-frequency induction welding: diagrammatic illustration and real process at Salzgitter Mannesmann Line Pipe GmbH, Germany. (Reproduced with permission of Salzgitter Mannesmann Line Pipe GmbH.)

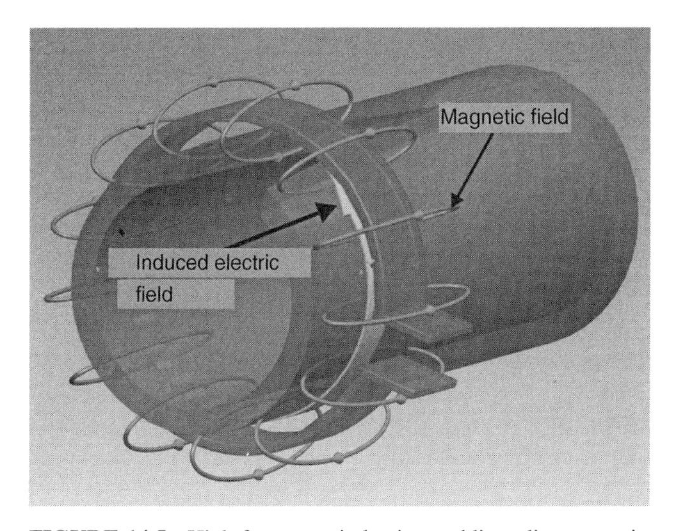


**FIGURE 14.5** High-frequency induction welding: diagrammatic illustration. (Reproduced with permission of Salzgitter Mannesmann Line Pipe GmbH.)

organization in general is necessary. However, at the end production steps can be retraced for all relevant production steps such as welding, hydrostatic testing, or dispatching.

### 14.3.1 Welding Line

With regard to welding in this context, the entire process chain from the pay-off reeling of the strip material up to the cutting of the endless pipe string in single pipes is covered. At first the ordered and foreseen pre-material has to fulfill the specified properties. These are described in relevant hot wide strip standards or, in most cases, in general, specified by in-house standards, agreed upon with the steel mill. For that purpose, the chemical analysis and, if applicable, the technological properties of the certificates are checked as well as the appearance, marking, width, and wall thickness of the material.

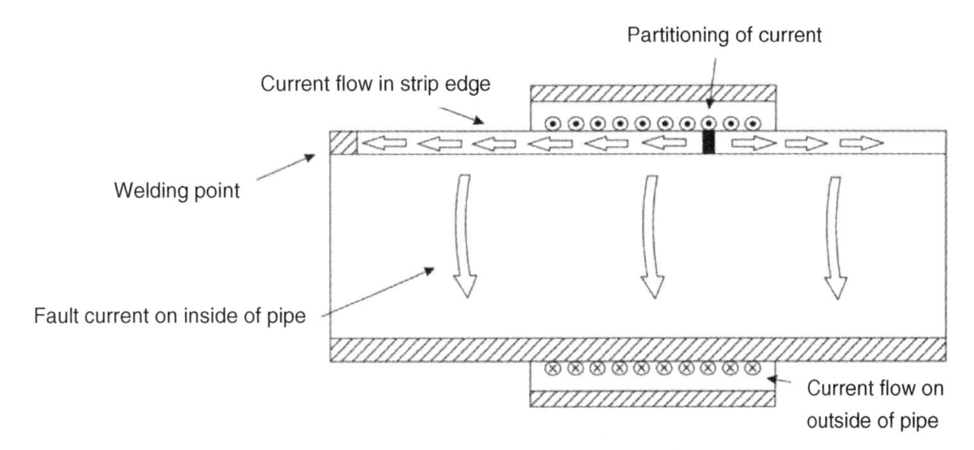


**FIGURE 14.6** High-frequency induction process [4]. (Reproduced with permission of Salzgitter Mannesmann Line Pipe GmbH.)

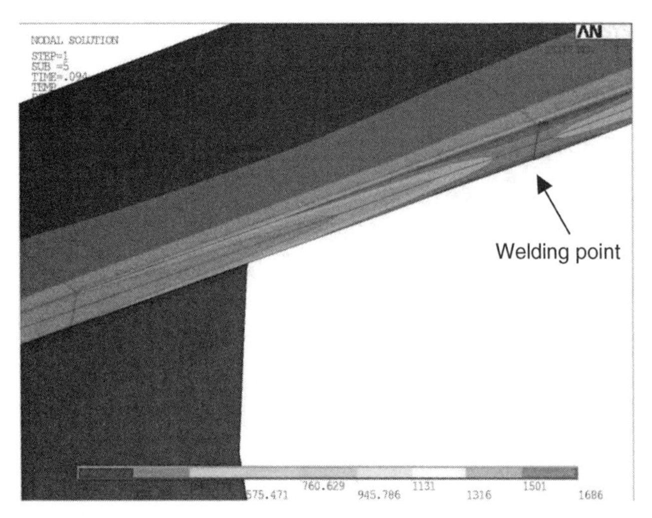


**FIGURE 14.7** Temperature distribution during high-frequency induction welding: FE-simulation [5]. (Reproduced with permission of Salzgitter Mannesmann Line Pipe GmbH.)

**FIGURE 14.9** Inductive heating of the longitudinal weld of HFI pipes at Salzgitter Mannesmann Line Pipe GmbH, Germany. (Reproduced with permission of Salzgitter Mannesmann Line Pipe GmbH.)

Before forming the strip material, the ultrasonic testing for laminations is performed, if required. This is usually performed according to the European standard ISO 10893-9 [6] or the equivalent North American Standard ASTM E213 or A578 [7,8]. The strip edges and strip body are tested with moving or fixed probes and an appropriate surface coverage according to standards and customers' specification.

Most up-to-date welding lines perform the first ultrasonic test of the weld in the welding line on the pipe string. Test sensitivities are dependent on the purpose as an intermediate process control or as a test for release purpose. Anyhow, a first ultrasonic test close to the welding and annealing makes sense in order to check the production parameters in case of given indications.

Furthermore, a first intermediate pipe dimension measurement immediately at the end of the welding line makes sense. This is done when the pipe is cut from the pipe string.

### 14.3.2 Finishing Line

The finishing line covers nondestructive testing (NDT) and destructive testing methods, both executed within the production process. Destructive testing performed in this context is the flattening test. The flattening test is performed on HFI pipes according to most specifications at the beginning and at

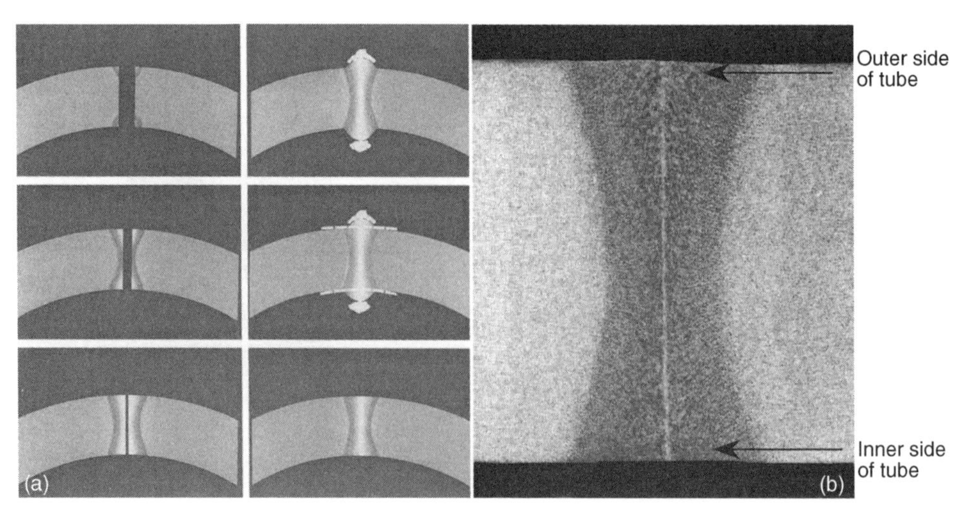


**FIGURE 14.8** (a) Diagrammatic illustration of the merging strip edges and the formation of the HFI weld [1]. (Reproduced with permission of Salzgitter Mannesmann Line Pipe GmbH). (b) HFI pressure weld in as-welded condition. (Reproduced with permission of Salzgitter Mannesmann Line Pipe GmbH.)

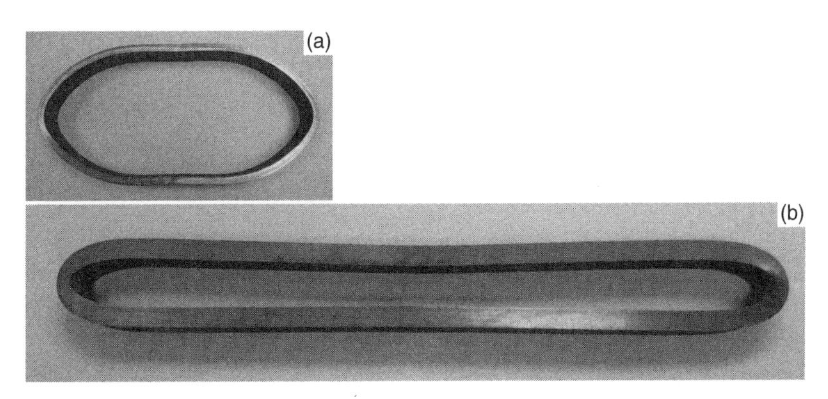


**FIGURE 14.10** Flattening test on HFI pipe at Salzgitter Mannesmann Line Pipe GmbH, Germany. Pipe dimensions: (a) Ø 219.1 mm × 7 mm; (b) Ø 610 mm × 25.4 mm. (Reproduced with permission of Salzgitter Mannesmann Line Pipe GmbH.)

the end of the coil with the weld in 12 o'clock and in 9 o'clock position, additionally before and after a weld stop singularly in 9 o'clock position [9,10]. This test is performed to prove the ductility of the base material as well as the ductility of the weld area [e.g. 11,12]. A flattening test on a 610.0 mm pipe with 25.4 mm wall thickness is shown in Figure 14.10.

Nondestructive tests performed on the finishing line are the geometrical measurements, for example, pipe size, ovality, wall thickness, drift, and so on. Additionally, the hydrostatic test is performed with a given test pressure, dependent on the relevant standard, pipe dimension, and grade. Typically, the test pressure is calculated to get a material load equivalent from 75 up to 90% of specified minimum yield strength (SMYS) referred to the nominal wall thickness. Alternatively, the specified minimum wall thickness may be used for calculation. In that case, the test pressure is usually determined to reach a hoop stress equivalent to 95% of SMYS. The weld seam during the test is located in 12 o'clock position and monitored by the operator. The purpose of the test is a leaking test, test of the weld seam, and a test to exclude material mix-up. The test pressure and the time are recorded in relation to the pipe.

Ultrasonic testing of the weld seam is performed according to the requirements of different standards over the entire pipe length. The European standard ISO 10893-11 [13] or the equivalent North American Standard ASTM E273 [14] is in most cases the applicable test standard for weld seam testing for longitudinal defects. Since the narrow weld seam is in longitudinal direction, testing for transversal defect is not common, but often possible, if required. Test sensitivities are typically in the range of N5 to N15 notch; alternatively or additionally, a drilled hole through thickness 3.2 or 1.6 mm in diameter may be used. While in many places in the world notches are used for reference standards, API/ISO/CSA allow use of holes and for others this is preferred. A typical test unit with moving test probes and pipe at rest is shown in Figure 14.11.

**FIGURE 14.11** Ultrasonic testing device for the longitudinal weld of HFI pipes at Salzgitter Mannesmann Line Pipe GmbH, Germany. (Reproduced with permission of Salzgitter Mannesmann Line Pipe GmbH.)

The third often required ultrasonic testing, besides strip testing for lamination and testing of the weld seam, is the circumferential testing of the pipe ends over a certain distance for lamination. This lamination test is performed to ensure that no problems during girth welding due to laminations occur.

Of course for special purpose or certain grades, such as some Oil Country Tubular Goods (OCTG) Q&T grades, additional NDT tests are sometimes required and performed for different reasons. In most cases, it is a question of full-body NDT or the usage of magnetic particle inspection (MPI). For full-body NDT, in most cases, eddy current or ultrasonic technology is used. Testing is performed for various defect types and sometimes additionally for wall thickness measurement. MPI is in most cases performed at the pipe ends on the bevel, if required.

**TABLE 14.1 Examples of Destructive Material Testing**

| Standard Tests | Tests for Special Purposes |
|---|---|
| • Flattening test | • CTOD test |
| • Charpy test | • HIC/SSC test |
| • Tensile test | • Strain-aging test |
| • Micro/macro examination | • Reeling test |
| • Hardness measurement | • DWT test |
| • Product analysis | • Ring-expansion test |
| | • Springback test |
| | • Burst test |
| | • Collapse test |
| | • Other nonstandardized special test according to customer requirements |

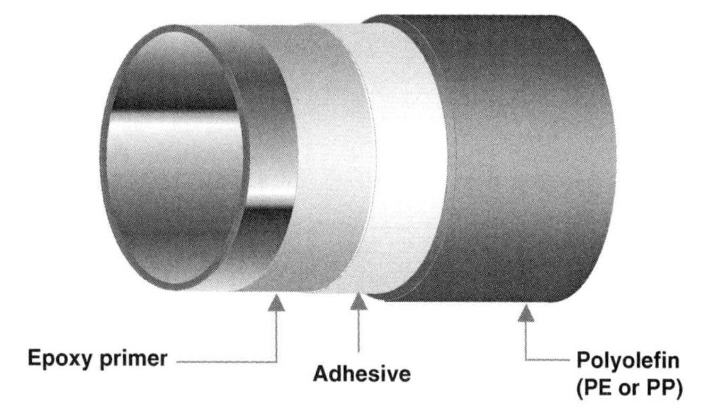


**FIGURE 14.12** Illustration of the three-layer coating. (Reproduced with permission of Salzgitter Mannesmann Line Pipe GmbH.)

### 14.3.3 Destructive Material Testing

Depending on the standard and specification, different destructive mechanical tests are required for different reasons. Depending on the pipe mill layout, some of them are performed online during production; for example, in the finishing line, others are performed in a mill integrated or external test house. A lot of the listed tests are applicable for the base material and the weld. However, some, such as the drop weight tear (DWT) test, are only applicable for the base material. The first column of Table 14.1 covers standard tests, which are more or less performed for a wide variety of pipe types to a different extent. The second column covers tests for special purposes, for example, sour service application (hydrogen-induced cracking (HIC)/sulfide stress cracking (SSC) test) or special offshore pipe laying technique, such as strain-aging test or reeling test. Of course, full-scale tests, such as burst test or other nonstandardized special tests, are also possible. However, these are not common on a daily basis and are in most cases very expensive and very time consuming.

## 14.4 RANGE OF GRADES AND DIMENSIONS

Hot wide strips are applied for (in most cases) the endless production process, either in the normalized (N) or in the thermomechanically (TM) rolled condition. Pipes with outside diameters of 20–660 mm (0.79–26 in.) and pipe wall thicknesses of 1.0–25.4 mm (0.04–1 in.) are produced using the HFW process. The wide manufacturing range of dimensions and line pipe steel grades up to API 5L X80M or casing up to API 5CT Q125 [15], respectively, are suitable for practically all requirements and applications. Of course additionally, pressure and boiler tubes, hollow sections, structural tubes, and others according to various international standards are furthermore produced via the use of the ERW process.

Different polymer coatings are used to ensure the highest possible level of pipeline safety against corrosion and/or mechanical damage, for an improved frictional resistance or for an enhanced abrasive resistance. The most common coating of the outside of the pipe is the three-layer polyethylene (PE) coating (Figures 14.12 and 14.13a), according to, for example, ISO 21809 [16] or CAN/CSA Z245 [17]. Further coatings such as polypropylene (PP), polyamide (PA), single- (maximum thickness of 700 μm) or double-layer epoxy resin coating (fusion bonded epoxy (FBE)) (Figure 14.13b), or (fiber) cement mortar (Figure 14.13c) are used, depending on the application and ambient temperature. On the inside, pipes can also be lined with a cement mortar lining, suitable for pipes for transportation of drinking water, or with a thin epoxy lining (flow coat) to minimize frictional resistance of gas line pipe.

## 14.5 TYPICAL FIELDS OF APPLICABILITY

Typical applications include precision steel tubes for automotive (e.g., airbags, shock absorbers, drive shafts) or industrial applications (for hydraulics and pneumatics, hydraulic cylinders); line pipe for transportation of oil, gas, drinking water, and sewage systems; tubes for machinery, mechanical engineering, and plant construction as well as oilfield tubes; pipes for long-distance heating systems; and structural tubes. One-third of the world OCTG consumption is provided by welded pipe and two-thirds are delivered in seamless conditions. Continuous improvements in welding, especially in the HFI technology, together with the comparably more costly production of seamless pipe have led to an increased usage of welded OCTG, mainly as casings [18]. According to API 5CT, the lower grades are manufactured without quenching and tempering, such as H40, K55, and N80 Type 1. Due to

**FIGURE 14.13** Coated pipes: (a) PE coating, (b) FBE coating, and (c) cement coating. (Reproduced with permission of Salzgitter Mannesmann Line Pipe GmbH.)

an increase in the technological capabilities, current market needs are calling for higher grades such as N80Q, L80, P110, or Q125 with additional requirements, for example, for the local underground gas storage:

- weldable;
- limited carbon content according to SEW 088;
- Charpy values at −20 °C for base material and weld;
- maximum hardness limitation of 23 HRC.

These grades have to be manufactured using a full-body quenching and tempering process.

## REFERENCES

1. Mannesmann HFI steel line pipe—longitudinally welded using the high-frequency induction process. Brochure, Salzgitter Mannesmann Line Pipe GmbH, 2002.
2. Brauer, H., Marewski, U., and Zimmermann, B. (2004) Development of HFIW line pipe for offshore applications. Proceedings of the 4th International Pipeline Technology Conference, May 9–13, 2004, Ostend, Belgium, Vol. 4, S. 1573/93.
3. Brauer, H., Löbbe, H., and Bick, M. (2010) HFI-welded pipes: where are the limits? Proceedings of the 8th International Pipeline Conference (IPC 2010), September 27–October 1, 2010, Calgary, Alberta, Canada.
4. Leßmann, H.J. (1994) Elektromagnetische und Thermische Vorgänge beim induktiven Längsnahtrohrschweißen. Elektrowärme International, Vulkan Verlag, Essen, 51(4), pp. 186–192.
5. Nikanorov, A., Baake, E., Brauer, H., and Weil, C. (2013) Approaches for numerical simulation of high frequency tube welding process. International Conference on Heating by Electromagnetic Sources (HES-13), May 22–24, 2013, Padua, Italy.
6. ISO (2011) ISO 10893-9: Non-destructive testing of steel tubes—Part 9: Automated ultrasonic testing for the detection of laminar imperfections in strip/plate used for the manufacture of welded steel tubes.
7. ASTM (2009) ASTM E213: Standard Practice for Ultrasonic Testing of Metal Pipe and Tubing.
8. ASTM (2012) ASTM A578: Standard Specification for Straight-Beam Ultrasonic Examination of Rolled Steel Plates for Special Applications.
9. API (2012) *API SPEC 5L: Specification for Line Pipe*, 45th ed., American Petroleum Institute.
10. ISO (2013) ISO 3183: Petroleum and natural gas industries—Steel pipe for pipeline transportation systems.
11. ASTM (2012) ASTM A370: Standard Test Methods and Definitions for Mechanical Testing of Steel Products.
12. ISO (1998) ISO 8492: Metallic materials—Tube—Flattening test.
13. ISO (2011) ISO 10893-11: Non-destructive testing of steel tubes—Part 11: Automated ultrasonic testing of the weld seam of welded steel tubes for the detection of longitudinal and/or transverse imperfections.

14. ASTM (2010) ASTM E273: Standard Practice for Ultrasonic Testing of the Weld Zone of Welded Pipe and Tubing.
15. API (2011) API SPEC 5CT: Specification for Casing and Tubing, 9th ed., American Petroleum Institute.
16. ISO (2011) ISO 21809-1: Petroleum and natural gas industries—External coatings for buried or submerged pipelines used in pipeline transportation systems—Part 1: Polyolefin coatings (3-layer PE and 3-layer PP).
17. Canadian Standards Association (2010) CSA Z245.21: External Fusion Bond Epoxy Coating for Steel Pipe/ External Polyethylene Coating for Pipe, Canadian Standards Association.
18. Brauer, H., Ehle, S., and Pinto, J. (2009) Progressive inductive quenching and tempering of HFI-welded oilfield tubular. Proceedings of the Pipeline Technology Conference, October 12–14, 2009, Ostend, Belgium.

# 15

# PIPE MANUFACTURE—SEAMLESS TUBE AND PIPE

ROLF KÜMMERLING AND KLAUS KRAEMER
*Vallourec & Mannesmann Deutschland GmbH, Düsseldorf, Germany*

## 15.1 THE ROLLING PROCESS

### 15.1.1 Introduction and History

The history of seamless steel tube starts with the Mannesmann brothers' invention of a process for piercing solid ingots by cross rolling, see Section 15.1.2. The invention was patented in 1882. The basic principle of cross-roll piercing and the most important components are shown in Figure 15.1.

The rolls arranged at an angle to each other and turning in the same direction force the billet between them into a helical motion. As it passes through the mill, it is pierced by a plug fixed to a freely rotating bar which is supported by a thrust block at the run-out end of the mill.

As opposed to rolling with flat or grooved rolls, in pierce rolling the contact surfaces between the billet and the rolls are small in relation to the total billet surface being worked. In addition, the forming operation mainly takes place circumferentially, while the material is actually intended to yield axially. Consequently, the hollows produced by pierce rolling, followed by size rolling to the finished diameter without an internal tool, were not marketable as tubes. Cross-roll piercing merely allowed the production of relatively thick-walled hollows called blooms.

This meant that the Mannesmann brothers had to develop an additional elongation process, which led to the three rolling stages in seamless tube production that are still standard practice today, namely, piercing, elongation, and final rolling (Figure 15.2).

The first production stage for nearly 90% of all seamless tubes is pierce rolling, and the final stage size rolling or stretch reducing, see Section 15.1.7. A large number of different longitudinal and cross rolling processes have become established for the elongation stage. Therefore, the type of elongation process is used to designate the entire production chain.

The first process to be employed for elongation was the pilger rolling process, which was developed by Mannesmann, see Section 15.1.3. The output of a pilger rolling mill is about 15–20 tubes/running hour.

In the 1920s, Ralph Stiefel introduced the plug rolling process for the elongation stage, see Section 15.1.4. Parallel to this, Stiefel significantly improved the elongation performance of the cross rolling process by using fixed guide shoes, while also developing various designs for cross rolling units. In terms of performance, Stiefel mills were far superior to pilger rolling mills and were increasingly replacing them, with an output of up to 120 pipes/running hour.

From about the middle of the twentieth century, control technology for electric drives had reached a stage of maturity that allowed the mandrel rolling process, which had actually been developed at the end of the nineteenth century, to be used highly cost effectively for the elongation stage with several consecutively arranged rolling stands, see Section 15.1.5. This mandrel rolling or continuous tube rolling process enabled the production of multiple lengths and achieved an output of up to 260 pipes/hour. Over the years, it has been further developed extremely successfully, and today nearly two-thirds of all seamless tubes are manufactured using this technology.

All elongation processes mentioned so far are longitudinal rolling processes. These can only compensate for the

*Oil and Gas Pipelines: Integrity and Safety Handbook*, First Edition. Edited by R. Winston Revie.

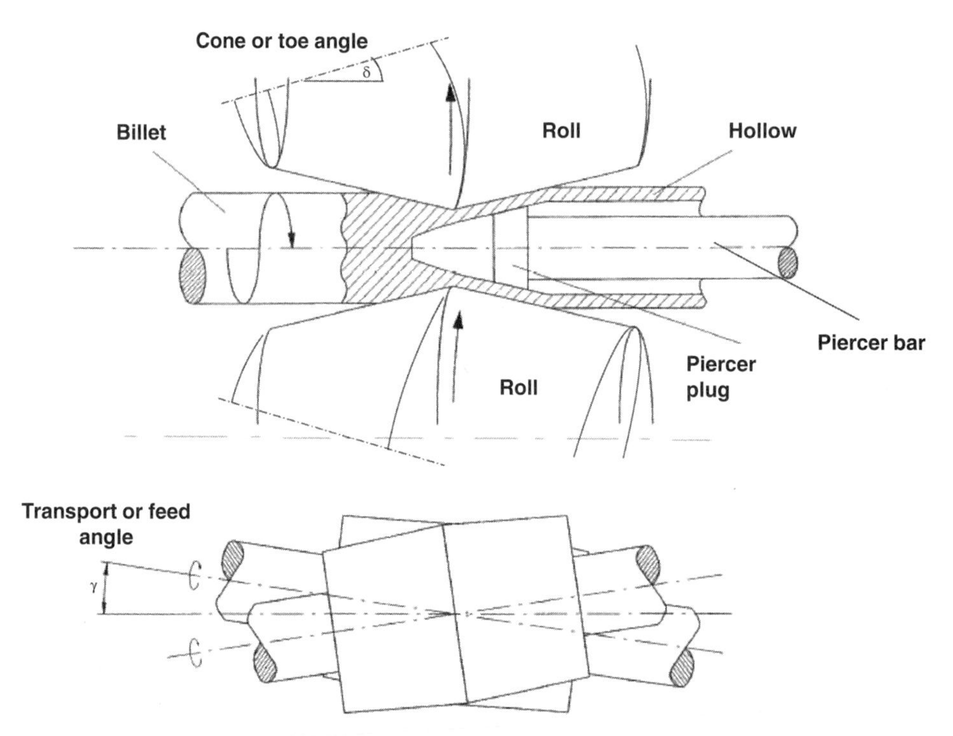


**FIGURE 15.1** Basic principle of cross-roll piercing. (Reproduced with permission of Vallourec & Mannesmann Deutschland GmbH.)

nonuniformity of the wall thickness resulting from the piercing process to some extent. Theoretically, cross rolling processes are better suited for this purpose that is why attempts have been made time and time again to employ them for the elongation stage. As regards performance, cross rolling—whether in an Assel, Diescher, or planetary cross rolling mill—is significantly inferior to longitudinal rolling. Besides, there are technical limitations—regarding the diameter-to-wall-thickness ratio in the case of Assel rolls and planetary rolls, and also, in the case of the Diescher process, the quality of the tube's inside surface [1].

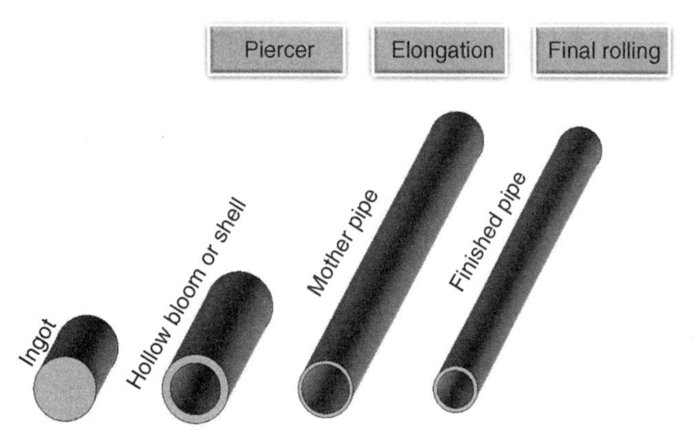


**FIGURE 15.2** Stages in the manufacture of seamless tubes. (Reproduced with permission of Vallourec & Mannesmann Deutschland GmbH.)

The most recent elongation process is tube forging on 4-hammer radial forging machines, see Section 15.1.6. Although here, too, there are limitations as to the diameter-to-wall-thickness ratio, and the output is even slightly below that of pilger rolling mills, the process yields substantial advantages regarding surface condition, wall thickness concentricity, and yield. Pipes produced with this technology are named Premium Forged Pipes (PFP©).

### 15.1.2 Cross Rolling Technology [2–4]

Cross rolling mills can be differentiated by the shape of their rolls—barrel type or conical—and by the guide elements they use, that is, fixed shoes or rotating Diescher discs. Furthermore, the roll arrangement may be horizontal (previously) or vertical (now). Today, the 15° cone-type cross-roll piercing mill with a vertical roll arrangement has become established as standard worldwide. The differences between barrel- and cone-type rolls can be seen in Figure 15.3.

Figure 15.4 illustrates the cases in which guide shoes are preferable over Diescher discs.

The stresses acting on the material during cross-roll piercing are extremely high. This applies particularly to the area before the tip of the piercing plug, where the tube material is subjected to alternating compression and tension; see Figure 15.5.

The result of these alternating stresses is the Mannesmann effect, that is, the billet core breaks open, and a hollow space

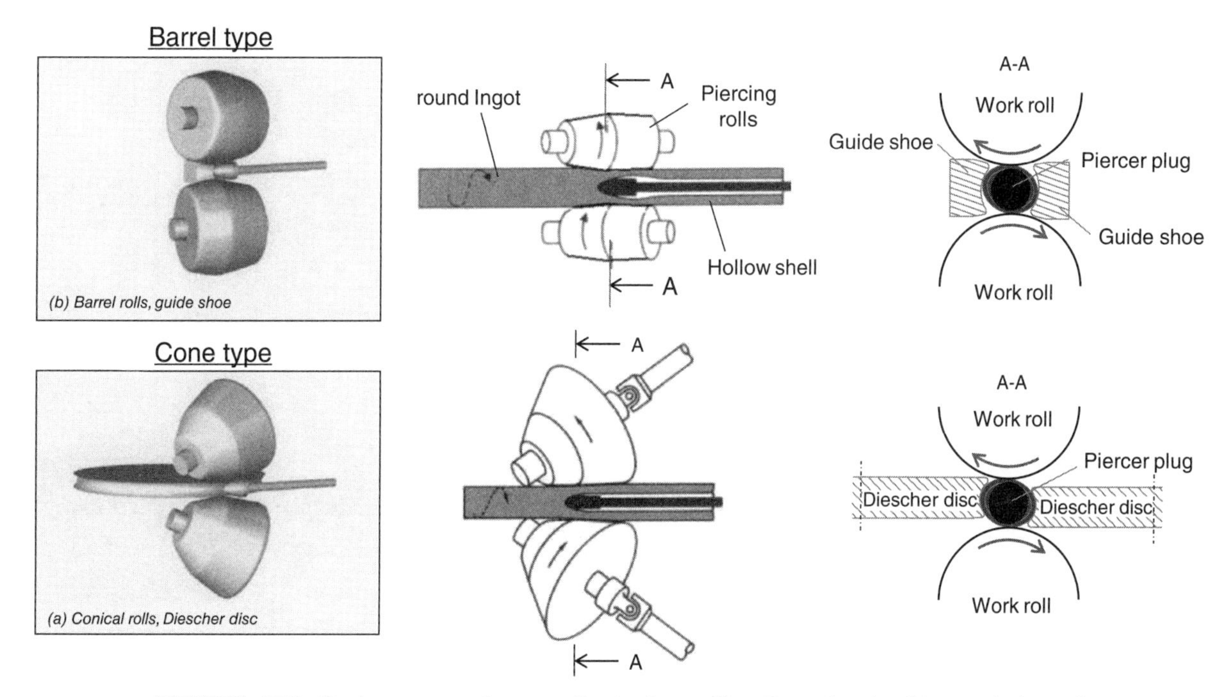


**FIGURE 15.3** Design types of cross-roll piercing mills. (Reproduced with permission of Vallourec & Mannesmann Deutschland GmbH.)

**Advantages and disadvantages**

| Criterion | Guide shoes | Diescher discs |
|---|---|---|
| Hollow shell length | - | + |
| Thinnest rollable hollow bloom wall | + | - |
| Expansion | + | - |
| Feed efficiency | - | + |
| Geometry change due to wear | - | + |
| Material stickings on guides with high-alloyed materials | + | - |
| Tool costs | - | + |
| Changing times due to tool wear | - | + |
| Changing times for size changes | + | - |

**FIGURE 15.4** Criteria for the use of Diescher discs or guide shoes in pierce rolling. (Reproduced with permission of Vallourec & Mannesmann Deutschland GmbH.)

forms quite naturally. Previously, this was considered necessary. Today, piercing is effected in front of the piercing plug with the billet closed, because the breaking of the core causes internal defects.

### 15.1.3 Pilger Rolling

Pilger rolling is an automated forging process. The rolling stock—that is, the pierced bloom—is manipulated by a feeder, which guides the forging mandrel plus bloom into the working pass between the open pilger rolls. As the rolls rotate against the rolling direction, they work the bloom and press the mandrel bar inside it backwards. As soon as the rolls open again, the feeder rotates the bloom through 90°, moves forward, and a new pilger rolling cycle is started, Figure 15.6.

The use of the pilger rolling process is justified to the present day, especially for larger tube sizes. For piercing, pilger rolling mills can be equipped with a piercing press or a cross rolling mill. Where ingots produced by stationary casting are predominantly used as the starting material, piercing presses combined with a downstream cross rolling mill are preferable, as they provide improved flexibility. Furthermore, it is advisable to reheat pilger-rolled tubes

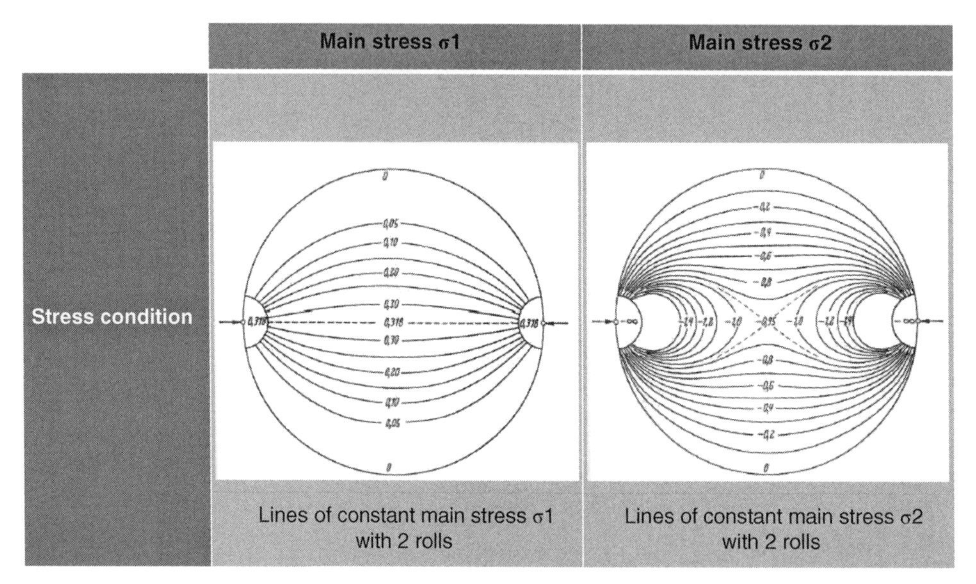


**FIGURE 15.5** Material Stresses in cross-roll piercing. (Reproduced with permission of Vallourec & Mannesmann Deutschland GmbH.)

before final rolling. This improves the diameter tolerances and permits the normalization of the material in the production flow of the rolling mill. A plant layout of this kind can be seen in Figure 15.7.

Shortcomings of the pilger rolling process compared with later developments include the quality of the outside surface (pilger humps) and the yield, because the tube end—the pilger head—cannot be rolled out, which means quite a bit of material must be cropped and is thus lost.

Pilger rolling mills can produce tubes to a specified tolerance on the inside diameter, which is important, for example, when tubes have to be welded together. The advisable size range for this process extends from diameters of 300 to 700 mm (12–28 in.). Its special strength lies in the range of medium to predominantly large wall thicknesses and its flexibility. In addition, there is hardly any limitation regarding materials. If the dimensions of the pilger rolling stand's delivery end accommodate this, lengths >20 m are producible. The annual output capacity of a single-stand pilger rolling mill can be up to 150 kt, depending on the size range and operating hours.

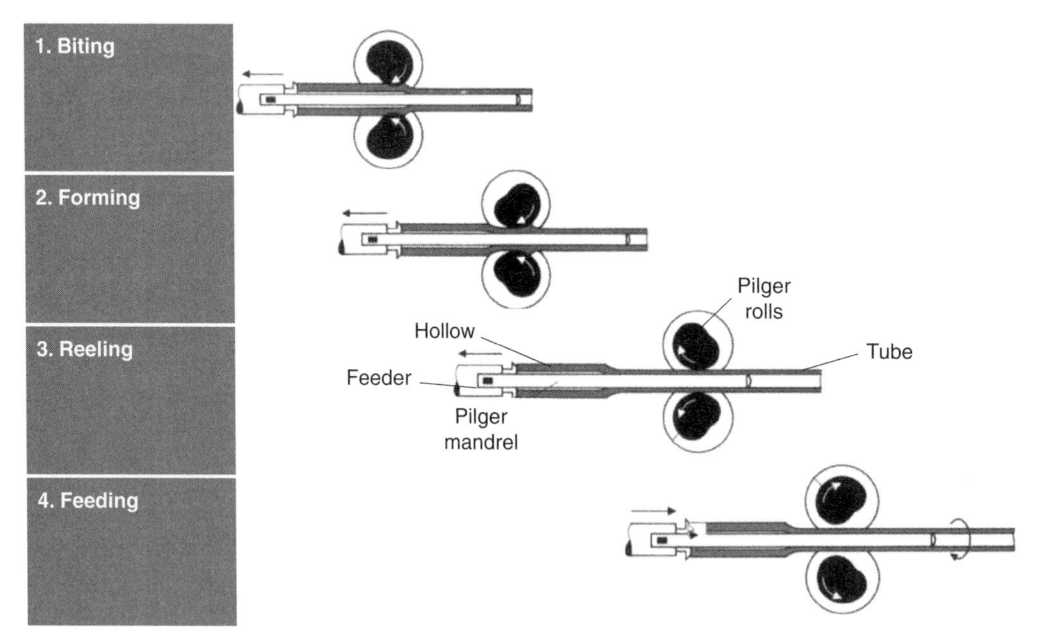


**FIGURE 15.6** The pilger rolling process. (Reproduced with permission of Vallourec & Mannesmann Deutschland GmbH.)

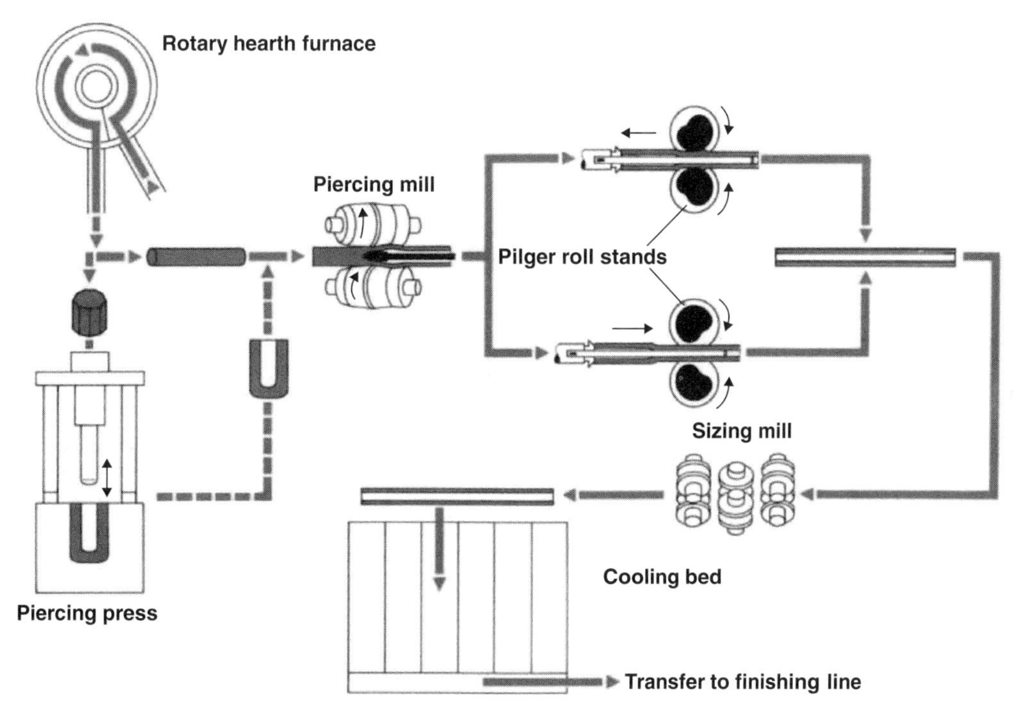


**FIGURE 15.7** Layout of a pilger rolling mill. (Reproduced with permission of Vallourec & Mannesmann Deutschland GmbH.)

### 15.1.4 Plug Rolling

In plug rolling, elongation takes place on a two-high rolling stand using a stationary plug lubricated with a salt–graphite mixture. After the first pass, the tube is returned to the pass entry end via stripper rolls and turned through 90°. Then a second roll pass is started using a plug with a larger diameter, see Figure 15.8.

The wall thickness reduction amounts to between 6 and 7 mm, of which about two-thirds is effected by the first pass.

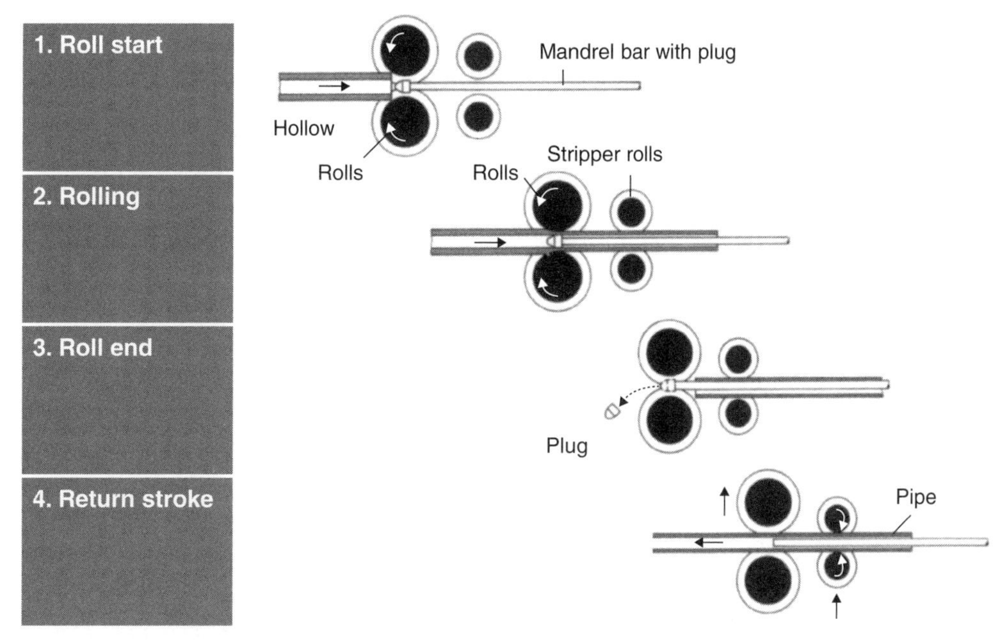


**FIGURE 15.8** The plug rolling process. (Reproduced with permission of Vallourec & Mannesmann Deutschland GmbH.)

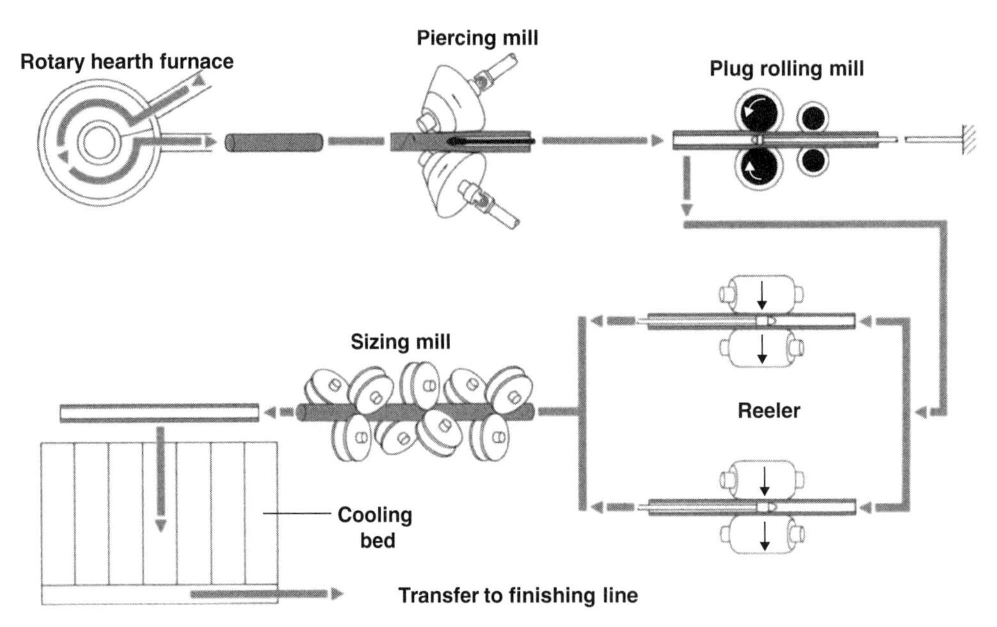


**FIGURE 15.9** Layout of a modern plug rolling mill. (Reproduced with permission of Vallourec & Mannesmann Deutschland GmbH.)

Any residual thickening of the tube wall in the area of the flanks and marks the plug may have left on the inside surface are smoothed out by subsequent reeling. To ensure that the reeling process can accommodate the cycle of the plug rolling mill, two reelers are operated in parallel.

Figure 15.9 shows the layout of a modern plug rolling mill.

Here, a single cross-roll stand with Diescher discs replaces the two consecutively arranged stands with guide shoes of the original plant configuration developed by Stiefel. The first of these served for piercing, and the second for preliminary elongation. Again, reheating of the tube is advisable, and it can—as in a pilger rolling mill—permit the normalization of the material in the production flow.

Plug rolling mills feature outstanding flexibility and achieve good wall thickness tolerances. The size range producible in a state of the art plug rolling mill covers outside diameters from about 180 to 406 mm (7–16 in.) and—exclusively—single lengths of up to 14.5 m. Depending on the size range and operating hours, annual output capacities of up to 450 kt are achieved. Materials up to and including 13% Cr steel can be processed. Thanks to the flexibility of plug rolling mills, there are no restrictions regarding the type of tubes producible, although they are less suitable for thick-walled tubes.

### 15.1.5 Mandrel Rolling

Mandrel mills—also known as continuous tube rolling mills—are the best performers among all tube-making plants. In a mandrel mill, several rolling stands are arranged in line. Wall thickness reduction is achieved by the rolls working together with a rolling mandrel that has been lubricated with graphite before insertion in the hollow bloom. Up until 2003, all mandrel mills were equipped with two-high stands. Since then, all the mills, but one, have been changed over to three rolls per stand. Depending on the number of rolls per stand, the rolls are arranged at 90° or 60° toward each other. Figure 15.10 illustrates the decisive reason for the superiority of the three-roll arrangement over the two-high stand.

Firstly, the differences in roll diameters and thus circumferential speeds from the roll base to the roll flank are significantly reduced, which improves the material flow. The risk of overstretching the wall is substantially reduced. And, secondly, the geometrical deviations in the roll gap resulting from the opening and closing of the rolls are halved. This is essential for ensuring a given reduction in wall thickness with a specified mandrel diameter. Thanks to the halved geometrical deviation, fewer mandrel sets with different diameters are required.

Besides the number of rolls per stand, mandrel rolling mills are distinguished by the type of mandrel guidance and by the method for stripping the tube off the mandrel.

Free-floating mandrels are inserted into the hollow bloom and the bloom with the mandrel inside it is then pushed into the first rolling stand via a roller table. Apart from acceleration forces, the resultant mandrel force must add up to zero. This means the friction forces in the direction of rolling and in the opposite direction must be in equilibrium. Since the rate of material flow increases from stand to stand, the mandrel speed during loading the rolling mill also increases.

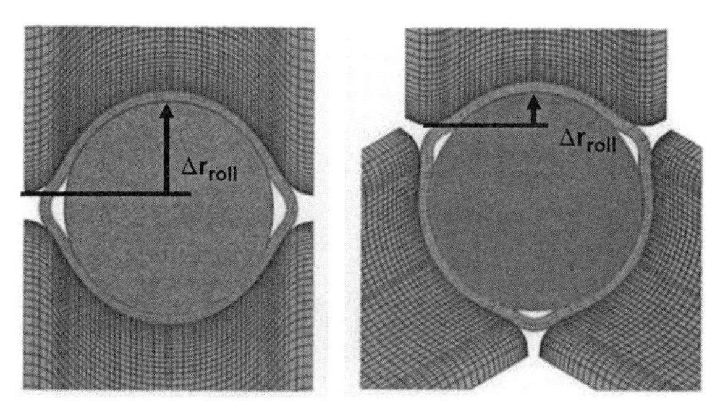


**FIGURE 15.10** Differences in roll diameters with 2-roll and 3-roll systems. (Reproduced with permission of Vallourec & Mannesmann Deutschland GmbH.)

During stationary rolling, it remains constant until it increases again when the rolling mill is emptied.

Variations in the forces that promote or brake longitudinal material flow have such a strong effect that the roll speed must be dynamically changed when rolling thin wall thicknesses. Another drawback is the great mandrel length required: for a rolled tube length of about 30 m, the mandrel must be about 27 m long. It should be pointed out in this context that a tube length of 30 m is not unusual for double lengths, as elongation by diameter reduction in the downstream sizing process is not always sufficient. Given the high resultant mandrel weights, the size range produced in free floating mandrel mills—also known as continuous tube rolling mills—is limited to diameters smaller than 180 mm (7 in.).

On completion of the rolling process, the continuously rolled tubes are stripped off the mandrel by a chain on a side conveyor. The mandrels are then cooled to between 80 and 120 °C in a water basin, lubricated (the lubricant should have dried by the start of rolling) and returned to the rolling process.

The reversal of friction forces between the mandrel and the tube material can be counteracted by feeding the mandrel at a controlled speed below that of the entry speed of the rolling stock into the first rolling stand. The mandrel is retained by a chain or an electromechanical retainer. In this way, the working section of the mandrel can be significantly shortened to between 12 and 13 m; see Figure 15.11.

In a variant of the retained mandrel rolling process referred to as semi-floating mandrel rolling, the mandrel is released at the end of rolling and withdrawn by a stripper chain on a side conveyor. Mills in which the tube is stripped off the mandrel in line by an extractor (i.e., three-stand sizing mill) are MPMs = multi-stand pipe mills if they have two-high stands. The manufacturers of mills with three-high stands call this rolling mill type PQF© = Premium Quality Finishing mill (SMS) or FQM© = Fine Quality Mill (Danieli).

The layout of such a plant is shown in Figure 15.12.

The advantages of mandrel mills—a high hourly throughput, excellent surface quality of the tubes, and a good yield—are opposed by drawbacks, including low flexibility regarding lot sizes and dimensional changes, and the high cost of mandrels in the case of retained mandrel mills.

Depending on operating hours and lot sizes, free-floating mandrel mills achieve annual output capacities of between 350 and 900 kt, thanks to their high hourly throughput. Three-high rolling mills can produce tube diameters up to 508 mm (20 in.), depending on the mill size. Here too, the annual capacity is at least 350 kt. A wide range of materials can be processed, up to and including 13% Cr steel. Although thin and thick-walled tubes can be produced in principle, the strengths of mandrel mills clearly lie with medium-wall tubes, and thus in the manufacture of oilfield tubes and line pipe in large lots.

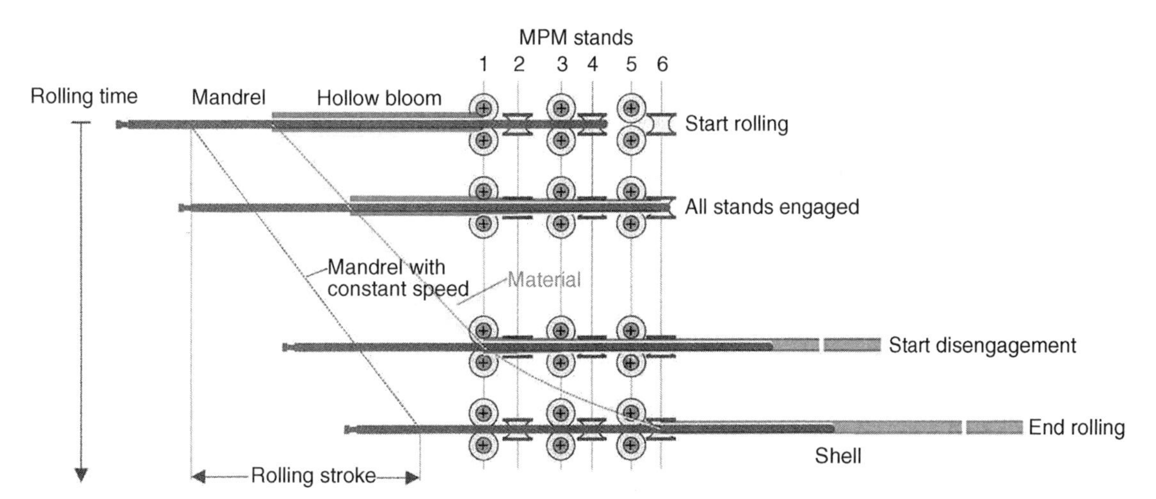


**FIGURE 15.11** Retained mandrel rolling process. (Reproduced with permission of Vallourec & Mannesmann Deutschland GmbH.)

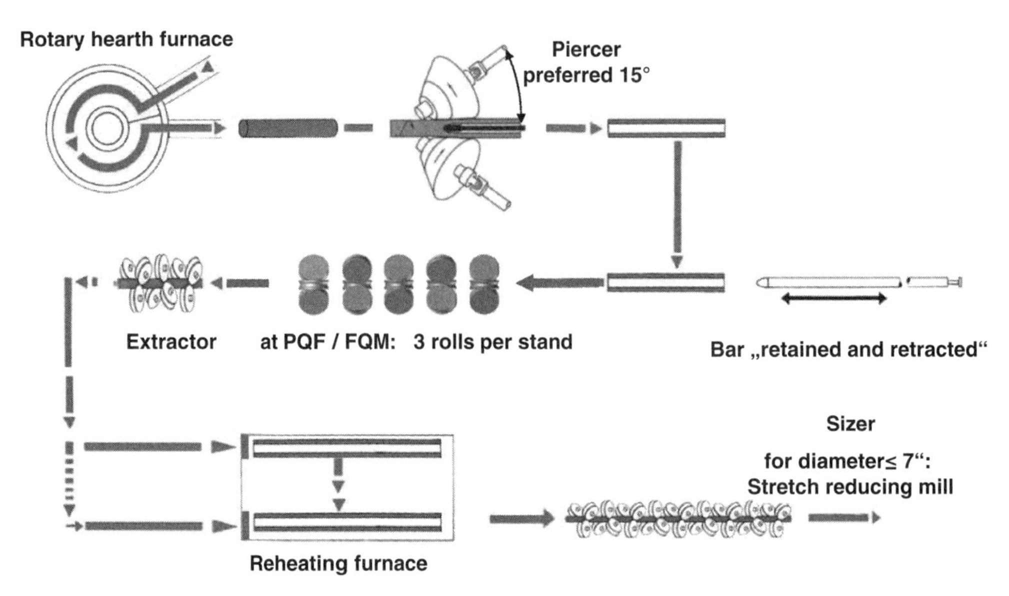


FIGURE 15.12 Layout of a three-high mandrel rolling mill. (Reproduced with permission of Vallourec & Mannesmann Deutschland GmbH.)

### 15.1.6 Forging

Pipe forging is affected by four synchronously operating forging hammers and a forging mandrel acting as internal tool. In addition, there are two manipulators—one at the inlet end and one at the outlet end—for handling the material during rolling. Subsequent size rolling can be dispensed with, see Figure 15.13 [5].

As opposed to all other current tube- and pipe-making processes, the final rolling stage in which the pipe receives its specified diameter can be omitted. For piercing, various processes can be employed, such as cross-roll piercing, piercing presses, or a combination of the two. The choice of the process depends on the size range, type of starting material and steel grade. Figure 15.14 shows a layout variant of a pipe forging plant.

Pipe forging offers many advantages. To start with, it is a forming process that involves very low stresses and is suitable for practically all steel grades. In addition, forged pipes not only possess outstanding geometrical accuracy but can also be made with stepped outside and/or inside diameters and with axially symmetrical cross sections. Furthermore, pipe forging plants are extremely flexible. One drawback is their low hourly throughput.

The process is used for pipe diameters from above 200 to 720 mm (8–28 in.), with wall thicknesses in the medium to extremely thick-walled range.

Radial forging technology of GFM started 1946

- Phase 1
  - Only manipulator A is pushing
- Phase 2
  - Manipulator B is pulling
  - Manipulator A is supporting
- Phase 3
  - Only manipulator B is pulling

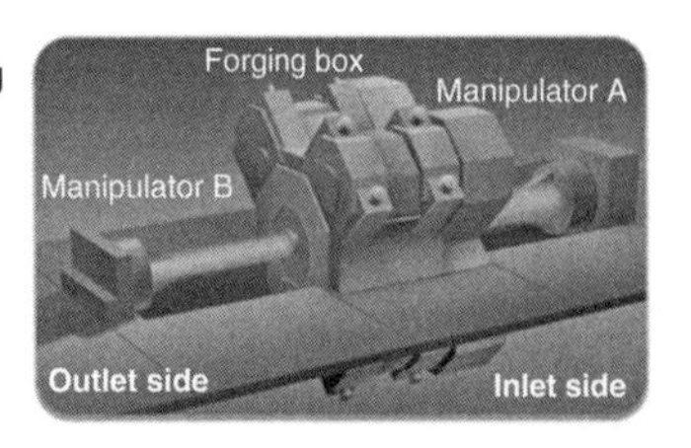


FIGURE 15.13 The pipe forging process. (Reproduced with permission of Vallourec & Mannesmann Deutschland GmbH.)

### 15.1.7 Size Rolling and Stretch Reducing

In size-rolling and stretch-reducing mills, tubes and pipes are rolled to their finished dimensions. There are mills with

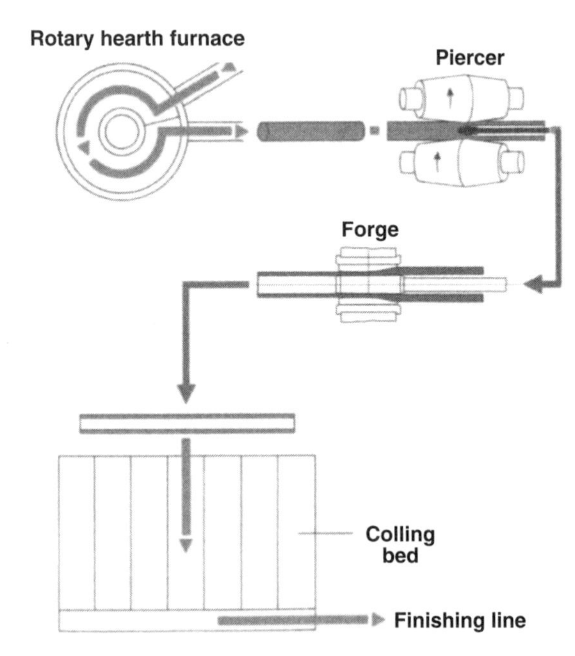


FIGURE 15.14 Layout of a pipe forging plant. (Reproduced with permission of Vallourec & Mannesmann Deutschland GmbH.)

**Sizing mill:**

- Slight reduction of the outside diameter (up to 25%)
- 2 or 3 rolls per stand
- Up to 12 stands
- Almost no traction between the stands

**Stretch-reducing mill:**

- Important reduction of the outside diameter (up to 84%)
- 3 rolls per stand
- Up to 30 stands
- Stretch between the stands

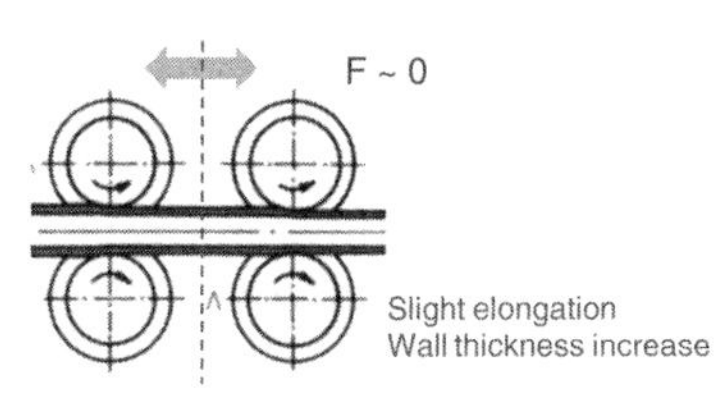


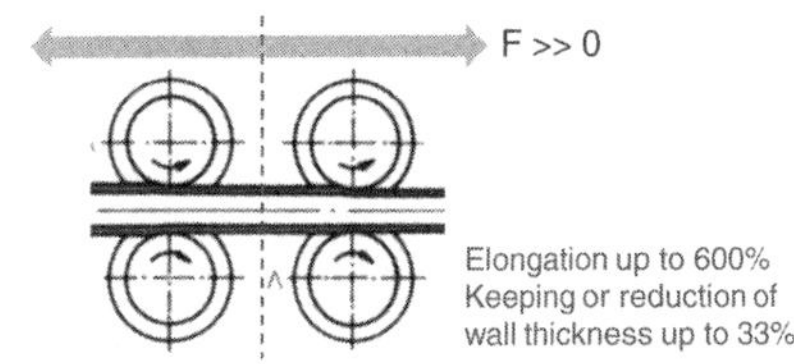


**FIGURE 15.15** Comparison of size rolling and stretch-reducing mills. (Reproduced with permission of Vallourec & Mannesmann Deutschland GmbH.)

two-high and three-high stands, which are arranged at a 90° or 60° offset to each other. Here too, the different circumferential speeds of the rolls across the rolling width have a negative effect on the material flow and roll wear. This is why, in new plants, the three-roll technology has completely replaced two-roll sizing, and four-roll technology is already waiting in the wings.

If the circumference is reduced, the rolling stock flows not only axially but also radially. The wall thickness is upset. In the case of large diameter reductions, this would mean the ingoing pipe would have to be rolled with a significantly thinner wall. Therefore stretch-reducing mills were developed which apply tension to the rolling stock between the individual rolling stands. One drawback here is that, when loading and emptying the mill, this tension has to be first built up and later relieved. The result is local thickening at the pipe ends, which significantly reduces the useful pipe length. However, this disadvantage can be avoided with appropriate speed control.

Figure 15.15 shows a comparison of size-rolling and stretch-reducing mills.

Stretch-reducing mills are chiefly used for rolling small tube diameters, while size-rolling mills are mainly used in plants producing outside diameters above 180 mm (7 in.).

## 15.2 FURTHER PROCESSING

Steel tube and pipe are subject to multiple demands under service conditions, which cannot always be met in the as-rolled condition. Accordingly, specific further processing is required after rolling.

### 15.2.1 Heat Treatment

To give steel its specific properties, in addition to determining its chemical composition, products made from it are in many cases given heat treatment after rolling. The effect of this process depends on the efficiency in achieving the desired crystal structures in the material with targeted heating and cooling in special heat treatment facilities. For line pipe of carbon steels, basically two treatment methods are employed: normalizing and quench-and-temper treatment.

A typical property of iron alloys is that they change their crystal structure as a function of temperature (allotropy) (Figure 15.16).

The main forming processes in the manufacture of seamless steel tubes take place at temperatures above 1000 °C. In this temperature range, the steel alloys typically used for line pipe have a face-centered cubic crystal structure (gamma iron or austenite). At temperatures below 700 °C, this crystal structure undergoes complete transformation to a body-centered cubic lattice (alpha iron or ferrite). This transformation frequently imposes a preferred orientation on the crystallites (banded microstructure), which can negatively affect the physical/technological properties of pipes.

In many cases, pipes are given a normalizing treatment to counteract this. For this purpose, the pipes are heated to austenitizing temperature and then allowed to cool in still air. This forced additional transformation of the crystal lattice

- $\alpha$-iron = ferrite cubically body centered T≤ 911 C°
- $\gamma$-iron = austenite cubically –face centered 911 °C < T < 1392°C

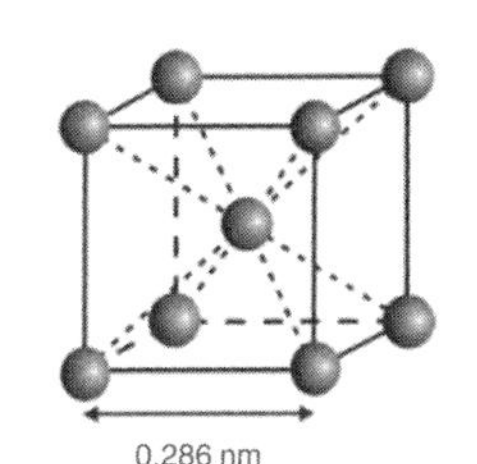


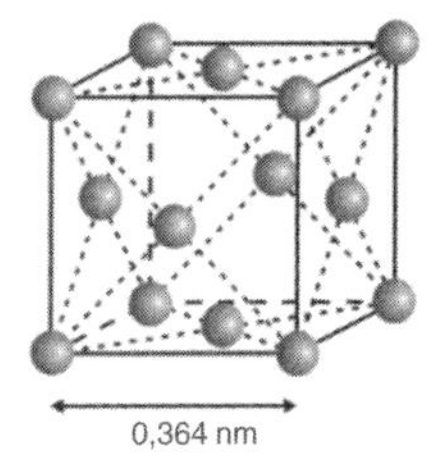


**FIGURE 15.16** Crystal structures of iron. (Reproduced with permission of Vallourec & Mannesmann Deutschland GmbH.)

FIGURE 15.17 Normalizing in a bogie hearth furnace. (Reproduced with permission of Vallourec & Mannesmann Deutschland GmbH.)

causes recrystallization, leading to the formation of a finer microstructure. Macroscopically, this microstructural change manifests itself in improved toughness properties and higher yield strength values. Depending on the plant configuration, this process can also be integrated in the hot forming process, for example, before the last forming stage. Figure 15.17 shows pipes being normalized in a bogie hearth furnace.

As a rule, pipes in higher strength steel grades are subjected to two-step heat treatment. This involves heating them to austenitizing temperature, followed by rapid cooling in a quenching medium. This is usually water, although steels with higher carbon contents are sometimes also quenched in oil or in a polymer-containing liquid, Figure 15.18.

Depending on the pipe wall thickness, quenching can, for example, be done in a basin. This should preferably have an

FIGURE 15.19 Quenching in a water basin. (Reproduced with permission of Vallourec & Mannesmann Deutschland GmbH.)

internal flushing device, to ensure fast cooling (Figure 15.19). Thin-walled tubes are usually quenched via a series of quenching rings or a ring-shaped arrangement of nozzles, which must be capable of applying the quenching medium at a sufficiently large flow rate to the axially travelling pipe. To heighten the quenching effect, an internal lance may also be used.

In another type of quenching system, nozzles arranged in a line along the tube axis spray water onto the surface of the axially stationary but rotating tube. In many cases, water ejected from an adjustable nozzle at the tube end flows through the tube, providing internal quenching along its length.

The objective of all these methods is to dissipate the heat stored in the tube material as quickly as possible, thus lowering the tube temperature.

This rapid temperature reduction impedes or prevents the diffusion processes taking place during recrystallization. Since the solubility of carbon is higher in a face-centered cubic austenite lattice than in the cubic body-centered lattice of ferrite, the carbon impeded in its diffusion movement causes distortion of the cubic body-centered lattice.

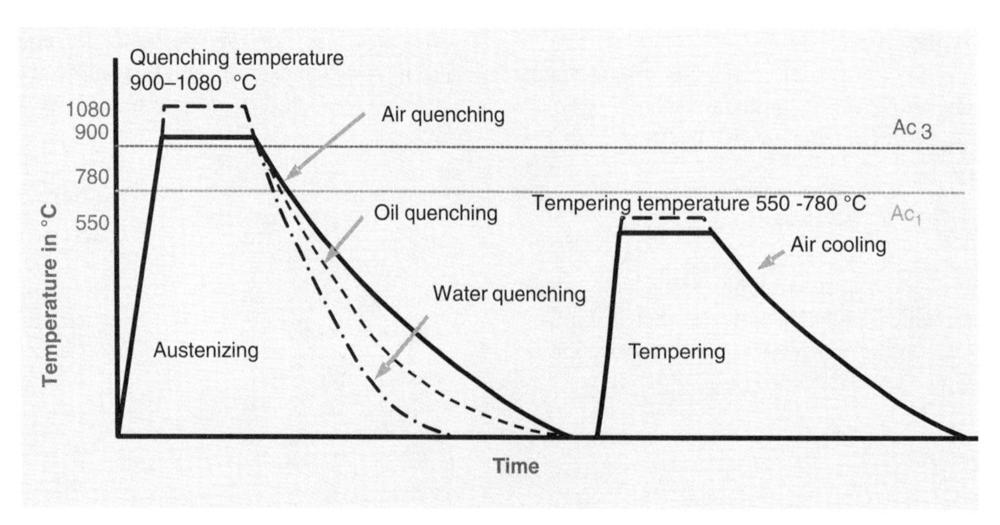


FIGURE 15.18 Schematic representation of heat treatment by hardening and tempering. (Reproduced with permission of Vallourec & Mannesmann Deutschland GmbH.)

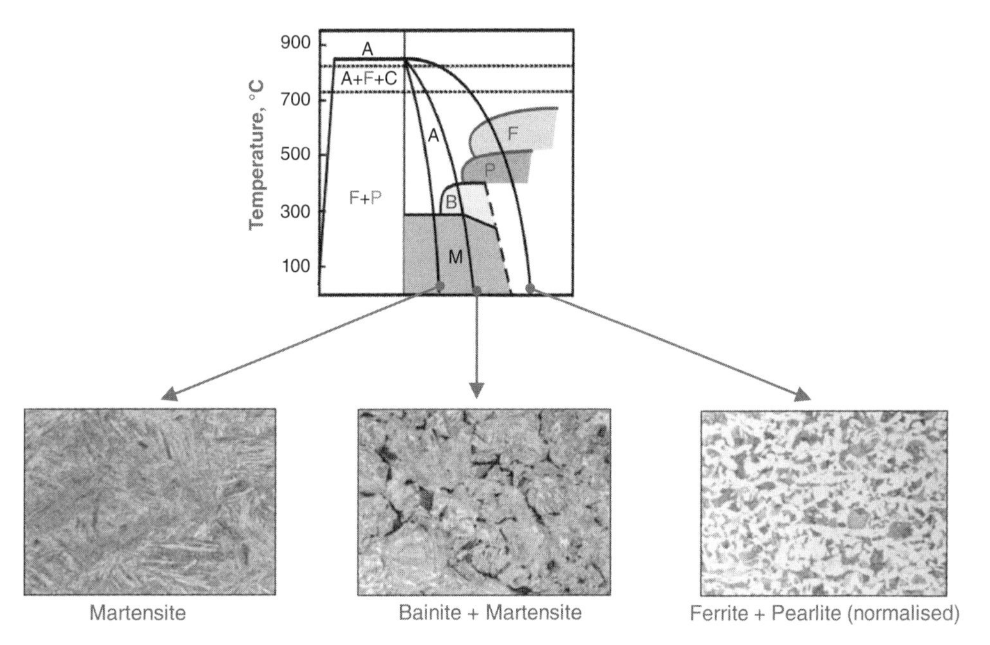


**FIGURE 15.20** Microstructure constituents as a function of cooling speed. (Reproduced with permission of Vallourec & Mannesmann Deutschland GmbH.)

Macroscopically, this leads to a substantial increase in yield strength, tensile strength and hardness. Since, an increase in steel's strength and hardness is usually accompanied by a reduction in its ductility, which restricts the service properties of the tubes; additional heat treatment is called for. For this purpose, the tubes are reheated to a temperature below the lattice transformation temperature and then allowed to cool in still air.

As a result, lattice tensions are reduced and the material's ductility is improved. The strength characteristics are still substantially higher than in the steel's non-heat-treated or normalized condition. The resultant microstructures are schematically illustrated in the following cooling diagrams, together with a logarithmic timeline (Figure 15.20).

The cooling curves shown in conjunction with the microstructures generated at the various cooling speeds (F = ferrite; P = pearlite; B = bainite; M = martensite) are typical of all steel alloys. The mechanical-technological properties that can be achieved depend on the knowledge and expertise in heat treating steels of the chemical compositions concerned.

### 15.2.2 Quality and In-Process Checks

The high demands placed on today's line pipe call for seamless control of the processes used. Ideally, these documented in-process checks are logged in a superordinate data acquisition system so they are available for analyses if and when required, together with the nondestructive and destructive tests specified by standards and customer specifications. Similarly, measurement results should be fed back to the process so as to ensure that the desired product properties are achieved within a narrow bandwidth.

Optimally, tests are coordinated according to various measurement principles, in order to achieve the highest possible level of test reliability. Here, a combination of electromagnetic techniques and a full-body ultrasonic test can be mentioned as an example. Traceability of measurement and test results to the individual heat treated pipe is the state of the art.

All the applicable delivery specifications demand that special data must be retained for a defined period after delivery of the products.

### 15.2.3 Finishing Lines

In the finishing lines, line pipes are prepared for delivery in accordance with the customer's specifications.

As a rule, this includes hydrostatic testing of the pipes with up to 90% yield strength utilization. The pressure level and holding time are documented for each test.

Pipe end sizing and the machining of welding bevels to the relevant specifications, as well as testing for laminations ensure the trouble-free welding of the delivered pipes.

On completion of all the magnetizing tests, the pipes are demagnetized in order to exclude impairment of the welding process due to residual magnetic flux leakage during pipelaying.

The final measurement and marking of the pipes mark the completion of the mill production process.

Then come the final inspection and release by the customer's representative (if specified) and mill experts. At the customer's request, the pipes can be provided with temporary corrosion protection and protective end caps before delivery.

## REFERENCES

1. Kümmerling, R. (1992) Comparison of various cross-rolling processes for elongation stages in the production of seamless tubes. Metallurgical Plant and Technology International 1/1992, pp. 48–57.
2. Bellmann, M. and Kümmerling, R. (1993) Kriterien für die umformtechnische Bewertung von Lochschrägwalzwerken für die Herstellung nahtloser Rohre. *Stahl und Eisen*, 113(8), 47–53.
3. Bellmann, M. and Kümmerling, R. (1993) Optimierung des Spreizwinkels von Lochschrägwalzwerken für die Herstellung nahtloser Rohre. *Stahl und Eisen*, 113(9), 111–117.
4. Lohmann, W. (1959) Spannungszustand und Verformungseffekt im Innern eines schräggewalzten vollen Blockes in einem Zweiwalzen- und einem Dreiwalzen-Schrägwalzwerk. Auszug aus P. Grüner, In: *Das Walzen von Hohlkörpern und das Kalibrieren von Werkzeugen zur Herstellung nahtloser Rohre*, Springer, 218–226.
5. Kümmerling, R. and Kneissler, A. (2012) Radial forging of seamless pipes. *Stahl und Eisen*, 132(10), 49–56.

# 16

# MAJOR STANDARDS FOR LINE PIPE MANUFACTURING AND TESTING

GERHARD KNAUF AND AXEL KULGEMEYER
*Salzgitter Mannesmann Forschung GmbH, Duisburg, Germany*

High quality and efficient line pipe manufacturing is based on standards and specifications that cover design, materials, and testing. Major international line pipe standards are briefly discussed in this chapter.

## 16.1 API SPEC 5L/ISO 3183

API SPEC 5L:2012-12: *Specification for Line Pipe* [1]

ISO 3183:2012: *Petroleum and natural gas industries—steel pipe for pipeline transportation systems* [2].

These two harmonized standards are the most basic standards for the line pipe material and provide requirements for seamless as well as welded steel line pipe to transport gas, water, and oil. They cover pipeline grades from A25 to X120. The product specification level PSL 1 provides a standard quality level for line pipe while PSL 2 has additional mandatory requirements for chemical composition, notch toughness, strength properties, and additional nondestructive testing. Special requirements, for example, for resistance to ductile fracture propagation in gas pipelines, for sour service, and for offshore service are included.

## 16.2 CSA Z662-11: *OIL AND GAS PIPELINE SYSTEMS*

This pipeline standard [3], a National Standard of Canada, provides guidance on the design, construction, operation, and maintenance of oil and gas pipeline systems and addresses requirements related to legislation, regulation, management systems, and technology. This includes guidance on the use of high-performance plastic piping, safety and loss management systems, integrity management program guidelines, incident reporting, and engineering assessments.

## 16.3 DNV-OS-F101-2012: *SUBMARINE PIPELINE SYSTEMS*

DNV-OS-F101 [4] is a line pipe standard for offshore applications and aims at ensuring that the concept development, design, construction, operation, and abandonment of pipeline systems are safe and conducted with due regard to public safety and the protection of the environment. With regard to pipe production, it provides requirements for materials engineering, material selection, and material specification, including format and characterization of material strength, requirements for nondestructive testing, supplementary requirements for fracture arrest properties, plastic deformation, dimensional tolerances and high utilization, sour service, welding, and corrosion control.

## 16.4 ISO 15156-1:2009: *PETROLEUM AND NATURAL GAS INDUSTRIES—MATERIALS FOR USE IN $H_2S$-CONTAINING ENVIRONMENTS IN OIL AND GAS PRODUCTION*

The standard ISO 15156 [5] gives important requirements and recommendations for the selection and qualification of carbon and low-alloy steels for service in equipment that is

*Oil and Gas Pipelines: Integrity and Safety Handbook*, First Edition. Edited by R. Winston Revie.

used in oil and natural gas production and natural gas treatment plants in $H_2S$-containing environments, where failure could pose a risk to the health and safety of the public and personnel or to the environment. It is equivalent to the NACE standard MR 0175. It can be applied to help avoid costly corrosion damage to the equipment itself. It supplements, but does not replace, the materials requirements of the appropriate design codes, standards, or regulations.

## 16.5 EFC PUBLICATION NUMBER 16, THIRD EDITION: *GUIDELINES ON MATERIALS REQUIREMENTS FOR CARBON AND LOW-ALLOY STEELS FOR $H_2S$-CONTAINING ENVIRONMENTS IN OIL AND GAS PRODUCTION*

This document [6] provides guidelines on the materials requirements for the safe application of carbon and low-alloy steels typically used in $H_2S$-containing environments in oil and gas production systems. It aims to be comprehensive in considering all possible types of cracking, which may result from exposure of such steels to $H_2S$, the conditions under which cracking may occur and appropriate materials requirements to prevent such cracks. In addition, the document recommends test methods for evaluating materials performance and particularly focuses on a fitness-for-purpose approach, whereby, the test conditions are selected to reflect realistic service conditions.

## 16.6 NACE TM0284 AND TM0177

NACE Standard TM0284-2011: Evaluation of Pipeline and Pressure Vessel Steels for Resistance to Hydrogen-Induced Cracking [7].

The occurrence of internal cracking in steel in the presence of aqueous hydrogen sulfide is known as hydrogen-induced cracking (HIC). This standard, NACE TM0284, addresses the testing of metals for HIC under standardized sour test conditions, for consistent evaluation of steel pipe. The primary purpose of this well-established standard is to facilitate conformity in testing so that data from different sources can be compared on a common basis.

NACE Standard TM0177-2005: *Laboratory Testing of Metals for Resistance to Sulfide Stress Cracking and Stress Corrosion Cracking in $H_2S$ Environments* [8].

Sulfide stress cracking (SSC) is defined as severe and sudden cracking and failure of steel in the presence of hydrogen sulfide ($H_2S$) and tensile stress. The standard NACE TM0177 addresses the testing of metals for resistance to sulfide stress cracking in aqueous environments containing hydrogen sulfide ($H_2S$) and contains methods for testing metals using tensile, bent-beam, C-ring, and double-cantilever-beam (DCB) test specimens. Test conditions and performance are described in detail and are the most commonly used in SSC testing.

## 16.7 ISO 10893-11—2011 NON-DESTRUCTIVE TESTING OF STEEL TUBES—PART 11: *AUTOMATED ULTRASONIC TESTING OF THE WELD SEAM OF WELDED STEEL TUBES FOR THE DETECTION OF LONGITUDINAL AND/OR TRANSVERSE IMPERFECTIONS*

This standard [9] specifies requirements for the automated ultrasonic testing of strip/plate in the manufacture of welded tubes for the detection of laminar imperfections. This testing is carried out in the pipe mill before or during pipe production.

## REFERENCES

1. API (2012) *API SPEC 5L:2012-12: Specification for Line Pipe*, American Petroleum Institute, Washington, DC.
2. ISO (2012) ISO 3183:2012: Petroleum and natural gas industries—steel pipe for pipeline transportation systems, ISO Central Secretariat, 1, ch. de la Voie-Creuse, CP 56, CH-1211 Geneva 20, Switzerland.
3. CSA (2011) *CSA Z662-11: Oil and Gas Pipeline Systems*, Canadian Standards Association, Mississauga, Ontario, Canada.
4. DNV (2007) DNV-OS-F101: Submarine Pipeline Systems, Det Norske Veritas AS, Oslo, Norway.
5. ISO (2009) ISO 15156-1:2009: Petroleum and natural gas industries, Materials for use in $H_2S$-containing environments in oil and gas production, ISO Central Secretariat, 1, ch. de la Voie-Creuse, CP 56, CH-1211 Geneva 20, Switzerland.
6. EFC (2009) Eliassen, S. and Smith, L. (editors), *EFC 16: Guidelines on Materials Requirements for Carbon and Low Alloy Steels for $H_2S$-Containing Environments in Oil and Gas Production*, 3rd ed., ISBN 978190654 0333, European Federation of Corrosion.
7. NACE (2011) *NACE TM0284-2011: Evaluation of Pipeline and Pressure Vessel Steels for Resistance to Hydrogen-Induced Cracking*, NACE International, Houston, TX.
8. NACE (2005) *NACE TM0177-2005: Laboratory Testing of Metals for Resistance to Sulfide Stress Cracking and Stress Corrosion Cracking in $H_2S$ Environments*, NACE International, Houston, TX.
9. ISO (2011) ISO 10893-11: 2011: Non-destructive testing of steel tubes—Part 11: Automated ultrasonic testing of the weld seam of welded steel tubes for the detection of longitudinal and/or transverse imperfections, ISO Central Secretariat, 1, ch. de la Voie-Creuse, CP 56, CH-1211 Geneva 20, Switzerland (2011-04).

# 17

# DESIGN OF STEELS FOR LARGE DIAMETER SOUR SERVICE PIPELINES

NOBUYUKI ISHIKAWA
*JFE Steel Corporation, Hiroshima, Japan*

## 17.1 INTRODUCTION

Linepipe material for sour gas service primarily needs to have strong resistance to the cracking caused by hydrogen invasion from the gas containing wet $H_2S$. Morphologies of the cracking caused in carbon steel or low alloyed steel under wet $H_2S$ environment are illustrated in Figure 17.1 [1]. Hydrogen-induced cracking (HIC) occurs without any applied stress, and there are several types of cracking morphologies, as shown in Figure 17.1a–g. Several terminologies have been used for describing these types of cracking, such as blister cracking, hydrogen-induced blister cracking, hydrogen-induced stepwise cracking, stepwise cracking, and so on. On the other hand, sulfide stress cracking (SSC) or sulfide stress corrosion cracking (SSCC), Figure 17.1h and i, occurs under the presence of a tensile stress field. The SSC has been investigated since the 1960s [2–4], and SSC testing method and guidelines for preventing SSC have been established in NACE MR 0175/ISO 15156-2 [5]. On the other hand, after the accident involving failure of a sour gas pipeline in the 1970s [6,7], extensive studies have been conducted to investigate the HIC mechanism and to improve material resistance [8–16]. Laboratory testing methods for evaluating the material resistance to HIC have been standardized in NACE TM 0284 [17] and EFC 16 [18]. In addition to the sour resistant properties, higher strength, toughness of base metal and weldments, and many other materials properties are required for the linepipe steels because of the demand for cost reduction in pipeline construction by applying higher grade linepipe and for expanding applications toward the harsh environmental fields such as deep water and cold climate. Material selection is one of the most important processes in the design of a sour gas pipeline. Since SSC is discussed in Chapter 24 of this handbook, HIC is the main subject of material design for sour gas pipelines in this chapter.

Reducing nonmetallic inclusions and center segregation is the basic measure to prevent initiation and propagation of HIC in linepipes used in wet $H_2S$ environments. Microstructural control of the base metal is also important for achieving both higher HIC resistance and other mechanical properties. In order to select appropriate linepipe steels and to develop the sour resistant steels used in severe environmental conditions, it is important to understand the behavior of HIC in the sour pipeline and the evaluation method for sour resistance properties, which are introduced in this chapter first. Then, material design concepts for controlling sour resistance properties in the production of the steels used in sour gas pipelines are summarized in this chapter.

## 17.2 HYDROGEN-INDUCED CRACKING OF LINEPIPE STEEL AND EVALUATION METHOD

### 17.2.1 Hydrogen-Induced Cracking in Full-Scale Test

Failure events caused in actual gas transmission pipelines in the early 1970s [5,6] initiated the extensive investigations for simulating HIC in laboratory specimens and full-scale tests [19–24]. One of the full-scale tests with pressurized wet

*Oil and Gas Pipelines: Integrity and Safety Handbook,* First Edition. Edited by R. Winston Revie.

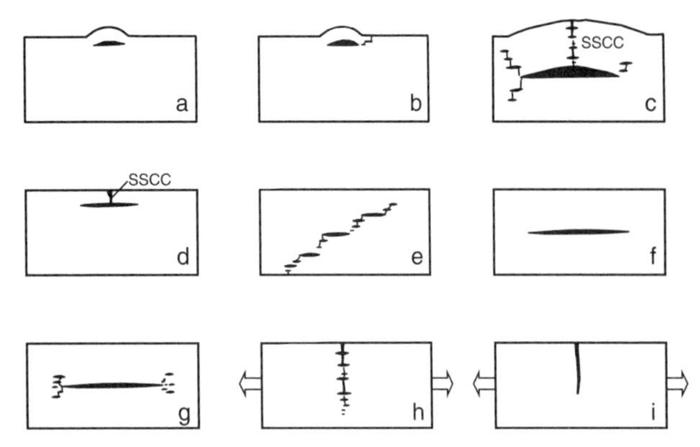


**FIGURE 17.1** Schematic illustration of $H_2S$ cracking morphology, (a) blister, (b) blister with stepwise cracking, (c) blister with stepwise cracking and SSCC, (d) blister with high-strength steel (usually accompanied with SSCC), (e) stepwise cracking, (f) straight cracking, (g) straight cracking with stepwise cracking at the crack tip, (h) SSCC in low-strength steel, (i) SSCC in high-strength steel [1].

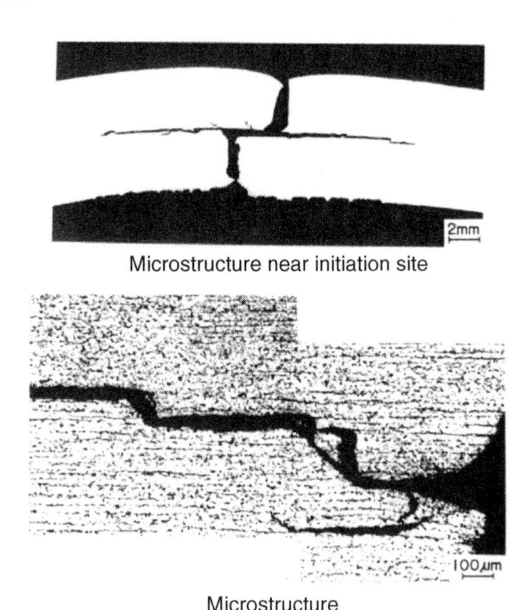


**FIGURE 17.2** Macrostructure and microstructure of the burst pipe in the full-scale test [20].

gas containing $H_2S$ was conducted by the HLP Committee of the Iron and Steel Institute of Japan [20]. Several large diameter pipes of strength grade up to X65 with different susceptibility to HIC were connected and installed in the test rig that circulated pressurized gas with brine in the bottom of pipes. Natural gas containing $H_2S$ gas was used for the test. Total gas pressure was 8.4 MPa and partial $H_2S$ pressure was up to 1.5 MPa. Susceptibility to HIC of the tested pipe was evaluated by the HIC tests with NACE TM0177 [25] solution (5% NaCl + 0.5% $CH_3COOH + H_2O + H_2S$ bubbling, currently the same as NACE TM0284 solution A), and HIC resistance was compared with the full-scale test result.

In this full-scale test, rupture of the pipe by HIC occurred in the pipe containing higher Mn and S content, which has higher susceptibility to HIC. Figure 17.2 shows the macrostructure and microstructure of the burst pipe. It was seen that the cracking was first initiated in the mid-thickness region of the pipe wall by forming stepwise cracking, and then the crack grew toward the surface of the pipe by combining with small cracks parallel to the wall. Figure 17.3 shows the comparison of the crack area ratio (CAR), measured by ultrasonic testing, in the laboratory test and full-scale test. In the full-scale test, immersion in the aqueous brine phase gave higher CAR. The pipes with lower HIC susceptibility, 20% or lower CAR, showed no HIC in the full-scale tests in both gaseous and aqueous regions. This result indicates that there is a good correlation between the HIC behavior in the full-scale test and the standardized HIC test. But, it was found that laboratory testing was much more conservative because the immersed samples are charged on 6 sides, while only one side is immersed in the full-scale test. Although NACE standard HIC test gave more conservative results than full-scale tests, which is same as other reports on the full-scale tests [22,24], the HIC test is capable of differentiating between highly susceptible and highly resistant materials.

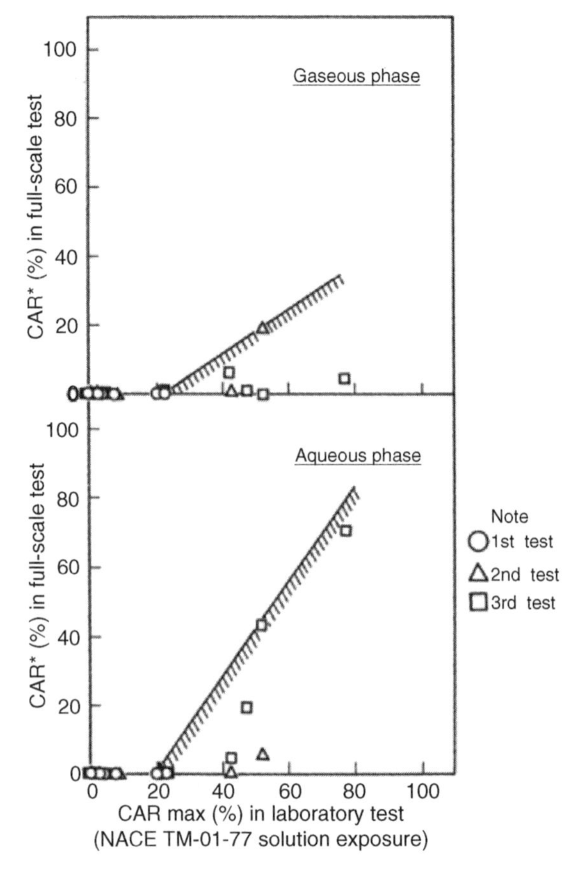


**FIGURE 17.3** Relation between small laboratory tests and full-scale tests [20].

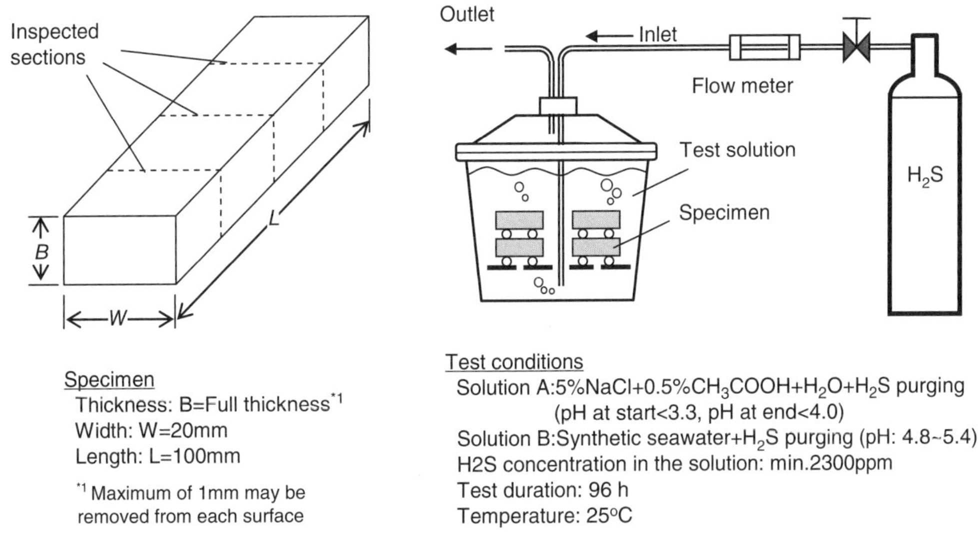


**FIGURE 17.4** Specimen and apparatus of HIC test by NACE TM 0284.

### 17.2.2 Standardized Laboratory Evaluation Method for HIC

Susceptibility to HIC of linepipe steel is generally evaluated in accordance with NACE TM 0284 [17]. This test method is not intended to reproduce the actual service condition of a sour gas pipeline, and the test result gives a relative measure of the susceptibility of materials under severe sour condition. Figure 17.4 shows the specimen and apparatus used for the HIC test. The specimen is machined, polished with emery paper, degreased by acetone, and then immersed in a test solution. Two types of the test solution are specified in NACE TM 0284. Solution A consists of 5.0% NaCl and 0.50% $CH_3COOH$ in distilled or deionized water, and then the solution is saturated with $H_2S$ gas. The value of pH at the start of test shall not exceed 3.3, and pH at the end of test shall not exceed 4.0. On the other hand, solution B is synthetic seawater saturated with $H_2S$ gas. The value of pH of solution B shall be within the range of 4.8–5.4 at the start and the end of test. The immersion time of the specimen is 96 h, and temperature of the test solution needs to be kept at 25 °C (±3 °C). After the immersion, three sections of the specimen are observed and crack dimensions are measured to calculate the crack sensitivity ratio (CSR), crack length ratio (CLR), and crack thickness ratio (CTR), as shown in Figure 17.5. Ultrasonic testing may be used to evaluate the CAR before cutting the sample. However, the use of ultrasonic testing for CRA measurement is not specified in the NACE standard.

### 17.2.3 Mechanisms of Hydrogen-Induced Cracking

The process of HIC in pipeline steels is explained in Figure 17.6 [26]. Hydrogen atoms are generated by the corrosion of steel, then hydrogen gas forms and bubbles off under normal cathodic reaction. In the $H_2S$ environment, the existence of hydrogen sulfide ion reduces the tendency to produce hydrogen gas and strongly promotes absorption of hydrogen atoms into the steel [27]. Hydrogen atoms diffuse into the steel and accumulate around the nonmetallic inclusions such as MnS and oxides. Then, recombination of hydrogen atoms to $H_2$ molecules builds up a heavy gas pressure in the interface between matrix and inclusions. Cracking initiates because of the tensile stress field caused by the hydrogen gas pressure, and the crack propagates

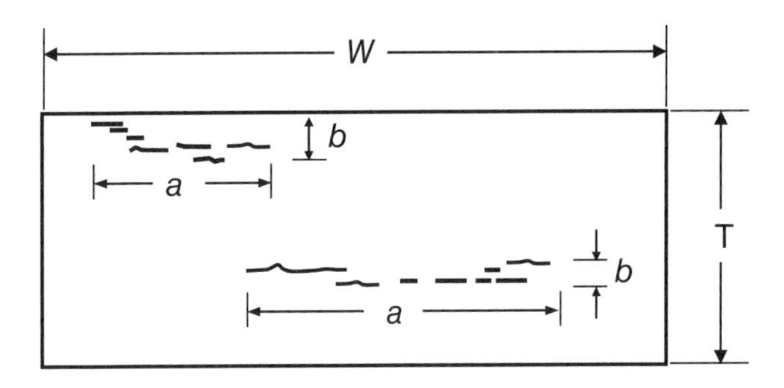


Crack sensitivity ratio $CSR = \frac{\Sigma(a \cdot b)}{W \cdot T} \times 100\%$

Crack length ratio $CLR = \frac{\Sigma a}{W} \times 100\%$

Crack thickness ratio $CTR = \frac{\Sigma b}{T} \times 100\%$

where: $a$=crack length
$b$=crack thickness
$W$=section width
$T$=specimen thickness

**FIGURE 17.5** Crack dimensions and calculation of CSR, SLR, and CTR.

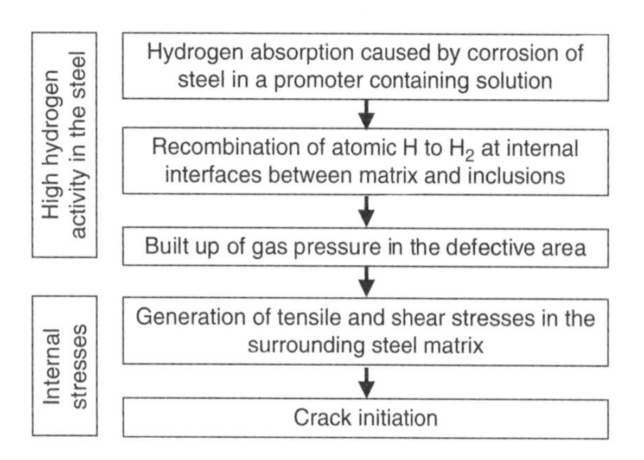


**FIGURE 17.6** Process of hydrogen-induced cracking. (Reproduced from Ref. [26] with permission of ASME.)

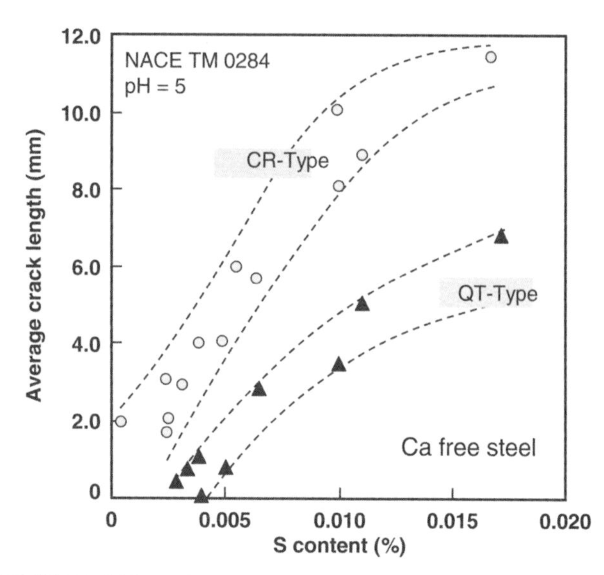


**FIGURE 17.8** Effect of sulfur content on the extent of HIC, (Adapted from [4].)

in the surrounding steel matrix. Therefore, controlling both inclusions and steel matrix is the key issue for preventing initiation and propagation of HIC.

## 17.3 MATERIAL DESIGN OF LINEPIPE STEEL FOR SOUR SERVICE

### 17.3.1 Effect of Nonmetallic Inclusions

Prevention of crack initiation is a primary issue for improving the material resistance to HIC. MnS inclusion is most harmful for HIC. Figure 17.7a shows a typical fracture surface of HIC of the steel that contains 0.0018% S and without Ca addition [28]. Elongated MnS inclusion acts as an initiator of HIC. Therefore, as a basic measure, sulfur content must be reduced for the steels used in sour environments. Figure 17.8 shows the effect of sulfur content on the extent of HIC in NACE TM 0284 solution with pH5 [4]. It is obvious that the resistance to HIC is strongly affected by sulfur contents; sulfur content should be reduced to a low level for the steels used in sour environments. Because of the progress in steelmaking process, modern steels for sour gas pipelines contains very low levels of sulfur, under 10 parts per million (ppm) [28].

Ca treatment is usually applied in the steelmaking process to prevent the formation of elongated MnS and to control the sulfide inclusions into spherical shapes [30,31]. However, there is an appropriate Ca content for preventing the cracking. An example of the fracture surface of steel with low sulfur content and Ca treatment is shown in Figure 17.7b. A cluster of Ca oxysulfide is seen on the fracture surface. If the Ca content is relatively high compared with the S content, excess Ca may form oxysulfides that can act as an initiator of HIC. A few parameters that represent effective Ca content to prevent HIC have been proposed for the material design for sour resistant linepipes [13,15,32,33]. [Ca]/[S], where [Ca] and [S] are contents of Ca and S in weight percent, is the simplest form of the parameter representing effective Ca treatment. The [Ca]/[S] value should be controlled within a narrow range in order to prevent the formation of both

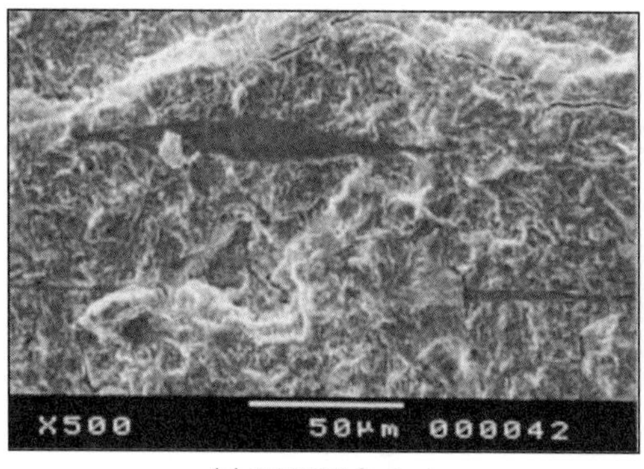


(a) 0.0018%S steel

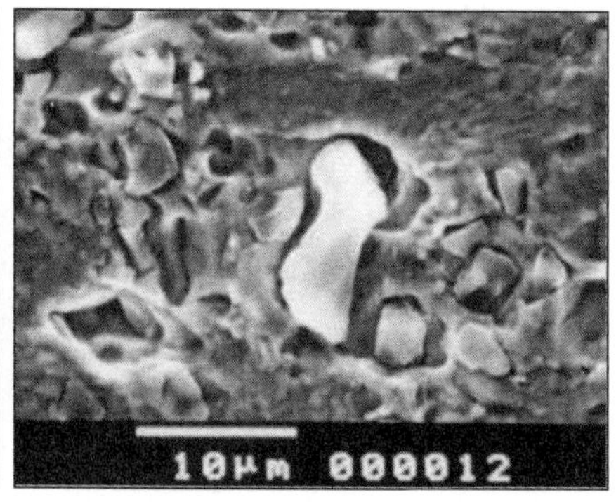


(b) 0.0003%S-0.0027Ca steel

**FIGURE 17.7** Fracture surfaces of typical HIC [28].

elongated MnS inclusions and Ca oxysulfides. The appropriate [Ca]/[S] range should depend on the steelmaking process, but it was pointed out that lower sulfur steel, such as [S] < 0.002%, gives wider [Ca]/[S] range [15,24]. The formation of Ca oxysulfide is strongly affected by the oxygen content in the steelmaking process. Therefore, further precise control of Ca, S, and O is necessary to achieve an excellent HIC resistance. Parameters consisting of Ca, S, and O contents are proposed, such as (atomic concentration ratio (ACR), {[Ca]–(0.18 + 130[Ca])[O]}/1.25[S]) [32,33] and ESSP (effective sulfide shape control parameter, [Ca] (1 − 124 [O])/1.25[S]) [13,16].

### 17.3.2 Effect of Center Segregation

The number of nonmetallic inclusions that act as crack initiator can be reduced by controlling S and Ca contents properly. It is impossible to reduce them completely, and HIC cannot be prevented if crack propagation easily occurs. Even for the steel plate produced from continuous cast slab, it is difficult to completely eliminate the center segregation that can cause centerline cracking in HIC test, as shown in Figure 17.9 [15]. Center segregation can occur during the solidification of molten steel because of the difference in the solubility of alloying elements in the liquid and solid phases of the steel. Alloying elements such as C and Mn tend to become enriched in the liquid phase, and the enriched region that remains in the steel forms a hard microstructure in the center segregation region, enhancing the propagation of cracking. Therefore, addition of C and Mn should be restricted to lower levels [34]. It was also shown that lower carbon content leads to lower Mn enrichment [24,35,36]. The Mn segregation ratio, the ratio of Mn concentration in the segregation zone to the average Mn content, can be maintained at about 1.5 for the steel containing 0.05% C or less, but Mn segregation ratio increases for steels with higher C content [24,37]. Phosphorus, P, is a harmful element that increases susceptibility to HIC in the center segregation zone since the segregation ratio of P is much higher than those of C and Mn, and P segregation causes increase of hardness and reduction of toughness [24,29,37]. Therefore, P content should also be restricted to lower levels.

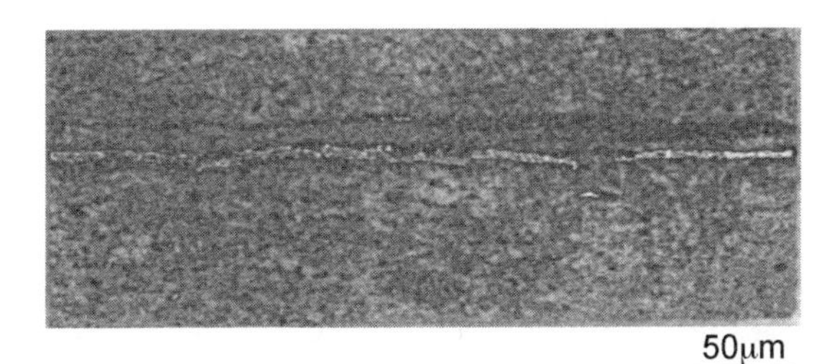


**FIGURE 17.9** Microstructure of centerline cracking [15].

### 17.3.3 Effect of Plate Manufacturing Condition

Modern steel plate for high-grade sour linepipe is produced by applying a controlled rolling and accelerated cooling process. Resistance to HIC is improved to a significant degree by fine bainitic microstructure [37–39]. During the slow cooling after hot rolling as-rolled or controlled-rolled steel, microscopic carbon distribution can occur by the transformation from austenite to ferrite–pearlite or bainite. Carbon-enriched region may turn into martensite–austenite constituent, so-called MA, and this hard second phase can increase susceptibility to HIC [38]. On the other hand, carbon distribution can be prevented by accelerated cooling, and the resulting fine bainitic microstructure shows excellent resistance to HIC.

However, microstructure of the controlled rolled and accelerated cooled steel is strongly affected by the plate rolling and cooling conditions that can, accordingly, change the HIC resistance [28,38,39]. Figure 17.10 shows examples of the microstructure of X65 linepipe steels produced with different cooling starting temperatures in plate rolling, which have the ferrite transformation temperature, $Ar_3$, of around 770 °C [28]. When the cooling starting temperature was near the $Ar_3$ temperature, the microstructure obtained was almost fully bainite or bainite and acicular ferrite. On the other hand, polygonal ferrite and bainite or pearlite microstructure was obtained when the cooling starting temperature was under $Ar_3$ temperature. Results of HIC tests are shown in Figure 17.11 [28]. Crack length ratio, in the HIC test was higher for the linepipe with lower cooling starting temperature. If the cooling starting temperature is lower than $Ar_3$ temperature, polygonal ferrite forms and this causes concentration of alloying element in austenite phase that will transform to bainite phase hardened with richer alloying composition. In this case, the difference in hardness of the ferrite and bainite phases becomes large, and cracking can easily propagate along the phase boundary. On the other hand, when the cooling starting temperature was near or above the $Ar_3$ temperature, almost no cracking was found. Therefore, a homogeneous bainitic microstructure is essential for improving crack resistance, and the cooling starting temperature in accelerated cooling process should be carefully controlled for high-strength sour linepipes.

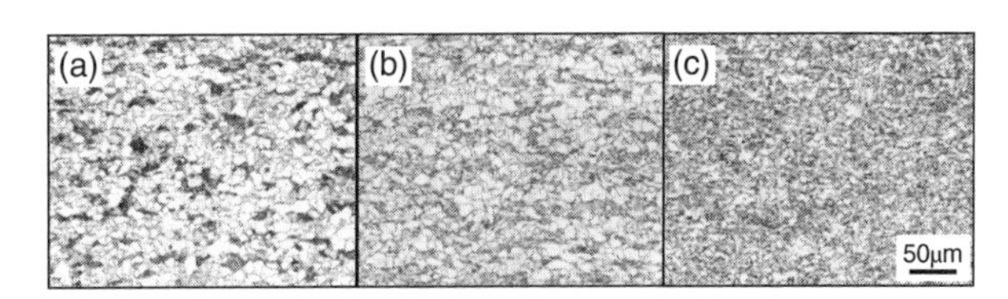


**FIGURE 17.10** Microstructure of X65 linepipe steel accelerated cooled from different temperature, (a) 760 °C, (b) 730 °C, and (c) 690 °C [28].

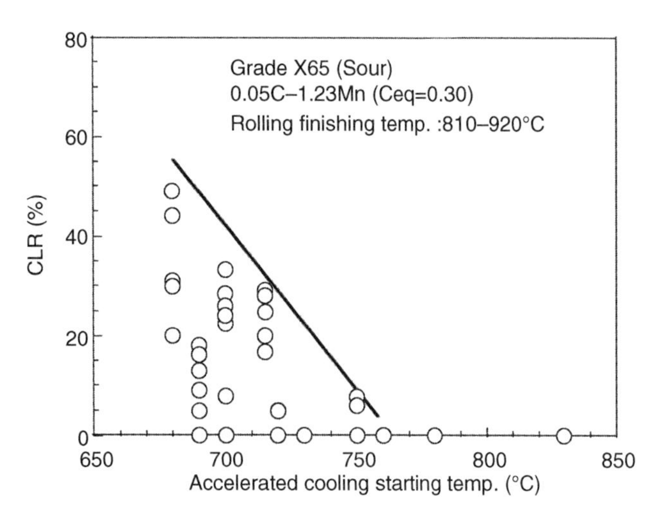


**FIGURE 17.11** Relation between accelerated cooling stop temperature and CLR [28].

The accelerated cooling stop temperature is also an important parameter to control. Same as the aforementioned, a hard second phase needs to be prevented for the HIC resistance, and there is the appropriate temperature range for the accelerated cooling stop temperature [38,39]. Higher cooling stop temperature may result in the formation of MA phase, as in the case of air cooled steel. On the other hand, strength of the steel is increased because of the increase of volume fraction of the harder transformed microstructure, which is susceptible to HIC. Therefore, plate manufacturing conditions should be carefully chosen for balancing the strength, toughness, and HIC resistance of the steel.

## REFERENCES


1. Ikeda, A. (1984) Hydrogen induced cracking of steels in wet hydrogen sulfide environment. *Tetsu-to-Hagane*, 70(8), 26–36.
2. Tresender, R.S. and Swanson, T.W. (1968) Factors in sulfide corrosion cracking of high strength steels. *Corrosion*, 37(2), 31–37.
3. Greer, J.B. (1977) Results of interlaboratory sulfide stress cracking using the NACE-1F-9 proposed test method. *Materials Performance*, 16(9), 9–19.
4. Taira, T., Tsukada, K., Kobayashi, Y., Inagaki, H., and Watanabe, T. (1981) Sulfide corrosion cracking of linepipe for sour gas service. *Corrosion*, 37(1), 5–15.
5. NACE MR0175/ISO 15156-2:2003, "Petroleum and natural gas industries—Materials for use in $H_2S$-containing environments in oil and gas production—Part 2: Cracking-resistant carbon and low alloy steels, and the use of cast irons," NACE International/ISO, 2009.
6. Bruno, T.V. and Hill, R.T. (1980) Stepwise cracking of pipeline steels—a review of the work of task group T-1F-20. NACE Conference Corrosion/80, 1980, Chicago, IL, USA, Paper No. 6.
7. Moore, E.M. and Warga, J.J. (1976) Factors influencing the hydrogen induced cracking sensitivity of pipeline steels. *Materials Performance*, 15(6), 17–23.
8. Miyoshi, E., Tanaka, T., Terasaki, F., and Ikeda, A. (1976) Hydrogen-induced cracking of steels under wet hydrogen sulfide environment. *Transactions of the ASME Series B*, 98, 1221–1230.
9. Kalwa, G., Pöpperling, R., Rommerswinkel, H.W., and Winkler, P.J. (1976) Steels with special properties for linepipes and structures in offshore application. Offshore North Sea (ONS) Conference, September, 21–24, 1976, Stavanger, Norway.
10. Iino, M. (1978) The extension of hydrogen blister-crack array in linepipe steels. *Metallurgical and Materials Transactions A*, 9, 1581–1590.
11. Inagaki, H., et al. (1977) Hydrogen induced cracking of line pipe for sour gas service. *Piping Engineering*, 14(4), 79.
12. Terasaki, F., et al. (1978) The hydrogen induced cracking susceptibility of various kinds of commercial rolled steels under wet hydrogen sulfide environment. *Sumitomo Metals*, 30(1), 40.
13. Nakasugi, H., Sugimura, S., Matsuda, H., and Murata, T. (1979) Development of new linepipe steels for sour service. *Nippon Steel Technical Report*, 297, 72–83.
14. Herbsleb, G., Poepperling, R.K., and Schwenk, W. (1981) Occurrence and prevention of hydrogen-induced stepwise cracking and stress corrosion cracking of pipeline steels. NACE Conference Corrosion/80, March 3–7, 1980, Chicago, IL, USA, Paper No. 9, see also: Corrosion NACE 37 (1981), No. 5, pp. 247–56.
15. Taira, T., Tsukada, K., Kobayashi, Y., Tanimura, M., Inagaki, H., and Seki, N. (1981) HIC and SSC resistance of line pipes for sour gas service—development of line pipes for sour gas service—part 1, *Nippon Kokan Technical Report Overseas*, 31, 1–13.
16. Jones, B.L. and Gray, J.M. (1993) Linepipe development toward improved hydrogen-induced cracking resistance. Proceedings of 12th International Conference of Offshore Mechanics and Arctic Engineering, Glasgow, Scotland, UK, 329–336.
17. NACE Standard TM0284-2011, Evaluation of Pipeline and Pressure Vessel Steels for Resistance to Hydrogen-Induced Cracking, NACE International, Houston, TX, USA, (2011).
18. European Federation of Corrosion. (2009) *Guidelines on Materials Requirements for Carbon and Low Alloy Steels for $H_2S$-Containing Environments in Oil and Gas Production*, 3rd ed., EFC Publication No. 16, Maney Publishing, Leeds, UK.
19. Taira, T., Kobayashi, Y., and Ichinose, H. (1982) Observation of cracks in line pipe by full size SSC test. *Nippon Kokan Technical Report*, 34, 14–26.
20. Hyodo, T., Iino, M., Ikeda, A., Kimura, M., and Shimizu, M. (1987) The hydrogen permeation and hydrogen-induced cracking behavior of linepipe in dynamic full scale tests. *Corrosion Science*, 27, 1077–1098.
21. Provou, Y. (1988) Full-scale and laboratory tests approach to $H_2S$-induced environmental fracture of pipeline steels and welds, Technical Steel Research, Report EUR 13514EN, Commission of the European Communities, 1988.

22. Christensen, C. and Hill, R.T. (1989) Full scale test of line pipe steels for sour service. NACE Conference Corrosion, New Orleans, LA, USA, Paper No. 89479.
23. Provou, Y., Bennett, C., Brown, A., Pöpperling, R., and Pontremoli, M. (1990) Testing linepipe steels under sour gas conditions—comparison of the results of full scale and laboratory tests. Pipeline Technology Conference, Oostende, Belgium 15–18.
24. Takeuchi, T., Kushida, T., Okaguchi, S., Yamamoto, A., and Miura, M. (2004) Development of high strength line pipe for sour service and full ring evaluation in sour environment. Proceedings of 23rd International Conference on Offshore Mechanics and Arctic Engineering, Halkidiki, Greece, 2004, Paper No. OM AE2004-51028.
25. NACE Standard TM0177-2009, Laboratory Testing of Metals for Resistance to Sulfide Stress Cracking and Stress Corrosion Cracking in $H_2S$ Environments, NACE International, Houston, TX, USA, (2009).
26. Liessem, A., Schwinn, V., Jansen, J.P., and Poepperling, R.K. (2002) Concepts and production results of heavy wall linepipe in grades up to X 70 for sour service. Proceedings of 4th International Pipeline Conference, Calgary, Canada, 2002, Paper No. IPC2002-27139. Published by ASME.
27. Burk, J.D. (1996) Hydrogen-induced cracking mechanisms, inspection, repair, and prevention, *SPE Production & Facilities*, 11(2), 49–53.
28. Ishikawa, N., Shinmiya, T., Endo, S., Wada, T. and Kondo, J. (2003) Recent development in high strength linepipe for sour environment. Proceedings of the 22nd International Conference on Offshore Mechanics and Arctic Engineering, Cancun, Mexico 2003, Paper No. OMAE2003-37065.
29. Kane, R.D. and Ribble, J.P. (1990) Corrosion of microalloyed steel weldments: HIC/sulfide stress cracking. Proceeding of Conference on the Metallurgy, Welding and Qualification of Microalloy (HSLA) Steel Weldments, Houston, TX, USA.
30. Luyckx, L. (1970) Sulphide shape control in low alloy steels. *Metallurgical Transactions*, 2, 3341.
31. Pircher, H. and Klapdar, W. (1975) Controlling inclusions in steels by injecting calcium in the ladle. *Microalloying*, 75, Union Carbide Corporation, New York, USA, 232.
32. Haida, O., Emi, T., Shiraishi, T., Fujiwara, A., and Sanbongi, K. (1978) Optimizing sulfide shape control in large HSLA steel ingots by treating the melt with calcium or rare earths. *Tetsu-to-Hagane*, 64, 1538–1547.
33. Amano, K., Kawabata, F., Kudo, J., Hatomura, T., and Kawauchi, Y. (1990) High strength steel line pipe with improved resistance to sulfide stress corrosion cracking for offshore use. Proceedings of 9th International Conference on Offshore Mechanics and Arctic Engineering, Houston, TX, USA, V, 21–26.
34. Matsumoto, K., Kobayashi, Y., Ume, K., Murakami, K., Taira, T., and Arikata, K. (1985) Effect of center segregation on hydrogen induced cracking susceptibility of high grade line pipe steels. NACE Conference Corrosion/85, Boston, MA, USA, Paper No. 239.
35. Ohtani, H. (1986) Development of low Pcm high grade linepipe for arctic service and sour environment. Proceeding of the Conference on Technology and Application of HSLA Steels, ASM, Philadelphia, PA, USA, 843.
36. Jacobi, H. (1999) Solidification structure and micro-segregation of unidirectionally solidified steels. *Steel Research*, 70, 362.
37. Ume, K. and Arikataq, K. (1986) High strength line pipe for sour gas service. *Nippon Kokan Technical Report*, 46, 106–108.
38. Tamehiro, H., et al. (1985) Effect of accelerated cooling after controlled rolling on the hydrogen induced cracking resistance of line pipe steel. *Transactions ISIJ*, 25, 982–988.
39. Kushida, T., Kudo, T., Komizo, Y., Hashimoto, T., and Nakatsuka, Y. (1988) Line pipe by accelerated controlled cooling process for sour service. *The Sumitomo Search*, 37, 83–92.

# 18

# PIPELINE WELDING FROM THE PERSPECTIVE OF SAFETY AND INTEGRITY*

DAVID DORLING[1] AND JAMES GIANETTO[2]
[1]*Starco Engineering, Calgary, Alberta, Canada*
[2]*CanmetMATERIALS, Natural Resources Canada, Hamilton, Ontario, Canada*

## 18.1 INTRODUCTION

A pipeline transportation system will most likely consist of gathering lines, a transmission system, and a distribution system. The gathering and distribution pipelines will be of a smaller diameter and will operate at lower pressures than the transmission line or lines.

The smaller diameter pipelines (<406-mm diameter) are, in general, welded with conventional welding processes, such as manual shielded metal arc welding (SMAW). Design and operating stresses are relatively low, and there should be little difficulty in meeting the strength and toughness requirements. The welding of these pipelines is, in most cases, adequately addressed in national pipeline standards.

For larger diameter transmission systems, material properties and subsequent weld metal property requirements may vary significantly. Economic considerations have resulted in the grade of pipe for long-distance pipeline construction increasing from X52 to the X80 in common use today. X100 and X120 pipe grades are in the later stages of development and implementation. This has been facilitated by advances in steelmaking and casting and the concurrent development of thermomechanically controlled processing (TMCP) of microalloyed steels.

For larger diameter pipelines, SMAW with cellulosic electrodes has also proven to be very effective and, as the pipeline industry has traditionally been an extremely conservative one and despite the fact that cross-country pipeline construction essentially involves the repetition, from 40 to 80 times/km, of the same joint, manual SMAW is still widely applied. The use of mechanized gas metal arc welding with narrow weld bevels considerably reduces the volume of welding required, and the productivity is considerably higher than that achievable with cellulosic SMAW electrodes. Since the early 1970s, mechanized, short-circuiting gas metal arc welding (GMAW-S) and, more recently, pulsed gas metal arc welding (GMAW-P) have been applied to large-diameter construction projects, both cross-country and, more particularly, offshore. The limitations of manual SMAW are strongly evident under the restrictive circumstances of lay-barge operation.

For traditional stress-based designs, minimum quality levels with respect to welding are incorporated into pipeline construction by way of regulatory requirements, codes, standards, and company specifications. Some pipeline systems, for example, onshore pipeline systems in areas of seismic activity or discontinuous permafrost and offshore pipelines laid by the reel-lay method, may be expected to experience higher longitudinal strains than the vast majority of existing pipelines, and this has implications with respect to the development, qualification, and application of the pipeline welding procedures.

Leaks and ruptures caused by defects introduced by welding are in the minority of pipeline incidents by far. Axial or longitudinal stresses from pressure loading are



*Oil and Gas Pipelines: Integrity and Safety Handbook*, First Edition. Edited by R. Winston Revie.

significantly lower in a completed pipeline than those in the circumferential or hoop direction. Discontinuities oriented in the axial direction, such as corrosion defects and stress corrosion cracks, have been a much bigger concern for pipeline owners and operators as potential failure of these axial-oriented discontinuities is driven by hoop stress. Pipeline incidents that do occur as a result of welding almost invariably arise not from a lack of knowledge but from failure to apply that knowledge.

## 18.2 CONSTRUCTION WELDING APPLICATIONS

The construction of a long-distance, cross-country pipeline is carried out by specialized crews generally using purpose-built equipment in a series of efficiently sequenced operations moving along the pipeline [1]. This is the "spread" method of pipeline construction that utilizes the principles of a production-line system; but, in the case of a pipeline, the product is static, and the workforce moves along the pipeline right of way. Welding is only one of the many operations and has to be undertaken in such a way as to maintain pace with and not impact the preceding and subsequent operations and crews.

The sequence of events on the main welding line or firing line of a submarine pipeline, pipe-lay vessel is practically identical to onshore pipeline construction. The main difference is that instead of each welding workstation moving along the pipeline, the joint being welded moves to successive welding stations as the vessel moves forward in steps corresponding to the length of the prefabricated pipe joint or pipe string [2].

A variety of conventional welding methods can and are being used for double-joint, mainline, tie-in, and repair welding. A number of less conventional methods are being actively pursued. The characteristics of the traditional welding processes, and some of the more innovative joining processes, are described by Dorling and Rothwell [3] and Jones and Hone [4].

### 18.2.1 Double-Joint Welding

The double-jointing process is the practice of welding two pipes together to form, for example, a single 24-m section from two 12-m pipe lengths. Double-joint welding is carried out prior to transporting to and stringing the pipe on the pipeline right of way. The advantage of double jointing is that half as many mainline welds are now required on the pipeline right of way. Double jointing is carried out at the pipe mill by the pipe manufacturer or at a pipe stockpile site by either the pipeline contractor or a double-jointing contractor. For submarine pipeline construction, double jointing of the pipe is carried out on board the pipe-lay vessel in very much the same way as it is done onshore [2].

As it is possible to rotate the double joint, a common practice is to use submerged arc welding (SAW), which has the advantage of high weld metal deposition rates with consistent weld bead profiles and properties for pipe materials to X80 for stress-based designs. SAW produces no spatter or visible arc and a relatively small amount of fumes. The process is very stable at high currents. The simplest form of the SAW process uses a single wire and a direct current (DC) power source; however, multi-torch and multi-wire systems are available [5].

Although not ideal from a productivity standpoint, double-joint welds can be completed using processes and procedures similar or identical to those qualified for mainline welding. They can be used in the 5G (fixed) position, as in an offshore pipeline spool base, or adapted for 1G (roll welding) as the pipe can be rotated. As with SAW, industrial multi-torch and multi-wire gas metal arc welding (or flux-cored arc welding) equipment are available if higher productivity is required.

### 18.2.2 Mainline Welding

For onshore pipeline construction, mainline welding refers to the welding of the pipeline in continuous lengths or sections between features such as roads, railway tracks, rivers, and other obstacles that may prevent the pipeline from being continuously installed in the trench. Completing this operation in a cost-effective manner relies on the continuous advance and optimum utilization of the welding crew, and any delays to this continuous process have to be minimized. Some joints will have to be left because they cannot be physically made at the time, others because the making of particular nonstandard welds, such as those required for heavy-wall pipe required at a crossing, will cause a disproportionate delay to the mainline production.

Once completed, the welds are inspected, coated, and the continuous lengths or sections of pipeline are lowered into the ditch using several side-boom tractors working together.

Two aspects of the mainline welding method determine the economics of pipeline construction:

1. The root-pass welding speed (and the root-pass percentage completion requirements) governs the overall productivity of the pipeline construction spread.
2. The fill-pass welding deposition governs the number of welding stations needed to maintain pace with the root pass.

Although the fill passes do not control the progress of pipeline construction, the efficiency of fill and cap processes determines the overall joint completion rates and the number of fill stations and pairs of welders required. They take on increasing significance as the pipe wall thickness and diameter increase. An additional factor to consider in the case of submarine pipeline construction is that the number of welding workstations on a lay barge is limited by the space available.

By correct choice of consumables and welding technique, the SMAW process can be used in all welding positions and will allow a wide range of property requirements to be met. It is thus an extremely versatile process; however, it is very dependent on the welder's manual skills for the attainment of defect-free welds having acceptable properties, and its productivity is inherently limited by its intermittent nature. SMAW with cellulosic stick electrodes is still widely used today for general pipeline applications, including mainline welding. Welds made using cellulosic electrodes can also be susceptible to hydrogen-assisted cracking, as will be discussed later.

Self-shielded, flux-cored arc welding (FCAW-S) wires are available for most pipe steels in common use today [6,7]. The products are specifically designed for pipeline welding with a vertical-down progression. They require no external gas shielding and, therefore, are ideally suited to manual welding on the pipeline right of way. Higher weld metal deposition rates can be achieved that, together with a higher duty cycle by virtue of the continuous nature of the consumable, results in a reduction in total welding time when compared with SMAW. Although attempts have been made to produce a wire for root bead welding, FCAW-S is primarily used for fill and cap passes and productivity in terms of welds per day, at best remains unchanged.

The continuous nature of the gas metal arc welding process and the virtual absence of slag covering on the completed weld lead to high productivity and a process ideally suited to mechanization and automation [8]. It is also inherently low hydrogen in nature. First used in 1969, the mechanized gas metal arc welding process has become the standard for major, large-diameter, cross-country pipelines in Canada and is increasingly being used worldwide.

Mechanized gas metal arc welding systems for mainline welding use lightweight tractors or "bugs" running on a band to align and carry the welding torches from the top to the bottom of the pipe. Small-diameter wires are used at a relatively high current (to give high metal deposition rates) with carbon dioxide or argon–carbon dioxide shielding gas mixtures. Welding progression is vertical down in a reduced-gap, compound bevel that is accurately machined on the pipe ends immediately ahead of the welding crew. The root bead is either completed with welding heads incorporated into the internal line-up clamp to produce a root bead from the inside of the pipe or all passes are completed externally, with the root bead being run on to a copper backing bar, which is incorporated into the internal line-up clamp. Shelters or "shacks" provide protection from the elements.

For mainline welding, mechanized gas metal arc welding increases the number of welds that can be achieved in production on a daily basis and significantly reduces the welding crew size required to complete these welds. It has been demonstrated to be capable of upwards of 140 defect-free welds per day under ideal conditions. Conventional, single-wire, mechanized GMAW-S was, up until recently, the most widely used mechanized process. However, the technology has now progressed to fully automated, higher productivity, multi-torch or multi-wire GMAW-P [9]. Most mechanized pipeline welding equipment suppliers have dual-torch welding systems that incorporate remote wire feeding and a pendant control and also the facility for through-arc sensing to guide the welding torch in the weld bevel. Multi-wire or tandem GMAW-P systems, which differ from conventional GMAW-S or GMAW-P as two welding wires are passed through the same welding torch, have been used in recent construction projects [10–12].

Efforts continue to be made to develop alternative methods of welding pipelines of all size ranges [13–16], with the aim of not only improving economics and efficiency, but also of providing weldment properties that are engineered for specific project requirements and reproducibly maintained in the field. The growing trend towards higher strength pipe steels, a fitness-for-purpose approach to weld flaw acceptance, and the application of strain-based pipeline designs places increasing importance on the second of these factors.

Lay-barge construction is most frequently used for submarine pipelines. The most common installation methods are S-lay, J-lay, and reeling. The methods, their history, and their application are described in some detail by Jensen [17].

With S-lay installation, the pipe is welded in the horizontal position, and the completed pipeline is laid on to the seabed over the stinger at the stern of the vessel. The stinger controls the curvature of the pipe from the horizontal to the inclined section. The pipeline takes an S shape from the vessel to the seabed. The welding of an offshore pipeline is, in many respects, not that much different from that of a cross-country pipeline. The root-pass welding speed governs the overall pace of construction. The wall thickness of an offshore pipeline is significantly greater than most onshore pipelines, and the number of workstations for filling and capping has to be adjusted within the confines of the space available to maintain pace with the root bead welding. Nevertheless, the same welding processes can be used on a lay barge as on a pipeline right of way.

In the J-lay method, the onshore prefabricated pipe lengths are welded on a vertical or near-vertical ramp and deployed almost vertically until it reaches the seabed where it assumes the J shape. The J-lay method avoids some of the difficulties of S-lay, such as tensile load forward thrust, and can be used in deeper waters. Onshore prefabrication of the quadruple- or hex joints used for J-lay installation can be made by mechanized, fixed-position GMAW-S or GMAW-P, or, as it is possible to rotate the joint, by SAW. On board the J-lay vessel, automated, vertical position GMAW-S or GMAW-P circumferential welding takes place in one welding station using between two and four welding heads in a carousel.

In the reeling method, the pipeline is installed from a large reel mounted on the pipe-lay vessel. Sections of the pipeline are welded at an onshore spool-base facility and spooled onto the reel. The vessel transports the pipeline to the installation site and deploys it to the seabed. Horizontal reels lay the pipe with an S-lay configuration; vertical reels most commonly deploy the pipeline in a J-lay configuration, but can also deploy in an S-lay mode.

### 18.2.3 Tie-In and Repair Welding

On a cross-country pipeline construction project, a tie-in is any joint that connects two pipeline sections at road and river crossings and at points along the right of way where access has been required to both sides (work side and ditch side) of the pipeline. A tie-in also connects the pipeline to prefabricated assemblies or to existing pipeline systems. These tie-in welds and, often the welds required to fabricate the short sections of pipelines to be inserted at these points, are, as mentioned earlier, not completed by the mainline welding crew in order to maintain optimum utilization of front-end welding personnel and equipment and maximize daily weld production.

Dedicated crews and equipment typically following some distance behind the mainline welding crew carry out tie-in and repair welds. Tie-in welds are normally performed with the pipe already in the trench. All operations are therefore carried out externally, and the accuracy of cutting, preparation, and alignment of the pipe ends prior to welding is critical.

For tie-in and repair welding in all but the higher strength pipe steels, the practice of using the SMAW process with cellulosic consumables throughout produces acceptable welds; nevertheless, as will be addressed in Section 18.5 of this chapter, the specification of welding procedures to avoid hydrogen-assisted cracking must take into account all the relevant field variables. For higher strength steels and heavier wall thicknesses, there is a greater susceptibility to hydrogen-assisted cracking, and the use of cellulosic electrodes requires even more care. Low-hydrogen, vertical-down SMAW consumables are available; however, while adequate for filling and capping, they tend to behave poorly for root bead welding when the fit up is less than ideal, and, consequently, productivity tends to be low. For up to X80 for stress-based designs, low-hydrogen, vertical-down SMAW is commonly used for hot, fill, and cap passes for both tie-in and repair with cellulosic SMAW for the root pass (i.e., a combination process) [18,19]. Traditional low-hydrogen, vertical-up SMAW consumables are available specifically for root bead applications and are frequently used, although it can prove challenging to master the techniques required.

For tie-in fill and cap passes, manual FCAW-S has been used for approximately 15 years [19]. Strength and toughness limitations have, in the past, restricted its application to X70 and below; however, consumables aimed at higher strength steels and low-temperature toughness applications are now available. Manual, gas-shielded, flux-cored arc welding (FCAW-G), with rutile cored wires, is being increasingly utilized for fill-pass applications in short sections of pipeline, as well as tie-in and repair welding for large-diameter pipeline projects where significant productivity improvements over SMAW can be realized [18]. The mechanization of FCAW-G for fill and cap passes offers the potential for further improvements in productivity, and mechanized FCAW-G has been used worldwide for a number of years for tie-in welding and the welding of short sections. The process is also seeing increasing use on mainline welding applications where flexibility of operation is more important than daily weld production [20,21]. Most of the pipeline-welding equipment manufacturers provide external welding systems for FCAW-G, and all are very capable for mainline, short-section, and tie-in welding.

Novel flux-cored arc welding wire formulations for the welding of higher strength pipe are becoming available. These new formulations result in outstanding weld metal strength and toughness while appearing to depart from the typical mechanisms of microstructural development [22,23].

Controlled GMAW-S is increasingly being used as an alternative root bead welding process, especially for higher strength pipe. The Lincoln Electric Company Surface Tension Transfer (STT™) [24], Miller Electric Regulated Metal Deposition (RMD™), and Fronius Cold Metal Transfer (CMT™) have been used for deposition of full-penetration root beads in conjunction with manual and mechanized FCAW-G on a number of pipeline projects [20,21,25]. The STT, RMD, and CMT process options are considered to be advantageous for the welding of higher strength steels because of their low-hydrogen potential and the ability to alter the deposited weld strength through selection of different welding wires and shielding gas compositions.

On a submarine pipeline, tie-in of pipeline segments and connection to surface facilities is completed by either welding or with mechanical connectors. The tie-in operations can be performed on board the pipe-lay vessel (in which case welding is preferred) or underwater. In the case of a tie-in welded underwater, hyperbaric welding is the more common method [2].

On the seabed, the pipeline segments lay parallel to each other and overlap. The ends of each segment are cut and a welding plug is then inserted into each end of the pipeline segments. The plugs are inflated for a perfect seal. The pipe ends are then beveled with a beveling machine to prepare them for welding. Pipe handling frames lift the pipeline ends and align them ready for welding. The welding habitat is then lowered and positioned precisely over the two pipe ends.

Compressed diving gas is used to push water out of the opening at the bottom of the habitat to make it a dry zone where the divers can work without diving equipment. The previously installed seals separate the water in the pipeline segments from the dry area of the welding habitat. Before welding begins, the beveled pipe ends are measured to ensure they meet specification. Divers, who are qualified welders, complete the welding or, in some cases, the entire process can be controlled remotely from on board the dive support vessel, in which case the divers monitor this process from within the welding habitat. The completed weld is inspected using ultrasonic inspection. If the weld meets the requirements, the welding habitat is lifted back aboard the vessel. The pipe handling frames lower the pipeline back to the seafloor, and the rest of the equipment is retrieved.

## 18.3 NONDESTRUCTIVE INSPECTION AND FLAW ASSESSMENT

For pipeline sections that will be designed and built in accordance with standard industry practices (in conventional pipeline areas where pipeline designs are stress based), the determination of girth welding inspection methods and weld flaw acceptance criteria is well defined. Flaw acceptance criteria can be derived directly from the workmanship-based acceptance standards contained in the governing code or standard or alternative weld acceptance standards based on engineering critical assessment (ECA) can be developed in accordance with the procedures and requirements also included in pipeline codes and standards, for example, the American Petroleum Institute (API) standard API 1104 Appendix A, Canadian Standards Association (CSA) standard CSA Z662 Annex K, and Det Norske Veritas (DNV) Offshore Standard DNV-OS-F101 Appendix A. Alternative weld acceptance standards based on ECA allow engineers to assess the suitability of welds containing flaws for intended service conditions and can provide a more relaxed defect allowance than the workmanship criteria, without sacrificing the integrity of the pipeline.

Historically, film-based radiography has been used to inspect pipeline girth welds, and most workmanship-based flaw acceptance criteria are founded on the assumption that weld inspection is carried out using radiography.

A source of radiation is placed on one side of the pipe, and the film is placed on the other side. X-and gamma rays can pass through the metal, and a two-dimensional, gray-scale image is obtained of the transmitted radiant energy. X-or gamma-ray absorption depends strongly on the material density and volumetric defects, which have either a higher or lower density (such as porosity or most slag inclusions). The radiation that reaches the film in a potential flaw area is different from the amount that impinges on the adjacent areas. This produces on the film a latent image of the flaw that, when the film is developed, can be seen as an "indication" of different photographic density from that of the image of the surrounding material.

Digital radiography is one of the newest forms of radiographic imaging. Digital radiographic images are captured using special phosphor screens containing micro-electronic sensors. Captured images can be digitally enhanced for increased detail and are easily archived, as they are digital files. Real-time radiography (RTR) is the latest application for inspecting pipelines, which lets electronic images to be captured and viewed in real time allowing cycle times of 4 min or less.

Unfortunately, radiography only provides quantitative information about the flaw length and lateral position across the weld. It does not provide information about flaw height and, unless favorably orientated, tight, crack-like defects can be missed. In addition, many cosmetic imperfections are repaired in the absence of key height information.

Ultrasonic-based inspection methods are well suited to find planar or crack-like flaws and techniques such as automated ultrasonic testing (AUT) are increasingly being used to inspect pipeline girth welds, especially those produced by mechanized welding systems using complex bevel geometries. AUT provides more information about the flaw, especially flaw height and through wall position; these dimensions have a significantly greater effect on integrity than flaw length alone. AUT is currently employed for the inspection of both cross-country and submarine pipelines. Systems are usually configured to perform a fine-increment, zone-focus inspection. A time-of-flight diffraction (TOFD) arrangement may also be used in parallel.

TransCanada Pipelines and RTD of the Netherlands developed the zone-focus approach in the mid-1980s following many years of field and laboratory experimentation [26]. This approach has been applied by TransCanada since 1989 and adopted worldwide for both onshore and offshore pipeline construction. The weld bevel is divided into discreet zones, typically 1–3 mm high. The number of zones is determined by the weld bevel configuration, wall thickness of the pipe, and welding procedure. The first AUT systems used arrays of transducers with individual transducers tailored for and dedicated to the portion of the bevel being interrogated and the type of defect expected. The sound field characteristics in terms of angle, focus, and focal size of beams generated by conventional transducers are fixed by the transducer itself (frequency, size, lens, wedge) and cannot be changed without a physical change of the transducer. The weld is divided into two virtual halves—an upstream side and a downstream side. Up- and downstream sides are inspected with mirror-like sets of transducers.

Phased-array ultrasonic inspection systems have recently been introduced [27,28] that can generate a range of ultrasound beams from the same transducer, controlled dynamically in real time by software. Instead of a single element, a phased-array system uses multiple elements in a common housing. Each element is connected to a separate pulser and time delay generator and is pulsed using a focal law. This focal law defines which elements are to be pulsed, with what time delay. By adjusting the time delays, normal beams, angled beams, focused or unfocused beams, shear waves or longitudinal waves can be generated.

For pipeline applications, a linear array of up to 128 elements in a single housing is used. The elements are pulsed in groups of 8–16. Pulsing a group of elements in sequence along the length of the transducer produces an electronic (linear) scan, which achieves the same objective as a raster scan. The beam is steered by delaying the pulsing of each element electronically at a set rate. Focusing is achieved by varying the rate of delay applied in sequence to the selected elements, while the actual size of the focus is determined by the number and size of the elements used. These electronic focusing and steering techniques, together with shaping the elements in the circumferential axis, result in a proper sound field being generated at the weld bevel.

Phased arrays are also capable of sectorial scanning, which can improve sizing accuracy for poorly oriented flaws. Sectorial scanning sequentially sweeps the beam electronically through a range of inspection angles.

TOFD has proven itself to be a particularly valuable tool for detecting hydrogen-assisted cracking, especially during the inspection of shielded metal arc welds. TOFD is a time-based method relying on the diffraction of ultrasonic waves from the tips of discontinuities. This technique uses a pair of longitudinal wave transducers, in a transmit–receive configuration, that have a refracted angle between 45° and 70°. When the wave is incident on a linear discontinuity, such as a crack, diffraction takes place at the extremities of the flaw in addition to the normal reflected wave. This diffracted energy is emitted across a wide angular range. The TOFD method also uses transmitted and reflected waves as references to place all other signals in their correct position in the pipe wall, through-thickness direction. The first reference is a lateral wave traveling directly between the transducers just under the scanning surface, and the second reference is the reflection from inside surface. An imperfection in the through-thickness direction produces signals that occur in the time interval between the reference signals, and the height of the imperfection is directly related to the time separation between these diffracted waves.

The approach of using a combination of fine-increment, zone-focus inspection with sectorial scanning and TOFD provides several different and separate assessments of any flaws present, significantly improving the probability of detection and confidence level achieved with respect to detection and sizing accuracy [12].

As the flaws can now be quantified based on their position within the weld and their vertical height and length, they can be readily assessed using an alternative weld acceptance standard based on engineering critical assessment [29–31]. This avoids the unnecessary repair of those imperfections that do not pass workmanship-based acceptance criteria and, on any rational basis, will have no effect on the integrity of the pipeline.

Work directed towards the treatment of weld discontinuities by engineering critical assessment has been in progress for almost three decades [32]. An approach to brittle fracture based on the crack-tip opening displacement (CTOD) test was developed in the UK over a number of years, and encapsulated in the then British Standards Institute (BSI) PD 6493 in 1980 (the revised PD 6493 is now published as British Standard Guide BS 7910). This basic methodology was adapted in Canada for use on pipeline girth welds and combined with a check for plastic collapse failure based on a relatively simple "missing ligament" model. A rather extensive series of full-scale tests was undertaken on large-diameter girth welds containing known defects and this, together with analysis of some comparable data from the literature, allowed the effective flaw size curves from PD 6493 to be adjusted to give a more acceptable range of safety factors (ratio between observed and predicted failure strains). It also confirmed the conservatism of the plastic collapse approach. Many girth weld ECA procedures have their roots in the CTOD design curve and PD 6493. By the mid-1980s, BSI 4515 Appendix H, API 1104 Appendix A, and CSA Z662 Appendix K (later Annex K) were added to their respective main documents.

In carrying out an ECA, one of the key inputs is the height of the imperfection; if this is known, detailed calculations of tolerable flaw lengths can be made. Alternative weld acceptance standards, which are based on the engineering design aspects of the pipeline, can therefore make full use of the relationship between applied stresses or strain, material toughness, and flaw size when ultrasonic inspection is performed. If radiographic techniques are used, then assumptions on the height of the imperfection have to be made. Assuming that the height of the imperfection is limited to one weld pass may be appropriate for volumetric flaws but is questionable for planar flaws as they may extend for more than one weld pass. As planar flaws are most significant in terms of structural integrity, the use of radiography would lead to a very conservative alternative weld acceptance standard, together with the likelihood that significant defects might not be detected anyway. The design of the ultrasonic approach gives specific information about the height of the imperfection in any region and, therefore, a detailed analysis can be carried out to specify allowable imperfection lengths, which can be read directly from the charts produced by the ultrasonic unit's data acquisition system.

From an overall quality and process control perspective, another considerable benefit of AUT is its ability to stay directly behind the welding crew, where it can facilitate immediate identification, and hence rapid feedback and correction, of any recurring welding problems. Radiographic inspection is typically delayed because of radiation exposure concerns for nearby personnel.

## 18.4 WELDING PROCEDURE AND WELDER QUALIFICATION

Qualified welding procedures are necessary to fulfill regulatory requirements and provide guidance to project engineers, pipeline contractors, and welders. The welding procedures, when incorporated in project-specific joining plans, provide essential guidance to the pipeline contactor with respect to welding crew setup and production targets. This will help maximize productivity with the appropriate combination of welding process, weld deposition, travel speed, and process efficiency (weld-to-weld time) while meeting defined weld quality requirements.

### 18.4.1 Welding Codes and Standards

For traditional stress-based designs, minimum quality levels are incorporated into pipeline construction by way of regulatory requirements, codes, standards, and company specifications.

For example, in the United States, it is first necessary to comply with Title 49 of the Code of Federal Regulations (CFR), Part 192 (49 CFR Part 192), and in particular Subpart E "Welding of Steel Pipelines." API standard API 1104 "Welding of Pipelines and Related Facilities" is incorporated by reference at 49 CFR. §§ 192.7, 192.225, 192.227, and 192.229. In Canada, it is necessary to comply with the National Energy Board (NEB) "Onshore Pipeline Regulations" (OPR-99). CSA standard CSA Z662 "Oil and gas pipeline systems" is incorporated in the NEB regulations and provides a technical basis for OPR-99 by setting out the minimum technical requirements for the design, construction, operation, and abandonment of pipelines.

Pipeline operating companies often develop their own general or project-specific welding requirements to be used in conjunction with regulatory requirements and national codes and standards.

### 18.4.2 Welding Procedures

Documented, qualified welding procedures are a mandatory requirement of pipeline codes and standards such as CSA Z662, API 1104, and associated regulations. These standards require that the welding procedures be qualified prior to the commencement of welding operations and that welds be deposited only by welders trained and qualified in the welding operation defined by the qualified welding procedure. The welding procedure is the primary tool for controlling quality and achieving the design requirements in welded construction.

As part of the quality assurance program, an overall welding procedure qualification protocol will need to encompass all of the combinations of pipe (grade, diameter, mill, etc.) and welding procedure (process, consumables, heat input, etc.), which will be encountered on the pipeline construction project. The codes and standards provide guidance and minimum requirements as to the number of and how the procedure qualification weld or welds are to be made, the inspection to be carried out, and the subsequent assessment of flaws. Also stipulated will be the types of mechanical tests to carried out, the number and dimensions of mechanical test specimens required, their location in the weld, and the minimum values to be achieved. When working with typical code-established, workmanship-based flaw acceptance criteria, only tensile and bend testing may be required to qualify a procedure. API 1104 and CSA Z662 also include nick-break testing. More stringent industry codes, or the operating company itself, may include Charpy V-notch toughness testing and hardness testing of the weld cross section. If engineering critical assessment-based, alternative flaw acceptance criteria are specified, fracture toughness testing such as CTOD testing may be stipulated. Standard ECA applicability normally covers demands up to only 0.5% strain.

A qualified welding procedure also provides the basis for producing welds with the required properties in an efficient manner without giving rise to defects. Company-specific welding procedure specifications (WPSs) may contain supplementary requirements that are applicable to the welding processes to be used and the project design requirements. Further assurance of quality is provided by the routine monitoring of welding parameters in the field and through visual and nondestructive inspections.

For pipeline facilities and assembly construction, it may be permissible to qualify welding procedures in accordance with the requirements of the American Society of Mechanical Engineers (ASME) Boiler and Pressure Vessel Code. For example, ASME BPV Section IX is required in Canada, and ASME BPV Section IX and API 1104 are required in the United States. This applies to welds that are

1. In stations;
2. Made at a manufacturing plant or fabrication shop remote from the final location of the weld;
3. Joining pipe to components or components to components.

### 18.4.3 Welding Procedure Specification

A WPS is mandatory and is the basis of the work instruction given to the welder. This work instruction may be in the form of a WPS data sheet. The WPS must reference the procedure qualification record(s) (PQR) on which it is based and provide sufficient information to produce the required weld in a controlled manner. It must include all the information to control any parameter that may adversely affect weld quality and the ranges of essential variables that the welder must work within to meet the qualification requirements of the applicable code or standard and any company- or project-specific supplementary specifications. Where a specific welding application requires the use of welding parameters outside of the range of these essential variables, the welding procedure must be re-qualified.

### 18.4.4 Procedure Qualification Record

A PQR is mandatory and is the critical record in the development and qualification of the welding procedure. It is used as the basis for the completion of the WPS and is a confirmation that the welding procedure to be used will comply with the requirements of the applicable code or standard and any company-specific supplementary specifications. A PQR may be used as the basis for more than one WPS.

This document serves as a record of all measured welding parameters used in the production of the qualification test weld and the results of required nondestructive and destructive testing.

### 18.4.5 Qualification of Welders

Welder training and qualification is a code requirement and must be completed with the welding procedures that will be used on the project. The pipeline codes in North America (CSA Z662 and API 1104) only require the welder to complete one weld (or segment of a weld) to demonstrate their skill; if the weld passes the required testing, the welder is considered qualified. However, with the increasing sophistication of pipeline welding processes, the welder must be given the opportunity and time to become fully acquainted with how the equipment works and what limits he or she has control over. For mechanized welding, past experience has shown that a minimum of three, half-circumferential, training welds will need to be completed before presenting the welder for testing.

## 18.5 HYDROGEN CONTROL IN WELDS AND THE PREVENTION OF HYDROGEN-ASSISTED CRACKING

Hydrogen-assisted cracking in welds was first recognized as a problem in the late 1930s and has been extensively studied since that time. Hydrogen cracking during fabrication of steel structures still occurs today but on a much-reduced level [33]. In fact, most of these continuing occurrences arise not because of a lack of industry understanding or knowledge but because of a lack of application of this knowledge.

Cracking can occur in the heat-affected zone (HAZ) of the pipe or fitting material at a weld root, weld toe, or in an underbead position. Cracking can also occur in the weld metal itself. Hydrogen-assisted cracking in pipeline girth welds is usually oriented along the weld length, although, in thicker wall pipe or components, it can occur transverse to the weld. The cracks may be buried or they may be surface breaking.

Four conditions, occurring simultaneously, must be met in order to cause hydrogen cracking: a sufficient amount of diffusible hydrogen, a susceptible microstructure, a tensile stress, and a temperature at which the sensitivity to hydrogen embrittlement is sufficient to cause cracking. The cracking is delayed in nature and can occur several minutes to many hours after welding. These aspects are covered in detail elsewhere [34].

Weld metal and HAZ hydrogen cracking can be avoided, even in higher strength pipe materials, through pipe and pipe mill qualification, and the appropriate construction welding procedures and quality surveillance in the field.

There are at least three factors influencing the risk of hydrogen cracking, which can be considered specific to girth welds in pipe. As mentioned earlier, SMAW with cellulosic stick electrodes is still widely used for general pipeline applications, including mainline welding. These electrodes are capable of producing welds with 60–80 ml $H_2$/100 g of deposited weld metal [33]. The electrodes are used with a vertical-down welding progression, and the heat input is characteristically low, primarily because of the high welding speed demanded by the "stovepipe" welding technique. The heat input may be typically 0.4–0.8 kJ/mm, and this will give very fast cooling rates. Heat input takes into account the collective effect of amperage, voltage, and travel speed on the thermal cycle of the weld. The third factor peculiar to cross-country, mainline construction is that the root pass can be strained as the internal line-up clamp is released, and the pipe is lifted and lowered to reposition the pipe on its supporting skids ready for the next length of pipe to be added. With high-strength pipe in particular, a cellulosic root bead can crack at that point in time, unless precautions are taken, as the weld is already saturated with hydrogen and further accumulation is therefore not required.

Incidences of delayed hydrogen-assisted cracking can occur. Cracking in these circumstances occurs in multi-pass welds and repairs to multi-pass welds, where the late application of a lower level of stress coupled with the steady buildup of hydrogen in the weld are contributing factors.

The susceptibility of the pipe material to hardening in the HAZ as a result of these rapid weld cooling rates is governed by its composition and often predicted by the use of carbon equivalent formulae. These formulae take into account the important elements that are known to have an effect on hardenability and the tendency to form crack-susceptible microstructures. The most widely used is the International Institute of Welding formula ($CE_{IIW}$) adapted from the original hardenability equation developed by Dearden and O'Neill [35]:

$$CE_{IIW} = C + Mn/6 + (Cr + Mo + V)/5 + (Cu + Ni)/15$$

Its calculation and use are described in detail by Bailey et al. [34].

However, hardenability in steel is not necessarily an indicator of HAZ hardness [35]. It describes how easily the HAZ becomes martensitic. A reduction in carbon produces a decrease in hardness of martensite formed by rapid cooling. Pipeline steels have traditionally used leaner alloying coupled with advanced processing routes, and it follows that a carbon equivalent formula that regards carbon as more important is preferable, especially for modern, low-carbon pipe steels. Carbon equivalent formulae of relevance to pipeline welding are discussed in detail in the paper by Yurioka [36].

With the widely used Pcm formula, the effect of carbon is given much more significance than that of other alloying elements, and Pcm is generally considered to be appropriate for the more modern, carbon-reduced, or microalloyed steels:

$$\begin{aligned} Pcm = {} & C + Si/30 + (Mn + Cu + Cr)/20 + Ni/60 \\ & + Mo/15 + V/10 + 5B \end{aligned}$$

The American Welding Society (AWS) method for determining the minimum necessary preheat to avoid HAZ cracking recommends Pcm to be used for steels with $C < 0.11\%$ and $CE_{IIW}$ for steels with $C \geq 0.11\%$. DNV-OS-F101 "Submarine Pipeline Systems" includes both $CE_{IIW}$ and Pcm and their use depends on pipe material "delivery condition."

A formula developed by Yurioka et al. as a weldability index for a range of low-carbon and carbon–manganese steels incorporates an interactive term for carbon and alloying elements [37]. The $CE_N$ formula approaches the values of $CE_{IIW}$ when applied to higher carbon steels and the values of Pcm when applied to lower carbon steels:

$$\begin{aligned} CE_N = {} & C + A(C)\{Si/24 + Mn/6 + Cu/15 + Ni/20 \\ & + (Cr + Mo + Nb + V)/5 + 5B\} \end{aligned}$$

$$\text{Where}: A(C) = 0.75 + 0.25 \tanh\{20(C - 0.12)\}$$

The CSA oil and gas pipeline code, CSA Z662 includes a form of the $CE_N$ formula that uses a look-up table for the interactive term for carbon.

Predictive methods are of only limited value in the context of pipeline girth welding. In addition, most of the predictive methods are intended for the prevention of hydrogen-assisted cracking in the HAZ. As mentioned in an earlier section, advanced steelmaking and TMCP technologies have reduced the reliance on carbon content and alloying levels as strengthening mechanisms. Low carbon levels and lean carbon equivalents provide improved weldability in modern pipeline steels and, for higher strength pipe steels in particular, the prevention of weld metal cracking may be the greater challenge.

To further reduce the risk of hydrogen cracking, procedural controls, such as the use of preheat, hot-pass techniques, and limited time delay between initial passes, can be incorporated into welding procedures. In addition, procedures can call for the retention of the line-up clamp until the root pass is complete. These controls are used in combination to raise and maintain the temperature above the critical temperature for hydrogen-assisted cracking, until at least the hot pass is complete. Graphical methods are available to help determine preheat temperatures [34], but they do not take into account those peculiarities of mainline construction referred to previously.

The application of preheat also allows hydrogen to diffuse out of the weld and, the proper use of preheating, including maintaining this temperature for all weld passes and not just the root bead, promotes slow cooling and hydrogen diffusion even after weld completion.

Dedicated crews and equipment, typically following some distance behind the mainline welding crew, carry out tie-in and repair welding and field fabrication of pipe-to-pipe or pipe-to-component welds. These welds do not experience the early application of stress of mainline construction, and it is delayed hydrogen-assisted cracking that has to be prevented. Maintaining preheat for all weld passes is especially beneficial in these circumstances.

It should be recognized that a given preheat may vary widely in its effectiveness as a result of changes in ambient temperature, humidity, and wind speed. For the same pipe steel welded in an Arctic winter or the middle of a tropical summer, although both will have the same cracking sensitivity, the latitude in the welding procedures may be quite different, and the same controls should not necessarily be applied to both projects.

HAZ hardness is often used as an indicator of the susceptibility of a microstructure to cracking during the evaluation of procedure qualification welds. The generally regarded notion that 350 HV is a hardness level below which hydrogen cracking is not expected dates back to the work by Dearden and O'Neill in the 1940s [35]. This concept was validated by Bailey in the early 1970s for welds with a diffusible hydrogen content of approximately 16 ml/100 g of deposited weld metal [38]. Bailey also noted that for welds with a diffusible hydrogen content of approximately 8 ml/100 g, the critical hardness, or the hardness level below which hydrogen cracking is not expected, was in the order of 400 HV.

The development and specification of welding procedures to avoid hydrogen-assisted cracking rely to a great extent on the ability and experience of the welding engineer to take into account all the relevant field variables. Although not always mandatory, it is a good practice to qualify welding procedures on pipe that will be used on the project under realistic field conditions. Procedure qualification should be performed on the highest carbon equivalent on which the procedure will be used. If construction experience is limited or additional verification is required of the efficacy of the selected and qualified welding procedures in the project pipe, there are physical tests that have been developed to help predict cracking in pipeline welding situations [39].

The Welding Institute of Canada (WIC) restraint test is most applicable to mainline welding, where the root is subjected to an early application of high stress. The test provides a good simulation of the critical factors, and the results have been correlated with full-scale test results and field performance. Operating companies have been known to include WIC testing in their pipe mill quality assurance programs in addition to limits on chemistry/carbon equivalent. The aspect of delayed hydrogen-assisted cracking of multi-pass welds in the HAZ or weld metal can be evaluated with a simple bend test. Both tests and their application are described in detail by Graville [39].

Mechanized gas metal arc welding, as a result of its intrinsically low hydrogen content, has a low susceptibility to hydrogen-assisted cracking when appropriate controls on preheat and inter-pass temperatures and cleanliness are adhered to. Similarly, shielded metal arc and flux-cored arc welding consumables designated, tested, and certified as low hydrogen are also less susceptible. To further reduce the risk of hydrogen cracking with the flux-shielded processes, procedural controls, such as the use of preheat, hot-pass techniques, and limited time delay between initial passes, should still be incorporated into welding procedures. In addition, in order to minimize the hydrogen potential of these consumables in production, a comprehensive consumable quality assurance program addressing storage, deployment, and utilization will need to be followed by the pipeline contractor.

## 18.6 IMPORTANT CONSIDERATIONS FOR QUALIFYING WELDING PROCEDURES TO A STRAIN-BASED DESIGN

A stress-based design is primarily focused on ensuring that pipeline wall thickness is appropriate to meet the design pressure and is typically inadequate for pipelines subjected to large ground deformations. With a strain-based design (SBD), the design loads and capacities are quantified in terms of longitudinal strain, and the pipelines are designed to sustain a prescribed level of plastic strain without rupture. Key components of SBD and their relationship are described by Wang et al. [40,41]. The strain demand, or applied strain, may be estimated from conditions causing the longitudinal strains, and compressive and tensile strain capacity may be determined from their respective models and/or experimental test data. The strain demand and strain capacity are then compared to determine if an event or condition is safe. The mechanical properties, dimensional tolerance, and features of the pipe and welds, especially flaw dimensions, can impact all of the key components of SBD to varying degrees.

The tensile strain capacity of a pipeline is controlled by the tensile strain capacity of the girth welds (the entire weld region, including the weld metal, fusion boundary and the HAZ). Girth welds tend to be the weakest link due to the likelihood of welding discontinuities and metallurgical and mechanical property changes from welding thermal cycles. Welding specifications will need to include supplemental requirements in terms of mechanical tests and their execution, essential variables and their control during construction, NDE requirements and weld acceptance standards, all targeted at ensuring adequate tensile strain capacity.

Hukle et al. and Newbury et al. [42–44] outline a comprehensive approach to qualify materials and welding procedures based on the need to view the pipeline as an integrated system composed of weld metal, HAZ, and pipe material. The procedure qualification protocol suggested encompasses all of the combinations of pipe (grade, mill, diameter, etc.) and welding procedure (process, consumables, heat input, etc.) that will be encountered on a pipeline construction project.

One of the most critical factors for SBD is to ensure consistent and adequate weld metal strength overmatch. To guarantee overmatch of the pipe by the weld metal, during the complete range of pipeline service loads, it is important that the weld metal maintains not only similar yield point behavior and high uniform elongation, but also the weld metal stress–strain curve should overmatch the line pipe stress–strain curve at all strains in the design strain region. To meet this requirement, a more consistent and reliable all-weld-metal tensile testing protocol was developed to better assess the strength of pipeline girth welds, especially those made with a narrow gap weld preparation [45]. The other material parameter that is important in SBD is the fracture toughness or fracture resistance. Not unexpectedly, as line pipe strength increases, it becomes more challenging to select welding consumables with adequate strength to meet the requirement of always providing a significant overmatch and adequate toughness for each and every pipeline welding application.

In terms of pipe properties, it is common in traditional stress-based design to only specify minimum properties required of the line pipe to be used. As long as these minimum values are met, the structure will satisfy the design and be sound. While these minimum properties are important in SBD, it is of equal importance to specify maximum properties in such a way so as to ensure adequate pipeline behavior under yielding conditions. Wang et al. outline some of the key considerations in terms of tensile properties and their measurement that need to be addressed in line pipe specifications for SBD applications [46].

Once pipe materials are defined for the project, welding processes, procedures, and consumables are selected for the various production welding applications. Where possible, emphasis should be placed on developing procedures that use low-hydrogen consumables. Welding procedures must then be qualified to ensure that the weld and HAZ will meet the strain capacity requirements of the design. As the welding heat input has a large effect on the size of the HAZ as well as the degree of properties change within weld and the HAZ, for each welding procedure required both a low and a high heat input weld may need to be produced and tested to determine the heat-input boundary limits, and the tolerance on individual welding parameters, for any given welding procedure [42–44]. To ensure that the mechanical properties of as-installed welds are consistent and strain capacity requirements are met, it will also be important to control and minimize the natural property variation, especially as pipe grade increases and the degree of weld overmatch becomes more challenging. Tighter control of welding parameters (e.g., heat input) and weld joint setup (e.g., root opening and land thickness) will be required than that currently practiced in industry.

Small-scale testing that is supplementary to the requirements for a stress-based design will be required for both strength and toughness [46–49]. Weld metal tensile and weld metal and HAZ toughness tests and test protocols specifically for welding procedure qualification for a SBD can be found in the work by Wang et al. [46,47], Cheng et al. [48], and Moore and Pisarski [49]. Tyson addresses fracture toughness testing for both stress-based and strain-based designs in his chapter in this handbook.

It is also important to monitor hardness in the weld and HAZ. If the hardness in this region drops significantly, strain localization can occur. If hardness increases significantly, problems with cracking can arise.

After small-scale testing of the weld and HAZ, the strain capacity must be confirmed using large- and/or full-scale testing. This can be achieved for some applications through curved wide-plate (CWP) testing of each weld procedure at the low and high heat inputs. CWP testing has been widely used for determining girth weld tensile strain capacity [42–44]; however, it does not account for the biaxial loading (pressure in combination with longitudinal service strains) [50,51] and girth weld high–low misalignment [52] in the installed pipeline. It would appear that the tensile strain capacity of a welded pipeline is best characterized by pressurized full-scale tests with tension or bending loading [53].

To simplify the process of SBD and to optimize welding procedure qualification and large- and/or full-scale testing requirements, strain capacity can be predicted through finite element analyses (FEA) or simplified equations based on input parameters such as pipe geometry, internal pressure, material properties, girth weld defect size, and high–low misalignment [41,54,55]. This technology can also be used to calculate critical girth weld defect size with target strain capacity as an input. In this way, strain capacity prediction technology can be used to calculate flaw acceptance criteria and, thus, for engineering critical assessments. Full-scale tests of welded pipelines are also essential to validate predictive tensile strain capacity methodologies.

Procedures for tensile strain design are included in CSA Z662, Annex C for onshore pipelines and DNV OS-F101 for submarine pipeline construction. These procedures are useful in certain circumstances but with limitations. DNV OS-F101 refers to DNV RP-F108 for specific guidance regarding testing and ECA procedures for pipeline girth welds subjected to cyclic plastic deformation, for example, during installation by the reeling method, as well as guidance for other situations with large plastic strains. A paper by Pisarski reviews this approach in some detail [56].

In terms of field welding, as with any pipeline-welding project, whether to a stress-based design or SBD, it is vital that welding procedures are thoroughly qualified to meet all the requirements stated in the governing standard and any supplementary requirements in the project welding specifications. As already mentioned, allowable parameter ranges for the essential welding variables may well be tighter than those required by code and must be strictly adhered to; however, it is also important to make sure the welding and construction requirements imposed by the design are realistic in terms of being achievable at a reasonable cost. For example, requirements for very tight fit-up tolerance or very small flaw size can provide higher strain capacity but, if taken to an extreme, they make field welding and inspection very difficult and can negatively impact construction schedule and costs.

## 18.7 WELDING ON IN-SERVICE PIPELINES

Welding on pipelines while they are in service is carried out to add a branch connection to an existing pipeline for the purpose of adding new supply, meeting new demand, or in order to isolate a portion of the line for maintenance reasons. Small-diameter connections, such as olets, may be added for measurement and recording purposes. Welding can also be used for the repair of corrosion or mechanical damage through the installation of a repair sleeve or through direct deposition of weld metal [57].

There are significant economic incentives for performing pipeline repair and maintenance without removing the pipeline from service. A shutdown involves revenue loss from the loss of pipeline throughput, in addition to that from the gas vented to the atmosphere. Since methane is a so called greenhouse gas, there are also environmental incentives for avoiding the venting of large quantities of gas into the atmosphere.

Full-encirclement, welded steel sleeves are widely used for the repair of pipe defects in onshore pipelines. There are two basic types of full-encirclement sleeves: Type A and Type B. Type A sleeves are for structural reinforcement only and do not need to be fillet welded to the in-service pipeline. Type B sleeves are also for structural reinforcement but can also contain a leak. As they are pressure containing, Type B sleeves must either have fillet welded ends or mechanical seals (widely used offshore) and are suitable for damage such as internal corrosion that may become a leak, or damage that is already leaking. A properly designed, fabricated, and installed steel sleeve can restore the strength of a defective piece of pipe to at least 100% of specified minimum yield strength (SMYS).

The sleeve should be designed to the same standard as the carrier pipe. It is acceptable to use a sleeve that is thicker or thinner than the carrier pipe and is of lesser or greater yield strength than the carrier pipe as long as the pressure-carrying capacity of the sleeve is at least equal to that of the carrier pipe. Many companies simply match the wall thickness and grade of the pipe material.

The Type B sleeve is installed by clamping two half shells to the pipeline, welding them together using full-penetration butt welds, and then fillet welding the sleeve ends to the pipeline.

The diameter of the sleeve is slightly greater than that of the carrier pipe so it fits over the carrier pipe. Usually, this point is ignored in the sleeve design even though it causes the sleeve to be slightly under-designed when made from the same material as the carrier pipe. The carrier pipe seam weld reinforcement is often removed; however, if material is removed from the sleeve for a groove to accommodate the carrier pipe seam weld or a backing strip for the sleeve side seam welds, the thickness of the sleeve should be greater than that of the carrier pipe by an amount that compensates for the material that is to be removed.

Pipeline repair by direct deposition of weld metal, or weld deposition repair, is an attractive alternative to the installation of full-encirclement sleeves or composite reinforcement for repair of wall loss defects on in-service pipelines [57,58]. This is especially true for wall loss in bend sections and fittings, where the installation of full-encirclement sleeves and composite reinforcement is difficult or impossible. Weld deposition repair is attractive because it is direct, relatively quick and inexpensive to apply, does not create additional corrosion concerns, and requires no additional materials beyond welding consumables.

Branch connections on operating pipelines and assemblies are, more often than not, installed by the hot-tapping process. In the case where a branch connection is smaller than nominal pipe size (NPS) 2, weldolets, sockolets, threadolets, elbowlets, or heavy couplings will be used depending on whether or not the branch connection will be subject to external static or vibration loads. For branch connections NPS 2 and larger to an operating pipeline or assembly, a full-branch or reduced-branch split tee is most commonly used. The split-tee configuration acts as both reinforcement and for pressure containment, and the ends are fillet welded to the carrier pipe. Alternatively, a pipe stub with full-encirclement saddle or a pipe stub with a full-encirclement sleeve and saddle can be utilized. As with Type A repair sleeves, the full encirclement is for structural reinforcement only; it is not a pressurized component and is not, therefore, welded to the carrier pipe. Stopple connections used to isolate a portion of the pipeline are also of a split-tee configuration.

Branch connections NPS 2 and larger to a new or decommissioned facility will be made cold using a welding tee, extruded header, or contoured insert fitting.

Particular care has to be taken when welding on an in-service pipeline for two reasons. First, the process of welding may lead to penetration of the pipe wall—a burn through—and result in the escape of the pressurized contents and the immediate shut down of the pipeline for repair. Second, the flowing contents of the pipeline can result in rapid cooling of the weld area and, in transformable steels, a high weld metal and/or HAZ hardness with a consequent risk of hydrogen-assisted cracking. This cracking, which may go undetected, can and has resulted in subsequent catastrophic failure. The unfortunate aspect of these two issues is that they work against each other; a low heat input and high rate of heat extraction results in reduced penetration but increased risk of hydrogen-assisted cracking.

Comprehensive, collaborative research and development programs have been undertaken over the last 20 or more years, largely under the Pipeline Research Council International (PRCI) sponsorship, to understand and address the key issues related to welding on in-service pipelines. These include the following:

1. Methods for predicting safe parameters for welding on in-service pipelines [58–61];
2. Procedures and guidelines for weld deposition repair [62–64];
3. Alternative processes and tools for welding on in-service pipelines [65,66];
4. Hardness limits for in-service welding [67–69].

These programs have led to the development of tools and guidelines that can assist the welding engineer in assessing the feasibility of a repair or hot tap at full line pressure and full flow, and in the development and qualification of the most appropriate welding approach and procedures.

## 18.8 PIPELINE INCIDENTS ARISING FROM WELDING DEFECTS AND RECENT INDUSTRY AND REGULATORY PREVENTATIVE ACTION

In any general review of pipeline welding, reference to literature on application of existing technologies and technology developments and their successful implementation is important to establish confidence in a process or approach; however, in the context of safety and integrity, an equally valuable source of information is an analysis of an unsuccessful application or component failure. Descriptions and analyses in the public technical literature of unfortunate incidents resulting from defective girth welds are sparse. Leaks and ruptures caused by defects introduced by welding are in the minority of pipeline incidents by far. Axial or longitudinal stresses from pressure loading are significantly lower in a completed pipeline than those in the circumferential or hoop direction. Discontinuities oriented in the axial direction, such as corrosion defects and stress corrosion cracks, have been a much bigger concern for pipeline owners and operators as potential failure of these axial-oriented discontinuities is driven by hoop stress.

Nevertheless, during 2008 and 2009, several newly constructed, large-diameter, higher strength gas and liquid pipelines in the United States experienced field hydrostatic test failures, in-service leaks, and in-service failures at girth welds. The U.S. Department of Transportation (DOT) Pipeline and Hazardous Materials Safety Administration (PHMSA) issued an Advisory Bulletin [70], which they summarized as follows:

> "PHMSA is issuing an advisory bulletin to notify owners and operators of recently constructed large-diameter natural gas pipeline and hazardous liquid pipeline systems of the potential for girth weld failures due to welding quality issues. Misalignment during welding of large-diameter line pipe may cause in-service leaks and ruptures at pressures well below 72 percent specified minimum yield strength (SMYS). PHMSA has reviewed several recent projects constructed in 2008 and 2009 with 20-inch or greater diameter, grade X70 and higher line pipe. Metallurgical testing results of failed girth welds in pipe wall thickness transitions have found pipe segments with line pipe weld misalignment, improper bevel and wall thickness transitions, and other improper welding practices that occurred during construction. A number of the failures were located in pipeline segments with concentrated external loading due to support and backfill issues. Owners and operators of recently constructed large diameter pipelines should evaluate these lines for potential girth weld failures due to misalignment and other issues by reviewing construction and operating records and conducting engineering reviews as necessary."

According to the supplementary information in the bulletin, post-incident metallurgical and mechanical tests and inspections of the line pipe, fittings, bends, and other appurtenances indicated pipe with weld misalignment, improper beveling of transitions, improper back welding, and improper support of the pipe and appurtenances. In some cases, pipe-end conditions did not meet the design and construction requirements of the applicable standards.

At a PHMSA Workshop on New Pipeline Construction Issues in 2009, it was revealed that most, if not all, of the failures identified previously were known to have been attributed, at least in part, to hydrogen-assisted cracking [71].

The PRCI contracted DNV to review these construction quality issues and to develop guidelines for the assurance of girth weld quality during pipeline construction and subsequent operation [72]. The report concludes that the incidents all involved production mainline and tie-in welds constructed using cellulosic-coated electrodes, particularly welds at wall-thickness transitions. None of the reported incidents involved welds completed solely with mechanized gas metal arc welding processes; however, repair welds made using cellulosic-coated electrodes in pipelines otherwise constructed using mechanized gas metal arc welding were included. Although full details of the operating company and PHMSA investigations are not disclosed, the presentation by Bauman [71] and the report by Bruce et al. [72] suggest that the main root cause of these failures was hydrogen-assisted cracking or a combination of hydrogen cracking and elevated stress situations and, for a small minority, elevated stress situations alone.

The PRCI report does present failure analyses of two girth weld hydrostatic test failures and six welds removed as part of the subsequent remedial actions from one 36 in. diameter, API 5LX70 construction project that serves to illustrate the key causes of the incidents of concern to PHMSA. With the permission and participation of the operating company concerned, the overall weld quality on this particular project was also reviewed and analyzed. Repair rates were particularly high, and the detailed analysis revealed many weld quality control malpractices. Many of the welds that were made were repaired by back welding, which was not part of the qualified welding procedure and was not approved by the company. Other aspects of the investigation raised concerns that welds were made without regard to the importance of following other critical aspects of the welding parameters (i.e., current, voltage, and travel speed) that are specified in the welding procedure. The required preheat was not being adequately or consistently applied, and many of the hydrogen cracks occurred in welds made during winter construction. Evidence also existed that proper care was not being exercised during field bending, ditching, and lowering-in activities.

Reported incidents arising from maintenance in-service welding on operating pipelines are also few and far between in the technical literature [73]. Two are known to have occurred in Canada. The first was attributed to delayed hydrogen cracking in the HAZ at the toe of a fillet weld made using cellulosic-coated electrodes between a full-encirclement repair sleeve and a pipeline containing flowing, pressurized products [74]. A study commissioned by API, which was stimulated in part by the severity of this particular incident, provided analyses of 15 of the 90 reported incidents in North America [75]. The following observations were made:

1. Hydrogen-induced cracking was the root cause;
2. Added stress causing failure some time after the welding has taken place;
3. API 5LX52 material;
4. High carbon equivalent material (0.44% and up, based on %C + %Mn/6);
5. The use of cellulosic-coated electrodes;
6. Welding with fluid in the pipeline; and
7. Circumferential breaks in the pipeline at the fillet weld around the ends of sleeves.

The second incident in Canada arising from in-service welding was subjected to a detailed, root-cause failure analysis, which was published [76] and provides one well-documented illustration where the aforementioned issues of hydrogen-assisted cracking, the influence of unanticipated tensile stresses and welding procedures not being followed have resulted in a costly failure.

In January 1992, TransCanada's NPS 36 Western Alberta Mainline ruptured at a location 100 km north of the city of Calgary. A fire at the adjoining James River Interchange Meter Station resulted from the rupture, rendering the exchange facility between the Alberta Eastern and Western systems inoperable. The cost of this failure in terms of the required repair and lost revenue was estimated at $9 million. The rupture initiated at a hot tap where a NPS 24 pipeline was tied in to the NPS 36 mainline. The Western Alberta System had been in operation since 1962 and was constructed of API 5LX52 pipe material, nominal dimensions 914-mm outside diameter by 10.3-mm wall thickness. The line had a maximum allowable operating pressure (MAOP) of 5826 kPa, which gave it a hoop stress of 71.9% SMYS. The pressure at the time of failure was 5392 kPa. The hot tap was installed in 1980 using the standard procedure at the time, which required the flow to be curtailed and the pressure reduced.

The metallurgical investigation concluded that the rupture originated at preexisting hydrogen cracking located at the toe of the hot tap stub weld on the NPS 36 carrier pipe. Two areas were identified with dimensions of approximately 66 and 68 mm in length separated by 32 mm and roughly 2 mm deep. Brittle fracture propagated in both directions consistent with the properties of the 1960s vintage pipe material. Viewed from the inside of the carrier pipe, the preexisting defect was located at approximately the 1 to 3 o'clock position. The appearance of the cap pass of the stub weld indicated that one side, from 6 to 12 o'clock, was welded using a vertical-up weave technique as required by the qualified welding procedure. However, on the side containing the preexisting defect, a stringer bead technique was used with a vertical-down progression from the 1 to 5 o'clock position at the weld toe onto the NPS 36 carrier pipe. Obviously, this area had not been welded in accordance with the specifications. Chemical analysis of the stringer bead was consistent with a low-hydrogen E8018-C2 electrode; in order to weld vertically down with a low-hydrogen electrode of this type, a very fast travel speed is unavoidable. Microhardness surveys of the HAZ and weld metal regions adjacent to the origin of failure revealed values in the range of 518–546 HV in the HAZ and 390–440 HV in the weld metal. Average hardness of the surrounding parent metal was 210 HV.

Although there was no doubt that the rupture originated at hydrogen cracking that occurred during the installation of the stub and the primary event was the noncompliant procedure used to weld the NPS 24 stub to the NPS 36 carrier pipe, a fracture assessment and stress analysis showed that this alone was insufficient to cause failure [77]. Additional construction at the James River Interchange Meter Station occurred during September 1991. Part of this construction included the installation of 150 m of NPS 30 piping onto the NPS 24 branch piping from the hot tap connection. The analysis showed that settlement of as little as 12–15 mm of this new construction was sufficient to cause the rupture. The settlement of the piping was believed to have occurred gradually over time; however, the stresses at the preexisting defect increased rapidly once a certain deflection was reached. During the few hours prior to the rupture, the internal pressure in the pipe was rising, although, as mentioned earlier, MAOP was never exceeded. Through-wall fracture occurred when the stresses on the preexisting defect, due to combined internal pressure and settlement, reached a critical value.

The PHMSA Advisory Bulletin and Workshop, the subsequent PRCI analysis and report, and the TransCanada failure investigation and subsequent publications serve to bring to notice some of the design, construction, operation, and maintenance issues that need to be considered before, during, and after construction and maintenance welding applications. In connections of the size and type under discussion, the pipe-to-pipe or pipe-to-fitting welds may have limited tolerance to undetected defects and to moments applied as a result of unanticipated stresses during construction or later, when the pipeline is in operation.

It is not possible to outline one all-encompassing strategy for mitigating hydrogen cracking in pipeline welding, as the most appropriate approach depends on the specific pipeline welding applications as well as local conditions such as ambient temperature, terrain, and so on. Nevertheless, the referenced PRCI report by Bruce et al. [72] does attempt to provide key measures for preventing girth weld failures caused by the occurrence of hydrogen cracking and is the most comprehensive document on the topic available today. There is no doubt that following the guidelines included in the document will result in a reduced likelihood of girth weld failures shortly after welding, during lowering-in, during hydrostatic testing, and in subsequent service.

As well as detailed strategies for managing hydrogen, the report includes procedures to control the longitudinal stresses imposed on girth welds during pipe lifting and lowering-in. These were developed as follows:

1. Idealization of the multiple side boom lifting and lowering-in process into simplified models.
2. Conducting FEA with the simplified models.
3. Combining the results of the simplified models into a single set of guidelines.

High–low misalignment is also addressed and guidance on limits provided together with consideration of the effects of unequal wall thickness, weld strength mismatch, and weld profile.

A separate document was developed as an appendix that is intended to provide welders and other field personnel with guidance as to how they can assist in the prevention of hydrogen cracking in pipeline girth welds. A video demonstration that shows the diffusion of hydrogen from welds made using different types of electrodes is also available from PRCI.

## APPENDIX 18.A: ABBREVIATIONS USED IN THIS CHAPTER

| Term | Definition |
|---|---|
| **Abbreviations for Units of Measurement** | |
| C | Celsius |
| g | Gram |
| HV | Vickers hardness |
| in | Inch |
| kJ | Kilojoules |
| km | Kilometre |
| kPa | Kilopascals |
| m | Metre |
| mL | Millilitre |
| mm | Millimetre |
| ° | Degree |
| **Abbreviations of Other Terms** | |
| API | American Petroleum Institute |
| Ar | Chemical element argon |
| ASME | American Society of Mechanical Engineers |
| AUT | Automated ultrasonic testing |
| AWS | American Welding Society |
| B | Chemical element boron |
| BSI | British Standards Institute |
| C | Chemical element carbon |
| CFR | Code of Federal Regulations |
| CMT | Cold metal transfer |
| Cr | Chemical element chromium |
| CSA | Canadian Standards Association |
| CTOD | Crack-tip opening displacement |
| Cu | Chemical element copper |
| CWP | Curved wide plate |
| DC | Direct current |
| DNV | Det Norske Veritas |
| DOT | US Department of Transportation |
| ECA | Engineering critical assessment |
| FCAW-S | Self-shielded flux cored arc welding |
| FCAW-G | Gas-shielded flux cored arc welding |
| GMAW-P | Pulsed gas metal arc welding |
| GMAW-S | Short circuit gas metal arc welding |
| HAZ | Heat-affected zone |
| MAOP | Maximum allowable operating pressure |
| Mn | Chemical element manganese |
| Mo | Chemical element molybdenum |
| NEB | National Energy Board |
| Ni | Chemical element nickel |
| NPS | Nominal pipe size |
| O | Chemical element oxygen |
| PHMSA | Pipeline and Hazardous Materials Safety Administration |
| PQR | Procedure qualification record |
| PRCI | Pipeline Research Council International |
| RMD | Regulated metal deposition |
| RTR | Real-time radiography |
| SBD | Strain-based design |
| SAW | Submerged arc welding |
| Si | Chemical element silicon |
| SMAW | Shielded metal arc welding |
| SMYS | Specified minimum yield strength |
| STT | Surface tension transfer |
| TMCP | Thermomechanically controlled processing |
| TOFD | Time-of-flight diffraction |
| V | Chemical element vanadium |
| WIC | Welding Institute of Canada |
| WPS | Welding procedure specification |

## APPENDIX 18.B: REGULATIONS, CODES, AND STANDARDS

This appendix lists the practices, codes, standards, specifications, and guidelines referred to in this chapter.

**API—American Petroleum Institute**

| Reference | Description |
|---|---|
| API 5L | Specification for line pipe |
| API 1104 | Welding of pipelines and related facilities |

**ASME—American Society of Mechanical Engineers**

| Reference | Description |
|---|---|
| ASME Section IX | BPVC section IX—qualification standard for welding and brazing procedures, welders, brazers, and welding and brazing operators |

**BSI—British Standards Institute**

| Reference | Description |
|---|---|
| BSI 4515 | Specification for welding of steel pipelines on land and offshore. Carbon and carbon manganese steel pipelines |
| PD 6493 | Guidance on methods for assessing the acceptability of flaws in fusion welded structures |

**CSA—Canadian Standards Association**

| Reference | Description |
|---|---|
| CSA Z662-11 | Oil and gas pipeline systems |

**DNV—Det Norske Veritas**

| Reference | Description |
|---|---|
| DNV-OS-F101 | Submarine pipeline systems |

**NEB—National Energy Board**

| Reference | Description |
|---|---|
| SOR/99-294 | Onshore pipeline regulations, 1999 |

**PHMSA—Pipeline and Hazardous Materials Safety Administration, Department of Transportation**

| Reference | Description |
|---|---|
| 49 CFR Part 192 Subpart E | Welding of steel in pipelines |

## REFERENCES


1. Hosmanek, M. (1984) Pipe Line Construction, University of Texas at Austin/Pipe Line Contractors Association, 1984.
2. Joining Methods—Technological Summaries. (2005) DNV Technical Report 2005–3394 to the Petroleum Safety Authority Norway, October 2005.
3. Dorling, D. and Rothwell, B. (1990) Field welding processes for pipeline construction. Proceedings of the Pipeline Technology Conference, October 15–18, 1990, Ostend, Belgium.
4. Jones, R.L. and Hone, P.N. (1990) Advances in techniques available for the girth welding of pipelines. Proceedings of the Pipeline Technology Conference, October 1990, Ostend, Belgium.
5. Orsini, T. and Gerbec, D. (2010) Improving Productivity with Submerged Arc Welding, ESAB Welding & Cutting Products Communication, May 2010.
6. Munz, R.P. and Narayanan, B. (2002) The welding of line pipe using the innershield process. Proceedings of WTIA International Conference on Pipeline Construction Technology, March 2002, Wollongong, Australia.
7. Hobrock, R. and Railling, D. (2012) Self-Shielded Flux-Cored Welding for Pipeline Applications, Oil and Gas Product News, June 2012.
8. Blackman, S.A. and Dorling, D.V. (1999) Capabilities and limitations of mechanized GMAW systems for transmission pipelines. Proceedings of WTIA/APIA/CRC-WSC International Conference on Weld Metal Hydrogen Cracking in Pipeline Girth Welds, March 1–2, 1999, Wollongong, Australia.
9. Blackman, S.A. and Dorling, D.V. (2000) Advanced welding processes for transmission pipelines. Proceedings of the Pipeline Technology Conference, May 21–24, 2000, Brugge, Belgium.
10. Blackman, S., Liratzis, T., Howard, R., Hudson, M., and Dorling, D. (2004) Recent tandem welding developments for pipeline girth welding. Proceedings of the 4th International Conference on Pipeline Technology, May 2004, Ostend, Belgium.
11. Glover, A., Horsley, D., Dorling, D., and Takehara, J. (2004) Construction and installation of X100 pipelines, Proceedings of the 5th International Pipeline Conference (IPC 2004), October 2004, Calgary, Alberta, Canada.
12. Zhou, J., Taylor, D., and Hodgkinson, D. (2008) Further large-scale implementation of advanced pipeline technologies. Proceedings of the 7th International Pipeline Conference (IPC 2008), September 2008, Calgary, Alberta, Canada.
13. Harris, I.D. and Norfolk, M.I. (2008) Hybrid laser/gas metal arc welding of high strength gas transmission pipelines. Proceedings of the 7th International Pipeline Conference (IPC 2008), September 2008, Calgary, Alberta, Canada.
14. Begg, D., Benyon, G., Hansen, E., Defalco, J., and Light, K. (2008) Development of a hybrid laser arc welding system for pipeline construction. Proceedings of the 7th International Pipeline Conference (IPC 2008), September 2008, Calgary, Alberta, Canada.
15. Pussegoda, L.N., Begg, D., Holdstock, R., Jodoin, A., Light, K., and Rondeau, D. (2010) Evaluating mechanical properties of hybrid laser arc girth welds. Proceedings of the 8th International Pipeline Conference (IPC 2010), September 2010, Calgary, Alberta, Canada.
16. Kumar, A., Fairchild, D.P., Macia, M.L., Anderson, T.D., Jin, H.W., Ayer, R., Ozekcin, A., and Mueller, R.R. (2011) Evaluation of economic incentives and weld properties for welding of steel pipelines using friction stir welding. Proceedings of the 21st International Offshore and Polar Engineering Conference, July 2011, Maui, Hawaii.
17. Jensen, G.A. (2010) Offshore pipelaying dynamics, Ph.D. Thesis, Norwegian University of Science and Technology, February 2010.

18. EWI/Microalloying International, TransCanada Pipelines and Miller/Hobart (2005) Development of best practice welding guidelines for X80 pipelines. Appendix B of Final Report for Research and Development Agreement DTRS56-04-T-0011, Project 152, U.S. Department of Transportation, Pipeline and Hazardous Materials Safety Administration, July 2005.
19. Glover, A.G., Horsley, D.J., and Dorling, D.V. (1999) High-strength steel becomes standard on Alberta Gas System. *Oil and Gas Journal*, 97, 44–50.
20. Widgery, D.J. (2002) Welding high strength steel pipelines—theory, practice and learning. Proceedings of WTIA International Conference on Pipeline Construction Technology, March 2002, Wollongong, Australia.
21. Bandera, G. (2010) Welding Well, World Pipelines, October 2010.
22. Narayanan, B.K., Soltis, P., McFadden, L., and Quintana, M. (2007) New process to girth weld pipe with a gasless technology. Proceedings of the 2007 Offshore Technology Conference, May 2007, Houston, TX.
23. Gerlich, A.P., Izadi, H., Bundy, J., and Mendez, P.F. (2014) Characterization of high-strength weld metal containing mg-bearing inclusions. *AWS Welding Journal*, 93, 15.
24. Stava, E.K. (1993) The surface-tension-transfer power source: a new, low-spatter arc welding machine. *AWS Welding Journal*, 72, 25–29.
25. Biery, N., Macia, M., Appleby, R., Fairchild, D., Hoyt, D., Dorling, D., and Horsley, D. (2006) Godin lake trial: X120 field welding. Proceedings of the 6th International Pipeline Conference (IPC 2006), September 2006, Calgary, Alberta, Canada.
26. Glover, A.G., Dorling, D.V., and Coote, R.I. (1988) Inspection and assessment of mechanized pipeline girth welds. Proceedings of TWI International Conference on Weld Failures, November 1988, London.
27. Moles, M., Dube, N., and Labbe, S. (2004) Special phased array applications for pipeline girth weld inspections. Proceedings of the 5th International Pipeline Conference (IPC 2004), September 2004, Calgary, Alberta, Canada.
28. Lozev, M., Spencer, R., Huang, T.-C., and Patel, P. (2006) Improved ultrasonic inspection and assessment methods for pipeline girth welds and repair welds. Final Report for Research and Development Agreement DTRS56-03-T-00126-Project 133, U.S. Department of Transportation, Pipeline and Hazardous Materials Safety Administration, August 2006.
29. Wang, Y.-Y., Lui, M., Horsley, D., and Bauman, G. (2006) A tiered approach to girth weld defect acceptance criteria for stress-based design of pipelines. Proceedings of the 6th International Pipeline Conference (IPC 2006), September 2006, Calgary, Alberta, Canada.
30. Pisarski, H. (2011) Assessment of flaws in pipe girth welds. Proceedings of CBMM-TMS International Conference on Welding of High Strength Pipeline Steels, November 2011, Araxá, Brasil.
31. Tyson, R.W., Xu, S., Ward, I., Duan, D.-M., and Horsley, D. (2012) ECA by failure assessment diagram. Proceedings of the 9th International Pipeline Conference (IPC 2012), September 2012, Calgary, Alberta, Canada.
32. Wang, Y.-Y. (2001) Compendium of updated pipeline girth weld ECA methodologies to support revisions to existing code practices. PRCI Report L51841, Pipeline Research Council International, September 2001.
33. Hart, P.H.M. (1999) Hydrogen cracking—its causes, costs and future occurrence. Proceedings of WTIA/APIA/CRC-WSC International Conference on Weld Metal Hydrogen Cracking in Pipeline Girth Welds, March 1999, Wollongong, Australia.
34. Bailey, N., Coe, F.R., Gooch, T.G., Hart, P.H.M., Jenkins, N., and Pargeter, R.J. (1993) *Welding Steels Without Hydrogen Cracking*, 2nd ed., The Welding Institute, Cambridge.
35. Dearden, J. and O'Neill, H. (1940) A guide to the selection and welding of low alloy structural steels. *Institute of Welding Transactions*, 3, 203–214.
36. Yurioka, N. (1999) Predictive methods for prevention and control of hydrogen assisted cracking. Proceedings of WTIA/APIA/CRC-WSC International Conference Weld Metal Hydrogen Cracking in Pipeline Girth Welds, March 1999, Wollongong, Australia.
37. Yurioka, N., Ohshita, S., and Tamehiro, H. (1981) Study on carbon equivalents to assess cold cracking tendency in steel welding. Proceedings AWRA Symposium on Pipeline Welding in the 80s, March 1981, Melbourne.
38. Bailey, N. (1970) Welding procedures for alloy steels. TWI Report Series, The Welding Institute, Abington Hall, Cambridge, July 1970.
39. Graville, B.A. (1995) Interpretive report on weldability tests for hydrogen cracking of higher strength steels and their potential for standardization, Welding Research Council Bulletin, No. 400, April 1995.
40. Wang, Y.-Yi., Liu, M., Zhang, F., Horsley, D., and Nanney, S. (2012) Multi-tier tensile strain models for strain-based design part 1—fundamental basis. Proceedings of the 9th International Pipeline Conference (IPC 2012), September 2012, Calgary, Alberta, Canada.
41. Wang, Y.-Yi., Liu, M., Song, Y., Stephens, M., Petersen, R., Gordon, J.R., and Horsley, D. (2011) Second generation models for strain-based design. Final Report for Research and development Agreement DTHP56-06-T00014-Project 201, U.S. Department of Transportation, Pipeline and Hazardous Materials Safety Administration, July 2011.
42. Hukle, M., Horn, A., Hoyt, D., and LeBleu, J. (2005) Girth weld qualification for high strain pipeline applications. Proceedings of the 24th International Conference on Offshore Mechanics and Arctic Engineering (OMAE 2005), June 2005, Halkidiki, Greece.
43. Hukle, M., Hoyt, D., LeBleu, J., Dwyer, J., and Horn, A. (2006) Qualification of welding procedures for ExxonMobil high strain pipelines. Proceedings of the 25th International Conference on Offshore Mechanics and Arctic Engineering (OMAE 2006), June 2006, Hamburg, Germany.
44. Newbury, B.D., Hukle, M.W., Crawford, M.D., and Lillig, D.B. (2007) Welding engineering for high strain pipelines. Proceedings of the 17th International Offshore and Polar Engineering Conference, July 2007, Lisbon, Portugal.

45. Gianetto, J.A., Bowker, J.T., Dorling, D.V., Taylor, D., Horsley, D., and Fiore, S.R. (2008) Overview of tensile and toughness testing protocols for assessment of X100 pipeline girth welds. Proceedings of the 7th International Pipeline Conference (IPC 2008), September 2008, Calgary, Alberta, Canada.
46. Wang, Y.-Yi., Zhou, H., Liu, M., Tyson, W., Gianetto, J., Weeks, T., Richards, M., and McColskey, J.D. (2011) Weld design, testing and assessment procedures for high strength pipelines. Report 277-S-01 for Research and development Agreement DTHP56-07-0001—Project 225, U.S. Department of Transportation, Pipeline and Hazardous Materials Safety Administration, December 2011.
47. Wang, Y.-Y., Liu, M., Tyson, W., Gianetto, J., and Horsley, D. (2010) Toughness considerations for strain-based designs. Proceedings of the 8th International Pipeline Conference (IPC 2010), September 2010, Calgary, Alberta, Canada.
48. Cheng, W., Tang, H., Gioielli, P.C., Minnaar, K., and Macia, M.L. (2009) Test methods for characterization of strain capacity: comparison of R-curves from SENT/CWP/FS tests. Proceedings of the Pipeline Technology Conference, October 2009, Ostend, Belgium.
49. Moore, P.L. and Pisarski, H.G. (2012) Validation of methods to determine CTOD from SENT specimens. Proceedings of the 22nd International Offshore and Polar Engineering Conference, July 2012, Rhodes, Greece.
50. Tyson, W.R., Shen, G., and Roy, G. (2007) Effect of biaxial stress on ECA of pipelines under strain-based design. Proceedings of the 17th International Offshore and Polar Engineering Conference, July 2007, Lisbon, Portugal.
51. Gioielli, P., Minnaar, K., Macia, M., and Kan, W. (2007) Large-scale testing methodology to measure the influence of pressure on tensile strain capacity of a pipeline. Proceedings of the 17th International Offshore and Polar Engineering Conference, July 1–6, 2007, Lisbon, Portugal.
52. Kibey, S., Minnaar, K., Issa, J., and Gioielli, P. (2008) Effect of misalignment on the tensile strain capacity of welded pipelines. Proceedings of 18th International Offshore and Polar Engineering Conference, July 6–11, 2008, Vancouver, BC, Canada.
53. Kibey, S.A., Lele, S.P., Tang, H., Krishnan, V.R., Wang, X., Macia, M.L., Fairchild, D.P., Cheng, W., Noecker, R., Wojtulewicz, P.J., Newbury, B., Kan, W.C., and Cook, M.F. (2011) Full-scale test observations for measurement of tensile strain capacity of welded pipelines. Proceedings of the 21st International Offshore and Polar Engineering Conference, July 19–24, 2011, Maui, Hawaii.
54. Kibey, S., Issa, J.A., Wang, X., and Minnaar, K. (2009) A simplified, parametric equation for prediction of tensile strain capacity of welded pipelines. Proceedings of the 5th Pipeline Technology Conference, October 2009, Ostend, Belgium.
55. Fairchild, D.P., Kibey, S.A., Tang, H., Krishnan, V.R., Wang, X., Macia, M.L., and Cheng, W. (2012) Continued advancements regarding capacity prediction of strain-based pipelines. Proceedings of 9th International Pipeline Conference (IPC 2012), September 2012, Calgary, Alberta, Canada.
56. Pisarski, H.G. (2011) Assessment of flaws in pipeline girth welds. Proceedings of the CBMM-TMS International Conference, November 2011, Araxá, Brasil.
57. Jaske, C.E., Hart, B.E., and Bruce, W.A. (2006) Updated pipeline repair manual. Final Report for Research and Development Agreement DTPH56-05-T-0006-Project 179, U.S. Department of Transportation, Pipeline and Hazardous Materials Safety Administration, August 2006.
58. Bruce, W.A., Swatzel, J.F., and Dorling, D.V. (2000) Direct weld deposition repair of pipeline defects. Proceedings of the 1st International Conference on Welding onto In-Service Petroleum Gas and Liquid Pipelines, March 2000, Wollongong, Australia.
59. Bubenik, T.A., Fischer, R.D., Whitacre, G.R., Jones, D.J., Kiefner, J.F., Cola, M.J., and Bruce, W.A. (1991) Investigation and prediction of cooling rates during pipeline maintenance welding. Final Report to American Petroleum Institute, December 1991.
60. Bruce, W.A., Li, V., and Citterberg, R. (2002) PRCI thermal analysis model for hot-tap welding—V 4.2. PRCI Report L51837, Pipeline Research Council International, May 2002.
61. Boring, M.A., Zhang, W., and Bruce, W.A. (2008) Improved burn-through prediction model for in-service welding applications. Proceedings of the 7th International Pipeline Conference (IPC 2008), September 2008, Calgary, Alberta, Canada.
62. Bruce, W.A., Mischler, H.D., and Kiefner, J.F. (1993) Repair of pipelines by direct deposition of weld metal. PRCI Report L51681, Pipeline Research Council International, June 1993.
63. Bruce, W.A., Holdren, R.L., and Kiefner, J.F. (1996) Repair of pipelines by direct deposition of weld metal—further studies. PRCI Report L51763, Pipeline Research Council International, November 1996.
64. Bruce, W.A. and Amend, W.E. (2009) Guidelines for pipeline repair by direct deposition of weld metal. Proceedings of WTIA/APIA Welded Pipeline Symposium, Welding Technology Institute of Australia, April 2009, Sydney.
65. Bruce, W.A. and Fiore, S.R. (2002) Alternative processes for welding on in-service pipelines. PRCI Report L51843, Pipeline Research Council International, March 2002.
66. Begg, D. (2009) Alternative welding processes for in-service welding. Final Report for Research and Development Agreement DTRS56-03-T-0010-Project 131, U.S. Department of Transportation, Pipeline and Hazardous Materials Safety Administration, April 2009.
67. Bruce, W.A. and Boring, M.A. (2005) Realistic hardness limits for in-service welding. PRCI Report L52242, Pipeline Research Council International, March 2005.
68. Bruce, W.A., Etheridge, B.C., and Arnett, V.R. (2009) Development of heat-affected zone hardness limits for in-service welding. Final Report for Research and Development Agreement DTPH56-07-T-000004-Project 216, U.S. Department of Transportation, Pipeline and Hazardous Materials Safety Administration, September 2009.
69. Bruce, W.A. and Etheridge, B.C. (2012) Further development of heat-affected zone hardness limits for in-service welding.

Proceedings of the 9th International Pipeline Conference (IPC 2012), September 2012, Calgary, Alberta, Canada.

70. PHMSA Advisory Bulletin (2010) Pipeline safety: girth weld quality issues due to improper transitioning, misalignment, and welding practices of large diameter line pipe, March 2010. Available at http://edocket.access.gpo.gov/2010/pdf/2010-6528.pdf.
71. Bauman, G. (2009) Construction Issues, Presentation at PHMSA Workshop on New Pipeline Construction Practices, Fort Worth, April 23, 2009. Available at http://www.regulations.gov/-!documentDetail;D=PHMSA-2009-0060-0012.
72. Bruce, W.A., Amend, W.E., Wang, Y-Y., and Zhou, H. (2013) Guidelines to address pipeline construction quality issues. Final Report to PRCI for Project MATH-5-1, Pipeline Research Council International, February 2013.
73. McHaney, J.H. and Bruce, W.A. (2000) Lessons to be learned from past in-service welding incidents. Proceedings of the 1st International Conference on Welding onto In-Service Petroleum Gas and Liquid Pipelines, March 2000, Wollongong, Australia.
74. National Energy Board (1986) In the matter of an accident on 19 February 1985 near Camrose, Alberta, on the pipeline system of Interprovincial Pipe Line Limited. National Energy Board Report, June 1986.
75. Bubenik, T.A., Fischer, R.D., Whitacre, G.R., Jones, D.J., Kiefner, J.F., Cola, M.J., and Bruce, W.A. (1991) Investigation and prediction of cooling rates during pipeline maintenance welding. Final Report to American Petroleum Institute, December 1991.
76. Chiovelli, S., Dorling, D.V., Glover, A.G., and Horsley, D.J. (1993) NPS 36 Western Alberta mainline rupture at James River Interchange. Proceedings of the 8th Symposium on Line Pipe Research, Houston, Pipeline Research Council International, September 1993.
77. Horsley, D.J., Nippard, F.E., and Pick, R.J. (1996) Finite element investigation of the NPS 36 Western Alberta Mainline rupture at the James River Interchange. Proceedings of the 9th Symposium on Pipeline Research, Houston, Pipeline Research Council International, September 1996.

# 19

# THE EFFECT OF INSTALLATION ON OFFSHORE PIPELINE INTEGRITY

ROBERT O'GRADY
*Wood Group Kenny, Galway, Ireland*

## 19.1 INTRODUCTION

The installation of offshore pipelines represents a considerable challenge that requires installation methods and procedures that are both safe and cost efficient. Successfully achieving these objectives requires an understanding of the effects of installation on pipeline integrity. This chapter discusses the following topics as a way of developing such an understanding:

1. Installation methods and pipeline behavior during installation
2. Critical factors governing installation
3. Installation analysis and design methodologies
4. Monitoring the installation process offshore
5. Implications of deeper water on installation

The discussion of these topics is covered in this chapter.

## 19.2 INSTALLATION METHODS AND PIPELINE BEHAVIOUR DURING INSTALLATION

The primary methods of rigid pipeline installation include S-lay, J-lay, and Reel-lay. Each of these methods has its own unique aspects from a pipe behavior point of view and so they are individually outlined here in turn, but first a general discussion on installation loading and failure modes is worthwhile. Note that other less adopted installation methods exist as well, such as tow out, but these have been omitted here for conciseness.

### 19.2.1 Pipeline Installation Loading and Failure Modes

Regardless of the method used there are generally four different regions of loading behavior through which a pipe joint passes during installation. Figure 19.1 highlights the four loading regions for the primary installation methods. Typically these regions are aligned in the same vertical plane and the first of them is known as the overbend.

The overbend represents the upper part of the suspended pipeline, which interacts with the installation vessel through various support and tensioning systems. This region encounters some level of hogging curvature, the magnitude of which depends on the installation method in question. For example, with S-lay there is typically a relatively high level of overbend curvature while for J-lay there is often only a small amount of such curvature. In fact, strictly speaking J-lay can sometimes be viewed as not having an overbend region at all, but for the sake of consistency a J-lay overbend is included in the discussion here.

Below the overbend there is an inflection region where the pipeline tends to straighten out before its curvature is reversed to sagging toward the bottom touchdown region of the pipeline, known as the sagbend. Once the pipeline passes through the sagbend and touches down, it then reaches the final seabed region where once again it straightens out.

*Oil and Gas Pipelines: Integrity and Safety Handbook,* First Edition. Edited by R. Winston Revie.

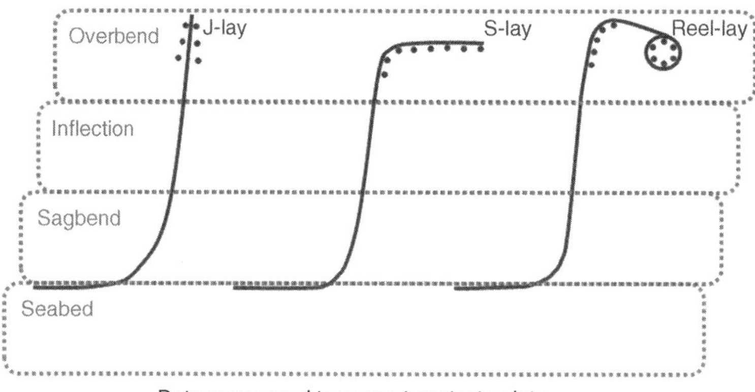


**FIGURE 19.1** Elevation views of primary installation methods with loading regions highlighted.

It is worth highlighting that the relative size, or length, of the respective loading regions varies depending on the installation method, the water depth in question and also the procedure being performed. For example, in shallow water the length of the inflection region shrinks considerably to the point that there can be an almost immediate transition between the overbend and sagbend. Also, as mentioned already, for J-lay the overbend region can shrink to zero length due to the lack of hogging curvature.

In addition to water pressure the two main types of loads encountered during installation are axial loads and in-plane bending moments. The magnitudes and directions of these two load types can vary across the different loading regions. For example, the in-plane bending moments are relatively insignificant in the inflection and seabed regions, which tend to be dominated by axial loads. On the other hand, in-plane bending moments are often dominant in the overbend and sagbend regions.

Out-of-plane bending moments sometimes occur as well, particularly when a pipeline is installed along a tightly curved route or when environmental loadings approach the pipeline and vessel in a somewhat lateral direction. In some more extreme circumstances other additional load types, such as shear point loads and distributed torque moments, also come into play and can influence pipeline behavior. Such circumstances are described later.

Given the primary load types discussed earlier, the governing failure mode for rigid pipelines during installation is typically local buckling with subsequent buckle propagation and so special attention needs to be given to this at the analysis and design phase of an installation project. This is particularly applicable to the bending dominated areas of the overbend and sagbend.

With dynamic movements of the vessel and pipeline, under environmental conditions, cyclic stresses also develop along the various regions and this then equates to fatigue damage. While fatigue damage represents another possible failure mode during installation, particularly in severe environments and deeper water where there is longer exposure to damage, it is more of a concern from the point of view of maintaining integrity over the life time of a pipeline. This means that fatigue damage during installation needs to be kept below a certain allowable threshold so as to provide adequate contingency for the fatigue damage that will inevitably be accumulated over the operational life span of a pipeline. Both fatigue damage and local buckling are discussed further in subsequent sections.

### 19.2.2 S-Lay Method

The S-lay method, as illustrated by Figure 19.1, involves welding pipe joints back to back to form an S-shaped line that runs from the installation vessel over a supporting stinger structure, down through the water column, and onto the seabed.

On the installation vessel the welding is done along a linear line of roller supports, known as the firing line, and the pipeline is held in place by a set of tensioners gripping the pipeline at intermittent points along the line. The vessel stinger is a truss type structure fitted with spaced roller beds that both guide and support a pipeline as it leaves the vessel during installation. The configuration of the roller beds is typically curved in profile in order to allow for the pipeline to gradually bend downwards as it enters the water column, thus minimizing the stress and strain in the pipe. Figure 19.2 shows a typical S-lay vessel spread.

The stinger can be either a fully rigid structure with a fixed rotation about a hinge point on the vessel or else it can be divided into a series of individual sections connected with multiple flexible hinges that allow for a certain level of rotation independent of the vessel.

In the latter multi-hinged case, the stinger sections usually have some level of inherent buoyancy so as to provide static uplift to the pipeline in shallow water scenarios where the departure angle of the pipeline, as it leaves the vessel, needs to be kept relatively small in order to minimize bending in the

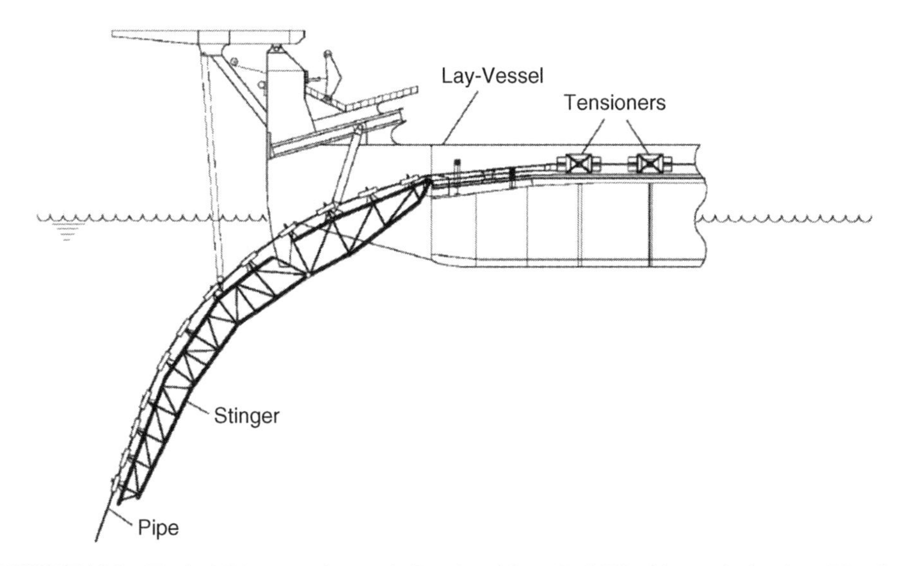


**FIGURE 19.2** Typical S-lay vessel spread. (Reprinted from Ref. [1] with permission from Elsevier.)

sagbend. The flexibility of the hinged sections also minimizes bending in the overbend during dynamic upwards movements of the vessel under wave loading. This overall configuration is known as a floating stinger and, as alluded to already, it is of use in shallow water scenarios.

The fully rigid stinger case mentioned previously is most applicable to moderate and deep water scenarios where the departure angle can be much larger without jeopardising the integrity of the sagbend.

The key benefit of S-lay is that it achieves a relatively high level of productivity with few restrictions on pipe diameter or wall thickness. Such productivity levels are possible due to the fact that the pipe joint welding is done simultaneously across multiple stations along the firing line, which typically runs the length of the installation vessel.

There is, however, a drawback to S-lay and that is the level of axial tension experienced by the pipeline and tensioner system on the vessel grows considerably with deeper water where the length of suspended pipeline in the inflection region inevitably increases. This then means that the radius of curvature for the stinger must be reduced in order to keep the suspended length, and so the tension, within necessary limits. This application of S-lay to deeper water is known as steep S-lay, given the almost vertical orientation of the pipeline as it leaves the stinger. Steep S-lay can in its own right be defined as another installation method, but for conciseness this is not done here.

Such a reduction in stinger radius during steep S-lay leads to direct increases in in-plane bending along the overbend. Larger bending in combination with the significant axial tension means that there is a greater potential for plastic strains developing along the pipeline over the stinger. This is because the bending moment capacity of pipe joints decreases in tandem with increases in tension. Also, because of the intermittent support of the spaced stinger roller beds, the larger tension causes concentrations of strain at the beds themselves.

To further elaborate on the aforementioned, under significant tension the pipe tends to straighten in the spans between roller beds, which in turn results in a higher curvature at the beds themselves so as to compensate for any straightening and to allow the pipeline to still follow the stinger. This concentration behavior also leads to increased shear loads from the roller beds, which if large enough induce permanent ovalization of the pipe cross section.

The overall net result of these tension effects is that pipelines often experience some level of plastic deformation as they pass over an S-lay stinger during deep water installation. This plastic deformation presents itself physically as residual hogging curvature and possibly ovalization.

The reversal of curvature in the sagbend region must overcome any residual hogging curvature from the overbend through both bending and twisting of the pipeline. This twisting behavior corresponds to a rotation about the pipeline longitudinal axis and is known as pipe roll. Such roll inevitably causes torque moments to build up and along the pipeline during installation. The level of pipe roll is dependent on a number of parameters, but one of the primary contributors is the magnitude of the residual curvature itself with larger curvature values generally resulting in greater levels of roll.

From a purely operational point of view, pipe roll is undesirable as it causes misalignment issues when installing in-line structures, such as tee pieces, along a particular pipeline. However, pipe roll, and indeed residual curvature, may also have implications for the bending response in the regions below the overbend. These implications are not well understood, but could indeed prove influential for future ultra-deep installation scenarios. More discussion on this topic is provided in a later section.

### 19.2.3 J-Lay Method

The J-lay method, as illustrated by Figure 19.1, also involves welding pipe joints back to back, but in this case the overall line forms a J-shaped profile starting at a steep angled tower on the installation vessel, down through the water column, and onto the seabed.

The J-lay tower is made up of a series of tensioners, clamps, and o-shaped supports whose diameters increase as the pipeline approaches the bottom of the tower so as to form a bell mouth effect that allows the pipe a controlled amount of free movement under dynamic conditions, thus preventing strain concentrations at the clamps further up the tower. Figure 19.3 shows as typical J-lay spread on an installation vessel. Given the relatively steep departure angle of the pipeline as it exits the tower, J-lay is only really applicable to deeper water scenarios where the sagbend is somewhat below the vessel.

Compared with S-lay, J-lay typically requires smaller axial tension loads to support the suspended pipeline in a given water depth and since the orientation of the load is almost vertical the vessel itself requires less thrust to hold position. In addition, with J-lay it is sometimes possible to simply support the pipeline using collars and a hang off table rather than using more complex tensioner systems. All of these aspects represent particular benefits of J-lay over S-lay.

In terms of pipeline deformation, J-lay imparts negligible overbend curvature when compared to S-lay and so the likelihood of continuous plastic strains developing is small. That being said, depending on the level of dynamic movement of the pipeline and vessel, due to environmental conditions, there is still the potential for local buckling, particularly in the sagbend, but also in the overbend. Fatigue damage is still a concern with J-lay as well.

The main drawback of J-lay is that because the tower is orientated almost vertically then it can only be of a finite length and so is usually only long enough for a small number of welding stations. This in turn means the productivity levels of J-lay are somewhat less than those of S-lay, meaning over time it can become a more expensive installation method. For this reason, J-lay is often deployed on the installation of shorter infield flow lines, whereas S-lay is used on the installation of longer export trunk lines.

### 19.2.4 Reel-Lay Method

The Reel-lay method fundamentally differs to both S-lay and J-lay in that the pipeline welding is done onshore in a fabrication yard prior to the installation project and then entire lengths of pipeline are wounded, or reeled, onto a carrousel, or reel, that is eventually deployed to the installation vessel. Once on the installation vessel the pipeline is reeled off the carrousel as it passes through a set of

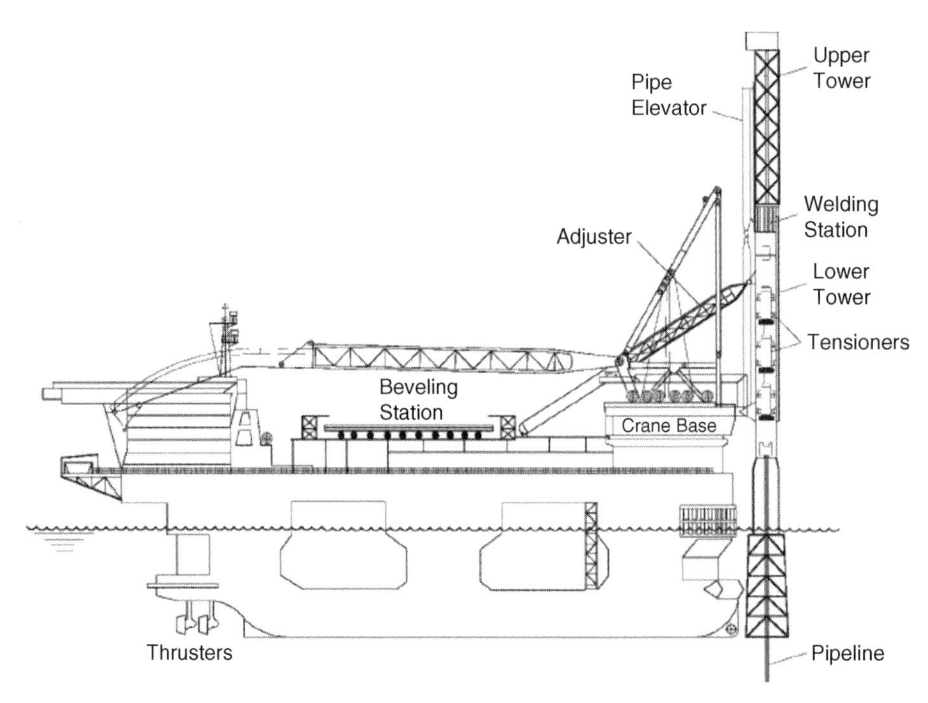


**FIGURE 19.3** Typical J-lay vessel spread. (Reprinted from Ref. [1] with permission from Elsevier.)

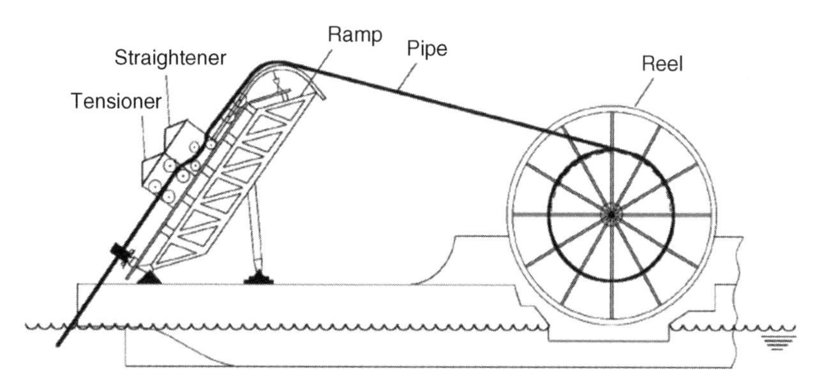


**FIGURE 19.4** Typical reel-lay vessel spread. (Reprinted from Ref. [1] with permission from Elsevier.)

straighteners, tensioners, and ramp supports before entering the water and adopting a sagbend catenary down to the seabed. Figures 19.1 and 19.4 illustrate this reeling off process.

Since pipelines are reeled off a carrousel at a relatively high speed, Reel-lay represents a very productive installation method. That being said, there are restrictions on the diameter and wall thickness of pipes that can be safely reeled onto a carrousel without jeopardising integrity. Also, due to the relatively small radius of the carrousels there is inevitably some level of plastic deformation imparted to a pipeline during the reeling on process and while the straighteners on the vessel are intended to remove residual curvature they often cannot eradicate all of the curvature before the pipe enters the water. As with S-lay, such residual curvature has implications for the regions below the overbend in terms of pipe roll and so on.

Ovalization can also occur during the reeling on process in addition to local buckling, which is itself a potential problem when the pipeline is deployed from the vessel during installation, as is the case with the other installation methods. The reeling on process limits the application of relatively thick coatings as well given the likelihood of crushing the coatings through excessive bending.

Given the speed at which a pipeline is installed during Reel-lay, the associated fatigue damage tends to be relatively minor when compared with the other installation methods. However, the relative velocity at which the Reel-lay installation vessel moves introduces additional behavior in terms of increased hydrodynamic loads acting on the pipeline.

## 19.3 CRITICAL FACTORS GOVERNING INSTALLATION

The factors governing the installation of a particular pipeline are a mixture of vessel-specific restrictions and pipeline integrity criteria. These factors are discussed over the following sections.

### 19.3.1 Vessel Restrictions

Any given installation vessel has an inherent set of restrictions that must be adhered to during an installation project, otherwise potential damage could be done to both the vessel and the pipeline. Sample vessel restrictions include

1. Tension capacity of pipeline tensioner systems;
2. Load capacity of supports;
3. Maximum separation between pipeline and supports at bottom of stinger, tower, or ramp;
4. Thrust capacity for DP vessels or mooring line capacity for barge vessels;
5. Maximum pay in/out velocities and accelerations of tensioner tracks.

Adherence to such restrictions is not only dependent on the static behavior of the suspended pipeline, but also on the dynamic response of the pipeline, and indeed the vessel, to environmental conditions such as waves and currents. Ensuring adherence means designing suitably safe installation procedures covering the entire installation project.

The design process is done prior to going offshore and involves predicting both static and dynamic behavior for a variety of procedures and a limiting set of environmental conditions. This generally requires the use of some numerical analysis method, such as finite element analysis.

Recommended installation analysis and design methodologies are discussed in a later section, but at this stage it is worth noting that the main deliverables from the analysis and design phase are a set of reports and tables that are used as a reference by vessel personnel to safely plan and monitor the installation procedures offshore.

### 19.3.2 Pipeline Integrity Criteria

The primary integrity criterion during installation is prevention of pipeline failure. As described in a previous section, the governing failure mode during installation is usually local buckling of the pipeline with possible subsequent buckle propagation. Prevention of such a failure again requires a suitable analysis and design process that covers the key aspects of the installation campaign both statically and dynamically. The design aspect is guided by checks outlined in industry accepted standards and recommended practices such as those developed by Det Norske Veritas or the American Petroleum Institute.

*19.3.2.1* ***Local Buckling Failure*** Local buckling is most likely to occur in either the overbend or sagbend regions. Typically, the overbend represents a safer area as it encounters relatively small external pressure and the corresponding displacement, or curvature, of the pipeline is somewhat controlled by the configuration of the stinger, tower, or ramp, which tends not to change much regardless of the level of loading. Should a buckle occur in the overbend it is relatively easy to address as only a small portion of pipeline needs to be cut back and annulus flooding is kept to a minimum; known as a dry buckle.

On the other hand, the sagbend is often the most susceptible region to local buckling given the likelihood of external pressure and the fact that the corresponding pipeline behavior is controlled purely by the level of loading encountered, which in turn can vary considerably, especially during a dynamic response. When a buckle occurs in the sagbend it can result in flooding of the pipeline, known as a wet buckle, which in turn can cause a dramatic increase in the tension load being carried by the tensioners and vessel. This larger load, coupled with the difficulty of cutting out a sagbend portion of the pipeline, makes a wet buckle highly undesirable.

The distinctive differences between overbend and sagbend buckling behavior are reflected in the checks within the design standards, which tend to be less strenuous, or conservative, for partially displacement controlled areas than for fully load controlled ones. With this in mind, care must be to taken when deciding where the transition point between the respective areas occurs. This transition point is generally somewhat above the actual bottom of the overbend region where the pipeline has a tendency to move quite a bit relative to the stinger, tower, or ramp supports during dynamic behavior and so is no longer displacement controlled, but is in fact load controlled.

To be conservative the transition point is often assumed to be the bottommost support where there is no lift off of the pipeline during dynamics. Any pipeline above this support is assumed to be partially displacement controlled while any pipeline below this support is assumed to be fully load controlled. Clearly, the identification of this support needs special consideration during the analysis and design phase. Note that in the case of a floating S-lay stinger where there is inherent flexibility and movement along the stinger the transition point can indeed move all the way up above the stinger to the fixed firing line vessel supports.

*19.3.2.2* ***Fatigue Damage*** Another important integrity criterion is the minimization of installation fatigue damage so as to maximize the possible operational life span of the pipeline. As alluded to already, fatigue damage is a dynamic phenomenon where movement along the pipeline generates cyclic stresses that in turn impart damage to the individual pipeline welds.

In a similar vain to local buckling, fatigue damage tends to be most prevalent in the overbend and sagbend regions, in particular where the pipeline comes into intermittent contact with support and seabed surfaces; something which inevitably generates periodic stress cycles. Note that supports in this context refer to the supports toward the bottom of the stinger, tower, or ramp, which come into and out of contact with the pipeline frequently during dynamic motion.

Given that the majority of fatigue damage is accumulated at a number of discrete points along the suspended pipeline, then it is important to ensure that the welds do not come to rest at these points for too long a duration. Therefore, a critical output from the analysis and design process is a maximum allowable standby time, when no new pipe is deployed, which if exceeded may result in the installation fatigue damage threshold being surpassed at a number of welds.

Note that the installation fatigue threshold is traditionally only a small percentage of the total fatigue damage a weld can encounter before failure. Again the selection of an exact installation threshold value is guided by industry standards and recommended practices.

*19.3.2.3* ***Damage to Coatings*** Prevention of damage to pipeline coatings can also be classified as pipeline integrity criterion during installation. Typical coating types include anticorrosion, concrete, and insulation. Each of these types serves an important purpose during the operational life of the pipeline, whether it is avoiding excessive corrosion, providing stability on the seabed, or maintaining necessary temperatures for production fluids.

If a coating is damaged, or compromised, then it can no longer fully fulfill its purpose and this can have significant impact on the pipelines operational performance and can even lead to pipeline failure over time; especially in the case of corrosion. Damage to coatings can be a result of crushing during excessive bending in overbend and sagbend or can arise from friction and impact loads during contact with overbend supports and the seabed.

*19.3.2.4* ***Additional Criteria*** Additional pipeline integrity criteria that also need to be satisfied during installation include

1. Limitation of plastic deformation and ovalization;
2. Achieving a desired level of residual bottom or lay tension;

3. Ensuring the pipeline is installed within a target installation corridor along the pipe route;
4. Correct positioning and orientation of in-line structures along the pipeline.

## 19.4 INSTALLATION ANALYSIS AND DESIGN METHODOLOGIES

The widely accepted approach to predicting what will happen to a given pipeline during installation is to analyze the process from a global perspective using a dedicated finite element analysis tool such as the **PipeLay** software package developed by Wood Group Kenny.

### 19.4.1 Global Installation Analysis

The term global here refers to the fact that the entire length of suspended pipeline is modeled using a string of suitable "pipe" finite elements, which is connected to both an installation vessel and a seabed. These global models and associated analysis tools usually allow for the following:

- Specification of various pipe section characteristics, including geometric properties, such as outer diameter and wall thickness, to capture weight and stiffness effects as well as hydrodynamic properties, such as drag coefficients, to account for water–particle interaction. Coatings on the pipe section can also be specified.
- Inclusion of nonlinear material properties to allow for deformation beyond the linear limit.
- Creation of secondary components such as cables, pull heads, in-line structures, and so on.
- Automated model meshing to ensure more refined element lengths in key areas such as the overbend and sagbend.
- Definition of elastic surfaces, representing both supports on the vessel and the mud line of the seabed, upon which the pipeline rests. The location of vessel supports is completely configurable and so allows for modeling of curved stinger profiles, and so on. The level of the seabed relative to the vessel is controllable through a water depth input and the seabed profile itself can if desired have an arbitrary shape. Also, if needs be, these elastic surfaces can incorporate a Coulomb friction model.
- Specification of both fixed and floating stingers.
- Automatic pressure loads acting on the "pipe" elements to account for increasing hydrostatic pressure when travelling from the water level down to the seabed.
- Creation of sea current profiles that influence the pipeline catenary through a corresponding hydrodynamic loading term.
- Automatic optimization of suspended pipeline shape to achieve a set of specified installation criteria.
- Both static and dynamic temporal analysis. In the case of the latter, the water surface profile incorporates a dynamic wave train that has its own set of water particle velocities and accelerations, which in turn interact with the pipeline through hydrodynamic effects. These wave profiles can be either regular sinusoids or else completely random in nature to reflect the behavior of actual offshore sea states. There are a number of spectrum types that can be used to define these random sea states. The selection of the correct spectrum type is dependent on the oceanographic data for the location where the installation is taking place. The same point applies to the selection of appropriate sea current profiles, as mentioned earlier.
- Dynamic first and second-order vessel motions arising from environmental loads. First-order motions are created by combining the predicted wave profile with response amplitude operators, or RAOs, for the vessel. Second-order motions represent vessel drift or DP generated motion and come directly from a user specified record or time history.
- Capturing the dynamic behavior of tensioner systems when compensating for vessel motion under wave conditions.
- Results postprocessing for fundamental parameters such as pipeline displacements, restoring forces, stresses, strains, as well as other more specific engineering parameters such as vessel loads, support reactions, fatigue damage, local buckling checks, and so on.

### 19.4.2 Methodologies

The preferred methodology used to analyze and design a complete installation project involves a series of steps, the first of which is to identify the key procedures that need to occur during the project. Examples of such procedures include

1. Initiation, which can be done using a variety of methods such as pull-in, dead man anchor, stab and hinge, and so on;
2. Normal lay across a variety of water depths;
3. Abandonment and recovery at various depths;
4. In-line structure installation, including pipeline end terminations, in-line tee assemblies, and so on;
5. Lifts using davit cable;
6. Transfers to other vessel;
7. Final laydown.

***19.4.2.1 Static Analysis*** Once the key procedures are known they are each analyzed statically, typically using a snapshot approach. This snapshot approach is summarized over the following points:

- Identify the governing stages in the procedure where the suspended pipeline length, shape, and response are going to change significantly. These stages reflect a certain level of pipe payout during initiation or a certain length of winch cable during abandonment or a certain water depth during normal lay, and so on.
- Create models and load cases for the governing stages and then through static analysis optimize the profile of the pipeline in each stage to achieve a set of primary installation criteria. These primary criteria are generally vessel-specific restrictions, such as tensioner tension, and once satisfied the analysis predictions should be reviewed to check that pipeline integrity criteria are also met. Refer to previous sections for details on vessel restrictions and integrity criteria.
- The optimization process is usually achieved by adjusting the position of the pipeline touchdown point relative to the vessel until it is in an optimum location that satisfies the criteria.
- If this initial optimization process fails to meet the necessary criteria then more considerable measures may be needed. For example, the angles of the stinger, tower, or ramp may need to be adapted or the individual support elevations may need fine-tuning, or indeed in very severe cases the actual installation vessel may need changing and even possibly the pipe wall thickness, although the window of opportunity to do this is often small to nonexistent.
- When the various installation criteria are satisfied across all stages, then a number of measured parameters from the optimized stage configurations are noted for use in installation analysis reports and tables, which will be later issued to offshore personnel who in turn try to achieve the same optimum parameters during the actual project.

***19.4.2.2 Dynamic Analysis*** When the static analysis cycle is complete, selective dynamic analysis is performed by taking the relevant optimized static snapshots and applying wave loadings to them in a time domain finite element solution. The term selective here refers to the fact that not necessarily all the snapshot stages are analyzed dynamically, but rather the ones that are of most concern from a pipeline integrity point of view. Clearly, the corresponding selection process requires good engineering judgment.

As mentioned before, the wave profile used in the dynamic analyses can either be a regular sinusoid or random in nature. The repetitive nature of the regular wave only requires short-duration analyses; however, it's not fully representative of actual offshore sea states and so only tends to be used for approximating extreme loading conditions.

On the other hand, the random wave is calculated from wave spectrums that are representative of offshore conditions and so in turn it provides more accurate predictions. That being said, due to the nature of random waves the corresponding analysis durations must be a number of hours long in order for the response statistics to reach steady state. This computational overhead with random wave simulation is one of the reasons for selective dynamic analyses. It is also worth noting that random wave analysis is a necessity for predicting realistic fatigue damage estimates.

The main goal of the dynamic analysis process is to determine a combined set of limiting values for the various wave parameters, such as height, direction, and period, which if encountered offshore could jeopardize pipeline integrity criteria as well as possibly causing vessel restrictions to be exceeded. The task of determining a governing set of wave parameters requires a sensitivity study where a load case matrix of different wave parameter combinations is analyzed to identify the respective governing values of the different parameters.

The extent of the dynamic load case matrix can sometimes be simplified by the fact that vessels tend to have a known governing wave heading and wave period that result in the worst dynamic motions; meaning only the limiting wave height needs to be determined. The use of short duration regular wave analysis to screen out insignificant load cases is also another method of reducing the load case matrix prior to performing time consuming random wave analyses. Note that it is important to exercise sound engineering judgment when adopting any simplification or reduction measures to the dynamic analysis process as the ability to use such measures safely is very much dependent on the scenarios under consideration.

Once the limiting wave parameters, or sea states, have been determined for the selected installation procedures and stages then they are noted in the installation analysis reports and tables so that vessel personnel can use them for monitoring and planning purposes offshore. Offshore monitoring is outlined further in a subsequent section.

Given the methodology described over the last number of paragraphs, considerable effort and expertise can be required to successfully complete the analysis and design phase of an installation project. For this reason, the industry is continually looking to develop new technology, particularly software, to help improve the ease at which analysis and design issues are addressed while also ensuring the necessary levels of detail and quality are maintained throughout. Wood Group Kenny PipeLay software is an example of such technology.

## 19.5 MONITORING THE INSTALLATION PROCESS OFFSHORE

Once the analysis and design phase is complete and the various installation procedures have been optimized and documented it is then possible to begin the offshore project. To guarantee safety and pipeline integrity during the project the entire installation process needs to be monitored throughout by vessel personnel.

### 19.5.1 Monitoring Process and Remedial Action

Offshore monitoring often involves comparing physical measurements of actual pipeline and vessel behavior against recommended parameters from the documented installation reports and tables. Typical parameters that are measured and compared include vessel position and motions, sea state variables, tensioner or winch tensions, support separations, support loads, touchdown point position, as well as possibly sea current profiles.

The various measurements require a mixture of different sensors and instrumentation most of which are located on the vessel, but in some cases they need to be deployed in the water. For example, an underwater ROV is necessary to record the touchdown point position. Given the variety of required measurements and corresponding instruments, considerable investments are being made in the IT systems on modern day installation vessels so that all the necessary information is relayed to the bridge and displayed in a clear and concise manner to the vessel management teams.

If while monitoring it becomes apparent that the measurements are deviating away from recommended values then remedial action may be necessary. In some cases these actions are relatively trivial, such as moving the vessel into a more optimum position to keep the pipeline on route or adjusting the suspended length slightly to ensure adequate tension levels are maintained. In other scenarios more significant action is required; particularly when the sea state variables are worsening and approaching their limiting values, in which case the pipeline may need to be abandoned on the seabed until there is an improvement in conditions.

### 19.5.2 Monitoring Analysis Software

The decision making and planning process behind remedial actions is traditionally based on the combined experience of the vessel management team as well as supporting documentation from the previous analysis and design phase. However, there is also a movement within the industry to complement this traditional process by performing live finite element simulations offshore on the vessel using real time measurements and forecast data. This is best done using specialized software, such as Wood Group Kenny's **OptiLay** package, that automatically acquires data from the various sensors and instrumentation before using it to create both static and dynamic analysis load cases, which in turn are applied to suitably selected models.

The autonomous nature of such software means that analysis predictions are provided at a high frequency and the fact that the load cases are based on actual offshore measurements, rather than assumed conditions, results in reduced conservatism when compared with onshore analysis. Both of these benefits can be amplified further when actual recorded vessel motions, due to wave loading, are used in the dynamic simulations rather than artificial motions generated from vessel response amplitude operator (RAOs) in combination with a random wave sea state, something which requires long-duration analyses and can be inherently conservative due to the RAOs themselves and the fact that all the wave energy tends to be concentrated in a single direction.

Taking advantage of the benefits from specialized onboard software improves offshore monitoring and planning by providing clarity on pipeline integrity for present and future conditions. Such clarity helps establish more optimum timeframes for performing key procedures, such as start-up, abandonment, recovery, and so on. Ultimately, this potentially allows for installation in more severe conditions that would be worse than the limiting conditions identified during the analysis and design phase, when an inherent amount of conservatism has to be used.

Another advantage of using such software on board the vessel is that it accurately determines and catalogues the fatigue damage incurred across an entire pipeline over the duration of the installation project. This catalogue of installation fatigue damage is inherently more realistic and less conservative than estimates derived onshore and so allows for improved fatigue life calculations, which can be beneficial to pipeline operators.

## 19.6 IMPLICATIONS OF DEEPER WATER ON INSTALLATION

The offshore oil and gas sector is continually moving into ever deeper water as it tries to gain access to new virgin fields. This movement to deeper water represents a strong growth area for the industry over the coming decades. To cater for this growth it is necessary to safely install pipelines at depths that would have been unthought-of in the recent past, for example, pipelines are now being installed in ultra-deep water depths in excess of 2000 m and indeed approaching 3000 m.

### 19.6.1 Increased Tension and Potential for Local Buckling

As mentioned already, the primary implication of deeper water installation is the growth in length of pipeline suspended from the installation vessel. This growth is encountered primarily along the inflection region between the overbend and sagbend, which is stretched out in deeper water, and this in turn results in increased axial tension loads acting on the pipeline and vessel.

Larger tensions make the selection of an installation vessel, and indeed installation method, more difficult as after a certain threshold the tension will exceed the capacities of many of the tensioner support systems on particular vessels. To help minimize increases in tension with depth, the inflection region is orientated at a relatively steep angle, so as to keep the suspended length down, but this then means a larger level of bending in the sagbend and, in the case of S-lay, the overbend.

The greater level of bending can have an impact on pipeline integrity as it introduces more potential for local buckling failure. In the sagbend, the additional water pressure with depth exacerbates the issue of local buckling even further and so the use of buckle arresters can be required in order to prevent catastrophic propagation of wet buckles.

### 19.6.2 Plastic Strains

Increased bending in the S-lay overbend in combination with the larger tension is also going to induce greater levels of plastic strain. As discussed in a previous section, this plastic strain manifests itself as residual hogging curvature, which in turn causes pipe roll as it passes through the sagbend.

Residual curvature and pipe roll can have implications for the response of the remaining regions along the suspended pipeline. For example, as the residual curvature is straightened in the tension dominated inflection region then inevitably a bending moment is generated where previously there was none. This is because any change in curvature, whether residual or not, requires a moment force.

Also, if the level of pipe roll is limited, for example, $<20°$, then as the residual curvature enters the sagbend it can cause an increase in bending moment as the sagging curvature is somewhat forced to counteract the residual hogging curvature. On the other hand, if the level of roll is significant, for example, approaching $90°$, then the residual curvature aligns itself with the sagbend curvature and therefore less bending moment is generated than in the case of no residual curvature.

Predicting such residual curvature and pipe roll implications is difficult and so to date they have generally been ignored or else assumed to be negligible in magnitude. However, with the continued growth of ultra-deep steep S-lay installation such an approach may need to be revised. The same point also applies to ovalization during installation whose magnitude increases with deeper water.

### 19.6.3 Prolonged Fatigue Exposure

Another effect of the longer suspended length in deep water is that the time taken for a pipe joint to reach the seabed grows considerably and so the joint is exposed to installation fatigue damage for a longer duration than in shallow water. This often means that there is not much contingency in terms of how long the installation vessel can go into standby mode, when no pipe is being deployed, before the design threshold for installation fatigue damage is exceeded in certain pipe joints suspended from the vessel. Note that this is more an issue for S-lay and J-lay rather than Reel-lay where the installation time is usually much shorter.

### 19.6.4 Design Implications

Given the relatively severe implications of deep water installation on pipeline integrity and vessel selection, the overall design of deep water pipelines is becoming ever more dependent on the possible installation method and associated procedures. This point particularly applies to wall thickness specification and material selection. For example, wall thickness in deeper water may need to be limited in order minimize the tension load on the installation vessel; however, at the same time the pipeline needs a greater load capacity to ensure integrity during the installation. These contradicting objectives can frequently only be satisfied through the selection of a higher grade of steel material, but even this solution eventually runs into problems when a certain threshold of material grade is reached.

In order to better address the issues of deeper water the industry is, and will have to continue, investing in newer generations of installation vessels, with more tension capacity, and so on, as well as other novel technologies and procedures that help improve the ease at which a given pipeline can be installed safely.

## REFERENCE

1. Kyriakides, S. and Corona, E. (2007) *Mechanics of Offshore Pipelines, Volume 1: Buckling and Collapse*, Elsevier, Oxford.

## BIBLIOGRAPHY

API (2000) *API 5L, Specification for Line pipe*, 42 ed., American Petroleum Institute, Washington.

Brooks, J., BP America, Inc., Cook, E.L., Consultant, and Hoose, J., INTEC Engineering. (2004) Installation of the mardi gras pipeline transportation system, Proceedings of Offshore Technology Conference, Houston, TX, 2004, OTC 16638.

BSI (1989) BS 8010, British Standard Code of Practice for Pipelines. British Standard Committees, London, England.

Bullock J.D., II, Geertse, E.M., and Landwehr, M.M., Subsea 7 Inc. (2011) Versatility in answering the challenge of deepwater field developments, Proceedings of Offshore Technology Conference, Houston, TX, 2011, OTC 21821.

Choi, H.S., Pusan National University and Jo, H.J., Korea Maritime University. (1999) Characteristics of ultra-deepwater pipelay analysis, Proceedings of Offshore Technology Conference,. Houston, TX, 1999, OTC 10710.

Det Norsk Veritas (2013) Submarine Pipeline Systems. DNV-OS-F101, October.

Dixon, M., Jackson, D., DeepSea Engineering & Management, and El-Chayeb, A., Stolt Offshore. (2003) Deepwater installation techniques for pipe-in-pipe systems incorporating plastic strains, Proceedings of Offshore Technology Conference,. Houston, TX, 2003, OTC 15373.

Endal, G., Ness, O.B., Verley, R., Holthe, K., and Remseth, S. (1995) Behaviour of offshore pipelines subjected to residual curvature during laying, Proceedings of the 14th International Conference on Offshore Mechanics and Arctic Engineering. pp. 513–523.

Endal, G., and Verley, R. (2000) Cyclic roll of large diameter pipeline during laying, Proceedings of the 19th International Conference on Offshore Mechanics and Arctic Engineering, New Orleans.

Gernon, G.O., Kenney, T.D., McDermott Incorporated, Harrison, G., Hudson Engineering, and Prescott, C.N., Kvaerner Earl & Wright, (1995) Installation of deepwater pipelines utilizing S-lay methods, Proceedings of Offshore Technology Conference,. Houston, TX, 1995, OTC 7843.

Grealish, F., Lang, D., Connolly, A., and Lane, M. Advances in contact modelling for simulation of deepwater pipeline installation. Proceedings of the Rio Pipeline Conference and Exposition, October 17–19, 2005, Rio de Janeiro, Brazil.

Hauch, S. and Bai, Y. (1998) Use of finite element analysis for local buckling design of pipelines. Proceedings of the 17th International Conference on Offshore Mechanics and Arctic Engineering. Lisbon.

Hauch, S. and Bai, Y. (1999) Bending moment capacity of pipes, Proceedings of the 18th International Conference on Offshore Mechanics and Arctic Engineering. Offshore Mechanical and Artic Engineering.

Heerema, E.P., Allseas Group S.A. (2005) Recent achievements and present trends in deepwater pipe-lay systems, Proceedings of Offshore Technology Conference,. Houston, TX, 2005, OTC 17627.

Ilie, D. and O'Grady, R. (2009) Fatigue analysis of staged pipelay operations, Proceedings of the 1st Floating Structures for Deepwater Operations Conference. 21–23 September, 2009, Glasgow, UK.

Kluwen, F., Allseas USA Inc. and Rijneveld, P., Allseas Engineering bv. (2007) Installation challenges/SCR/in-line tees, Proceedings of Offshore Technology Conference,. Houston, TX, 2007, OTC 19060.

Kopp, F., Light, B.D., Preli, T.A., Rao, V.S., and Stingl, K.H., Shell international E&P—EP projects. (2004) Design and installation of the na kika export pipelines, flowlines and risers, Proceedings of Offshore Technology Conference,. Houston, TX, 2004, OTC 16703.

Macara, J.C., Shell Philippines Exploration B.V. (2002) Malampaya deep water gas pipeline and flowlines: technical and engineering challenges faced in the execution of the malampaya pipeline scope, Proceedings of Offshore Technology Conference, Houston, TX, 2002, OTC 14040.

Mckinnon, C., J P Kenny Ltd, Staines, Middlesex, UK. (1999) Design, material and installation considerations for ultra deepwater pipelines, Proceedings of Offshore Europe Conference, SPE 56910. Aberdeen, Scotland, 1999.

O'Grady, R., Bakkenes, H.R., Lang, D., and Connaire, A. (2008) Advancements in response prediction methods for deep water pipe-in-pipe flowline installation, Proceedings of the Offshore Technology Conference. May 5–6, 2008, Houston, TX.

O'Grady, R. and Harte, A. (2013) Localised assessment of pipeline integrity during ultra-deep S-lay installation. *Journal of Ocean Engineering*, 68, 27–37.

O'Grady, R., Ilie, D., and Lane, M. (2009) A novel approach to pipeline tensioner modeling, Proceedings of the Rio Pipeline Conference and Exposition. September 22–24, 2009, Rio de Janeiro, Brazil.

O'Grady, R., Lang, D., and Lane, M. (2010) An improved approach to modelling in-line structure installation, Proceedings of the 5th International Offshore Pipeline Forum. October 20–21, 2010, Houston, TX.

Perinet, D., Acergy France and Frazer, I., Acergy UK. (2008) Strain criteria for deep water pipe laying operations, Proceedings of Offshore Technology Conference, Houston, TX, 2008, OTC 19329.

Perinet, D., and Frazer, I., Acergy. (2007) J-lay and steep S-lay: complementary tools for ultradeep water, Proceedings of Offshore Technology Conference,. Houston, TX, 2007, OTC 18669.

PipeLay User Manual. (2014) MCS Kenny Software. Version 3.1, March 2014, Galway, Ireland.

Pipeline Installation and Fatigue Monitoring System Operators Procedures Manual. (2012) MCS Kenny Software. Version 2.1, August, Galway, Ireland.

Rashdi, K.R., PETRONAS Carigali Sdn. Bhd., Sainal, M.R., PETRONAS, and Yusoff, M.N.M., PETRONAS Carigali Sdn. (2008) Samarang pipeline-replacement project: application of an innovative approach for shallow-water pipeline installation, Proceedings of Society of Petroleum Engineers Annual Technical Conference and Exhibition,. 2008, Denver, Colorado, USA, SPE 114888,

Wincheski, R., Bertrand, C., SIEP/ Dampman, B., Pegasus International/ Eisenhauer, D., MMI. (2002) Brutus export pipelines—improvement opportunities and challenges in deepwater pipeline installation, Proceedings of Offshore Technology Conference, Houston, TX, 2002, OTC 13994.

Yun, H.D., Peek, R.R., PasLay, P.R., and Kopp, F.F. (2004) Loading history effects for deep-water S-lay of pipelines. *Journal of Offshore Mechanics and Artic Engineering*, 126, 156–163.

# PART III

# THREATS TO INTEGRITY AND SAFETY

# 20

# EXTERNAL CORROSION OF PIPELINES IN SOIL

Homero Castaneda and Omar Rosas

*Chemical and Biomolecular Engineering Department, National Center for Research and Education in Corrosion and Materials Performance, The University of Akron, Akron, OH, USA*

## 20.1 INTRODUCTION

Current technologies for detecting and monitoring the state of external corrosion of oil and gas pipelines consider the system as electrochemical or electrical in nature. Modeling of external corrosion has become an important tool to implement the current preventative technologies and is used to support analyses of the information acquired from each survey or inspection method. Critical factors incorporated into modeling help to characterize not only the position of external corrosion but also the time-dependent effect of the dissolution mechanism. These factors range from macro- to microscale in nature. This chapter summarizes the description, quantification, and influence of critical factors for external corrosion of buried pipelines. The classification of the factors is based on the multiscale approach for onshore pipelines, considering the soil/pipeline system as an electrochemical cell.

## 20.2 BACKGROUND

Pipelines are exposed to and interacting with various environmental elements causing integrity damage. The soil, oxygen in the air, rainfall, temperature, moisture, and electrolytes suspended in the atmosphere can interact with the pipeline and cause it to corrode. Corrosion is a time-dependent threat in pipelines exposed to underground conditions and is one of the prime causes of pipeline failure. Soil, as defined by Rim-Rukeh et al. [1], is a complex material consisting of a discontinuous and heterogeneous environment that is constituted by an organic solid phase, liquid water phase, air, and other gas phases. It is estimated that between two and four pipeline failures occur every month in the United States, with the most common failure mechanism being localized corrosion [2,3].

In soils, water and gas occupy the spaces between solid particles, and these spaces can constitute as much as half the volume of dry soil. Some of the water in soil is bound to mineral surfaces, whereas, bulk water can flow through porous soil. Fluid flow through soil is controlled by the permeability of the soil, which, in turn, depends on the size distribution of the solid particles in the soil. For example, good drainage can occur in coarse-grained sand, and atmospheric oxygen can penetrate to a depth greater than that which would be reached, for example, in a fine-grained soil high in clay. Capillary action in fine-grained soil can draw water up, keeping the soil saturated with water, preventing drainage, retarding evaporation, and restricting oxygen access from the atmosphere to the pipeline.

The electrochemical corrosion processes taking place on metal surfaces in soils occur in groundwater that is in contact with the metallic structures. Damage evolution of the pipe coating, of the coating/steel interface, and of the steel beneath the coating is based on initiation conditions and time-dependent parameters that produce and influence various processes. Both, the soil and the climate influence the groundwater composition and the factors that control the damage/failure process.

The damage evolution process of coated underground pipelines can be defined in four stages as illustrated in Figure 20.1. Stage I is the initiation, which comprises the

*Oil and Gas Pipelines: Integrity and Safety Handbook,* First Edition. Edited by R. Winston Revie.

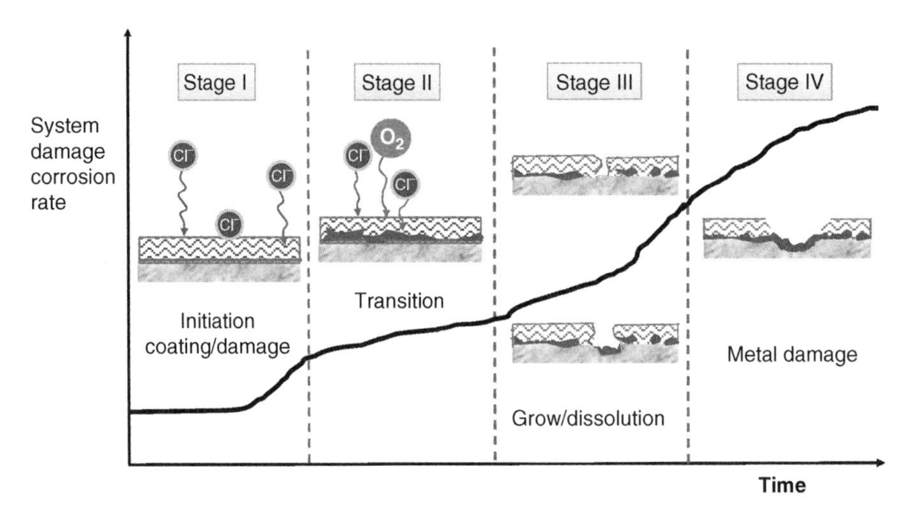


**FIGURE 20.1** Damage evolution stages for buried pipeline/coating systems.

transport of species towards and within the pipe coating under cathodic protection (CP) conditions. Stage II includes the activation of the steel substrate and corrosion product formation, during which any change in volume or physical modification of the pipe coating is a natural consequence. A holiday, which is a defect in the pipe coating, is an indication for stage III, where the environment is in direct contact with the steel substrate. Stage IV, the final stage, is the failure or metal loss leading to a leak or rupture of the pipe. The parameters of each stage that are critical to the performance of the corrosion control and protection system are explored in this chapter.

Pipelines are expected to function reliably and continuously over several decades without failure or extended damage from corrosion. Various parameters in the soil can affect pipeline damage, for example, pH, moisture content, resistivity, ion content, redox potential, and type of soil, as well as the oxygen content and microbes present in the soil [4]. The different components that form the electrochemical cell affect the damage evolution and will influence, directly or indirectly, each of the damage evolution stages. The electrochemical and transport processes at the metal/coating interface are influenced by the water uptake on the metal surface as well as the geometry and chemistry of the film in addition to the oxide layer formation on the surface and the interactions with the soil. The external interactions with the soil are electrical and mechanical in nature and include stray currents and telluric effects. The microenvironments due to oxygen, moisture, and other species migrating through the soil to the metal surface control these macro metal/moisture environments and influence corrosion. The macro environment will further influence the soil condition and corrosivity [5]. A multi-scale concept becomes important for the critical factors for external corrosion in buried onshore pipelines.

## 20.3 CRITICAL FACTORS OF SOIL CORROSIVITY THAT AFFECT PIPELINES

Based on a multi-scale concept, various factors affect the aggressivity of the soil, which, in turn, influences the integrity and reliability of a pipeline. Most often, the corrosivity of the soil determines the mode of protection offered to a pipeline as well as the survey technology used, and the current estimate of reliability. Several important variables have been identified that have an influence on corrosion rates and damage evolution in soils; these include the type of soil, amount of water, degree of aeration, pH, redox potential, resistivity, soluble ionic species (salts and organics), and microbiological activity. The complexity of soil is pictured in Figure 20.2. Some variables affecting pipeline corrosivity are discussed in the following section.

### 20.3.1 Soil Types and Resistivity

Historically, resistivity has often been used as a broad indicator of soil corrosivity. Because ionic current flow is associated with transport mechanisms leading to corrosion, high soil resistivity will arguably slow down the transport kinetics for each stage in the damage evolution process. Liu et al. [7] found that soil resistivity was the most important factor contributing to pipeline damage. The resistivity value of the soil decreases with the increase of the moisture content in the soil [8]. Obtaining a good correlation between resistivity observed in the field and in the laboratory has yet to be achieved [9]. Soil resistivity generally decreases with increasing water content and the concentration of ionic species. It is by no means the only parameter affecting the risk of corrosion damage; a high soil resistivity alone will not guarantee the absence of metallic and coating damage.

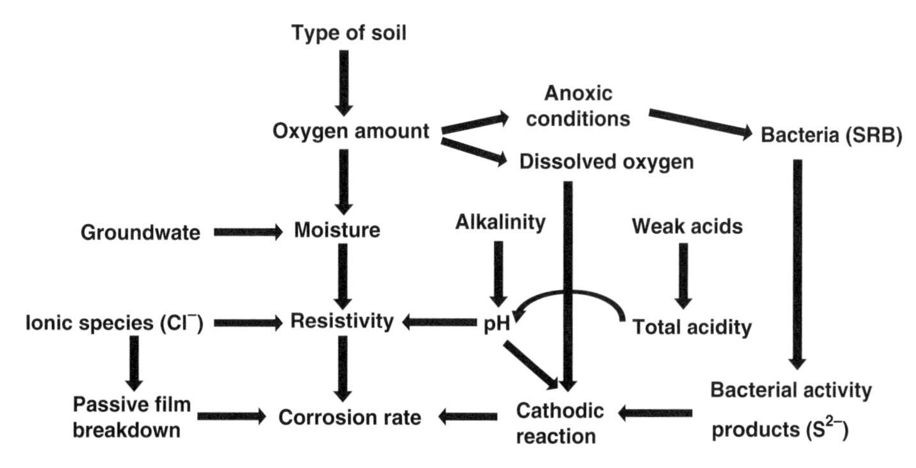


**FIGURE 20.2** Variables affecting corrosion in soil. MIC effects are presented only for sulfate-reducing bacteria. (Adapted from Ref. [6].)

Variations in soil resistivity along the length of a pipeline are highly undesirable because they will lead to the formation of macro corrosion cells. Therefore, for pipelines, the merit of a corrosion risk classification based on an absolute value of soil resistivity is not robust. Soil resistivity can be measured by the so-called Wenner four-pin technique as illustrated in Figure 20.3 or, more recently, by 3D electromagnetic survey technologies. The latter allow measurements in a convenient manner and at different soil depths. Another option for soil resistivity measurements is the so-called soil box method, whereby, a sample is taken during excavation. Preferably, the sampling will be performed in the immediate vicinity of the pipeline (e.g., in a pipe trench). Table 20.1 shows the effect of soil type and moisture content on the corrosive behavior of the soil.

### 20.3.2 Water Coverage Due to Vapor Transportation and Drainage

Water in liquid form represents the essential electrolyte required for electrochemical corrosion reactions. In soils, a distinction is made between saturated and unsaturated water flow/accumulation. Unsaturated water flow refers to the movement of water from wet areas toward dry soil areas. Saturated water flow is dependent on pore size and distribution, texture, structure, and organic matter. High moisture content facilitates faster transport of ions between the pipeline surface and the soil, causing the soil to be more corrosive. Water movement in soil can occur by the following mechanisms: gravity, capillary action, osmotic pressure (from dissolved species), and electrostatic interaction with soil particles. The water-holding capacity of a soil is strongly dependent on its texture; coarse sands retain very little water while fine clay soils store water to a high degree. Gupta et al. [13] observed that the maximum corrosivity of steel occurs when the moisture content is 65% of its water-holding capacity and termed this as the "critical moisture-holding content." Moreover, moisture content is not the only factor that is found to control the corrosion of ferrous metals. Neale et al. [14] observed that depending on the soil type, the saturation of water in the soil may vary from as low as 0.5% for sand to as high as 217% for bentonite clay. In addition, it

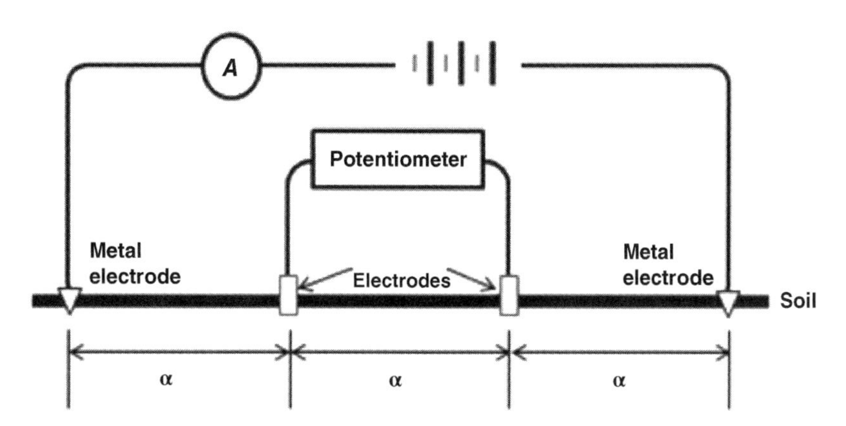


**FIGURE 20.3** Wenner four-pin technique for measuring soil resistivity. (Adapted from Ref. [12].)

**TABLE 20.1 Soil Types/Resistivity and their Effect on the External Corrosion Rate of Bare Carbon Steel with no Cathodic Protection [10,11]**

| Soil Type | Soil Resistivity (ohm-cm) | Moisture | Corrosion Rate (mm/year) |
|---|---|---|---|
| Muskeg/sloughs/free water accumulations | <500 | Always wet | Very corrosive (>1.0) |
| Loams/clays | 500–2000 | Mainly wet | Corrosive to moderately corrosive (0.5–1.0) |
| Gravels, sandy | 2000–10,000 | Mainly dry | Mildly corrosive (0.2–0.5) |
| Arid, sandy | >10,000 | Always dry | Noncorrosive (<0.2) |

has also been reported that low moisture content favors localized corrosion [5]. Castaneda et al. [15] created drainage maps for water accumulation at different depths. The spatial distribution of precipitation, evapo-transpiration, and drainage was used to map water accumulation, as illustrated in Figure 20.4.

### 20.3.3 pH of Soils

Alkaline soil conditions can be corrosive or lead to threats to steels under certain conditions. This is corroborated by thermodynamic data for Fe–$H_2O$ equilibrium. Moreover, pipelines at high pH (more than 11 or depending upon the iron concentration, temperature, and ionic species in solution) can produce alkali and hydrogen formation producing localized conditions for active corrosion, coating delamination, or hydrogen embrittlement [16]. Acidic environments, however, are detrimental to steel pipelines. Acid acts as a depolarizing agent and causes difficulty in the polarization of the pipeline along the protective potential. Thus, a higher current density is required to maintain CP in the acidic area. Table 20.2 shows, in general, the effect of pH on the corrosion of buried steel pipelines [17], although factors other than pH can also be important and must be considered. Stress corrosion cracking (SCC) is a corrosion mode influenced by the pH of the soil. Near neutral SCC and alkaline SCC include two different forms of attack that are produced by different mechanisms affecting steel pipelines. The former produces transgranular attack; the latter includes intergranular attack.

### 20.3.4 Chlorides and Sulfates in Soils

The chlorides and sulfates present in the soil can also influence the corrosivity of the soil and damage evolution of the pipeline/soil system. Their contribution to the corrosion process is significant, and at high concentrations, they may cause severe corrosion in steel pipelines. Chlorides promote corrosion due to their conductive nature. In addition, they inhibit the formation of a passive oxide film, thus, promoting localized corrosion. Moreover, the presence of chloride tends to decrease soil resistivity [1]. Chloride ions can originate from brackish groundwater, mine drilling shafts, and from human activities, such as the de-icing of

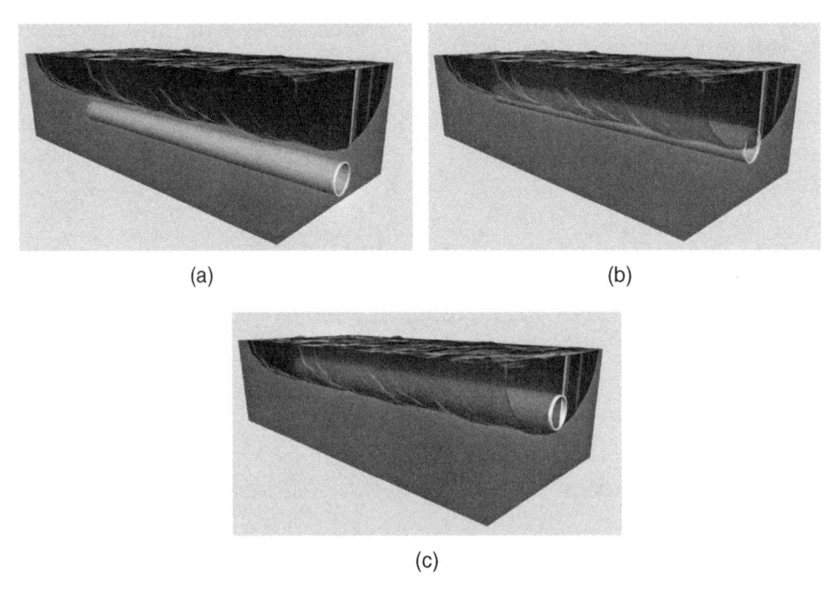


**FIGURE 20.4** Varying quantities of water covering the entire pipe: (a) no coverage, (b) partial coverage, and (c) total coverage.

**TABLE 20.2 Effect of pH on the Corrosion of Buried Steel Pipelines[a]**

| pH | Relative Corrosivity |
|---|---|
| <5.5 | Severe |
| 5.5–6.5 | Moderate |
| 6.5–7.5 | Low |
| >7.5 | Very low |

[a]*Note*: Factors other than pH (e.g., resistivity and moisture) can also be important and must be considered.

*Source*: Adapted from Ref. [17].

roads. The concentration of chlorides can also vary with the season. Data have shown that there is a systematic decrease in soil resistivity with an increase in both chlorides and sulfates [18]. Table 20.3 shows the effect of chlorides and sulfates on the corrosion of steel pipelines.

### 20.3.5 Differential Aeration Corrosion Cells

Differential cell corrosion is the most common mechanism that causes localized corrosion of pipelines. The soil corrosivity in underground pipelines can be influenced by the level of oxygen concentration at different regions of the pipeline, creating differential cell aeration, as well as by any differences in soil chemistry or pipeline surface [6]. The region of higher oxygen concentration acts as a cathode, whereas, the oxygen-deficient region acts as an anode. Differential cell corrosion can be autocatalytic in nature. The chemical and electrochemical reactions and ion migration tend to generate conditions favorable to the continuation of the electrochemical cell [6]. Figure 20.5 depicts a pipeline passing through two different types of soil. The potential of the pipeline in the clay soil is more anodic compared with the potential of the pipeline in the dry sandy soil, as a result, a corrosion cell is generated. The pipeline in the clay soil is corroded, whereas, the pipeline in the sandy soil is protected. Differential corrosion cells can also be generated by differences in

**TABLE 20.3 Effect of Chlorides and Sulfates on the Corrosion of Buried Steel Pipelines**

| Concentration (ppm) | Relative Corrosivity |
|---|---|
| **Chloride** | |
| >5000 | Severe |
| 1500–5000 | Considerable |
| 500–1500 | Moderate |
| <500 | Low |
| **Sulfate** | |
| >10,000 | Severe |
| 1500–10,000 | Considerable |
| 150–1500 | Moderate |
| 1–150 | Negligible |

*Source*: Adapted from Ref. [10].

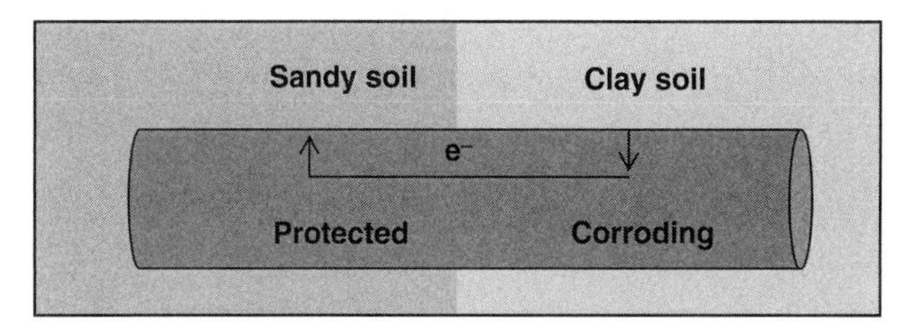


**FIGURE 20.5** A differential corrosion cell generated due to differences in the soil type.

moisture content in soil or differences in oxygen diffusion through the soil.

### 20.3.6 Microorganisms in Soils

The presence of bacteria and fungi in the periphery of the pipeline can change the corrosion behavior of the pipeline. This is also known as microbial-induced corrosion (MIC) [19–22]. This type of corrosion is estimated to contribute up to 20–30% of the external corrosion of pipelines [23]. Certain bacteria can exist in the absence of oxygen (anaerobically) at the surface of the pipeline and electrochemically reduce any sulfates present therein while consuming hydrogen during the process. A hydrogen deficiency will depolarize the steel at the cathodic areas, resulting in more rapid degradation of the metal/pipeline. Thus, the corrosion of steel pipelines is accelerated in the presence of sulfate-reducing bacteria (SRB) because they create an environment that supports a more rapid attack. The bacterial activity causes the open circuit potential (OCP) of the anodic areas to be more negative. The attack is aggressive in the presence of bacterial communities with acid-producing bacteria causing the most aggressive corrosion of underground pipelines [23]. A higher than normal potential is used during CPof such pipelines. An additional −0.1 V in addition to the conventional −0.85 V copper sulfate electrode (CSE) is suggested under this condition [10,24,25].

### 20.3.7 Redox Potential

The redox potential measures the degree of soil aeration. It is also an indicator of whether the soil can sustain SRB. A high redox potential value indicates higher oxygen content and vice versa. SRB proliferate at low redox potentials [26,27]. It has also been observed that redox potential and resistivity are better indicators of soil corrosiveness than moisture content [17].

## 20.4 IDENTIFYING CORROSIVE ENVIRONMENTS

Ideally, any highly corrosive environments along a proposed pipeline route would be identified before pipeline installation. Various American Society for Testing and Materials

(ASTM) standard methods can be used to test for soil resistivity and pH. ASTM G57 is one such test for field measurement of soil resistivity. Other ASTM test methods include ASTM G51-95(2102) for measuring the pH of soil, ASTM G200-09 for the measurement of oxidation–reduction potential (ORP) of soil, and ASTM G162-99(2010) for conducting and evaluating laboratory corrosion tests in soils [28–31].

## 20.5 CATHODIC PROTECTION AND STRAY CURRENTS

After identifying the risk or likelihood for corrosion of a buried pipeline in a certain environment by evaluating the factors involved in this process, different methodologies have to be followed to either prevent or counterattack the effect of corrosion. The main methods for corrosion mitigation on underground pipelines are coatings and CP (discussed in Chapters 30 and 32, respectively).

On a cathodically protected pipe, the coating reduces the surface area of exposed steel on the pipeline, thereby reducing the current necessary to cathodically protect the steel. CP is used to reduce the corrosion rate of a metal surface by making it the cathode of an electrochemical cell.

The principle for CP is illustrated in Figure 20.6 [6]. The electrons are supplied to the pipeline by using an ac source coupled to a rectifier and an anode. In the case of a coated pipeline, current flows to the areas where the coating is defective. An electron current flows along the electric cables connecting the anode to the cathode, and ionic current flows in the soil between the anode and cathode to complete the circuit.

The primary goal of CP is lowering the potential of the buried structure. As long as the cathodic current increases, the potential of the metal decreases; for each increment that the potential of the metal is reduced, the current requirements tend to increase exponentially.

A poor CP design may be a cause of damage to the pipeline; some of the key indicators are the bad current distribution, interferences, or stray currents. Stray currents are currents flowing in the electrolyte (soil) from external sources not associated with the CP system. Corrosion damage to an underground structure caused by a CP system on another structure is a form of stray current corrosion that is commonly called "interference." Stray current damage is most commonly associated with impressed current CP systems. A foreign pipeline or other metallic structure can form a second resistance path where the current distribution can be deviated from the coverage areas that the pipeline requires to complete the protection system.

Figure 20.7 illustrates a foreign pipeline passing through a zone of positive soil potentials (area of influence) and be covered by impressed current system. If we have a pipeline crossing the one selected for protection at a more remote location. The positive soil potentials will force the foreign pipeline to pick up current at points within the area of influence. This current must then complete the electrical circuit and return to the negative terminal of the DC power source. The figure illustrates this by showing the current path distribution flowing along the foreign line (non-original protected) toward the point where the two lines cross and then leaving the foreign line in the vicinity of the crossing. This current then flows into the protected pipeline and returns to the rectifier. The area where the current leaves the foreign line in the vicinity of the crossing becomes a potential site for corrosion in the foreign pipeline. Usually, a small amount of current will flow along the foreign pipeline in the opposite direction from the ground bed area. This is indicated as final current path in the figure. This current will leave the foreign pipeline at remote locations, usually in areas of relatively low

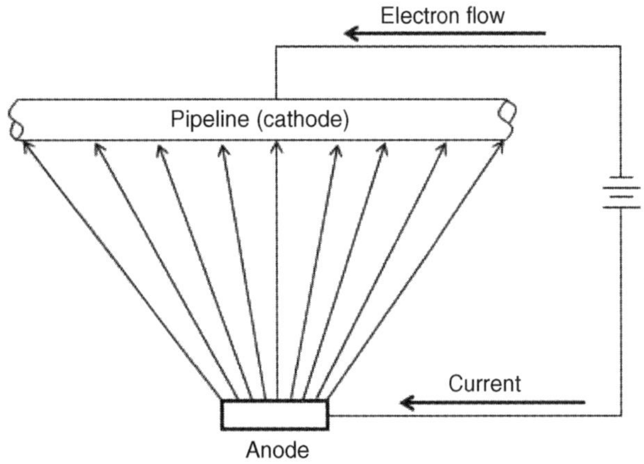

Electron flow
Pipeline (cathode)
Current
Anode

FIGURE 20.6 Current distribution during the cathodic protection of a pipeline. (Adapted from Ref. [6].)

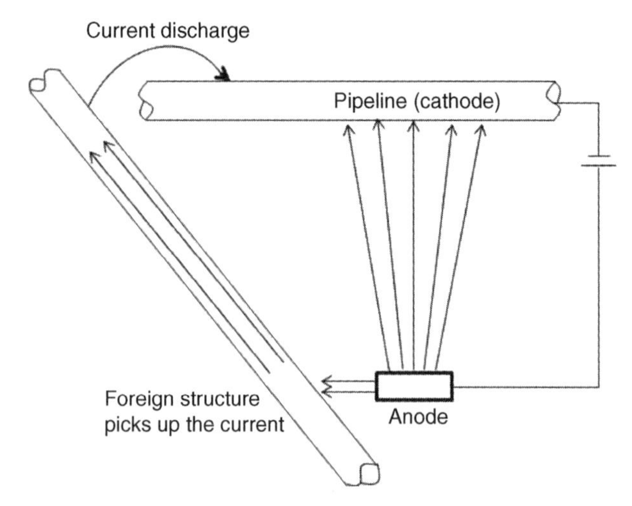


FIGURE 20.7 Anodic and cathodic interference example. (Adapted from Ref. [6].)

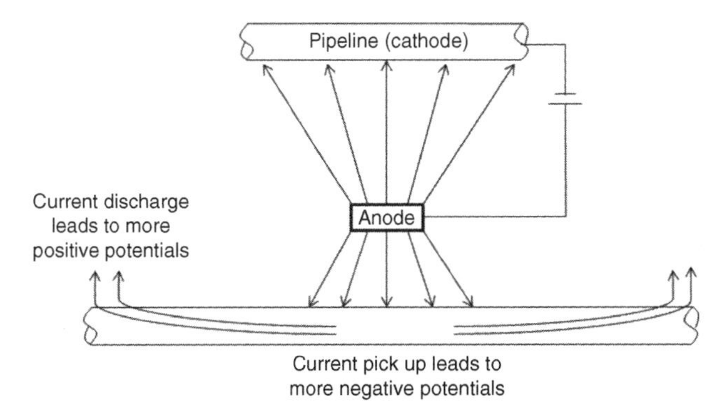


**FIGURE 20.8** Anodic interference example. (Adapted from Ref. [6].)

soil resistivity. The severity of the effect is largely a function of the impressed voltage and the proximity of the foreign pipeline to the anodic bed. Where the impressed voltage is high and the foreign pipeline is close to it, current forced onto the foreign pipeline tends to be high and can cause damage. In such instances, the foreign pipeline represents a potential threat leading to weak integrity. Once the source is identified, the current path can be distributed to correct the problem. A better electrical balance can produce suitable current distribution and avoid the high-current drainage at the potential corrosion location; connecting a resistance between the two metallic conductors (protected and foreign pipelines) the magnitude of the electrical passive element (resistance) can be calculated in order to distribute the current drain from specific location to higher areas. CP design (or redesign) can also be considered as a preventive or control action, the location of the anodic bed, addition of passive elements (such as resistors), and more efficient current pathways in the electrical configuration are some examples of remediation actions [10].

Another condition of a poor CP design is depicted in Figure 20.8. This figure illustrates a condition where a foreign structure approaches a CP anodic bed but does not cross the protected pipeline. In this case, the foreign pipeline is forced to pick up current in the area of anode influence and this current will flow along the foreign structure away from the anode. This stray current leaves the foreign structure in remote areas with low soil resistivity to flow to the protected pipeline and then back to the rectifier to complete the circuit. In this way, there may be many areas of current discharge and damage to the foreign structure rather than a single discharge area, as it was described in the previous case. Corrective actions may include the use of bond cables from the foreign structure to the negative terminal of the rectifier or the installation of CP systems on the foreign structure to reverse the flow of current. As in the preceding case, current pickup by the foreign pipeline may be so intense that correction of the condition may not be practical if the anode is too close to the foreign pipeline In areas of high soil resistivity, where relatively high-voltage rectifiers may be used, the area of influence surrounding an anodic bed may extend for several hundred feet. Tests must be made to assure that neighboring structures will not be damaged or that the possibility of damage can be corrected [10].

## REFERENCES

1. Rim-Rukeh, A. and Awatefe, J.K. (2006) Investigation of soil corrosivity in the corrosion of low carbon steel pipe in soil environment. *Journal of Applied Sciences Research*, 2, 466–469.
2. Kirmeyer, G.J., Richards, W., Weston, R.F., and Smith, C.D. (1994) *An Assessment of the Water Distribution System and Associated Research Needs*, American Water Works Research Foundation, Denver.
3. Pipeline & Hazardous Materials Safety Administration (PHMSA) (2015) *Incident Data Base*, PHMSA, Washington, DC. Available at http://www.phmsa.dot.gov/pipeline/library/datastatistics/pipelineincidenttrends.
4. Li, S.Y., Jung, S., Park, K., Lee, S.M., and Kim, Y.G. (2007) Kinetic study on corrosion of steel in soil environments using electrical resistance sensor technique. *Materials Chemistry and Physics*, 103, 9–13.
5. Cole, I. and Marney, D. (2012) The science of pipe corrosion: a review of the literature on the corrosion of ferrous metals in soils. *Corrosion Science*, 58, 5–16.
6. Roberge, P. (2000) *Handbook of Corrosion Engineering*, McGraw-Hill, New York, NY.
7. Liu, Z., Sadiq, R., Rajani, B., and Najjaran, H. (2009) Exploring the relationship between soil properties and deterioration of metallic pipes using predictive data mining methods. *Journal of Computing in Civil Engineering*, 24, 289–301.
8. Norhazilan, M., Nordin, Y., Lim, K., Siti, R., Safuan, A., and Norhamimi, M. (2012) Relationship between soil properties and corrosion of carbon steel. *Journal of Applied Sciences Research*, 8, 1739–1747.
9. Ferreira, C.A.M., Ponciano, J.A.C., Vaitsman, D.S., and Pérez, D.V. (2007) Evaluation of the corrosivity of the soil through its chemical composition. *Science of the Total Environment*, 388, 250–255.
10. Peabody, A.W. (2001) *Peabody's Control of Pipeline Corrosion*, NACE International, Houston, TX.
11. Romanoff, M. (1957) Underground corrosion. *National Bureau Of Standards Circular*, 579, 227; see also: Richard E. Ricker, Analysis of pipeline steel corrosion data from NBS (NIST) studies conducted between 1922-1940 and relevance to pipeline management, NISTIR 7415, NIST, Gaithersburg, MD, May 2, 2007.
12. Revie, R.W. and Uhlig, H.H. (2008) Corrosion and corrosion control. In: *An Introduction to Corrosion Science and Engineering*, 4th ed., John Wiley & Sons, Inc., Hoboken, NJ.
13. Gupta, S. and Gupta, B. (1979) The critical soil moisture content in the underground corrosion of mild steel. *Corrosion Science*, 19, 171–178.
14. Neale, C.N., Hughes, J.B., and Ward, C. (2000) Impacts of unsaturated zone properties on oxygen transport and aquifer reaeration. *Ground Water*, 38, 784–794.

15. Castaneda, H., Zhang, L., Rahimi, A., Rivera, H., Martinez, L., and Arya, F. (2013) Dynamic macro modeling for the soil characteristics as complement for pre-evaluation step in the ECDA method. Proceedings of CORROSION/2013, NACE Paper.
16. Evans, U.F., Asamudo, N.U., Wansah, J., and Okore, S.K. (2012) A study of corrosion potential of the Nigerian coastal plain sands using numerical corrosivity scale. *International Review of Physics*, 6(5), 417.
17. Arzola, S., Palomar-Pardave, M., and Genesca, J. (2003) Effect of resistivity on the corrosion mechanism of mild steel in sodium sulfate solutions. *Journal of Applied Electrochemistry*, 33, 1223–1231.
18. United States Salinity Laboratory Staff (1954) Determination of the properties of saline and alkali soils. In: Richards, L.A. (editor), *Diagnosis and Improvement of Saline and Alkali Soils, Agriculture Handbook No. 60*, US Department of Agriculture, Washington, DC, pp. 7–33.
19. Sherar, B., Power, I., Keech, P., Mitlin, S., Southam, G., and Shoesmith, D. (2011) Characterizing the effect of carbon steel exposure in sulfide containing solutions to microbially induced corrosion. *Corrosion Science*, 53, 955–960.
20. Iwona, B.B. and Gaylarde, C.C. (1999) Recent advances in the study of biocorrosion: an overview. *Revista de Microbiologia*, 30, 177–190.
21. Little, B., Wagner, P., and Mansfeld, F. (1991) Microbiologically influenced corrosion of metals and alloys. *International Materials Reviews*, 36, 253–272.
22. Videla, H.A. and Herrera, L.K. (2005) Microbiologically influenced corrosion: looking to the future. *International Microbiology*, 8, 169–180.
23. Koch, G.H., Brongers, M.P.H., Thompson, N.G., Virmani, Y.P., and Payer, J.H. (2001) Corrosion Cost and Prevention Strategies in the United States, Appendix BB Defense. Report No. FHWA-RD-01-156.
24. NACE (1999) *Materials Performance: MP*, National Association of Corrosion Engineers. Houston, TX, Vol. 38, Issues 7–12.
25. Hamilton, W.A. and Lee, W. (1995) Biocorrosion. In: Barton, L.L. (editor), *Sulfate-Reducing Bacteria, Biotechnology Handbooks*, Vol. 8, Springer Science+Business Media, New York, pp. 243–264.
26. Schwermer, C.U., Lavik, G., Abed, R.M.M., Dunsmore, B., Ferdelman, T.G., Stoodley, P., Gieseke, A., and De Beer, D. (2008) Impact of nitrate on the structure and function of bacterial biofilm communities in pipelines used for injection of seawater into oil fields. *Applied and Environmental Microbiology*, 74, 2841–2851.
27. Hamilton, W. (1985) Sulphate-reducing bacteria and anaerobic corrosion. *Annual Reviews in Microbiology*, 39, 195–217.
28. ASTM (2003) *ASTM Standard G57: Standard Test Method for Field Measurement of Soil Resistivity Using the Wenner Four-Electrode Method*, ASTM International, West Conshohocken, PA, DOI: 10.1520/G0057-06R12, Available at www.astm.org.
29. ASTM (2003) *ASTM Standard G51: Standard Test Method for Measuring pH of Soil for Use in Corrosion Testing*, ASTM International, West Conshohocken, PA, DOI: 10.1520/G0051-95R12, Available at www.astm.org.
30. ASTM (2003) *ASTM Standard G200: Standard Test Method for Measurement of Oxidation-Reduction Potential (ORP) of Soil*, ASTM International, West Conshohocken, PA, DOI: 10.1520/G0200-09, Available at www.astm.org.
31. ASTM (2003) *ASTM Standard G2162: Standard Practice for Conducting and Evaluating Laboratory Corrosions Tests in Soils*, ASTM International, West Conshohocken, PA, DOI: 10.1520/G0162-99R10, Available at www.astm.org.

# 21

# TELLURIC INFLUENCE ON PIPELINES*

David H. Boteler and Larisa Trichtchenko
*Earth Sciences Sector, Natural Resources Canada, Ottawa, Ontario, Canada*

## 21.1 INTRODUCTION

Telluric currents are the electric currents produced in the ground by natural geomagnetic field variations. One of the first experiments to record telluric currents was carried out in 1862 [1]. The currents and corresponding electric fields are primarily induced by changes in the earth's magnetic field, which are usually caused by interactions of the solar disturbances propagating and impacting the earth's natural electromagnetic environment. Recordings of these magnetic field variations and electric fields have been widely used since 1950s for mineral exploration [2].

In pipelines, telluric currents are responsible for variations in pipe-to-soil potentials (PSP) that interfere with pipeline surveys and might contribute to pipeline corrosion [3,4]. Evaluating the telluric influence on a pipeline is difficult due to its irregular nature. This is nicely described by Shapka [5]:

> "Pipeline potentials would remain unchanged for several weeks, than begin fluctuating anywhere from 100 millivolts to 15 volts. Further investigation showed that these disturbances correlated closely to variations in the Earth's magnetic field."

Telluric variations have a continuous frequency spectrum, as distinct from the single frequency variations due to AC interference. As well, PSP variations due to tellurics are different for different locations along the pipeline, which can be confusing when doing close interval pipeline survey.

To help understand all these features of telluric currents, this chapter provides a review of the literature followed by an explanation of the geomagnetic sources of telluric activity, the impacts of earth's deep conductivity structures, a pipeline's response to telluric electric fields and the methodology for assessing telluric effects, and mitigation/compensation of telluric effects. To conclude, we examine the knowledge gaps and open questions about telluric effects.

## 21.2 REVIEW OF THE EXISTING KNOWLEDGE ON PIPELINE-TELLURIC INTERFERENCE

The first report of telluric currents in pipelines comes from Varley in 1873 who commented that currents on a telegraph cable in London appeared to be associated with telluric currents in nearby gas pipelines [6,7]. However, studies focused on telluric current effects on pipelines have been reported for just over half a century. Early work on a 1128-mile pipeline in Canada noted the existence of stray currents that could not be accounted for by any of the usual sources and indicated a possible association with magnetic disturbances [8]. The variations of PSP or currents along the pipeline associated with telluric activity have been reported from nearly every part of the world, for example, New Zealand [9–11], Africa [12], Australia [13–16], Europe [17–30], USA [31–40], Argentina [41–44], Canada [45–47] and not only on pipelines buried in the ground, but on the seafloor as well [48].

The telluric influence has represented an ongoing problem for engineers setting up cathodic protection systems and the variations in PSP produced by telluric currents often make pipeline surveys difficult [9,35,49]. The most significant telluric currents are associated with large geomagnetic



*Oil and Gas Pipelines: Integrity and Safety Handbook,* First Edition. Edited by R. Winston Revie.

disturbances [50,51], although tidally induced effects have also been reported [11]. Early attempts to supress the telluric currents effects include the pipeline grounding, investigated in Norway [52]. Later observations in Northern Canada by Seager showed how the amplitude of telluric PSP fluctuations is affected by the introduction of an insulating flange into the pipeline [53]. Experimental techniques have also been developed for correcting close interval potential surveys [39,54,55] and pipeline test station surveys [56], although they can require knowledge of telluric effects to use them appropriately [57].

Construction of the Alaska pipeline in the high-latitude region noted for enhanced geomagnetic and telluric activity prompted more dedicated quantitative studies and numerical modeling. For the Alaska pipeline, Campbell [58] calculated the telluric currents that could be produced in the pipeline and their dependence on geophysical parameters such as the earth conductivity structure and frequency spectrum of the geomagnetic disturbances. Numerical modeling of the effects of telluric currents on pipelines have also been made for pipelines in Finland [26], Argentina [59], Nigeria [60]. All the aforementioned modeling has been done based on the assumption that the pipeline is a single infinitely long conductor. Inclusion of the coating properties was made by modeling the pipeline as multiple concentric cylinders [61,62]. Pirjola and Lehtinen [63] modeled a single pipeline with discrete ground connections. The modeling was extended to include the continuous high-resistance connection to ground through a pipeline coating by the application of distributed source transmission line (DSTL) theory. This was based on the methods developed for AC interference [64–67], adapted for telluric modeling [68–70] and continues to be the most widely used method due to its versatility [46,71–77].

Development of numerical modeling of telluric effects on pipelines has provided the theoretical foundation for understanding the telluric observations and provided tools to aid in the design of cathodic protection systems [78–80]. The telluric influence is now often considered as part of the pipeline design process [81–83]. Further explanations of the methodology for assessing telluric effects on a pipeline are presented in the section on mitigation.

Greater awareness of telluric effects has prompted several large-scale investigations, such as the simultaneous measurements of the telluric variations of the PSPs in multiple locations along a pipeline made in different countries [84], as well as the detailed Pipeline Research Council International(PRCI) review of the existing knowledge on the telluric interference with pipelines [85].

Telluric effects on pipelines are part of a wider field of geomagnetic effects on ground infrastructure [7,86–89] that dates back to the last century and the early days of the telegraph [90]. In a major magnetic storm in 1859 telegraph services around the world were disrupted with many accounts of messages being interrupted during the disturbance [91]. Subsequent telluric disturbances produced problems and in 1921 telluric currents started fires at several telegraph stations in Sweden [92]. As technology changed, telluric effects were observed on other systems. Submarine phone cables were affected [93] and a major storm of August 4, 1972 produced an outage of the L-4 phone cable system in the Midwestern United States [94]. Telluric currents also occur in power systems where they are referred to as geomagnetically induced currents (GIC). The GIC induced in transmission lines flow to ground at substations causing saturation of power transformers leading to a variety of problems with power system operation [95,96]. On 13 March 1989, one of the biggest magnetic storms of the last century sent electric currents surging through power systems in North America and Northern Europe [97]. The result was equipment and lines tripped out of service, burnt-out transformers, and the collapse of the Hydro-Québec power system, leaving the 6 million residents of Québec without power for over 9 h [98]. This all has stimulated research into the impacts of natural electromagnetic environment on ground infrastructure and has significantly contributed to advancing our understanding of telluric effects on pipelines [99–103].

## 21.3 GEOMAGNETIC SOURCES OF TELLURIC ACTIVITY

The geomagnetic field variations that produce telluric activity have their origin with processes on the Sun [104]. The electromagnetic and particle radiation from the Sun affects the earth's magnetic environment in a variety of ways [105]. The regular electromagnetic radiation (light) heats the dayside part of the earth's atmosphere, producing vertical convection that drives electric currents in the ionosphere (conductive top layer of the earth's atmosphere). These electric currents create a magnetic field on the dayside of the earth that is superimposed on the earth's internally generated magnetic field. A site on the earth's surface is carried by the earth's rotation into this extra magnetic field in the morning and out again in the evening. This produces the normal magnetic field variation seen on quiet days. Figure 21.1 shows an example of the geomagnetic diurnal variation in Norway and simultaneous recordings of PSP on a nearby pipeline [25].

Irregular periods of solar activity produce disturbances such as coronal mass ejections (CMEs) and high-speed streams that travel from the Sun, through interplanetary space (solar wind), to the earth. These disturbances interact with the earth's magnetic field creating the geomagnetic storms that enhance telluric currents. Figure 21.2 shows an example of a solar disturbance reaching the earth, seen by the sudden changes in the interplanetary magnetic field and the solar wind speed in the top two panels. The next two panels show the magnetic disturbance and telluric electric field observed

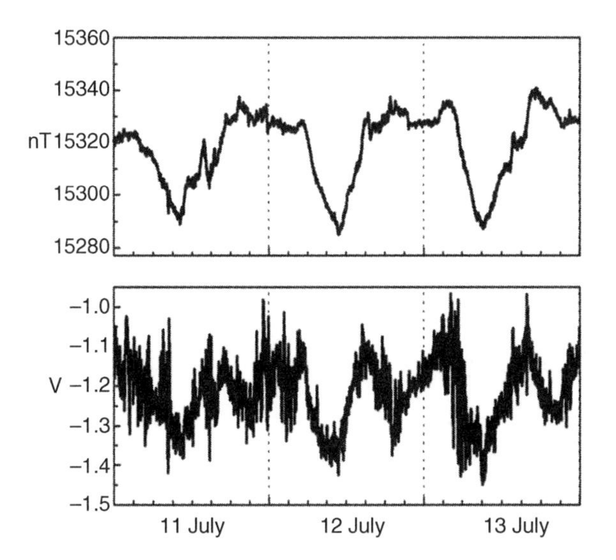


**FIGURE 21.1** Examples of simultaneous geomagnetic (top graph) and PSP variations. Daily regular variations, recorded in Norway for 3 days.

on the ground. The bottom panel shows recordings of the resulting PSP variations on a pipeline.

The sizes of geomagnetic field variations associated with particular types of solar phenomena are not the same all over the earth. The diurnal variation is largest near the equator, where the currents driven by solar heating are greatest, and decreases slowly with increasing latitude. In contrast the irregular geomagnetic field variations and geomagnetic storms produced by high-speed streams and CMEs are most intense at high latitudes (above 65 degrees). It should be noted that the same processes that give rise to the geomagnetic disturbances also produce the aurora and the "auroral zone," the region where aurora are most frequently seen, is also a region of high telluric activity. Mid-latitudes will experience a mixture of these auroral and equatorial disturbances. Early accounts of telluric activity were generally from high-latitude pipelines, but the use of higher resistance coatings on modern pipelines (see Section 21.6.3) has increased the amplitude of telluric PSP fluctuations and means that they can now be seen on pipelines at any latitude.

Figure 21.3 shows an example of PSP recordings made on an Australian pipeline during a geomagnetic disturbance [106]. Ap and Dst are global geomagnetic activity indices showing the worldwide level of disturbance in the geomagnetic field. The middle panel shows the rate of change of the magnetic field, *dH*/*dt*, recorded at Canberra, the magnetic observatory closest to the pipeline location. It can be seen that the size of

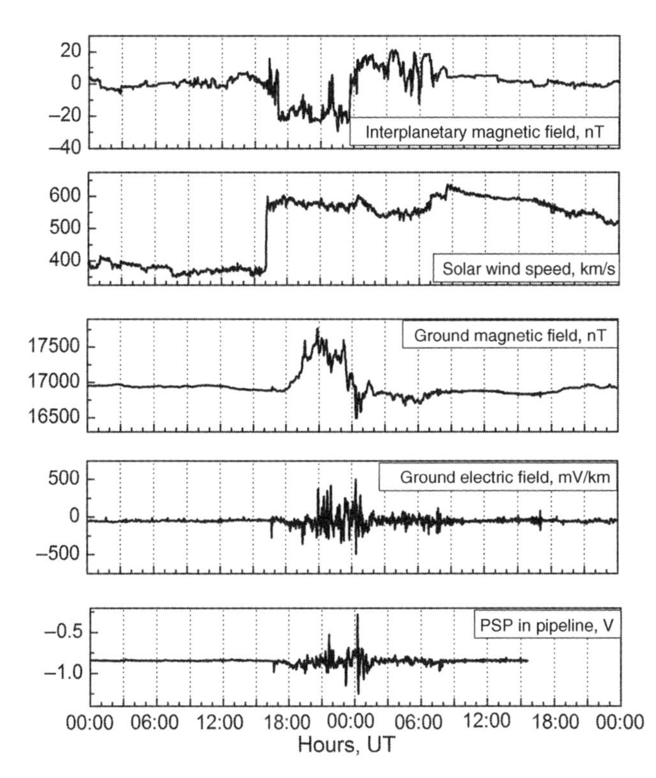


**FIGURE 21.2** Chain of events on 6–7 April 2000 recorded on 6 April showing large changes in magnetic field and solar wind speed and resulting magnetic storm on the ground. Recordings of the ground electric field in Ontario and telluric currents in Maritime pipeline had a maximum around midnight on 6 April.

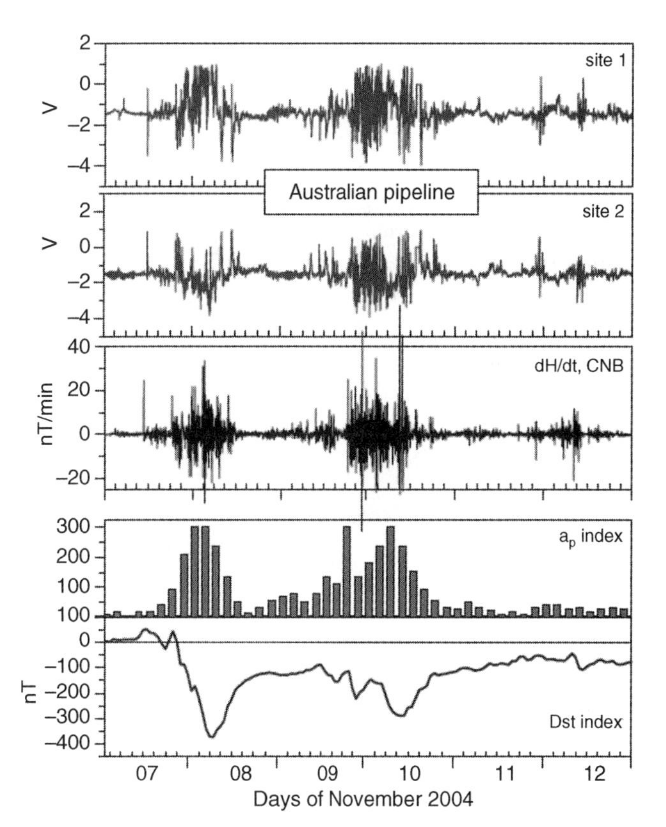


**FIGURE 21.3** Pipeline voltage at two sites on an Australian pipeline and rate of change of the magnetic field recorded at the Canberra magnetic observatory and magnetic activity indices ap and Dst during a magnetic disturbance in November 2004.

these variations follow the general trend of the global activity indices. The PSP variations at two sites on a nearby pipeline are closely related to the rate of change of the local magnetic field.

Thus, in understanding the influence of telluric currents on PSP variations, it is important to compare these variations with nearby geomagnetic recordings if available or with the variations of the global geomagnetic activity [107–109].

## 21.4 EARTH RESISTIVITY INFLUENCE ON TELLURIC ACTIVITY

The size of telluric electric fields produced by the geomagnetic field variations depends on the rate of change of the magnetic field variations and the deep earth resistivity structure of the region traversed by the pipeline [2,107]. Geomagnetic field variations at low frequencies (mHz) penetrate tens to hundreds of kilometers into the earth and the resistivity down to these depths needs to be taken into account to calculate the electric fields at the earth's surface.

The Earth is comprised of a number of layers as shown schematically in Figure 21.4. At the earth's surface the resistivity depends on the rock type with sedimentary rocks being less resistive than granitic igneous rocks. Below the surface layer, the crust of the earth is resistive but at greater depths the increasing temperature and pressure causes a decrease in resistivity.

Higher resistivities cause larger telluric electric fields to be produced during geomagnetic disturbances, while the lower resistivity rocks have lower telluric electric field values. Thus, a pipeline passing through regions with different resistivities will experience larger telluric currents in the region with the higher resistivity [43,59]. Changes in soil resistivity at the surface can also affect the discharge of telluric currents from a pipeline [39].

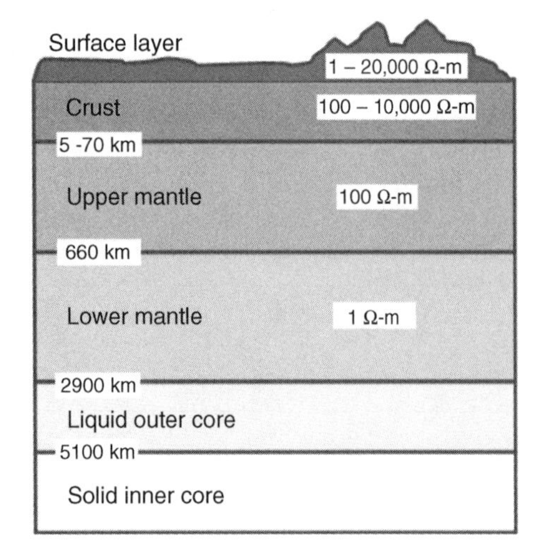


**FIGURE 21.4** Internal structure of the earth, showing main divisions of crust, mantle, and core and associated resistivity values.

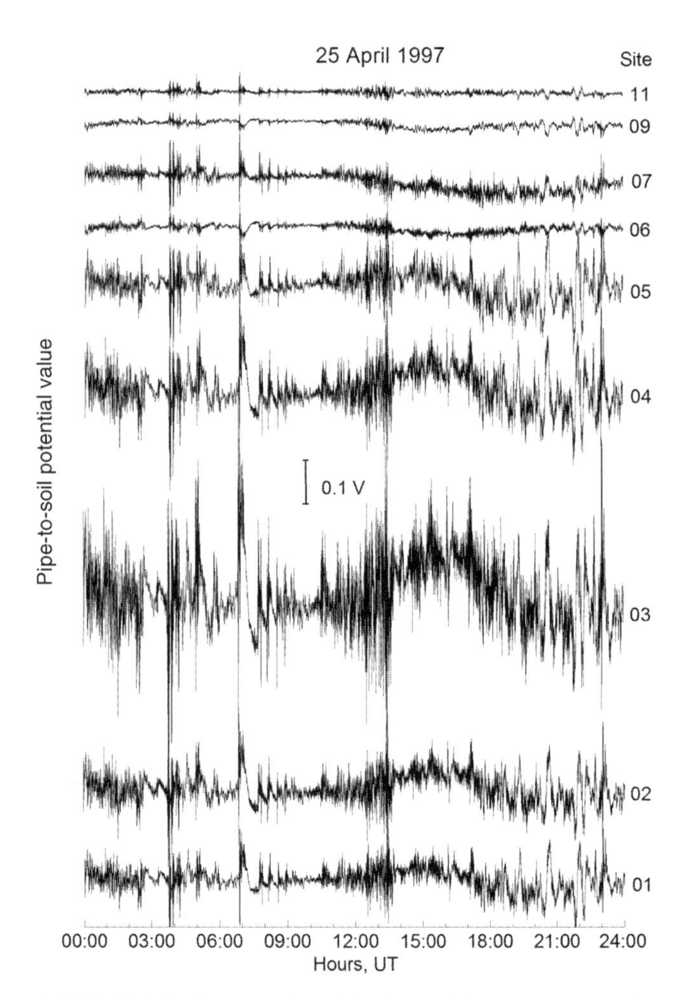


**FIGURE 21.5** The stackplot of the PSP variations simultaneously recorded on different sites along the pipeline. Amplitude changes from site to site show possible amplification of PSP due to ground conductivity contrast [84].

Boundaries between regions of different resistivity, such as at the interface between geological terrains also produce localized enhancements of the electric fields that can affect pipelines [25,110–113]. For example, Figure 21.5 shows the changes in amplitude of PSP variations at sites along a 150 km stretch of pipeline. These could not be explained by the pipeline structure, but coincided with earth resistivity changes associated with a geological boundary [84]. The resistivity boundary effect is also significant at a coastline because of the large contrast between the resistivity of the land (100–10,000 ohm-m) and seawater (0.25 ohm-m) [85,114].

## 21.5 PIPELINE RESPONSE TO TELLURIC ELECTRIC FIELDS

The telluric electric fields produced by geomagnetic field variations will drive telluric currents along a pipeline, as well as in the ground. The "potential drop" produced by these currents flowing on and off the pipeline is the cause of the telluric PSP variations. During the course of a geomagnetic disturbance, the telluric electric field will reverse direction many times causing a similar change in direction of the telluric current resulting in many fluctuations of the PSP on the pipeline as shown in Figure 21.6 [84]. It shows that the PSP variations at the southern end (KP 862) of the pipeline are out of phase (oppositely directed) from the PSP variations in the northern part of the pipeline consistent with the telluric currents generally flowing on and off the pipeline at opposite ends.

The size of PSP variations produced by telluric currents flowing on and off a pipeline depends on the resistance to ground from the pipeline steel. In the absence of any ground beds, this is determined by the resistance through the pipeline coating: the more resistive the coating, the larger the telluric PSP fluctuations that are produced. In fact it is the trend over the last 40 years for use of higher resistance coatings that has caused telluric PSP variations to be larger on modern pipelines. The amplitude of the PSP fluctuation generally decreases with distance from the end of the pipeline. This falloff is characterized by the "adjustment distance" determined by the resistance along the pipeline steel and the conductance through the coating to ground and typically has values of many tens of kilometers [68]. The decrease in coating conductance has not only increased the size of the telluric PSP fluctuations but has also increased the adjustment distance meaning that sizeable telluric PSP fluctuations are seen further along the pipeline.

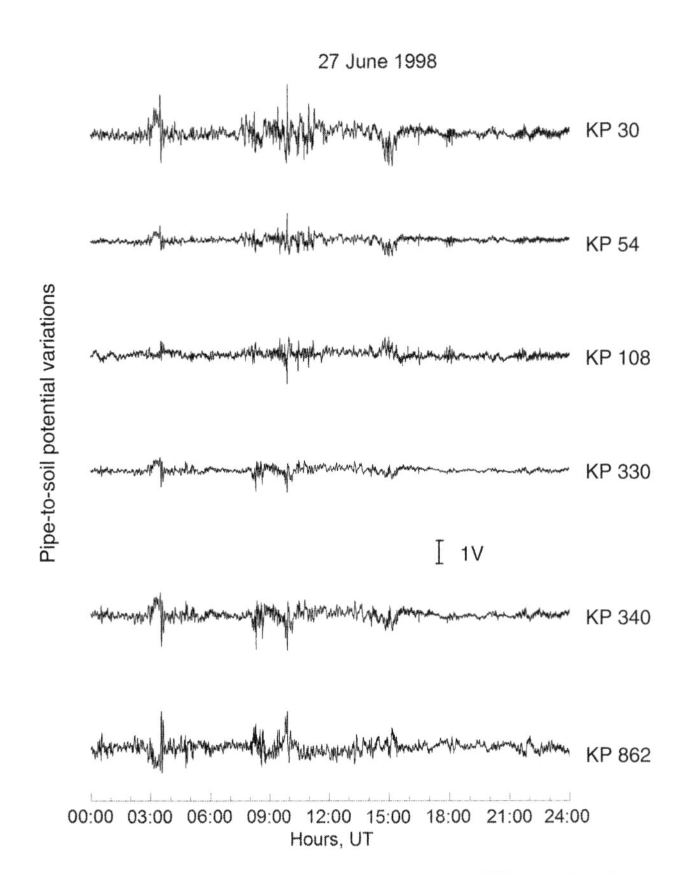


**FIGURE 21.6** Simultaneous PSP recordings at different sites along the pipeline, with the top being at one end of the pipeline (KP 30) and the bottom one is at the another end of the pipeline (KP 862).

In practice, many specific things about the pipeline structure or the electric fields can affect the telluric current flow along the pipeline affecting the size and location of the telluric PSP fluctuations that occur. If an insulating flange is inserted into the pipeline it effectively creates new end points where the telluric current flows on or off the pipeline, thus creating extra sites where telluric PSP fluctuations occur. In fact, any pipeline feature that alters the flow of telluric currents along the pipeline will give rise to the currents flowing on or off the pipeline producing the associated telluric PSP fluctuations. Changes in direction (bends) of a pipeline or changes from one to two pipelines (such as sometimes occurs at a compressor station on a gas pipeline) are places where larger telluric PSP fluctuations may occur [115].

## 21.6 TELLURIC HAZARD ASSESSMENT

The size of telluric PSP variations produced during a geomagnetic disturbance depends on (i) the amplitude of the geomagnetic field variations, (ii) the conductivity structure of the earth, and (iii) the pipeline response, as already discussed. To assess the telluric hazard to a pipeline requires knowledge about all three factors.

### 21.6.1 Geomagnetic Activity

Geomagnetic disturbances come in many sizes with many small ones, a lesser number of medium disturbances, and a few very large disturbances. A good guide to the geomagnetic activity to which a pipeline will be exposed in the future can be provided by analyzing a representative sample of past activity. Recordings of geomagnetic disturbances are made at magnetic observatories around the world, and data back to 1991 are available from Intermagnet (www.intermagnet.org) [Earlier data may also be obtained by contacting the institutes operating the magnetic observatories.] By analyzing data from an observatory in the vicinity of a pipeline, it is possible to determine the rate of occurrence of different levels of geomagnetic activity that will be experienced by the pipeline.

As an example, the results of the statistical analysis made of the magnetic activity in different latitudinal zones of Canada represented by data from four geomagnetic observatories for the year 2002 [116] are presented in Figure 21.7. Observatories at Cambridge Bay (CBB), Yellowknife (YKC), and Meanook (MEA) are located in the auroral zone, and Ottawa (OTT) in the subauroral zone. The range of the magnetic field variations in each hour is used as a measure of the size of the geomagnetic

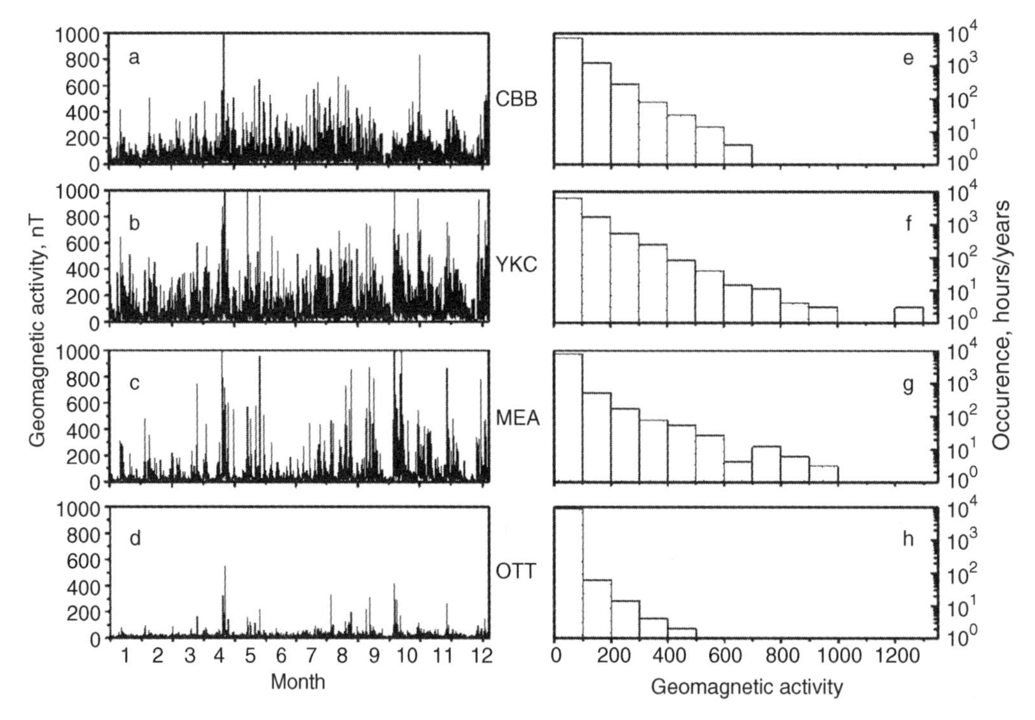


**FIGURE 21.7** Annual magnetic activity in different latitudinal zones of Canada represented by data from four geomagnetic observatories for the year 2002. Observatories at Cambridge Bay CBB, YKC and MEA are located in the auroral zone, and OTT in the subauroral zone. (a)–(d) Variations of hourly range values (in nanoTesla) during the year; (e)–(h) number of hours per year below certain level of geomagnetic activity.

activity. Figures 21.7a–d show the variations of hourly range values (in nanoTesla) during the year while Figures 21.7e–h show the number of hours per year with different levels of geomagnetic activity.

### 21.6.2 Earth Conductivity Structure

Earth conductivity varies according to rock type and with the changing conditions at increasing depths within the earth. This variation of the conductivity with depth can be modeled using a layered model of the earth. Examples of such models are shown in Figure 21.8.

Earth models such as in Figure 21.8 can be used to calculate the transfer function of the earth. This provides the relationship between the surface electric field and the magnetic field variations as a function of frequency. Any time series of magnetic field data can then be decomposed into its frequency components, each multiplied by the corresponding transfer function value to give the electric field frequency components. The summation of these components then gives the electric field variations for the specified interval [107].

Electric fields calculated from archived magnetic field data can be used to construct electric field statistics in the same way as done for the magnetic data [81,116,117]. Figure 21.9 shows the occurrence of electric fields calculated using the magnetic data from the Yellowknife observatory and the earth models in Figure 21.8. These results show how often the telluric electric fields exceed specified values and provide values to use as inputs to a pipeline model.

### 21.6.3 Pipeline Response

The next step of the assessment process is the modeling of the pipeline. This can be done in a manner similar to that used for studying AC induction in pipelines [64–67]. Each pipeline section is defined by its electrical properties (series impedance along the pipeline determined by the resistivity and cross-sectional area of the pipeline steel, and parallel admittance given by the conductance (1/resistance) to ground through the pipeline coating) [68]. The telluric electric field is represented by a voltage source in each section. Multiple pipeline sections can be combined using a network modeling approach to provide models for a complete pipeline network [70,77].

The calculated electric fields are then used as inputs to the model to obtain the PSP across the pipeline. Figure 21.10 shows recordings of PSP variations on pipeline produced by telluric currents (left panel) and modeled using a calculated telluric electric field and an example of pipeline [107].

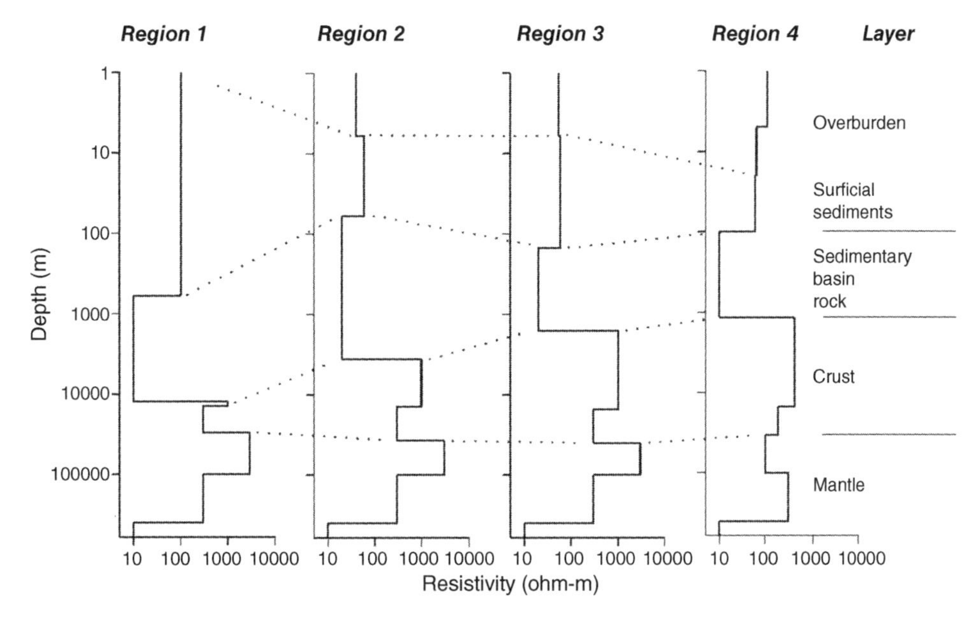


**FIGURE 21.8** One dimensional models of earth conductivity for four regions in Northern Canada. Dashed lines separate the various earth layers. Logarithmic scales [81,116,117].

These results demonstrate where in the pipeline network the largest telluric PSP will occur and, combined with the E field statistics of Figure 21.9, can show how often specific PSP levels will be exceeded. This provides the pipeline designer/operator with information about the telluric influence on the pipeline and can be used to examine mitigation options if required.

## 21.7 MITIGATION/COMPENSATION OF TELLURIC EFFECTS

NACE report "CEA 54276" in 1988 says

"Telluric effects can give rise to large fluctuations in measured potential values, sometimes of the order of several volts

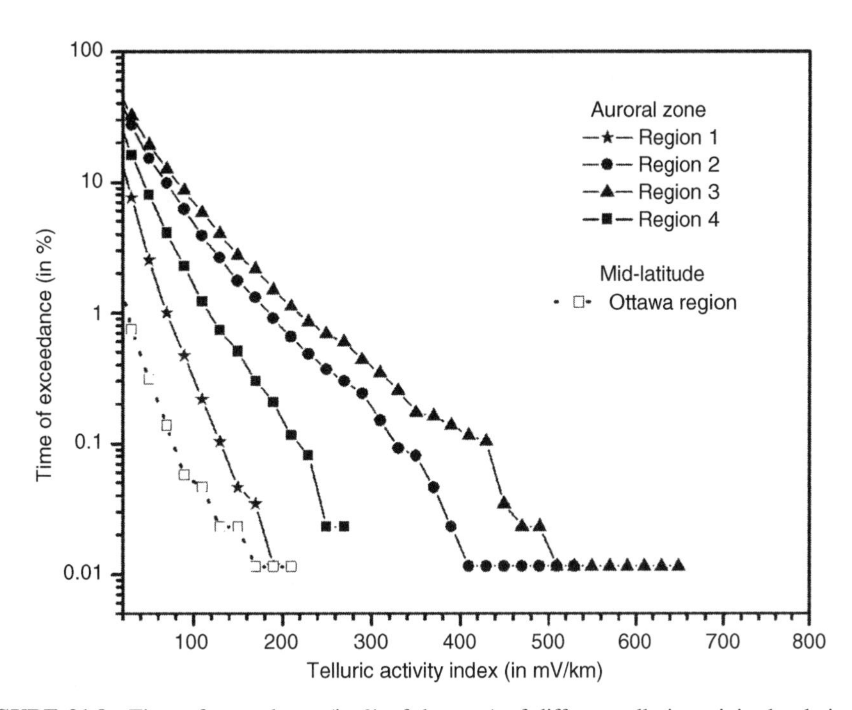


**FIGURE 21.9** Time of exceedance (in % of the year) of different telluric activity levels in four different conductivity regions of the auroral zone and the same characteristic for the mid-latitude zone (Ottawa) [116].

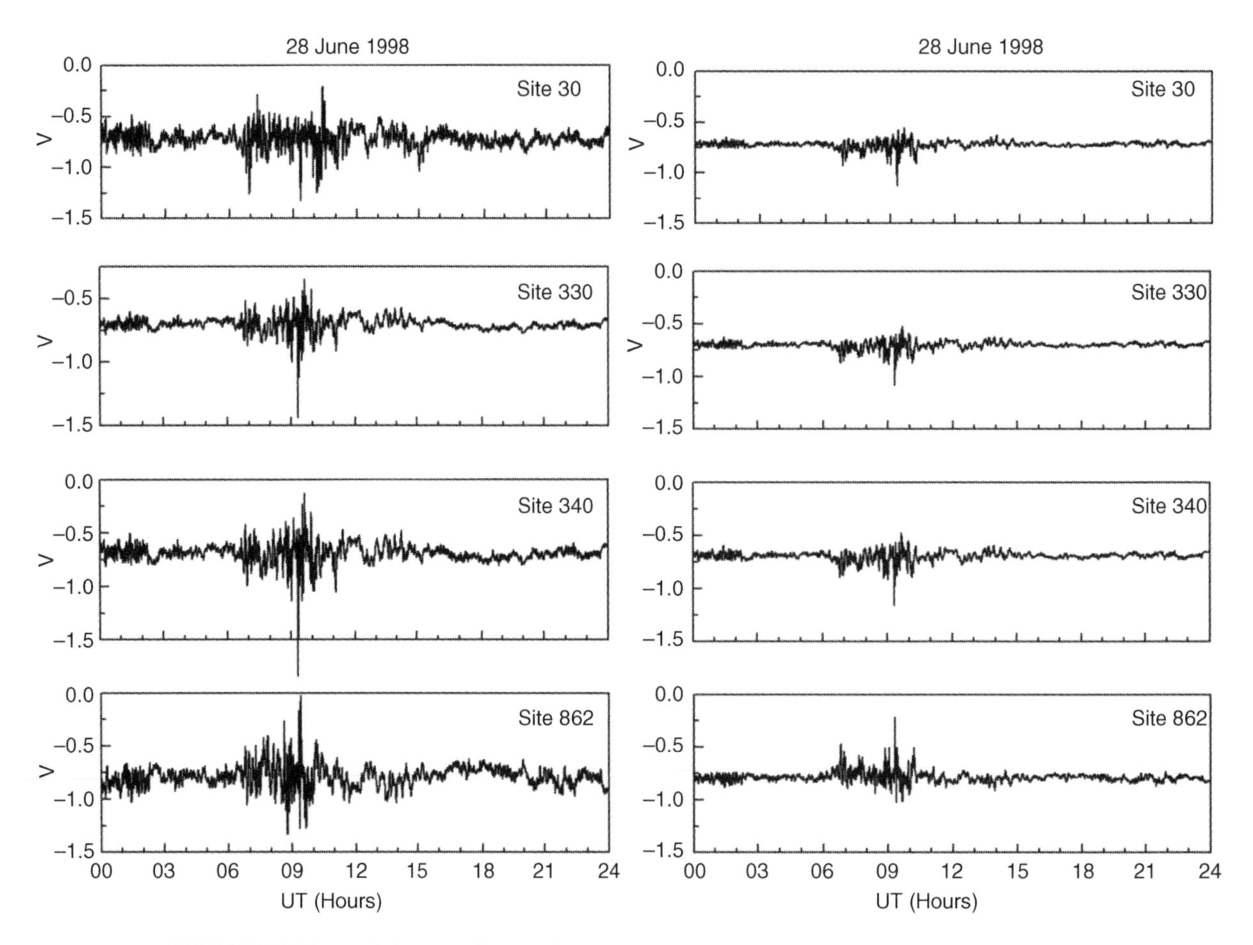


**FIGURE 21.10** Telluric recordings (left) and simulations (right) showing the PSP variations at different sites along a pipeline for the same day.

positive or negative. In such cases [ . . . ] specialist advice is sought either to monitor and suitably correct the measured values or to choose a survey period when geomagnetic activity is at a minimum."

Views on the most appropriate action to take to mitigate telluric effects have evolved over time as understanding has improved. An early suggestion for mitigation of telluric effects was to install insulating flanges to block the flow of currents along the pipeline [35]. However, as illustrated in Figure 21.11, it is now recognized that disrupting the telluric current flow along the pipeline forces it to flow through the coating to ground, increasing the PSP variations for the pipeline on either side of the flange.

The strategy adopted in the last 20 years is to remove obstacles to the flow of telluric currents by removing or jumpering around insulating flanges and providing good connections to ground, that is, a path for telluric currents to easily flow on and off the pipeline without passing through the pipeline coating. As well, ground beds are also placed at other places along the pipeline, such as bends [4,85]. The effect of good grounding on reducing telluric PSP variations can be seen in Figure 21.12.

Another strategy is to use potential-controlled rectifiers as shown in Figure 21.13 [45,79]. Here the pipeline modeling can be useful in determining the range of variations that must be controlled. Information from the modeling about where the telluric fluctuations change sign should also be used to guide placement of potential-controlled rectifiers otherwise there is a danger that telluric compensation in one part of a pipeline may "overflow" into a neighboring part where it would actually increase the PSP variations.

Telluric fluctuations on pipelines may make it difficult to obtain reliable potential survey data. In some cases, it may be sufficient to wait for geomagnetic quiet times to conduct pipeline surveys. [Long-term forecasts of geomagnetic activity can help with survey planning.] However, in other cases it can be necessary to take active steps to compensate the survey data for telluric effects. This has been done by using a reference site to provide a telluric level that can be subtracted from the survey data [57,118]. For the Alaska pipeline measurements of the telluric current along the

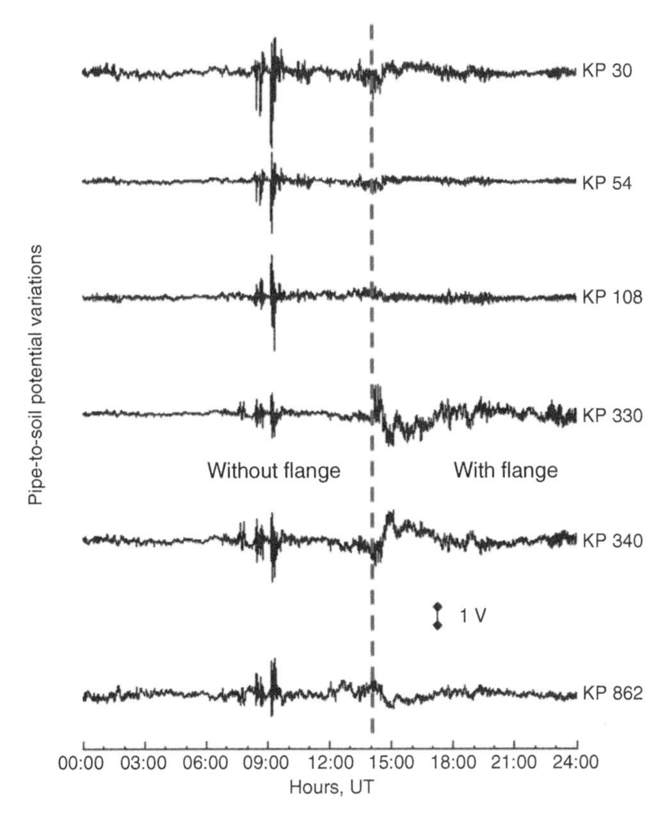


**FIGURE 21.11** PSP recordings without a flange (i.e., jumper cable connected around the flange) and with a flange (jumper cable disconnected) in a pipeline. The dashed line shows the time when the jumper cable was disconnected.

pipeline were used [56], again with the purpose of subtracting off the telluric fluctuations in the survey data. Finally, online model calculations can be made using geomagnetic data to calculate the telluric PSP variations occurring during the time of a survey [117,119].

## 21.8 KNOWLEDGE GAPS/OPEN QUESTIONS

Above we have shown the methodology for assessing the telluric hazard to a pipeline, but this needs the appropriate input data: for example, geomagnetic data and earth conductivity structure. Geomagnetic data are available from many magnetic observatories around the world (see www .intermagnet,org). The proximity of an observatory to a pipeline varies considerably. Even in places such as North America with a reasonably good coverage of magnetic observatories, the observatory recordings may not be sufficient to map the changes in magnetic activity along a pipeline route. In such cases, temporary installation of magnetometers along the pipeline route could be considered.

Earth conductivity models to use for the electric field calculations can be constructed from magnetotelluric (MT) survey results published in the geophysical literature. Such earth models are also required for studies of geomagnetic effects on power systems and this has led to the production of compendiums of earth models that can be used for both pipeline and power system studies [120,121]. These earth models are one-dimensional (1D) models, that is, they only account for the variation of conductivity with depth, while, as mentioned earlier, lateral variations in conductivity also influence the electric fields experienced by pipelines study of such effects requires the use of 2D or 3D modeling techniques [122].

Telluric PSP variations are a concern both because of the impacts on the PSP recordings, which make it difficult to identify the real reason for PSP anomalies, and because of their possible contribution to corrosion [4]. There are definitely not enough measurements done of the corrosion rates due to the telluric activity [28,29], which makes it difficult to evaluate the contribution of telluric activity in the overall corrosion process. Like other stray currents, the variable nature of the telluric shifts of PSP will produce less effect on corrosion rates than a DC shift [123,124]. In this regard, their effect may be similar to AC corrosion. The chapter of AC corrosion states that it is necessary to understand the interplay of the AC and DC levels on the pipeline in order to assess the corrosion rates. Similar investigations need to be undertaken for telluric activity.

## 21.9 SUMMARY

Telluric PSP variations have been observed on pipelines in many parts of the world and are a concern because interference to pipeline surveys and for creating conditions that may contribute to corrosion of the pipeline.

The telluric activity originates from the activity on the Sun that causes geomagnetic disturbances on the earth. The telluric electric fields enhanced due to these disturbances depend on the amplitude and frequency content of the geomagnetic field variations and on the deep earth conductivity structure.

Pipeline response to telluric electric fields is controlled by the electrical properties of the pipeline: the series impedance of the steel and the parallel admittance through the coating. Use of higher resistance coatings on modern pipelines has increased the size of telluric PSP variations that occur making telluric effects visible on pipelines all over the world.

If telluric influence is suspected on a pipeline then a telluric hazard assessment can be made taking account of the geomagnetic activity, earth structure, and pipeline characteristics to determine how often specified PSP levels will be exceeded in different parts of the pipeline.

Mitigation now focuses on allowing telluric currents to flow along the pipeline and providing good ground connections to drain currents off the pipeline at the ends, bends, and so on. Comparison with reference recordings can be used to correct survey data.

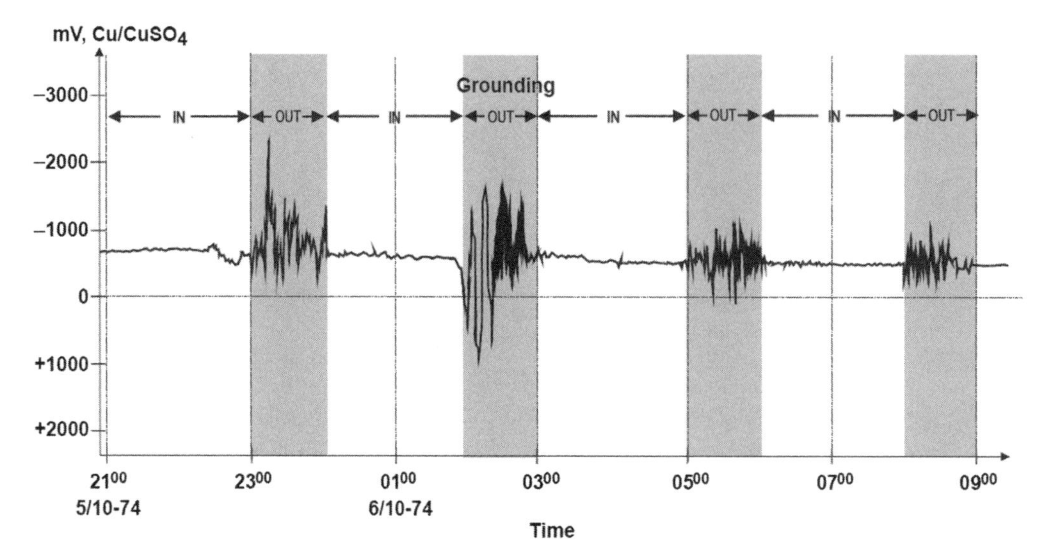


FIGURE 21.12 PSP for times with and without ground connections [85]. (Redrawn from Ref. [52]. Reprinted with the permission of Pipeline Research Council International, Inc., all rights reserved.)

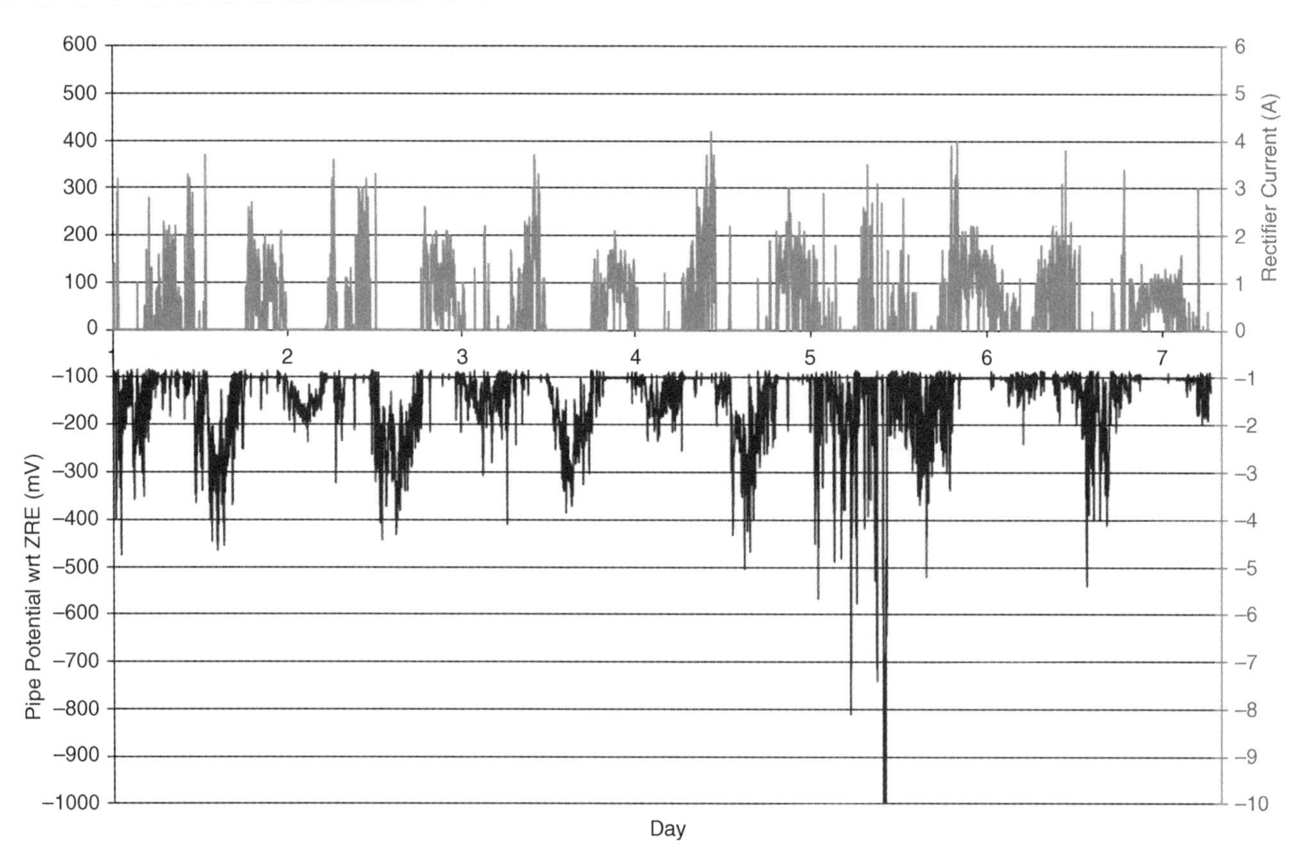


FIGURE 21.13 Pipe-to-soil potential and rectifier current time variations for impressed current system operating in potential control mode. (From Ref. [85]. Reprinted with the permission of Pipeline Research Council International, Inc., all rights reserved.)

Substantial work is required to properly evaluate the impact of telluric PSP fluctuations on pipeline corrosion and possible disbondment of coatings.

## ACKNOWLEDGMENTS

The authors have benefitted from many useful discussions with R.A. Gummow. Funding from the Panel for Energy Research and Development (PERD) has contributed to a number of studies that have helped to develop the understanding of telluric currents.

## REFERENCES

1. Lamont, J.V. (1862) *Der Erdstrom und der Zusammen desselben mit dem Erdmagnetismus*, Leopold-Voss-Verlag, Leipzig und Muenchen.
2. Kaufman, A.A. and Keller, G.V. (1981) *The Magnetotelluric Sounding Method, Methods in Geochemistry and Geophysics*, Elsevier, Amsterdam, Vol. 15, p. 595.
3. Peabody, A.W. (2001) *Peabody's Control of Pipeline Corrosion*, 2nd ed., NACE International, Houston, TX.
4. Gummow, R.A. (2002) GIC effects on pipeline corrosion and corrosion control systems. *Journal of Atmospheric and Solar Terrestrial Physics*, 64, 1755–1764.
5. Shapka, R.S. (1993) Geomagnetic effects on a modern pipeline system. Proceedings Solar-Terrestrial Predictions Workshop, May 18–22, Vol 1, 1993, Ottawa, Canada. pp. 163–170.
6. Varley, C.F. (1873) Discussion of a few papers on Earth currents. *Journal of the Society of Telegraph Engineers*, 2, 111–114.
7. Lanzerotti, L.J. and Gregori, G.P. (1986) Telluric currents: the natural environment and interactions with man-made systems. In: *The Earth's Electrical Environment*, National Academy Press, Washington, DC, pp. 232–257.
8. Allison, N.J. and Huddleston, W.M.E. (1952) Extraneous currents noted on large transmission pipe line system. *CORROSION*, 8 (1), 1.
9. Procter, T.G. (1974) Experience with telluric current interference in the cathodic protection of a buried pipeline in New Zealand. *Materials Performance*, 13 (6), 24.
10. Procter, T.G. (1975) Pipeline telluric current interference as one phase of a wider interdisciplinary technological problem. *Materials Performance*, 14 (8), 24–28.
11. Boteler, D.H. and Cookson, M.J. (1986) Telluric currents and their effects on pipelines in the cook strait region of New Zealand. *Materials Performance*, 25 (3), 27–32.
12. Barker, R.H. and Skinner, N.J. (1980) The flow of electric currents of telluric origin in a long metal pipeline and their effect in relation to corrosion control. *Materials Performance*, 19 (2), 25–28.
13. McCoy, J. (1989) Cathodic protection on the Dampier to Perth pipeline—Australia. *Materials Performance*, 28 (2), 16–20.
14. Martin, B.A. (1993) Telluric effects on a buried pipeline. *CORROSION*, 49 (4), 343–350.
15. Martin, B.A. (1993) Telluric effects on an equatorial pipeline in Papua New Guinea, Proceedings of the Australasian Corrosion Association Conference 33, Paper No. 40, November 1993.
16. Martin, B.A. (1994) Cathodic protection of a remote river pipeline. *Materials Performance*, 33 (3), 12–15.
17. Brooks, D.R. (1983) Responding to the challenge in engineering. *Gas Engineering and Management*, 23, 355–367.
18. Brasse, H. and Junge, A. (1984) The influence of geomagnetic variations on pipelines and an application for large-scale magnetotelluric depth sounding. *Journal of Geophysics*, 55, 31–36.
19. Camitz, G., Edwall, H.-E., and Marbe, A. (1997) The influence of telluric currents on the cathodic protection of natural gas pipelines. Proceedings of the CEOCOR 4th International Conference, 1997, Sector A, Vienna, Austria. pp. 125–135.
20. Edwall, H.-E. and Boteler, D. (2001) Studies of telluric currents on pipelines in Southern Sweden. Proceedings, CORROSION 2001, NACE, March 11–16, Houston, TX, Paper No. 01315.
21. Hesjevik, S.M. and Birketveit, O. (2001) Telluric current on short gas pipelines in Norway—risk of corrosion on buried gas pipelines. Proceedings, CORROSION 2001, NACE, March 11–16, Houston, TX, Paper No. 313.
22. Hejda, P. and Bochníček, J. (2005) Geomagnetically induced pipe-to-soil voltages in the Czech oil pipelines during October-November 2003. *Annals of Geophysics*, 23, 3089–3093.
23. Pulkkinen, A., Viljanen, A., Pajunpää, K., and Pirjola, R. (2001) Recordings and occurrence of geomagnetically induced currents in the Finnish natural gas pipeline network. *Journal of Applied Geophysics*, 48, 219–231.
24. Pirjola, R., Pulkkinen, A., and Viljanen, A. (2003) Studies of space weather effects on the Finnish natural gas pipeline and on the Finnish high voltage power system. *Advances in Space Research*, 31 (4), 795–805.
25. Trichtchenko, L., Boteler, D., Hesjevik, S.M., and Birketveit, O. (2001) The production of telluric currents in Norway. Proceedings, CORROSION 2001, NACE International, March 11–16, 2001, Houston, TX, Paper No. 01314.
26. Viljanen, A. (1989) Geomagnetically induced currents in the Finnish natural gas pipeline. *Geophysica*, 25(1 and 2), 135–159.
27. Viljanen, A., Pulkkinen, A., Pirjola, R., Pajunpää, K., Posio, P., and Koistinen, A. (2006) Recordings of geomagnetically induced currents and a nowcasting service of the Finnish natural gas pipeline system. *Space Weather*, 4 (10), 1–9.
28. Kioupis, N. and Maroulis K. (2006) AC-corrosion detection on electrical resistance probes connected to a natural gas transmission pipeline. Proceedings of the 8th International Conference Pipeline Rehabilitation & Maintenance, September 11–15, 2006, Istanbul, Turkey.
29. Kioupis, N. (2011) Corrosion of buried pipelines detected on ER-probes: comparison with MFL ILI results. Proceedings, CEOCOR conference.

30. Demirel, B. and Yalcin, H. (2013) Investigation of the telluric effects arising along the cathodically protected natural gas pipeline between Karadeniz Eregli and Duzce. *Turkish Journal of Electrical Engineering and Computer*, 21, 758–765.
31. Gideon, D.N., Hopper, A.T., and McClure, G.M. (1968) Earth current effects on buried pipelines. Analysis of Ohio and Vancouver Field Tests. American Gas Association Catalogue No. L30510.
32. Gideon, D.N., Hopper, A.T., and Thompson, R.E. (1969) Earth current effects on buried pipelines analysis of observations of telluric gradients and their effects. American Gas Association Catalogue No. L30570.
33. Gideon, D.N., Hopper, A.T., and Thompson, R.E. (1970) Earth current effects on buried pipelines—analysis of observations of telluric gradients and their effects, AGA Project PR-3-41. p. 72.
34. Gideon, D.N. (1971) Telluric current effects on buried pipelines. *Materials Protection and Performance*, 10 (7), 5–10.
35. Peabody, A.W. (1979) Corrosion aspects of arctic pipelines. *Materials Performance*, 18 (5), 27.
36. Merritt, R.P. (1979) Measurement of the auroral-induced current in the TransAlaska pipeline. *Transactions of the ASME*, 101, 156–158.
37. Campbell, W.H. (1980) Observation of electric currents in the Alaska oil pipeline resulting from auroral electrojet current sources. *Geophysical Journal of the Royal Astronomical Society*, 61 (2), 437–449.
38. Campbell, W.H. and Zimmerman, J.E. (1980) Induced electric currents in the Alaska oil pipeline measured by gradient, fluxgate, and squid magnetometers. *IEEE Transactions on Geoscience and Remote Sensing*, GE-18(3), 244–250.
39. Smart, A.L. (1982) The TransAlaska pipeline—potential measurements and telluric current. *IEEE Transactions on Industry Applications*, 1A-18(5), 557–567.
40. Sackinger, W.M. (1991) The relationship of telluric currents to the corrosion of warm Arctic pipelines. Proceedings of International Arctic Technology Conference, May 29–31, Society of Petroleum Engineers (SPE 22099), Anchorage, Alaska. pp. 361–366.
41. Favetto, A. and Osella, A. (1999) Numerical simulation of currents induced by geomagnetic storms on buried pipelines: an application to the Tierra Del Fuego, Argentina, gas Transmission route. *IEEE Transactions on Geoscience and Remote Sensing*, 37 (1), 614–619.
42. Osella, A., Favetto, A., and Lopez, E. (1999) Corrosion rates of buried pipelines caused by Geomagnetic storms. *CORROSION*, 55 (7), 699–705.
43. Osella, A. and Favetto, A. (2000) Effects of soil resistivity on currents induced on pipelines. *Journal of Applied Geophysics*, 44 (4), 303–312.
44. Osella, A., Martinelli, P., Favetto, A., and López, E. (2002) Induction effects of 2-D structures on buried pipelines. *IEEE Transactions on Geoscience and Remote Sensing*, 40 (1), 197–205.
45. Seager, W.H. (1991) Adverse telluric effects on northern pipelines. Presented at the International Arctic Technology Conference, May 29–31, 1991, Society of Petroleum Engineers (SPE 22178), Anchorage, Alaska.
46. Boteler, D. and Trichtchenko, L. (2001) Observations of telluric currents in Canadian pipelines. CORROSION 2001, NACE International, Paper No. 01316.
47. Gummow, R.A. (2001) Telluric current effects on corrosion and corrosion control systems on pipelines in cold climates. Proceedings NACE N.W. Area Conference, Anchorage, Alaska.
48. Weldon, C.P., Schultz, A., and Ling, T.S. (1983) Telluric current effects on cathodic protection potential measurements on Subsea pipelines. *Materials Performance*, 22 (8), 43–47.
49. Russell, G.I. and Nelson, L.B. (1954) Extrinsic line current fluctuations seriously restrict progress of coating conductance surveys on large trunk line. *CORROSION*, 10, 400.
50. Campbell, W.H. (1986) An interpretation of induced electric currents in long pipelines caused by natural geomagnetic sources of the upper atmosphere. *Surveys in Geophysics*, 8, 239–259, D. Reidel Publishing Company.
51. Hessler, V.P. (1976) Causes, recording techniques, and characteristics of telluric currents. *Materials Performance*, 15 (4), 38.
52. Henriksen, J.F., Elvik, R., and Granasen, L. (1978) Telluric current corrosion on buried pipelines. Proceedings of the 8th Scandinavian Corrosion Congress, Theory and Praxis at Corrosion Prevention, Volume II, Helsinki, Alaska. pp. 167–176.
53. Boteler, D.H. and Seager, W.H. (1998) Telluric currents: a meeting of theory and observation. *CORROSION*, 54 (9), 751–755.
54. Nicholson, P. (2003) Stray and telluric current correction of close interval potential survey data, presented at EUROCORR 2003, Budapest, Hungary.
55. Nicholson, J.P. (2003) Pipeline integrity, World Pipelines.
56. Degerstedt, R.M., Kennelley, K.J., Lara, P.F., and Moghissi, O.C. (1995) Acquiring "telluric-nulled" pipe-to-soil potentials on the Trans Alaska Pipeline. CORROSION 1995, NACE International, Paper No. 345, pp. 1–26.
57. Carlson, L., Dorman, B., and Place, T. (2004) Telluric compensation for pipeline test station survey on the Alliance pipeline system. Proceedings of the International Pipeline Conference (IPC 2004), October, 2004, Calgary, Canada, Paper IPC04-0762, p231-241. pp. 4–8.
58. Campbell, W.H. (1978) Induction of auroral zone electric currents within the Alaska pipeline. *Pure and Applied Geophysics*, 116 (6), 1143–1173.
59. Osella, A., Favetto, A., and Lopez, E. (1998) Currents induced by geomagnetic storms on buried pipelines as a cause of corrosion. *Journal of Applied Geophysics*, 38 (3), 219–233.
60. Ogunade, S.O. (1986) Induced electromagnetic fields in oil pipelines under electrojet current sources. *Physics of the Earth and Planetary Interiors*, 43, 307–315.
61. Trichtchenko, L. and Boteler, D.H. (2001) Specification of geomagnetically induced electric fields and currents in

pipelines. *Journal of Geophysical Research*, 106 (A10), 21039–21048.

62. Trichtchenko, L., Boteler, D.H., and Larocca, P. (2004) Modeling the effect of the electromagnetic environment on pipelines, Geological Survey of Canada, Open File 4826.
63. Pirjola, R. and Lehtinen, M. (1985) Currents produced in the Finnish 400 kV power transmission grid and in the Finnish natural gas pipeline by geomagnetically-induced electric fields. *Annales Geophysicae*, 3 (4), 485–491.
64. Taflove, A. and Dabkowski, J. (1979) Prediction method for buried pipeline voltages due to 60 Hz AC inductive coupling. *IEEE Transaction on Power Apparatus and Systems*, PAS-98, 780–794.
65. Dawalibi, F. and Southey, R.D. (1989) Analysis of electrical interference from power lines to gas pipelines part I: computation methods. *IEEE Transactions on Power Delivery*, 4 (3), 1840–1846.
66. Dawalibi, F. and Southey, R.D. (1990) Analysis of electrical interference from power lines to gas pipelines part I: computation methods. *IEEE Transactions on Power Delivery*, 5 (1), 415–421.
67. Boteler, D.H. and Trichtchenko, L. (2005) A common theoretical framework for AC and telluric interference on pipelines. Proceedings, CORROSION 2005, NACE International, Houston, TX, Paper No. 05614.
68. Boteler, D.H. (1997) Distributed source transmission line theory for active terminations. Proceedings of the 1997 Zurich EMC Symposium, URSI Supplement, February 18–20, ETH, Zurich. pp. 401–408.
69. Boteler, D.H., Seager, W.H., Hohansson, C., and Harde, C. (1998) Telluric current effects on long and short pipelines. CORROSION 1998 NACE International, Paper No. 363. pp. 1–12.
70. Boteler, D.H. (2013) A new versatile model of geomagnetic induction of telluric currents in Pipelines, *Geophysical Journal International*, 193, 98–109.
71. Boteler, D.H. (2000) Geomagnetic effects on the pipe-to-soil potential of a continental pipeline. *Advances in Space Research*, 26 (1), 15–20.
72. Trichtchenko, L. (2004) Modeling electromagnetic induction in pipelines. Proceeding of NACE CORROSION 2004, Paper No. 04212.
73. Trichtchenko, L. and Boteler, D.H. (2002) Modelling of geomagnetic induction in pipelines. *Annales Geophysicae*, 20, 1063–1072, SRef-ID: 1432-0576/ag/2002-20-1063.
74. Pulkkinen, A., Pirjola, R., Boteler, D., Viljanen, A., and Yegorov, I. (2001b) Modelling of space weather effects on pipelines. *Journal of Applied Geophysics*, 48, 233–256.
75. Trichtchenko, L., Boteler, D.H., and Fernberg, P. (2008) Space weather services for pipeline operations. Proceedings ASTRO, 2008, CASI, April 29–May 1, 2008, Montreal, Canada.
76. Boteler, D.H., Trichtchenko, L., Blais, C., and Pirjola, R. (2013a) Development of a telluric simulator. Proceedings CORROSION 2013, NACE International, March, 2013, Orlando, United States.
77. Boteler, D.H., Trichtchenko, L., and Edwall, H.-E. (2013b) Telluric effects on pipelines. Proceedings of the CEOCOR Symposium, June 6–7, 2013b, Florence, Italy.
78. Boteler, D.H., Gummow, R.A., and Rix, B.C. (1999) Evaluation of telluric current effects on the Maritimes and Northeast Pipeline. NACE International Northern Area Eastern Conference, October 24, 1999, Ottawa, Canada, Paper No. 8A, 3.
79. Rix, B.C., Boteler, D., and Gummow, R.A. (2001) Telluric current considerations in the CP design for the Maritimes and Northeast Pipeline. CORROSION 2001, NACE International, March 11–16, 2001, Houston, TX, Paper No. 01317.
80. Rix, B.C. and Boteler, D.H. (2001) GIC effects on pipeline cathodic protection systems. *Ocean Resources*, 19 (8), 58–61.
81. Trichtchenko, L. and Fernberg, P. (2012) 16284 Assessment of Telluric Activity in Mackenzie Valley Area, Geological Survey of Canada Open File 7143, 127 p. doi: 10.495/291562.
82. Trichtchenko, L., Fernberg, P., and Harrison, M. (2012) Assessment of telluric activity in the area of proposed Alaska Highway Pipeline. Geological Survey of Canada Open File 7142, 111pp. doi: 10.495/291561.
83. Trichtchenko, L. (2012) Assessment of telluric activity in the area of the proposed Alaska Highway pipeline. Proceedings of CORROSION 2012, March, 2012, Salt Lake City, United States, Paper No. 0001192.
84. Boteler, D.H. and Trichtchenko, L. (2000) International study of telluric current effects on pipelines. Final Report, GSC Open File 3050.
85. Gummow, R., Boteler, D.H., and Trichtchenko, L. (2002) Telluric and ocean current effects on buried pipelines and their cathodic protection systems. Report for Pipeline Research Council International, Catalog No. L51909.
86. Lanzerotti, L.J. (1979) Geomagnetic influences on man-made systems. *Journal of Atmospheric and Terrestrial Physics*, 41, 787–796.
87. Lanzerotti, L.J., Kennel, C.F., and Parker, E.N. (editors), (1979) *Impacts of Ionospheric/magnetospheric Processes on Terrestrial Science and Technology, Solar Systems Plasma Physics*, North-Holland Publishing Company, Vol. III, pp. 319–363.
88. Lanzerotti, L.J. (1983) Geomagnetic induction effects in ground-based systems. *Space Science Reviews*, 34, 347–356.
89. Boteler, D.H., Pirjola, R.J., and Nevanlinna, H. (1998) The effects of geomagnetic disturbances on electrical systems at the Earth's surface. *Advances in Space Research*, 22, 17–27.
90. Prescott, G.B. (1866) *History, Theory and Practice of the Electric Telegraph*, Ticknor and Fields, Boston.
91. Boteler, D.H. (2006) The super storms of August/September 1859 and their effects on the telegraph system. *Advances in Space Research*, 38, 159–172.
92. Karsberg, A., Swedenborg, G., and Wyke, K. (1959) The influences of earth magnetic currents on telecommunication lines, In: *Tele*, English ed., Televerket (Swedish Telecom), Stockholm, pp. 1–21.
93. Anderson, C.W. (1978) Magnetic storms and cable communications. In: Kennel, C.F., Lanzerotti, L.J., and Parker, E.N.

(editors), *Solar System Plasma Physics*, North-Holland, Amsterdam.

94. Anderson, C.W., Lanzerotti, L.J., and Maclennan, C.G. (1974) Outage of the L-4 system and the geomagnetic disturbances of August 4, 1972. *Bell System Technical Journal*, 53, 1817–1837.
95. Molinski, T.S. (2002) Why utilities respect geomagnetically induced currents. *Journal of Atmospheric and Solar-Terrestrial Physics*, 64, 1765–1778.
96. Kappenman, J.G. (2007) Geomagnetic disturbances and impacts upon power system operation. In: Grigsby, L.L. (editor), *The Electric Power Engineering Handbook*, 2nd ed., CRC Press/IEEE Press, Chapter 16, pp. 16-1–16-22.
97. Allen, J., Frank, L., Sauer, H., and Reiff, P. (1989) Effects of the March 1989 solar activity. *EOS Transactions AGU*, 70, 1479.
98. Bolduc, L. (2002) GIC observations and studies in the Hydro-Québec power system. *Journal of Atmospheric and Solar-Terrestrial Physics*, 64, 1793–1802.
99. Boteler, D.H. (1991) Prediction of extreme disturbances with applications to geomagnetic effects on pipelines and power systems. Proceedings of the Solar-Terrestrial Predictions Workshop, Leura, Australia. pp. 53–68.
100. Boteler, D.H. and Pirjola, R.J. (1997) Nature of the geoelectric field associated with GIC in long conductors such as power systems, pipelines, and phone cables. Proceeding of the Beijing EMC Symposium, May, 1997. pp. 68–71.
101. Boteler, D.H. (2003) Geomagnetic hazards to conducting networks. *Natural Hazards*, 28(2–3), 537–561.
102. Pirjola, R., Viljanen, A., Pulkkinen, A., and Amm, O. (2000) Space weather risk in power systems and pipelines, Physics and Chemistry of the Earth, part C: solar. *Terrestrial and Planetary Science*, 25 (4), 333–337.
103. Trichtchenko, L. and Boteler, D.H. (2003) Effects of natural geomagnetic variations on power systems and pipelines. Proceedings of the EMC Conference, September, 2003, St Petersburg, Russia.
104. Moldwin, M. (2008) *Introduction to Space Weather*, Cambridge University Press, p. 134.
105. Campbell, W.H. (2003) *Introduction to Geomagnetic Fields*, 2nd ed., Cambridge University Press, p. 352.
106. Trichtchenko, L., Zhukov, A., van der Linden, R., Stankov, S.M., Jakowski, N., Stanislawska, I., Juchnikowski, G., Wilkinson, P., Patterson, G., and Thomson, A.W.P. (2007) November 2004 space weather events: real time observations and forecasts. *Space Weather*, 5 (6), S06001.
107. Trichtchenko, L. and Boteler, D.H. (2002) Modeling of geomagnetic induction in pipelines. *Annales Geophysicae*, 20, 1063–1072.
108. Trichtchenko, L. and Boteler, D.H. (2004) Modeling geomagnetically induced currents using geomagnetic indices and data. *IEEE Transactions on Plasma Science*, 32 (4), 1459–1467.
109. Marshall, R.A., Waters, C.L., and Sciffer, M.D. (2010) Spectral analysis of pipe-to-soil potentials with variations of the Earth's magnetic field in the Australian region. *Space Weather*, 8 (5), 1–13.
110. Boteler, D.H., Trichtchenko, L., and Samson, C. (2003) Investigation of earth conductivity influence on pipe-to-soil potentials. Proceedings, NACE Northern Region, Eastern Conference, September 15–17, 2003, Ottawa, Canada.
111. Fernberg, P.A., Samson, C., Boteler, D.H., Trichtchenko, L., and Larocca, P. (2007) Earth conductivity structures and their effects on geomagnetic induction in pipelines. *Annales Geophysicae*, 25 (1), 207–218.
112. Osella, A., Martinelli, P., Favetto, A., and López, E. (2002) Induction effects of 2-D structures on buried pipelines. *IEEE Transactions on Geoscience Remote Sensing*, 40 (1), 197–205.
113. Trichtchenko, L. (2005) Influence of surface conductivity contrasts on the currents and fields induced in buried pipelines by sources of variable frequencies. Proceedings of the NACE CORROSION/2005, April 2005, Paper No. 05615.
114. Pirjola, R. (2013) Practical model applicable to investigating the coast effect on the geoelectric field in connection with studies of geomagnetically induced currents. *Advances in Applied Physics*, 1 (1), 9–28.
115. Boteler, D.H. (2007) Assessing pipeline vulnerability to telluric currents. Proceedings of CORROSION 2007, NACE, March 2007, Houston, TX, Paper No. 07686.
116. Fernberg, P.A., Trichtchenko, L., Boteler, D.H., and McKee, L. (2007) Telluric hazard assessment for northern pipelines. Proceedings of CORROSION 2007, NACE International, March, 2007, Houston, TX, Paper No. 07654.
117. Trichtchenko, L., Fernberg, P., and Harrison, M. (2010) Use of geomagnetic data for evaluation of telluric effects on pipelines. Proceedings of CORROSION 2010, NACE International, 2010, San Antonio, Houston, TX, Paper No. 14262.
118. Place, T.D. and Sneath, T.O. (2001) Practical telluric compensation for pipeline close-interval surveys. *Materials Performance*, 40, 22–27.
119. Trichtchenko, L., Boteler, D.H., and Fernberg, P. Space weather services for pipeline operation. Proceedings, ASTRO 2008, April 29–May 1, 2008, Montreal, Canada.
120. Ferguson, I.J. and Odwar, H.D. (1997) Earth conductvity models. In: Boteler, D.H., Boutilier, S., Wong, A.K., Bui-Van, Q., Hajagos, D., Swatek, D., Leonard, R., Hughes, I.J., Ferguson, I.J., and Odwar, H.D. (editors), *Geomagnetically Induced Currents: Geomagnetic Hazard Assessment, Phase II*, Geological Survey of Canada, Open File No.3420.
121. Adam, A., Pracser, E., and Wesztergom, V. (2012) Estimation of the electric resistivity distribution (EURHOM) in the European lithosphere in the frame of the EURISGIC WP2 project. *Acta Geodaetica et Geophysica Hungarica*, 47, 377–387.
122. Börner, R.-U. (2010) Numerical modelling in geo-electromagnetics: advances and challenges. *Surveys in Geophysics*, 31, 225–245.
123. McCollum, B. and Ahlborn, G.H. (1916) Influence of Frequency of Alternating or Infrequently Reversed Current on Electrolytic Corrosion, National Bureau of Standards Tech, Paper No. 72.
124. López, E., Osella, A., and Martino, L. (2006) Controlled experiments to study corrosion effects due to external varying fields in embedded pipelines. *Corrosion Science*, 48 (2), 389–403

# 22

# MECHANICAL DAMAGE IN PIPELINES: A REVIEW OF THE METHODS AND IMPROVEMENTS IN CHARACTERIZATION, EVALUATION, AND MITIGATION

Ming Gao and Ravi Krishnamurthy
*Blade Energy Partners, Houston, TX, USA*

## 22.1 INTRODUCTION

Pipelines can be mechanically damaged by external force from third-party intrusion, contact with rocks in the backfill, or by settlement onto rocks. Mechanical damage typically includes pipe coating damage, dent(s) in the pipe, and gouge(s); all are localized or confined to some portion of the pipe along the pipe's length. Since coating damage is often coincident with dent and/or gouge, mechanical damage is commonly divided into two categories: dents and gouges (which are defects in the pipe wall that serve as failure initiation sites).

Dents typically result from a purely radial displacement. A pipe impinging on a rock may result in a dent. If the pipe slides on the rock, a dent with a gouge may result. Also, third-party mechanical damage caused during construction and excavation is a common cause of a dent and a dent with a gouge. A gouge normally results in a highly deformed, work hardened surface layer and may involve metal removal. Additionally, mechanical damage in the form of a crack could instantly occur on the external surface (OD) if the sliding contact causes extreme heating of the material and if there is any re-rounding upon removal of the indenting force [1–3]. Cracking may also instantly occur on the internal surface (ID) of the pipe due to denting strains exceeding the strain limit [1]. Mechanical damage to pipelines from outside forces can result in either immediate or delayed failure.

A majority of the anomalies caused by outside forces do not have dire consequences. However, a few prominent pipeline failures have been attributed to mechanical damage [4]. While dents are common, failures from dents alone, without additional surface mechanical damage such as scratches, gouges, and cracks, are relatively rare. Dents with additional surface mechanical damage have resulted in immediate failure approximately 80% of the time [5]. In the remainder of mechanical damage events, damage was not severe enough to cause immediate failure. However, it may lead to delayed failure if the internal pressure is raised sufficiently, or if corrosion or cracking develops in the damaged material, or if there is pressure-cycle fatigue. Table 22.1 shows the total number of reportable incidents in the United States from all cases and the number of incidents from mechanical damage in a total of 460,000 miles of gas and liquid pipelines from 1985 through 2003.

Statistical data showed that mechanical damage is one of the major threats for pipeline integrity. As illustrated in Figure 22.1, 33.7% of the serious incidents on all types of pipelines from 1994 to 2013 were caused by mechanical damage during excavation, which is more than by any other single cause [6].

Significant efforts have been made over the past 30 years by the Office of Pipeline Safety (OPS) of the Pipeline Hazardous Material Safety Administration (PHMSA) of the U.S. Department of Transportation, the pipeline industry, and stakeholder organizations with regard to the reduction of serious pipeline incidents. The efforts include (1) increasing public awareness of the risks of excavation in pipeline corridors and prevention, (2) investing in research to detect

*Oil and Gas Pipelines: Integrity and Safety Handbook*, First Edition. Edited by R. Winston Revie.

**TABLE 22.1 Analysis of Reported Mechanical Damage Incidents in the USA 1985–2003 [4]**

| | Total Number of Reportable Incidents from all Causes 1985 through 2003 | Number of Immediate Incidents from Mechanical Damage | Number of Delayed Incidents from Mechanical Damage | Ratio of Immediate to Delayed |
|---|---|---|---|---|
| 300,000 Miles of natural gas transmission and gathering pipelines | 1583 | 440 (28% of total) | 49 (4% of total) | 9 to 1 |
| 160,000 Miles of liquid petroleum pipelines | 3366 | 724 (21% of total) | 153 (5% of total) | 5 to 1 |

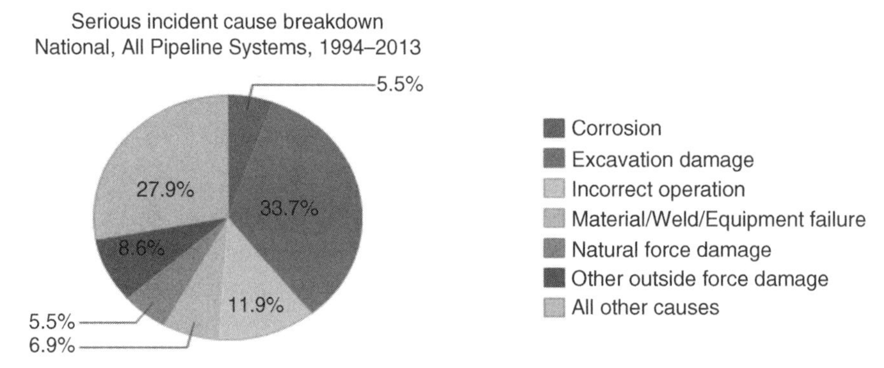


**FIGURE 22.1** Serious incident cause breakdown 1994–2013, indicating 33.7% caused by excavation damage [6].

mechanical damage using in-line inspection (ILI), (3) improving evaluation of the severity of mechanical damage, and (4) developing mitigation measures [7]. Such significant efforts have greatly reduced serious incidents over time. Figure 22.2 shows a decreasing trend in serious pipeline incidents caused by excavation damage for the 20-year period from 1988 to 2008. This is the result of a broad array of initiatives at all levels designed to engage all stakeholders in efforts to reduce the risk of damage to underground facilities [6].

In this chapter, the current status of ILI technologies for characterization of mechanical damage, mainly dents and dents with gouges/cracks, is reviewed. State-of-the-art technologies for in-ditch assessment of mechanical damage and validation of ILI tool performance are then summarized. Methods to assess the severity of mechanical damage, and associated regulatory and industry guidelines, are presented. Finally, a brief review of the prevention and mitigation of mechanical damage is given.

## 22.2 CURRENT STATUS OF IN-LINE INSPECTION (ILI) TECHNOLOGIES FOR MECHANICAL DAMAGE CHARACTERIZATION [8,9]

Technologies and associated ILI tools identified of having potential for mechanical damage detection and discrimination can be categorized as one of the three types: dimensional (calipers), magnetic flux leakage (MFL) and ultrasonic [8–10]. Dimensional measurement technology (mainly calipers) directly measures the deviation from the circular form of the pipe wall and has been used for detecting, locating, and sizing dents, wrinkles, cold bends, and so on. This technology is generally regarded as providing the most accurate results for sizing dents and wrinkles at specified detection thresholds (e.g., 2% OD for dents), but it is not capable of detecting other defects associated with dents such as corrosion, cracks, and gouges. Therefore, dimensional technology in conjunction with other technologies, such as MFL technology, is utilized to reveal the severity of mechanical damage defects.

Magnetic flux leakage (MFL) ILI technology, using both axial and circumferential fields, is capable of detecting some mechanical damage because it has been shown to be sensitive to the associated geometric and magnetic changes [11]. The magnetic signal indicating mechanical damage is driven primarily by geometric changes (local metal loss or deformed metal and denting). Other parts of the signal are mainly associated with changes in the magnetic properties resulting from stresses, strains, and metallurgical changes [12].

Ultrasonic technologies have also been used to detect dents and cracks. Identification of dents by ultrasonic tools (UT)s is feasible; however, dent sizing is inconsistent. Despite a high probability for detection of cracks in plain pipe, the probability of detection (POD) of cracks within

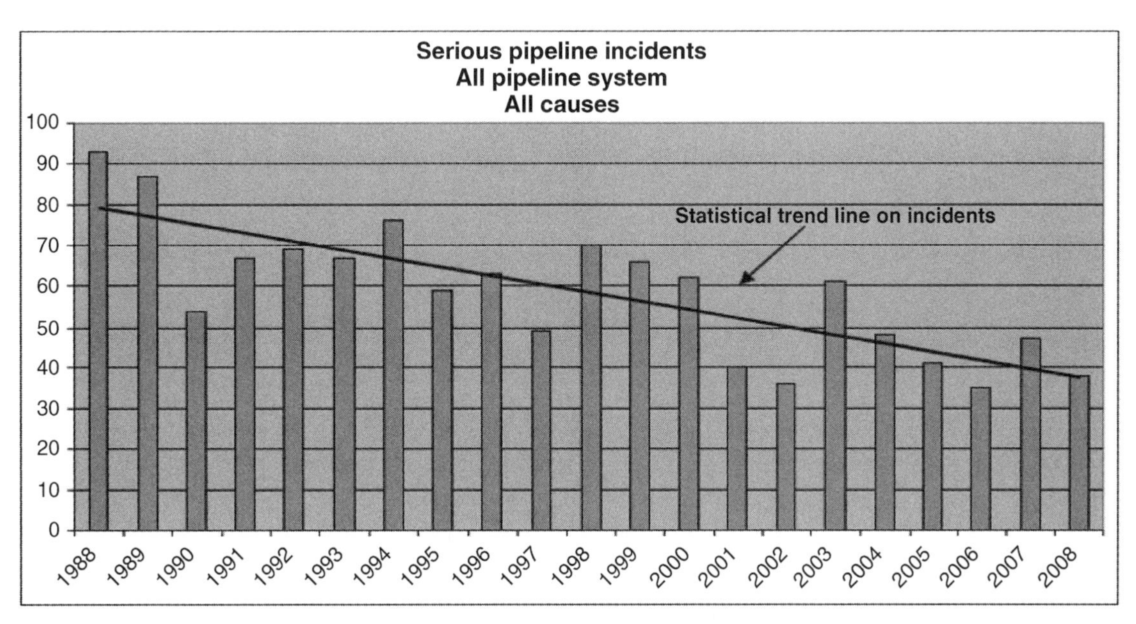


**FIGURE 22.2** Serious pipeline incidents caused by excavation damage—all pipeline systems (onshore and offshore). (*Source:* DOT/PHMSA Pipeline Incident Data [6].)

dents decreases with increasing size of the dents. Sizing of cracks within dents by ILI is not feasible with any degree of reliability. Additionally, UT requires a liquid medium; consequently, operational limitations exist for application in gas pipelines. To overcome this limitation, electromagnetic acoustic transducer (EMAT) technology is being developed for gas pipelines. However, the EMAT commercialization is just initiated within the last several years for detecting cracks and dent with cracks in pipelines [8,9].

### 22.2.1 Geometry (Caliper) Sensing Technologies

Multiple geometry ILI tools are available that can provide various levels of deformation details. These tools can generally be grouped into two categories: single-channel tools and multiple-channel tools [13].

Single-channel tools only provide the minimum pipeline diameter and distance traveled, which are useful for new construction or line segments that have never been inspected. They offer simple operation, low inspection cost, ability to pass large reductions of pipe diameter, and rapid analysis of the geometrical data. However, they offer little information on the orientation and geometrical profiles of the deformation, which are required for performing strain and stress calculation from the results [13].

In contrast, multiple-channel ILI tools provide detailed information on the orientation, length, width, maximum depth, and geometrical profile of the deformation, which can be used for deformation severity evaluation based on strain calculation. They are also reliable for locating and characterizing deformation anomalies. Therefore, a multiple-channel ILI tool is recommended for use in an integrity management plan for mechanical damage [13].

Three types of multiple-channel geometry sensing technologies are used to detect and characterize plain dents in the pipe wall:

- Direct Arm Measurement (caliper type)
- Direct Arm Measurement with electromagnetic (EM) proximity sensors
- Indirect Electromagnetic Measurement (caliper type)

Direct Arm Measurement sensing technologies are generally employed in tools known as caliper. All Direct Arm Measurement Calipers (DAMC) employ multiple mechanical arms (fingers) for dent characterization, which measure the interior of the pipe geometry by contacting the inner surface of the pipe and are often referred to as "multichannel" calipers. Each arm is equipped with a sensor that measures the angle of the arm (such as Hall effect or eddy current transducers). These multichannel tools provide data, such as depth, length, width, location, and orientation of the deformed area. Various proprietary mechanical designs are incorporated to address issues associated with sampling intervals, inspection speed excursions, sensor bounce, vibration, and sensor coverage of the internal pipe surface.

Some DAMC technologies were reported to be augmented with EM sensors at the ends of the mechanical arms. These EM sensors provide an additional data stream for analyses that address potential sensor liftoff, which occurs due to tool speed/sensor inertia, deformation geometry, interior cleanliness of the pipe, and other conditions.

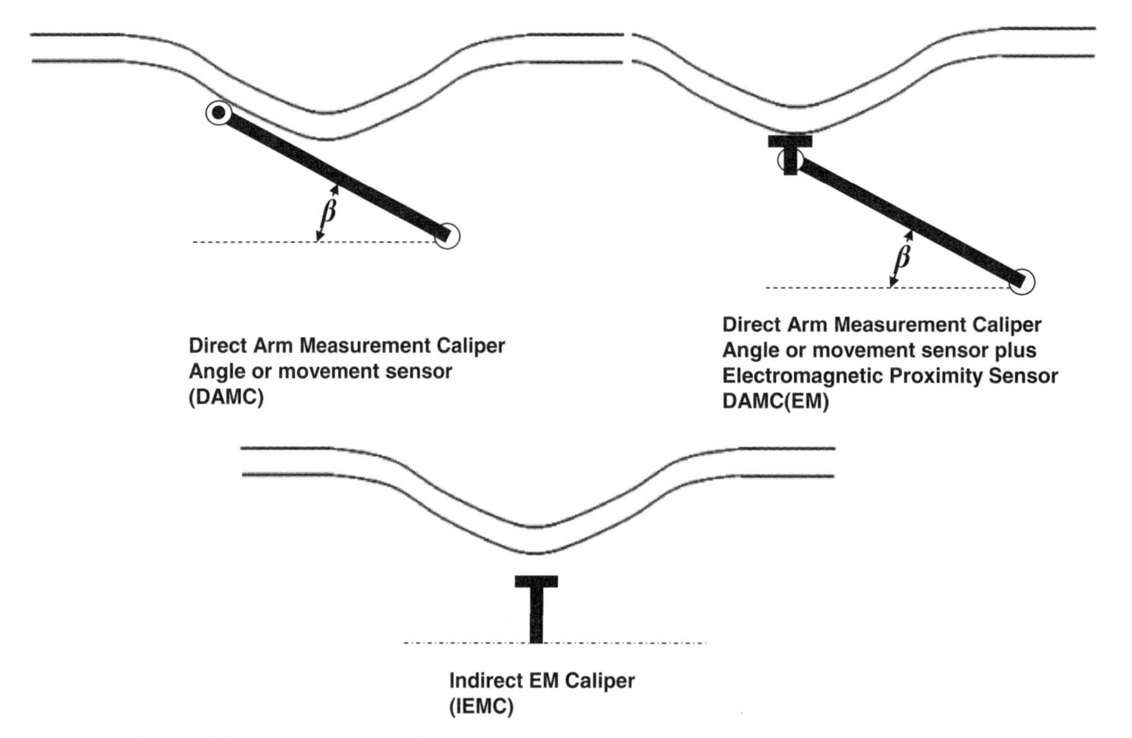


**FIGURE 22.3** Three types of caliper sensor technology for dent characterization [8,9]. (Reproduced from Ref. [9]. The full copy of the report can be found at www.prci.org, ©Pipeline Research Council International, Inc. PRCI bears no responsibility for the interpretation, utilization or implementation of any information contained in the report.)

Indirect Electromagnetic Calipers (IEMC) were also identified for deformation measurements. This type of sensor does not rely on direct contact with the pipe wall. EM sensors (eddy current) are mounted on a fixed diameter ring, with a diameter sufficient to allow for passage of the tool through a minimum expected bore, and gauge the distance between the ring and pipe wall. Limited validation data have been made available for this technology.

Figure 22.3 shows the three types of caliper sensing technology for detection, discrimination, and sizing of plain dents.

The POD and sizing performance for all deformation technologies are greatly influenced by the circumferential coverage of the sensing area by the caliper sensors [14]. Data sampling rate and contact area of individual sensors also affect the performance.

Design and spacing of the sensing arms determine the circumferential resolution of a caliper tool. Narrow arms provide greater resolution of the contour. Conventional resolution direct arm calipers are typically designed with the same number of arms as the nominal pipe size (NPS), such as 12 arms or 20 arms for NPS 12 or 20 pipelines, respectively. The circumferential spacing between the sensing arms is approximately 3.14 in. when the number of sensors equals the NPS [13]. Higher resolution caliper tools are designed with narrower spacing as small as ¾ in., but typically 1–2 in. Caliper tools have also been designed with paddles. Radius-tipped paddles provide more coverage of the inside surface than narrow arms but tend to average the width of the deformation. Conversely, narrow arms provide greater resolution of the deformation contour, but do not record deformation that passes between adjacent arm sensors.

### 22.2.2 Coincident Damage Sensing (Dent with Metal Loss) Technologies

Coincident damage for the purpose of this chapter is classified as corrosion, gouges, and cracks associated with dents. Damage from gouges was broadly categorized as localized damage due to moved or removed metal [15].

MFL and UT tools in combination with deformation tools are utilized to detect and discriminate localized mechanical damage in the form of corrosion, gouge, and gouges with crack, within the deformation The sensitivity to deformation detection depends on the type and size of sensor, and the type of sensor assembly.

MFL, using either axial or circumferential magnetic fields, is capable of detecting dents and dents with metal loss (DMLs) [11]. MFL tools do respond to any sharp portions of the deformation by the sensors "lifting off" from the inside pipe surface. This creates a magnetic field change, causing a signal response. However, the signal shape for a typical MFL

**TABLE 22.2 Categorization of In-line Inspection Technologies for Mechanical Damage Characterization [8,9]**

| Generic Technology Key | Deformation Sensing Technology | Coincident Sensing Technology | Inspection Process |
|---|---|---|---|
| Caliper + MFL (axial) | Direct Arm Measurement Caliper (DAMC) | MFL (axial magnetic field) | Two separate runs |
| Caliper + MFL (circumferential) | Direct Arm Measurement Caliper (DAMC) | MFL (circumferential magnetic field) | Two separate runs |
| Caliper + MFL (axial) Combo | Direct Arm Measurement Caliper (DAMC) | MFL (axial magnetic field) | One single tool, run once |
| Caliper (EM) + MFL (axial) | Direct Arm Measurement Caliper (DAMC (EM)) with electromagnetic sensor | MFL (axial magnetic field) | Two separate runs |
| Caliper (EM) + MFL (axial) Combo | Direct Arm Measurement Caliper (DAMC (EM)) with electromagnetic sensor | MFL (axial magnetic field) | One single tool, run once |
| MFL (3-axis) | Triaxial MFL Hall-3, ID/OD EM | Triaxial MFL Hall-3, ID/OD EM | One single tool, run once |



signal from a dent is fundamentally different than that seen from metal loss. Furthermore, an MFL signal from a cold worked region will generally exhibit a signal shape different from that of metal loss. This enables MFL to detect and discriminate the metal loss and cold work within the dent. Proper analysis of the MFL raw data is the key to gaining information about the deformation anomalies. In all cases of MFL technology, except one, that is, triaxial MFL technology [8,9], assessment of deformation coincident with metal loss requires an integrated analysis of caliper-based deformation data in conjunction with MFL data.

Generally, conventional MFL tools alone, except the triaxial MFL tools, are not considered to be reliable for identifying and sizing dents. Therefore, in the majority of cases, the detection and discrimination of dents and DMLs rely on the integration and analysis of multiple data streams. Often these data originate from multiple ILI tools or from tool technologies that have been mechanically combined into a single tool. For multiple ILI tools, mechanical damage is inspected with deformation (caliper) tools and MFL tool separately. For the combined technologies, a single Combo tool runs once.

For multiple data streams, automated analysis of ILI signals from calipers and metal loss from MFL tools is a normal practice for inspection vendors. However, the integration of data from multiple tools and analysis of these data streams for the purpose of mechanical damage detection and discrimination rely heavily on expert analysis and interpretation.

As for the triaxial MFL technology [9], since it utilizes triaxis Hall effect sensors along with magnetic models to incorporate both stress and geometry effects to interpret MFL signals from dents, this technology does not require a caliper data stream to characterize deformation. The performance of the triaxial MFL tools for detecting and discriminating deformation with coincident metal loss was evaluated with limited field data [9] and is reported later in this chapter.

Table 22.2 summarizes these technologies in their generic category description, which are used by the pipeline industry for characterization of mechanical damage. In the table, the technologies for crack detection, UT and EMAT, are not included but they are similar to MFL and can be combined with caliper but run separately. So far there is no Combo type of tool for dents with crack detection.

### 22.2.3 Capabilities and Performance of the Current In-Line-Inspection Technologies for Detection, Discrimination, and Sizing of Mechanical Damage

The capabilities and performance of the current mechanical damage technologies have been evaluated under a project "Investigate Fundamentals and Performance of Current Inspection Technologies for Mechanical Damage Detection" funded by Pipeline Research Council International (PRCI) [9,16]. Since this evaluation provided the latest and most complete information on the capabilities and performance of the current mechanical damage technologies and were validated using extensive excavation data collected from both ILI vendors and pipeline operators, the results have been used in this chapter.

The capabilities of the current deployed ILI-based technologies for detecting, discriminating, and sizing dents and DMLs are summarized in Table 22.3. Clearly, geometry (caliper) ILI tools alone are capable of detecting and sizing dents but incapable of detecting and sizing metal loss in the deformation. For detecting metal loss coincident with deformation, the combined technologies, or the triaxial MFL technology, should be used. However, their capability in discrimination of metal loss between corrosion and gouge/crack is limited (high uncertainty and low confidence), if not impossible.

***22.2.3.1 Performance Measures*** The performance of any ILI technologies can be measured with the following four

**TABLE 22.3 Capabilities of Current In-Line Inspection Technologies to Detect and Discriminate Dents and Dents with Metal Loss Condition [9,16]**

| | Dent | | | | Metal Loss | | | | |
|---|---|---|---|---|---|---|---|---|---|
| Generic Technology Type | Detection | Depth | Length | Width | Detection | Depth | Length | Width | Gouge/Corrosion Discrimination |
| Geometry sensing technologies | Y | Y | Y | Y | Y | N | N | N | N |
| Caliper + MFL | Y | Y | Y | Y | Y | Y | Y | Y | N or limited |
| MFL (triaxial) | Y | Y | Y | Y | Y | Y | Y | Y | Limited |



parameters: probability of detection (POD), probability of identification (POI), probability of false call (POFC), and sizing tolerance, in accordance with the procedures recommended by API 1163 ILI [17], and using binomial probability distribution methods [18,19]. The parameters are defined as follows:

- **POD:** The probability (or, certainty) that the number of anomalies (e.g., dents) successfully detected by the ILI tool with respect to the total number of anomalies of the same type (dents) on the pipeline. Usually, the pipeline industry adopts POD = 80% (of the time) with a confidence level of 95%, indicating with which the POD is satisfied.
- **POI:** The probability (or, certainty) that the number of anomalies (e.g., gouge) is successfully classified and/or characterized by the ILI tool from the anomalies having similar ILI signal characteristics (e.g., corrosion, crack) on the pipeline. Usually, POI = 80% (of the time) at a confidence level of 95% is adopted by the pipeline industry.
- **POFC:** It is a closely associated measure with the POD and is the frequency that the tool falsely reports an anomaly where no anomaly exists. POFC is defined as a probability (or, the certainty) of a nonexisting feature being reported as a feature by the ILI tool.
- **Sizing tolerance:** It is a measure of an ILI technology's ability to predict an anomaly's dimensions, typically, the depth, length, and width. API 1163 indicates that the sizing accuracy shall include the following three parameters when using a small sample size of excavations for tool performance validation:
  - A tolerance, for example, within ±10% of wall thickness.
  - A certainty, that is, the probability that a reported anomaly depth is within the specified tolerance.
  - A confidence level indicating the confidence with which the tolerance and certainty levels are satisfied.

The methods used for calculating POD, POI, POFC, and sizing tolerance are the binomial probability distribution and confidence interval methods in accordance with API 1163 [18–20]. For confidence interval methods, the Clopper–Pearson method known as the "exact" confidence interval for binomial probability distribution is preferred because this method provides the most conservative intervals, $P_L$, and $P_U$, that is, the lower and upper ends of the interval, respectively, for a given confidence level. The details of the binomial distribution analysis and Clopper–Pearson confidence interval methods are given elsewhere [18–20].

In the following section, the performance of current ILI technologies for characterization of dents and DMLs in terms of POD, POI, POFC, and sizing is summarized and reviewed [8,9,16].

***22.2.3.2 Dent: POD, POI, POFC, and Sizing Performance*** It is known that the probabilities of detection and identification of dents for depth >2.0% OD in pipelines are very high and the POFC is very low for all types of generic technologies listed in Table 22.3. POD, POI, and POFC were not the concerns. Only the sizing performance was evaluated and reported [9,19].

For dent sizing, geometry sensing technologies (calipers) are generally regarded as providing the most accurate results at specified detection thresholds because they directly measure the deviation from a circular form of the pipe wall. Other technologies, for example, MFL and UT/EMAT alone are not capable of sizing dents with acceptable certainty and confidence, except triaxial MFL.

Two data sources have been used for evaluation of the depth sizing performance: one from ILI vendors and the other from pipeline operators. Some of the data from ILI vendors were obtained from pull-through tests without rebounding and re-rounding of dents errors. "Rebound" refers to the reduction in the dent depth and change in shape due to elastic unloading that occurs when the indenter is removed from the pipe and "re-rounding" refers to the change in dent depth and shape due to change in internal pressure. Therefore, the vendors' data, obtained through pull-through testing, represents the sizing accuracy that can be achieved by the ILI tools; however, the data from operators were affected by the conditions of ILI run and re-rounding and rebounding due to excavation and pressure reduction, in particular, for bottom side of constrained dents. Another source of large error in operator's data was variability in in-ditch protocol for dent measurement. Hence, the operators' data represent the current level of tool performance influenced by various conditions in addition to the error from the ILI tool itself.

**TABLE 22.4 Dent Depth Sizing Tolerance Evaluation**

| | Tolerance for Certainty = 0.8 at 95% Confidence Level | | | |
|---|---|---|---|---|
| Technology Type | Sample Size for Validation | Limits of Detection (% OD) | Binomial Distribution Analysis (% OD) | Clopper–Pearson Certainty Interval Method (% OD) |
| DAMC (EM) | 130[a] | 0.5 | ±1.10 | ±1.22 |
| DAMC (EM) | 20[b] | 0.5 | ±0.74 | ±0.74 |
| DAMC (EM) | 15[c] | 0.5 | ±0.51 | ±0.78 |
| MFL (3-axis) | 273[d] | 2.0 | ±0.78 | ±0.80 |

[a]Direct examination of the dents from NPS 16 pipeline.
[b]Direct examination of the dents from multiple ILI.
[c]Laboratory pull.
[d]Dent depth predicted by MFL (3-axis) and validated against multiple tool size ILI by caliper.
(Reproduced from Ref. [9]. The full copy of the report can be found at www.prci.org, ©Pipeline Research Council International, Inc. PRCI bears no responsibility for the interpretation, utilization or implementation of any information contained in the report.)

***Dent Sizing Performance—Vendors' Data:*** A total of 438 depth sizing data points collected from ILI vendors are evaluated in Table 22.4. The estimated tolerance varies from ±0.51% to ±1.10% OD with 80% certainty at a confidence level of 95% evaluated using the binomial method. The tolerance was slightly large when the Clopper and Pearson confidence interval method was used. The results showed the tolerance to be more or less close to ±1.0% OD, or better. This tolerance value probably represents the best that can be obtained from the current ILI technologies for dent depth measurement.

***Dent Sizing Performance—Operators' Data:*** A total of 359 data points were collected from pipeline operators. Overall, the data scatter was larger and consequently a larger tolerance than that of vendors' data. The results are summarized in Table 22.5.

Among the 359 data points, 103 data points were collected in 2006, and the other 256 data points were collected in 2009 [8,16]. During the earlier time of the study (2006), the data collected from operators showed significant scatter compared with the validation populations obtained from the vendors. The depth tolerance was in the range between ±1.6 and ±4.48% OD. Progress was made in reducing the data scatter during the later time (2009) due to the better quality and more complete information of the data gathered, which allowed filtering the top-side and bottom-side dents

**TABLE 22.5 Dent Depth Sizing Performance Evaluation (LOD—Limit of Detection)**

| | | | | | Mid-Range Size Caliper Specifications | | Validation (% OD) | |
|---|---|---|---|---|---|---|---|---|
| Technologies for Dents | Data Source | # of Dent Features Evaluated | | | LOD (% OD) | Depth Tolerance (% OD) | Binomial Distribution Method | C–P Confidence Interval Method |
| 1 DAMC(EM) | Phase I | | 28 | | 0.5 | ±1.00 | ±1.60 | ±1.60 |
| 2 Caliper (EM) + MFL Combo | Phase I | | 58 | | 0.5 | ±1.00 | ±4.48 | ±4.65 |
| 3 Caliper + MFL Combo | Phase I | | 17 | | 1.0 | ±0.50 | ±1.63 | ±1.63 |
| 4 Caliper + MFL Combo | Phase II | All | 135 | Dents | 1.00 | ±0.50 | ±1.07 | ±1.08 |
| | | Top side | 25 | Dents | 1.00 | ±0.50 | ±1.00 | ±1.20 |
| | | Bottom side | 110 | Dents, uncorrected | 2.00 | ±0.50 | ±1.10 | ±1.10 |
| | | Bottom side | 110 | Dents, corrected* | 2.00 | ±0.50 | ±1.32 | ±1.32 |
| 5 MFL (3-axis) | Phase II | All | 26 | Dents | 0.50 | ±1.00 | ±2.40 | ±2.88 |
| | | Top side | 6 | Insufficient for analysis | 0.50 | ±1.00 | NA | NA |
| | | Bottom side | 20 | Uncorrected | 0.50 | ±1.00 | ±2.88 | ±3.00 |
| | | Bottom side | 20 | Corrected* | 0.50 | ±1.00 | ±0.95 | ±1.00 |
| 6 Caliper + MFL | Phase II | All | 95 | Dents | 0.60 | ±0.80 | ±5.30 | ±5.34 |
| | | Top side | 1 | Insufficient for analysis | 0.60 | ±0.80 | NA | NA |
| | | Bottom side | 94 | Uncorrected | 0.60 | ±0.80 | ±5.34 | ±5.34 |
| | | Bottom side | 94 | Corrected* | 0.60 | ±0.80 | ±1.51 | ±1.52 |

*Depth corrected for spring back using EPRG recommended equation [21].

from the total populations. For the bottom-side dents, attempts have been made to apply corrections for re-rounding and rebounding using Equation 22.1 recommended by the European Pipeline Research Group (EPRG) [21]:

$$\frac{H_r}{D} = 0.436\frac{H_p}{D} - 0.282 \qquad (22.1)$$

where $D$ = pipe diameter, $H_r$ = residual dent depth measured upon excavation at reduced pressure and unrestrained condition, $H_p$ = maximum dent depth during indentation as measured by ILI under restrained condition, and $\frac{H}{D}$ is expressed in %.

Application of Equation 22.1 to the bottom-side dents could predict the residual dent depth upon excavation when the indenter would be removed. The calculated residual depth is then used to compare with the in-ditch validation measurements. As shown in Table 22.5, with this correction, the overall scatter bands are narrowed to the range between ±1 and ±1.5% OD.

***Sizing Accuracy—Pull Test Data:*** It has long been recognized that the difficulties in validation of ILI sizing performance for dents are associated with multiple sources of error, mainly due to resolution (sensor spacing) of the ILI tool, rebounding, re-rounding, and in-ditch measurement errors. The dent sizing correlations between ILI and in-ditch measurements collected from pipeline operators usually exhibit a large scatter. Consequently, a laboratory pull test was performed with a high-resolution caliper and validated using the LaserScan 3D technology, an improved and more accurate method for nondestructive examination (NDE) measurement (see Section 22.3) [19]. The purpose of the pull test is not only to eliminate the errors resulting from re-rounding and rebounding, but also to minimize uncertainties in in-ditch direct measurements.

The pull test was conducted on a 30 in. OD × 0.342 in. wall × 10 ft long test piece that contained 10 dents with varying depths (0.5–6% OD nominal depth) made by Battelle Memorial Institute. Figure 22.4 is a sketch of the 10 dents and the LaserScan image of the deformed pipe test piece.

The ILI tool used for the pull test was a high-resolution DAMC (EM) caliper with a total of 72 sensors (sensor spacing = 1.3 in.). This ILI tool was repeatedly pulled 12 times with a speed between 0.5 and 1.8 m/s with the majority of tests at 1 m/s to provide statistically significant sizing results. A total of 120 data points (12 data points/dent) were generated.

For direct measurement, a commercially available laser profilometer, Nvision 3D scanner, was used. This has a spatial resolution of 0.1 mm and a depth accuracy of 0.1 mm, which is considered to be sufficient for dent profiling. The laser scans were done on the outer surface. A one-on-one comparison of ILI and LaserScan dent profile methods was used. The dent depth is directly measured from the overlapped profiles with the same reference points. Figure 22.5 illustrates how the measurements and comparisons were made.

The evaluation of ILI tool performance was performed using the linear regression best fit (maximum depth) against the laser profilometry validation measurements. Figure 22.6 is the graph of the ILI dent depth prediction versus LaserScan measurement.

From Figure 22.6, it is seen that the linear correlation between ILI and LaserScan data is very good with the $R^2$ value = 0.9889. The slope of the regression line is 0.9673, that is, close to one. There is a small shift of the regression line from the unity line, indicating that the ILI tool underestimates the dent depth by 0.20% OD. The reason for this shift is unclear. It may be partially attributed to the difference in measurements from external and internal surfaces and requires further investigation.

The linear regression analysis showed that the tool tolerance from the 12 sets of repeated test data is ±0.20% OD at 80% confidence level and ±0.31% OD at 95% confidence level. This compares well with the vendor's specification for this caliper technology: ±0.5% OD at 95% confidence level.

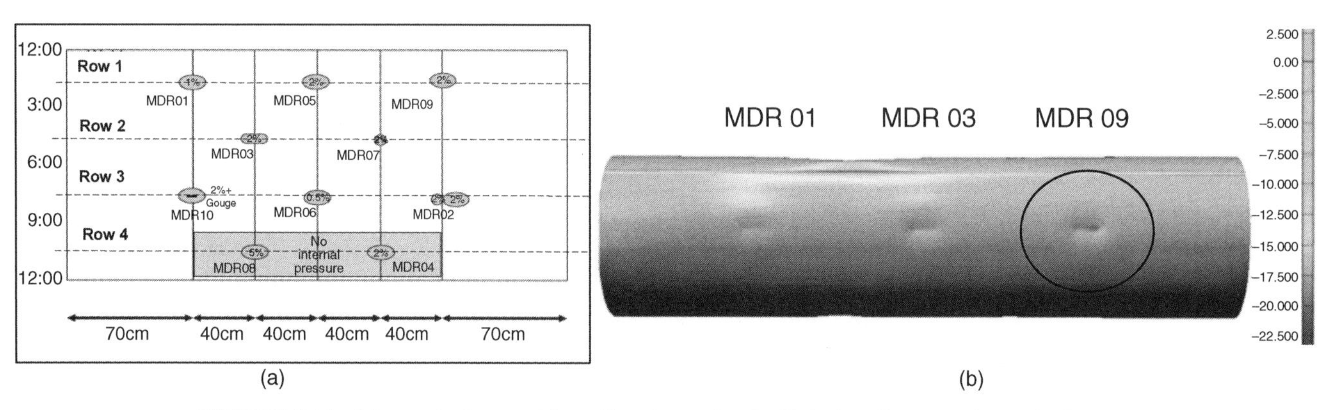


**FIGURE 22.4** Man-made dents and LaserScan image for the 30-inch pull test specimen: (a) layout of 10 dents on a 10 ft pipe and (b) laser-scanned image showing three dents on row 1.

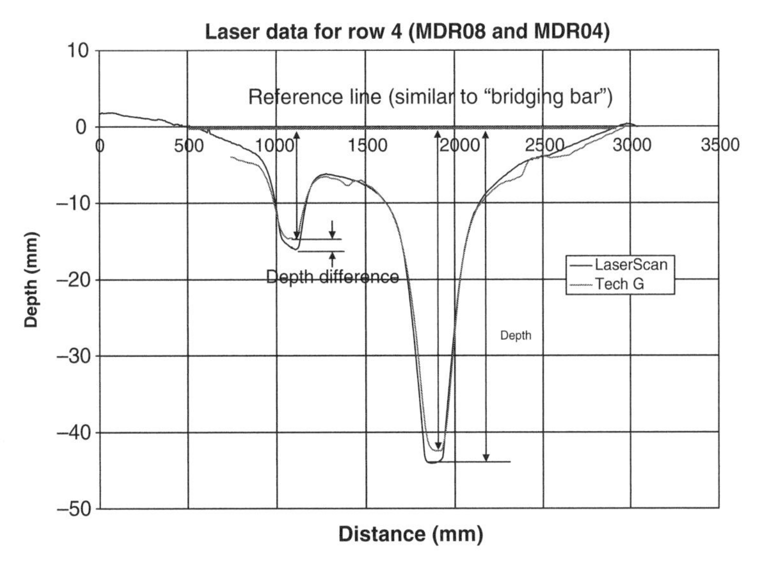


**FIGURE 22.5** An example of one-on-one comparison of profile between ILI and laser.

These results appear to be promising. The data suggest that, without re-rounding and rebounding, and the minimized NDE measurement error, the ILI tool's "inherent capacity" can be evaluated. The evaluated results in Figure 22.6 can serve as a basis for assessing the ILI errors.

***Sizing Accuracy—Length and Width:*** Similar evaluations were conducted on dent length and width sizing. The ILI vendors all reported that their technologies had the capability to measure length and width; however, there are no claimed or expected performance specifications on length and width from the ILI vendors. Table 22.6 shows the dent length and width sizing performance for two ILI technologies evaluated using the binomial distribution and Clopper–Pearson confidence interval technique.

From the table, it seems that the length and width tolerances are in the range 15–20 in., which is quite large. Historically, the main parameters used to determine severity of mechanical damage are (a) the nature of the mechanical damage (such as a plain dent, a dent with gouges, cracks, welds, etc.) and (b) the depth of the dent.

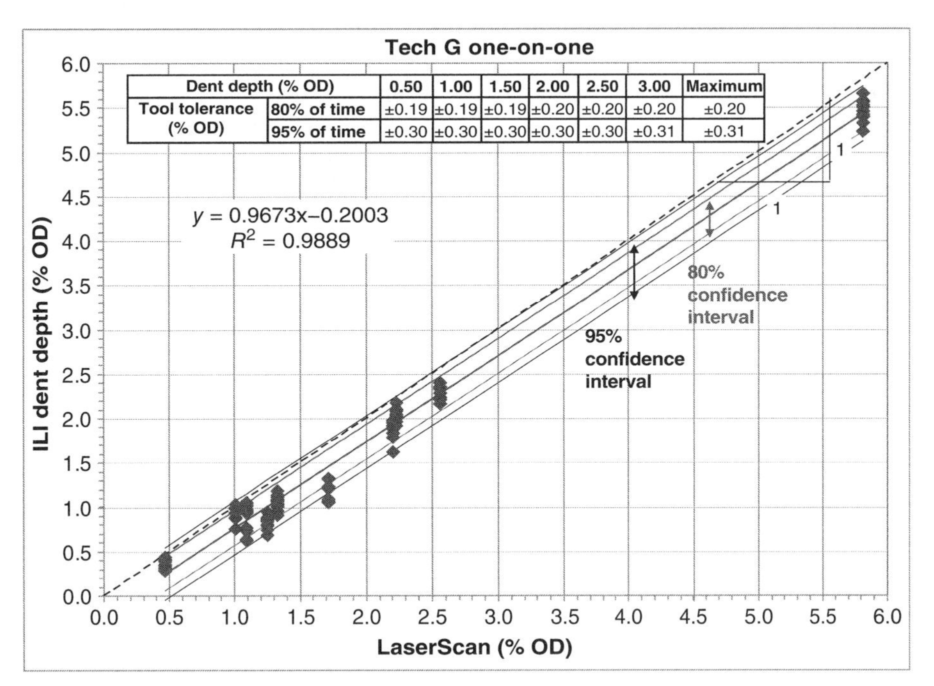


**FIGURE 22.6** Dent depth validation of a high-resolution DAMC (EM) ILI tool using a one-on-one profile comparison method, 12 times repeating pull tests.

**TABLE 22.6 Dent Length and Width Performance for the Data Collected in 2009**

| Technologies for Dents Sizing | Data Source | # of Dent Features Evaluated | Validation of Length and Width Tolerance at 80% Certainty and 95% Confidence Level (in.) | |
|---|---|---|---|---|
| | | | Binomial Distribution Method | Clopper–Pearson Confidence Interval Method |
| Length Assessment | | | | |
| Caliper + MFL | Phase II | 28 | ±17.00 | ±17.00 |
| MFL (3-axis) | Phase II | 12* | ±14.70 | NA* |
| Width Assessment | | | | |
| Caliper + MFL | Phase II | 23 | ±15.00 | ±21.00 |
| MFL (3-axis) | Phase II | 12* | ±20.00 | NA* |

**Note:* for Clopper-Person confidence interval method, the minimum required excavation number for validation is 14 for a certainty of at least 0.8 at 95% confidence. No tolerance values can be obtained for the validation excavation number 12 [19,20].

In fact, since depth is the only geometry parameter in DOT guidelines (49 Code of Federal Regulations (CFR) 192 and 195) for evaluation of disposition of dents [16], there is no claimed performance from ILI vendors on length and width, and very limited data are available for evaluation.

Clearly, the vendor-supplied data indicated a higher performance level than the operators' data. The causes for this discrepancy are debatable, but it is likely that the vender's data were gathered under more controlled conditions, and the data represent the best effort for the technologies while the operators' data came from routine tool runs and did not have the benefit of special analysis by the vendors. As mentioned previously, the operators' data would reflect multiple sources of error, including (a) variability inherent in the pipe, most notably, variability in shape and size of deformations due to changes in internal pressure (re-rounding) and changes in external confining forces (rebounding), (b) errors from the ILI tool itself, and (c) errors from in-ditch validation measurements.

#### *22.2.3.3 Dents with Metal Loss: POD, POI, POFC and Sizing Performance*

Since metal loss (such as corrosion, gouges, grooves, and/or cracks on the dent) causes stress concentration and is one of the major threats to pipeline integrity, evaluation of the capabilities and performance of ILI technologies in detecting, discriminating, and sizing metal loss is particularly important.

A total of 718 samples were used for evaluating the tool performance for characterization of DML [9,16]. The results showed that the overall performance of the current technologies is POD = 83.5%, POI = 91.2%, and POFC = 29.5% at a 95% confidence level using the binomial distribution method and POD in the range of 83.4–89.9%, POI = 91.4–96.4%, and POFC = 36.6–44.3% using the Clopper–Pearson confidence interval method.

From these results, the performance of ILI tools for detecting and discriminating DML is acceptable in accordance with the pipeline industry commonly acceptable criteria for POD and POI = 80% at 95% confidence level. However, the probability of a false call, 29.5%, is high, and could have resulted in many unnecessary excavations. Tables 22.7–22.9 summarize the results as reported by the study [9,16].

However, the ILI tool capabilities for detection and identification of DML vary greatly with individual technologies. The POD can be as high as 94.7% but can be below 80%. The same pattern can be seen for POI; out of 10, six are higher than 80% and four are below 80% with the lowest 47.3%.

Since no detailed information is available about the conditions for ILI runs and excavations, it is not possible to make any conclusions about which technologies are better than the others. The only conclusion that can be made from the available information is that high values of POD and POI have been achieved and potentially are achievable for all the technologies.

***Sizing Accuracy for Coincident Metal Loss*:** Only a small amount of data, that is, 59 coincident metal loss features are available for depth sizing evaluation. Among them, 56 were collected in 2003 while only 3 were obtained in 2009. A depth tolerance of ±12% wt is determined for 80% certainty at 95% confidence level. This is slightly higher but generally consistent with the MFL tool tolerance for corrosion, that is, ±10% wt for 80% certainty at 95% confidence level. Table 22.10 is a summary of the evaluation.

### 22.2.4 Closing Remarks

From the aforementioned review, the following remarks may be made on the overall performance of ILI tools:

(a) The capabilities of current ILI technologies for detecting and discriminating dents and DMLs are acceptable based on a POD and POI level of 80% at 95% confidence. However, the POFC is high at around 29.5%. Some ILI vendors stress the importance of manual data interpretation by a subject matter expert (SME) in discriminating metal loss within dents as a significant factor affecting POFC.

**TABLE 22.7 POD of the Current ILI-Based Technologies for Dents with Metal Loss (DML)**

| | Technologies for Dents and Dents with Metal Loss | Project Phase Data Source | Total Investigations | ILI Reported DML (Exc. Based on ILI Call) | Correct Calls (Including Type of DML) | True Calls (Not Including Type of DML) | Number of Missed DML | Number of False Calls | POD | | | | | |
|---|---|---|---|---|---|---|---|---|---|---|---|---|---|---|
| | | | | | | | | | Proportion (%) | | | POD at 95% Confidence Level | | |
| | | | | | | | | | | $n$ | $x$ | Binomial p, at 95% Conf for given $x$, $n$ (%) | Confidence Interval (%) Lower | Upper |
| 1 | Caliper (EM) + MFL Combo | Phase I (vendor) | 138 | 82 | 27 | 27 | 18 | 55 | 60 | 45 | 27 | 49.0 | 46.7 | 72.3 |
| 2 | Caliper (EM) + MFL Combo | Phase I (vendor) | 34 | 26 | 25 | 25 | 0 | 1 | 100 | 25 | 25 | 89.0 | 88.7 | 100.0 |
| 3 | Caliper + MFL Combo | Phase I (vendor) | 61 | 58 | 52 | 49 | 3 | 6 | 95 | 55 | 52 | 89.0 | 86.5 | 98.5 |
| 4 | Caliper (EM) + MFL | Phase I (vendor) | 26 | 23 | 20 | 16 | 3 | 3 | 87 | 23 | 20 | 75.0 | 69.6 | 96.3 |
| 5 | Caliper + MFL | Phase I (operator) | 27 | 8 | 6 | 6 | 3 | 2 | 67 | 9 | 6 | 45.0 | 34.5 | 90.2 |
| 6 | Caliper + MFL | Phase I (operator) | 114 | 37 | 31 | 30 | 5 | 6 | 86 | 36 | 31 | 76.0 | 73.0 | 94.4 |
| 7 | Caliper + MFL (circumferential magnet) | Phase I (operator) | 63 | 56 | 26 | 22 | 7 | 30 | 79 | 33 | 26 | 67.0 | 63.8 | 89.6 |
| 8 | Caliper + MFL Combo | Phase II (operator) | 135 | 135 | 53 | 53 | 0 | 82 | 100 | 53 | 53 | 94.7 | 94.5 | 100.0 |
| 9 | Caliper + MFL | Phase II (operator) | 94 | 36 | 33 | 33 | 0 | 3 | 100 | 33 | 33 | >91.3 | 91.3 | 100.0 |
| All | Current ILI (MFL based) MD technologies | Combined all (1–12) | 719 | 466 | 278 | 262 | 42 | 188 | 87 | 320 | 278 | 83.5 | 83.4 | 89.9 |

**TABLE 22.8 POI of the Current ILI-Based Technologies for Dents with Metal Loss (DML)**

| | Technologies for Dents and Dents with Metal Loss | Project Phase Data Source | Total Investigations | ILI Reported DML (Exc. Based on ILI Call) | Correct Calls (Including Type of DML) | True Calls (Not Including Type of DML) | Number of Missed DML | Number of False Calls | Dent with Metal Loss Performance | | | | | |
|---|---|---|---|---|---|---|---|---|---|---|---|---|---|---|
| | | | | | | | | | Proportion (%) | | | POI — POI at 95% Confidence Level | | |
| | | | | | | | | | | $n$ | $x$ | Binomial, $p$, at 95% Conf for given $x$, $n$ (%) | Confidence Interval (%) Lower | Upper |
| 1 | Caliper (EM) + MFL Combo | Phase I (vendor) | 138 | 82 | 27 | 27 | 18 | 55 | 100.0 | 27 | 27 | 89.9 | 89.5 | 100.0 |
| 2 | Caliper (EM) + MFL Combo | Phase I (vendor) | 34 | 26 | 25 | 25 | 0 | 1 | 100.0 | 25 | 25 | 89.1 | 88.7 | 100.0 |
| 3 | Caliper + MFL Combo | Phase I (vendor) | 61 | 58 | 52 | 49 | 3 | 6 | 94.2 | 52 | 49 | 88.4 | 85.8 | 98.4 |
| 4 | Caliper (EM) + MFL | Phase I (vendor) | 26 | 23 | 20 | 16 | 3 | 3 | 80.0 | 20 | 16 | 66.0 | 59.9 | 92.9 |
| 5 | Caliper + MFL | Phase I (operator) | 27 | 8 | 6 | 6 | 3 | 2 | 100.0 | 6 | 6 | 65.1 | 60.7 | 100.0 |
| 6 | Caliper + MFL | Phase I (operator) | 114 | 37 | 31 | 30 | 5 | 6 | 96.8 | 31 | 30 | 90.8 | 85.6 | 99.8 |
| 7 | Caliper + MFL (circumferential magnet) | Phase I (operator) | 63 | 56 | 26 | 22 | 7 | 30 | 84.6 | 26 | 22 | 72.7 | 68.2 | 94.6 |
| 8 | Caliper + MFL Combo | Phase II (operator) | 135 | 135 | 53 | 53 | 0 | 82 | 100.0 | 53 | 53 | 94.5 | 94.5 | 100.0 |
| 9 | Caliper + MFL | Phase II (operator) | 94 | 36 | 33 | 33 | 0 | 3 | 100.0 | 33 | 33 | >91.3 | 91.3 | 100.0 |
| All | Current ILI (MFL based) MD technologies | Combined all (1–12) | 719 | 466 | 278 | 262 | 42 | 188 | 94.2 | 278 | 262 | 91.2 | 91.4 | 96.4 |

**TABLE 22.9 POFC of the Current ILI-Based Technologies for Dents with Metal Loss (DML)**

| | Technologies for Dents and Dents with Metal Loss | Project Phase Data Source | Total Investigations | ILI Reported DML (Exc. Based on ILI Call) | Correct Calls (Including Type of DML) | True Calls (Not Including Type of DML) | Number of Missed DML | Number of False Calls | POFC | | | | | |
|---|---|---|---|---|---|---|---|---|---|---|---|---|---|---|
| | | | | | | | | | | | | POFC at 95% Confidence Level | | |
| | | | | | | | | | | $n$ | $x$ | Binomial, $p$, @ 95% Conf for given $x$, $n$ (%) | Confidence Interval (%) Lower | Upper |
| 1 | Caliper (EM) + MFL Combo | Phase I (vendor) | 138 | 82 | 27 | 27 | 18 | 55 | 67.1% | 82 | 55 | 75.6 | 57.6 | 75.6 |
| 2 | Caliper (EM) + MFL Combo | Phase I (vendor) | 34 | 26 | 25 | 25 | 0 | 1 | 3.8% | 26 | 1 | 17.0 | 0.2 | 17.0 |
| 3 | Caliper + MFL Combo | Phase I (vendor) | 61 | 58 | 52 | 49 | 3 | 6 | 10.3% | 58 | 6 | 19.4 | 4.6 | 19.4 |
| 4 | Caliper (EM) + MFL | Phase I (vendor) | 26 | 23 | 20 | 16 | 3 | 3 | 13.0% | 23 | 3 | 30.4 | 3.7 | 30.4 |
| 5 | Caliper + MFL | Phase I (operator) | 27 | 8 | 6 | 6 | 3 | 2 | 25.0% | 8 | 2 | 60.0 | 4.6 | 60.0 |
| 6 | Caliper + MFL | Phase I (operator) | 114 | 37 | 31 | 30 | 5 | 6 | 16.2% | 37 | 6 | 29.5 | 7.3 | 29.5 |
| 7 | Caliper + MFL (circumferential magnet) | Phase I (operator) | 63 | 56 | 26 | 22 | 7 | 30 | 53.6% | 56 | 30 | 65.1 | 41.8 | 65.1 |
| 8 | Caliper + MFL Combo | Phase II (operator) | 135 | 135 | 53 | 53 | 0 | 82 | 60.7% | 135 | 82 | 67.8 | 53.3 | 67.8 |
| 9 | Caliper + MFL | Phase II (operator) | 94 | 36 | 33 | 33 | 0 | 3 | 8.3% | 36 | 3 | 2.4 | 2.3 | 20.2 |
| All | Current ILI (MFL based) MD technologies | Combined all (1–12) | 719 | 466 | 278 | 262 | 42 | 188 | 40.3% | 466 | 188 | 29.5 | 36.6 | 44.2 |

**TABLE 22.10 Coincident Metal Loss Sizing Performance**

| Technologies for Dents with ML | Data Source | # of Dent Features Evaluated | Validation at 80% Certainty and 95% Confidence Level, (%wt) Binomial Distribution Method | Clopper–Pearson Confidence Interval Method |
|---|---|---|---|---|
| DAMC (EM), DAMC + MFL, DAMC (EM) + MFL Combo, DAMC_MFL (circ), | Phase I and phase II | Phase I 56,and phase II 3 | ±12.0 | ±13.0 |

(b) The capability for detecting and discriminating dents and DMLs varies greatly with individual technologies; however, high values of POD and POI have been achieved by some of the current technologies.

(c) The capability of an ILI tool to discriminate metal loss between corrosion and gouges was not evaluated because there was insufficient data to validate the vendors' claims. However, one technology (MFL plus caliper type) claimed a success rate between 50 and 90% for identification of gouges but, as pointed out by those vendors, the capabilities are subject to limitations suggesting that the confidence level is low. Additional research would be required to address this issue.

(d) Applications of ultrasonic sensor technologies have been reported, but the available data for validation are not sufficient. Therefore, capabilities of ultrasonic sensor technologies, including EMAT for detecting and discriminating coincident metal loss and cracks, are not evaluated and will be a subject for future study.

(e) One of sources for the scatter in pipeline operator data was the lack of appropriate in-ditch inspection tools and inconsistent protocols for measurement of dents and DMLs. Developing an improved in-ditch measurement technology and well-defined measurement protocols will improve in-ditch data. This will be discussed in the next section.

## 22.3 IMPROVED TECHNOLOGIES FOR IN-DITCH MECHANICAL DAMAGE CHARACTERIZATION

Standard and high-resolution ILI caliper tools are used to characterize and profile the dents along a pipeline. The profile data can be used to determine the severity of a dent using depth-based and strain-based assessment. However, in assessing the performance of ILI caliper tools, one of the major factors causing errors or biases (other than the ILI error) in validation is the variability of the in-ditch manual measurements. Conventional technology comprising a simple profile gage or straight edge method has been used to characterize dent profiles. These traditional practices have been time consuming, tedious, and operator dependent and have produced wide scatter in data and inconsistent results for ILI performance evaluation [22]. Inaccuracy and insufficiency of in-ditch data made interpretation of depth and strain values a difficult task. Moreover, for dents associated with other anomalies such as gouge or corrosion, conventional in-ditch methods fail to precisely profile the associated anomalies, which again lead to incorrect estimation of the severity of the dent and make managing pipeline integrity a daunting task.

With the advantage of advanced laser 3D scan technologies that have been in use in the aerospace and automotive industries for reverse engineering, styling, design and analysis, and in medical applications like generation of 3D digital profiles from body parts, an in-ditch portable 3D LaserScanner for profiling dents and dents with associated anomalies was developed [23,24]. This improved measurement tool has made in-ditch measurement of mechanical damage not only significantly easier and more accurate, but also essentially operator independent. Mitigation decisions can now be made in real time based on the measured depth and strain values of a dent. It is noted that the 3D LaserScan profiling technologies have been tested in the past to characterize and profile dents. However, the application of laser tools was limited because they could not readily provide precise digitized profiles with a reference frame. Now, the technical barriers have been overcome.

More recently, a structured-light 3D scanner system was developed [24–26], and has been successfully applied to in-ditch measurement of corrosion in pipelines [27]. Potentially, this technology can be used for in-ditch mechanical damage measurement [28] and will provide opportunities to the pipeline operators and NDE inspection service companies to select technologies that best fit their needs. However, since the structured-light 3D scanner system for in-ditch mechanical damage measurement is still under development, only the 3D LaserScan technology is discussed.

### 22.3.1 In-Ditch Laserscan Technology

The 3D LaserScan system consists of two major components, that is, hardware (scanner) and software. The scanner hardware includes automatic surface generation. When the pipe surface is scanned, the curves and laser lines are recorded by the scanner and an optimization loop is initiated between the LaserScan and the data collected. The output of the scan is not a point cloud but an optimized surface. This is an improvement over several non-portable conventional laser-scanning devices, where getting a precise reference profile geometry is a tough task.

This optimized surface is then processed by the system's proprietary software. The software models the scanned undeformed pipe surface to best fit a perfect circular cross section of the cylinder, and then calculates the depth profile of the dent as the difference or displacement between the scanned 3D surface and the undeformed cylinder surface. Each of the depth points on the surface will have a calculated cylindrical coordinate. Once the profile is generated, the axial and circumferential profile can be extracted and the maximum depth and deformation strain can be calculated.

***22.3.1.1 Hardware: 3D LaserScanner*** Figure 22.7 shows an overview of the LaserScan system hardware. The system consists of a scanner with built-in camera and a laptop computer and cables. The scanner is portable and light, weighing approximately two pounds.

**FIGURE 22.7** An overview of the LaserScan system hardware. (a) Scanner with computer; (b) scanning a 16 in. OD pipe; (c) dent with positioning targets.

**TABLE 22.11 Specifications of Commercially Available LaserScanners**

| Technical Specifications | Model 1 | Model 2 |
|---|---|---|
| Weight | 1.25 kg (2.75 lb) | 980 g (2.1 lb) |
| Dimensions | $172 \times 260 \times 216$ mm$^3$ ($6.75 \times 10.2 \times 8.5$ in$^3$) | $160 \times 260 \times 210$ mm$^3$ ($6.25 \times 10.2 \times 8.2$ in$^3$) |
| Measurements | 25,000 measures/s | 18,000 measures/s |
| Laser class | II (eye safe) | II (eye safe) |
| Resolution in *Z* axis | 0.05 mm (0.002 in.) | 0.1 mm (0.004 in.) |
| Accuracy | Up to 40 μm (0.0016 in.) | Up to 50 μm (0.002 in.) |
| ISO | 21 μm + 100 μm/m | 20 μm + 200 μm/m |
| Depth of field | 30 cm (12 in) | 30 cm (12 in) |

Table 22.11 shows two specifications of the LaserScanner commercially available. The single point accuracy is up to 40 and 50 μm, and the resolution in the *Z* axis (i.e., depth direction) is 0.05 and 0.1 mm, depending on the model of the scanner.

The LaserScanner has auto-positioning stereo vision. The camera sees the patterns of the positioning targets that are applied on the surface to be scanned, and by triangulation, the scanner is able to determine its position relative to the targets (Figure 22.8).

The scanner can scan all non-shiny surfaces with minimal surface preparation. However, for shiny surfaces, the surface should be covered with an opaque powder. The positioning targets should be placed in and around the area of interest with a random distance of 3–4 in. from each other. Once the positioning targets are placed, the LaserScanner can be used to scan the surface with the positioning targets.

The accuracy, reliability, and repeatability of the scanner tool were tested by measuring the step wedge with nominal dimensions of 0.1 in. per step. Figure 22.9 shows the step wedge and dimensions as measured using Model 1 Laser-Scanner by a specialist and a volunteer. The difference is in the order of few thousandths of an inch, which is acceptable for in-ditch dent measurement. The specialist-measured step wedge is within the tool accuracy specification of 0.0016 in.

The LaserScanner was used to characterize typical pipe dents. Figure 22.10 shows the pipe used in this work, containing five dents with one of them, that is, dent 1, containing a gouge.

***22.3.1.2 Data Processing*** Once the area of interest is scanned, the scanning system digitally resembles the original surface of the pipe, dents, and associated defects. Figure 22.11 is the resembled pipe, showing the dents and dents with gouge, and a zoomed gouge with the measured length, width, and depth.

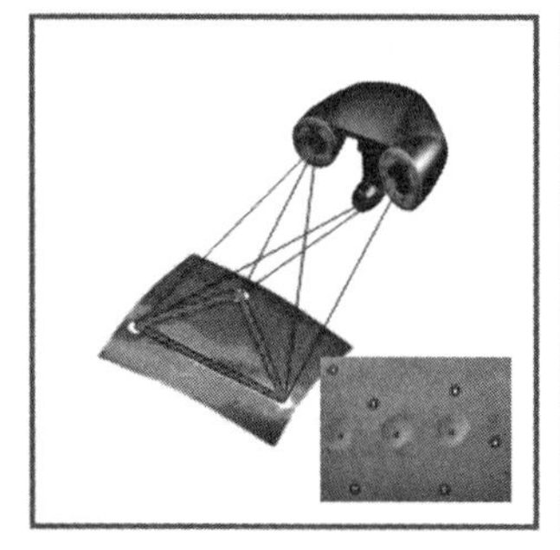
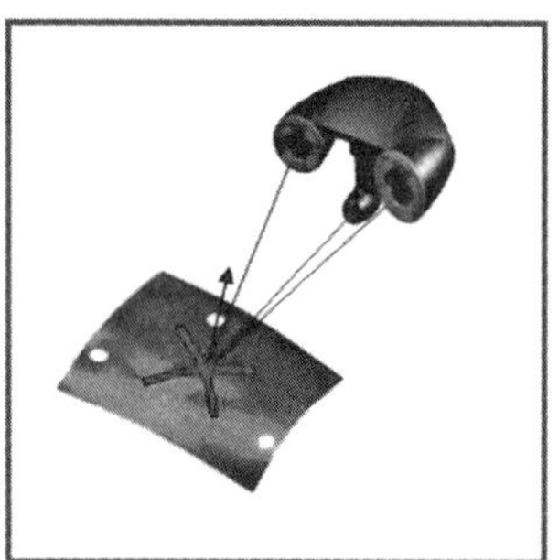
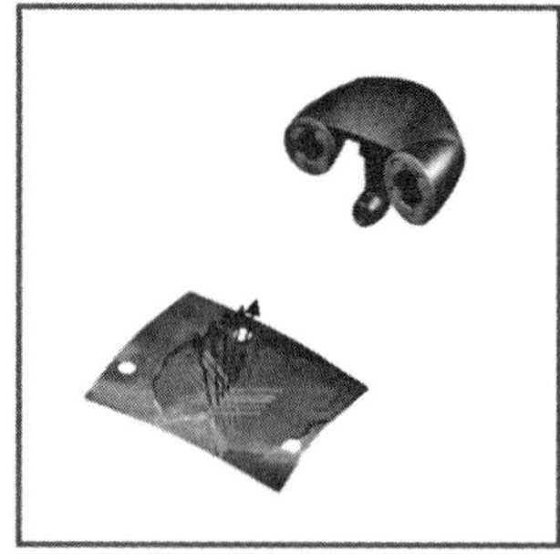

**FIGURE 22.8** (a) Scanner positions itself by triangulation, (b) cameras read the laser cross deformations, and (c) the scanner moves to generate a 3D model.

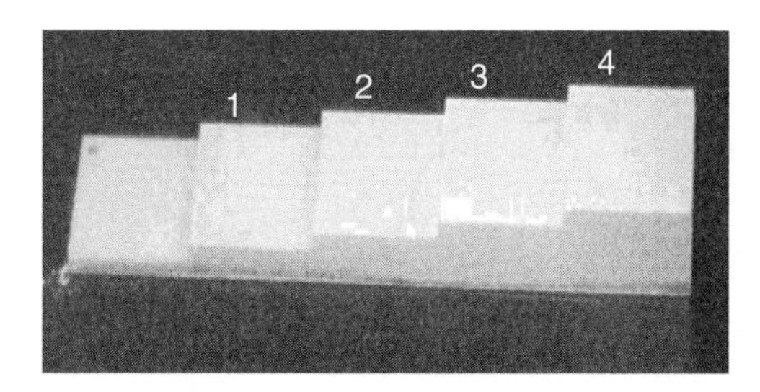


| Step # | Depth (in.) | Specialist Measured depth (in.) | Volunteer Measured depth (in.) |
|---|---|---|---|
| 1 | 0.1000 | 0.1005 | 0.1014 |
| 2 | 0.2000 | 0.2008 | 0.2039 |
| 3 | 0.3000 | 0.3008 | 0.3046 |
| 4 | 0.4000 | 0.4000 | 0.4049 |

**FIGURE 22.9** Comparison between nominal and measured step depths by a specialist and volunteer, (a) step wedge and (b) scanned data.

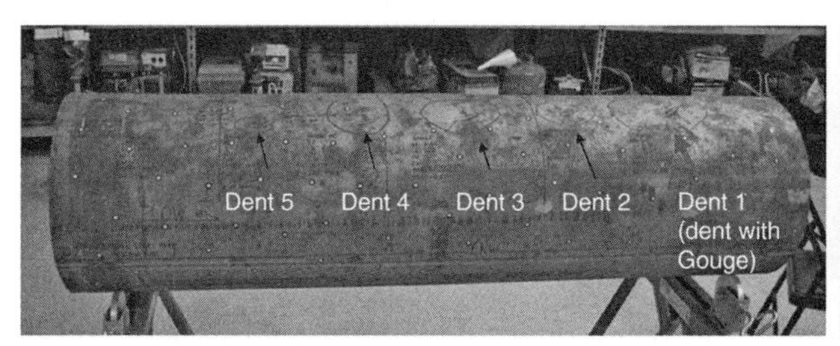

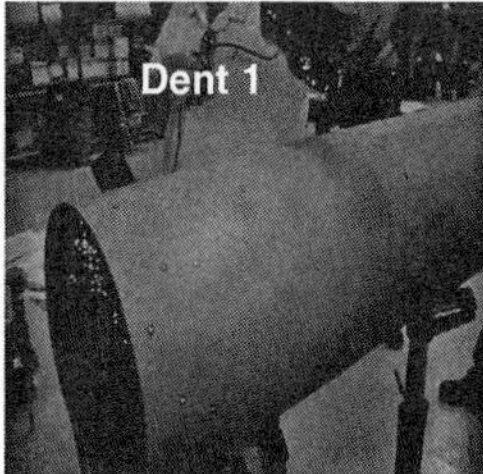

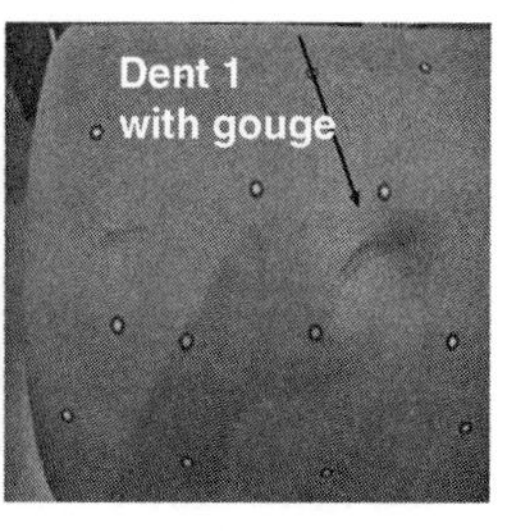


**FIGURE 22.10** Pipe with five dents. Dent 1 is associated with a gouge. The small dots on the pipe surface are positioning targets.

These 3D image data are further processed with the system's proprietary software, which extracts and converts each of the points on the scanned pipe surface into a cylindrical coordinate system. The figure next to the table in Figure 22.12 shows a Point P($r$, $\theta$, $z$) on the scanned pipe surface, where $r$ and $\theta$ are the point P($r$, $\theta$, $z$) projection onto the polar-coordinate $r$–$\theta$ plane, and $z$ is the coordinate of the point on the $z$ axis, that is, the center line of the pipe. The numbers inside the table in Figure 22.12 are the $r$ values of the data points. For undeformed pipe, $r = r_o$, that is, the radius of the pipe. In the deformed or dented area, the r value shows the change in $r_o$ due to deformation.

Once the scanned surface data are extracted and saved in the cylindrical coordinate system format (Figure 22.12), the axial and circumferential profiles of a dent can be readily extracted and plotted for any given value of $z$ and $\theta$. Figure 22.13 shows 180 axial profiles of a dent (i.e., dent 1, Figure 22.11) around the circumference of the pipe from 0° to 360° with an increment of 2° from $\theta = 0°$ (reference). The resolution of the profile is 2 mm (i.e., $z = 2$ mm increment for each point, Figure 22.12). The deepest point of the dent is located at the axial distance of $z = 146$ mm at $\theta = 86°$. As a common practice for both ILI and in-ditch manual measurement, the axial profile passing

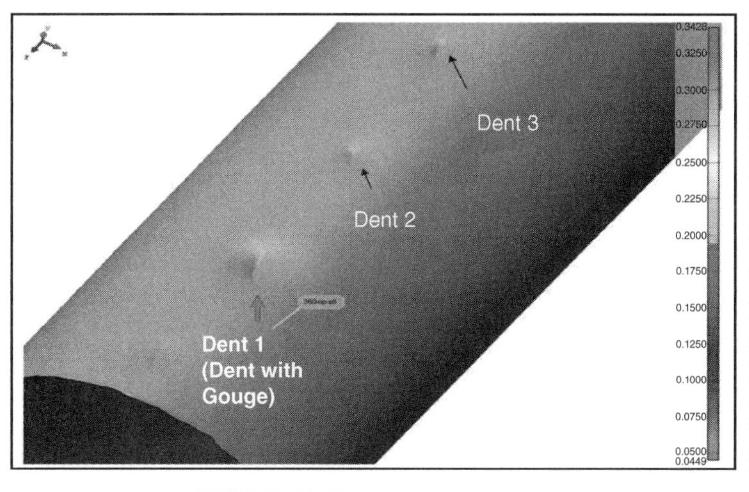

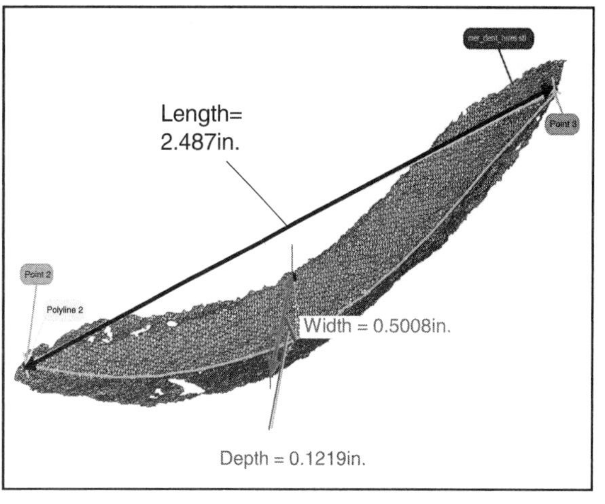


**FIGURE 22.11** (a) Scanned image of the dented pipe surface and (b) gouge profile with depth, length, and width measured by data processing. The scanned pipe is that shown in Figure 22.10.

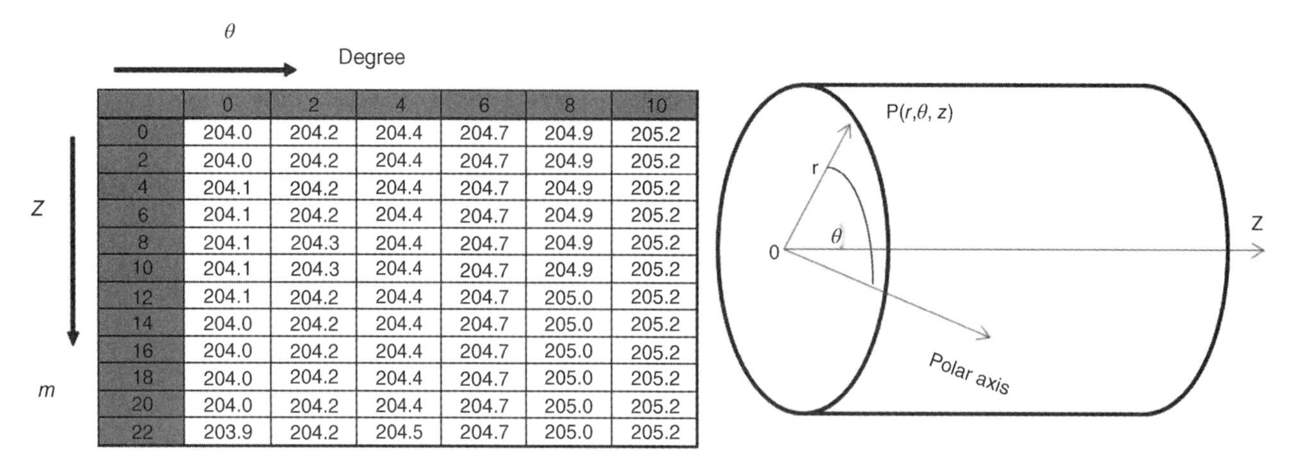


| | 0 | 2 | 4 | 6 | 8 | 10 |
|---|---|---|---|---|---|---|
| 0 | 204.0 | 204.2 | 204.4 | 204.7 | 204.9 | 205.2 |
| 2 | 204.0 | 204.2 | 204.4 | 204.7 | 204.9 | 205.2 |
| 4 | 204.1 | 204.2 | 204.4 | 204.7 | 204.9 | 205.2 |
| 6 | 204.1 | 204.2 | 204.4 | 204.7 | 204.9 | 205.2 |
| 8 | 204.1 | 204.3 | 204.4 | 204.7 | 204.9 | 205.2 |
| 10 | 204.1 | 204.3 | 204.4 | 204.7 | 204.9 | 205.2 |
| 12 | 204.1 | 204.2 | 204.4 | 204.7 | 205.0 | 205.2 |
| 14 | 204.0 | 204.2 | 204.4 | 204.7 | 205.0 | 205.2 |
| 16 | 204.0 | 204.2 | 204.4 | 204.7 | 205.0 | 205.2 |
| 18 | 204.0 | 204.2 | 204.4 | 204.7 | 205.0 | 205.2 |
| 20 | 204.0 | 204.2 | 204.4 | 204.7 | 205.0 | 205.2 |
| 22 | 203.9 | 204.2 | 204.5 | 204.7 | 205.0 | 205.2 |

**FIGURE 22.12** Digitized dent profile data ($r$ values) in cylindrical coordinates.

through the deepest point is used to determine the depth and length of the dent.

It is noted that in the profile next to the deepest point at $\theta = 84°$, there is a depression near the tip, indicating the presence of a gouge near $z = 126$ mm, which is consistent with Dent 1 as described.

Similarly, any circumferential profile for any given $z$ value can readily be extracted and plotted. Figure 22.14a is a circumferential profile at $z = 146$ mm, which passes through the deepest point of the dent. This profile can be used to determine the circumferential angular span and width of the dent. Figure 22.14b-1 and b-2 is the circumferential profile at $z = 126$ mm, which passes through the gouge in the dent. Again, a depression near the tip of the profile is noticeable due to the presence of the gouge.

Once the profile is generated, the axial and circumferential data can be used to quantify the dent depth, length, and width and to calculate the deformation strains at each point in the dent using strain calculation models [12,29–32].

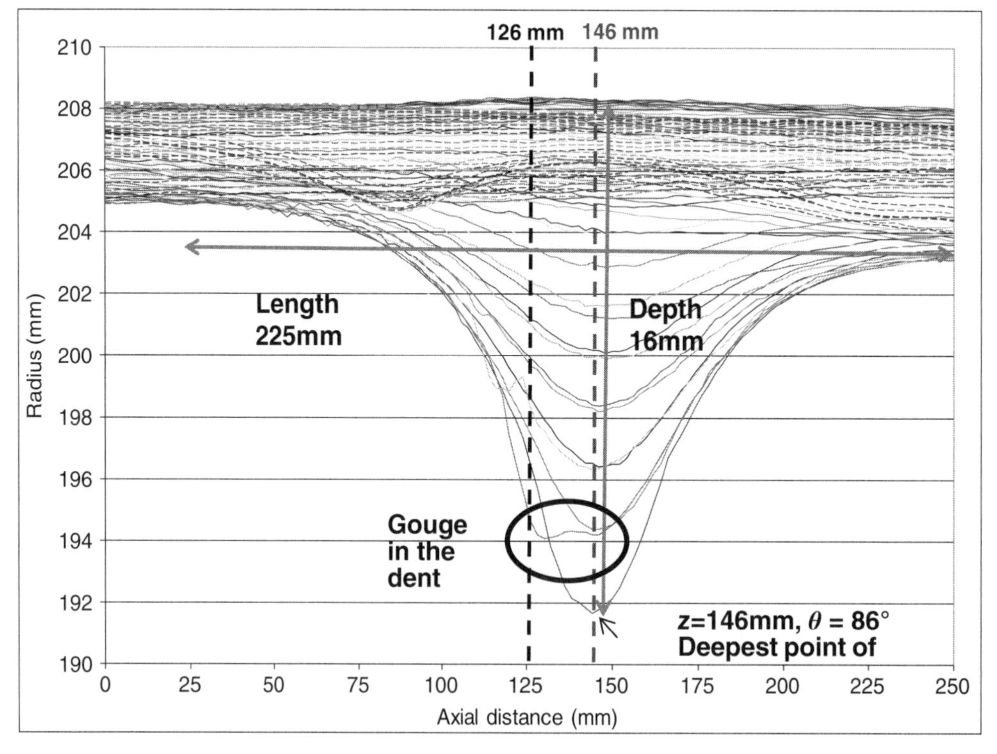


**FIGURE 22.13** 180 axial profiles of dent 1. Each profile corresponds to a specific $\theta$ value. One of the profiles next to the deepest profile has a depression suggesting that it passed through a gouge.

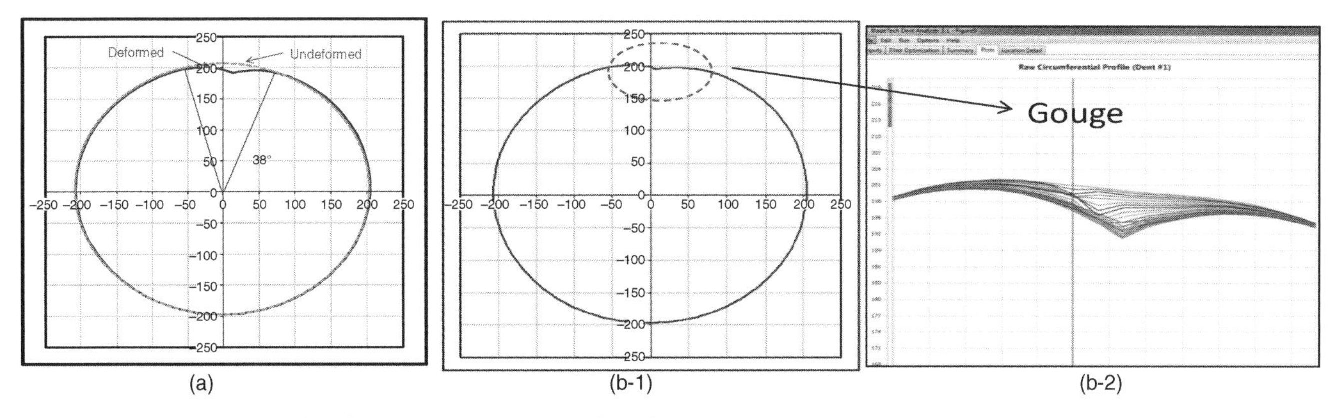


**FIGURE 22.14** Two circumferential profiles of Dent 1: (a) at $z = 146$ mm where the deepest point of the dent is located, (b-1) at $z = 126$ mm where the gouge is located, and (b-2) closeup of (b-1) showing gouge.

### 22.3.2 Application of the State-of-the-Art In-Ditch Measurement Technology

The benefits of using the LaserScan in-ditch measurement technology are obvious, and may be described in the following three aspects.

***22.3.2.1 Calibration of Depth-Based ILI Technologies for Mechanical Damage Evaluation [22]*** Usually, ILI tools are calibrated for mechanical damage evaluation using *pull-through* tests. These tests use pipe samples with premade dents and are performed at ambient pressure without the influence of rebounding and re-rounding. The calibration is generally limited to the maximum depth of the deformation (dent) and associated metal loss such as gouge and corrosion, which are manually measured with the traditional bridging bar and profile gauge technologies. The LaserScan system can provide 3D information of deformation, including axial and circumferential profiles, and is less dependent on technician skill than bridging bar.

It is known that the standard and high-resolution multiple-channel caliper tools can provide dent axial and circumferential profile data with the same format and the same cylindrical coordinate system as used by the 3D scanner. Therefore, this makes it possible to perform one-on-one comparison of the profiles between ILI and 3D scanner data, which could provide the most accurate data (at least one order of magnitude higher than ILI) for calibration of ILI tools. The data shown in Figures 22.5 and 22.6 (Section 22.2.3.2) may be used for the comparison. The tool tolerance can be accurately and reliably determined.

***22.3.2.2 Better Evaluation of the Depth-Based Tool Performance for ILI Runs*** For caliper and caliper–MFL tool ILI runs, evaluation of ILI sizing performance requires manually measuring the maximum depth of the dent and coincident metal loss such as corrosion and gouge. This is generally a quite difficult and tedious job, in particular, for bottom-side dents in a very confined space. The errors from in-ditch measurement may be significant, which adds additional uncertainties for ILI sizing performance evaluation. With the aid of this improved technology, not only can intensive labor be reduced, but also in-ditch measurement errors can be minimized. Moreover, differentiation of the effect of re-rounding and rebounding from tool performance itself becomes possible.

Figure 22.15 is a plot of 31 depth data points of bottom-side dents measured in-ditch with a 3D scanner against those reported by ILI caliper tool. In spite of the large error bands ($R^2 = 0.51$), the overall correlation looks good. The slope of the regression line is 1.08, that is, close to 1. The shift of the regression line from the ideal 3D ILI 1:1 ratio line (i.e., the dashed line) is about 0.42% OD. This systematic error is mainly caused by dent re-rounding and rebounding and can be corrected for integrity management of dents.

Using the 1:1 ratio dashed line as a reference, 19 points are above the dashed line, indicating that the depths measured in-ditch by 3D scanner are shallower than those reported by ILI. This is mainly caused by the effect on depth of re-rounding and rebounding by removing rocks underneath the dents. Six points are essentially on the dashed line. For these, there might be a possibility that the dents were not constrained even though they are on the bottom side; in this case, little of the re-bounding effect would be expected. Six points fall below the dashed line, and for these, the depths measured in-ditch by the 3D scanner is deeper than those reported by ILI. This reflects the limited resolution/error in the ILI tool, where it might not have captured the deepest point of the dents.

It should be noted that the ILI tool measures the inner surface while the in-ditch 3D scanner measures the outside surface of the pipeline. This could cause a discrepancy in a dent profile between these two types of measurement;

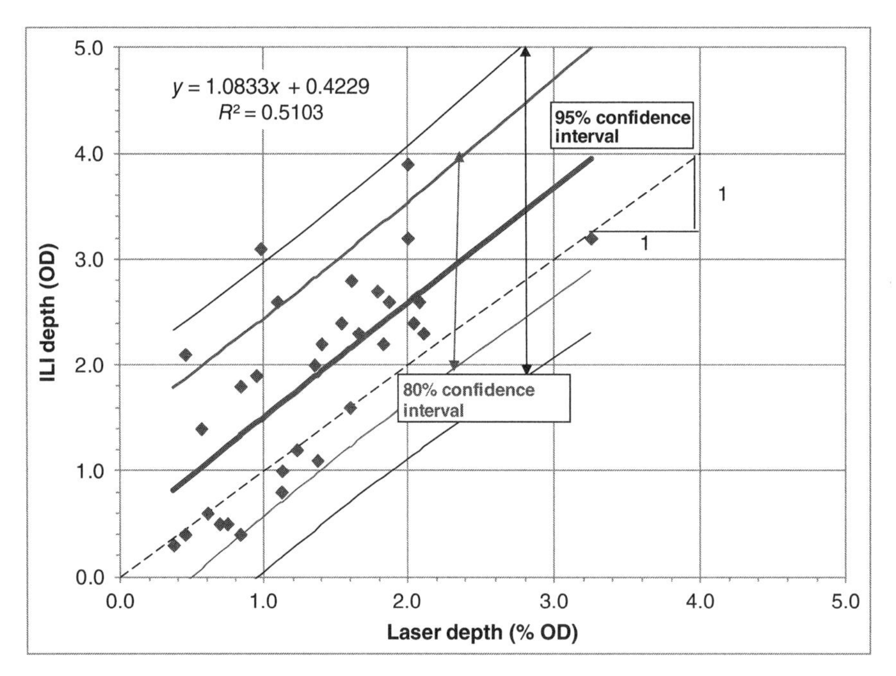


**FIGURE 22.15** Dent depth correlation between ILI reported and 3D LaserScanner.

however, the associated error is relatively small, in the order of wall thinning due to deformation.

***22.3.2.3 Better Evaluation of Strain-Based Tool Performance for ILI Runs*** The severity of the dents has traditionally been defined by depth [5,7,13]. Practical experience has shown that the use of depth alone can result in both unnecessary excavations due to many deep dents and ovalities that are not necessarily harmful, and also may miss moderate or shallow dents that are in fact severe due to their overall size and sharpness [5,31,33]. Since pipeline mechanical damage is related to strain, strain levels derived from the dent profile may offer another measure of dent severity [32]. Therefore, ASME B31.8 (American Society of Mechanical Engineers) for gas pipelines affecting high-consequence areas offers an option to use strain-based assessment methods, including nonmandatory equations to estimate component and total strains. The nonmandatory equations use the dent geometrical parameters (axial and circumferential radius of curvature R1 and R2, and depth/length ratio) to compute the bending and membrane strain components and total strain. Since some of the equations recommended in early editions of ASME B31.8 are incorrect, a major modification was made by Lukasiewicz et al. [29–31] and has been accepted for use [1,2,22,23,34]. A more detailed discussion is given in Section 22.4.2.2 of this chapter.

It is known that both ILI and 3D LaserScan technology are able to provide dent 3D geometry and detailed axial and circumferential profiles. Strains at each of the points in the dent can be calculated using the *local* deformation geometry at that point. The location and magnitude of the highest strain in the dent can be identified, which may not be the same as identified by the dent's global geometry.

Since the LaserScan technology provides much higher resolution than an ILI tool, the calculated strains would have higher accuracy than from ILI. Therefore, the 3D LaserScan technology can be used to correct the strains obtained by the ILI tool data.

All strain components and total strain are computed at every point of a dent using its local deformation geometry from axial and circumferential profiles [35]. The raw displacement data are first filtered and smoothed with a fast Fourier transform (FFT) low pass and Gaussian smooth filters to minimize both electronic noise and false readings from surface irregularities. The software then uses piecewise parametric cubic interpolation to calculate the radius of curvature and the bending strain components, and the finite difference in length before and after deformation to calculate the membrane strain components. This will be further discussed in Section 22.4.2 of this chapter.

Figure 22.16 is a comparison of the strain profile and maximum total strain reported using the standard ILI caliper and the LaserScanner using the proprietary software, showing that there is a factor of 2 difference between these two technologies. Since the caliper used for ILI is a low-resolution tool equipped with only 18 sensors for a 26 in OD pipeline, the sensor spacing is about 4.5 in,; the difference in the computed strain value between ILI and LaserScan may be mainly attributed to the low resolution of the caliper tool. Statistical analysis showed that a correction factor of 2.3

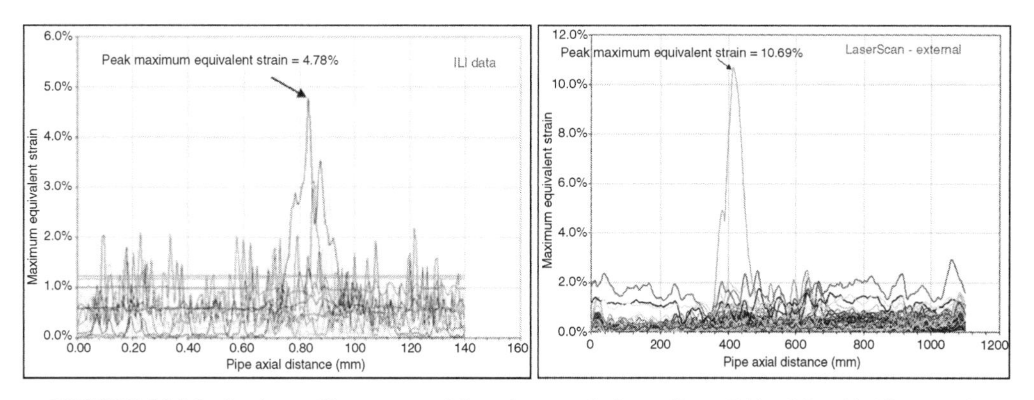


**FIGURE 22.16** Strain profiles computed from low-resolution caliper (26 in. OD with 18 sensors) and LaserScanner, showing a 5.91% difference in maximum equivalent strain. (a) Standard caliper; (b) LaserScanner.

should be applied to the ILI strain data for integrity assessment of the dent severity.

For a caliper with higher resolution, the difference in computed strain between caliper and LaserScanner should be smaller. Figure 22.17 shows 14 strain data points computed from the ILI and LaserScan data. The ILI caliper has 48 sensors in a 24 in OD pipeline, that is, the sensor spacing is 1.75 in,, which is about one-third of the previous example. The correlation in strain between ILI and LaserScan appears to be good, with the correlation coefficient $R^2 = 0.80$.

Using the correlation in Figure 22.17, the following regression equation may be applied to dents reported from ILI for correction at the 95% confidence level:

$$\text{Strain\%}_{\text{Laser}} = 1.3605 \times (\text{Strain\%}_{\text{ILI}}) - 0.0989 \pm 2.06 \quad (22.2)$$

The implication of these examples is that using a high-resolution tool such as the Laserscan as a reference and the ILI tool strain estimation can be corrected. Then, the adjusted ILI tool strain estimation can be utilized to rank the dent's based on the strain severity and consequently prioritizes for further investigation.

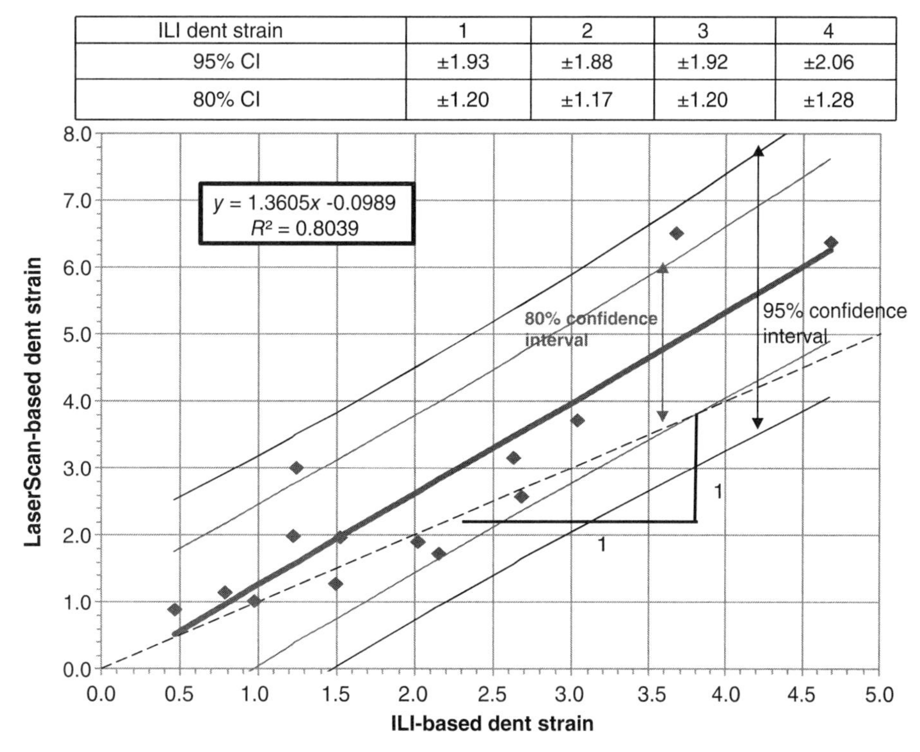

| ILI dent strain | 1 | 2 | 3 | 4 |
|---|---|---|---|---|
| 95% CI | ±1.93 | ±1.88 | ±1.92 | ±2.06 |
| 80% CI | ±1.20 | ±1.17 | ±1.20 | ±1.28 |



**FIGURE 22.17** Strain based on ILI and on LaserScan data.

## 22.4 ASSESSMENT OF THE SEVERITY OF MECHANICAL DAMAGE

This section provides an overview of mechanical damage assessment guidelines, including guidelines that utilize strain-based methods.

### 22.4.1 Regulatory and Industry Standard Guidance

Comprehensive reviews of the regulations and industry guidance for managing various forms of mechanical damage have been published by a special Task Group [5,36] and DOT OPS [13]. The first review provided by the special Task Group [5,36] served as a basis for the new criteria in the version of ASME B31.8S-2003 for prioritization and repair of mechanical damage in gas pipelines. These criteria are still included in its newer and current editions of B31.8S [32]. The second review issued by DOT OPS further provided an overview of the potential effects of dents on the integrity of both gas and liquid pipelines [13], as well as guidance for prioritization and repair as provided by the regulations, industry standards, and recommended practices. The regulations and industry standards operative in the United States pertinent to the assessment of mechanical damage, including dents, are as follows:

**For gas pipelines**

- 49 CFR 192 Transportation of Natural and Other Gas by Pipeline: Minimum Federal Safety Standards
- ASME B31.8-1995 Gas Transmission and Distribution Systems (1995)
- ASME B31.8-2003 Gas Transmission and Distribution Systems (2003)
- ASME B31.8-2007 Gas Transmission and Distribution Systems (2007)
- ASME B31.8-2010 Gas Transmission and Distribution Systems (2010)
- ASME B31.8-2012 Gas Transmission and Distribution Systems (2012)

**For liquid pipelines**

- 49 CFR 195 Transportation of Hazardous Liquids by Pipeline
- ASME B31.4-1998 Pipeline Transportation Systems for Liquids Hydrocarbons and Other Liquids (1998)
- ASME B31.4-2002 Pipeline Transportation Systems for Liquids Hydrocarbons and Other Liquids (2002)
- ASME B31.4-2006 Pipeline Transportation Systems for Liquids Hydrocarbons and Other Liquids (2006)
- ASME B31.4-2009 Pipeline Transportation Systems for Liquids Hydrocarbons and Other Liquids (2009)
- ASME B31.4-2012 Pipeline Transportation Systems for Liquids Hydrocarbons and Other Liquids (2012)
- API Publication 1156 Effect of Smooth and Rock Dents on Liquid Petroleum Pipelines (1997)
- API Publication 1160 Managing System Integrity for Hazardous Liquids Pipelines (2001)

Both reviews indicate that the main parameters used to determine severity of mechanical damage are the

- Nature of the mechanical damage (such as plain dents, dents with gouges, cracks, etc.);
- Depth of dent, which is expressed as a percentage of the pipe diameter.

In fact, depth is the only dent geometry parameter currently mentioned in 49 CFR 192 and 195 for the evaluation of dent disposition. Table 22.12 is a summary of 49 CFR 192 and 195 on the disposition of plain dents and dents associated with other defects for gas and liquid pipelines [37,38]. The table shows the timescale requirements for operators to take prompt remediation action based on the dent conditions discovered through integrity assessment.

*For gas pipelines*, 49 CFR 192 places dent conditions into three categories with respect to urgency—immediate, 1 year, and monitored—as follows:

- Immediate: A dent with any indication of metal loss, cracking, or a stress riser falls into the "immediate repair" category. To maintain safety, an operator must, as soon as possible after receiving the ILI report without excavation verification, temporarily reduce the operating pressure or shut down the pipeline until all immediate conditions are repaired.
- One year: (i) A smooth dent with depth at least 6% of the nominal diameter (OD) and located in the upper 2/3 of the pipe, or (ii) a dent with depth 2% OD that affects pipe curvature at a girth weld or longitudinal seam weld. The operator must take action to remediate within 1 year of the discovery of the condition.
- Monitored: (i) A dent with depth >6% OD located in the lower 1/3 of the pipe, (ii) a dent with depth >6% OD located in the upper 2/3 of the pipe for which engineering analyses of the dent demonstrate that critical strain levels are not exceeded, or (iii) a dent with a depth >2% OD that affects pipe curvature at a girth weld or a longitudinal seam weld, and for which engineering analyses of the dent and girth weld or seam weld demonstrate that critical strains are not exceeded. These analyses must consider weld properties. An operator does not have to schedule remediation, but must record and monitor the conditions during subsequent risk

**TABLE 22.12 Summary of 49 CFR 192 and 49 CFR 195 Regarding Dents**

| Anomaly | 49 CFR 192 Condition | 49 CFR 195 Condition |
|---|---|---|
| A dent that has any indication of metal loss, crack, or a stress riser | Immediate | Upper 2/3 of the pipe—immediate<br>Lower 1/3 of the pipe—60 day |
| A dent with depth >6% of the nominal pipe diameter | Upper 2/3 of the pipe—1 year<br>Upper 2/3 of the pipe—monitored[a]<br>Lower 1/3 of the pipe—monitored[a] | Upper 2/3 of the pipe—immediate<br>Lower 1/3 of the pipe—180 day |
| A dent with a depth >3% of the nominal pipe diameter on the upper 1/3 of the pipe | Not defined | 60 day |
| A dent with a depth >2% of the nominal pipe diameter on the upper 2/3 of the pipe | Not defined | 180 day |
| A dent with depth >2% of the nominal pipe diameter that affects pipe curvature at girth weld or at a longitudinal weld. | 1 year<br>Monitored[b] | 180 day |

[a]Engineering analysis of the dent demonstrates that the critical strain levels are not exceeded.
[b]Engineering analysis of the dent and girth or seam weld demonstrates that the critical strain levels are not exceeded. This analysis must consider the weld properties.

assessments and integrity assessments for any change that may require remediation. *For liquid pipelines* (49 CFR 195), the acceptance/rejection criteria are also based on the nature of dents and their depth. However, differences can be found in timescale requirements for an operator to take actions to address the integrity issues. These are summarized in Table 22.12 and may be described as follows:

- As for liquid pipelines, 49 CFR 195 places dent conditions into three repair categories in timescale: immediate repair, 60-day condition, and 180-day condition. The respective conditions for gas pipelines are immediate, 1 year, and monitored, respectively. There is no "monitored" condition for liquid pipelines.
- A dent located in the upper 2/3 of the pipe and associated with a stress riser, such as a crack, or with a depth >6% OD, falls into the immediate repair category. However, it becomes a 60-day or 180-day condition if it is located in the lower 1/3 of the pipe.
- A dent located in the upper 2/3 of the pipe with depth >3 or 2% OD are 60-day or 180-day conditions, respectively, for a liquid pipeline. There are no corresponding requirements for a gas pipeline.
- A dent with depth >2% OD that affects pipe curvature at a girth weld or at a longitudinal seam weld is a 180-day condition but is a 1-year condition for a gas pipeline.

From the aforementioned review, it is seen that 49 CFR 192 and 49 CFR 195 both use "depth-based" criteria but differ on the disposition of anomalies. 49 CFR 192 places anomalies into one of the three categories: immediate repair, 1 year, and monitored conditions; while 49 CFR 195 defines immediate repair, 60-day, and 180-day conditions.

Current 49 CFR 192 and 49 CFR 195 documents both incorporate ASME codes B31.8 (2007) and B31.4 (2002) by referencing their Repair Procedures (Paragraph 851) and "Deposition of Defects" (Paragraph 451.6.2, B31.4). However, the 2007 edition of ASME B31.8 and the 2002 edition of B31.4 have been replaced by new editions, B31.8 (2012) and B31.4 (2012). Even though the new editions of the ASME codes are not currently referenced by 49 CFR 192 and 49 CFR 195, the requirements recommended by the new editions are considered to be aligned and compatible with 49 CFR 192 and 49 CFR 195.

It is noted that B31.8 (2007) (Paragraph 851.41and Appendix R) and its newer editions [32] provide the option to use strain criteria and assess corrosion features in dents using remaining strength criteria for corroded dents. For example, 6% strain in pipe bodies and 4% strain in welds are acceptable for plain and rock dents. This provides operators with safe alternatives, which are particularly important for features located in areas that are difficult to access, such as river crossings. For strain calculation, B31.8 (2007) and its newer editions provide nonmandatory formulas (B31.8, Appendix R) and allow others from the open literature or derived by a qualified engineer [5,13,36]. There are no similar options in B31.4 (2002) and its newer additions for liquid pipelines. However, from static behavior of the dents, the respective strain-based criteria in B31.8 (2007) and its newer editions may be applicable to liquid pipelines.

Table 22.13 is a summary of the acceptance/rejection criteria of B31.8S (2003) [36] that have remained the same in its newest edition (2012) for mechanical damage in gas pipelines.

**TABLE 22.13 Criteria of Mechanical Damage for Gas Pipelines[a]**

| Feature | Failure Mode | Safe Limit | Mitigation |
|---|---|---|---|
| Plain, unrestrained dents (anywhere on OD) | None | 6% OD; deeper up to 6% strain in pipe body. 4% strain in welds | Repair coating if excavated: monitor for corrosion if not excavated |
| Rock dents | Corrosion, SCC | | |
| External mechanical damage, gouges, scrapes, cracks, or SCC in dents | Rupture<br>Low cycle fatigue | None | Cut out<br>Sleeve<br>Grind out |
| Internal (pig passage) mechanical damage | None | None | None |
| Dents affecting ductile girth welds or seam welds | Fatigue | 2% OD; deeper per analysis, subject to 4% max strain limit | |
| Dents on acetylene welds and brittle seams | Brittle fracture | None | Cut out |
| Dents with metal-loss corrosion | Rupture | 6% OD and metal loss per corrosion criterion | Sleeve |
| Dents with grind repair | Rupture | 4% OD and metal loss per grind criterion | |

(Reproduced from Ref. [36] with permission of ASME.)

[a]Operating at hoop stress levels at or above 30% of the specified minimum yield strength.

In Table 22.14, Dawson et al. [39] have summarized the international code guidance and recommended practices relevant to the assessment of dents in pipelines. It is noticeable that all but ASME B31.8 (2003) are based on simple depth and dent nature criteria. For plain dents, most codes adopt a 6% OD criterion with some exceptions, except that the Pipeline Defect Assessment Manual (PDAM) allows 7% OD for unconstrained and 10% OD for constrained plain dents.

### 22.4.2 Strain-Based Assessment Methods

As described, ASME B31.8 (2003) and its newer editions [32] offer the option to use strain-based methods in assessing the severity of mechanical damage. In this section, the strain-based assessment methods are reviewed. A new approach that combines strain-based severity assessment with MFL signal recognition to discriminate DMLs among *corrosion, gouge, and crack in the dent* is also presented. Strain-based case studies are also included here.

***22.4.2.1 Strain Calculation Equations Recommended by ASME B31.8*** So far there are no standard methods for the calculation of dent strains. Appendix R of ASME B31.8 (2003) and its newer edition provide nonmandatory equations for computing three strain components, namely, two bending strains (longitudinal and circumferential) and one membrane strain (axial) using the dent's global geometry. Formulae for total strain calculation are provided using these three component strains. ASME B31.8 does not provide a formula for computing circumferential membrane strain, assumed to be compressive.

The following are the six nonmandatory equations recommended by ASME B31.8S. Figure 22.18 defines the parameters used in the equations. Circumferential bending strain ($\varepsilon_1$),

$$\varepsilon_1 = \frac{t}{2}\left(\frac{1}{R_o} - \frac{1}{R_1}\right) \tag{22.3}$$

Longitudinal bending strain ($\varepsilon_2$),

$$\varepsilon_2 = \frac{-t}{2R_2} \tag{22.4}$$

Extensional strain ($\varepsilon_3$),

$$\varepsilon_3 = \frac{1}{2}\left(\frac{d}{L}\right)^2 \tag{22.5}$$

Total strain on inside pipe surface ($\varepsilon_i$) and on outside pipe surface ($\varepsilon_o$),

$$\begin{aligned} \varepsilon_i &= \left[\varepsilon_1^2 - \varepsilon_1(\varepsilon_2 + \varepsilon_3) + (\varepsilon_2 + \varepsilon_3)^2\right]^{\frac{1}{2}} \\ \varepsilon_o &= \left[\varepsilon_1^2 + \varepsilon_1(-\varepsilon_2 + \varepsilon_3) + (-\varepsilon_2 + \varepsilon_3)^2\right]^{\frac{1}{2}} \end{aligned} \tag{22.6}$$

**TABLE 22.14 Summary of Published Guidance on the Assessment of Dents in Pipelines**

| Published Guidance | Top of Line Dents (8–4 o'clock) | | | | Bottom of Line Dents (4–8 o'clock) | | | |
|---|---|---|---|---|---|---|---|---|
| | Plain Dent | Dents with Cracks/Gouges | Dents at Welds | Dents with Corrosion | Plain Dent | Dents with Cracks/Gouges | Dents at Welds | Dents with Corrosion |
| CSA Z662-03 (2005) | Up to 6% OD | Not allowed | Up to 2% for >NPS 12 in. or up to 6 mm for <NPS 12 in. | As per ASME B31.G up to max depth of 40%wt | As for top of line dents | | | |
| AS2885.3 (2001) | Up to 6% OD | Not allowed | Not allowed | Detailed assessment allowed | As for top of line dents | | | |
| ASME B31.8[a] (2003) | Up to 6% OD or unlimited if strain <6% | Not allowed | Up to 2% OD or unlimited if strain <4% for ductile welds (no safe limit for brittle welds) | As per ASME B31.G limits | As for top of line dents | | | |
| ASME B31.4[b] (2004) | Up to 6% OD | Not allowed | Not allowed | External corrosion <87.5% RWT required for design, that is, <12.5%wt. Internal corrosion as per ASME B31.G | As for top of line dents | | | |
| API 1160[b] (2001) | Up to 2% OD for 12[a] NPS (6.35 mm < 12 in. NPS) | Not allowed | Not allowed | Not allowed | Up to 6% OD | Not Allowed | Investigate/mitigate within 6 months | Investigate/mitigate within 6 months |
| PDAM (2003) | Up to 7% OD (unconstrained) Up to 10% OD (constrained) | Method provided | Not allowed | Not allowed | Up to 6% OD (unconstrained) | Method provided | Not allowed | Not allowed |
| DOT Liquid Rule (Part 195)[b] (2000) | 1. Up to 6% OD (immediate condition) 2. Up to 3% OD for NPS 12[a] or >6.35 mm for NPS <12 in. (60-day condition) 3. Up to 2% OD (180-day condition) | Not allowed (immediate condition) | Up to 2% OD (180 day condition) | Not allowed | Up to 6% OD | Not allowed (60-day condition) | Up to 2% OD (if > 180-day condition) | Not allowed (60-day condition) |
| DOT Gas Rule (Part 192)[a] (2001) | 1. Up to 6% OD for 12 in. NPS or 12.7 mm for <12 in. NPS (1-year condition) 2. Monitor dents >6% OD with acceptable strain levels | Not allowed (immediate condition) | 1. Up to 2% for NPS 12[a] or up to 6.35 mm for <NPS 12 in. (1-year condition) 2. Monitor dents >2% OD with acceptable strain levels | Not allowed | Monitor dents >6% OD | Not allowed | 1. Up to 2% for NPS 12[a] or up to 6.35 mm for <NPS 12 in. (1-year condition) 2. Monitor dents >2% OD with acceptable strain levels | Not allowed |

[a]Only relevant to gas pipelines.
[b]Only relevant to liquid pipelines.
(Reproduced from Ref. [39] with permission of ASME.))

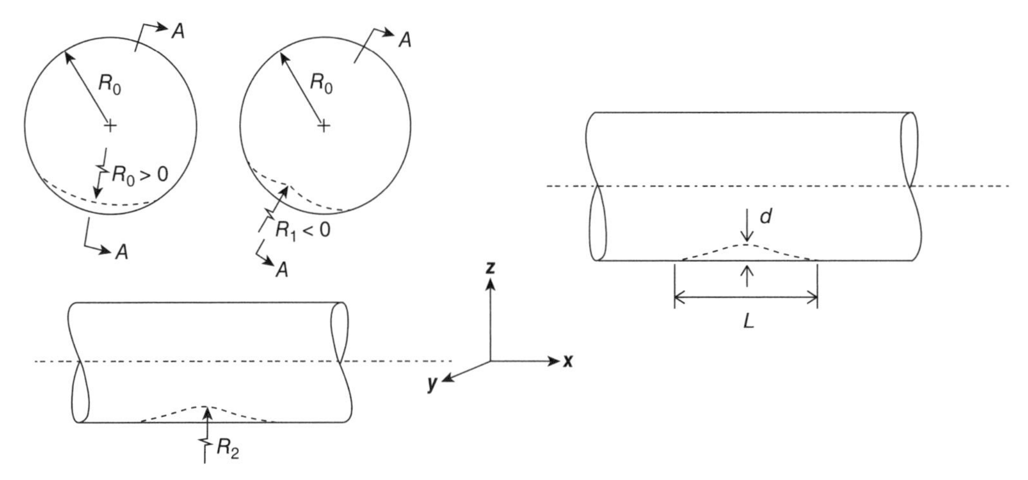


**FIGURE 22.18** Dent geometry and definitions of the parameters used in ASME B31.8 2003 and its newer editions.

The overall maximum strain, $\varepsilon_{max}$, defined by the following equation, is then compared with acceptance/rejection criteria recommended by ASME B31.8 (2003):

$$\varepsilon_{max} = \text{Max}[\varepsilon_i, \varepsilon_o] \tag{22.7}$$

The nomenclature in Equations 22.3–22.7 is that used in ASME B31.8, that is,

$t$ = Wall thickness
$R_o$ = Initial pipe surface radius
$R_1$ = Radius of curvature in transverse plane, negative for reentrant dents
$R_2$ = Radius of curvature in longitudinal plane, negative for reentrant dents
$L$ = Dent length
$d$ = Dent depth

From Equation 22.6, the total effective strain on the pipe surface can be further deduced using $\varepsilon_x = \pm\varepsilon_2 + \varepsilon_3$, $\varepsilon_y = \pm\varepsilon_1$ [13,41]:

$$\varepsilon_{eff} = \sqrt{\varepsilon_x^2 - \varepsilon_x\varepsilon_y + \varepsilon_y^2} \tag{22.8}$$

where $x$ and $y$ refer to the longitudinal and circumferential axes of the pipe as defined in Figure 22.18, and the positive and negative signs in the definitions of $\varepsilon_x$ and $\varepsilon_y$ refer to the inner and outer wall surface, respectively. Again, Equation 22.7 is used to determine the overall maximum strain. The subscripts in Equation 22.8 refer to the strain directions shown in Figure 22.18.

Finally, the calculated maximum total strain is compared with the acceptance/rejection criteria to determine if the dent satisfies fitness for service (FFS). As described previously, 6% strain in pipe bodies and 4% strain in welds are acceptable for plain and rock dents.

It is noted that Equations 22.3–22.4 can be used to calculate dent strain components based on either global or local geometry using ILI and field inspection data [5,40]. When using local dent geometry, piecewise interpolation methods, such as Bessel cubic interpolation or fourth-order B-spline [41,42], should be used to approximate the local dent profile and estimate the radius of curvature in both longitudinal ($\theta$) and circumferential ($z$) directions:

$$\kappa = \frac{r^2 - rr'' + (r')^2}{[r^2 - (r')^2]^{3/2}} \tag{22.9}$$

where $\kappa$ = radius of curvature; $r$ = radius; r′ = first derivative; r″ = second derivative; $z$ (longitudinal) and $\theta$ (circumferential) are the coordinates in the cylindrical coordinate system as defined in Figure 22.12.

The local axial membrane strain may be estimated using the local difference in length before and after deformation [41]:

$$\varepsilon_{axial,i} = \frac{(L_{s,i} - L_{o,i})}{L_{o,i}} \tag{22.10}$$

where $L_{s,i}$ is the local deformed length and $L_{o,i}$ is its original length in the longitudinal direction at point $i$ (coordinates $r$, $\theta$, $z$) in the dent.

By using Equations 22.9 and 22.10, the total strain at each point of the dent can be computed using the dent profile as reported by caliper ILI, and the location and magnitude of the maximum strain in the dent can be identified. The accuracy of the computed results depends on the resolution of the ILI tool. High-resolution tools yield better accuracy than low-resolution tools. This localized computing method has been called a "point-to-point" method for convenience [1,2,22,31].

Equations 22.3 and 22.4 for the bending component calculation were developed using thin-plate and shell theory [43]. However, Equation 22.5, the global axial membrane strain, was empirically established against a limited number of finite element analyses (FEA) [13]. For the total effective strains, that is, Equation 22.8, there are no documented explanations in the public domain regarding how they were actually established. Baker [13] indicated that Equation 22.8 can be seen to be either the effective strain in plane strain $e_{effective} = \frac{2}{3}\sqrt{e_1^2 - e_1 e_2 + e_2^2}$ where $e_1 = \varepsilon_1$, $e_2 = \varepsilon_2 + \varepsilon_3$, $e_3 = 0$ without having a preceded constant 2/3, or can be thought of as analogous to von Mises stress for a plane stress condition, $\sigma_{effective} = \sqrt{\sigma_1^2 - \sigma_1\sigma_2 + \sigma_2^2}$, where $\sigma_3 = 0$.

As noted by various authors [29–31,34], Equation 22.8 implied that the radial strain $e_3 = 0$ for thin-wall pipes is incorrect because the stress and strain in the radial direction cannot both be zero [31,34]. It is also incorrect to assume that the von Mises thin-wall strain formula is analogous to the von Mises stress for a plane stress condition because it is not supported by plastic strain theory [44].

Lukasiewicz et al. [29,45] further pointed out that the accuracy of Equation 22.5, from ASME B31.8 for estimation of global longitudinal membrane strain, is extremely poor. A more rigorous derivation showed that Equation 22.5 underestimates axial membrane strain by a factor of 4; the correct equation should be $\varepsilon_3 = 2\left(\frac{d}{L}\right)^2$ [31].

Lukasiewicz et al. [29] also indicated that ASME B31.8 (2003) only calculates longitudinal membrane strain, which neglects circumferential membrane and shear strains. The finite element method (FEM) analysis of actual dents has shown that these strain components can be similar in magnitude to the longitudinal strain [45].

In the following section, improved equations for computing dent strains are discussed.

### 22.4.2.2 *Improved Strain Calculation Methods*

*Lukasiewicz Method* Lukasiewicz et al. [29] and Czyz et al. [30] proposed an alternative method that combines mathematical algorithms and FEA for dent strain calculation. The authors indicate that strains in a pipe wall consist of two main components: longitudinal and circumferential, each of which can be further separated into bending and membrane strains. The membrane strain is constant through the wall, while the bending component changes linearly from the inner to the outer surface. Figure 22.19 shows these strain components.

The respective longitudinal and circumferential bending strains, that is, $\varepsilon_x^b$ and $\varepsilon_y^b$, can then be calculated from the curvature of the radial displacement $w$ in the axial $x$ and circumferential $y$ directions (see Figure 22.20).

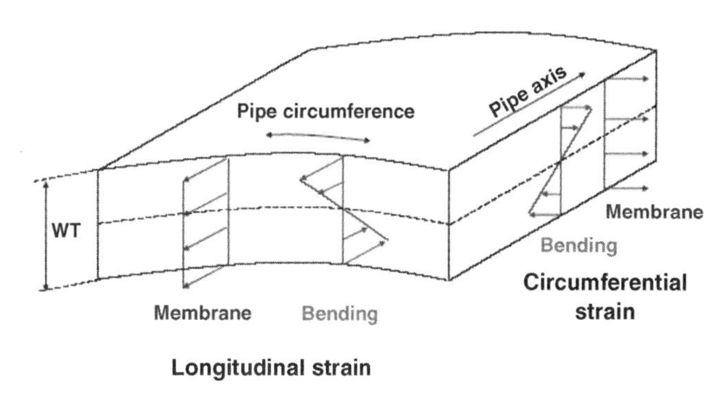


**FIGURE 22.19** Strain components in a pipe wall [29,30]. (Reproduced from Ref. [29] with permission of ASME.)

$$\varepsilon_x^b = z\frac{\partial^2 w}{\partial x^2}, \quad \varepsilon_y^b = z\frac{\partial^2 w}{\partial y^2} \tag{22.11}$$

The bending strain is maximum on the pipe surface, that is, $z = \pm t/2$ where $t$ is the pipe wall thickness. The strains at the outer surface are as follows:

$$\varepsilon_x^b = \frac{t}{2}\frac{\partial^2 w}{\partial x^2}, \quad \varepsilon_y^b = \frac{t}{2}\frac{\partial^2 w}{\partial y^2} \tag{22.12}$$

On the inner surface, they have an opposite sign.

These authors [29,30] further showed that calculation of the bending components is fairly straightforward and can be done directly from the dent displacement measured by an in-line high-resolution caliper tool. The in-line caliper measures the pipe wall deflection $w$ in the $Z$ direction. A fourth-order B-spline approach may be employed with FEM simulation to calculate curvature of dents from caliper data, as demonstrated by Noronha et al. [42].

The main difficulty in strain-based methods is in determining membrane strains in the dented region. At present, no exact (analytical) solutions are available for calculation of membrane strains of the dented region in the pipe [29]. However, the membrane strains may be calculated using the relationships between membrane strain and displacement

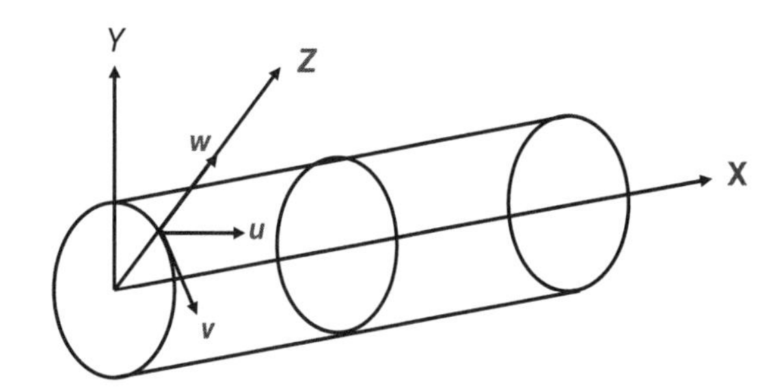


**FIGURE 22.20** Coordinate system ($x$, $y$) of the pipe and displacement components ($u$, $w$). (Reproduced from Ref. [29] with permission of ASME.)

for large deformations of a cylindrical shell [45]:

$$\varepsilon_x^m = \frac{\partial u}{\partial x} + \frac{1}{2}\left(\frac{\partial w}{\partial x}\right)^2 + \varepsilon_x^o$$
$$\varepsilon_y^m = \frac{\partial v}{\partial y} - \frac{w}{R} + \frac{1}{2}\left(\frac{\partial w}{\partial y}\right)^2 + \varepsilon_y^o \qquad (22.13)$$
$$\gamma_{yx} = \frac{\partial u}{\partial y} + \frac{\partial v}{\partial x} + \left[\frac{\partial w}{\partial x}\right]\left[\frac{\partial w}{\partial y}\right]$$

where,

$\varepsilon_x^m$ and $\varepsilon_y^m$ are the membrane strains in the axial ($x$) and the circumferential ($y$) directions, respectively;
$\gamma_{xy}$ is the shear strain in the plane $x$, $y$;
$R$ is the mean radius of the pipe;
$\varepsilon_x^o$ and $\varepsilon_y^o$ are the initial strains due to pressure in the pipe, thermal expansion, and so on.

These equations are solved numerically for the unknown displacements $u$ and $v$ (see Figure 22.20) using two-dimensional (2D) FEA. The normal displacement $w$ is the only measured variable, preferably obtained using a high-resolution geometry tool.

The computed membrane strains can be superimposed on the bending components, thereby producing maximum values of strain in the axial and circumferential directions:

$$\varepsilon_x = \varepsilon_x^m \pm \overline{\varepsilon}_x^b$$
$$\varepsilon_y = \varepsilon_y^m \pm \overline{\varepsilon}_y^b$$

where the positive and negative signs refer to the outer and inner wall surfaces, respectively.

The maximum equivalent strain (both on the inner and outer surface) in the dented area of the pipe is then given by

$$\varepsilon_{eq} = \frac{2}{\sqrt{3}}\sqrt{\varepsilon_x^2 + \varepsilon_x\varepsilon_y + \varepsilon_y^2} \qquad (22.14)$$

or, more precisely,

$$\varepsilon_{eq} = \frac{2}{\sqrt{3}}\sqrt{\varepsilon_x^2 + \varepsilon_x\varepsilon_y + \varepsilon_y^2 + \gamma_{xy}^2/2} \qquad (22.15)$$

Equations 22.14 and 22.15 are derived directly from plastic strain theory [43], with an incompressibility condition in the plastic regime $\varepsilon_x + \varepsilon_y + \varepsilon_z = 0$. Note that $\varepsilon_z$ cannot be assumed to be identical to zero. The pipe wall behaves as a thin shell and therefore it is in plane stress (not plane strain). The through-thickness strain is in fact the sum of the two in-plane components. Equation 22.14 is a simplified version of Equation 22.15 by neglecting shear strain $\gamma_{xy}$, which is generally small and insignificant.

The combined mathematical algorithm and 2D FEA model allow the calculation of all the strain components based on the measurement of the radial deformation of a pipe with a high-resolution geometry in-line tool. The proposed method is likely to provide a practical tool for assessment of strain in dents. However, it still requires incorporating a 2D FEA model. For most pipeline operators, this is still impractical.

There are two major differences between Equation 22.8 (ASME B31.8 recommended) and Equation 22.14 (Lukasiewicz et al. suggested): (a) a preceding constant ($2/\sqrt{3}$) and (b) a sign difference in the mid-term $\varepsilon_x\varepsilon_y$. These differences could result in significant discrepancy in the calculated effective strain value. Czyz et al. [30] demonstrated that the difference in effective strain can be more than 50% with the assumption that $\gamma_{xy}^2/2 = 0$. Further comparison [31] was made using the three-case data provided by Baker [13]. Again, the comparison assumes $\gamma_{xy} = 0$. The results showed that ASME B31.8 Equation 22.8 underestimates the effective strain by a factor of up to 1.8 and 2 for external and inner surfaces, respectively.

*Blade Energy Partners Simplified Model [31,35,46,47]* Blade Energy Partners developed a simplified model for local strain calculation of mechanical deformation in the form of dents and bulges on a point-to-point basis. The simplified model adopted the following:

- ASME B31.8 Equations 22.3 and 22.4 [13] to calculate bending strain components,

$$\varepsilon_1 = \frac{t}{2}\left(\frac{1}{R_o} - \frac{1}{R_1}\right) \text{ and } \varepsilon_2 = \frac{-t}{2R_2}$$

- Rosenfeld's local axial membrane strain Equation 22.10 [41] using the finite difference method,

$$\varepsilon_{axial,i} = \frac{(L_{s,i} - L_{o,i})}{L_{o,i}}$$

- Finite difference method to calculate local circumferential membrane strain,

$$\varepsilon_{circumferential,i} = \frac{(\mathrm{ARC_{s,i}} - \mathrm{ARC_{o,i}})}{\mathrm{ARC_{o,i}}} \qquad (22.16)$$

where $\mathrm{ARC_{s,i}}$ is the local deformed arc length and $\mathrm{ARC_{o,i}}$ is the original pipe arc length (perfect circle) at point $i$ (i.e., at $r_i$, $\theta_i$, $z_i$) in the dent.

- Lukasiewicz et al.'s total strain model [29,30] for total or equivalent strain calculation, Equation 22.14

$$\varepsilon_{eq} = \frac{2}{\sqrt{3}}\sqrt{\varepsilon_x^2 + \varepsilon_x\varepsilon_y + \varepsilon_y^2}$$

**TABLE 22.15 A Comparison of the Computed Strains Between FEA and Blade Simplified Model**

| | Maximum Equivalent Strain | | Percent of error (%) |
|---|---|---|---|
| Dent Depth (% OD) | Blade's Dent Analyzer Tool (%) | FEA (%) | Error = (Blade Strain − FEA Strain)/FEA Strain |
| **3.0%** | **11.9** | **11.6** | **2.6** |
| 6.0% | 22.2 | 25.1 | **−11.4** |
| 9.0% | 24.6 | 26.5 | **−7.3** |
| 12.0% | 29.6 | 29.7 | **−0.3** |
| 15.0% | 30.1 | 31.3 | **−3.9** |
| Rebound | 31.5 | 31.3 | **0.5** |
| | **Average error** | | **−3.3** |
| | **Std deviation** | | **5.3** |

The Blade model takes the correct equations from ASME B31.8S for bending strain calculation but replaces ASME's incorrect equation for total strain with Lukasiewicz's equation that was derived directly from plastic strain theory. Further, a finite difference method is applied to calculate and include both longitudinal and circumferential membrane strains. The Blade method greatly improves the prediction accuracy over ASME B31.8S but is simpler than the Lukasiewicz et al. method.

More than 14 FEA runs with various pipe sizes and dent depths have been performed to validate the simplified model. Table 22.15 shows a comparison of the total strain between FEA and the simplified method. The pipe used for the validation has a size of 34 in. OD, 0.375 in. wall, and steel grade X52. The indenter is a spherical ball with 2.5 in. diameter made of hardened bearing steel (aircraft grade E52100 Alloy steel). A high-resolution LaserScanner was used for dent profiling. The scanned data were extracted and exported to the strain analyzer by using Blade's strain equations.

The results showed that the agreement between FEA and Blade strain is good, with an average error = −3.3%, minimum error = −0.3%, and maximum error = −11.4% with standard deviation 5.3%. In general, the Blade method provides a slightly lower strain compared with FEA, by about 5% of the FEA strain (see Table 22.15). The FEA verification demonstrates the effectiveness of the model.

### 22.4.3 A Combined Approach to Evaluate Dent with Metal Loss

Pipelines constructed in mountainous or rocky terrain are vulnerable to damage such as denting. ILI of these pipelines often reports thousands of dent features. Some of them are associated with corrosion, gouges, and/or cracks. Current ILI technologies are incapable, or have limited capability, of discriminating between corrosion and cracks/gouges. Hence, ILI vendors generally report dents with associated metal loss without distinguishing the type of loss. Since a dent associated with corrosion can be assessed separately (as per ASME B31.8 guidelines), while a dent with cracks/gouges could pose an immediate threat to pipeline integrity [32], it is important to discriminate between dents with corrosion and those with cracks/gouges.

In two PRCI projects, namely, MD 1-2 [9,16] and MD1-4 [22,23], an approach was proposed by Wang et. at [1,48] for identifying dents with cracks and discriminating dents with cracks/gouges from DMLs.

The approach combines two criteria: strain severity and MFL signal characteristics. The strain severity-based criterion is used to assess the severity of mechanical damage, such as susceptibility to cracking. Only those dents with strains meeting or exceeding the strain criterion are considered as candidates for containing gouges/cracks and are investigated further using the MFL signal characteristics criterion. Actually, meeting the strain criterion is necessary for causing cracking/gouging regardless of whether the dent is a plain dent or a DML. However, for dents that do not meet the strain severity criterion but are associated with metal loss, MFL signals would also be reviewed to validate that the metal loss is not associated with cracks and/or gouges.

The MFL signal characteristics criterion is used to determine (1) if the candidate dents are indeed plain dents without any suspicious MFL signal association and (2) if the MFL signals reported as metal loss are associated with corrosion or cracking/gouging. Actually, the MFL criterion is a sufficient and complementary condition for discriminating dents with gouges/cracks from plain dents and DMLs.

Limited data showed that this combined approach is adequate to effectively identify dents with gouges/cracks from thousands of ILI-reported dents. The following are the procedures used:

- Apply a prescreening criterion such as dent aspect ratio to the large number of dents reported by ILI making an inference about the severity of the dent and identify candidate dents for full strain analysis.
- Perform full strain analysis on the identified candidate dents and apply the strain severity criterion to identify

candidate dents for further investigation using MFL signal characteristics.

- Apply MFL signal characteristics criterion to the candidate dents and (1) determine if there is an MFL signal associated with the dent but not reported by the ILI vendor because it is below the reporting threshold, (2) check if the MFL signal (both reported and not reported) is an indication of metal loss, and (3) classify the metal loss into cracks/gouges or corrosion.
- Finally, make a dig selection based on the aforementioned procedures and proceed with field investigation and validation.

***22.4.3.1 Dent Strain Severity-Based Criterion*** A dent represents permanent damage to a pipeline by local plastic deformation. Dent severity and its susceptibility to cracking can be assessed using a plastic damage criterion. Dents with high strain could be potentially associated with cracks.

The combined approach adopts the ductile failure damage indicator (DFDI) as the severity criterion for identifying candidate dents that may contain cracks/gouges for further investigation. The detailed discussion of DFDI can be found in Refs [49–51] and Section 22.5.1.1 of this chapter. A simplified upper-bound DFDI equation is as follows:

$$\mathrm{DFDI}_{\mathrm{upperbound}} = \frac{1.65\varepsilon_{eq}}{\varepsilon_o} \quad (22.17)$$

where $\varepsilon_{eq}$ is the equivalent strain calculated using the ILI dent profile and $\varepsilon_o$ is the critical strain of the material. The critical strain (true strain) of the pipe material is measured from material testing and is usually in the range of 0.3–0.5.

In general, dents are susceptible to cracking when DFDI $\geq 1$. For conservatism and screening purposes, a DFDI value of 0.6, that is, DFDI $\geq 0.6$, is suggested as the severity criterion to determine if the MFL criterion should be applied for the dent of interest.

***22.4.3.2 MFL Signal Criterion*** The second criterion of the combined approach is to characterize the MFL signals, in particular, the triaxial MFL signals, for the respective candidate dents.

It is known that MFL ILI technology is capable of detecting crack signal(s) from a dent [9,52,53]. However, when the signal(s) is below the reportable threshold, the dent may be reported as a plain dent. In other cases, signals are reportable, yet the ILI tool is incapable of differentiating or discriminating cracking/gouging from corrosion according to signal(s) alone, and therefore ILI vendors commonly report them as metal loss. Sometimes, due to lack of experience, data analysts may ignore or dismiss the signals even though they are visible or reportable [1,48].

Whenever a dent is found to be associated with metal loss, a careful review of the MFL signal characteristics is required. The steps involved in the MFL signal criterion to differentiate the metal loss between cracks, gouges, and corrosion are as follows:

- Determine how many metal loss features are within the dent region, that is, only one isolated metal loss signal or many clustered features.
- If only a single strong MFL signal is found and it is located at the dent apex or spot of highest strain in the dent, then the metal loss feature is most likely a crack.
- If a strong metal loss signal is located at the dent apex or spot of highest strain but is surrounded by general shallower metal loss features, then it is probably either a gouge or a crack.
- If there is a strong metal loss feature signal oriented circumferentially and located at the dent apex or spot of highest strain, then it is probably a gouge.
- If many general metal loss signals clustered within the dent area are found, then it is most likely corrosion.

Although this procedure appears to be quite simple and straightforward, it requires experience in analyzing and discriminating MFL signal data and is still under development.

***22.4.3.3 Illustration of the Effectiveness of the Approach*** A case study was reported by Wang et. at. [1,48] to illustrate the effectiveness of the approach utilizing three ILIs with a MFL/caliper Combo tool. The inspection reported a total of 6361 dents. First, a quick screening was performed based on the strain value of the dent, which identified 150 dents with high strain values. Then, the upper-bound DFDI values were calculated for these dents and ranked in a descending order. A DFDI limit of 0.6 was set to determine those dents that required a detailed review of MFL signals. This limit was chosen based on a conservative material critical strain as outlined earlier and will be fine-tuned in the future based on experience from excavations and feedback. Finally, a total of 15 dents were identified for careful examination of MFL signals. In addition, other dents reported by ILI as DML were also checked with the MFL criterion. At the end, seven dents were predicted to be associated with cracks/gouges. Another eight dents were predicted not to be associated with cracks/gouges. All 15 dents were excavated, verified in-ditch, and mitigated with appropriate methods as specified in the industry code. Table 22.16 is a summary of the 15 cases studied. The excavation results show excellent agreement between the predictions and in-ditch findings with a 100% success rate. Even though the approach is still in the development stage, its effectiveness is obvious and shows promise to eventually become a powerful tool for safe and cost-effective integrity management of mechanical damage.

**TABLE 22.16 Summary of Case Study—Prediction versus Excavation Findings for 15 Dents**

| Case | $\varepsilon_{eq}$ (%) | Upper-bound DFDI | Prediction | Excavation | Prediction–Excavation |
|---|---|---|---|---|---|
| 1 | 10.20 | 0.6 | Possible crack | Through wall crack | Positive–positive |
| 2 | 15.00 | 0.8 | Possible crack | Through wall crack | Positive–positive |
| 3 | 16.90 | 0.9 | Possible crack and gouge | Gouge | Positive–positive |
| 4 | 13.00 | 0.7 | Low confidence of possible crack | ID crack | Positive–positive |
| 5 | 31.00 | 1.7 | Crack with MD | MD | Positive–positive |
| 6 | 17.60 | 1 | Possible crack | ID/OD crack | Positive–positive |
| 7 | 11.50 | 0.6 | Possible crack | Through wall crack | Positive–positive |
| 8 | 6.80 | 0.4 | No crack | No crack but 6% "ML" [a] caused by sitting on rock | Negative–negative |
| 9 | 7.10 | 0.4 | No crack | No crack but 16.4% "ML" [a] caused by sitting on rock | Negative–negative |
| 10 | 10.50 | 0.6 | No crack | No crack but 12% "ML" [a] caused by sitting on rock | Negative–negative |
| 11 | 9.90 | 0.5 | No crack | No crack but 16% "ML" [a] caused by sitting on rock | Negative–negative |
| 12 | 5.00 | 0.3 | No crack | No crack but 36% corrosion | Negative–negative |
| 13 | 9.00 | 0.5 | No crack | No crack but 10% "ML" [a] caused by sitting on rock | Negative–negative |
| 14 | 7.80 | 0.4 | No crack | No crack but 15% corrosion | Negative–negative |
| 15 | 3.60 | 0.2 | No crack | No crack but 37% corrosion | Negative–negative |

[a] "ML" here refers to impression(s) caused by contacting irregular-shaped rock and is not a gouge caused by third-party damage.

### 22.4.4 Fatigue Assessment of Dents

The methods and associated criteria described in Tables 22.12 and 22.13 are for assessing static dent behavior only. This is adequate for constrained dents because they cannot re-round under internal pressure. American Petroleum Institute (API) 1156 and its addendum [54] conclude, based on full size fatigue tests, that if dents are constrained, then the time required for the number of cycles necessary for failure is likely to be greater than the normally expected life of the pipeline. Therefore, there is usually little concern for fatigue of constrained plain dents, except in aggressive pressure cycles over a long period of time. The implication of these results is that, for liquids pipelines, considering not digging up rock dents at all may not be a bad idea in order to avoid fatigue in service [5,36]. However, long-term corrosion control and monitoring of dents should be addressed [5,36,54].

An important exception to this is associated with double-centered dents with a flattened or saddle-shaped area between them. The flattened area between two dent centers is susceptible to pressure-cycle fatigue because it is effectively unconstrained and flexes readily in response to pressure cycles. Hence, API 1156 recommends that dents having space between centers of one pipe diameter or less should be candidates for excavation and be repaired. This rule is primarily applicable to liquid lines because no failure in gas pipelines due to fatigue has been observed or reported.

For unconstrained plain dents, the major concern is pressure-cycle-induced fatigue. This occurs due to high local bending stresses associated with operating pressure fluctuations [54]. An empirical method was proposed by the EPRG for predicting the fatigue life for an unconstrained plain dent. The EPRG method may be the "best" semiempirical approach in terms of the quality-of-fit to full-scale test data but could be very conservative [55,56]. More work on fatigue models has been done in the past decade [34,63]; however, applications of those models are still in development.

Fatigue is a main concern for liquid pipelines but not for gas pipelines because their operating pressure cycles are not as aggressive as those of liquid lines. PDAM provides a review of plain dent fatigue assessment models [55].

***22.4.4.1 EPRG Dent Fatigue Analysis Method (Stress Based)*** An empirical method developed by the EPRG for predicting the fatigue life of an unconstrained plain dent may be considered as the best method in terms of quality-of-fit to published full-scale test data [21,57]. A comparison between the predictions made using this model and the published full-scale test data of rings and vessels containing plain dents is shown in Figure 22.21 [55,56].

The model in DIN 2413 [58] calculates the fatigue life of a plain dent using the fatigue life of plain pipe from an *S–N* curve for submerged arc welded long seam pipe, modified for the stress concentration due to the dent.

As indicated in [55], the following points should be borne in mind when applying the fatigue model:

- The stress concentration factor ($K_s$) is not dimensionless; the model is derived in SI units, and modifications would be required if a different system of units were used.
- The EPRG model does not account directly for the re-rounding behavior that has been observed in full-scale tests.

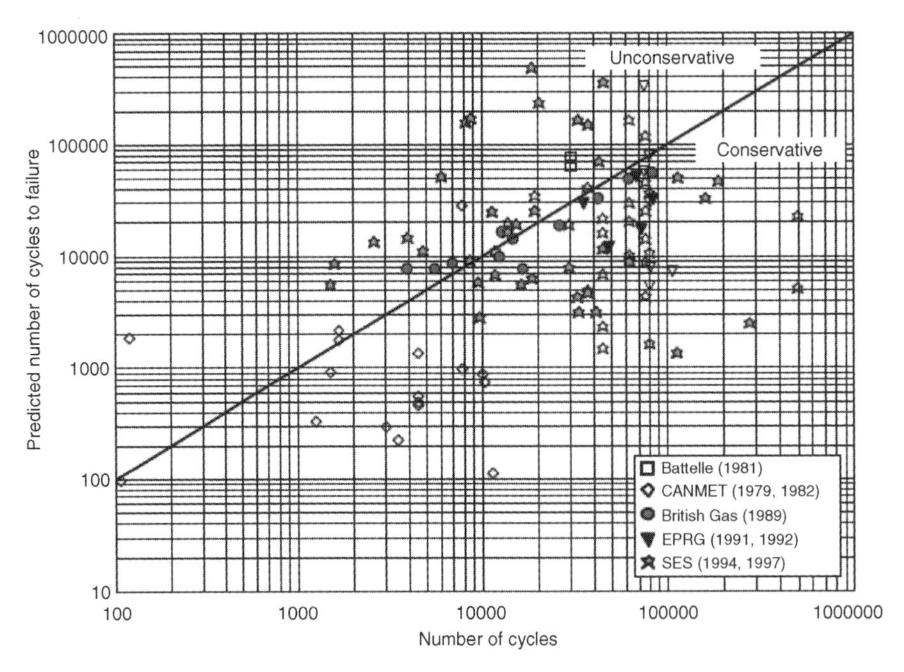


**FIGURE 22.21** Predictions of the fatigue life of a plain dent made using the EPRG model.

- The EPRG model does not predict the experimentally observed effect of mean pressure on the fatigue life of a plain dent, in which a longer fatigue life is observed for a higher mean pressure over the same stress range because of permanent re-rounding.
- There is experimental evidence to indicate that the dent shape has an effect on the fatigue strength of the dent, but insufficient results from which to draw firm trends.
- The model is based on tests in which the damage was introduced at zero pressure. The depth of dent is measured after spring-back and at zero pressure. Therefore, a correction must be made for dents introduced and measured at pressure. An empirical re-rounding correction factor was given by EPRG, see Equation 22.22 [21,55].
- The model is not directly applicable to high-strain/low-cycle fatigue behavior, that is, reverse plasticity. However, it is noted that the reduction in fatigue life due to the cyclic plasticity associated with incremental re-rounding in the first few cycles is implicitly taken into account in this empirical model.

The fatigue life of an unconstrained plain dent, $N$ (cycles), is given by the following equations:

$$N = 1000\left[\frac{(\sigma_U - 50)}{2\sigma_A K_s}\right]^{-4.292} \qquad (22.18)$$

where:

$$2\sigma_A = \sigma_U\left[B\left(4 + B^2\right)^{0.5} - B^2\right] \qquad (22.19)$$

$$B = \frac{\frac{\sigma_a}{\sigma_U}}{\left[1 - \left(\frac{\sigma_{max}-\sigma_a}{\sigma_U}\right)\right]^{0.5}} = \frac{\frac{\sigma_a}{\sigma_U}}{\left[1 - \frac{\sigma_a}{\sigma_U}\left(\frac{1+R}{1-R}\right)\right]^{0.5}} \qquad (22.20)$$

$$R = \frac{\sigma_{min}}{\sigma_{max}}$$

$$K_d = H_o\frac{t}{D} \quad K_s = 2.87\sqrt{K_d} \qquad (22.21)$$

$$H_o = 1.43H_r \qquad (22.22)$$

The parameters in the equations are defined as follows:

| | |
|---|---|
| $t$ | pipe wall thickness (mm) |
| $D$ | outside diameter of pipe (mm) |
| $H_o$ | dent depth (mm) |
| $H_r$ | dent depth measured at pressure (mm) |
| $N$ | number of cycles to failure |
| $R$ | ratio of minimum to maximum stress in stress cycle |
| $\sigma_U$ | ultimate tensile strength (N/mm$^2$) |
| $2\sigma_a$ | nominal cyclic hoop stress range (N/mm$^2$) ($=\Delta\sigma = \sigma_{max} - \sigma_{min}$) |
| $2\sigma_A$ | equivalent cyclic hoop stress range corresponding to $R=0$ (N/mm$^2$) |
| $\sigma_{max}$ | maximum hoop stress in stress cycle (N/mm$^2$) |
| $\sigma_{min}$ | minimum hoop stress in stress cycle (N/mm$^2$) |

Other semiempirical or empirical methods for predicting the fatigue life of a plain dent subject to cyclic pressure loading have also been proposed, including models API 1156 [33,58], AGA [59], Rosenfeld [60,61], Hope et al. [62], and Alexander [63]. Detailed Information can be found from these reports and technical papers. However, from a practical point of view, the EPRG model was established on more than 45 full-scale fatigue tests and may be considered as the "best" method in terms of the quality-of-fit to full-scale test data [21,57].

*Factor to Account for Model Uncertainty*

PLAIN DENTS Equations 22.18–22.22 are recommended by EPRG to calculate fatigue life for non-constrained plain dents. However, there is no factor to account for the model uncertainty. The uncertainty factor on the predicted fatigue life of a plain dent was derived by linear regression analysis on all of the relevant full-scale fatigue tests in which the dent failed due to fatigue. The model at a 95% confidence was estimated to correspond to an uncertainty factor of 13.3. This means that, by reduction of the computed fatigue life $N$ from Equation 22.18 by a factor of 13.3, the confidence level of the reduced fatigue life being $N/13.3$ is 95%.

DENTS ON WELDS The fatigue life of a weld is less than that of the parent plate because of stress concentrations, local material defects, and imperfections within the weld. Consequently, the fatigue life of a dented weld will be lower than that of an equivalent plain dent. The PDAM proposed [55,56] that the fatigue life of an otherwise sound weld in a smooth dent can be estimated by reducing the predicted fatigue life of an equivalent plain dent by a ***factor of ten***. The proposed limit is applicable only if the maximum depth of the dent (as measured at zero pressure after spring back ($H_o$)) is limited to 3.9% of the pipe diameter (approximately 2.7% as measured at pressure ($H_r$)).

DENTS WITH GOUGES A gouge is a stress riser that would significantly decrease fatigue life. The PDAM recommended [55,56] that the fatigue life of a smooth dent containing a gouge can be estimated by reducing the predicted fatigue life of an equivalent plain dent by a factor of ***one hundred***. The proposed limit is applicable only if (i) the maximum depth of the dent is limited to 4.0% of the pipe diameter (as measured at zero pressure after spring back ($H_o$)) (approximately 2.8% as measured at pressure ($H_r$)), (ii) the maximum depth of the gouge is limited to 20% of the wall thickness, and (iii) the upper shelf Charpy V-notch impact energy for 2/3 thickness specimen size is at least 47 J (35 ft-lbf).

***22.4.4.2 Strain-Based Fatigue Life*** It should be noted that the EPRG fatigue model [21,55–57] and other empirical and semiempirical fatigue models [33,58–63] are stress based. It is also possible to estimate fatigue life using strain-based models; for example, Francini and Yoosef-Ghodsi [34] suggested using the strain fatigue life equation that has been determined for a wide variety of metals [64]:

$$\Delta\varepsilon = 3.5\frac{\sigma_{ult}}{E}N^{-0.12} + e_o N^{-0.6} \qquad (22.23)$$

where $\Delta\varepsilon$ = cyclic strain range ($\varepsilon_{max} - \varepsilon_{min}$), $\sigma_{ult}$ = ultimate strength of the pipe, $E$ = elastic modulus, $N$ = number of cycles to failure, and $e_o$ = true strain to failure.

The difficulty with the strain life equation is that the ultimate tensile strength or true strain to failure is unknown in most cases. Francini and Yoosef-Ghodsi [34] suggested assuming that the ultimate tensile strength is the specified minimum tensile strength (SMTS) and the true strain to failure is the minimum elongation ($e_f$) for the grade of pipe, then a conservative estimate of the strain life for the measured dent strain can be made. The approximate strain fatigue life equation becomes

$$\Delta\varepsilon = 3.5\frac{\text{SMTS}}{E}N^{-0.12} + e_f N^{-0.6} \qquad (22.24)$$

Equation 22.24 can be solved numerically for $N$.

Rosenfeld [65] also introduced a generic strain curve that is a lower bound, except for low-alloy steel data from Stout and Pense [66], and is given by the expression:

$$\Delta\varepsilon/2 = 0.004(2N_f)^{-0.11} + 0.35\text{E}(2N_f)^{-0.54} \qquad (22.25)$$

where $\Delta\varepsilon/2$ is the alternating strain amplitude and $2N_f$ is the number of complete load cycles to failure. Furthermore, a simple strain fatigue equation by Markl [67] was also introduced for computing cycles to failure based on the cyclic strain range:

$$N = \left[0.01655/(\Delta\varepsilon/2)\right]^5 \qquad (22.26)$$

In general, the cyclic strain range for a dent is not known. FEA may be used to determine the cyclic strain range with internal operating pressure data. Although some applications have been made, more work needs to be done both theoretically and experimentally.

## 22.5 MITIGATION AND REPAIR

In Section 22.4.1, there is a detailed discussion on conditions for mitigation in accordance with dent severity in terms of the nature of mechanical damage (such as plain dents, dents with gouges/cracks, etc.) and the depth of a dent. Table 22.12 in Section 22.4.1 provides a summary of 49 CFR 192 and 195 on the disposition of plain dents and dents associated with other defects for gas and liquid pipelines [37,38]. The table shows the timescale requirements for operators to take prompt remediation action based on the dent conditions discovered through integrity assessment. Table 22.14 provides a comparison of international code guidance and recommended practices relevant to the assessment of dents in pipelines.

For strain-based assessment, ASME B31.8 (Paragraph 851.41and Appendix R) provides the option to use strain-based criteria and sets a 6% strain in pipe bodies and 4% strain in welds as the acceptance limits for plain and rock dents, as seen in Table 22.13, Section 22.4.1. These mitigation criteria are based on the nonmandatory equations recommended in B31.8 for strain calculation. It has been shown that one of these equations (namely, the total or equivalent strain equation) is incorrect, which underestimates dent total strain by a factor of up to 1.8–2.0 [29–31]. It is also known that the 6% strain limit for plain dents was based on engineering judgment [5,41] and was not rigorously justified. Recent progress in improvement of the dent strain calculation methods as presented in Sections 22.4.2 and 22.4.3 has raised a question about the appropriateness of ASME B31.8 strain limits. Many dents that were originally accepted by ASME B31.8 have now become unacceptable and are required to be repaired. The 6% strain limit for plain dents appears to be overly conservative and inappropriate when using more accurate dent strain calculation methods. Alternative strain limits need to be established based on the material's critical strain for failure for safe and cost-effective management of mechanical damage. In the following sections, alternative strain-based dent severity criteria are summarized and reviewed.

### 22.5.1 Improved Strain-Based Dent Severity Criteria – Alternatives

To date, there are three alternative criteria that have been proposed for mitigation of mechanical damage.

***22.5.1.1 Criterion Based on Ductile Failure Damage Indicator (DFDI)*** In the metal forming and solid expansion of tubular applications, the DFDI is adopted as one of the ductile failure criteria. The DFDI criterion is derived from the concept that ductile failure results from initiation, growth, and coalescence of voids on a microscale, and formation of cracks during large plastic deformation. Hancock and Mackenzie [51] in the mid-70s followed Rice et al.'s work [50] and proposed a reference failure strain, $\varepsilon_f$, that is, a strain limit for ductile failure, which can be expressed by stress triaxiality, $\frac{\sigma_m}{\sigma_{eq}}$, and material critical strain, $\varepsilon_o$:

$$\varepsilon_f = 1.65\varepsilon_o \exp\left(-\frac{3}{2}\frac{\sigma_m}{\sigma_{eq}}\right) \tag{22.27}$$

$$\sigma_m = \tfrac{1}{3}(\sigma_1 + \sigma_2 + \sigma_3) \tag{22.28}$$

$$\sigma_{eq} = \frac{1}{\sqrt{2}}\sqrt{(\sigma_1 - \sigma_2)^2 + (\sigma_2 - \sigma_3)^2 + (\sigma_3 - \sigma_1)^2} \tag{22.29}$$

where $\sigma_m$ = mean stress of three principal stresses in a triaxial stress field; $\sigma_{eq}$ = von Mises stress; and $\sigma_1$, $\sigma_2$, and $\sigma_3$ are principal stresses in the directions 1, 2, and 3, respectively. The ratio $\sigma_m/\sigma_{eq}$ represents the triaxiality of the stress field, and $\varepsilon_o$ = critical strain of the material, a material property measured by uniaxial tension testing and usually in the range of 0.3–0.5 for typical pipeline steels.

The DFDI, $D$, reflects the total plastic damage:

$$D = \int_0^{\varepsilon_{eq}} \frac{d\varepsilon_{eq}}{1.65\varepsilon_o \exp\left(-\frac{3}{2}\frac{\sigma_m}{\sigma_{eq}}\right)} \tag{22.30}$$

In Equation 22.30, $\varepsilon_{eq}$ is the equivalent strain, that is, $\varepsilon_{eq} = \frac{2}{\sqrt{3}}\sqrt{\varepsilon_x^2 + \varepsilon_x\varepsilon_y + \varepsilon_y^2}$ (i.e., Equation 22.14 in Section 22.4.2.2), and $D$ is the DFDI ranging from 0 (undamaged) to 1 (cracking). By definition, ductile failure or failure of the dent (cracking) will occur when $D > 1$. In order to calculate the DFDI, FEA should be conducted to extract three principal stresses and the equivalent plastic strain at every node in the dent. However, it is not practical to conduct FEA for all reported dents in a pipeline. Most recently, the present authors [1,2,48,49] developed simplified DFDI equations that do not use FEA results. The following are the simplified DFDI upper- and lower-bound equations:

$$\text{DFDI}_{\text{upper bound}} = \frac{\varepsilon_{eq}}{\left(\frac{\varepsilon_o}{1.65}\right)} \tag{22.31}$$

$$\text{DFDI}_{\text{lower bound}} = \frac{\varepsilon_{eq}}{\varepsilon_o} \tag{22.32}$$

The upper-bound and lower-bound DFDI are calculated using the maximum equivalent strain of the dent (calculated from 3D dent profile [23,46]), provided the critical strain of the material is known. The critical strain in Equation 22.32 is the critical *true* strain (not critical engineering strain) of the pipe material, which can be measured using a uniaxial tensile test with a specially equipped measuring system that performs real-time monitoring and recording of the specimen cross section up to necking and final rupture.

As demonstrated in Section 22.4.3.1, this criterion has been applied to the development of the combined approach for discrimination of metal loss between corrosion, cracking, and gouging, and effectively used for management of mechanical damage, including failure analysis [1,2,47,48].

***22.5.1.2 Criterion Based on Strain Limit Damage (SLD)*** The ASME Boiler & Pressure Vessel Code [68] Section VIII, Division 3 recommends a strain limit damage

(SLD) criterion using elastic-plastic FEA to estimate the accumulated plastic damage in pressure vessel components. Additionally, Section VIII also contains approximations for material properties based on specified minimum reduction in area and elongation to failure. These material properties are incorporated into the equations given ahead. The total SLD, $D_{\varepsilon t}$, is the total accumulated damage and is given by Equation 22.33. $D_{\varepsilon t} > 1$ indicates the limit state for the structure to carry no further loads (failure condition).

$$D_{\varepsilon t} = D_{\varepsilon form} + \sum_{k=1}^{M} D_{\varepsilon,k} \leq 1.0 \qquad (22.33)$$

where,

$$D_{\varepsilon,k} = \frac{\Delta\varepsilon_{peq,k}}{\varepsilon_{L,k}} \qquad (22.34)$$

$$\varepsilon_{L,k} = \varepsilon_{Lu}\left(e^{\frac{-m_5}{1+m_2}\left(\frac{\sigma_{1,k}+\sigma_{2,k}+\sigma_{3,k}}{3\sigma_{e,k}}-\frac{1}{3}\right)}\right) \qquad (22.35)$$

In these equations, $D_{\varepsilon,k}$ = damage occurring during the $k$th load increment, $D_{\varepsilon form}$ = damage occurring during forming, $\Delta\varepsilon_{peq,k}$ = change in total equivalent plastic strain during the $k$th load increment, $\varepsilon_{L,k}$ = maximum permitted local total equivalent plastic strain at the $k$th load increment, $\varepsilon_{L,u}$ = maximum of $m_2$, $m_3$, and $m_4$ where $m_2$, $m_3$, $m_4$ are coefficients calculated using specified minimum material property as per the Table KD-230 of ASME Boiler & Pressure Vessel Code, Section VIII Division 3 (2010), $m_5$ = value listed in Table KD-230, ($\sigma_{1,k}$, $\sigma_{2,k}$, $\sigma_{3,k}$) = principal stresses in the 1, 2, 3 directions, respectively, at a point of interest for the $k$th load increment, and $\sigma_{e,k}$, = von Mises equivalent stress at a point of interest for the $k$th load increment.

The advantage of the SLD criterion is that it utilizes the specified minimum reduction area and elongation to failure, which does not require measurement of actual material properties, including critical strain.

The SLD criterion is developed for validating pressure vessel design and is conservative with a built-in safety factor. The calculated SLD value is always larger than the DFDI value under the same loading conditions.

The SLD method requires FEA to extract the stress and strain parameters at every node in the dent. In general, FEA is not practical for evaluating the large number of dents in a typical pipeline assessment. Simplified SLD equations that estimate upper- and lower-bound SLD values for practical use [48,49] are

$$\text{SLD}_{\text{upper bound}} = \frac{\varepsilon_{eq}}{0.2248} \qquad (22.36)$$

$$\text{SLD}_{\text{lower bound}} = \frac{\varepsilon_{eq}}{0.4308} \qquad (22.37)$$

***22.5.1.3 Specified Minimum Elongation Criterion*** Francini and Yoosef-Ghodsi [34] reviewed the current ASME B31.8 strain equations and proposed an alternative strain limit for plain dents using specified minimum elongation of the pipe steel grade. The proposed limit of the equivalent strain ($\varepsilon_{eq}$) in a plain dent is

$$\varepsilon_{eq} \leq \frac{\varepsilon_f}{\text{SF}} \approx \frac{e_f}{\text{SF}} \qquad (22.38)$$

where $\varepsilon_f$ and $e_f$ are the true fracture strain and the specified minimum elongation to failure, respectively, and SF is the safety factor.

Francini and Yoosef-Ghodsi recommended the following simple criterion as the alternative to the current ASME B31.8 6% strain limit: *"If the plain dent is not associated with a weld, a dent with a calculated equivalent strain less than one-half of the specified minimum elongation for the pipe steel grade is considered benign."* Using this minimum elongation limit criterion, the alternate strain limit is 9–12% for typical line pipe steels (i.e., typical elongation to failure 18–24%).

It is noted that both the minimum specified elongation criterion and the SLD criterion are developed to assess fitness for purpose, not for failure prediction. For prediction of susceptibility to cracking, critical engineering assessment, or failure analysis, the actual material properties should be used, in particular, the true strain for failure. Since both the minimum specified elongation and SLD criteria utilize typical lower-bound material properties, their application for these purposes is limited. DFDI may be the better approach for critical engineering assessment of mechanical damage.

Since strain-based criteria are still under development, more work needs to be done. However, the aforementioned strain-based criteria can be used as a measure of dent severity for assessing the need for mitigation.

### 22.5.2 Repair

Depending on the severity and type of mechanical damage, several remediation options are available [7,13]:

- Recoating
- Grinding out of a scrape or gouge to create a smooth contour (with limitations)
- Steel reinforcement sleeve repair
- Steel pressure-containing sleeve repair
- Composite wrap repair (with limitations)
- Hot tap (with limitations)
- Pipe replacement

A repair can be considered temporary or permanent. A temporary repair is one that is to be removed within a period typically specified by the pipeline operator's written procedures. Temporary repairs are sometimes implemented in order to maintain continuous service and are used when the operator plans to return later to complete a more comprehensive repair, such as a pipe replacement. Any repair that is intended to restore the pipeline to service for a period >5 years, without a requirement for reevaluation, should be considered permanent [7,69].

All repairs carry limitations or add requirements, and the guidance for repair selection is reviewed and summarized by Baker [7,13] and may be found in more detail in the Repair Manual [69] issued by the PRCI as well as in applicable industry consensus standards, such as ASME B31.4 or B31.8.

## 22.6 CONTINUING CHALLENGES

Effectively and accurately characterizing and evaluating mechanical damage is a long-standing challenge for governmental and professional organizations, researchers, ILI vendors, and operators in the pipeline industry. Significant progress has been made during recent years, and serious incidents caused by mechanical damage are decreasing, although mechanical damage still ranks as the highest single cause of failure [6]. The reviews by Baker [7,13] identified gaps related to prevention, detection and characterization, and assessment/mitigation. He further pointed out areas for further research to address critical needs, and the need for guidelines and codes that ensure safe operation around the world.

In this chapter, the current status of ILI technologies, improved in-ditch tools for mechanical damage characterization, and strain-based assessment models and mitigation criteria are reviewed and summarized. Though the examples showed the effectiveness of state-of-the-art approaches and technologies for identification, prioritization, and integrity management of significant dents and dents with cracks and gouges, they are still under development, facing new challenges and continuing to improve.

It has been shown that the 3D in-ditch LaserScan is an accurate and valued tool that can measure a dent profile with an accuracy that is at least one order of magnitude better than ILI geometry tools. The reliable data measured by the 3D in-ditch LaserScanner can be used for calibration and validation of ILI vendors' pull-through tests and ILI run data, in particular, for high-resolution geometry tools. The calibration is no longer limited to depth but can be done for the overall profile and associated strains. However, the disadvantage of the tool is its current high cost. As indicated previously in Section 22.3.1, a structured-light 3D scanner system has recently been developed [24–26] and is commercially available at a much lower cost and has been successfully applied for corrosion in-ditch measurement [27]. Potentially, this technology can be used for in-ditch mechanical damage measurement, but accelerated research and development for this technology could provide opportunities to pipeline operators and NDE inspection service companies to select technologies that best fit their needs from a wider range of options.

The ILI tool sizing performance reviewed in this chapter was based on in-ditch measurement data, mainly of dents, using traditional technologies such as bridge bar or profile gage, and so on. Errors were large due to inconsistency of in-field measurement protocols. With the improved 3D in-ditch measurement technology, the evaluation of errors, both for ILI pull-through tests and actual ILI runs, can be greatly reduced by one-on-one comparison of ILI and in-field dent profiles. Reevaluation of the ILI geometry tool performance in terms of sizing tolerance in parallel with the application of improved in-ditch measurement tool is desirable.

Moreover, current assessment of re-rounding and rebounding is, by and large, limited to depth. A more accurate assessment would be based on the change in dent profile, including length, width, and strains in addition to depth, and could provide a more comprehensive picture of the effect of re-rounding and rebounding.

The current ASME code for strain-based mitigation criteria for gas pipelines sets a 6% strain limit for plain dents and 4% limit for dents with welds; however, alternative criteria have been proposed and are in use by various operators. A thorough review of the various criteria is needed, including their advantages, limitations, and gaps for further study. Guidelines for using strain-based assessment of mechanical damage are critical for future ASME codes and government regulations for the purpose of integrity management of mechanical damage.

Finally, the fatigue analyses methods based on depth and *S–N* curves, such as EPRG equations, require a large safety factor to ensure 95% accurate results because of the large scatter in full-scale fatigue data. Strain-based fatigue methods have been proposed using equations from other industries and text books [34,61,65–67]; however, their advantages, disadvantages, and limitations for remaining life calculations have not been rigorously evaluated and further investigation is needed.

## REFERENCES


1. Wang, R., Kania, R., Arumugam, U., and Gao, M. (2012) A combined approach to characterization of dent with metal loss. Proceedings of the 9th International Pipeline Conference (IPC 2012), September 24–28, 2012, Calgary, Alberta, Canada, Paper No. 90499.
2. Arumugam, U., Gao, M., and Wang, R. (2012) Root cause analysis of dent with crack: a case study. Proceedings of the 9th

International Pipeline Conference (IPC 2012), September 24–28, 2012, Calgary, Alberta, Canada, Paper No. 90504.

3. Katchmar, P.J. (2012) Failure Investigation Report TransCanada/Bison Pipeline, November 7, 2012. Available at http://phmsa.dot.gov/pipeline/library/failure-reports.
4. Kiefner, J., Kolovich, C., and Kolovich, K. (2006) Mechanical damage technical workshop. PHMSA/DOT, February 28, 2006, Houston, TX. Available at http://primis.phmsa.dot.gov/rd/mtg_022806.htm.
5. Rosenfeld, M.J. (2001) Proposed new guidelines for ASME B31.8 on assessment of dents and mechanical damage, GRI, May 2001.
6. DOT/PHMSA (2014) Pipeline Significant Incident files. Available at http://primis.phmsa.dot.gov/comm/DamagePrevention.htm.
7. Baker, M., Jr. (2009) Mechanical damage study. Final Report, PHMSA TTO 16 DTRS56-02-D-70036, April, 2009.
8. Gao, M. and Krishnamurthy, R. (2010) Investigate performance of current in-line inspection technologies for dent and dent associated with metal loss damage detection. Proceedings of the 8th International Pipeline Conference (IPC 2010), September 27–October 1, 2010, Calgary, Alberta, Canada, Paper No. 31409.
9. McNealy, R., Gao, M., McCann, R., Cazenave, P., and Krishnamurthy, R. (2008) Investigate fundamentals and performance improvements of current in-line-inspection technologies for mechanical damage detection, Phase I report. PRCI Report, PR-328-063502, DTPH56-T-000016, ECT#204, 2008.
10. Davis, R.J. and Nestleroth, J.B. (1999) *Pipeline Mechanical Damage Characterization by Multiple Magnetization Level Decoupling, Review of Progress in Quantitative Nondestructive Evaluation*, Plenum Press, New York, NY, Vol. 18.
11. Panetta, P.D., et al. (2001) Mechanical damage characterization in pipelines. Pacific Northwest National Laboratory Report PNNL-SA-35467, October 2001, Department of Energy, DE-AC06-76RLO1830, USA.
12. Clapham, L., Babbar, V., and Rubinshteyn, A. (2006) Understanding magnetic flux leakage signals from dents. Proceedings of the 6th International Pipeline Conference (IPC 2006), September 25–29, 2006, Calgary, Alberta, Canada, Paper No. 10043.
13. Baker, M., Jr. (2004) Dent study. Final Report, TTO Number 10 DTRS56-02-D-77036, November 2004.
14. Beuker, T. and Rahe, F. (2005) High Quality Smart Pig Inspection of Dents, Compliant with the U.S. Code of Federal Regulations, Oil and Gas Processing Review, 2005.
15. Bubenik, T.A., et al. (2000) In-line-inspection technologies for mechanical damage and SCC in pipeline. Final Report, DOT DTR56-96-C-0010, June, 2000.
16. Gao, M. and Krishnamurthy, R. (2009) Investigate fundamentals and performance improvements of current in-line-inspection technologies for mechanical damage detection, Phase II report. PRCI Report, PR-328-063502, DTPH56-T-000016, ECT#204, 2009.
17. API (2005) *API 1163: In-Line Inspection Systems Qualification Standard*, 1st ed., American Petroleum Institute, Washington, DC.
18. Desjardins, G., Reed, M., and Nickle, R. ILI Performance Verification and Assessment Using Statistical Hypothesis.
19. McCann, R., McNealy, R., and Gao, M. (2007) In-line inspection performance verification. Proceedings of CORROSION/2007, NACE International, Nashville, TN, Paper No. 0713.
20. McCann, R., McNealy, R., and Gao, M. (2008) In-line inspection performance, II: validation sampling. Proceedings of CORROSION/2008, NACE International, Paper No. 1177.
21. Corder, I. and Chatain, P. (1995) EPRG recommendations for the assessment of the resistance of pipelines to external damage. Proceeding of the EPRG/PRC 10th Biennial Joint Technical Meeting on Line Pipe Research, April, 1995, Cambridge, UK.
22. Gao, M. (2009) Evaluation report: LaserScan technology for in-ditch mechanical damage characterization for MD1-4. PRCI Report, PR-328-073511, June, 2009.
23. Arumugam, U., Tandon, S., Gao, M., et al. (2010) Portable LaserScan for in-ditch profiling and strain analysis: methodology and application development. Proceeding of the 8th International Pipeline Conference (IPC 2010), September 27–October 1, 2010, Calgary, Alberta, Canada, Paper No. 31336.
24. Fechteler, P., Eisert, P., and Rurainsky, J. (2007) Fast and high resolution 3D face scanning. Proceedings of the IEEE International Conference on Image Processing (ICIP 2007), September 16–19, 2007, San Antonio, TX.
25. Gupta, M., Agrawal, A., Veeraraghavan, A., and Narasimhan, S.G. (2011) Structured light 3D scanning in the presence of global illumination. Proceeding of the IEEE Computer Vision and Patter Recognition (CVPR), June 21–23, 2011, Colorado Springs, USA.
26. Liu, K., Wang, Y., Hao, Q., and Hassebrook, L. (2010) Dual-frequency pattern scheme for high-speed 3-D shape measurement. *Optics Express*, 18, 5229–5244.
27. Cazenave, P., Tinacos, K., Gao, M., Kania, R., and Wang, R. (2014) Evaluation of new in-ditch methods for measurement and assessment of external corrosion. Proceedings of the 10th International Pipeline Conference (IPC 2014), September 29–October 3, 2014, Calgary, Alberta, Canada, Paper No. 33578.
28. Seikowave: "Applications of 3D Scanner", http://www.seikowave.com.
29. Lukasiewicz, S.A., Czyz, J.A., Sun, C., and Adeeb, S. (2006) Calculation of strains in dents based on high resolution in-line caliper survey. Proceedings of the 6th International Pipeline Conference (IPC 2006), September 25–29, 2006, Calgary, Alberta, Canada, published by ASME, Paper No. 10101.
30. Czyz, J.A., Lukasiewicz, S.A., and Adeeb, S. (2008) Calculating dent strain, Pipeline and Gas Technology, January/February 2008.
31. Gao, M., McNealy, R., Krishnamurthy, R., and Colquhoun, I. (2008) Strain-based models for dent assessment—a review. Proceedings of the 7th International Pipeline Conference (IPC 2008), Calgary, Alberta, Canada, Paper No. IPC2008-64565.
32. American Society of Mechanical Engineering (ASME) Gas Transmission and Distribution Piping System ASME B31.8, Editions of 2003, 2007, 2010, and 2012.

33. Kiefner, J.F. and Alexander, C.R. (1999) Effect of Smooth and Rock Dents on Liquid Petroleum Pipelines (Phase 2), Addendum to API Publication 1156, October 1999.
34. Francini, B.R. and Yoosef-Ghodsi, N. (2008) Development of a model for predicting the severity of pipeline damage identified by in-line inspection. Pipeline Research Council International (PRCI) Report, PR-218-063511-B, Final Report No. 08-124, December 2008.
35. Blade Energy Partners (2008) A Point-to-Point Dent Strain Analysis Tool Using ILI and 3D LaserScan Reported Dent Profiles, User Instruction Manual, 2008.
36. Rosenfeld, M., Pepper, J., and Leewis, K. (2002) Basis of the new criteria in ASME B31.8 for prioritization and repair of mechanical damage. Proceedings of the 4th International Pipeline Conference (IPC 2002), September 29, 2002. Published by ASME, Paper No. 27122.
37. ANON (2014) What actions must be taken to address integrity issues? Transportation of Natural and Other Gas by Pipeline: Minimum Federal Safety Standards, Section 192.933, US Government Code of Federal Regulation, 49 CRF Part 192, March 2014.
38. ANON (2014) Pipeline integrity management in high consequence areas, US Government Code of Federal Regulation, 49 CFR 195.452, March 2014.
39. Dawson, S.J., Russell, A., and Patterson, A. (2006) Emerging techniques for enhanced assessment and analysis of dents. Proceedings of the 6th International Pipeline Conference (IPC 2006), September 25–29, 2006, Calgary, Canada, Paper No. 10264. Published by ASME.
40. Warman, D.J., Johnston, D., Mackenzie, J.D., Rapp, S., and Travers, B. (2006) Management of pipeline dents and mechanical damage in gas pipelines. Proceedings of the 6th International Pipeline Conference (IPC 2006), September 25–29, 2006, Calgary, Canada, Paper No. 10407.
41. Rosenfeld, M.J., Porter, P.C., and Cox, J.A. (1998) Strain estimation using Vetco deformation tool data. Proceedings of the ASME 2nd International Pipeline Conference, 1998, Volume 1, Calgary.
42. Noronha, D.B., Martins, R., Jacob, B., and Souza, E. (2005) The use of B-splines in the assessment of strain levels associated with plain dents. Proceedings of Rio Pipeline Conference and Exposition, October, 2005, Rio de Janeiro, Brazil, Paper No. IBP 1245_05.
43. Gibson, J.E. (1980) *Thin Shells: Computing and Theory*, Pergamon Press, Oxford, NY.
44. Dieter, G.E. (1976) *Mechanical Metallurgy*, McGraw-Hill, New York, NY.
45. Lukasiewicz, S.A., Czyz, J.A., Sun, C., and Adeeb, S. (2006) Calculation of strains in dents based on high resolution in-line caliper survey. Proceedings of the 6th International Pipeline Conference (IPC 2006), September 25–29, 2006, Calgary, Canada, Paper No. 10101.
46. Arumugram, U., Kendrick, D., Limon-Tapia, S., and Gao, M. (2010) An approach for evaluation and prioritizing dents for remediation as reported by ILI tools. Proceedings of the 8th International Pipeline Conference (IPC 2010), September 27–October 1, 2010, Calgary, Canada, Paper No. 31401.
47. Tinacos, K., Cazenave, P., Gao, M., Krishnamurth, R., Seman, D., and Joes, C. (2012) ILI based dent screening and strain based analysis. Proceedings of the 9th International Pipeline Conference (IPC 2012), September 24–28, 2012, Calgary, Canada, Paper No. 90484.
48. Wang, R., Kania, R., Arumugam, U., and Gao, M. (2013) Characterization of plastic strain damage and MFL signals for discrimination of crack/gouge/corrosion in dents—a combined approach. Proceedings of the 19th Biennial Joint Technical Meeting on Pipeline Research, April 29–May 3, 2013, Sydney, Australia, Paper No. 2.
49. Arumugram, U., Gao, M., Wang, R., Kania, R., and Katz, D. (2013) Application of plastic strain model to characterize dent with crack. Proceedings of CORROSION/2013, NACE International, Orlando, FL, Paper No. 002858.
50. Rice, J.R. and Tracey, D.M. (1969) On the ductile enlargement of voids in tri-axial stress fields. *Journal of the Mechanics and Physics of Solids*, 17, 201–217.
51. Hancock, J.W. and MacKenzie, A.C. (1976) On the mechanisms of ductile failure in high-strength steels subjected to multi-axial stress-states. *Journal of the Mechanics and Physics of Solids*, 24, 147–169.
52. Miller, S. and Sander, F. (2006) Advances in feature identification using tri-axial MFL sensor technology. Proceedings of the 6th International Pipeline Conference (IPC 2006), September 25–29, 2006, Calgary, Canada, Paper No. 10327.
53. Vanessa, I.S., Ellis, C., and Wilkie, G. (2006) Characterization of mechanical damage through use of the tri-axial magnetic flux leakage technology. Proceedings of the 6th International Pipeline Conference (IPC 2006), September 25–29, 2006, Calgary, Canada, Paper No. 10454.
54. API (1997) *API 1156: Effect of Smooth and Rock Dents on Liquid Petroleum Pipelines*, API Publication, Washington, DC.
55. Cosham, A. and Hopkins, P. (2003) The pipeline defect assessment manual (PDAM). Report to the PDAM Joint Industry Project, May, 2003.
56. Cosham, A. and Hopkins, P. (2003) The effect of dents in pipelines—guidance in the pipeline defect assessment manual. Proceedings of ICPVT-10, July 7–10, 2003, Austria.
57. Roovers, P., Bood, R., Galli, M., Marewski, U., Steiner, M., and Zaréa, M. (2000) EPRG methods for assessing the tolerance and resistance of pipelines to external damage. Proceedings of the 3rd International Pipeline Technology Conference, R. Denys, Ed., Elsevier Science, Volume II, May 21–24, 2000, Brugge, Belgium, pp. 405–425.
58. Anon (2013) Deutsche Norm, Design of steel pressure pipes, DIN 2413 Part 1, October 1993.
59. Owler, J.R., Alexander, C.R., Kovach, P.J., and Connelly, L.M. (1994) Cyclic pressure fatigue life of pipelines with plain dents, dents with gouges, and dents with welds, AGA Pipeline Research Committee, Report PR-201-927 and PR-201-9324, June 1994.
60. Rosenfeld, M.J. (1998) Investigations of dent rerounding behaviour, Volume 1, Proceedings of the 2nd International

Pipeline Conference (IPC 1998), Calgary, Canada, American Society of Mechanical Engineers, 1998, pp. 299–307.

61. Rosenfeld, M.J. and Kiefner, J. (2006) Basics of metal fatigue in natural gas pipeline system—a primer for gas pipeline operators. PRCI Report, Catalog No. L52270, 2006.
62. Hope, A.D., Voermans, C.M., Lasts, S., and Twaddle, B.R. (1995) Mechanical damage of pipelines: prediction of fatigue properties. Proceedings of the 2nd International Pipeline Technology Conference, September 11–15, 1995, Oostende, Belgium, R. Denys, Ed., Elsevier, 1995.
63. Alexander, C. and Jorritsma, E. (2010) A systematic approach evaluation dent severity in a liquid transmission pipeline system. Proceedings of the 8th International Pipeline Conference (IPC 2010), September 27–October 1, 2010, Calgary, Canada, Paper No. IPC2010-31538.
64. Frost, N.E., Marsh, J.K., and Pook, L.P. (1974) *Metal Fatigue*, Oxford University Press, Oxford, UK.
65. Rosenfeld, M.J., Hart D.H. and Zulfiqar, N. (2008) Acceptance Criteria for Mild Ripples in Pipeline Field Bends, Report on Contract PR-218-9925, Final Report No.08-081, Pipeline Research Council International (PRCI), August, 2008.
66. Stout, R.D. and Pense, A.W. (1965) Effect of composition and microstructure on the low cycle fatigue strength of structural steels. *Journal of Basic Engineering*, 87, 269–274, Paper No. 64-Met-9,
67. Markl, A.R.C. (1952) Fatigue tests of piping components. *Transactions of the ASME*, 74, pp. 77–78.
68. ASME (2010) Alternative Rules for Construction of High Pressure Vessels ASME Boiler & Pressure Vessel Code, Section VIII, Division 3.
69. Jaske, C.E., Hart, B.O., and Bruce, W.A. (2006) Pipeline repair manual. Report on Contract PR-186-0324, Catalog No. L52047, Pipeline Research Council International (PRCI), Houston, TX, August, 2006.

# 23

# PROGRESSION OF PITTING CORROSION AND STRUCTURAL RELIABILITY OF WELDED STEEL PIPELINES

ROBERT E. MELCHERS
*Centre for Infrastructure Performance and Reliability, The University of Newcastle, Newcastle, NSW, Australia*

## 23.1 INTRODUCTION

Pitting is an important form of corrosion that often is responsible for the perforation of physical infrastructure such as pipelines, tanks, and vessels. It may occur for a variety of metal construction materials and for a variety of exposure conditions. In this chapter, discussion is limited to structural and low-alloy steels such as those commonly employed in the offshore oil industry, for ships and for coastal infrastructure. It is also limited to pipelines and so on exposed to sea and other waters. Generally such exposure produces rather aggressive corrosion, including pitting. Herein, the development of analytical models to represent the progression of pit depth with increased exposure time and as a function of influencing factors is outlined. Such models obviously must be of sufficient simplicity to be used in practice. The development is built on a very brief review of the extensive literature and research on nucleation, initiation, and early growth of pits, although for infrastructure applications these early phenomena are of lesser interest. The model for the progression of pitting corrosion described herein is based on reasonable assumptions from corrosion science and also is based on field observations. There is only very limited discussion of this topic in the corrosion science or engineering literature. A summary is given of the principal factors known to influence pitting corrosion, including steel composition and water quality.

Since the risk associated with corrosion, and in particular, the risk associated with pitting and perforation, such as for pipe or vessel walls, is of interest in most practical applications, a short overview is also given of relevant theory and techniques for the assessment of structural reliability. This sets the scene for an examination of the uncertainty associated with maximum pit depth and the application of extreme value (EV) distributions, a tool used for over 50 years for this purpose, but as will be seen, open to new interpretations when examined in the light of better understanding of the development of pit depth with increased exposure time. To complete this chapter, short descriptions are given of some research efforts to estimate the external corrosion and pitting of steel pipelines, and its extension to the external corrosion of welds in marine immersion conditions. Also, a brief discussion is given of the important problem of the internal corrosion of steel pipelines used for water injection systems, noting that this also may have relevance to the severe corrosion sometime observed in oil pipelines. However, the corrosion of oil production pipelines is not considered specifically. Typically such pipelines are subject to carbon dioxide ($CO_2$) corrosion. By comparison, this is typically much less aggressive except where water (usually present in recovered oil) is present. In this case, the most severe (and often very aggressive) corrosion occurs at the bottom of (near) horizontal pipelines and is attributed to water separation to the bottom of the pipe [1]. It follows that for these pipelines too, pitting in seawater conditions is of interest.

While in principle protective coatings and cathodic protection should provide adequate protection against corrosion, in

*Oil and Gas Pipelines: Integrity and Safety Handbook,* First Edition. Edited by R. Winston Revie.

practice this is not always the case or cannot be applied. There are many cases in practice of older steel infrastructure near or in seawater showing signs of corrosion, despite protective coatings or cathodic protection. Where wear or erosion also is involved, such as for coal or iron ore bulk cargoes for ships, protective coatings last only a very short time and cathodic protection is ineffective. Similarly, for mooring chains in the offshore industry [2] and for water injection pipelines [3], protective coatings generally are not effective and cathodic protection has been found to be problematic.

## 23.2 ASSET MANAGEMENT AND PREDICTION

Asset managers and engineers often are faced with the need to make decisions about the technical adequacy or safety as well as the possible future performance and safety of such infrastructure [4]. Predicting the future rate of progression of corrosion, and particularly, possible future pitting corrosion is of interest. As argued elsewhere, the most appropriate approach for predicting future pitting corrosion is through the development of rational, mathematical, and probabilistic models calibrated to real-world data and observations [5]. As noted, these models should aim to represent the pitting process at a level suitable for engineering and managerial asset management decisions, and to reflect the effect of time and other influencing factors.

It may be helpful to review, briefly, why models for the progression of pit depth with time are useful for prediction of future corrosion, and also, why they are useful for interpreting observations. The basic ideas are illustrated in Figure 23.1 [6]. Consider first the case in which it is desired to estimate the longer-term rate of corrosion or pitting shown by $r_s$ in Figure 23.1a when all that is available is an observation of corrosion $c(t_i)$ (or pit depth $d(t_i)$) at time $t_i$. The usual approach is to estimate the rate of future corrosion by $c(t_i)/t_i$ shown as the "apparent" corrosion rate in Figure 23.1a. However, a better estimate of the long-term corrosion rate, given by $r_s$ (and also its intercept $c_s$ at $t = 0$), can be obtained if the theoretical model is known, as shown in Figure 23.1b. Note that the model, particularly if properly calibrated to actual field data obtained in past experiments or from field experience, strengthens the prediction since it also brings to bear information and knowledge contained in the model. In turn, this means that the best models are those based both on theoretical concepts as well as accumulated past experience. Such models differ in concept and in predictive capability from models based only on data. In addition, if multiple observations are available, it is also possible to make estimates of uncertainty or variability in corrosion losses or maximum pit depth, as shown schematically in Figure 23.1c. Such estimates are important for constructing the high-quality models necessary for modern whole-of-life asset decision making processes [4]. As argued previously [5], development of such models requires a combination of scientific understanding of corrosion processes and sound approaches to mathematical modeling.

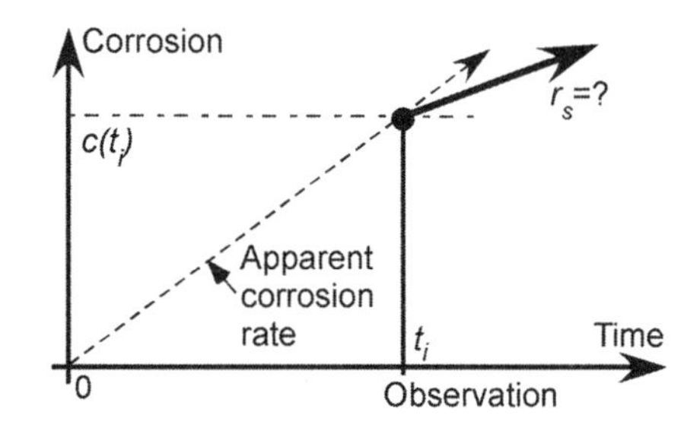


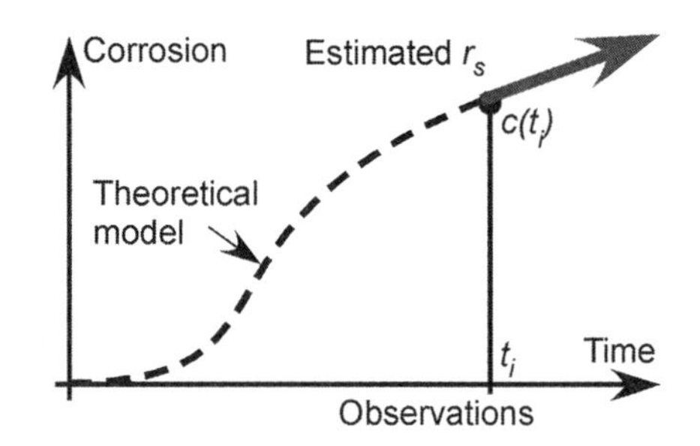


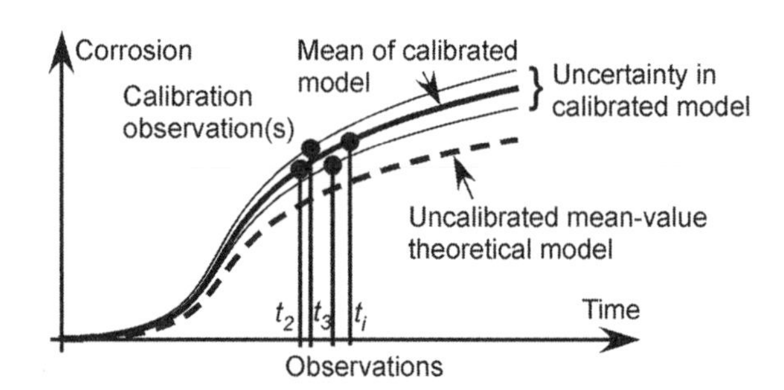


**FIGURE 23.1** (a) Corrosion loss observation $c(t_i)$ at time $t_i$ and the estimated apparent corrosion rate compared with the better estimate $r_s$, (b) estimate of $r_s$ based on a known theoretical model, and (c) showing variability estimated from multiple observations.

## 23.3 PITTING

### 23.3.1 Terminology

Differences of opinion arise sometimes about the use of the term "pitting." Some electrochemists would prefer to reserve the term for the highly localized corrosion seen only on metals that have passive films, such as stainless steels and aluminum, and therefore, should not be applied to mild and low-alloy steels. In this view, the existence of a passive film is a necessary condition for pitting [7]. Herein, the more classical view is adopted, namely, that the term also can be applied to metals without passive films, or weak films, including mild and low-alloy steels. This is entirely consistent with major corrosion researchers and authors such as Evans [8], Butler et al. [9], Uhlig and Revie [10], Szklarska-Smialowska [11], Sharland and Tasker [12], Shrier [13], and

Mercer and Lumbard [14], all of whom use the term pitting in relation to mild and low-alloy steels.

### 23.3.2 Initiation and Nucleation of Pits

Although electrochemistry lies at the heart of corrosion processes, much of the discussion to follow in this chapter deals with engineering application aspects and for that purpose, detailed discussion of electrochemistry usually is not necessary, in a similar way that atomic particle physics is not necessary for the understanding of the behavior of a steel bridge. However, to set the scene for an understanding of pitting corrosion a little background is required.

Corrosion may be considered as the (electrochemical) transfer of material (steel) away from the original location to somewhere else—usually deposited elsewhere as oxidized iron (i.e., rusts) but also lost into the surrounding environment (usually water or moisture). This process is accompanied by the physical transport of electrons (i.e., energy) released from the sites where metal is lost (the anodes) to those sites (the cathodes) where metal is deposited and oxidized, usually by oxygen (although other electron acceptors may be involved). In contrast, in purely chemical reactions, there is no distinction between anodic and cathodic sites and the electrons involved in the reactions simply move from one electron orbit to another [15].

Typically there are numerous anodic and cathodic sites on a surface subject to corrosion and these may be very close to each other (nanometers) or much further apart, provided there is electrical conductivity between them. In active corrosion the conductive paths must include both the steel and water (or other electrolyte) in series. The driving force (electrical potential) for the corrosion current is the result of very small differences in the surface topography and grain structure of a steel surface. This results in very small (electrical) potential differences. These provide the initial driving force for electron flow. For metals such as stainless steel and aluminum, which have a so-called passive film, this stage is also known as pit "nucleation" in the pitting corrosion literature [7,16].

Mild steel does not have a significant passive film and the process is slightly different, which, for passive films, has received by far the most attention in the corrosion science literature [17]. However, an important mechanism for pitting, including for mild steel, is the likely involvement of sulfide inclusions that are almost invariably present in all commercial steels [18], something recognized many years ago [19–21]. In both cases, the overall effect is much the same—the surface corrodes at localized (anodic) areas under overall readily available oxygen conditions, with the corrosion products (the oxidized metal) forming mainly at and near the cathodic regions. The process is driven by differences in localized electrical potential and produces localized corrosion at a microscale. Whereas, in metals with passive films there usually is some delay before the passive film is ruptures and pitting commences, pitting for mild and structural steels commences almost immediately, as demonstrated by careful experimental work [9].

### 23.3.3 Development of Pitting

In theory, the rate of corrosion, and thus, also of pitting, can be expressed through the rate of flow of electrons, either over larger areas (as for general corrosion) or on small areas (as for pitting). As in any electrical circuit, the rate of electron flow that will occur (and thus, the theoretical corrosion rate kinetics) depends on the resistivity of the path (water and steel) through which electron flow (or current) occurs. This mechanism is employed for electrochemical measurement on some types of laboratory experiments [16] and then used sometimes to explain the rate of corrosion under particular circumstances. However, for most practical situations this is largely theoretical. In practice, most corrosion processes in wet environments are rate controlled by the rate at which the critical diffusion process can occur. Candidate diffusion processes include the diffusion of ferrous irons out of the localized corrosion regions and later out of the pits and also the diffusion of oxygen from the external environment toward the cathodic sites. In actual corrosion pitting, one of the diffusion processes usually is the rate-limiting step, controlling the rate of corrosion, since typically the diffusion processes are much slower than the kinetic reactions (electron or corrosion current flows [16]. As will be seen, this distinction is crucial in modeling the corrosion process under realistic conditions.

Actual observations of the progression of corrosion of mild or structural steels in seawater support these theoretical concepts. Usually, already within hours of first exposure, very small areas of localized corrosion or pits can be seen to have developed on a steel surface. These localized regions grow quickly in depth, reaching perhaps up to some 100 microns [9], within days–weeks of exposure and then tend to stop growing in depth but become wider with further exposure [22]. It has also been observed that many early pits stop growing soon after formation and are "overtaken" by others [11].

Figure 23.2 shows a schematic view of the development of pitting with increased exposure time. In particular, it shows that the initial pits or regions of localized corrosion stop growing in depth but amalgamate to form shallow depressions and that later new pitting develops on the depression surfaces. The result is the formation of a series of depressions and a range of pit depths and sizes. This shows that the growth of pit depth is not a continuous process, at least for longer exposures. Figure 23.3 shows some microscope photographs (at the same scale) of the progression of pitting [23]. This pattern of behavior for pit growth and development contrasts with the conventional wisdom, which

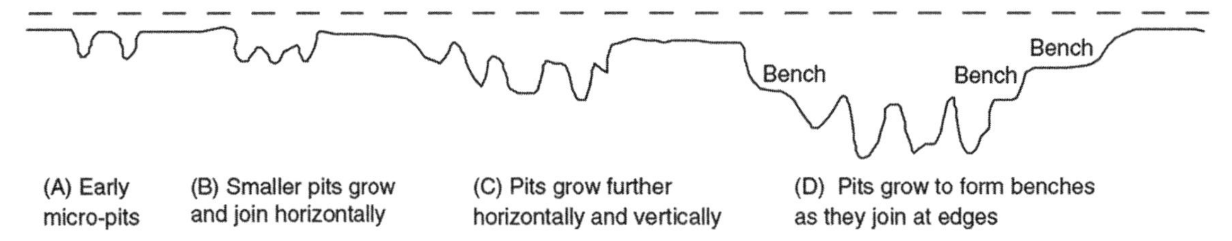
Bench
Bench
Bench
(A) Early micro-pits
(B) Smaller pits grow and join horizontally
(C) Pits grow further horizontally and vertically
(D) Pits grow to form benches as they join at edges

**FIGURE 23.2** Development of pitting as a function of time, showing initial pitting broadening out to form a plateau on which new pits then form [23].

assumes a continuous single functional process for pit depth development.

### 23.3.4 Biological Influences

Real seawaters are more than "salt water" and typically contain a wide range of chemicals and biological materials. For metals exposed to real seawater invariably there will also be a colonization of the steel surface by biofilms [24]. These can provide environments suitable for colonization by microorganisms. In turn, these may provide electron acceptors alternative to oxygen, and thus, contribute to the corrosion process. There is a wealth of laboratory observations of the contribution of microbiologically influenced corrosion (MIC) to initial and short-term corrosion losses and to early pitting [24]. Usually, these experiments are conducted in solutions with artificially high levels of bacterial culture and high supplies of nutrients necessary for microorganism metabolism [25]. There also is evidence in field exposure conditions of increased mass losses for waters with elevated levels of nutrients, for both short-term [26,27] and long-term exposures [28], including in brackish and fresh waters [29,30]. In addition, more recent controlled experimental observations of steel surfaces exposed to normal coastal seawater show that the presence of microorganisms has a strong effect on localized corrosion and pitting when compared with the surfaces resulting from corrosion in essentially sterile seawater and that this effect continues for many years [31]. It follows that microbiological factors can have an important influence on the severity of pitting corrosion.

Although Figures 23.2 and 23.3 show the development of pitting under seawater exposure conditions, it is important to note that similar observations have been made in laboratory experiments using triply distilled water and at elevated water temperatures (70 °C) [14]. This means that this type of corrosion behavior is not confined just to seawater exposures and is not necessarily related to the effect of microorganisms, although as discussed earlier, they can be involved [32].

### 23.3.5 Trends in Corrosion with Time

It now should be clear that the corrosion of a steel surface in various waters, including seawater, is a complex, changing phenomenon. This means that any microscopic examination of a corroded steel surface will reveal only a "snapshot" of a complex mix of larger and smaller pits as well as regions that may not yet be unaffected by corrosion. It also shows that so-called uniform or general corrosion, widely used in corrosion studies, is actually an idealization of a more complex situation. However, in practical applications it is still a convenient concept and is readily ascertained from changes in mass loss of nominally identical coupons exposed for different periods of time. This is much easier to measure than pit depths or the form of an undulated surface consisting of pits and depressions (e.g., Figure 23.3). Examples of corrosion mass losses as functions of exposure period are shown in Figure 23.4, for four different marine exposure environments, including tidal and atmospheric exposures [33,34]. Similar trends have been interpreted from published data for many other exposure sites [35].

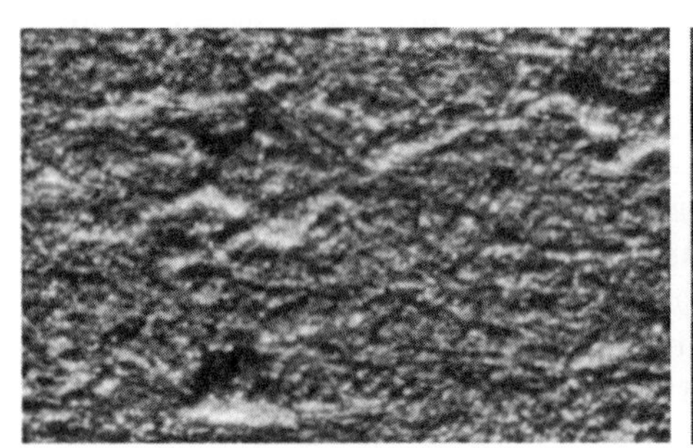
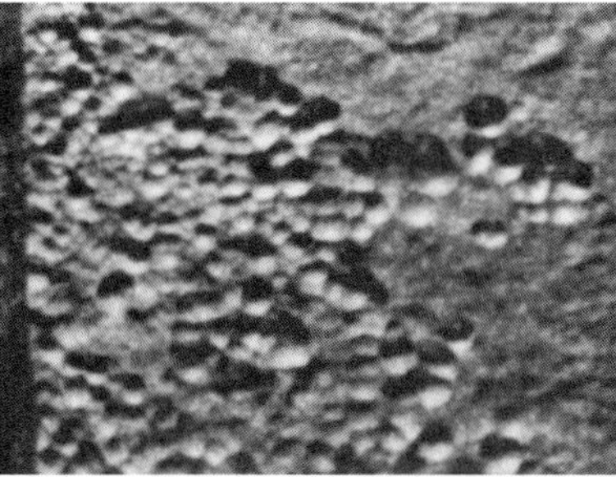

**FIGURE 23.3** Successive views of pitted surface of steel coupons. (Height of images approximately 1 mm) [23].

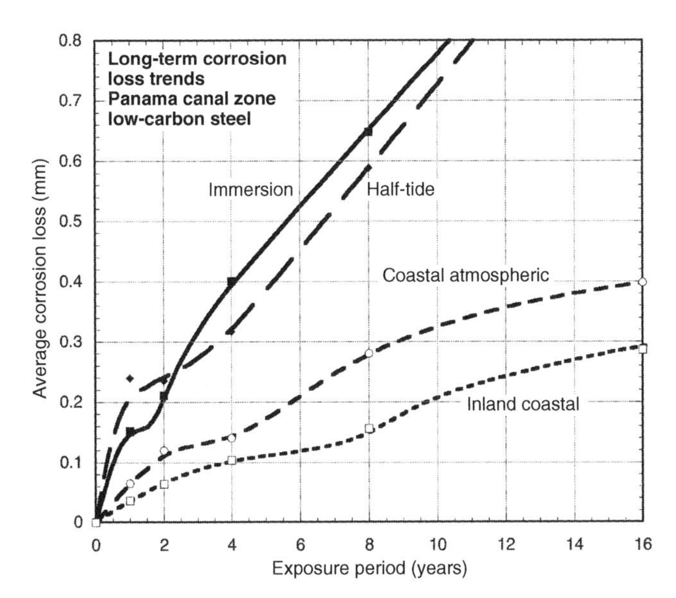


**FIGURE 23.4** Corrosion as measured by mass loss for different exposure zones in the Panama Canal Zone, based on reported data with trends curves added. (Data from Refs [33,34].)

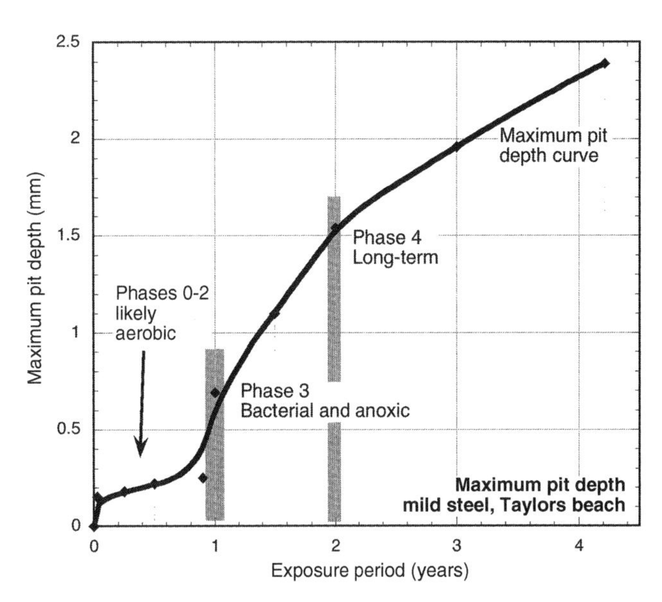


**FIGURE 23.5** Typical growth of pit depth with increased exposure time as inferred from observations of mild steel coupons exposed to coastal seawater [36].

Maximum pit depths for the sites shown in Figure 23.4 have been reported. However, the data trends are inconsistent, sometimes showing pit depths at some exposure periods being (much) lower than what was observed for the preceding period. This cannot be true in actual exposure conditions but is the result of the experimental difficulty of measuring actual (absolute) pit depths relative to the original surface. Usually, all that can be done is to measure the pit depth relative to some part of the remaining surface and then to add an allowance, based on mass loss to estimate the absolute pit depth. This attempts to allow for the plateaus shown in Figure 23.2. For maximum pit depth, Figure 23.5 shows measurements of pit depths corrected to produce estimates of absolute pit depth for mild steel exposed to coastal seawater for up to 4 years.

It is seen that pits develop very soon after first exposure to a maximum pit depth that then changes little for some time (about 8 months in Figure 23.5) after which there is a gradual increase followed by a very considerable increase in pit depth and this continues, at a gradually reducing rate, for several years. Overall, the pattern (but not the relative intensities) is similar to the trends shown in Figure 23.4 for "general" corrosion. However, it will be seen that the increase in the rate of corrosion after time $t_a$ is much greater than that for mass losses shown in Figure 23.4, reflecting the concentrated corrosion inside pits. Similar trends for maximum pit depth have been observed for many other field trials [36].

Because of the close links between pitting and general corrosion outlined earlier, it is not surprising that Figure 23.5 shows the same general characteristic functional form as seen for general corrosion loss as a function of time (Figure 23.4), although as noted, it is much more severe in phases 3 and 4 for pitting than it is for general corrosion. This functional form can be considered as "bimodal." It forms the basis for the approach to modeling the progression of pitting corrosion with increased exposure time. It also has been found to be useful in explaining various practical observations for corrosion loss and for pit depth.

## 23.4 MODEL FOR LONG-TERM GROWTH IN PIT DEPTH

The characteristic functional form for the progression of corrosion and for maximum pit depth seen in Figures 23.4 and 23.5 has been found to occur in many data sets [35,36] and led directly to postulating the bimodal model shown in Figure 23.6, first developed for mass loss as a function of exposure period. It consists of a number of sequential phases, 0–4, each representing a different corrosion rate-controlling mechanism. Modes 0–2 occur before and modes 3 and 4 after time $t_a$ (Figures 23.5 and 23.6).

The corrosion rate-controlling mechanisms in phases in Figures 23.5 and 23.6 may be summarized as follows. Immediately upon first exposure, the corrosion loss trend consists of a very short phase 0. During this time, corrosion initiates and also, for seawater, the metal surface is colonized by biofilm and then by microorganisms. In phase 1, the rate of corrosion is controlled by the rate of diffusion of oxygen from the water or moisture immediately adjacent to the metal surface ("concentration control") while in phase 2 the corrosion rate is controlled by the rate of oxygen diffusion through the increasing thickness of corrosion products on the metal surface. This produces the characteristic attenuation of the rate of corrosion. Eventually, at around $t_a$, it becomes

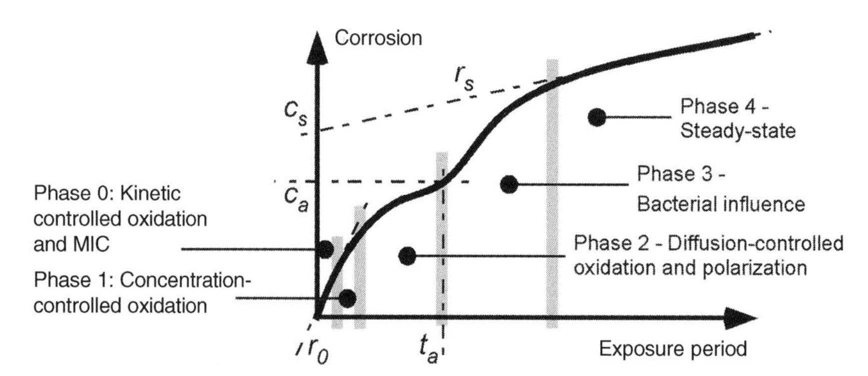


**FIGURE 23.6** Bimodal model for marine corrosion loss (and maximum pit depth) as a function of exposure period. Parameter $t_a$ defines the transition between the modes.

possible for anoxic condition to govern the rate of corrosion, through aggressive pitting corrosion under the rust products or through microbiological activity within the rust layers adjacent to the metal surface [37]. This is phase 3. Phase 4 represents long-term, probably steady state, corrosion condition. The bimodal trend is not confined to the corrosion loss of mild steel in seawater. Without going into detail, it can be stated that it has been found to be relevant for mild steel in fresh- and brackish waters too, and for a range of steel compositions, including some stainless steels. Details are available in the literature.

For pitting, Figure 23.5 shows very clearly that the rate of growth of maximum pit depth in phases 3 and 4 is very much greater than in phase 2 and also that the pits are much deeper than in phase 2, particularly, when compared with the bimodal trend for mass loss (Figure 23.4). As noted, in part this reflects the fact that the deepest pits occur over relatively small areas of a steel surface and that the electron currents that are essential in the corrosion process are concentrated there. In part, also, it is the result of the mechanisms involved in pitting in phase 3 being much more aggressive than earlier, being the result of an autocatalytic reaction process of steel that can occur under the localized anoxic conditions [18]. Such conditions will exist under the corrosion products and as part of the highly irregular corroded surface formed earlier during corrosion in phase 2. In addition, the highly nonhomogeneous nature of the corroded surface also permits, under the anoxic conditions that will prevail, the activity of obligatory anaerobic bacteria such as the sulfate-reducing bacteria, to flourish during phase 3 and add to the severity of local corrosion [38], provided of course that nutrients are available. Experimental evidence for this mechanism is available [31] and evidence from field trials shows correlation between corrosion and pitting and nutrient concentration in the water [27,29,39]. A more detailed description of the reactions and mechanisms involved is available [32].

Mathematical formulations exist for the most important phases in Figure 23.6 [35,40]. Also, the bimodal model has been calibrated to a range of field observations reported in the literature using the results reported for independent field investigations. For example, Figure 23.7 shows the calibrated variation of the model parameter $t_a$ with an average seawater temperature [35].

It is evident that the progression of pit depth $d(t)$ over extended periods of exposure time $t$ does not follow a simple constant "corrosion rate," implying a linear function. It also does not follow, except perhaps for short exposures, the so-called power law, widely used to represent the progression of corrosion [41] as well as the development of pit depth [42]. Typical it has the form

$$\begin{aligned} d(t) &= A.(t - t_i)^B \quad t \geq t_i \\ &= 0 \qquad\qquad\quad t < t_i \end{aligned} \tag{23.1}$$

where, $d(t)$ is a characteristic dimension such as pit depth or diameter and $t_i$ is the time to pit initiation. $A$ and $B$ are constants obtained from curve fitting (1) to experimental data. The power law, or the simplified expression $d(t) = A.t^B$, is relevant for relative periods of exposure, usually much

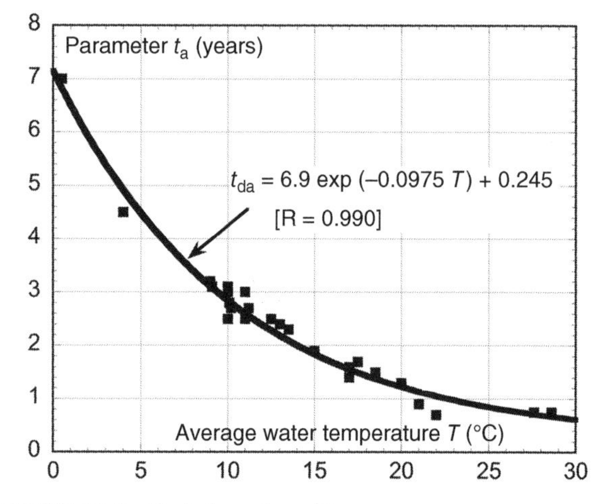


**FIGURE 23.7** Variation of $t_a$ with average seawater temperature based on worldwide data [35].

<1 year and more usually measured only in hours or days. Moreover, the calibrations have been based on laboratory observations [11]. As an approximation Equation (23.1) can be used to represent the progression of maximum pit depth for phases 0–2, but as seen in Figure 23.5 (and much other data besides) the corrosion mechanisms in these phases are very different from those in phases 3 and 4, as clearly seen in the bimodal characteristic. Representing the whole of the data set such as in Figure 23.5 by Equation (23.1) loses the subtleties of the various rate-controlling processes. As will become clear when the statistics of maximum pit depth are considered ahead, this can have considerable ramifications for predicting the likely future maximum pit depths and their uncertainty.

## 23.5 FACTORS INFLUENCING MAXIMUM PIT DEPTH DEVELOPMENT

Apart from structural size, orientation, steel composition, and grain structure, a considerable number of environmental factors may influence the pitting corrosion of structural (carbon) steels. For example, for seawater corrosion, Table 23.1 summarizes the known effects [43]. Schematic representations of the effect of three of the most important effects, nutrient levels in the water, water temperature, and concentration of dissolved oxygen in the water on corrosion, and hence, on pitting, are shown in Figure 23.8. There is insufficient space for a review, but details of these and other effects have been described in the literature.

## 23.6 STRUCTURAL RELIABILITY

### 23.6.1 Formulation

For pipelines, both structural strength capability and containment are important design and operational conditions. Modern trends are to move away from "factors of safety" and to consider instead risks or reliability [1]. The reliability of pipelines can be presented as a conventional structural

**TABLE 23.1 Factors Known to Influence Corrosion of Steel in Seawater**

| Factor | Importance | Factor | Importance |
|---|---|---|---|
| Bacteria/nutrients | High | Pollutants | Varies |
| Biomass | Low | Temperature | High |
| Oxygen supply | Short term | Pressure | No |
| Carbon dioxide | Little | Suspended solids | No |
| Salinity | Not by itself | Wave action | High |
| pH | High | Water velocity | High |
| Carbonate solubility | Low | | |

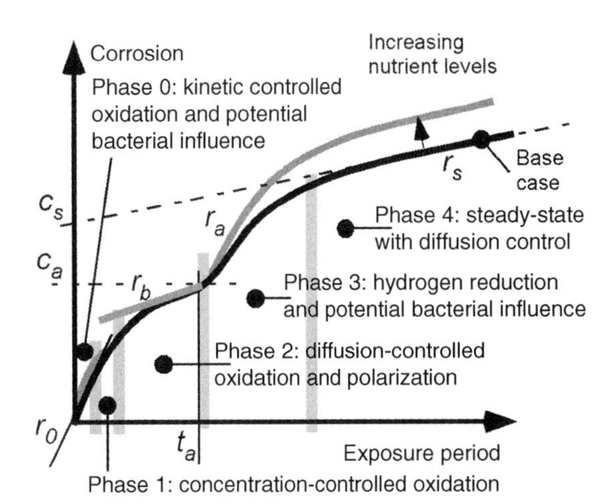


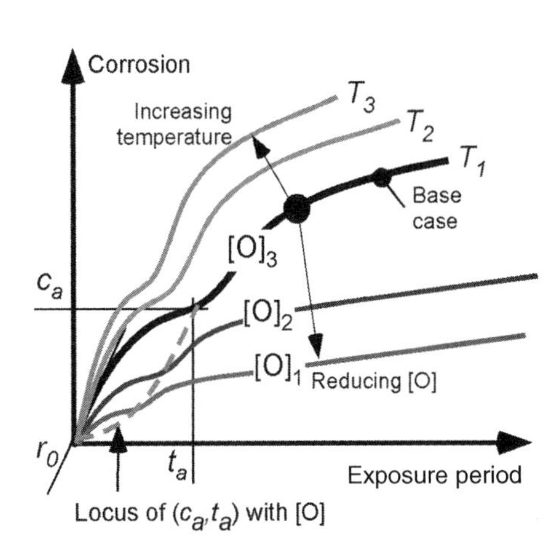


**FIGURE 23.8** Influence of nutrients to support microbiological activity, average water temperature and average oxygen concentration on the "base case" bimodal model for corrosion of mild steel in seawater [37].

reliability problem and represented in the space of both loading and resistance and time $t$ as shown in Figure 23.9. It shows a stochastic loading process system $Q(t)$ and a time-dependent, monotonically decreasing random variable resistance $R(t)$. The joint probability density at any time $t$ of $Q(t)$ and $R(t)$ is denoted $f_{QR}(t)$. Both $Q(t)$ and $R(t)$ can comprise multiple components to represent more complex systems. Of interest is the probability of failure $p_f(t)$ as a function of time $t$. This is a function of the decreasing nature of the resistance or load capacity $R(t)$ of the system.

At any time $t$ conventional structural systems reliability theory defines the probability of failure as [44]

$$p_f|t = \iint_{D_f} f_Q(x).f_R(x)dx \tag{23.2}$$

where, $f_Q(t)$ is the (conditional) probability density function (PDF) for $Q$ and similarly for $R$ at $t$. The failure domain $D_f$

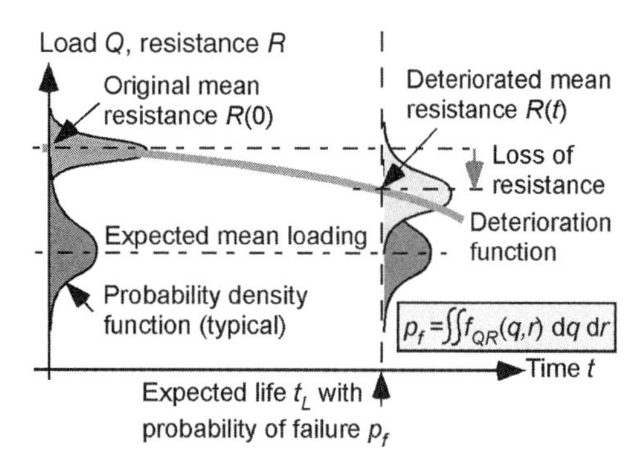


**FIGURE 23.9** Schematic relationship between loading process $Q(t)$, deteriorating resistance $R(t)$ and the time-dependent probability of failure $p_f(t)$.

may be defined through a limit state function (or performance function):

$$G(X) = G(R, Q) = R - Q < 0 \qquad (23.3)$$

where, $X$ collects all the relevant random and other variables, which, for the simplest case shown here, are simply $Q$ and $R$. If one or more of the components of $X$ is time dependent, $G(X)$ also is time dependent. Stochastic variables may be included in a variety of ways but often the simplest approach is to consider them using a combination rule such as Turkstra's [44].

### 23.6.2 Failure Conditions

For any structural reliability problem the crucial matter for definition is the limit state function (or functions) $G(X)$ describing the condition(s) considered to constitute failure of the system. For pressure pipelines supported by soil or within soil or in water this may include [1,45] bursting, bending moment and/or shear load limitations, crushing all with or without corrosion of the pipe wall leading to unacceptable leakage, or simply to corrosion of the pipe wall without other influences. Here, only the latter case will be considered.

The limit state function for wall perforation through pitting corrosion can be considered as the event when the deepest pit is equal to or exceeds the local wall thickness $D$. In practice, it is likely to be a random variable with spatially varying properties. Ignoring these for the present as second-order effects, the limit state function for the case of the deepest pit first penetrating the wall thickness $D$ can be written as

$$G(X, t) = R(t) - Q(t) < 0 = [D - d_{\max}(t)] < 0 \qquad (23.4)$$

where, $d_{max}(t)$ is the maximum pit depth at time $t$. It is best treated as a random variable. Its statistical properties are required to determine (4) and thus to evaluate (2). Since $d_{max}(t)$ is a maximum over many individual pit depths, EV theory provides an appropriate approach to estimating its statistical properties. In the following, attention is given to determining the statistical properties of $d_{max}(t)$ since once these are obtained the structural reliability can be carried out. The details of the calculation procedures are well-established [44] and are not considered further herein.

If failure of the pipeline can be defined only through the probability of pipe-wall perforation by pitting, without consideration of any possible failure mechanism and without consideration of uncertainties in wall thickness $D$, a simpler approach can be adopted. In this case, the probability of failure for any time interval $t$ from zero becomes

$$p_f(t) = P[D - d_{\max}(t)] < 0 \qquad (23.5)$$

## 23.7 EXTREME VALUE ANALYSIS FOR MAXIMUM PIT DEPTH

### 23.7.1 The Gumbel Distribution

As noted earlier, in applications, the maximum pit depth that may occur at some point in time is of interest. There is a long history of assuming that the probability distribution describing the uncertainty in maximum pit depth can be represented by an EV distribution. In fact, pitting corrosion usually is considered one of the prime applications of EV analysis [46]. The EV distribution usually assigned to maximum pit depth is the Gumbel distribution for maxima, following the early work of Aziz [47]. In some cases, other distributions with long "tails" such as the lognormal, Weibull, or Pearson 3 have been used, but unlike the Gumbel distribution, there is no underlying theoretical basis for these choices—their use is purely empirical.

To ascertain whether a set of maximum pit depths is consistent with the Gumbel distribution it is convenient to use a so-called Gumbel plot. This is a graph of cumulative probability against (in this case) maximum pit depth, with the cumulative probability axis distorted in such a way that data that are truly Gumbel distributed plot as a straight, sloping line. Lines with greater slopes denote a greater variability in the data. (The concept may be compared directly with so-called normal probability plots.) Figure 23.10 gives an example of a Gumbel plot.

In Figure 23.10, the right vertical axis shows the conventional cumulative probability (i.e., the probability that the variable of interest is less than the value on the horizontal axis). The left vertical axis shows the standardized variable $w$, defined as $w = (y - u)\alpha$ where $y$ is the maximum pit depth having cumulative distribution function (CDF) $F_Y(y)$ and PDF $F_Y(y)$ each defined by

$$F_Y(y) = F_W[(y - u)\alpha] \text{ with } F_W(w) = \exp(-e^{-w}) \qquad (23.6)$$

$$f_Y(y) = \alpha f_W[(y - u)\alpha] \qquad (23.7)$$

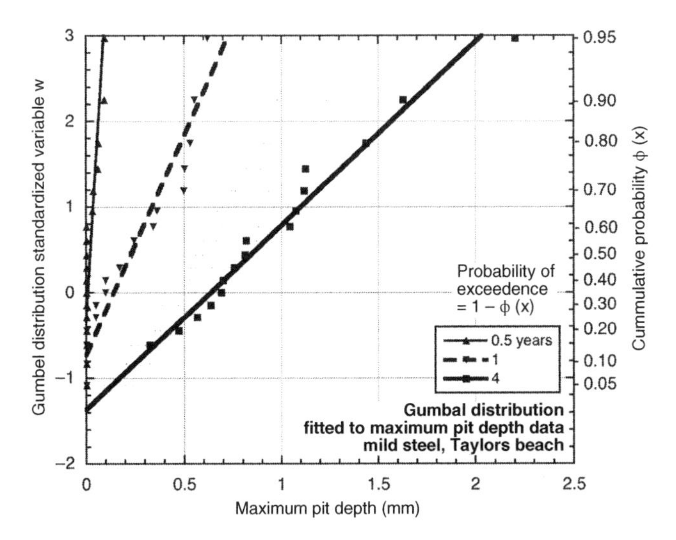


**FIGURE 23.10** Gumbel plot with maximum pit depth data at various exposure periods for 18 surfaces mild steel plate coupons continuously immersed in Pacific Ocean seawater [50].

In the language of EV theory, here, $u$ and $\alpha$ are the "mode" and "slope" of the Gumbel distribution, respectively. They are related to the mean $\mu_Y$ and standard deviation $\sigma_Y$ through $\mu_Y = u + 1.1396/\alpha$ and $\sigma_Y = 0.40825\pi/\alpha$.

Standard techniques are available to estimate the values of $F_Y(y)$ for individual pit depths in a data set so as to plot the data set on a Gumbel plot. The data sets for different exposure periods shown on Figure 23.10 were treated using the rank-order method [46]. In each case shown in Figure 23.10, a straight line can be fitted through the data sets. This indicates that each data set could be considered consistent with, or representable by, the conventional Gumbel distribution. Also, the slopes of the lines (cf. $\alpha$ in Equations 23.6 and 23.7) increase with longer exposure, indicating greater variability in pit depth with increasing exposure period. In the conventional thinking, once the Gumbel distribution has been fitted to the data, it can be used to estimate probability of exceedence $[=1 - \phi(y)]$ of a given depth of pit for longer exposures and for greater pit depths, simply by extrapolation of the upper (right hand) end of each Gumbel line [46].

### 23.7.2 Dependence between Pit Depths

The approach described previously has a long history [47] and has been applied for relatively shallow pits or, similarly, for relatively short-term exposures. Ultimately, it is based on the assumption that each data set is homogeneous, that is, each data point is an independent sample drawn from the same (statistical) population. Closer examination of the pit depth data plotted in Figure 23.10 shows that they do not fit particularly well to the fitted straight line. This may be dismissed as natural variability in data [48] and indeed this is possible. But examination of pitting behavior and pit characteristics against the underlying assumptions for EV theory reveals a different and more likely possibility.

The phases in the bimodal model (Figures 23.6 and 23.8) imply changes in the nature of pitting process from one phase to the next and in particular between the first and the second mode of the bimodal model, that is, between phases 0–2 and 3–4. At the very least, this means that the pit depths generated in the first mode are not from the same statistical population as the pit depths in the second mode. This may be true also for the pit populations generated by each of the phases. Thus, for pit data extending over extended periods of exposure time, that is, with some pit already in phases 3 or 4, there are (at least) two different statistical populations for each data set. EV theory using a particular EV distribution requires all data to come from the one population [46].

EV theory is based on the assumption that each sample observation is an independent statistical event or outcome from the (same) underlying distribution. For pitting, it is very unlikely that the pit depths are statistically independent. This can be expected since the corrosive environment is common to the whole steel surface and the surface itself is largely homogeneous except for microscopic crystal structure and its associated defects, with the latter tending to become irrelevant as corrosion proceeds. Data on correlation between pit depths is scarce, and only recently has it been shown that there is indeed a high degree of correlation for pit depth on larger steel plate surfaces [49]. Fortunately, a degree of dependence between events or outcomes can be tolerated in EV analysis. This occurs under the assumption of so-called asymptotic independence, meaning that the maximum depth pits may be assumed independent as the total number of pits increases to a large number [46]. Presumably, this is because the very deepest pits are considered likely to occur in non-adjacent areas.

### 23.7.3 EV Distribution for Deep Pits

In practice, the pit depths of most interest are the deepest of the deepest and this implies, for extended periods of exposure, pits in phase 3 and 4 (Figure 23.5). If these data, which are all in the second mode, are examined on the Gumbel plot, it is seen that the data set has a distinct curvilinear aspect. Indeed, if the corresponding pit depth data are plotted on a Frechet plot (Figure 23.11), it is seen that they are a reasonable fit to the Frechet lines. Theoretical justification for the use of the Frechet EV distribution has been proposed based on the sharing of nutrients for MIC in phases 3 and 4 [50], although other reasons also are both possible and likely. Irrespective of the theoretical aspects, empirically the Frechet EV distribution appears a better fit to the deep pit data than the Gumbel EV distribution. When the Frechet lines are plotted back on the Gumbel plot they appear as curved lines above the locus AA that is meant to represent the region for phases 3 and 4 of the model (Figure 23.12).

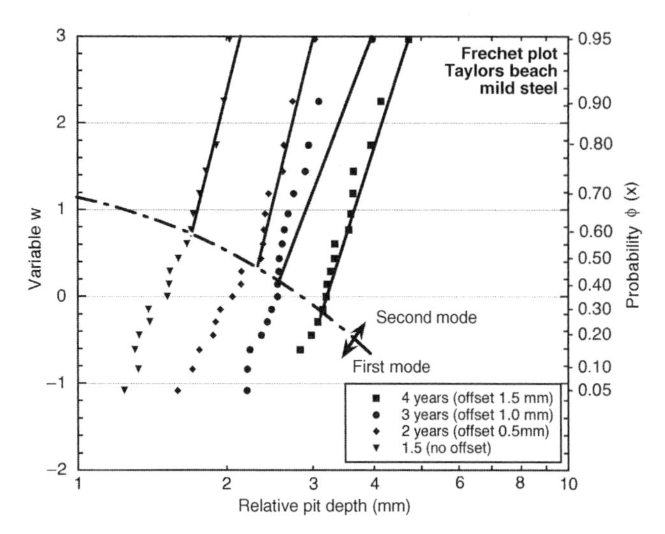


**FIGURE 23.11** Weibull plot showing that the data for the deepest pits in the longer-term exposures tend to be consistent with the Weibull extreme value distribution [50]. The first and second modes refer to these modes in the model for progression of maximum pit depth in Figure 23.5.

### 23.7.4 Implications for Reliability Analysis

Figure 23.12 shows that for exposures >1.5 years and for the deeper pits (i.e., those above and to the right of the line AA), the curved lines representing the Frechet EV distribution are considerably at variance with the straight lines fitted through the data and representing the Gumbel distribution. In particular, it is seen that for a given cumulative probability the deepest pits are much greater for the Frechet compared with the Gumbel distribution. This shows clearly that for the prediction of occurrence of deeper pits still, as could be obtained by extrapolation of the curves shown, the corresponding pits will be much deeper than would be predicted using the Gumbel distribution. Obviously, this can have important practical implications.

## 23.8 PITTING AT WELDS

### 23.8.1 Short-Term Exposures

Welds on steel welded pipelines can be longitudinal or circumferential or both. Typically, buried pipelines are cathodically protected but this may not be the case for steel pipelines used in marine conditions in the oil industry, for an example. In this case, the general and the pitting corrosion at welds is of interest [1,51]. It is known that welding process causes the generation of a heat-affected zone (HAZ) immediately adjacent to the weld itself. The weld may be a hot fusion weld or performed using a weld material different from the parent (pipe) metal (PM). It has been observed that corrosion in the HAZ usually is more severe than that in the weld zone (WZ) and that typically this is more severe than for the patent metal [52]. However, recent observations indicate a more complex situation (Figure 23.13). Field experiments using multiple coupons cut from the longitudinal welded region of industry-standard API Spec 5L X56 longitudinal seam welded pipe (186.3 mm diameter) and exposed to seawater over an extended period show that the relativity changes as corrosion progresses (Table 23.2) [53]. Moreover, the trends show consistency with the trend for deepest pits on plates and with the bimodal model for maximum pit depth, but also appear to be somewhat more complex (Figures 23.5 and 23.6).

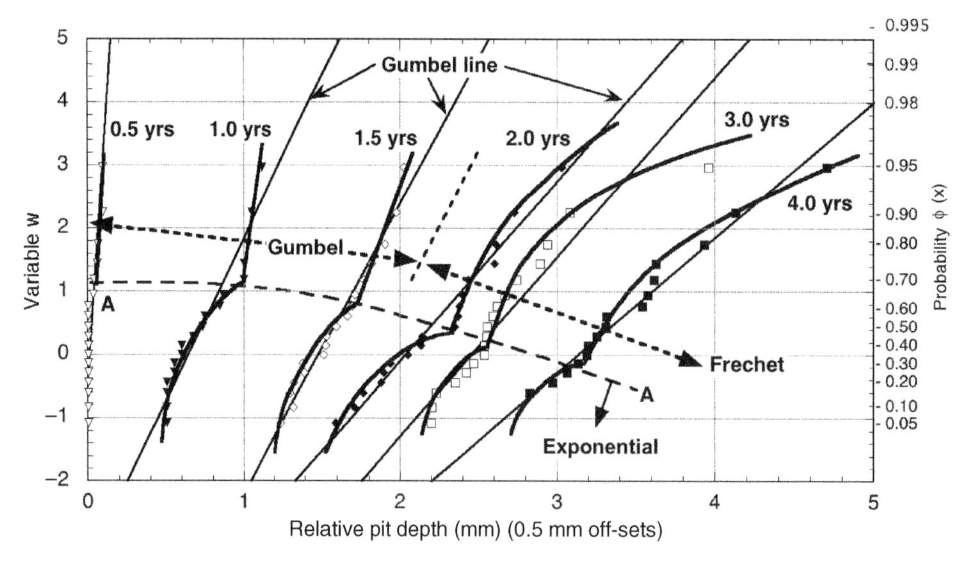


**FIGURE 23.12** Gumbel plot of Figure 12.10 with each data set is offset by 0.5 mm for clarity [50]. The straight light lines through the data are the Gumbel trends, as shown. The upper curved lines (above AA) correspond to the Frechet trends in Figure 23.11 for exposures after about 1.5 years.

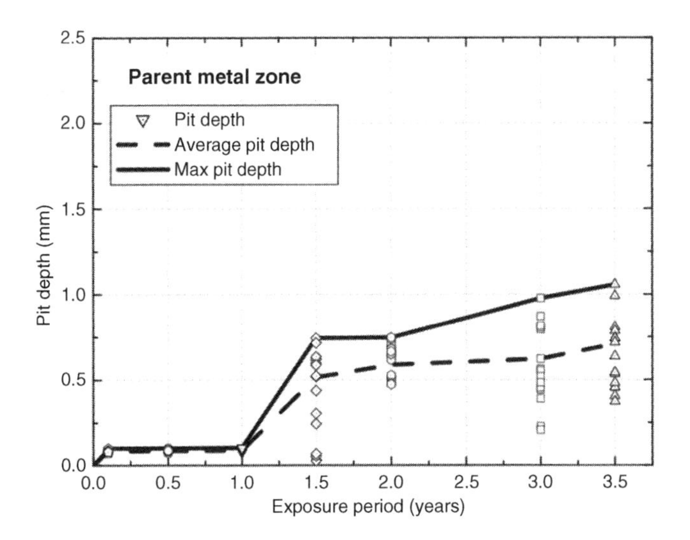


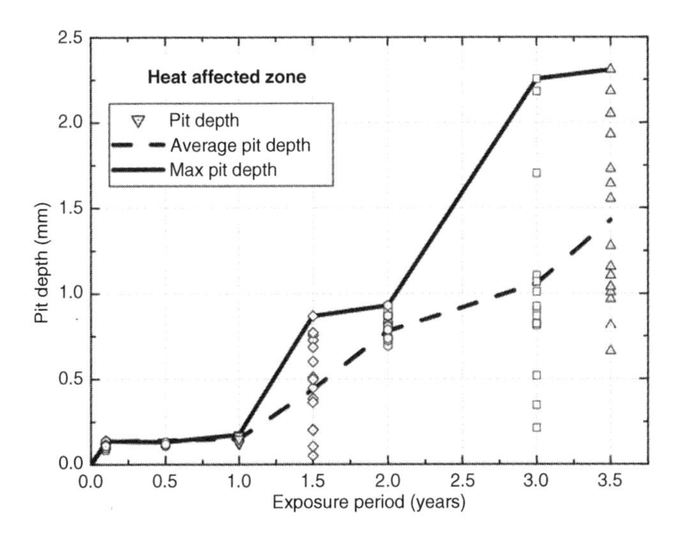


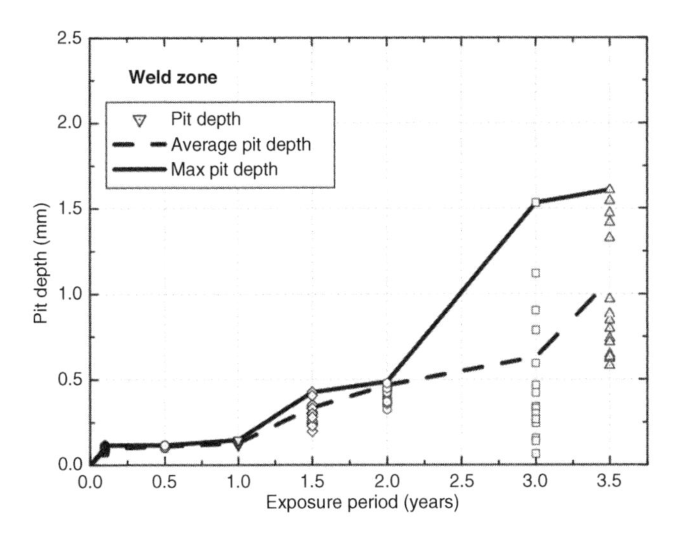


**FIGURE 23.13** Progression of maximum pit depths measured on coupons recovered at the times shown, for each of (a) parent metal, (b) heat-affected zone, and (c) weld zone [54].

**TABLE 23.2 Maximum Pit Depth (mm) in Different Zones After Various Exposure Periods [54]**

| Exposure period (years) | 1 | 1.5 | 3.5 | 33 |
|---|---|---|---|---|
| Parent metal | 0.10 | 0.75 | 1.1 | 4.9 |
| Heat-affected zone | 0.12 | 0.85 | 2.3 | 7.3 |
| Welds zone | 0.10 | 0.40 | 1.6 | 5.1 |

### 23.8.2 Estimates of Long-Term Pitting Development

For pipelines in practice, much longer exposure periods are usual and the information in Figure 23.13 is likely to be of limited practical interest. Pit depth data for welds on pipelines after much longer exposures would be useful. Such information is not readily available. Instead, data for maximum pit depth for welded steel tubular piling exposed to seawater conditions in Newcastle harbor for some 33 years has been reported [54]. The piling was removed because of excessive corrosion at around the low water mark. It is known that corrosion at lower levels, that is, in the immersion zone, is reasonably uniform [8]. In this case, the original thickness of the steel could be estimated quite accurately from the part of the pile exposed in the atmosphere since atmospheric corrosion losses are very much lower than those in the immersion and tidal zones.

Figure 23.14 shows the pit depth data at 33years as well as the data up to 3.5 years for the HAZ. Also shown is the constructed trend for maximum pit depth with time, based on the assumption that the differences in welding and in weld and steel composition between the pipeline steel and the 33-year-old steel tubular were sufficiently smaller to be neglected in a first estimate. There is support for this assumption [8,55]. Table 23.2 shows the maximum pit depths at 33 years compared with those for much lesser exposure periods.

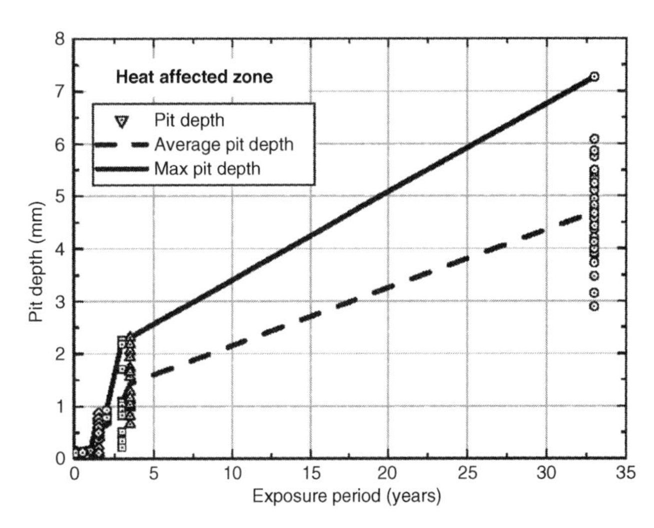


**FIGURE 23.14** Maximum pit depth data at 33 years exposure for welded steel piling in the immersion zone plotted together with data for welded steel pipeline coupons (Figure 23.13b) [54].

It is clear from Figure 23.14 that an average maximum pitting corrosion rate, as would be obtained by dividing the maximum pit depth at 33 years by 33, considerably overestimates the rate of pitting after the first few years. As before, a better estimate of longer-term pit depth development $d(t)$ as a function of exposure time $t$ would be obtained from $d(t) = c_s + r_s.t$ consistent with parameters $c_s$ and $r_s$ in Figures 23.6 and 23.8a. This simplifies the early pit depth development but is more realistic for longer-term exposures than the "average" corrosion rate.

### 23.8.3 EV Statistics for Weld Pit Depth

The statistics of maximum pit depths for each of the three WZs was analyzed using EV theory and using the 15 deepest pit depth observations for each zone at each recovery time point. Figure 23.15 shows the results for the HAZ on a Gumbel plot, with best fit Gumbel lines fitted. As before (see Figure 23.12), it is clear that there is a consistent departure from the best fit Gumbel lines. It is easily shown that this is consistent with the deeper pits following a Frechet EV distribution (not shown).

## 23.9 CASE STUDY—WATER INJECTION PIPELINES

Water injection is a technique to increase the yield of oil fields, usually mature ones, in which water, usually seawater or produced water and sometimes aquifer water is injected into the reservoir. Internal corrosion is a major issue for these pipelines, with potentially severe economic and environmental consequences should failure occur [56]. Depending on the oil field, the pipelines may be 10–20 km long and 200–300 mm in diameter with wall thickness around 40–50 mm. Pipelines typically are constructed from carbon steel. Current industry practice is that cathodic protection is not used. Stainless steel pipelines are not common and have had mixed success [57].

Typically, the injection water is degassed before being pumped under high pressure (around 200 bar) into the pipeline. In all cases, the water is de-aerated, nominally to <20 ppb oxygen concentration. In some cases, oxygen scavengers, scale inhibitors, and corrosion inhibitors are added. This suggests that aggressive corrosion should not be expected but field experience sometimes shows otherwise. Quite different corrosion patterns have been observed. Some pipelines show only moderate corrosion and pitting around the circumference while others show severe corrosion at the so-called 6 o'clock position, that is at the lower level of the inside of the pipeline. Others still have corrosion mainly at the circumferential WZ in the form of severe pitting. Figure 23.16 shows a short extract from a trace generated by intelligent pigging and evidence that in this case severe pitting has occurred at the WZ.

Several causatives have been proposed, including MIC, but there is no clear consensus [3,57]. To combat MIC typically biocides are added to the water at the upstream end of the pipeline, in various dosing regimes. Also, some operators have the pipes pigged, periodically or otherwise. Pigging in this context is the mechanical cleaning of the internal surfaces of the pipe using a multiple set of brushes in a train that is forced down the pipeline. So-called intelligent pigs also measure the internal surfaces geometry from which pitting and corrosion losses can be estimated.

The causes of the different pitting corrosion patterns that have been observed for water injection pipelines are currently under investigation [58]. Similarly, statistical analysis of the

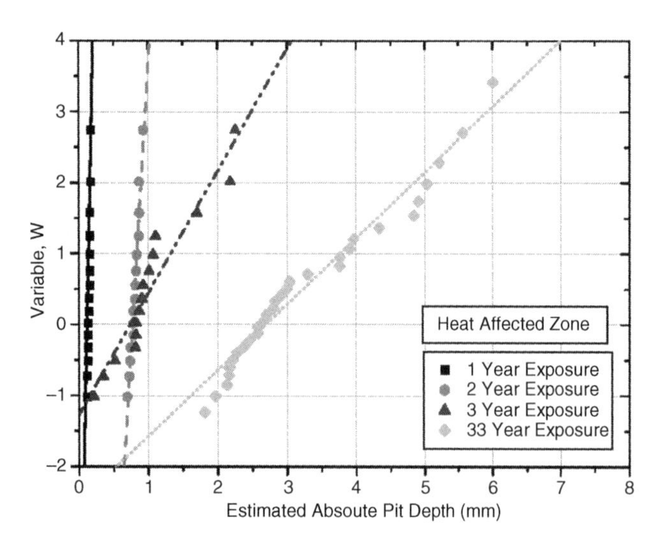


FIGURE 23.15 Gumbel plot showing the data sets for the 15 deepest pit depth readings for each zone and a straight (Gumbel) line fitted to each to represent the best fit Gumbel distribution [54].

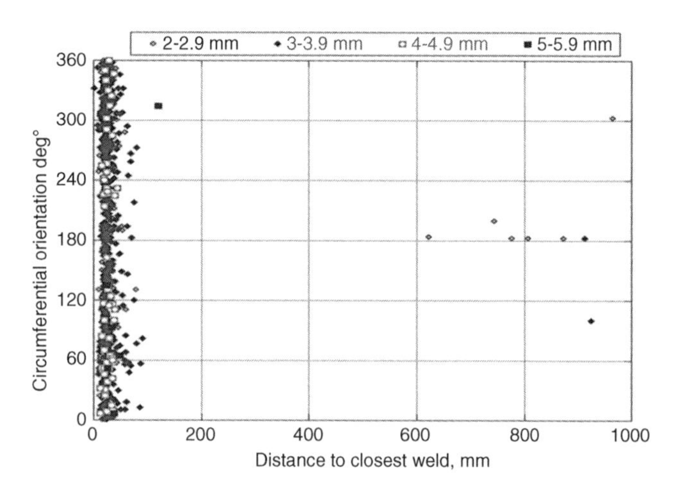


FIGURE 23.16 A 1 m length of the pigging trace report for a pipeline in the North Sea, showing severe pitting corrosion within the weld zone [58].

maximum pit depths for the WZs and elsewhere also are under investigation, together with the trends for pit depth progression. It can be expected, however, that the trends discussed earlier for pitting of plates and pile and external welds exposed to seawater will be revealed also for the internal corrosion of these carbon steel pipelines conveying seawater or produced and aquifer water. The main difference is likely to be that the early "oxic" pitting corrosion phases (0–2) will be much curtailed as a result of deaeration but otherwise the anoxic conditions potentially may lead to MIC if the nutrient supply is sufficient.

## 23.10 CONCLUDING REMARKS

Several implications for pipeline reliability arise from the aforementioned descriptions and analyze. The first is that a simple "corrosion rate," as widely used in industry and in corrosion estimates is not a sufficient model for the actual progression of pit depth development and therefore not a sufficient model for the prediction of likely future pit depth. For deeper pitting, the parameters $r_s$ and $c_s$ (Figure 23.6) should be identified and used to describe the mean value increase of maximum pit depth for longer-term corrosion.

Second, for simple estimation of the likely future probability of a given maximum pit depth being reached, the Frechet distribution should be considered as an appropriate EV distribution, since, for the deepest pits, this is likely to provide a greater probability of a given maximum value of pit depth being exceeded. The fact that the Gumbel EV distribution is not necessarily sufficient was foreshadowed already by Aziz [47] but ignored for over 50 years subsequently. Largely, this must be attributed to the Gumbel distribution often being seen as the "natural" EV distribution to represent the "greatest" or "deepest," whereas, the Frechet distribution is a derived result [50]. It is not an obvious choice from empirical data alone. This highlights one of the reasons for developing models based on theory—empirical models can only reflect the data and therefore do not add new or additional information.

More generally, seawater is a complex mix of many different chemicals and usually is considered one of the most hostile environments for structural steel. For that reason, a good understanding of the factors involved in the corrosion, and in particular pitting, is essential in managing the performance and continued safety of steel infrastructure exposed to marine conditions. The present chapter has outlined some of the current outcomes of research in this area, focusing mainly on engineering rather than corrosion science investigations. Largely, this is driven by the needs of industry and by the need to develop tools and techniques that can be applied in the offshore, shipping, and coastal environment engineering industries. As noted, these are enhanced when based on theory rather than data alone.

## ACKNOWLEDGMENTS

Much of the research reported herein was and continues to be supported financially by the Australian Research Council and various schemes, including the ARC Professorial Fellowship scheme.

## REFERENCES

1. Palmer, A.C. and King, R.A. (2008) *Subsea Pipeline Engineering*, 2nd ed., PennWell Corporation, Tulsa, OK.
2. Fontaine, E., Potts, A.E., Kai-tung, M., Arredondo, A.L., and Melchers, R.E. (2012) SCORCH JIP: examination and testing of severely-corroded mooring chains from West Africa, Proceedings of Offshore Technology Conference, April 30–May 3, 2012, Houston, TX, Paper OTC 23012.
3. Heidersbach, K.L. and van Roodselaar, A.C. (2012) Understanding, preventing, and identification of microbial induced erosion-corrosion (channelling) in water injection pipelines. Proceedings of NACE International Corrosion Conference and Expo 2012, Paper No. C2012-0001221.
4. Paik, J.K. and Melchers, R.E. (editors) (2008) *Condition Assessment of Aged Structures*, Woodhead Publishing Ltd., Cambridge, UK.
5. Melchers, R.E. (2005) The effect of corrosion on the structural reliability of steel offshore structures. *Corrosion Science*, 47 (10), 2391–2410.
6. Melchers, R.E. (2013) Long-term corrosion of cast irons and steel in marine and atmospheric environments. *Corrosion Science*, 68, 1–292.
7. Burstein, G.T., Liu, C., Souto, R.M., and Vines, S.P. (2004) Origins of pitting corrosion. *Corrosion Engineering, Science and Technology*, 39 (1), 25–30.
8. Evans, U.R. (1960) *The Corrosion and Oxidation of Metals: Principals and Practical Applications*, St. Martin's Press Inc Publisher, Edward Arnold Ltd., New York, NY.
9. Butler, G., Stretton, P., and Beynon, J.G. (1972) Initiation and growth of pits on high-purity iron and its alloys with chromium and copper in neutral chloride solutions. *British Corrosion Journal*, 7 (7), 168–173.
10. Uhlig, H.H. and Revie, R.W. (1985) *Corrosion and Corrosion Control*, 3rd ed., John Wiley & Sons, Inc., New York, NY.
11. Szklarska-Smialowska, Z. (1986) *Pitting Corrosion of Metals*, NACE International, Houston, TX.
12. Sharland, S.M. and Tasker, P.W. (1988) A mathematical model of crevice and pitting corrosion—I. The physical model. *Corrosion Science*, 28 (6), 603–620.
13. Shreir, L.L. (1994) Localised corrosion. In: Shrier, L.L., Jarman, R.A., and Burstein, G.T. (editors), *Corrosion*, 3rd ed., Butterworth-Heinemann, Oxford, UK, Vol. 1, pp. 1:151–1:212.
14. Mercer, A.D. and Lumbard, E.A. (1995) Corrosion of mild steel in water. *British Corrosion Journal*, 30, 43–55.
15. Brown, T.L. and Lemay, H.E. (1981) *Chemistry: The Central Science*, 2nd ed., Prentice-Hall, Englewood Cliffs, NJ.

16. Jones, D. (1996) *Principles and Prevention of Corrosion*, 2nd ed., Prentice-Hall, Upper Saddle River, NJ.
17. Burstein, G.T. (1994) Passivity and localised corrosion. In: Shrier, L.L., Jarman, R.A., and Burstein, G.T. (editors), *Corrosion*, 3rd ed., Butterworth-Heinemann, Oxford, UK, Vol. 1, pp. 1:118–1:150.
18. Wranglen, G. (1974) Pitting and sulfphide inclusions in steel. *Corrosion Science*, 14, 331–349.
19. Homer, C.E. (1934) The influence of non-metallic inclusions on the corrosion of mild steel. Second Report of the Corrosion Committee, Iron and Steel Institute, London, 225–240.
20. Mears, R.B. (1935) Metallurgical factors influencing the probability of corrosion of iron and steel. Third Report of the Corrosion Committee, Iron and Steel Institute, London, 1935, 111–112.
21. Evans, U.R. (1968) *The Corrosion and Oxidation of Metals, First Supplementary Volume*, Edward Arnold, London, UK, p. 193.
22. Phull, B.S., Pikul, S.J., and Kain, R.M. (1997) Seawater corrosivity around the world: results from five years of testing. In: Kain, R.M. and Young, W.T. (editors), *Corrosion Testing in Natural Waters: Second Volume, ASTM STP 1300*, American Society for Testing and Materials, Philadelphia, PA, pp. 34–73.
23. Jeffrey, R. and Melchers, R.E. (2007) The changing topography of corroding mild steel surfaces in seawater. *Corrosion Science*, 49, 2270–2288.
24. Little, B.J. and Lee, J. (2007) *Microbiologically Influenced Corrosion*, Wiley-Interscience, New York, NY.
25. Lee, W., Lewandowski, Z., Nielsen, P.H., and Hamilton, W.A. (1995) Role of sulfate-reducing bacteria in corrosion of mild steel: a review. *Biofouling*, 8, 165–194.
26. Cragnolino, G. and Tuovinen, O.H. (1984) The role of sulphate-reducing and sulphur-oxidizing bacteria on the localized corrosion of iron-base alloys—a review. *International Biodeterioration*, 20 (1), 9–26.
27. Melchers, R.E. (2007) The influence of seawater nutrient content on the early immersion corrosion of mild steel—1 empirical observations. *CORROSION (NACE)*, 63 (1), 318–329.
28. Jeffrey, R. and Melchers, R.E. (2003) Bacteriological influence in the development of iron sulphide species in marine immersion environments. *Corrosion Science*, 45 (4), 693–714.
29. Melchers, R.E. (2005) Effect of nutrient-based water pollution on the corrosion of mild steel in marine immersion conditions. *CORROSION*, 61, 237–245.
30. Melchers, R.E. (2007) The effects of water pollution on the immersion corrosion of mild and low alloy steels. *Corrosion Science*, 49 (8), 3149–3167.
31. Jeffrey, R. and Melchers, R.E. (2010) The effect of microbiological involvement on the topography of corroding mild steel in coastal seawater. Proceedings of NACE Conference, San Antonio, TX, Technical Symposium TEG 187X, Paper No. 10224.
32. Melchers, R.E. (2012) The relative influence of microbiological and abiotic processes in modelling longer-term marine corrosion of steel. Proceedings, Eurocorr2012, Istanbul.
33. Southwell, C.R., Forgeson, B.W., and Alexander, A.L. (1958) Corrosion of metals in tropical environments—part 2—atmospheric corrosion of ten structural steels. *CORROSION*, 14 (9), 53–59.
34. Forgeson, B.W., Southwell, C.R., and Alexander, A.L. (1960) Corrosion of metals in tropical environments—part 3—underwater corrosion of ten structural steels. *CORROSION*, 16 (3), 105t–114t.
35. Melchers, R.E. (2003) Modeling of marine immersion corrosion for mild and low alloy steels—part 1: phenomenological model. *CORROSION*, 59 (4), 319–334.
36. Melchers, R.E. (2004a) Pitting corrosion of mild steel in marine immersion environment—1: maximum pit depth. *CORROSION (NACE)*, 60 (9), 824–836.
37. Melchers, R.E. and Jeffrey, R. (2008) The critical involvement of anaerobic bacterial activity in modelling the corrosion behaviour of mild steel in marine environments. *Electrochimica Acta*, 54, 80–85.
38. Hamilton, W.A. (1985) Sulphate-reducing bacteria and anaerobic corrosion. *Annual Review of Microbiology*, 39, 195–217.
39. Melchers, R.E. and Jeffrey, R. (2012) Corrosion of long vertical steel strips in the marine tidal zone and implications for ALWC. *Corrosion Science*, 6, 26–36.
40. Melchers, R.E. and Wells, P.A. (2006) Models for the anaerobic phases of marine immersion corrosion. *Corrosion Science*, 48 (7), 1791–1811.
41. Tidblad, J., Mikailov, A.A., and Kucera, V. (2000) Application of a model for prediction of atmospheric corrosion in tropical environments, In: *Marine Corrosion in Tropical Environments, ASTM STP 1399*, American Society for Testing and Materials, West Conshohocken, PA, pp. 18–32.
42. Engelhardt, G. and Macdonald, D.D. (1998) Deterministic prediction of pit depth distribution. *CORROSION*, 54, 469–479.
43. Schumacher, M. (editor) (1979) *Seawater Corrosion Handbook*, Noyes Data Corporation, Park Ridge, NJ.
44. Melchers, R.E. (1998) *Structural Reliability Analysis and Prediction*, John Wiley & Sons, Inc., Chichester, UK.
45. Ahammed, M. and Melchers, R.E. (1997) Probabilistic analysis of underground pipelines subject to combined stresses and corrosion. *Engineering Structures*, 19 (12), 988–994.
46. Galambos, J. (1987) *The Asymptotic Theory of Extreme Order Statistics*, 2nd ed., Krieger, Malabar, FL.
47. Aziz, P.M. (1956) Application of the statistical theory of extreme values to the analysis of maximum pit depth data for aluminum. *CORROSION (NACE)*, 12 (10), 495t–506t.
48. Melchers, R.E. (2007) Reply to discussion: statistical characterization of pitting corrosion—part1 1: data analysis and part 2: probabilistic modeling for maximum pit depth, A Valor, D Rivas, F Caleyo and JM Hallen. *CORROSION (NACE)*, 63 (2), 112–113.
49. Melchers, R.E., Ahammed, M., Jeffrey, J., and Simundic, G. (2010) Statistical characterization of corroded steel surfaces. *Marine Structures*, 23, 274–287.
50. Melchers, R.E. (2008) Extreme value statistics and long-term marine pitting corrosion of steel. *Probabilistic Engineering Mechanics*, 23, 482–488.

51. Bai, Y. and Bai, Q. (2005) *Subsea Pipelines and Risers*, Elsevier, Oxford, UK.
52. Eid, N.M.A. (1990) Localized corrosion at welds in structural and stainless steel under marine conditions part 1. Proceedings of the International Offshore Mechanics and Artic Engineering Symposium, 3, pp. 647–657.
53. Chaves, I.A. and Melchers, R.E. (2011) Pitting corrosion in pipeline steel welds zones. *Corrosion Science*, 53, 4026–4032.
54. Chaves, I.A. and Melchers, R.E. (2012) External corrosion of carbon steel pipeline weld zones. Proceedings of ISOPE, Rhodes, Paper TPC-0628.
55. Melchers, R.E. (2005c) Effect of alloying on maximum depth of pits in mild steel in marine immersion environments. *CORROSION (NACE)*, 61 (4), 355–363.
56. Latifi, L., Berry, M., and Maxwell, S. (2007) Mitigation of microbiologically influenced corrosion in water injection flowlines. Proceedings of NACE International Corrosion 2007 Conference and Expo, March11–15, Nashville, Tennessee, Paper No. 07511.
57. Stott, J.F.D. (2012) Implementation of nitrate treatment for reservoir souring control: complexities and pitfalls. Proceedings of SPE International Conference and Exhibition on Oilfield Corrosion, May 28–29, 2012, Aberdeen, UK.
58. Comanescu, I., Melchers, R.E., and Taxen, C. (2012) Correlation between MIC and water quality, pigging frequency, and biocide dosing in oil field water injection pipelines. Proceedings of Eurocorr2012, Istanbul, RSP2, Oil and Gas.

# 24

# SULFIDE STRESS CRACKING

RUSSELL D. KANE
*iCorrosion LLC, Houston, TX, USA*

## 24.1 INTRODUCTION

Sulfide stress cracking (SSC) is just one of several forms of hydrogen-related deterioration that can result in pipelines, facilities piping and equipment, and oil country tubulars from exposure to sour (wet $H_2S$) service conditions. These are environments that involve exposure to aqueous environments containing hydrogen sulfide ($H_2S$) as found in oil and gas production, gas and chemical processing, and petroleum refining. As a result of this multiplicity of sour cracking mechanisms, it is critical to understand the differences between SSC and the other forms of wet $H_2S$ cracking. While the parameters that control all wet $H_2S$ cracking mechanisms are generally the same, the influence of specific environmental and metallurgical parameters may vary depending on the cracking mechanisms under consideration.

SSC is particularly insidious form of wet $H_2S$ cracking that can lead to embrittlement with loss of ductility and brittle fracture, formation of multiple propagating crack fronts, and sudden catastrophic failure. Under severe situations with susceptible materials such as high strength or hardness steels, SSC can result in high crack growth rates of between $10^{-8}$ and $10^{-5}$ mm/s for low-alloy steels [1], where $10^{-5}$ mm/sec was demonstrated for hard welds (>HRC 30) in carbon steel [2].

## 24.2 WHAT IS SULFIDE STRESS CRACKING?

SSC is a form of hydrogen stress cracking (HSC), also referred to as hydrogen embrittlement cracking (HEC). It typically occurs in high-strength steels [3] and localized hard zones in the pipe [4], particularly, in weldments and the heat-affected zone (HAZ). The proclivity for SSC in the area of the weld results from the fast cooling rates commonly applied in joining of pipelines and the associated metallurgical transformations these cooling rates produce in the steel. This can be in either the longitudinal or helical seam weld made during manufacturing of the pipe or in the girth (circumferential) welds made during laying and fabricating the pipeline in the field. SSC can also occur in high-strength steels and other materials with ultimate tensile strength and hardness that are in excess of accepted maximum values. This is an important concept in prevention of SSC. While most engineers focus on obtaining materials that meet minimum strength or hardness requirements, the materials that usually result in failure by SSC are those on the maximum end of the specification that are often left unchecked.

## 24.3 BASICS OF SULFIDE STRESS CRACKING IN PIPELINES

SSC is produced in a susceptible material by the simultaneous action of residual and/or applied tensile stresses and internal hydrogen atoms produced by corrosion. In the case of steel, corrosion occurs under generally acidic conditions ($pH < 7$) by the sulfide corrosion reactions shown as follows [5]:

$$\text{Anode}: Fe \rightarrow Fe^{+2} + 2e\text{-}$$

*Oil and Gas Pipelines: Integrity and Safety Handbook*, First Edition. Edited by R. Winston Revie.

$$\text{Cathode}: H_2S + H_2O \rightarrow H^+ + HS^- + H_2O$$

$$HS^- + H_2O \rightarrow H^+ + S^{-2} + H_2O$$

$$\text{Net reaction}: Fe + H_2S + H_2O \rightarrow FeS + 2H^0$$

In sulfide corrosion, FeS is a solid compound that is insoluble in aqueous solutions and commonly precipitates on exposed metal surfaces as a scale. In practice, as shown above, iron sulfide corrosion products are based on FeS, but can take other forms such as $FeS_2$, $Fe_7S_8$, or $Fe_8S_9$ depending on the pH, $H_2S$ partial pressure, and oxidizing potential of the environment. As a consequence, sulfide corrosion products can vary in their morphology and the degree of their protectiveness.

As also shown in the aforementioned equations, one of the natural by-products of the sulfide corrosion reactions is atomic hydrogen ($H^0$) that forms on the surface of the material at local cathodic sites. Atomic hydrogen is a very small atom (with a diameter just over 1 Angstrom) and under the right conditions can be adsorbed onto the metal where it can readily diffuse even at near room temperature.

In most acidic corrosive environments, the vast majority of corrosion-generated atomic hydrogen recombines on the metal surface to form molecular hydrogen gas ($H_2$), which harmlessly bubbles off before it has a chance to be absorbed into the metal. However, in the presence of sulfur species such as $H_2S$ the kinetics of recombination of atomic hydrogen to molecular hydrogen can be significantly retarded resulting in increased absorption of atomic hydrogen by the metal. This characteristic of sour environments results because sulfur is a highly effective hydrogen recombination "poison", which acts similarly to species such as Sn, Pb, Sb, and P, which slow recombination, and thus, increase the efficiency of hydrogen charging into the metal.

It is accepted that SSC is a form of HSC in most steels. The role of $H_2S$ in this phenomenon is generally considered to be two-fold: (1) it increases the rate of corrosion of steel in aqueous solutions through its property like $CO_2$ of being an acid gas, where it has the ability to form a weak acid solution when dissolved in water (see Figure 24.1 [6]) and (2) it poisons the hydrogen recombination/evolution reaction as discussed previously leading to higher than normal levels of hydrogen flux in steel. Consequently, $H_2S$ results in an increase in the severity of hydrogen charging and HSC versus that observed in similar acid solutions without $H_2S$.

A comprehensive mechanistic SSC study was conducted to identify specifically the mechanism of SSC in steel. Tests were conducted both with and without $H_2S$ under conditions of applied potential to produce equivalent hydrogen permeation currents in both environments. As shown in Figure 24.2, susceptibility to SSC was found to be directly related to the amount of hydrogen available, not the presence or absence of $H_2S$. In an $H_2S$-containing environment, the absorption of atomic hydrogen into the metal is enhanced by the poisoning effect of S-containing species indicated previously. Therefore, hydrogen charging conditions of a particular magnitude were produced more readily in the $H_2S$-containing environment [7].

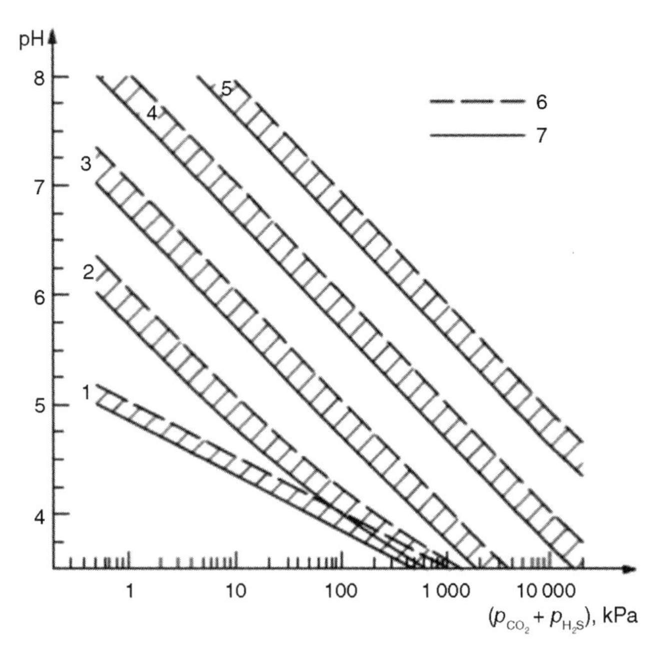


**FIGURE 24.1** Solution pH versus acid gas ($H_2S + CO_2$) partial pressure also showing the effect of bicarbonate that is commonly present in the aqueous solution; bicarbonate levels: 1–0 meq/l, 2–0.1 meq/l, 3–1 meq/l, 4–10 meq/l, 5–100 meg/l; temperature: 6–100 °C, 7–20 °C [6]. (From ©NACE International 2009.)

Once in the steel, atomic hydrogen readily diffuses at near ambient temperatures to sites of high internal stress (i.e., grain boundaries, inclusions, and regions of triaxial stress at internal or surface stress concentrations). In the presence

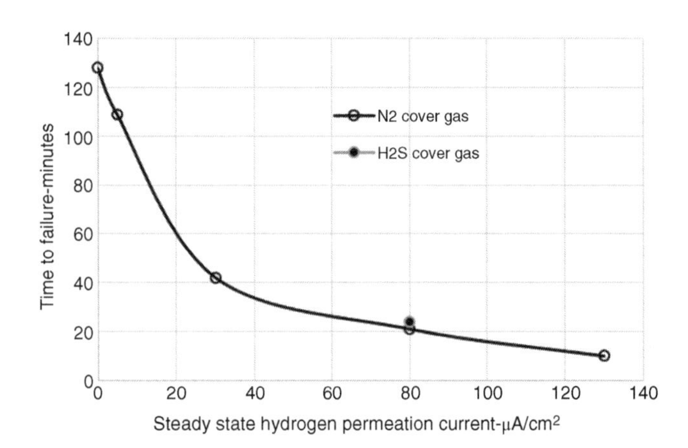


**FIGURE 24.2** HSC susceptibility of C–Cr–Mo low-alloy steel (HRC 32) versus hydrogen permeation current for cathodic charging in solutions purged with nitrogen compared with a solution with $H_2S$ purge [7]. (From ©NACE International 1984.)

of tensile stress, these locations become sites for embrittlement and the initiation of SSC.

## 24.4 COMPARISON OF SSC TO OTHER SOUR CRACKING MECHANISMS

As mentioned previously, SSC is one of the many environmental cracking mechanisms resulting in pipeline steels exposed to aqueous sour service environments. The other mechanisms are discussed elsewhere in this book and include: hydrogen blistering, hydrogen-induced cracking (HIC), stress-oriented hydrogen-induced cracking (SOHIC) and stress corrosion cracking (SCC). Table 24.1 provides a useful comparison of the features of these cracking processes.

One of the interesting aspects in the literature on SSC is the reference to two categories of SSC; referred to as SSC Type I and Type II starting in 1979 [8] and more recently reviewed [9].

Type II is the most common form of SSC, and as mentioned previously, herein, it is associated with HSC mechanisms found in the results of laboratory studies and field performance. Failures are often associated with transgranular cleavage in moderate strength steels (550–690 MPa YS; HV 248–300) or intergranular fracture in higher strength/hardness steels (>690 MPa YS; HV>300).

In contrast, as described in the literature [9], Type I SSC involves two stages: the first being the initiation of small blister cracks (HIC) oriented parallel to the applied stress and rolling direction while the second stage is the linking together of the blister cracks by short segments of transgranular cleavage by an HEC mechanism. This latter type of SSC has been typically observed in lower strength steels (<550 MPa YS; <HV 248) and is the same as described and defined in Table 24.1 as SOHIC. As can be seen from this description, it is a hybrid mechanism involving both HIC (due to hydrogen pressure build up) and SCC (HSC embrittlement process) in plate steels. It is important to realize that it

**TABLE 24.1 Forms of Environmental Cracking in Aqueous Sour (wet $H_2S$) Service**

| Cracking Mechanism | Phenomenon | Materials | Stress | Method to Prevent |
|---|---|---|---|---|
| SSC | Embrittlement by atomic hydrogen at areas of high internal or surface tensile stress (e.g., grain boundaries, notches, etc.) | Usually associated with high strength (>550–690 MPa YS) steels or high hardness (>HRC 22) areas; can occur at lower HRC values if cold work is present | Tensile stress required | Limiting YS or HRC; steels with accelerated cooling or QT better than NT or hot rolled; limit design stress |
| Blistering | Accumulation of atomic hydrogen at weak internal interfaces (e.g., inclusions) and the formation of hydrogen gas pressure, which inflates blisters | Lower strength pipeline steels (<550 MPa YS; <HRC22) with centerline segregation or high impurity levels resulting in massive sulfide inclusions | No applied stress required | Use of steels with lower impurities (especially sulfur) |
| HIC | The link up of small to moderate sized internal hydrogen blisters in the form of stepwise blister cracks | Low strength pipeline steels (<550 MPa YS; <HRC 22) with intermediate sulfur levels especially made with extensive hot rolling | No applied stress required | Use of steels with lower impurities ($S < 0.006\%$; $P < 0.015\%$) with Ca-treatment for sulfide shape control |
| SOHIC | The linear through thickness array of small blister cracking that link by brittle cleavage under the influence of high applied or residual tensile stress | Low strength and often in low sulfur, HIC-resistant steels (<550 MPa YS; <HRC 22) with low-sulfur levels | Usually requires high-hydrogen charging conditions and high-residual/applied tensile stress; exacerbated by severe notch or weld toe geometries | Use of HIC resistant steels with ultralow sulfur <0.002 with Ca-treatment (caution not to over Ca-treatment); QT > NT > hot rolled. Reduce stress concentrations and hydrogen charging severity; replace with CRA-clad steel |
| SCC | Brittle environmental cracks that form and propagate from the action of local anodic sites and tensile stress | In Ni-containing steels in aqueous sour service; at higher temperatures when $H_2S$ combined with chloride results in SCC in stainless and Ni-base alloys | Tensile stress required | Keep Ni content of steels <1%; alloy stainless materials with sufficient Cr, Ni, Mo, N for resistance to local corrosion, pitting, SCC |

can also be observed in pipe made from rolled plate, particularly, if high residual tensile stress or mechanical stresses co-exist with severe service conditions of high hydrogen charging.

## 24.5 INFLUENCE OF ENVIRONMENTAL VARIABLES ON SSC

### 24.5.1 Availability of Liquid Water

One of the premises on which the occurrence of SSC is the requirement for the presence of a conductive electrolyte that can sustain electrochemical corrosion reactions. In many cases, one of two conditions prevail in oil and gas environment service: (1) there is an oil wetting condition, whereby, water is excluded from the metal surface, or (2) water is contained totally in the vapor phase with no liquid (free) water present in the system for the temperature and pressure that exist. Under these two conditions, there is effectively no corrosion, and thus, no possibility of sulfide corrosion or SSC. In some systems (e.g., highly processed hydrocarbon streams), where the environment is precisely controlled, the aforementioned conditions may allow the use of metals and alloys that would normally be as risk of SSC. However, if conditions are more variable and can change leading to a reversal in the water wetting tendencies on the metal surface from condensation or varying ratios of water and hydrocarbon, it is important to have selected materials that can handle this situation without suffering the ill effects of SSC.

### 24.5.2 pH and $H_2S$ Partial Pressure

In most service environments involving exposure to $H_2S$ in the presence of free liquid water, tendencies toward SSC of pipeline steels generally increases with decreasing pH of the aqueous environment between pH 6.5 and 3. The reason for this behavior is the increased availability of hydrogen ions in aqueous media with lower pH values. It has also been observed that SSC of pipeline steels increases with increasing $H_2S$ partial pressure (to a point) a situation that results in a depression in pH, which increases the hydrogen flux into the steel, and thus, promotes SSC. High $CO_2$ partial pressure in combination with $H_2S$ also helps to depress the pH of the aqueous environment that further increases hydrogen ion availability and the associated hydrogen flux (see Figure 24.3) [10].

As also shown in Figure 24.3, at still higher $H_2S$ partial pressures, hydrogen permeation rates can actually decrease from the peak levels seen at lower $H_2S$ partial pressures. This effect results in cases where a more stable sulfide film is formed on the steel surface that reduces the higher hydrogen charging rates observed at lower $H_2S$ partial pressures. Therefore, it is important to assess not just $H_2S$ partial

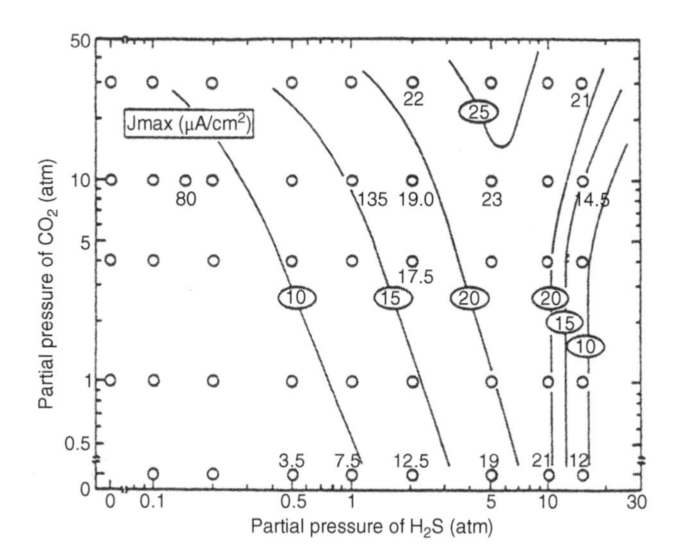


**FIGURE 24.3** Peak hydrogen permeation rate ($J_{max}$) values indicated for various conditions of $H_2S$ and $CO_2$ partial pressures (in synthetic seawater) [10]. Note: 1 atm = 102 kPa. (From ©NACE International 1996.)

pressure, but it should be reviewed in combination with other variables. Standard laboratory tests have been developed for SSC, which evaluate materials for cracking in solutions adjusted to pH values in the range of 3–5.5 with buffering agents such as bicarbonate or acetate. While many tests are run at ambient pressure with 15 psia $H_2S$ partial pressure, $H_2S$ partial pressures in tests are now routinely varied from 0.1 to over 100 kPa to simulate intended service conditions. For more information on testing for SSC, see NACE MR0175/ISO 15156 Part 2 [6] and NACE TM0177 methods A—D [11].

Figure 24.4 shows commonly encountered sour service conditions as depicted in NACE MR0175/ISO15156 Part 2. This figure is commonly known as the "$H_2S$ serviceability diagram" and demonstrates the strong relationship between pH and $H_2S$ partial pressure when assessing SSC in service conditions with region 3 being more severe than region 2, which in turn is more severe than regions 1 and 0. To evaluate resistance to SSC of metals with laboratory tests, this industry standard requires applied tensile stress values in the range of 80–100% Actual Yield Strength (AYS) using uniaxial tension specimens, C-rings or four point bend specimens. Commonly used solutions are the standard NACE TM0177 Solution A (5% NaCl, 0.5% acetic acid saturated with $H_2S$) at a nominal pH 3 or a simulated service environment of 5% NaCl + 0.4% CH3COONa with the pH adjusted between 3.5 and 5.5 to required value from service conditions.

Hydrogen permeation in steels is highly temperature-dependent phenomenon and, as such, so are SSC tendencies. As shown in Figure 24.5, SSC susceptibility in steels reaches a maximum (i.e., minimum time to failure) around room temperature (23–25 °C) [12]. This interesting but well

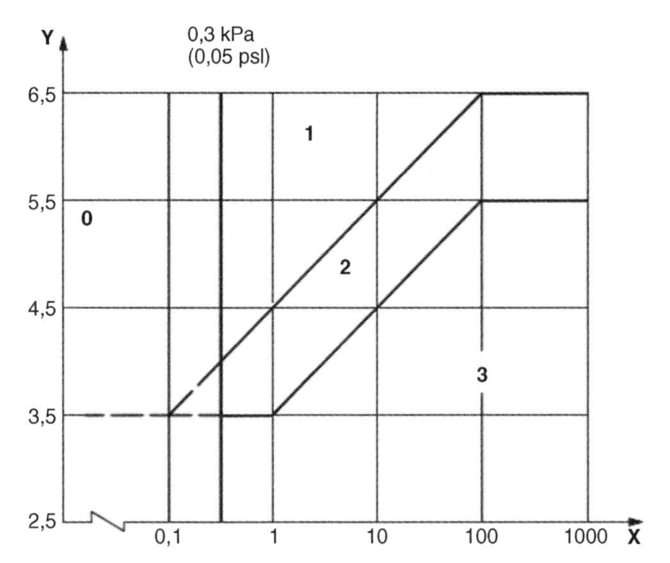


**FIGURE 24.4** $H_2S$ Serviceability Diagram (pH versus $H_2S$ partial pressure in kPa) shows regions of environmental severity with respect to the SSC of carbon and low-alloy steels; regions 0–3 show generally increasing severity of SSC [6]. (From ©NACE International 2009.)

documented trend results from an interplay between the high diffusion rate of atomic hydrogen at room temperature with its ability to still be trapped at sites in the steel where it can concentrate in the metal lattice affected by tensile stress. SSC tends to decreases with both decreasing and increasing temperature. At lower temperatures, production in atomic hydrogen decreases and the mobility of atomic hydrogen

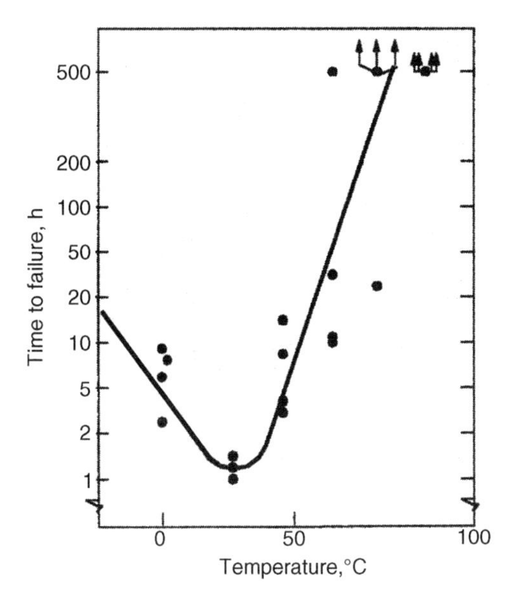


**FIGURE 24.5** Effect of environment temperature on SSC susceptibility in terms of time-to-failure for C–Mn steel. Note: Maximum susceptibility at near room temperature (23–25 °C) [12]. (©NACE International 1972.)

decreases, which impedes diffusion. At higher temperatures, thermal energy more easily overcomes the trapping energy for hydrogen where SSC initiates, thus, producing the decreasing SSC susceptibility at elevated temperature.

## 24.6 INFLUENCE OF METALLURGICAL VARIABLES ON SSC IN STEELS

The major variable used to predict SSC susceptibility and to mitigate this form of environmental cracking is the material strength as characterized in hardness for steels and most other engineering alloys. Hardness commonly serves as a measure of strength through its correlation with the ultimate tensile strength. Susceptibility to SSC generally increases with hardness, although at times the actual yield strength of the material may be alternatively specified. For selection of steels for use in sour service conditions, NACE MR0175/ISO15156 Part 2 uses Rockwell B and C scale hardness values (HRB/HRC) for specifying the allowable parent metal conditions for sour service while using Vickers hardness (HV5 or HV10) for indicating acceptable conditions associated with welds and applied overlay materials.

For unrestricted use in sour service conditions, steels with a maximum hardness of 22 HRC are generally acceptable so long as they contain <1% Ni as an alloying element and do not have extensive cold working, which is usually limited between 5 and 15% strain depending on the steel and product form. However, it is important to realize that this does not necessary indicate that these soft steels provide acceptable resistance to hydrogen blistering, HIC or SOHIC (Type I SSC). Separate considerations for these cracking phenomena are provided in Part 2 of the NACE MR0175/ISO15156 and should be reviewed (Also, see Chapter 17 in this book for further information.)

Allowance is made for some higher hardness steels up to 26 HRC provided that they are heat treated as tubular components made of Cr–Mo low-alloy steels and quenched and tempered to achieve complete transformation to martensite on quenching. Still higher hardness steels can be utilized if evaluated under laboratory sour conditions as per the requirements necessary to ballot this material condition into this standard (i.e., data from three commercial heats at maximum hardness with triplicate specimens per heat as discuss in NACE MR0175/ISO15156 Part 2 Annex B).

For steel weldments, extensive laboratory tests show that isolated hard zones in parent metal, weld metal, and the HAZ region can be detrimental for SSC resistance [4,13,14] and that parent metal composition, metallurgical processing and selection of weld consumables and procedures should be selected to minimize this problem. As with base metals, welds can be deemed acceptable provided hardness testing for a welding procedure qualification is carried out; however, the hardness method is Vickers HV 10 g or HV 5 g methods.

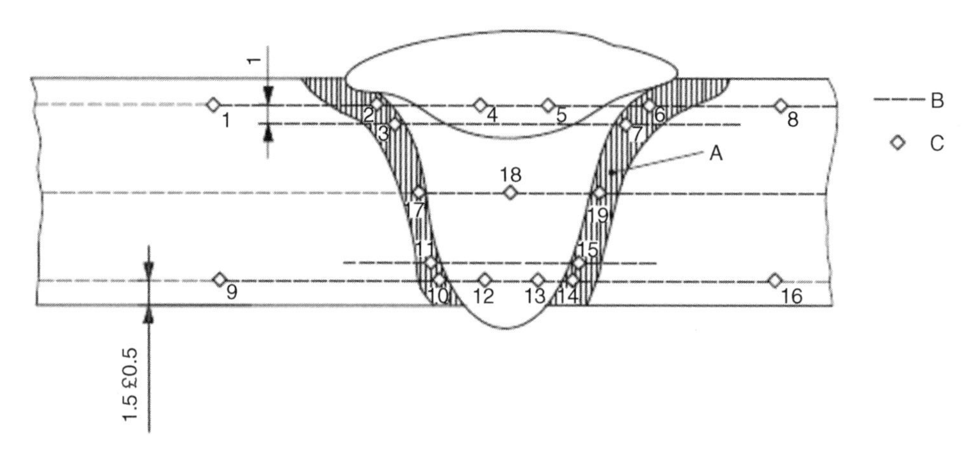


Key
A weld heat-affected zone (visible after etching)
B lines of survey
C hardness impressions: impressions 2, 3, 6, 7, 10, 11, 14, 15, 17 and 19 should be entirely within the heat-affected zone and located as close as possible to the fusion boundary between the weld metal and the heat-affected zone.

The top line of survey should be positioned so that impressions 2 and 6 coincide with the heat-affected zone of the final run or change of profile of the fusion line associated with the final run.

**FIGURE 24.6** Allowable 16-point HV hardness test configuration for butt weld from NACE MR0175/ISO15156 Part 2 [6]. Note: Dimensions in millimeters. (From ©NACE International 2009.)

Acceptable steel welds are those with a maximum hardness of up to HV 250 for the root and mid-passes and if the external surface of the pipe is not affected by a sour environment, then a cap pass hardness limit, in some cases, can be up to HV 275.

A typical HV hardness test arrangement for a butt weld configuration is shown in Figure 24.6 from NACE MR0175/ISO15156 Part 2 [6]. Alternative arrangements are available for other weld configurations in this standard. This NACE standard also allows the HRC hardness method to be used for welding procedure qualification if the design stress does not exceed 2/3 of specified minimum yield strength (SMYS) and the welding procedure specification includes post-weld heat treatment require the agreement of the equipment user.

Extensive laboratory testing has shown the general relationship between $H_2S$ partial pressure and acceptable weld hardness. Figure 24.7 shows that the NACE/ISO HV 250 limit for weldments and base metal is justified based on laboratory test data conducted at 15 psia $H_2S$ partial pressure [4]. However, this figure also shows that welds with higher hardness values may be serviceable in environments with lower $H_2S$ partial pressures with the hardness limits at 5 kPa being about HV 300 and the limit for HV350 being near 0.5 kPa. However, the present NACE/ISO standard for sour service requires procedure qualification and $H_2S$ testing for establishing hardness/$H_2S$ limits for conditions such as these that are not currently in this standard.

Carbon and low-alloy steels subjected to cold deforming by rolling, cold forging, or other manufacturing process that results in permanent strains >5% are commonly thermally stress-relieved with a minimum temperature of 595 °C when a final maximum hardness of 22 HRC or HV250 is obtained. However, based on service experience there are exceptions with some cold-worked line pipe fittings are acceptable with up to 15% cold strain if the hardness does not exceed 190 Brinell hardness (HBW) [6].

## 24.7 USE OF CORROSION-RESISTANT ALLOYS TO RESIST SSC

Stainless steels and Ni-base alloys offer great potential in engineering design by eliminating the dependence on corrosion inhibitors. However, it must be realized that these materials can, in some cases, act very differently than steels and are subject to a different set of engineering limitations particularly in sour service conditions involving exposure to aqueous $H_2S$ environments. In addition, to justify the cost of corrosion-resistant alloys (CRAs) for many applications, they must also have longer service life and higher strength capabilities in sour service than conventional steels. As such, these materials must be able to overcome the detrimental effects of $H_2S$ and $CO_2$ gas, sulfur compounds, and concentrated brine solutions without susceptibility to environmental cracking. To obtain high strength combined with corrosion resistance, there has been increased use of materials containing high levels of chromium, nickel, and molybdenum (See Table 24.2) [5].

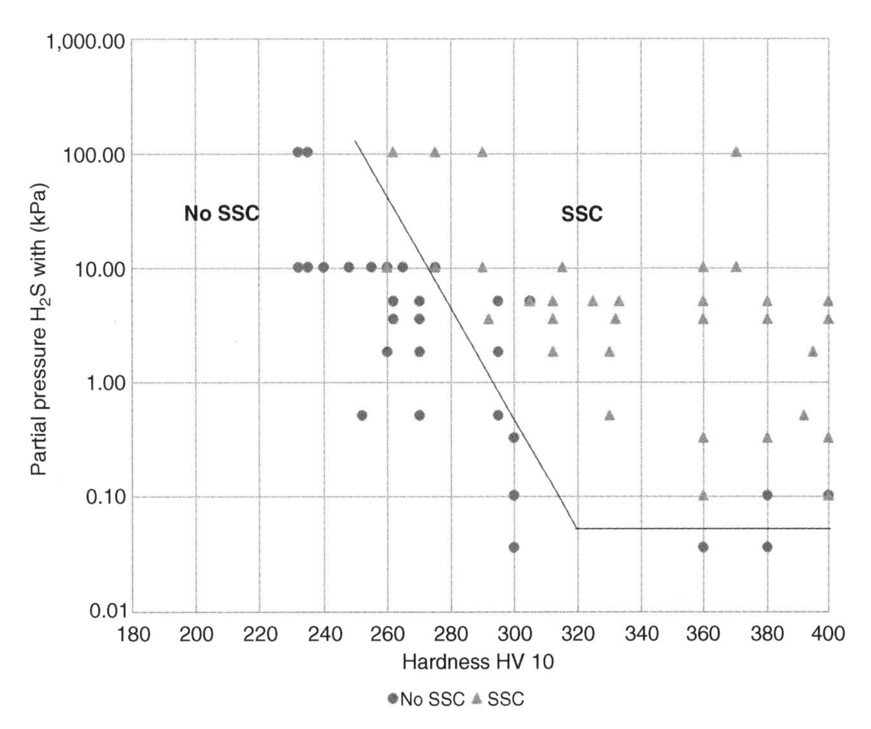


**FIGURE 24.7** Relationship between hardness and $H_2S$ partial pressure defining serviceability regions for steel weldments showing SSC and no SSC behavior [4]. (From ©NACE International 1981.)

There are four classes of CRA materials based on Fe–Ni–Cr–Mo compositions: (1) martensitic stainless steels, (2) duplex stainless steels, (3) high-ahoy austenitic stainless steels, and (4) Ni-base alloys. Typically, conventional austenitic stainless (AISI 3XX series) steels do not have adequate strength or resistance from SCC for use in many $H_2S$ applications and exhibit chloride SCC at 60 °C and higher. Each classification has different ranges of composition, microstructure, strengthening mechanism, and performance in corrosive production environments. The guidelines for the use of these materials in sour service is governed by NACE MR0175/ISO15156 Part 3 [15] that provides material hardness limits and acceptable metallurgical conditions for service conditions based on specific limitations of $H_2S$ partial pressure, pH, chloride concentration, and the presence/absence of elemental sulfur.

While the first step up in CRA materials are 9–13Cr alloys, martensitic stainless steels for $H_2S$ service have been developed with improved corrosion resistance and mechanical properties versus the conventional 12–13% Cr grades. For example, high-strength tubular grades of martensitic stainless steels with yield strengths up to 760 MPa have been developed with nominally 13% Cr and are used in the quenched and tempered condition. These materials contain 13–17% Cr, 4–6% Ni and 1.5–2% Mo for added resistance to localized corrosion and SSC superior to straight 12–13Cr alloys. Maximum susceptibility to SSC in martensitic stainless steels typically occurs near room temperature with decreasing cracking susceptibility with increasing service temperature. However, for grades that contain Ni, there is an additional need to consider environmental cracking from SCC usually occurring at local anodic sites and pits. Serviceability limits for these materials are commonly in the range of 1–10 kPa $H_2S$ at pH 3.5–4.5 with materials available in the 650–860 MPa minimum yield strength grades at up to HRC 28–30. Selected SSC data for such martensitic stainless steels (with $H_2S$ and pH limits) are shown in Figure 24.8 [16].

Duplex stainless steels are a mixture of two microstructural constituents, ferrite and austenite. They can be produced in both cast and wrought conditions and typically have 22–25% Cr with yield strengths of 350–620 MPa when in the cast or annealed conditions and up to 900 MPa when cold worked during processing. These steels derive their mechanical properties and microstructure from a balanced chemical composition that contains alloying additions of Cr, Ni, Mo, (sometimes W), and N. This results in a material with a corrosion resistance comparable to or superior to those of the martensitic stainless that is also greater than conventional austenitic stainless steels.

NACE MR0175/ISO15156 Part 3 uses the alloy pitting resistance equivalent number (PREN = Cr + 3.3(Mo + 0.5 W) + 16 N) for evaluating these alloys for serviceability in sour environments with PREN values in the ranges of 30–40 and 40–45. Duplex stainless steels generally show serviceability with $H_2S$ partial pressures in the range 2–20 kPa that varies with the chloride content of the environment and PREN, cold

**TABLE 24.2 Common Corrosion Resistant Alloys (CRAs) Used in Sour Service [5]**

Stainless Steels

| Designation | Nominal Compositions (wt)% | | | | | | Comments |
|---|---|---|---|---|---|---|---|
| | Fe | Ni | Cr | Mo | N | Other | |
| *Martensitic SS* | | | | | | | |
| 13Cr | Bal | – | 13 | – | – | – | Low cost/resists $CO_2$ corrosion |
| Modified 13Cr[a] | Bal | 5.0 | 13 | 2.0 | – | – | improved corrosion resistance (esp. in $H_2S$) and strength over 13 Cr |
| 15Cr | Bal | 1.5 | 14.5 | 0.5 | – | – | Improved corrosion resistance over 13Cr |
| Duplex SS | | | | | | | |
| 18Cr[a] | Bal | 4.5 | 18.5 | 2.5 | 00.7 | – | Lower cost than 22/25 Cr alloy |
| 22Cr | Bal | 5.5 | 22.0 | 3.0 | 0.10 | – | resists $CO_2$ and low/mod $H_2S$ |
| 25Cr | Bal | 6.0 | 25.0 | 3.5 | 0.20 | W | higher corrosion and SCC resistance than 22 Cr esp w/added Mo and W |
| *High Alloy Austenitic SS* | | | | | | | |
| Alloy 28 | Bal | 31.0 | 27.0 | 3.5 | – | – | Higher resistance to $CO_2$, mod $H_2S$ and Cr |
| Alloy 254[a] | Bal | 18.0 | 20.0 | 6.0 | 0.20 | – | Low Ni and Cr content but high Mo and N for pitting resistance |
| Alloy 904[a] | Bal | 25.0 | 21.0 | 4.5 | – | – | Low Ni and Cr content but high Mo and N for pitting resistance |
| Alloy 6XN[a] | Bal | 24.0 | 21.0 | 6.5 | 0.20 | – | Low Ni and Cr content but high Mo and N for pitting resistance |

Nickel-Base Alloys

| Designation | Nominal Composition (wt)% | | | | | | Comments |
|---|---|---|---|---|---|---|---|
| | Fe | Ni | Cr | Mo | N | Other | |
| *Cold Worked* | | | | | | | |
| 2535[a] | Bal | 38.0 | 25.0 | 3.0 | – | – | Lower nickel content than 825, higher, higher pitting resistance |
| 825 | Bal | 42.0 | 21.0 | 3.0 | – | – | Good resistance to $CO_2$/high $H_2S$ and $Cl^-$/moderate temperature |
| G-30[a] | 20 | Bal | 22.0 | 7.0 | – | – | Good resistance to $CO_2$/high $H_2S$ and $Cl^-$/high temperature |
| 2550[a] | 20 | 50.0 | 25.0 | 6.0 | – | 2.5W | Better resistance than G3 to $CO_2$/high $H_2S$ and $Cl^-$ |
| G-50 | 17 | 50.0 | 20.0 | 9.0 | – | – | Better resistance than G3 to $CO_2$/high $H_2S$ and $Cl^-$ |
| C-22[a] | 4.0 | Bal | 21.0 | 13.5 | – | 3.0 W | Alternative to C-276 in some environments |
| C-276 | 5.5 | Bal | 15.0 | 16.0 | – | 3.5 W | Excellent resistance to $CO_2$/high $H_2S$ and $Cl^-$/very high temperature and sulfur |
| *Precipitation Hardened* | | | | | | | |
| 925[a] | Bal | 42.0 | 21.0 | 3.0 | – | 2.0 Ti | Lower Ni and cost option to alloy 718 |
| 718 | Bal | 52.0 | 19.0 | 3.0 | – | 1.0 Ti/5.0 Cb | Good resistance to $CO_2$/mod $H_2S$ and $Cl^-$ |
| 725[a] | 8.0 | Bal | 21.0 | 8.0 | – | 3.4 | Better pitting and SCC resistance than 718 |
| 625 plus[a] | 5.0 | Bal | 21.0 | 8.0 | – | 1.5 Ti/3.5 Cb | Better pitting and SCC resistance than 718 |

Source: ©NACE International 1998.
[a]Most commonly used of the CRA materials listed.

work used in processing and hardness of the material up to HRC 36. Maximum susceptibility to environmental cracking in duplex stainless steels typically occurs at intermediate temperatures generally found in the range 60–120 °C, where both HEC and SSC mechanisms are operable.

Highly alloyed stainless and Ni alloys are for $H_2S$ service applications primarily for use as high-strength tubular materials and ancillary pressure control equipment. They can be cold worked in the range of 30–50% cold reduction to strength levels in tubular material between 750 and 1,000 MPa yield strength. Alternatively, for other components, such as valves and specialized equipment that often require more complex shapes or welding, precipitation-hardened Ni-base alloys, which obtain their strength via aging heat treatments, are available with nearly the same strength levels.

Per NACE MR0175/ISO15156 Part 3, high-alloy austenitic stainless steels at up to HRC 35 are serviceable under conditions of 100–690 kPa $H_2S$ partial pressure that varies with PREN and total alloy composition, along with chloride content and maximum service temperatures in the range of 60–171 °C. By comparison, Ni-base alloys varying in the performance based on alloy composition and strength, and environmental variable of $H_2S$ partial pressure, chloride and exposure to elemental sulfur. Precipitation-hardened Ni-base

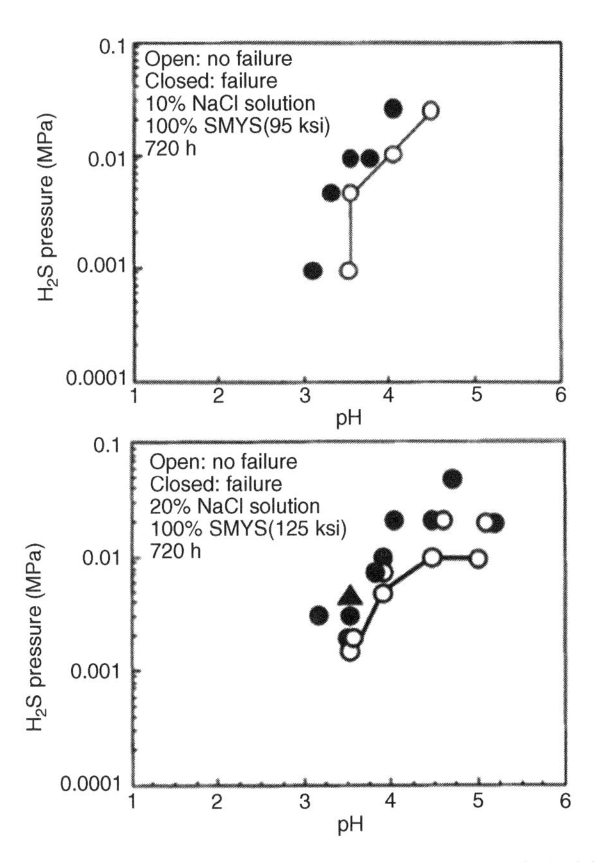


**FIGURE 24.8** $H_2S$ and pH limits for SSC in a modified 13Cr (Ni–Mo) alloy for 655 MPa (above) and 863 MPa (below) specified minimum yield strength conditions in 10–20% NaCl solutions, respectively. Note: 1 MPa = 1000 kPa; 100 ksi = 690 MPa) [16].

alloys with hardness values in the range HRC 35–44 have limits of $H_2S$ partial pressure in the range 200–3400 kPa with chlorides up to 180,000 mg/l, maximum (in some cases) and maximum service temperatures varying from 135 to 232 °C. By comparison, the cold worked Ni-based alloys show service limits from 200 to 6900 kPa $H_2S$ at temperatures ranging from 132–232 °C with unrestricted chloride levels.

High-temperature sour service environments, when found in combination with elemental sulfur, appear to accelerate anodic SCC in much the same way that oxidizing agents increase the severity of SCC in non-sulfide containing environments. The most highly alloyed materials (with minimum level of 14.5% Cr, 52% Ni + Co, 12% Cr) are utilized for service conditions that include exposure to elemental sulfur in combination with up to 6900 kPa H2 S with some alloys available up to HRC 45.

It must be kept in mind that when corrosion-resistant alloys (martensitic, duplex, and austenitic stainless steels and Ni-base alloys) are used for sour service applications in the welded condition or as weld overlays, the limits for SCC may vary from those provided herein based on the parent metal composition and metallurgical condition, along with the weld consumables and procedures used. For such situations, the applicable sections of NACE MR0175/ISO15156 Part 3 need to be consulted for specific advice for appropriate guidelines, qualification, and specification. These $H_2S$ serviceability limits may need to be determined through procedure qualification and SSC testing per methods described in this document and the ancillary SSC test methods that it references.

## REFERENCES


1. Iwadate, T. et al., (1988) Hydrogen effect on remaining life of hydroprocessing reactors. *CORROSION*, 44, 103–112 NACE International.
2. Cayard, M.S., Kane, R.D., Kaley, L., and Prager, M. (1994) Research report on Characterization and Monitoring of Cracking in Wet $H_2S$ Service, Publication 939, American Petroleum Institute, (Oct. 1994) P10.
3. Kane, R.D. and Greer, J.B. (1977) Sulfide stress cracking of high-strength steels in laboratory and oilfield environments. *Journal of Petroleum Technology*, 28, 1483–1488.
4. Omar, A.A., Kane, R.D., and Boyd, W.K. (1981) *Factors Affecting the Sulfide Stress Cracking Resistance of Steel Weldments*, Corrosion/81, Paper No. 186, NACE International, Houston, TX.
5. Kane, R.D. and Cayard, M.S. (March 1998) *Roles of $H_2S$ in the Behavior of Engineering Alloys: A Review of Literature and Experience*, Corrosion/98, Paper No. 274, NACE International, Houston, TX.
6. ANSI/NACE D (2009) *ANSI/NACE MR0175/ISO 15156: Part 2, Petroleum and Natural Gas Industries—Materials for Use in $H_2S$-Containing Environments in Oil and Gas Production—Part 2: Cracking-Resistant Carbon and Low-Alloy Steels, and the Use of Cast Irons.* NACE International, Houston, TX.
7. Berkowitz, B.J. and Henbaum, F.H. (1984) AT The role of hydrogen in sulfide stress cracking of low alloy steels. *CORROSION*, 40 (5), 240–245.
8. Kaneko, T., et al. (1979) Proceedings of NACE Middle East Corrosion Conference, Bahrain, p. 932.
9. Takahasi, A. and Ogawa, H. (1996) Influence of microhardness and inclusion on stress oriented hydrogen induced cracking of line pipe steels. *ISIJ International*, 36 (3), 334–340.
10. Kimura, M., et al. (1996) Effect of environmental factors on hydrogen permeation in linepipe steel. In: Kane, R.D., Horvath, R.J., and Cayard, M.S. (editors), *Wet $H_2S$ Cracking of Carbon Steels and Weldments*, NACE International, Houston, TX, p. 536.
11. NACE *NACE TM0177: Laboratory Testing of Metals for Resistance to Sulfide Stress Cracking and Stress Corrosion Cracking in $H_2S$ Environments, Methods A–D*, NACE International, Houston, TX.
12. Townsend, H.E. Jr. (1972) Hydrogen sulfide stress corrosion cracking of high strength steel wire. *CORROSION*, NACE International, Houston, TX, 28 (1), 39–46.
13. Kotecki, D.J. and Howden, D.G. (1996) Weld cracking in a wet sulfide environment, Proc. American Petroleum Institute,

In: Kane, R.D., Horvath, R.J., and Cayard, M.S. (editors), *Wet $H_2S$ Cracking of Carbon Steels and Weldments*, NACE International, Houston, TX, p. 337.

14. Lawson, V.B., et al. (1996) Pipeline failures in the grizzly valley sour gas gathering system. Corrosion/87, Paper No. 52, In: Kane, R.D., Horvath, R.J., and Cayard, M.S. (editors), *Wet $H_2S$ Cracking of Carbon Steels and Weldments*, NACE International, Houston, TX, p. 415.
15. ANSI/NACE (2009) *ANSI/NACE MR0175/ISO 15156—Part 2: Petroleum and Natural Gas Industries—Materials for Use in H2S-Containing Environments in Oil and Gas Production—Part 3: Cracking-Resistant CRAs (Corrosion Resistant Alloys) and other Alloys*, NACE International, Houston, TX.
16. Kimura, M., Tamari, T., and Shimamoto, K. (2006) High Cr stainless steel OCTG with high strength and superior corrosion resistance, JFE Technical Report, No. 7.

# 25

# STRESS CORROSION CRACKING OF STEEL EQUIPMENT IN ETHANOL SERVICE

Russell D. Kane
*iCorrosion LLC, Houston, TX, USA*

## 25.1 INTRODUCTION

Research and field experience with ethanol stress corrosion cracking (eSCC) has evolved a better understanding of the fundamentals of this environmental cracking phenomenon and its implications in fuel grade ethanol (FGE) operations (e.g., facilities piping, equipment, tanks, and pipelines) worldwide. The process of identifying and understanding this phenomenon began in 2003 with a meeting of representatives from oil companies, fuel ethanol producers, and technical organizations when eSCC was first identified in steel tanks, piping and associated equipment used to handle and store FGE currently provided under specifications given in ASTM D4806 [1] and other international standards.

The dilemma presented in 2003 was to explain the cause and key parameters behind the unexpected failures in steel equipment exposed to FGE and to identify its potential impact on the supply chain for fuel ethanol. The initial effort involved a review of service experience and the published literature and culminated in a research report [2] published by the American Petroleum Institute (API) and a supporting technical paper [3] that summarized the initial findings. Subsequently, updates to API 939D were issued in 2007 and 2013 as the research, literature review and field surveys progressed [4]. The results of this work were utilized for the preparation of guidelines for the identification, repair, and mitigation of eSCC in steel equipment used in fuel ethanol service, which were originally issued in 2008 and updated in 2013 [5].

The major findings of this work indicated that the failures in ethanol handling equipment, tanks, and piping were the result of eSCC and are as follows:

- In some cases, eSCC has produced failures in <1 year, but in other cases may take years.
- Stress appeared to be an important variable in the propensity for eSCC with most reported cases occurring in welds and components that had high residual, applied and/or dynamic stresses.
- Based on a literature review and preliminary laboratory studies, the factors that increased corrosivity of fuel ethanol appeared to be increased water content, decreased pHe, and other factors, including the presence of sulfur, sulfate, and chloride concentration. However, none of these variables appeared to correlate with severity of eSCC.
- Based on a review of preliminary laboratory studies, eSCC of steel was produced in samples of commercial FGE within the limits of ASTM D4806, including some that did not contain denaturants or inhibitors (allowable and widely used per this specification).

## 25.2 FACTORS AFFECTING SUSCEPTIBILITY TO ETHANOL SCC

In eSCC, as in all situations involving environmental cracking, there are three primary variables that come together to

*Oil and Gas Pipelines: Integrity and Safety Handbook,* First Edition. Edited by R. Winston Revie.

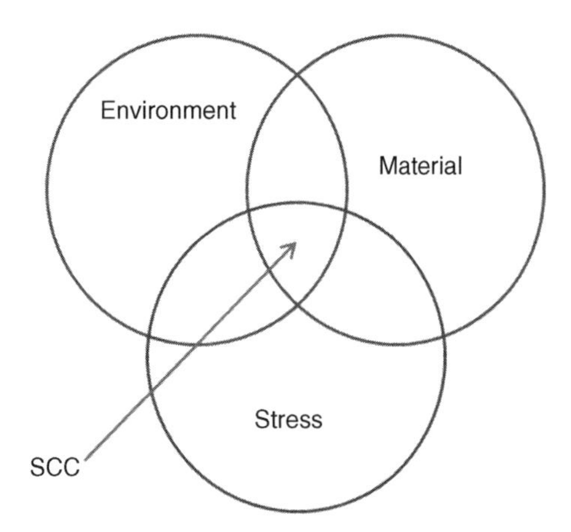


**FIGURE 25.1** Schematic representation of the three primary factors causing environmental cracking (such as stress corrosion cracking).

define the severity of the cracking response. They are as follows: (a) environment, (b) metallurgical condition, and (c) stress. This is shown schematically in Figure 25.1.

The situation for eSCC is similar with specific parameters in each area having greater/lesser roles in the cracking process. The major parameters in each category are summarized as defined by laboratory and/or field studies.

### 25.2.1 Environmental Variables in FGE

In the Unites States, the chemical makeup of FGE is defined by ASTM D4806, which is shown in Table 25.1. Per this specification, FGE contains up to 1 vol% water, up to 56 mg/l acidity as acetic acid, and (per its most recent revision in 2009) the level of inorganic chlorides reduced to 8 mg/l maximum from 32 mg/l in the earlier versions of this standard. The pHe (analogous to pH in aqueous solutions, but with neutral in ethanolic environments being around 8.5) ranges from 6.5 to 9. Within the specification limits of ASTM D4806, none of the variables have been found to conclusively correlate with the occurrence of eSCC in field failures. However, certain environmental variables have been found through laboratory testing to control susceptibility to eSCC of steel.

***25.2.1.1 Oxygen*** FGE is handled without procedures designed to exclude ambient air, which contains approximately 20 mol% oxygen. Therefore, FGE should be considered to contain substantial dissolved oxygen and it can have levels up to 60–80 parts per million (ppm) dissolved oxygen when saturated with ambient air. The results of laboratory studies have shown that the susceptibility to eSCC was only found in aerated ethanolic media, not in deaerated solutions [6].

Recent laboratory studies show that the level of aeration needed to sustain stress corrosion cracking (SCC) in ethanolic solutions is only a small percentage of ambient air saturation. Below this level of aeration, there appears to be no susceptibility to SCC. However, since most field equipment used for handling and storage of FGE are not designed to remove or exclude ambient air, it is presently difficult to utilize this key variable for control of eSCC in facilities. Additionally, in a pipeline, dissolved oxygen can be consumed by corrosion reactions with the inner surface of the steel producing de facto deaeration. However, with FGE, there exists the combination for high dissolved oxygen content (up to 10 times that in water) and very low general corrosion rates (generally 0.01–0.1 mpy with occasional excursions to 1 mpy under highly turbulent and aerated conditions). These low corrosion rates may not substantially reduce the concentration of dissolved oxygen in FGE inside a pipeline and excluded from ambient air, which maintains the potential for eSCC in pipeline applications. It appears from limited measurements that most systems handling FGE are in excess of 50% of ambient air saturation, thus providing ample aeration to promote eSCC initiation and support crack propagation in steel equipment.

***25.2.1.2 Corrosion Potential*** Laboratory tests have shown that steel can assume a range of corrosion potentials

**TABLE 25.1 Fuel Grade Ethanol Specification per ASTM D4806**

| Property | Units | Specification | ASTM Designation |
|---|---|---|---|
| Ethanol | %v min | 92.1 | D5501 |
| Methanol | %v max | 0.5 | – |
| Solvent-washed gum | mg/100 ml max | 5.0 | D381 |
| Water content | %v max | 1.0 | E203 |
| Denaturant content | %v min | 1.96 | D4806 |
| | %v max | 5.00 | |
| Inorganic chloride content | ppm (mg/l) max | 10 (8) | E512 |
| Copper content | mg/kg max | 0.1 | D1688 |
| Acidity as acetic acid | %m (mg/l) | 0.007 (56) | D1613 |
| pHe | – | 6.5–9.0 | D6423 |
| Appearance | Visibly free of suspended or precipitated contaminants (e.g., clear and bright) | | |

in FGE and synthetic FGE (SFGE) solutions depending on the composition of solution and the level of aeration. The range of corrosion potential for eSCC of steel is at potentials between 0.0 and 350 mV (Ag/AgCl ethanol reference electrode), which are usually only obtained as a result of aeration of the solution [4,5].

Understanding and possibly controlling the corrosion potential of steel in ethanolic solutions, through chemical additions or galvanic interaction, is an area of ongoing development as is the ability to conveniently and accurately measure potential in operating equipment in fuel ethanol service.

*25.2.1.3* ***Impurities*** Studies have shown that eSCC is affected only to a limited degree for most of the constituents of FGE in the allowable range cited in ASTM D4806. These variables include: pHe, acetic acid, methanol, and water contents. The only chemical variable defined in ASTM D4806 that appears to have an effect is the presence of chloride ions. As indicated in Figure 25.2, laboratory tests have shown that inorganic chloride (within the ASTM limit) causes increased eSCC severity in terms of crack density and velocity when evaluated using the slow strain rate testing technique [7]. Presently, the 2009 version of this standard reduced the level of inorganic chlorides to 10 ppm (8 mg/l).

This may suggest that inorganic chlorides may be a factor in eSCC field failures. However, limited field examinations performed, thus far of inorganic chloride levels in FGE lading or tank analyses have not shown a conclusive correlation in cases where eSCC has taken place. Most analyses examined show concentrations of chlorides to be below 2 mg/l. That said, there are many potential sources of extremely low level chloride contamination that include seawater, road salts, and cleaning agents to name a few.

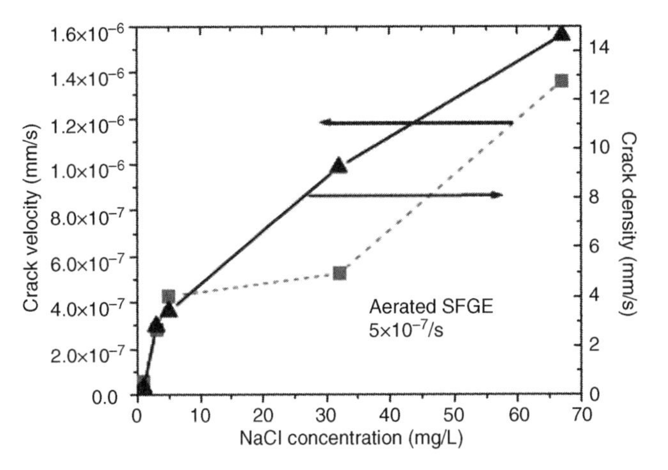


**FIGURE 25.2** Influence of NaCl concentration on crack velocity and density for eSCC [7]. (From © NACE International 2010.)

### 25.2.2 Metallurgical Variables

By far, the typical material utilized for transport, handling, and storage of FGE is steel.

*25.2.2.1* ***Steel Grade/Composition*** Carbon steels used in fuel ethanol service for piping and tanks include those classified under ASTM A36, ASTM A53, ASTM A106, and ASTM A516-70. Steels used in existing pipelines often include those defined under API 5L (e.g., X-42 and X-52) and higher strength grades (e.g., X-60, X-70, and X-80) proposed for new construction. Based on surveys of field experience and laboratory testing programs, susceptibility to eSCC is not limited to any particular steel grade for conventional steels with ferritic/pearlitic microstructures. Numerous studies have indicated that eSCC in these steels initiates at hard nonmetallic (alumina and silicate) inclusions, particularly areas that have undergone yielding as a result of combined residual and applied stresses at or beyond the engineering yield strength of the material. Cast and wrought seamless steels had slightly lower eSCC crack growth rates, but all showed susceptibility to environmental cracking. There is some tendency for low-carbon pipeline steels to have a somewhat lower susceptibility to eSCC than the higher carbon plate steels when tested using slow strain rate methods in laboratory formulated simulated FGE solutions [2–4,8].

Additionally, from the survey of field data, all grades of steel already mentioned have been shown to exhibit SCC under certain circumstances and, consequently no relationship has been identified between steel grade and susceptibility to eSCC based on field experience. Laboratory studies are in progress currently to examine the eSCC resistance of advanced steels with lower carbon content and processed by thermo-mechanical controlled processing (TMCP) techniques, which have predominantly ferritic and bainitic microstructures.

Most field failures and leaks resulting from eSCC in steel equipment has been found to occur in non-post weld heat treated (non-PWHT) welded equipment. This cracking does not appear to be related to the weld metal or the heat-affected zone (HAZ) hardness of the weldments. Rather cracking is usually initiated in the base metal adjacent to the weldment associated with the point of maximum residual tensile stress around the weldment. PWHT is usually recommended for increasing resistance to eSCC in welded equipment, where possible.

### 25.2.3 Mechanical Variables

*25.2.3.1* ***Tensile Stress*** The association of eSCC with the base metal adjacent to the weldment and regions of high tensile residual stress in non-PWHT welds indicates a strong relationship between eSCC and locally high levels of tensile stress in the steel. As such, no eSCC failures have thus far been reported in welded steels having undergone post weld heat treated (PWHT). The current guidelines for PWHT for prevention of eSCC as found in API Tech. Bulletin 939E involve heating to 611 °C (1150 °F) minimum and holding at this temperature for 1 h per 25 mm of metal thickness, or fraction thereof, with a 1 h minimum hold time. For cases where PWHT is not possible, ethanol immersion-resistant coatings have been utilized that act as a barrier between the steel and FGE.

Based on the aforementioned experience, high levels of tensile stress appear necessary to initiate eSCC. Fracture mechanics tests also indicate a similar association between eSCC and a high level of stress is required for its initiation. Based on laboratory fracture mechanics tests, the stress intensity for eSCC of steels appears to be in the range of about 35 to slightly over 50 ksi(in.)$^{-1/2}$ in steel for environmental conditions supporting eSCC [9,10]. This range appears to apply for wet milled US corn-based FGE and various SFGE blended in the laboratory to the ASTM D4806 specification (See Figure 25.3).

*25.2.3.2 **Plastic Strain and Dynamic Stress*** eSCC has also been found to occur in non-welded steels, which have been plastically deformed (and which may have high tensile residual stresses) in small shop built tanks and pipelines. Two occurrences of eSCC in pipelines have been documented in API surveys and include segments of pipelines involving a field bending and ancillary plant equipment used to handle FGE, which was cold worked during fabrication [5]. These observations suggest that either the high tensile residual stresses or plastic deformation (or both) support eSCC.

Laboratory test data also supports the association between plastic strain, high tensile stresses and eSCC. Examination of slow strain rate tests that involve straining up to and beyond the material yield strength also show that eSCC is limited to conditions of high stress at or beyond the yield strength of the material. Such tests show a reduction of the plastic deformation in a slow tension test, which results from the exposure of steel to FGE [4]. Interestingly, these studies also show that statically stressed specimens commonly used for laboratory SCC evaluation (e.g., tensile, C-ring, and U-bend geometries) have not shown the ability to reliably initiate eSCC, even when stressed beyond their yield strength. This suggests that dynamic or cyclic loading of components may accelerate eSCC in service.

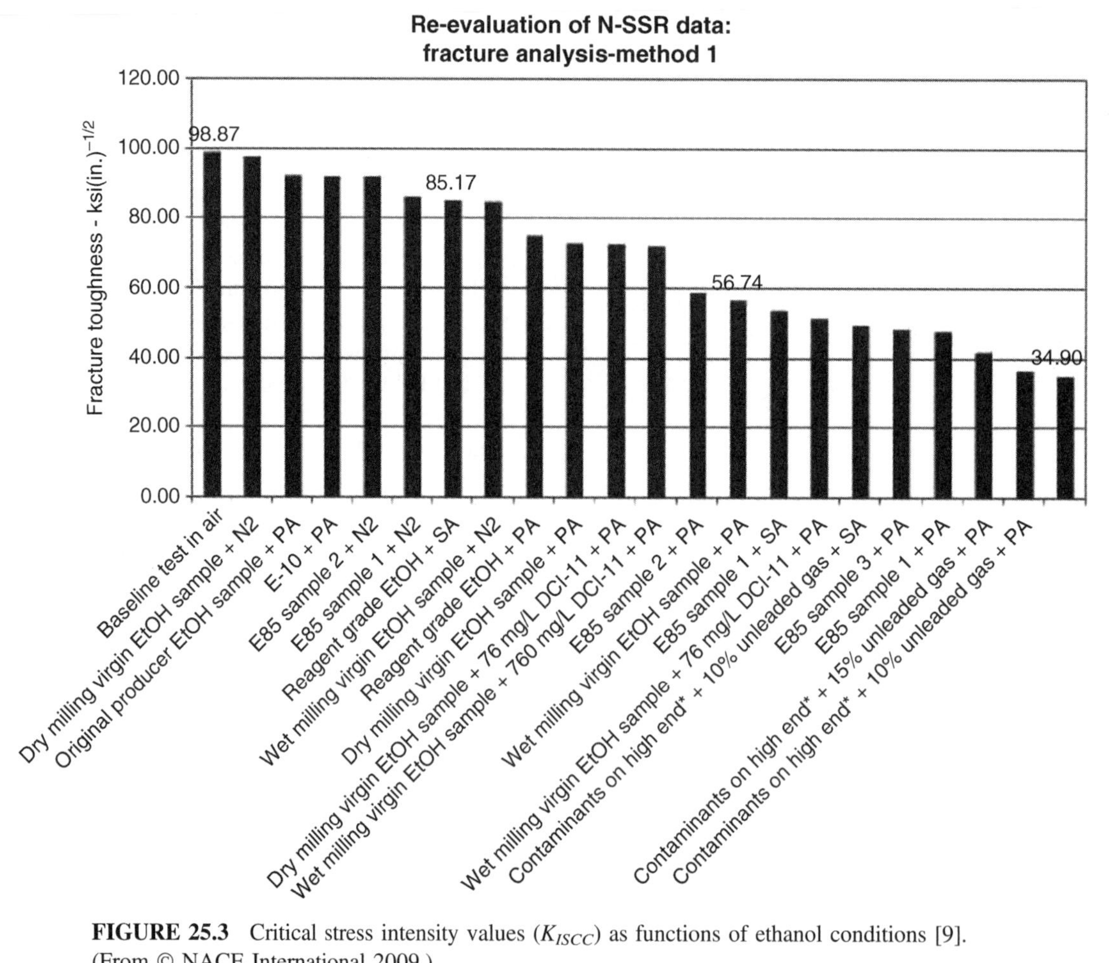


**FIGURE 25.3** Critical stress intensity values ($K_{ISCC}$) as functions of ethanol conditions [9]. (From © NACE International 2009.)

## 25.3 OCCURRENCES AND CONSEQUENCES OF eSCC

A total of 31 cases of eSCC have been reported in the API survey efforts conducted from 2003 through 2010 [5]. To date, none of the cases of eSCC have been reported in steel equipment at ethanol manufacturing facilities; all have been in the mid-stream and downstream portions of the FGE distribution system. A listing of the documented instances of SCC from these API surveys and the relative percentages of applications they represent are given as follows:

- Facilities piping/fittings—24%
- Tank floor plates seam welds—22%
- Tank floor/sidewall fillet welds—9%
- Tank sidewall 1st course butt weld—9%
- Tank floating roof seam welds—9%
- Tank roof springs—6%
- Facilities piping/supports—6%
- Ancillary handling equipment—6%
- Tank nozzle—3%
- Shop built tank in E85—3%
- Pipeline—3%

Obviously, not all cases of eSCC can be, or have been, documented. It does show, however, that facilities piping and tanks in mid- and downstream ethanol storage and handling operations are main locations of concern for eSCC failures with only two cases of documented eSCC in pipeline service.

There has also been a lone eSCC failure report in a small shop-built steel tank used to hold an E85 gasoline blend (85% ethanol). However, in this later case, the tank was severely cold worked during fabrication and was previously used in waste oil service where it suffered pitting damage. In this case, eSCC initiated in the region that had experienced both cold working and pitting damage.

Another potential aspect of the issue of eSCC may relate to the source of ethanol. It should be kept in mind that fuel ethanol in the United States is not sold water free. Denatured ethanol with <0.5 vol% water is considered "anhydrous ethanol." Ethanol with higher water content (up to 6 vol%) is usually referred to as "hydrated ethanol." eSCC has occurred in FGE made to ASTM D4806 specifications with up to 1 vol % water and commonly ranges from 0.5 to 1.0 vol% water and is mostly derived from corn as its bio-feedstock (either wet or dry milled processes). Anhydrous ethanol with about 0.1 vol% water has very low propensity for eSCC of steel. Hydrated ethanol is uncommon in the United States but has been used as a fuel in Brazil. Brazil also makes anhydrous ethanol, which is used for blending into gasoline. Both anhydrous and hydrated grades of fuel ethanol made in Brazil are made from sugarcane and the amount of each product manufactured fluctuates with the price. Recent accounts indicate that the Brazilian market for fuel ethanol is dominated by anhydrous ethanol. There have been no documented experiences with eSCC in Brazil with either hydrated or anhydrous ethanol despite over 20 years of research and use in Brazil. Laboratory SSR tests also show differences in eSCC severity based on ethanol feedstock and processing. Laboratory tests have also showed that steels tested in Brazilian anhydrous FGE samples (made from sugar cane feedstock) showed lower susceptibility ($K_{Iscc}$ values from 55 to 68 ksi(in.)$^{-1/2}$ shown in Figure 25.4) [11] for eSCC than

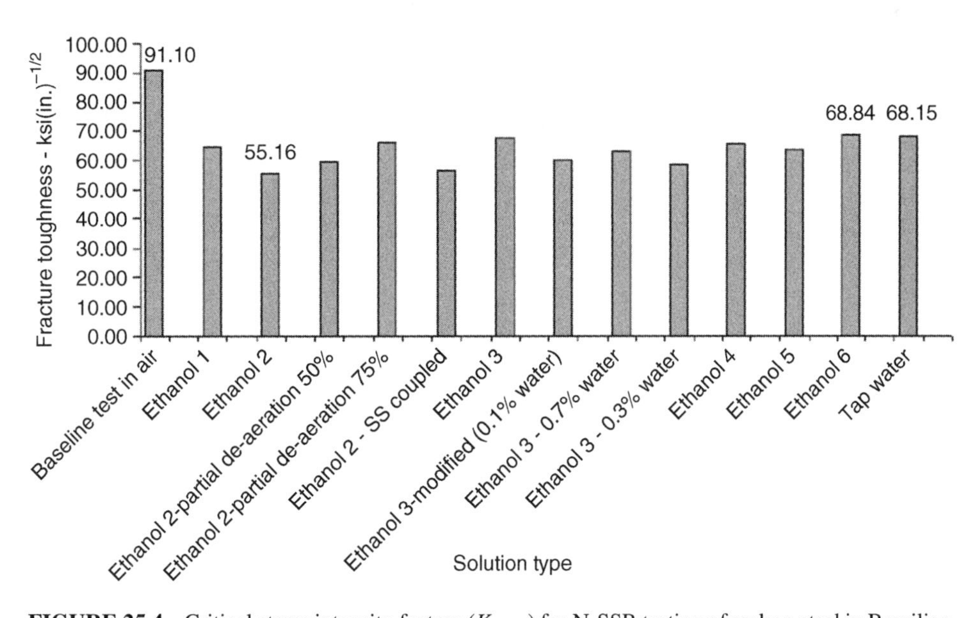


**FIGURE 25.4** Critical stress intensity factors ($K_{ISCC}$) for N-SSR testing of carbon steel in Brazilian FGE [11]. (From © NACE International 2010.)

US wet milled corn ($K_{Iscc}$ values from 40 to 45 ksi(in.)$^{-1/2}$ in Figure 25.3) [9].

## 25.4 GUIDELINES FOR IDENTIFICATION, MITIGATION, AND REPAIR of eSCC

### 25.4.1 Identification

The cracks produced by eSCC are typically branched and mostly intergranular (see Figure 25.5) [10]. However, they can also be transgranular, or mixed mode particularly if produced by laboratory slow strain rate tests that are characteristically high stress and employ post-yield straining to failure. This figure shows regions of intergranular cracking indicated by the arrows. While other environmental cracking mechanisms can cause predominantly intergranular cracking, none have been identified to be active in FGE or ethanol fuel blends.

Visual inspection of equipment for SCC in many systems (including FGE) is difficult since cracks are tight and often filled with corrosion products. To confirm eSCC in steel equipment a metallurgical analysis is recommended that includes metallographic sectioning and examination for confirmation of cracking mode by comparison to examples of eSCC provided in API 939E.

### 25.4.2 Inspection

Development of the guidelines for inspection for eSCC cited in API 939E, it was complicated since in most cases, the procedures used by many companies were not yet formalized. However, due to the similarity of eSCC to other forms of environmental cracking (especially, amine SCC), an attempt was made to combine documented procedures for eSCC inspection, which were combined with those provided in other API standards. For example, dry fluorescent magnetic testing (DFMT) used for tank inspection is not recommended for inspection of tight, corrosion product-filled cracks produced by eSCC. Wet fluorescent magnetic testing (WFMT) is typically more sensitive and commonly used method for detecting surface-connected cracks, small discontinuities, and environmental cracking, and has been used with success for eSCC detection.

Shear wave ultrasonic testing (UT) methods have also been used for detection of SCC in equipment not amenable to internal inspection by WFMT. It is a useful technique where through thickness propagation of cracking needs to be quantified. Other methods that may be applied to eSCC inspection include alternating current field measurement (ACFM), which is an electromagnetic technique that can be used to detect and size surface-breaking cracks in ferromagnetic materials such as steel and may be applied through thin surface coatings and corrosion products. Thus far, there has been no documented field experience with the eddy current (EC) method to detect eSCC, but it appears that it might prove useful in certain situations, if needed.

In a recent occurrence of eSCC in a short pipeline carrying FGE, the cracking location was identified prior to leakage or

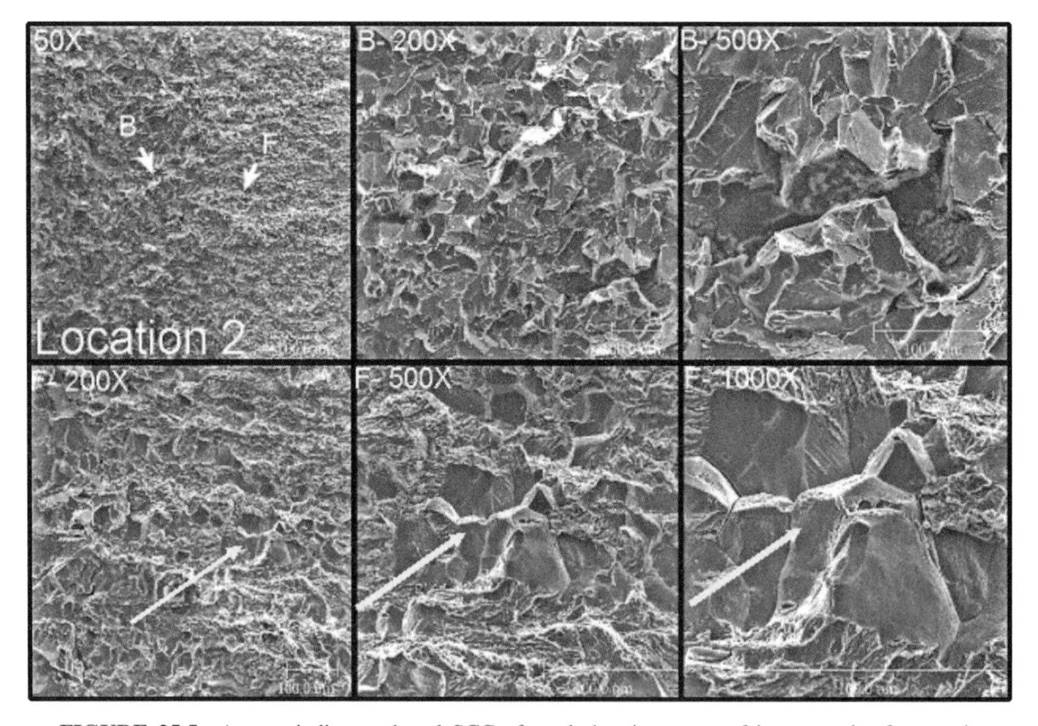


**FIGURE 25.5** Arrows indicate ethanol SCC of steel showing areas of intergranular fracture in Area F. Area B is the overload fracture (mostly quasi-cleavage) produced after SCC produced a critical size defect. (Photos by R. D. Kane.)

failure by using hydrostatic testing. While this was the first reported case involving hydrostatic testing to identify eSCC, similar procedures are common in pipeline systems and are used to evaluate pipeline segments that can suffer external cracking due to high pH carbonate–bicarbonate SCC or near-neutral SCC. Pipeline operators have suggested that UT shear wave sensors may also be used for in-line inspection (ILI) of pipelines for eSCC; however, the sensors need to be configured depending on the orientation of the cracks produced by SCC. Presently, most UT ILI tools are configured for longitudinal cracking, whereas the two cases of ethanol SCC found in pipelines, thus far were cracks oriented circumferentially produced as a result of the residual tensile stress from bending. In cases where ILI crack inspections involve possible cracks in both longitudinal and circumferential orientations, multiple passes of ILI tools may be necessary [5].

### 25.4.3 Mitigation

The most commonly used approaches to mitigation of eSCC in steel equipment appear to fall into two categories: (a) reduce mechanical and/or residual tensile stresses in components, and (b) use of a barrier coating between the internal surface of steel equipment and FGE.

***25.4.3.1 Stress Reduction and Post Weld Heat Treatment*** Locations of high mechanical and residual tensile stresses are prime sites for initiation of eSCC. They potentially combine several factors key to cracking initiation and growth:

- Plasticity that tends to provide for local breakdown of the passive film.
- High tensile stress that acts as a driving force for initiation and crack growth.

Therefore, efforts should be taken to minimize high (and highly localized) tensile stresses and plastic strains if at all possible. High mechanical tensile stresses and stress concentrations can occur as a result of subsidence, lap-seam welds, and poor component fit-up. High (near yield point) residual tensile stresses are usually present in non-PWHT welds.

For new weld construction and repairs in FGE service, PWHT has been used to reduce susceptibility to eSCC through lowering of tensile residual stresses. These procedures have been mainly used in applications involving piping and equipment in ethanol storage and blending facilities where eSCC has been observed or is anticipated. It is recognized that under most circumstances for storage tanks and possibly pipelines, thermal stress relief procedures are impracticable or not possible, and coatings or other measures are the primary eSCC mitigation technique.

***25.4.3.2 Coatings*** The selective use of coatings has been successful in tanks exposed to FGE. They are used primarily on the bottom, lower wall, and roof components. Techniques for the use of coating and identification of ethanol resistant coatings are given in API 939E. However, FGE and gasoline–ethanol blends are incompatible with many types of polymeric coatings. Therefore, selection of coatings for FGE should be based on long-term compatibility of the coatings (and specific formulations) with ethanolic environments *under conditions of full immersion* obtained from either laboratory or field tests or field service experience. Examples of the coating types often cited for this service include those composed of (a) novolac epoxy or (b) epoxy phenolic resins. It is best to check with the coating manufacturer to confirm chemical resistance to ethanol immersion.

***25.4.3.3 Other Potential Mitigation Techniques*** Other potential mitigation techniques have been identified in recent laboratory research studies on eSCC of steels. As mentioned previously, the presence of dissolved oxygen in FGE is important to maintaining conditions conductive for eSCC initiation and growth. Therefore, deaeration, while not widely used in industrial handling and storage of FGE, can be a technique for reducing susceptibility to eSCC of steel. Deaeration can be attained by use of inert gas purging, mechanical deaeration, or by the addition of chemical scavengers for dissolved oxygen. Alternatively, addition of chemical agents that increase the alkalinity of the ethanolic solution or provide inhibitive benefits specifically for eSCC can also be used (see Figure 25.6) [6]. It needs to be stressed that, while these have been shown to work in the laboratory, such chemicals need to be evaluated prior to use in service for

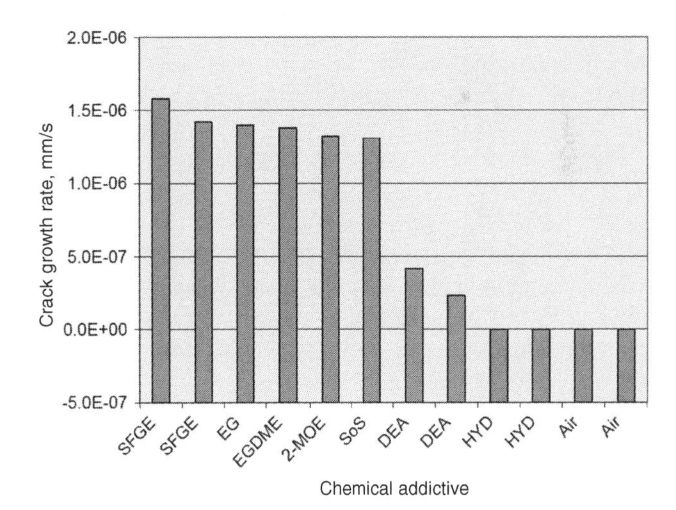


**FIGURE 25.6** Crack growth rate for ethanol SCC in SSR tests performed in SFGE compared with SFGE with chemical additives intended to reduce eSCC. (Air refers to baseline SSR tests performed in air with no SCC.) Sodium sulfite (SoS)—oxygen scavenger, ethylene glycol dimethyl ether (EGDME)—inhibitor, 2-metyoxyethanol (2-MOE)—inhibitor, diethanolamine (DEA)—inhibitor/pH modifier, hydrazine (HYD)—oxygen scavenger, and ethylene glycol (EG) [6]. (From © NACE International 2008.)

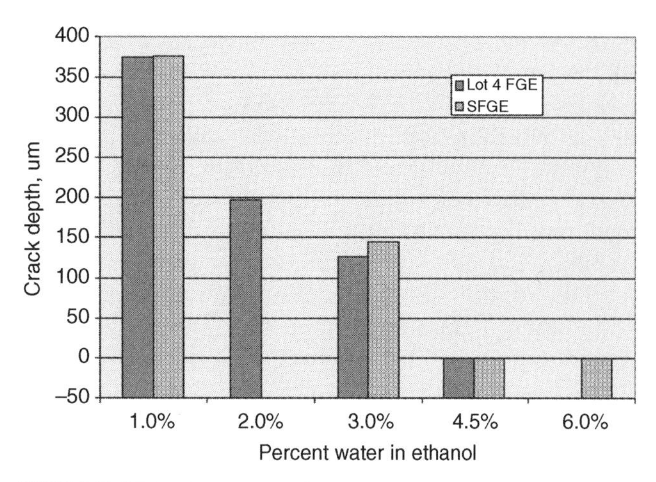


**FIGURE 25.7** Crack depth in SSR tests as a function of water content for a SFGE and one lot of corn-based FGE [12]. (From © NACE International 2010.)

their efficacy in mitigating eSCC and for conformance with end use product specifications, and for any unintended issues that they may cause in the ethanol and fuel distribution systems, association processes, or in the final fuel blend used in combustion engines.

One other potential avenue for eSCC mitigation is based on the reduced severity of ethanol SCC water concentrations in the ethanol in excess of the 1 vol% maximum amount specified in ASTM D4806 (See Figure 25.7) [12]. As shown in this figure, eSCC susceptibility in steels decreases rapidly in the range 1–4 vol% in terms of depth of SCC crack penetration in laboratory slow strain rate tests. One potential issue with this approach is the separation of a water phase in gasoline–ethanol blends at low temperature unless the water is removed from the ethanol before blending.

## 25.5 PATH FORWARD

Much of the initial effort in understanding eSCC came from the petroleum refining sector that operates facilities used to store FGE and blend it into conventional gasoline blends (currently up to 10% ethanol). This initial effort focused on literature review, laboratory research, and field studies to define an environmental cracking process that reportedly had already caused a substantial number of field failures. It was also used to uncover key variables that may be incorporated into fabrication and/or operating practices, which could ultimately reduce the occurrence of or eliminate eSCC.

More recently, the bulk of the work on eSCC has been aimed at eSCC as a potential damage mechanism in the pipeline distribution system for FGE and ethanol/fuel blends. The number of eSCC failures reported in pipelines has been very limited (only two documented at the time of this publication). However, as the supply of ethanol being transported via pipelines grows, the concern for the potential disruption of pipeline operations from eSCC is the obvious reason for the more recent studies.

As such, many of the issues discussed herein, which can be used to minimize susceptibility to eSCC (e.g., deaeration, inhibitors, and chemical treatments), are of keen interest. The current investigations on eSCC have shifted emphasis from issues affecting refining storage and blending facilities to assessing the potential for eSCC (and methods of mitigation, if present) in the pipeline distribution system (including cyclic loading characteristic of liquids pipelines and corrosion fatigue) and the impact of MIC [13,14] in equipment handing FGE and ethanol/fuel blends.

## REFERENCES

1. ASTM D4806 (2009) *Standard Specification for Denatured Fuel Ethanol for Blending with Gasolines for Use as Automotive Spark-Ignition Engine Fuel*, ASTM International, West Conshohocken, PA.
2. API (2003) *Publication 939D, Stress Corrosion Cracking of Carbon Steel in Fuel Grade Ethanol: Review and Survey*, American Petroleum Institute, Washington, DC, November 2003.
3. Kane, R.D., Maldonado, J.G., and Klein, L.J. (2004) Stress corrosion cracking in fuel ethanol: a newly recognized phenomenon. Proceedings of CORROSION/2004, NACE International, March 2004, Houston, TX, Paper No. 04543.
4. API (2007) *Technical Report 939D, Stress Corrosion Cracking of Carbon Steel in Fuel-Grade Ethanol: Review, Experience Survey, Field Monitoring, and Laboratory Testing*, 2nd ed., American Petroleum Institute, Washington, DC, Updated 2007 and 2013.
5. API (2008) *Technical Bulletin 939E, Identification, Repair, and Mitigation of Cracking of Steel Equipment in Fuel Ethanol Service*, 1st ed., American Petroleum Institute, Washington, DC, November 2008 and 2013.
6. Beavers, J.A., et al. (2008) Prevention of internal SCC in ethanol pipelines. Proceedings of CORROSION/2008, NACE International, March 2008, Houston, TX, Paper No. 08153.
7. Lou, X., et al. (2010) Understanding the stress corrosion cracking of X-65 pipeline steel in fuel-grade ethanol. Proceedings of CORROSION/2010, NACE International, 2010, Houston, TX, Paper No. 10073.
8. Beavers, J.A., et al. (2009) Effects of steel microstructure and ethanol-gasoline blend ratio. Proceedings of CORROSION/2009, NACE International, March 2009, Houston, TX, Paper No. 09532.
9. Venkatesh, A. and Kane, R.D. (2009) Fracture analysis of slow strain rate test for stress corrosion cracking. Proceedings of CORROSION/2009, NACE International, March 2009, Houston, TX, Paper No. 09296.
10. McIntyre, D.R., et al. (2009) SCC behavior of steel in fuel ethanol and butanol. Proceedings of CORROSION/2009, NACE International, 2009, Houston, TX, Paper No. 09530.

11. Venkatesh, A., et al. (2010) Evaluation of stress corrosion cracking behavior of steel in multiple ethanol environments. Proceedings of CORROSION/2010, NACE International, March 2010, Houston, TX, Paper No. 10077.
12. Beavers, J., et al. (2010) Recent progress in understanding and mitigating SCC of ethanol pipelines. Proceedings of CORROSION/2010, NACE International, March 2010, Houston, TX, Paper No. 10072.
13. Sowards, J.W., et al. (2011) Effect of ethanol fuel and microbiologically influenced corrosion on the fatigue crack growth behavior of pipeline steels. Proceedings of DOD Corrosion Conference (La Quinta, CA), NACE International, 2011, Houston, TX.
14. Jain, L. (2013) Microbiologically influenced corrosion of linepipe steels in ethanol and acetic acid solutions. Proceedings of CORROSION/2013, NACE International, March 2013, Houston, TX, Paper No. 2013–2250.

## BIBLIOGRAPHY OF ADDITIONAL eSCC PAPERS

### Laboratory Techniques, Environmental Variables and eSCC Crack Propagation

Gui, F., et al. (2010) Effect of ethanol composition on the SCC susceptibility of carbon steel. Proceedings of CORROSION/2010, NACE International, March 2010, Houston, TX, Paper No. 10075.

Lou, X., Yang, D., and Singh, P.M. (2009) Effect of ethanol chemistry on stress corrosion cracking of carbon steel in fuel-grade ethanol. *CORROSION*, 65(12), 785–797.

Sridhar, N., Price, K., Buckingham, J., and Dante, J. (2006) Stress corrosion cracking of carbon steel in ethanol. *CORROSION*, 62(8), 687–702.

Torkkeli, J., et al. (2011) Evaluation of ost-weld heat treatment as a method to prevent stress corrosion cracking of carbon steel in ethanol by notched constant tensile load testing. Proceedings of the European Corrosion Congress, EUROCORR 2011, September 4–8, 2011, Stockholm, Sweden.

Xiaoyuan, L. and Singh P.M. (2010) Role of water, acetic acid and chloride on corrosion and pitting behaviour of carbon steel in fuel-grade ethanol. *Corrosion Science*, 52(7), 2303–2315.

### Measurement of Oxygen in FGE

Lou, X., Yang, D., and Singh, P.M. (2009) Effect of ethanol chemistry on stress corrosion cracking of carbon steel in fuel-grade ethanol. *CORROSION*, 65(12), 785–797.

Sridhar, N., Price, K., Buckingham, J., and Dante, J. (2006) Stress corrosion cracking of carbon steel in ethanol. *CORROSION*, 62(8), 687–702.

# 26

# AC CORROSION

LARS VENDELBO NIELSEN
*MetriCorr ApS, Glostrup, Denmark*

## 26.1 INTRODUCTION

Alternating current (AC) corrosion can be defined as corrosion caused or influenced by AC. If the AC component is removed or limited below a certain level, the corrosion will be diminished or completely mitigated. Virtually always, the actual AC corrosion case is equally under the influence of direct current (DC). This means that in addition to mitigation by limiting the AC component, AC corrosion can be almost completely controlled by carefully adjusting the DC component—in terms of the cathodic protection (CP) condition.

AC corrosion in buried pipelines has increasingly become a concern as shared right-of-ways with high-voltage power lines have become common practice in the so-called energy corridors (Figure 26.1). Proximity with AC-powered traction systems is another source of AC. The AC is induced in the pipeline through extended length of parallelism between the pipeline and the interfering power source. Figure 26.2 shows schematically an arrangement of the conductors (L1, L2, and L3) carrying the current in a high-voltage power line as well as the neutral earth wire (or top wire), which is often also included in the tower arrangement primarily for fault and lightning protection. The AC current in each of the conductors (L1, L2, and L3) creates a magnetic field that induces AC in the pipeline like a secondary coil in a transformer. Since the current in the phase conductors reach their instantaneous peak values at one-third of a cycle from each other (phase shift 120°), the net resulting magnetic field that influences the pipeline will be significantly reduced compared with the effect of one single conductor. In fact, in principle, if the pipeline was located completely symmetrical relative to the position of the conductors, the net induction would be zero. The neutral earth wire also exhibits the magnetic field; hence AC will be induced in the earth wire. An alternative path for the AC may be created through the earth points of these wires. If the pipeline is grounded close to the earth points, a stray AC current interference may result.

The AC corrosion attach is usually occurring at small rather than large coating defects and forms a characteristic circular or hemispheric attach as if initiated in a small point and growing laterally in all directions resulting in a ball-shaped morphology (Figure 26.3) with corrosion rates that can reach several millimeters per year in severe cases.

This chapter has been divided into two parts. The first part provides an overview of the current understanding of AC corrosion and the parameters behind it. The second part provides engineering practices for the mitigation and monitoring of AC corrosion.

## 26.2 BASIC UNDERSTANDING

AC corrosion is influenced by both AC and DC parameters, which are correlated through the spread resistance. Following is a short presentation of the parameters needed to understand AC corrosion:

- **AC voltage** ($U_{AC}$) is the AC voltage that can be measured between the pipeline and remote earth. It is the driving force for AC current exchanged between the pipe and the earth though a coating defect—which may give rise to corrosion—or through grounding devices,

*Oil and Gas Pipelines: Integrity and Safety Handbook*, First Edition. Edited by R. Winston Revie.

**FIGURE 26.1** Shared corridor—high-voltage power lines and underground pipeline indicated by test station.

including galvanic anodes that may have been installed for mitigation purposes. The AC voltage will change along the length of the pipeline since the induced voltage depends on characteristics of the pipeline, characteristics of the interfering AC power system, as well as the geometrical and geographical alignment. The AC voltage often also changes with time primarily because of intermittent conditions in the AC power system, for instance, because household power consumption is different during daytime and nighttime.

- **AC current density** ($J_{AC}$) is commonly expressed in A/m$^2$. In terms of AC corrosion, this parameter most often refers to the AC current density in a coating defect or in a coupon or probe used to simulate a coating defect of a certain area.
- **ON potential** ($E_{ON}$) is the DC potential measured between the pipeline and a reference electrode placed in the adjacent soil while the CP current is flowing. This potential includes IR drops in the soil.
- **IR-free potential** ($E_{IR\ free}$): Other terms that are essentially used for the same quantity are polarized potential and OFF potential. IR-free potential and polarized potential may be defined as the pipe-to-soil DC potential without the IR-drop error included. OFF potential is indicating the potential measured without the CP current flowing and is therefore supposed to compensate for the IR drop caused by the CP current.
- **DC current density** ($J_{DC}$) is commonly expressed in A/m$^2$. In terms of AC corrosion, this parameter most often refers to the DC current density in a coating defect or in a coupon or probe used to simulate a coating defect of a certain area.
- **Spread resistance** ($R_S$) is commonly expressed in $\Omega\cdot$m$^2$. In terms of AC corrosion, this parameter most often refers to the resistance from the pipe to earth through a specific coating defect with known area, or the resistance from a coupon or probe with a known exposed area to earth. The spread resistance multiplied by the DC current density constitutes the IR drop

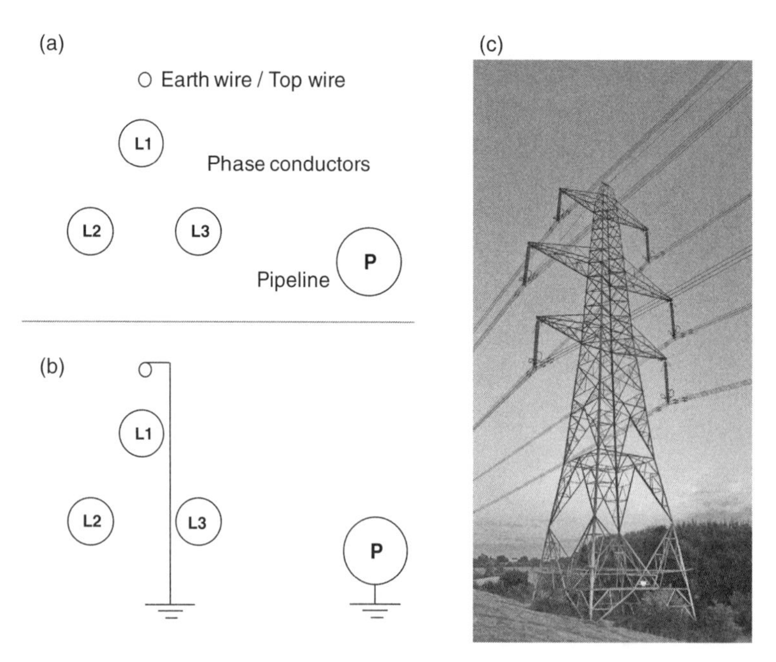


**FIGURE 26.2** Illustration of conductors in a high-voltage power line relative to a pipeline. (a) Conductors L1, L2, and L3 and an earth wire on top. (b) Earth wire grounded next to pipeline grounding. (c) Pylon carrying earth wires and two L1, L2, L3 systems.

**FIGURE 26.3** Examples of AC corrosion developed from small coating defects in the pipeline. (Upper left: Courtesy of CeoCor (Comité d'étude de la Corrosion et de la Protection des Canalisation), A.C. Corrosion on Cathodically Protected Pipelines—Guidelines for risk assessment and mitigation measures—2001 [1]. Upper and lower right: Photographs courtesy of Roger Ellis, Pipeline Manager, Shell UK Ltd. Lower left: Courtesy of PTT Plc, Thailand.)

mentioned earlier, and the spread resistance is (approximately) the proportionality factor between AC voltage and AC current density.

The spread resistance is a key parameter in the AC corrosion process as further demonstrated in Section 26.2.1.

### 26.2.1 The Spread Resistance

The spread resistance plays a key role in the AC corrosion process since in many ways it ties together the DC and AC parameters. In the following, some observations and features of the spread resistance are given.

Typically, the AC voltage and AC current density exchanged between the pipe and the soil through a circular coating defect with area ($A$) and diameter ($d$) are described as proportional through the relation:

$$U_{AC}(\mathrm{V}) = \frac{\rho_{soil}(\Omega \cdot \mathrm{m})}{2 \times d(\mathrm{m})} \times A(\mathrm{m}^2) \times J_{AC}(\mathrm{A/m}^2) \qquad (26.1)$$

In this relation, $\rho_{soil}$ is the soil resistivity. The spread resistance is the proportionality factor between the AC voltage and AC current density and may in this case be described as

$$R_S(\Omega \cdot \mathrm{m}^2) = \frac{\rho_{soil}(\Omega \cdot \mathrm{m})}{2 \times d(\mathrm{m})} \times A(\mathrm{m}^2) \qquad (26.2)$$

Hence, according to this equation the spread resistance depends both on the magnitude (size) of the coating defect and on the soil resistivity.

This is true only with some modifications as follows.

***26.2.1.1 The Spread Resistance Dependency on DC Current Density*** The CP of the bare steel at a coating defect will affect the soil chemistry at the close proximity of the steel surface—particularly through the production of hydroxyl ions from water by the CP current—see Figure 26.4.

The electrochemical processes resulting from CP may influence the resistivity of the soil close to the coating defect in three different ways:

1. Increase the spread resistance—leading to decrease in the resulting AC current density at a constant AC voltage in accordance with Equations 26.1 and 26.2.

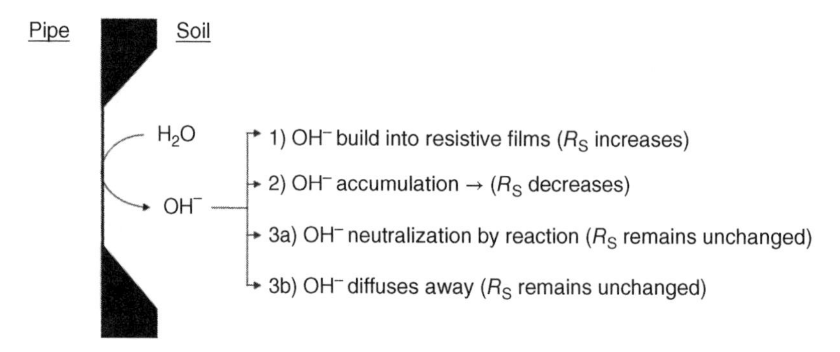


**FIGURE 26.4** Schematic illustration of the modification of the spread resistance due to the production of hydroxyl ions through the electrochemical cathodic protection processes.

2. Decrease the spread resistance—leading to an increase in the resulting AC current density at a constant AC voltage in accordance with Equations 26.1 and 26.2.
3. No influence.

*Case 1* (increase of the spread resistance). Due to the CP processes, alkalinity ($OH^-$) is produced at the bare steel surface at the coating defect. If the soil contains earth alkaline cations such as $Ca^{2+}$ and $Mg^{2+}$, the alkalinity production will lead to precipitation of calcium- and/or magnesium hydroxides. This scale forming process will form resistive layers on the steel surface, and the spread resistance will increase regardless of a constant soil resistivity in the bulk. The process may increase the spread resistance by 2 orders of magnitude.

An example is shown in Figure 26.5 [2]. The development of the spread resistance is shown throughout approximately 70 h in an environment with a mixture of scale forming earth alkaline cations ($Ca^{2+}$ and $Mg^{2+}$) This development is compared with the development in a similar but non-scale forming environment where earth alkaline cations ($Ca^{2+}$ and $Mg^{2+}$) were replaced with $Na^+$ equivalents. In both cases, the DC ON potential was −850 mV CSE (copper–copper sulfate electrode) whereas the AC voltage was 20 V. In the scale forming environment, the spread resistance increases by a factor of 20 (and continues its increase), whereas the spread resistance in the non-scale forming environment is constant over the period.

*Case 2* (decrease of the spread resistance). This relates to the accumulation of the hydroxyl ions ($OH^-$) produced by the CP processes. If produced in sufficiently large amounts—with a certain reaction rate—hydroxyl ions will concentrate at the surface and contribute to the conductivity of the soil close to the steel surface. Besides an increase in pH, the accumulation will lower the spread resistance and cause

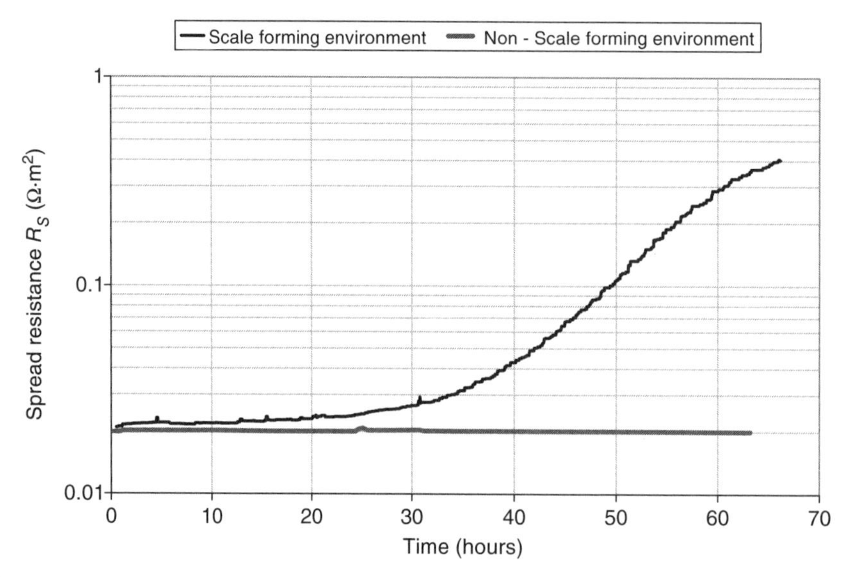


**FIGURE 26.5** Development of the spread resistance in a scale-forming environment versus a non-scale-forming environment under mild cathodic protection (−850 mV CSE).

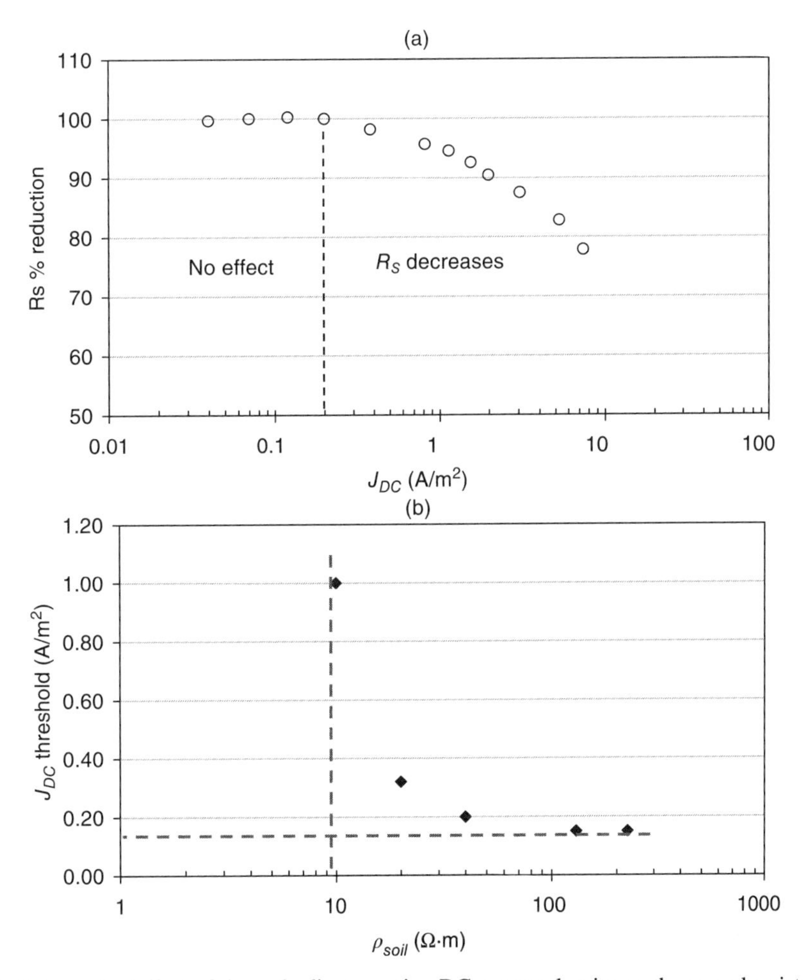


**FIGURE 26.6** (a) Effect of the cathodic protection DC current density on the spread resistance—illustrating a threshold DC current density above which the spread resistance decreases. (b) Threshold cathodic DC current density above which the spread resistance decreases—as a function of the soil resistivity.

increase in the AC current density at constant level of AC voltage according to Equations 26.1 and 26.2.

Figure 26.6a illustrates an example established in a soil box with a chemical composition having a non-scale forming capability. A coupon having area of 1 $cm^2$ was increasingly polarized in the environment and the spread resistance measured simultaneously. The decrease in spread resistance is observed to occur when the cathodic DC current density exceeds a threshold of 0.15 $A/m^2$ in this environment.

In a similar manner, the spread resistance in soil boxes with various soil resistivities (various concentrations of non-scaling solutions based on same chemical substances) was recorded as a function of the applied CP current density. The threshold CP current densities were defined where the spread resistance began decreasing. This threshold level versus the soil resistivity is shown in Figure 26.6b [3]. As can be observed, such simple experiments illustrate how the CP current density can take control of the environment close by the coating defect and in particular, above a certain threshold current density, gradually decrease the spread resistance. This threshold depends on the actual soil resistivity. The higher the resistivity, the lower the level of CP needs to be before the resistivity close to the coating defect is affected. In conclusion, Equation 26.1 must be used with care and never substitute real measurements.

*Case 3* (spread resistance remains unchanged). This relates to the situation where the hydroxyl ions ($OH^-$) are not produced sufficiently fast by the CP processes to accumulate at the surface at the coating defect (below the threshold CP current density).

#### 26.2.1.2 The Spread Resistance Dependency on Coating Fault Size and Geometry

As already indicated through Equations 26.1 and 26.2, the spread resistance value depends on the diameter of the coating fault. Substituting $A = \pi\cdot(½d)^2$ in Equation 26.2 gives

$$R_S(\Omega\cdot\mathrm{m}^2) = \frac{\rho_{soil}(\Omega\cdot\mathrm{m})\times d(\mathrm{m})\times\pi}{8} \quad (26.3)$$

As observed from this, the spread resistance (in $\Omega\cdot\mathrm{m}^2$) increases with increasing area of the coating defect. Consequently, small coating defects are more susceptible to AC corrosion because the spread resistance is lower compared with larger defects, and the AC current density produced at small defects will be higher than the AC current density produced at larger defects—given the same level of AC voltage.

This above is a simplification and relates to a circular defect.

Figure 26.7 illustrates further the geometry around a circular coating defect. Besides the diameter ($d$) describing the size of the defect, one must consider the effects on the spread resistance of the thickness ($t$) of the coating and one must consider the angle between the exposed metal surface and the coating. Further, it should be considered how a scratch behaves in comparison with a circular coating defect. These questions have been studied by mathematical numerical simulations, which afterwards have been partly checked in controlled practical experiments [4]. Some essential results are shown in Figure 26.8. Generally, the spread resistance is illustrated for a constant soil resistivity. Upper figures relate to conditions where the coating is very thin (0.001 mm) and with no practical influence of the spread resistance. Lower figures relate to conditions where the coating is significant (3 mm thick). Left figures describe calculated actual values, whereas right figures describe indexed values where index 100 is the case of a circular defect. Besides the circular defect, calculations were made on rectangular defects with increasing length ($L$) to width ($w$) ratios (1:1, 10:1, 100:1, and 1000:1) virtually describing the transition from a circular defect to a scratch. The calculations have been made with the area of the coating defect as the primary parameter.

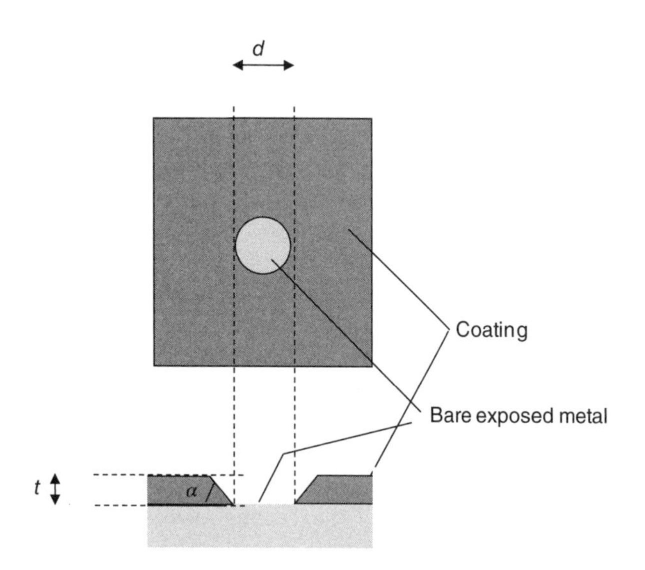


**FIGURE 26.7** Illustrations of the geometrical parameters related to a circular coating defect. Besides diameter ($d$) and area ($A$) of the coating defect, the coating thickness ($t$) and coating contact angle ($\alpha$) with the metal surface are also considered [3].

The upper figures show that as long as the coating is thin, and therefore have no effect on the spread resistance, the circular defect will have the highest spread resistance (practically identical to the square (1:1)), whereas increasing length to width ratio provides lower and lower spread resistances. The spread resistance relating to the scratch-like defects—length to width ratio 1000:1—will be approximate only 20% of the spread resistance relating to the circular defect. Hence, the scratch-like defect will produce five times higher AC current density than the circular one.

The lower figures take into account the geometrical limitations provided by a coating with a defined thickness (3 mm). It is understandable that the spread resistance of a pin hole defect in a 3 mm coating will be conducted primarily by the narrow gab in the coating rather than the actual area of the bare steel surface. The figures show how the coating becomes predominant at small defects and that the spread resistance for all geometries practically converts toward the same value when the coating defect is small (0.01 mm).

In conclusion to this it is clear that the geometry of the coating defect conducts the susceptibility to AC corrosion; small defects are more likely to suffer from AC corrosion than larger defects, and scratch-like defects are more susceptible than circular or square defects.

The combined effect of coating fault geometry and influences of DC current on the chemistry at the coating defect remains to be an area with lack of research.

### 26.2.2 The Effect of AC on DC Polarization

Another phenomenon being important in the understanding of AC corrosion is the effect of AC on DC polarization behavior. Most often the AC will depolarize the electrode kinetics, which implies in practice that the DC current demand for maintaining a certain CP potential will be increased with increasing AC. This tendency can be readily observed in practice by controlling the pipeline ON potential and register the current demand alongside the pipe-to-soil AC voltage. In the laboratory, the effect can be demonstrated for example in a soil box experiment where a working electrode is kept at a certain ON potential with superimposed AC at different levels. Figure 26.9 gives an example of this—clearly illustrating that AC causes depolarization of both the anodic and the cathodic electrode kinetics.

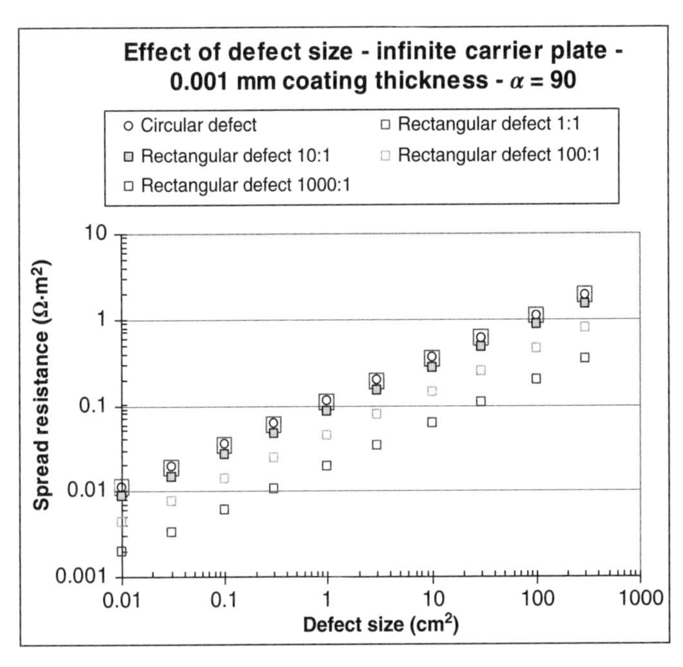


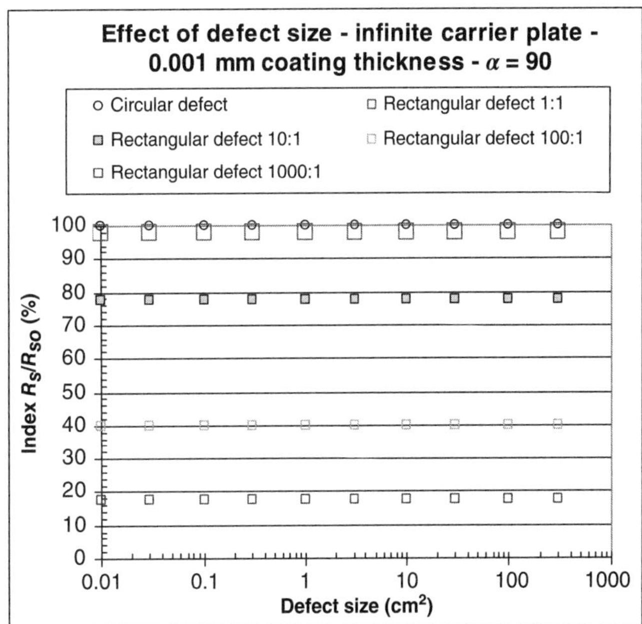


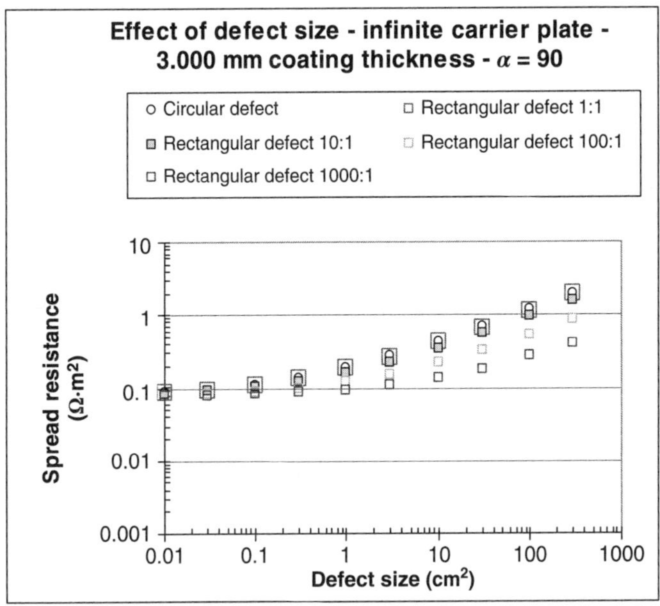


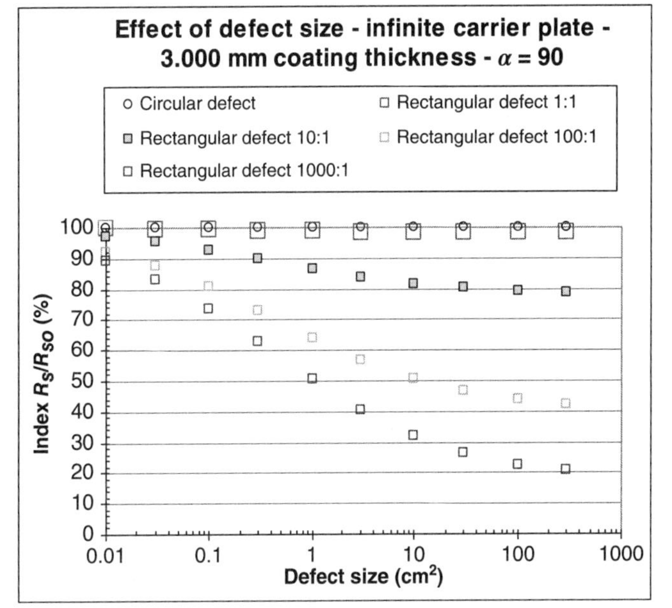


**FIGURE 26.8** Effects of the coating defect size and geometry on the coating defect spread resistance. Data are mainly calculated with random experimental checks [3].

The effect of AC on the IR-free DC potential measurement is a further challenge as illustrated in Figure 26.10. The figure shows three cathodic polarization curves—in all three cases represented by the ON potential versus CP current density and the IR-free potential versus CP current density. The difference between the figures is the superimposed AC—0.5, 4, and 10 V. Besides showing the aforementioned effect that AC causes a depolarization of the DC kinetics, it can also be observed that as the AC increases, the IR-free polarization curve starts acting in an odd manner, in the sense that increased CP current actually leads to more anodic IR-free DC potentials. The reason for this behavior remains so far unsolved, but the phenomenon is readily reproduced and has been reported independently from different sources (e.g., Büchler [5]). Figure 26.11 quantifies the effect, which in this book is referred to as the Faradaic rectification.

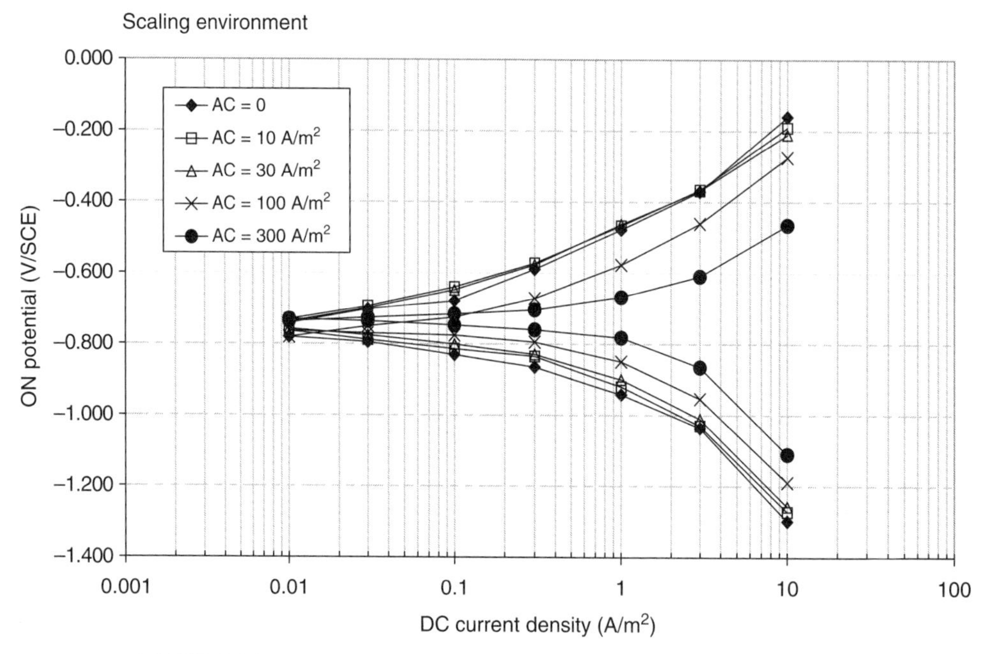


**FIGURE 26.9** Effects of AC current densities on DC polarization curves.

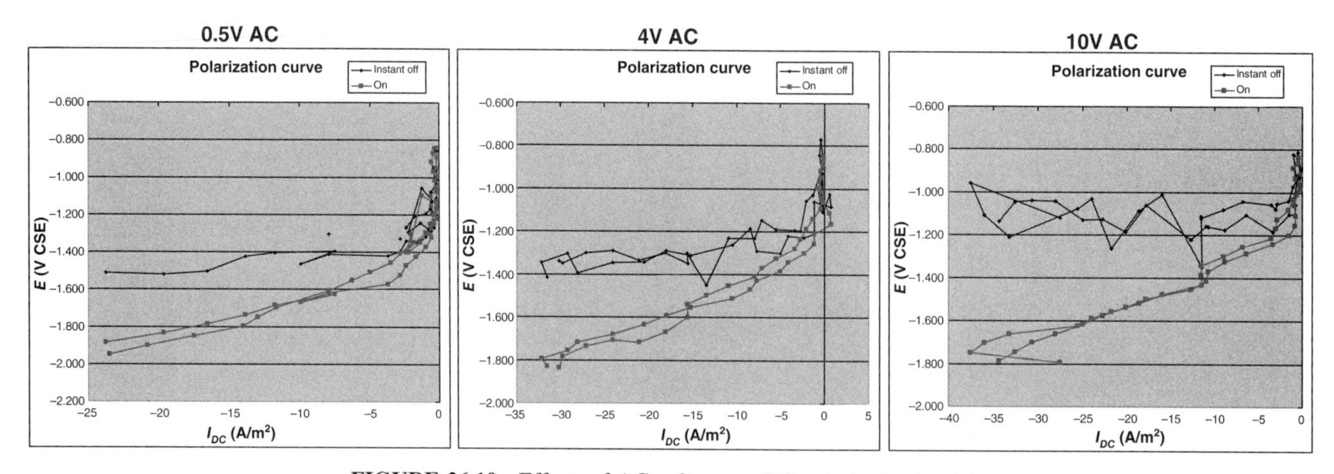


**FIGURE 26.10** Effects of AC voltage on DC polarized potentials.

### 26.2.3 The Vicious Circle of AC Corrosion—Mechanistic Approach

A proposed mechanism of AC corrosion has been based on reactions and stability domains illustrated in the Pourbaix diagram—Figure 26.12. During the anodic (positive) half wave of an AC cycle, the bare metal is oxidized and forms an oxide film—such as $Fe_3O_4$. During the cathodic (negative) half wave of the AC cycle, this oxide film is reduced and converted into a non-protective rust layer—such as $Fe(OH)_2$. In the next anodic cycle, a new oxide film grows that is converted into a larger amount of rust and so on; see Figure 26.13 [5,6,7].

This may be a true mechanism as long as the following prerequisites—found in any laboratory experiment and in any field detection of AC corrosion—are observed (Figure 26.14) with reference to the previous illustrations:

- Induced AC;
- A small coating defect; and
- Excessive CP.

The induced AC is the driving force for the process, whereas the smaller coating defects as well as the excessive CP facilitate a low spread resistance. The lower the spread

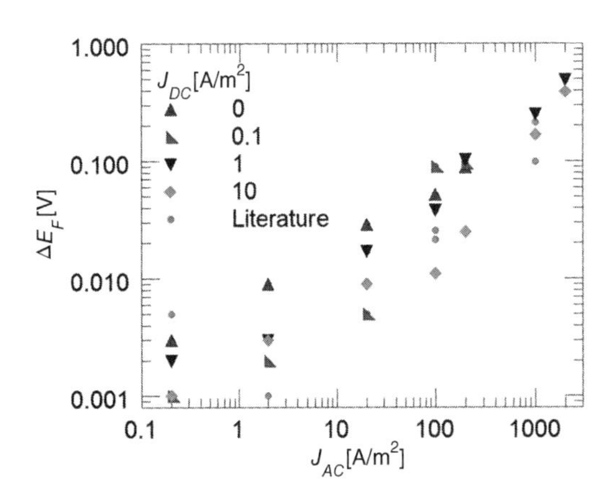


**FIGURE 26.11** Effects of AC voltage on DC polarized potentials [5]. $\Delta E_F$ is explained as the Faradaic rectification caused by the AC. (Courtesy of Dr. Markus. Büchler, GSK Switzerland.)

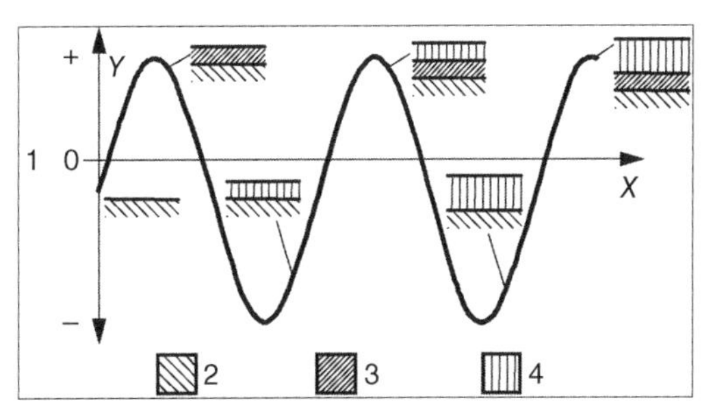


**FIGURE 26.13** Schematic description of an AC corrosion process. (Courtesy of Dr. Markus. Büchler, GSK Switzerland.)

**Key**

| | | | |
|---|---|---|---|
| 1 | A.C. current present on a coating defect | 2 | Metal |
| 3 | Passive film (e.g., $Fe_3O_4$) | 4 | Iron hydroxide (e.g., $Fe(OH)_2$) |
| *X* | Time | *Y* | Current |

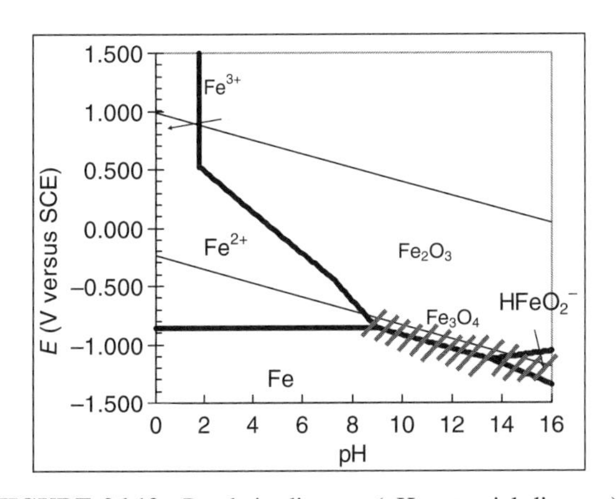


**FIGURE 26.12** Pourbaix diagram (pH–potential diagram).

resistance, the higher the AC current density at a certain level of AC voltage, since AC voltage, spread resistance, and AC current density are correlated simply through Ohm's law.

If the AC current density is high enough, the DC current density (for the CP process) will increase due to depolarization effects of the AC. The net result is that the cathodic DC current increases—i.e., the CP current at the coating defect becomes even more excessive. Such excessive CP current produces alkalinity/hydroxyl ions ($OH^-$) nearby the coating defect by either of these reactions:

$$2H_2O + 2e^- \rightarrow H_2 + 2OH^-$$

$$O_2 + 2H_2O + 4e^- \rightarrow 4OH^-$$

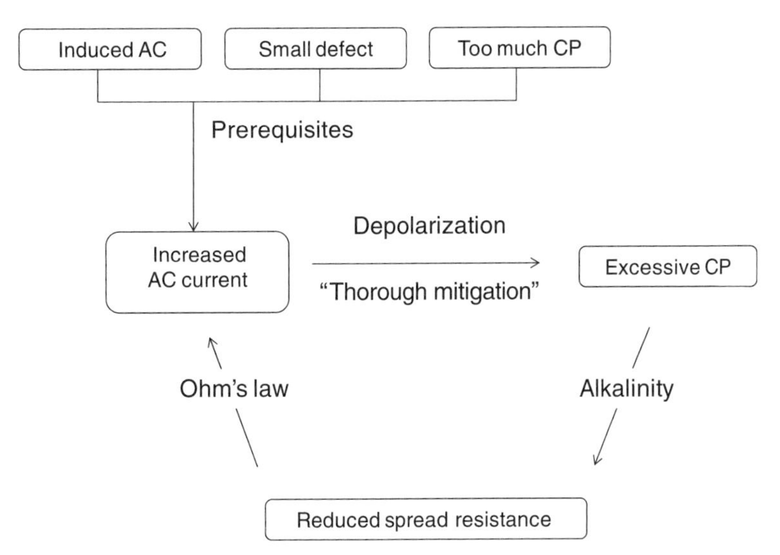


**FIGURE 26.14** The vicious circle of AC corrosion.

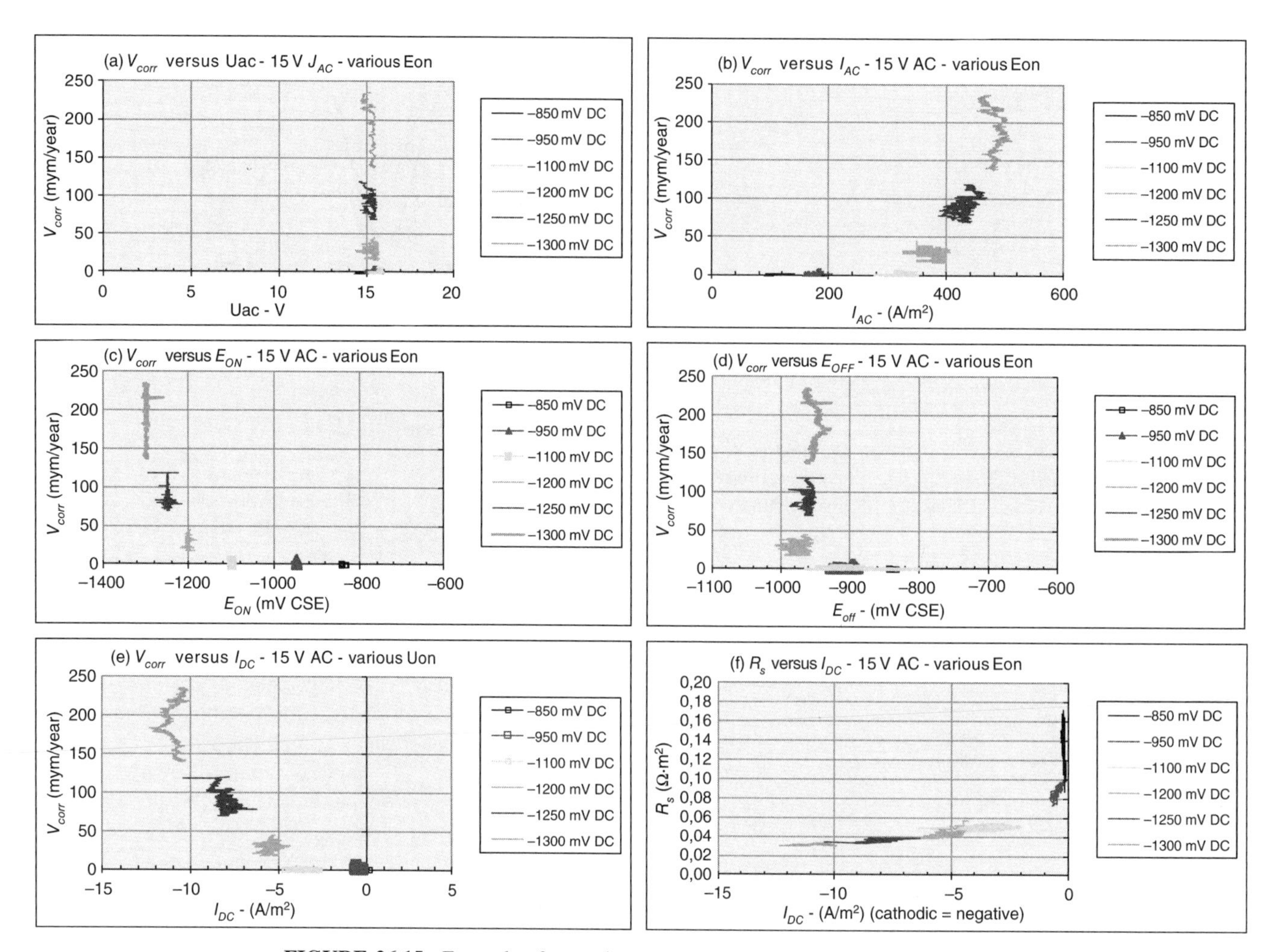


**FIGURE 26.15** Example of corrosion rate versus various electrical parameters.

Regardless of which of these reactions is taking place, the net production of hydroxyl ions is achieved nearby the coating defect, and this can result in alkalization (pH raise) and a significant lowering of the spread resistance. In turn, the lowering of the spread resistance results in ever higher levels of AC current density, which again depolarizes the CP kinetics, produces more alkalinity, raises pH further, lowers spread resistance further, increases AC current density further, and so on. A vicious circle has been produced (Figure 26.14). Eventually, the pH at the coating defect may reach such high values that the high-pH free corrosion area ($HFeO_2^-$) in the Pourbaix diagram (Figure 26.12) will lead to an open wound of corrosion. Corrosion rates in excess of 10 mm/year have been observed.

Figure 26.15 illustrates some interesting observations that follow from the "vicious circle" scenario. The results have been produced in a laboratory soil box environment with a non-scaling electrolyte embedded in purified quartz sand. At a constant 15 V AC, six different levels of CP (ON potentials) were maintained for some weeks to study the development of the corrosion rate measured by electrical resistance (ER) probes and the various electrical parameters associated.

Figure 26.15a shows the corrosion rate displayed as a function of the AC voltage—which was held constant in all experiments. As observed, corrosion can be created or kept at zero depending on the CP level. The AC voltage itself therefore is not a valid indicator of corrosion risk.

Figure 26.15b shows the corrosion rate displayed as a function of the AC current density. What can be deduced from this is that despite a constant AC voltage, the AC current density changes from approximately 100 A/m²—at low CP level—up to 500 A/m² at higher CP levels. The corrosion rate may be tied to the AC current density.

Figure 26.15c shows the corrosion rate displayed as a function of the DC ON potential. Clearly, the corrosion rate increases the more CP is applied.

Figure 26.15d shows the corrosion rate displayed as a function of the IR compensated DC potential. This graphic illustrates above all the Faradaic rectification effect discussed in Figures 26.10 and 26.11. Increasing the CP level (more negative ON potential) below −1200 mV (ON) results in less negative IR-free potentials.

Figure 26.15e shows the corrosion rate displayed as a function of the CP DC current density (note that the negative sign indicates cathodic current). Clearly, corrosion rate increases when CP current density increases.

Figure 26.15f shows the spread resistance as a function of the DC current density. This again illustrates the spread resistance as being a very central parameter in AC corrosion. The graph shows that the CP DC current density is able to lower the spread resistance by a factor of 5–6 in this case. In turn, the AC current density is increased by the same factor of 5–6 (Figure 26.15b).

Getting out of the vicious circle can be managed, as will be discussed hereafter.

## 26.3 AC CORROSION RISK ASSESSMENT AND MANAGEMENT

Standards and guidelines on how to assess the risk of AC corrosion are currently being developed, both under the auspices of NACE, ISO, and CEN. The latter—EN 15280:2013—*Evaluation of AC corrosion likelihood of buried pipelines applicable to cathodically protected pipelines*—published in September 2013 [8]—takes into account the up-to-date understanding of AC corrosion as presented in Section 26.2, and therefore, will be the backbone of this section on risk assessment and management.

The development of this particular standard has to a large degree been based on established monitoring in selected test stations in which corrosion rates on ER corrosion probes have been compared with the electrical quantities ($U_{AC}$, $J_{AC}$, $E_{ON}$, $E_{OFF}$, $J_{DC}$) as previously defined [5,6,9].

### 26.3.1 Criteria

EN 15280 operates with a set of normative criteria based on AC/DC current densities or direct measurements of corrosion rates on coupons or probes and a set of informative criteria based on AC voltage and DC ON potentials (ON potential approach). Since the conditions vary from case to case, it is acknowledged that a single threshold value cannot be applied. Previously, a 30 A/m$^2$ AC current limit measured on coupons has been applied as a rule of thumb, but recent observations both regarding thresholds and regarding qualitative understanding like the vicious circuit of AC corrosion have demonstrated that such simple criterion is not applicable.

### 26.3.2 Current Criteria

The following is based closely on section 7 in EN 15280:2013 [8].

The design, installation, and maintenance of the CP system shall ensure that the levels of AC voltage do not cause AC corrosion. This is achieved by reducing the AC voltage on the pipeline and the current densities (in coupons) as follows:

- As a *first* step, the AC voltage on the pipeline should be decreased to a target value that should be 15 V or less. This value is measured as an average over a representative period (example given is 24 h).
- As a *second* step, effective AC corrosion mitigation can be achieved by complying with (usual) CP criteria in EN 12954 Table 1 [10] (see footnote[1]), and maintaining the AC current density (rms) over a representative period of time (example given is 24 h) to be lower than 30 A/m$^2$ on a 1 cm$^2$ coupon or probe;

  or
- maintaining the average CP current density over a representative period of time (example given is 24 h) lower than 1 A/cm$^2$ on a 1 cm$^2$ coupon or probe if the AC current density (rms) is higher than 30 A/m$^2$;

  or
- maintaining the ratio between AC current density and DC current density <5 over a representative period of time (example given is 24 h).

These current density criteria have been illustrated in Figure 26.16.

As a normative alternative to the previously mentioned list of criteria, it is also allowed to demonstrate effective AC corrosion mitigation by corrosion rate measurements (ER probes, weight loss coupons).

***26.3.2.1 ON Potential Approach*** As an informative alternative the annex E of EN 15280:2013 suggests an ON-potential approach having criteria as follows:

The CP criteria in EN 12954 Table 1 [10] (same footnote as mentioned earlier), should be respected.

1) First scenario: "more negative" CP level. In this case, one of the three parameters given ahead, in order of priority, can be applied:
   - The following formula should be satisfied: $\frac{U_{AC}}{|E_{ON}|-1.2} < 3$

     Note that in this case, it is important to ensure that there is no corrosion risk due to cathodic disbondment and no adverse effect caused by hydrogen evolution.

     or
   - AC current density <30 A/m$^2$.

     or
   - ratio between AC and DC current density <3 if AC current density is >30 A/m$^2$.

[1] This is essentially the CP criteria −850 mV versus CSE under normal conditions, −950 mV CSE under anaerobic conditions, −750 mV CSE in sandy aerated soil, and −650 mV in sandy aerated soil with resistivity higher than 1000 ?om.

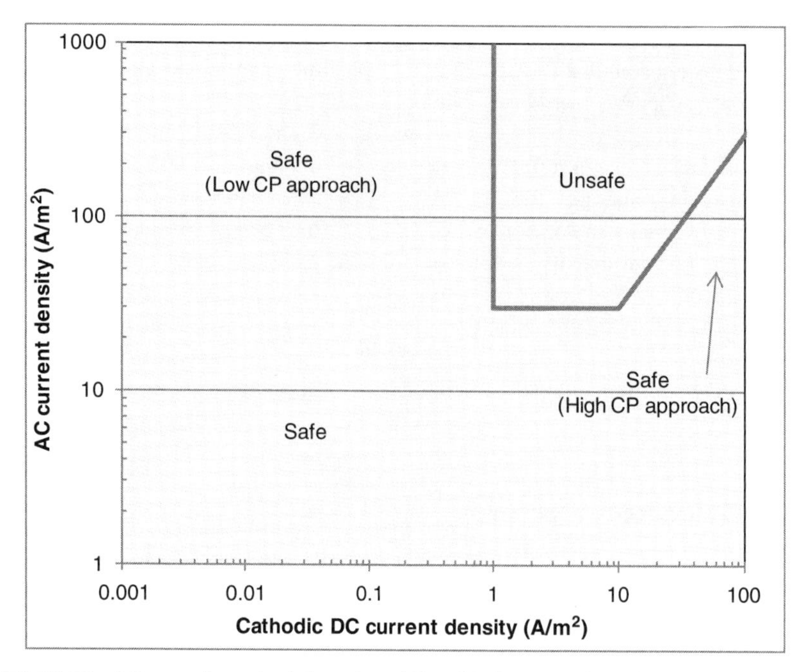


**FIGURE 26.16** AC corrosion criteria based on AC and DC current densities measured on installed coupons. Criteria are taken from EN 15280:2013 [8].

2) Second scenario: "less negative" CP level. In this case, one of the three parameters given ahead, in order of priority, can be used:
   - Average $U_{AC} < 15$ V if the average $E_{ON}$ is more positive than $-1.2$ V CSE;
   
     or
   - average AC current density less than 30 A/m$^2$;
   
     or
   - CP average current density <1 A/m$^2$ if average AC current density is >30 A/m$^2$.

AC voltage and DC potential should be measured using the same reference electrode placed in remote earth.

Figure 26.17 illustrates these guidelines for AC corrosion risk assessment.

### 26.3.3 Mitigation Measures

Methods for the mitigation of AC corrosion address both the AC component and the adjustment and control of the CP level. The mitigation should be incorporated both in the design and construction phase of a pipeline, as well as in the operational stage.

For the assessment of the level of induced voltage, modern computer software programs are available that are capable of performing detailed calculations and modeling [11,12]. For preparation of AC calculations and recommendations for optimized earth electrode positions, some basic information is required regarding the pipeline of matter and regarding the high-voltage power line(s) paralleled with the pipeline as follows:

*Pipeline input data*

- Diameter
- Wall thickness
- Length
- Depth
- Coating and preferable average coating resistance
- Mapping of pipeline in GPS coordinates
- Installed earth electrodes (location and resistance to earth)
- Location of isolation devices
- Soil resistivity
- Number of transformer/rectifier (T/Rs) (optional—nice to know)
- Approximate on potential (optional—nice to know)
- Approximate off potential (optional—nice to know)
- Total CP current (optional—nice to know)

*High-voltage power line system input data—for each interfering line*

- Nominal voltage (e.g., 500 kV, 230 kV, etc.);
- Normal load/maximum current;
- Earth fault current;

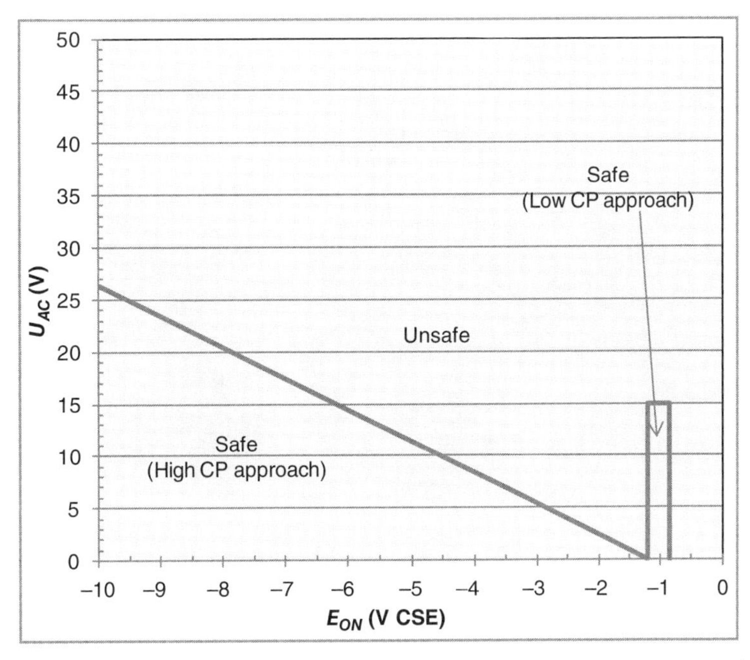


**FIGURE 26.17** AC corrosion criteria based on AC voltage and DC ON potentials. Criteria are taken from EN 15280:2013—informative annex E [8].

- Location of all conductors on the pylons, for example, indicated on a pylon drawing—for illustration see Figure 26.18;
- Distance between the conductors and earth and the individual distance between the conductors; see Figure 26.18;
- Mean sag between two pylons; see Figure 26.18;
- Location of pylons (GPS coordinates) where the power line route is changing and where transposition of the phases is made;
- Conductor's geometry;
- Earth wire/cable shield information;
- Electrical parameters for phase and earth conductors.

*Output data*

For the pipeline the calculations and iterative modeling should as a minimum lead to (fault conditions excluded) the following:

- Maximum AC voltage along the AC pipeline in a normal operation condition—without earth electrodes.

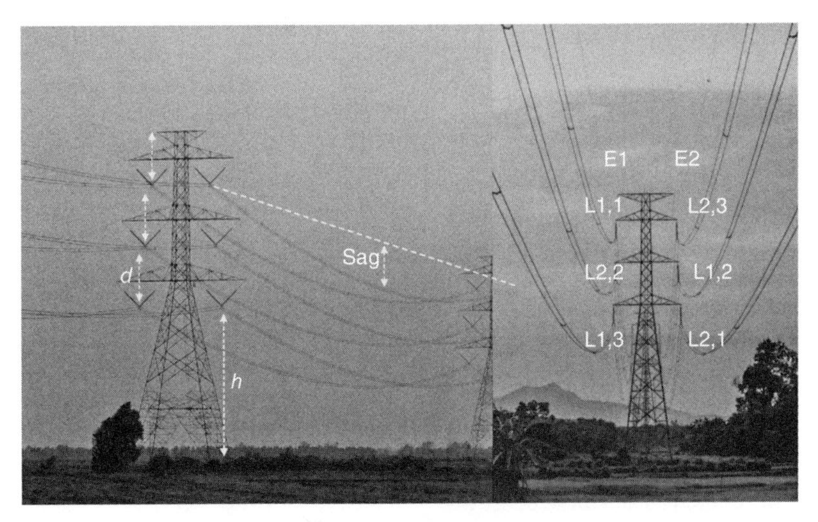


**FIGURE 26.18** Illustration of information from pylons required for AC voltage modeling.

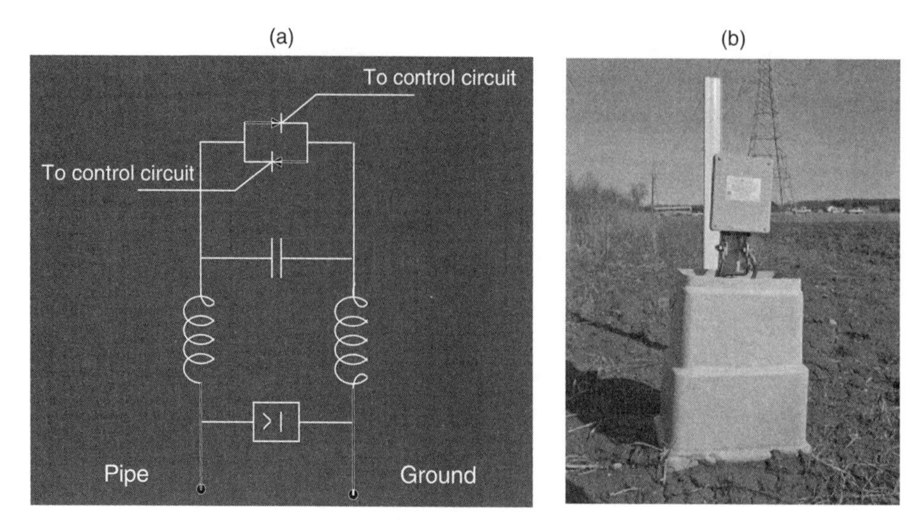


**FIGURE 26.19** (a) Electrical components circuitry typical of some AC discharge devices. (Courtesy of MetriCorr.) (b) Polarization Cell Replacement (PCR) installed at an AC interfered location. (Courtesy of Dairyland Electrical Industries.)

- Recommendation for an optimized earth electrode position in order to minimize AC voltage on the pipeline.
- Maximum AC voltage along the pipeline in a normal operation condition—with recommended earth electrodes established

In the design stage, the installation of isolating joints should be considered. Since the length of parallelism between an electrically continuous pipeline and a high-voltage power line is a determining factor for the overall induced voltage, the shortening of the electrically continuous sections of pipe by inserting isolating joints will decrease the level of induced AC. Of course, the installation of isolating joints also creates discontinuity in terms of DC, and therefore, must be considered in view of the CP design and position of CP sources (rectifiers and galvanic anodes).

The level of induction on a pipeline located in urban areas as opposed to rural areas may be moderated through the existence of other metallic constructions such as other pipe works, and so on. Such presence of other metallic subjects will reduce the induction in the pipeline caused by the so-called civilization factor. A systematic way of introducing a civilization factor is by installing mitigation wires along the pipeline. These are wires installed in close proximity to the pipeline. The drawback of these wires is their adverse effect on coating defect surveys.

Since the AC interference mainly depends on the proximity and parallel routing of the pipeline and high-voltage power lines, it is evident that mitigation of AC induction can also be achieved by increasing the distance between the pipeline and the high-voltage power lines. Typically, this is not practical because of the desire to keep these constructions within narrow energy corridors. Another factor that can be introduced in the construction stage is how to arrange the phase conductors on the pylons (Figure 26.18) in order to minimize AC induction, or ultimately to arrange high-voltage power transportation in cables where the distance between the power conductors is very small compared with the distance between the individual conductors and the pipeline.

Mitigation measures most typically imply the establishment of earth electrodes, which drains AC from the pipeline to the adjacent soil and into "far earth." The earth electrodes are typically copper rods, zinc ribbons, and so on. It is essential that the resistance to earth is as low as possible to facilitate the AC drainage. The earth electrodes can be directly connected to the pipeline, or they can be connected through AC discharge devices (DC decoupling devices, polarization cells, polarization cell replacements, etc.); see Figure 26.19. These devices essentially block the DC while creating a low-resistance AC path. The advantage of inserting these devices is that a direct connection implies a possible increased demand of CP, or that the earth electrodes are consumed if they act as sacrificial anodes and thereby over time will increase the resistance to earth and become less effective. Requirements for the AC discharge devices are that they need to be completely neutral and should not interfere with the applied CP system. A common AC discharge component is an electrolytic capacitor, which can be installed along with other electronic components (spark gabs, thyristors, and diodes) to protect from transients and so on; see Figure 26.19.

Figure 26.20 shows an example of simple AC induction modeling for a short pipeline (approximately 20 km) that is paralleled with a high-voltage power line. Quite typically, the induced voltage has maximum values in the end points and

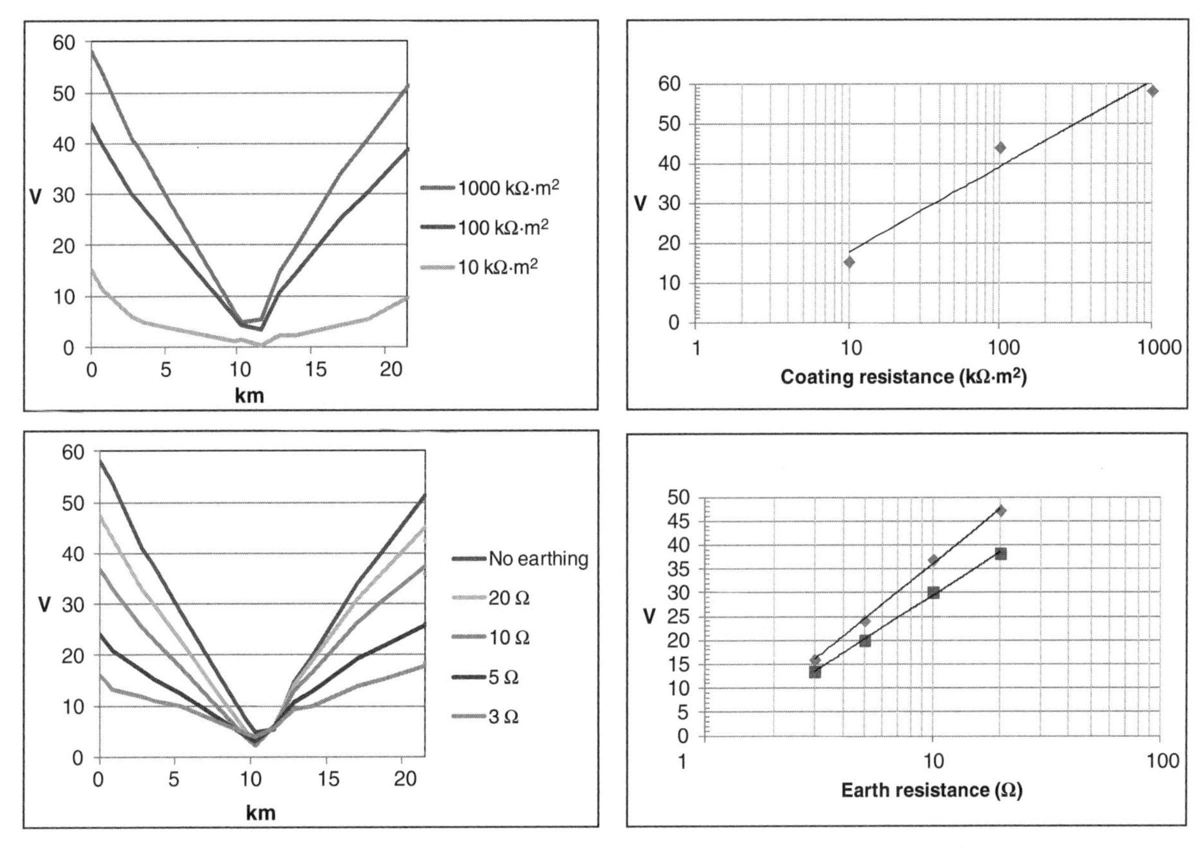


**FIGURE 26.20** Examples of modeled AC voltage data and effects of coating resistance and resistance from earth electrode to earth.

minimum value close to the midpoint of the pipeline. The upper figures illustrate the significant effect of the coating resistance on the induced AC. The higher the coating resistance, the higher the AC voltage induced. In other words, modern efficient coatings introduce higher induced AC, and therefore, constitute an increased risk of AC corrosion in tiny coating defects. This must be encountered when repair of coating defects are considered as a means for minimizing AC corrosion risk in areas deemed to be high-risk areas.

The lower illustrations in Figure 26.20 illustrate the effect of the resistance to earth of two earth electrodes installed in the end points of the pipeline. As observed, the resistance from the pipe to earth through the earth electrode is a crucial factor for the effectiveness of the AC drainage. Sometimes low resistances are costly in high-resistive areas—ultimately such as rocky terrains.

As discussed previously, the mitigation of AC is one way—and a first step—of the AC corrosion mitigation procedure and the first step to get out of the vicious circle of AC corrosion. The second step, which is sometimes (often) the easiest and most cost effective, is to control the level of CP applied to the pipeline. It must apply to usual criteria but not too much more. Therefore, in theory and in the simple case, AC corrosion may be controlled by simply adjusting the CP level.

CP conditions may, however, vary quite a lot from one pipeline to another. The current requirement may be in the order of amperes per kilometer for pipelines with a very poor coating quality ranging all the way down to microamperes per kilometer for pipelines with a very good coating quality (a very high coating resistance).

When the coating quality is very good, and the CP current consumption therefore very low, the pipeline current is low as well, and therefore, only very small IR drops in the pipe occur. Under these circumstances, it is quite easy to maintain a homogeneous CP potential throughout the pipeline length. If there is no complicating circumstance like DC interference, it is therefore easily obtainable to control the DC potential in a narrow range—on one hand complying with the usual CP potential criterion—and on the other hand make sure that no excessive CP is present that could initiate AC corrosion.

On the other hand, if the coating quality is very poor and the CP current requirement is very high, the pipeline current is likely to be very high, and large IR drops in the pipeline may occur. Under these circumstances, both conditions of excessive CP nearby CP sources and low, and perhaps

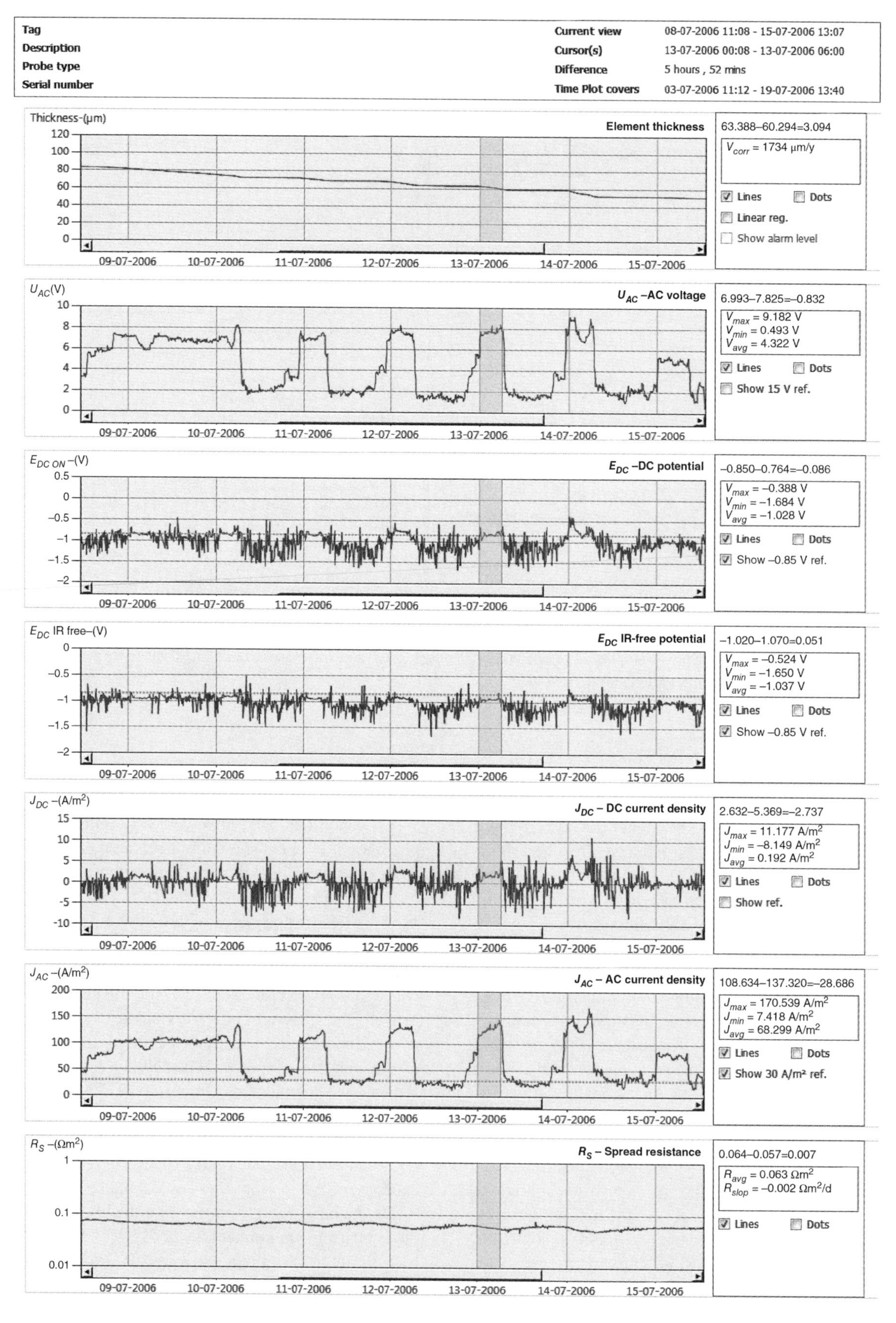


**FIGURE 26.21** Simultaneously measured corrosion rates (ER probes) and the electrical data involved in AC corrosion risk assessment.

inadequate, CP in mid-positions between CP sources can be present on the same pipeline. Further complications with DC interference can even be present. Even though these kinds of pipelines are normally well grounded due to the low average coating resistance, and therefore, AC induction level can be moderate, it can be difficult to maintain conditions for safe operation in terms of AC corrosion.

### 26.3.4 Monitoring and Management

To document the safe operation of pipelines with regard to AC corrosion (as well as other types of corrosion), it should be considered to establish monitoring in selected test stations along the pipeline. For the evaluation of AC corrosion the installation of coupons or probes is a necessary requisite. The coupons or probes should be installed anywhere the criteria for AC corrosion likelihood needs to be evaluated. The coupon or probe should gimmick a small coating defect—for instance 1 $cm^2$.

AC modeling procedures may be applied in connection with the selection of sites, which may include the following:

- Sites where the AC voltage is unacceptably high.
- Sites with low soil resistivity.

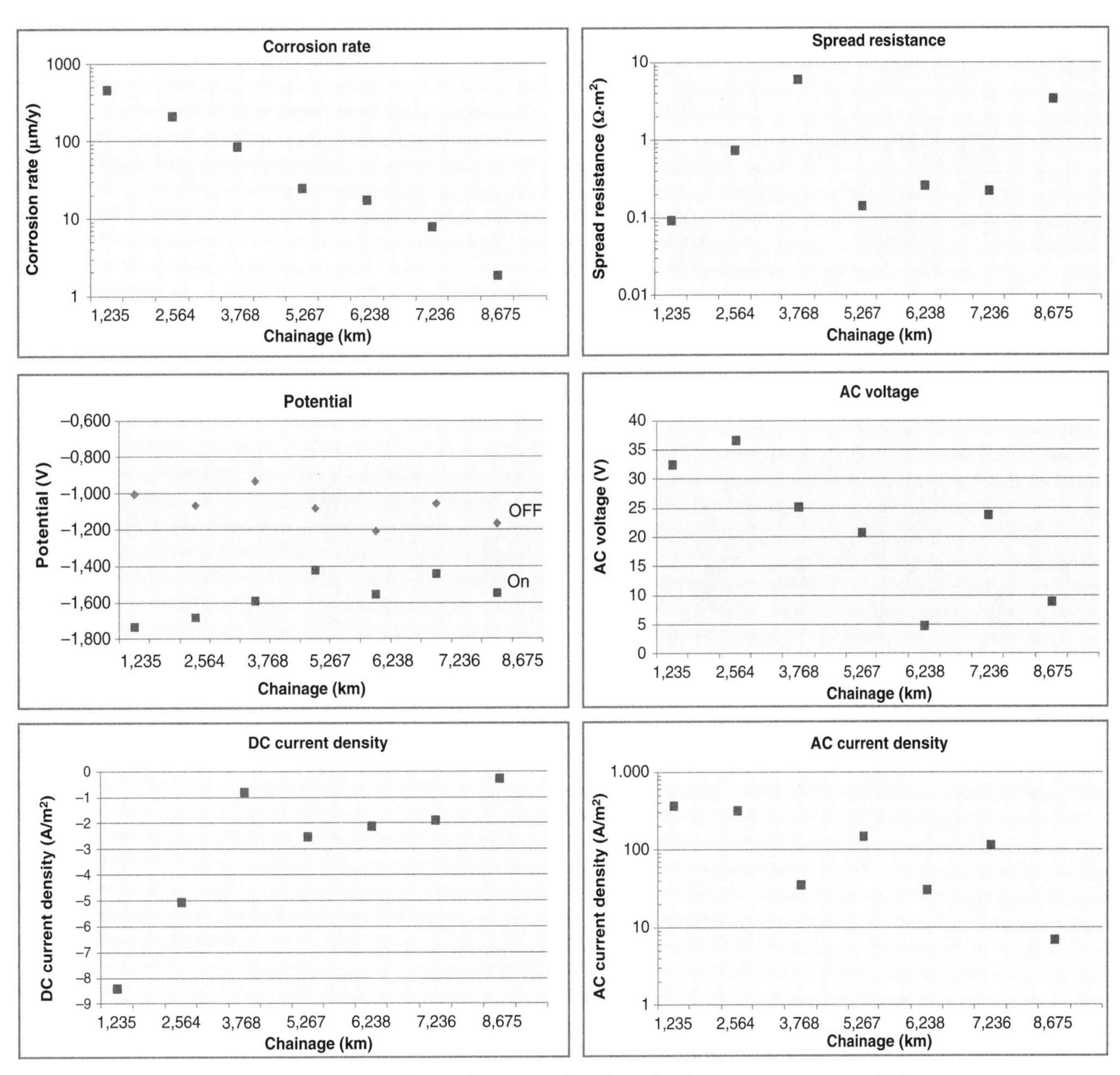


**FIGURE 26.22** Example of corrosion rate and various electrical parameters measured along a pipeline.

- Sites where (AC) corrosion has previously been experienced.
- Sites where DC interference (anodic and/or cathodic) is present
- Sites where AC has been of concern for the CP operator. The experience of the operator has high value.
- Sites with excessive CP.
- Sites with low level of CP.
- Sites where instant off measurements are difficult to obtain.

Other points may also be considered. Some operators prefer to make a statistical test before selecting a higher number of test stations for systematic monitoring.

During installation of the coupons or probes special care has to be taken to make sure that the coupon is embedded in the very same soil as the pipeline.

The selection of parameters to be monitored must reflect the analysis that must be performed to ensure safe operation—according to Section 26.3.1. Figure 26.21 shows an example where all optional parameters are measured simultaneously, that is, both corrosion rate measured on ER probes, AC and DC current densities, DC potentials (ON potential and IR compensated potential), as well as spread resistance. Such concept allows for analyzing under which electrical conditions corrosion takes place and subsequently adjust these conditions and make sure the corrected conditions are maintaining safe operation. Figure 26.22 shows comparable data from different pipeline locations at which all relevant parameters have been measured simultaneously.

Figures 26.23 and 26.24 show corrosion rate data obtained on 31 pipelines compared with the criteria given in Section 26.3.1. Figure 26.23 shows data based on AC and DC current densities while Figure 26.24 is based on AC voltage and DC ON potential. As observed from these figures and especially Figure 26.24, the threshold parameters are not always very well in correlation with corrosion rates, and corrosion rate measurements could be considered.

As an allowed option, it is therefore suggested to simplify AC corrosion risk assessment by using the corrosion rate approach and to optimize the electrical parameters only if corrosion is unacceptably high.

A simple scheme for AC corrosion evaluation has been suggested in Figure 26.25.

Risk assessment may be necessary for

- New constructions of pipelines,
- Construction of new interference sources, which comprise both AC sources like overhead power lines and DC sources such as DC traction systems, interfering rectifiers, and so on;
- Existing pipelines.

For new constructions of either pipelines or interfering sources, the analysis should include calculations using the computerized tools for predicting the interference levels. The same tools should be used for designing the mitigation. For old pipelines, it could be a pragmatic solution to allow a judgment based on the track record of corrosion incidents. The mitigation strategy could be "no action" if the pipeline

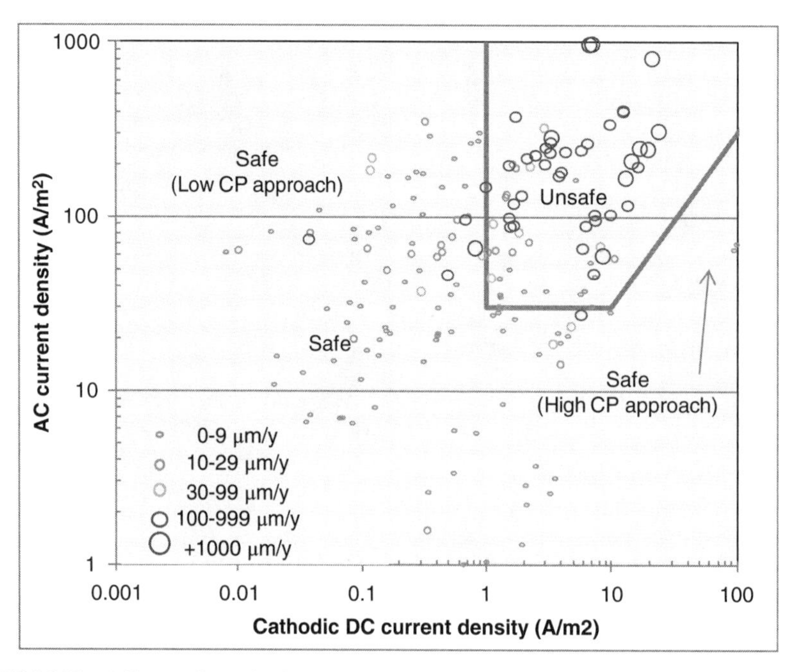


**FIGURE 26.23** AC corrosion criteria based on AC and DC current densities measured on installed coupons—with illustration of corrosion rates measured simultaneously [9].

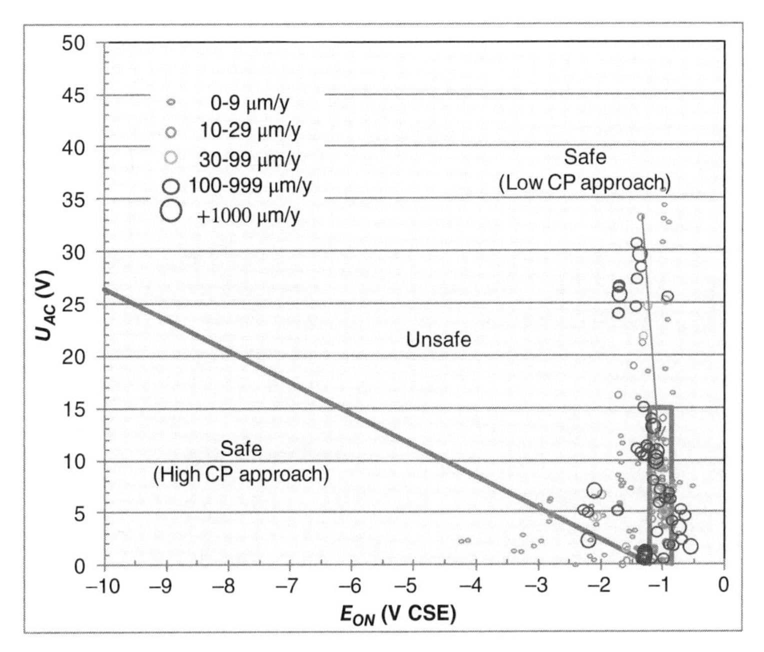


**FIGURE 26.24** AC corrosion criteria based on AC voltage and DC ON potentials with corrosion rates measured simultaneously [9].

has been present for a long time and no corrosion incident has ever occurred.

The monitoring strategy could be based on the electrical fingerprints as allowed in the EN 15280 standard, or it could be based on the corrosion rate measurements strategy as illustrated previously. If, for instance, the monitoring strategy involves measuring the AC/DC current on coupons and it shows that it is impossible to mitigate AC to below acceptable limits (unsafe condition), then the monitoring strategy could be changed to the corrosion rate strategy for the demonstration of a safe condition.

A direct way to the demonstration of a safe operation would imply application of the corrosion rate strategy—which is operational and fully allowed according to the new standard.

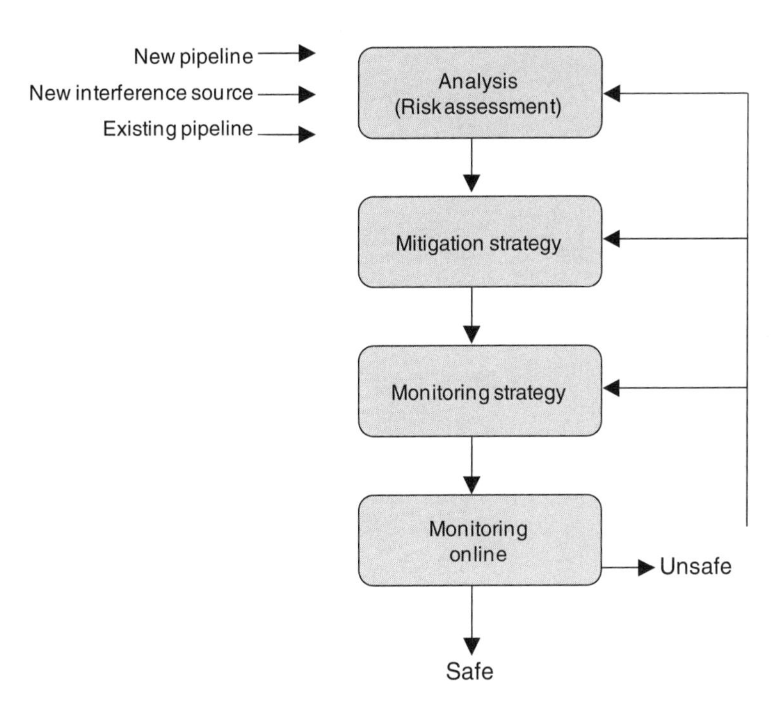


**FIGURE 26.25** A suggested scheme for AC corrosion assessment.

## REFERENCES

1. A.C. corrosion on cathodically protected pipelines: guidelines for risk assessment and mitigation measures, CeoCor Booklet, (2001)
2. Nielsen, L.V. and Cohn, P. (2000) AC corrosion and electrical equivalent diagrams. Proceedings of the 5th International Congress, Bruxelles, Belgium, CEOCOR.
3. Nielsen, L.V. (2010) Effect of coating defect size, coating defect geometry, and cathodic polarization on spread resistance—consequences in relation to AC corrosion monitoring. Proceedings of the CEOCOR International Congress, Editor: CEOCOR, c/o C.I.B.E., Brussels, Belgium.
4. Bossche, B. (2010) Development of a time dependant numerical model for the quantification of AC corrosion phenomena. Proceedings of the CEOCOR International Congress, Editor: CEOCOR, c/o C.I.B.E., Brussels, Belgium.
5. Büchler, M. (2013) Determining the A.C. corrosion risk of pipelines based on coupon measurements. Proceedings of the CEOCOR International Congress, Editor: CEOCOR, c/o C.I. B.E., Brussels, Belgium.
6. Büchler, M. (2012) Effect of variation of cathodic protection level over time on the a.c. corrosion process. Proceedings of the CEOCOR International Congress, Editor: CEOCOR, c/o C.I.B.E., Brussels, Belgium.
7. Büchler, M., Voûte, C.-H., and Joos, D. (2011) Field investigation of A.C. corrosion. Proceedings of the CEOCOR International Congress, Editor: CEOCOR, c/o C.I.B.E., Brussels, Belgium.
8. EN 15280 (2013) Evaluation of AC corrosion likelihood of buried pipelines applicable to cathodically protected pipelines.
9. Nielsen, L.V. (2014) A.C. influenced corrosion in pipelines: cost effective assessments. Proceedings of the 15th Middle East Corrosion Conference and Exhibition. February 2–5, 2014, Manama, Bahrain.
10. EN 12954 Cathodic protection of buried or immersed metallic structures—general principles and application for pipelines.
11. Bortels, L. Baeté, C., and Dewilde, J.-M. (2012) Accurate modelling and trouble shooting of AC interference problems on pipelines. Proceedings of CORROSION/2012, NACE International, Houston, TX, Paper No. 0001664.
12. AC predictive and mitigation techniques. Final Report, for Corrosion Supervisory Committee, PRC International, (1999).

## BIBLIOGRAPHY

### Standards and Recommended Practices

ANSI/NACE RP0104 The use of coupons for cathodic protection monitoring applications.

EN 13509 (2003) Cathodic protection measurement techniques.

EN 50443 Effects of electromagnetic interference on pipelines caused by high voltage AC electric traction systems and/or high voltage AC power supply systems.

Guide on the influence of high voltage AC power systems on metallic pipelines, CIGRE Technical Brochure No 95, 1995.

NACE SP0177 Mitigation of alternating current and lightning effects on metallic structures and corrosion control systems.

NACE SP0169 Control of external corrosion of underground or submerged metallic piping systems.

NACE TM0497 (2012) Measurement techniques related to criteria for cathodic protection on underground or submerged metallic piping systems.

### Books and Reports

NACE (2010) *AC Corrosion State-of-the-Art: Corrosion Rate, Mechanism, and Mitigation Requirements*, NACE International, Publication 35110.

Von Baeckmann, W., Schenk, W., and Prinz, W. (editors) (1997) *Handbook of Cathodic Corrosion Protection, Theory and Practice of Electrochemical Protection Processes*, 3rd ed., Gulf Publishing Company, Houston, TX.

### Papers/Conference Proceedings

Baeté, C. (2013) Simulation of AC corrosion on pipelines under AC and DC interference. Proceedings of the CEOCOR International Congress, Editor: CEOCOR, c/o C.I.B.E., Brussels, Belgium.

Bertocci, U. (1979) AC induced corrosion. The effect of an alternating voltage on electrodes under charge-transfer control. *CORROSION*, 35(5), 211–215.

Bette, U. (2010) Effect of coupon geometry on the lateral current distribution. Proceedings of the CEOCOR International Congress, Editor: CEOCOR, c/o C.I.B.E., Brussels, Belgium.

Bette, U. (2012) Results of measurements on ER coupons with 16 bit digital oscilloscope. Proceedings of the CEOCOR International Congress, Editor: CEOCOR, c/o C.I.B.E., Brussels, Belgium.

Blotzky, K. (2010) CP measurements on severely high voltage polluted pipelines. Proceedings of the CEOCOR International Congress, Editor: CEOCOR, c/o C.I.B.E., Brussels, Belgium.

Bolzoni, F., Goidanich, S., Lazzari, L., Ormellese, M., and Pedeferri, M.P. (2004) Laboratory testing on the influence of alternated current on steel corrosion. Proceedings of CORROSION/2004, NACE International, Houston, TX, Paper No. 208.

Bosh, R.W. and Bogaerts, W.F. (1998) A theoretical study of AC-induced corrosion considering diffusion phenomena. *Corrosion Science*, 40(2/3), 323–336.

Boteler, D.H. (2013) A.C. interference in Sweden. Proceedings of the CEOCOR International Congress, Editor: CEOCOR, c/o C.I. B.E., Brussels, Belgium.

Bruckner, W. (1964) The Effects of 60 Cycle Alternating Current on the Corrosion of Steels and Other Metals Buried in Soils, University of Illinois, Technical Bulletin No. 470, November 1964.

Büchler, M., Stalder, F., and Schöneich, H.-G. (2005) Eine neue elektrochemische Methode für die Ermittlung von Wechselstromkorrosion, *3R International*, 44, 396.

Büchler, M., Schöneich, H.-G., and Stalder, F. (2005) Discussion of criteria to assess the alternating current corrosion risk of cathodically protected pipelines. Joint Technical Meeting on Pipeline Research, Proceedings Volume Paper No. 26, PRCI.

Büchler, M., Voûte, C.-V., and Schöneich, H.-G. (2007) Evaluation of the effect of cathodic protection levels on the A.C. corrosion on pipelines. EUROCORR Conference Proceedings.

Büchler, M., Voûte, C.-H., and Schöneich, H.-G. (2008) Diskussion des Wechsel strom korrosion sme chanismus auf kathodisch geschätzten Leitungen: Die Auswirkung des kathodischen Schutzniveaus. *3R International*, 47, 304.

Büchler, M., Voûte, C.-H., and Schöneich, H.-G., (2008) Discussion of the mechanism of A.C.-corrosion of cathodically protected pipelines: the effect of the cathodic protection level. Proceedings of the CEOCOR International Congress, Editor: CEOCOR, c/o C.I.B.E., Brussels, Belgium.

Büchler, M. and Schöneich, H.-G. (2009) Investigation of alternating current corrosion of cathodically protected pipelines: development of a detection method, mitigation measures, and a model for the mechanism. *CORROSION*, 65, 578–586.

Büchler, M. and Voûte, C.-H. (2009) Wechselstromkorrosion an kathodisch geschätzten Rohrleitungen: neue Erkenntnisse zum Mechanismus. *DVGW energie, wasser-praxis*, 6, 18.

Büchler, M., Voûte, C.-H., and Schöneich, H.-G. (2009) Kritische Einflussgrössen auf die Wechselstromkorrosion: Die Bedeutung der Fehlstellengeometrie. *3R International*, 8, 324.

Büchler, M. (2010) Cathodic protection: a general discussion of the involved processes and their consequences for threshold values. Proceedings of the CEOCOR International Congress, Editor: CEOCOR, c/o C.I.B.E., Brussels, Belgium.

Büchler, M. (2010) Kathodischer Korrosionsschutz: Diskussion der grundsätzlichen Mechanismen und deren Auswirkung auf Grenzwerte. *3R International*, 49, 342.

Chin, D.T. and Fu, T.W. (1979) Corrosion by alternating current: a study of the anodic polarization of mild steel in $Na_2SO_4$ solution. *CORROSION*, 35(11), 514.

Chin, D.T. and Venkatesh, S. (1979) A study of alternating voltage modulation on the polarization of mild Steel. *Journal of Electrochemical Society*, 126(11) 1908–1913.

Chin, D.T. and Sachdev, P. (1983) Corrosion by alternating current: polarization of mild steel in neutral electrolytes. *Journal of Electrochemical Society*, 130(8), 1714–1718.

Dabkowski, J. and Taflove, A. (1978) Mutual design considerations for overhead ac transmission lines and gas transmission pipelines. Volume one: engineering analysis, particularly section 7 Pipeline system susceptibility, PRCI Project PR-132-80, Catalog No. L51278e, September 1978.

Di Biase, L. (1995) Corrosion due to alternating current on metallic buried pipelines: background and perspectives. Proceedings of the 1995 II National Conference, held November 21–22, 1995, Milano, Italy: APCE(10), 1996, pp. 6–10.

Di Biase, L. (2000) Corrosion due to alternating current on metallic buried pipelines: background and perspectives. Proceedings of 5th International Congress, Committee on the Study of Pipe Corrosion and Protection, Bruxelles, Belgium, CEOCOR.

Dibiase, L., et al. (2011) A.C corrosion 1989–2011: 22 years of researches and experience in Europe. Proceedings of the CeoCor Compendium, Editor: CEOCOR, c/o C.I.B.E., Brussels, Belgium.

Edwall, H.-E. and Malmborn, K. (2011) Practical case of measures to reduce the stray current impact on a 40 km long natural gas pipeline caused by intersecting High Voltage transmission lines. Proceedings of the CEOCOR International Congress, Editor. CEOCOR, c/o C.I.B.E., Brussels, Belgium.

Fernandes, S.Z., Mehendale, S.G., and Venkatachalam, S. (1980) Influence of frequency of alternating current on the electrochemical dissolution of mild steel and nickel. *Journal of Applied Electrochemistry*, 10(5), 649–654.

Floyd, R.D. (2004) Testing and mitigation of AC corrosion on 8″ line: a field study. Proceedings of CORROSION/2004, NACE International, Houston, TX, Paper No. 210.

Frazier, M.J. (1994) Induced AC Influence on Pipeline Corrosion and Coating Disbondment, GRI, Project No. A381, December 1994.

Freiman, L.I. and Yunovich, M. (1991) Special behavior of steel cathode in soil and protection assessment of underground pipe with a buried coupon. *Protection of Metals*, 27(3), 437–447.

Fu, A.Q. and Cheng, Y.F. (2010) Corrosion of pipeline steel in the presence of alternating current and the new CP recommendation. Proceedings of the 8th International Pipeline Conference (IPC 2010), Paper IPC2010-31658.

Funk, D. and Schoeneich, H.G. (2002) Problems with coupons when assessing the AC-corrosion risk of pipelines. *3R International*, 41(10–11), 54.

Gilroy, D.E. (2003) AC interference—important issues for cross country pipelines. Proceedings of CORROSION/2003, NACE International, Houston, TX, Paper No. 699.

Goidanich, S., Lazzari, L., Ormellese, M., and Pedeferri, M.P. (2005) Influence of AC on carbon steel corrosion in simulated soil conditions. Proceedings of 16th International Corrosion Congress, NACE International, September 19–24, 2005, Beijing, China, Paper No. 04-03.

Goidanich, S., Lazzari, L., Ormellese, M., and Pedeferri, M.P. (2006) Effect of AC on cathodic protection of carbon steel in simulated soil conditions. Proceedings of EUROCORR 2006, Maastricht, the Netherlands: EFC(8), Paper No. 279.

Goidanich, S., Lazzari, L., and Ormellese, M. (2010) AC corrosion—part 1: effects on overpotentials of anodic and cathodic processes. *Corrosion Science*, 52, 491–497.

Goidanich, S., Lazzari, L., and Ormellese, M. (2010) AC corrosion—part 2: parameters influencing corrosion rate. *Corrosion Science*, 52, 916–922.

Gomila, A. (2012) DC current density control at steel coupons against AC corrosion risk. Proceedings of the CEOCOR International Congress, Editor: CEOCOR, c/o C.I.B.E., Brussels, Belgium.

Goran, C. (2000) Alternating current corrosion on cathodically protected steel in soil—a long term field investigation. Proceedings of the 5th International Congress, Bruxelles, Belgium, CEOCOR.

Gregoor, R. and Pourbaix, A. (2003) Detection of AC corrosion. *3R International*. 42(6), 289–395.

Gummow, R., Wakelin, R., and Segall, S. (1998) AC corrosion—a new challenge to pipeline integrity. Proceedings of CORROSION/1998, NACE International, Houston, TX, Paper No. 566.

Gummow, R.A. (1999) Cathodically protection considerations for pipelines with ac mitigation facilities. PRCI Contract PR-262-9809, Catalog L51800e, January 1999.

Hamlin, A.W. (1986) Alternating current corrosion. *Materials Performance*, 25(1) 55.

Helm, G., Helm, T., Heinzen, H., and Schwenk, W. (1993) Investigation of corrosion of cathodically protected steel subjected to alternating currents. *3R International*, 32(5), 246.

Hosokawa, Y., Kajiyama, F., and Nakamura, Y. (2002) New CP criteria for elimination the risks of AC corrosion and over-protection on cathodically protected pipelines. Proceedings of CORROSION/2002, NACE International, Houston, TX, Paper No. 02111.

Ibrahim, I., Meyer, M., Tribollet, B., Takenouti, H., Joiret, S., Fontaine, S., and Schoneich, H.G. (2008) On the mechanism of AC assisted corrosion of buried pipelines and its CP mitigation. Proceedings of ASME International Pipeline Conference (IPC 2008), Paper IPC2008-64380.

Jones, D. (1978) Effect of alternating current on corrosion of low alloy and carbon steels. *CORROSION*, 24(12), 428–433.

Juetner, K., Reitz, M., Schaefer, S., and Schoeneich, H. (1998) Rotating ring-disk studies on the impact of superimposed large signal AC currents on the cathodic protection of steel. *Electrochemical Methods in Corrosion Research VI, Materials Science Forum*, 289–292(2), 107.

Kajiyama, F. and Nakamura, Y. (2010) Development of an advanced instrumentation for assessing the AC corrosion risk of buried pipelines. Proceedings of CORROSION/2010, NACE International, March 14–18, 2010, San Antonio, TX, Paper No. 10104

Kajiyama, F. (2013) Methodology for evaluation of a.c. corrosion risk using coupon d.c. and a.c. current densities. Proceedings of the CEOCOR International Congress, Editor: CEOCOR, c/o C.I.B.E., Brussels, Belgium.

Kioupis, N. (2011) Corrosion of buried pipelines detected on ER-probes. Proceedings of the CEOCOR International Congress, Editor: CEOCOR, c/o C.I.B.E., Brussels, Belgium.

Kroon, D.H. (2014) AC interference and AC corrosion of pipelines. Proceedings of the 15th Middle East Corrosion Conference and Exhibition, February 2–5, 2014, Manama, Bahrain.

Lalvani, S.B. and Lin, X.A. (1994) A theoretical approach for predicting AC-induced corrosion. *Corrosion Science*, 36(6), 1039–1046.

Lalvani, S.B. and Lin, X. (1996) A revised model for predicting corrosion of materials induced by alternating voltages. *Corrosion Science*, 38(10), 1709–1719.

Lazzari, L., Goidanich, S., Ormellese, M., and Pedeferri, M.P. (2005) Influence of AC on corrosion kinetics for carbon steel, zinc and copper. Proceedings of CORROSION/2005, NACE International, Houston, TX, Paper No. 189.

Linhardt, P. and Ball, G. (2006) AC corrosion: results from laboratory investigations and from a failure analysis. Proceedings of CORROSION/2006, NACE International, Houston, TX, Paper No. 160.

Martinez, S. (2013) Estimation of AC interference on pipelines running parallel to railway line prior to changing traction from 3 kV DC to 25 kV AC and design of mitigation. Proceedings of the CEOCOR International Congress, Editor: CEOCOR, c/o C.I.B. E., Brussels, Belgium.

Mateo, M.L., Fernandez Otero, T., and Schiffrin, D.J. (1990) Mechanism of enhancement of the corrosion of steel by alternating currents and electrocatalytic properties of cycled steel surfaces. *Journal of Applied Electrochemistry*, 20(1), 26–31.

McCollum, B. and Ahlborn, G. (1916) Influence of frequency of alternating or infrequently reversed current on electrolytic corrosion, Technologic papers of the Bureau of Standards, No. 72, August 15, 1916.

Mengarini, G. (1891) Electrolysis by alternating currents. *Electrical World*, 16(6), 96.

Nielsen, L.V. and Cohn, P. (2003) AC corrosion in pipelines. Field experiences from a highly corrosive test site using ER corrosivity probes. Proceedings of the CEOCOR.

Nielsen, L.V., Nielsen, K.V., Baumgarten, B., Breuning-Madsen, H., Cohn, P., and Rosenberg, H. (2004) AC induced corrosion in pipelines: detection, characterization and mitigation. Proceedings of CORROSION/2004, NACE International, Houston, TX, Paper No. 04211.

Nielsen, L.V., Baumgarten, B., and Cohn, P. (2004) On-site measurements of AC induced corrosion: effect of AC and DC parameters. Proceedings of the CEOCOR International Congress, Editor: CEOCOR, c/o C.I.B.E., Brussels, Belgium.

Nielsen, L.V., Baumgarten, B., and Cohn, P. (2004) Investigating AC and DC stray current corrosion. Proceedings of the 7th International Congress 2004, Groniue, Holland: CEOCOR.

Nielsen, L.V. and Galsgaard, F. (2005) Sensor technology for on-line monitoring of AC induced corrosion along pipelines. Proceedings of CORROSION/2005, NACE International, Houston, TX, Paper No. 375.

Nielsen, L.V. (2005) Role of alkalization in AC induced corrosion of pipelines and consequences hereof in relation to CP requirements. Proceedings of CORROSION/2005, NACE International, Houston, TX, Paper No. 05188.

Nielsen, L.V., Baumgarten, B., and Cohn, P. (2005) Investigating AC and DC stray current corrosion. Proceedings of the CEOCOR International Congress, Editor: CEOCOR, c/o C.I.B.E., Brussels, Belgium.

Nielsen, L.V., Baumgarten, B., and Cohn, P. (2006) A field study of line currents and corrosion rate measurements in a pipeline critically interfered with AC and DC stray currents. Proceedings of the CEOCOR.

Nielsen, L.V. (2011) Considerations on measurements and measurement techniques under AC/DC interference conditions. Proceedings of the CEOCOR International Congress, Editor: CEOCOR, c/o C.I.B.E., Brussels, Belgium.

Nielsen, L.V. and Petersen, M.B. (2011) Monitoring AC corrosion in buried pipelines. Proceedings of the DoD Conference, July 2011.

Nielsen, L.V. (2012) Possible temperature effects on a.c. corrosion and a.c. corrosion monitoring. Proceedings of the CEOCOR International Congress, Editor: CEOCOR, c/o C.I.B.E., Brussels, Belgium.

Ormellese, M., Lazzari, L., Goidanich, S., and Sesia, V. (2008) CP criteria assessment in the presence of ac interference. Proceedings of CORROSION/2008, NACE International, Houston, TX, Paper No. 8064,

Ormellese, M. (2013) A proposal of ac corrosion mechanisms of carbon steel in cathodic protection conditions. Proceedings of CORROSION/2013, NACE International, 02457 2013 CP.

Panossian, Z., Abud, S., Almeida, N., et al. (2009) Effect of alternating current by high power lines voltage and electric transmission systems in pipelines corrosion. Proceedings of CORROSION/2009, NACE International, Paper No. 09541.

Panossian, Z., et al. (2010) A new thermodynamic criterion and a new field methodology to verify the probability of AC corrosion in buried pipeline. Proceedings of CORROSION/2010, NACE International, March 14–18, 2010, San Antonio, TX, Paper No. 10115.

Pookote, S. and Chin, D.T. (1978) Effect of alternating current on the underground corrosion of steels. *Materials Performance*, 17(3), 9.

Pourbaix, A., Carpentiers, P., and Gregoor, R. (2000) Detection and assessment of alternating current corrosion. *Materials Performance*, 38(3), 34–39.

Prinz, W. (1993) Alternating current corrosion of cathodically protected pipelines. Proceedings of the 1992 International Gas Research Conference, November 16–19, 1992 (Government Institutes Inc., Rockville, MD: 1993).

Qiu, W., Pagano, M., Zhang, G., and Lalvani, S.B. (1995) A periodic voltage modulation effect on the corrosion of Cu-Ni alloy. *CORROSION*, 37(1), 97–110.

Quang, K.V., Brindel, F., Laslaz, G., and Buttoudin, R. (1983) Pitting mechanism of aluminium in hydrochloric acid under alternating current. *Journal of Electrochemical Society*, 130(6), 1248–1252.

Radeka, R., Zorovic, D., and Barisin, D. (1980) Influence of frequency of alternating current on corrosion of steel in seawater. *Anti-Corrosion Methods and Materials*, 27(4), 13.

Ragault, I. (1998) AC corrosion induced by V.H.V electrical lines on polyethylene coated steel gas pipelines. Proceedings of CORROSION/1998, NACE International, Houston, TX, Paper No. 557.

Ruedisueli, R.L., Hager, H.E., and Sandwith, C.J. (1987) An application of a state-of-the-art corrosion measurement system to a study of the effects of alternating current on corrosion. *CORROSION*, 43(6), 331–338.

Sandberg, B. (2010) A new corrosion probe. Proceedings of the CEOCOR International Congress, Editor: CEOCOR, c/o C.I.B. E., Brussels, Belgium.

Sandberg, B. (2011) AC stray current due to ohmic interference. Proceedings of the CEOCOR International Congress, Editor: CEOCOR, c/o C.I.B.E., Brussels, Belgium.

Schoeneich, H.G. (2004) Research addresses high voltage interference, AC corrosion risk for cathodically protected pipelines. *Oil and Gas Journal*, 102(7), 56–63.

Simon, P.D. (2007) Dynamic nature of HVAC induced current density on collocated pipelines. Proceedings of CORROSION/2007, NACE International, Houston, TX, Paper No. 650.

Simon, P. (2007) Case histories where ac assisted corrosion was identified below the 15 VAC safety threshold. Proceedings of NACE Eastern Area Conference, October 8, 2007.

Song, H., Kim, Y., Lee, S., Kho, Y., and Park, Y. (2002) Competition of AC and DC current in AC corrosion under cathodic protection. Proceedings of CORROSION/2002, NACE International, Houston, TX, Paper No. 02117

Song, H., Kim, Y., Lee, S., Kho, Y., and Park, Y. (2002) Competition of AC and DC current in AC corrosion under cathodic protection. Proceedings of CORROSION/2002, NACE International, Houston, TX, Paper No. 117.

Stalder, F. (2000) Influence of soil composition on the spread resistance and of ac corrosion on cathodically protected coupons. Proceedings of 5th International Congress, Committee on the Study of Pipe Corrosion and Protection, Bruxelles, Belgium, CEOCOR.

Stalder, F. (2002) AC corrosion of cathodically protected pipelines. Guidelines for risk assessment and mitigation measures, Annex N.5-4. Proceedings of the 4th International Congress 2002, Groniue, Holland: CEOCOR.

Tan, T.C. and Chin, D.T. (1989) Effect of alternating voltage on the pitting of aluminium in nitrate, sulfate and chloride solutions. *CORROSION*, 45(12), 984–989.

Wakelin, R., Gummow, R., and Segall, S. (1998) AC corrosion—case histories, test procedures, and mitigation. Proceedings of CORROSION/1998, NACE International, Houston, TX, Paper No. 565.

Williams, J. (1966) Corrosion of metals under the influence of alternating current. *Materials Protection*, 5(2), 52–53.

Yunovich, M. and Thompson, N.G. (2004) AC corrosion: corrosion rate and mitigation requirements. Proceedings of CORROSION/2004, NACE International, Houston, TX, Paper No. 206.

Yunovich, M. and Thompson, N.G. (2004) AC corrosion: mechanism and proposed model. Proceedings of International Pipeline Conference 2004 (IPC 2004), October 4–8, 2004, Materials Park, OH: ASME(7), Paper No. IPC04-0574.

Yunovich, M. and Thompson, N.G. (2004) AC corrosion: mechanism and proposed model. Proceedings of the 5th Biennial International Pipeline Conference (IPC 2004), October 4–8, 2004, Calgary, Alberta, Canada, Paper IPC2004-0574.

# 27

# MICROBIOLOGICALLY INFLUENCED CORROSION

Brenda J. Little and Jason S. Lee
*Naval Research Laboratory, Stennis Space Center, Mississippi, USA*

## 27.1 INTRODUCTION

Low alloy steel pipelines, used to transport crude oil, petroleum products, and natural gas, are located in a variety of microbiologically active environments, including belowground in soils and undersea [1]. The U.S. Department of Transportation (DOT) Office of Pipeline Safety has complied statistics for pipeline releases, including oil and gas, from 2002 to 2011 [2]. Over that period of time approximately 34% of all releases were attributed to corrosion. National Association of Corrosion Engineers (NACE) International [3] estimated the cost of corrosion for onshore gas and liquid transmission pipelines was $7 billion. However, there are no specific statistics related to microbiologically influenced corrosion (MIC) of low alloy steel pipelines. Russian investigators [4] estimated that 30% of the corrosion damage in equipment used for oil exploration and production was directly attributable to MIC.

The term MIC is used to designate corrosion due to the presence and activities of microorganisms, that is, those organisms that cannot be seen individually with the unaided human eye. Causative microorganisms are from all three main branches of evolutionary descent, that is, bacteria, archaea (methanogens), and eukaryota (fungi). The list of microorganisms involved in MIC and the mechanisms by which they influence corrosion is continuously growing. Mechanisms are the result of specific metal/microbe/electrolyte interactions. Corrosion is directly related to oxidation (anode) and reduction (cathode) reactions and microbial processes require one- and two-electron transfers (either oxidation or reduction reactions). Microorganisms can accelerate rates of partial reactions in corrosion processes or shift the mechanism for corrosion. MIC can involve a conversion of a protective metal oxide to a less protective layer (e.g., a sulfide) or removal of the oxide layer, for example, by metal oxide reduction or acid-production. Microorganisms can produce localized attack, including pitting, dealloying, galvanic corrosion, stress corrosion cracking, and hydrogen embrittlement. Microorganisms can also produce non-tenacious corrosion products, for example, sulfides that are easily detached by mechanical sheer, resulting in enhanced erosion corrosion. However, microorganisms do not produce a unique corrosion morphology that distinguishes MIC from abiotically produced corrosion.

Discussion in the following sections will be limited to MIC of low alloy steels, for example, carbon steel. The main alloying element in carbon steel is carbon and its mechanical properties depend on the percentage of carbon. Carbon content has little effect on the general corrosion resistance [5]. Low-carbon steel contains approximately 0.05–0.3 wt% carbon and mild steel, 0.3–0.6% carbon. In referencing the work of others in this chapter, the alloy terminology used in the original work will be maintained.

Both internal and external oil and gas pipeline surfaces can be affected by MIC. DOT statistics [2] suggest that 7 and 16% of releases of crude oil in the United States are due to external and internal corrosion, respectively. Information in this chapter related to internal MIC will be limited to petroleum-based hydrocarbon fuels in low-alloy steel piping. The focus of most testing, monitoring, and research related to MIC in the oil and gas industry for internal and external pipeline surfaces is on sulfate-reducing bacteria (SRB) [6]. Consequently, any discussion of causative microorganisms, in this chapter, will be dominated by references to SRB. The

*Oil and Gas Pipelines: Integrity and Safety Handbook,* First Edition. Edited by R. Winston Revie.

significance of other causative microorganisms will be acknowledged and discussed.

## 27.2 REQUIREMENTS FOR MICROBIAL GROWTH

Microorganisms have developed several strategies for survival in natural environments: (1) spore formation (2) biofilm formation (3) dwarf cells, and (4) a viable, but non-culturable state. Many microorganisms produce spores that are resistant to temperature, acids, alcohols, disinfectants, drying, freezing, and other adverse conditions. Spores may remain viable for hundreds of years and can germinate when conditions become favorable. However, there is a difference between survival and growth. MIC requires growth of the causative organisms and growth requires water, electron acceptors/donors, and nutrients. The potential for MIC is determined by the availability of these essentials and any proposed mechanism must account for their availability.

### 27.2.1 Water

Liquid water is needed for all forms of life. Microbial interaction, distribution, and growth with oil and gas are limited by water availability. Microbial growth in hydrocarbons is concentrated at oil/water interfaces, that is, emulsified water, and separate water phases. The volume of water required for microbial growth in hydrocarbon fuels is extremely small. Since water is a product of the microbial mineralization of organic substrates, it is possible for *in situ* microbial mineralization of a hydrocarbon to generate a water phase that can be used for further proliferation.

### 27.2.2 Electron Donors and Acceptors

Microorganisms obtain energy through electron transfer processes. Not all electron donors and acceptors are water soluble. Electrogenic bacteria are capable of moving electrons to and from solid materials. Petroleum hydrocarbons, organic matter, reduced inorganic compounds, molecular hydrogen, and iron can act as electron donors, which release electrons during cellular respiration. Electrons are then channeled to electron acceptors. During this process the electron donor is oxidized and the electron acceptor is reduced. Microorganisms can use a variety of electron acceptors for respiration in dissimilatory reactions, that is, the acceptors are not assimilated. In aerobic respiration, energy is derived when electrons are transferred to oxygen, the terminal electron acceptor. In anaerobic respiration, a variety of organic and inorganic compounds may be used as terminal electron acceptors, including sulfate, carbon dioxide, nitrate, nitrite, $Cr^{+6}$, $Fe^{+3}$, and $Mn^{+4}$. There is specificity among anaerobes for particular electron acceptors; bacteria are routinely grouped based on the terminal electron acceptor in anaerobic respiration, for example, sulfate-, nitrate-, and metal-reducing bacteria.

Facultative anaerobic bacteria can use oxygen or other electron acceptors. Obligate anaerobic microorganisms cannot tolerate oxygen for growth and survival. Obligate anaerobic bacteria and archaea are, however, routinely isolated from oxygenated environments associated with particles, crevices, and most importantly, in association with aerobic and facultative bacteria that effectively remove oxygen from the immediate vicinity of the anaerobe.

SRB are a group of ubiquitous, diverse anaerobes that use sulfate as the terminal electron acceptor, producing hydrogen sulfide ($H_2S$). Several SRB can also reduce nitrate, sulfite, or thiosulfate. Under specific conditions, some SRB can accept electrons directly from iron and transfer the electrons for sulfate reduction. Enning et al. [7] demonstrated direct uptake of electrons from iron through a semiconductive ferrous sulfide corrosion crust. Many archaea can also produce sulfides. The inclusive term for all sulfide-producing microorganisms is sulfide-producing prokaryotes (SPP).

Several corrosion mechanisms have been attributed to SPP, including cathodic depolarization by the enzyme dehydrogenase, anodic depolarization, production of iron sulfides, release of exopolymers capable of binding metal ions, sulfide-induced stress corrosion cracking and hydrogen-induced cracking or blistering [8]. During corrosion of carbon steel influenced by SPP, a thin (approximately 1 μm), adherent layer of mackinawite [$(Fe,Ni)_9S_8$)] is formed. If the ferrous ion concentration is high, mackinawite and green rust 2, a complex ferrosoferric oxyhydroxide will form. Under some circumstances, green rust 2, unstable in the presence of oxygen, can be an electron acceptor for SRB [9]. Once electrical contact is established between corrosion products and carbon steel, the carbon steel behaves as an anode and electron transfer occurs through the iron sulfide. In the absence of oxygen, the metabolic activity of SPP causes accumulation of $H_2S$ near metal surfaces. At low ferrous ion concentrations, adherent and temporarily protective films of iron sulfides form on low-alloy steel surfaces with a consequent reduction in the corrosion rate. High rates of SPP-induced corrosion of carbon steel are maintained only when the concentrations of ferrous ions are high.

In the absence of oxygen, sulfides, from whatever source, react with carbon steel to from a layer of iron sulfide that prevents further reaction, that is, diminution of corrosion. Aggressive SPP corrosion of low alloy steel has been reported in the presence of dissolved oxygen. Hardy and Bown [10] investigated the weight loss of mild steel exposed to successive aeration-deaeration shifts. In their experiments the highest corrosion rates were observed during periods of aeration. In laboratory seawater/hydrocarbon fuel incubations, Aktas et al. [11] demonstrated that there was minimal sulfate reduction and no corrosion of carbon steel in the total absence of oxygen. Aggressive corrosion was observed when low levels of dissolved oxygen (<100 parts per billion) were present in the seawater. Hamilton [12] reviewed mechanisms for MIC and concluded that oxygen was the terminal electron acceptor in many MIC reactions. Following this logic, when SPP are involved in corrosion sulfate could serve as the terminal electron acceptor in respiration, but oxygen will be the terminal electron acceptor in the corrosion reaction.

### 27.2.3 Nutrients

Waters with suitable forms of carbon, nitrogen, phosphorus, and sulfur are required to support microbial growth. Hydrocarbons can be degraded under aerobic and anaerobic conditions to provide a carbon source for microbial assimilation [13–18]. Aerobic biodegradation of hydrocarbons is faster than anaerobic degradation. Rates depend on the specific electron acceptors used in the process ($O_2 > NO^{3-} > Fe^{3+} > SO_4^{2-} > CO_2$). As a practical matter, carbon availability is not typically the main constraint to crude oil degradation. Low concentrations of assimilable forms of nitrogen and phosphorus can limit hydrocarbon biodegradation.

## 27.3 INTERNAL CORROSION

Pipelines are classified by function. Gathering pipelines collect products from sources, such as wells, tankers or other pipelines. They move products to storage or processing facilities. Transmission pipelines transport liquids or natural gas over longer distances. These pipelines lines deliver crude oil to refineries or refined products to markets. Distribution pipelines move products to customers.

Pitting is the typical type of internal corrosion in pipelines, both isolated pits and overlapping ones [19]. Internal corrosion due to MIC is directly related to the biodegradability of the contents, water, and electron acceptors/donors. Some microorganisms are naturally occurring in hydrocarbon fuels; others are introduced from air or water. Susceptibility of hydrocarbons to microbial degradation can be generally ranked as follows: linear alkanes > branched alkanes > small aromatics > polyaromatics > cyclic alkanes. Some compounds, such as the high molecular weight polycyclic aromatic hydrocarbons, may not be degraded. Walker and Colwell [20] concluded that bacteria showed decreasing abilities to degrade alkanes with increasing chain length.

The sulfur content of crude oils is a particular concern from a MIC perspective because SRB could use oxidized sulfur compounds, including sulfate, as electron acceptors to produce $H_2S$. However, in past surveys, sulfur content did not correlate to $H_2S$ content [21]. Most of the sulfur in crude oils is organic sulfur in heterocyclic ring structures, for example polycyclic saturated carbonaceous ring structures. Gogoi and Bezbaruah [22] concluded, ". . . most prevalent naturally occurring microorganisms do not effectively breakdown sulfur-bearing heterocycles," suggesting that these compounds are not readily biodegradable.

### 27.3.1 Production

SPP-related MIC of carbon steel, used in oil production, has been reported around the world. The petroleum production environment is particularly suitable for the activities of SRB because it handles large volumes of water from underground reservoirs, which contain nutrients [23]. Ciaraldi et al. [24] concluded that the factors influencing MIC in production lines in the Gulf of Suez were low flow velocities, deposit accumulations, water flooding and increased levels of bacteria. El-Raghy et al. [25] reported that pipelines used to transport El-Morgan field crude in the Gulf of Suez lost 75% of their original wall thickness due to the activities of bacteria, particularly SRB.

### 27.3.2 Transmission

Petroleum transmission lines are less susceptible than production lines to MIC because oxygen, water, and sediment are removed to specified limits. For example, in Canada, the National Energy Board (NEB) requires that crude transmission pipelines cannot accept a product that contains more than 0.5% basic sediment and water (BS&W). Lillebo et al. [26] demonstrated that growth of SRB was inhibited in crude oils containing <0.5% water. In the United States, the Federal Energy Regulatory Commission (FERC) allows BS&W levels of 1.0%. At these low levels, the water in crude oil exists as a microemulsion, resulting in carbon steel surfaces being oil-wetted and corrosion is negligible. BS&W values are averaged readings, meaning that it is possible to have slugging events that are not detected. Despite the potential differences in the water content, Friesen et al. [27] demonstrated there was no direct relationship between water content of crudes and corrosivity. Furthermore, internal corrosion has been observed in crude pipelines with <0.5% BS&W at locations where water can accumulate [28]. Unintentional introduction of water or oxygen into crude oils increases the likelihood for MIC.

Papavinasam [29] concluded that "bulk crude oil may indirectly affect the corrosion by influencing the locations where water accumulates, by influencing the type of emulsion, by impacting the wettability of phases on the steel surface and by supplying chemicals that can partition into the water phase." Accumulation of water depends on inclination of the pipe, the flow velocity and the cleanliness of the pipeline. Water solubility increases with hydrocarbon molecular weight [30]. However, industrial experience indicates that heavier crudes, while high in water content, are less corrosive owing to their elevated viscosity and resulting low conductivity ($<10^{-7}$ S/cm) [30]. Asphaltenes and resins in heavy crudes act as surfactants to stabilize water-in-oil emulsions.

Crudes carry water-wetted particles that can drop out at locations downstream of over-bends. Sludge deposits concentrate water from oil at the pipeline surface shifting the oil-wet surface to a water-wet surface. Sludge deposits are combinations of hydrocarbons, sand, clay, corrosion products, and biomass that can reach 50% water by weight. Mosher et al. [28] demonstrated that high bacterial activity and/or water content in the sludge alone did not produce corrosive conditions over a 3-month period. Analyses of pipeline deposits obtained from pigging operations indicated a range of particle sizes with diameters from 44 to 400 μm. Most of the solids were fine particles of silica sand and iron minerals. Larger sand particles were uniformly coated with very fine clay surrounded by a film of water. Under low flow conditions, these particles precipitate and form a sludge deposit.

## 27.4 TESTING

### 27.4.1 A Review of Testing Procedures

One of the first attempts to quantify microorganisms related to MIC in oil and gas systems was published in 1975 by the American Petroleum Institute (API) [31]. "*Recommended Practice (RP) 38 Biological Analysis of Subsurface Injection Waters*" describes liquid media for cultivation of SRB and heterotrophic bacteria. The RP states that the presence of SRB is "a potential problem." It further states, "The extent of the problem will depend upon additional evidence . . ."

In 1990, the Gas Research Institute (GRI) published "*MIC: Methods of Detection in the Field*" [32]. The GRI guide, "designed to help gas industry personnel determine whether or not the corrosion occurring at a particular site is MIC . . ." was the first to emphasize the importance of acid-producing bacteria (APB) to the corrosion of carbon steel gas pipelines. The guide also specified localized corrosion morphologies that were suggestive of MIC, including cup-type, scooped-out hemispherical pits, and striation lines. The guide provided a numerical rating for predicting the probability that MIC had occurred based on two parameters, that is, the number of bacteria and the characteristics of pit morphology. The guide did not suggest using either parameter independently to diagnose MIC. The guide was not meant to be a predictive tool. Unintended consequences of the guide [32] were the proliferation of liquid media test kits, a strong reliance on numbers of particular types of bacteria to diagnose and predict MIC, and an over interpretation of pit morphology to diagnose MIC [33].

In 2004, NACE International Standard Test Method TM0194-2004 "*Field Monitoring of Bacterial Growth in Oil and Gas Systems*" [34] provided sampling procedures for planktonic and sessile bacteria, a lactate-based culture medium, and a serial dilution to extinction methodology for enumerating SRB. Vials that turned black due to formation of iron sulfide within a 28-day period were scored as positive for SRB. Additionally, the time required for blackening was suggested as a measure of the "strength (i.e., activity) of the growing culture." There is an acknowledgment within TM0194-2204 that the lactate-based medium cannot be used to grow SRB requiring other carbon sources, for example, acetate, propionate, or butyrate.

Several attempts have been made to improve liquid culture media used for the detection of SRB. A complex medium was developed containing multiple carbon sources that could be degraded to both acetate and lactate [35]. In comparison tests, the complex medium produced higher counts of SRB from waters and surface deposits among five commercially available media [36]. Jhobalia et al. [37] developed an agar-based culture medium for accelerating the growth of SRB. The authors noted that over the sulfate concentration range from 1.93 to 6.50 g/l, SRB grew best at the lowest concentration. Cowan [38] developed a rapid culture technique for SRB based on rehydration of dried nutrients with water from the system under investigation. The author claimed that using system water reduced the acclimation period for microorganisms, ensuring that the culture medium had the same salinity as the system water used to prepare the inoculum. Cowan [38] reported quantification of SRB within 1–7 days.

The distinct advantage of culturing techniques to detect specific microorganisms is that low numbers of cells grow to easily detectable higher numbers in the proper culture medium. Under all circumstances though, culture techniques underestimate the organisms in a natural population [39,40]. Kaeberlein et al. [41] suggested that 99% of microorganisms from the environment resist cultivation in the laboratory. A major problem in assessing microorganisms from natural environments is that viable microorganisms can enter into a non-culturable state [42]. Another problem is that culture media cannot approximate the complexity of a natural environment. Growth media tend to be strain specific. As previously mentioned, lactate-based media sustain the growth of lactate-oxidizers, but not acetate-oxidizing bacteria. Incubating at one specific temperature is further selective. Zhu et al. [43] demonstrated dramatic changes in the microbial population from a gas pipeline after samples were introduced into liquid culture media. For example, using culture techniques SRB dominated the microflora in most pipeline samples. However, using culture-independent quantitative polymeric chain reaction (qPCR) techniques they found that methanogens were more abundant in most pipeline fluid samples than denitrifying bacteria and that SRB were the least abundant bacteria. Similarly, Romero et al. [44] used molecular monitoring to identify bacterial populations in a seawater injection system. They found that some bacteria present in small amounts in the original waters were enriched in the culture process.

It is well established that the microbial constituents in the sessile population (attached to the surface) are different from those of planktonic population (passively floating). Wrangham and Summer [45] used metagenomic analyses of planktonic and sessile samples from three different geographical locations to demonstrate that the planktonic population was not representative of the sessile population from the same location. They reported, ". . . planktonic and sessile populations from the same location may be as different from each other as they are to samples obtained from other locations." Similarly, Larsen et al. [46] used molecular microbiological methods (MMM) to demonstrate that the bacteria and archaea in scale and produced water were "somewhat different" from each other.

### 27.4.2 Current Procedures

More recent test methods, for example, NACE International Standard Test Method TM0212-2012 "*Detection, Testing, and Evaluation of MIC on Internal Surfaces of Pipelines*" [47] acknowledge that many types of microorganisms, including archaea, can contribute to MIC. In Section 7.2.4, the method clearly indicates that the type of medium used in liquid culture techniques determines, to a large extent, the numbers and types of microorganisms that grow. In addition to liquid culture, the document describes other techniques to identify microorganisms, including microscopy, adenosine triphosphate photometry, hydrogenase measurements, adenosine phosphosulfate reductase, 4′,6-diamidino-2-phenylindole (DAPI), and MMM. MMM include qPCR, fluorescence *in situ* hybridization (FISH), denaturing gradient gel electrophoresis (DGGE), and clone library building. The advantages and disadvantages for each test have been described in detail elsewhere [8,48,49]. TM0212-2012 stresses the need to collect microbiological, operational, and chemical data from corroded sites and to compare with similar types of data collected from areas that are not corroded.

Alabbas et al. [50] reported that DGGE was an ineffective method for fingerprinting DNA, specifically DNA from sour crude oil and seawater injection pipelines, because it is difficult to reproduce among different users and the information is visual, that is, there are no databases for comparative purposes. Investigators have used other approaches to describe microbial populations in petroleum reservoirs. Guan et al. [51,52] used phylogenetic analyses of gene fragments of the dissimilatory sulfite reductase (dsr) gene that encodes for the key enzymes in the anaerobic dissimilatory respiration of sulfate. The dsr gene is present in all SPP. Their investigation demonstrated the diversity of SPP that could potentially be involved in reservoir souring and corrosion.

Sequencing can provide the order of nucleotides in DNA or RNA. DNA sequencing can be used to determine the sequence of individual genes, gene clusters, chromosomes, or full genomes. Wang et al. [53] compared the results from pyrosequencing data and clone library searches to estimate bacterial diversity in aqueous and oil phases from a water-flooded petroleum reservoir. Pyrosequencing is a method that involves extracting DNA, suspending it in a fluid, breaking it apart using chemiluminescent enzymatic reactions, and using a high resolution camera to infer its makeup. In molecular biology a library is a collection of DNA fragments. The term can refer to a population of organisms. Using both pyrosequesncing and clone library approaches, Wang et al. [53] determined that at a high phylogenic level, the predominant bacteria detected by the two methods were identical. However, they reported, ". . . pyrosequencing allowed the detection of "more rare bacterial species than the clone library method."

Prediction of MIC in gas and oil carbon steel pipelines has been unreliable because of uncertainties in the time to pit initiation and the rate of propagation. There have been attempts to predict MIC based on corrosivity factors. For example, Pots et al. [54] considered SRB the major contributor to MIC and determined that the following parameters influenced SRB activity: water, pH of the water, salinity, temperature, and nutrients, for example, sulfate, total carbon, nitrogen, and C:N ratios. Each parameter was given a rating factor (F) based on their influence. In addition, the operational history of the pipeline was reviewed, for example, duration of periods of stagnation. Sooknah et al. [55,56] used a similar approach to develop an internal pitting corrosion model that predicts susceptibility to MIC (Figure 27.1). Use of this type of model requires a thorough understanding of the specific system to which it is applied.

Risk-based inspection programs that include MMM have been designed and are being tested [57,58]. Larsen et al. [59] developed a model that estimates corrosion risk and time-before-pit initiation using qPCR enumeration of MIC-causing microorganisms and reverse transcript qPCR as a measure of cellular activity. Larsen et al. [59] used the approach during an inspection of two pipelines in the North Sea to develop a strategy for remediation.

### 27.4.3 Monitoring

Many techniques claim to monitor MIC, however, none has been accepted as an oil and gas industry standard or as a RP by ASTM or NACE International. NACE Standard TM0194-2004 "*Field Monitoring of Bacterial Growth in Oil and Gas Systems*" [34] describes a standard test method for monitoring growth of microorganisms and evaluating the effectiveness of control chemicals, but does not relate directly to corrosion. The major limitation for MIC monitoring programs has been the inability to relate microorganisms to corrosion in real time. Some techniques can detect a specific modification in the system due to the presence and activities

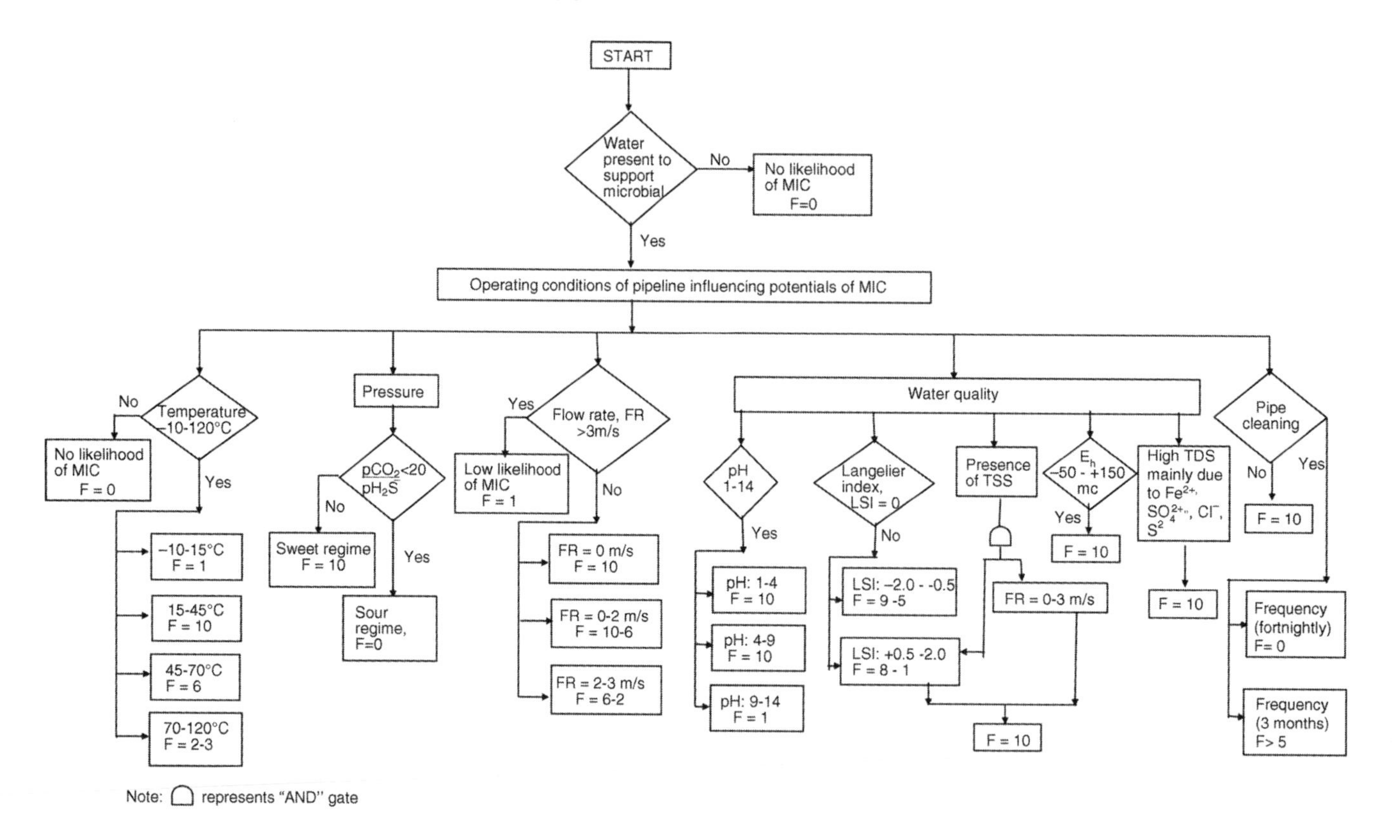


**FIGURE 27.1** Internal pipeline MIC—risk assessment flowchart (© NACE International 2007 [56].)

of microorganisms (e.g., heat transfer resistance, fluid friction resistance, galvanic current) and assume something about the corrosion. Others measure some electrochemical parameter (e.g., polarization resistance, electrochemical noise) and assume something about microbial activities. Either experience or knowledge of a particular operating system can be an effective monitoring tool, especially for evaluating a treatment regime (biocides or corrosion inhibitors).

There have been approaches to derive real-time microbiological data. Chattoraj et al. [60] suggested that fluorogenic bioreproters could be used to determine total microbial contamination/activity on line and in real time. Fluorogenic bioreporters are compounds that undergo a change in their fluorescent signal after interaction with microorganisms. Chattoraj et al. [60] demonstrated that fluorescence, monitored with a fluorometer, before and after addition of the bioreporter, provided a ratio related to microbial activity. Several investigators have attempted to simultaneously quantify corrosion and some property related to microorganisms. For example, Haile et al. [61] developed a four-probe sensor for simultaneously monitoring corrosion rate and sulfide oxidase to detect sulfides. The probe has been demonstrated in the laboratory.

The application of MMM techniques for monitoring microbial populations has been suggested [62,63]. Hoffmann et al. [64] concluded, ". . . there are no 'off the shelf' solutions and standardized methods will ultimately be required if comparative data are to be generated across the industry."

### 27.4.4 Control

Standard mitigation procedures for controlling internal corrosion in oil and gas pipelines include physical removal of deposits (pigging) and chemical treatments that include non-oxidizing biocides (e.g., glutaraldehyde, quaternary amines, and tetrakis (hydroxymethyl) phosphonium sulfate). Biocides kill or slow the growth of microorganisms by a number of mechanisms, including protein cross-linking, disrupting cell membranes, or by inhibiting a vital process (e.g., synthesis or respiration) [65]. Maxwell and Campbell [66] used the approach developed by Pots et al. [54] described earlier to predict the risk of MIC in oil transportation lines. Maxwell and Campbell [66] concluded, ". . . frequency of pigging—which will have no biocidal effect—is predicted to provide the greatest mitigation of MIC compared to the possible bacteriostatic effect of high salinity and the bactericidal effect of biocide additions."

Harris et al. [67] used maximum pitting rate to evaluate the impact of film forming corrosion inhibitors as a means of controlling MIC of carbon steel in produced waters. Film forming inhibitors (e.g., quaternary amines) are frequently used to control $CO_2$ corrosion and can be used in combination with biocides (e.g., glutaraldehyde) intended to reduce numbers of microorganisms. In addition, some corrosion inhibitors contain toxic components. Harris et al. [67] indicated the following: (1) MIC pitting rates increased in the presence of some corrosion inhibitors, presumably because of the biodegradability of the inhibitor, (2) some microorganisms developed a resistance to supposedly toxic inhibitors, (3) MIC control was not related to toxicity, that is, a less toxic inhibitor provided better MIC control than a more toxic one, and (4) severe corrosion was not related to numbers of SRB.

Similarly, Campbell et al. [68] suggested that evaluating MIC control or mitigation by monitoring a decrease in bacterial numbers is inaccurate and misleading. In their experiments, they concluded that biocides can injure cells in a biofilm, rather than kill the cells. Recovering SRB had significantly reduced doubling times compared with SRB that had not been exposed to biocides.

Ciaraldi et al. [24] described the following tactics to prevent MIC in oil production lines and equipment:

- "Scale inhibition and paraffin dispersal/dissolution;
- Periodic chemical and mechanical cleaning;
- Use of fiberglass linings, coatings, and sacrificial anodes in vessel and piping bottoms;
- Reroutings of production to increase flow velocities;
- Resizing of needed replacement piping to increase flow velocities;
- Rotation of piping to extend life (damaged areas relocated from 6 o'clock position);
- Refurbishment pigging of pipelines (i.e., multiple pig runs with increased aggressiveness to restore to near bare-metal condition; experience has shown that in many cases 50–100 pig runs are required);
- Routine, periodic, and aggressive maintenance pigging of refurbished pipelines;
- Routine, batch biocide treatments (when possible, immediately following pigging, mechanical, or chemical cleaning); and
- Improved monitoring with coupons, gas analyses, bacterial culturing of liquid, deposit and pigging debris samples, and increased UT surveys/smart pig inspections." UT refers to ultrasonic testing.

### 27.4.5 Alter Potential Electron Acceptors to Inhibit Specific Groups of Bacteria

Biocides control MIC by decreasing the microbial population, whereas control by manipulation of electron acceptor relies on growth stimulation or retardation of specific microbial populations. Both removal and addition of electron acceptors have been used as a means of controlling microbial populations and MIC in seawater injection systems where seawater is injected into oil reservoirs to maintain pressure. In these applications, oxygen is removed to minimize corrosion. However, in the anaerobic environment, growth of SRB is encouraged. The concentration of sulfate in seawater is typically >2.0 g/l. Rizk et al. [69] used nanofiltration to reduce sulfate in seawater from 2.6 to 0.05 g/l. In laboratory studies they demonstrated that the amount of $H_2S$ in the seawater was a direct function of the amount of sulfate in the water. The authors discussed the implication for corrosion, but did not make corrosion measurements. In contrast, Jhobalia et al. [37] demonstrated that high-sulfate concentration in a laboratory medium (increase from 1.93 to 6.5 g/l) could inhibit growth of *Desulfovibrio desulfuricans* and the corrosion rate of mild steel. Addition of sulfate was presented as a "biochemical approach" for dealing with MIC [37], but has not been tested as a practical control strategy for MIC. The authors hypothesized that the observation was due to increasing toxicity of sulfate toward SRB metabolism or sulfate reduction.

Laboratory and field experiments have demonstrated that nitrate, an alternative electron acceptor, treatment can be an effective replacement for biocide treatment to reduce the sulfide production by SRB [70,71] a process known as biocompetitive exclusion. The addition of nitrate can induce a shift in the dominant population from SRB to nitrate-reducing bacteria (NRB). NRB reduce nitrate to $N_2$ with several possible intermediate by-products, including nitrite and ammonium. One motivation for nitrate injection is to prevent souring due to SRB. The other is to reduce corrosion risks.

Nitrate treatment was implemented on an oil platform in the North Sea (Veslefrikk) [72]. The change from glutaraldehyde treatment to nitrate resulted in a dramatic change in the bacterial community. The SRB population decreased and the numbers of NRB increased. After 4 months of nitrate addition the activity of SRB in the biofilm was markedly reduced as measured with radiorespirometry and an enrichment of NRB was measured. After 32 months of nitrate treatment, SRB numbers were reduced 20,000 fold and SRB activity was reduced 50-fold. Corrosion measurements decreased from 0.7 to 0.2 mm/year. Similar applications have been made to reduce souring [73,74]. Gullfaks platforms have been treated with nitrate to reduce $H_2S$ production [75]. The authors also observed a 1000-fold reduction in SRB numbers and a 10- to 20-fold reduction in sulfate respiration activity and a 50% reduction in corrosion.

Voordouw et al. [76] and Hubert et al. [77] demonstrated a nitrate-reducing, sulfide-oxidizing bacterium capable of reducing nitrate to nitrite, nitrous oxide, or nitrogen and oxidizing sulfide to sulfate or sulfur. The stoichiometry of the reactions catalyzed by the organism depended on the ratio of sulfide to nitrate. Dunsmore et al. [78] isolated an organism from a Danish North Sea oilfield water injection system that had been continuously treated with nitrate since the start of the injection. This species, an SRB, could reduce nitrate and produce ammonium in the presence of sulfate, increasing the likelihood of corrosion. Hubert et al. [79] demonstrated that both nitrate and nitrite were effective treatments for decreasing sulfide concentrations. The required dose depended on the concentration of oil organics used as the energy source by the microbial community. Sunde et al. [75] suggested that reservoir characteristics and nutrient availability have a significant impact on the effectiveness of nitrate injection.

There are several potential mechanisms for the observed inhibition of SRB due to addition of nitrate. Microbial nitrate reduction produces more energy than sulfate reduction. It follows that if both nitrate and sulfate are present, nitrate will be the preferred electron acceptor. Toxic reaction products from the reduction of nitrate to $N_2$, for example, nitrite, may inhibit SRB. A shift in the redox potential in the system may also inhibit SRB. As a consequence of nitrate reduction, the redox potential will likely increase, producing unfavorable conditions for sulfate reduction. Nitrite may also act as a scavenger for $H_2S$. When competing for the same carbon source, nitrate-utilizing bacteria out-compete SRB. This argument is valid only in carbon-limited waters.

The success of nitrate addition relies on a population of nitrate-utilizing bacteria in the system. Hubert et al. [77] suggested that bioaugmentation, in which *ex situ* grown microorganisms could be injected with the nitrate if indigenous NRB were lacking. Despite the possibility of bioaugmentation, there are several reports of failures. Bouchez et al. [80] attempted to inoculate a nitrifying sequencing batch reactor with an aerobic denitrifying bacterium. The added bacterium disappeared after 2 days. Similarly, Hubert et al. [79] reported that introduction of microorganisms into natural communities was difficult. Microorganisms other than SRB and NRB may be involved in subterranean nitrogen cycling. For example, in an anaerobic environment, ammonium-oxidizing (anammox) bacteria can convert ammonium and nitrite into $N_2$. Li et al. [81] have identified five genera of anammox bacteria in high-temperature petroleum reservoirs.

In summary, nitrate addition and sulfate removal/reduction both attempt to control MIC caused by SRB. The long-term consequences on the microbial populations and MIC are unknown. There are limited data indicating that nitrate injection is ineffective at slowing $H_2S$ production where souring has already occurred [82].

## 27.5 EXTERNAL CORROSION

As is the case with internal corrosion, the potential for MIC impacting the outer surfaces of buried or submerged pipelines is controlled by availability of water, electron acceptors/donors and nutrients in the environment. All line pipe is externally coated (e.g., asphalts, polyolefin tapes, and fusion bonded epoxies) and can be further protected with cathodic protection (CP). Coatings isolate the pipe exterior surface from the environment. CP is the application of current to a pipeline, overriding local anodes and making the entire pipeline surface a cathode. Coatings reduce the exposed area of pipeline surfaces, making CP economically feasible.

NACE International Standard Practice (SP) SP0169-2007 (formerly RP0169-2002) "*Control of External Corrosion on Underground or Submerged Metallic Piping Systems*" [83] describes the ". . . procedures and practices for achieving effective control of external corrosion . . . ." The SP lists the following conditions in which CP is ineffective "elevated temperature, disbonded coatings, thermal insulating coatings, shielding, bacterial attack, and unusual contaminants in the electrolyte." Microorganisms can affect CP in several ways. Some microorganisms are attracted to the areas surrounding a cathodically polarized pipe, that is, microbial abundance, activity and diversity in saturated soils, seawater or sediments could increase as a result of CP. If microorganisms compromise the coating, the potential required to prevent corrosion will be more negative than the −850 mV versus saturated copper/copper sulfate reference electrode recommended in SP0169-2007. Barlo and Berry [84] confirmed that the criterion for CP of buried pipelines was valid in concept; however, the actual protection potential varied with environment. Microorganisms can increase the kinetics of corrosion reactions, necessitating a current increase to maintain a specific potential. When sufficient CP levels exist, corrosion is mitigated even in the presence of bacteria. The difficulty is determining the adequate protection potential.

### 27.5.1 Buried Pipelines

One of the earliest reports of MIC [85] identified SRB as the cause of external corrosion of iron pipe failures in sulfate-rich soils. SRB are still the organisms that are most frequently cited as causing external corrosion on pipe surfaces. Soil corrosivity increases with moisture, salts, and dissolved oxygen concentration. Jack et al. [86] reported that MIC was responsible for 27% of all corrosion deposits on the exterior of line pipe in a survey of Nova Gas Transmission Ltd. (Calgary, Alberta) pipe lines. Peabody [87] indicated that soil moisture and numbers of bacteria were greater in back fill material than in undisturbed soil adjacent to a pipeline. Backfill was less consolidated and allowed greater penetration of moisture and oxygen.

Microorganisms can degrade adhesives and coatings exposing metal. Pope and Morris [88] indicated that almost all cases of MIC on external surfaces were associated with disbonded coatings or other areas shielded from cathodic protection. Abedi et al. [89] reported the failure of an oil transmission line due to external corrosion. In their study a polyethylene tape coating on the exterior of the pipeline became loose, exposing the pipe surface to wet soil. The failure was attributed to SRB and stress corrosion cracking. With adequate cathodic protection fusion bonded epoxy coatings allow for cathodic protection of the pipe surface should the coating bond fail [90] because they do not shield the cathodic current. NACE International Standard TMO106-2006 "*Detection, Testing, and Evaluation of MIC on External Surfaces of Buried Pipelines*" [91] is meant to identify MIC after it has occurred and does not describe a predictive methodology.

### 27.5.2 Submerged Pipelines

DOT [19] reports, "although salt water is more corrosive than most soil environments, cases of significant external corrosion on offshore pipelines are extremely rare." The report concludes that control of external corrosion in the offshore environment has been "mastered" because of the homogeneity of the offshore environment.

## 27.6 CONCLUSIONS

Enumeration of microorganisms by any of the available methods (culture- or molecular based) cannot be used to independently diagnose MIC in the field, predict risk due to MIC or evaluate of biocide efficacy to prevent MIC. In all cases there must be some accompanying measure of corrosion. Current MIC corrosion risk assessment models attempt to assign corrosivity factors to operational and environmental parameters. The weight attributed to each factor is arbitrary and requires a complete understanding of the system. Much of the information included in this chapter relates to processes that have been evaluated at 40 °C or less. The present standards and methodology will not be adequate for evaluation of MIC in deeper, hotter reservoirs. Future testing may require different methodologies (i.e., higher pressure vessels and flow cells) and new standards to measure and monitor corrosion.

## ACKNOWLEDGMENTS

NRL Publication no. NRL/BC/7303—14-2045. Funding support from NRL Base Program.

## REFERENCES

1. Hopkins, P. (2007) Oil and Gas Pipelines: Yesterday and Today. American Society of Mechanical Engineers. Available at http://www.penspen.com/Downloads/Papers/Documents/OilandGasPipelines.pdf.
2. Keener, B. (2012) Hazardous Liquids Accident Data with emphasis on Crude Oil and Internal Corrosion. Transportation Research Board. Available at http://onlinepubs.trb.org/onlinepubs/dilbit/keener102312.pdf.
3. Koch, G.H., Brongers, M.P.H., Thompson, N.G., Virmani, Y.P., and Payer, J.H. (2002) Corrosion Costs and Preventive Strategies in the United States. Publication No. FHWA-01-156. Available at http://www.nace.org/Publications/Cost-of-Corrosion-Study/.
4. Agaev, N.M., Smorodina, A.E., Allakhverdova, A.V., Efendizade, S.M., Kosargina, V.G., and Popov, A.A. (1986) Prevention of corrosion and environmental protection. *VNIIOENG (Russian)*, 7 (59), 1–63.
5. Osarolube, E., Owate, I.O., and Oforka, N.C. (2008) Corrosion behaviour of mild and high carbon steels in various acidic media. *Scientific Research and Essays*, 3 (6), 224–228.
6. Donham, J.E. (1987) Corrosion in petroleum production operations. In: Davis, J.R. (editor), *ASM Handbook, Corrosion, J. R. Davis.*, ASM International, Materials Park, OH, Vol. 13, pp. 1232–1235.
7. Enning, D., Venzlaff, H., Garrelfs, J., Dinh, H.T., Meyer, V., Mayrhofer, K., Hassel, A.W., Stratmann, M., and Widdel, F. (2012) Marine sulfate-reducing bacteria cause serious corrosion of iron under electroconductive biogenic mineral crust. *Environmental Microbiology*, 14 (7), 1772–1787.
8. Little, B.J. and Lee, J. S. (2007) *Microbiologically Influenced Corrosion*, John Wiley & Sons, Inc., Hoboken, NJ.
9. Zegeye, A., Huguet, L., Abdelmoula, M., Carteret, C., Mullet, M., and Jorand, F. (2007) Biogenic hydroxysulfate green rust, a potential electron acceptor for SRB activity. *Geochimica et Cosmochimica Acta*, 71 (22), 5450–5462.
10. Hardy, J.A. and Bown, J.L. (1984) The corrosion of mild steel by biogenic sulfide films exposed to air. *CORROSION*, 42 (12), 650–654.
11. Aktas, D.F., Lee, J.S., Little, B.J., Duncan, K.E., Perez-Ibarra, B.M., and Suflita, J.M. (2013) Effects of oxygen on biodegradation of fuels in a corroding environment. *International Biodeterioration & Biodegradation*, 81, 114–126.
12. Hamilton, W.A. (2003) Microbially influenced corrosion as a model system for the study of metal microbe interactions: a unifying electron transfer hypothesis. *Biofouling*, 19 (1), 65–76.
13. Rueter, P., Rabus, R., Wilkes, H., Aeckersberg, F., Rainey, F.A., Jannasch, H.W., and Widdel, F. (1994) Anaerobic oxidation of hydrocarbons in crude-oil by new types of sulfate-reducing bacteria. *Nature*, 372 (6505), 455–458.
14. Aitken, C.M., Jones, D.M., and Larter, S.R. (2004) Anaerobic hydrocarbon biodegradation in deep subsurface oil reservoirs. *Nature*, 431 (7006), 291–294.
15. Atlas, R.M. (1981) Microbial degradation of petroleum hydrocarbons: an environmental perspective. *Microbiological Reviews*, 45 (1), 180–209.
16. Atlas, R.M. and Bartha, R. (1992) Hydrocarbon biodegradation and oil-spill bioremediation. *Advances in Microbial Ecology*, 12, 287–338.
17. Das, N. and Chandran, P. (2011) Microbial degradation of petroleum hydrocarbon contaminants: an overview. *Biotechnology Research International*, 2011, 13.

18. Fritsche, W. and Hofrichter, M. (2000) Aerobic degradation by microorganisms. In: Klein, J. (editor), *Biotechnology, Enviromental Processes II*, Weinheim, Germany, Wiley-VCH, Vol. 11b, pp. 146–155.

19. Baker, M. and Fessler, R.R., (2008) Pipeline Corrosion. U.S. Department of Transportation, Pipeline and Hazardous Materials Safety Administration, Office of Pipeline Safety, Final Report DTRS56-02-D-70036. Available at https://primis.phmsa.dot.gov/gasimp/docs/FinalReport_PipelineCorrosion.pdf.

20. Walker, J.D. and Colwell, R.R. (1975) Some effects of petroleum on estuarine and marine microorganisms. *Canadian Journal of Microbiology*, 21 (3), 305–313.

21. Been, J. and Zhou, J. (2012) Corrosivity of dilbit and conventional crude oil in transportation pipelines, Proceedings of 2012 NACE Northern Area Eastern Conference: Corrosivity of Crude Oil Under Pipeline Operating Conditions. Houston, TX, NACE International, Paper No. 2012-03.

22. Gogoi, B.K. and Bezbaruah, R.L. (2002) Microbial degradation of sulfur compounds present in coal and petroleum. In: Stapleton, R.D. and Singh, V.P. (editors), *Biotransformations: Bioremediation Technology for Health and Environment: Progress in Industrial Microbiology*, Elsevier, Amsterdam, Vol. 36, pp. 427–456.

23. Mora-Mendoza, J.L., Garcia-Esquivel, R., Padilla-Viveros, A.A., Martinez, L., Martinez-Bautista, M., Angeles-Ch, C., and Flores, O. (2001) Study of internal MIC in pipelines of sour gas mixed with formation waters. Proceedings of CORROSION/2001, NACE International, Houston, TX, Paper No. 01246.

24. Ciaraldi, S.W., Ghazal, H.H., Abou Shadey, T.H., El-Leil, H.A., and El-Raghy, S.M. (1999) Progress in combating microbiologically induced corrosion in oil production. Proceedings of CORROSION/99, NACE International, Houston, TX, Paper No. 181.

25. El-Raghy, S.M., Wood, B., Aduleil, H., Weare, R., and Saleh, M. (1998) Microbiologically influenced corrosion in mature oil fields—a case study in El-Morgan field in the Gulf of Suez. Proceedings ofCORROSION/98, NACE International, Houston, TX, Paper No. 279.

26. Lillebo, B.L.P., Torsvik, T., Sunde, E., and Hornnes, H.K. (2010) Effect of water content on the grwoth of SRB in crude oil. Proceedings of CORROSION/2010, NACE International. Houston, TX, Paper No. 10206.

27. Friesen, W.I., Perovici, S., Donini, J.C., and Revie, R.W. (2012) Relative corrosivities of crude oils from oil transmission pipelines. Proceedings of 2012 NACE Northern Area Eastern Conference: Corrosivity of Crude Oil Under Pipeline Operating Conditions, Houston, TX, NACE International, Paper No. 2012-08.

28. Mosher, M., Crozier, B., Mosher, W., Been, J., Tsaprailis, H., Place, T., and Holm, M. (2012) Development of laboratory and pilot scale facilities for the evaluation of sludge corrosivity in crude oil pipelines, Proceedings of 2012 NACE Northern Area Eastern Conference: Corrosivity of Crude Oil Under Pipeline Operating Conditions, Houston, TX, NACE International, Paper No. 2012-07.

29. Papavinasam, S. (2012) Managing internal corrosion and public perception of oil transmission pipelines. Proceedings of 2012 NACE Northern Area Eastern Conference: Corrosivity of Crude Oil Under Pipeline Operating Conditions, Houston, TX, NACE International, Paper No. 2012-10.

30. Craig, B. (1998) Predicting the conductivity of water-in-oil solutions as a means to estimate corrosiveness. *CORROSION*, 54 (8), 657–662.

31. American Petroleum Institute (1975) API RP 38: API Recommended Practice for Biological Analysis of Subsurface Injection Waters. Dallas, TX,

32. Pope, D.H. (1990) *GRI-90/0299 Field Guide: Microbiologically Influenced Corrosion (MIC): Methods of Detection in the Field*, Gas Research Institute, Chicago, IL.

33. Little, B.J., Lee, J.S., and Ray, R.I. (2006) Diagnosing microbiologically influenced corrosion: a state-of-the-art review. *CORROSION*, 62 (11), 1006–1017.

34. NACE International (2004) NACE Standard TM0194-2004: Field Monitoring of Bacterial Growth in Oil and Gas Systems. Houston, TX,.

35. Scott, P.J.B., and Davies, M. (1999) Monitoring bacteria in waters: the consequences of incompetence. *Materials Performance*, 38 (7), 40–43.

36. Scott, P.J.B., and Davies, M. (1992) Survey of field kits for sulfate-reducing bacteria. *Materials Performance*, 31 (5), 64–68.

37. Jhobalia, C.M., Hu, A., Gu, T., and Nesic, S. (2005) Biochemical engineering approaches to MIC. Proceedings of CORROSION/2005, NACE International. Houston, TX, Paper No. 05500.

38. Cowan, J.K. (2005) Rapid enumeration of sulfate-reducing bacteria. Proceedings of CORROSION/2005, NACE International, Houston, TX, Paper No. 05485.

39. Giovannoni, S.J., Britschgi, T.B., Moyer, C.L., and Field, K.G. (1990) Genetic diversity in Sargasso Sea bacterioplankton. *Nature*, 344, 60–63.

40. Ward, D.M., Ferris, M.J. Nold, S.C. and Bateson, M.M. (1998) A natural view of microbial diversity within hot spring cyanobacterial mat communities. *Microbiology and Molecular Biology Reviews*, 62 (4), 1353–1370.

41. Kaeberlein, T., Lewis, K., and Epstein, S.S. (2002) Isoltaing "uncultivable" microorganisms in pure culture in a simulated natural environment. *Science*, 249, 1127–1129.

42. Roszak, D.B. and Colwell, R.R. (1987) Survival strategies of bacteria in the natural environment. *Microbiological Reviews*, 51 (3), 365–379.

43. Zhu, X., Ayala, A., Modi, H. and Kilbane, J.J. (2005) Application of quantitative, real-time PCR in monitoring microbilogically influenced corrosion (MIC) in gas pipelines, Proceedings of CORROSION/2005, NACE International. Houston, TX, Paper No. 05493.

44. Romero, J.M., Velazquez, E. Villalobos, G., Amaya, M., and Le Borgne, S. (2005) Genetic monitoring of bacterial populations in a sewater injection system. Indetification of biocide resistant bacteria and study of their corrosive effect. Proceedings of CORROSION/2005, NACE International. Paper No. 05483:

45. Wrangham, J.B. and Summer, E.J. (2013) Planktonic microbial population profiles do not accurately represent same location sessile population profiles. Proceedings of CORROSION/2013, NACE International. Houston, TX, Paper No. 2248.

46. Larsen, J., Skovhus, T.L., Saunders, A.M., Hoejris, B., and Agerbaek, M. (2008) Molecular identification of MIC bacteria from scale and produced water: similarities and differences. Proceedings of CORROSION/2008, NACE International. Houston, TX, Paper No. 08652.

47. NACE International (2012) NACE Standard TM0212-2012: Detection, Testing, and Evaluation of Microbiologically Influenced Corrosion on Internal Surfaces of Pipelines. Houston, TX.

48. Whitby, C. and Skovhus, T.L. (editors) (2011) Applied Microbiology and Molecular Biology in Oilfield Systems. *Proceedings from the International Symposium on Applied Microbiology and Molecular Biology in Oil Systems (ISMOS-2), 2009*, Springer, London.

49. Juhler, S., Skovhus, T.L., and Vance, I. (2012) A Practical Evaluation of 21st Century Microbiological Techniques for the Upstream Oil and Gas Industry, Energy Institute.

50. Alabbas, F.M., Kakopovbia, A., Mishra, B., Williamson, C., Spear, J.R., and Olson, D.L. (2013) Microbial community associated with corrosion products collected from sour oil crude and seawater injection pipelines, Proceedings of CORROSION/2013, NACE International. Houston, TX, Paper No. 2248.

51. Guan, J., Xia, L.P., Wang, L.Y., Liu, J.F., Gu, J.D., and Mu, B.Z. (2013) Diversity and distribution of sulfate-reducing bacteria in four petroleum reservoirs detected by using 16S rRNA and dsrAB genes. *International Biodeterioration & Biodegradation*, 76, 58–66.

52. Guan, J., Zhang, B.-L., Mbadinga, S., Liu, J.-F., Gu, J.-D., and Mu, B.-Z. (2014) Functional genes (dsr) approach reveals similar sulphidogenic prokaryotes diversity but different structure in saline waters from corroding high temperature petroleum reservoirs. *Applied Microbiology and Biotechnology*, 98 (4), 1871–1882.

53. Wang, L.Y., Ke, W.J., Sun, X.B., Liu, J.F., Gu, J.D., and Mu, B.-Z. (2014) Comparison of bacterial community in aqueous and oil phases of water-flooded petroleum reservoirs using pyrosequencing and clone library approaches. *Applied Microbiology and Biotechnology*, 1–13.

54. Pots, B.F.M., Randy C.J., Rippon, I.J., Simon Thomas, M.J.J., Kapusta, S.D., Girgis, M.M., and Whitman, T. (2002) Improvements on De Waard-Milliams corrosion prediction and applications to corrosion management corrosion. Proceedings of CORROSION/2002, NACE International, Houston, TX, Paper No. 02235.

55. Sooknah, R., Papavinasam, S., and Revie, R.W. (2008) Validation of a predictive model for microbiologically influenced corrosion. Proceedings of CORROSION/2008, NACE International, Houston, TX, Paper No. 08503.

56. Sooknah, R., Papavinasam, S., Revie, R.W., and De Romero, M.F. (2007) Modeling the occurrence of microbiologically influenced corrosion. Proceedings of CORROSION/2007, NACE International, Houston, TX, Paper no. 07515.

57. Jensen, M., Lundgaard, T., Jensen, J., and Skovhus, T.L. (2013) Improving risk-based inspection with molecular microbiological methods. Proceedings of CORROSION/2013, NACE International; Houston, TX, Paper No. 2247.

58. Sorensen, K.B., Juhler, S., Thomsen, T.R., and Larsen, J. (2012) Cost efficient MIC management system based on molecular microbiological methods. Proceedings of CORROSION/2012, NACE International, Houston, TX, C2012-000111.

59. Larsen, J., Sorensen, J.V., Juhler, S., and Pedersen, D.S. (2013) The application of molecular microbiological methods for early warning of MIC in pipelines. Proceedings of CORROSION/2013, NACE International, Houston, TX, Paper No. 2029.

60. Chattoraj, M., Fehr, M.J., Hatch, S.R., and Allain, E.J. (2002) Online measurement and control of microbiological activity in industrial waters systems. *Materials Performance*, 41 (4), 40–45.

61. Haile, T., Papavinasam, S., and Gould, D. (2013) Evaluation of a portable online 4-probe sensor for simultaneous monitoring of corrosion rate and sulfate reducing bacteria activity. Proceedings of CORROSION/2013, NACE International, Houston, TX, Paper No. 2148.

62. Zhu, X., Lubeck, J., Lowe, K., Daram, A., and Kilbane, J.J., (2004) Improved method for monitoring microbial communities in gas pipelines. Proceedings of CORROSION/2004, NACE International, Houston, TX, Paper No. 04592.

63. Larsen, J., Skovhus, T.L., Thomsen, T.R., and Nielsen, P.H. (2006) Bacterial diversity study applying novel molecular methods on halfdan produced waters. CORROSION/2006, NACE International, Houston, TX, Paper No. 06668.

64. Hoffmann, H., Devine, C., and Maxwell, S. (2007) Application of molecular microbiology techniques as tools for monitoring oil field bacteria. Proceedings of CORROSION/2007, NACE International, Houston, TX, Paper No. 07508.

65. Dickinson, W.H., Campbell, S., and Turk, V. (2012) Tank reactor studies of biocide performance and mitigation of deadleg corrosion. Proceedings of CORROSION/2012, NACE International, Houston, TX, Paper No. C2012-0001196.

66. Maxwell, S. and Campbell, S. (2006) Monitoring the mitigation of MIC risk in pipelines, Proceedings of CORROSION/2006, NACE International, Houston, TX, Paper No. 06662.

67. Harris, J.B., Webb, R., and Jenneman, G. (2010) Evaluating corrosion inhibitors as a means to control MIC in produced water, Proceedings of CORROSION/2010, NACE International, Houston, TX, Paper No. 10256.

68. Campbell, S., Duggleby, A., and Johnson, A. (2011) Conventional application of biocides may lead to bacterial cell injury rather than bacterial kill within a biofilm. Proceedings of CORROSION/2011, NACE International, Houston, TX, Paper no. 11234.

69. Rizk, T.Y., Stott, J.F.D., Eden, D.A., Davis, R.A., McElhiney, J.E., and Di Iorio, C. (1998) The effects of desulphated seawater injection on microbiological hydrogen sulphide generation and

inplication for corrosion control. Proceedings of CORROSION/98, NACE International, Houston, TX, Paper No. 287.

70. Larsen, J. and Skovhus, T.L. (2011) Problems caused by microbes and treatment strategies the effect of nitrate injection in oil reservoirs—experience with nitrate injection in the Halfdan oilfield. In: Whitby, C. and Skovhus, T.L. (editors), *Applied Microbiology and Molecular Biology in Oilfield Systems: Proceedings from the International Symposium on Applied Microbiology and Molecular Biology in Oil Systems (ISMOS-2), 2009*, Springer, London, pp. 109–115.

71. Telang, A.J., Ebert, S., Foght, J.M., Westlake, D.W.S., Jenneman, G.E., Gevertz, D., and Voordouw, G. (1997) Effect of nitrate injection on the microbial community in an oil field as monitored by reverse sample genome probing. *Applied and Environmental Microbiology*, 63 (5), 1785–1793.

72. Thorstenson, T., Bodtker, G., Sunde, E., and Beeder, J. (2002) Biocide replacement by nitrate in sea water injection systems. Proceedings of CORROSION/2002, NACE International, Houston, TX, Paper No. 02033.

73. Larsen, J. (2002) Downhole nitrate applications to control sulfate reducting bateria activity and reservoir souring. Proceedings of CORROSION/2002, NACE International, Houston, TX, Paper No. 02025.

74. Larsen, J., Malene, R.H., and Zwolle, S. (2004) Prevention of reservoir souring in the Halfdan field by nitrate injection. Proceedings of CORROSION/2004, NACE International, Houston, TX, Paper No. 04761.

75. Sunde, E., Lillebo, B.L.P., Bodtker, G., Torsvik, T., and Thorstenson, T. (2004) $H_2S$ inhibition by nitrate injection on Gullfaks field. Proceedings of CORROSION/2004, NACE International, Houston, TX, Paper No. 04760.

76. Voordouw, G., Nemati, M., and Jenneman, G.E. (2002) Use of nitrate-reducing, sulfur-oxidizing bacteria to reduce souring in oil fields: interactions with SRB and effects on corrosion. Proceedings of CORROSION/2002, NACE International, Houston, TX, Paper No. 02033.

77. Hubert, C., Arensdorf, J., Voordouw, G., and Jenneman, G.E. (2006) Control of souring through a novel class of bacteria that oxidize sulfide as well as oil organics with nitrate. Proceedings of CORROSION/2006, NACE International. Houston, TX, Paper No. 06669.

78. Dunsmore, B.C., Whitfield, T.B., Lawson, P.A., and Collins, M.D. (2004) Corrosion by sulfate-reducing bacteria that utilize nitrate. Proceedings of CORROSION/2004, NACE International, Houston, TX, Paper No. 04763.

79. Hubert, C., Voordouw, G., Nemati, M., and Jenneman, G. (2004) Is souring and corrosion by sulfate-reducing bacteria in oil fields reduced more efficiently by nitrate or nitrite? Proceedings of CORROSION/2004, NACE International, Houston, TX, Paper No. 04762.

80. Bouchez, T., Patureau, D., Dabert, P., Juretschko, S., Dore, J., Delgenes, P., Moletta R., and Wagner, M., (2000) Ecological study of a bioaugmentation failure. *Environmental Microbiology*, 2 (2), 179–190.

81. Li, H., Chen, S., Mu, B.Z., and Gu, J.D. (2010) Molecular detection of anaerobic ammonium-oxidizing (anammox) bacteria in high-temperature petroleum reservoirs. *Microbial Ecology*, 60 (4), 771–783.

82. Mitchell, A.F., Anfindsen, H., Liengen, T., and Molid, S. (2012) Experience of molecular monitoring techniques in upstream oil and gas operations, Proceedings of CORROSION/2012, NACE International, Houston, TX, C2012-0001756.

83. NACE International (2007) NACE Standard Practive SP0169-2007: Control of External Corrosion on Underground or Submerged Metallic Piping Systems. Houston, TX.

84. Barlo, T.J. and Berry, W.E. (1984) An assessment of the current criteria for cathodic protection of buried steel pipelines. *Materials Performance*, 23 (9), 9–16.

85. von Wolzogen Kuhr, C.A.H. and van der Vlugt, L.S. (1934) The graphitization of cast iron as an electrobiochemical process in anaerobic soils. *Water*, 18 (16), 147–165.

86. Jack, T.R., van Boven, G., Wilmot, M., and Worthingham, R.G. (1996) Evaluating performance of coatings exposed to biologically active solis. *Materials Performance*, 35 (3), 39–45.

87. Peabody, A.W. (2001) *Peabody's Control of Pipeline Corrosion*, NACE International, Houston, TX.

88. Pope, D.H. and Morris, A. (1995) Some experiences with microbiologically influenced corrosion of pipelines. *Materials Performance*, 34 (5), 23–28.

89. Abedi, S.S., Abdolmaleki, A. and Adibi, N. (2007) Failure analysis of SCC and SRB induced cracking of a transmission oil products pipeline. *Engineering Failure Analysis*, 14 (1), 250–261.

90. Norsworthy, R. (2006) Fusion bonded epoxy—a field proven fail safe coating system. Proceedings of CORROSION/2006, NACE International, Houston, TX, Paper No. 06044.

91. NACE International (2006) NACE Standard Test Method TM0106-2006: Detection, Testing and Evaluation of Microbiologically Influenced Corrosion (MIC) on External Surfaces of Buried Pipes. Houston, TX.

# 28

# EROSION–CORROSION IN OIL AND GAS PIPELINES*

SIAMACK A. SHIRAZI,[1] BRENTON S. MCLAURY,[1] JOHN R. SHADLEY,[1] KENNETH P. ROBERTS,[1]
EDMUND F. RYBICKI,[1] HERNAN E. RINCON,[2] SHOKROLLAH HASSANI,[3]
FAISAL M. AL-MUTAHAR,[4] AND GUSAI H. AL-AITHAN[4]

[1]*Erosion/Corrosion Research Center, The University of Tulsa, Tulsa, OK, USA*
[2]*ConocoPhillips Company, Houston, TX, USA*
[3]*BP America Inc., Houston, TX, USA*
[4]*Saudi Aramco, Dhahran, Saudi Arabia*

## 28.1 INTRODUCTION

Erosion–corrosion has many definitions. In all definitions of erosion–corrosion, there is a mechanical component—erosion, and an electrochemical component—corrosion, the two components working together to degrade metallic piping, tubing, and many other types of equipment.

One of the most costly problems facing oil and gas production is related to material performance. According to a 2-year study conducted by the U.S. Federal Highway Administration (FHWA) on the cost of corrosion, the annual direct cost of corrosion in the oil and gas exploration and production industry is as high as $1.4 billion [1]. Also, the Department of Transportation Pipeline and Hazardous Materials Safety Administration (PHMSA) in significant incident files from 2012 reported corrosion to be the cause of 17.9% of all pipeline reported incidents recorded from 1992 to 2011 [2].

Erosion–corrosion is not as common as other corrosion-related issues faced by the oil and gas industry, but it has been reported to be amongst the five main failure causes with a 9% incidence of erosion–corrosion related failures, behind pitting (12%), preferential weld corrosion (18%), $H_2S$ related (18%), and $CO_2$ corrosion (28%). Thus, the understanding of erosion–corrosion in the oil and gas industry is indeed, quite relevant [3].

In most cases, the rate of metal loss is greater for the combined processes than for either erosion or corrosion working alone. And there are often significant interactions between erosion and corrosion. One useful way of defining the participation of the main components and their interactions in the total erosion–corrosion process is shown in Equation 28.1.

$$\begin{aligned}\text{Erosion-Corrosion Rate} &= \text{Erosion Rate} + \text{Corrosion Rate}\\ &\quad + \text{Effect of erosion on the corrosion rate}\\ &\quad + \text{Effect of corrosion on the erosion rate}\end{aligned} \tag{28.1}$$

In Equation 28.1, “erosion rate” refers to the mechanical erosion rate of the metal that would occur under the given conditions in the absence of an electrochemical corrosion component, and “corrosion rate” refers to the corrosion rate of the metal that would occur under the given conditions in the absence of a mechanical erosion component. For the two interaction terms in Equation 28.1, “effect of erosion on the corrosion rate” refers to the increment, positive or negative, that the corrosion rate undergoes when the full effect of the mechanical erosion component is considered, and “effect of corrosion on the erosion rate is the increment, positive or

* Research for this chapter has been done at the Erosion/Corrosion Research Center, Tulsa, Oklahoma as part of the MS and/or PhD programs.

*Oil and Gas Pipelines: Integrity and Safety Handbook,* First Edition. Edited by R. Winston Revie.

negative, that the erosion rate undergoes when the full effect of the electrochemical component is considered. The term "synergy" is often used to describe cases for which the sum of the interaction terms $\neq 0.0$, that is, when the erosion–corrosion rate given in Equation 28.1 is different from simply the sum of the erosion rate and corrosion rate.

In oil and gas pipelines, the erodent is usually sand or another solid particle that is produced along with the production fluids. The most prevalent corrosion agent in oil and gas production and transportation is carbon dioxide gas dissolved in produced water. The presence of $H_2S$ gas, even in small quantities, is a major factor in the degradation of piping in production, gathering, and transportation systems, but is not so much associated with erosion–corrosion as is $CO_2$.

In many cases, the effect of corrosion on erosion rate term in Equation 28.1 is small compared with the rates of the other erosion–corrosion components and can be neglected. In some erosion–corrosion systems, however, the effect of erosion on corrosion rate can be very substantial, for example when erosive forces strip away a corrosion product layer that would otherwise provide corrosion protection. In these cases, the combined erosion and corrosion metal loss rate is larger than the sum of the loss rates from erosion and corrosion acting individually, and often very much larger. Because the synergistic effect can be quite large, indeed, much larger than either the erosion or corrosion component, erosion–corrosion research has favored systems involving a significant interaction term; and, usually this interaction term is the effect of erosion on corrosion rate.

In oil and gas pipelines, the erodent can be various produced particles or liquid drops entrained in a gas carrier fluid. However, the most common, and most destructive erodent is sand; so, the mechanical partner in oilfield erosion–corrosion is most often sand erosion. In these cases, what is ultimately important is the so-called erosivity. Erosivity refers to the *severity* of the erosion conditions. Erosivity factors do not include properties of materials, but they include properties of the sand and carrier fluids, sand rate, flow conditions, geometry, among others. For example, increasing the sand rate would generally increase the erosivity. Because of the importance of solid particle erosion to understanding erosion–corrosion, the section of this chapter following the present introductory section focuses on the factors influencing solid particle erosion and on models developed to predict erosion rate. Following the section on solid particle erosion, three erosion–corrosion processes are discussed, each involving an effect of erosion on corrosion that can result in an erosion–corrosion rate significantly greater than the sum of the erosion and corrosion rates taken separately.

The first of these three erosion–corrosion systems, for which the effect of erosion on corrosion rate can be strong, involves erosion of a protective scale. In $CO_2$-saturated systems, bare metal corrosion rates of carbon steel can be more than 800 mpy (20 mm/year). However, under certain environmental conditions, an iron carbonate ($FeCO_3$) scale can form on metal surfaces that can reduce corrosion rates to a few mpy (0.1–0.2 mm/year) [4]. Environmental conditions for which protective iron carbonate scale is likely to form include pH > 5.0, temperature >150 °F (66 °C), plus a significant concentration of $Fe^{2+}$ ions. When sand is entrained in the flow, impingement of the sand particles onto the iron carbonate scale can prevent the scale from forming, or removing it completely or partially. As a result of this erosion of the scale, corrosion rates rise significantly or possibly return to bare metal values.

A second example of a potentially very substantial effect of erosion on corrosion rate involves corrosion-resistant alloys (CRAs). CRAs represent a class of iron-based materials that exhibit resistance to corrosion in many environments. Typically, a CRA will have a significant chromium alloying component and possibly many other alloying components as well. Most materials in this class work well in oil and gas production and transportation applications. When exposed to a corrosive environment, an oxide layer forms on the surface of the CRA forming a barrier to the reactant species. In most particle-free oilfield fluids, the oxide layer is able to keep corrosion rates of the CRA at very low values. However, if sand or another solid particle is entrained, impingement of the particle onto the CRA can damage or partially remove the oxide layer, and increased corrosion rates may result.

Erosion rate also can affect corrosion rate strongly in systems that are protected by a chemical inhibitor. The inhibitor in a liquid carrier fluid is injected into the system at the upstream end of the segment of piping that is to be protected. Typically, the inhibitor molecule attaches itself to the metal piping surface through adsorption. At a high enough inhibitor concentration, a barrier is created that sharply reduces the cathodic reactant penetration to the metal surface, keeping corrosion rates low. However, even at moderately high-flow velocities, entrained sand particles impinging on the metal surfaces can mechanically dislodge the inhibitor molecules from the surface, thereby exposing the bare metal, which leads to high corrosion rates.

While much has been done to better understand erosion–corrosion and identify conditions where it will occur, it is important to note that attention to erosion–corrosion is relatively new when compared with the attention directed at understanding and controlling erosion and corrosion individually. Research on erosion and corrosion has helped build the foundation for the work on erosion–corrosion. The following sections introduce the reader to current testing and modeling approaches to better understand and mitigate metal loss in pipelines due to solid particle erosion and erosion–corrosion.

## 28.2 SOLID PARTICLE EROSION

During the production of oil and gas, solid particles are often present in the production fluids. The particles can be of sand or carbonate, which is released from the reservoir or a material such as magnetite resulting from scale. The particles are a detrimental by-product of production and serve no favorable purpose. One potentially catastrophic result of the presence of particles is solid particle erosion, which can be defined as the mechanical removal of wall material by the impact of solid particles. Several removal mechanisms have been proposed with three of the most common being cutting, fatigue, and brittle fracture. The authors have examined scanning electron microscope (SEM) images from specimens eroded by sand for several materials used in the oil and gas industry such as carbon steels, stainless steels, titanium, and other materials such as aluminum, and the eroded surfaces look similar and provide insight into the typical erosion mechanism process. An SEM image of 316 stainless steel eroded by silica flour with a nominal diameter of approximately 20 microns is shown in Figure 28.1. Particles impinge the surface creating craters pushing material to the edges of the craters creating lips of material that are removed through subsequent impacts. Note that impingement of irregularly-shaped 20 micron particles onto a steel surface can result in surface scratches and craters that are approximately 2–3 microns in length or diameter. Figure 28.2 shows a schematic of this process. However, removal of grains through fracture along grain boundaries has also been observed.

Solid particle erosion is complex and not well understood. This is primarily due to the large number of factors that affect erosion and the difficulty in isolating the effect of a single factor. Considering particles flowing in a single-phase fluid,

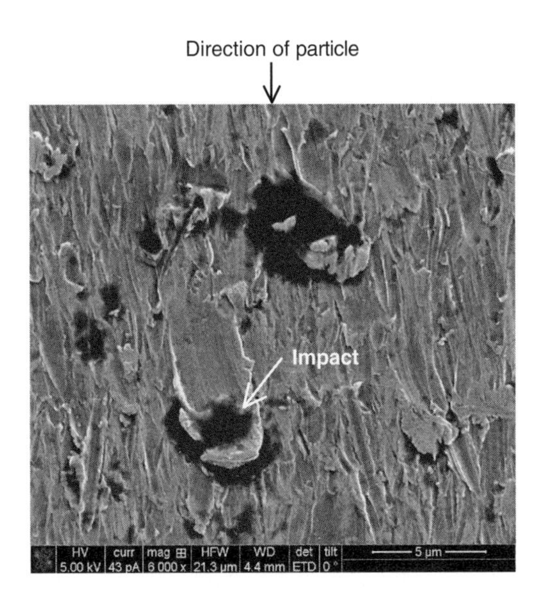


**FIGURE 28.1** SEM image of 316 stainless steel eroded by silica flour.

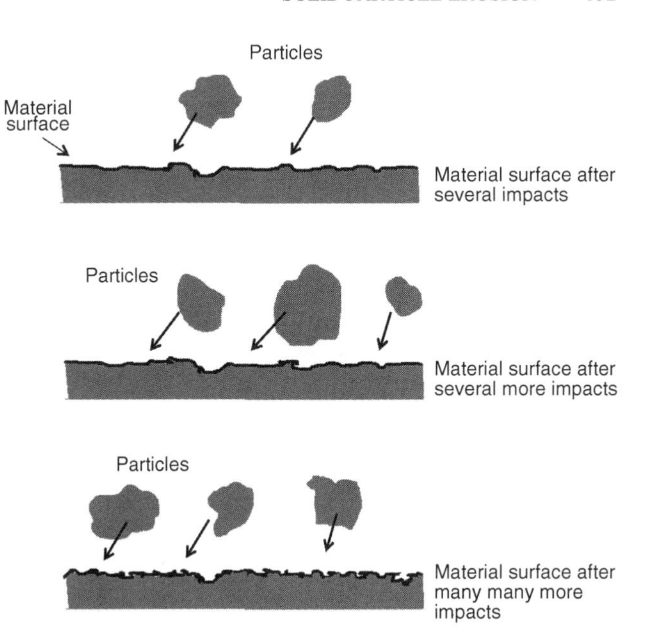


**FIGURE 28.2** Schematic of erosion mechanism commonly observed for materials used in oil and gas industry.

particle properties such as size, shape, density, and hardness; fluid properties such as density and viscosity; and pipe wall material such as density all affect the severity of erosion. The problem becomes substantially more complex when particles flow in a multiphase carrier fluid due to the wide range of possible flow structures. Meng and Ludema [5] performed a literature study of wear equations and found 182 acceptable for analysis; 28 of which were specific to solid particle erosion. There were 33 different parameters in the 28 equations with an average of five parameters in an equation. This demonstrates the level of incongruence among researchers.

Even though solid particle erosion is not fully understood, it is still necessary to estimate erosion rates or at least predict production velocities that would not cause excessive erosion. Historically, American Petroleum Institute recommended practice 14E (API RP 14E) [6] has been used as a guide to prevent significant erosion damage. This guideline provides an equation to calculate an "erosional" or a threshold velocity, presumably a flow velocity that is safe to operate. The recommended practice states that

*"The following procedure for establishing an "erosional velocity" can be used where no specific information as to the erosive/corrosive properties of the fluid is available. The velocity above which erosion may occur can be determined by the following empirical equation:*

$$V_e = \frac{c}{\sqrt{\rho}} \tag{28.2}$$

where:

$V_e$ = Fluid erosional velocity, ft/s

$c$ = Empirical constant

$\rho$ = Gas/liquid mixture density at flowing pressure and temperature, lbs/ft$^3$.

*Industry experience to date indicates that for solids-free fluids values of* $c = 100$ *for continuous service and* $c = 125$ *for intermittent service are conservative*." The recommended practice adds "If solids production is anticipated, fluid velocities should be significantly reduced. Different values of "*c*" may be used where specific application studies have shown them to be appropriate."

Unfortunately, this equation only considers one factor, the density of the medium, and does not consider many other factors that can contribute to erosion/corrosion in multiphase flow pipelines. It is obvious that Equation 28.2 is very simple and easy to use; but, factors such as environmental conditions, fluid properties, flow geometry, type of metal, temperature, pH, sand production rate and size distribution, and flow composition are not accounted for. The only physical variable accounted for in Equation 28.2 is the fluid density. The equation suggests that the limiting velocity could be higher when the fluid density is lower. This does not agree with experimental observations for sand erosion or even for erosion caused by liquid droplets, because sand and liquid droplets entrained in gases with lower densities will cause higher erosion rates than liquids with higher densities and viscosities. Therefore, the form of Equation 28.2 does not seem to be appropriate for many oilfield situations, including those involving sand production. Moreover, the basis for the development of the API RP 14E is not very clear. Salama [7] provides a review of many ideas and suggestions for the basis of the equation. Jordan [8] and McLaury and Shirazi [9] have compared the threshold velocities that were predicted by the API RP 14E to field data for $c = 100$ and concluded that the API equation is not conservative enough at high gas velocities and very low liquid rates. Recognizing the complexities involved in determining a suitable threshold velocity for controlling erosion and erosion/corrosion, many investigators are developing data and models for predicting erosion–corrosion.

The API RP 14E equation with slightly different values of $c$ is also proposed in ISO 13703 [10]. The origin of the expression is unclear but it is unlikely that the original intent was to provide a criterion for solid particle erosion. However, this became the calculation of choice for oil and gas producers for many years probably due to ease of use and lack of other models, but it is apparent that many factors are not considered in the equation. In an attempt to overcome the shortcomings of API RP 14E many companies developed tables of $c$ values to use under various operating conditions. As a replacement for API RP 14E, other researchers developed other approaches to predict erosion or determine an erosional velocity specifically for solid particle erosion. Salama and Venkatesh [11] proposed an equation to predict erosion rate as a function of sand rate, fluid velocity, material hardness, and pipe diameter. They further showed that the expression could be rearranged to provide an equation for erosional velocity for steel pipes with an allowable penetration rate of 10 mils/year as shown in Equation 28.3,

$$V = \frac{4d}{\sqrt{W}} \tag{28.3}$$

where $V$ is the erosional velocity in ft/s, $d$ is the pipe diameter in inches, and $W$ is the sand rate in barrels/month. The authors do note that this assumes the particles impact the pipe wall at velocities similar to the flow velocity, which may only be true for gases at low densities. Bourgoyne [12] also developed expressions for erosion rate prediction based on his experimental work. Equation 28.4 is his equation for penetration rate for dry gas or mist flow,

$$h = F_e \frac{\rho_a q_a}{\rho_s A} \left( \frac{V_{sg}}{100 F_g} \right)^2 \tag{28.4}$$

where $h$ is the erosion rate in m/s, $F_e$ is a specific erosion factor (kg/kg) based on experimental results, $\rho_a$ is particle density, $\rho_s$ is the density of wall material, $q_a$ is the volume flow of particles in m$^3$/s, $A$ is the cross-sectional area of the pipe in m$^2$, $V_{sg}$ is the superficial gas velocity in m/s, and $F_g$ is the fractional volume of gas. A similar equation is also provided for liquid continuous flows replacing $V_{sg}$ and $F_g$ with $V_{sl}$, superficial liquid velocity, and $F_l$, fractional volume of liquid. Svedeman and Arnold [13] rearranged Bourgoyne's equation to provide an erosional velocity assuming an acceptable erosion rate of 0.13 mm/year (5 mils/year) as shown in Equation 28.5,

$$V_e = K_s \frac{d}{\sqrt{Q_s}} \tag{28.5}$$

where $V_e$ is the erosional velocity in ft/s, $K_s$ is a fitting erosion constant, $d$ is pipe diameter in inches, and $Q_s$ is the volumetric flow of solids in ft$^3$/day. $K_s$ accounts for several factors such as geometry type and material and phases present such as gas–sand, liquid–sand, or gas–liquid–sand. Salama [7] extended his work in erosion to develop an expression for erosion rate for particles in gas–liquid flows as a function of sand rate, mixture velocity, particle diameter, pipe diameter, mixture density, and an empirical value that accounts for geometry type and material. Assuming a sand size of 250 microns and an allowable erosion rate of 0.1 mm/year (4 mils/year), he rearranged the expression to solve for the erosional velocity in m/s as shown in Equation 28.6,

$$V_e = S \frac{D\sqrt{\rho_m}}{\sqrt{W}} \tag{28.6}$$

where $S$ is a geometry dependent constant, $D$ is the pipe diameter in mm, $\rho_m$ is the fluid mixture density in kg/m$^3$, and $W$ is the volumetric sand rate in kg/day.

The models discussed to this point either assumed that particles impact at velocities close to the flow velocity or required empirical coefficients based on data, so applicability is extremely limited. In order to develop a more general approach, the particle impact velocity must be modeled. Currently, the most sophisticated approach to calculate erosion is the use of computational fluid dynamics (CFD). The larger, commercially available CFD programs have frameworks in place to predict erosion. The most common use of CFD for erosion prediction is an Eulerian–Lagrangian approach. The carrier fluid flow field is determined using an Eulerian method; then individual particle paths are determined using a Lagrangian approach. Particle impact information, such as location, speed, and angle, are extracted during particle tracking and applied in an erosion equation. However, an Eulerian–Eulerian approach to predict erosion is also used where particles are treated as another phase instead of discrete entities. The CFD approach allows for calculation of erosion in complex three-dimensional geometries and provides a surface contour of erosion. Figure 28.3 displays a predicted erosion contour plot for a bend carrying liquid and sand particles. The areas shown in red in the downstream section of the bend indicate the regions of maximum erosion. The maximum erosion for this specific condition is along the surface of the inner radius due to strong secondary flows that redirect particles to this location.

CFD users must perform computational grid refinement studies to insure that the flow field and particle motion are independent of the grid. For the Eulerian–Lagrangian approach, particle number studies must also be performed to insure that a sufficient number of particles are examined to obtain statistically representative results.

Another important factor for CFD simulations is an appropriate erosion equation. Erosion equations are typically determined experimentally and ideally are performed with

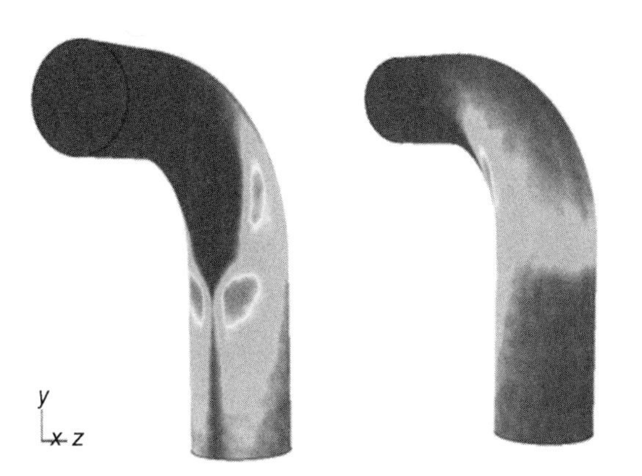


**FIGURE 28.3** CFD prediction of erosion in a bend carrying liquid and sand particles.

the particles and wall material of interest. Data are commonly collected using a direct impingement (blasting) device at a variety of impact angles as described in ASTM G-76 [14]. In order to control impact speed and angle, testing is normally performed in air. The particle rate in the tests must be low enough to avoid concentration effects such as particle–particle interaction or particles shielding the material being eroded. It is also mandatory that the particle velocity (if possible, impact velocity) be measured, since the particle impact velocity can be significantly different than the average fluid velocity leaving the nozzle of the testing device. Zhang et al. [15] provide details on two forms of erosion equations commonly used in CFD simulations. The first form was developed by the Erosion/Corrosion Research Center (E/CRC) and is shown in Equations 28.7 and 28.8.

$$\mathrm{ER} = C(\mathrm{BH})^{-0.59} F_s V_p^n F(\theta) \tag{28.7}$$

$$F(\theta) = \sum_{i=1}^{z} A_i \theta^i \tag{28.8}$$

ER is the erosion ratio in terms of mass loss of materials per mass of sand, $F_s$ is the sand sharpness factor, $V_p$ is the particle impact velocity, and $F(\theta)$ is the angle function. $n = 2.41$ is suggested by E/CRC but others have suggested values between 1.5 and 3. $C$ and $A$ are empirical coefficients. BH is the Brinell hardness of the wall material. It is important to note that a general function of erosion with material hardness is not seen experimentally except within carbon steels for which this function was originally developed. So, for other materials, $C(\mathrm{BH})^{-0.59}$ can be combined into one empirical coefficient. The second form was developed by Oka and is shown in Equations 28.9–28.11.

$$E(\theta) = g(\theta) E_{90} \tag{28.9}$$

$$g(\theta) = (\sin\theta)^{n1}(1 + Hv(1 - \sin\theta))^{n2} \tag{28.10}$$

$$E_{90} = K(\mathrm{Hv})^{k1} \left(\frac{V_p}{V'}\right)^{k2} \left(\frac{D_p}{D'}\right)^{k3} \tag{28.11}$$

$E(\theta)$ is the volume loss of material divided by the mass of sand, $E_{90}$ is the erosion at the normal impact angle; $Hv$ is the Vickers hardness of the wall material; $V_p$ and $D_p$ are the particle impact speed and diameter; $V'$ and $D'$ are the reference impact speed and reference particle diameter; $n1$, $n2$, and $k2$ are empirical functions of wall hardness; and $K$, $k1$, and $k2$ are empirical constants. It is apparent that there are several empirical constants contained in the equations to account for specific conditions used in the testing, but there are a couple important similarities. Knowledge of the particle impact speed and angle are imperative.

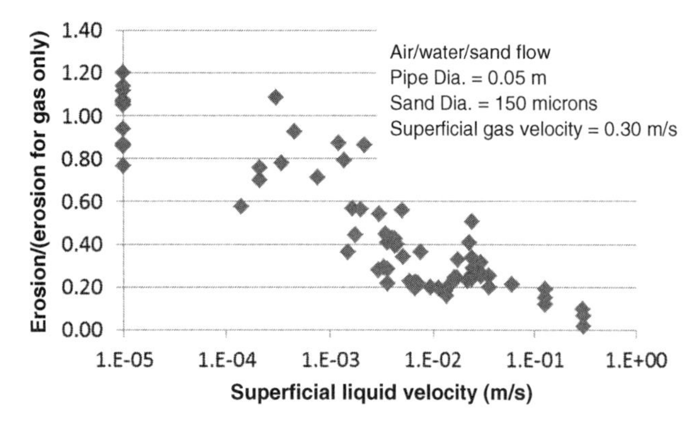


**FIGURE 28.4** Effect of liquid rate on erosion for a gas dominant flow.

While CFD is an incredibly useful erosion prediction tool, there are disadvantages. Access to CFD programs must be available, and users must have a level of expertise. Additionally, significant progress has been made in the simulation of gas–liquid flows, but simulation of particles in gas–liquid flows still needs substantial development. So other simpler models or guidelines have been developed. One guideline is provided in DNV RP O51 [16]. This guideline provides erosion prediction in various geometries and states that the models presented are based on experiments and modeling efforts but warns that the data were generated at low pressures for relatively small pipe diameters. It appears that CFD was one of the tools used to generate the proposed models and for gas–liquid flows mixture properties were used in the simulations. Due to difficulties of simulated gas–liquid flows, this is a common approach but often does not capture the effect of flow regime on erosion. Consider the erosion data shown in Figure 28.4 that is normalized to the gas only condition, where erosion is shown as a function of superficial liquid velocity with constant superficial gas velocity. Superficial velocity is defined as the volumetric flow of that phase divided by the cross-sectional area of the pipe. The maximum erosion occurs for the gas only condition and then drops with the addition of liquid until the flow becomes annular, and then there is a slight increase then decrease with further increase in liquid rate. The increase and then decrease with liquid rate can be explained by examining the liquid film and the entrainment of liquid. As the liquid rate increases in this region, both the liquid film and the entrainment of liquid (and particles) increase. Initially, the increased entrainment is the dominant factor, so erosion increases since more particles are entrained in the gas core where velocities are higher. When the liquid rate increases further, the increased film thickness becomes the dominant factor, and the particle impact velocities reduce due to the increased film thickness. This behavior cannot be simulated assuming mixture properties. For low liquid rates, the predicted erosion values actually increase slightly from the gas only conditions when applying mixture properties.

Another tool for erosion prediction is a program developed by the E/CRC at the University of Tulsa called Sand Production Pipe Saver (SPPS). This model has its origins in the early 1990s but has remained under continued development. The goal of the model is to predict a representative particle impact velocity for a set of conditions, and then use this information in an erosion ratio equation to predict erosion. A simple one-dimensional particle tracking routine is used to predict the particle velocity as it passes through a one-dimensional fluid velocity field as the particle approaches the wall. When the particle is a radius away from the wall, the particle tracking routine is stopped and the particle velocity at that position is the impact velocity [17]. The length for which the particle tracking is performed is a function of the geometry type and size. Treating a three-dimensional flow in a one-dimensional manner has its limitations especially for small particles and liquids, so a two-dimensional approach was also developed [18]. For gas–liquid flows, a similar approach to a single-phase carrier fluid is applied for flow patterns that are considered homogeneous, including churn and dispersed bubble. For annular [19] and slug flows [20], specific models are applied that account for the multiphase flow characteristics such as liquid film thickness and entrainment fraction for annular flow. Figure 28.5 shows the normalized predicted results using DNV RP O51 and SPPS for the data shown in Figure 28.4.

SPPS and DNV RP O501 can also be used to predict flow rates that result in an allowable penetration rate. In this manner, results from SPPS can be compared with API RP 14E. Figure 28.6 displays such a comparison. In this figure, operating to the left of the curves is considered safe and operating to the right of the curves results in excessive erosion. Figure 28.6 illustrates a danger in applying API RP 14E. At low liquid rates, the allowable gas rate predicted by API RP 14E is too large. As the liquid rate reduces, the flow approaches gas only which is a very erosive condition. It is important to mention that the API RP 14E results are

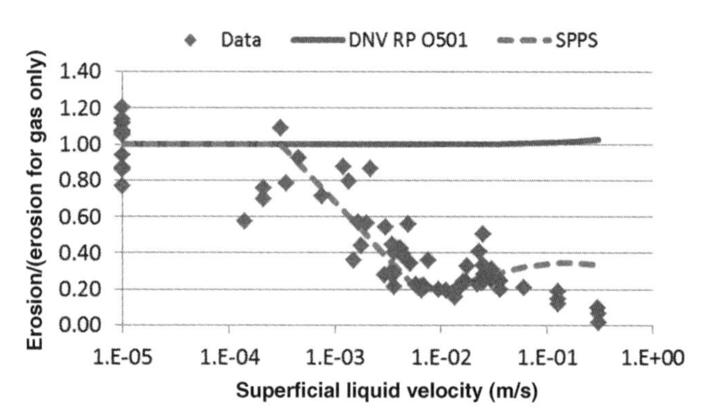


**FIGURE 28.5** DNV RP O51 and SPPS prediction trends of erosion for gas flow with low liquid loading.

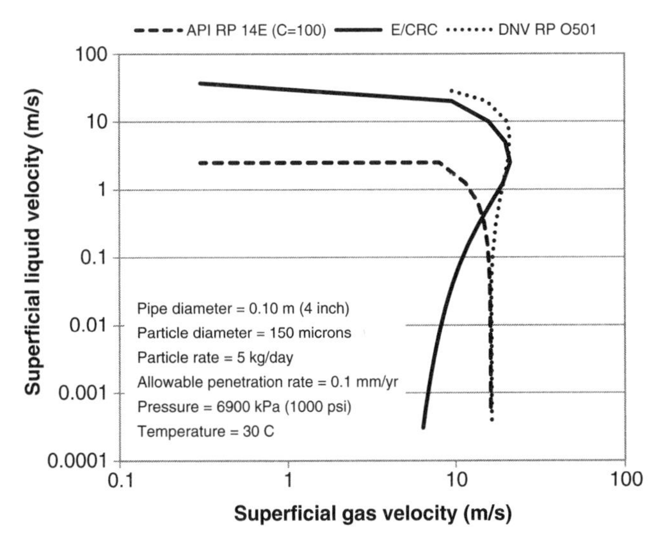


**FIGURE 28.6** Comparison of allowable flow rates in an elbow using API RP 14E, SPPS, and DNV RP O501.

independent of pipe diameter, sand size and rate, and allowable penetration rate. However, SPPS can be used to predict variations in threshold flow rates with these parameters. Figure 28.7 demonstrates the effect of sand rate using SPPS. The other conditions are the same as those used in Figure 28.6.

## 28.3 EROSION–CORROSION OF CARBON STEEL PIPING IN A $CO_2$ ENVIRONMENT WITH SAND

Under some oil and gas production conditions involving a $CO_2$ environment, an $FeCO_3$ (iron carbonate) scale can form on carbon steel piping surfaces and provide considerable corrosion protection to the pipeline. But, if sand is entrained in the production fluids, sand particles impinging the walls of the piping and pipe fittings can cause erosion of the protective $FeCO_3$ scale at particle impingement points and permit corrosion at bare metal rates. This type of erosion–corrosion is a particularly menacing form of metal loss because it can be more severe than either erosion or corrosion acting alone. Whether erosion–corrosion will be a problem in a particular oil and gas production situation depends on environmental (relating to corrosion) and mechanical (relating to erosion) factors.

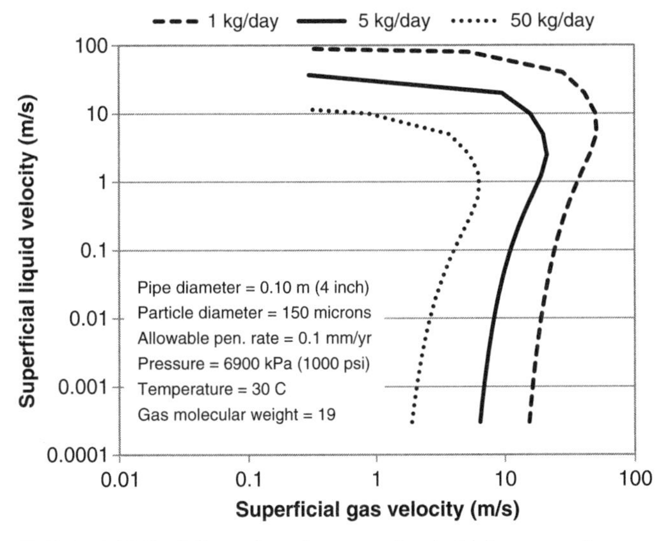


**FIGURE 28.7** Effect of sand rate on threshold flow rates in an elbow using SPPS.

For an iron carbonate scale to form, Equation 28.12 has to be satisfied at the surface of the metal

$$[Fe^{2+}][CO_3^{2-}] > K_{sp} \tag{28.12}$$

where $[Fe^{2+}]$ and $[CO_3^{2-}]$ are the molar concentrations of iron and carbonate ions, respectively, at the metal surface, and $K_{sp}$ is the solubility product of $FeCO_3$. This condition usually requires the temperature to be above 150 °F (66 °C), the pH to be above 5.0, and a supersaturation of $Fe^{2+}$ ions. Iron carbonate scales formed at temperatures near or above 200 °F (93 °C) are nonporous and tough. Scales formed at these temperatures may be only a few micrometers in thickness and still provide excellent protection against corrosion. Scales formed at lower temperatures tend to be more porous and to exhibit cracking and possibly separation from the metal substrate. But, they can be much thicker than the higher temperature scales. Corrosion protection is typically lower for the lower temperature scales, but in some cases still significant.

With the introduction of sand into the flow stream, erosion of the scale ensues. The scale may be thinned by the erosion, in which case some corrosion protection may be lost. At the same time, $FeCO_3$ scale is precipitating with the effect of growing the scale. Eventually, steady state conditions are reached at which the scale deposition rate is equal to the erosion rate. Of course, for extremely high erosivities, the scale may be removed completely, in which case the corrosion rates return to bare metal values.

Among the factors that determine the erosion rate of the scale are the flow velocity, flow geometry, sand properties and sand rate, and erosion resistance of the scale. Factors that determine the deposition rate of the scale are the temperature and the supersaturation factor, SS, at the metal surface. The higher the value of SS, the higher the precipitation rate. The SS is given in Equation 28.13.

$$SS = \frac{[Fe^{2+}][CO_3^{2-}]}{K_{sp}} - 1.0 \tag{28.13}$$

Flow velocity is a parameter affecting both scale formation and scale reduction through dissolution and erosion. In many oilfield situations, flow velocity can be controlled to some

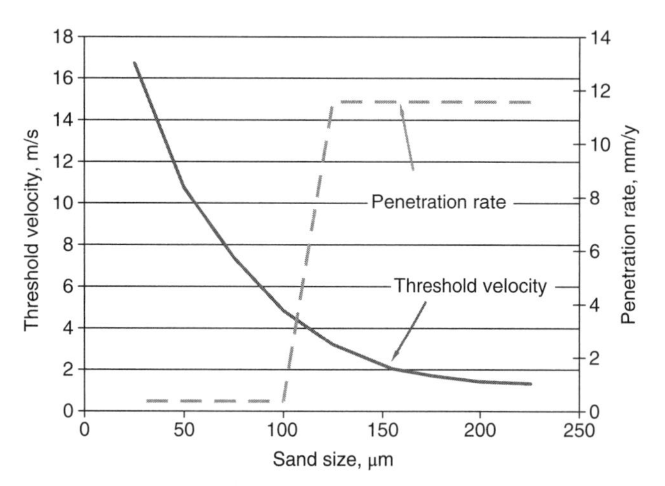


**FIGURE 28.8** Threshold velocity and corrosion penetration rate (without considering the erosion rate) as a function of sand particle size. Test conditions were as follows: 205 kPa $CO_2$, 74 °C, 5.5 pH, 10 ppm $Fe^{2+}$, 5.0 kg/d sand production in a 76.2 mm diameter elbow, 3.8 m/s flow velocity.

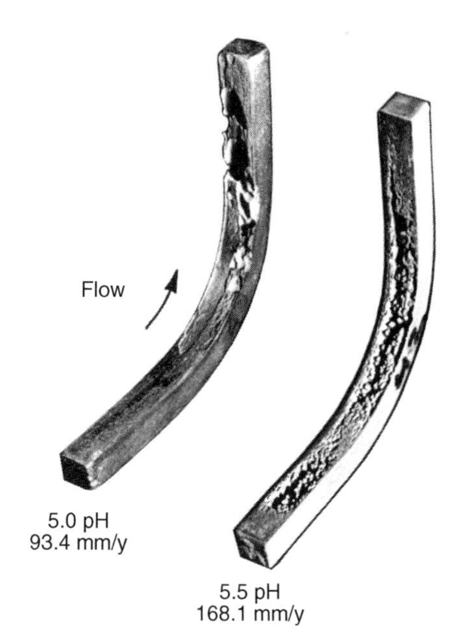


**FIGURE 28.9** Examples of severe pitting in elbow specimens for cases slightly above the threshold velocity where iron carbonate scale was removed at sand particle impingement points. For this geometry and other conditions tested, pitting most often occurred when pH was in the range 5.0 to 6.0. Test conditions were as follows: 80 °C, 450 kPa $pp_{CO2}$, 0.5 wt% sand, 155 μm sand, 20 ppm $Fe^{2+}$, 16 mm diameter elbow, 2% NaCl solution, 5.0 m/s flow rate at pH 5.5, 2.0 m/s flow rate at pH 5.0.

extent by operational procedures. Shadley et al. [21] used a solid particle erosion model developed by Shirazi, et al. [22] and results of erosion–corrosion testing in a laboratory flow loop to develop a semiempirical prediction model for erosion–corrosion in a carbon steel elbow in a $CO_2$ environment with sand. The model provided an estimate of the "threshold velocity," that is, the minimum flow velocity that would remove the scale for the given erosion and corrosion conditions. Figure 28.8 shows an example of how the threshold velocity and corrosion rate would change with increasing sand size. Threshold velocity falls sharply with sand size initially, then begins to level off for larger sand sizes. For a flow velocity of 3.8 m/s, the corrosion rates in this scenario would be very low for sand sizes <100 μm because an iron carbonate scale is providing good corrosion protection. Above 100 μm, the protective iron carbonate scale would be eroded away, eventually leading to bare metal corrosion rates for larger sand sizes.

One unexpected, but important outcome of the laboratory work behind the threshold velocity model was the finding that for flow velocities just above the threshold velocity, where the protective scale was intact in most places, but completely removed at sand impingement points, pitting was observed; and, in most cases the pitting was quite severe. Figure 28.9 shows a photograph of two such specimens. These specimens formed the inner surface of the outside radius of elbows through which the sand-laden fluid flowed. The arrow in the photograph indicates the direction of flow. Often, at test end, iron carbonate scale would be observed on the upstream end of the elbow; but, at the downstream end, the scale would be partially or completely removed by impinging sand.

This finding defined three regions for corrosion behavior under scale-forming conditions: (1) low corrosion rates for flow velocities below the threshold velocity; (2) bare metal corrosion rates for flow velocities well above the threshold velocity; and (3) potential for severe pitting for flow velocities just above the threshold velocity.

## 28.4 EROSION–CORROSION MODELING AND CHARACTERIZATION OF IRON CARBONATE EROSIVITY

$CO_2$ (sweet) corrosion is one of the major causes of tubing and piping failure in the oil and gas industry. In 2002, the total cost of failures due to corrosion was estimated to be approximately $7 billion annually in the United States [1]. Knowledge of the damage risks is the first step in a program for controlling the costs of corrosion. Anticipating where corrosion is likely to happen, estimating its severity, and implementing suitable corrosion controls have become increasingly important in all phases of oil and gas extraction and transportation. Acquiring knowledge of the damage risks is the first step in an erosion–corrosion control program. And a model to predict where erosion–corrosion damage may occur and how severe the damage may be is an important tool toward achieving that objective.

A model for an erosion–corrosion system in which iron carbonate scales provide significant corrosion protection must compare the rate at which iron carbonate scales are being deposited on the metal surface to the rate which iron carbonate scales are being eroded from the metal surface. The rate at which iron carbonate scales are being eroded away depends on the erosivity of the system and the erosion resistance of the scale. The factors affecting the erosivity include sand properties, sand rate, carrier fluid properties, flow rate, and others. There are many factors that contribute to the formation and physical properties of iron carbonate scale. These are discussed in the following sections.

### 28.4.1 $CO_2$ Partial Pressure

Increasing $CO_2$ partial pressure increases the concentration of carbonate ions both in the bulk and at the metal surface due to increased corrosion rate, which in turn increases supersaturation. When supersaturation exceeds 0.0, $FeCO_3$ precipitation rate is proportional to supersaturation.

### 28.4.2 pH

The pH of the solution influences the formation of iron carbonate scale in two ways: First, $FeCO_3$ scales do not readily form when pH is much lower than about 5.0 [23]; and second, the scales formed at high pH are likely more protective and more erosion resistant than scales formed at lower pH.

### 28.4.3 Temperature

When the temperature is lower than about 140 °F (60 °C), iron carbonate scales may not be observed due to the high solubility of iron carbonate [23]. For conditions for which supersaturation exceeds 0.0, precipitation rate of the iron carbonate increases with increasing temperature. The higher the temperature, the more dense and protective is the scale.

### 28.4.4 Flow Velocity

Shear stresses impressed on the metal surface by high-velocity flows may mechanically disturb the formation of iron carbonate scale. Furthermore, high flow velocity may reduce the supersaturation level at the metal surface through increased mass transfer of iron and carbonate ions away from the metal surface.

### 28.4.5 Supersaturation

Supersaturation increases by two ways, either increasing the concentrations of ferrous ions and carbonate ions at the metal surface, or by decreasing the solubility product, $K_{sp}$. The solubility product is given by Equation 28.14 [24]. The solubility product increases with increasing absolute temperature, $T$, and solution ionic strength, $I$

$$\log K_{SP} = 59.3498 - 0.041377T - \frac{2.1963}{T} + 24.5724 \log T + 2.5181\sqrt{I} - 0.657I$$
$$I = \frac{1}{2}\sum_{i=1}^{n} 2c_i z_i^2 \tag{28.14}$$

where $c_i$ are the concentrations of the various species in the aqueous solution in (mol/l), and $z_i$ are ion charges. The relations between the solubility product, and temperatures for different ionic strengths are shown in Figure 28.10.

Based on the aforementioned discussion, scale is formed and deposited on the steel surface at different rates. Several models were developed for predicting scale precipitation and deposition rate [25–28]. The most influential factors controlling $FeCO_3$ deposition rate (and supersaturation) are temperature and solubility product.

### 28.4.6 Erosion of Scale

When environmental conditions are right for forming iron carbonate scale ($FeCO_3$), the scale will deposit onto the metal surface. The scale microstructure and mechanical properties vary dependent upon the conditions under, which the scale was formed [29,30]. For example, scale formed at 190 °F, 10 ft/s, pH 6.4, 2 wt% NaCl, 1900 parts per million (ppm) $NaHCO_3$ eroded five times faster than bare mild steel tested under the same conditions of temperature, velocity, and erosivity. However, scale formed at 150 °F, 10 ft/s, pH 6.24, 2 wt% NaCl, 1900 ppm $NaHCO_3$ eroded 15 times faster than bare mild steel. The erosion rate was determined by the difference in weight and volume before and after the erosion. Test results are shown in Figure 28.11.

### 28.4.7 Erosion–Corrosion

In the erosion–corrosion process, two different mechanisms contribute to the metal loss of the pipe or tubing. The first one

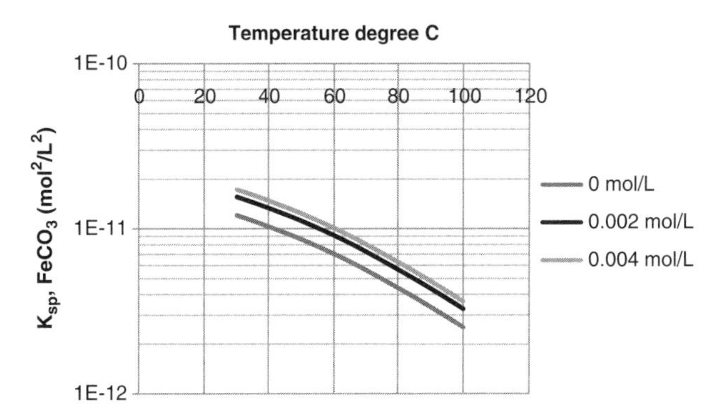


**FIGURE 28.10** The relation between the solubility of iron carbonate at different temperatures and ionic strength 0, 0.002, and 0.004 mol/l.

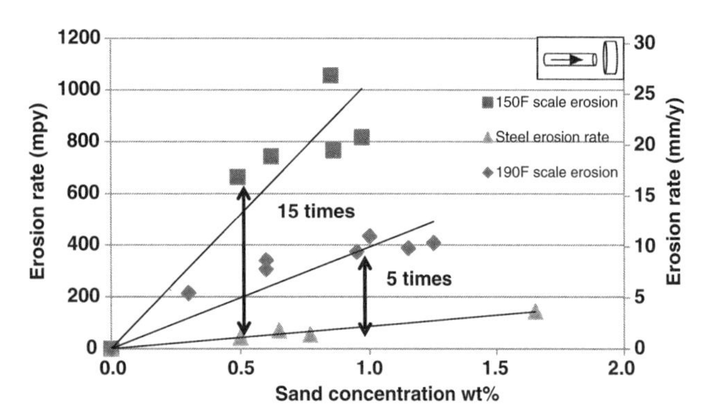


**FIGURE 28.11** Scale erosion rates and bare G10180 mild steel erosion rate. (Adapted from Ref. [31].)

is through the anodic reaction in the corrosion process. The second mechanism is the removal of the metal surface as a result of erosion by sand particle impingement. The sand particles in the carrier fluid impact the pipe surface causing chipping and gouging. Under scale-forming conditions, iron carbonate scale may cover the metal surface. In this case, it is the $FeCO_3$ scale that is eroded, and the corrosion that is taking place is merely replacing scale at places where the scale has been thinned due to the erosion.

Figure 28.12 shows the results of many tests, in which $FeCO_3$ scales were formed, then sand was entrained in the flow stream. The figure shows that when sand erosion rates were zero or low, the corrosion rate (corrosion part of erosion–corrosion) was very low. However, when erosion rates were increased, corrosion rates increased sharply. In some cases, the iron carbonate scale was completely removed allowing corrosion to take place at bare metal rates.

### 28.4.8 Erosion–Corrosion Model Development

Al-Mutahar et al. [32] developed a model to predict steady state erosion–corrosion rates. A flowchart of the erosion–corrosion model procedure is shown in Figure 28.13. A sand erosion prediction model [32] is used to estimate the erosion rate of iron carbonate scale. A $CO_2$ corrosion rate prediction model [32] was modified to estimate the corrosion rate under scale-forming conditions. A third model was developed that combined the erosion and corrosion models and computed the corrosion rate under steady conditions. The sequence of the computational process can be summarized in the following steps for thermodynamic conditions that are favorable for iron carbonate scale formation:

i) Compute the erosion rate of the scale and the corrosion rate for a scale thickness of zero (bare metal conditions)
ii) Compare the formation rate of the iron carbonate scale to the erosion rate calculated in step (i).
iii) On the other hand, if the formation rate of iron carbonate scale is smaller than the erosion rate, then it is assumed that the scale cannot maintain itself and erosion–corrosion rate equals the summation of the corrosion rate and erosion rate of bare metal.
iv) On the other hand, if the formation rate of the iron carbonate scale is greater than the erosion rate, the scale thickness, $h$, is increased incrementally until the rate of scale formation is equal to the erosion rate of the scale. At this point, the scale thickness is considered maintainable and equal to the steady state scale thickness $h'$. Once steady state conditions are reached, the corrosion rate is computed for an iron carbonate scale of steady state thickness, $h'$, and the erosion rate of the underlying metal is set equal to zero.

As a further illustration of how the erosion–corrosion model works, a sample calculation is presented in this section for a multiphase flow in a 90° elbow. For the case described in Table 28.1, the $CO_2$ corrosion model predicts that scale

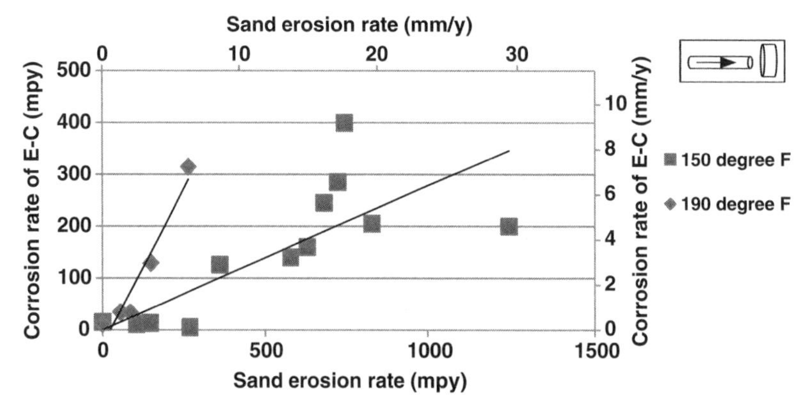


**FIGURE 28.12** Corrosion part of erosion–corrosion process showing increase with increase of erosivity at two different testing conditions. (Adapted from Ref. [31].)

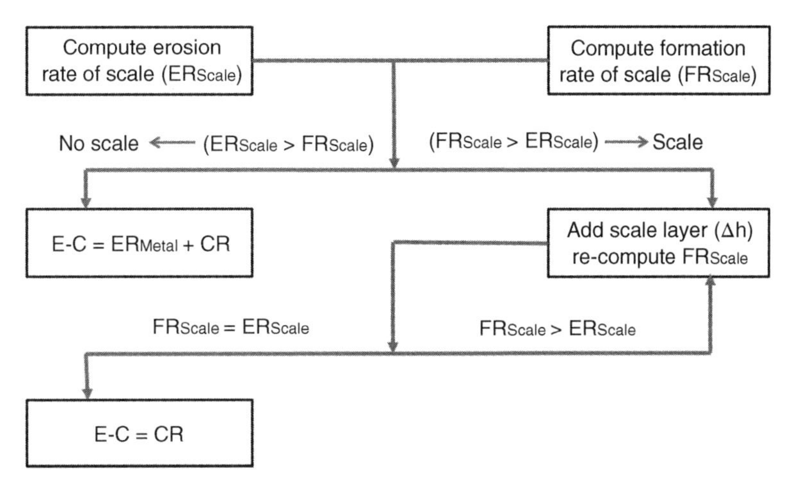


**FIGURE 28.13** Flow chart of erosion–corrosion model computational procedure. (Adapted from Ref. [32].)

**TABLE 28.1 Flow Conditions for Multiphase Case**

| | |
|---|---|
| Geometry | 90° Elbow |
| Diameter | 89 mm |
| Temperature | 90 °C |
| $p_{CO2}$ | 4.5 bars |
| $V_{sl}$ | 3.0 m/s |
| $V_{sg}$ | 10.0 m/s |
| pH | 5.9 |
| $[Fe^{2+}]$ | 10 ppm |
| NaCl | 3 wt% |
| Sand rate | 36 kg/day |
| Sand size | 150 μm |
| Sand shape | Semi rounded |

was likely to form under these thermodynamic conditions. The erosion rate of the scale was then predicted using the erosion rate prediction model.

Figure 28.14 shows the erosion rate of the iron carbonate scale computed for a range of superficial gas velocities.

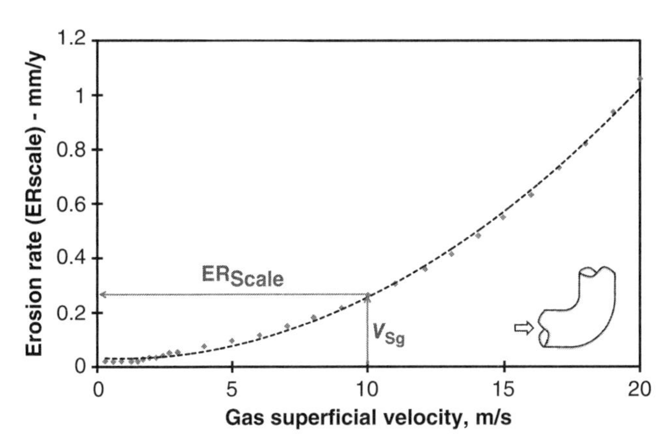


**FIGURE 28.14** Prediction of the erosion rate of the scale for the multiphase case described in Table 28.1. (Adapted from Ref. [32].)

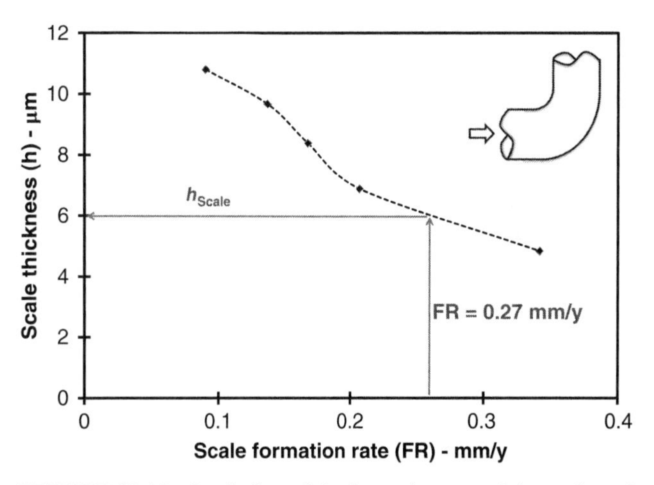


**FIGURE 28.15** Prediction of the formation rate of the scale and steady-state scale thickness for the multiphase case described in Table 28.1. (Adapted from Ref. [32].)

Next, the steady state scale thickness is calculated. The steady state scale thickness is the thickness at which the formation rate of the iron carbonate scale equals the erosion rate of the scale. Figure 28.15 shows the formation rate of the scale computed over a range of scale thicknesses. For a formation rate of the scale equal to 0.27 mm/year (equal to the predicted erosion rate), the steady state scale thickness is predicted to be 6 microns. Finally, the corrosion rate is calculated for the steady state scale thickness of 6 microns, which is much thicker than stainless steel oxide, which is <10 nm. Figure 28.16 shows the corrosion rate computed over a range of scale thicknesses. For a 6 micron scale thickness, the predicted material corrosion rate is 1 mm/year (≈40 mpy).

Further details of the model development are described by Al-Mutahar et al. [32] and Al-Aithan [31].

## 28.5 EROSION–CORROSION OF CORROSION-RESISTANT ALLOYS

In general terms, CRAs are defined by the Specification API 5CRA/ISO 13680:2010 as “alloy intended to be resistant to general and localized corrosion and/or environmental cracking in environments that are corrosive to carbon and low-alloy steels.” This definition is very broad, including the alloy families of martensitic, duplex, austenitic, and precipitation-hardened stainless steels and Ni alloys as well as titanium alloys. Most of these alloy groups have excellent resistance to $CO_2$ corrosion but they may suffer from other damage mechanisms such as pitting corrosion by oxidants, environmental assisted cracking, and other depassivation-related damage mechanisms, including erosion–corrosion.

For some applications, it may be more economical in the long term to use CRA instead of the more commonly used carbon steel with inhibitors. Recently, the industry has witnessed the increasing use of CRAs in oil and gas production tubing. In a sweet environment, the most widely used CRA for downhole tubular application is martensitic stainless steel 13Cr-L80 and its modified types. The main reason for its broad usage is its excellent corrosion resistance to flow-assisted $CO_2$ corrosion. In fact, the effects of high gas velocity on carbon steels in carbon dioxide environments are reduced to a low level or virtually eliminated by alloying with 12% or more chromium [33]. Also, the relatively low cost of 13Cr stainless steel compared with alloys such as Duplex stainless steel or more costly Ni Austenitic alloys makes the martensitic stainless steel a very attractive option.

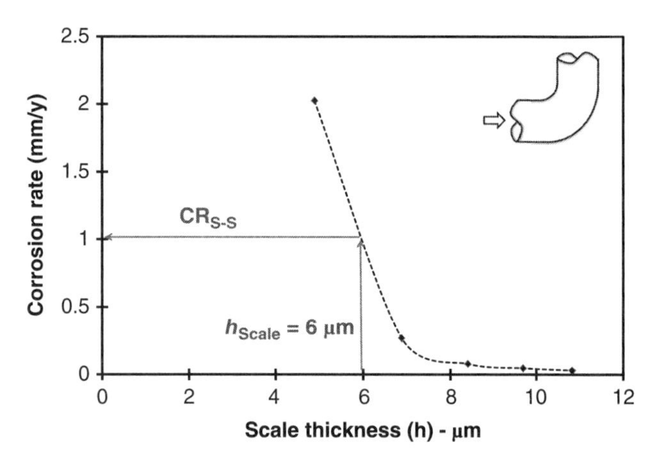


FIGURE 28.16 Prediction of the steady-state corrosion rate for the multiphase case described in Table 28.1. (Adapted from Ref. [32].)

### 28.5.1 Erosion–Corrosion of Carbon Steels versus CRAs

Some studies have been conducted on 13Cr to explore the interactions between sand particle erosion and $CO_2$ corrosion in a flowing two-phase (water/$CO_2$ gas) system. Erosion corrosion resistance of the 13Cr alloy family may be significantly better than carbon steel under certain conditions (see Figure 28.17 [34]).

At low erosivity conditions such as low velocities, single flow liquids with sand concentrations as high as 5–7%, 13Cr has shown erosion–corrosion rates within industry acceptable limits for long-term service (as low as 5 mpy). At similar

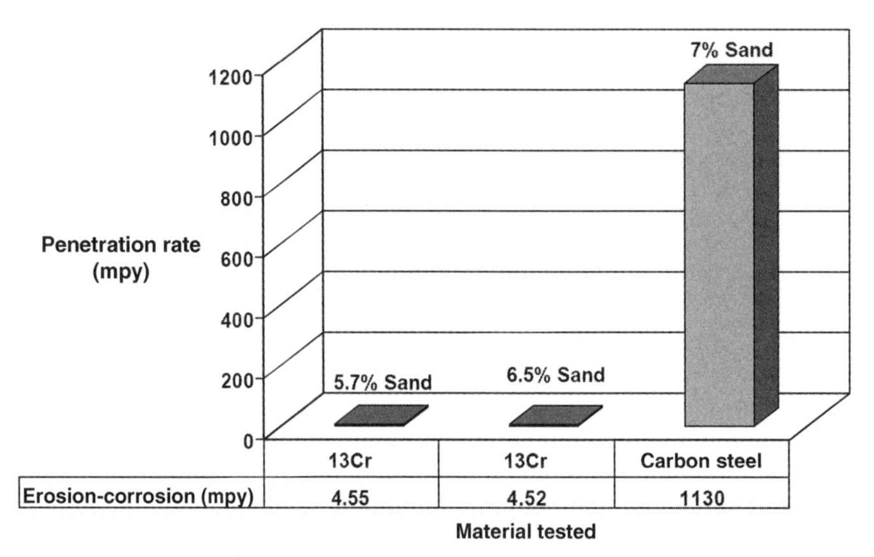


FIGURE 28.17 Comparison between the erosion–corrosion penetration rate for carbon steel and 13Cr in single-phase liquid conditions [34].

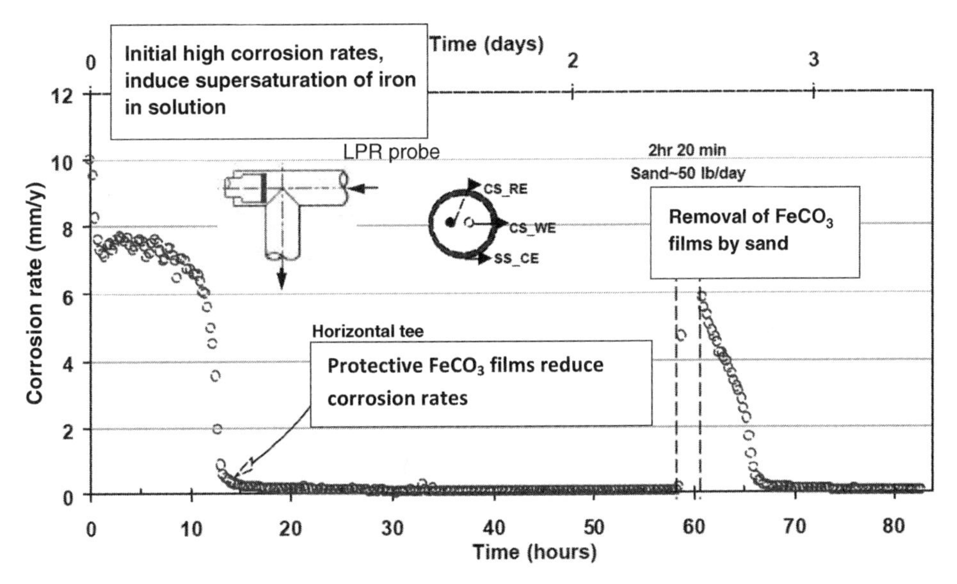


**FIGURE 28.18** Corrosion rate (LPR on horizontal to down plugged tee) for carbon steel under scale-forming conditions ($V_{sg}$ = 10 ft/s (3.1 m/s), $V_{sl}$ = 3.5 ft/s (1.1 m/s), $T$ = 195 °F (91 °C), pH 6.2, initial $[Fe^{2+}]$ = 10 ppm). (Adapted from Ref. [35].)

erosivity conditions, carbon steel may experience erosion–corrosion rates much higher than bare metal $CO_2$ corrosion rates and as high as one inch of wall thickness lost per year. In such conditions, erosion–corrosion rates are dominated by the corrosion component of the damage mechanism. Even at low erosivity conditions, removal of pseudo-protective iron carbonate layers on carbon steel by the mechanical action of solid particle impingements can increase metal loss rates to rates higher than bare metal corrosion rates (see Figure 28.18 [35]).

At similar conditions the 13Cr is still able to either maintain its passivating film or to repassivate fast enough (re-healing of the Cr–Fe oxide layer) to hold the total metal loss rate to within acceptable limits. This remains true for multiphase flows involving $CO_2$ and sand as long as the erosivity of the sand-bearing fluid is relatively low. Figure 28.19 [36] shows the similarity between the pure erosion rates and erosion–corrosion rates of 13Cr exposed to multiphase flow conditions at background (very low) sand concentrations. Note that the 95% confidence intervals of the total metal loss rate based on five measurements for both pure erosion (distilled water, sand, and inert $N_2$) and erosion–corrosion ($CO_2$-saturated brine and sand) overlap to a large extent. At these flow rates and sand concentration combinations (erosivities), both pure erosion and erosion–corrosion remain low and well within acceptable long-term penetration rate limits.

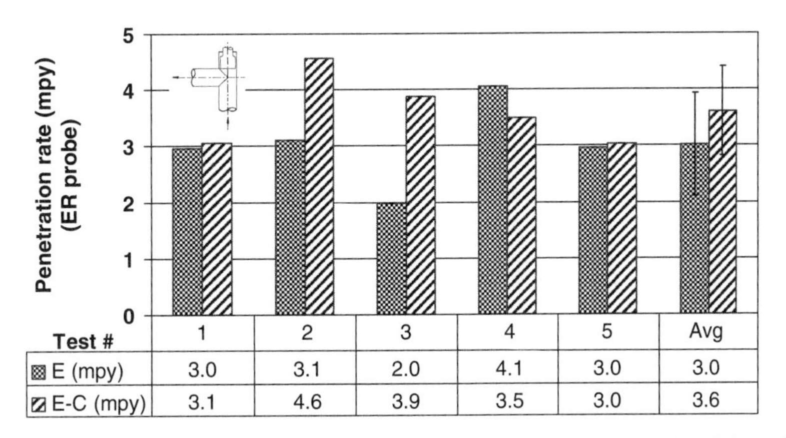


| Test # | 1 | 2 | 3 | 4 | 5 | Avg |
|---|---|---|---|---|---|---|
| E (mpy) | 3.0 | 3.1 | 2.0 | 4.1 | 3.0 | 3.0 |
| E-C (mpy) | 3.1 | 4.6 | 3.9 | 3.5 | 3.0 | 3.6 |

**FIGURE 28.19** Penetration rates for pure erosion and erosion–corrosion tests of 13Cr in multiphase flow at 76 °F (24.4 °C). The error symbols for the average of the data represent the 95% confidence interval on the mean value. (Adapted from Ref. [36].)

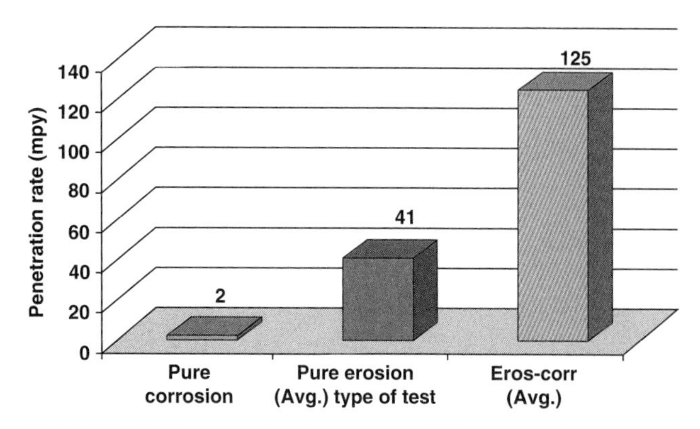


**FIGURE 28.20** Annular flow: $V_{SG}=97$ ft/s; $V_{SL}=0.2$ ft/s sand rate ~48 lb/day; $N_2$ or $CO_2$, $T=76$ °F; 13Cr. (Adapted from Ref. [36].)

### 28.5.2 Erosion–Corrosion with CRAs under High Erosivity Conditions

When erosivity conditions are severe enough, that is, either flow rates are too high or solid particle concentrations are high, the removal of passive films in some CRAs could overcome the ability of the passive film to reform enabling large erosion–corrosion rates. When the erosion–corrosion rates are larger than the pure erosion rates and the pure corrosion rates added as if they were taking place separately, the system is said to behave synergistically, that is, the whole is greater than the sum of the parts. Figure 28.20 illustrates this concept for 13Cr L80 exposed to high flow rates combined with high sand production rates at room temperatures. At lower pH and higher temperatures, the effect would likely be larger [36]. Other researchers have found similar behaviors in other martensitic stainless steel tested under different sets of conditions [37]. This mechanism has been of concern to the industry. Erosion–corrosion damage of this type and the likelihood of failure can be controlled by limiting flow rates (but, with great economic impact due to limiting of the production rate), or by better down hole sand control measures or through appropriate materials selection if the potential problem can be recognized in the early stages of the project.

Other, higher alloyed CRAs, such as super martensitic stainless steel, have shown reduced synergistic effects of erosion–corrosion as compared with the 13 Cr-L80; and at similar conditions, higher grade alloys such as cold worked duplex stainless steels can completely curb the synergistic effect at severe erosivity conditions for pure erosion rates as high as 100 mpy. Off course at erosion rates this high, it would likely be necessary to control flow rates, and or solids concentrations to reduce the pure erosion rates to acceptable limits as well [38] (see Figure 28.21).

On a side note, it is worth taking notice from this figure of the proximity of the values of pure erosion rates for 13Cr-L80, Super 13Cr-95, and 22Cr-110 in spite of their differences in Rockwell C hardness values (22 HRC, 28 HRC, and 36 HRC, respectively). Hence, although the erosion–corrosion resistances of different CRA grades could be significantly different, their pure erosion resistance is very similar. Therefore, it is not recommended to anticipate better resistance to erosion due to higher hardness when working within this range of engineering alloys. In fact, the typically higher mechanical properties of the CRAs has justified the use of thinner piping for a given mechanical design requirement. A thinner wall permits less reaction time to avoid loss of containment should a severe erosion event occur due to temporary production of sand, thus increasing the risk of failure. These considerations should be taken into account, while designing high production facilities or service with risk of sand production.

Significantly greater hardness may yield significant reductions in erosion rates as in the use of Co, Ni, and tungsten carbides alloys in the form of intermetallic laves phase alloys, hard facing alloys, or carbide-strengthened alloys, hence their success when use as inserts in choke valves in high flow or sandy services, or drilling tools in severe wearing service [39]. The effect of hardness on erosion resistance alloys has also been proved, but once again, even large

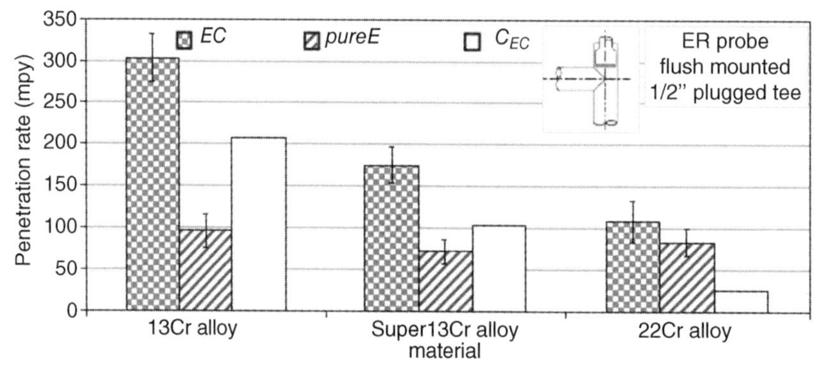


**FIGURE 28.21** Erosion–corrosion (EC), pure erosion (*pureE*), and corrosion component of the erosion–corrosion ($C_{EC}$) penetration rates at 150 °F. (Adapted from Ref. [38].)

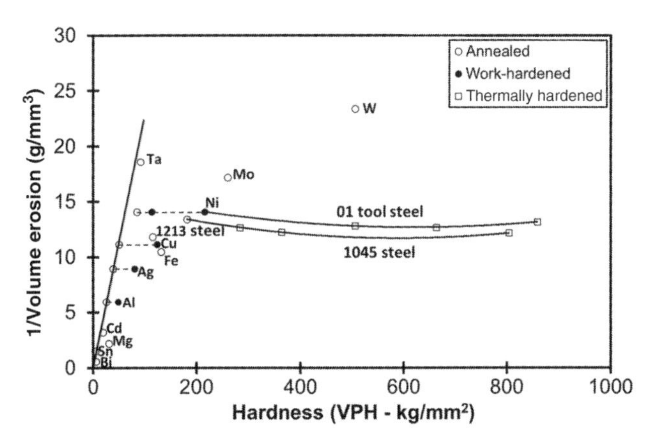


FIGURE 28.22 Erosion resistance (1/volume eroded in mm$^3$) as a function of Vickers hardness for several pure metals and steels with diverse work hardening and heat treatments. (Adapted from Ref. [40]. Data from Finnie, I., Wolak, J. and Kabil, Y., 1967, Journal of Materials, 2, 682–700.)

changes in hardness for low-alloy steels with diverse heat treatments have shown little effect on erosion resistance [40] (see Figure 28.22).

### 28.5.3 Repassivation of CRAs

A laboratory method previously introduced by McMahon and Martin [41] was also used by Rincon [36] to study the effect of pH and temperature on the repassivation behavior of CRAs in CO2-saturated brine. The laboratory method was conducted in a glass cell. Data were collected based on an electrochemical technique. The working electrode (CRA) is scratched, while the potential and current are recorded. After some simple mathematical calculation, Rincon et al. [38] show that thickness loss and the repassivation time can be estimated. The scratch made on a well-defined area of the working electrode may be interpreted as similar to the removal of the passive layer due to a single particle impingement; and, it is believed that this approach can be used to study the effect of erosion on the corrosion response. With scratch tests, the mechanical process making the scratch and the repassivation process of a CRA can be viewed as nearly separate events. The sand erosion process bears some similarity to a simple scratch test except that mechanical and electrochemical processes involved are taking place simultaneously and continuously. The goal of investigating the simple scratch test was to shed some light on the erosion–corrosion behavior of CRAs. The effects of pH and temperature on the repassivation of 13Cr, Super 13Cr, and 22Cr were examined in previous research periods.

For a typical scratch test of an active passive alloy, the anodic current is very high immediately after making the scratch. How fast the current decays with time depends on the healing rate of the oxide film, which depends on the alloy grade as well as environmental conditions such as

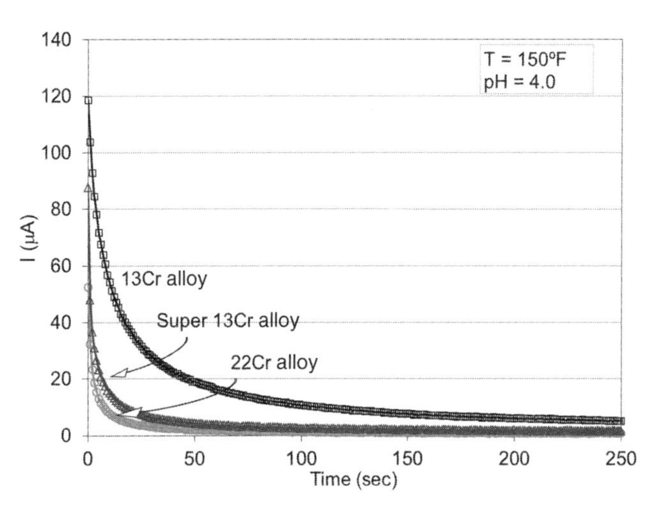


FIGURE 28.23 Decay of the anodic current for three different alloys. (Adapted from Ref. [38].)

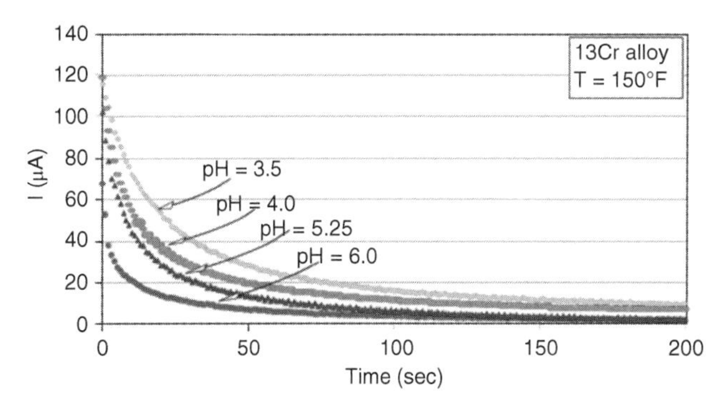


FIGURE 28.24 Effect of pH on the decay of the anodic current after scratching the surface of a 13Cr Alloy [42].

corrosive agent, pH, temperature among others (see Figures 28.23–28.25).

Figure 28.23 shows a comparison of current responses among the three CRAs at a temperature of 150 °F. Scratch tests made at 76 and 200 °F indicate that the ranking of the alloys is the same regardless of the temperature. For a given pH and temperature, the current response, thus the corrosion

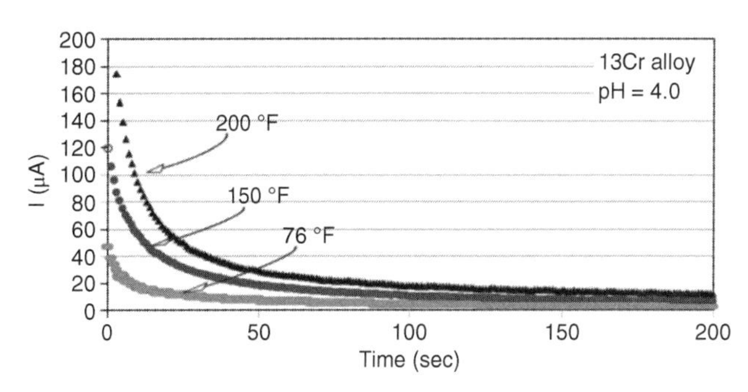


FIGURE 28.25 Effect of temperature on the decay of the anodic current after scratching the surface of a 13Cr Alloy [42].

rate, for Super 13Cr is much lower than the one for 13Cr but not as low as for 22Cr. As expected the Super 13Cr is performing better than 13Cr, but not as well as 22Cr.

Figure 28.24 shows the current decays for 13Cr at 76 °F for different pH conditions. A couple of trends can be seen in this figure. The current decays at faster rates at higher pHs as indicated by the greater (more negative) initial slopes shown for pH 5.25 and 6 as compared with lower pH (3.5 or 4). Also, the value of the current observed is lower for higher pH (5.25 or 6) than for lower pH (3.5 or 4) over the entire interval of 200 s shown in the figure. Since the corrosion rates are proportional to the current, these results suggest very high corrosion rates immediately after making the scratch and exposing the bare metal to the corrosive solution. Then the corrosion rates decay with time, while the healing process takes place on the metallic surface. The passivated state is achieved sooner for higher pH. For higher temperatures, namely, 150 and 200 °F, similar trends are observed (results were presented by Rincon et al. [34,36]). However, the initial currents obtained immediately after making the scratch for these temperatures are much higher than the current obtained for 76 °F.

Figure 28.25 shows the comparison of the current decays at different temperatures; 76, 150, and 200 °F at pH 4. As expected, at higher temperatures higher healing rates were found as suggested by the steeper slopes shown by the current versus time curves at 200 °F. However, the higher temperature always causes a current that is initially higher and remains higher over the entire test period. As a result, at higher temperatures, it takes longer time for the current to return to its original level.

It has been shown that the anodic current decay over an initial repassivation period can be characterized as approximately following the second-order behavior expressed by Equation 28.15:

$$\frac{dI}{dt} = -I^2 m \qquad (28.15)$$

where $I$ is the measured current, $t$ is time, and $m$ is a constant that depends on material and environmental conditions.

Integration of Equation 28.15 yields a convenient form for characterizing the healing process (repassivation process). The first integration yields

$$\frac{1}{I} = \frac{1}{I_0} + mt \qquad (28.16)$$

where $I_0$ is the initial current at $t = 0$.

Equation 28.17 can be used to compare the repassivation responses of a particular alloy under different environmental conditions, or to compare the repassivation responses of several alloys exposed to a particular environment. The comparison of material B (or environment B) to material A (or environment A) can be done according to the thickness loss (TL) accumulated by the time the corrosion rate reduces to (near) zero as follows:

$$\mathrm{TLR}_{B/A} = \frac{\mathrm{TL_B}}{\mathrm{TL_A}} = \lim I^* \rightarrow 0 \frac{(C/m_B) \times \ln(I_B/I^*)}{(C/m_A) \times \ln(I_A/I^*)} = \frac{m_A}{m_B} \qquad (28.17)$$

where the subscripts A and B refer to materials (or environments) A and B, TLRB/A is the ratio of thickness losses between the two material (or environments), $C$ is a units conversion factor, and $I^*$ is the value of the current at a given time lapse, $t^*$ after the scratch has been made. This means that knowing $m$ and $I_0$, one can predict the time required for the corrosion current to return to the value of interest, $I^*$. If $I^*$ is set to correspond to a corrosion rate of for instance 1 mpy, Equation 28.17 can be used to estimate the cumulative metal loss experienced by the alloy during the repassivation process between the time of the scratch and the time the corrosion rate returns to 1 mpy. Taking the limit of the thickness loss ratio as the final current $I^*$ approaches zero yields the thickness loss ratio simply equal to $m_A/m_B$. $I_A$ and $I_B$ in Equation 28.17 refer to the initial ($t = 0$) corrosion currents for material (or environment) A and material (or environment) B, respectively. The second order model was found to be a very good approximation to describe the repassivation process of stainless steels such as 13Cr and 22Cr under all pH and temperature combinations for a $CO_2$-saturated solution for at least the first 200 s after the scratch.

Figure 28.26 shows the cumulative thickness losses from current responses shown in Figure 28.25. As expected, and proved elsewhere ([34,36]) with flowloop testing data, thickness losses are shown to be higher for higher temperatures. The same trends are obtained for pH 3.5, 5.25, and 6.0, but results are not shown in this chapter.

Table 28.2 shows flow loop test values and predicted values of the corrosion component ($C_{e\text{–}c}$) of the erosion–corrosion penetration rates for 3 alloys tested at similar conditions (76 °F (24.4 °C), pH 4). The predicted values were obtained by means of Equation 28.17 where material A is 13Cr and material B is S13Cr or 22Cr. As can be seen, loop test and predicted $C_{e\text{–}c}$ values match very well for both S13Cr and 22Cr, suggesting that scratch data as used in Equation 28.17 may be useful to predict erosion–corrosion penetration rates provided that data from at least one loop test is available for similar erosivity and environmental conditions. Similar analysis has been done for other temperatures and pHs using Equation 28.17 and scratch test slopes, $m$, indicating reasonable agreement between flow loop test values and predicted values.

The testing predicting methodology already discussed has been used as an effective and efficient procedure for investigating the erosion–corrosion behavior of CRAs in oilfield environments. The main idea was to reduce the need for expensive and time consuming loop tests by using a simplified scratch test.

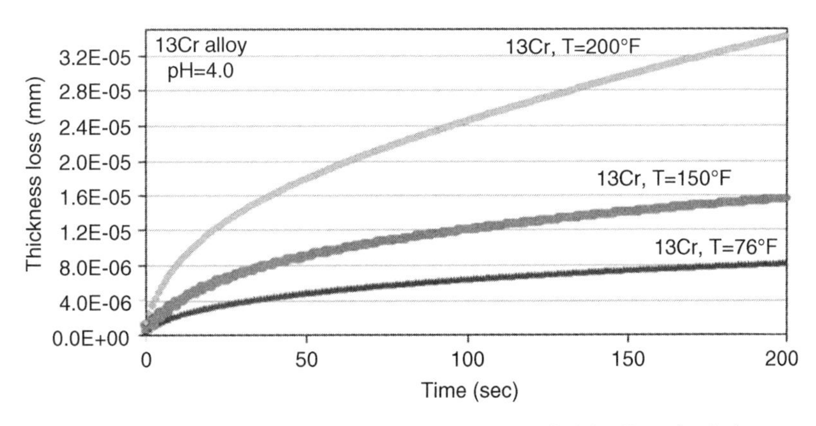


**FIGURE 28.26** Comparison of the cumulative thickness loss of 13Cr alloy after being scratched at pH = 4.0 [42].

**TABLE 28.2 Loop Test Values and Predicted Values of the Corrosion Component ($C_{e\text{–}c}$) of the Erosion–Corrosion Penetration Rates for 3 Alloys Tested at Similar Conditions (76 °F (24.4 °C), pH 4) [38]**

| Alloy | Loot test $C_{e\text{–}c}$ rate[a] $C_{e\text{–}c} = EC - E$ (mpy) | Scratch test data $TLR_{B/A} = (m_{13Cr}/m_{CRA})$ | Predicted $C_{e\text{–}c}$ rate $C_{e\text{–}c,CRA} = TLR_{B/A}\, C_{EC,13Cr}$ (mpy) |
|---|---|---|---|
| 13Cr | 99 | 1 | (Reference) |
| Super13Cr | 47 | 0.40 | 40 |
| 22Cr | 29 | 0.22 | 22 |

[a]Obtained by subtracting the erosion rate from the erosion–corrosion rate.

A novel procedure to predict erosion–corrosion of CRAs using the second order model and scratch test along with solid particle impinging data generated from CFD simulations was proposed by Rincon [42]. Figure 28.27 shows the comparison between experimental data and predicted values for erosion–corrosion rates of three CRAs. The figure shows how a model [42] based on Equations 28.15, 28.16, and 28.17 was able to grasp the trends for the erosion–corrosion behavior of the three alloys with respect to CRA type and temperature. These results are encouraging, especially when considering the complexity of the phenomenon under study and the large number of variables taken into account in the model.

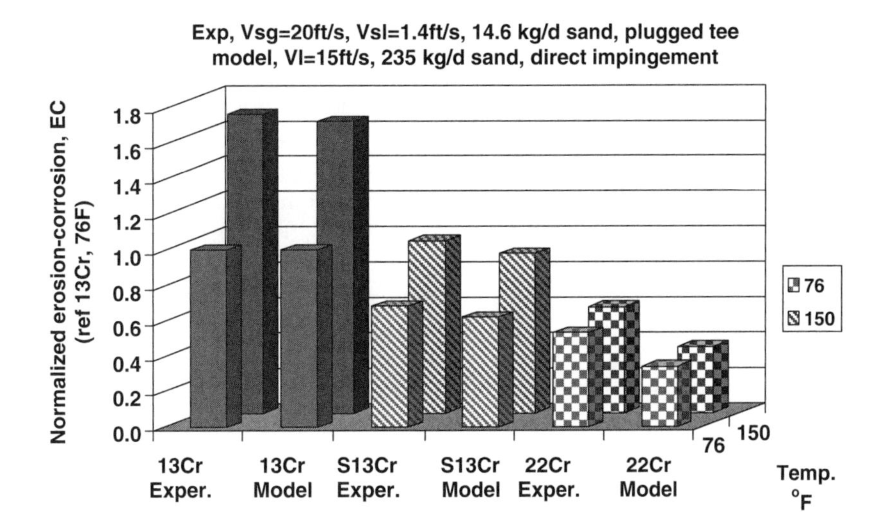


**FIGURE 28.27** Comparison between experimental data and predicted trends of the erosion–corrosion, EC of the three CRAs at two different temperatures. (All data normalized to 13Cr at room temperature value [42].)

### 28.5.4 Effect of Microstructure and Crystallography on Erosion-Corrosion

The previous section showed the breakdown of the total erosion–corrosion weight lost as the summation of 4 different components. Changes in exposure conditions, such as flow velocity, sand rate, and alloy grade with different chemistry or microstructure could affect each component differently shifting the damage mechanism dominance and, therefore, the total waste loss rate. This phenomenon was observed by He et al. [43] when studying the erosion–corrosion behavior of 4 different stainless steels with a diverse range of chemistry, microstructure, and microhardness. Under the experimental conditions of this study, the corrosion component was a small contributor to the total erosion–corrosion weight loss. The mechanical removal action of the solid particle impingements produced most of the weight loss. Thus, the microstructure and respective microhardness proved to be significant factors in the total erosion–corrosion resistance, shifting the ranking of the alloys with changes in the acidity of the media and flow velocities.

Aribo et al. [44] conducted studies to compare the erosion–corrosion performance of lean duplex stainless steels, (UNS S32101) and austenitic steel 304 stainless Steel (UNS S30403). Lean duplex steels appear to have better resistance to erosion and erosion–corrosion damage than austenitic stainless steels when exposed to flows having velocities of 15 m/s and containing 500 mg/l of sand. The generality of this trend is still subject to verification by consideration of a broader experimental matrix. XRD spectra analysis for fresh surfaces and sand-flow exposed surfaces for these two alloys demonstrated that sand impingement was able to induce enough stress to transform some of the austenite volume fraction to an $\alpha$-martensite, causing an increase in localized hardness on both austenitic and the lean duplex stainless steel alloys. The extension of the microstructure phase transformation seems to be related to the erosion and erosion resistance of these alloys and their relative ranking with regard to erosion–corrosion resistance. More research in this area is needed to improve the prediction of erosion–corrosion rates of CRAs.

### 28.5.5 Summary

At high erosivity conditions, a synergistic effect between erosion and corrosion was confirmed in three CRAs (13Cr-L80, Super13Cr-95, and 22Cr-110) since the total metal loss rate was shown to be higher than the sum of the rates of corrosion (without sand) and erosion measured separately. It appears that under the high sand rate conditions tested, the erosivity is severe enough to damage the passive layer, thereby enabling the raised corrosion rate.

Synergism seems to occur for the three alloys; however, the degree of synergism is quite different for the three alloys and may not be significant for 22Cr under field conditions where erosivities are typically much lower than those occurring in the small bore loop used in this testing.

For the others two alloys, 13Cr and Super13Cr, especially at high temperatures, if the erosion rate of the passive film is great enough, then an accelerated erosion–corrosion process may take place with a significant contribution from a corrosion component. In most cases, there is likely a competition between the protective film removal due to mechanical erosion and the protective film healing. If erosivity conditions are severe enough, base metal can also be removed by the mechanical erosion component. A mechanistic model of this competition would need to account for film characteristics, the concentration and distribution of sand, the flow pattern and flow velocities, flow geometry, and environmental factors.

Predictions of the corrosion component of erosion–corrosion based on scratch test data compared well to test results from a multiphase gas/liquid flow loop for the three CRAs at high erosivity conditions. Second-order behavior appears to be an appropriate and useful model to represent the repassivation process of CRAs. A procedure to predict penetration rates for erosion–corrosion conditions was developed based on the Second- order model behavior observed for the re-healing process of the passive film of CRAs under tested conditions. Good agreement between the actual and predicted penetration rates was found.

## 28.6 CHEMICAL INHIBITION OF EROSION–CORROSION

Application of chemical inhibitors is one of the most common and successful practices for controlling corrosion and erosion–corrosion in the oil and gas industry. Considerable research in the oil and gas industry has been devoted to the development of inhibitor systems that can reduce corrosion rates in production equipment to acceptable levels. There are many different types of inhibitors that are used for the oil and gas industry applications. Imidazoline-based inhibitor is commonly used by oil and gas companies for sweet corrosion applications. Several commercial inhibitors have been developed by chemical companies to mitigate corrosion especially for $CO_2$ and $H_2S$ systems. Some of these inhibitors are claimed to be useful for erosion–corrosion mitigation as well. There are several parameters affecting the inhibitor performance, including pH of solution, flow velocity and flow regime, temperature, sand erosion, oil/water emulsion, metal cations, chlorides, type of metal, and electrolyte [45–55].

The effect of sand erosion on inhibitor performance is an important area of concern in oil and gas production because inhibitors that are effective in a $CO_2$-saturated static brine, are not necessarily effective under erosion–corrosion conditions [56]. Sand production also raises the inhibitor concentration needed for optimum inhibitor effectiveness [48]. In this section, the effect of sand erosion on inhibitor performance and also prediction of inhibited erosion–corrosion rate as a function of environmental conditions will be discussed in detail.

### 28.6.1 Effect of Sand Erosion on Chemical Inhibition

Reduction of chemical inhibition performance by sand erosion has been reported by several authors [54,55,57–60]. Interactions of sand particles with chemical inhibitors in oil and gas wells depend on different parameters, including inhibitor characteristics, sand type, sand size, and presence of oil in the system. In general, this interaction has been explained by mechanical removal of inhibitor from the surface and also by reducing the effective inhibitor concentration in the system. McMahon et al. 2005 [61] experimentally showed that large amounts of corrosion inhibitor can be lost from the bulk solution by adsorption of inhibitor onto the surfaces of produced sand particles. This can reduce the concentration of the inhibitor that is available to protect steel surfaces. However, this effect only becomes significant for high concentrations of sand (>1000 ppm) and small particle sizes (<10 μm diameter), especially for oil-wetted particles [61]. On the other hand, Ramachandran et al. [62] showed that the effect of inhibitor loss due to sand adsorption is small for industrial corrosion inhibitors used under field conditions. Tandon et al. [57] showed inhibitor protection provided by concentrations of 100 ppm and above appeared not to be affected much by the addition of small amounts of sand under non-scale-forming conditions. However, for some inhibitor concentrations, increasing sand concentration adversely affected the ability of the inhibitor to maintain optimum protection over extended periods of time [57].

Using an inhibitor can change the erosion–corrosion behavior. Shadley et al. [63,64] showed that using an effective inhibitor can increase the threshold velocity for removal of an iron carbonate scale in a carbon steel elbow in $CO_2$-saturated brine solution. The velocity limit below which scale can be maintained is defined as the "threshold velocity." Working above the threshold velocity, pitting or uniform corrosion can occur. At velocities below the threshold velocity, a protective scale can form on pipe walls, and metal loss rates are typically low. The upward shift of threshold velocity provides a much larger scale formation region and, thus greater flexibility in piping system design and operation.

Tummala et al. [56] studied two inhibitor treatment scenarios involving preexisting iron carbonate scale: In the first scenario, sand was added to the system one day after adding inhibitor; and, in the second scenario, inhibitor was added to the system one day after adding sand. In both scenarios, adding sand increased the corrosion rate and reduced the inhibitor performance but final corrosion rates were the same—about 0.5 mm/year. Chemical inhibitor added to the system before or after sand entrainment shows the same efficiency. However, inhibitor efficiency is reduced because of the sand interaction.

### 28.6.2 Modeling and Prediction of Inhibited Erosion–Corrosion

The need to be able to predict inhibitor performance under sand production conditions is particularly acute when the wells are deep or offshore because of the difficulty in running coupon tests to evaluate the inhibitor efficiency at critical points throughout a system, which might involve a number of wells flowing into a single header. Further, it would be highly desirable to be able to make decisions on the design of the well given advanced knowledge of the inhibition options and their predicted effectiveness.

Crossland et al. [65] suggested a correlation for the prediction of the likelihood of erosion–corrosion inhibition success in the oil and gas industry as a function of environmental factors such as temperature, shear stress, and total dissolved solids Equation 28.18. This correlation is a statistical correlation based on experimental and field data over a wide range of production conditions, but it does not include the effects of sand production. They defined four different regions for inhibitor performance, which can be used for the estimation of the amount of the inhibitor needed to get the corrosion rate lower than 0.1 mm/year (Table 28.3) [65].

$$\text{ILSS} = \frac{\text{Temp}(^{\circ}\text{C})}{40} + \frac{\text{Shear stress(Pa)}}{240} + \frac{\text{TDS(ppm)}}{125000} + \frac{\text{Predicted} - \text{C.R.}\left(\frac{\text{mm}}{\text{yr}}\right)}{10} \qquad (28.18)$$

where ILSS is inhibitor likelihood success score and TDS is total dissolved solids.

**TABLE 28.3 Summary of Inhibition Risk Categories and Guidance [63]**

| Category | ILSS | Description |
|---|---|---|
| 1 | ILLS < 2.5 | Corrosion inhibition success is highly likely (inhibitor concentration up to 50 ppm) |
| 2 | 2.5 < ILSS < 4 | Corrosion inhibition success is expected (inhibitor concentration up to 100 ppm) |
| 3 | 4 < ILSS < 5.5 | Corrosion inhibition is challenging (inhibitor concentration up to 300 ppm) |
| 4 | 5.5 < ILSS | Corrosion inhibition may not be viable (inhibitor concentration more than 400 ppm) |

Al Zawai et al. [66] used the response surface method design to model the erosion–corrosion rate as a function of sand load, flow velocity, and inhibitor concentration. They conducted the erosion–corrosion experiment using a jet impingement geometry for three different sand loads (SL = 150, 500, and 1500 ppm), flow velocities ($V$ = 7, 10, and 20 m/s), and inhibitor concentrations (CI = 0, 100, and 150 ppm) in order to develop an empirical model for erosion–corrosion prediction. Equation 28.19 is an empirical correlation for prediction of erosion–corrosion rate in inhibited system based on total weight loss (TWL).

$$\mathrm{TWL} = 31.3 + 51.3 \times \mathrm{SL} + 17 \times V - 11 \times \mathrm{CI} + 4.06 \times \mathrm{SL} \times V + 0.58 \times \mathrm{SL} \times \mathrm{CI} \tag{28.19}$$

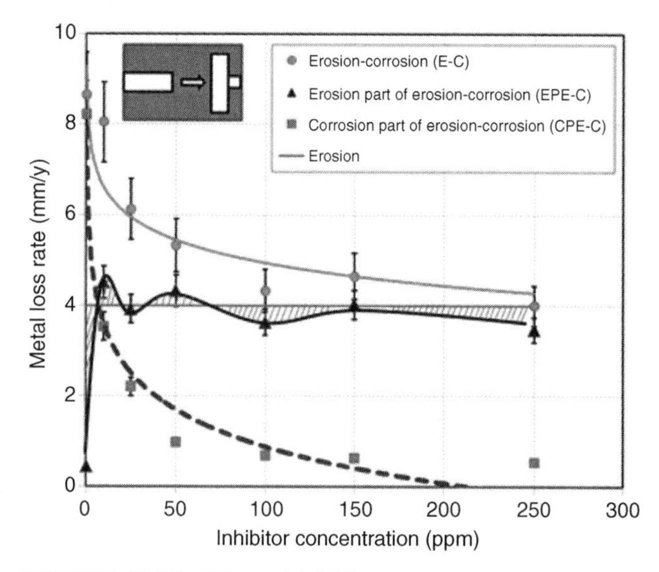


**FIGURE 28.28** Effect of inhibitor concentration on erosion–corrosion rate and its components at $T$ = 57 °C, pH = 4.8, $pCO_2$ = 2.4 bar and target material is X65 carbon steel [54].

Hassani et al. [54,55] developed a methodology for prediction of inhibited erosion–corrosion rate, which is a combination of sand erosion model, $CO_2$ corrosion model, and modified inhibitor adsorption isotherm. There are several factors involved in this model, including; sand erosion parameters (sand size, sand shape, sand density, sand concentration, flow velocity, geometry, flow regime, fluid density, and viscosity) and $CO_2$ corrosion parameters (temperature, pH, $CO_2$ pressure, flow velocity, and regime).

Erosion–corrosion consists of two parts: erosion part of erosion–corrosion and corrosion part of erosion–corrosion. The erosion part can be predicted by predicting pure erosion and erosion affected by corrosion. The corrosion part also has two components, including pure corrosion and corrosion affected by erosion. Corrosion can affect erosion by changing the surface roughness and also by forming corrosion products on the surface, which may have a different erosion resistance in comparison with base metal. Erosion can also affect the corrosion behavior by removing the corrosion products from the surface, removing corrosion inhibitor from the surface and also changing the surface morphology. Erosion–corrosion, involving solid particles, is a complicated phenomenon and many factors affect erosion–corrosion mechanisms and the synergistic effects that occur between the erosion and corrosion of mild steels. Addition of inhibitors adds another order of complexity to the phenomenon. However, in order to make engineering predictions of erosion–corrosion with inhibitors, some simplifying assumptions have to be made.

Hassani et al. [54,55] experimentally showed that for $CO_2$ corrosion of carbon steel, erosion affected by corrosion can be neglected in comparison with total erosion–corrosion rate (Hatched area in Figure 28.28). Harvey et al. [67] also neglected this term in the erosion–corrosion prediction modeling. Using this assumption, erosion part of erosion–corrosion is equal to the pure erosion rate, which can be predicted using pure erosion prediction models. The corrosion part can be predicted by combining the pure $CO_2$ corrosion prediction model with a modified inhibitor adsorption isotherm to predict the corrosion part of the erosion–corrosion. Pure $CO_2$ corrosion rate can be predicted using the available mechanistic models. However, corrosion affected by erosion was predicted using a modified inhibitor adsorption isotherm. The Frumkin inhibitor adsorption isotherm was modified by including an erosivity term in the adsorption/desorption constant (Eq. 28.20).

$$\left[K_{\mathrm{a/d}}(\mathrm{ER}, \Delta T)\right] C_{\mathrm{inh}} = \left(\frac{\theta}{1-\theta}\right) e^{-f\theta} \tag{28.20}$$

where $K_{\mathrm{a/d}}$ is adsorption/desorption constant, ER is erosion rate, $\Delta T$ is temperature change, $C_{\mathrm{inh}}$ is inhibitor concentration, $\theta$ is surface coverage by inhibitor, and $f$ is an adjustable parameter accounting for the interaction between adsorbed inhibitor molecules. The adsorption/desorption constant is the best term for including the effect of erosivity since it is a function of entropy of the system. Entropy is an expression for randomness and disorder in the system and sand erosion causes randomness and disorder in the inhibitor adsorption/desorption process. Therefore, total erosion–corrosion rate can be predicted by predicting the pure erosion rate, pure $CO_2$ corrosion rate, and effect of sand erosion on inhibitor performance (i.e., modified inhibitor adsorption isotherm). Figure 28.29 shows a comparison between experimental and predicted erosion–corrosion rates as a function of inhibitor concentration and temperature when erosivity of the system (erosion rate of mild steel) is about 4 mm/year.

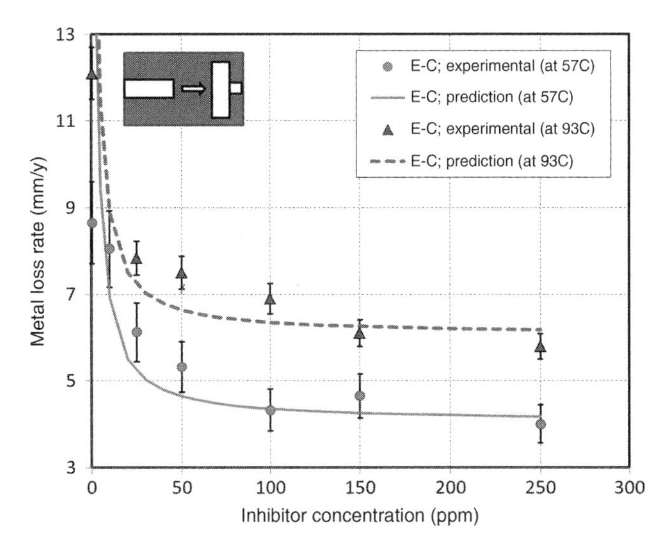


**FIGURE 28.29** Comparison between experimental data and prediction of total erosion–corrosion rate as a function of inhibitor concentration at $T=57$ and 93 °C, pH = 4.8, $pCO_2=2.4$ bar and target material is X65 carbon steel [54].

## 28.7 SUMMARY AND CONCLUSIONS

The oil and gas production industry has long recognized that corrosion and erosion can damage flow lines, elbows, chokes, valves, headers and other flow line, and production equipment. More recently, the process of erosion–corrosion has received increased attention as a damaging process. Erosion–corrosion process involves both erosion and corrosion and is inherently more complex than erosion or corrosion. There are several conditions where erosion–corrosion can cause severe damage to metallic flow components. One condition occurs after a protective corrosion scale has been formed on a carbon steel, and sand erosion removes the scale in a localized area to bare metal. The localized area of bare metal acts like a small anode and the scale area acts like a large cathode, resulting in severe pitting, see Figure 28.9. Another erosion–corrosion condition occurs with CRAs, which have a thin protective oxidation layer. Sand erosion can remove this layer and again expose bare metal to cause increased corrosion rates. A third condition is when inhibitors are used and the sand production is severe enough to damage the inhibitor protective coating. Tests show corrosion rates increase for this condition. Much has been done the past two decades to determine procedures for dealing with erosion–corrosion. Threshold velocities have been determined identifying the flow velocities above which pitting is a risk. Models and experimental tests to predict erosion of metals for a variety of flow regimes, flow geometries, particle sizes, and fluid properties have been or are being developed. These tools, coupled with corrosion models and flow analysis, are being applied to erosion–corrosion problems. Studies on the erosion removal of corrosion scales have been successful, which is an important step toward developing possible procedures for predicting erosion–corrosion-induced pitting. While important steps have been taken and progress has been made, there are still challenges to meet in raising the level of understanding of erosion–corrosion, before it reaches the levels of application that corrosion or erosion have attained.

## ACKNOWLEDGMENTS

The authors wish to acknowledge the support of over 50 oil and gas and service companies that have supported the E/CRC since 1982, and many project monitors from these companies who have contributed to the research of the E/CRC.

The authors especially acknowledge Mr. Edward Bowers, Senior Technician of the E/CRC for his ideas, design, construction, and maintenance of all 12 of the flow loops and other test equipment in the E/CRC and for guiding students in their research.

## REFERENCES

1. Koch, G.H., Brongers, M.P.H., Thompson, N.G., Virmani, Y.P., and Payer, J.H. (2002) *Corrosion Costs and Preventive Strategies in the United States*, NACE International, Publication NO. FHWA-RD-01-156.
2. PHMSA (2002) PHMSA Website: Significant Incidents Files, June 29, 2012.
3. Kermani, M.B. and Harrop, D. (1996) The impact of corrosion on the oil and gas industry. *SPE Production Facilities*, 11(3), 186–190.
4. Shadley, J.R., Shirazi, S.A., Dayalan, E., and Rybicki, E.F. (1996) Erosion-corrosion of a carbon steel elbow in a carbon-dioxide environment. *CORROSION*, 52(9), 714–723.
5. Meng, H.C. and Ludema, K.C. (1995) Wear models and predictive equations: their form and content. *Wear*, 181–183, 443–457.
6. American Petroleum Institute (1991) *API RP 14E: Recommended Practice for Design and Installation of Offshore Production Platform Piping Systems*, 5th ed., American Petroleum Institute, Washington, DC.
7. Salama, M.M. (1998) An alternative to API RP 14E erosion velocity limits for sand laden fluids. Proceedings of Offshore Technology Conference (OTC), Houston, TX, Paper No. OTC-8898.
8. Jordan, K. (1998) Erosion in multiphase production in oil and gas. Proceedings of NACE Corrosion 1998 Conference & Expo, Paper No. 58.
9. McLaury, B.S. and Shirazi, S.A. (2000) An alternate method to API RP 14E for predicting solids erosion in multiphase flow. *ASME Journal of Energy Resources Technology*, 122(3), 115–122.
10. International Standard Organization (2000) *ISO 13703, Petroleum and Natural Gas Industries—Design and Installation of Piping Systems*. 1st ed., International Standard Organization.

11. Salama, M.M. and Venkatesh, E.S. (1983) Evaluation of API RP 14E erosional velocity limitations. Proceedings of the 15th Offshore Technology Conference (OTC), Houston, TX, Paper No. OTC-4485.
12. Bourgoyne, A.T., Jr. (1989) Experimental study of erosion in diverter systems due to sand production. Proceedings of SPE/IADC Drilling Conference, LA, Paper No. SPE/IADC-18716.
13. Svedeman, S.J. and Arnold, K.E. (1993) Criteria for sizing multiphase flow lines for erosion/corrosive service. Proceedings of the 68th SPE Annual Fall Technical Conference, Houston, TX, Paper No. SPE-26569.
14. ASTM (2004) ASTM G76-04: Standard Test Method for Conducting Erosion Tests by Solid Particle Impingement Using Gas Jets.
15. Zhang, Y., Reuterfors, E.P., McLaury, B.S., Shirazi, S.A., and Rybicki, E.F. (2007) Comparison of computed and measured particle velocities and erosion in water and air flows. *Wear*, 263 (1–6), Special Issue, 330–338.
16. Det Norske Veritas (2007) Recommended Practice, Erosive Wear in Piping Systems, DNV RP O501. Revision 4.2, 2007.
17. Shirazi, S.A., McLaury, B.S., Shadley, J.R., and Rybicki, E.F. (1995) A procedure to predict solid particle erosion in elbows and tees. *Journal of Pressure Vessel Technology*, 117(1), 45–52.
18. Zhang, Y., McLaury, B.S., Shirazi, S.A., and Rybicki, E.F. (2011) Predicting sand erosion using a two-dimensional mechanistic model. Proceedings of the NACE Corrosion 2011 Conference & Expo, Paper No. 11243.
19. McLaury, B.S., Shirazi, S.A., Mazumder, Q.H., and Viswanathan, V. (2006) Effect of upstream pipe orientation on erosion in bends for annular flow. Proceedings of the NACE Corrosion 2006 Conference & Expo, Paper No. 06572.
20. Throneberry, J.M. Zhang, Y., McLaury, B.S., Shirazi, S.A., and Rybicki, E.F. (2010) Solid-particle erosion in slug flow. Proceeding of the SPE Annual Technical Conference, Florence, Italy, Paper No. SPE 135402.
21. Shadley, J.R., Shirazi, S.A., Dayalan, E., Ismail, M., and Rybicki, E.F. (1996) Erosion-corrosion of a carbon steel elbow in a $CO_2$ environment. *CORROSION*, 52(9), 714–723.
22. Shirazi, S.A., McLaury, B.S., Shadley, J.R., and Rybicki, E.F. (1995) Generalization of the API RP 14E Guideline for Erosive Services. *Journal of Petroleum Technology*, 47, 693–698.
23. Schmitt, G. and Hörstemeier, M. (2006) Fundamental aspects of $CO_2$ metal loss corrosion—part II: influence of different parameters on $CO_2$ corrosion mechanisms. Proceedings of CORROSION/2006, NACE International, Houston, TX, Paper No. 06112.
24. Sun, W., Nešić, S., and Woollam, R. (2009) The effect of temperature and ionic strength on iron carbonate ($FeCO_3$) solubility limit. *Corrosion Science*, 51(6), 1273–1276.
25. Johnson, M. and Tomson, M.B. (1991) Ferrous carbonate precipitation kinetics and its impact on $CO_2$ corrosion. Proceedings of CORROSION/1991, NACE International, Houston, TX, Paper No. 268.
26. Van Hunnik, E., Pots, B., and Hendriksen, E. (1996) The formation of protective $FeCO_3$ corrosion product layers in $CO_2$ corrosion. Proceedings of CORROSION/1996, NACE International, Houston, TX, Paper No. 6.
27. Yean, S., Al Saiari, H., Kan, A.T., and Tomson, M.B. (2008) Ferrous carbonate nucleation and inhibition. Proceedings of the SPE International Oilfield Scale Conference, Aberdeen, Scotland, UK: Society of Petroleum Engineers Paper No. SPE 114124.
28. Sun, W. and Nešić, S. (2008) Kinetics of corrosion layer formation: part 1-iron carbonate layers in carbon dioxide corrosion. *CORROSION*, 64(4), 336–843.
29. Schmitt, G., Mueller, M., Papenfuss, M., and Strobel-Effertz, E. (1999) Understanding localized $CO_2$ corrosion of carbon steel from physical properties of iron carbonate scales. Proceedings of CORROSION/1999, NACE International, Houston, TX, Paper No. 38.
30. Akbar, A., Hu, X., Wang, C., and Neville, A. (2012) The influence of flow rate, sand and inhibitor on iron carbonate scales under erosion-corrosion conditions using a submerged impingement jet. *CORROSION*, 4, 3125–3137, NACE International.
31. Al-Aithan, G., Al-Mutahar, F.M., Shadley, J.R., Shirazi, S.A., Rybicki, E.F., and Roberts, K.P. (2014) A mechanistic erosion-corrosion model for predicting iron carbonate ($FeCO_3$) scale thickness in a $CO_2$, environment with sand. *CORROSION*, 1, NACE International, Houston, TX.
32. Al-Mutahar, F.M., Roberts, K.P., Shirazi, S.A., Rybicki, E.F., and Shadley, J.R. (2012) Modeling and experiments of $FeCO_3$ scale growth and removal for erosion-corrosion conditions. *CORROSION*, 1, 683–701, NACE International, Houston, TX.
33. Van Der Horst, J.M.A. and Sloan, C.R. (1973) ASTM STP 567: Erosion Corrosion of Finned Heat Exchanger Tubes, Erosion, Wear and Interfaces with Corrosion. American Society for Testing and Material, Philadelphia, United States. pp. 18–29.
34. Rincon, H. (2001) MSME: Erosion-Corrosion Phenomena of the 13Cr Alloy in Flows Containing Sand Particles. Master of Science Thesis in Mechanical Engineering, The University of Tulsa, Tulsa, OK.
35. Rincon, H., Shadley, J.R., Rybicki, E.F., and Roberts K.P. (2006) Erosion-corrosion of carbon steel in $CO_2$ saturated multiphase flows containing sand. Proceedings of CORROSION/2006, NACE International, Houston, TX, Paper No. 06590.
36. Rincon, H., Shadley, J.R., and Rybicki, E.F. (2005) Erosion corrosion phenomena of 13Cr at low sand rate levels. Proceedings of CORROSION/2006, NACE International, Houston, TX, Paper No. 05291.
37. Mori, G., Vogl, T., Haberl, J., and Havlik, W. (2010) Erosion corrosion and synergistic effects under high velocity multiphase conditions. Proceedings of CORROSION/2010, NACE International, Houston, TX, Paper No. 10382.
38. Rincon, H., Shadley, J.R., Rybicki, E.F., and Roberts K.P. (2008) Erosion-corrosion of corrosion resistant alloys used in the oil and gas industry. Proceedings of CORROSION/2008, NACE International, Houston, TX, Paper No. 08571.
39. Raghu, D. and Wu, J. (1997) Recent developments in wear and corrosion resistant alloys for the oil and gas industry. Proceedings of CORROSION/1997, NACE International, Houston, TX, Paper No 16.
40. Hutchings, I.M. (1983) *Monograph on the Erosion of Materials by Solid Particle Impact*, MTI Publication, No. 10.

41. McMahon, A.J. and Martin, J.W. (2004) Simulation tests on the effect of mechanical damage or acid cleaning on CRA's used for oil/gas production well tubulars. Proceedings of CORROSION/2004, NACE International, Houston, TX, Paper No. 4127.
42. Rincon, H. (2006) Testing and prediction of erosion-corrosion for resistant corrosion alloys used in the oil and gas production, Ph.D. dissertation, The University of Tulsa, Tulsa, OK.
43. He, D.D., Jiang, X.X., Li, S.Z., and Guan, H.R. (2005) Erosion-corrosion of stainless steels in aqueous slurries—a quantitative estimation of synergistic effects. *CORROSION*, 61(1), 30–36.
44. Aribo, S., Bryant, M., Hu, X., and Neville, A. (2013) Erosion-corrosion of lean duplex stainless steel (Uns S32101) in a $CO_2$-saturated oilfield brine. Proceedings of CORROSION/2013, NACE International, Houston, TX, Paper No. 2382.
45. Shadley, J.R. and Palacios, C.A. (1990) Effects of flow velocity on inhibitors used in oil and gas production. 22nd Annual OTC Proceedings, Vol. 2, Paper OTC 6281. pp. 107–113.
46. Ochoa, N., Moran, F., Pébère, N., and Tribollet, B. (2005) Influence of flow on the corrosion inhibition of carbon steel by fatty amines in association with phosphonocarboxylic acid salts. *Corrosion Science*, 47(3), 593–604.
47. Liu, X., Okafor, P.C., and Zheng, Y.G. (2009) The inhibition of $CO_2$ corrosion of N80 mild steel in single liquid phase and liquid/particle two-phase flow by aminoethyl imidazoline derivatives. *Corrosion Science*, 51(4), 744–751.
48. Jiang, X., Zheng, Y.G., and Ke, W. (2005) Effect of flow velocity and entrained sand on inhibition performances of two inhibitors for $CO_2$ corrosion of N80 steel in 3% NaCl solution. *Corrosion Science*, 47(11), 2636–2658.
49. Menendez, C.M., Weghorn, S.J., Ahn, Y.S., Jenkins, A., and Mok, W.Y. (2005) Electrochemical evaluations of high shear corrosion inhibitors using jet impingement. Proceedings of CORROSION/2005, NACE International, Houston, TX, Paper No. 05331.
50. Schmitt, G., Werner, C., and Schöning, M.J. (2002) Micro-electrochemical efficiency evaluation of inhibitors for $CO_2$ corrosion of carbon steel under high shear stress gradients. Proceedings of CORROSION/2002, NACE International, Houston, TX, Paper No. 280.
51. Bartos, M. and Watson, J.D. (2000) Oilfield corrosion inhibition under extremely high shear conditions. Proceedings of CORROSION/2000, NACE International, Houston, TX, Paper No. 00068.
52. Swidzinski, M., Fu, B., Taggart, A., and Jepson, W.P. (2000) Corrosion inhibition of wet gas pipelines under high gas and liquid velocities. Proceedings of CORROSION/2000, NACE International, Houston, TX, Paper No. 00070.
53. Seal, S., Jepson, W.P., and Anders, S. (2002) The effect of flow on the corrosion product layer in presence of corrosion inhibitors. Proceedings of CORROSION/2002, NACE International, Houston, TX, Paper No. 02277.
54. Hassani, S., Roberts, K.P., Shirazi, S.A., Shadley, J.R., Rybicki, E.F., and Joia, C. (2012) Characterization and prediction of chemical inhibition performance for erosion-corrosion conditions in sweet oil and gas production. *CORROSION*, 68 (10), 885–896.
55. Hassani, S., Roberts, K.P., Shirazi, S.A., Shadley, J.R., Rybicki, E.F., and Joia, C. (2013) A new approach for predicting inhibited erosion-corrosion in $CO_2$-saturated oil/brine flow condition. *SPE Production and Operations*, 28. doi: 10.2118/155136-PA http://dx.doi.org/10.2118/155136-PA.
56. Tummala, K., Roberts, K.P., Shadley, J.R., Rybicki, E.F., and Shirazi, S.A. (2009) Effect of sand production and flow velocity on corrosion inhibition under scale forming conditions. Proceedings of CORROSION/2009, NACE International, Houston, TX, Paper No. 09474.
57. Neville, A., Wang, C., Ramachandran, S., and Jovancicevic, V. (2003) Erosion-corrosion mitigation using chemicals. Proceedings of CORROSION/2003, NACE International, Houston, TX, Paper No. 03319.
58. Tandon, M., Shadley, J.R., Rybicki, E.F., Roberts, K.P., Ramachandran, S., and Jovancicevic, V. (2006) Flow loop studies of inhibition of erosion-corrosion in $CO_2$ environments with sand. Proceedings of CORROSION/2006, NACE International, Houston, TX, Paper No. 06597.
59. Dave, K., Shadley, J.R., Rybicki, E.F., Roberts, K.P., Ramachandran, S., and Jovancicevic, V. (2008) Effect of a corrosion inhibitor for oil and gas wells when sand is produced. Proceedings of CORROSION/2008, NACE International, Houston, TX, Paper No. 08570.
60. Hu, X., Alzawai, K., Gnanavelu, A., Neville, A., Wang, C., Crossland, A., and Martin, J. (2011) Assessing the effect of corrosion inhibitor on erosion-corrosion of API-5L-X65 in multi-phase jet impingement conditions. *Wear*, 271(9–10), 1432–1437.
61. McMahon, A.J., Martin, J.W., and Harris, L. (2005) Effects of sand and interfacial adsorption loss on corrosion inhibitor efficiency. Proceedings of CORROSION/2005, NACE International, Houston, TX, Paper No. 05274.
62. Ramachandran, S., Ahn, Y.S., Bartrip, K.A., Jovancicevic, V. and Bassett J. (2005) Further advances in the development of erosion corrosion inhibitors. Proceedings of CORROSION/2005, NACE International, Houston, TX, Paper No. 05292.
63. Shadley, J.R., Shirazi, S.A., Dayalan, E., and Rybicki, E.F. (1996) Velocity guidelines for preventing pitting of carbon steel piping when the flowing medium contains $CO_2$ and sand. Proceedings of CORROSION/1996, NACE International, Houston, TX, Paper No. 15.
64. Shadley, J.R., Shirazi, S.A., Dayalan, E., and Rybicki, E.F. (1998) Prediction of erosion-corrosion penetration rate in a $CO_2$ environment with sand, *CORROSION*, 54(12), 972–978.
65. Crossland, A., Vera, J. Woollam, R.C., Turgoose, S., Palmer, J., and John, G. (2011) Corrosion inhibitor efficiency limits and key factors. Proceedings of CORROSION/2011, NACE International, Houston, TX, Paper No. 11062.
66. Alzawai, K., Hu, X., Neville, A., Roshan, R., Crossland, A., and Matrtin, J. (2011) Predicting the performance of API 5L-X65 in $CO_2$ erosion-corrosion environments using design of experiments. Proceedings of CORROSION/2011, NACE International, Houston, TX, Paper No. 11240.
67. Harvey, T.J., Wharton, J.A., and Wood, R.J.K. (2007) Development of synergy model for erosion-corrosion of carbon steel in a slurry pot. *Tribology: Materials, Surfaces and Interfaces*, 1, 33–47.

# 29

# BLACK POWDER IN GAS TRANSMISSION PIPELINES

Abdelmounam M. Sherik
*Research and Development Center, Saudi Aramco, Dhahran, Saudi Arabia*

## 29.1 INTRODUCTION

Black powder is a worldwide phenomenon experienced by most, if not all, transmission gas pipeline operators with internally uncoated lines [1–7]. In the gas industry, the term "black powder" is a color-descriptive term loosely used to describe a blackish material that collects in gas transmission pipelines. Black powder can be found in several forms such as dry fine powder or wet and has a tar-like appearance. Figures 29.1 and 29.2 show dry and tar-like black powder, respectively, collected at scraper receiving doors of gas transmission lines. Black powder has been reported in recently commissioned as well as older gas transmission pipelines.

Case histories showed that large quantities of black powder can be generated inside pipelines. For example, 3600 kg (8000 lb) of black powder were removed from a 400 mm (16 in.) × 80 km (50 miles) gas pipeline in Houston and 500 kg (1111 lb) of black powder were removed from a 914 mm (36 in.) Greek sales gas transmission pipeline extending for a distance of 12 km [8].

Black powder could have major adverse effects on customers by contaminating the customer gas supply leading to interruption of customer's operations and/or poor quality of products in which the gas is used as feed stock. It could also negatively affect gas pipeline operations. For example, it can lead to instrument scraping delays, reduced in-line inspection (ILI) accuracy, erosion of pressure control valves as well as flow reduction [1–7]. Figures 29.3 and 29.4 show cage and casing of a control valve, which were eroded by black powder in 45 days of service.

Finally, black powder could present a major health and environmental hazard. This is because some black powder is contaminated with mercury and naturally occurring radioactive materials (NORM) [4,9]. Iron sulfides are also potentially pyrophoric. The presence of these hazardous substances in black powder necessitates special procedures for handling and disposal of the removed black powder.

Black powder mainly consists of various forms of corrosion products, namely, iron sulfide, iron oxide, and iron carbonate, mechanically mixed, or chemically combined with any number of contaminants such as salts, sand, liquid hydrocarbons, and metal debris [1–7]. These products have common characteristics: They are all relatively high in specific gravity (sp.gr. 4–5.1), abrasive and are typically difficult to remove in cleaning operations [10].

Different gas pipeline operators report different compositions for the black powder removed from their pipelines. For example, some gas operators determined their black powder as being predominantly iron sulfides [1,6,7] others report the complete absence of iron sulfides, but the presence of iron oxides and hydroxides [2–5,8,9], while others report a combination of all of these products (iron sulfides, iron carbonates, and iron oxides) [10].

All types of black powder have one common source: they are formed inside the sales gas pipelines as a result of corrosion of the internal walls of these lines.

There are several removal and prevention methods available to gas pipeline operators for mitigating the formation and managing the impact of black powder [1,2,4]. Typically, there is no one best solution to control black powder. For example, in the case of existing uncoated pipelines, strict

*Oil and Gas Pipelines: Integrity and Safety Handbook,* First Edition. Edited by R. Winston Revie.

**FIGURE 29.1** Very fine black powder collected at scraper door receiver of a sales gas pipeline. (Courtesy of Saudi Aramco.)

adherence to the transmission gas product specifications would ensure elimination of condensed water and in turn the formation of black powder. However, because of process upsets, excess moisture may enter the line grid leading to potential condensation and internal corrosion. Therefore, the optimum black powder management practice usually consists of a combination of several control methods that must include moisture control coupled with downstream removal methods such as filtration. In contrast, in the case of new pipelines, internal coatings primarily used for drag reduction provide a cost effective and economical solution for the prevention of black powder.

**FIGURE 29.2** Tar-like black powder collected at scraper door receiver of a sales gas pipeline. (Courtesy of Saudi Aramco.)

**FIGURE 29.3** A control valve cage that was eroded by black powder in 45 days of service. (Courtesy of Saudi Aramco.)

**FIGURE 29.4** A control valve casing that was eroded by black powder in 45 days of service. (Courtesy of Saudi Aramco.)

In the next sections, several technical and operational aspects related to black powder sources, compositions, properties, formation mechanisms, and management methods will be presented and discussed.

## 29.2 INTERNAL CORROSION OF GAS TRANSMISSION PIPELINES

In typical transmission gas production, triethylene glycol (TEG) is used to remove water from the produced gas to reduce its dew point below the lowest temperature found in transportation operations. In the Middle East, and in similar temperature regions (e.g., the Southern United States), transmission gas is dried to a maximum water content of 7 lbs/mmscf. In contrast, in colder regions such as Canada and northern Europe, transmission product gas specifications are set to a maximum water content of 4 lbs/mmscf. Because of this drying process, it is often perceived that the gas is "dry" and therefore no water condensation and subsequent internal corrosion are expected [11,12]. However, field observations show that even under "dry" transmission gas conditions, general internal corrosion is widely encountered leading to what is known in the industry as the black powder phenomenon. The reason for internal corrosion in nominally "dry" gas pipelines can be attributed to the fact that it is not normally possible to completely eliminate water from pipelines [2,4,6,11,13]. For example, water vapor can potentially condense on the inner walls of the pipeline due to high dew points resulting from the presence of hygroscopic salts deposited on pipeline internal surfaces [14]. These salts can pick up water from the gas (even though it meets the 7 or 4 lb/mmscf specifications) and promote corrosion. The source of salts is normally the original pipeline hydrotest process. Corrosion attributed to the presence of salts is expected to decrease rapidly with increasing service life of the pipeline as these salts are continually removed during cleaning operations of the lines. Another potential source of condensed water in "dry" gas is operational upsets in the dehydration process, which lead to excess moisture in the line grid. Yet another source of condensed water is the water absorbed with TEG carryovers or co-condenses with TEG vapor that normally enters the line with the gas [12,15]. This leads to the condensation of TEG–water homogenous solutions that are high in TEG content, typically in the range of 95–99.5 wt% TEG [11,12]. For example, HYSYS simulation using the operating conditions of a gas treating plant, with 7 lb/mmscf water specification, showed that the composition of the condensed TEG–water solution is 97.66 wt% TEG and 2.34 wt% water [15]. Because TEG is heavy, viscous, and has very low vapor pressure, it collects on the internal walls of the pipelines and becomes mixed with various residuals such as salts, corrosion products, sand, and other contaminants. This mixture forms a layer of black sludge under which corrosion could occur. Recent laboratory studies have shown, although this sludge will not lead to general corrosion of the pipeline steel under simulated transmission gas conditions, it could lead to pitting corrosion [16].

Corrosion due to $H_2S$, $CO_2$, and $O_2$ in sales gas pipelines has well established mechanisms. The following are simplified electrochemical reactions that describe these corrosion processes and their respective corrosion products. It is important to note that in all of these electrochemical reactions, condensed water is a necessary condition for these reactions to proceed.

### 29.2.1 Siderite-$FeCO_3$ ($CO_2$ Corrosion)

The source of siderite–$FeCO_3$ corrosion product is the chemical reaction of dissolved $CO_2$ in condensed water producing carbonic acid, which in turn reacts directly with steel to produce $FeCO_3$, in accordance with these reactions [3,17].

$$H_2O\,(\text{condensed water}) + CO_2\,(\text{in gas}) \rightarrow H_2CO_3\,(\text{carbonic acid})$$
$$H_2CO_3 + Fe\,(\text{pipeline steel}) \rightarrow FeCO_3 + H_2$$

Carbon dioxide ($CO_2$) is a naturally occurring constituent of natural gas; this is in contrast to oxygen, which could ingress through leaks at low pressure points throughout the pipeline systems [6].

### 29.2.2 Iron Sulfides ($H_2S$ Corrosion)

Iron sulfides (FeS) corrosion products are usually formed from $H_2S$ reacting directly with the steel wall of the pipeline as per the following reactions.

$$H_2O\,(\text{condensed water}) + H_2S\,(\text{in gas}) \rightarrow H_3O^+ + HS^- HS^- + Fe\,(\text{pipeline steel}) \rightarrow FeS + H_2$$

Hydrogen sulfide ($H_2S$) can also be a naturally occurring constituent of natural gas or alternatively, produced by sulfate-reducing bacteria (SRBs) [6]. These anaerobic bacteria use the reduction of sulfate as a source of energy and oxygen, in accordance with reactions such as

$$2H^+ + SO_4^{-2} + CH_4 \rightarrow H_2S + CO_2 + 2H_2O$$

It is important to note that condensed water is a prerequisite for these bacteria to thrive and multiply and as such, the aforementioned reaction cannot occur in the absence of water. Because of this, typical transmission gas product specification specifies a maximum moisture content limit of 7 lbs water/mmscf (0.112 mg/l).

### 29.2.3 Iron Oxides ($O_2$ Oxidation)

Oxygen is not a natural constituent of sales gas but it gets introduced during gas treating operations and gas transportation. For example, oxygen could ingress through leaks at low pressure points throughout the pipeline systems, or is inevitably introduced due to the use of technical grade nitrogen gas in blanketing of TEG storage tanks. Technical grade nitrogen normally contains 3 wt% oxygen. Pipeline scraping operations are also another intermittent source of oxygen ingress. A 1988 survey of 44 natural gas transmission pipeline companies in North America indicated that their gas quality specifications allowed maximum $O_2$ concentrations ranging from 0.01 to 0.1 mol% with typical value of 0.02 mol% [13,18]. It has been shown that oxygen content of approximately 0.001 mol% has little effect on steel corrosion in the presence of stagnant water inside gas transmission pipelines, while 0.01 mol% produces fairly high corrosion rates. As a general rule of thumb, it is recommended that transmission pipelines should consider limiting maximum oxygen concentrations to 10 ppmv (0.001 mol%) [13,18].

In cyclical wet–dry low dissolved oxygen environments, as is the case in sales gas, iron oxides are usually formed by the direct oxidation of pipeline steel walls, in accordance with the following reactions [17,18].

$$2Fe + H_2O\,(\text{condensed water}) + 3/2\,O_2 \rightarrow 2\alpha, \beta \text{ or } \gamma\text{-}FeO(OH)$$

In waters containing low concentrations of dissolved oxygen as is the case in sales gas environment, the $\gamma$-FeO(OH) is unstable and will quickly transform to magnetite–$Fe_3O_4$ and water by the following reaction [17].

$$8\,\gamma\text{-}FeO(OH) + Fe \rightarrow 3Fe_3O_4 + 4H_2O\,(\text{condensed water})$$

But if the water is nearly saturated with dissolved oxygen, then hematite ($Fe_2O_3$) is often present [17].

Alternatively, iron oxides may be formed due to microbiologically induced corrosion (MIC) resulting from acid-producing bacteria (APB) or iron-oxidizing bacteria (IOB). Once again, condensed water is a prerequisite for these bacteria to thrive and multiply and as such, MIC cannot occur in the absence of water [2,3,6].

## 29.3 ANALYSIS TECHNIQUES

The importance of analysis and identification of the organic and inorganic composition and particle size range of black powder cannot be overemphasized. Accurate analysis of black powder is important in many aspects. For example, accurate analysis is essential in selecting the proper chemical cleaning procedure and chemistry, identifying the corrosion mechanism and develop appropriate mitigation methods, and also in developing a proper disposal strategy. In analyzing black powder, black powder samples are normally collected at scraper receiving doors as a result of pigging operations. The collected samples are subjected to various tests to determine their organic and inorganic composition. Results of the analysis of these samples are as good as the quality of the analyzed samples and therefore proper sample collection and preservation is essential for accurate analysis. Black powder samples must be collected immediately in clean glass bottles containing argon gas or mineral silicon oil. This is to avoid any possible oxidation of the corrosion products such as iron sulfides (in case the generated black powder is FeS based). If bacterial presence is expected, then the collection should be made in sterilized glass bottles.

However, and due to the large pipeline networks and their presence, oftentimes, in remote regions away from well-equipped testing laboratories, it is a common practice in the oil and gas industry to test corrosion products using simple techniques such as the use of acids or magnets to test for certain corrosion products. But these methods can be completely inaccurate and therefore often misleading. For example, a drop of acid on a corrosion product could make this corrosion product to effervesce, which suggests that this product could be either iron carbonate ($FeCO_3$) or calcium carbonate ($CaCO_3$). These two products have completely different formation mechanisms and sources, where the former is a corrosion product and the latter is a scale. Not knowing positively the product being tested leads to erroneous conclusions and mitigation actions. Likewise, the different forms of iron sulfide react differently with acid spot test. For example, makinawite is highly reactive with hydrochloric acid evolving $H_2S$ gas. But the other forms of iron sulfides such as pyrrhotite, troilite, marcasite, and pyrite are less reactive with HCl and much less or no $H_2S$ is evolved during the short time of an acid spot test. Therefore, an acid spot test failing to make the product evolve $H_2S$ does not guarantee that a corrosion product is free of sulfides. Similarly, if a product was found to be magnetic when using magnetic spot tests, it is often assumed that the product is magnetite ($Fe_3O_4$). But greigite ($Fe_3S_4$) is also magnetic and has black color as $Fe_3O_4$ and the two compounds are formed under entirely different circumstances.

More recently energy dispersive X-ray (EDX) technique has been used to analyze the composition of black powder. However, this technique provides only elemental analysis and not compound compositional analysis. For example, in black powder samples that contain various compounds of iron oxide (e.g., $Fe_3O_4$, FeO(OH), $Fe_2O_3$), EDX technique is unable to ascertain which of these compounds is present, but it can only inform the engineer that the elements Fe and O exist in the analyzed black powder sample. It is left to the engineer's judgment to infer from this elemental information which compounds are present, which is an impossible task. X-ray diffraction (XRD) is the correct technique that should be used to positively identify the compounds of the collected black powder samples. XRD technique requires small (usually 1 g) of black powder and produces full compound compositional analysis of the sample analyzed. With the correct identification of the corrosion products, the electrochemical reactions given in Sections 29.2.1–29.2.3 can be used to arrive at the source of these corrosion products and develop the appropriate preventative strategy.

Wet tar-like or slurry-like black powder samples collected from transmission gas pipelines should also be analyzed for their organic content using Fourier transformed infrared (FTIR), thermal gas analysis (TGA), and gas chromatography/mass spectroscopy (GC/MS) techniques. These analyses will show type and levels of organic-based liquids found in the analyzed samples. The liquid phases can be a mixture of TEG and hydrocarbon-based liquids such as compressor oils. The presence of such hydrocarbon liquids in black powder is indicative of an operational problem such as excessive TEG carry over and compressor leaks.

Also, collected black powder should be analyzed for its bacterial content, more specifically for sulfur-reducing bacteria (SRB), APB, and IOB using conventional cultivation media (for live bacteria). Conventional cultivation media methods can only detect live bacteria; however, DNA analysis techniques can be used to detect bacteria that are dead as a result of poor sample preservation. Bacterial analyses, particularly DNA analysis, are warranted especially if localized corrosion attacks with morphologies that are characteristic of microbial-induced corrosion (MIC) are observed on the internal walls of the lines.

Determination of black powder particulate size and size distribution is important and essential for designing the appropriate filtration system. SEM or optical microscopic techniques are most commonly used to determine particulate size on black powder samples collected from filters or at scraper receiving doors. These microscopy techniques will show that black powder is composed of fine and jagged micron-sized particles. However, when the same sample is subjected to analysis using XRD technique it shows that the average particle size of black powder is in the nanometer range (<100 nm) [2,3]. The apparent discrepancy in particle size measurement as determined by scanning electron microscope (SEM) versus XRD techniques can be attributed to agglomeration of the individual black powder particles. Whereas SEM measures the agglomerated particle size, XRD measures the average size of individual (non-agglomerated) particles. The question is which particle size the engineer should use for better design of filtration systems or estimating the erosion rate of control valves. For this reason, one should measure the particle size of black powder particles in movement with the gas. There are several *in situ* techniques such as isokinetic or laser-based sampling techniques [5].

## 29.4 COMPOSITION AND SOURCES OF BLACK POWDER

Table 29.1 shows typical XRD analysis results of black powder samples collected from transmission gas and sour gas pipelines in the Middle East. As can be seen from this table, the black powder collected from onshore transmission gas pipelines was predominantly iron oxides with minor amounts of iron carbonate. No iron sulfides were detected in these powders [3,8]. In case of subsea offshore transmission gas pipelines, iron sulfides are normally detected. This difference in composition is primarily due to oxygen ingress into onshore lines but the lack of oxygen ingress in subsea lines. Oxygen will react with $H_2S$ in the gas producing elemental sulfur and water [2,19]. In contrast, black powder collected from sour gas pipelines showed, as expected, iron sulfides to be the predominant species with minor amounts of iron hydroxides and iron carbonate.

**TABLE 29.2 Compounds in Black Powder and Their Respective Potential Sources**

| Constituent | Potential Sources |
|---|---|
| **$Fe_3O_4$** | 1. Low dissolved oxygen-induced corrosion<br>2. Conversion of γ-FeOOH<br>3. Bacterial-induced corrosion (APB, IOB)<br>4. Conversion of $FeCO_3$ and FeS (inside pipeline) due to oxygen ingress. Minor<br>5. Mill scale (minor and limited to newly commissioned pipelines) |
| **α-FeOOH** | Low dissolved oxygen-induced corrosion |
| **Iron Sulfides** | $H_2S$-induced corrosion<br>1. Chemical source<br>2. Bacterial source (SRB) |
| **Siderite—$FeCO_3$** | $CO_2$ corrosion |
| **Elemental sulfur** | Oxygen ingress leading to scavenging of $H_2S$ to elemental sulfur and water |

Table 29.2 shows the different compounds found in black powder samples and their potential sources. Comparison of results presented in Table 29.1 and Table 29.2 clearly shows that there could be several potential sources for the black powder collected from transmission gas pipelines.

Table 29.3 shows typical transmission gas quality specification for a major oil and gas producer in the Middle East. From the moisture content values and composition of the gas (see Tables 29.1 and 29.2), it is believed that internal corrosion due to low dissolved oxygen corrosion is the main source of black powder in transmission gas lines that experience oxygen ingress. The *in situ* conversion of primary

**TABLE 29.1 Major and Minor Corrosion Products in Black Powder Collected from Sales Gas Transmission and Sour Gas Flow Pipelines**

| Pipeline | Major Corrosion Products | Minor Corrosion Products |
|---|---|---|
| **Onshore sales gas pipelines** (oxygen ingress) | $Fe_3O_4$, and α-FeOOH | $FeCO_3$ |
| **Subsea sales gas pipelines** (no oxygen ingress) | Iron sulfides | |
| **Onshore wet sour gas pipelines** | Iron sulfides | $Fe_3O_4$, α-and γ-FeOOH and $FeCO_3$ |

**TABLE 29.3 Typical Sales Gas Product Specification and Measured Levels for Components That are Deemed Critical to Internal Corrosion of Sales Gas Pipelines**

| Component | Company Sales Gas Specification (Maximum Limits) | Measured Levels |
|---|---|---|
| $H_2S$ (ppm) | 16 | 1–2 |
| $O_2$ (mol%) | Not tested (assumed to be 0.0) | 0.01–0.03 |
| $CO_2$ (mol%) | 3 | <1.0 |
| Moisture, mg/l (l b/mmscf) | 0.112 (7) | 0.12 (7.5)–0.55 (35) |

corrosion products such as FeS and $FeCO_3$ to secondary iron oxide corrosion products as well as MIC and degradation of mill scale are expected to be minor contributors to the formation of black powder in sales gas pipelines.

In the case of black powder collected from sour gas pipelines, $H_2S$-induced corrosion, as expected, is the main contributor. The presence of minor amounts of $Fe_3O_4$, α-and γ-FeOOH indicates oxygen ingress into these lines leading to direct corrosion and/or conversion of small amounts of FeS to these iron oxide species.

Elemental sulfur in the range of 3–14 wt% was detected in some black powder samples by XRD technique. In contrast, X-ray florescence (XRF) analysis detected sulfur in all black powder samples. The fact that XRD technique did not detect sulfur in all black powder samples whereas XRF did indicates that the sulfur detected by XRF is not all elemental but combined in an amorphous compound that is not detectable by XRD.

The small amounts of elemental sulfur, detected by XRD, most likely have originated from the oxidation, inside the pipeline by oxygen ingress, of small amounts of $H_2S$ and/or FeS as per the following reactions:

$$2H_2S + O_2 \rightarrow 2H_2O + 2S\ \text{(elemental)}$$
$$3FeS + 2O_2 \rightarrow Fe_3O_4 + 3S\ \text{(elemental)}$$

XRF analysis revealed the presence of minor contaminants such as salts, sand, metal debris (from eroded equipment), and sulfur. These minor contaminants make up approximately 15–20 wt% of black powder. Mercury in the range 0.5–2 wt% was also detected in some black powder samples by XRF technique but not by XRD.

## 29.5 PHYSICAL AND MECHANICAL PROPERTIES

Figures 29.5 and 29.6 show typical SEM images of black powder particles collected at a scraper receiving door. It is clear from these images that black powder is composed of fine and jagged micron-sized particles. XRD analysis further showed that the average particle size of black powder is in the nanometer range (<100 nm). The apparent discrepancy in particle size measurement as determined by SEM versus XRD techniques can be attributed to agglomeration of the individual black powder particles. Whereas, SEM measures

**FIGURE 29.5** Image of black powder particles as taken by SEM.

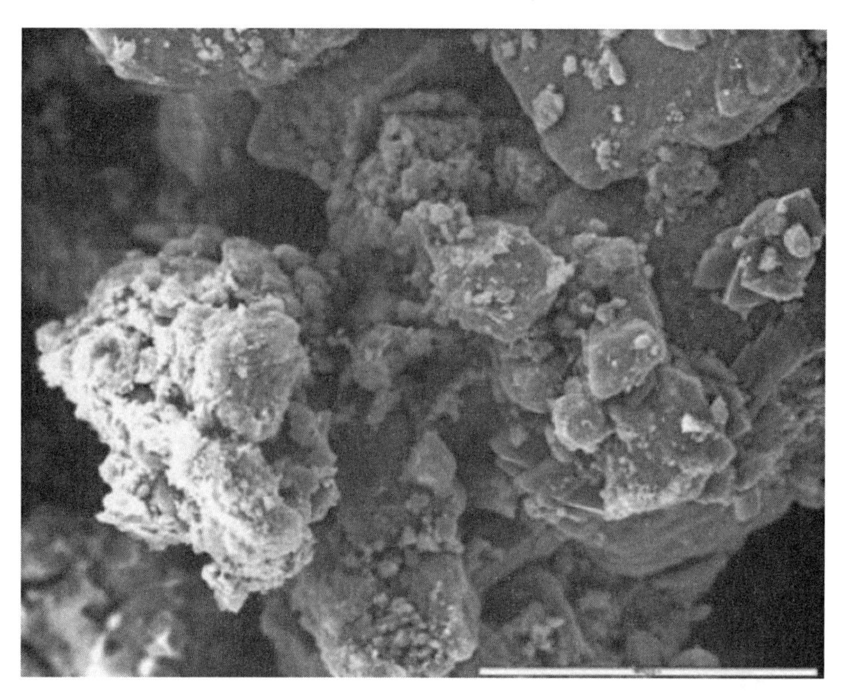

**FIGURE 29.6** High-magnification SEM images of black powder particles.

the agglomerated particle size, XRD measures the average size of individual (non-agglomerated) particles.

Microhardness measurements conducted on pipeline steel and black powder particles revealed average Vickers hardness values of 179 ± 10 VHN (87 ± 2 Rb) and 498 ± 62 VHN (49 ± 4 Rc), respectively. Hardness measurements conducted on individual black powder particles using the nanoindentation technique revealed hardness values ranging from 32 Rc to 64 Rc. From these measurements, it can be concluded that (1) both hardness measurements techniques provide comparable hardness values for black powder, and (2) black powder is significantly harder than carbon steel (the hardness of black powder is in the Rockwell C range whereas the pipeline steel is significantly much softer and has hardness in the Rockwell B range). The large standard deviation in hardness measurements using the nano-indentation technique is attributed to the fact that the hardness indents were made into individual particles that could be of differing compositions (iron oxides or iron carbonates), which are known to have different hardness values.

## 29.6 IMPACTS ON OPERATIONS

Black powder has several major impacts on transmission gas pipeline operations:

1. Black powder delays ILI runs and/or lowers the accuracy of the data gathered by the ILI tools. Thus, timely revalidation of transmission gas pipelines could become an issue.
2. Back powder reduces gas flow and therefore decreases operational efficiency leading to higher costs.
3. Black powder erodes carbon steel pressure control valves as quickly as 45 days. For example, Saudi Aramco has retro-fitted control valves with trims designed to resist erosion, such as solid tungsten carbide for valve cages and tungsten carbide inserts for valve plugs and seat rings.
4. Black powder erodes or clogs metering instruments leading to deviations in measured transmitted gas to customer.
5. Black powder increases operational expenses because of required cleaning, handling, and disposal operations. The latter will become more significant if black powder contains NORM or naturally occurring mercury (NOM).
6. Finally, black powder contaminates the customer gas supply and could lead to frequent replacement of customers' cartridge filter elements.

## 29.7 BLACK POWDER MANAGEMENT METHODS

Generally, pipeline companies practice various methods to manage and control black powder in the gas line grid.

Management methods can broadly be divided into two categories: (a) removal and (b) prevention methods.

### 29.7.1 Removal Methods

Removal methods consist of many techniques such as the use of filters and separators to capture black powder particles and also the use of pigging operations to mechanically or chemically clean a pipeline. Cleaning (mechanically or chemically) of gas lines is performed mainly for the following reasons:

a. Increase operational efficiency.
b. Facilitate effective corrosion inspection.
c. Increase operational safety.
d. Alter operational parameters.

Regardless of which cleaning method is used, when is a pipeline considered clean is the most difficult question. The level of cleanliness and the cleaning approach is governed by the reason for cleaning.

- ILI
- Corrosion control
- Increased flow efficiency

Cleaning to restore ILI accuracy is the most stringent and requires the highest level of cleanliness.

When a cleaning plan is implemented, waste disposal must be considered up front due to the large quantities of solids and solvents used in the operation. This is especially important if the removed black powder is contaminated with NORM or NOM.

***29.7.1.1 Mechanical Cleaning*** A significant portion of any pipeline cleaning job is performed by mechanical scrapers used to perform the cleaning job. Mechanical pigs are commonly deployed into a pipeline to scrape debris from the pipeline wall and remove black powder. In cases where black powder is not a major problem, this cleaning method may suffice to keep the pipeline in fairly clean condition, However, this cleaning method gives poor results when the black powder problem becomes major (large quantities). At one company with serious black powder formation, mechanical scraping operations were made every 3 months. However, this scraping frequency was found to further exacerbate the problem (corrosion) by exposing fresh steel surface for under specification "dry" gas. The frequency was dropped to once per year. The use of advanced mechanical cleaning tools to physically remove black powder can further improve the removal efficiency of the pigging operations. These advanced designs include bidirectional disks, brushes, magnets, front-end jet nozzles, and supporting wheels. However, in line with serious black powder formation, mechanical cleaning is typically not effective in removing the black powder. In this case, a combined approach of mechanical and chemical cleaning is used.

***29.7.1.2 Chemical Cleaning*** The addition of a chemical cleaner greatly improves the performance of the mechanical pigging operations. The proper analogy to consider is that the chemical solution acts as the liquid soap and the scraper acts as the brush. Chemical cleaners are used to

- Enhance the penetration into the solids deposits;
- Soften and loosen the deposits;
- Lift and carry the solids out of the line.

Chemical cleaning is more effective than mechanical cleaning but it is significantly more expensive.

There are several chemical cleaning agents used for the removal of black powder from transmission gas pipelines. Gel and surfactant cleaning are the most common solutions used. Gel shows excellent capability to carry large amounts of solids, but in situations where cleaning has to be done online, dealing with large slugs of gel becomes problematic. Also, removal of gel residues from the pipeline needs extra attention. Surfactant cleaning has a proven record in removing black powder. These chemicals can be dissolved in diesel or organic solvents (dissolution in water should be avoided to ensure the pipeline is not exposed to water). Surfactants will have the ability to penetrate contaminants and lower the surface tension properties of the pipeline leading to the removal of large amounts of black powder.

***29.7.1.3 Separators*** The use of separators and cyclones is based on the principle of centrifugal force. The black powder-laden gas passes through these devices and the black powder particles are physically knocked out of the gas stream to the walls of the separator where they fall and are collected at the bottom in a collection hub. This removal method is effective only if the concentration of solid particles is relatively high and if the particle size is relatively large (larger than 8–10 μm).

Filters are usually cartridge filters placed downstream of the gas pipeline to protect control valves and customers' equipment. The design and size of these filters will depend on the amount of black powder, its particle size, and hardness.

***29.7.1.4 Cyclo-Filters*** Cyclo-filters combine the best features of cyclones and filters in a two-stage removal process. The first stage of the removal is achieved by the cyclone, which knocks out black powder particles >8–10 μm; the second stage of cartridge filters removes the finer black powder particles.

Each of the aforementioned removal methods can be applied separately or in combination. For example, mechanical cleaning by instrument scraping can be combined with installation of filters downstream closest to the customer. This combination ensures that the scraped black powder gets filtered out from the gas supply before reaching the customer.

Although the removal approach is successful in protecting downstream operations, including the customers from the impacts of black powder, the removal methods have several common drawbacks: (a) they are after-the-fact treatments, that is, they do not address the root cause of black powder formation, (b) these methods are not a one-time solution but require frequent applications, (c) multiple location installations are most often necessary as in the case of filters and cyclones, (d) these methods constitute an additional cost to the ongoing operational costs of gas transport systems, and (e) subsequent handling and disposal procedures and processes are required. The handling and disposal procedures could become challenging and costly if the black powder contains health and environmentally hazardous materials such as mercury and NORM.

### 29.7.2 Prevention Methods

This management philosophy has at its core the belief that internal corrosion of transmission gas pipelines is the source of black powder. As such, these methods are based on preventing corrosion from occurring. These methods are as follows:

***29.7.2.1 Internal Coatings*** These are organic coatings such as high solids solvent based epoxy polyamine films. These coatings are applied at the pipe mill to protect the internal surfaces of pipelines from corrosion during storage. Nowadays, they are typically used for reducing drag; however, prevention of black powder formation would be an added benefit. These coatings are typically applied with a thickness range of 2–3 mils (50–80 μm) to cover pipe roughness (Ry5 = 30 μm). They have been in use for the last 60 years and are used in over 300,000 km of pipelines worldwide. They have demonstrated very good ageing properties (e.g., no degradation after 30 years exposure to sales gas) and are designed to withstand hydrotest (1 year), pigging operations, chemicals, including methanol and TEG, and fast decompression. International standards (API 5 L 2 and ISO 15741) cover the specification of internal coating for gas pipelines. As per the ISO 15741 Standard, the coating should be subjected as follows:

- Resistance to neutral salt spray (480 h);
- Resistance to water immersion (480 h);
- Resistance to chemicals (168 h);
- Cyclohexane, 95% diethylene glycol (DEG) in water, hexane, methanol, toluene, lubrication oil;
- Resistance to gas variation (240 h);
- Resistance to hydraulic blistering (24 h).

Internal coatings are considered a cost effective means of prevention of black powder in new transmission gas pipelines. However, they are very difficult to apply and are not cost effective for existing pipelines, particularly buried pipelines. In the case of applying these coatings to a new pipeline segment that is to be connected to an existing uncoated network, it is important that a filtering system is installed upstream of the start of the new segment. This is to filter out the black powder which will damage (erode) the internal coating in the new line if not removed.

***29.7.2.2 Moisture Control*** Elimination of water condensation in the pipeline, particularly internally uncoated pipelines, is the most critical step in preventing black powder formation in a gas grid. This can be achieved by improving the efficiency of the gas dehydration process to ensure dry gas, as per company specification, in the pipeline. Appropriately sized TEG dehydration units coupled with the installation of appropriately sized refrigeration and knockout drum units upstream and downstream of TEG dehydrator units, respectively, will help ensure drier gas entering the gas lines. Controlling and minimizing process upsets, such as water carry over (i.e., sales gas with water levels in excess of 7 lb/mmscf), is also important in limiting water moisture in the pipeline. The use of appropriately sized, designed and maintained molecular sieves, and chillers might be an expensive CAPEX, but it would ensure drier gas.

In internally bare transmission gas pipelines, strict adherence to the transmission gas quality specifications would ensure elimination of condensed water and in turn the prevention of black powder formation. However, because of process upsets, excess moisture may inevitably enter the line grid leading to potential condensation and internal corrosion. Moisture control in a gas grid with multiple connected gas treating plants is especially challenging because of the potential additive effect of process upsets. For example, in a gas network connected to seven treating plants, a 3-day upset per year in each plant results in a potential accumulative 21 days of moisture condensation.

***29.7.2.3 Commissioning Practices*** This concept involves the improvement of the required hydrotesting procedure. A major oil and gas producer in the Middle East revised its standards and construction practices to prevent black powder from being generated during new pipeline installation. Air was no longer allowed during the hydrotest water drying operations but flash drying with methanol or nitrogen gas should be used instead. Also, chemical cleaning was made a mandatory requirement in sales gas and product pipelines prior to commissioning to ensure a clean pipeline prior to starting operations. The use of sweet water with biocides and corrosion inhibitors will ensure no corrosion takes place during hydrotest wait in periods. If sweet water is not readily available in the field, as in many Middle East regions, then freshwater slugs can be used between pigs to wash the line and remove salt water.

## 29.8 GUIDANCE ON HANDLING AND DISPOSAL OF BLACK POWDER

NORM, principally the radionuclides lead-210 (Pb-210) and polonium-210 (Po-210), and NOM have been identified in black powder found in many transmission gas lines. Naturally occurring radionuclides are abundant in the earth's crust and originate from the presence of uranium or thorium. As gas, oil or water is produced from ground formations some radionuclides, such as radon in gas and radium in water, can migrate to surface facilities and create NORM or NOM as scales or deposits. The decay scheme for the most abundant uranium radionuclide, $^{238}U$, is shown in Figure 29.7.

Exposure or uncontrolled release of NORM and NOM is considered to represent potential exposure hazards to workers and the community. The importance of obtaining reliable data on these contaminants is considered essential for making informed decisions on the development of hazard evaluation and environmental control strategies. For this reason, a major oil and gas producer in the Middle East set up a NORM and NOM mapping program of its sales gas transmission lines. Approximately 30 locations throughout the sales gas network were identified for sample collection. It is believed that these locations will address any geographical, temporal, or seasonal variations in radioactivity concentrations. A minimum of three samples were collected from each location separated by a long period of time (months) to ensure seasonal variations. This program was believed to provide the company with credible and representative data. The radionuclides $^{210}Pb$ and $^{210}Po$ were selected for the study due to their relatively long half-lives (22 years and 138 days, respectively) and confirmed radiotoxicity. Mercury, observed as a contaminant in produced gas and also an environmental concern, was included in the program. The analysis of the Pb-210 and Po-210 required extensive method development to attain satisfactory analyte recoveries from the black powder matrix. The main problem was to overcome the chemical interference of high levels of iron in the black powder acid extracts. The Po-210 method involved the deposition of polonium from dilute acidic solutions onto a silver disc, with the Po-210 determined using alpha spectrometry [20–22]. Two methods were identified for the determination of Pb-210, which decays by low-energy beta and low-energy gamma (46.5 keV) emissions. The method of choice for the Pb-210 analysis involved liquid scintillation counting, which was found to be suitable for the low levels encountered and provided acceptable sample throughputs. Exemption levels of 0.20 Bq/g were established for each radionuclide of concern (Pb-210 and Po-210). Exemption levels are defined as radiation activity concentration levels below which a material is considered exempt from control as NORM. These exemption levels for radionuclides of concern in sales gas pipelines are detailed in Table 29.4.

The results of this mapping campaign showed that over 70% of the samples collected indicate levels of Pb-210 and Po-210 above the exemption level, and therefore would be regarded as having enhanced levels of NORM.

**TABLE 29.4 Exemption Levels for Pb-210 and Po-210**

| Radionuclide | Exemption Level (Bq/g) |
|---|---|
| Lead-210 | 0.2 |
| Polonium-210 | 0.2 |

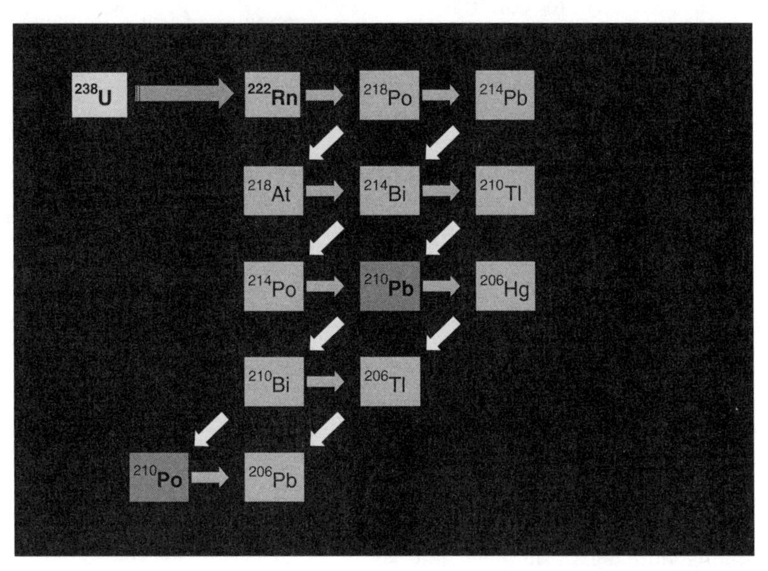


**FIGURE 29.7** $^{238}U$ natural decay scheme highlighting the origin of $^{210}Pb$ and $^{210}Po$.

NOM analysis was conducted using the toxicity characteristic leaching procedure (TCLP) as per US Environmental Protection Agency Method 1310 was used to establish whether a solid waste is hazardous, based on specified limits for selected metals. The TCLP data also indicate the relative environmental mobility of selected metals. Black powder samples were extracted with an amount of extraction fluid equal to 20 times the weight of the sample. The extraction fluid employed was determined based on the alkalinity of the solid phase of the waste. Following the extraction, the liquid TCLP extract was separated from the solid phase by filtration through a 0.6–0.8 nm glass fiber filter.

Exemption levels below which no action is required have been established for mercury. These exemption levels are as follow: total Hg of 0.4 mg/kg and TCLP Hg of 0.2 mg/l.

### 29.8.1 Workers Protection and Contamination Control

Pipeline pigging activities have been identified as an operation where workers can come into direct contact with potentially enhanced levels of NORM and where radioactive contamination can be spread in the immediate and surrounding work area. Information is provided in the following paragraphs on ways to ensure workers' protection and control the spread of contamination.

***29.8.1.1 Workers Protection*** Personnel required to work with NORM must be trained in the associated hazards.

- All NORM operations shall be covered by a safe system of work, which shall identify the hazards and highlight the precautions to be taken.
- Any item or area with detectable levels of loose NORM contamination shall be subject to radiological controls.
- Appropriate personal protective equipment (PPE) shall be worn, which may include but not be restricted to
  - Tyvek coveralls.
  - Neoprene, PVC, or NBR gloves.
  - Half-face respirators with HEPA cartridges; these must be fit tested.
  - Quarter-face HEPA disposable respirators.
- Eating, drinking, smoking, and chewing gum are not allowed in work areas where potential NORM contamination exists.
- Only essential personnel shall be allowed in the work areas where potential NORM contamination exists.
- Personnel shall thoroughly wash up with copious quantities of soap and water, after working with contaminated equipment, and before eating, drinking, or smoking, and at the end of the workday.

#### *29.8.1.2 Contamination Control*

- All NORM operations shall be carried out in a manner that prevents the spread of NORM contamination and minimizes the potential for workers to be exposed to NORM.
- NORM operations shall only be undertaken in areas that are clearly demarcated and access is restricted to those directly involved in the operations.

Waste debris from scraping activities is required to be contained and stored in suitable receptacles pending its NORM status being determined.

## 29.9 SUMMARY

- Black powder is a worldwide phenomenon experienced by most, if not all, gas pipeline operators. Black powder is regenerative and is formed inside natural gas pipelines as a result of corrosion of the internal walls of the pipeline. More specifically, black powder forms through chemical reactions of iron (Fe) present in ferrous pipeline steel with condensed water containing oxygen ($O_2$), hydrogen sulfide ($H_2S$), and carbon dioxide ($CO_2$). These chemical species are benign in dry sales gas, but can become corrosive when dissolved in water moisture.
- Dehydration process upsets and inefficient gas dehydration due to under design of TEG contactors are the source for condensed water in "dry" transmission gas lines causing black powder formation.
- In transmission gas transmission pipelines that experience oxygen contamination, black powder is composed mainly of iron hydroxides and iron oxides with minor amounts of iron carbonates.
- In transmission gas pipelines that experience no oxygen ingress, such as deep subsea gas lines, black powder is mostly composed of iron sulfides.
- Contaminants such as sand, dirt, hydrocarbons, elemental sulfur, and metal debris typically make up 20 wt% of black powder. The jagged shape and high hardness of black powder make it very erosive to many engineering materials such as pipeline control valves.
- Black powder is a complex phenomenon that is strongly affected by a multitude of operational parameters and network design that, in many cases, are unique to a specific operator, and sometimes, to a specific line within the same operator's network. Therefore, generalized conclusions and a "one-solution-fits-all" scenario might not be valid for all situations.
- There are several removal and prevention methods available to gas operators for mitigating the formation

and managing the impact of black powder. Typically, there is no one solution to control black powder.

- In the case of existing uncoated pipelines, strict adherence to a transmission gas moisture content standard would ensure elimination of condensed water and in turn the formation of black powder. However, because of process upsets, excess moisture may enter the line grid leading to potential condensation and internal corrosion. The best black powder management practice usually consists of a combination of several control methods that must include moisture control coupled with downstream removal methods.
- In the case of new pipelines, organic solvent-based internal coatings primarily used for drag reduction provide a cost effective and economical method for the prevention of black powder.
- Some black powder may contain NORM and NOM, which necessitate special handling and disposal procedures.

## ACKNOWLEDGMENTS

The author would like to thank Saudi Aramco for the permission to publish this work.

## REFERENCES

1. Baldwin, R.M. (1998) Black powder in the gas industry—sources, characteristics and treatment. GMRC, Report No. TA97-4, May 1998.
2. Sherik, A., Zaidi, S., Tuzan, E., and Perez, J. (2008) Black powder in gas transmission systems. Proceedings of CORROSION/2008, NACE International, Paper No. 08415.
3. Sherik, A., Perez, J., Jutaily, S., and AbdulHadi, A. (2007) Composition, sources and formation mechanisms of black powder in gas transmission pipelines. Proceedings of EUROCORR 2007, September 9–13, 2007, Freiburg, Germany.
4. Sherik, A. (2008) Black Powder: study examines sources, makeup in dry gas system. *Oil and Gas Journal*, 106, 54–59.
5. Sherik, A. (2012) Corrosion of sales gas pipelines. Proceedings of EUROCORR Conference, September, 2012, Istanbul, Turkey.
6. Baldwin, R.M. (1998) Here are procedures for handling persistent black powder contamination. *Oil and Gas Journal*, 96(43), 51–58.
7. Baldwin, R.M. (1998) Black powder control starts locally, works back to source. *Pipe Line & Gas Industry*, 82(4), 81–87.
8. Tsochatzidis, N.A. and Maroullis, K.E. (2007) Methods help remove black powder from gas pipelines. *Oil and Gas Journal*, 105, 52–58.
9. Godoy, J.M., Carvalho, F., Cordilha, A., Matta, L.E., and Godoy, M.L. (2005) (210) Pb content in natural gas pipeline residues ("black-powder") and its correlation with the chemical composition. *Journal of Environmental Radioactivity*, 83, 101–111.
10. Arrington, S. (2006) Pipeline debris removal requires extensive planning. *Pipeline and Gas Journal*, 233(11), 77–87.
11. Gartland, O. (2003) Internal corrosion in dry gas pipelines during upsets. PRCI Report No. 8292, March 2003.
12. Smart, J. and Roberts, R. (2006) Possible glycol corrosion in nominally dry gas pipelines. Proceedings of CORROSION/2006, NACE International, Paper No. 06442.
13. Sridhar, N., Dunn, D.S., Anderko, A.M., Lencka, M.M., and Schutt, H.U. (2001) Effects of water and gas compositions on the internal corrosion of gas pipelines—modeling and experimental studies. *CORROSION*, 57(3), 221–235.
14. Kolts, J. (2004) Design for internal corrosion resistance of sales gas pipelines, Proceedings of EUROCORR 2004, September 12–16, 2004, Nice, France.
15. Sherik, A.M., Lewis, A.L., Rasheed, A.H., and Jabran, A.A. (2010) Effect of triethylene glycol on corrosion of carbon steel in $H_2S$, $CO_2$, and $O_2$ environments. Proceedings of CORROSION/2010, NACE International, Paper No. 10188.
16. Sherik, A.M., Rasheed, A., and Jabran, A. (2011) Effect of sales gas pipelines gas pipelines black powder sludge on the corrosion of carbon steel. Proceedings of CORROSION/2011, NACE International, Paper No. 11087.
17. Craig, B. (2002) Corrosion product analysis—a road map to corrosion in oil and gas production. *Materials Performance*, 41, 56–58.
18. Lyle, F.F. (1997) Carbon dioxide/hydrogen sulfide corrosion under wet low-flow gas pipeline conditions in the presence of bicarbonate, chloride and oxygen. PRCI Final Report PR-15-9313.
19. Sherik, A. and Davis, B. (2009) Thermodynamic analysis of formation of black powder in sales gas pipelines. Proceedings of CORROSION/2009, NACE International, Paper No.09560.
20. Flynn, W.W. (1968) The determination of low levels of 210Po in the environmental materials. *Analytica Chimica Acta*, 43, 221–227.
21. Holtxman, R.B. (1987) The determination of $^{210}Pb$ and $^{210}Po$ in biological and environmental materials. *Journal of Radioanalytical and Nuclear Chemistry*, 115, 59–70.
22. Jia, G., et al. (2001) Determination of $^{210}Pb$ and $^{210}Po$ in mineral and biological environmental sample. *Journal of Radioanalytical and Nuclear Chemistry*, 247, 491–499.

# PART IV

# PROTECTION

# 30

# EXTERNAL COATINGS

Doug Waslen
*Sherwood Park, Alberta, Canada*

## 30.1 INTRODUCTION AND BACKGROUND

Pipeline coatings are critical to the long-term reliability of pipelines. External corrosion prevention coatings will be discussed in this chapter. The history of pipeline coatings, modern high-performance coatings, coating properties and performance, coating selection, and application techniques will be presented.

Pipeline coatings are the first line of defense against soil side corrosion, which means that proper selection and quality application are key to pipeline integrity. Where coatings become damaged or degraded, cathodic protection can (usually) minimize corrosion at these sites. Pipeline coatings are primarily designed to isolate the pipeline from the soil (electrolyte). This isolation is achieved by the coating acting as a dielectric and moisture barrier. The more effective the isolation, the less opportunity for corrosion and less cathodic protection current is required.

Pipeline coatings have been used since the 1940s and have seen many evolutions in terms of formulation and application. Early on, wax tapes and hot applied asphalt and coal tar coatings were applied "over the ditch." Asphalt and coal tar were often applied by bucket and hand and sometimes covered with asbestos or fiberglass wrap. Field-applied tape coatings evolved in the 1950s and 1960s and were anticipated to be a step change in technology for corrosion prevention; however, years later their degradation has become one of the biggest challenges for continued safe pipeline operation due to under film corrosion. Nevertheless, tape coatings did eliminate the health hazards associated with the application of coal tar and asphalt. Plant-applied coatings became commonplace in the early 1970s with extruded polyethylene and fusion bond epoxy (FBE) powder coatings obsoleting most field applied coating (with the exception of girth welds). These coatings remain state of the art and are considered high performance by many standard associations. Multilayer and composite coatings combine epoxy and polyethylene (extruded or powdered) technologies resulting in a high-performance system that enhances the mechanical damage limitations associated with FBE. Higher applied costs accompany these multilayer and composite coatings but that is often balanced by better mechanical properties and lower cathodic protection demand. Figure 30.1 shows the timeline of coating development.

## 30.2 COATING PERFORMANCE

Coating performance is achieved through matching coating properties to assessed needs, identifying applicable industry standards, documenting application requirements and quality assurance in application.

### 30.2.1 Needs Assessment

Each pipeline project is unique, and the coating requirements should be evaluated individually for each project. Pipeline diameter/long seam type, construction and operating temperature, soil conditions, project economics, and coating product availability all contribute to coating selection.

***30.2.1.1 Pipe Diameter/Long Seam Type*** Generally, pipes less than nominal pipe size (NPS 16) (sometimes <NPS 24) are electric resistance welded (ERW) or seamless

*Oil and Gas Pipelines: Integrity and Safety Handbook,* First Edition. Edited by R. Winston Revie.

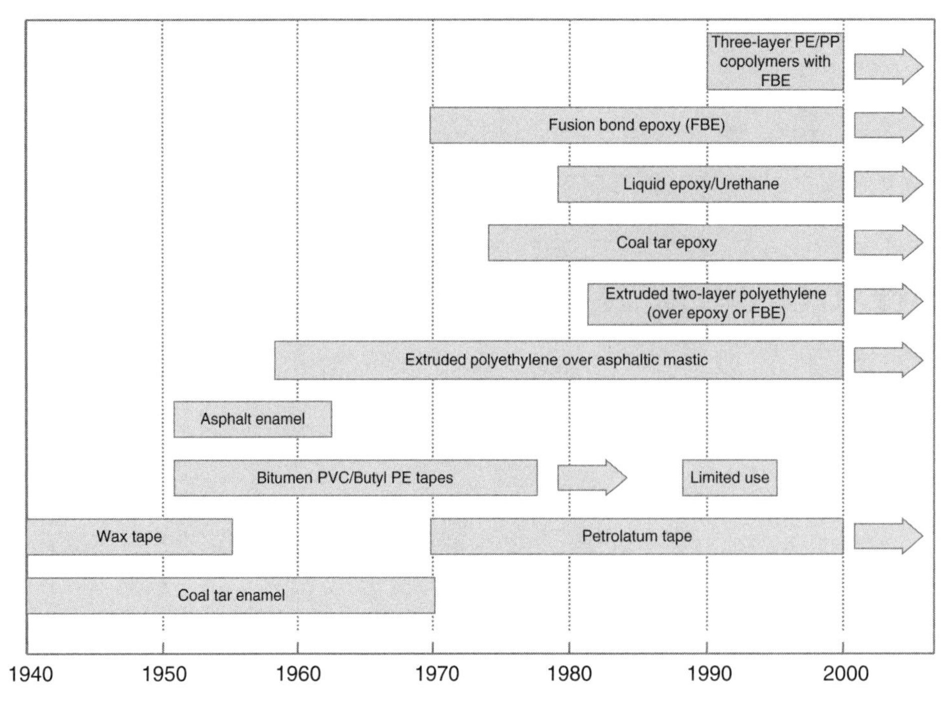


**FIGURE 30.1** Pipeline coating historical timeline.

and provide the most alternatives for plant coatings due to the lack of a raised weld seam. Raised long seams associated with larger pipes have historically prohibited the use of extruded coatings as they tend not to conform to the sharp long seam weld contour and may result in a conduit for water to migrate under the coating. Typically, hard film coatings such as FBE, liquid epoxy, or liquid urethanes conform well to raised welds, and therefore, multilayer coating systems that use epoxy primers reduce the risk of this scenario.

***30.2.1.2 Construction/Operating Temperatures*** The ambient temperature during construction needs to be considered. Low temperatures may

- Contribute to cracking of polyolefin coatings (water migration);
- Reduce the impact resistance of coatings (a concern during backfill);
- Result in incomplete cure of liquid coatings.

High surface temperatures due to ultraviolet (UV) heating can

- Result in irregular cure of liquid field-applied coatings (a concern during backfill and penetration damage during operation);
- Cause mastic flow of adhesive-based coatings (loss of adhesion).

Maximum pipeline operating temperatures must never exceed coating temperature rating. It is best practice to select coatings that have temperature ratings 5–10 °C above the maximum operating temperature. Coating degradation due to temperature is a function of time at temperature; so long-term small exceedances may be just as detrimental as short-term large exceedances.

***30.2.1.3 Soil Conditions*** Soil conditions must be considered as they can cause significant coating damage due to penetration, impact, and soil stress. Coating impact and penetration resistance at various temperatures are commonly reported by the manufacturer on data sheets or can be determined in the laboratory. Soil stress resistance is more difficult to assess; however, there are laboratory shear tests that can be used to compare coatings. Generally, clay soils can impart high shear force on the coating, and therefore, hard film coatings (FBE, liquid epoxy, etc.) are preferred. Soft mastic backed coatings can wrinkle and disbond as a result of high soil stress.

***30.2.1.4 Project Economics*** Consideration should be given to several coating alternatives that meet the project requirements. Higher integrity coatings generally have a higher initial cost but have lower repair rates during construction, lower cathodic protection demand, and somewhat longer service life. However, quantifying the economic benefit of higher performance can be subjective. Typically, coating cost is a relatively small portion of the project cost, and pipeline economic lives are 30–50 years so higher performance coatings can often be justified.

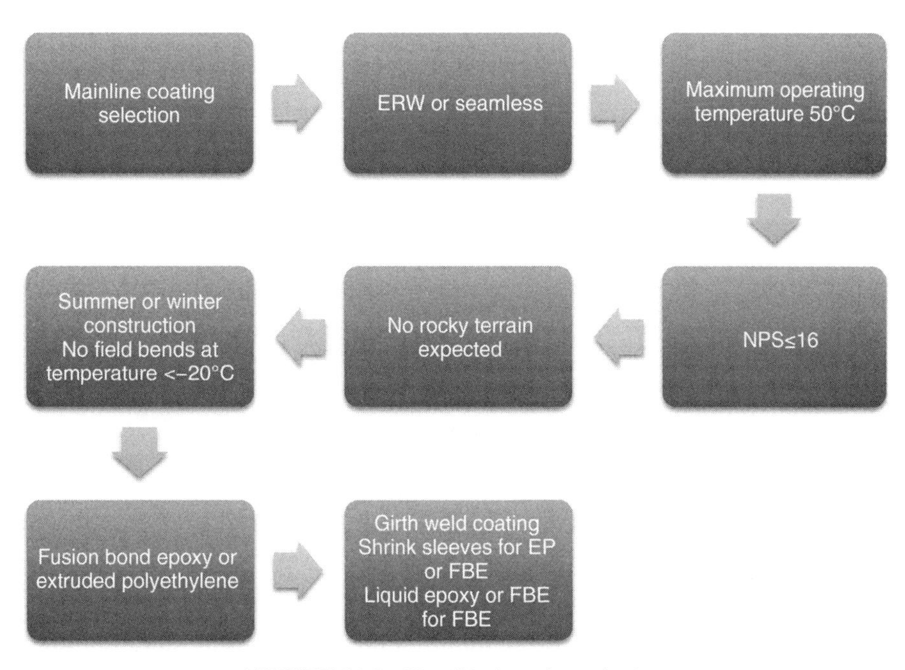


**FIGURE 30.2** Simplified coating selection.

*30.2.1.5* ***Product Availability*** High-performance coatings such as extruded polyethylene, FBE, multilayer, and composite coatings are widely available in North America. Depending on the location and the unique demands of the project, specialty coatings may only be available from certain applicators. Often, pipe coating applicators are located nearby the pipe manufacturer, and therefore, it may be possible to reduce transportation costs by utilizing local relationships. Prequalifying the coating applicator is recommended to verify the plant capability and quality.

A simple example of the coating selection process is provided in Figure 30.2.

## 30.3 PRODUCT TESTING

### 30.3.1 Cathodic Disbondment Resistance

Coatings must resist the application of cathodic protection. Unfortunately during service, cathodic protection creates strain on the coating at defect locations. Cathodic protection is necessary at these locations to prevent external corrosion. Therefore, the resistance of the coating to disbondment from cathodic protection can be an indicator of the life span. Common American Society for Testing and Materials (ASTM) and Canadian Standards Association (CSA) test methods can be used to compare the disbondment resistance of various coatings (ASTM G95-07, CSA Z245.20/21). Keep in mind, however, that the test parameters are orders of magnitude more severe than in-service conditions. Cathodic disbondment tests can be used for qualification and quality assurance testing during coating application. Examples of cathodic disbondments are shown in Figure 30.3.

### 30.3.2 Adhesion

As a method to compare coating performance, adhesion tests can be utilized. Hot water soak, peel, pull-off, and bend tests are commonly used (ASTM D4541, CSA Z245.20/21). Water soak is simply an immersion test with preexisting coating defects to observe loss of adhesion. Peel tests can be used to determine relative adhesive strength for coatings, such as extruded polyethylene and tape. Pull-off tests make dry adhesion comparison simple; however, many pipe coatings have adhesion strength greater than the test maximum. Bend tests determine the flexibility of the coating at various temperatures but also indicate the adhesion strength. Water soak adhesion tests (Figure 30.4) are often used for coating qualification and application quality assurance.

Adhesion tests can also be used to evaluate the compatibility between the mainline coating, girth weld, and assembly and repair coatings.

### 30.3.3 Flexibility

During pipeline construction, the pipeline is field bent to contour to the ground surface to maintain constant depth of cover. Coatings have bend limitations (degree of bend over a length of pipe) and become more restrictive as temperature decreases. Laboratory tests can be conducted to determine the failure point of the coating at various temperatures and bend angles (CSA Z245.20/21).

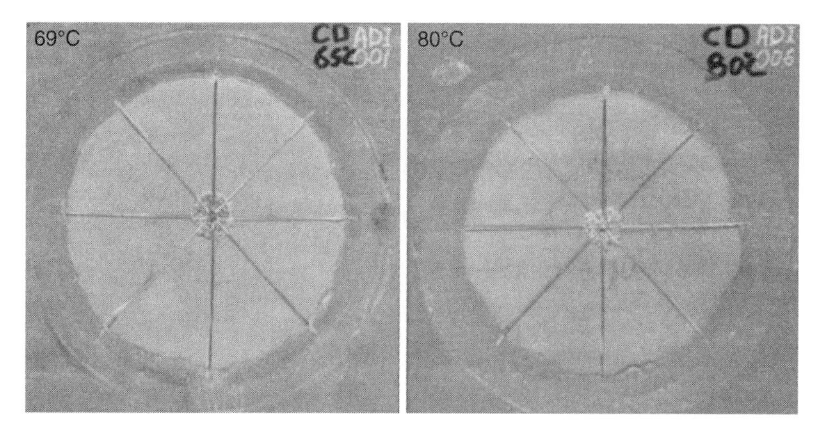


**FIGURE 30.3** Cathodic disbondment panel—after test.

### 30.3.4 Aging

Electrochemical impedance spectroscopy (EIS) can be used to determine the rate of deterioration of a coating subjected to a stress and can also be used to compare the rate of deterioration of several coatings [1]. The test involves the application of an AC current to the sample and measures resistance and capacitance of the coating. These properties help to evaluate the aging rate of the coating. Aged coating from pipeline failures or cutouts can be assessed and compared to new coating results to estimate the degradation as a function of time.

### 30.3.5 Temperature Rating

Several test methods should be used to estimate the temperature limits for the coating. Impact tests at cold temperatures can identify limitations for handling and construction. Hot water adhesion and high-temperature cathodic disbondment tests can be conducted at various temperatures to identify a relative limit for operating temperature (ASTM G95-07, CSA Z245.20/21). As mentioned, any temperature limitation of the coating should exceed the maximum pipeline operating temperature by at least 5 °C. Consideration should also be given to future operational conditions that may result in increased temperatures (new product shipments, etc.). Temperature effects associated with UV heating of the coating during construction should be considered.

### 30.3.6 Damage Resistance

The coating must withstand damage associated with handling and installation. This may include

- Repeated loading on and off of trains or trucks;
- Storage on skids;
- Handling during construction using rollers and slings;
- Insertion into bored or directionally drilled crossings.

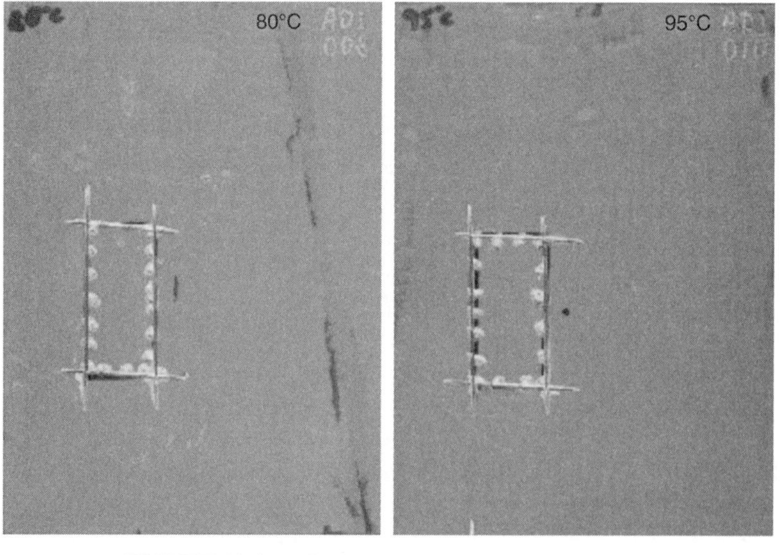


**FIGURE 30.4** Adhesion water soak panel—after test.

The challenge associated with assessing damage resistance is establishing a criterion. Typically, several coatings are evaluated and acceptance is based on a "reasonable" criterion that provides the design engineer with a few choices to select from. Some test methods (i.e., CSA Z245.20/21) have impact criteria established based on practical limits. However, these limits may not be appropriate for all situations. Abrasion resistance of various coatings can be compared using a standard test like Taber Abrasion (ASTM D4060-10). Penetration resistance can be determined by measuring the degree of penetration of the coating during 48-h static loading (CSA Z245.20/21). Hardness can be measured using any of the standard hardness methods. Typically for hard film coatings, the Shore D scale is utilized (ASTM D2240).

### 30.3.7 Cure

Hardness is typically used to evaluate the degree of cure. Simple fingernail impression practices can be used to assess cure of field-applied coatings. Sequential laboratory tests should be conducted to determine the cure curve for various preheat temperatures and ambient temperatures. This will ensure that the correct preheat is chosen for the ambient temperatures on the day of application.

### 30.3.8 Electrical Isolation

Dielectric tests can be used to compare the coating resistance to electrical current (ASTM D149). This property indicates the ability of the coating to isolate the pipe from the environment; the higher the dielectric resistance, the better the isolation.

## 30.4 STANDARDS AND APPLICATION SPECIFICATION

Today, there is guidance provided by standard associations for the application of pipeline coatings; however, there is little guidance for coating selection. These standards should be consulted to ensure compliance (where necessary) and to guide application requirements. For example, National Association of Corrosion Engineers (NACE) International, CSA, and ISO have many industry-accepted standards for the specification, application, and inspection of coatings (CSA Z245.20/21, NACE RP0394). A written performance-based application specification should be developed prior to request for proposal, likely during the engineering design stage.

The application specification should be performance based, meaning that it should refer to the applicator's specification where necessary, but should specify desired end results. For example, the specification should include the following:

1. Coating products permitted
2. Surface cleanliness and preparation requirements
3. Minimum coating thickness requirements
4. Minimum inspection requirements and frequencies
5. Test sample requirements, pass criteria, and retesting protocol
6. Documentation requirements

### 30.4.1 Quality Assurance

Assuring quality during application is actually the responsibility of the applicator; however, the owner's application specification should identify any additional inspections, tests, or checks that the applicator shall perform and documentation requirements. The applicator's quality and test plan should be reviewed prior to application and accepted or modified as necessary. As a further requirement, a third-party audit requirement can be specified to ensure that the applicator is performing the necessary checks. If the owner has not used the applicator previously, it is recommended that the application mill be inspected and qualified prior to use. Also, if the coating has not been used previously or if there have been coating formulation changes, the coating should undergo qualification testing using laboratory-prepared panels and/or application samples.

## 30.5 FIELD-APPLIED COATINGS

Field-applied coating selection and application (Figure 30.5) must be considered along with the mainline plant-applied

**FIGURE 30.5** Field spray epoxy—rehabilitation excavation.

coating to ensure compatibility with the parent coating, to identify application constraints (temperature, humidity, surface preparation, etc.), and to ensure worker training and qualifications. Field-applied coatings are subject to the most application variables, and therefore, require the utmost care and attention. Most coating failures (up to 80%) are due to improper application and not the fault of the coating itself. Surface preparation errors, lack of worker training, and improper raw material storage are most often causes of coating failure. Similarly to the plant-applied coatings, the same rigor of selection applies; however, more prescriptive application specifications, inspection, and worker training are required. Typically, the coating supplier or manufacturer can provide on-site application training for the pipeline construction company or ideally a separate coating applicator can be contracted. The coating supplier can certify the workers trained and only those certified shall be permitted to apply the coating.

Documentation of the products applied, the atmospheric conditions, the worker's name, date, and so on is required, just as it is during pipe joining/welding, to ensure traceability should integrity problems surface during operation.

## 30.6 COATING TYPES AND APPLICATION

High-performance coatings include extruded olefins, powder epoxy coatings, liquid epoxy, liquid urethane coatings, and many other specialty coatings. Girth weld and assembly coatings include petrolatum products, liquid epoxy/urethane, mastic tapes, two- and three-layer heat shrinkable sleeves, and powder epoxy.

### 30.6.1 Fusion Bond Epoxy

FBE is a plant- or field-applied coating that is applied in powder form at very high application temperatures (>225 °C). The coating cures almost immediately and generally conforms well to raised welds (long seam and girth welds). It is usually applied to a minimum thickness of 12 mils (300 μm) unless it is used as a primer coat as a part of a multilayer system. New construction of a pipeline with FBE coating is illustrated, before backfilling, in Figure 30.6.

FBE can be sensitive to surface cleanliness, and therefore, acid washing and testing for salt contamination are conducted prior to application. A strong process and quality management are critical to successful performance. A typical booth of the type used for in-plant application of FBE coatings is shown in Figure 30.7.

FBE is fairly susceptible to handling and construction damage; however, it is very resistant to disbondment from cathodic protection and penetration damage. It can be applied in a shop to induction bends and piping assemblies. FBE is the only coating that does not inhibit effective cathodic protection transmission. It is often referred to as a fail-safe coating meaning that even when it disbonds, corrosion processes remain controlled by cathodic protection. There are some formulations of fusion bond powder that have enhanced abrasion and impact resistance. Typically, these powders are applied in a dual-layer system over conventional FBE.

**FIGURE 30.6** Fusion bond epoxy—new construction prior to backfill.

### 30.6.2 Extruded Olefins

Extruded coatings include various olefins, typically polyethylene and polypropylene. They can be applied over an extruded adhesive or over liquid or FBE primer with an interlayer adhesive (multilayer). Adhesive and jacket thickness and uniformity are key to performance. Extruded coatings have high resistance to soil stress, impact, and

**FIGURE 30.7** Fusion bond epoxy powder booth—plant application.

**FIGURE 30.8** Extruded polyethylene coating—plant applied. (Reproduced with permission of ShawCor Ltd.)

penetration damage and good resistance to cathodic protection. It provides an excellent moisture barrier and high dielectric strength. A pipe with extruded polyethylene coating is illustrated in Figure 30.8.

### 30.6.3 Liquid Epoxy and Urethane

A pipe with shop-applied liquid epoxy coating is illustrated in Figure 30.9. Liquid coatings are often used for abrasion resistance either on their own or as an overcoat on FBE, typically. They can also be shop spray applied for coating of piping assemblies, valves, and field applied to girth welds. These coatings are highly resistant to impact and abrasion and commonly used in directionally drilled crossings and for pipelines in rocky backfill. Liquid coatings can be sensitive to surface preparation, and therefore, the surface must be free of oil, grease, salts, and so on. Solvent washing should be used to ensure cleanliness. Minimum surface profile of 3 mils (76 μm) is typically required to ensure optimum performance. Coating thickness ranges from 20 mils (500 μm) to 60 mils (1500 μm) depending on the purpose.

### 30.6.4 Composite Coatings

Composite coatings differ from multilayer coatings in that they are applied in powder form that results in an integrated film, which is not dependent on an adhesive layer. The most common composite coating comprises a fusion bond powder primer, a powder interlayer followed by a powdered polyethylene topcoat. This type of coating has little potential for under film corrosion because the cohesive bond between components exceeds the adhesive bond to the pipe surface. These coatings are extremely resistance to all forces, including mechanical damage, and have high dielectric strength.

### 30.6.5 Girth Weld Coatings

As mentioned, it is much more difficult to achieve quality application in the field; therefore, modern pipeline construction techniques aim to limit the amount of field-applied coating. Generally, coatings on girth welds represent the majority of field-applied coatings. The best practice is to match the girth weld coating to the parent coating. For example, if the parent coating is epoxy, then the girth weld coating should be epoxy. Or if the parent coating is polyethylene, the girth weld coating should be a polyethylene shrink sleeve. Shrink sleeves are applied to an ambient surface and heated using a torch or induction heater to shrink the sleeve to the pipe surface. Epoxy coatings can be applied to an ambient surface if the necessary cure temperature can be maintained. Often, the pipe surface is preheated to accelerate cure (to speed the time to backfill) or to achieve the necessary surface temperature during cure (during cold temperatures). Field application of a liquid epoxy coating is illustrated in Figure 30.10.

**FIGURE 30.9** Liquid epoxy shop applied for directional drill.

**FIGURE 30.10** Liquid epoxy field application for girth welds.

### 30.6.6 Specialty Coatings

Specialty coatings include those designed for temperature extremes, irregular shapes, and unique application conditions (i.e., wet surface). Liquid epoxy and urethane coatings typically require application temperatures above 10 °C for the duration of the cure otherwise their properties are significantly compromised. Often during pipeline construction and maintenance, coatings need to be applied at temperatures as cold as −30 °C. Vinyl ester-based coatings can be applied at temperatures as low as −20 °C. Overall, these coatings have slightly lower performance compared to conventional epoxies so they should only be used when necessary. Alternative approaches include hoarding and heating the ditch and pipe to allow conventional coatings to be applied in a controlled environment.

Irregular shapes such as valves, fittings, flanges, and induction bends can create coating challenges. If shop coating these components using liquid coatings is not available, field-applied petrolatum coatings, hand-applied liquid coatings, or some tape-style coatings can be considered because of their ease of application and conformability. Trial application is recommended to ensure user-friendly application and to achieve the desired conformability. A petrolatum tape coating shop applied to a valve is shown in Figure 30.11.

Most coatings have maximum operating temperatures between 45 and 80 °C. High-temperature tapes, epoxies, and polypropylene can withstand temperatures up to 110 °C. Pipeline design temperatures are sometimes higher than this for service such as hot bitumen pipelines, sometimes as high as 150 °C. Specialty FBEs have dry service rating as high as 150 °C. A dry service rating means that the corrosion coating must be isolated from the environment. Typically, high-temperature pipelines are insulated (polyurethane foam) to maintain flow temperatures and also to lessen the thermal gradient in the soil. Furthermore, the foam is covered with polyethylene or polypropylene to protect it from degradation due to water exposure and ingress.

**FIGURE 30.11** Petrolatum tape coating shop applied to valve.

**FIGURE 30.12** Tape coating during integrity excavation—note petrolatum coating as a transition to original coating.

Sometimes it is necessary to apply coating to a wet substrate either due to groundwater or humidity. Petrolatum products can be used on a wet surface because they displace water as they are applied. Some rigid wrap products and epoxies can be applied underwater.

### 30.6.7 Repair Coatings

Most pipelines undergo continuous integrity management programs that inspect and repair the pipeline to ensure safe reliable long-term operation. These integrity programs involve excavation, coating removal, steel inspection, repair (if necessary), and re-coating. It is critical that the new coating is properly selected and applied to ensure that there isn't a need to re-excavate the area in the future as a result of on-going corrosion. Furthermore, the workers must be trained in the proper application and limitations of the repair/rehabilitation coating(s). Several coatings may be required to achieve good results. For example, if the existing coating is coal tar enamel, a transition coating should be considered that would bond to the repair coating as well as the existing coating (see Figure 30.12).

## REFERENCE

1. Papavinasam, S., Attard, M., and Revie, R.W. (2009) Electrochemical impedance spectroscopy measurement during cathodic disbondment experiment of pipeline coatings. *Journal of ASTM International*, 6(3), Published Online 20 March, 2009. doi: 10.1520/JAI101246

# 31

# THERMOPLASTIC LINERS FOR OILFIELD PIPELINES

JIM MASON
*Mason Materials Development, LLC, Birdsboro, PA, USA*

## 31.1 INTRODUCTION

Internal corrosion of oilfield pipelines has been an operating concern since the beginning of the industry. Of the many solutions adopted by the industry few have the durability and reliability of the thermoplastic liner, which was introduced to oil and gas more than 30 years ago. In the intervening years the technology has matured, with advanced materials and designs delivering high reliability and durability to pipelines from 50 mm (2 in.) to over 1000 mm (39 in.) in diameter.

Thermoplastic liners consist of a thermoplastic pipe that is inserted into the metallic host pipe, and terminated at each end with a corrosion-resistant coupling such as a thermoplastic lined flange or a corrosion resistant alloy welded connection (Figures 31.1 and 31.2). For onshore application the flanged connection is generally preferred due to low cost and ease of installation, inspection, and maintenance. Thermoplastic liners are also used for offshore, subsea pipelines. In that case welded, flangeless, connections are normally used.

All thermoplastic liner materials are somewhat permeable to small molecules such as water, $CO_2$, and $H_2S$, which can cause corrosion. The liner works as a corrosion management system because a protective corrosion scale is formed on the interior pipe wall behind the liner shortly after going into service. The liner is a physical barrier that protects the scale layer from damage by the pipeline contents. As long as this protective scale layer remains intact, the corrosion rate from that time on is essentially negligible, and experience is that this is normally the case.

One of the best reasons to use a liner is that it provides a double-containment benefit. If the liner fails, the pipeline contents are still contained. A repair can be completed on a nonemergency basis and without a pipeline spill. Liners are a preferred corrosion management solution in cases where the long-term reliability of chemical corrosion inhibition systems is in doubt, or inhibitor consumption rate is so high that it becomes more expensive than a liner over the lifetime of the pipeline, or when there is a significant erosion process in the pipeline caused by abrasive contents such as slurries or high-speed flowstreams with entrained sand. Liners are almost always less expensive than specialty alloy steel pipes and, because the liner is an extruded pipe with a wall thickness of at least several millimeters, it provides a pinhole-free barrier—a benefit that high-performance internal pipeline coatings cannot guarantee.

Liners are used in upstream pipelines transporting corrosive fluids such as produced water, oil well multiphase, and sweet or sour gas multiphase. In most applications, the liner retains a smooth interior surface over its entire service life, contributing to improved flow characteristics compared with aging, corroding steel.

## 31.2 CODES AND STANDARDS

There is only one international standard that addresses thermoplastic liners for oilfield pipelines. NACE International Recommended Practice RP0304, "Design, Installation, and Operation of Thermoplastic Liners for Oilfield Pipelines" [1] was first issued in 2004 and has recently undergone a major revision to include more complete design guidance.

The Canadian Standards Association (CSA) has developed and regularly updates a comprehensive standard for

*Oil and Gas Pipelines: Integrity and Safety Handbook,* First Edition. Edited by R. Winston Revie.

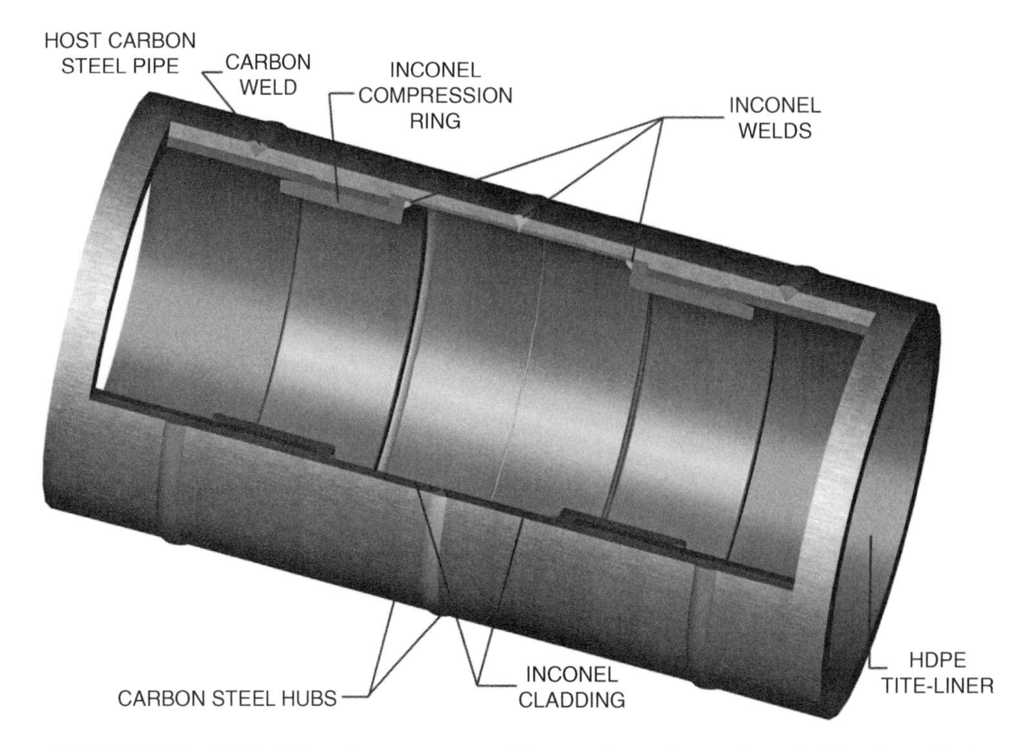


**FIGURE 31.1** Welded flangeless connector with corrosion-resistant alloy cladding in unlined section.(Illustration courtesy of United Pipeline Systems, Durango, Colorado, USA.)

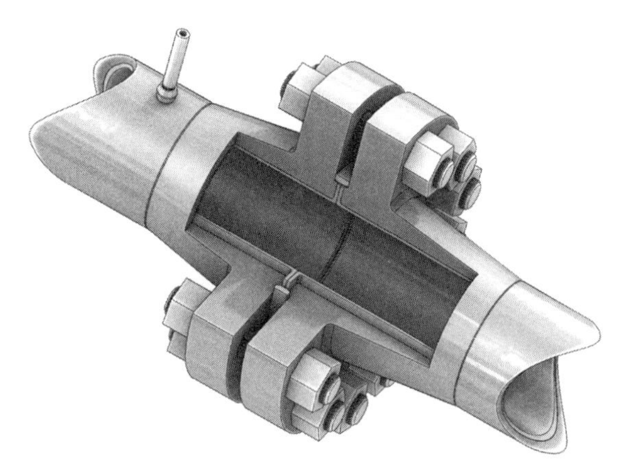

**FIGURE 31.2** Fully lined flanged connection showing annulus vent riser at upper left.(Illustration courtesy of United Pipeline Systems, Durango, Colorado, USA.)

thermoplastic-lined oilfield pipelines. Clause 13 of Canadian Standard Z-662 "Oil and Gas Pipeline Systems" [2] provides guidance on design and operation of thermoplastic-lined oilfield pipelines. The standard is based on decades of operating experience with thermoplastic lined systems in Canada. Compliance with Z-662 is mandatory in Canada.

In addition to international and national standards, some oil companies have developed their own internal standard practices or specifications to which thermoplastic lined pipelines must comply. Access to these proprietary standards by contractors and suppliers is on a case-by-case basis at the discretion of the oil company.

## 31.3 THE INSTALLATION PROCESS

Liners can be installed in new pipelines to provide corrosion protection from the start, or in existing pipelines to rehabilitate a corroding pipeline and prevent the progress of existing corrosion damage. There can be minor differences in the installation procedure for liner intended for new pipelines compared with installation of liner for rehabilitation purposes. The differences are largely in how the pipeline is prepared to receive the liner.

New pipelines can be designed to accommodate the liner by a using large radius swept bends of at least 15X nominal pipeline diameter instead of elbows and small radius bends. The access points for liner installation can be planned and left uncovered after the rest of the pipeline is buried to facilitate access for liner installation. New pipelines normally do not need internal cleaning to remove corrosion product debris as is the case for pipelines being rehabilitated. The same types of end fittings are suitable for new and rehabilitation projects. Both flanged and flangeless designs are available (Figures 31.1 and 31.2).

In rehabilitation projects, before liner insertion the host pipeline must be prepared to receive the liner. This usually involves excavating the host pipe at convenient access points, cutting the pipe and attaching an end fitting suitable for use as a liner termination to create a pipeline segment to be lined. In existing pipelines, elbows and short-radius bends must be cut out and replaced with swept bends with a radius of at least 15X the nominal pipeline diameter, Different installation techniques may have somewhat larger minimum radius requirements. The installer should always be consulted in this matter.

For both new and rehabilitation projects the segment length depends on the pipeline topography. Changes in direction increase the total frictional load and therefore the pulling force necessary to insert the liner, so the installation contractor will choose the access points to limit the insertion forces to a safe level below the damage threshold of the liner. Depending on local regulations the segment might be hydro tested to ensure that the host pipe is capable of sustaining the design pressures of the lined pipeline. Any deficiencies in pressure capacity of the steel host pipe must be resolved before liner insertion.

At one end of the segment the contractor positions an instrumented winch with an accurate length counter and a load cell to measure pulling force. At this point the contractor may run a sizing pig through the segment to test for any interior intrusions such as abnormally large weld roots, different ID repair sections, pipeline dents, kinks, and so on. If any are found, they must be fixed before liner insertion. The interior of the pipeline must be free of debris and loose corrosion products before insertion. In the case where the liner is being installed in a pipeline that has been in use and may have corrosion products on the internal surfaces, a wire brush pig is often used for this basic cleaning task. Some oil companies require treatment of the interior of the host pipe with a suitable corrosion inhibitor before liner insertion in both new and rehabilitation projects. Finally, the pulling cable is passed to the other end of the segment using a flexible-cup pig and air pressure so that the liner can be pulled in.

The next step is preparation of the liner. Liner pipe is normally delivered to the job site in lengths that are shorter than the length of the steel segment to be lined. These shorter lengths are joined together using a process known as butt fusion in which the ends of two segments to be joined are melted in a controlled manner and then pressed together under carefully controlled conditions until the polymer solidifies again. The process is repeated to join enough liner segments for insertion into the host pipe. For high-density polyethylene (HDPE) [3] and polyamide 11 (PA11) [4] there are standardized procedures for this process, although most installers optimize and fully qualify their own procedures to accommodate environmental and operational realities of the field.

A specially designed pull head is attached to the liner. The pull head is then connected to a pulling cable via a swivel to prevent torque loads from transferring to the liner from the cable. If any special diameter reduction devices are to be used, the cable passes through these devices before being attached to the pull head. The liner is then pulled through the diameter reduction devices and into the host pipe. The ends of the liner are then terminated in the end fitting. For the flange fitting it is necessary to fuse on a flange adapter made from the same type of polymer as the liner.

After the liner is inserted it is normal to do an integrity check using compressed air. This may be done on a segment-by-segment basis, or the whole pipeline can be tested as one unit. During this pressure test the annulus vents at each flange are monitored for excess flow from the annulus between the liner and host pipe, which would indicate a breach in the liner that must be repaired before commissioning. Some jurisdictions require a full hydro test before the lined pipeline can be certified for use. Before burial, the vents are connected to riser pipes terminated aboveground to a venting control system, which might be as simple as a hand-operated valve at the end of a small diameter riser pipe. The vent ports of several adjacent lined segments may be connected together with jumpers and vented through a valve at the downstream end of the chain. More sophisticated venting systems may include automatic vent valves that open at a certain pressure differential and pass the gases into a manifold that runs along the pipeline path. A benefit of the hand-operated valve is that an inspector can operate the valve periodically and observe the pressure and flow of gas from the vent. The presence of an unexpectedly high pressure or flow volume may signal a liner failure.

## 31.4 IMPORTANT MECHANICAL DESIGN ASPECTS

The lined pipeline system is a composite system in which the host pipe bears all of the same mechanical stresses as it would in an unlined system. It is important to remember that the liner is not designed to be a load bearing component. Its sole function is to separate the host pipe from the corrosive pipeline contents. Liners are normally designed to be just thick enough to facilitate proper fabrication and installation, plus any added thickness needed to address resistance to liner collapse.

Thermoplastic polymers are not perfect gas barriers. Because of this, gases in the pipeline contents will permeate through the liner into the space, known as the annulus, between the liner and the host pipe, eventually reaching the same partial pressure in the annulus as in the pipeline contents. If the pipeline is depressurized with a high gas pressure in the annulus it is possible that the liner will collapse to accommodate the expanding annulus gases. Collapse of the liner is generally considered to be a catastrophic event requiring replacement of the liner. To prevent annulus gas pressure buildup, the annulus must be vented continuously or periodically through the venting systems described earlier. The pressure differential across the liner wall required to cause radial buckling is known as the critical buckling pressure, $P_{crit}$, which can be approximated for an unconstrained liner by

$$P_{crit} = 0.2334\,E(t/r)^2$$

Where $E$ is the young's modulus of the liner material, $t$ is the thickness, and $r$ is the average radius of the liner.

Liners can be installed in three stress states: loose, neutral, and tight. Each type offers advantages and disadvantages to system design and performance. When discussing this topic it is essential to remember that thermoplastic liner materials are generally elastoplastic. In this context, it means that they exhibit elastic behavior over the range of strain from zero to the tensile yield point, after which they exhibit plastic, irreversible, deformation. They also exhibit creep behavior. When held under load at strains below the yield point for long periods of time, the material undergoes partial stress relaxation leading to some unrecoverable strain after the load is removed.

In the case of loose fitting liners, commonly called "slip-lining," at the time of insertion the outside diameter of the liner is smaller than the inside diameter of the host pipe when the liner is not pressurized. On pressurization the loose liner expands to come into contact with the interior surface of the host pipe to transfer the hoop stress to the host pipe. When held under pressure, especially at elevated temperatures, the liner may stress relieve and remain in close contact with the steel host pipe after the pressure is removed. But usually, under field installation conditions, and especially in colder climates, loose liners may move slightly away from the wall of the host pipe when not under pressure because the deformation during the initial pressurization is not sufficient to cause fully plastic deformation at the strain required for contact with the host pipe inner wall, and/or the time allowed at high temperature was not sufficient to allow complete stress relaxation, or the pressurization water temperature was not sufficiently high to accomplish the stress relaxation job in the time allowed. Over time creep, stress relaxation, thermal expansion, and swelling due to absorption of hydrocarbons at the operating conditions may reduce this tendency, but it is unlikely be completely eliminated.

Neutral fitting liners are those in which the initial outside diameter of the liner is the same as the inside diameter of the host pipe. For insertion it is usually necessary to temporarily reduce the diameter if the liner slightly through mechanical means, especially for long segment insertions. Axial loading during pull-in prevents reversion to the initial diameter. When insertion is complete and the axial load is removed the liner reverts to its initial diameter, to come in contact with the host pipe interior wall with no residual stress in the liner material.

Tight-fitting liners are designed to have an initial, preinstallation, outside diameter that is slightly larger than the inside diameter of the host pipe. The liner must be reduced in diameter by mechanical means to be smaller than the inside diameter of the host pipe for insertion. Axial loading during the insertion prevents reversion during the insertion process. When the load is removed the liner will elastically recover until it makes contact with the host pipe wall. At this point the liner is still in radial compression with a reaction force that is pushing out against the host pipe wall.

Liners in gas and multiphase services are more prone to collapse due to annulus pressure buildup than liners used in water injection service, which does not contain gas to permeate into the annulus. Liners in hydrocarbon service undergo changes in stiffness and dimension as they absorb hydrocarbons from the flowing pipeline contents. Designing the liner to be collapse resistant requires knowledge of the modulus of the polymer at equilibrium in the service conditions and at the temperature of service. These characteristics can be determined through laboratory studies and are very useful to liner designers [5]. The following equation can be used to calculate the collapse pressure $P_{crit}$ of a tight fitting liner [6].

$$P_{crit} = E\left(\frac{t}{r}\right)^2\left(2.334 - 0.0385\varepsilon_{swell}\frac{r}{t}\right)$$

where

$P_{crit}$ = Critical collapse pressure (bar)
$E$ = Polymer Young's modulus (MPa)
$T$ = Liner wall thickness (mm)
$r$ = Average radius of the liner
$\varepsilon_{swell}$ = Volumetric swell of the polymer (%)

Neutral and tight fitting liners are installed using special diameter reduction equipment so that during insertion the OD of the liner is less than the ID of the host pipe. The principal methods of diameter reduction are conical die reduction [7] and roller-box reduction [8]. In both cases, the diameter of the liner is reduced while under axial tension by passing through a device that concentrates compressive hoop stress along a narrow zone of the circumference of the liner pipe. As the diameter is reduced, the liner pipe downstream of the diameter reduction device gets longer as would be predicted by Poisson's ratio. Because the deformation is less than the yield strain the liner is able to recover nearly 100% of its initial diameter when the axial load is removed.

The implications of installed stress state (tightness) on liner performance are profound, affecting how the liner should be installed, operated, and the likely failure modes. Tightness can also affect the maximum lined segment length. Depending on the service type (gas, liquid, multiphase, abrasive slurry, etc.) liner tightness affects, for example, venting frequency, permissible pressure cycle range, and pigging procedures. Understanding the dynamics of the liner after installation is critical to successful operation of a lined system.

Tightness affects the resistance of the liner to collapse or radial bucking. The critical buckling pressure is the differential pressure across the liner wall that results in radial instability, leading to buckling and collapse of the liner. Several studies have concluded that for a given liner wall thickness, tighter liners have a higher critical buckling pressure than looser liners [9–12]. These studies also confirm that thicker liners have a higher buckling pressure than thinner liners for the same tightness.

Because thermoplastic materials have a much larger thermal expansion coefficient than metals, the plastic liner will try to lengthen much more than the steel host pipe when there is a temperature increase. This can happen when the liner is installed in winter at low ambient temperature and operated at a much higher temperature. Let's consider the case of a HDPE liner installed at −20 °C and operated at 60 °C. The coefficient of linear thermal expansion of HDPE is $116 \times 10^{-6}$ m/m/°C ($2.13 \times 10^{-4}$ ft/ft/°F), which is about 9X more than the coefficient for steel. Using these temperatures we calculate that the 80 °C temperature change will result in 9.28 mm/m (0.114 in./ft) length change. Over the length of a 500 m (1640 ft) liner segment the total unconstrained length change is 4.64 m (15.2 ft), while the steel will only lengthen about 0.5 m (1.6 ft). Because the ends of the liner are constrained, the resulting stress in the plastic material can result in axial buckling, also known as "accordion buckling," which is one of the principal failure modes for liners. Tight fitting liners are much more resistant to this failure mode because of the large frictional resistance to axial movement when the liner is in radial compressive contact with the host pipe. The temperature increase increases tightness, too. Neutral fitting liners have some frictional resistance to axial movement, and loose fitting liners have much less. During installation, expert contractors will often add some "stretch" to the liner before capturing the ends with fittings at the low installation temperature. This elongation, fully within the elastic range of the polymer, offsets some of the anticipated thermal expansion because as the liner increases in temperature and gets longer, the elastic stress stored during the stretch is relieved, resulting in a near-zero axial loading at operating temperatures while maintaining radial compression of the liner.

## 31.5 LINER MATERIALS

In principle, any thermoplastic material can be used as a liner for oilfield pipelines. Polymer material properties are often reported at standard laboratory temperature, around 23 °C, and without any chemical exposure conditioning prior to property measurement. These properties are significantly affected by temperature, and may be dramatically affected by chemical exposure. All key properties should be reported at or above the design temperature of the pipeline. There are several important criteria that affect the material choice:

- The material must be able to be processed to make the liner pipe of the specified diameter and wall thickness.
- The material must be able to be installed by the installation method offered by the installation contractor.
- The material should be sufficiently well characterized that the resistance to the service environment is well quantified.
- The permeability coefficients for the principal gases in a gas or multiphase system should be known.
- The short- and long-term mechanical properties should be well characterized.

Because of its relatively low cost, widespread availability, ease of production and installation, and long history of service in oil and gas applications, HDPE is the most frequently used thermoplastic for oilfield pipeline liners. The basic engineering grades of HDPE, designated PE-3708, PE-3710, PE-4708, and PE-4710 in the American Society for Testing and Materials (ASTM) classification system [13], or PE 100 in the International Organization for Standardization (ISO) classification system [14] are the preferred grades for liner applications, although medium density materials such as PE-2406 and PE80 may be suitable especially in water applications. HDPE can be affected significantly by hydrocarbons in the service environment. The effects include swelling and loss of stiffness, but these effects can be accounted for when designing a liner if the effects are quantitatively known [15]. The effect is more severe at higher temperatures. Hydrocarbon absorption causes reduction in modulus, which reduces the critical buckling pressure. The attendant swelling increases the liner tightness, partially offsetting the loss of modulus until the point that the internal stress in the swollen, softened HDPE is relieved by buckling of the liner at some random mechanical defect point. Even a minor defect can be the initiation site. Even so, at temperatures up to about 65 °C, HDPE is a good choice for most liners in hydrocarbon transport pipelines. For water transport pipelines the upper temperature limit depends on the amount of hydrocarbons in the water. Modern bimodal HDPE materials, such as PE100 and PE4710, have been used in water lines up to 80 °C successfully with proper design and operation.

In the case that the operating conditions are too severe for HDPE there are alternative materials available. PA11 has been used in hot, sour hydrocarbon, and multiphase pipelines good results in appropriate service conditions [16]. PA11 is has been used as the pressure barrier, essentially a liner, in offshore flexible pipes that comply with API Spec 17J [17] for more than 25 years. Polyamides are condensation polymers that can be degraded by water under the right conditions. Acid conditions accelerate the process. The lifetime of the PA11 material can be predicted with reasonable accuracy using the methodologies of API 17 TR2 [18]. Because the grade of PA11 used for liners is plasticized, it is subject to damage when large amounts of methanol are used in the pipeline [19]. The methanol extracts the plasticizer and accelerates degradation of the polymer. Polyamide 12 is chemically similar to PA11 and can be considered equivalent in the application. Other engineering polymers such as polyvinylidene fluoride (PVDF) and polyetherether ketone (PEEK) have also been studied from use as liners in oilfield pipelines [20]. Laboratory data indicate that these materials are highly promising for field use.

## 31.6 OPERATING A PIPELINE WITH A LINER

Pipelines that are fitted with liners are operated differently compared with unlined systems. The differences are related almost entirely to preventing liner damage such as collapse and axial buckling. When liners are new there are always some residual gas and/or liquid bubbles trapped in the annulus. This is a normal consequence of the installation process. As the lined pipeline is operated, these bubbles will move downstream until they reach a liner termination. This volume must be removed periodically by operating the vents located at each segment termination point. Over time these fluids will be purged from the annulus. Once the annulus is purged of gases and liquids, unless the pipeline is going to be used in gas or multiphase service, the vents need not be operated except occasionally to check liner integrity.

When there is gas in the pipeline during normal operation, some gas will permeate through the liner and eventually accumulate at a downstream termination. This permeated gas must be vented off or there will be a greatly increased risk of liner collapse and loss of protection. Expert installers normally recommend a venting schedule based on their knowledge of the material and pipeline service conditions. Some jurisdictions may require specific vent inspection intervals. A general rule is that venting should be done frequently during the first few months, perhaps starting weekly, then biweekly, then monthly, and as the amount of annulus gas vented declines at subsequent venting operations, the venting frequency can be reduced. Nevertheless, checking the events periodically is a critical inspection technique to verify liner integrity.

Operators must be aware of the likely vent fluid composition. It may not be the same as the pipeline contents because of the differential permeation rates of the different species. Annulus gas analysis may not show the same concentration ratios of gases that are in the pipeline. For example, if $H_2S$ is the fastest gas permeant, it will grow in concentration in the annulus faster than more benign gases, potentially creating an increased hazard during the venting operation. Hazardous conditions could exist if the gases vented are highly toxic such as with $H_2S$. It may be necessary to use scrubbing equipment or a self-contained breathing apparatus when manually venting an annulus with a potential inhalation hazard such as sour gas pipelines. It is also important to recognize that venting has to be done even in bad weather and plans made to account for this. Venting systems have been known to plug because of solid debris in the annulus that moves to the vent, or because of ice plugs. If there is a potential for liquid ice plugs in the vent riser, it may be necessary to heat-trace the riser for cold weather operations.

Lined pipelines should never be allowed to go under vacuum without first operating all the annulus vents to remove accumulated gas while the pipeline is under full line pressure, then taking special care that all the vent valves are completely shut. Since atmospheric pressure is greater than internal pressure when the pipeline goes under vacuum, the result will be liner collapse as air rushes through open vents into the annulus and the flexible liner complies with the pressure differential. This is usually a fatal condition for the liner. In upset conditions where advance venting is not possible, the liner should be inspected by operating each vent valve after return to stable operation. If there has been a liner breach there will be a steady flow of fluid at the vent.

A successful design approach to deal efficiently with permeated annulus gases is the grooved liner concept. In this design, the outer surface of this tight-fitting liner has small axial grooves that facilitate transport of gases to the vents [21]. This may be the preferred solution in cases when the amount of permeated gas is expected to be large or when it is imperative to prevent liner collapse and the ability to effectively vent permeated gases from the annulus of a conventional tight fitting liner is in doubt. To accommodate the grooves the liner can be designed slightly thicker so as not to compromise the collapse resistance. In this design venting is critical. The cumulative volume of the grooves is very large compared with volume normally found in the annulus of a tight fitting liner system. This large volume at high pressure is more than capable of collapsing the liner in the event that the pipeline is depressurized while the annulus is at high pressure.

## 31.7 LINED PIPELINE SYSTEMS—APPLICATION EXAMPLES

These examples are intended to illustrate success and failures in lined pipeline systems and not necessarily to illustrate leading edge, innovative liner use. Because it can take years for concepts to be proven good or bad, it is most useful in a review to highlight what has been shown to work well, and what has been shown to have problems.

### 31.7.1 Liners in Hydrocarbon Flow Lines

Petroleum Development Oman LLC (PDO) has been operating PE liners since 1988, first with water injection pipelines, and eventually with hydrocarbon flow lines. Experiences in PDO were reported by de Mul, et al. [22]. Their first PE lined crude product pipeline was installed in 1992. They reported that they have used PE liners in over 150 water and crude product pipelines in diameters up to 864 mm (34 in.). The liner design was established using properties of the PE liner material at equilibrium conditions of hydrocarbon absorption at the operating temperature. Their liners are designed to resist collapse, but are not so thick that they are inherently collapse resistant against full line pressure in the annulus. This was found to be too conservative as long as annulus pressure was properly managed by operating the vents at regular intervals. They reported that the minimum specified collapse pressure for their designs is 3 barg (43.5 psi). The upper temperature limit for PE lined pipelines was established at 60 °C using the best PE materials available at the time. For water lines with PE liners 70 °C was the upper temperature limit. In practice, PDO's experience with crude flow lines ranges from 10 to 90 barg (145–1350 psi) operating pressure and 30 to 50 °C operating temperature with very good outcomes provided that appropriate attention was paid to good field practices such as venting.

### 31.7.2 Grooved PE Liners

Kinder Morgan chose to use a grooved PE liner in 29 km (18 miles) of their gathering system in the Yates Field in west Texas, USA [23]. That asset already was using more than 13,533 lineal meters (44,400 linear feet) of PE lined steel in diameters from 152 to 406 mm (6–16 in.). For the new, grooved liner design pipelines, the diameters ranged from 102 to 610 mm (4–24 in.). The grooved design was chosen because the produced fluid was from a reservoir that was under tertiary oil recovery using high-pressure $CO_2$ injection. The produced fluid was a hot, highly corrosive, mixture of oil, water, gas, and $CO_2$. The decision to use a grooved liner instead of a typical tight-fitting liner was that the amount of $CO_2$ permeating throughout the PE liner was projected to be large at the production temperatures.

### 31.7.3 Liners in a Reeled, Water Injection Pipeline

A North Sea Operator used a PE lined, reeled, water injection pipeline for 13 years before a sample was removed for testing and analysis during a scheduled shut down to modify the subsea production system. The liner was inserted as a tight fitting liner at an onshore spool base using an established diameter reduction technique. Extensive testing and analyses were conducted on the specimen and reported in 2010 [24]. Lined steel segments of 472 m (1548 ft) length were joined using a welded, flangeless connector to comprise a single length to be deployed offshore. Comparison of the recovered liner polymer with retained, unexposed polymer of the same pipe lot showed that the liner had not stress-relaxed during service and had retained the interference fit of the initial installed condition after reeling onto a reel-lay barge, unreeling at the installation site, and 13 years of service. Pull-out testing of the recovered sample showed that the liner remained securely captured in the segment end fitting joint up to the yield stress of the polymer. Corrosion on the internal surface of the steel pipe that was protected by the liner was superficial. Metallographic analysis showed that the worst case corrosion was 1.6% (280 μm or 0.011 in.) of the nominal wall thickness of 17 mm (0.67 in.). The majority of the steel surface was not corroded and was covered by cracked mill scale.

### 31.7.4 Liners in Sour Gas and Gas Condensate Pipelines

BP America, then operating in the 1990s as BP Amoco in Canada, operated a network of highly corrosive sour gas and gas condensate pipelines in Alberta, Canada [25]. A 101 mm (4 in.) carbon steel pipeline transported gas condensate, sour gas, and salt water from the well to a separator facility about a mile away. Operating temperature was 65 °C and operating pressure was 41.4 barg (600 psig). HDPE liners with a tight design were installed by a reputable and experienced contractor. After 6 months of normal operation and frequent venting, it became impossible to adequately remove pressure from the annulus, indicating a likely failure. When the line was excavated for a repair it was discovered that the line had sustained a major collapse failure and loss of containment. A new HDPE liner was installed with shorter lined segments that should make it easier to control annulus pressure and move permeated gas to the vents, and a more neutral fit that would be more forgiving as the liner material became swollen by hydrocarbon absorption. This liner lasted approximately 18 months before failure. A third HDPE liner was installed using the same design. An economic value calculation indicated that the liner could be replaced every 18 months and be less costly and more reliable than an inhibition program. When the third liner was scheduled for replacement engineers concluded that a liner made from PA11 offered the possibility for a much longer lifetime in service. PA11 had been used extensively in offshore flexible risers and flow lines as the fluid containment barrier for more than a decade. Through trials in the laboratory and the field it was concluded that PA11 could be installed using the same techniques as for HDPE, but with minor modifications. The operator reported that annulus vent pressures rose much more slowly than for any of the HDPE liners. This is because PA11 has a much lower permeation rate for hydrocarbons than HDPE, and is not swollen by liquid hydrocarbons. Swelling is responsible for a large increase in permeation rates in HDPE compared with unexposed material. Samples of liner were recovered during normal maintenance by removal of a test spool located in line with the production fluids. Analysis revealed excellent retention of properties after more than 3 years being exposed to the production environment [26]. The operator also deployed PA11 liners in several other high-temperature sour gas pipelines with good success and meaningful learnings.

### 31.7.5 PA11 Liners in Sour Gas Pipelines

Shell Canada Limited installed tens of kilometers of PA11 liner in 152 and 203 mm (6 and 8 in.) pipelines in a highly sour gas field in southern Alberta, Canada in 2000 [27]. Due to the higher cost of PA11 compared with PE, a thinner liner wall was chosen. To accommodate long segment pull lengths of up to 800 m (2625 ft) a slightly loose design was chosen. This design choice was based on laboratory evidence that the PA11 material, being much more fatigue resistant and ductile than HDPE, would be collapse tolerant—the idea that multiple cycles of partial collapse and re-rounding under pressure cause by line pressure fluctuations while gas was in the annulus would not damage the liner. Operating pressure was 7000 kPa (1015 psi). For about a year after installation the liners performed as expected, but eventually it became difficult to control annulus pressures. In 2002, it was discovered that liner breaches were happening. The causes were multiple. The most general learnings from this experience are that loose designs are not a good idea for high-pressure gas. They allow too much axial motion due to flowing gas pressure and can result in axial collapse and loss of containment. The worst case scenario for polymer properties changes with injected fluids absorption must be considered; softening and swelling with absorption or shrinkage and stiffening with extraction of plasticizing species. Most of the liners installed in the original construction were removed and replaced with a collapse resistant grooved HDPE liner.

Shell encountered problems after replacing the PA11 liners [28,29]. The steel pipeline failed due to internal corrosion after just under 4 years of service. There was no evidence of liner failure prior to the steel pipeline failure. A thorough investigation concluded that the failure was caused by internal corrosion that had reduced the steel thickness to about 10% of the original thickness in the area of the failure. Although in pipeline segments located away from the failure site had fully intact iron sulfide scales, it appears that the protective iron sulfide scale did not cover the steel surface completely in the area of the failure, leaving patches unprotected and susceptible to corrosion. Between the PE liner and the steel pipe, they found a paste-like substance consisting of an iron sulfide corrosion product, small amounts of formic acid, and an acidic liquid, which they identified as mostly methanol. Methanol was used as a hydrate inhibitor in normal pipeline operations and under the conditions of the subject pipeline was vaporized downstream of the injection site. This increased the permeation rate of methanol through the HDPE liner. Since, HDPE is not a perfect barrier to methanol some methanol permeated through the liner and came to reside in the annulus. Formic acid from downhole treatment chemical mixtures also can permeate the PE liner. They concluded that methanol and water, by themselves, were corrosive in the conditions of the annulus and that the presence of formic acid accelerated the corrosion rate in the absence of a fully intact protective scale. Shell has changed the operating practice to address this issue and it has proven successful.

## REFERENCES


1. NACE, NACE Recommended Practice RP0304, Design, Installation, and Operation of Thermoplastic Liners for Oilfield Pipelines. NACE International, Houston, TX.
2. CSA, CSA Standard Z-662, Oil and Gas Pipeline Systems. Canadian Standards Association, Mississauga, Ontario, Canada.
3. Plastics Pipe Institute (2012) TR-33, Generic butt fusion joining procedure for field joining of polyethylene pipe. Irving, Texas, 2012.
4. Plastics Pipe Institute (2008) TR-45, Butt fusion joining procedure for field joining of polyamide-11 (PA-11) pipe. Irving, Texas, 2008.
5. Baron, J., Szklarz, K., and MacLeod, L. (2000) Nonmetallic liners for gas/condensate pipelines. Corrosion 2000 Proceedings, NACE International Meeting, March 2000, Orlando, Paper No. 00789.
6. Brogden, S., Lu, L., Dowe, A., Messina, N., and Robinson, J. (2012) The use of engineering polymers for internal corrosion protection of hydrocarbon pipelines. Proceedings of MERL Oilfield Engineering with Polymers 2012, October 2012, London.
7. U.S. Patents 5,048,174 and 5,839,475. U.S. Patent and Trademark Office (USTPO), USPTO Contact Center (UCC), Crystal Plaza 3, Room 2C02, P.O. Box 1450, Alexandria, VA 22313-1450.
8. Canadian Patent 1,241,262. Canadian Intellectual Property Office (CIPO), Client Services Centre, Place du Portage I, 50 Victoria Street, Room C-229, Gatineau, QC K1A 0C9 Canada.
9. Wang S.S., Mason, J., and Yu, T.P. (2000) Effect of tightness on thermoplastic liner collapse resistance. Corrosion 2000 Proceedings, NACE International Meeting, March 2000, Orlando, Paper No. 00789.
10. Frost, S.R., Korsunsky, A.M., Wu, Y.S., and Gibson, A.G. (2000) 3-D modeling of liner collapse, Corrosion 2000 Proceedings, NACE International Meeting, March 2000, Orlando, Paper No. 00789.
11. Gerard, F., Dang, P., and Mason, J. (2004) Design analysis of polyamide-11 liners: operating higher than the critical buckling

pressure. Corrosion 2004 Proceedings, NACE International Meeting, March 2004, New Orleans, Paper No. 04712.

12. Lo, K.H. and Zhang, J.Q. (1994) Collapse resistance modeling of encased pipes, In: Eckstein, D. (editor), *Buried Plastic Pipe Technology: ASTM STP 1222*, American Society for Testing and Materials, Philadelphia, Vol. 2, pp. 97–110.
13. ASTM, ASTM D3350, *Standard Specification for Polyethylene Plastic Pipe and Fittings Materials*, ASTM International, West Conshohocken, PA.
14. ISO, 12162:1995 *Thermoplastics Materials for Pipes and Fittings for Pressure Applications—Classification and Designation—Overall Service (design) Coefficient*, International Organization for Standardization (ISO), Geneva, Switzerland.
15. Baron, J., Szklarz, K., and MacLeod, L. (2000) Nonmetallic liners for gas/condensate pipelines. Corrosion 2000 Proceedings, NACE International Meeting, March 2000, Orlando, Paper No. 00789.
16. Mason, J. (1999) Pipeline liner material wears well in tests of field specimens. *Oil and Gas Journal* 97(42), 76–82.
17. API (2008) API Specification 17J, Specification for Unbonded Flexible Pipe. American Petroleum Institute, Washington, DC. July 2008.
18. API (2003) The ageing of PA11 in flexible pipes. API Technical Report 17 TR2, American Petroleum Institute, Washington, DC June 2003.
19. Szklarz, K. and Baron, J. (2004) Learnings from thermoplastic liner failures in sour gas pipeline service and replacement liner design and installation. Proceedings of NACE International Meeting, March 2004, New Orleans, LA, Paper No. 04702.
20. Brogden, S., Lu, L., Dowe, A., Messina, N., and Robinson, J. (2012) The use of engineering polymers for internal corrosion protection of hydrocarbon pipelines, Proceedings of MERL Oilfield Engineering with Polymers 2012, October 2012, London.
21. Taylor, J., Groves, S., and Melve, B. (2000) Effective annular venting of thermoplastic liners for added value and benefit, Corrosion 2000 Proceedings, NACE International Meeting, March 2000, Orlando, Paper No. 00784.
22. de Mul, L.M., Gerretsen, J.H., and van de Haterd, A.W. (2000) Experiences with polyethylene lined pipeline systems in Oman. Corrosion 2000 Proceedings, NACE International Meeting, March 2000, Orlando, FL, Paper No. 00788.
23. Schmitz, J. (2010) High density polyethylene liners for high temperature and gaseous applications. Proceedings of 13th Middle East Corrosion Conference and Exhibition, February, 2010, Manama, Bahrain, Paper No. 10073.
24. Brogden, S. and Fryer, M. (2010) Effectiveness of Polymer Lined Pipelines for Subsea Water Injection Service, September 2010, London, MERL Oilfield Engineering with Polymers 2010.
25. Lebsack, D. (2000) Operator tests nylon-based liner on highly corrosion gas pipe line. *Pipeline and Gas Journal*, 83, 77–80.
26. Mason, J. (1999) Pipeline liner materials wears well in test of field specimens. *Oil and Gas Journal*, 97(42), 76–82.
27. Szklarz, K.E. and Baron, J.J. (2004) Learnings from thermoplastic liner failures in sour gas pipeline service and replacement liner design and installation. Corrosion 2004 Proceedings, NACE International Meeting, March 2004, New Orleans, LA.
28. Simon, L., MacDonald, R., and Goerz, K. (2010) Corrosion failure in a lined sour gas pipeline—part 1: case history of incident. Proceedings on the NACE Northern Western Area Conference, February, 2010, Calgary, Alberta, Canada.
29. Macdonald, R., Simon, L., Goerz, K., and Girgis, M. (2010) Corrosion failure in a lined sour gas pipeline—part 2: role of methanol in corrosion behind liner. Proceedings on the NACE Northern Western Area Conference, February, 2010, Calgary, Alberta, Canada.

# 32

# CATHODIC PROTECTION

Sarah Leeds and John Leeds
*DC Voltage Gradient Technology & Supply Ltd., Wigan, UK*

## 32.1 INTRODUCTION

Cathodic protection (CP) is a very important technique for the mitigation of corrosion on buried or immersed structures such as pipelines. It involves an electrical current being applied to make the pipeline act as the cathode in a simulated electrochemical cell, hence, prevent the corrosion reaction from occurring at the pipe surface. The pipeline being protected must be exposed to an electrolyte such as buried in moist soil or seawater. CP is more advantageously used to protect coated structures because CP will only be required to protect uncoated portions of the pipeline and intact coatings will act as insulators reducing the amount of current required.

If CP is not properly applied it can have detrimental effects on the coating and ultimately the pipeline being protected.

There are two methods for applying CP, namely, sacrificial anode cathodic protection system and impressed cathodic protection system. Sacrificial anode CP system is where an active metal is used as an anode (e.g., aluminum or zinc or magnesium, etc.) and the potential between the active metal and the pipeline being protected, forces the protective current to flow to areas of bare steel exposed on the buried pipeline. In an impressed CP system, an external power source forces the protective current to flow from an immersed anode to the pipeline to be protected through the conductive soil.

It is important to make sure that the CP system is designed, installed, monitored, and maintained properly to ensure that the CP system will last the life span of the design.

## 32.2 HISTORICAL FOUNDATION OF CATHODIC PROTECTION

Knowledge of CP dates back to the Imperial Roman times of 27 B.C.–385 A.D [1]. However, the science of CP was first established in 1824, when Sir Humphrey Davy [1] made a presentation to the Royal Society of London based on his experimental findings of protecting copper (which was used on ships bottoms to stop marine fouling) against corrosion from seawater through the use of cast iron anodes [2].

In 1834, Faraday discovered the quantitative connection between electrodeposition, corrosion, and electrical current, which lead to the scientific foundation of electrolysis and CP principles [3,4]. Subsequently, without any knowledge of Davy's work, Frischen, who was the Inspector of Telegraphs in Germany, reported to a meeting in Hanover in 1865, the results of experiments carried out over a long period looking into the protection of wrought iron, the most widely used metal at that time. Frischen connected zinc to iron for protection against seawater and was able to establish that effective protection of iron was due to the influence of galvanic electricity [1,5].

In 1902, Cohen was the first person who successfully applied CP using an impressed current system. In 1906, Geppert, the manager of the public works in Karlsruhe, Germany, constructed the first CP system for pipelines using a direct current generator with a 10 V/12 A capacity to protect 300 m of gas and water pipelines close to a tramline [1].

*Oil and Gas Pipelines: Integrity and Safety Handbook,* First Edition. Edited by R. Winston Revie.

In 1906, Deutscher Verein des Gas und Wasserfachs (DVG) instigated a basic investigation into the scientific fundamentals of CP, which was carried out by Haber and Goldschmidt [6]. They recognized that CP and stray current corrosion were based on electrochemical phenomenon. Haber utilized a non-polarized zinc sulfate electrode for potential measurement. Two years later, McCollum used the first copper sulfate electrode, which has since become the standard reference electrode for potential measurements of land based systems [1].

Between 1910 and 1918, Bauer and Vogel in Berlin determined the current density required for CP [7]. In 1920, it was found that the Rhineland cable near Hanover, Germany, was damaged by the formation of corrosion couples in geologically mixed soil strata. This was the first time in Germany that zinc plates were built into the cable shafts in order to protect the sheathing metal from corrosion [1].

In the 1920s, introduction of safe, reliable welds allowed the laying of continuously welded pipelines and other structures making CP practically possible. However, this was not entirely achieved as engineers who built the pipelines had limited understanding of electrochemical protection, regarding it as a "black art" and even the electrical engineers underestimated the costs of the processes and the dangers that an applied CP current can have on other local pipelines. This resulted instead to improvements in the pipeline coating resistance to aggressive soils to reduce the effects of stray current corrosion. The lack of understanding and the limited interest in CP can be further acknowledged by the limited amount of information on CP in Evans' first book on corrosion, which was published in 1924 [1].

From 1928 onwards, development of CP as a practical application for pipelines was carried out in the United States. Kuhn, who is the undisputed "father of pipeline CP," installed the first CP rectifier on a long distance gas pipeline in New Orleans, United States in 1928. Kuhn found that the protection of plain cast pipes, which had poorly conducting joints, did not extend beyond individual pipe joints, so he added more rectifiers. Through a series of experiments, Kuhn found that the polarized potential of −850 mV against a saturated copper/copper sulfate reference electrode (CSE) provided sufficient protection against corrosion. Kuhn reported what are now regarded as simplified results of his experiments to the Washington Conference for Corrosion Protection of the National Bureau of Standards in 1928. Kuhn's findings formed the basis of modern CP technology. He was the first person who proposed that a polarized (OFF) protection potential should be −850 mV (CSE) as the value to be used for minimizing corrosion, which is a criterion still used today [8,9]. The polarized potential is referred to as the IR free potential or the instant OFF potential.

The theoretical basis of CP was independently established through the work of Hoar [10] in the United Kingdom and Mears and Brown [11,12] in the USA. The Mears and Brown work on CP is more widely acknowledged and their definition for obtaining complete protection is used most often in the CP industry; "in order that protection may be complete, the potential of the cathode area must be polarized to a potential equal to or more anodic than the open circuit potential of anodes." Jones [12] in 1971 further substantiated the work of Mears and Brown by showing that the electrode kinetic theory could be applied to improve the understanding of CP. Jones claimed that the kinetics of the anodic dissolution reaction could be used to determine the cathodic polarization required for adequate protection.

CP of pipelines in Europe began in Belgium and was widely applied later than in the United States. From 1932 de Brouwer applied CP to the supply pipelines of Distrigaz in Brussels, Belgium. In 1939, there were more than 500 cathodically protected pipelines in Russia, which were mostly utilizing galvanic anodes. CP utilizing galvanic anodes was carried out in Great Britain and Germany after 1940 and 1950, respectively [1].

From humble beginnings, CP has grown to have many uses in marine; underground structures; water storage tanks; oil, gas, and water pipelines; oil platform support's; and many other facilities exposed to a corrosive environment. CP had become second to protective coatings, as the most widely used method for limiting the corrosion deterioration of metallic structures in contact with any forms of corrosive environments.

## 32.3 FUNDAMENTALS OF CATHODIC PROTECTION

### 32.3.1 Mechanism of Cathodic Protection

It is generally assumed that the corrosion rate of a metal is controlled by the diffusion of oxygen dissolved in the bulk electrolyte solution to the metal surface. Figure 32.1 details a typical Evans diagram of steel in neutral aerated water. It can be seen that the corrosion rate of steel or the maximum consumption of electrons is limited by the rate of oxygen reduction. If no current is applied then electron supply will be due to the dissolution of iron, which will then be equal to $i_{corr}$. However, under application of CP, where the power source supplies the electrons this will ultimately lead to a maximum applied current, $i_{corr}$, which will then be equal to the limiting current density, $i_{Lim}$. Applying more current ultimately leads to the reduction of water, which causes the generation of hydrogen gas [13].

CP is a technique used to mitigate corrosion of a metallic structure. It involves applying a potential in such a way that the current flows from the anode through the surrounding electrolyte (moistened soil or seawater) to the structure surface. In concept, it is thought that if a sufficiently negative potential is applied, then corrosion of the metal structure should be mitigated.

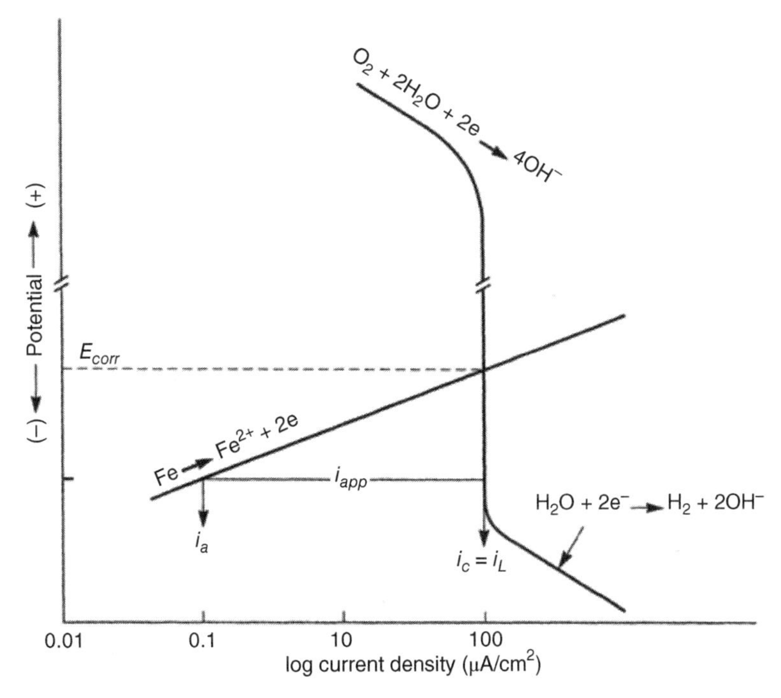


**FIGURE 32.1** CP by impressed current density, $i_{app}$ for steel in neutral aerated water [13]. (JONES, DENNY A., PRINCIPLES AND PREVENTION OF CORROSION, 2nd Edition, © 1996, p. 444, Reprinted with permission of Pearson Education, Inc., Upper Saddle River, NJ.)

Application of CP causes two electrochemical reactions to occur; reduction of oxygen and reduction of water. The first reaction occurring as potentials are made more negative is the reduction of oxygen dissolved in the electrolyte (moist soil or seawater) to form hydroxide ions (cathodic reaction 1).

Cathodic reaction (1) $^{1}/_{2}O_2 + H_2O + 2e^- \rightarrow 2OH^-$

Reduction of oxygen is limited by the amount of available oxygen, which is typically 8 ppm dissolved in water (moist soil or seawater) at ambient temperature [14]. The limiting current density of this reaction is diffusion controlled. It has been found that when reduction of oxygen is the dominant reaction and occurs at applied potentials less negative than −1.17 V (CSE), then the typical pH of the metal surface is never more alkaline than 10.57 [15]. This is thought to be because there is such a small concentration of hydroxyl ions generated at the metal surface that the pH is limited because of the transfer of hydroxyl ions away from the metal surface.

The second reaction that occurs is reduction of water (cathodic reaction 2). This reaction normally occurs as the applied potential is made more negative than −1.17 V (CSE). This reaction is not controlled by any limiting current density but is controlled by diffusion and migration of ions to and from the metal surface.

Cathodic reaction (2) $2H_2O + 2e^- \rightarrow H_2 + 2OH^-$

It has been found that as the applied potential was made more negative, the pH at the metal surface becomes progressively more alkaline. This can be explained by more hydroxide ions being retained on the metal surface when the surface films forming are becoming thicker, generating a more alkaline surface pH.

The mixed-potential theory, as described in Figure 32.2 introduced in 1957, is the theory most thought today to explain the mechanism of cathodic protection. The mixed-potential theory is thought to be accomplished by polarizing the metal from its corrosion potential ($E_{corr}$) to its redox/reversible potential ($E_{o,a}$). The amount of cathodic polarization depends on the amount of applied current, which is the difference between the anodic current at a particular potential. It can be seen that the anodic and cathodic species in a particular environment are combined. When cathodically polarizing the metal surface at some set potential, the point of $\varepsilon_c$ on Figure 32.2 represents the amount of potential needed to be applied in order for there to be a potential change causing $E_{corr}$ to be shifted cathodically to its reversible or equilibrium potential ($E^o_{Fe/Fe^{2+}}$) of its local anodes. It is assumed that this is because at this potential, the anodic dissolution rate will equal the deposition rate, such that $i_a = i_{ac} = i_{oa}$. As in Figure 32.2, it can be seen that to reduce the corrosion rate from $i_{corr}$ to $i_a$, the iron must be polarized for $\varepsilon_c$ (potential change from $E_{corr}$), which should be in the order of 0.1 V (CSE). This is the basis for the 100 mV shift, which is one of the three National Association of Corrosion

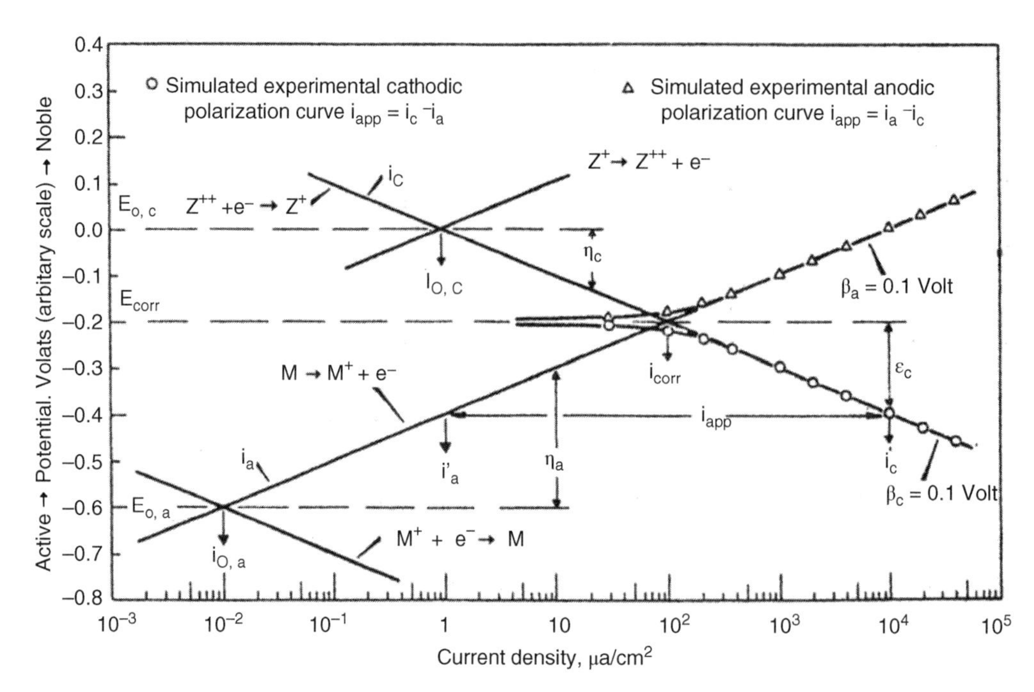


**FIGURE 32.2** Anodic and cathodic tafel curves for corrosion of a metal, M. Explanation for the mechanism of CP as based on the mixed-potential theory. (Adapted from D. A. Jones, Corrosion 28(11), p. 42, 1972 © NACE International, 1972.)

Engineers (NACE) criterions used as a guide for the reduction in corrosion and outlined in NACE SP0169-2007 [16].

### 32.3.2 E-pH Pourbaix Diagram

A Pourbaix diagram can be used to determine the propensity of a metal such as iron to corrode in solutions of varying pH and potential. The Pourbaix diagram can also be used to determine the region of immunity, which is the domain where iron is thermodynamically stable. The Pourbaix diagram can also determine the region of passivation, which is the domain where iron is covered by films of either $Fe_2O_3$ or $Fe_3O_4$ dependent on the electrolyte solution. The addition of anions to the electrolyte solution can have an effect on the thermodynamics of the metal reactions. In the case of addition of chlorides to the electrolyte solution, it can break down the films formed, which could lead to localized attack of the metal.

When constructing the Pourbaix diagram for the Fe–$H_2O$ system, such as that constructed in Figure 32.3, it is important to define the conditions and it is assumed that temperature is 298 K (25 °C) and pressure of 1 atm. The solid species were considered to be Fe, $Fe_3O_4$, and $Fe_2$O3 under these conditions. The Pourbaix diagram was plotted using a pH range between −2 and +16 and an electrode potential range between −1.8 and +2.2 $V_{SHE}$. If the temperature and pressure are changed, a new Pourbaix diagram would have to be redrawn to fit these new conditions.

Another factor to take into consideration when constructing a Pourbaix diagram for the Fe–$H_2O$ system is to define what ionic activity will represent "significant corrosion". To determine what is considered to be corrosion relies on the amount of metal dissolution that is permissible under the defined conditions. It has been suggested that when the Pourbaix diagram such as in Figure 32.3 was constructed then a solubility of $10^{-6}$ g atoms of soluble ion (iron) per liter was chosen to represent "no corrosion for all practical purposes" for the Fe–$H_2O$ system [23].

The *E*–pH Pourbaix diagram can be used to explain how CP works. Pourbaix suggested that CP can work by either lowering the metal (iron) potential into the region of immunity or by placing it in the region of passivation (Figure 32.3). Pourbaix suggested that in order to cathodically polarize iron into the region of passivity, it should be possible by anodically protecting or by passivating by alkalization. Pourbaix believed that in order to polarize iron into the region of passivity then the protection potential should be below the breakdown or pitting potential, so as to avoid localized corrosion or to avoid the growth or exiting pits. Figure 32.3 suggests that passivation of iron is only possible if the pH of the electrolyte is between 10 and 13 [17].

A Pourbaix diagram can be used to predict how iron in the form of steel works to suppress the corrosion reaction when CP is applied. When CP is applied to a metal such as a pipeline it is energized and has the effect of making the potential more negative. This causes the steel (iron)

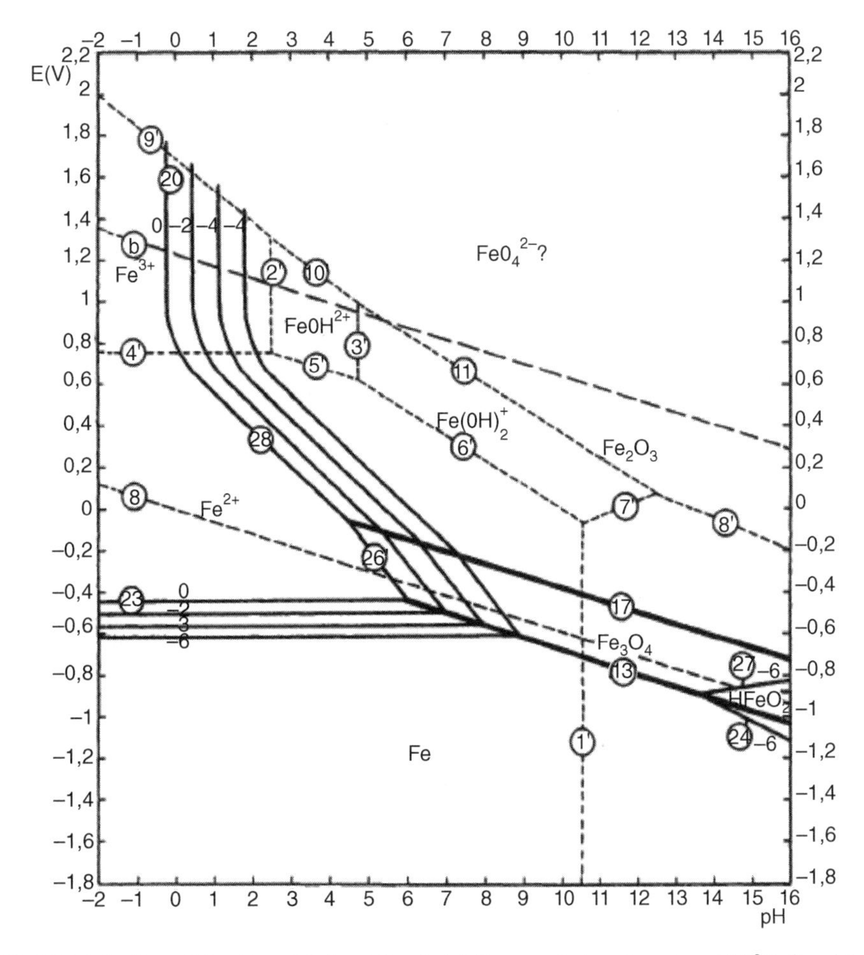


**FIGURE 32.3** Potential–pH equilibrium (Pourbaix) diagram for iron in water at 25 °C. Considering Fe, $Fe_3O_4$, and $Fe_2O_3$ as solid substances [17].

potential to move from its native potential to a more negative potential. This causes the steel (iron) potential to move so that it is in the region of immunity. In practice this is not as simple as CP generates alkalinity due to the cathodic reaction with the production of hydroxyl ions. The effect of the cathodic-generated alkalinity is to move the potential not only in a negative direction but also towards a higher pH found in the region of passivity, which occurs before reaching immunity.

Pourbaix recommended the following conditions to ensure CP by placing iron into the region of immunity as detailed in Table 32.1 [17].

It should be noted that the values detailed in Table 32.1 assume that the activity of iron (Fe) is $10^{-6}$ g.ion/l and would represent no corrosion. Theoretically, the conditions in Table 32.1 are possible; however, this may not be so simple practically. The application of CP could mean that corrosion of iron does not occur due to immunity but rather due to "cathodic passivity," When determining the domain regions as set out in the Pourbaix diagram all the conditions need to be adjusted for determining whether a metal will corrode or not.

**TABLE 32.1 CP Conditions for Immunity**

| | Potential | |
|---|---|---|
| pH | Volt versus. Standard Hydrogen Electrode (S.H.E.) | Volt versus. Copper-Saturated $CuSO_4$ Electrode (C.S.E.) |
| Below 10 | E0 = −0.62 | E0 = −0.94 |
| Between 10 and 13 | E0 = −0.08 to 0.059 pH | E0 = −0.40 to 0.059 pH |
| Above 13 | E0 = +0.31 to 0.088 pH | E0 = −0.01 to 0.088 pH |

*Source:* Data on pH and voltage versus. S.H.E. from Ref. [17].

Data used for drawing *E*–pH Pourbaix diagrams for different metals detailed in the "Atlas of Electrochemical Equilibria in Aqueous Solutions" written by Marcel Pourbaix has been improved over the years through new research. Access to this new research can be found in the CEBELCOR website: www.cebelcor.org.

## 32.4 HOW CATHODIC PROTECTION IS APPLIED

CP can be achieved by using buried or immersed sacrificial anodes connected electrically to a structure/pipeline, which causes the corrosion potential to be lowered. The galvanic series or electromotive force series (emf) (Table 32.2) is used to determine the most suitable metal to be selected as an anode. Examples are magnesium, aluminium, or zinc as good anode metals. The other way to achieve CP is by an impressed current system where a source of external current is used in conjunction with an anode ground bed to suppress the corrosion.

The choice between an impressed current and sacrificial anode system for a particular situation generally relates to one of economics, with the sacrificial anodes being more economical for smaller systems, while impressed current systems are more suited to larger installations. It also depends on environmental conditions of where the CP system will be utilized as sacrificial anodes are more suitable to a lower resistivity environment.

### 32.4.1 Sacrificial Anode Cathodic Protection System

CP can be achieved by burying or immersing an electrode called a sacrificial or galvanic anode that is more electronegative than the metal, such as a steel pipeline to be protected, and then connecting it into the circuit.

The current available from a sacrificial anode can be limited by the relatively low driving potential available (difference in the potential between the anode and structure such as pipeline, when referred to a common reference electrode) and the size that can be conveniently handled. Because of this, it is necessary to ensure a uniform distribution of the anodes over the structure

**TABLE 32.2 Electromotive Force Series of Metals**

| Metal | Volts[a] |
|---|---|
| Magnesium | −2.38 |
| Aluminum | −1.66 |
| Zinc | −0.76 |
| Iron | −0.44 |
| Tin | −0.14 |
| Lead | −0.13 |
| Hydrogen | 0.00 |
| Copper | +0.34 to +0.52 |
| Silver | +0.80 |
| Platinum | +1.20 |
| Gold | +1.42 to +1.68 |

[a]Half-cell potential in a one-molar solution of its own salts, measured with respect to hydrogen reference electrode at 25 °C

*Source:* Data from Ref. [13]. L. L. Shrier, R. A. Jarman, and G. T. Burstein, Corrosion, 3rd edition, Butterworth-Heinemann, Volume 1, 1994. Also NACE Basic Corrosion Course.

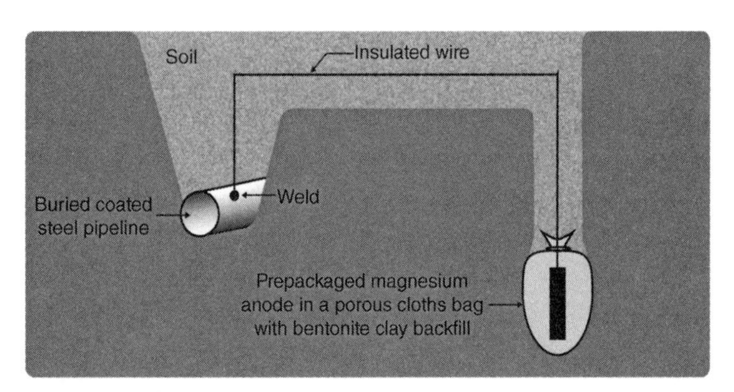


**FIGURE 32.4** Sacrificial anode CP system using magnesium as the anode. (Adapted from Ref. [23].)

unless the structure is well coated when generally only a few anodes are required [18].

Anodes may be buried or immersed at a distance from the surface of the metal to be protected and connected to it by means of an insulated conductor or alternatively may be directly attached to the surface of the structure such as pipeline. Direct attachment to the pipeline can be achieved satisfactorily if the electrolyte resistivity is low. However, when the electrolyte resistivity is high, adequate spacing of 5 or 10 m should be provided between the anode and pipeline to ensure that the available current is uniformly distributed. Figure 32.4 details a sacrificial anode CP system.

Using sacrificial anodes as a means of protection involves the formation of a galvanic couple by connecting the pipeline to be protected to a more active material (i.e., the sacrificial anode) so the structure, sacrificial anode, and corrodant forms an electrochemical cell.

With the sacrificial anode method of protection, the maximum voltage that can be applied is equal to the emf of the electrochemical cell formed, whereas, in the case of an impressed current arrangement, any applied voltage in theory may be utilized.

The fixed driving voltage can have a number of implications:

- The limited voltage means limited current, which means only a limited area of pipeline can be protected, that is, limited spread of current. On small structures, sacrificial anodes are economical. On larger structures, sacrificial anodes may be uneconomic. Some notable examples to this exception are offshore oil rigs and very well coated pipelines.
- The limited current available from sacrificial anodes reduces possible interference problems.
- Galvanic anodes located in high-resistivity soils will provide only limited current due to a high circuit resistance.

### 32.4.2 Sacrificial Anode Design

The current output of a sacrificial anode depends upon the resistance of the anode to the electrolyte, and the driving potential between the anode and the structure to be protected.

From this, the output of the anode is determined by Ohms law:

$$I = \frac{V}{R}$$

where,

$I$ = Anode output,
$V$ = Driving potential of circuit, and
$R$ = Resistance of the anode to the electrolyte.

The resistance $R$ of the anode to the electrolyte is given by the formula:

$$R = P\frac{0.0627}{L}X\left(\ln\frac{4L}{r} - 1\right)$$

where,

$P$ = Electrolyte resistivity
$L$ = Anode length
$R$ = Anode radius

This formula is based on a simplification of Dwight's formula for calculation of the resistances to ground [19].

From the simplified Dwight formula, it is evident that the resistance to electrolyte of an anode depends upon the electrolyte of an anode, which depends upon the electrolyte resistivity and the size and shape of the anode. For a given volume of anode material used, for example, in the form of a cylindrical rod, the resistance decreases if the length is increased. Thus, for electrolytes of high resistivity (e.g., most soils) anodes are generally in the form of rods or relatively slender castings, or in very high-resistivity surroundings, as extruded strip or ribbon. In seawater, which has a low resistivity, anodes generally are in the form of heavy castings or in extreme areas approximately spherical castings to minimize size. Anodes buried in soil are generally surrounded with a chemical backfill, which has a lower resistivity than the soil to reduce the circuit resistance and also to provide a suitable environment to ensure uniform corrosion of the anode material.

Sacrificial anodes generally contain steel inserts, which maintain electrical continuity and mechanical strength of the anode material toward the end of its effective life. Where electrical cables are connected to the anodes to provide attachment to the structure, the cables are attached to the steel insert, with the cable insert connection being protected by a tar epoxy coating [1,3,4,18].

**TABLE 32.3 Typical Sacrificial Anode Characteristics**

| Alloy | Environment | Potential (versus $Cu/CuSO_4$) (V) | Practical Consumption Rate (kg/A—Year) |
|---|---|---|---|
| Magnesium | Soil | −1.50 | 8.5 |
| Zinc | Seawater | −1.10 | 12 |
| Zinc | Soil | −1.10 | 12 |
| Aluminum alloys Al-Zn-Hg | Seawater | −1.10 | 3.5 (approx) |
| Aluminum alloys Al–Zn–In | Seawater | −1.15 | 3.7 (approx) |

*Source:* Data from Ref. [18] and European/CEM Meeting on CP.

### 32.4.3 Anode Material

The alloys that are used for sacrificial anodes are universally based on magnesium, zinc, or aluminum. The general performance of each of the galvanic anode types vary with the operating conditions such as electrolyte resistivity, temperature, and operating current density. Typical operating results for sacrificial anodes can be found in Table 32.3.

Anode backfill is widely used to surround sacrificial anodes that are buried in earth. The backfill provides a uniform environment to promote anode consumption and maximum efficiency. It also isolates the anode material from direct contact with the earth, which prevents negative effects from soil minerals that might build up high-resistance films on the anode surface. Examples of backfill used for magnesium sacrificial anodes are 75% hydrated gypsum, 20% bentonite clay, and 5% sodium sulfate. A typical mixture utilized with zinc anodes consists of 50% plaster of Paris and 50% bentonite clay.

### 32.4.4 Impressed Current System

Impressed current systems are like sacrificial anode systems in the sense that electrical currents are used to control corrosion flow from one or more anodes to the structure to be protected. An impressed current system, however, requires an external direct current source, such as a rectifier typically called transformer/rectifier (T/R) to impress current through an anode, which provides a very low corrosion rate. This controls the magnitude and direction of the current flow used for CP.

The most important characteristic of materials utilized for impressed current anode groundbeds is a low consumption rate, which in most instances is related to current density at the surface of the anode. Most often the anodes utilized are made from low consumption alloys such as graphite, high-silicon cast iron, platinum, and ceramics (e.g., Si–Fe, Pb–Sb, Sb–Ag, or Pt–Ti). Graphite and high-silicon cast iron are commonly used for buried installations. High-silicon cast

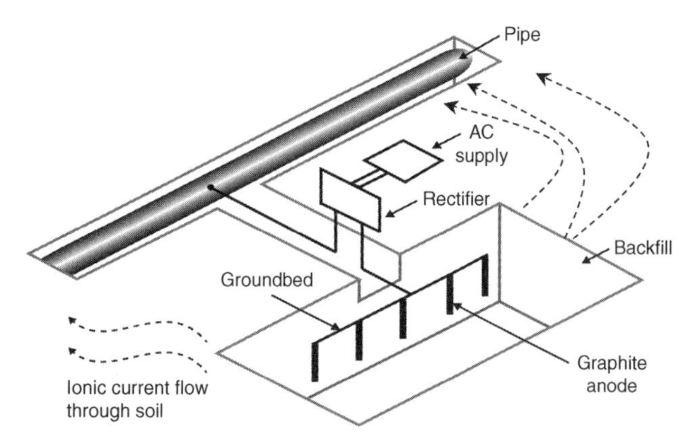


**FIGURE 32.5** Impressed current system for a buried pipeline.

iron, platinum, and ceramics are commonly used in fresh water and seawater applications. Chlorine or oxygen is normally evolved at the inert impressed current anodes.

Figure 32.5 details a general impressed current CP system. The figure shows how a group of anodes are installed remotely (typically 100 feet or more) from the pipeline and usually spaced on 10–30 ft centers and make up the groundbed that comes with back fill. The rectifier, typically called T/R, supplies the power source but power sources can also be supplied by solar cells, engine generator sets, wind powered generators, and turbine generators.

Groundbeds are not really providing the driving power for the CP system, but instead an external source of direct electric current is connected between the structure to be protected and the groundbed anodes. The positive terminal of the power source is required to be connected to the groundbed, which will be forced to discharge as much current as possible. If a mistake is made and the positive terminal is connected to the structure then the structure will become the anode and will actively corrode.

There are several main factors that should be taken into consideration when selecting a site to be utilized as a groundbed site:

Soil resistivity: One of the most important factors when determining a location to install a groundbed in. The number of anodes required for a groundbed, the DC voltage output of the transformer/rectifier and the AC power costs will be directly influenced by soil resistivity. Theoretically, it is assumed that the lower the soil resistivity, the fewer the number of anodes will be required, and hence, the lower DC voltage output will be required. Thus, ideally a location that has the lowest soil resistivity would be considered as the best groundbed site.

Soil moisture: Soil moisture should be taken into consideration at the proposed anode depth. Generally, the resistivity of soil decreases as the moisture of soil increases until the soil becomes saturated. Anodes should ideally be installed below the water table or in areas where there is high moisture (swamps or ditches). The season when the soil moisture was analyzed should be taken into consideration as it can directly affect the moisture content and water table effect could influence the performance of the groundbed. Electroosmosis also affects the moisture content of the soil surrounding the anodes. This effect is seen under situations when the DC current is discharged to the soil from the anode causing the current to push away the water from surface of the anode causing an increase in the groundbed resistance. In locations where the soil consists of high clay content, any removal of moisture from the clay could reduce the conductivity of the soil causing the groundbed to become ineffective.

Interference with foreign structures: A thorough analysis of the surrounding environment of a proposed groundbed site should be taken into consideration to make sure that there are no other metallic structures in the local vicinity. If there are any other metallic structures then structures such as pipelines, underground metallic sheathed electrical cables, or well casings could be subjected to stray current interference and the groundbeds should be located well away from these. If in the unlikelihood that groundbeds should be located near to other metallic structures then thorough studies should be carried out to determine the effect of any interference and this should be carried out before the groundbed has been installed.

The power supply availability, accessibility to the groundbed for testing and maintenance, and any third-party effect such as vandalism of the groundbed should be taken into consideration when selecting a grounbed site.

Groundbeds are forced to discharge current and will corrode, thus, it is important to provide anode material that is consumed at relatively low rates. Materials such as scrap steel pipe, rail, rod, and so on, may be used but they tend to be consumed at the rate of 44 kg per ampere/year (one ampere flowing for 1 year). Thus, a large amount of material will be required for a desired operating life. There are various anodes that are available for use but the design life should be calculated to determine whether they are suitable. Those include graphite, platinized anodes, mixed metal oxide coated anodes, and so on.

The type of anode groundbed used is dependent on the design. Anode groundbeds are either located close to the structure being protected or remotely, so that the current can be distribute over the entire structure to be protected. Anode groundbeds are either local distribute anode groundbed system, vertical distributed anode groundbed system, horizontal distributed anode groundbed system, deep anode groundbed, or remote anode groundbed system. The location of the anode groundbed is relative to the structure being protected and other structures in the vicinity are very important in minimizing shielding effects.

Typically, coke or graphite back fill are used to surround anodes used in impressed current CP systems. Figures 32.6 and 32.7 detail CP systems utilizing anodes in soil and water conditions [1,3,4].

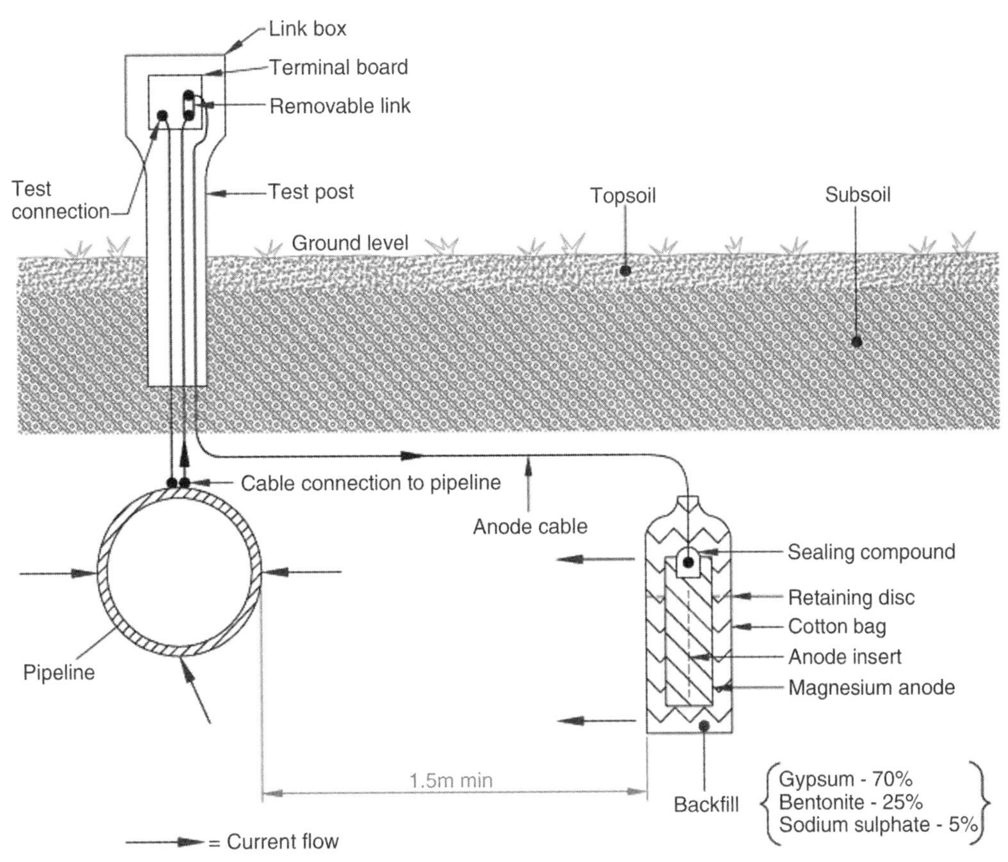


**FIGURE 32.6** Impressed current CP system for a buried pipeline in soil.

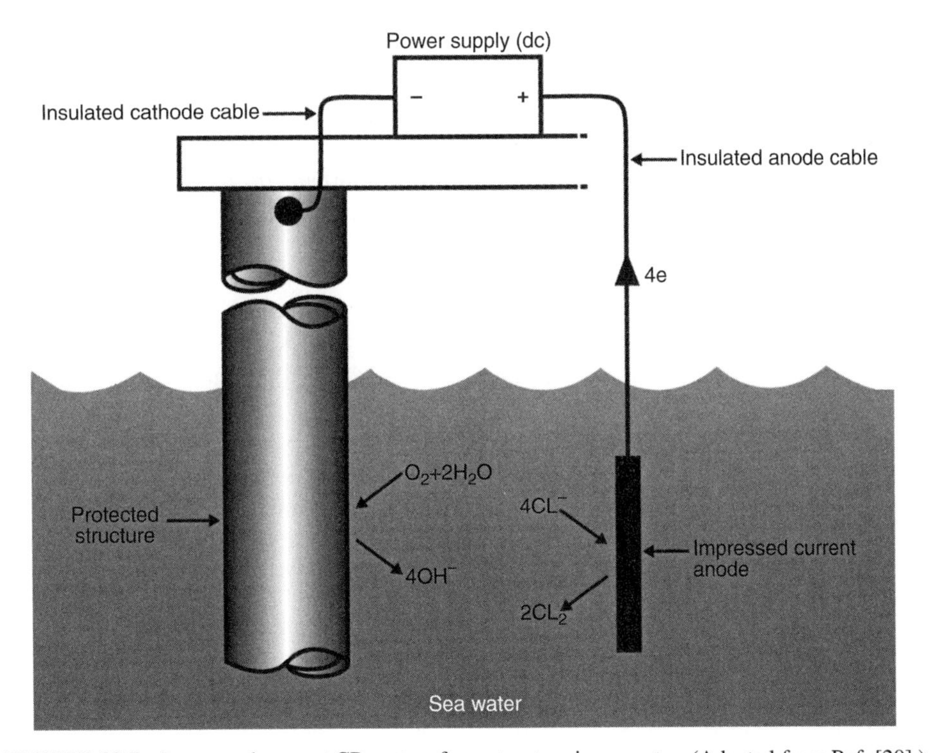


**FIGURE 32.7** Impressed current CP system for a structure in seawater. (Adapted from Ref. [20].)

### 32.4.5 Sacrificial Anode versus Impressed Current Systems

The major difference between a sacrificial anode and impressed current CP system that will determine the type of system chosen is the difference in the driving voltage attainable, and therefore, the current output. A sacrificial anode system has a driving voltage limited by the emf of the galvanic cell set up between the anode and the structure; for example, −1.5 V for a magnesium anode and steel structure; whereas, an impressed current system is free from this restriction and different design techniques are employed to take advantage of this. If protecting a large structure such as a pipeline, then utilizing sacrificial anodes would require a large number of them distributed along the pipeline, which would involve a multitude of electrical connections and a lot of installation work. While an impressed current CP system being utilized to protect the same pipeline would require a smaller number of anodes, each fed with a greater current, though it would require a higher driving voltage of about 20 V to achieve protection. An undesirable feature of protecting the pipeline this way is that certain sections of the pipeline may be over-protected in achieving complete protection in other parts of the pipeline. This is always seen on a pipeline where the relative nearness of the anode and groundbed and buildup of voltage in the pipe metal causes local over protection. Often a pipeline is over protected from one anode position because overall protection can be more easily obtained from one anode than from a number of smaller anodes, which could cause a more even protection [18].

Generally, any decision taken to install one type of CP system over another requires careful consideration. Generally, a compromise is used and may include a combination of both systems. Parameters that may affect the decision of which CP system to use are

(a) *Coating effectiveness:* The more effective the protective coating, the lower will be the CP current requirements. The use of sacrificial anodes tends to be on smaller surface areas with a high current density.

(b) *Size of protected structure:* It is generally more economical to protect a large structure by impressed current. A large structure may be protected by one impressed current system; however, if a sacrificial anode system is used, it would require a large number of anodes. In such a system, the capital cost of the impressed system may be less than the sacrificial anode system and may have a higher maintenance cost associated with anode replacement. With small structures, the capital cost of sacrificial anode protection may be considerably less than that of providing a power supply, a transformer rectifier, and an anode system. In such circumstances, maintenance costs would vary little between impressed current and sacrificial anode systems.

(c) *Soil resistivity:* Sacrificial anodes have a fixed low driving voltage with respect to the protected metal. In high-resistivity soils, such low driving voltages cannot produce sufficient current output from the anode. Therefore, in high-resistivity soils, the use of an impressed current system may be more economic.

Sacrificial anodes are infrequently used in soils where the resistivity is in excess of 75 ohm-m.

(d) *Interference:* A large impressed current system can cause voltage gradients in the soil sufficiently high for interference to affect foreign structures, especially in the vicinity of the anode and of the structure. The sacrificial anode system can minimize such problems.

(e) *Availability of anode sites:* In urban areas, where the structure is located under concrete or bitumen, it may not be economic to excavate through such surfaces to install sacrificial anodes, reinstate the surface, and then repeat this activity should the anode need replacing. It may be cheaper to install an impressed current system in a large open area, for example, a park that will protect a long length of such a structure. The reverse may also be true. If there are no suitable impressed current sites because of lack of space, and so on, then a sacrificial anode system will have to be used. In some areas, the lack of available power may prohibit the use of an impressed current system.

(f) *Stray current:* A sacrificial anode system tends to earth the structure. This can be an advantage in overcoming stray current effects—both stray earth currents (from operation of DC traction systems) and induced currents (as a result of paralleling high-voltage electrical transmission lines). Caution should be exercised to ensure, in stray current areas, which sacrificial anodes do not cause corrosion at a distant discharge point through current pick-up. Impressed current systems are not subject to this problem [18].

(g) *Control:* An impressed current system has the advantage of greater flexibility and control. As one system protects so much structure, there is flexibility of location because all the current is flowing back to one source and problems such as coating defects and underground short-circuits are more readily traced.

However, a structure protected by sacrificial anodes generally requires less frequent monitoring and is self-controlling.

See Table 32.4 for a summary of the comparison between sacrificial anode and impressed current CP systems.

**TABLE 32.4 Summary of the Comparison between Sacrificial Anode and Impressed Current Systems**

| Sacrificial Anodes | Impressed Current System |
|---|---|
| 1. They are independent of any source of electrical power | 1. Requires mains supply or other source of electrical power |
| 2. Their usefulness is generally restricted to the protection of well-coated structures or the provision of local protection, because of the limited current that is economically available | 2. Can be applied to a wide range of structures, including, if necessary, large, uncoated structures |
| 3. Their use may be impracticable except with soils or waters with low resistivity | 3. Use is less restricted by the resistivity of the soil or water |
| 4. They are relatively simple to install; additions may be made until the desired effect is obtained | 4. Needs careful design, although the ease with which output may be adjusted allows unforeseen or changing conditions to be catered for |
| 5. Inspection involves testing with portable instruments, at each anode or between adjacent pairs of anodes | 5. Needs inspection at relatively few positions; instrumentation at points of supply can generally be placed where it is easily reached |
| 6. They may be required at a large number of positions. Their life varies with the conditions so that replacements may be required at different intervals of time at different parts of the system | 6. Requires generally a small number of anodes |
| 7. They are less likely to affect any neighboring structures because the output at any one point is low | 7. Requires the effects on other structures that are near the groundbed of the protected structures to be assessed but interaction is often easily corrected, if necessary |
| 8. Their output cannot be controlled but there is a tendency for their current to be self-adjusting because if conditions change such that the metal to be protected becomes less negative, driving the emf, and hence current, increases. It is possible, by selection of material, to ensure that the metal cannot reach a potential that is sufficiently negative to damage the paint coating | 8. Requires relatively simple controls and can be made automatic to maintain potentials within close limits despite wide variations of conditions. Since, emf used is generally higher than with sacrificial anodes the possible effects of ineffective control or incorrect adjustment, for example, damage to paint work or coatings are greater |
| 9. Their bulkiness may restrict flow and/or cause turbulence and restrict access in circulating water systems | 9. Requires high integrity of insulation on connections to the positive side of the rectifier, which are in contact with the soil or water: otherwise they will be severely corroded |
| 10. They may be bolted or welded directly to the surface to be protected, thus, avoiding the need to perforate the metal of structure to be protected internally, and so on | 10. Requires the polarity to be checked during commissioning because misconnection, so that polarity is reversed, can accelerate corrosion |
| 11. They cannot be misconnected so polarity is reversed | |

## 32.5 DESIGN PRINCIPLES OF CATHODIC PROTECTION

Methods for designing CP systems can be rather complex but basically it can be split up into three different questions:

1. How much current is required?
2. What is the most economical way of supplying current?
3. How is the protective current distributed over the structure?

### 32.5.1 Current Requirement for a Cathodic Protection System

Usually, the amount of current supplied to a CP system or on a specific exposed area is based on past experience or on actual measurements, which have been taken at that particular CP system location. The current required to protect a new structure is calculated according to its overall exposed area. Another method of calculating the current requirement is that the structure to be protected should be preinstalled. Then, electrical measurements should be performed directly to determine the current requirement for protection of that particular structure.

### 32.5.2 What is the Most Economical Way for Supplying Current?

There are two different ways of supplying a CP system and each has its own requirement for how to provide the estimated current.

*32.5.2.1 Sacrificial Anode Systems* If sacrificial anodes are chosen as the way of supplying CP then the type of anode material must be chosen and the number of anodes to be used must be calculated. The anode life can be calculated to determine the most suitable anodes to use. The type and size of anode material used can be adjusted accordingly to ensure that it produces the required current for the desired anode life for the most economic cost.

The type of anode material and the number and sizes of anodes required depends on the corrosivity of the environment, which can be obtained through the resistivity of the electrolyte. The resistivity can then be used to calculate the output from a single anode of a particular size and type of material used. The number of anodes then required for the CP system is calculated by dividing the total current required by the output from a single anode.

The anode life can be calculated by determining the amount of active material that is present in the anode and the current provided. Thus the weight of the anode material being consumed is proportional to the amount of current being provided.

*32.5.2.2* ***Impressed Current Systems*** If an impressed current system is chosen then the system is designed as a simple direct current circuit. The most important variables to this circuit are the resistances of the components making up the system, in particular, anode resistance. The anode resistance is dependent on the number and size of the anodes used and the resistivity of the environment. It is more desirable to have an overall low resistance as this will reduce the amount of voltage required to provide the required current to run the CP system. For a given current requirement, a lower overall resistance and resultant lower voltage will cause lower power costs and make a more economic system.

For underground anode installations an electrically conductive material is utilized such as coke breeze and is placed around the buried anodes. This electrically conductive material has the effect of reducing the electrical resistance by increasing their effective size. For further reduction of the resistance of the anodes it will require the installation of more anodes or larger sized anodes. If there is a requirement for more or larger sized anodes then the expense of this must be balanced against the increased costs of the T/R and power costs associated with higher voltages.

As with sacrificial anode systems, the factors involved in supplying an impressed current system must also be balanced in the overall design of the CP system to ensure that the cost is minimized but while still maintaining an overall effective CP system.

### 32.5.3 How Is the Protective Current Distributed over the Structure?

This final question for the design of a CP system depends on the system configuration, be it for a sacrificial anode CP system or an impressed current CP system.

Several different real case scenarios are examined for the better understanding of how the current is distributed over the whole structure CP system.

(a) For the protection of a pipeline utilizing sacrificial anodes, a resistive shunt was placed in the cable from the anode to the pipeline. This allowed the current to be easily measured and the potentials along the pipeline were measured to determine whether the CP system was functioning properly. The NACE book, "Cathodic Protection Survey Procedures, Second Edition" by Holtsbaum details procedures for measuring the current utilizing a resistive shunt. Particular consideration should be given to Chapters 3 and 4, which explain how shunts are used to measure the current and outlines the process for determining pipeline current measurements [21].

(b) For the protection of a buried pipeline utilizing an impressed CP system, anodes were placed at a greater distance away from the pipeline than in the previous case of a sacrificial anode system as detailed in a). It is a preference for anodes to be located at a remote location away from the pipeline when one anode groundbed is used to protect a very long section of pipeline because it allows a more even distribution of the current along the pipeline. T/Rs supplying the power supply for the CP system are typically located close to the anode groundbed because it minimizes the length of cable required between the anode and the T/R, which could suffer rapid anodic corrosion if the insulation of the cable is damaged. Often this type of failure of the impressed current CP system is due to damage to the cable between the anode and T/R [18].

## 32.6 PROTECTIVE COATINGS AND CATHODIC PROTECTION

The function of a coating is to reduce the area of metal exposed to the electrolyte, which enables the current density required for CP to be greatly reduced. Even a coating in a poor condition can reduce its current requirement by approximately 80–90%. A reduction in current required will reduce the cost and extend the life of the CP system. Also, it can reduce the undesirable side effects such as stray currents that occur when large amounts of current are required.

For both seawater and buried pipeline conditions, a protective coating used in conjunction with CP can be far more economical way of applying CP as coatings will act as the premier method for stopping corrosion.

The basic requirements of pipeline coatings to be summarized are as follows:

(1) High electrical resistance.

(2) Provide continuous coatings, that is, there should be minimal defects (holidays).

(3) Resistant to chemical or bacterial action.

**TABLE 32.5 Coating Resistance against Current Requirement [22]**

| Effective Coating Resistance (ohm for 1 average $ft^2$) | Current Required (A) |
|---|---|
| Bare pipe | 500 |
| 10,000 | 14.91 |
| 25,000 | 5.96 |
| 50,000 | 2.98 |
| 100,000 | 1.49 |
| 500,000 | 0.29 |
| 1,000,000 | 0.15 |
| 5,000,000 | 0.03 |
| Perfect coating | 0.000058 |

*Source:* © NACE International 2003.

(4) To withstand all temperature variations to which it may be subjected.
(5) Adhere strongly to the surface to be protected.
(6) Possess satisfactory ageing characteristic and adequate mechanical strength.
(7) Ability to resist abrasion and mechanical damage during installation.

Materials used for pipeline coatings range from fusion bonded epoxy, extruded polythene, asphaltic enamel coal tar or and plastic tapes. The choice of the coating type utilized depends on a large number of factors, although the final aim should be to achieve the best overall economy in the combined cost of protected structure and the initial and running cost of the protection schemes. It is generally accepted that the highest quality coatings are provided by the coal tar enamels and the extruded polythenes [22,23].

The importance of coating resistance in terms of reducing the current required for CP is given by the following table, which details the protective current requirement for ten mile of 36 in. diameter pipe with varying quality of protective coating (Table 32.5).

CP can have beneficial and adverse effects on the performance of protective coatings depending on the particular situation. Some of these effects are listed in the following section:

### 32.6.1 Beneficial Effects of Cathodic Protection Used in Conjunction with Coatings

- Holidays present in the coating and coating deterioration that occurs with time do not ultimately result in local metal loss.
- Since corrosion at holidays and defects is controlled, there is likely to be no undercutting of the coating at these areas.

### 32.6.2 Adverse Effects of Cathodic Protection Used in Conjunction with Coatings

If the CP system is not properly designed and operated or if the coating system is not properly selected to be compatible with the CP system then three problems may occur:

1. Deterioration of coatings if they are not resistant to alkaline conditions, which occur due to CP. As a result of CP more hydroxyl ions will form on the metal surface causing the protected structure to be more alkaline in nature. If the coating is not resistant to alkalinity it can be damaged and fail. Oil-based paints, such as alkyds, are particularly susceptible to this type of damage due to increased alkalinity as a direct effect of applying CP. This sort of damage can occur under normal CP operation.
2. Disbonding of coatings as an effect of the generation of hydrogen at the metal/coating interface where there will be excessive protective current. This can cause disbondment of the coating and premature failure of the coating used. Disbondment can be prevented if the CP system is properly designed to limit excessive evolution of hydrogen.
3. Deterioration of the coating due to electroendosmosis. Electroendosmosis (Figure 32.8) can cause a failure of the coating system due to the coating not being resistant to this form of coating damage. The less permeable the coatings such as epoxies then the more resistant to electroendoosmosis than the more permeable coating such as oleoresinous phenolics.

Specific tests can be found in various standards such as American Society for Testing and Materials (ASTM) such as G8, G9, G19, G42, G80, and G95, which can be used to evaluate the resistance of coatings to damages that can be caused by alkalinity, cathodic disbondment, and electroendosmosis [24–30].

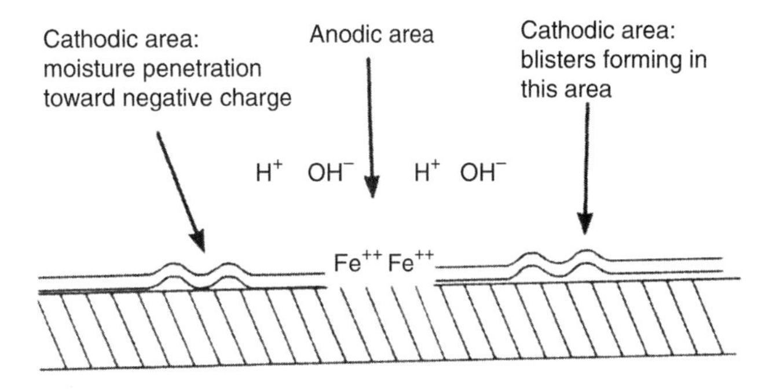


**FIGURE 32.8** Mechanism of electroendosmosis [24].

## 32.7 MONITORING CATHODIC PROTECTION SYSTEMS

All CP systems require routine monitoring to make sure that the systems are running efficiently and effectively to prevent corrosion and subsequent metal loss from occurring, which could be very costly or cause environmental damage or be harmful to human life. Most often CP systems require a periodic inspection and subsequent maintenance to keep optimum performance of the system.

Sacrificial anode system inspection requirements are fairly minimal. Typically, they require some sort of inspection program every set time interval, which can be 6 monthly to yearly. A typical inspection of the system requires electrical measurements of the structure at several points within the system every set time interval. In most cases, test stations are the locations for performing the measurements and are usually installed as a permanent part of the CP system. Within, the test stations are usually connection wires for testing the system that are typically located within an enclosure. The potential readings recorded at the test stations can be used to indicate whether there is a problem with the CP system. Analyzing the potential readings can be used to assess whether there is a need for a more thorough survey consisting of potential measurements taken at locations along the structure and the current output measurement of the anodes of that particular CP's system. The most common problem associated with a sacrificial anode system is the consumption of the anode and the breaking of the wires from the anode to the structure.

Impressed current CP system requires more frequent inspections. Typical inspection plans involve inspections every month of T/R to determine the amount of current and voltage being provided to the CP system. Potential measurements will then be taken of the structure between every 6 months and every 12 months. This will determine if there is a problem within the CP system and what maintenance is required and if any other tests are required.

There are several different survey techniques that are used to monitor CP systems to determine its functionality. Standard test point and annual compliance surveys are used to monitor CP performance. The direct current voltage gradient (DCVG) is used to locate coating faults in buried pipelines and determine the severity of the coating faults. The close interval potential surveys (CIPS) are used to monitor the efficiency of the CP system. Soil resistivity is used to identify corrosive soil areas where the pipeline is located and determine the likelihood of metal loss at these areas.

### 32.7.1 Commissioning of Cathodic Protection System

Commissioning is used by a CP engineer to verify that the CP system has been installed correctly as that outlined during the design of the system. The CP system should be started, then verification of the CP effectiveness should be investigated, and adjustments should be made if required in order to ensure that CP is applied to the whole pipeline.

During initial stages of commissioning of a CP system, before the CP system has been switched on, all electrical installations should be carried out by the appropriate electrical engineer according to the local regulations. Prior to energizing the CP system, all the circuits, T/R, electrical drainage stations, isolation joints, earthing devices, metallic casings, monitoring test stations, coupons, and working electrodes should be checked to verify that they have been installed correctly and are functioning correctly. Potential measurements should be made of the corrosion potential of all structures on the test stations to verify these. Resistance measurements should be measured at remote earth for groundbed or sacrificial anode locations and resistance between the pipeline being protected and the groundbed. The influence of an AC or DC sources towards the pipeline should be investigated. The anode to electrolyte potential of sacrificial anodes should be measured. The pipe to electrolyte potential of any neighboring foreign pipelines should be investigated [31].

During the energizing of the CP system it involves the following procedures:

1. Switch on the CP system such as an impressed current station and check that it is functioning properly.
2. Adjust any settings at the stations such as T/R in order to ensure that the potential design requirements are met.
3. After energizing of the CP system the current should be adjusted a little at a time until the potential at the drain point reaches it limiting critical potential. The T/R should be set at this setting until the pipeline has polarized.
4. After connection of all sacrificial anodes as per the design; these should all be checked that they are functioning properly [31].

### 32.7.2 Monitoring Test Stations (Test Points)

Pipelines generally have monitoring facilities installed along the pipeline route to check that CP is being supplied to all parts of the pipeline. Monitoring test stations are usually installed above the pipeline but with connection to the pipeline. Within the test station, there should be at least two separate cables that are connected directly to the pipeline. Each cable within the test station should be color coded to identify what they are.

Typically, pipe-to-soil potentials, current and possible interference are monitored at test stations. Test stations are usually installed on pipelines at a distance no longer than 3 km and in urban and industrial locations they are installed at intervals no longer than 1 km [31]

Monitoring test stations are installed at crossings, isolating joints, at connections to earthing systems, metallic casings, bond connections to other pipelines or structures, and other connections that have coupons or groundings attached to them.

Monitoring test stations should also be installed on pipelines where the pipeline crosses other pipelines, where they cross major roads or dykes, where pipelines cross railways and rivers and where the pipeline runs close to other structures. Ideally, if several pipelines run in parallel such as in a corridor but not in the same trench then each pipeline should be provided with separate potential monitoring facilities.

Pipelines should be bonded where a pipeline crosses another pipeline. The bonding should consist of two separate cables, each connected to each of the individual pipeline, which should terminate within the test station and have a facility to install direct or resistive bonds when needed.

At Impressed current stations, the T/R output voltage should be checked, the protection current output should be checked and the ON potential at the drain point should be checked.

Sacrificial anodes should be checked to verify the protective output of each sacrificial anode and the equipotential of each bond current.

At the test stations the ON potentials should be carried out at the extremities of the pipeline. The ON potential and current should be measured for any coupons installed at the test stations. The pipe to electrolyte of neighboring pipelines should be measured to assess any influence these may have on the pipeline of interest. The ON potential should be measured to determine any electrical isolation of the pipeline at the isolating joints, metal casings, any reinforced concrete sections, and at any earthing systems. The ON potential and current should be measured at the bonding sections on the pipeline. The AC voltage should be measured on the pipeline to determine any likelihood of interference.

The methods for carrying out the measurements that are required to be taken should be carried out according to local standards and specifications such as NACE TM0497:2012 Item No. 21231 [32] or in accordance with the procedures set out in the very comprehensive NACE book, Cathodic Protection Survey Procedures, Second edition by Holtsbaum [21].

### 32.7.3 Annual Compliance Surveys

Inspection and monitoring of CP systems should be carried out at regular intervals and in some countries it is legal requirements to carry out these surveys to determine that the protection criteria is being fulfilled and to determine any deficiencies with the CP system. There are routine measurements and checks that should be taken out annually or as specified legal requirements or determined in standards. Table 32.6 details some routine annual inspections and the frequency at which these measurements should be taken.

The frequency of carrying out the measurements can be increased in areas where there is stray current interference or at critical sites and where there are foreign connections.

If CP systems are monitored by remote monitoring then any equipment malfunctions can be easily determined from the first available data transmission than measurements carried out monthly or annually.

Information regarding the methods for carrying out these measurements can be found in NACE standards such as NACE Standard TM0497:2012 or from the NACE book, Cathodic Protection Survey Procedures, Second edition by Holtsbaum [21].

### 32.7.4 Direct Current Voltage Gradient Surveys—DCVG

In CP when current flows through the resistive soil to the bare steel exposed at coating faults in the protective coating, a voltage gradient is generated in the soil. The larger the coating fault the greater the current flow, and hence, voltage gradient and this is utilized in the DCVG technique to prioritize coating faults for repair. The voltage gradient is observed by measuring the out of balance between two copper sulfate electrodes utilizing a specially designed millivoltmeter. When the two electrodes are placed 1–1.5 m apart on the soil in the voltage gradient from a coating fault, one electrode adopts a more positive potential than the other, which allows the direction of current flow to be established and coating fault to be located. To simplify the coating fault location interpretation, the applied CP is separated from all other DC influences such as tellurics, DC traction, and so on, by pulsing the local CP ON and OFF in an unsymmetrical fashion. The pulsing DC can be from the pipeline CP system itself, or from an independent source such as a portable DC generator or batteries utilizing a temporary groundbed and impressed on top of a pipelines existing CP system.

In surveying a pipeline, the operator walks the pipeline route testing for a pulsing voltage gradient at regular intervals. As a coating fault is approached, the operator will observe that the millivoltmeter needle begins to respond to the pulse, pointing in the direction of current flow that is toward the coating defect.

When the coating fault is passed the needle direction completely reverses and slowly decreases as the surveyor moves away from the coating fault. By retracing to the coating fault, a position of the electrodes can be found where the needle shows no deflection in either direction (a null). The coating fault is then sited midway between the two electrodes. This procedure is then repeated at right angles to the first set of observations, and where the two midway positions cross is the epicenter of the voltage gradient. This is usually directly above the coating fault.

In order to determine various characteristics about a coating fault, such as severity, shape, corrosion behavior,

**TABLE 32.6 Routine Compliance Surveys and Frequency of Measurements**

| Item | Type of Measurement | Time Schedule of Measurement |
|---|---|---|
| Impressed current test station | Check visually the T/R for any problems and measure the output voltage and current | 1–3 months |
| Impressed current test station | Carry out a function check of the impressed current station to make sure it is working adequately. This involves checking the T/R unit, permanent reference electrodes, groundbed, earthing system, and calibration of equipment used for measurements. Then measure the output voltage and current | 1–3 years |
| Drainage station | Visually check the drainage point unit and then measure the drain point potential and current | 1 month |
| Drainage station | Carry out a detailed function check of the drainage station. Verify that the permanent reference electrodes are working. Measure the efficiency of the diodes and their protection devices. Check the setting of the resistors and check the calibration of instruments used. Measure the drain point potential and current | Annually |
| Connections to foreign pipelines—resistive or direct bond | Measure the current flow. Carry out a detailed functional check of the device and measure the current flow | Annually |
| Bonding devices and grounding systems | Measure the electrical continuity | At least annually |
| Safety and protection devices that include AC decoupling devices | Measure the settings and function | Annually |
| Permanent reference electrodes | Check validity of reference electrode by checking with a known standard reference electrode | Annually |
| Test stations | Measure the (VOFF) OFF pipe-to-soil potential[a] | Annually |
| Sacrificial anode stations | Check the sacrificial anode stations visually. Measure pipe–to-soil potential | Annually |
| Sacrificial anode stations | Carry out a detailed inspection of the sacrificial anode station. Check resistor settings, efficiency of the bonding connection, verify that permanent reference electrodes are working properly and measure the pipe to soil potential | 3 years—but can be sooner if required |

[a]Where stray current may influence VOFF potential measurements then alternative measurement techniques maybe required.
*Source:* Data compiled from Ref. [31].

and so on, various electrical measurements around the epicenter and from epicenter to remote earth are made for interpretation.

The beauty of the DCVG technique is that it utilizes the existing pipeline CP system at its normal setting wherever possible, and hence, the results reflect the intimate interaction between the protective coating degradation and the effectiveness of CP at individual coating faults. This is very powerful information in the fight against corrosion.

The DCVG technique can be applied in city streets, process plant and refineries, across rivers and estuaries, swamps, parallel pipeline systems, to gas, oil, chemical, and water pipelines. Also, DCVG can be used to evaluate the protective coating on telephone and power cables.

Figure 32.9 details the response of the analog DCVG meter as an operator is surveying the pipeline. Figure 32.10 details an actual DCVG operator surveying above a pipeline. Examples of coating faults that can be located using DCVG are illustrated in Figure 32.11.

### 32.7.5 %IR Severity

All buried pipelines and structures have coating damage and coating faults occur in different shapes and sizes. The common factor in the variety of damage is that all of the coating faults will have a voltage gradient epicenter for the bared steel, which is in contact with corrosive soil and under CP.

The evidence from excavating many thousands of coating faults, has led to two main conclusions: (1) The need to understand severity and its repair importance, which provides information enabling %IR severity assessment to be made. (2) More often than not, the usual case is to find that two types of coating provide protection against the electrochemical corrosion reaction, the organic (paint) and inorganic (surface films) coatings.

The general thought is that the corrosion opportunity at every coating fault can be contained by an effective CP system. This assumption is a delicate balance and can be upset as the

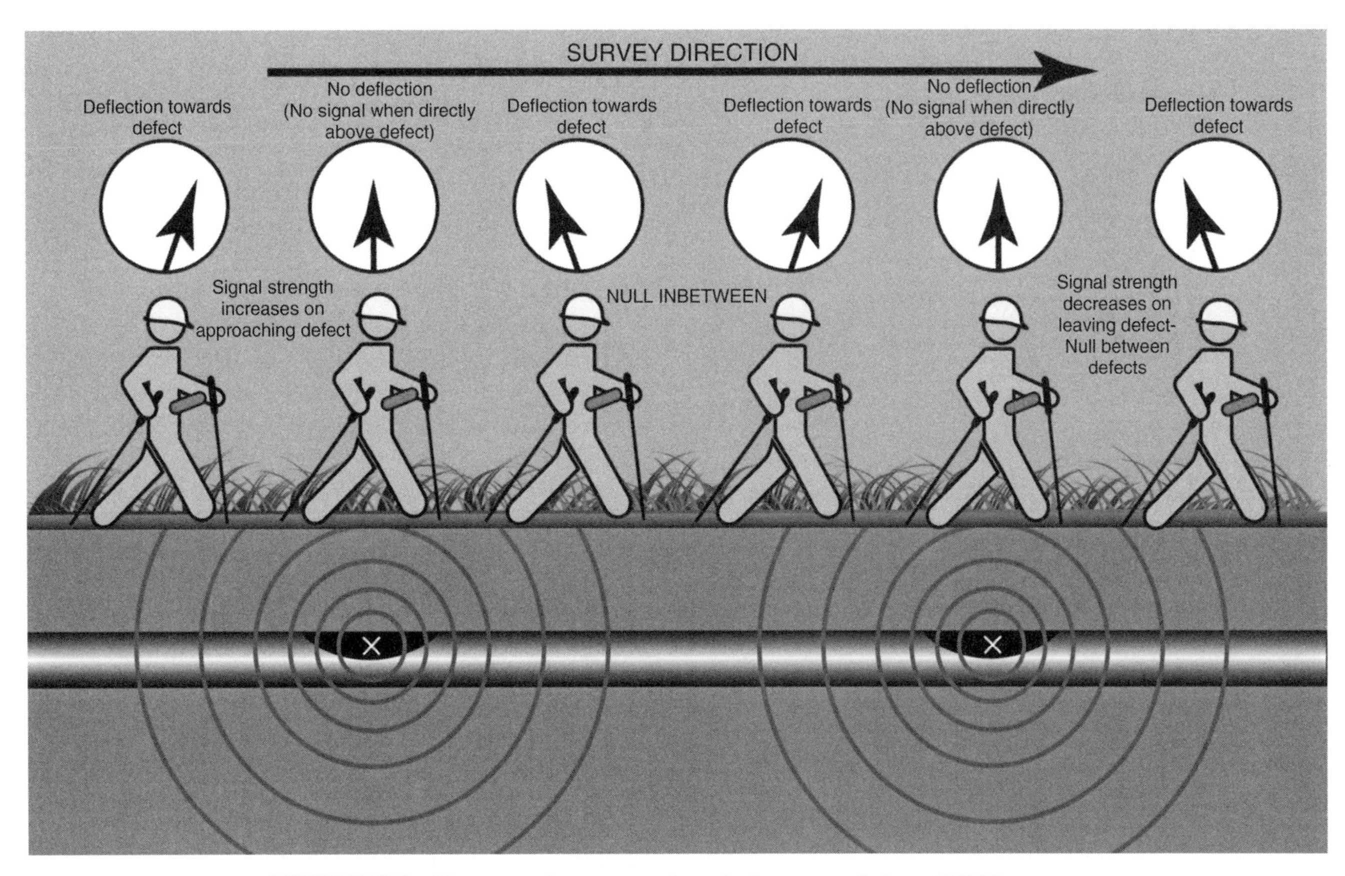


**FIGURE 32.9** Diagrammatic representation of what occurs during a DCVG survey.

severity of the coating fault increases, while the CP system becomes more and more ineffective in its protection.

The DCVG survey technique can accurately "fingerprint a pipeline" by locating the epicenter of any coating faults. From electrical measurements, the severity of the coating fault can be assessed and then graded into four categories, which are based on practical experience. The importance of each category will prioritize any immediate or future refurbishment of the coating and any rehabilitation of the pipeline and its CP system.

The relative severity of a coating fault is expressed by the term %IR, which is calculated using the following formula:

Coating fault severity (%IR)

$$= \frac{\text{Fault epicenter to remote Earth}}{\text{Calculated pipe to remote Earth}} \times 100$$

In a shorter form,

$$\%\text{IR} = \frac{\text{OL}_{\text{RE}}}{P_{RE}} \times 100$$

Calculation of the pipe to remote earth potential (*P*/RE) is an important figure needed to calculate the severity (%IR) value for a coating fault. To be able to calculate the severity (%IR) of coating faults, it is necessary to know the distance of coating faults and the DCVG signal strengths at test posts either side of the sector being surveyed.

The pipe to remote earth potential (*P*/RE) is calculated as follows: -

$$P_{RE} = S_1 - \frac{d_X(S_1 - S_2)}{D_2 - D_1}$$

where,

$S_1$ = Signal at upstream test post: in example = 800 mV

$d_x$ = Distance between upstream test post and defect: in example = 400 m

$D_2$ = Distance of downstream test post: in example = 1000 m

$D_1$ = Distance of upstream test post: in example = 0 m

$S_2$ = Signal at downstream test post: in example = 300 mV

**FIGURE 32.10** Surveyor carrying out a DCVG survey on a buried pipeline.

In the example given in Figure 32.12, the severity of a coating fault (%IR) is calculated as follows:

Overline to remote earth from Figure 32.12 is 130 mV.

Pipe to remote earth is calculated using the formulae given previously and using the values given in Figure 32.12; then

$$\%\text{IR} = \frac{130 \times 100}{600} = 21.7\%$$

What does this figure of 21.7% mean?

Because the severity of a coating fault is calculated from electrical measurements of the voltage gradient, the results obtained are dependent upon the current flowing to each coating fault and soil/film resistivity. The current flowing depends upon how large the coating fault is, how well

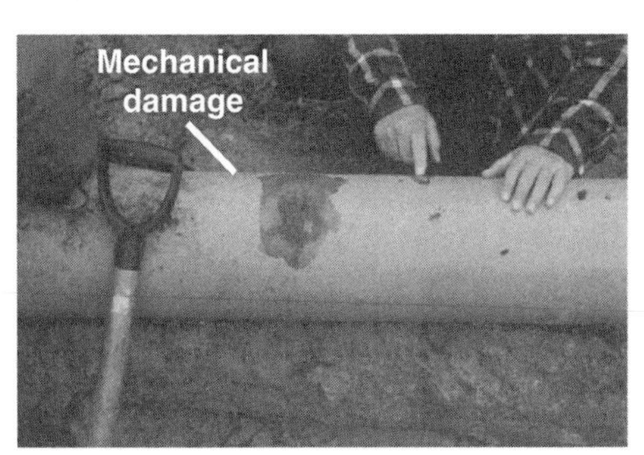


Coating defect observed in an epoxy coating.

Rock damage caused by laying pipeline in backfill without any shielding, such as rock shield or rock guard, to prevent damage to pipe caused by rocks.

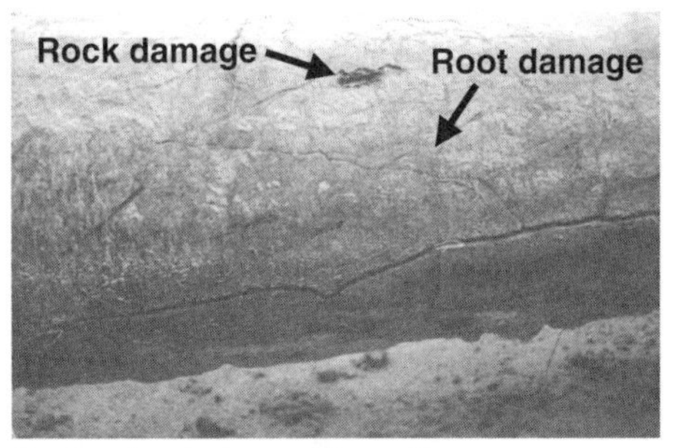


Root and rock damage to coating

Mechanical damage from digger from installation of pipe or replacing backfill

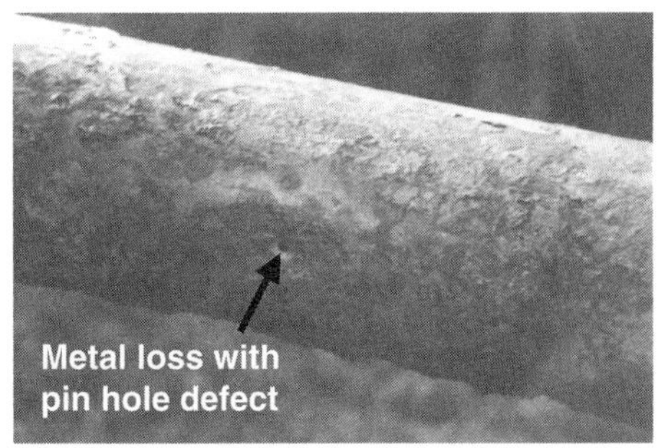


Pipe corroding due to poor cathodic protection application

**FIGURE 32.11** Examples of coating defects that can be located by DCVG.

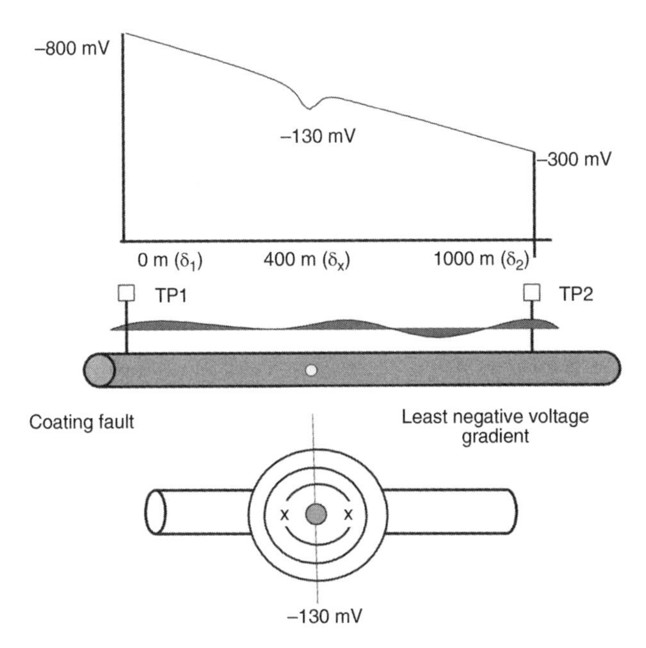


**FIGURE 32.12** %IR severity of the defect.

protective films of magnetite have grown on the exposed steel surface, and how close the coating fault is to the CP drain point. Resistivity affects current flow, and hence, voltage gradient. By resistivity we mean a combination of soil resistivity and the resistivity of films of corrosion product on the exposed steel surface. The latter resistivity can be quite large, that is, 100,000 ohm/cm$^2$ of surface area. Ideally, all measurements should be converted to a standard soil resistivity; we have chosen 5000 ohm-cm.

This has allowed coating faults to be graded to determine those that should be repaired from those that it is possible to live with. Please note on new pipelines most of the %IR values will be very small. The pipeline operator will determine if all small %IR defects should be repaired or not.

### 32.7.6 Coating Fault Gradient

The coating fault grading is

| | |
|---|---|
| 0–15%IR: | Characterized as small coating faults. Such coating faults can usually be left unrepaired provided the pipeline CP is good and there are not too many small coating faults in close proximity. |
| 15–35% IR: | Characterized as medium coating faults. These coating faults may need repair usually within normal maintenance activities. |
| 35–70% IR: | Characterized as medium–large coating faults. These coating faults need to be excavated for inspection and repair in order to fix what could be considered a significant coating fault. |
| 70–100% IR: | Characterized as large/important coating faults. These coating faults should be excavated early for inspection and repair. |

The aforementioned characterizations of coating faults are only one input but a very important input to the excavation and repair decision. Other important factors are shape and method of coating failure, corrosion behavior, soil pH and resistivity, presence of hydrogen sulfide in the soil, operating temperature, age, coating type, pipe-to-soil potentials, leak and metal loss history, and so on.

### 32.7.7 Close Interval Potential Surveys - CIPS/CIS

The theory behind the CIPS is to measure pipe-to-soil potential along the whole pipeline by taking readings at regular intervals (1 m). A CSE is connected to a test point and positioned in the ground directly over the pipeline to obtain a pipe to soil potential profile for the entire pipeline. Such potentials are then interpreted according to general criteria originating largely from one of the NACE criteria, for example, SP0169-2007, as to what constitutes the potential divide between protection and under protection.

Typically, the pipeline right-of-way should have been identified either before carrying out a CIPS survey or one operator walks ahead of the CIPS operator and locates the pipeline using a pipe locator.

CIPS surveys are often quite demanding as it requires the operator to walk the entire right of way. The copper wire can also be broken many times along the pipeline when the operator encounters obstacles such as road crossings, fences, and so on.

One important consideration of a CIPS survey is the ohmic (IR) drop error that is included in the potential measurement when the CP system is operational. Most often instant OFF measurements are made to correct for the IR drop. However, it is not always easy to make IR-free measurements because some currents cannot be easily interrupted such as sacrificial anodes, which are directly bonded to the structure, foreign T/R, stray currents, telluric currents, and long line cells. Interrupters are generally used to switch the CP system ON and OFF for a limited amount of time, which ensures that depolarization of the pipeline occurs gradually due to the accumulative effect of switching the CP OFF. Typically all T/R along the pipeline under inspection are synchronized to switch ON and OFF at the same time. The CP system should be switched ON and OFF in such a way that it avoids capacitive spikes that occur shortly after the current is interrupted as this may hide the instant OFF potential. Utilizing data loggers that have been synchronized with the interrupters will avoid this problem and make sure that the data are recorded at the correct intervals. It is a preference to have some sort of GPS system as a distance measurement for greater accuracy to pin point where data have been taken.

Figure 32.13 details how CIPS measurements should be taken. Figure 32.14 shows a surveyor carrying out a CIPS survey.

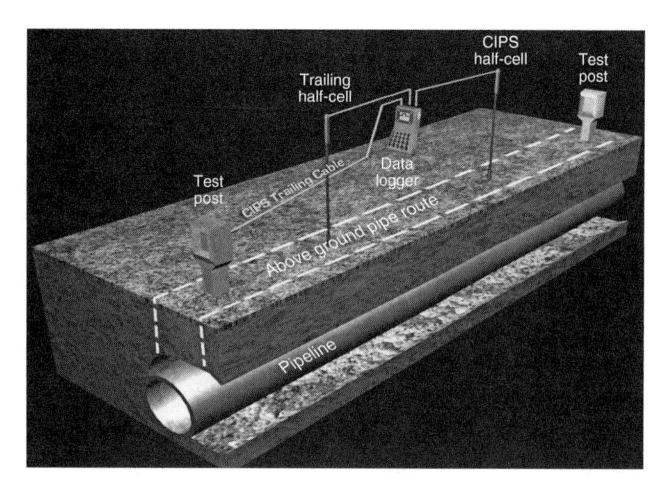


**FIGURE 32.13** Diagrammatic representation of CIPS/CIS method.

**FIGURE 32.14** Surveyor carrying out a CIPS/CIS survey on a buried pipeline.

An example of the sorts of graphs that are produced during a CIPS pipe–to-soil potential survey is detailed in Figure 32.15.

### 32.7.8 Soil Resistivity

Soil resistivity is an inverse function of soil moisture and the concentration of soluble salts available. Generally, most soil and rock minerals are highly resistive. Current flows primarily through moisture-filled pore spaces and cracks in soils and rocks. It can be assumed that soil resistivity must, therefore, be predominantly controlled by pore spaces, cracks, and the amount and composition of water filling the pore spaces.

Soil resistivity has been historically used as a broad indicator of soil corrosivity. Ionic current flow is associated with soil corrosion reactions, so it can be assumed that high soil resistivity should slow down the corrosion reactions. Soil resistivity generally decreases with increasing water content

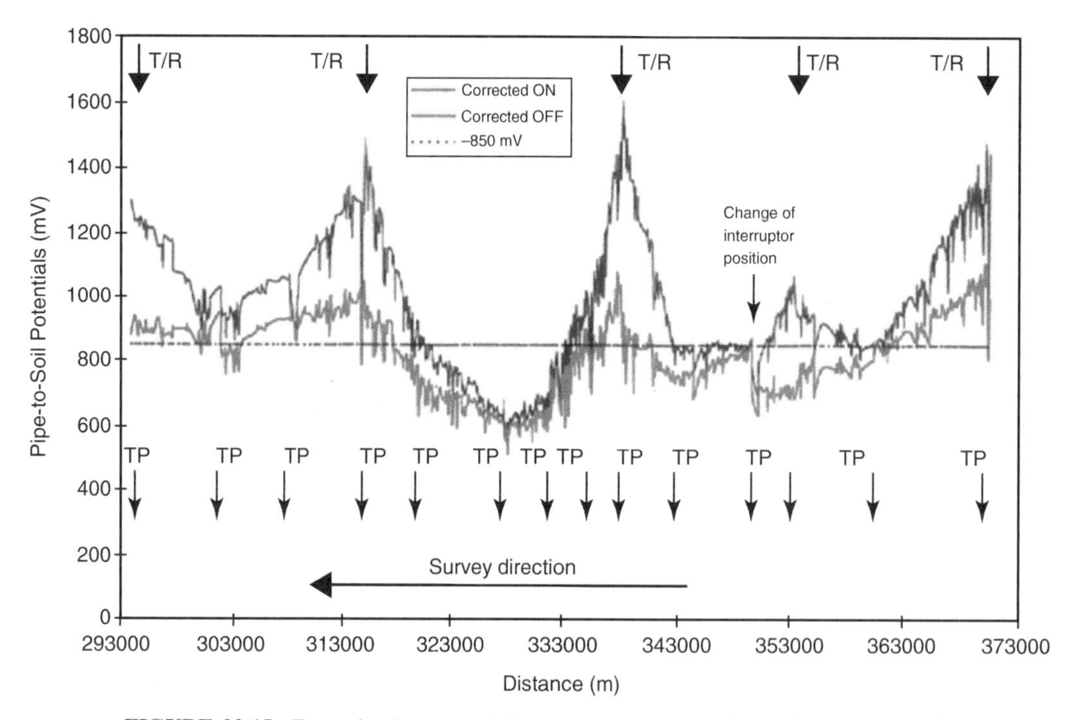


**FIGURE 32.15** Example of exponentially corrected coating defect epicenter potentials.

**TABLE 32.7 Relationship between Soil Resistivity and Steel Corrodibility [34]**

| Resistivity Range (ohm-cm) | Steel Corrodibility | Years to Penetrate Sheet Steel |
|---|---|---|
| <500 | Very severe | 1–5 |
| 500–1000 | Severe | 5–10 |
| 1000–2000 | Severe to moderate | 10–15 |
| 2000–5000 | Moderate to slight | 15–20 |
| 5000–10,000 | Slight | >20 |
| 10,000–1,000,000 | Slight to none | |

and concentration of ionic species. A high soil resistivity alone does not always guarantee an absence of any corrosive attack. Variations in the soil resistivity along a pipeline can cause problems leading to the formation of macro corrosion cells. Therefore, for metallic structures such as pipelines, the merit of a corrosion risk based on absolute values of soil resistivity is limited. Soil resistivity has different effects on different metals. For example, cast iron corrodes only slightly in soils with resistivities above 1000 ohm-cm [33]. Table 32.7 details the relationship between soil resistivity and corrosion [34].

There are two different methods for obtaining soil resistivity, the four-pin (Wenner) method and an electromagnetic method. The four-pin method involves the use of four metal probes that are driven into the ground along a straight line, equidistant from each other. Soil resistivity is then derived from the voltage drop between the center pair of pins with the current flowing between the two outer pins. This assumes that the measured resistivity will be a measure of the hemispherical volume of the earth probe by the central pins. An alternating current is applied from the soil resistance meter causing the current to flow through the soil, between pins C1 and C2 and the voltage or potential, $\Delta E$, is measured between pins P1 and P2 (Figure 32.17). Resistivity of the soil calculated and read off the instrument meter, but obeys the following formula:

$$P = 2\Pi a \frac{\Delta E}{I}$$

where,

$P$ = Soil resistivity (ohm.cm)
$a$ = Distance between the probe (cm)
$\frac{\Delta E}{I}$ = Soil resistance (ohms), from instrument reading
$\Pi$ = 3.1416

The resistivity values obtained from the four-pin method represent an average resistivity of soil to a depth equal to the pin spacing. Resistance measurements are typically performed to a depth equal to that of the buried structure under evaluation. The depth is typically 0.5–2.5 m [34].

If whilst taking soil resistivity measurements and the line of soil pins are placed parallel to a bare underground pipe or some other metallic structure then the presence of the bare metal may cause the soil resistivity values to be lower than actually thought to be. This is because a portion of test current will flow along the metallic structure rather than through the soil, so these types of measurements should be avoided. The four-pin soil resistivity measurements could be misleading unless it is remembered that the soil resistivity calculated with each additional depth increment is averaged in the test with that of all the soil in the layers above. The four pin can also only be taken at selected sections and a full profile of the pipeline right of way cannot be made. Figure 32.16 details the four-pin soil resistivity method.

The electromagnetic soil resistivity instrument is used to measure soil conductivity, which is the reciprocal of soil resistivity. In fact, two parameters are measured at the same time, the soil conductivity for locating changes in soil resistivity and the in-phase component of the vertical magnetic field that permits the detection of buried metal objects. This method gives a continuous profile of the full pipeline right-of-way. The method is accurate, easy to use, faster, and simpler than the four-pin technique. Electromagnetic methods should not be carried out when switching the CP system ON and OFF. Figure 32.17 details the electromagnetic soil resistivity equipment.

### 32.7.9 Corrosion Coupons

Corrosion coupons are used as a method to take IR free potential measurements. A corrosion coupon is typically a small piece of metal that must have the same metallurgical detail and surface finish as the actual structure it is connected to. It is electrically connected to the test station. It is thought that the potential of the coupon should closely match the potential of an exposed part of the structure located in the vicinity of the coupon. Disconnecting the coupon from the structure at the test station, then an instant OFF potential can be made on the coupon without having to interrupt any of the other current sources. It is doubtful if these measurements are IR free as any voltage drop occurring within the electrolyte in the distance between the reference electrode and the coupon surface will be incorporated into the measurement. The reference electrode would have to be placed as close as possible to the coupon to minimize this error.

Any corrosion buildup will affect the results of a coupon and the environmental conditions of the coupon must exactly match that of the structure under test, such as temperature, soil conditions, pH and oxygen concentration, and so on.

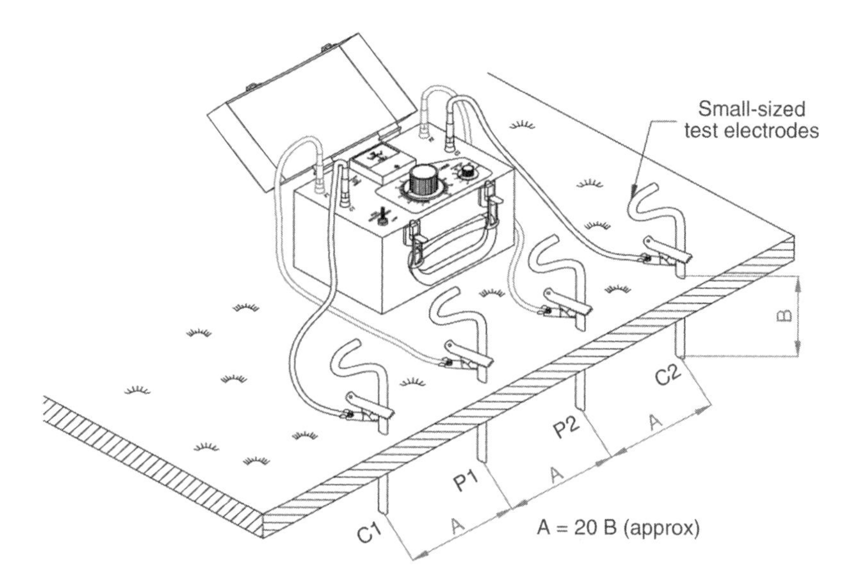


**FIGURE 32.16** Four-pin soil resistivity.

**FIGURE 32.17** Electromagnetic soil resistivity.

Measurement of the current flow can be calculated using a shunt resistor in the bond wire. Corrosion rates should also be possible to calculate from coupons.

## 32.8 CATHODIC PROTECTION CRITERIA

Sacrificial anode and impressed current CP systems are electrical circuits and so electrical measurements should be made to evaluate their activity. Therefore, in order to monitor the effectiveness of CP, it is important to be able to measure a relatively simple parameter such as structure/electrolyte potential, and compare this value to a particular standard.

Corrosion ceases and CP has been achieved when there is net flow of current onto all structure surfaces.

Mears and Brown postulated that when a metallic structure such as a pipeline is polarized to a half cell potential then the surface will have a corrosion rate of zero. At the half-cell potential, the rate of the deposition reaction, reverse reaction, $Fe^{2+} + 2e^{-} \rightarrow Fe$, is the same as dissolution reaction, forward reaction, $Fe^{-} \rightarrow Fe^{2+} + 2e$ and the net reaction is zero. The half-cell potential of $Fe^{-} \rightarrow Fe^{2+} + 2e$ in neutral electrolytes can be calculated using the Nernst equation

$$e^{2+}_{Fe/Fe} = -0.44 + 0.59 \log(Fe^{2+})^2$$

where,

activity of the ferrous ions $(Fe^{2+})$ = solubility product/$(OH^{-})2$, then according $Fe(OH)_2 = Fe^{2+} + 2OH^{-}$. The solubility product is $1.8 \times 10^{-15}$ in neutral solutions and $(OH^{-}) = ½ (Fe^{2+})$. The calculated potential will be −0.59 V (SHE), which corresponds to −0.90 V (CSE), which is not that far from the empirical value of −0.85 V, which is used worldwide as the accepted criterion for prevention of corrosion [13].

There are three criteria that have been accepted over the years by corrosion engineers as a guide for the divide between corrosion and protection. The most common criteria that have been accepted by NACE International include a negative (cathodic) potential of at least 850 mV with the CP applied, a negative polarized potential of at least 850 mV relative to saturated copper/copper sulfate reference electrode and a minimum of 100 mV cathodic polarization between the structure surface and a stable reference electrode contacting the electrolyte [16,31]:

### 32.8.1 − 850 mV versus $Cu/CuSO_4$ with the Cathodic Protection Current Applied Criterion

This is one of the most widely used criterions for determining if an acceptable degree of CP has been attained on a buried or submerged metallic structure. In the case of a steel structure, an acceptable degree of protection is achieved when at least a −850 mV potential difference exists between the structure and a saturated $Cu/CuSO_4$ reference electrode contacting the soil directly above and as close to the pipe as possible.

It is assumed that if a protection potential of −850 mV or more negative is applied with respect to a $Cu/CuSO_4$ reference placed in contact with the soil directly over or adjacent to the steel structure would then be an indication that adequate protection has been achieved at the point on the structure as long as any voltage drops (IR drops) across the electrolyte or through the metallic return path are taken into account. The wording for this particular criterion does state that voltage drops other than those across the structure-to-electrolyte boundary must be considered. Consideration is then defined as

- Measuring or calculating the voltage drop(s).
- Reviewing the historical performance of the CP system.
- Evaluating the physical and electrical characteristics of the pipe and its environment.
- Determining whether or not there is a physical evidence of corrosion.

Voltage drops can be reduced by placing the reference electrodes as close to the pipe surface as possible. Voltage drops can also be eliminated by interrupting all of the sources of the CP system DC current supply and measuring the ON and OFF potentials. The instantaneous OFF potential should be free of voltage drop error. Comparison of the two potentials and noting the difference between the two will indicate the approximate voltage drop included in the potential measurement when the measurement is made with the protective current applied.

Voltage drops are more prevalent in the vicinity of an anode bed or in areas where dynamic stray currents are present. Usually, high-resistivity soil will also cause high-voltage drops. In areas where dynamic stray currents are present, an instantaneous reading may be impossible to accurately record, and therefore, meaningless. The use of voltage recording instruments is recommended as to obtain potential values over a 24-h period, or at least during the known period of highest stray current activity.

The −850 mV criterion with CP applied is one that is almost always used in areas with large amounts of dynamic stray current activity. It is generally accepted that if the potential of the structure to reference electrode remains more negative than −850 mV at all times, even if there are substantial fluctuations in potential with time, then the pipe can be considered protected at that particular test point. It is possible that the amount of test points surveyed and the frequency at which those surveys are performed may have to increase in heavy stray current areas due to ever changing conditions.

***32.8.1.1 Limitations of the − 850 mV Potential to a Cu/ $CuSO_4$ Reference Electrode with Cathodic Protection Applied Criterion*** A limitation of this criterion for insuring total protection is that the potentials can vary widely from one area of the underground structure to another as a result of coating damage, interference effects, and so on. This suggests the possibilities of potentials less negative than −850 mV in sections of the pipe between two consecutive test readings. Also the voltage drop component in the potential measurement has to be considered unless the instant OFF or polarized potential of −850 mV is used. In many cases, removal of the voltage drop component or testing for polarized potential may be difficult, particularly when it is not possible to remove and/or interrupt all sources of current for CP systems.

More negative potentials than −850 mV can of course be maintained at all test points, which could insure a greater degree of protection between test points. However, economics is a factor to consider, higher potentials may result in higher power consumption. Also, in this situation, care has to be taken to insure that polarized potentials of the pipeline does not reach a value at which coating damage could occur due to hydrogen evolution at the surface of the pipeline.

Although this accepted criterion can be and is widely used for all types of structures because of its straightforward approach and simplicity, its most economical use is in the case of coated structures. Use of this criterion to protect an old bare structure could require substantially higher protective current than if one or the other criterion is used.

Another limitation of this criterion is due to the requirement that the potential reading be taken with the reference electrode contacting the electrolyte directly over or adjacent to the pipeline to minimize voltage drop. In cases of river crossings, road crossings, and so on, where the electrode cannot be properly placed, an additional criterion must be used.

### 32.8.2 Polarized Potential of −850 mV Measured to a $Cu/CuSO_4$ Reference Electrode Criterion

This method considers the voltage drop (IR) across the electrolyte by measuring the structure potential with all current sources switched off. The potential then measured is the polarized potential of the structure at a particular test point. A polarized potential has been defined as the potential across the structure/electrolyte interface, which is the sum of the corrosion potential and the cathodic polarization. The difference in the potential between the static or native potential and the instant OFF potential is the amount of polarization that has occurred due to the operation of the CP system. The definition of polarization is the deviation from the corrosion potential of an electrode resulting from the flow of current between the electrode and the electrolyte [16,31,32].

*32.8.2.1 Limitations of Polarized Potential of −850 mV Measured to a $Cu/CuSO_4$ Reference Electrode Criterion* Industry standard for protection is to have instant polarized potentials between −850 and −1050 mV. Polarized potentials more negative than −1100 mV should be avoided to minimize the possibility of cathodic disbondment of the coating. Standards such as NACE SP0169-2007 and European standards point out that polarized potentials that result in an excessive generation of hydrogen should be avoided on all metals, particularly, higher strength steel, certain grades of stainless steel, titanium, aluminum alloys, and prestressed concrete pipes [16,31,32].

### 32.8.3 100 mV Polarization Criterion

The 100 mV polarization criterion, like the −850 mV polarized potential criterion, is based on the development of polarization. This causes the structure to exhibit a more negative value with respect to a $Cu/CuSO_4$ reference electrode, than in its native state.

Measurement of the polarization shift can be determined by either measuring its formation or decay. Determination of the amount of polarization is normally made during the polarization decay (positive shift) period subsequent to de-energizing the CP system, or in the case of sacrificial anodes, when they are disconnected. When the CP system is de-energized, there is an immediate rapid drop of potential. The polarization decay does not include this drop. The potential of the structure after this recorded initial rapid drop is then used as a basis for determining the polarization shift. To obtain the total polarization shift, the final potential after polarization decay has taken place, is measured and subtracted from the potential read immediately after the CP system has been turned off or disconnected. If this total shift is 100 mV or more then the structure is considered to be cathodically protected.

There is, however, a drawback to this method and it is the time that is required for full polarization, which could take from several hours to days for a coated structure to several weeks for a bare structure. This could make this method very time consuming and may also cause the structure to be unprotected for an extended period of time. Normally, the bulk of the depolarization will take place in the initial phase of the polarization decay; therefore, it may not be necessary to wait for the full decay period, except in those cases in which the total actual polarization shift of the structure is relatively close to 100 mV. If during the early phase of polarization decay measurement, the potential drops to 100 mV or more, then there is no real need (unless the actual value is desired) to wait for the further depolarization. If on the other hand, the potential drop in the initial phase of the decay period is only on the order of 50–60 mV, then it may be doubtful that the 100 mV shift will occur. In this case, a determination should be made as to whether a longer wait for total depolarization is required and justifiable.

To determine the formation of polarization on a pipeline, it is first necessary to obtain the static or native potentials (before the application of CP current) on the structure at a sufficient number of test locations. Once the CP system has been energized and the structure has had time to polarize, then these potential measurements are repeated with the current source interrupted. The amount of polarization formation can then be determined by comparing the static/natural potential with the instant OFF potential [16,31,32].

*32.8.3.1 Applications of the 100 mV Polarization Criterion* In a lot of cases, the cost of power required to achieve the 100 mV polarization decay criterion may be less than that required to meet the −850 mV potential with CP applied criterion or the −850 mV polarized potential. Further, the polarized potential corresponding to 100 mV of polarization may be less negative than −850 mV. This criterion takes into account only the polarization film on the pipe surface and is independent of any voltage (IR) drops in the soil or at the structure electrolyte interface. This criterion should not be used in areas subject to stray currents. The results obtained under these conditions could be misleading. Corrosion engineers must determine the effectiveness of this criterion in areas of stray current activity.

The 100 mV polarization criterion is mostly used on poorly coated or bare structures and in some instances could be useful in large pipe networks such as in compressor and regulating stations where the cost of a CP system to achieve either a −850 mV "on" potential or a −850 mV polarized potential may be prohibitive.

This criterion is often used for installations where it is impractical to meet either of the −850 mV criteria.

In piping networks, where new pipe is coupled to old pipe, it may be good practice to use the −850 mV polarized potential criterion for the new pipes and the 100 mV polarization criterion for older pipes [16,31,32].

#### 32.8.3.2 Limitations of the 100 mV Polarization Criterion

The limitations for the 100 mV polarization criterion are

- Interrupting all DC current sources is not always possible.
- Stray current potential variations can make this criterion difficult to measure.
- If dissimilar metals are coupled in the system, protection may not be complete on the anodic metal.
- The only additional limitation is related to the time required for the pipe to depolarize. In some cases, adequate time may not be available to monitor the polarization decay of the pipe to the point where the criterion could have been met. [31,32]

There is one further criterion that can be used to monitor the CP system, the net current flow criterion.

### 32.8.4 Net Current Flow Criterion

This criterion is used to cover the situations for bare or ineffectively coated pipelines where it appears that long-line corrosion activity is the primary concern.

This criterion is based on the premise that if the net current at any point in a structure is flowing from the electrolyte to the structure, there cannot be any corrosion current discharging from the structure to the electrolyte at that point. The principle is based on first locating points of active corrosion and then measuring current flow to or from the structure.

Some pipeline companies use the side drain method for application of this criterion on the basis that if the polarity of the voltage readings on each side of the structure indicates current flow toward the structure, then the structure is receiving protective current. If the reference electrode located over the pipe is positive with respect to the other two reference electrodes, then the current is discharging from the pipe to the electrolyte and corrosion is taking place.

The aforementioned statement is considered correct as long as there are not outside sources of influence such as other pipelines or other gradient sources such as stray currents. In cases where other gradient sources exist, the results could be misleading. The results might be questionable in areas with high-resistivity surface soil for deeply buried pipelines, or where local corrosion cells exist.

In cases where there are sources of outside gradients it may be difficult to evaluate the results of these tests properly. NACE SP0169 and NACE TM0497-2012 provide further information about the net current flow criterion [16,32].

#### 32.8.4.1 Limitations of the Net Current Flow Criterion

The use of this criterion is to be avoided in all areas of stray current activity because potential variations could interfere with its use. Also, extreme care must be taken in common pipeline corridors because other gradient sources may exist that could result in erroneous or misleading measurements.

Even though the results indicate a net current flow towards the pipe/structure, that net current flow is indicative only of what is happening at the specific point of test and does not represent what may be happening at other points on the pipeline/structure. Pipeline companies that use the side drain method for application of this criterion normally require that tests be conducted at close intervals (2–20 ft along the pipeline) [16,31,32].

### 32.8.5 Use of Criteria

There may arise a situation where a single criterion cannot be used to evaluate the effectiveness of a CP system and it may be necessary to employ a combination of criteria.

There will also be situations when the application of any of the criteria listed may be insufficient to achieve protection. This is often in cases where there is the presence of sulfide, bacteria, elevated temperatures, acid environments, and/or dissimilar metals increase the amount of current required for protection. At other times, values less negative than those listed may be sufficient. An example of this is when pipelines are encased in concrete or dry or aerated high-resistivity soil.

In comparison to criteria utilized in North America, European standards detail the environmental conditions where specific criteria should be utilized. For example, for steels of the type typically used in pipelines, in soils, and waters under aerobic conditions at temperatures <40 °C in environments of resistivity >1000 ohm-m, the protection potential is listed as −0.65 V versus $Cu/CuSO_4$ [31]. Under the same conditions, but with resistivity between 100 and 1000 ohm-m, the protection potential is listed as −0.75 V versus $Cu/CuSO_4$ [31]. For full details, the reader should consult the latest edition of European Standard [31].

Table 32.8 details the minimum potentials required to prevent corrosion of iron and steel from occurring as based on a European standard [31].

**TABLE 32.8 Minimum Potentials of Iron and Steel for CP as Based on a European Standard [31]**

| Environmental Condition | Free Corrosion Potential Range (V) | Protection Potential (V) | Limiting Critical Potential (V) |
|---|---|---|---|
| Soils and waters in all conditions except those hereunder described | −0.65 to −0.40 | −0.85 | To prevent hydrogen embrittlement on high strength steel non-alloyed and low-alloyed steels (>550 N.mm$^2$), the critical limit potential shall be documented experimentally |
| Soils and waters at 40 °C < $T$ < 60 °C | – | For temperatures 40 °C ≤ $T$ ≤ 60 °C then the protection potential may be interpolated linearly between the potential value determined for 40 °C (−0.65, −0.75, −0.85, or −0.95 V) and the potential value for 60 °C (−0.95 V) | |
| Soils and waters at $T$ > 60 °C | −0.8 to −0.50 | −0.95 | |
| Soils and waters in aerobic conditions $T$ < 40 °C with 100 < $P$ < 1000 ohm-m | −0.50 to −0.30 | −0.75 | |
| Soils and waters in aerobic conditions at $T$ < 40 °C with $P$ > 1000 ohm-m | −0.40 to −0.20 | −0.65 | |
| Soils and waters in anaerobic conditions with sulfate-reducing bacteria activity | −0.80 to −0.65 | −0.95 | |

*Source:* Permission to reproduce extracts from British Standards is granted by BSI. British Standards can be obtained in PDF or hard copy formats from the BSI online shop: www.bsigroup.com/Shop or by contacting BSI Customer Services for hard copies only: Tel: +44 (0) 845 086 9001, Email: cservices@bsigroup.com.

## REFERENCES


1. Von Baeckmann, W., Schwenk, W., and Prinz, W. (1997) *Handbook of Cathodic Corrosion Protection: Theory and Practice of Electrochemical Protection Processes*, 3rd ed., Gulf Professional Publishing, Houston, TX, ISBN: 0-88415-056-9.
2. Davy, H. (1824) On the corrosion of copper sheeting by sea water; and on their methods of preventing this effect; and on their applications to ships of war and other ships. *Philosophical Transactions of the Royal Society*. 114, 151–158.
3. Shrier, L.L., Jarman, R.A., and Burstein, G.T. *Corrosion*, 1st, 2nd, and 3rd eds., Butterworth-Heinemann, Oxford, UK. ISBN: 0-7506-1077-8-2000.
4. Fontana, M.G. and Greene, N.D. (1978) *Corrosion Engineering*, 2nd ed., McGraw-Hill, New York, NY. ISBN: 0-07-021461-1, 1978.
5. Frischen, C. (1857) *Zeitschrift des Architechten-und Ingenieur-Vereins fur das Königreich von Hannover*, 3, 14.
6. Haber, F. and Goldschimdt, F. (1908) *Zeitschrift für Elektrochemia*, 12, 49.
7. Vogel, O. and Bauer, O. (1908) gwf 63: p. 172, 1920; 68: p. 683, 1908.
8. Kuhn, R.J. (1928) Galvanic currents on cast iron pipe. *Bureau of Standard*, 73B75.
9. Kuhn, R.J. (1958) The early history of cathodic protection in the United States: some recollections. *Corrosion Prevention and Control*, 5, 46.
10. Hoar, T.P. (1938) The electrochemistry of protective metallic coatings. *Journal of the Electrochemical Society*, 33.
11. Mears, R.B. and Brown, R.H. (1938) A theory of cathodic protection. *Transactions of Electrochemical Society*, 74, 519–531.
12. Jones, D.A. (1971) The application of electrode kinetics to the theory and practice of cathodic protection. *Corrosion Science*, 11, 439.
13. Jones, D.A. (1996) *Principles and Prevention of Corrosion*, 2nd ed., Prentice Hall, Upper Saddle River, NJ. ISBN: 0-13-359993-0.
14. Humble, R.A. (1948) *Corrosion*, 4(7), 358.
15. Evans, U.R. (1981) *An Introduction to Metallic Corrosion*, 3rd ed., Edward Arnold Ltd., London, UK. ISBN: 0-7131-2758-9.
16. NACE (2007) NACE Standard SP0169-2007: Control of External Corrosion on Underground or Submerged Metallic Piping, Item No. 21001.
17. Pourbaix, M. (1974) *Atlas of Electrochemical Equilibria in Aqueous Solutions*, NACE International, Houston, TX and CEBELCOR, Brussels, BE. Library of Congress Catalog Card No.: 65–11670.
18. Morgan, J. (1993) *Cathodic Protection*, 2nd ed., NACE International, Houston, TX ISBN: 0-915567-28-8,
19. McAllister, E.W. (2009) *Pipeline Rules of thumb Handbook*, Gulf Professional Publishing, 7th revised edition, ISBN: 978-1-85617-500-5.
20. The Marine Technology Directorate Ltd. (1990) *Design and Operational Guidance on Cathodic Protection of Offshore Structures, Subsea Installations and Pipelines*, MTD Ltd. Publication, ISBN: 1-870553-04-7.
21. Holtsbaum, W.B. (2012) *Cathodic Protection Survey Procedures*, 2nd ed., NACE International, ISBN: 1575902524.

22. Kehr, J.A. (2003) *Fusion-Bonded Epoxy (FBE): A Foundation for Pipeline Corrosion Protection*, NACE International, Houston, TX. ISBN: 1-57590-148-X.
23. Revie, R.W. (editor) (2011) *Uhlig Corrosion Handbook*, 3rd ed., John Wiley & Sons, Inc., New York, ISBN: 978–0470080320.
24. Drisko, R.W. (1998) *Corrosion and Coatings. An Introduction to Corrosion for Coatings Personnel*, Society for Protective Coatings, Pittsburgh, PA. SSPC 98-08; ISBN: 1-889060-24-0.
25. ASTM (2010) ASTM G8-96: Test Method for Cathodic Disbonding of Pipeline Coatings.
26. ASTM (2013) ASTM G9-07: Test Method for Water Penetration into Pipeline Coatings.
27. ASTM (2004) ASTM G19-04: Test Method for Disbonding Characteristics of Pipeline Coatings by Direct Soil Burial.
28. ASTM (2012) ASTM G42-11: Test Methods for Cathodic Disbondment of Pipeline Coatings Subjected for Elevated Temperatures.
29. ASTM (1998) ASTM G80-88: Test Method for Specific Cathodic Disbondment of Pipeline Coatings.
30. ASTM (2013) ASTM G95-07: Test Method for Cathodic Disbondment of Test of Pipeline Coatings (Attached Cell Method).
31. ISO (2003) *Petroleum and Natural Gas Industries-Cathodic Protection of Pipeline Transportation Systems—Part 1: On-Land Pipelines*, International Organization for Standardization, London, UK. ISO/FDIS 15589-1.
32. NACE (2012) *Measurement Techniques Related to Criteria for cathodic protection on underground or Submerged Metallic Piping System: NACE Standard TM0497-2012*, NACE International, Houston, TX. Item No.: 21231.
33. Benson, R.C. (2002) Materials Performance, C January 2002: p. 28.
34. Bradford, S.A. (2004) In: Bringas, J.E. (editor), *Practical Handbook of Corrosion Control in Soils*, CASTI Publishing Inc., Edmonton, AB. and ASM International, Geauga County, OH. ISBN: 1-894038-48-7-2004.

# PART V

# INSPECTION AND MONITORING

# 33

# DIRECT ASSESSMENT

John A. Beavers, Lynsay A. Bensman, and Angel R. Kowalski
*Det Norske Veritas (U.S.A.), Inc., Dublin, OH, USA*

## 33.1 INTRODUCTION

There are three common techniques used for assessing the integrity of pipelines subject to time-dependent threats: hydrostatic testing, in-line inspection (ILI), and direct assessment (DA). Each methodology has its strengths and weaknesses and a combination of these methodologies typically provides the most effective integrity management.

Hydrostatic testing is the oldest technology used to confirm the integrity of pipelines and prevent future failures. The pipelines are pressure tested with water at pressures significantly higher than the operating pressure in order to remove near critical flaws. Any flaws remaining in the pipeline will be smaller than the critical size at normal operating pressures and, given that the flaw-grow rates are reasonably low, will thus require some time to grow to a critical size. The operator must estimate the growth rate of the flaws remaining in the pipeline in order to establish the retest interval. This growth rate can be estimated based on the prior hydrostatic test history of the pipeline, laboratory research, or the results obtained from ILI or DA. While hydrostatic testing has been effective in reducing service failures, it has a number of limitations. Very few, if any, flaws are removed and the pipeline must be taken out of service for testing. The removed flaws are those that are sufficiently large to cause a rupture at the hydrostatic test pressure. In dry climates, obtaining adequate sources of water can be a challenge while freezing of the water can be an issue in winter months and northern climates. In some pipelines, the addition of water introduces the threat of internal corrosion where previously it did not exist. For liquid petroleum pipelines and some gas pipelines, the water must be treated prior to discharge to prevent the introduction of contaminated water into the environment. A pressure reversal also may occur where the failure pressure after the test is lowered by the hydrostatic test due to growth of flaws during the test.

ILI is another technique used to manage time-dependent integrity threats on operating pipelines. There is a long history of using magnetic flux leakage (MFL) and, to a lesser extent, ultrasonic tools to address internal and external corrosion threats on pipelines. Over the past 20 years, there has been significant work performed in the development of ultrasonic crack detection tools. These ILI technologies have a big advantage over hydrostatic testing in that they are capable of detecting flaws that are much smaller than the critical flaws that could fail a hydrostatic test, resulting in longer reinspection intervals than hydrostatic testing. ILI also generally does not require that the pipeline be taken out of service and problems associated with obtaining and handling large volumes of water. However, the pipeline must be capable of accommodating the ILI tools. This requires the presence of appropriate launchers and receivers and the absence of components in the pipeline that would obstruct a tool, such as high angle bends, diameter changes, and certain types of valves. The single biggest issue with ILI technologies is that there is a finite probability that a near-critical feature will not be detected or will be detected and improperly sized. The technologies for detection of corrosion flaws are more mature and therefore less likely to miss a near critical defect than those for detection of crack-like defects.

Because of these detection and sizing issues, one or more validation methods are typically employed. In some cases, pipeline operators will validate the detection and sizing

*Oil and Gas Pipelines: Integrity and Safety Handbook,* First Edition. Edited by R. Winston Revie.

capabilities of the tool by means of a dig program, where the reported flaw dimensions are compared with the dimensions measured in the field. In other cases, pipeline operators perform confirmatory hydrostatic tests on portions of their system to demonstrate the performance of the ILI tool. Pipeline operators may be faced with a situation where a large number (thousands) of features are found by ILI on a segment of a pipeline. They can use elements of DA, such as those used in the pre-assessment and indirect inspection steps, in conjunction with sizing of the ILI features, to identify which features are most likely to be an integrity threat and should be excavated. These DA elements also are used to prioritize pipeline systems for ILI.

The limitations of hydrostatic testing and ILI have increased interest within the pipeline community in the development of alternatives, such as DA, to these integrity management tools. Elements of DA have always been used in the management of pipeline integrity but the DA procedures have only recently been formalized. DA is somewhat of a misnomer in that the direct inspection or assessment of the pipe is only one, the third, element of the DA process. For the most common DA procedures, there are three other elements or steps: (1) pre-assessment, (2) indirect inspections, and (4) post assessment. The details of each step can differ significantly for each different type of DA.

Formal DA procedures have been developed to address external and internal corrosion threats to both gas and liquid petroleum pipelines. This chapter summarizes these procedures.

## 33.2 EXTERNAL CORROSION DA (ECDA)

External corrosion of buried onshore ferrous piping has been identified as a serious threat to the structural integrity around the world. In the United States, a congressional funded research program conducted between 1999 and 2001 [1] determined that the corrosion-related cost to the transmission pipeline industry was approximately $5.4–$8.6 billion annually.

The external corrosion direct assessment (ECDA) methodology was developed to assist pipeline operators when evaluating the impact of external corrosion on the structural integrity of buried piping. Pipeline operators have historically managed external corrosion using some of the ECDA tools and techniques. Often, data from aboveground inspection tools have been used to identify areas that may be experiencing external corrosion. The ECDA process takes this practice several steps forward and integrates information on a pipeline's physical characteristics and operating history with data from multiple field examinations and pipe surface evaluations to provide a more comprehensive integrity evaluation with respect to external corrosion.

### 33.2.1 Overview of Technique/Standard

ECDA is a four-step process that combines pre-assessment, indirect inspections, direct examinations, and post assessment to evaluate the likelihood of external corrosion on a pipeline. The first recommended practice was issued in 2002 (NACE Standard RP-0502-2002 [2]). The standard was reaffirmed in 2008 [3] without significant changes and republished again in 2010 (NACE—ANSI Standard Practice SP0502-2010 [4]).

The pre-assessment step has three main objectives: (1) determine whether ECDA is feasible for the pipeline to be evaluated, (2) select the indirect inspection tools, and (3) identify the ECDA regions. The feasibility of the methodology may be affected by availability of data, conditions for which indirect inspections are not available or cannot be collected, or sections of pipe where visual inspection of the pipe may not be practical or feasible. An ECDA region is a section or sections of the pipeline segment that have similar physical characteristics, corrosion histories, expected future corrosion conditions, and in which the same indirect inspection tools are used.

Indirect inspection covers aboveground inspections that help identify and define the coating condition, other anomalies, and areas where corrosion activity may be occurring. The inspection tools include, but are not necessarily limited to, close interval potential survey (CIS), direct current voltage gradient (DCVG), and alternating current voltage gradient (ACVG). An example of a technician performing DCVG measurements to locate coating defects is shown in Figure 33.1. The indirect inspection requires at least two complementary tools to be used over the entire length of each ECDA region, although it is wise to complete as many as are practical. These complementary tools improve detection

**FIGURE 33.1** DCVG measurements being performed to locate coating anomalies on an underground pipeline.

**FIGURE 33.2** Direct examination of a pipe section. Ultrasonic wall loss measurements are being performed at the center of each grid square.

reliability under the wide variety of conditions that may be encountered along a pipeline segment. After the tests are completed, indications are identified and aligned for comparison. The criterion to establish a classification is determined and the severity of each indication is classified as severe, moderate, or minor.

The direct examination step determines which indications from indirect inspections are most severe, and data are collected to assess corrosion activity. Direct examination requires excavations to expose the pipe surface so that measurements can be made on the pipeline and in the immediate surrounding environment. Direct examination of a pipe segment for external corrosion is shown in Figure 33.2. A minimum of one direct examination per ECDA region is required (two per ECDA region when ECDA is applied for the first time on the pipeline segment) regardless of the results of the indirect inspections and pre-assessment steps. During the direct examination, defects other than external corrosion may be found and should be assessed according to applicable standards and codes. The pipeline operator evaluates or calculates the remaining strength at locations where corrosion defects are found.

During post assessment, analyses of data collected from the previous three steps are conducted to assess the effectiveness of the ECDA, identify existing root causes of external metal loss identified and determine reassessment intervals. The reassessment interval may be defined as half of the remaining life of the piping system on the basis of remaining indications not mitigated during the process. The estimate of remaining life is based on conservative growth rates and conservative growth periods.

### 33.2.2 Strengths

Some of the advantages of ECDA are as follows: (1) it is not an intrusive method as it does require interruption of the pipeline operation, (2) it is an alternative assessment methodology for pipelines that cannot be inspected by ILI, (3) it identifies the direct cause of external metal loss due to corrosion and (4) it may be used as a complementary assessment to improve results of ILI and hydrostatic testing.

### 33.2.3 Limitations

Some of the limitations of ECDA are as follows: (1) the effectiveness and feasibility of DA is based on the amount and quality of relevant data, (2) it may not be feasible for pipelines susceptible to electrical shielding, (3) it may not be cost effective when indirect inspection is limited by external conditions that require additional activities such as drilling, and (4) excavation of the pipeline for direct examination may not be practical or feasible due to associated risks.

### 33.2.4 Status of Standard

Task Group TG-041 of NACE International currently is reviewing and updating NACE SP0502-2010 based on industry experience with the standard over the past 10 years, changes to referenced standards, and new findings from applied research.

### 33.2.5 Context of Technique/Standard in Integrity Management

As already discussed, the task group working on the standard is incorporating lessons learned from experience to improve the effectiveness of the methodology. One aspect being analyzed at the present time is the incorporation of new guidelines to assess the integrity of pipeline sections located in steel casings. There is also interest in providing more guidance to pipeline operators on how to address the ECDA feasibility when required data are not available.

### 33.2.6 Where ECDA Technique Is Headed

Future ECDA improvements will most likely come from innovation in technologies associated with predictive capabilities of indirect inspections and also with overcoming existing limitations related to electrical shielding and cased pipe sections.

## 33.3 STRESS CORROSION CRACKING DA (SCCDA)

The first recognized stress corrosion cracking (SCC) failure of an underground pipeline occurred in 1965 in Natchitoches,

Louisiana [5]. The SCC associated with this failure was intergranular in nature (the cracks followed the grain boundaries in the metal) and is referred to as high-pH or classical SCC. A second form of external SCC was discovered in the 1980s [6]. This form of SCC is referred to as near-neutral pH SCC and is transgranular (the crack path is through the grains in the metal). Both forms of cracking are associated with crack colonies that may contain up to thousands of cracks. The cracks most typically are axially oriented but may be at other orientations depending on the nature of the stresses in the pipelines. In some cases, growth and interlinking of the stress corrosion cracks produce flaws of sufficient size to cause leaks or ruptures of the pipelines at normal operating pressures.

### 33.3.1 Overview of Technique/Standard

The first recommended practice for Stress Corrosion Cracking Direct Assessment (SCCDA) was issued in 2004 (NACE Standard RP0204-2004) [7]. The standard was reaffirmed in 2008 without significant changes [8]. SCCDA is a structured process that is intended to assist pipeline companies in assessing the extent of SCC on buried pipelines and thus contribute to their efforts to improve safety by reducing the impact of external SCC on pipeline integrity. The term is somewhat of a misnomer in that the process is much more extensive than simply examining the pipeline for evidence of SCC. SCCDA is a four-step process; pre-assessment, indirect inspections, direct examinations, and post assessment.

The pre-assessment step involves the collection of existing information on the pipeline that can be used to assess the likelihood that the pipeline is susceptible to SCC, prioritize sections of the pipeline for examination, and select dig sites. In the indirect inspection step, aboveground or other types of measurements are conducted to supplement the data obtained from the pre-assessment. This could include CIS survey data, coating fault survey data, or geological information. In the direct examination step, the pipe is examined to assess the presence, extent, type, and severity of SCC. An example of a colony of stress corrosion cracks found during direct examination is given in Figure 33.3. Other information, such as coating condition, soil information, or electrolyte composition, can also be collected. In the post-assessment step, remedial actions, if required, are identified and prioritized, reassessment intervals are defined, and the effectiveness of SCCDA is evaluated.

Table 33.1 [9] summarizes important factors that should be initially considered in ranking susceptible valve segments of a pipeline and in the selection of dig sites. The information in this table is based on laboratory and field research performed over the past 50 years on SCC. The first recommended practice for SCCDA RP0204-2004 [7], adopted the criteria found in the 2004 version of ASME B31.8S [10] for the selection of pipeline segments that have the highest likelihood of containing SCC colonies. In the case of high-pH SCC, the initial selection of susceptible segments is based on five factors: operating temperature (>100 °F), operating stress (>60% of the specified minimum yield strength), distance from compressor station (<20 miles), pipeline age (>10 years), and coating type (other than fusion bonded epoxy (FBE)) [10,11]. The 2004 version of ASME B31.8S [10] only addressed high-pH SCC. SP0204-2008 [7,8] used the same criteria for selection of near neutral pH SCC susceptible segments but removed the temperature criterion. The 2010 version of ASME B31.8S [11] excludes both the temperature criterion and the distance criterion for

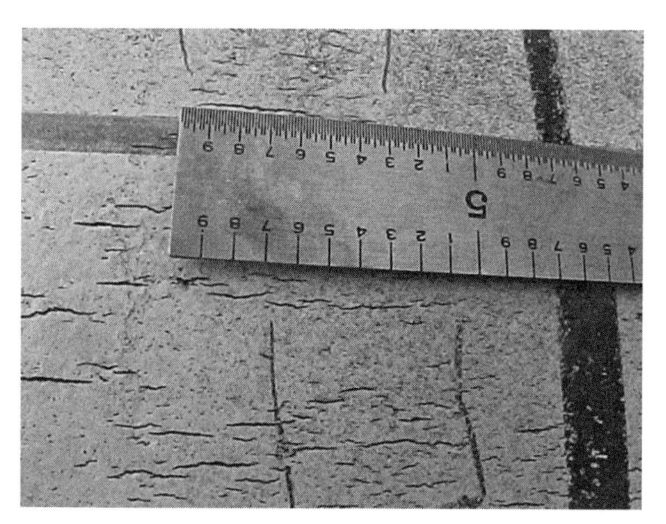

**FIGURE 33.3** Colony of stress corrosion cracks found during the DA step of SCCDA. The cracks were revealed by magnetic particle inspection of the external surface of the pipe.

**TABLE 33.1 Important Factors for Consideration in Ranking Susceptible Segments and in Site Selection for SCC [9]**

| | Ranking | |
|---|---|---|
| Factor | High-pH SCC | Near-Neutral pH SCC |
| Pipe temperature | A | C |
| Pipeline age | A | A |
| Soil properties | B | A |
| Maximum stress | A | A |
| Stress fluctuations | B | A |
| Stress concentrators | B | A |
| Surface condition | A | A |
| Pipe coating | A | A |
| Girth weld coatings | A | A |
| Pipe manufacturer | C | B |
| Cathodic protection related parameters | B | B |

A = major factor; B = factor having some effect; C = minor effect.

near-neutral pH SCC. The distance parameter is somewhat redundant in that it is generally thought to incorporate the stress effect, as well as the temperature effect, in the case of high-pH SCC.

Other types of information on the pipeline can be used for the selection of dig sites in the chosen segments. Again, much of this information is based on the applied research previously performed and includes factors such as topography, drainage, and soil type (near-neutral pH SCC), the magnitude and frequency of cyclic pressure fluctuations, the specific coating type, including the girth weld coating, surface preparation for the coating, coating condition, the pipe to soil potentials, and prior history on the pipeline. For near-neutral pH SCC, the pipe manufacturer may also be an important consideration.

### 33.3.2 Strengths

For complicated pipeline systems, the pre-assessment phase of SCCDA can be used to eliminate SCC as a significant integrity threat, allowing operators to focus their efforts on portions of the system where SCC is likely to occur. Where it has been determined that SCC is a serious threat, carrying out SCCDA is complementary to ILI and hydrostatic testing. As already discussed, it can be used to prioritize segments for ILI or hydrostatic testing. For validation of an ILI run, SCCDA can be used to identify which ILI indications are most likely to be due to SCC and require excavation.

### 33.3.3 Limitations

Appendix A3 of ASME B31.8S [10,11] (both versions) states that each pipeline segment should be assessed for risk for the possible threat of SCC if all of the relevant criteria are present, implying that SCC is not an integrity issue for segments that do not meet these criteria. Batte [12] reviewed the service experience of the gas transmission pipeline industry from the mid-1960s through 2010. It was found that between 80 and 90% of SCC failures met all of the relevant criteria; thereby, demonstrating the importance of these parameters in the SCC process. On the other hand, this is not adequate performance for an integrity management program that relies solely on SCCDA. The authors of NACE RP0204-2004 [7] recognized this issue and included wording in the standard to state the following. "It is recognized that these screening factors will identify a substantial percentage of the susceptible locations, but not necessarily all of them." In other words, the criteria are great for selecting the most susceptible sites for SCC but will not find all of them. Therefore, a significant limitation of SCCDA is that it presently is not capable of reliably identifying the location or locations of the most severe SCC on a pipeline segment. Perhaps, the predictive capability of SCCDA will improve with better techniques and or understanding of the phenomenon. However, at the present time, it is not always a suitable replacement for hydrostatic testing or ILI. The pre-assessment step of SCCDA may indicate that a particular pipeline segment is not likely to be susceptible to SCC and; therefore, other threats are of a more immediate concern. An example would be a newer pipeline with an FBE coating. At the other extreme, ILI, hydrostatic testing, or even pipe replacement may be warranted if extensive, severe, SCC is found. On this basis, SCCDA currently is better suited as a threat evaluation tool than for integrity verification.

### 33.3.4 Status of Standard

Task Group TG-273 of NACE International currently is reviewing and updating NACE RP0204-2008 based on industry experience with the standard over the past 10 years, changes to referenced standards, and new findings from applied research.

### 33.3.5 Context of Technique/Standard in Integrity Management

As already discussed, SCCDA, in its current form, is better suited as a threat evaluation tool than for integrity verification because it presently is not capable of reliably identifying the location or locations of the most severe SCC on a pipeline segment. For pipelines where SCC has been identified as a major threat, SCCDA can be used to prioritize segments for ILI or hydrostatic testing and to assist in the selection of dig sites for verification of ILI.

### 33.3.6 Where SCCDA Technique Is Headed

It is unlikely that, in the near future, technology will change to the extent that SCCDA will be capable of locating the most severe or all SCC on a pipeline segment. This is primarily the result of the inability of the indirect tools used to address the issue of the variation in SCC susceptibility of individual pipeline segments. These tools typically are used to look for evidence of coating damage or cathodic protection levels. Major changes to the standard will likely address recent changes to the American Society of Mechanical Engineers (ASME) [11] and Canadian Energy Pipeline Association (CEPA) [13] recommended practices with respect to severity categories. In the earlier versions of these documents [10,14], there were two SCC severity categories (significant and not significant) and the number of these categories was expanded in the later documents.

## 33.4 INTERNAL CORROSION DA (ICDA)

All pipeline systems can be susceptible to internal corrosion. Even "dry" gas transmission lines can suffer from internal corrosion as a result of upsets at upstream gas processing facilities or the transportation of gas that does not meet quality specifications designed to prevent water condensation. The concept of internal corrosion direct assessment (ICDA) was first formalized in 2002 [15] and was developed based on existing technologies, namely modeling techniques, to evaluate and assess internal corrosion. The original concept behind ICDA for gas transmission lines was that examination of locations where water is expected to accumulate allows conclusions to be drawn regarding the remaining portions of a pipeline.

### 33.4.1 Overview of Technique/Standard

The first NACE ICDA standard for dry gas ICDA (DG-ICDA) was issued in 2006 [16], followed by liquid petroleum ICDA (LP-ICDA) in 2008 [17], and wet gas ICDA (WG-ICDA) in 2010 [18].

The pre-assessment step involves the collection of data that can be used to determine the feasibility of performing ICDA, identify locations of potential water input, and perform flow modeling and corrosion rate modeling/factor analysis. The indirect inspection step involves performing flow modeling in order to determine the location of expected water and/or solids accumulation as well as corrosion rate modeling/factor analysis in order to select locations for examination. In the direct inspection step, which is often referred to as the detailed examination step for ICDA, the pipe is inspected for the presence, extent, and severity of internal corrosion. Non-destructive testing methods are utilized as the internal surface of the pipe cannot be visually examined while the pipeline is in service. In the post-assessment step, remedial actions, if required, are identified and prioritized, reassessment intervals are defined, and the effectiveness of ICDA is evaluated.

The particulars for how the indirect inspection step is executed are different for each ICDA method and are discussed as follows.

### 33.4.2 Dry Gas ICDA

Dry gas ICDA operates under the premise that examination of the furthest downstream locations where water is expected to accumulate allows inferences to be made regarding the integrity of the remaining downstream sections of the pipeline. In the indirect inspection step, basic flow modeling is performed to determine the critical inclination angle past which water is not expected to flow. At pipeline inclinations less than the critical angle, the forces exerted by gas movement overcome the force of gravity acting on the water and cause the water to be transported through the pipe. At inclinations greater than or equal to the critical angle, the force exerted by gravity overcomes the force exerted by gas flow and the water is expected to "holdup" or accumulate. The highest velocity experienced by the pipeline (or other combination of velocity and pressure that yields the greatest critical inclination angle) is used to select the furthest downstream location to which water could have been transported. Two locations downstream from this point are also inspected as confirmation that there is no corrosion downstream from the furthest location where water could have been transported. If internal corrosion is found downstream of the initial site, then either a higher velocity has occurred than the velocity utilized in the critical angle calculation or the pipeline operated outside of a "normally dry" environment at some point historically. In order to account for periods of lower flow, upstream (or subregion) inspections are also performed. If internal corrosion is found at any of the inspection locations, additional inspections are required. Thus, this method is not applicable to pipelines on which extensive internal corrosion is present.

### 33.4.3 Wet Gas ICDA

Wet gas ICDA operates under the premise that the pipeline is continuously wet and therefore focuses on the differentiation in predicted wall loss due to internal corrosion. In the indirect inspection phase, flow modeling is performed in order to identify where different flow regimes are expected to occur and liquid hold-up volumes are estimated for locations where accumulation is predicted. Corrosion rate modeling is then performed to estimate the corrosion wall loss that has occurred during the service life. Multiple scenarios may need to be modeled in order to account for different internal environments that existed during the service life. The number of inspections performed is then determined based on the total length of pipeline being assessed and the predicted wall loss. Success of the WG-ICDA process is determined by comparison of the results found during the direct examination stage versus what was predicted during the indirect inspection step.

### 33.4.4 Liquid Petroleum ICDA

Similar to DG-ICDA, liquid petroleum ICDA is intended to identify locations along the pipeline where water may accumulate. Liquid flow modeling is utilized to determine whether or not water will remain entrained within the liquid hydrocarbon product being transported or will separate out into a separate layer (this separation is referred to as stratified flow). Solids flow modeling is also performed to identify whether solids accumulation is expected to occur. Factor analysis, which considers how different factors impact the corrosion rate and thus corrosion distribution within the pipeline, are used to identify which accumulation locations are most likely to experience internal corrosion. Alternatively, corrosion rate modeling may be utilized to prioritize locations. Inspections are performed at locations where water and solids accumulation is predicted. Again, similar to DG-ICDA, additional inspections are required if internal corrosion is found. Thus, this method may not be applicable to pipelines in which extensive internal corrosion is present.

### 33.4.5 Strengths

Both DG-ICDA and LP-ICDA are best suited for long pipelines with limited numbers of inputs because direct examination requirements are set per region (and new regions are set for each input location). For networks or systems with multiple inputs, DG- or LP-ICDA can quickly become economically unfeasible to perform.

ICDA techniques can be incorporated into overall internal corrosion management programs in order to identify pipelines that are more susceptible to internal corrosion or to select locations for installation of monitoring devices.

### 33.4.6 Limitations

Historical information is crucial to the success of ICDA. Multiple data inputs are needed to perform flow modeling and corrosion rate modeling (where needed). The modeling is intended to be performed on the most severe conditions experienced from an internal corrosion standpoint or for a sufficient number of conditions experienced to fully represent the history of the segment. While it is possible to collect data regarding current conditions; for older pipelines, these data may not sufficiently represent historical conditions. In such cases, it is possible to draw erroneous conclusions based on the limited number of direct examinations performed. Additionally, solids models are less mature than liquid flow models; thus, it is more difficult to accurately predict locations where solids accumulation will occur. One final note from a corrosion rate modeling perspective; widely accepted models that predict the rate and location where microbiologically influenced corrosion (MIC) is expected to occur have not yet been developed. Because MIC is known to be a prevalent internal corrosion threat, which can cause very high internal corrosion rates, this can be a significant limitation for systems in which MIC is occurring.

### 33.4.7 Status of Standards

An additional standard for multiphase flow ICDA (MP-ICDA) is currently under development. Once this standard is complete, the entire suite of internal operating environments will have been addressed. There are National Association of Corrosion Engineers (NACE) task groups working on maintaining and revising, as necessary, the standard practices governing the other three ICDA methodologies.

### 33.4.8 Context of Technique/Standard in Integrity Management

ICDA can be used as a stand-alone integrity management technique, and is routinely used as such, mainly in non-piggable pipelines. However, it is also an excellent tool to use in conjunction with ILI in order to better understand why internal corrosion is occurring at certain locations.

### 33.4.9 Where ICDA Technique Is Headed

ICDA usage continues to grow, especially outside the United States. As users gain experience with the techniques, improvements will undoubtedly continue to be made in the process. Additionally, as improvements continue to be made for both flow and corrosion rate modeling techniques, the ICDA methodologies will continue to increase in accuracy.

## REFERENCES


1. Koch, G.H., Brongers, M.P.H., Thompson, N.G., Virmani, Y.P., and Payer, J.H. (2002) FHWA-RD-01-156: Corrosion Cost and Preventive Strategies in the United States, U.S. Department of Transportation, Federal Highway Administration, March 2002.
2. NACE (2002) NACE Standard Recommended Practice RP-0502-2002: Pipeline External Corrosion Direct Assessment Methodology, Item # 21097.
3. NACE (2008) NACE Standard Recommended Practice RP-0502-2008: Pipeline External Corrosion Direct Assessment Methodology, Item # 21097.
4. NACE (2010) SP0502-2010: Pipeline External Corrosion Direct Assessment Methodology, NACE—ANSI Standard Practice, Item # 21097.
5. Bartol, J.A., Wells, C.H., and Wei, R.P. (1977) Environmentally induced cracking of natural gas and liquid pipelines. U.S. Department of Transportation, Report No. DOT/MTB/OPSPO-78/01, December 1977.
6. Delanty, B.S. and O'Beirne, J. (1991) Low-pH stress corrosion cracking. Proceedings of the 18th World Gas Conference, IGU/C6-91, Berlin, July 8–11, 1991, International Gas Union, Paris, France.
7. NACE (2004) NACE Standard Recommended Practice RP 0204–2004: Stress Corrosion Cracking (SCC) Direct Assessment, Item # 21104.
8. NACE (2008) NACE Standard Practice SP0204-2008: Stress Corrosion Cracking Direct Assessment Methodology, September 2008.
9. Beavers, J.A. (2014) 2013 Frank Newman Speller Award Lecture: Integrity Management of Natural Gas and Petroleum Pipelines Subject to Stress Corrosion Cracking, Corrosion, January 2014, 70(1), 3–18.

10. ASME (2004) ASME B 31.8S-2004: Managing System Integrity of Gas Pipelines. The American Society of Mechanical Engineers, Appendix A-3.
11. ASME (2010) ASME B31.8S-2010: Managing System Integrity of Gas Pipelines. The American Society of Mechanical Engineers, Appendix A-3.
12. Batte, A.D., Fessler, R.R., Marr, J.E., and Rapp, S.C. (2012) A new joint industry project addressing the integrity management of SCC in gas transmission pipelines. Proceedings of Pipeline Pigging and Integrity Management Conference, Clarion Technical Conferences, February 6–9, 2012, Houston, TX.
13. CEPA (2007) *Stress Corrosion Cracking Recommended Practices*, 2nd ed., Canadian Energy Pipeline Association (CEPA), Calgary, Alberta, Canada.
14. CEPA (1997) *Stress Corrosion Cracking Recommended Practices*, Canadian Energy Pipeline Association (CEPA), Calgary, Alberta, Canada.
15. Moghissi, O., Norris, L., Dusek, P., Cookingham, B., and Sridhar, N. (2002) Internal Corrosion Direct Assessment of Gas Transmission Pipelines—Methodology, Gas Technology Institute, Contract No. 8329, April 2002.
16. NACE (2006) NACE Standard Practice SP0206-2006: Internal Corrosion Direct Assessment for Normally Dry Natural Gas Pipelines.
17. NACE (2008) NACE Standard Practice SP0208-2008: Liquid Petroleum Internal Corrosion Direct Assessment.
18. NACE (2010) NACE Standard Practice SP0110-2010: Wet gas internal corrosion direct assessment.

# 34

# INTERNAL CORROSION MONITORING USING COUPONS AND ER PROBES

## A Practical Focus on the Most Commonly Used, Cost-Effective Monitoring Techniques

DANIEL E. POWELL
*Williams, Tulsa, OK, USA*

### 34.1 INTRODUCTION—CORROSION MONITORING USING COUPONS AND ER PROBES

The focus for this chapter is to provide an overview of the primary, most commonly used methodologies for monitoring potential corrosion within the interior of pipelines transporting crude oil, hydrocarbon liquids, or natural gas from production wellheads to the processing facilities, and ultimately to the customers. This monitoring enables us to detect potential threats of internal corrosion, as well as measure the effectiveness of any remediation measures, such as the application of corrosion inhibitor or biocide treatments.

There are numerous books, reports, and standards that highlight corrosion monitoring as a tool for assessing and managing the integrity of pipelines. These include "Corrosion Control in Petroleum Production" [1], "Uhlig's Corrosion Handbook (Third Edition)" [2], "Field Guide for Investigating Internal Corrosion of Pipelines" [3], "NACE Internal Corrosion for Pipelines" [4] manual, "NACE Internal Corrosion for Pipelines—Advanced Course" [5] manual, the internal corrosion direct assessment methodologies [6–8], as well as "Preparation, Installation, Analysis, and Interpretation of Corrosion Coupons in Oilfield Operations" [9], and "Techniques for Monitoring Corrosion and Related Parameters in Field Applications (3T199)" [10], which provides an overview of a number of monitoring techniques, providing the strengths and limitations of each. Our focus in this chapter is on the most commonly used and lowest cost monitoring techniques that are used worldwide, including those in remote locations, where telecommunications or electrical power may be limited.

This chapter will focus on the use of mass loss coupons and electrical resistance (ER) corrosion probes, as those are the most commonly used monitoring techniques. Although there may be specialized applications, such as the use of highly polished EM coupons that are used for evaluating and monitoring the effectiveness of mitigative measures, using electron microscopy, such analysis is very labor intensive and expensive, and as such is primarily undertaken for specialized diagnostics or analyses. Again, the reader is referred to papers that address this specialized topic [11–14].

Although linear polarization resistance (LPR) corrosion monitoring is a common technique, our discussions in Section 34.2.2.7 will be somewhat abridged for two reasons. First, note that the internal corrosion monitoring discussed in this chapter is intended for the assessment of corrosivity within pipelines primarily transporting hydrocarbons, including crude oil, natural gas liquids, and natural gas, with relatively small proportions of produced or condensed water. Documents that discuss corrosion monitoring within pipeline systems transporting hydrocarbon have typically noted a limitation of LPR probes for this application [1,10], citing

*Oil and Gas Pipelines: Integrity and Safety Handbook*, First Edition. Edited by R. Winston Revie.

"the need for a sufficiently conductive medium precludes the use of this technique for many applications in oil and gas, refinery, chemical, and other low-conductivity applications [10]." Another limitation associated with such an electrochemical technique is the "bridging of electrodes with conductive deposits causes short circuits that affect results and can render them invalid until the electrodes are cleaned [10]." Due to these perceived limitations, LPR probes are predominantly used in water handling systems for the crude oil and natural gas production and processing systems (including waterflood systems) and are ideal for such applications.

There are other, fine corrosion monitoring techniques, such as electrochemical noise (ECN or EN) [2,10]. However, such monitoring systems require elaborate telecommunications systems, sophisticated data management, and skilled personnel to properly evaluate the data. Although EN monitoring systems have been successfully implemented at a few facilities, this technique has not been widely adopted. Instead, mass loss coupons and ER probes are still the most commonly used monitoring techniques used throughout the crude oil and natural gas industry for detecting and managing potential internal corrosion.

This chapter has limited discussions regarding the use of ultrasonic thickness measurements as a method to measure internal corrosion rates. NACE International Publication 3T199 "Techniques for Monitoring Corrosion and Related Parameters in Field Applications" [10] cites that "the most commonly used techniques are mass loss coupons, ER probes, linear polarization probes, and ultrasonics." The ultrasonic thickness measurements serve as a direct, nonintrusive technique for measuring metal thickness and calculating corrosion rates. In Section 3.1.1, this NACE publication states that ultrasonics "is normally used for corrosion prediction over long periods of time and is not suited for monitoring real-time corrosion rate changes." This publication also states in their Section 3.2.2 and 3.2.3 for flaw detection and flaw sizing, respectively, that ultrasonics "does not quantify as a corrosion monitoring technique [10]." A recently published paper from NACE Corrosion 2012 reviewed "Nonintrusive Techniques for Monitoring Internal Corrosion of Oil and Gas Pipelines" concluded that "ultrasonic handheld appears to be the most ideal nonintrusive technique currently," and noted a limitation associated with liquid couplant drying out [15]. This paper also pointed out that the handheld ultrasonic testing (UT) can reliably measure pits with an error of $\pm$0.01 in. ($\pm$0.25 mm), and is capable of detecting the location of a defect or pit. Thus, based on the NACE Publication 3T199 and the published paper [15], the ultrasonic wall thickness measurement serves as an excellent tool for inspecting pipe wall and assessing the integrity of pipelines, but cannot offer the precision over a short period of time for closely monitoring corrosion rates and assessing the effectiveness of remedial measures, such as chemical treatments.

We will also have only limited discussions regarding the use of in-line inspection (ILI) data from intelligent pig runs through pipelines. First, a separate chapter, Chapter 35, will discuss ILIs. Second, although many people may view ILI as a monitoring technique, it is actually an inspection technique, and does not have the same sensitivity as monitoring instrumentation placed within the pipeline. ILI runs through individual pipelines are typically several years apart and there are also limitation in the precision of individual measurements versus those from coupons or electronic probes placed within a system. (This will be discussed later in this chapter.) Due to the limited precision and long duration between ILI runs, results should not be used as a basis for controlling remedial measures, for example, chemical treatments, when other internal corrosion monitoring methodologies, such as coupons or electronic probes are available. Nevertheless, corrosion specialists should review any and all ILI data as part of the overall integrity assessment and evaluating potential internal corrosion.

Again, the purpose of this chapter is to provide the reader with a practical perspective for the most commonly used, most cost-effective approach to detecting and controlling potential internal corrosion in crude oil and natural gas gathering systems, processing facilities, and the piping to the customers. The key is to place the monitoring instrumentation along the path where any free water would be swept. We will then present a strategy by which these monitoring results are used to trend corrosion rates and "drive" remedial measures, such as corrosion inhibitor or biocide treatments. Finally, we will present an overview of the relative ability of mass loss coupons or electronic probes to detect active corrosion, versus inspection techniques, such as ultrasonic wall thickness measurements (UT), radiography (radiographic testing (RT)), or magnetic flux leakage (MFL) techniques. It will be apparent that mass loss coupons or electronic probes placed within the process systems offer the best option for detecting and controlling potential internal corrosion, and thus helping to maintain the system integrity.

### 34.1.1 Corrosion—A Definition

Corrosion is the deterioration of a material, usually a metal, that results from a reaction with its environment [7]. As applied to the oil and gas industry, *corrosion* is the unwanted deterioration of the piping or process equipment associated with (1) crude oil or natural gas gathering systems, (2) the production or processing facilities, any water or gas injection systems used to enhance production, (3) the export or sales lines, and (4) distribution systems. If the corrosion processes continues untreated, the cumulative effect could initiate sufficient deterioration that would affect the integrity of the piping or process equipment, and possibly result in failure.

### 34.1.2 Corrosion and Use of Coupons and ER Probes as Integrity Management Tools

In the following sections, we will present an overview of internal corrosion monitoring, based on the use of corrosion coupons or electrical probes (primarily ER) probes for detecting corrosion. (These are the most commonly used monitoring tools per NACE 3T199 [10] for pipelines transporting liquid or gaseous hydrocarbons.) Strengths and limitations will be noted. Next, we will present the placement of the corrosion monitoring within the process streams and in particular the pipelines. "Placement" addresses not only the axial position within a pipeline, but also the optimum orientation or azimuth for installing corrosion monitoring instrumentation. We will then present a strategy for applying chemical treatments, primarily corrosion inhibitors, such that corrosion rates can be kept low, and the designed service life can be achieved. Next, we will compare the relative sensitivities of nondestructive inspection techniques, such as ILIs, radiography, and ultrasonic inspections versus corrosion coupons or electronic probes to detect the onset of corrosive conditions. Finally, we will have a brief discussion of fluid sample analysis, which should be used as an independent check to verify or confirm internal corrosion monitoring results. Collectively, this approach should provide the reader with a comprehensive overview of internal corrosion monitoring, and how to use those results to maintain system integrity.

## 34.2 CORROSION COUPONS AND ELECTRICAL RESISTANCE CORROSION PROBES

Internal corrosion monitoring is the tool used by corrosion professionals to detect the onset of corrosive conditions. The discussions will be focused on monitoring corrosion rates for crude oil, hydrocarbon liquids, and natural gas in gathering pipeline, production facilities, and sales or export/transmission pipelines to the customers. Our focus will be on the use of mass loss coupons and ER corrosion probes that are placed within pipelines. These are the primary "work horses" of the oil and gas industry. However, the monitoring techniques discussed can also be used in other industries, such as the chemical industry, water processing, etc. In Section 34.6, we will briefly discuss fluid sampling analysis, which complements the metal coupon and electronic probe measurements of corrosion rates, and should provide independent confirmation of the onset of internal corrosion.

The most obvious question is "Why does the oil and gas industry place such a priority on the use of metal coupons and ER probes, as opposed to the use of (a) nonintrusive techniques, such as ultrasonic wall thickness measurements or radiography, (b) comparatively modern monitoring techniques, such as EN or ECN, or, (c) or others, such as the use of EM coupons?" First, we need to define "intrusive" and "nonintrusive." Please refer to Section 34.8, which provides definitions from NACE's 3T199 [10]. "Intrusive" monitoring requires the monitoring coupon or electronic probe to be placed directly within the test environment, for example, within the pipeline. In contrast, "nonintrusive" monitoring is a measurement from the outside of the pipe or vessel, such as a measurement of the remaining thickness of the pipe wall. Generally, corrosion engineers prefer monitoring results from coupons or probes placed directly within the test environment or process stream. However, if those results are not available, monitoring results from nonintrusive techniques would become the primary analytical tool.

Here are a few considerations that highlight why mass loss coupons and electronic probes are the most widely used monitoring tools—particularly for remote locations or when cost considerations may limit monitoring options:

- (a) The nonintrusive monitoring techniques such as ultrasonic (UT), radiography (RT), and do not have the sensitivities that are offered by intrusive monitoring. We will discuss the relative sensitivities of different monitoring techniques in Section 34.5. Corrosion coupons and ER probes offer the best, low-cost option for monitoring and early detection for the onset of corrosion. Their use allows appropriate remedial measures to be quickly implemented before significant damage occurs. Corrosion coupons are used to measure general and localized (pitting) corrosion rates typically in 60, 90, or 180 days exposure cycles. ER probes provide general corrosion rates in near real time, based on comparison of the present and previous readings.
- EN or ECN monitoring techniques require sophisticated telecommunication systems, which may not be practical for monitoring at remote locations. When ER probes are used to make their measurements, the calculated corrosion rates are dependent on the previous reading. The effects from any corrosion, which may have occurred between measurements, will be detected by the measured change in resistance. In contrast, each EN or ECN measurement is independent of the previous reading. Thus, in order to detect any corrosion-related events, the time between readings must be shortened to seconds—not hours. (The same limitation also applies to LPR measurements.) It is only through the employment of sophisticated telecommunication systems and an extensive collection of data that corrosion professionals will be assured that EN or ECN techniques will have detected any corrosion-related event. Since many gathering system, pipelines, or cross-country transmission lines are well away from production stations or compressor stations, the installation of the infrastructure to support corrosion monitoring could be an expensive proposition. Consequently, coupons or ER probes,

coupled with remote data collectors, are favored by many pipeline operators.

- EM coupons are highly polished coupons that are removed from the system and are examined with scanning electron microscopy. This helps characterize the corrosion mechanisms and the possible initiation of any pits, but is a labor-intensive technique [12–14]. As such, EM coupons are primarily utilized for special diagnostics.

Consider mass loss metal coupons first.

### 34.2.1 Metal Coupons

The vast majority of pipelines found in gathering systems, processing facilities, or transmission pipelines are constructed of carbon steel. Accordingly, one of the most commonly used internal corrosion monitoring technique is based on inserting carbon steel mass loss coupons into the process stream for a set period of time, for example, 90 or 180 days, and then removing and cleaning them to remove any corrosion products. The reduction in mass over the exposure period is used to determine the internal corrosion rate. If pits are observed within the metal coupons at the end of the exposure period, then a pitting rate would be based on the depth of the deepest pit and the exposure period [4,9].

(If we are to monitor corrosion rates within a section of a gathering pipeline or piping within a production facility, which has a different metallurgy, the corrosion monitoring should be based on that metallurgy. For example, coupons used to monitor a pipeline constructed with duplex stainless steel should be made from the duplex stainless steel. The companies that provide metal coupons or electrical probes for corrosion monitoring, will typically offer to fabricate coupons or electronic probes using whatever base material is required to match system metallurgy. However, there is likely to be a premium charge for the customization.)

Figure 34.1 illustrates the variety of metal coupons that are commercially available.

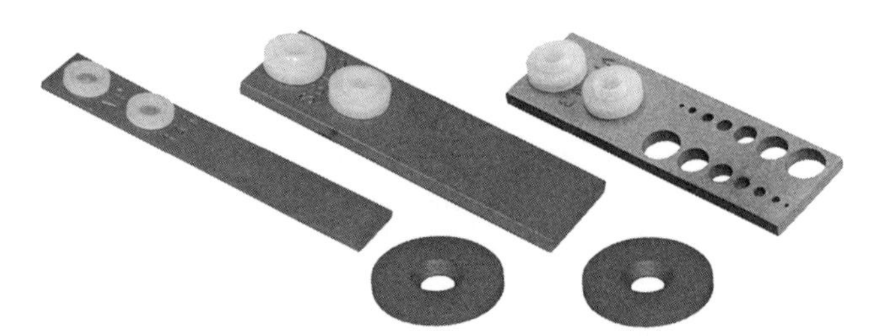

**FIGURE 34.1** Manufacturers offer a variety of styles of metal coupons for monitoring internal corrosion. *The coupon on the right is used to detect scale depositions.* (Photo, courtesy of Rohrback Cosasco Systems.)

***34.2.1.1 Metal Coupons—Determining General Corrosion Rates*** It is fairly straightforward to determine general corrosion rates from metal coupons removed from the system. We will start with a known initial mass for the coupon. The coupon will also have a known geometry, whether it has a rectangular shape, a cylindrical shape, or appears like a round washer. The initial dimensions of the metal coupon will determine the total surface area exposed to the environment being monitored. The density of the material, which is the mass divided by volume, is known. The date or time the metal coupon was installed within the system, and the date or time that the metal coupon was removed from the system, are also known. Finally, we measure the weight of the coupon after it has been removed from the system. The weight of most interested in is the final weight, after any deposition and corrosion products have been removed. (It may also be useful to document the appearance of the coupon as soon as possible after the coupon is removed from service, including the color and depth of any deposits or corrosion products.) From the already mentioned data, we can then calculate the corrosion rate.

NACE RP-0775–2005 (or the most current year) provides several equations for determining corrosion rates [9]. The basic formula for the corrosion rate in mils per year (mpy) or in mm/year is

$$\mathrm{CR} = F\times[W/(A\times T\times D)] \quad \text{mpy or mm/year}^1 \qquad (34.1)$$

where

$F$ = Conversion factor, depending on units of input and whether output is in mpy or mm/year
$W$ = Mass loss in grams (g)
$A$ = Initial exposed surface area ($mm^2$ or $in.^2$)
$T$ = Exposure period in days
$D$ = Density of the metal coupon (typically $g/cm^3$)

To report corrosion rates in mpy, use

$$\mathrm{CR}=(2.227\times 10^4)*[W/(A\times T\times D)] \quad \text{mpy [9]}^2 \qquad (34.1a)$$

where

$W$ = Mass loss in grams (g)
$A$ = Initial exposed surface area ($in.^2$)
$T$ = Exposure period in days
$D$ = Density of the metal coupon ($g/cm^3$)

(Note that 1000 mils equal 1 in.)

[1] © NACE International 2005.
[2] © NACE International 2005.

To report corrosion rates in mm/year, use

$$\text{CR}=(3.65 \times 10^5)*[W/(A \times T \times D)]\ \text{mm/year}\ [9]^3 \quad (34.1\text{b})$$

where

$W$ = Mass loss in grams (g)
$A$ = Initial exposed surface area ($mm^2$)
$T$ = Exposure period in days
$D$ = Density of the metal coupon ($g/cm^3$)

Note that corrosion rates, which could be reported in mm/year, may also be reported in μm/year:

$$1\ \text{mm/year} = 1000\ \mu\text{m/year} \quad \text{and} \quad 25.4\ \mu\text{m/year} = 1\ \text{mpy}$$

The factors in Equations 34.1a and 34.1b are derived from the conversion factors, such as inches to centimeters, millimeters to centimeters, inches to mils, and days to years, to generate corrosion rates in mpy or mm/year. For calculating general corrosion of coupons, we make the fundamental assumption that the exposed surface area remains constant, and the only thing that changes is the relative thickness of the coupon.

Consider as an example, a carbon steel coupon, which had a 0.0510 g mass loss over a 92-day exposure. Also, note that the steel has a density of 7.86 $g/cm^3$. Assume an exposed surface area of 3.40 $in.^2$. For this example, we want to determine corrosion rate in mpy, where 1000 mils = 1 in. Here's an example of how the corrosion rate is calculated

$$\text{CR (mpy)} = 2.227 \times 10^4 \times \frac{0.051\ \text{g}}{3.40\ \text{in}^2 \times 92\ \text{days} \times 7.86\ \text{g/cm}^3} = 0.46\ \text{mpy}^4 \quad (34.2)$$

In this example, the calculated corrosion rate is 0.462 mpy, which is well below a generally accepted maximum acceptable corrosion rate of 1.0 mpy, as per NACE RP-0775-2005 [9].

#### 34.2.1.2 *Metal Coupons—Determining Pitting Corrosion Rates*

When corrosion coupons are placed within pipelines, it is possible that pits may form on the metal coupons, just like they could form on the base metal of the pipe. If left unchecked, pits in pipelines could develop and grow to the point of penetrating through the wall thickness, and could result in a leak. Accordingly, an inherent point of internal corrosion monitoring is to check the metal coupons for evidence of pit formation. Note that pits could form for many different reasons, whether from microbiological results, such as the action of sulfate-reducing bacteria (SRB) or acid-producing bacteria (APB), oxygen ingress, $H_2S$, or perhaps from the presence of carbon dioxide, which reacts with water to form carbonic acid.

As in the case of determining general corrosion rates, the first step is to clean the coupons by removing any corrosion products from the surface of the metal coupons. Then, the depth of individual pits can be measured either through the use of a light microscope or by using a needle gauge and comparing that reading to one taken on unaffected base metal. There are also developing technologies that are just being introduced in 2013, which use laser or three-dimensional structured light to profile pits. These new pit depth/profile technologies are expected to become more widely used in future applications due to the reduced time required for each assessment [16].

Although the depth of the deepest pit is of the greatest interest, it should also be noted whether the pit was an isolated pit, or if there are numerous pits in close proximity, for example, a network of pits.

Unlike laboratory work, the analysis of pitting observed on metal coupons must rely upon some basic assumptions. We know when the coupon was installed, when it was removed, and we also know the depth of the deepest of all the pits. However, we do not know the specific date the pit initiated. Hence, determination of a pitting corrosion rate has to be made, based on the date of installation.

#### 34.2.1.3 *Categorizing General and Pitting Corrosion Rates*

NACE RP-0775-2005 presents a listing of general corrosion rates and pitting corrosion rates, and categorizes them as "low," "medium," "high," and "severe" [9]. These are presented in Table 34.1.

We use this table as a guideline for defining acceptable and unacceptable corrosion rates. We would take the worst, highest general or pitting corrosion rate for any individual corrosion coupon, and use that value to categorize the observed internal corrosion. If the categorization is above acceptable limits, that would be used as a basis for adjusting chemical treatments or other mitigative measures. (These are discussed in Section 34.4.)

Before leaving this topic, we should note that the maximum acceptable corrosion rate is really based on the remaining corrosion allowance of a pipeline and the projected service life. Pipelines are designed to have a certain wall thickness, which is required for pressure containment. This portion of the pipe wall thickness is based on the maximum allowable operating pressure (MAOP). The additional portion of the wall thickness is the allowance for corrosion. Corrosion professionals use this additional wall thickness (that beyond the thickness required for pressure containment) and the projected remaining service life to determine the maximum acceptable corrosion rate. A pipeline, which has experienced very little corrosion throughout most of its service life, may have sufficient additional wall thickness

**TABLE 34.1 Categorizing Carbon Steel Corrosion Rates for Oil Production Systems [9]**

| | Average Corrosion Rate | | Maximum Pitting Rate | |
|---|---|---|---|---|
| | mm/year | mpy | mm/year | mpy |
| Low | <0.025 | <1.0 | <0.13 | <5.0 |
| Moderate | 0.025–0.12 | 1.0–4.9 | 0.13–0.20 | 5.0–7.9 |
| High | 0.13–0.25 | 5.0–10 | 0.21–0.38 | 8.0–15 |
| Severe | >0.25 | >10 | >0.38 | >15 |

mm/year = millimeters per year, mpy = mils per year (1000 mils = 1 in.).
*Source:* NACE RP-0775-2005.

(beyond that thickness required for pressure containment) to allow the acceptable corrosion rates to exceed the 1.0 mpy (0.025 mm/year) guideline late in service life. In contrast, a pipeline, which has operated under more severe conditions, and has only a small remaining corrosion allowance, might have a maximum acceptable corrosion rate below the 1.0 mpy (0.025 mm/year) guideline late in the service life for that particular pipeline.

[Another option for pipeline and corrosion engineers to consider is reducing the MAOP for the section of pipe. That would reduce the portion of the pipe wall thickness required for pressure containment, but would increase the portion of the pipe wall thickness for the corrosion allowance. However, this option should only be used very judiciously.]

### 34.2.2 Electrical Resistance Probes

ER probes are very valuable tools for measuring internal corrosion rates without having to physically remove the sensing element from the process stream. The corrosion control professional can simply go to the location of the ER probe and make an electronic connection to the probe or remote data logger, and record or download the data, which would then be taken to an office for assessing the corrosion rate trends. Figure 34.2 illustrates several styles of ER probes used to monitor internal corrosion within the field, and will be discussed in more detail momentarily. Figure 34.3 illustrates how corrosion rate measurement data can be retrieved "from the field" through a handheld data logger. (Intrinsic Safe Data Loggers are commercially available for collecting or downloading data in hazardous environments.)

***34.2.2.1 ER Probes Determine Corrosion Rates by Measuring Changes in Resistance*** All ER probes use measurements of the resistance of the sensing unit to quantify the internal corrosion rates. Internal corrosion mechanisms cause the resistance of the probe's sensing element to change (increase). Hence, the internal corrosion rate can be determined from successive measurements of the resistance of a wire loop, strip, or cylindrical element style ER probes versus time. Note that the "raw data" from ER probes reflect the cumulative general corrosion on the probe's sensing element, including the effects from any short duration transient event that may have occurred between readings. Thus, like metal coupons, ER probe measurements reflect cumulative corrosion, and the consequences of any transient corrosive events, which might occur between readings, will not be missed.

Note that ER probes are designed to measure general corrosion rates, rather than pitting corrosion rates. If pits form on the sensing element, it will be obvious that this data is invalid, since successive readings typically follow an

**FIGURE 34.2** Various corrosion probe designs. (Photo, courtesy of Rohrback Cosasco Systems.)

FIGURE 34.3 ER probe connected to remote data collector. Note the cable connected to a probe located on top of this pipe. The cable connects to a remote data collector, which can be programmed to measure the corrosion rate at regular intervals, such as every 15 min, hourly, daily, weekly, and so on. When the hand held device is connected to the remote data collector, the archived data can be retrieved for analysis. Such a data collection system is beneficial for remote locations typical of the oil and gas industry. However, for monitoring corrosion rates near compressor stations or processing facilities, it may be more beneficial to connect the electronic probes to the facility's SCADA systems. (Photo, courtesy of Rohrback Cosasco Systems.)

exponential curve. When such behavior is observed, the ER probe should be replaced.

Temperature effects are minimized by a second, protected element within all ER probes. Readings from the protected element are compared to those from the exposed element, thereby providing the basis for calculating the "true" metal loss and corrosion rate. Note that probe design is critical for obtaining valid monitoring results, and details can be found in technical literature provided by the probe manufacturer.

If electrical probes are covered by conductive corrosion products or deposits, the probes would not be able to accurately measure corrosion rates. (Presumably, the pipe walls would be in a similar condition.)

The following sections describe the most common types of ER probes. These are the wire loop, tubular type, tube type, spiral type, and flush type probes. Each has a different surface area, sensitivity, and effective probe lifetime. Figure 34.4 depicts the appearance of several styles of ER corrosion probes.

Probe manufacturers often use a letter to indicate the different types of probes, such as W for wire, T for tubular or tube, F for flush probe, and so on. The numerical value following the letter designation indicates the relative thickness of the sensing element, which indirectly provides the effective service lifetime of the probe.

As an example, a T10 probe would be a tubular type probe, which would have an effective lifespan of ½ the probe's rated element thickness, which in this example would be 5 mils (127 μm) corrosion. Thus, if the corrosive environment was 5 mpy (127 μm/year), the probe could theoretically be used for up to 1 year. However, if the corrosion rate was 20 mpy (508 μm/year), the effective life of the probe would

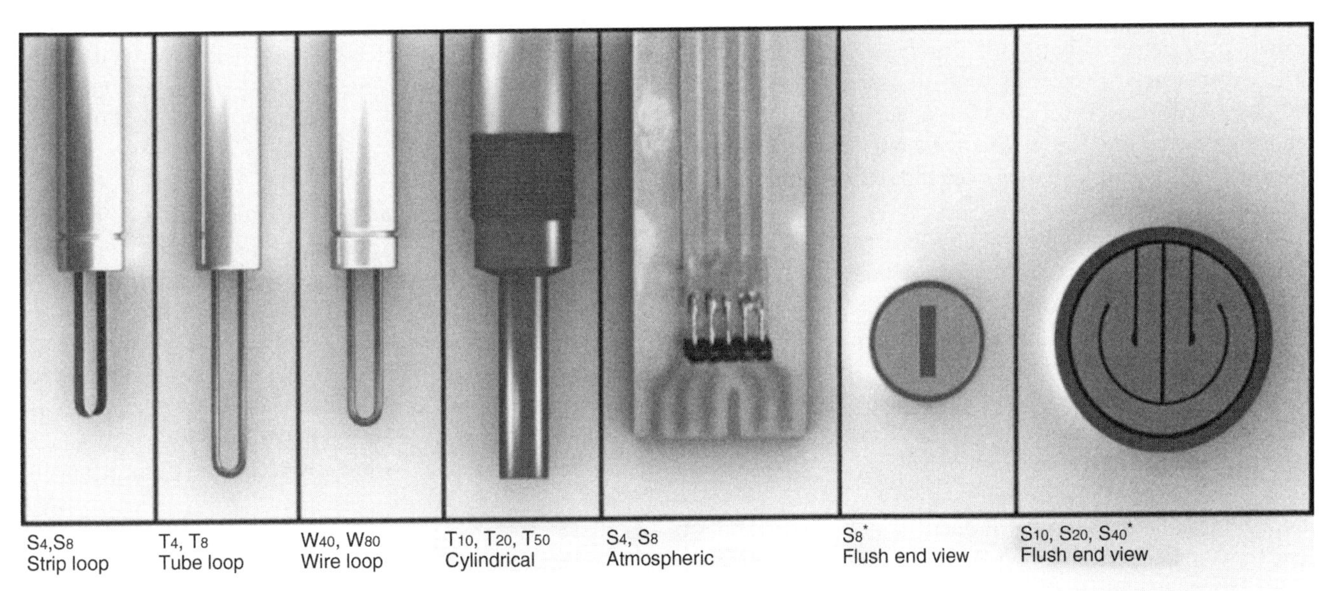


FIGURE 34.4 Various styles of ER probes. (Illustration courtesy of Rohrback Cosasco Systems.)

be only 90 days. Consult with the probe manufacturer to obtain the most current data regarding each of their probes, and select the most appropriate probe having a reasonable service life, while at the same time having adequate sensitivity for detecting corrosion.

*34.2.2.2* ***Wire Loop ER Probes*** Wire loop ER probes are commonly used due to their high sensitivity. Basically, they appear to have what appears to be a simple wire loop extending from the end of the probe into the system being monitored. Wire loop probes appear like the probe on the left in Figure 34.4. These probes provide a high sensitivity for detecting the onset of corrosive conditions, and are very useful when first assessing corrosion rates within a new pipeline or when conducting performance tests, such as for the selection of new corrosion inhibitor products. However, the temperature stability of wire loop ER probes can be relatively poor, and the probe elements are subjected to premature failure (due to fatigue) under high flow rates or when particulates such as sand or formation fines are transported through the pipeline and strike the sensing element. Velocity shields are recommended to help protect the sensing element of wire loop ER probes.

Wire loop ER probes can be used in small diameter pipelines, or can be located near pipe walls. A variety of sizes for the wire elements are available, each offering a trade-off between sensitivity and the service life of the probe. Probe response time charts are available from the probe manufacturers. As an example, those charts would show that for a 10 mil (254 μm) span wire loop probe, it would take approximately 100 h or approximately 4 days in order to detect a 10 mpy (254 μm/year) corrosion rate. Those charts also show that it would take approximately 8 days to detect a 5 mpy (127 μm/year) change in corrosion rate, and a bit over a month to detect a 1 mpy (25.4 μm/year) corrosion rate. These response times are very short compared to the typical 90-day exposure cycle for mass loss coupons. That provides an insight regarding why wire loop are often used when first establishing baseline corrosion rates and setting the chemical treatment rates.

Note that wire loop ER probes are susceptible to iron sulfide bridging, which may occur within sour systems. "Bridging" refers to the deposition of iron sulfide, which creates temporary low-resistance pathways across the deposition ("bridge"). When the deposition is on the sensing element (here a wire loop ER probe), the new electrical path will render measurements invalid. Hence, wire loop ER probes should not be used in sour service.

*34.2.2.3* ***Tubular Loop ER Probes*** Tubular loop probes are very similar to wire loop ER probes. They use capillary tubes instead of wires as the sensing element; see Figure 34.4. The sensing element is quite weak but is very sensitive due to the 0.004 in. (101.6 μm) wall thickness of the capillary tube. Tubular loop ER probes are normally used with flow detectors, and are placed in low shear rate environments.

Like wire loop ER probes, tubular ER probes are prone to premature failure due to fatigue at the base of the element, where the capillary tube extends from the end of the probe. Velocity shields may help reduce fatigue and premature failure of the capillary tubes due to fluid flow.

Tubular loop ER probes are also subjected to sulfide bridging if the probe is installed in sour service. Hence, they are not recommended for that service.

Since tubular ER probes are able to detect corrosion rates as small as 1 mpy (25.4 μm/year) in 10–12 days, they are also commonly used when first measuring the corrosion rates within a pipeline. They may also be used when conducting performance tests for selecting new corrosion inhibitor products.

*34.2.2.4* ***Tube or Cylindrical ER Probes*** Tube or cylindrical ER probes have a large surface area, and a stout design. They appear like tubes or cylinders, as depicted in Figure 34.4. These stout probes have a high resistance to premature failure, and are designed for environments having potentially high corrosion rates, including locations where there may be significant corrosion or erosion. Tube ER probes would be the probe of choice for oil production pipelines having high flow rates. The tube or cylinder shaped ER probes are also less susceptible to sulfide bridging, and can be used for monitoring corrosion in sour service. They should have relatively long service lives when placed in environments having relatively low corrosion rates.

The tube type ER probes appear like solid cylinders and are available in various element sizes, such as 0.01 in. (254 μm), 0.02 in. (508 μm), and 0.05 in. (1270 μm) elements for effective element thickness of 0.005 in. (127 μm), 0.010 in. (254 μm), and 0.025 in. (635 μm). It would require a T10 probe (0.010 in. (254 μm) effective element thickness) approximately 25 days to detect a 1 mpy (25.4 μm/year) corrosion rate.

*34.2.2.5* ***Flush Mounted ER Probes*** Flush mounted probes are designed for placement into pipelines or vessels without protruding beyond the wall thickness. The two probes on the right side of Figure 34.4 depict a top view of two different designs for the sensing elements of flush mounted ER probes. Note that unlike intrusive probes, which extend part way into the diameter of a pipeline, flush mounted probes are even with the pipe wall, and can be left within pipelines, while the pipeline is mechanically cleaned by pigs.

There are advantages and disadvantages with each design. The size of the sensing elements may be restricted, due to the size of the access fitting. The most common flush mounted probes are the F5 and F10 models, which offer between 5 and 10 mil (127 and 254 μm) range or span for cumulative corrosion measurements. (Thicker elements have been made, but have been less effective for the corrosion rates normally encountered in the oil and gas industry.)

For most oil and gas applications, the optimum flush mounted probe would be a flush mounted probe having a 5 mil (127 μm) range or span, which is oriented at 6:00 o'clock on horizontal section of piping. It would require approximately 10–12 days to detect a change, based on a 2 mpy (50.8 μm/year) corrosion rate.

When used in sour service, flush mounted ER probes, which have a sensing element with a spiral or complex design, are susceptible to sulfide bridging. If the sensing element is in the shape of a rectangle, it will have a very short resistance path, and a very low resistance. However, probes with that design have a high "noise" level, which means a longer response time is needed to separate the true signal from "noise." Consequently, regardless of the design of the probe element, flush mounted ER probes are not recommended for monitoring within sour service conditions. Instead, it would be more advantageous to use high-resolution ER techniques for such a specialized application.

#### *34.2.2.6 High Sensitivity or High-Resolution ER Probes*

High-resolution ER probes (Figure 34.5) are "enhanced" ER probes, which provide metal loss data that can be converted to corrosion rates. These probes use a similar method of measurement compared to the conventional ER probe, but provide significantly improved compensation for temperature changes. Modifications to instrumentation and probe design have also contributed to significant improvements in electronic noise reduction and enhanced measurement resolution.

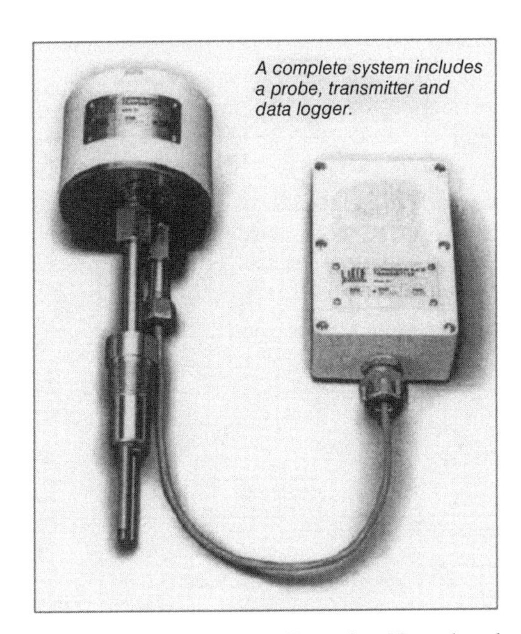


**FIGURE 34.5** High-resolution ER probe. Note the electronic package above the corrosion probe's sensing element. This helps to reduce/enhance the signal to noise ratio, improving the resolution. (Photo, courtesy of Rohrback Cosasco Systems.)

High-resolution ER probes are commercially available. The probes must be continuously monitored, using either a data logger, or an online data collection system. The higher resolution allows trend information to be collected in a shorter period. High-resolution ER probes have perform well in sour service conditions (2–10% $H_2S$), where iron sulfide bridging could occur, and most conventional probes would be inoperable.

According to product literature, the sensitivity of high-resolution ER probes can be as low as $2 \times 10^{-7}$ mm (0.2 μm = 0.0079 mils). The main application for these probes would be in oil and gas gathering systems, where corrosion rates would be expected to be approximately 1–3 mpy (25–76 μm/year).

#### *34.2.2.7 Linear Polarization Resistance (LPR) Probes*

LPR probes use a totally different technique for measuring corrosion rates than is used by the ER type probes. Unlike the ER probe, which uses two or more successive readings of the probe's resistance to calculate a corrosion rate, each LPR measurement of the corrosion rate is independent of previous readings. LPR readings are essentially "real-time" measurements of corrosion and cannot detect any transient corrosive conditions, which may have occurred between LPR readings. Consequently, the time between readings needs to be very much shorter than for ER probe readings in order to be able to detect corrosion associated with transient conditions.

There are two basic designs of the LPR probe, namely, the three-electrode and two-electrode designs. Figure 34.6 depicts a two element LPR corrosion probe. For each type of probe, a fixed potential difference is established between the electrodes, and the resulting current is measured. At the low potential differences used in this polarization technique, the corrosion rate is proportional to the corrosion current. (A detailed explanation of linear polarization techniques is available from LPR manufacturers.)

The three-electrode LPR probe used to be superior to the two-electrode probe, since the three-electrode LPR probe

**FIGURE 34.6** The two-electrode LPR probe. (Photo, courtesy of Rohrback Cosasco Systems.)

could better compensate for IR drops in low-conductivity fluids. However, for the last 20 years, instruments have been commercially available that compensate for IR drop by using a DC measurement plus high-frequency AC compensation. Hence, the issue of using either a two- or three-electrode LPR probe is no longer relevant.

LPR techniques should be used for measuring general corrosion—not pitting corrosion. The LPR technique requires conductive solutions around the electrodes, and works best when the solution is hydrocarbon free [10]. LPR probes have historically not been recommended over ER probes for measuring corrosion rates in oil or gas systems, which are predominantly hydrocarbon based, and have low water cuts [10]. The hydrocarbons tend to film or coat the electrodes, lowering the electrical conductivity. Also, if the sensing elements are coated with iron sulfide (FeS), there could be a bridging between the elements, which could also distort or invalidate the results. Consequently, there should be limited usage of LPR probes in sour environments [10]. However, LPR probes are great for water applications, such as water processing or water injection systems.

## 34.3 PLACING CORROSION MONITORING COUPONS OR ELECTRONIC PROBES WITHIN PIPELINES

"Garbage in, garbage out" is a commonly used adage to emphasize the importance for obtaining good—valid—data. It also applies to the collection of internal corrosion monitoring data. Many pipeline and facility designers make the assumption that the flow through a pipeline is a homogeneous mixture. When access fittings for internal corrosion monitoring coupons or electronic probes are placed within systems, they are often placed at convenient locations—not at locations or orientations, where internal corrosion is most likely to occur. In this section we will describe the strategies for placing the internal monitoring weight loss coupons or electronic probes within pipelines. First, we are going to discuss where along the length of a pipeline the monitoring should be placed, and then, we will discuss the orientation of that monitoring.

### 34.3.1 Placement of the Corrosion Monitoring Point on a Pipeline

Consider a major trunk line or pipeline segment having no other pipelines flowing into or out of the pipeline. Liquids or gas entering the pipeline will also exit the pipeline. Thus, we have a system, such as depicted in Figure 34.7. For this illustration, we have a starting point, where the trunk line departs from a pad containing numerous producing wells, and then flows to a crude oil or natural gas processing facility.

Two monitoring points are illustrated in Figure 34.7. One monitoring location is at the beginning of the line and just upstream of the chemical injection point. This location could be used for obtaining baseline internal corrosion rates. The second location is at the end of the pipeline just upstream of the gathering facility. The second is the more important of the two monitoring points, as it provides the internal corrosion rate for the pipeline, and provides a measure for the effectiveness of corrosion inhibitors in controlling (minimizing) internal corrosion.

The strategy is that regardless of how corrosive the liquids or gas entering the pipeline, *the corrosion inhibitors injected at the start of the pipeline should be adjusted to a sufficient rate in order to attain acceptable corrosion rates at the end of the pipeline*. The point is to ensure there is sufficient chemical remaining within the process stream at the end of the pipeline to protect the base metals, and that it had not been consumed upstream of that monitoring point.

Let's consider a real world example from Southeast Asia. Production from a number of wells on one offshore platform fed a pipeline that transported wet (saturated) gas to a second downstream platform. Figure 34.8 illustrates the monitoring point on the downstream platform, and represents the end of that pipeline, before the production was comingled with any other production. Note that the internal corrosion monitoring probe was installed through an existing fitting on the piping at this downstream platform.

Besides for being at the end of the pipeline, the monitoring point was also in a (relative) low point on the process stream, where any free water (produced or condensed) could accumulate. Thus, the monitoring would be at a point where water could accumulate, and internal corrosion is more likely to occur.

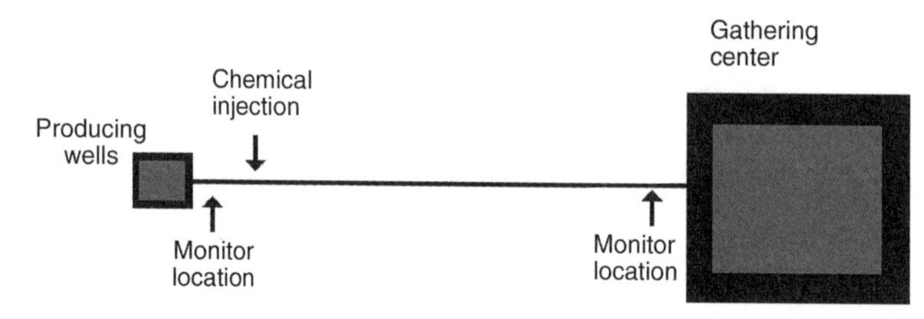


**FIGURE 34.7** Ideal location for monitoring internal corrosion [17,19]. (© NACE International 2001, 2002.)

**FIGURE 34.8** Location of internal corrosion monitoring near the end of one subsea pipeline [17,18]. (© NACE International 2001.)

Before discussing the monitoring results, we should bring to the reader's attention the internal corrosion direct assessment methodologies [6–8]. Modeling is a key to knowing where any liquids could accumulate in pipelines transporting natural gas. These "hold up" points are ideal locations for monitoring any potential internal corrosion, provided they are accessible. (The reader is referred to the NACE direct assessment standard practices and models.) For the piping depicted in Figure 34.8, we note that the gas flow came from a subsea pipeline that ran to the base of the offshore platform, followed by a vertical riser, a short horizontal run, and then a short dip to the horizontal section depicted in the figure. Thus, the line had an obvious low point, which was easily accessible and where any liquids would accumulate. As such, it was an obvious monitoring point.

Figure 34.9 presents a graph from an ER corrosion probe installed at this downstream location and used to measure the corrosion rates. The initial slope illustrates the baseline corrosion rates. The vertical line on the left marks the time when corrosion inhibitors were first injected. (The inhibitors were applied continuously, rather than in batch treatments.) Note how the corrosion rates almost immediately dropped to zero. Finally, the vertical line on the right illustrates the time when the corrosion inhibitor injection was stopped. Note how the low corrosion rates continued for a number of days, before returning to the baseline corrosion rate. This illustrates the film persistency of the corrosion inhibitor and how protection continues for a brief period of time—even if there is a short interruption in the treatments, such as a failure of the pumps or a logistical disruption.

The example mentioned earlier shows how internal corrosion monitoring can be used for a single line segment, whether it is part of an upstream gathering system or part of a crude oil, hazardous liquids, or a natural gas transmission pipeline [17,18]. Let's now look at the placement of internal corrosion monitoring devices within a gathering system having several laterals.

Next, consider the gathering system illustrated in Figure 34.10 [17,19]. We have production at three different sets of wellheads or drill site pads. Production from two drill site pads are first mixed together, and then production from the third drill site pad is added to the process stream. The combined production then flows to the crude oil and natural gas processing facility, identified as a gathering center in the Figure 34.10.

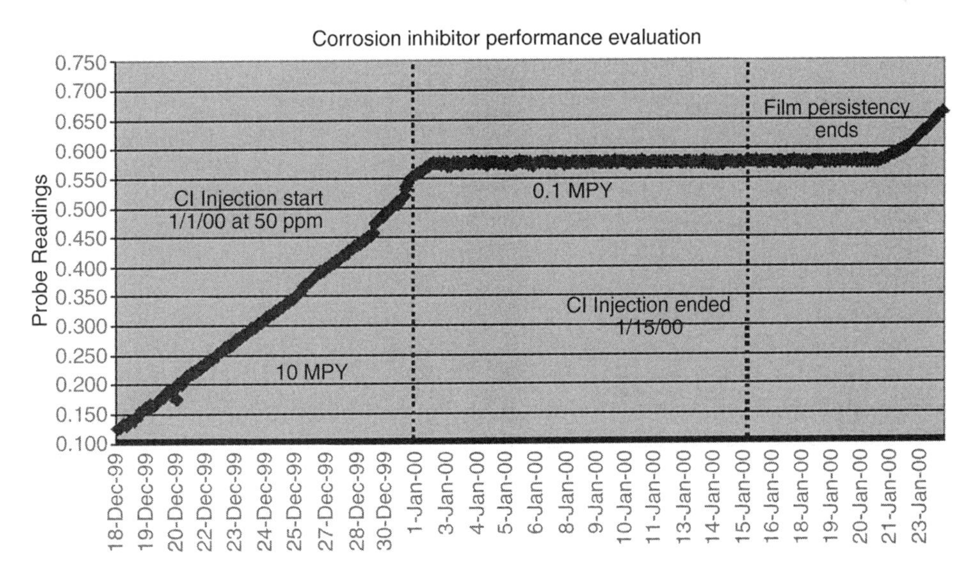


**FIGURE 34.9** Monitoring results from an ER probe on a pipeline and how they are used [17,18]. (© NACE International 2001.)

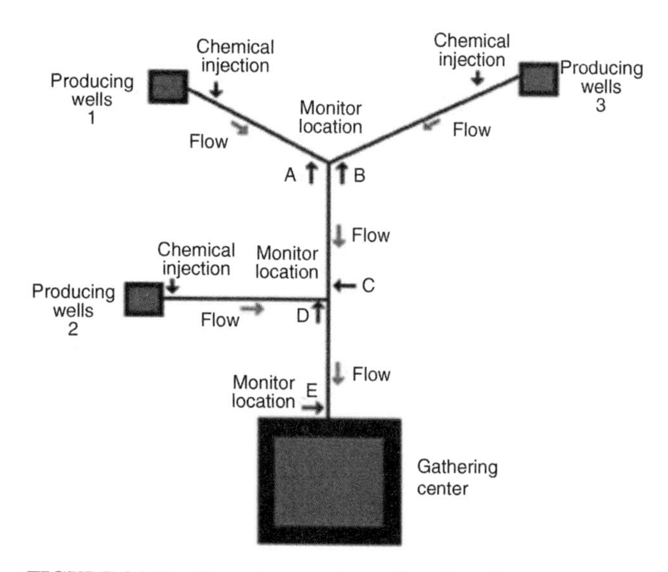


FIGURE 34.10 Corrosion monitoring locations for a three-phase gathering system (gas, oil, water) [17,19]. (© NACE International 2001, 2002.)

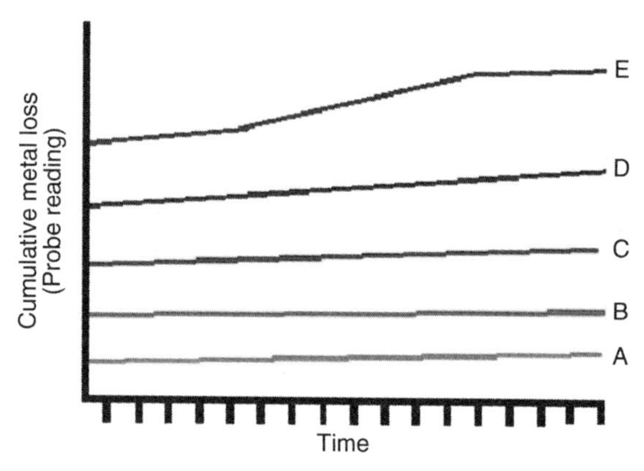


FIGURE 34.11 Overlaying results from multiple monitoring points to ensure acceptable corrosion rates and to optimize chemical treatments.

The production stream from producing wells locations 1 and 3 near the top of Figure 34.10 will flow downward to a mixing point. We should install monitoring points near the end of each of these two laterals, just upstream of the point where the production mixes, and the monitoring points are depicted as points A and B in the figure. Note the location for the chemical injection for the producing wells at 1 and 3. If corrosion monitoring rates measured at points A or B are above acceptable limits, we would adjust (increase) the treatment rates for producing wells at 1 or 3 (or both) until acceptable corrosion rates are obtained for each line. There would be no change to the chemical treatment rates, if the corrosion rates were already at or below acceptable limits.

Next, we look at monitoring results from the combined flow from 1 and 3. This is monitored at point C. If the corrosion rate from the combined flow is at or below acceptable limits, we would make no changes to the chemical treatments for the production from producing wells at 1 or 3. However, if the corrosion rates were above the acceptable limits, we would increase the treatments at either 1 or 3 or both until acceptable rates are obtained at points A, B, and C.

Let's now consider flow from producing wells 2. These are monitored at point D. As before, if the monitoring results are not acceptable at point D, we would increase treatments for producing wells at location 2 until acceptable corrosion rates are achieved at point D.

Finally, we look at monitoring results at point E. This is just upstream of the processing facility or gathering center, as depicted in Figure 34.10. If acceptable corrosion rates are obtained at point E, no additional adjustments are needed at this time. However, if the corrosion rates are above acceptable limits, then the corrosion inhibitor treatment rates should be increased at injection points 1, 2, and/or 3 until acceptable corrosion rates are obtained for all monitoring points, that is, A, B, C, D, and E. Although an iterative process may be required to "line out" the chemical treatment injection rates the first time, once established, the chemical injection rates should be much easier to maintain.

Figure 34.11 presents an overlay of corrosion monitoring results from ER probes located on each pipeline segment for the gathering system depicted in Figure 34.10. The goal is to look at results from each of the corrosion probes and balance the treatments, so that satisfactory corrosion rates are obtained for all the monitoring locations. Since the corrosion mechanisms are dynamic and subject to change, the monitoring should continue throughout the operational lifetime of the pipeline.

The graph in Figure 34.11 illustrates the proper placement of internal corrosion monitoring points within a pipeline and how to use these points for adjusting chemical treatment rates for controlling internal corrosion. The underlying assumption is that if you inject corrosion inhibitor at the beginning of the pipeline and have protection at the end of the pipeline segment, then intermediate points would have received adequate corrosion inhibitor treatments to protect the pipeline. We have seen cases where the application of corrosion inhibitors may provide more than sufficient protection at the beginning of the pipeline, but the inhibitor is consumed, and insufficient protection may be provided as the fluids flow through the second half of the pipeline. That is why the key monitoring points need to be placed as close to the end of the pipeline as practical. The monitoring points should also be located where produced or condensed water would be swept or could accumulate.

A second point is that for initially establishing the required chemical treatment rates, we relied upon results from ER probes within the pipeline—not metal coupons. As described in Section 34.2.2, ER probes are very sensitive, and readings can be collected at short intervals, such as hourly or daily without removing the ER probes from the pipeline. In contrast, metal coupons are left within the pipelines for longer intervals, such as 90 days. If the chemical treatments are inadequate for protecting the pipelines, we need to know that immediately, such that remedial measures can be implemented in a timely manner—well before corrosion can be measured by ultrasonic or radiographic inspection techniques. Our goal is to control the internal corrosion mechanisms, such that the pipelines can be operated safely and in an environmentally responsible manner.

### 34.3.2 Orientation of the Corrosion Monitoring Coupons or Electronic Probes within a Pipeline

Earlier, we noted that many pipeline designers placed corrosion monitoring access fittings at locations, which are most convenient for personnel who remove and replace corrosion coupons or electronic probes. Actually, the orientation of corrosion monitoring within a pipeline is crucial for obtaining good, valid corrosion monitoring results. In this section, we share our experiences in selecting the proper orientation for installing corrosion monitoring access fitting, such that the metal coupons or ER corrosion probes will be positioned to monitor internal corrosion along the path where the most corrosive electrolytes would either be swept or "pool," that is, collect.

First, imagine a pipeline transporting produced fluids consisting of a combination of crude oil, natural gas, and water. Determine the velocities, based on the production rates for each phase of the fluid and take into account the internal diameter of the pipe. Also, look at the local elevation changes to determine if the critical angles could be exceeded, which would allow liquids to accumulate [6]. Unless you're in a high-velocity flow regime, there will be some stratification, in which the different phases would separate, based on relative density. The natural gas, being the least dense, would be on top. Produced or condensed water, being the densest, would be swept along the bottom of the pipeline, and the crude oil, being less dense than water, would be swept on top of the water.

Since produced water is the electrolyte that is essential for the corrosion mechanisms, the key to successful monitoring is to place the corrosion coupon or the ER probe within that phase. Try to install the corrosion monitoring flush with the bottom of the pipeline, where any liquid water would most likely be swept. The rationale is illustrated in Figure 34.12 [2,17,18].

Figure 34.12a illustrates how a corrosion monitoring coupon or ER probe oriented at the 3:00 o'clock orientation could miss the mark and not be able to detect the onset of corrosive conditions, whereas Figure 34.12b depicts corrosion monitoring on the bottom of the pipeline, which places the coupon or probe within the fluid (electrolyte), where internal corrosion is most likely to occur.

Although not as desirable as monitoring installed at 6:00 o'clock, coupons and ER probes are frequently installed from the top of the pipeline and across the diameter of the pipeline [1,2,20]. Be sure that the rod holding the coupon or probe

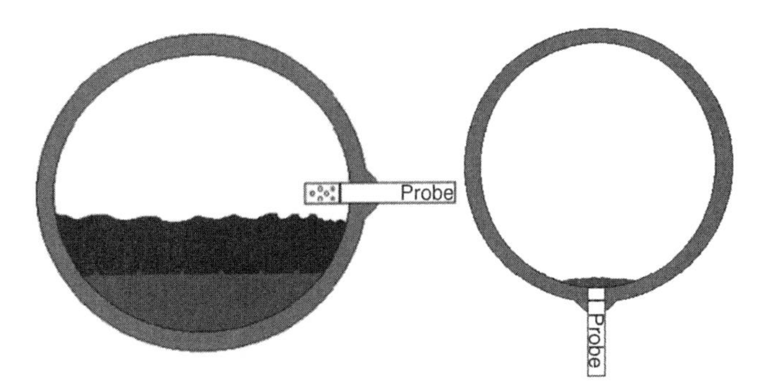


**FIGURE 34.12** (a) Probe at 3:00 or 9:00 o'clock [2,17,18]. This probe is not immersed within the most corrosive phase and will not yield valid results. (b) Probe at 6:00 o'clock [2,17,18]. Probe is properly positioned to monitor corrosion rates for liquids swept through pipe. (© NACE International 2001.)

is not in the flow path of any pigs that may pass through the pipeline. Also, ensure the coupon or ER probe is fully wetted by any liquids flowing through the pipeline. Further, make sure that any flush mounted coupon or ER probe that is installed across the diameter of a pipeline is not so close to the interior bottom of the pipe as to create an artificial crevice.

Before ending this particular section, we should note that the 6:00 o'clock orientation is ideal for pipelines transporting clean fluids—those containing virtually no solids or sediments, such as sand or formation fines. It is the recommended orientation for transmission pipelines. However, the 6:00 orientation on a horizontal section of pipe would not be the location of choice for pipelines that sweeps at least some hard solids, for example, sand, through the pipeline. The concern is that a portion of those hard solids would fall into and gall the threads of an access fitting when the corrosion coupon or ER probe is changed out. When the presence of such hard solids or sands are suspected, a better location for installing internal corrosion monitoring would be on the outside radius of a 90° bend, where the flow transitions from vertical to horizontal. The flow would continuously sweep solids away from the fitting, rather than allowing those solids to accumulate. (It could also be a very expensive operation to "blow down" and depressurize a section of a pipeline to repair an access fitting.)

Figure 34.13 illustrates one such access fitting, which was installed in a compressor station. Any solids or liquids that are transported through the pipeline, whether by regular production of pigging operations, would tend to be swept away from a monitoring point located on the extrados (outside radius) of a 90° bend from vertical to horizontal.

## 34.4 MONITORING RESULTS "DRIVE" CHEMICAL TREATMENT AND MAINTENANCE PIGGING PROGRAMS

Figure 34.14 illustrates the methodology behind the key role of using internal corrosion monitoring results to "drive" any

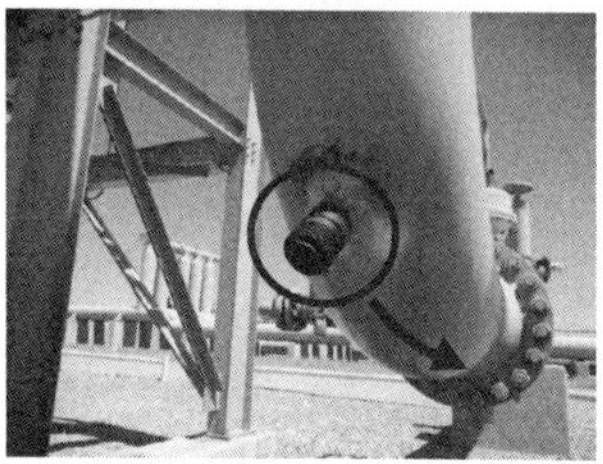

**FIGURE 34.13** Location of monitoring access fitting on extrados of 90° bend. If sand or sediment is present within your pipeline and could accumulate above an access fitting at 6:00 o'clock orientation, instead consider installing that access fitting upstream, at a 90° bend where the flow would wash away the solids and keep the threads clean.

necessary changes in the chemical treatment program. This will be an iterative process of monitoring and adjusting the chemical treatment rates throughout the service life of any production or transmission pipeline. We should expect to adjust chemical treatment rates when new wells or production is added to the process stream of a gathering system, as well as making changes when producing wells or production streams are taken out of service. We may also need to make changes to the treatment programs within gathering systems throughout field life in response to changing conditions in the reservoir, such as if/when the reservoir "sours," due to microbiological activity.

First, we set up ranges or steps for our chemical treatments. For continuous treatments, they could be steps, such as 10, 20, 40, 60, 80, 100 parts per million (ppm), and so on. For batch applications of chemical products, the steps could be values such as 5 gallons (or liters), 10 gallons (or liters), 20 gallons (or liters), and so on. We could also set up steps for batch treatments, based on the frequency of application. For example, we could treat daily, every other day, weekly, monthly, and so on.

The key for establishing the series of chemical treatment steps is that they are developed to be site specific, and would be based on historical knowledge of the relative corrosivity of the fluids at that particular location or site, and concentrations of treatments that appear to be effective in controlling the corrosion. For example, if we have a corrosive environment, which has required 100 ppm corrosion inhibitor, the

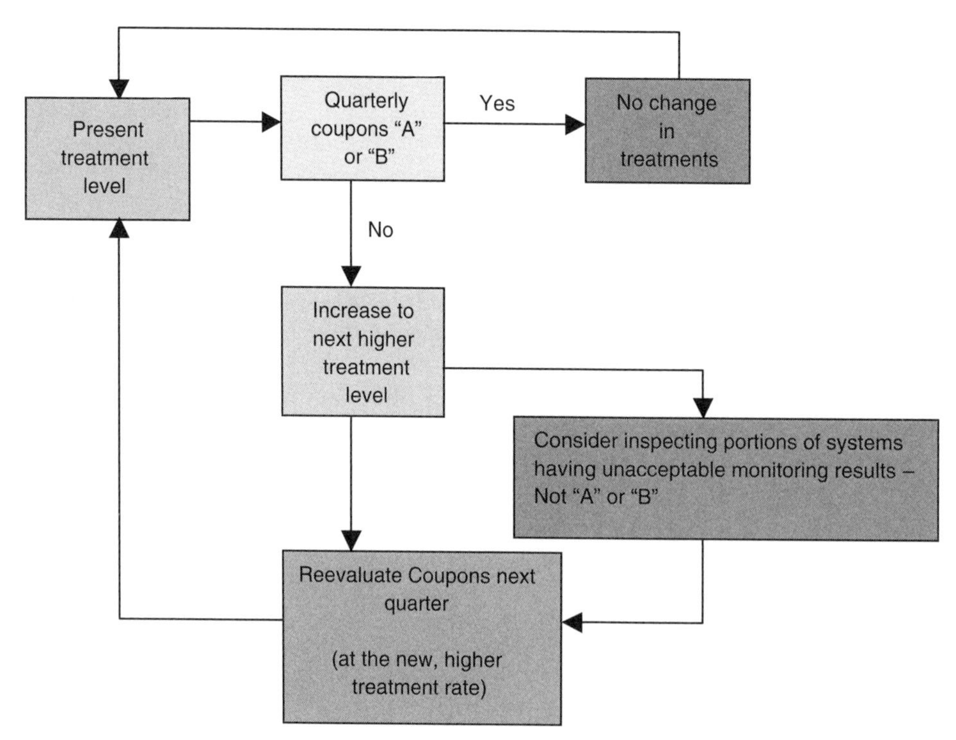


**FIGURE 34.14** Logic diagram for increasing inhibitor treatments based on monitoring results [21]. (© NACE International 2004.)

treatment rate steps would be built around those concentrations, such as 80, 100, 120, 140 ppm, and so on.

Second, establish a series of steps or levels for the observed internal corrosion rates. One such sequence is illustrated within the most current version of NACE RP-0775-2005. This Standard Practice document, which is entitled "Preparation, Installation, Analysis, and Interpretation of Corrosion Coupons in Oilfield Operations," includes a Table 2, which lists ranges for "low," "moderate," "high," and "severe" corrosion rates based in both mm/year and mpy [9]. (This table is duplicated as Table 34.1 in Section 34.2.1.3.) Since this table provides ranges for both general corrosion rates, as well as pitting corrosion rates, we would categorize, based on the more severe of the general or pitting corrosion rates.

Note that the name of the ranges that NACE used in the RP-0775 is not the key. Instead, it's the process of establishing the steps. Instead of the four ranges used by NACE, we could have adopted a system using five or six ranges, such as "A," "B," "C," "D," and "E," or "slight," "mild," "moderate," "significant," "severe," and "very severe." Also, note that the "A," "low," or "slight" descriptions, that is, the lowest of the corrosion rate ranges, must correspond to rates that enable the designed service life to be achieved.

The ranges used within the NACE Standard Practice and any similar set of ranges for corrosion rates are based on the design and corrosion allowance for the pipeline transporting crude oil or natural gas, whether it is part of a gathering system or a transmission line. The corrosion allowance is the additional or remaining portion of the thickness of the pipe wall beyond that required for pressure containment. The maximum acceptable average corrosion rate would then be the remaining wall thickness, divided by the remaining service life (typically in years). If we have a 2 mm corrosion allowance and a 40 year service life, the maximum acceptable corrosion rate would be 50 μm/year (2 mm = 2000 μm and ((2000 μm)/(40 years)) = 50 μm/year.) [In English units, the 2 mm = 78.7 mils, where 1000 mils = 1 in. Thus, 78.7 mils/40 years = 1.97 mpy would be the maximum allowable average corrosion rate while still achieving the desired service life.] Note that if we have corrosion rates in excess of the maximum allowable average corrosion rate for one or several years, the maximum allowable average corrosion rates in subsequent years would need to be reduced to compensate for the periods of higher corrosion rates. Alternately, the desired service life could be proportionally reduced.

Now refer to Figure 34.14. Look at the results from the internal corrosion monitoring and match the results to the range of values for different corrosion rates. For purposes of illustration, we will use the letters to define ranges of corrosion rates. If the results from the internal corrosion monitoring are characterized as "A" or "B," we will typically make no change in the chemical treatments for that particular pipeline. However, if the result from the coupon monitoring was "C", "D," or "E," we would follow the arrow downward, which tells us to increase the chemical treatments to the next higher injection rate. Thus, for example, if we had a "C" rated coupon and had been treating at 40 ppm, we would increase the corrosion inhibitor treatment rate to 60 ppm, which is the next higher step or treatment rate. (However, if we had a "D" or "E" rated coupon, which looked pretty severe, we might increase the treatment rate to 80 or 100 ppm. The first priority is protecting the integrity of the pipeline system. Later, we can optimize the treatments.)

Note the sidebar box, which suggests that if elevated corrosion rates are observed, consideration should be given to inspections, such as radiography, ultrasonic techniques, or fluid sample analyses. The purpose would be to independently confirm the observed corrosion rates.

The next step in the process is to reevaluate the monitoring results at the end of the next exposure cycle, which for coupons is typically 90 days later. The regular exposure cycles provides the opportunity to confirm the effectiveness of the treatments and determine whether any additional changes or updates are warranted.

Please also note that, although the logic diagram in Figure 34.14 is based on coupon monitoring results, general corrosion results from ER corrosion probes could also have been used.

Internal corrosion monitoring is an iterative process, requiring a constant review of new monitoring data. We need to confirm the effectiveness of the present chemical treatment rates (including the case when no chemical products are being applied) and to document the effectiveness of increases in the treatment rates. As necessary, it may also be appropriate to make an additional step change or increase in the treatment rate.

## 34.5 RELATIVE SENSITIVITIES OF NDT VERSUS INTERNAL CORROSION MONITORING TECHNIQUES

Production and management personnel will often assume that results from successive ILIs of pipelines, using MFL, or nondestructive testing (NDT), such as ultrasonic wall thickness measurements or radiography, will provide an equivalent understanding of the conditions within a pipeline, and could be obtained through internal corrosion monitoring. Indeed, corrosion professionals should evaluate all available NDT data to glean any new information regarding the integrity or operation of pipelines, including results from NDT or ILIs of pipelines. However, results from MFL-based ILIs or nonintrusive NDT do not offer the same precision in measurements for determining corrosion rates that is offered by corrosion coupons or ER corrosion probes. In the subsequent sections we will emphasize the value of NDT for integrity assessments, while, the more sensitive internal corrosion monitoring techniques should be used for controlling corrosion rates.

### 34.5.1 Precision of UT, RT, or MFL Nondestructive Inspection Techniques

Numerous techniques are available for nondestructive inspections of pipelines or piping within facilities. These are primarily used for locating and providing a measurement of general corrosion, as evidenced by the reduction of wall thickness, or the depth of any pits. The axial and circumferential dimensions associated with pipeline indications (anomalies) are used when assessing the integrity of a particular section of pipe, and determining whether the pipeline can continue to operate safely at the present MAOP. Ultrasonic inspection (UT) provides measurements of the remaining wall thickness or the depth of individual pits and are based on direct ultrasonic measurements at the location of interest. Radiographs (RT) also provide measurements of the remaining wall thickness or the depths of localized pits, and are based on subtle changes in the density of the radiographic images. MFL is still another technique for identifying the location of general corrosion or localized pitting. MFL ILI vehicles or "smart pigs" are the most widely known application of this technology, and are used to inspect pipelines. However, there are MFL applications, such as tools for inspecting the tubes within heat exchangers. In this technique, magnetic fields saturate the base metal, but the magnetic fields will be distorted by any anomalies, allowing them to be detected.

The UT, RT, and MFL nondestructive inspection techniques are excellent tools when conducting pipeline integrity assessments. However, as we shall see, these techniques are not sufficiently precise for quickly detecting any active internal corrosion and verifying the effectiveness of any adjustments in the corrosion inhibitor treatment rates. Let's look at the reported accuracy of the UT, RT, and MFL techniques, and then compare the relative sensitivities of these techniques to those available through corrosion coupons or ER probes, which are placed within the process systems.

- The accuracy of ultrasonic (UT) wall thickness measurements is typically reported as ±10 mils (0.010 in.) (254 μm), although many UT inspectors state they can detect changes as small as ±5 mils (0.005 in.) (127 μm) [15].
- Likewise, radiographers typically report pipeline anomalies to the nearest ±10 mils (254 μm), while some report to the nearest ±5 mils (127 μm), the same as for ultrasonic inspections.
- MFL is a great technique for quickly inspecting a pipeline or tubing within a heat exchanger. However, its accuracy is not the same as that available from ultrasonic or radiographic techniques. The accuracy of MFL results are reported as a percent of wall thickness, with 5% wall thickness being typical by the ILI companies. If, for example the thickness of pipe wall was 0.375 in. (9.525 mm) and 5% accuracy was claimed, the actual resolution would be $0.375 \times 0.05 = 0.019$ in., which is 19 mils (483 μm). Thus, MFL is a great tool for screening pipeline integrity and finding anomalies, but cannot offer the same precision as the UT or RT techniques.

Let's assume an internal corrosion rate of 10 mpy (254 μm/year). If we use the nominal 10 mils (254 μm) resolution for UT or RT, it would take one year for UT or RT to be able to detect the corrosion, plus one additional year to confirm the trend. That's 2 years before we could detect and confirm the corrosion using RT or UT techniques. We note that any metal, which is lost from the pipe wall during that time, cannot be placed back onto the pipeline, and undetected corrosion can shorten the effective service life of pipelines.

Next consider the relative sensitivity of MFL ILIs. As noted previously, the precision of the readings is most typically based on 5% of the nominal wall thickness. Thus, if we have a nominal wall thickness of 0.375 in. (9.525 mm), the best resolution we could have would be approximately 19 mils (483 μm). It would take 2 years to be even able to detect the onset of corrosion and another 2 years to be able to confirm the trend. Thus, MFL ILIs are excellent for detecting dents or areas of corrosion, and should be used as an integral part of pipeline integrity assessments. However, since MFL data at best only offers a precision of 5% of nominal wall thickness, it does not have the sensitivity required for detecting changes in pre-existing indications within a timely manner. As such ILI data should not be used for controlling corrosion mitigation programs, such as the application of corrosion inhibitors, when other corrosion control measures are available.

We now look at the time response available through corrosion coupons or ER probes.

### 34.5.2 Typical Exposure Periods for Coupons or ER Probes to Detect Active Corrosion

First, consider corrosion rate coupons. When corrosion coupons are first placed within a system, the initial corrosion rates will be fairly high, and then will rapidly drop. The initial high corrosion rates are due to the metal coupon initially having a large surface area, which is free of any corrosion products or oxide. However, as the corrosion products develop on the exterior surface of the coupons, those products will slow subsequent reactions, and result in slower corrosion rates. Corrosion professionals within the oil and gas industry have recognized this, and as a consequence, a general "rule of thumb" for this industry is to leave the metal coupons within the systems for a minimum of 30–60 days before removal, with most monitoring based on "standard" 90-day exposure cycles.

Next consider a wire loop ER probe within the same system. As described in Section 34.2.2.2, it would take approximately 4 days to detect a 10 mpy (254 μm/year) corrosion rate and approximately 8 days to detect a 5 mpy (127 μm/year) corrosion rate. High-resolution ER probes also have similar abilities to quickly detect corrosion rates, while offering a longer service life.

### 34.5.3 Relative Time for Coupons, ER Probes, or Inspection Techniques to Detect Active Corrosion

Based on the previous discussion of the most widely used inspection and monitoring techniques, it is apparent that wire loop ER probes are preferred by many corrosion professionals, since they offer quick detection (as little as 4 days) for the onset of corrosion conditions. The wire loop ER probes are also the tool of choice for adjusting corrosion inhibitor treatment rates. However, these probes will typically have a shorter lifetime than ER probes having cylindrical or flush mounted probe elements. Hence, the choice of probe element may be based on a balance between sensitivity and probe life.

Metal coupons require brief exposure cycles (30–60 days) for detecting and confirming internal corrosion rates. However, coupons are much more cost-effective than ER probes, and, consequently, are "work horses" for routine monitoring of oil and gas production and transmission systems, particularly where operations have stabilized.

UT and RT are also valuable tools, but their primary role should be that of routine inspections and integrity assessments, since they do not offer sensitivities for detecting corrosive conditions, that are comparable to those of corrosion coupons or electronic probes placed within the systems. Likewise, MFL techniques are excellent tool for inspecting pipelines, but again, do not have sufficiently sensitive for detecting the onset of corrosive conditions and being used to control mitigative measures, such as corrosion inhibitor treatments.

Please note, however, that this recommendation to rely upon ER probes and corrosion coupons as the primary, most cost-effective monitoring technique rather than nondestructive inspection techniques, such as UT, RT, or MFL inspections is subject to revision if or when new inspection technologies are developed and become commercially available.

Table 34.2 summarizes the discussions already mentioned and makes the point that whereas ultrasonic and radiographic inspections are very valuable tools for integrity assessments, they are not adequate tools for measuring corrosion rates and initiating remedial measures. Metal coupons are far better, but typically require 30–60 days to establish the initial surface oxides on the coupons and establish a baseline corrosion rate. However, wire loop ER probes can detect a 10 mpy (254 μm/year) corrosion rate in as few as 4 days. Such prompt detection of the onset of corrosive conditions enables corrosion professionals to rapidly initiate mitigative measures to quickly bring corrosion rates to acceptable levels and maintain the projected service life for the pipeline assets.

**TABLE 34.2 Relative Time for Monitoring and Inspection Techniques to Detect 10 mpy Corrosion**

| Monitoring or Inspection Technique | Time to Detect and Confirm 10 mpy (254 μm/year) Corrosion Rate |
|---|---|
| MFL | 2–3 years |
| Ultrasonic or radiography | 1–2 years |
| Metal coupon | 30–60 days |
| Most sensitive wire loop ER probe | 4 days |

J.B. Mathieu also presented this same concept in a paper entitled "On-Line Monitoring Techniques," which is in Proceedings of the NACE Middle East Conference in Bahrain in 1994 [22]. A chart from his paper (Figure 34.15) shows the relative response time for various monitoring techniques to be able to detect corrosion, which is occurring at 10 mpy (254 μm/year).

Note in Figure 34.15 that MRTV is MicroVertilog, which is a MFL technique. Its relative responsiveness would be comparable to that from MFL inspections.

## 34.6 FLUID SAMPLE ANALYSIS TO COMPLEMENT AND VERIFY MONITORING RESULTS

The final component of an internal corrosion monitoring program would be the collection and analysis of fluid samples from the process stream being monitored. The purpose is to collect data that independently support and corroborate the results from internal corrosion monitoring.

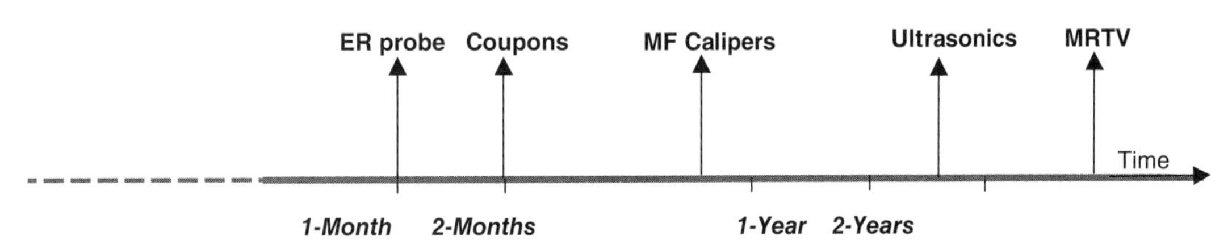


**FIGURE 34.15** Relative responsiveness of monitoring and inspection techniques [22]. (© NACE International 1994.)

If the monitoring is for a gas transmission pipeline, the sample analysis could be as simple as routine gas analysis, which includes a measure of the moisture content of that gas. (Water is an electrolyte that is necessary to support corrosion mechanisms.) The analysis could also include checks to quantify the ppm of hydrogen sulfide or perhaps the mole percent of carbon dioxide. Again, the parameters, which are monitored, are those components or parameters that may pose a potential threat to the pipeline integrity.

If the monitoring is for a crude oil and or natural gas gathering system, the analysis would be focused on the chemical species that pose a potential threat to the integrity of that pipeline. This could mean serial dilution tests to quantify populations of SRB, APB, pH of any produced water, the concentration of dissolved $H_2S$ within produced or condensed water, the ppm of $H_2S$ within the gaseous phase, the mole percent $CO_2$, the volume and character of any material removed from a pipeline during maintenance (cleaning) pig runs, and so on. The fluid sample analyses could also include measures of the concentration of biocides, corrosion inhibitors, scale inhibitors, and so on, the residuals, which could tell whether sufficient chemical product has been applied to treat the entire pipeline. Note that corrosion inhibitor products are frequently combinations of several active components. However, residual measurements may "key" on only one of these components. If the "key" component has been depleted but the other components are still providing adequate protection, it is possible to have a false indication for depletion of the inhibitor. Hence, review monitoring results along with other data such as fluid sample analyses, production data, and so on, to get a consistent picture of the effectiveness of the corrosion control and mitigation measures.

## 34.7 SUMMARY

This chapter has focused upon corrosion monitoring within oil and gas pipelines or processing facilities. The primary workhorse and most cost-effective approach for routine internal corrosion monitoring is through the use of metal mass loss coupons or ER corrosion probes that are inserted into the process stream. Mass loss coupons provide general corrosion rates, as well as pitting corrosion rates, and typical exposure periods are 90 days. There are a variety of different styles of ER corrosion probes. Like coupons, the ER probes can be installed to be even with the pipe wall (flush mounted) or intruding into the process stream. There is a tradeoff between probe life and sensitivity, and the particular choice should be based on knowledge regarding the potential corrosivity of the system being monitored.

We also noted that there are other types of monitoring available. As opposed to mass loss coupons, we could use highly polished coupons, which are assessed with scanning electron microscopes after the exposure cycle. EM coupons are excellent for specialized diagnostics, but this time-intensive option is rarely used for routine monitoring. Besides ER probes, there are also options for LPR or EN or ECN probes. A limitation of LPR or EN technology is that, unlike coupons or ER probes, each reading is independent of all previous readings. Hence, to reduce the chance that a transient corrosion event occurs between readings and is missed, the frequency of readings may be increased. That may require extensive telecommunication systems, and consequently may not be appropriate for remote locations, which may not have power sources, or mountains/hills that would block telephone signals.

We also presented an overview of where corrosion monitoring probes or coupons should be placed within pipelines or process streams. Any chemical treatments, such as corrosion inhibitors, should be applied at the start of a pipeline segment, and the monitoring should be located near the end of pipeline segments. The monitoring near the end of the pipeline ensures there is adequate corrosion inhibitor to effectively treat the pipeline, and that the corrosion inhibitor has not been fully consumed. Modeling is also highly recommended to identify locations along the pipelines, where liquids may accumulate, and where internal corrosion would be most likely to occur. These are ideal locations for placing internal corrosion monitoring coupons or probes. Remember that the corrosion mechanisms within pipelines require an electrolyte, for example, any produced or condensed water, and hence, we need to place the sensing element of a corrosion probe or metal coupon within that fluid to measure the corrosion rates.

The treatment rates should be adjusted, so that satisfactory performance, for example, corrosion control, is attained near the end of the pipeline. Additionally, the monitoring coupon or electronic probe should be oriented to ensure it is wetted by the most corrosive phase being swept through the pipeline, and this is typically the path the produced or condensed water travels.

We also compared the relative ability for MFL, UT, and RT nondestructive inspection techniques to detect a 10 mpy (254 μm/year) internal corrosion rate, versus metal coupons or ER probes that are installed within a pipeline. It could take 2–3 years for MFL inspections to be able to detect what UT or RT inspections can detect in 1–2 years and metal coupons could detect in 2–3 months, or the most sensitive wire loop ER probes could detect in 4–8 days. Since it is not possible to put metal back onto a pipe wall, the best option for being able to control internal corrosion and minimize any corrosion is to monitor the systems with a combination of sensitive ER probes and metal coupons. Remedial measures would then be initiated when internal corrosion rates rise above previously defined acceptable limits.

## 34.8 DEFINITIONS OF CORROSION MONITORING TERMS FROM NACE 3T199 [10]. © NACE INTERNATIONAL 1999

**Direct Measurement**: Describes measurements of metal loss or corrosion rate.

**Indirect Measurement**: Describes measurement of any parameters that may influence or are influenced by, metal loss or corrosion.

**In-Line Monitoring**: Refers to installation of monitoring equipment directly in the bulk fluid of the process, but data acquisition requires extraction of probes or process shutdown for analysis, for example, mass loss coupons.

**Intrusive Monitoring**: Requires penetration through the pipe or vessel wall to gain access to the interior of the equipment.

**Nonintrusive Monitoring**: Monitoring from the outside of the pipe or vessel wall without having to gain access to the interior of the equipment.

**On-Line Monitoring**: Refers to installation of monitoring equipment for continuous measurement of metal loss, corrosion rate, or other parameters in an operating system; data are obtained without removing the monitoring device.

**Off-Line Monitoring**: Refers to monitoring methods in which the sample is taken for subsequent analysis.

**Real-Time Measurements**: Refers to measurements that can detect changes in the parameter under investigation essentially as they occur. The changes can be detected because the technique is sufficiently sensitive and can be made continuously or with sufficient frequency to follow the changes in the parameter. (This is particularly relevant for many of the measurement techniques that require a significant cycle time.)

**Sample**: A small quantity of a fluid that has been removed from the process stream for testing purposes.

**Side (Slip) Stream**: A bypass loop or a direct outlet from the process stream. (Fluid may or may not be at the same velocity, temperature, and pressure as the main process stream.)

## REFERENCES


1. Byers, Harry G. (1999) *Corrosion Control in Petrochemical Production*, TPC Publication 5, 2nd ed., NACE International, Houston, TX.
2. Roberge, P.R. (2011) Corrosion monitoring, Part IV. *Uhlig's Corrosion Handbook*, 3rd ed., John Wiley & Sons, Inc., New York, NY.
3. Eckert, R. (2003) *Field Guide for Investigating Internal Corrosion of Pipelines*, NACE International, Houston, TX.
4. Internal Corrosion for Pipelines—Course Manual, NACE International, 2008.
5. Internal Corrosion for Pipelines—Advanced Course Manual, NACE International, 2009.
6. NACE (2006) *NACE SP0206-2006, Internal Corrosion Direct Assessment Methodology for Pipelines Carrying Normally Dry Natural Gas (DG-ICDA)*, NACE International, Houston, TX.
7. NACE (2008) *NACE SP0208-2008, Internal Corrosion Direct Assessment Methodology for Liquid Petroleum Pipelines*, NACE International, Houston TX.
8. NACE (2006) *NACE SP0106-2006, Control of Internal Corrosion in Steel Pipelines and Piping Systems*, NACE International, Houston, TX.
9. NACE (2005) *NACE RP0775-2005, Preparation, Installation, Analysis, and Interpretation of Corrosion Coupons in Oilfield Operations*, NACE International, Houston, TX.
10. NACE (1999) NACE International Publication 3T199 Techniques for Monitoring Corrosion and Related Parameters in Field Applications, NACE, Houston, TX, December.
11. Eckert, R.B. and Cookingham, B.A. (2002) Field use proves program for managing internal corrosion in wet-gas systems. *Oil and Gas Journal*, 100(3), 48.
12. Eckert, R., Aldrich, H., and Pound, B.G. (2004) Biotic pit initiation on pipeline steel in the presence of sulfate reducing bacteria. NACE Corrosion 2004, Paper No. 04590.
13. Eckert, R.B., Aldrich, H.C., Edwards, C.A., and Cookingham, B.A. (2003) Microscopic differentiation of internal corrosion initiation mechanisms in natural gas pipeline systems. NACE Corrosion 2003, Paper No. 03544.
14. Zintel, T.P., Kostuck, D.A., and Cookingham, B.A. (2003) Evaluation of chemical treatments in natural gas systems vs. MIC and other forms of internal corrosion using carbon steel coupons. NACE Corrosion 2003, Paper No. 03574.
15. Papavinasam, S., Doiron, A., Attard, M., Demoz, A., and Rahimi, P. (2012) Non-intrusive techniques to monitor internal corrosion of oil and gas pipelines. NACE Corrosion 2012, Paper No. 0001261.
16. Pikas, J. (2014) Understanding 3D structured light. NACE Corrosion 2014, Paper C2014-3792.
17. Powell, D.E., Ma'Ruf, D.I., and Rahman, I.Y. (2001) Methodology for establishing near real-time corrosion monitoring capabilities and for conducting field tests of corrosion inhibitors. Proceedings of NACE Northern Area Conference, February, 2001, Anchorage, Alaska, Paper OilGas 02.
18. Powell, D.E., Ma'Ruf, D.I., and Rahman, I.Y. (2001) Monitoring internal corrosion and inhibitor effectiveness in gathering and production piping, using corrosion probes. NACE Materials Performance, 40(8), 50.
19. Powell, D.E., Ma'Ruf, D.I., and Rahman, I.Y. (2002) Field testing corrosion inhibitors in an oil and gas gathering system. NACE Materials Performance, 41(8), 42.
20. Eden, D.C., Cayard, M.S., Kintz, J.D., Schrecengost, R.A., Breen, B.P., and Kramer, E. (2003) Making credible corrosion measurements—real corrosion, real time, Proceedings of NACE Corrosion 2003, NACE International. Paper No. 03376.
21. Powell, D. and Islam, M., (2014) Monitoring experiences—contrasting the practices of two crude oil producers. Proceedings of NACE Corrosion/04, NACE International. Paper No. 04441.
22. Mathieu, J.B. (1994) On-line monitoring techniques. Proceedings of NACE Middle East Conference, 1994, Bahrain.

# 35

# IN-LINE INSPECTION (ILI) ("INTELLIGENT PIGGING")

Neb I. Uzelac
*Neb Uzelac Consulting Inc., Toronto, Ontario, Canada*

## 35.1 INTRODUCTION

In-line inspection (ILI, also known as smart or intelligent pigging) has become an indispensable tool in pipeline integrity management, to establish the actual condition of the pipeline. The term ILI pertains to running devices equipped with some form of nondestructive testing (NDT) technology through the interior of pipelines in order to detect various types of defects and anomalies. In most applications, ILI tools are autonomous, free swimming devices that are propelled through pipelines by the flow of the fluid, and there is no external influencing the operation, controls, processing, and recording. Exceptions are tethered tools, typically for shorter pipeline sections and sections out of service, ILI tools that are connected by an umbilical with control and data storage done online from the outside.

Intelligent pigging started in the late sixties when the first tools aimed at detecting corrosion, utilizing magnetism, were developed. ILI has evolved in the meantime with additional types of tools, their wider applicability and improved reliability, accompanied by an overall increase in acceptance within the industry.

This chapter will introduce ILI in general terms, its place within pipeline integrity management, describe running tools, give a nomenclature of tools based on their inspection purpose, and discuss verification and usage of ILI data.

Since ILI has been introduced and increasingly in the recent years, there have been many contributions and case studies presented at industry conferences and in journals with a wealth of information pertaining to ILI methods, performance, specifics, limitations, and altogether very helpful when one is addressing ILI-related issues. A list of conferences and journals that cover the topics is given in Section "Bibliography".

## 35.2 PLACE OF ILI IN PIPELINE INTEGRITY MANAGEMENT

Knowing the condition of a pipeline is an essential prerequisite for being able to devise an appropriate set of measures and activities to manage its integrity. There are numerous parameters defining the condition of a pipeline, most notably changes to its original, if known, state. Those range from precise pipeline route, changes to it due to unforeseen causes, mechanical damage such as dents, corrosion, or other types of metal loss, and cracks, all of which can develop over the lifetime of a pipeline. In addition, there could be defects that were present from laying of the pipeline unaccounted for that might develop over time and compromise its integrity, such as manufacturing and weld defects. Additionally, given the age of much of the pipeline network, there are pipelines for which there are not sufficient, incorrect, or no data at all.

The fact that a buried pipeline is best accessible from the inside facilitated development of methods to look for increasingly many of potential flaws by having tools travel through the interior of the pipeline. This of course, providing that the pipeline is "piggable," meaning that there are no obstacles for an ILI tool passing through it.

*Oil and Gas Pipelines: Integrity and Safety Handbook,* First Edition. Edited by R. Winston Revie.

Once a baseline inspection of a pipeline with clearly defined goals is performed, not only is its condition (presence of corrosion, cracks, etc.) captured in time, but repeating the inspections enables monitoring growth of detected defects allowing prioritization of mitigation measures and gauging their efficiency.

ILI has been widely used throughout the world since the 1990s, and the first documents related to ILI were published by The European Pipeline Operators Forum (POF) in 1998, Specifications and requirements for intelligent pig inspection of pipelines [1], and NACE in 2000, SOTA 35100, In-Line Nondestructive Inspection of Pipelines [2], both since reissued. The goal of the latter document was to introduce existing ILI technologies and give an overview of activities involved with running tools, both form organizational and technical point of view.

Since the introduction in the USA of Federal Rules for Pipeline Integrity Management in High Consequence Areas [3], the need arose for recommendations and more specific guidelines, also for utilizing ILI, and several documents were developed as a result:

- API 1160, Managing System Integrity for Hazardous Liquid Pipelines. This consensus standard explains the role of ILI within pipeline integrity management [4].
- API 1163, ILI Systems Qualification Standards, umbrella standard to be used with and complement the companion standards listed below, developed to establish an approach that will consistently qualify the equipment, people, and processes utilized in the ILI industry [5].
- NACE SP0102-2010 (formerly RP 0102), Standard Practice, ILI of Pipelines, covering operational issues, activities involved in planning, organizing, and execution of an ILI project [6].
- ASNT ILI PQ, ILI Personnel Qualification and Certification, laying out requirements for personnel involved in handling ILI tools and performing data analysis [7].

## 35.3 RUNNING ILI TOOLS

Running ILI tools is typically discussed between pipeline operators and ILI service providers ahead of the time in order to optimize the inspection. An indispensable step is filling out a pipeline questionnaire, a list of questions regarding characteristics of a pipeline section to be inspected, such that all potential restrictions can be identified on time and addressed. A sample questionnaire is included in Appendix 35.A of this chapter.

In order for an ILI to be successful, the following steps need to be fulfilled:

- Chose the proper inspection technology for the goal of inspection. If operators want to find corrosion on the pipeline for example, they must choose an inspection technology that provides sufficient performance in detecting, discriminating, and sizing of corrosion defects. Tool technologies are discussed in Section 35.4.
- A tool, with the right inspection technology, has to be chosen that matches the characteristics of the pipeline section where it will be utilized, its size internal diameter (ID), wall thickness, potential restrictions (e.g., bends), proper launchers and receivers, type of fluid, to name just a few. Complete list of characteristics will be listed in a pipeline questionnaire. Some of ILI technologies operate in both gas and liquids, some not, pressure, temperature, speed of the fluid, all may play a role and restrict usage of some of the tools, or tools have to be built specifically for those conditions. Proper compatibility assessment of ILI tools will help avoid damage during inspections or, even worse, tools getting stuck in the pipeline.
- An inspection run with properly chosen ILI technology and tool matching the characteristic of the pipeline section still has to be successfully performed, which entails pipeline preparation (cleaning, or even some modifications), gauging, launching, tool tracking (might include pipeline route surveying), assuring the proper operating regime, and finally receiving.
- After the inspection run, recorded data have to be properly analyzed and an inspection report issued that lists all the detected features, which the inspection technology is capable of detecting. Depending on the type of inspection technology and complexity of the tool, types of analyses greatly vary and are ILI service provider specific and proprietary.
- In many cases, verification excavations are being performed to confirm the findings of ILI and very often they reveal systematic errors and assist in their correction.

Guidance in performing all of those steps is provided in the above listed standards, [5], [6] and [7].

### 35.3.1 Tool Type Selection

Goal and objectives of an inspection have to be analyzed and matched with capabilities of ILI technologies regarding their ability to detect, locate, identify, and size defects adequately well to match the purpose of inspection and intended defect assessment methods. API 1163 provides guidance in understanding tool performance and how it should be specified, presented and verified [5].

### 35.3.2 Making Sure the Tool Fits the Pipeline

Chosen ILI tool has to be able to match the characteristics of the pipeline (see pipeline questionnaire in Appendix 35.A), its diameter, wall thickness, type of welds, and to pass all physical restrictions that might exist in the pipeline: valves, bends, tees, off-takes, river crossings (change of wall thickness and, hence, ID), and so on. In addition, the type of fluid has to be considered, its chemical suitability (aggressiveness), but also as operation of some of the technologies depend on type of liquid, cleanliness (generally, debris impacts operation of all ILI tools), speed of inspection controlled by flow of fluid, as for each ILI technology there is an optimal range, and so on. Majority of ILI runs are performed in operating pipelines, to minimize disruption, but there is also the option of using tethered tools and crawler devices for those that are not. Running those tools is different, but all that will be said in the section on ILI tool technologies and their performance will still apply, provided that they can be applied under that configuration and regime of operation.

Launchers and receivers, additions to a pipeline section through which tools are inserted and retrieved from the pipe, have to fulfill certain requirements regarding configuration and size, depending on size and type (liquids or gas) of the pipeline, and ILI tool type. There are also elaborate procedures to make sure that launching and receiving is made safely (e.g., explosion protection) and effectively, all of which has to be covered in operational procedures [6].

### 35.3.3 Conducting the Survey

Performing a successful ILI run, after having considered characteristics of the pipeline section and compatibility of ILI tools, entails a range of activities that have to be well planned and executed. It is a joint effort by the pipeline operator and the ILI service provider and involves activities from launching the tool, its tracking, controlling the inspection speed, and receiving the tool [5], [6], [7].

Each run is specific, and it is essential that the operator and the ILI provider discuss all the necessary steps before the run, as described above.

## 35.4 TYPES OF ILI TOOLS AND THEIR PURPOSE

ILI tools can be categorized based on the main purpose of inspection and further breakdown is based on underlying basic principles of operation, which in turn define their characteristics and limitations.

ILI has started its foray into mainstream pipeline integrity management in the 1990s. Originally, tools were used for looking at corrosion and deformation, but there has been an evolution of ILI technologies in both looking at different types of features and improving detection, identification, and sizing capabilities. With it came a higher level of understanding and acceptance within the industry, but also more stringent demands on performance and an improved set of requirements to define the performance.

POF, The European Pipeline Operators Forum, started in the late 1990s, to work on “Specifications and requirements for intelligent pig inspection of pipelines” [1], suggesting what needed to be included in ILI reports. Very importantly, they introduced the concepts of:

- *POD:* probability of detection, the probability that the ILI system will detect a defect of a certain type and larger than a minimum specified size.
- *POI:* probability of identification, the probability that the ILI system will properly identify detected features.
- *Sizing capabilities:* defined with a tolerance (e.g., ±0.5 mm) a probability (e.g., 80%) and confidence level (e.g., 95%).

These characterizations have been accepted and ILI technologies and tools have been characterized by following those guidelines. Each of the ILI service provider companies (pigging vendors) describe the performance of their tools by ILI technologies using those qualifiers in their performance specifications. In addition, for each of the ILI tools there will be a tool specification listing its physical and operational characteristics [1], [5].

This section describes ILI technologies and divides them into general categories based on the main purpose of inspection. There might be an overlap in detection capabilities, for example, a corrosion detection tool may be able to detect dents, but the tool technologies will be organized by their primary purpose.

Bidirectional tools: by and large, ILI tools are deployed to go with the flow in pipelines, from launcher to receiver, moved by pressure differential building on their cups, designed to withstand pressure from the back. In some cases, however, tools have to be able to move in both directions (bidirectionally), if there are no receivers, for example, so they have to be configured accordingly.

### 35.4.1 Geometry (Deformation) Tools

Geometry tools, also known as deformation or caliper tools measure the bore of the pipe and so can detect deviations from its nominal, circular shape such as ovalities, dents and wrinkles. Typical applications for geometry tools are:

- In acceptance of new pipelines to detect potential anomalies during installation, like denting during backfill;
- Verify passage for other, more complex, ILI tools;
- Detect mechanical and third-party damage.

**FIGURE 35.1** Geometry Tools Mechanical: (a) Baker Hughes [8] (Photo courtesy of Baker Hughes.) (b) Lin Scan [9] (Photo Courtesy of LIN SCAN.); Electronic: (c) ROSEN [10]; and Combined Mechanical and Electronic: (d) ROSEN [10].

***35.4.1.1 Principle of Operation*** These tools utilize mechanical arms, electromagnetic methods or both, with the mechanical geometry tool, being prevalent in the industry. Mechanical arms are mounted on the tool body with their free end sliding along the internal surface of the pipe wall as the tool moves through the pipeline. There are different configurations of mechanical arms, some have wheels at the end, and some are mounted underneath cups; see Figure 35.1. The number of mechanical arms defines the circumferential resolution and the sensitivity to the angle of deflection the accuracy with which the deformations are being recorded. Each arm has to be recorded individually for circumferential position of deformations to be revealed. Those tools generally have the highest collapsibility of all ILI tools, such that they can be used to verify passage for other, more complex, tools.

There are also tools based on eddy current and tools combining mechanical and eddy current method for increased accuracy; see Figure 35.1.

***35.4.1.2 Types of Detectable Features*** Geometry tools can be used to detect deviations from the circular internal shape of the pipe such as

- Dents
- Ovalities
- Wrinkles
- Buckles
- Installations, for example, vales, off-takes
- ID changes
- Girth welds

***35.4.1.3 General Performance Characteristics*** These tools can be operated

- In gas and liquids lines;
- As stand-alone tools or combined with other ILI technologies.

### 35.4.2 Mapping/GPS Tools

Mapping tools are used for mapping pipelines, that is, running them enables establishing accurate centerlines of pipelines in geographic information system (GIS) coordinates and generating precise pipeline route documentation.

***35.4.2.1 Principle of Operation*** Mapping or inertial guidance tools use gyroscopes and accelerometers to detect changes in movement as the tool travels through the pipeline. It is necessary to know the spatial location for the starting and finishing points (launchers and receivers) and the tool will map the pipeline route in between. As these tools typically have a time-dependent drift, it is important to set intermediate

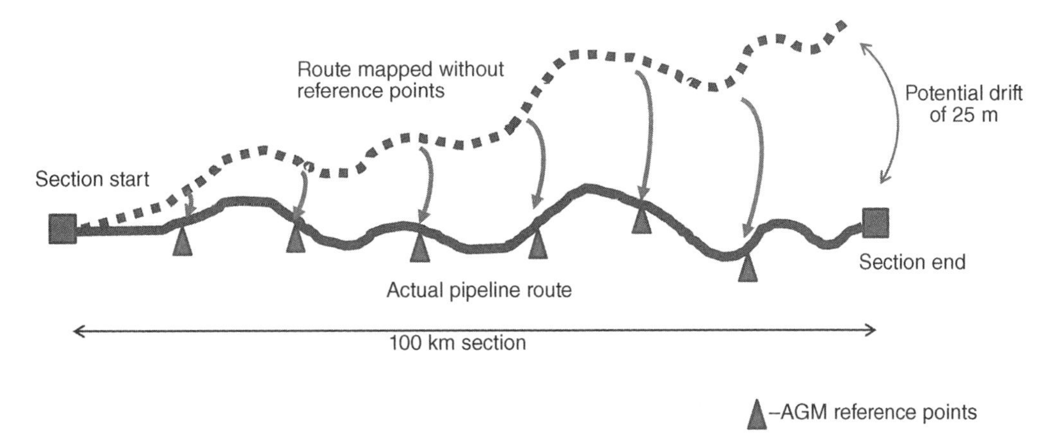


**FIGURE 35.2** Mapping tool's inherent drift is corrected using reference points provided with high accuracy GPS location.

reference points for correcting the ensuing error. The spatial locations of starting, ending, and the reference points are best established using GPS systems, such that the resulting data for the location of the pipeline can be achieved with high accuracy; see Figure 35.2. Above ground markers (AGMs) that sense passage of ILI tools are positioned at those reference points and their exact GIS position is subsequently correlated with the ILI tool's distance recording.

These tools are used for mapping of newly laid pipelines or pipelines for which there is no proper documentation. They are also used to detect movement of pipe, like after landslides or earthquakes, if information exists about the pipeline route from before the event to which the results of mapping run can be compared. The results allow for calculating large-scale stresses that the pipeline is submitted to by being moved.

***35.4.2.2 Types of Detectable Features*** Tools record the route of the pipeline and yield the location of the centerline of the pipeline in GIS coordinates, that is, longitude, latitude, and elevation, including locations of girth welds. Established route of a pipeline can be used for tying in all relevant pipeline information into a database and overlaying of mapping results over areal pictures and maps provides a most enhanced overview of the pipeline systems.

#### 35.4.2.3 General Performance Characteristics

- Can be run as standalone tools or on more complex tools in conjunction with other ILI technologies, most often with geometry, but increasingly also with metal loss inspection.
- Operate in gas and liquids pipelines.
- Accuracy of established centerline points is dependent on number and accuracy of external reference points.

### 35.4.3 Metal Loss Tools

Metal loss tools, also called corrosion tools, are the most widely used application of ILI. They are used to detect and size internal and external metal loss in the pipe wall, as caused, for example, by corrosion, erosion or third-party damage. Precise locating and sizing of metal loss allows for estimating the pressure carrying capacity of the pipe diminished by the reduced wall thickness enabling prioritizing of maintenance and repair measures.

There are different mechanisms causing metal loss in pipeline wall resulting in different shapes, locations, and sizes. The principle of operation and configuration of ILI tools for metal loss detection will affect their suitability to detect, identify, and size different types of metal loss and hence, the associated POD, POI, and sizing capabilities. It is important to understand the NDT technologies that the tools utilize in order to be able to assess their performance on different types of metal loss defects.

***35.4.3.1 Magnetic Flux Leakage (MFL) Tools*** When this technology was started in the 1960s, it was the very first to be applied for ILI inspection for metal loss and it is still the most widely used.

Principle of operation: Pipeline wall is magnetized such that there is a uniform distribution of magnetic flux in the ferromagnetic steel, which will be disturbed if there are changes in the cross section of the pipe wall in the direction of magnetization, such as caused by metal loss; see Figure 35.3. MFL tools carry strong permanent magnets configured to magnetize the pipe wall to saturation, such that each reduction in the cross section of the pipe wall, as is the case with metal loss, will cause magnetic flux to "leak" or spread outside of the pipe wall. This flux leakage will be detected by sensors mounted on the tool sliding along the interior pipe wall, and will provide the basis for

Axial pipe magnetization

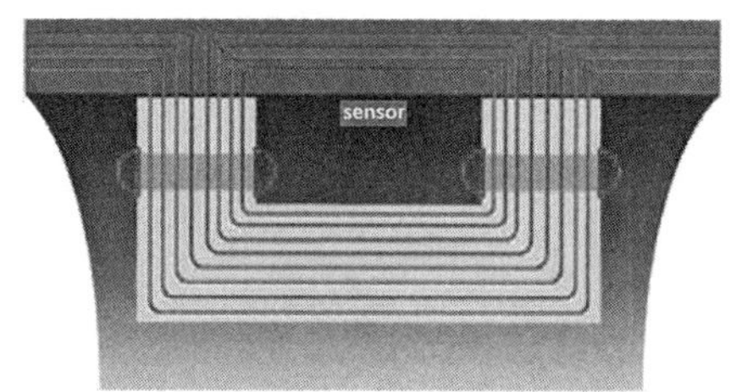


Flux distribution in pipe without defects

Flux distribution in pipe with an internal defect

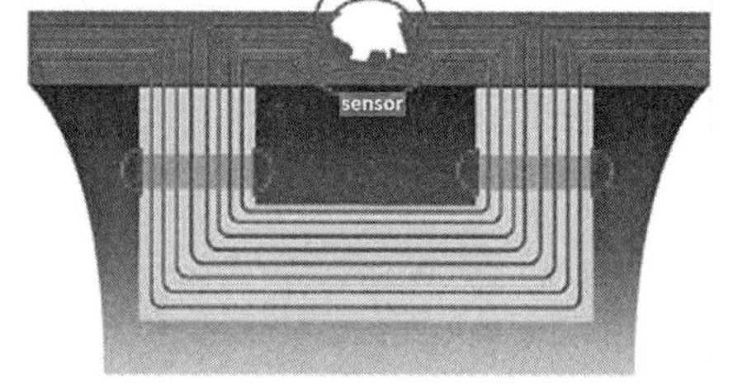


Flux distribution in pipe with an external defect

**FIGURE 35.3** Simplified schematics of axial pipe magnetization in which a sensor at the internal surface of the pipe detects magnetic flux leakage associated with internal and external metal loss. (From Ref. [11].)

estimating the amount of missing metal (metal loss). Hall element sensors are typically used for measuring the magnitude of flux leakage; in older tools, coils were used. The number of sensors around the circumference of the pipe define circumferential resolution and the sampling distance as the tool moves, the axial resolution of the tools, both affecting the type and minimum size of detectable defects.

Even though the principle is the same, there are different designs and implementations of MFL, which all affect operational characteristics and inspection performance of the tools. Also, additional sensors are required in order to distinguish between external and internal defect location (ID/OD discrimination) in most cases mounted on a separate tool body, trailing the main magnetizing unit. Figure 35.4 shows some of the tools.

Each metal loss type and its geometry will generate a specific pattern of flux leakage, and after an inspection run the shape of recorded flux leakage will be analyzed to infer the metal loss that caused it. Over the decades of utilizing this technology, algorithms have been developed and expertise built up to improve detection and sizing capabilities of this technology.

Because of differences in the ways that the tools are built that influence their mechanical performance and detection capabilities, labels "low" and "high resolution" have historically been used to denote their capabilities. It is, however, prudent to look at tool specifications rather than relying on such vague terms when looking for the right tool for one's application.

MFL metal loss tools typically magnetize the pipe in axial direction, but there are also some that magnetize it in circumferential and recently also helical direction; see next section. Both these versions were devised in order to overcome the difficulty with detecting and characterizing defects that are narrow in the direction of magnetization, like narrow axial defects for an axially magnetizing tool.

Types of detectable features: MFL ILI tools are primarily geared toward detecting internal and external metal loss, such as

- General corrosion
- Pitting
- Erosion
- Mechanically induced metal loss

The technology will also detect

- Girth welds and installations (valves, off-takes, etc.);
- Dents and bends.

Note: As with any ILI technology, POD, POI, and sizing capabilities have to be looked at for individual tool configurations and regarding specific types of defects.

General performance characteristics:

- Operate equally well in gas and liquids pipelines.
- Due to the requirement of sufficient magnetization to saturate the pipe wall, which is increasingly difficult as

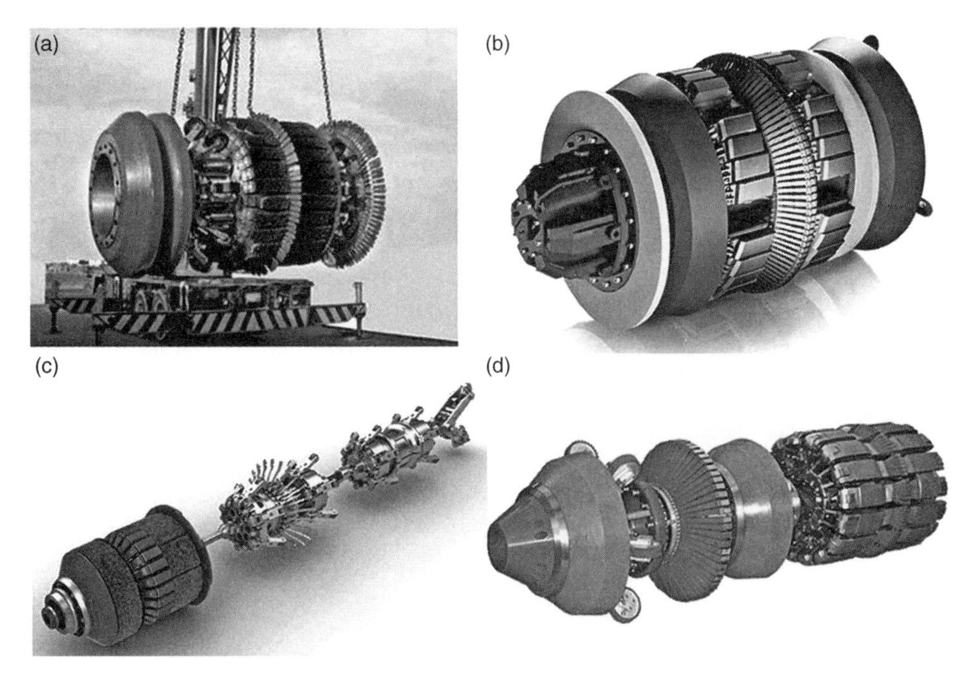


**FIGURE 35.4** MFL metal loss tools, (a) LIN SCAN [9] (Photo courtesy of LIN SCAN.); (b) ROSEN [10] (Photo courtesy of ROSEN.); (c) GE PII [12] (Photo courtesy of GE PII.); (d) Baker Hughes [8] (Photo courtesy of Baker Hughes.)

the wall thickness of the pipe increases and pipe diameter decreases, MFL tools have an upper wall thickness limit for proper operation.

- Full specifications can be typically achieved up to inspection speeds of 3–4 m/s (7–9 mph). For inspecting at higher speeds, like in high speed gas pipelines, tools have to be capable of bypassing gas and should be fitted, or combined with, a speed controlling unit.
- Calibration might be required prior to runs to improve detection and sizing capabilities.
- Cleanliness of the pipe plays a role in that it can cause sensor lift-off and erroneous readings.

Note: Generally, cleanliness is required for any ILI, some more than others, but it needs to be strived for!

***35.4.3.2 Circumferential Magnetic Flux Leakage Tools*** In the second half of the 1990s, tools came out with MFL applied circumferentially (transverse field, or transverse magnetization) in order to inspect for narrow axially oriented defects, in most cases associated with long seam welds in pipelines.

Principle of operation: Axially magnetizing MFL, as used in metal loss tools, is less effective on longitudinally oriented very narrow metal loss as they don't obstruct axial flux sufficiently for reliable detection and sizing; see illustration of the effect that shape of defects have on MFL in Figure 35.5. But if the magnetization is oriented circumferentially, long and narrow axially oriented defects will present themselves as more of an obstacle and promote flux leakage; see Figure 35.6. Similar results were also reported for helical magnetization, where the field is oriented spirally.

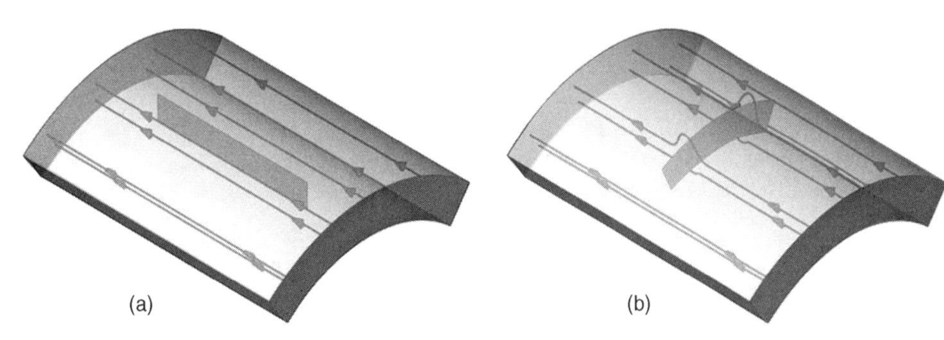


**FIGURE 35.5** Narrow defects aligned (a) with the direction of magnetization do not disrupt flux leakage as much as those oriented perpendicularly (b).

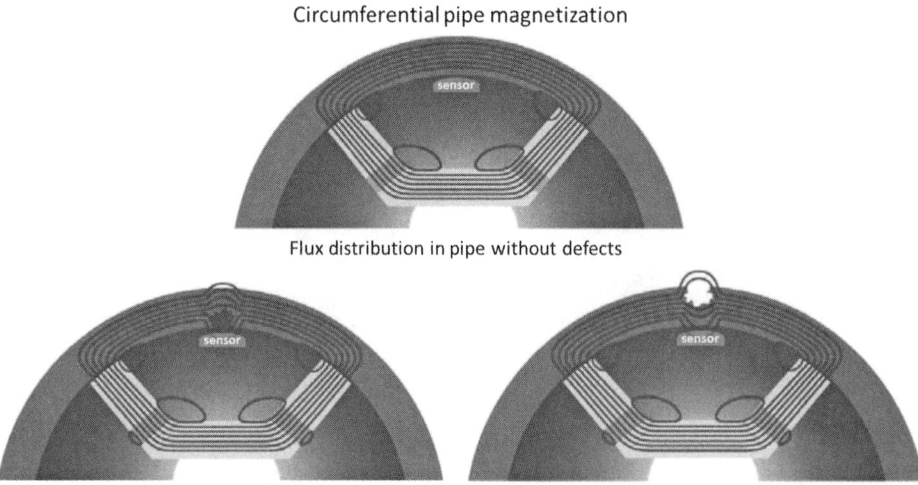


**FIGURE 35.6** Circumferential pipe magnetization in which a sensor inside the internal surface of the pipe detects magnetic flux leakage associated with internal and external metal loss, including narrow axial defects such as axial crack-like defects. (From Ref. [11].)

Magnetizing the pipe circumferentially results in different levels of magnetization compared with axial MFL and as such has some attributes that make it different in performance, not just the types of defects that can be detected. There are also several configurations of magnetizers on tools presently available, each of which has its own performance characteristics.

Types of detectable features:

- Axially oriented narrow defects, like narrow axial external corrosion (NAEC) associated with longitudinal welds.

General performance characteristics:

- Operate readily in gas and liquids pipelines.
- There has to be a certain opening to a crack for sufficient POD.
- Magnetization levels are generally not as high as with axial MFL and magnetic properties of steel and stresses, for example, residual stresses, influence tool performance.
- Inspection speed influences inspection results more than it does with axial MFL.
- Combining circumferential MFL with axial MFL tools improves the characterization ability for defects traditionally not suitable for axial MFL, for example, long narrow defects.

***35.4.3.3 Ultrasonic Testing (UT) Tools*** Ultrasonic testing (UT) is the other major technology utilized for metal loss inspection. UT has been used in other NDT applications, but was not introduced to ILI until the late 1980s.

Principle of operation: UT is based on direct and linear wall thickness measurement, the same principle that is used in the manual (or automated) testing in-the-ditch. Ultrasonic piezoelectric transducers are mounted on a sensor carrier perpendicularly to the wall emitting pulses and recording incoming reflections (pulse-echo) (Figure 35.7). By timing the reflections from the inner and outer surfaces of the pipe wall (time of flight) and knowing the speed of sound in the coupling medium and the pipe steel, the distance of the transducer to the inner surface of the pipe wall and the wall thickness can be deduced. Deploying a sufficient number of transducers to cover the full circumference of the pipe and timing the measurement of each transducer as it moves down the pipeline, the tool yields a grid of wall thickness measurements with an axial resolution defined by the shot-to-shot distance and circumferential resolution given by [circumferential] sensor to sensor distance. Figure 35.8 illustrates UT transducers deployed on a sensor carrier of a UT metal loss tool, and Figure 35.9 shows pictures of some of the tools.

This method requires the transducers (sensors), and by extension the tool, to be immersed in a homogeneous liquid with suitable ultrasonic properties for this type of measurement, as most of the light and heavy crude oils and refined products being pumped in liquids lines are.

Types of detectable features: UT metal loss tools detect changes in the pipe wall thickness and so detect external and internal metal loss, such as:

- General corrosion
- Pitting

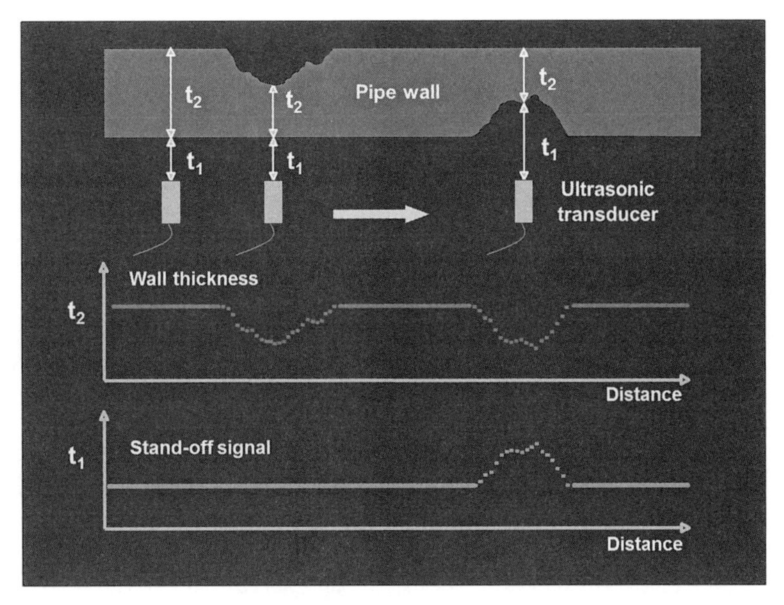


**FIGURE 35.7** Ultrasonic wall thickness measurement, principle of operation of UT metal loss tools.

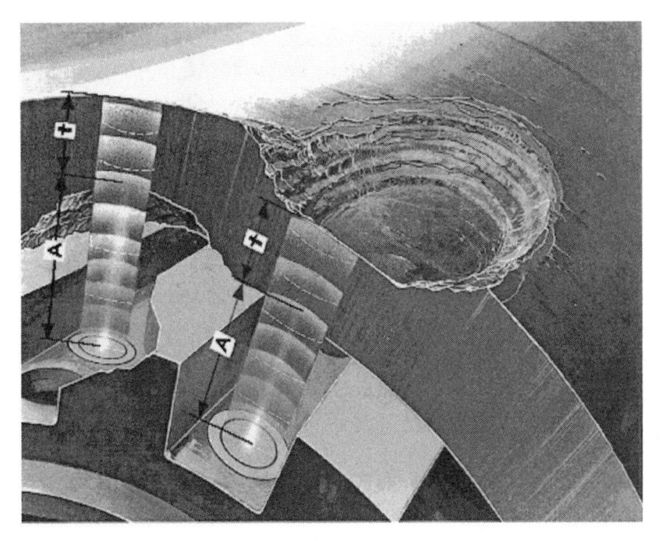

**FIGURE 35.8** UT transducers uniformly deployed to cover the full circumference of the pipe wall and measure metal loss.

- Erosion
- Mechanically induced

The technology will also detect

- Girth and long seam welds and installation (valves, off-takes, etc.);
- Dents and bends;
- Laminations;
- Hydrogen-induced cracking (HIC) and blisters.

Note: As with any ILI technology, specific POD, POI and sizing capabilities have to be looked at for individual tool configurations and regarding specific types of defects. For example, feasibility of detecting small pits will depend on the spatial resolution of UT sensors and might or might not be available in a specific pipe size.

General performance characteristics:

- This type of tool cannot operate in gas or multiphase environment as they require a suitable and homogeneous liquid as a couplant for ultrasound.
- Cleanliness of the pipeline can be critical, because excessive layers of wax, for example, can block the ultrasound from penetrating the wall causing loss of measurement.
- Speed of inspection does not affect the operation of the tool itself, but surpassing certain value induces an increase in axial shot-to-shot distance and hence reduced axial measurement resolution. This also means that the minimum detectable defect size increases accordingly. Depending on configuration, this speed is typically in the range of 1–2.5 m/s (2.2–5.6 mph).
- Increased wall thickness can become an issue only as it might require slower inspection (due to increased time required for the ultrasonic pulse to get across the pipe wall), but there is a limit in the remaining wall thickness, that is, no inspection is possible in pipe walls thinner than a minimum specified (~2 mm (0.08 in.)).

### 35.4.4 Crack Detection

Cracks in pipelines have been found to lead to ruptures and are among the most difficult types of defects to detect and even more difficult to size. The awareness about cracks in

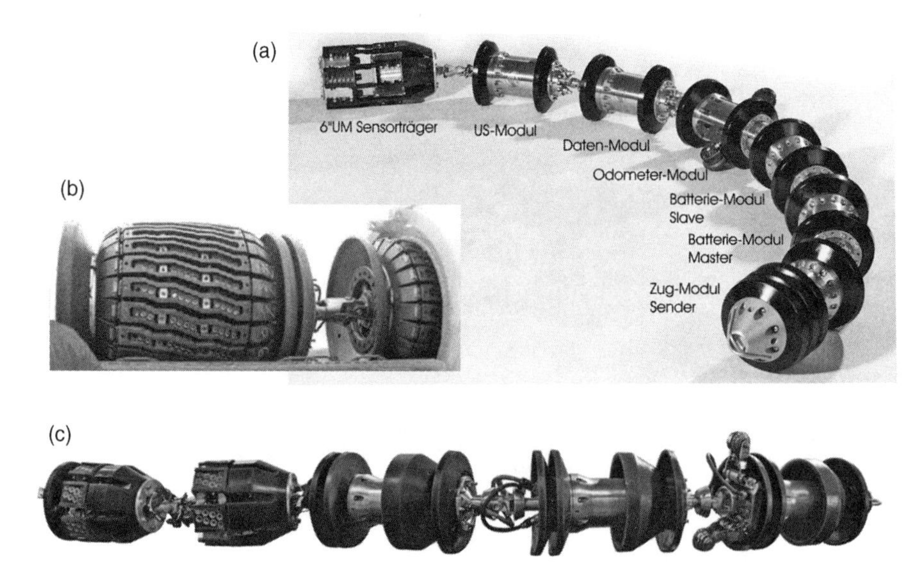


**FIGURE 35.9** UT metal loss tools: (a) NDT Global (Courtesy of NDT Global [13]), (b) Baker Hughes (with permission © 2014 Baker Hughes [8]), (c) LIN SCAN (Photo courtesy of LIN SCAN. [9])

pipelines got a boost in Canada after the National Energy Board (NEB) Public Inquiry in 1996, prompted by several pipeline ruptures and it wasn't until the early 2000s that the operators and regulators in the United States started openly talking about cracks, especially stress corrosion cracking (SCC).

Generally cracks, even very tight ones, cause strong reflections for ultrasound hitting them perpendicularly, but it is the questions how to get an ultrasound pulse going in a pipe wall. In the late 1980s, an approach was tried with liquid filled wheels providing coupling between the UT transducers and pipe wall in gas pipelines. In 1995, a liquid coupled tool was brought to the market and has remained the most reliable and accurate crack detection tool for liquids pipelines [14]. In the late 1990s, electromagnetic acoustic transducers (EMATs) were first introduced and continue to evolve as the method of choice for crack detection in gas pipelines. As cracks and crack-like defects have huge impact on pipeline integrity, other methods have also been tried to detect them, most notably using circumferential MFL, which, however, hasn't shown much success on tight cracks.

Since the introduction of all the technologies, there have been many inspections and also publications of tool performances and case studies reported at major industry conferences and in pipeline journals; see Section "Bibliography".

***35.4.4.1 Liquid Coupled UT*** Principle of operation: Liquid coupled ultrasonic crack detection is based on corner reflection of ultrasound travelling in a direction perpendicular to the crack orientation. Piezoelectric transducers are mounted on a sensor carrier of the ILI tool oriented at an angle to the surface of the pipe, such that, based on the differences in propagation speed of ultrasound in the liquid filling the interior of the pipeline and the pipe wall (Snell's law), a shear wave propagating under 45° in the pipe wall is generated (Figure 35.10). Similarly to UT wall thickness measurement, pulse-echo method is being utilized where reflections are timed to derive location of reflectors, internal, external, and mid-wall.

Since, due to hoop stress in pipelines, the prevailing orientation of cracking is axial, the first tools developed to use this technology were for detecting axial cracks and hence the propagation of ultrasound was set to be circumferential; see Figure 35.11a depicting a sensor carrier of a liquid coupled UT crack detection tools aimed at axial cracks. Types of detectable features:

- Cracks and crack-like features;
- SCC;
- Fatigue cracks;
- Seam-weld imperfections such as lack of fusion, toe cracks, etc.;
- Narrow axial corrosion.

Circumferentially or spirally oriented cracks can be detected as well, using this technology, but the sensor carrier has to be modified such that ultrasound propagates perpendicularly to the cracks; see Figure 35.11b.

Other types of tight crack-like defects, like surface breaking laminations or stress oriented hydrogen induced cracking (SOHIC), can be detected with this technology too, but it is important to know their orientation in order to configure the sensor carrier, i.e. mount UT transducers accordingly.

Tools configured for axial crack detection are shown in Figure 35.12

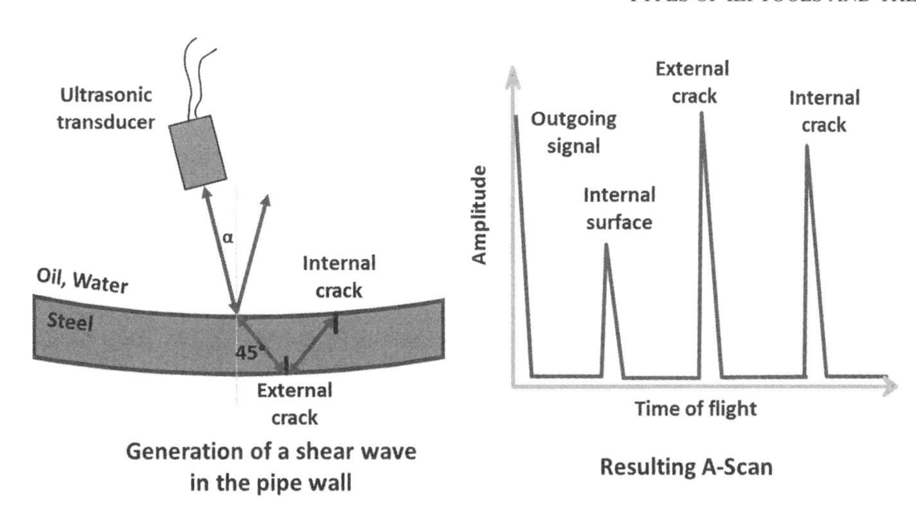


**FIGURE 35.10** Principle of operation of a liquid coupled ultrasonic crack detection tool.

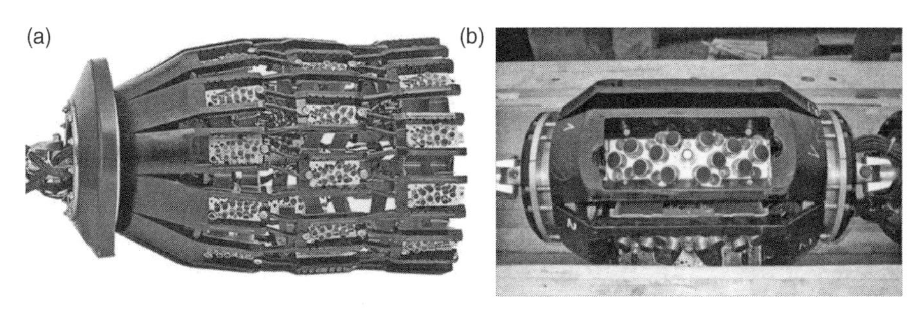


**FIGURE 35.11** Ultrasonic liquid coupled crack detection sensor carriers for axial (a) and circumferential cracks (b). (Courtesy of NDT Global [13].)

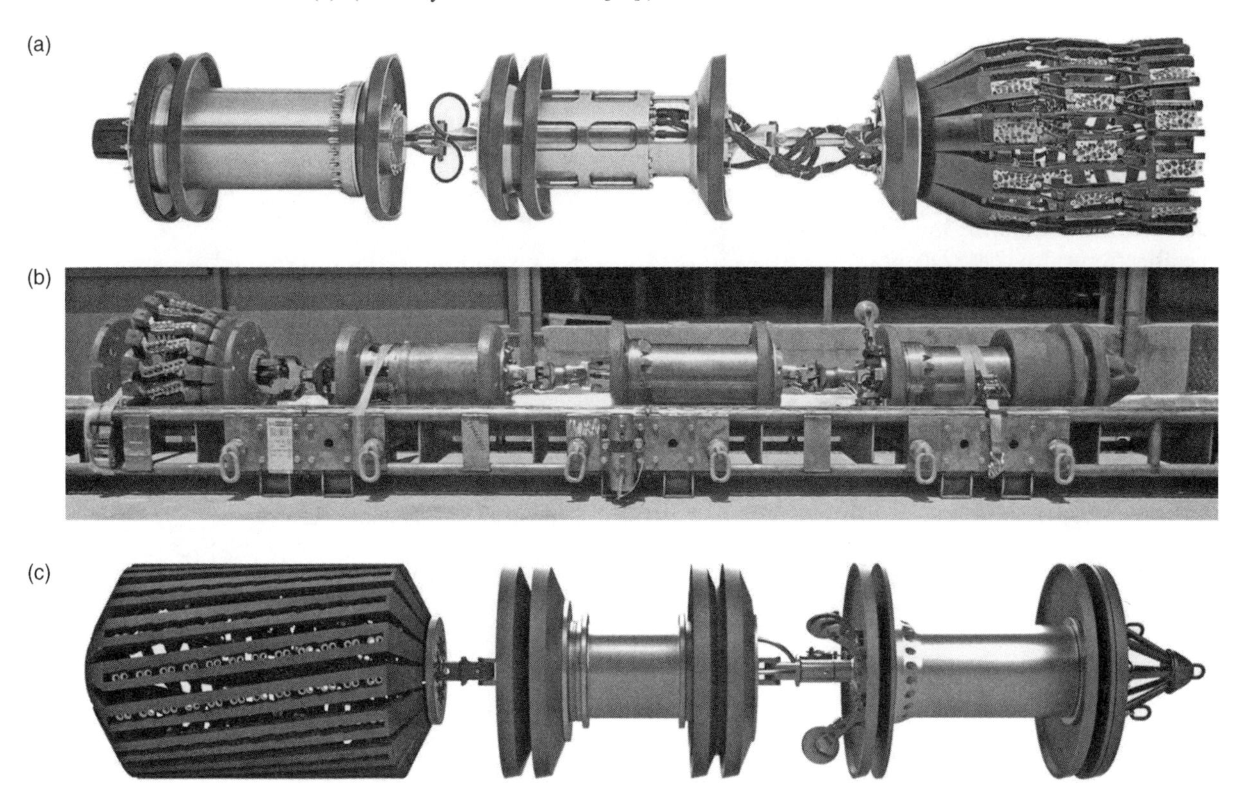


**FIGURE 35.12** UT crack detection tools: (a) NDT Global [13] (Courtesy of NDT Global); (b) Baker Hughes [8] (with permission © 2014 Baker Hughes); (c) GE PII [12]. (Courtesy of GE PII.)

General performance characteristics:

- Operate only in suitable liquids.
- Gas pipelines can be inspected using a liquid batch, where the tool is immersed in a suitable liquid being pumped through the pipeline.
- Cleanliness is equally important as with UT metal loss tools, layers of wax or other type of debris in pipelines may block or attenuate the ultrasound and cause loss of measurement.

*35.4.4.2 Electromagnetic Acoustic Transducers (EMAT)* Principle of operation: Ultrasonics has established itself as the most reliable method for crack detection, but generating ultrasound on the tool requires a medium to couple ultrasonic energy into the wall. EMAT overcomes this in that the ultrasound is generated in the pipe wall itself through electromagnetic interaction between the sensor on the tool and pipe wall. The sensor comprises a coil in a magnetic field where a pulsing current in the coil generates vibrations in the pipe wall generating ultrasound (Figure 35.13), utilizing magnetostriction or eddy current mechanisms. The configuration of the sensor, that is, the layout of the magnetic field, shape of the coil, and characteristics of the current, dictates which mode of ultrasound will be generated in the pipe wall, which in turn determines its suitability for crack detection. Generally, guided waves are generated to propagate circumferentially through the pipe wall, sensitive to axial cracks and sensors are configured for detecting both ensuing reflection and transmission, see Figure 35.14.

Types of detectable features: Tools presently on the market (some of them shown in Figure 35.15) utilize slightly different EMAT configurations, but are all geared toward detecting axial cracks and crack-like defects, such as SCC, fatigue cracks, and long seam weld imperfections.

General performance characteristics:

- Can operate in gas pipelines;
- Can detect coating disbondment, as type of coating and its bonding affect the attenuation of ultrasound.

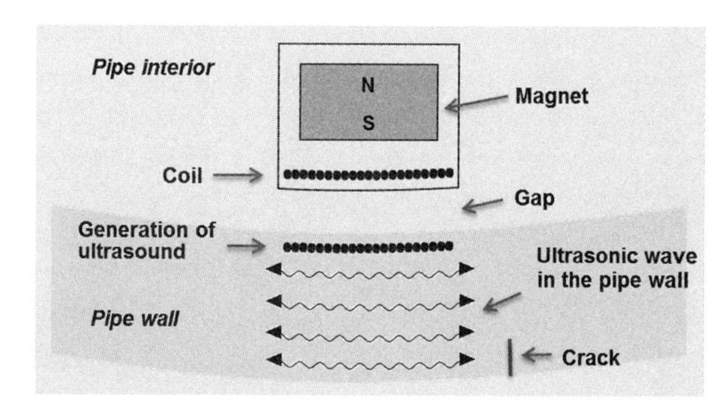


**FIGURE 35.13** EMAT principle of operation, generating ultrasound in pipe steel without liquid coupling.

## 35.4.5 Other

As said before, ILI has evolved not only in performance of existing ILI technologies, but also in development of new applications and combination of existing ones.

*35.4.5.1 Combined Technology Tools* The performance of each of ILI technologies described so far has improved over the years, with advancements in mechanical and electronic engineering along with progress in capabilities of data processing, but also with increasing usage and expertise that come with it. The need for operational practicality and ability to build more into tools than previously possible, has led ILI companies to combine different ILI technologies into their tools. The types of technologies that can be combined will depend on pipe size, section length, and types of restrictions present. Generally, the smaller the pipe diameter, the less space is available, and tools get longer, same with longer sections as additional batteries are needed to cover the energy consumption that increases accordingly.

Most common combinations are

- Mapping and geometry;
- Metal loss, geometry and mapping (Figure 35.16);
- Metal loss and crack detection;
- Axial and circumferential MFL.

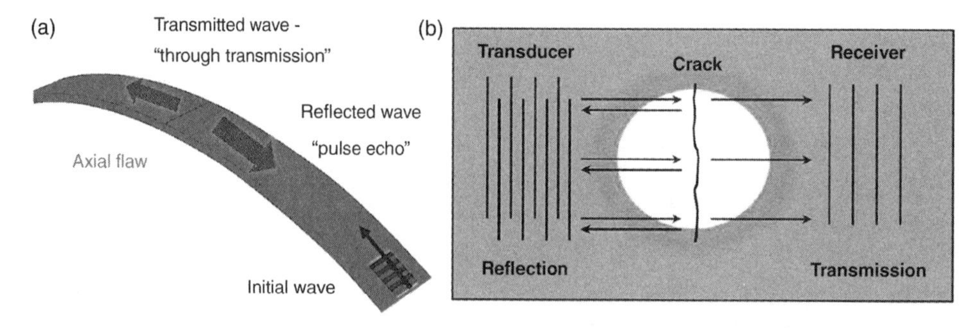


**FIGURE 35.14** EMAT sensors operating in reflection and transmission, (a) from Ref. [15]. (Published by ASME, reproduced with permission.) (b) From Ref. [16]. (Published by ASME, reproduced with permission.)

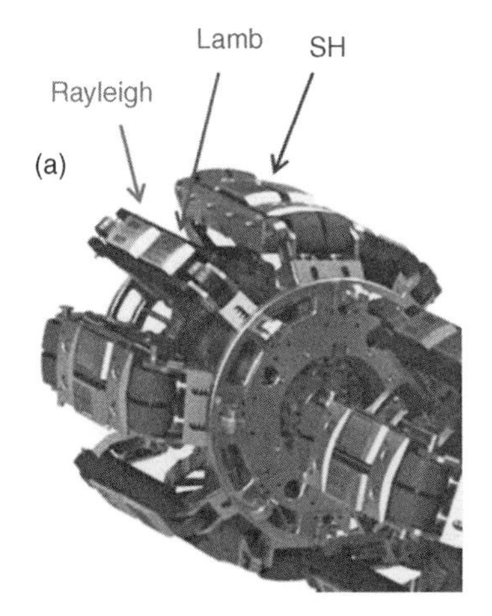

**FIGURE 35.15** EMAT tools: (a) From Ref. [15]. (Published by ASME, reproduced with permission.) (b) From Ref. [16]. (Published by ASME, reproduced with permission.)

***35.4.5.2 Detection of Mechanical Damage*** MFL as it is used most commonly for metal loss inspection, namely saturating the pipe wall, isn't sensitive to changes in material properties, for example, cold working after denting or gauging, as it isn't to remnant magnetization. But if the magnetization of the pipe wall is far lower, changes in magnetic properties, caused by changes in material properties, start affecting magnetic flux leakage and recorded signals. By combining two unites with MFL on the same ILI tool, one saturating the wall of the pipe and the other not and combining the results, it is possible to deduce and detect parts in the pipe wall that have undergone cold working, or generally changes in material properties like mechanical damage caused by third parties; see example of such a tool in Figure 35.17.

***35.4.5.3 Cathodic Protection Current Measurement*** One of the recent additions to the family of ILI technologies is measurement of currents induced by cathodic protection systems to the pipeline. The tool measures the voltage drop in the pipe created by cathodic protection current the characteristics of which can be traced back to the condition of cathodic protection in effect. Figure 35.18 shows such a tool.

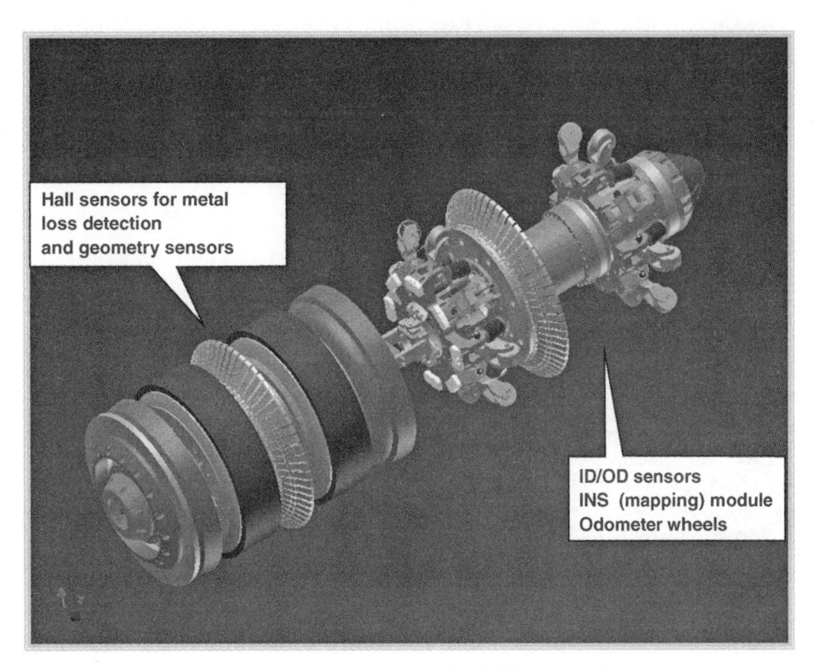


**FIGURE 35.16** Drawing of a combination tool with MFL metal loss, geometry and mapping capabilities.

**FIGURE 35.17** Multiple data set MFL tool [17] (© NACE International 2006.)

***35.4.5.4 Leak Detection*** ILI for leak detection has been considered for many years, but has never really been developed as a mainstream ILI application. Tools available in the market have, for the most part, utilized acoustic sensors to detect and locate the acoustic noise associated with the leaking fluid; see Figure 35.19. There have also been attempts to measure pressure differentials on tools as they pass by a leak [18].

## 35.5 UTILIZING ILI DATA/VERIFICATION

One of the main objectives of ILI is to obtain data for estimating the burst pressure of a pipeline containing defects. A successful inspection will detect anomalies above certain size, properly identify and size them, such that they can be fed into appropriate defect assessment algorithms. Since there are different types and levels of assessment algorithms, it is important to match their requirements with adequate ILI data. For example, an algorithm might require just the length and maximum depth of a defect, whereas another, more advanced, requires the knowledge of its profile, so then the ILI technology has to be utilized that can deliver that profile. Verifications of results from ILI are essential not only to confirm the findings, but also provide valuable feedback to ILI service providers that helps them improve their tools and data analysis. The API 1163 standard [5] gives recommendations on conducting verifications.

**FIGURE 35.18** Tool for inspecting cathodic protection current and influences of external AC and DC current sources. (Photo courtesy of Baker Hughes [8].)

**FIGURE 35.19** Leak detection tools (a) GOTTSBERG leak detection [19]. (Reproduced with permission of Gottsberg Leak Detection GmbH & Co. KG.) (b) (SmartBall® Leak Detection © 2014 Pure Technologies Ltd.) [20].

## 35.6 INTEGRATING ILI DATA

Versatility of information pertaining to a pipeline section or system calls for integrating it in order to maximize the insight into pipeline integrity issues. The information gained from ILI should be viewed in conjunction with all the other knowledge, such as information on soil, coating conditions, operational regime, to name just a few. As an example, when analyzing the distribution of corrosion in a section of a pipeline it is essential to look at those facts to fully understand what caused the distribution to be like that which, in turn, will facilitate devising most effective mitigation measures.

## APPENDIX 35.A[1]: SAMPLE PIPELINE INSPECTION QUESTIONNAIRE (NONMANDATORY)

| **Company Name** | |
|---|---|
| **Completed by** | |
| Name | Fax |
| Office phone | Date |
| **Checked by** | |
| Name | Fax |
| Office phone | Date |

| **Site information** | |
|---|---|
| Pipeline name | |
| Line length (km) (mi) | Line OD (mm) (in.) |
| Launch site | L station # |
| Launch phone | Receive phone |
| Receive site | R station # |
| Base location | Base station # |
| Base shipping address | |
| Base contact | Base phone |
| Type of inspection required: SCC MFL Dent Profile clean | |
| Dummy tool required? | Locator required? |
| Pipeline alignment maps available? | |

| **Product Details** | |
|---|---|
| Product type | $H_2O$ content |
| Wax content | Slackline? |
| $CO_2$ content | Hazardous? |
| $H_2S$ content | Protective equipment? |

[1] SP0102-2010/In-Line Inspection of Pipelines © NACE International 2010. All rights reserved by NACE. Reprinted with permission. NACE standards are revised periodically. Users are cautioned to obtain the latest edition; information in an outdated version of the standard may not be accurate.

| Type of flow: laminar turbulent two-phase (Transitional) | | | |
|---|---|---|---|
| Flow property: liquid gas two-phase (Both) | | | |
| Will the line be isolated? | Constant velocity? | | |
| Flow rate controllable? | | | |
| **Line Conditions** | Min. | Normal | Max. |
| Launch pressure (kPa) | | | |
| (psi) | | | |
| Launch velocity (km/h) | | | |
| (mph) | | | |
| Launch flow rate ($m^3$/d) | | | |
| (MMcfd) | | | |
| Launch temperature (°C) | | | |
| (°F) | | | |
| Receive pressure (kPa) | | | |
| (psi) | | | |
| Receive velocity (km/h) | | | |
| (mph) | | | |
| Receive flow rate ($m^3$/d) | | | |
| (MMcfd) | | | |
| Receive temperature (°C) | | | |
| (°F) | | | |

Note: These values are recorded for regular line conditions. Pressures and velocities vary during the pig run.

| **Pipe Details** | |
|---|---|
| Last inspection year | MAOP |
| Design pressure | Type of cleaning pig |
| Cleaning program? | Frequency |
| Known/suspected damage | |
| Relevant historical data | |

| **Pipeline Conditions** | |
|---|---|
| Year of construction | Sphere tees installed? |
| Pipe cover depth max. min. | Type of pipe cover? |
| Are there high-voltage lines in the vicinity of the pipeline? Where? | |
| Insulating flanges in the pipe? Where? | |

| R.O.W. access (road, air, etc.) |
|---|
| Does pipe have hot taps? |
| Relevant historical data |

| **Pipe Features** | Yes | No | | Yes | No |
|---|---|---|---|---|---|
| Does the pipeline contain the following features? | | | | | |
| Thread and collar couplings | | | Chill rings | | |
| Bell and spigot couplings | | | Hydrocouples | | |
| Stepped hydrocouples | | | Stopple tees | | |
| Nontransitioned wall-thickness changes | | | Wye fittings | | |
| | | | Miter joints | | |
| Corrosion sampling points | | | Acetylene welds | | |
| Internal probes | | | Vortex breakers | | |

| **Trap Details** (Figure 35.A1) | | Launch Length | Receive Length |
|---|---|---|---|
| A | Closure to reducer (m/ft) | | |
| B | Closure to trap valve (m/ft) | | |
| C | Closure to bridle CL (mm/in.) | | |
| d | Pipeline diameter (mm/in) | | |
| d′ | Pipeline internal diameter (mm/in) | | |
| D | Overbore (NPS # or mm/in.) | | |
| D′ | Overbore internal diameter (mm/in.) | | |
| E | Axial clearance (m/ft) | | |
| F | Reducer length (mm/in.) | | |
| F′ | Reducer wall thickness (mm/in.) | | |

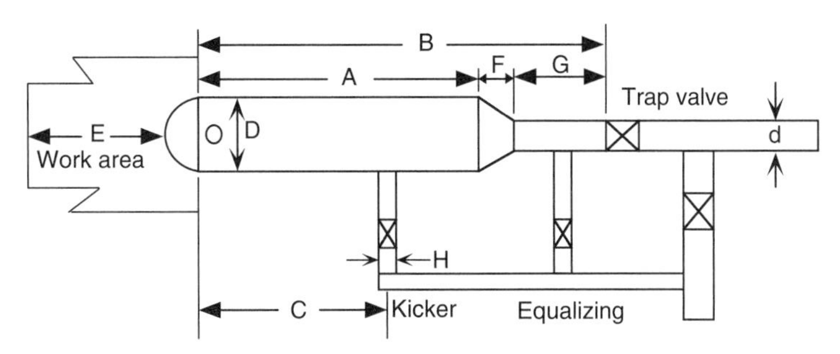


**FIGURE 35.A1** Plan view of a generic pig trap.

| | | | |
|---|---|---|---|
| G | Reducer to valve (m/ft) | | |
| H | Kicker line (NPS # or mm/in.) | | |
| **Trap Conditions** | | Launcher | Receiver |
| Orientation | | | |
| Type/internal diameter of trap valve (mm/in) | | | |
| Centerline height of trap | | | |
| (Aboveground) (mm/in.) | | | |
| Is hoist available? | | Yes No<br>Capacity<br>Lift height | Yes No<br>Capacity<br>Lift height |
| Is trap equipped with:<br>Pig-sig?<br>Sphere tee?<br>Coupons or internal fittings? | | | |
| Trap closure type | | | |
| Trap pressure rating (kPa/psi) | | | |
| Concentric or eccentric reducer? | | | |
| Workshop near trap? | | | |
| Access limitations (due to available area or ground conditions)? | | | |
| AC power at trap site? | | | |
| Intrinsic safe area, level? | | | |
| Site drawings available? | | | |

| **Pipe Information** | | | | | |
|---|---|---|---|---|---|
| Nominal Wall Thickness of Pipe (mm [in.]) | Length of Each Wall Thickness (km [mi]) | Pipe Weld Type | Pipe Grade (MPa/psi) | Mill | OD (mm [in.]) |
| | | | | | |
| Total length= | | | | | |
| **Repair History** | | | | | |
| Nominal Wall Thickness and Grade of Pipe (mm/MPa [ksi/in.]) | Length of Each Wall Thickness (km [mi]) | Start Chainage | End Chainage | Comments | Date of Repair |
| | | | | | |
| Total length= | | | | | |

| Bends | | | | | |
|---|---|---|---|---|---|
| Type | Chainage of Bend (km [mi]) | Angle (degrees) | Bending Radius | Minimum Bore (mm [in.]) | Comments |
| | | | | | |

| Tees/Off Takes/Branches | | | | | |
|---|---|---|---|---|---|
| Type (Forges, Stopple, etc.) | Chainage of T/OT/B (km [mi]) | O'clock Position | Max. Off-Take Diameter (mm [in.]) | Barred or Unbarred | Comments |
| | | | | | |
| | | | | | |
| | | | | | |
| | | | | | |

| Valves | | | | |
|---|---|---|---|---|
| Type | Chainage of Valve (km [mi]) | Manufacturer | Model | Minimum Bore (mm [in.]) |
| | | | | |

| Diameter Changes | | | | | |
|---|---|---|---|---|---|
| Type of Reducer | Chainage of Diameter Change (km [mi]) | Upstream Diameter (mm [in.]) | Downstream Diameter (mm [in.]) | Diameter Transition Length (mm [in.]) | Comments |
| | | | | | |

| **Coatings**<br>(If concrete coated, is there any magnetic content?) | |
|---|---|
| Internal | |
| External | |

| **Aboveground References**<br>Can any of the following be located from aboveground for references? | | | |
|---|---|---|---|
| Line valves | | Large bends | |
| CP connections | | Off tees | |
| Major WT changes | | Sleeves | |
| Anodes | | Casings | |
| Girth welds | | Insulation flanges | |

| **Known Metal Loss Information** | |
|---|---|
| Internal | |
| External | |
| Mechanical damage | |
| Other | |

| **Special Attention** |
|---|
| |
| |
| |
| |
| |

| **Comments** |
|---|
| |
| |
| |
| |
| |

**Completed by**________________________________________

| Name | Signature | Date |
|---|---|---|

**Checked by** ________________________________________

| Name | Signature | Date |
|---|---|---|

**Updated by**________________________________________

| Name | Signature | Date |
|---|---|---|

## REFERENCES


1. Pipeline Operators Forum (2009) Specifications and requirements for intelligent pig inspection of pipelines. s.l. Available at http://www.pipelineoperators.org/.
2. NACE (2010) In-Line Inspection of Pipelines, Houston, TX: NACE International, Publication 35100 (2012).
3. US DOT (1999) 49 CFR 195.452—Pipeline integrity management in high consequence areas, Washington, DC.
4. API (2001) API 1160: Managing System Integrity for Hazardous Liquid Pipelines. s.l., American Petroleum Institute, 2001.
5. API (2005) API 1163: In-line Inspection Systems Qualification Standard. s.l., American Petroleum Institute.
6. NACE (2010) NACE SP0102-2010: In-Line Inspection of Pipelines, Houston, TX: NACE International, 2010.
7. ASNT ILI PQ, ILI Personnel Qualification and Certification, The American Society for Nondestructive Testing, 2010.
8. Baker Hughes Available at http://www.bakerhughes.com/products-and-services/process-and-pipeline-services/pipeline-services.
9. LIN SCAN Available at http://www.linscaninspection.com/.
10. ROSEN Available at http://www.roseninspection.net/.
11. ROSEN Group—Products & Services Product Brochure. s.l. Available at www.roseninspection.net.
12. GE PII Available at http://www.ge-energy.com/products_and_services/services/pipeline_integrity_services/.
13. NDT Global Available at http://www.ndt-global.com/solutions/tools/ultrasonic-tools.html.
14. Willems, H.H., Barbian, O.A., and Uzelac, N.I. (1996) Internal insepction device for detection of longitudinal cracks in oil and gas pipelines. Proceedings of ASME International Pipeline Conference, June 9–14, 1996, Calgary, AL: ASME.
15. Sutherland, J., Tappert, S., Kania, R., Kashammer, K., Marr, J., Mann, A., Rosca, G., and Garth, C. (2012) The role of effective collaboration in the advancement of EMAT inline inspection technology for pipeline integrity management. Proceedings of International Pipeline Conference (IPC 2012), September 24–28, 2012, Calgary, AB, Paper No. IPC2012-90021, a case study, ASME.
16. Kania, R., Klein, S., Marr, J., Rosca, G., San E., Riverol, J., Ruda, R., Jansing, N., Beuker, T., Ronsky N.D., and Weber R. (2012) Validation of EMAT technology for gas pipeline crack inspection. Proceedings of International Pipeline Conference (IPC 2012), September 24–28, 2012,, Calgary, AB: ASME. Paper No. IPC2012-9240.
17. Simek, J. (2006) Magnetic flux leakage and multiple data set. Proceedings of NACE Central Area Conference, October 2006, Houston, TX: NACE International.
18. Camerini, D.A., et al. (2003) Pig Detector de Vazamentos em Oleoductos. Proceedings of Rio Pipeline 2003 Conference & Exposition, September 24–26, 2003, Rio Janeiro: ASME.
19. Gottsberg Leak Detection GmbH & Co. KG Available at http://www.leak-detection.de/.
20. Pure Technologies Available at http://www.puretechltd.com/products/smartball/smartball_oil_gas.shtml.


## BIBLIOGRAPHY: JOURNALS, CONFERENCES AND OTHER SOURCES WITH ILI RELATED CONTENT

### Journals


- 3R-International English ed., http://www.di-verlag.de
- European Oil & Gas Magazine, http://www.oilandgaspublisher.de
- Journal of Pipeline Engineering, http://www.gs-press.com.au/print/journal_of_pipeline_engineering
- Oil & Gas Journal, http://www.ogj.com/index.html
- Pipeline & Gas Journal, http://www.pipelineandgasjournal.com/
- Pipelines International, http://pipelinesinternational.com/
- PetroMin Pipeliner, http://www.safan.com
- Scandinavian Oil & Gas Magazine, http://www.scandoil.com

**Conferences:**

- International Pipeline Conference IPC, ASME http://www.internationalpipelineconference.com/
- International Pipeline Technology Conference, http://pipelinetechnology2013.com/home
- NACE Corrosion Conference, NACE http://events.nace.org/conferences/c2013/president.asp
- Pipeline Pigging and Integrity Management, http://www.clarion.org/ppim/ppim12/
- Pipeline Technology Conference, EITEP—Euro Institute for Information and Technology Transfer in Environmental Protection GmbH, http://www.eitep.de/sendy/w/XoHjc67ml J0lLTPsc5P6Nw/hxe1iB4F5JZV4jkg93MSiw/oNjc1RTwkFnHffWd9720dQ
- Rio Pipeline, ASME http://www.ibp.org.br/main.asp?ViewID=%7B5D99112A-2BE3-4F29-8955-E36E50029744%7D&Lang ID=en
- The Best Practices in Pipeline Operations & Integrity Management, Clarion, http://www.clarion.org/bestpractices/bahrain2013/index.php
- Unpiggable Pipeline Solutions Forum, Clarion, http://www.clarion.org

**Other Sources:**

- Inline Inspection Association, http://www.iliassociation.org/
- NACE International, www.nace.org
- PHMSA, DOT https://primis.phmsa.dot.gov
- Pigging Products & Services Association, http://www.ppsa-online.com/
- PRCI - Pipeline Research Council International, http://www.prci.org/

# 36

# EDDY CURRENT TESTING IN PIPELINE INSPECTION

Konrad Reber
*Innospection Germany GmbH, Stutensee, Germany*

## 36.1 STANDARD EDDY CURRENT TESTING

### 36.1.1 Introduction

The basis for eddy current testing (ECT) was laid by the discoveries of electromagnetic effects in the nineteenth century. The protagonists were Michael Faraday (1791–1867) for the discovery of the interaction between magnetic fields and electrical currents and Leon Foucault (1819–1868) for the discovery of eddy currents in metal objects. Only the sound electromagnetic theory developed by James Clerk Maxwell (1831–1879) allows us today to engineer circuits and probes for inspection. The tribute for the first application of eddy currents is given to David Edward Hughes for metal sorting. Friedrich Foerster (1908–1999) developed the first devices for ECT as we know them today. Eddy current developed into a major branch of nondestructive testing (NDT). Although it is limited to conductive materials, the possibility to automate the scanning and signal analysis has increased its importance since it was introduced. As will be described in the following section, its role in pipeline inspection is minor today, although recent developments suggest that this is currently changing.

### 36.1.2 How Eddy Current Testing (ECT) Works

The basic principle of ECT consists of measuring the impedance of an electrical coil. The impedance is its complex resistance, which also depends on a property called inductance. The inductance is a physical property of a coil that describes how a magnetic field is generated, when an electrical current is sent through it. Electromagnetic theory shows that the effect of inductance leads to a phase shift and amplitude change in the current, when an AC voltage is applied to the coil. The inductance of a coil can, thus, be measured by measuring the AC resistance (impedance) of the coil. It depends on the size of the coil, the number of turns, and most importantly on the electromagnetic properties of all materials in the vicinity of the coil. A coil in the vicinity of a metallic surface will have a different impedance as compared to the same coil in air. Moreover, a coil close to a metallic surface will have a different impedance depending on whether the metallic surface is flawless or it has a crack. This is the basis of ECT. To understand the abilities and limitations of ECT better, few more details are required.

The details of the set-up are shown in Figures 36.1 and 36.2. An AC current flows through a coil. An AC magnetic field is generated around the coil. If the coil is in air, the magnetic field only depends on the nature of the coil (its inductance). If the coil is brought closer to a metallic surface, the magnetic field will penetrate into the metal piece. According to the law of electromagnetic induction, an AC magnetic field will lead to the generation of currents in the surface. These currents generate a magnetic field that, according to Lenz's law, is opposed to the initial field. It weakens the original field, and thus, influences the inductance of the coil. Anything that changes the nature of the eddy currents in the metal can, thus, be detected by a change in the impedance. Naturally, the vicinity of the coil to the surface and the presence of a crack in the metal are such influences on the eddy currents. The deeper the crack, the more it will impede the flow of currents in the surface and the higher will be the effect.

It lies in the nature of electromagnetic induction that the currents flow in the surface only (skin-effect). The degree to

*Oil and Gas Pipelines: Integrity and Safety Handbook*, First Edition. Edited by R. Winston Revie.

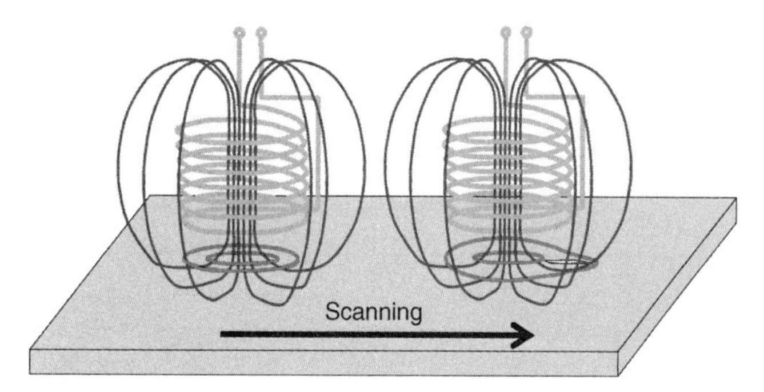


**FIGURE 36.1** The concept of eddy current testing.

which they will penetrate into the bulk of the material depends on the material properties. Electromagnetic theory shows that the higher the frequency of the currents and the higher the conductivity and permeability of the metal, the more the currents will be confined to the surface. Standard eddy current techniques employ a frequency of a few kHz to MHz. This means that stainless steel tubes of a few mm wall thickness can be inspected while on carbon steel tubes, only surface effects are detected. The effect of limited penetration depth is shown in Figure 36.2b.

The coils used for inspection vary considerably in size, shape, and design. Either single coils (absolute coil) or two coils in a bridge mode (differential coil) can be used. The latter makes the arrangement very sensitive to sudden changes while the first is more suitable for the detection of gradual changes in material properties. Also, coils can be arranged in a send–receive setup. In general, it can be said that the smaller the defect to be detected, the smaller the coil should be and the closer it should be to the surface.

The instrumentation itself then consists of a scanner that holds sensors in a predefined distance to the surface (liftoff) and allows moving them smoothly in a certain direction. The electronics carry out a phase-sensitive detection to determine the complex impedance. The value of the impedance is shown in the impedance plane (Figure 36.2a). As the sensor moves over a surface, the locus point of the impedance changes.

### 36.1.3 Limitations for Pipeline Inspection

Eddy current sensing technology has widespread application in the aviation industry. Here, the detection of cracks in riveted joints is of paramount importance. As the material to be tested is aluminum and the sheets are rather thin, ECT has been the method of choice for many years now. Also, ECT is applied in all kinds of component testing. Another application is material sorting, which makes use of the fact that different metals have varying conductivity values.

The application of ECT in the oil and gas industry and, in particular, for pipeline inspection is so far of minor importance. The task in pipeline inspection consists of finding cracks and corrosion in comparatively thick-walled pipe, which are always produced from low-carbon ferritic steel (exceptions are described in the following section). Also, far-side defects need to be detected. For internal inspection (pigging), external corrosion defects are of interest. With the limitation in penetration depth, all of these tasks cannot be achieved with standard eddy current technology.

In summary, it can be said that standard ECT in pipeline inspection is rather used to complement other techniques. The most often applied technique consists of measuring the stand-off of a probe from a surface. This allows for profiling the near-side surface.

Altogether, three methods can be identified that use an alteration of the standard eddy current idea and that are

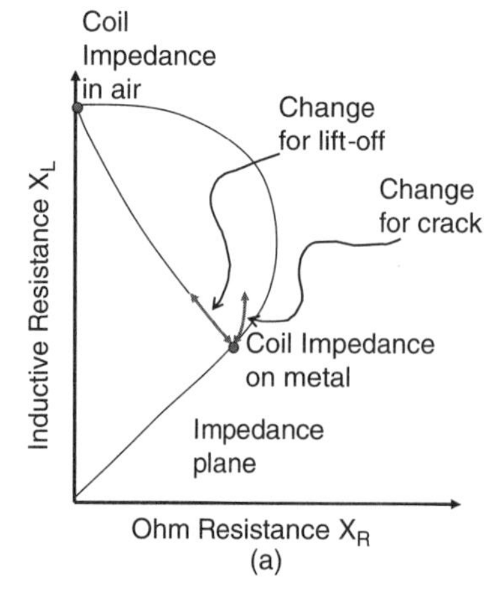


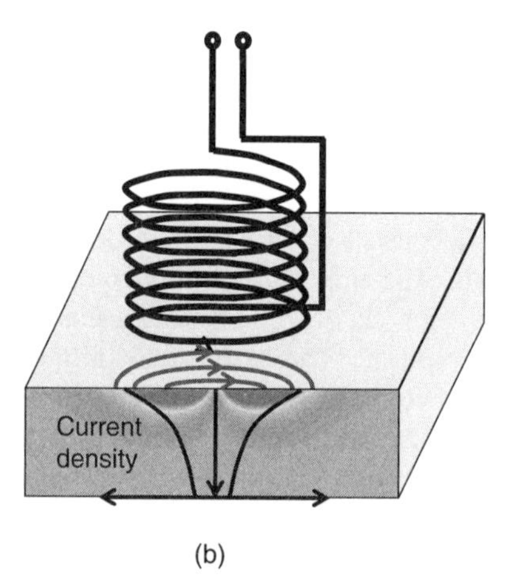


**FIGURE 36.2** (a) The change of the locus point in the impedance plane will indicate defects and (b) the depth of penetration of eddy currents into the metal is limited.

frequently used in the oil and gas industry. It is a matter of viewpoint whether these methods should be called eddy current technology or not. Nevertheless, all of these techniques employ an AC electromagnetic method and are commonly called "eddy current" techniques. These methods comprise remote field eddy current testing (RFEC) and the closely related low-frequency eddy current testing (LFET), pulsed eddy current technology (PEC) and magnetic eddy current testing (MEC™ or SLOFEC™).

## 36.2 ENHANCED EDDY CURRENT TESTING

### 36.2.1 Remote Field Eddy Current Testing (RFEC)

The main idea about RFEC is that it uses very low frequencies. Very low, in this context, means frequencies in the range of 10–50 Hz. At these frequencies, currents and magnetic fields are no longer confined to the surface. If an exciter coil is placed inside a ferritic tube, a characteristic distribution of eddy currents and magnetic fields will be established. The tube conducts these fields along the axis. If detecting coils are placed at a certain distance, the received signal will depend on the quality of the transmission path. Metal loss defects or cracks impeding the flow of eddy currents or magnetic flux will influence the signal. The setup is shown Figure 36.3.

The detector coils need to be placed at a certain distance to avoid any direct coupling with the exciter coil. This coined the name "remote field." In the early days of RFEC, the working principle was not so evident, but has been explored and developed through experimental studies. Today, the interaction principle is well understood [1]. The key is the distribution of the magnetic field. The magnetisation is in the axial direction of the tube or pipe. This reduces the interaction with defects oriented along the axis of the pipe, such as axial cracks. Magnetic flux leakage (MFL) suffers from the same drawback.

Closely related to RFEC, is a technique called low frequency eddy current (LFET). Here also, low-frequency magnetization processes are used. The technology is used for external inspection of piping. The typical cylindrical symmetry is not present, which determines the spacing between the sending and receiving coil. Special send–receive configurations are required [2]. External pipe and tank floor scanners based on this technology are available.

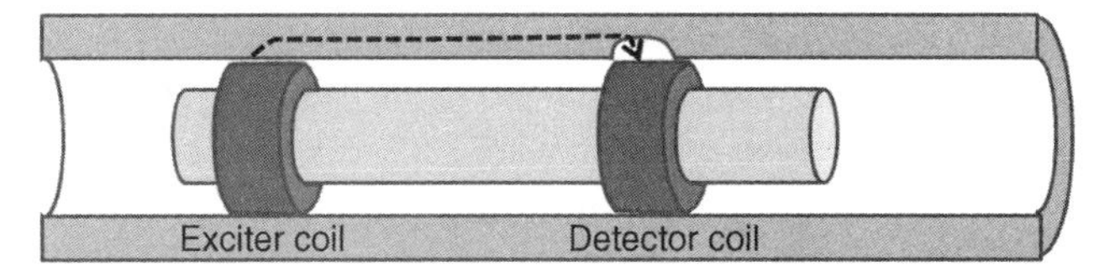


**FIGURE 36.3** The idea of remote field eddy current testing.

### 36.2.2 Pulsed Eddy Current (PEC) Testing

The development of the PEC and its application to the oil and gas industry was driven by Shell Global Solutions [3]. The idea is to use a short pulse of current and investigate the response over time. A pulse contains a broad spectrum of frequencies. The lower frequency range will lead to eddy currents propagating through the pipe wall. The speed of this propagation is given by material properties and can be described as a diffusion process. Due to the limit in wall thickness, the response will show a change in signal, once the eddy currents have diffused through the wall. The principle is shown in Figure 36.4. Figure 36.4a shows a steep, broadband, excitation pulse. The resulting eddy currents diffuse through the wall and lead to a delayed response in the coil. Figure 36.4b shows the response on a logarithmic scale. The response originates from deeper and deeper layers of the metal with increasing time. The limitation of the metal sheath thickness shows up as a sudden change in the decay rate of the response. The greater the wall thickness, the later this changeover will occur.

The advantage and difference to the other mentioned techniques is that it can deliver absolute wall thickness readings. It is, thus, suitable for a monitoring application, where a stationary sensor measures wall thickness over a long period of time. A drawback is that because of the use of a broader frequency band, a single measurement takes too much time to use PEC on a scanning tool. The sensor needs to be at rest

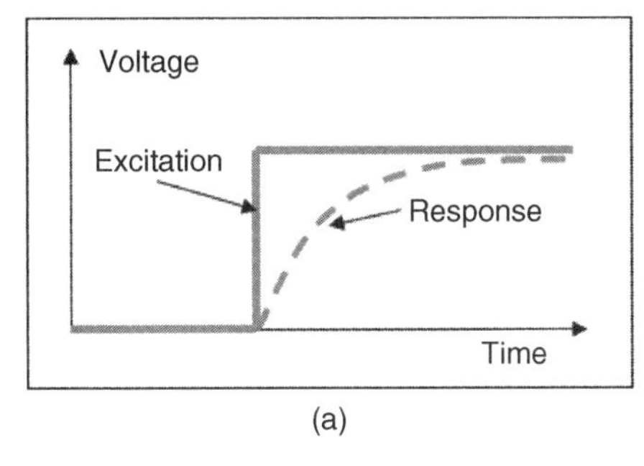


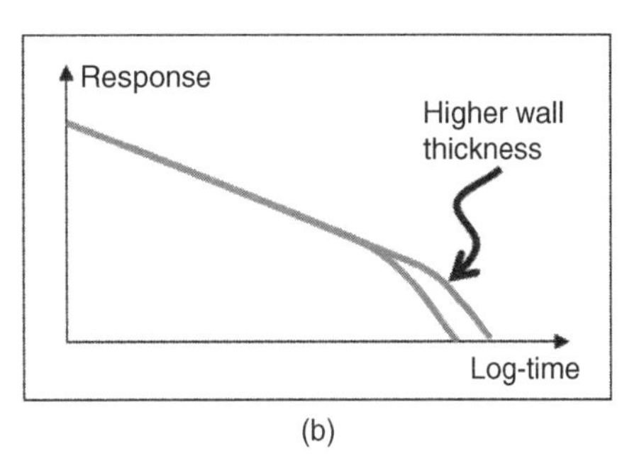


**FIGURE 36.4** In pulsed eddy current testing, the decay of the step response is analyzed.

when a measurement is done. The available sensors are made such that larger liftoff values are permissible for the measurement. A measurement through several mm of coating is possible. At the same time, the measured wall thickness is an average over the active area of the sensor. Today, service companies can obtain licenses to use this technology. Refer to Section 36.3.3 for applications.

### 36.2.3 Magnetic Eddy Current Testing (SLOFEC or MEC)

The inability to use standard eddy current techniques on ferritic pipe material is due to its high values of permeability. This limits the depth of penetration and the obtained information to the near-side surface only. Also, far-side defects can influence the near-side response of eddy current sensors. If a pipe wall is magnetized, the magnetic flux distribution will be distorted by the presence of a crack-like or metal loss defect. The distortion of the magnetic flux leads to a higher level of magnetization on the near side of the pipe wall. This concept is shown in Figure 36.5.

The different level of magnetization can be detected with a regular eddy current sensor. At the same time, near-side defects are detected by the regular eddy current effect. In the impedance plane, the two effects can easily be distinguished by analysis of the signal phase.

It is important to understand the difference between MEC and MFL. The setups of the two methods look quite similar. However, MFL requires a leakage field to be measured. The method senses on the surfaces of the pipe wall. For a repeatable signal in MFL, the magnetizing field requires a rather high level [4]. A value of about 10 kA/m is considered a minimum. In MEC, a field of 3 kA/m is usually sufficient. The requirement for a minimum magnetization level is due to the need to discriminate other permeability related signals. Pipeline stresses or material inhomogeneities may be an interesting aspect for inspection. However, a clear distinction of metal loss defects is required.

## 36.3 APPLICATIONS FOR PIPELINE INSPECTION

### 36.3.1 Standard EC Applications

In the early days of in-line inspection (ILI), a need for the detection of internal cracks was deemed to be necessary. Pipelines under cyclic loading may suffer from fatigue. Fatigue cracks may develop from the internal surface as tensile stresses in cylindrical pressure vessels are slightly higher on the inside of the wall. A pig has been developed to find internal cracks using eddy current-based sensing technology [5]. This tool employed a special PEC-technology. The purpose of the pulsation here was to limit power consumption and data storage requirements. The project was discontinued mainly because the need to find internal cracking today is considered to be less important as compared to finding stress corrosion cracking (SCC) on the external pipe surface. Also, for liquid pipelines, a method based on pulse-echo ultrasonic testing (UT) technology is available. Today there is no ILI tool based on eddy-current technology alone.

Eddy current sensing is, however, employed in regular ILI to complement other technologies. In regular ILI using MFL, a special sensor is required to determine whether a signal originates from the near or the far-side of a pipe wall. Traditionally, a magnetostatic reluctance measurement was used. The problem was that this measurement interferes with the strong magnets of the standard MFL measurement. In this case, an ILI provider included eddy current sensors to measure lift-off and metal loss at the position of an internal

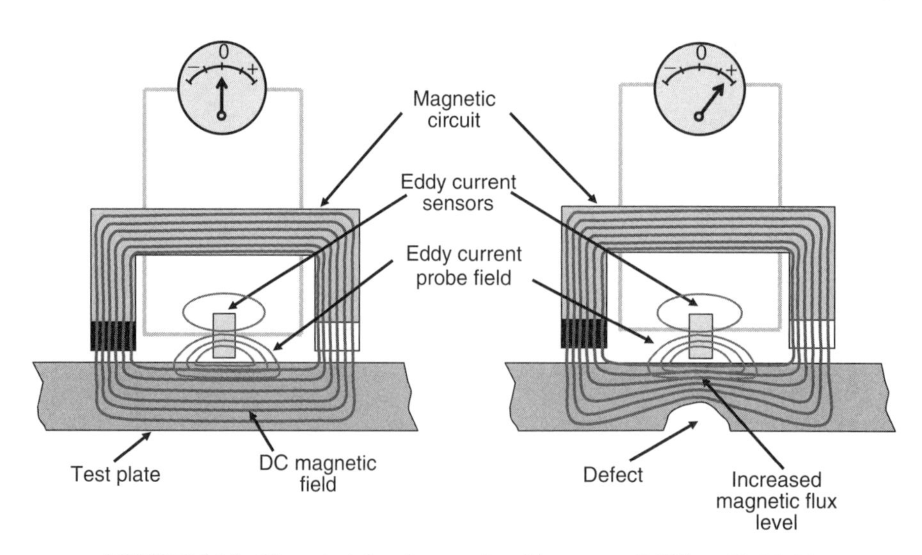


**FIGURE 36.5** The principle of magnetic eddy current (MEC or SLOFEC).

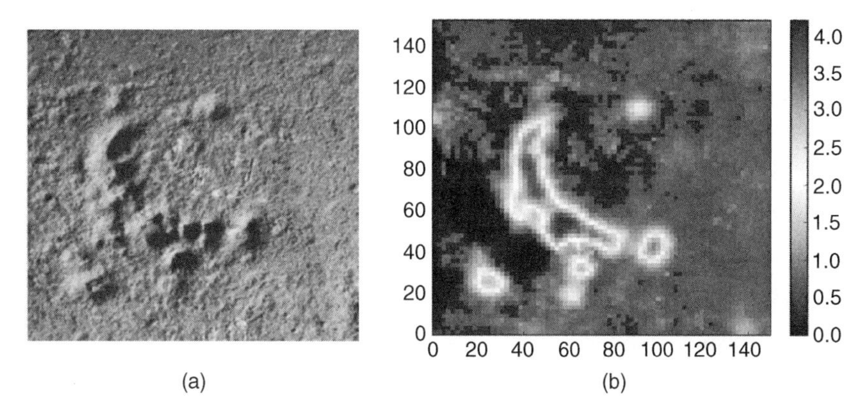


**FIGURE 36.6** Shallow internal corrosion measured with eddy current sensors. (Reproduced with permission of ROSEN.)

defect [6]. This type of corrosion is sometimes called shallow internal corrosion (SIC). If no signal is found, the potential corrosion site is external. However, more than simple yes/no information can be derived. A calibration allows for converting the signal to the corrosion depth. Figure 36.6 shows a near-side corrosion defect. Figure 36.6a shows the picture of the defect and Figure 36.6b shows the quantitative measurement by an ILI tool. A similar technology is used by a different vendor to complement an electromagnetic acoustic transducer (EMAT) sensing technology [7].

For external inspection of pipelines, the mapping of general corrosion is important. This is often done to verify the results of ILIs. Before making a decision on the severity of a corrosion site, a complete mapping of the area is done. One commercial system is based on sheets consisting of an eight-by-eight array of eddy current sensors [8]. The sheet is conformable to allow placing it onto any pipe radius. Through lift-off measurement, a profile of the defect under the array is obtained rapidly. An advanced form of this system has been given the name meandering winding magnetometer (MWM$^{TM}$) [9].

### 36.3.2 Remote Field and Low Frequency Testing

Remote field is an internal pipe or tube inspection method. It is a common method for tube inspection. Applications in the pipeline industry were initially found in well inspection [10]. ILI tools based on remote field technology were also developed [11]. The main goal had been to find external crack-like indications in gas pipelines as this has been a defect type not addressed by MFL and UT inspection tools. The drawback for this technology has been that the inspection speed is limited. For ILI tools, a speed of at least approximately 2 m/s is desirable. This, however, is not achievable due to the low frequency. Today, inspection pigs based on RFEC are not used in gas pipelines. However, the same technology is successfully employed in liquid, mainly water pipelines [12].

### 36.3.3 Pulsed Eddy Current Applications

The PEC method has several applications. One key application is absolute wall thickness measurement in piping, in particular, offshore piping. Most electromagnetic methods are very sensitive to localized corrosion defects. Gradual wall thickness thinning is more difficult to detect. In addition, PEC is ideal for monitoring defects [13]. A monitoring process means that a sensor is affixed to a certain position on the pipe. No scanning is performed, but the sensor reading is taken from that position continually. Hence, PEC sensors can be placed where problems like erosion are expected, such as potentially in elbows. PEC is also used for scanning surfaces, in particular, when UT is not feasible or difficult to apply. Figure 36.7 shows a device for scanning subsea pipelines using PEC [14].

Another application is the inspection of piping under insulation (CUI for corrosion under insulation) [15]. One form of this technology is offered under the trade name insulated component testing (INCOTEST$^{TM}$). Since PEC sensors can be designed such that they measure over relatively large liftoff, they can measure through heat insulation. Also, PEC can measure through a thin layer of aluminum shield as the low-frequency content is mainly responsible for the wall thickness reading.

### 36.3.4 Magnetic Eddy Current Testing (MEC, SLOFEC)

The combined eddy current technique in combination with DC magnetic field was developed into a method also for piping and pipelines under the trade name SLOFEC or MEC [16]. This technology is available in the form of handheld pipe scanners or cable operated tools for external and internal pipe inspection, respectively. A typical application is shown in Figure 36.8.

Concerning the application of handheld pipe scanners, the method is often used for the verification of ILI results.

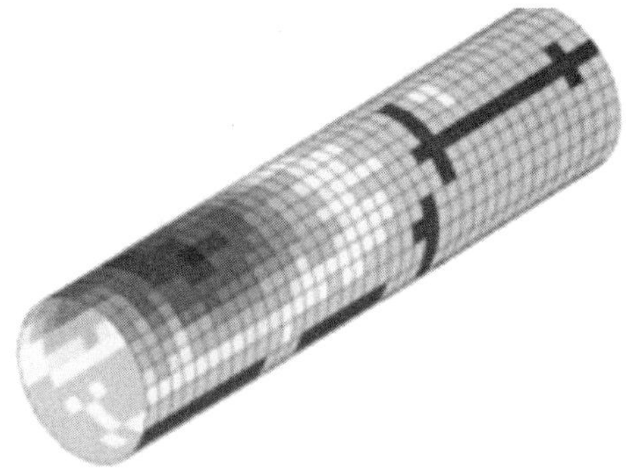

**FIGURE 36.7** (a) An inspection tool for the external inspection of subsea pipelines using PEC and (b) depiction of results of the same (Both reproduced with permission of Impresub.)

**FIGURE 36.8** Using SLOFEC to verify results of in-line inspection. (Reproduced with permission of Innospection.)

**FIGURE 36.9** Using MEC-HUG™ to externally inspect a flexible riser. (Reproduced with permission of Innospection.)

Typically, ILI tools can point to the location of a defect to within 10 cm. Nevertheless, in the case of an in-the-ditch verification, a larger area of coating needs to be removed for the verification by, for instance, handheld UT. SLOFEC allows locating the defect under the coating, and cutting the coating can be avoided or at least reduced to a smaller size.

As the required magnetization can be supplied by permanent magnets, the technology is also used for subsea pipeline inspection, where the power for electromagnets is not available. The technology is mounted on a crawler system and inspects pipelines or other tubular or flat structures. It is deployed either by an remotely operated vehicle (ROV) or from topside with an umbilical. Whereas, this technology can be used to inspect bare steel piping, it is particularly advantageous when an inspection through a coating of polymer or cladding of Monel or stainless steel is required. Also, inspections of flexible risers are carried out with this technique [17]. The deployment of a tool for the inspection of a flexible riser is shown in Figure 36.9.

## REFERENCES


1. Palanisamy, R. and Lord, W. (1980) Finite element analysis of eddy current phenomena. *Materials Evaluation*, 38(10), 39–43.
2. Sun, Y.-S. (1999) Electromagnetic field-focussing remote field eddy current probe system and method for inspecting anomalies in conducting plates. U.S. Patent 6.002.251, December 14, 1999.
3. Crouzen, P.C.N. (2003) Shell Global Solutions, Monitoring wall thickness. U.S. Patent 0130322 A1, December 18, 2003.
4. Jansen, H.J., van de Camp, P.B., and Geerding, M. (1994) Magnetisation as a key parameter of magnetic flux leakage pigs for pipeline inspection. *Insight*, 36(9), 672–677.
5. Bach, G. (1995) Pipetronix, First experiences with the detection of cracks in long-distance onshore pipelines. *Material and Corrosion*, 46, 418–421.
6. Stawicki, O., Ahlbrink, R., and Schroeer, K. (2009) Rosen Shallow internal corrosion sensor technology for heavy pipe

wall inspection. Proceedings of PPSA Seminar, Aberdeen, 2009.

7. Beller, M., Barbian, A., Thielager, N., and Willems, H. (2009) A new in-line inspection tool for the quantitative wall thickness measurement of gas pipelines: first results, NDT Systems and Services. Proceedings of the 4th Pipeline Technology Conference, Hannover, 2009.
8. Patrick, A. (2005) Clock Spring Company, ILI tool validation—feature assessment and mapping. Proceedings of PPSA Seminar, Aberdeen, 2005.
9. Washabaugh, A. (2013) Jentek Sensors, Inspection of pipelines using high-resolution MWM and MR-MWM-arrays. Proceedings of API Pipeline Conference, San Diego, 2013.
10. Schmidt, T.R. (1961) The casing inspection tool—an instrument for the *in-situ* detection of external casing corrosion in oil wells. *CORROSION*, 17, 81–85.
11. Atherton, D. (1995) Remote field eddy current inspection. *IEEE Transactions on Magnetics*, 31(6), 4142–4147.
12. Russel, D. and Shen, V. (2008) Increased use of remote field technology for in-line inspection of pipelines. Proceedings of the 17th World Conference on Nondestructive Testing, Shanghai, China, 2008.
13. Crouzen, P. and Munns, I. (2006) Pulsed eddy current testing corrosion monitoring in refineries and oil production facilities—experience at Shell. Proceedings of ECNDT, Shell Global Solutions, Berlin, 2006.
14. Slomp, E., Bertelli, M., and Milijanovic, D., IDMC Overseas FZE. (2012) Subsea PEC: the NDT diverless robotic system for the corrosion inspection by the PEC method. Proceedings of ADIPEC, Abu Dhabi, 2012.
15. Bouma, T., de Raad, J.A., Robers, M., and Wassink, C. (2000) Condition monitoring of inaccessible piping. Proceedings of the 15th WCNDT, Rome, 2000, Applus RTD.
16. Boenisch, A., Kontrolltechnik, Dijkstra, F.H., and de Raad, J. A. (2000) RTD, Magnetic flux and SLOFEC inspection of thick walled components. Proceedings of WCNDT, Rome, 2000.
17. Reber, K. and Boenisch, A. (2011) Innospection, Inspection of flexible riser pipe with MEC-FIT™. Proceedings of PPSA Seminar, Aberdeen, 2011.

# 37

# UNPIGGABLE PIPELINES

Tom Steinvoorte
*ROSEN Europe, Oldenzaal, The Netherlands*

## 37.1 INTRODUCTION

Pipelines are proven to be the safest way to transport and distribute gaseous and liquid hydrocarbons. Regular inspection is required to maintain that reputation. Today, a large part of the pipeline systems can be inspected with "standard" in-line inspection (ILI) tools. Since first use in the early 1960s, the ILI tools have been significantly developed and improved. Challenges of the early days of pipeline inspection, such as high speed in gas lines, tight 1.5D bends, dual diameter, heavy wall thickness, high pressure and temperature, multi-diameter, long distances, and so on, have been resolved and modern ILI tools are able to cope with these challenges. Figure 37.1 shows the "unpiggable timeline."

It is believed that to date still up to 40% of the world's oil and gas pipelines cannot be inspected with existing tools and procedures and as mentioned previously, reasons why pipelines are classified as unpiggable are numerous.

To operators these unpiggable pipelines are equally important to the overall integrity of the pipeline system, and suitable inspection solutions are therefore required. Although alternatives such as direct assessment and spot checks using infield, nondestructive testing exist, the most valuable information can only be obtained from the inside of the pipeline using ILI devices.

This chapter will discuss the definition of an unpiggable pipeline, the challenges related to their inspection, ways of making unpiggable pipelines piggable, recent technology advances, and the process of selecting a suitable inspection solution.

As a result of the increasing focus on the inspection of these unpiggable pipelines many case studies and contributions have been presented at industry conferences and seminars. A list of relevant topics is presented in Appendix A.

### 37.1.1 What Is an Unpiggable Pipeline?

Although the word "unpiggable" does not exist in dictionaries, the word is frequently used in the pipeline industry. The word "unpiggable" is typically used for pipelines that do not allow an ILI tool to be run from start to finish. In the process of identifying a solution for an unpiggable pipeline, it first has to be identified why the pipeline is considered "unpiggable." The reasons why pipelines cannot be pigged are numerous but can be categorized as follows:

- Pipeline access

  A prerequisite to an ILI is that ILI tools should be able to enter and exit the pipeline. Conventional inspection approaches require launchers and receivers that enable the ILI tools to access and exit the pipelines.
- Mechanical pipeline design

  The pipeline contains installations, such as plug valves, 90° miter bends, dead ends, off-takes, and so on, that cannot be passed with conventional ILI tools. Figure 37.2 shows typical challenges.
- Pipeline medium and operating conditions

  The pressure and/or flow in the pipeline is not sufficient, or too high, to propel the ILI tool through the pipeline within its specified operating velocity. Aggressive media may also pose a challenge for the ILI tool to navigate the pipeline.

*Oil and Gas Pipelines: Integrity and Safety Handbook,* First Edition. Edited by R. Winston Revie.

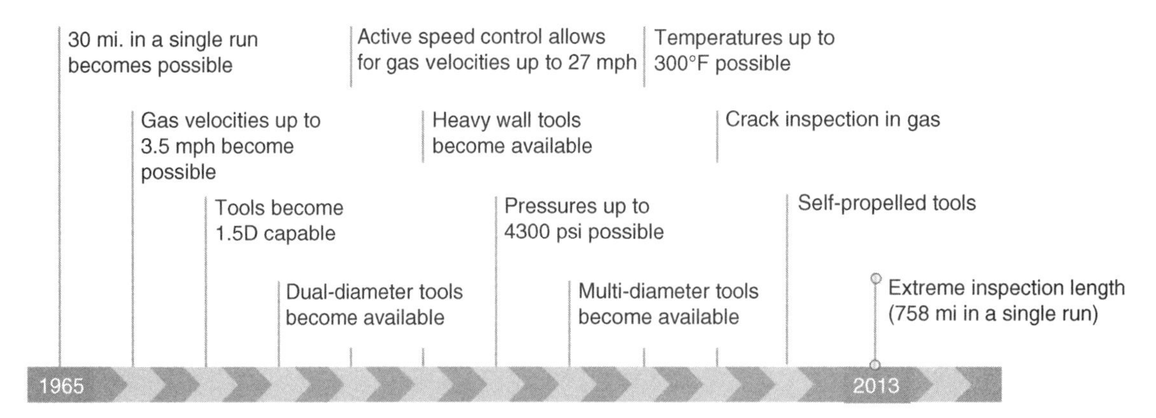


FIGURE 37.1 Unpiggable timeline.

Not only the individual categories, but also the combination of these has to be considered. For example, for multi-diameter, low flow and thick wall gas pipelines magnetic flux leakage (MFL) tools are available. The combination of the three challenges may, however, require development of a new ILI tool or inspection is perhaps not even possible at all.

***37.1.1.1 "Unpiggable," "Challenging," or "Difficult to Pig?"*** In the recent past, there has been an increasing focus on providing solutions for the population of these unpiggable pipelines and it has been recognized that the word "unpiggable" is not appropriate. Any unpiggable pipeline can be transformed into a piggable pipeline by modification of the pipeline and/or its operating conditions; if there are no

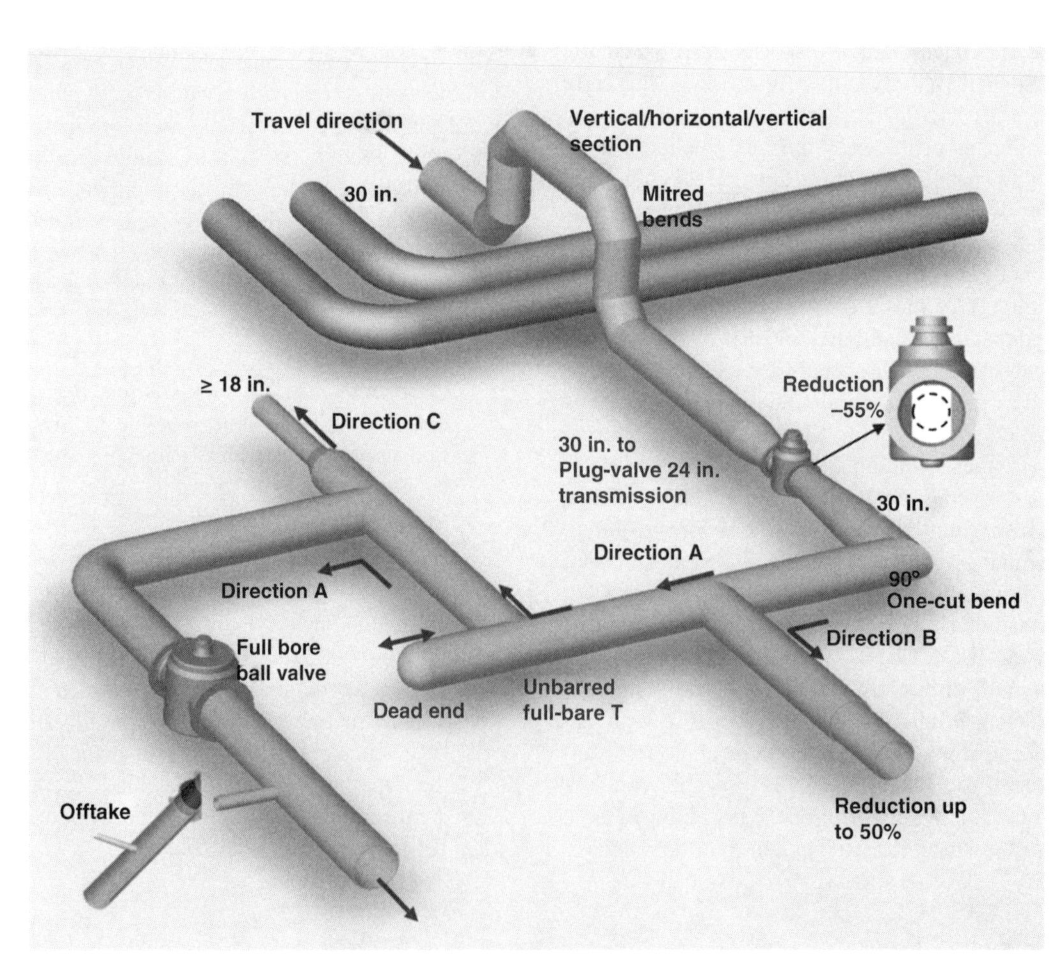


FIGURE 37.2 Unpiggable installations.

traps, they can be installed; is a pipeline considered unpiggable due to for example, a one-cut 90° miter bend—this can be changed out; unfavorable operating conditions can be adjusted to the needs of the inspections tools. The pigging of one pipeline may just require more effort and investment than another one. Therefore, the industry now refers to these pipelines as "challenging" or "difficult to pig," rather than "unpiggable."

*37.1.1.2* ***Continued Development*** As already discussed, the development of solutions for unpiggable pipelines is a function of time and market need. In response to the increasing demand, ILI vendors have developed and will continue to develop innovative solutions and procedures for the inspection of challenging pipelines.

A challenging pipeline inspection today can be a standard pipeline inspection tomorrow.

*37.1.1.3* ***Definition*** Considering the aforementioned contributing aspects we now come to the definition: A pipeline is defined "unpiggable, or challenging" if inspection with current technologies and procedures is considered not feasible.

### 37.1.2 The Main Challenges

*Documentation/Knowledge* Contrary to piggable pipelines, challenging pipelines were often not designed to be pigged. As a result, documentation is often incomplete or not available at all. Pre-inspection cleaning and gauging may not be possible either, so important questions related to bore restrictions and cleanliness are left unanswered. Consequentially, these unknown conditions need to be considered in the overall inspection approach.

*Access* Not designed to be pigged often means that there is no access point available for a tool to enter/exit the pipeline. Temporary or permanent traps or (modified) spool pieces can be considered. Just a pipeline opening in explosion safe conditions may already be sufficient, for example, for self-propelled solutions.

Modifications often require the pipeline to be taken out of service. Not only the pipeline modification itself, but also the consequence of the pipeline taken out of service (e.g., production deferment) need to be considered.

*Measurement* The selection of a suitable inspection technology is not only depending on the objective of the inspection. Also, the conditions of the pipeline have to be considered; the technology may have to cope with the (often unknown) conditions in the pipeline. Especially for self-propelled solutions, friction forces and power consumption need to be considered as well.

*Motion Platform* Once a suitable technology has been selected, this technology has to be brought to the pipe wall where it needs to do its intended job. Conventional pigging uses the medium to propel the tool/technology. When this is not feasible, other solutions, such as tethers or crawlers, have to be considered.

*Cleanliness* A clean pipeline is a prerequisite for a successful inspection of a challenging pipeline for several reasons. To achieve optimum measuring results the inspection tools require a fairly clean pipe surface (MFL, electromagnetic acoustic transducer (EMAT), eddy current (EC), remote field eddy current (RFEC)), or clean surface (ultrasonic testing (UT)). Debris may also influence the run behavior of the ILI tool, thus potentially affecting the data quality (e.g., speed excursions in low-flow low-pressure gas pipelines). Larger amounts of (unexpected) debris could even prevent the tool from reaching the end of the pipeline (the tool becomes lodged).

*Risk Mitigation* Whereas the risks associated to running conventional ILI tools are well understood and accepted, this is not necessarily the case for unconventional inspection solutions.

Although operators often would like to use the most recent and advanced technologies, they often also hesitate "to be the first."

The question "what if the system fails" has to be answered and well-designed fail safe mechanisms have to be in place in case the risk of a tool being lodged in a pipeline cannot be accepted. Extensive testing may also be required to demonstrate the systems capability in a controlled environment.

*Combined Challenges* One of the biggest challenges with difficult to inspect pipelines is the unique combination of the specific factors that make each of the pipelines an inspection challenge.

## 37.2 CHALLENGING PIPELINE INSPECTION APPROACH

As with every pipeline inspection, the first step is to complete the pipeline questionnaire. Additional information, for example, configuration of bends, elevation profile, possibility to change operation conditions, or to take the pipeline out of service, is typically required to support the selection process.

Once the objective of the inspection is known, the challenges are identified and the boundary conditions are known, there are various options that need to be considered to prepare for the actual inspection. Figure 37.3 shows the basic approach.

### 37.2.1 Pipeline Modification

To inspect an unpiggable pipeline from the inside the first step that should be considered is to modify the pipeline (e.g.,

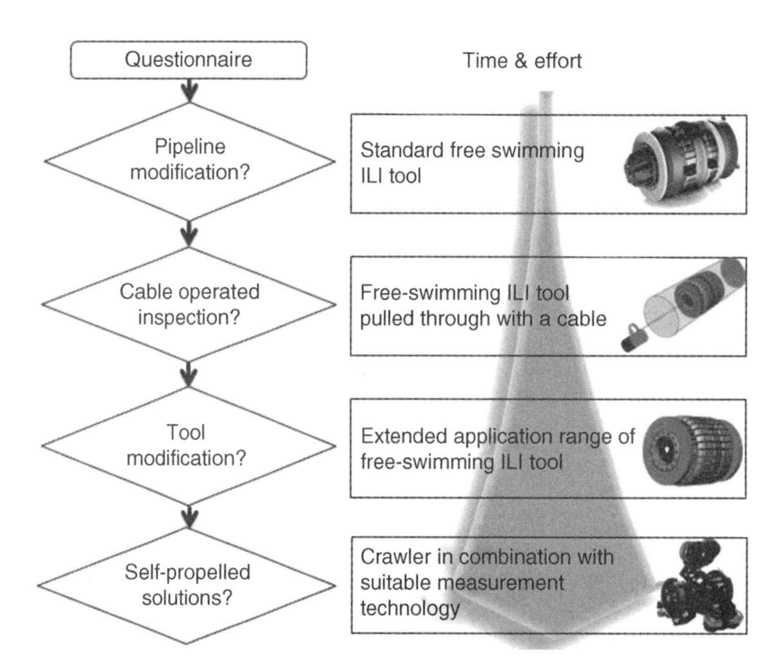


**FIGURE 37.3** Challenging pipeline inspection approach.

creating flow or pressure, exchange fittings or tight bends preventing an ILI tool to pass or installing pig traps).

### 37.2.2 Cable Operated Inspection

An alternative approach to inspection lines with insufficient pressure or flow, or for pipelines that do not have traps, is to pull an ILI tool through the line by using a cable and winch. This requires the pipeline to be taken out of service and can only be used for shorter sections of pipeline. Vendors specify inspection length capability of up to 2.5 mile (4 km) but presence of bends significantly reduces the practical distance that can be covered. This is especially the case for MFL technology because of the drag resulting from the magnets. The individual capabilities depend on the amount, location, orientation and radius of bends, friction force of the selected technology in the pipeline, total length to be inspected, and amount of debris expected and require detailed assessment.

A cable operated inspection approach can be used both unidirectionally and bidirectionally. For a bidirectional application, the tools can be pumped in with, for example, air and pulled back via the cable.

The cable operated inspection is also commonly referred to as wire line inspection.

### 37.2.3 Modification of Existing Tools

If a pipeline modification is not feasible for the intended application then the next step is to assess the feasibility of modifying an existing tool.

Examples include multi-diameter tools, bidirectional tools, and low friction tools that are capable of inspection low pressure gas pipelines. Section 4 provides an overview of recent developments in ILI tools.

### 37.2.4 Self-Propelled Inspection

If the inspection of these pipelines is not feasible with a free swimming or cabled inspection tool, even after modification of piping or tools, then self-propelled solutions are required.

In this case the running of pre-inspection cleaning and gauging tools is typically not possible. As a part of the inspection approach the cleanliness of the pipeline has to be considered and alternative cleaning solutions may be required. High speed flushing with, for example, diesel to dissolve paraffin or to remove sediments can be considered. Hydro jetting may be possible as well.

In case up front cleaning is not possible, the inspection solution has to cope with a certain amount of debris; inspection without cleaning.

Other challenges relate to the geometry of the pipeline as well as the operational conditions during the inspection. Slippery surface may affect traction for drive wheels and inclination, wall thickness changes, bends, and off takes need to be considered as well.

Due to the variety of challenges the optimal solution for unpiggable pipelines often is not a single tool, but rather a project by project-driven tailored solution.

A detailed upfront assessment is required to identify the specific conditions that the inspection device is expected to encounter. The unique nature of the specific factors that make each of the pipelines an inspection challenge often requires a solution tailored to the specific needs. A wide range of solutions is required to satisfy the markets needs but pigging vendors often hesitate to invest in them because the application range for of each of these solutions might be limited.

Self-propelled solutions are available nevertheless but the specific requirements limit each application to a smaller group of pipelines matching the conditions required for these solutions. Section 4 provides an overview of available solutions.

### 37.2.5 Selection Process

Once one or more suitable inspection solutions are identified, the most suitable approach has to be selected. Aspects to consider in the selection process include

- Objective/measurement quality—what is the expected performance of the inspection solution and can the risk of a reduced specification be accepted (if applicable).
- Timing—is sufficient time available to develop solutions that are not yet available.
- Overall cost—what is the overall cost to execute the inspection.
- Repeat inspections—how does the cost of a repeat inspection of a tailored solution compare to modifying the pipeline in a way that conventional tools can be run.
- Risk—commercial, legal, and operational risks have to be considered.
- Similar inspection requirements—existence of similar challenging sections will help financially justify development of tailored solutions.
- ILI cost versus replacement—how does the cost of the inspection and the expected lifetime extension compare to pipeline replacement.

## 37.3 FREE SWIMMING ILI TOOLS FOR CHALLENGING PIPELINE INSPECTIONS

Since the early days of pigging, ILI tools have been significantly developed and improved. This chapter will discuss several of the more recent developments in ILI tools.

**FIGURE 37.4** Bidirectional geometry tool [2]. Figure 3 in "Multidisciplinary Pipeline Inspection Project, development of in-line ultrasonic Inspection tool to be able to pass multiple mitre bends and 1D back to back bends in one inspection run," by L.J. Gruitroij, A. Hak Industrial Services B.V., The Netherlands, © 2013 Pigging Products and Services Association.

### 37.3.1 Bidirectional Inspection

For those pipelines that have only access from one end, special bidirectional ILI tools have been developed. Geometry, Circumferential MFL, (CMFL) and UT wall thickness measurement (UT WM) tools are readily available (see Figures 37.4–37.6, and also Ref. [1]). Other technologies such as EMAT or UT crack detection (UT CD) can also be developed for bidirectional applications and their availability will depend on market demand.

For pipelines requiring a bidirectional inspection, specific procedures have to be developed to bring the ILI tool to the end point and back to the entry point. The propulsion methods described ahead are required to be carefully reviewed on a project-by-project basis. If a propulsion method has been selected for the application of a bidirectional tool, procedures have to be developed to ensure the ILI tool traverses the entire section of pipe to be inspected to achieve the inspection objectives.

- Reversed flow

  For a free-swimming bidirectional tool, the easiest way of navigating through a pipeline is by pumping the

**FIGURE 37.5** Bidirectional UT geometry and WM ILI tool. (Adapted from Ref. [3].)

**FIGURE 37.6** Bidirectional MFL ILI tool. (Adapted from Ref. [4].)

tool to the end point, stop it and then reverse the flow in the pipeline system to propel the tool back to the launching facilities. This can be done if a parallel line is available by simply connecting these via flexible hoses or by installing a temporary pump spread on the other side to push the product back in the opposite direction.

- Push against product pressure

  Short sections of gas pipelines can be inspected by pushing the tool against the product pressure into the pipeline up to the end point. The remaining gas volume behind the tool is then used to send the tool back to the entry point by controlled pressure release at the launching facilities.
- Gravity based

  For the inspection of vertical sections a gravity based approach can be used. The tool is pushed down the pipeline using the hydrostatic pressure of a storage tank and pumped back by permanently available or temporarily installed pumps. Typical applications include loading lines and branch connections in storage tank farms.
- Push in–pull out

  Another approach of deploying a bidirectional tool is to push the tool into the pipeline by using flow (e.g., nitrogen or air) and then recovering it by pulling it out using a winch and a cable. Figure 37.7 shows a setup.

### 37.3.2 ILI Tools for Launch Valve Operation

Pipelines with a need for frequent maintenance cleaning are sometimes fitted with ball valves, which have the option to launch a cleaning tool. Figure 37.8 shows such a valve.

This kind of ball valve offers a cost-efficient solution because the pigging operation requires significantly less effort compared with cleaning using conventional pig traps.

The cleaning tool is inserted into the valve while the valve is closed and the door can be opened. After the door is closed, the valve is open and the tool is launched with the flow. The next downstream ball valve has the same functionality and is used as a receiver. At one end of the ball valve, a steel mesh is installed to stop the tool. The ball of the valve can be rotated

**FIGURE 37.7** (a) Tool loading and (b) launcher setup. (Adapted from Ref. [5].)

**FIGURE 37.8** Pig launch valve [5].

by 360°; hence, it can be used as launcher and receiver (not applicable to all pig valves).

If only one valve is available, the cleaning can be done with a bidirectional cleaning tool. The tool is launched as already discussed. Once the end of the pipeline has been reached the flow is reversed and the tool is pumped back to the ball valve.

In the past, these ball valves have been used for cleaning only. With the new small and short bidirectional ILI tools, these lines can now also be inspected. MFL metal loss and geometry inspection services are available. Figure 37.9a shows a 10 in. MFL tool and Figure 37.9b shows the tool being inserted in the valve.

### 37.3.3 Low-Pressure Inspection of Gas Pipelines

The main challenge in inspecting low-pressure gas pipelines are speed excursions that are usually caused by obstacles, such as excessive girth welds, bends, dents, wall thickness changes, and debris. The tool stops at this obstacle, as the pressure (force) is not high enough to move the tool over it. After a certain time, the pressure behind the tool builds up and forces the tool over the obstacle. In some cases, tool speeds of over 67 mph (30 m/s) are reached, whereas most MFL ILI tools have a maximum speed specification of 11 mph (5 m/s). Furthermore, after the tool starts to move again, the head pressure in front of the tool is typically low, therefore, the high speed phase is elongated.

When exceeding the target tool speed the magnetic field is affected in a way that makes it difficult to correctly size or even identify corrosion features. Furthermore, the resolution is impaired.

To reduce the speed excursions in low-pressure gas pipelines conventional MFL tool performance can be improved by selecting a friction reducing cup/disc configuration. Dedicated low pressure tools also feature friction reducing magnetizers in combination a reduced tool length and weight. Figure 37.10 shows a selection guideline for low-pressure gas pipelines. With dedicated tools pipelines with gas flow below 0.5 m/s can nowadays be inspected as well. A detailed assessment of the pipeline geometry and operating conditions is typically required to better estimate the tool's performance in the given conditions.

### 37.3.4 Multi-Diameter Inspection

Another challenge for pipeline ILI is multi-diameter pipelines. Most of the multi-diameter pipelines have not been designed as such, but over decades operators decided to connect two or more pipelines of different diameters to fulfill their system requirements. Also, conscious decisions have been made to build multi-diameter pipelines.

Over the past years, several multi-diameter tools have been developed, built, and successfully deployed in a multitude of projects. Figure 37.11 shows multi-diameter MFL tools; see also Ref. [6].

## 37.4 SELF-PROPELLED INSPECTION SOLUTIONS

One of the most significant challenges for self-propelled ILI solutions is the required pulling force to propel the

**FIGURE 37.9** (a) 10 in. MFL tool for launch valve and (b) tool loading. (Adapted from Ref. [5].)

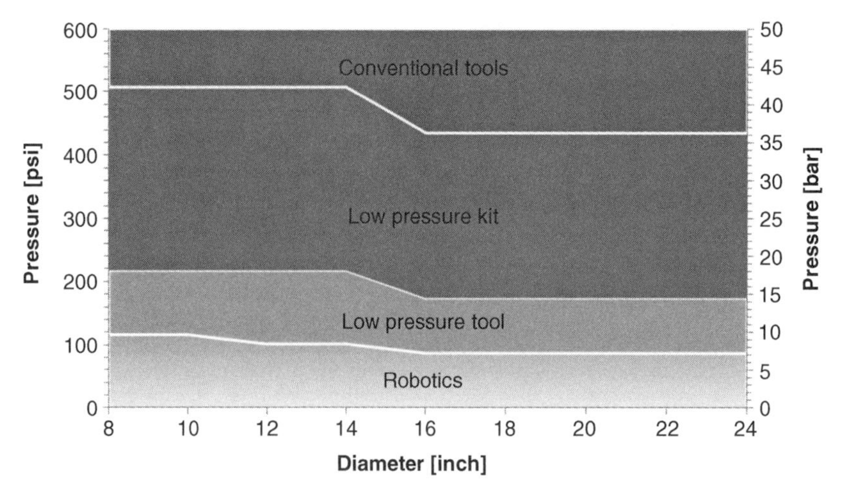


**FIGURE 37.10** Selection guideline for low-presure gas pipelines.

**FIGURE 37.11** Multi-diameter ILI tool. (Adapted from Ref. [7].)

measurement unit and cable (if present) while coping with ID restrictions, wall thickness changes, bends, debris, and so on. Traditionally, power is provided by a tether that connects the device to an external power source. The tether is furthermore used for data transfer that allows for external data storage and online monitoring and control of the inspection device. The third function of the tether is to pull the tool back to its entry point, either as a part of the inspection approach or as a fail-safe system.

For tethered operations the friction forces resulting from bends need to be considered as well. A detailed assessment considering the combination of individual bend radius, angle, and location of each bend is required. As a rule of thumb, tethered operators typically specify a maximum cumulative bend opening of 270°. Although larger bend openings can be negotiated, they will limit the possibility to pull the tool back.

Furthermore, the inspection length in combination with the power consumption needs to be considered while the pipeline trajectory, geometry and configuration play a role in power consumption as well (e.g., vertical climb, longer distances require thicker cables to minimize the power loss over the cable). Thicker cables result in return to higher power consumption because additional force is required to pull the cable.

If the application of a tether is not feasible than onboard power supply can be considered (e.g., batteries). To extend the inspection range for battery powered crawlers online charging systems and onboard power generators using turbines are being developed.

### 37.4.1 UT-Based Crawlers

As mentioned previously, a low friction measurement technology is preferred. The most common technology applied on self-propelled inspection tools is therefore UT technology [8,9].

To reduce the drag, neutrally buoyant cables are used. Winches with cable length of up to 15 km are available. Presence of bends may, however, significantly reduce the practical range.

### 37.4.2 MFL-Based Crawlers

For pipelines that cannot be cleaned or pipelines that do not have a liquid coupling, MFL is often preferred. Also, the type of defects (e.g., pitting) to be detected may require MFL. MFL-based crawlers are shown in Figures 37.12 and 37.13.

**FIGURE 37.12** 10–14 in. MFL crawler. (Adapted from Ref. [10].)

**FIGURE 37.13** Multi trotter MFL crawler. (Adapted from Ref. [11].)

The bulky design of the MFL measurement principle is, however, less favorable, especially considering the passage capability that is desired to better cope with the often unknown conditions in a pipeline. Also, the friction forces associated with MFL technology make this technology more difficult to apply. The forces required to move the inspection unit are significantly higher compared to propelling UT units. This is especially valid if the device is also required to overcome wall thickness changes, bends, dents, or debris.

To (partly) overcome these challenges low friction magnetizers and/or magnetizers with low saturation levels are applied and more powerful crawlers have been developed.

To avoid the limitations of a tether, MFL crawlers typically carry their own batteries, and the tether is used for control and monitoring only. Wireless control or autonomous systems are available if the operation does not allow for a tether. The practical range highly depends on the specific pipeline conditions.

### 37.4.3 Other

Next to the two mainstream ILI technologies MFL and UT that have been addressed in the previous chapters, a variety of crawler solutions and measurement technologies have become available in recent years. Several of them are described hereafter.

Saturated low-frequency eddy current (SLOFEC) is an internal and external corrosion screening corrosion technique. shallow external corrosion (SIC) technology is used for detection of internal corrosion up to 10 mm depth. The low friction and low power consumption of these eddy current solutions are favorable for crawler applications [12].

EMAT for wall thickness measurement is used in combination with a crawler and rotating sensor arms. Figure 37.14 shows a tool.

Pipeline intervention crawlers are also made available to the market as an independent “delivery system.” These crawlers, based on a powerful brush drive technology, are able to transport existing measurement technologies such as MFL and UT. Figure 37.15 shows the brush drive.

The Helix is a motion platform that uses a helical drive for the deployment of the measurement technology as well as for the axial movement of the inspection device. Figure 37.16 shows a tool.

**FIGURE 37.14** EMAT crawler. [13] (Image courtesy of Diakont).

**FIGURE 37.15** Modular brush drive. (Adapted from Ref. [14]. Printed by permission of SIG Ltd. All rights reserved.)

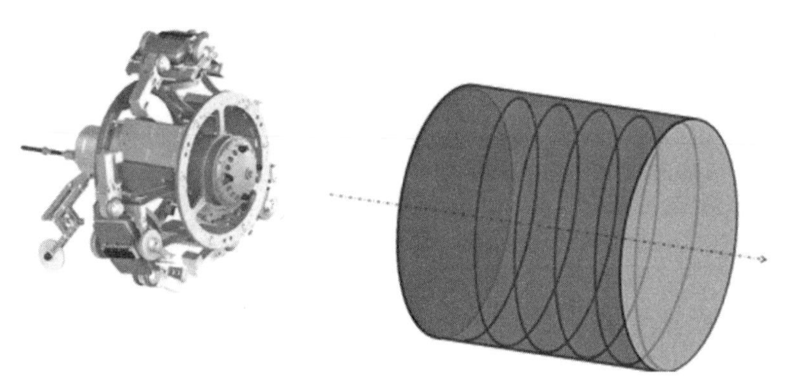

**FIGURE 37.16** RoHelix crawler. (Adapted from Ref. [11].)

## REFERENCES


1. 3P Services Available at http://ppsa-online.com/papers/09-Aberdeen/2009-06-Hostage.pdf.
2. A. Hak Industrial Services Available at http://ppsa-online.com/papers/13-Aberdeen/2013-02-AHak-paper.pdf.
3. Quest Integrity Group Available at http://www.questintegrity.com/assets/PDFs/Case%20Study%20-%20InVista%20Inspection%20Wharf%20Line.pdf.
4. ROSEN Group Available at http://www.rosen-group.com/global/solutions/services/service/rocorr-mfl-a_bidi.html.
5. Steinvoorte, T., et al. (2013) Tailored ILI services—delivering solutions for challenging pipelines. Proceedings of the Rio Pipeline Conference & Exhibition, 2013.
6. PII Pipeline Solutions Available at http://www.ge-energy.com/content/multimedia/_files/photos/PII_SmartScan-660x450.jpg.
7. ROSEN Group Available at http://www.rosen-group.com/global/solutions/services/service/multi-diameter-technology.html.
8. APPLUS RTD Available at http://advantec-is.com/hp_wp/wordpress/wp-content/uploads/10-16-PIT-Tool.jpg.
9. Oceaneering Available at http://www.oceaneering.com/oceandocuments/brochures/inspection/INS%20-%20Pipescan%20-%20Tehered%20Pipeline%20Solution.pdf.
10. Pipetel Technologies Inc. Available at http://www.pipetelone.com/explorer_10-14.html.
11. Steinvoorte, T. (2013) Practical Solutions for Challenging pipelines. PPSA seminar, 2013.
12. GE Industrial Solutions Available at http://pipelineandgasjournal.com/commercial-robotic-inspection-solution-unpiggable-pipelines.
13. Diakont Available at http://www.diakont.com/solutions/ili/robotic-in-line-inspection.
14. Pipecrawlers Available at http://www.pipecrawlers.com/pipeline-crawler-english.pdf.

## BIBLIOGRAPHY: SOURCES OF ADDITIONAL INFORMATION

Journals, conferences, and other sources with content related to unpiggable pipeline inspections.

Oil & Gas Journal Available at http://www.ogj.com/index.html.

Online forum Difficult and Unpiggable Pipeline Inspection Solutions Available at http://www.linkedin.com/groups?home=&gid=3356338&trk=anet_ug_hm.

Pipeline & Gas Journal Available at http://www.pipelineandgasjournal.com.

Pipelines International Available at http://pipelinesinternational.com.

The Australian Pipeliner Available at http://Pipeliner.com.au.

Unpiggable pipeline solutions forum Available at http://www.clarion.org/UPS_Forum/ups2013.

# 38

# IN-THE-DITCH PIPELINE INSPECTION

Greg Zinter
*Applus RTD, Edmonton, Alberta, Canada*

## 38.1 OVERVIEW

There are two primary and complementary types of in-service pipeline inspection—internal and external. Internal inspection is conducted by a range of inline inspection (ILI) tools and is discussed in Chapter 35.

External inspection is conducted in the ditch on the outside surface of a pipe that has been excavated (which is the usual case, as most pipelines are buried) and exposed.

In-the-ditch inspection begins with locating and safely digging up a pipeline in a known specific location. The criteria for selecting a location are usually based on the detection of a potentially dangerous anomaly in a pipeline by an ILI tool.

This chapter discusses the application of nondestructive examination (NDE) techniques and technologies in the external inspection of in-service pipelines. Specifically, the chapter focuses on the following:

- The role of in-the-ditch pipeline inspection and the pipeline integrity program;
- Types of coating typically encountered in a pipeline integrity excavation;
- Types of anomalies that threaten pipeline integrity;
- Types of NDE measurement technologies and methods;
- The interrelation of ILI and NDE data;
- NDE in-the-ditch procedures;
- NDE data gathered and its management, use and reporting;
- Pipeline repair;
- Technological development for in-the-ditch inspection.

## 38.2 INTRODUCTION TO NONDESTRUCTIVE EXAMINATION OF PIPELINES

NDE is a critical part of any pipeline integrity program. The term NDE is interchangeable with nondestructive testing (NDT). American Society for Nondestructive Testing (ASNT) defines NDE as "the process of inspecting, testing, or evaluating materials, components, or differences in characteristics without destroying the serviceability of the part or the system." For pipeline inspection, NDE is using an appropriate inspection technology to examine and assess a pipe and pipe anomalies without destroying it. Most commonly, this examination occurs when a pipeline is in service.

In this chapter, NDE refers to a "direct" measurement technology that is conducted from the outside of the pipe. In this examination method, a technician physically examines a pipe with a variety of techniques that begin with a visual assessment. Direct measurement and evaluation provides the most accurate and reliable data. By comparison, ILI is an inferred measurement where a tool with a specific inspection technology, such as magnetism or ultrasonics, is inserted into a pipeline. This ILI tool travels down the pipeline and takes measurements that require some type of post processing to provide useable data. Technically, ILI also fits the aforementioned definition of NDE.

## 38.3 NDE AND A PIPELINE INTEGRITY PROGRAM

It is important to understand that the two reasons to conduct NDE on a pipeline are safety and verification of ILI data.

*Oil and Gas Pipelines: Integrity and Safety Handbook,* First Edition. Edited by R. Winston Revie.

Both of these reasons directly support the primary objective of a well-designed pipeline integrity program, which is to support the safe and economical operation of a pipeline over its entire life cycle.

### 38.3.1 Safety

A pipeline is designed with a factor of safety to allow safe operation without ruptures or leaks. This factor of safety is an engineered value that is well understood and it provides a buffer between the operating pressure of a pipeline and the failure pressure of the pipe material. The results of NDE provide data that can be used to calculate the remaining strength of a pipeline in a location where a change in the material characteristics results in a change of the design pressure of a pipeline.

This information can be used immediately to confirm the protection of the excavation and assessment crew working on the pipeline. A pipeline operator will lower the pressure of a pipeline to ensure that there is an appropriate factor of safety to protect the crew. A typical excavation pressure is the rupture pressure calculated from the information from the ILI inspection multiplied by a safety factor. If this value is greater than the maximum operating pressure, the operator will apply a safety factor to the highest actual operating pressure during a given period of time prior to the excavation. By verifying the size of the defect, this initial assessment will provide confirmation of a safe operating pressure.

The larger safety perspective is to ensure the pipeline is safe to operate at the operating pressures. The NDE data will lead to an engineering calculation that will dictate the appropriate repair method. The repair might require the installation of a reinforcement sleeve to reestablish the design safety factor. Regardless of whether or not a sleeve is installed, the final steps in every assessment excavation are to recoat and bury the pipe.

### 38.3.2 Verification and Advancement of Technology

NDE data are always compared with ILI data. This comparison verifies the stated accuracy of the tool, which either confirms the results of the ILI tool, or identifies inaccuracies with the tool or data interpretation. Any inaccuracies might result in the immediate review and adjustment of the entire ILI data log, result in a complete reanalysis of the integrity of a pipeline, or lead to further development of the tool or the analysis to improve the ILI tool's accuracy.

## 38.4 PIPELINE COATINGS

Most anomalies that threaten the integrity of a pipeline are associated with a defect in the external pipeline coating. For this reason, some background understanding of pipe coatings is helpful.

Regardless of the type of coating, the most common root causes of coating failures are as follows:

- Improper surface preparation;
- Poor application;
- Interaction between coatings and immediate environment (soil stresses and interaction with soil chemistry);
- Mechanical damage;
- Extreme operating conditions;
- Age.

For most large-diameter hydrocarbon pipelines, there are three primary types of coating.

### 38.4.1 Asphalt or Coal Tar Enamel

While these are different coating types, they are often confused with each other when they are first encountered on an integrity dig. In both cases, there is a polymer—either coal tar or asphalt—a fibrous binder, and some solvents to help the polymer flow onto the pipe during application. In some cases, an outer kraft paper wrapper may be used.

Both types of coating are hot applied to the pipe. As the hot mixture cools and the solvents evaporate, the material hardens and forms a durable protective layer on the outside of the pipe. Historically, these coatings come with significant environmental and safety concerns due to the hazardous nature of the petroleum solvents and the asbestos, which was sometimes used as a binder and an insulator. They were most commonly used on large diameter hydrocarbon pipelines until the late 1960s. Over time, these coatings dry out and become quite friable. They are very effective coatings as long as they are not disturbed. These coatings will also degrade rapidly when exposed to air. The exposed coatings that will not be removed as part of the inspection (i.e., at either end of the excavation) should be supported and protected to prevent further degradation.

Some typical problems related to asphalt or coal tar epoxy coatings are as follows:

- Disbondment can occur over time (consider that this coating has not been widely used on hydrocarbon pipelines for over 40 years) and the coating will fall away from the pipe as soon as the surrounding soil is removed.
- Irregular, circumferential cracks or disbondment may be caused by cold bending at low temperatures or improper adhesion.
- Mechanical/impact damage can cause loss of adhesion at and/or around the point of damage.
- Soil loading: This can cause irregular cracks on the crown of large diameter pipes.

### 38.4.2 Tape Wrap

Tape wrap coatings were a functional improvement over coal tar or asphalt-based coatings because they were less hazardous products and easier to apply in the field. The tape is usually a plastic film of polyvinyl-chloride (PVC) or polyethylene (PE) with a self-adhesive backing. Once the pipe surface has been blasted and primed, the tape is spiral wrapped on the pipe. It is common to use a layer of mastic to bond the tape to the pipe. The performance of the tape wrap coating is directly related to proper surface prep and application procedures. The tape can be hot or cold wrapped.

Disbondment is a common issue, particularly with cold wrap applied tape or when the operating temperature of a pipeline has increased significantly. This disbondment leads to two problems: First, the pipe is exposed to a corrosive environment and, second, the disbonded tape effectively shields the pipe from cathodic protection. Tape coatings replaced coal tar or asphalt enamel coatings and were most commonly used from the late 1960s to the middle 1970s.

Some typical problems associated with tape wrap coatings are as follows:

- Loss of adhesion at overlap is caused by poor surface prep and/or improper overlapping.
- Sagging or wrinkling is caused when the coating is too thick and heavy or there are external stresses caused by the soil settlement or movement.
- Tenting/bridging is caused when the tape is pulled over a high spot like a weld.

### 38.4.3 Fusion Bonded Epoxy (FBE)

Fusion Bonded Epoxy (FBE) coatings have become the most popular coating systems on large diameter hydrocarbon pipelines in North America. Electrically charged epoxy powder is sprayed onto the heated, grounded pipe joint. The powder melts and fuses with the steel. Because FBE is applied in a shop environment, quality control of the surface preparation and application process is easily managed. The result is a very durable and impermeable hard shell coating.

There are variations on FBE where a multilayered coating is applied for greater protection and abrasion resistance.

Typical problems encountered with FBE coatings include:

- Blisters are caused by moisture between the pipe surface and the coating when the pipe is heated for any reason. The moisture can be present during application or it may migrate through the coating by osmosis. As long as the blister is intact, the coating is still effective.
- Cathodic disbondment can occur where there is excessive CP current.
- Cracking can occur during by cold bending at low temperatures or by overcuring of the coating.
- Inclusions will occur when material is trapped under the coating layer at the time of application.

There are other types of coatings such as extruded polythene jacket, or yellow jacket (YJ), wax-based coatings or fiberglass. An extra protective layer, such as "rock guard," is used to wrap pipes where the soil is rocky.

Concrete weights are also placed over pipelines in wet environments to counteract buoyancy.

## 38.5 TYPES OF ANOMALIES

### 38.5.1 Introduction

Technically, an anomaly is a deviation from the norm. For pipelines, it is an imperfection in the steel, such as a crack or an area of corrosion. Anomalies are also known as features, defects, or indications. Most anomalies represent a threat to pipeline integrity because they can grow and further compromise pipe strength during the operation of the pipeline.

Much has been written about the type of pipeline anomalies and the way they are categorized. For the purposes of this discussion, the categories have been based on those described in pipelinealias.com [1]. Note that it is common to encounter a complex pipeline anomaly or feature where two or more anomaly types exist in one location.

### 38.5.2 Volumetric

Volumetric anomalies in the pipe wall are caused by a change in the volume of material in the pipe wall compared with the adjacent nominal wall. The most common types of volumetric anomalies are as follows:

*Corrosion*—as this is the most common reason for metal loss on pipelines, it is also one of the most common reasons for pipeline failures. External corrosion will occur when there is a break in the coating and the cathodic protection system is not able to prevent the electrochemical reaction between the electrolyte in the soil and the pipeline steel. Internal corrosion can be caused by microbial activity or erosion from particulates that are entrained in the fluid.

Corrosion can be spread over a general area or be concentrated in a pit, or both conditions can exist. The biggest influence to the size of an external corrosion anomaly is the type of coating and the size of the coating defect. Internal corrosion can be managed by regular pipeline cleaning and injected corrosion inhibitors. Figure 38.1 shows a typical deep external corrosion feature.

*Scratches and gouges*—these can occur during construction, either when the pipeline is first built, during subsequent excavations for inspection, maintenance or modification, or when a third party is excavating the right-of-way. These

**FIGURE 38.1** Typical external corrosion feature. (Reproduced with permission of Applus RTD.)

anomalies occur because of a piece of equipment hitting the pipe. This is another very common cause of pipeline failure. Figure 38.2 shows a typical gouge in a dent.

*Manufacturing-related volumetric anomalies*—these are defects that originate in the mill and can include general variation in wall thickness (either more or less than nominal), inclusions, mill slug, or trim marks from trimming the long seam. These anomalies are usually present during the original hydrotest, and do not always grow with pipeline usage. As a result, volumetric anomalies represent less of a threat than corrosion, or scratches and gouges.

*Construction-related volumetric anomalies*—this can include arc burns or puddle welds from poor girth weld practices, grind marks, or handling damage. These are also less of a threat to the integrity of the pipeline as they are usually present during the post-construction hydrotest.

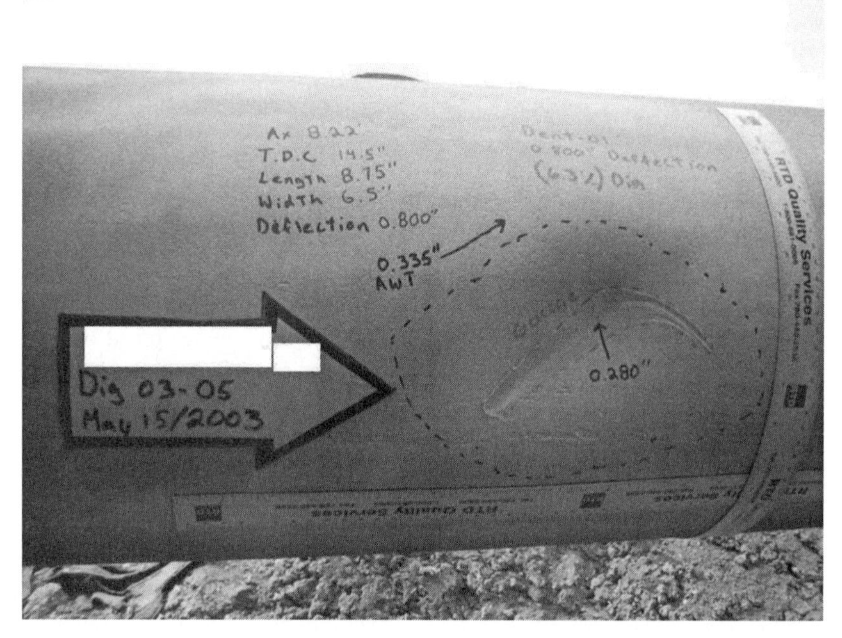


**FIGURE 38.2** Gouge and dent. (Reproduced with permission of Applus RTD.)

### 38.5.3 Planar

A planar anomaly "is formed by separation of previously continuous material. Planar anomalies are two dimensional and do not significantly change the total cross-sectional area of the pipe wall" [1]. The most critical anomalies in this category are those that break the surface of the pipe. The most common types of planar anomalies are as follows:

*Cracks*—cracks, like corrosion, will grow in pipelines that are in service, unless they are found and repaired. Cracks form due to environmental conditions, such as stress corrosion cracking (SCC) or elevated hydrogen concentrations, or manufacturing processes (hook cracks). Cracks are often found in association with corrosion or gouges. Given the complexity of detection and the consequences of an undetected feature, the industry relies on in-the-ditch assessment to continually improve ILI technology that is used to find cracks. Figure 38.3 shows an SCC colony.

*Laminations*—these can occur in the pipe body or on the pipe surface. Unlike cracks, laminations that are not surface breaking do not normally grow with time. If a mid-wall lamination is found in an integrity assessment, it was probably present when the pipeline was constructed and pressure tested. Regardless, some operators choose to install sleeves for these defects to manage the pipeline risk.

### 38.5.4 Geometric

Geometric anomalies result from a change in the shape of the pipe. The most common cause in this category is mechanical damage. These anomalies include the following:

*Dents*—these occur when an external object interacts with the pipe to cause a deformation. Dents are different from scratches and gouges in that a dent is a larger deformation of the pipe wall. The most common cause is from excavation equipment, or rocks in the bottom of the pipeline ditch. Dents often contain gouges or scratches. Figure 38.2 shows a gouge inside a large dent, which was probably caused by an excavator.

*Ovality*—this occurs due to excessive external stress on the pipeline that usually acts at right angles to the pipe surface. This stress might cause the pipe to deform into an oval-shaped cross section or ovality. The design strength of a pipeline assumes the pipe is circular, and any change in this shape will change the ultimate design strength.

*Buckle*—this is also caused by excessive external stresses. These stresses are usually aligned in the axial direction and are often caused by geotechnical movements on slopes or at river crossings. A buckle means that the pipe has yielded past its elastic limit and the original design strength is compromised.

Any one of these anomaly types can compromise the design strength of a pipeline. It is common for a number of anomaly types to coexist in the same feature. For example, corrosion and cracking are often found together on some pipe types. This factor makes an assessment much more complex and more critical, and increases the probability of a pipeline failure.

## 38.6 NDE MEASUREMENT TECHNOLOGIES

A variety of measurement technologies are available for in-the-ditch pipeline integrity assessments. It is important to be prepared with the appropriate technology for the type of anomaly that will be assessed. Table 38.1 is a summary of these technologies, with a list of the most appropriate use.

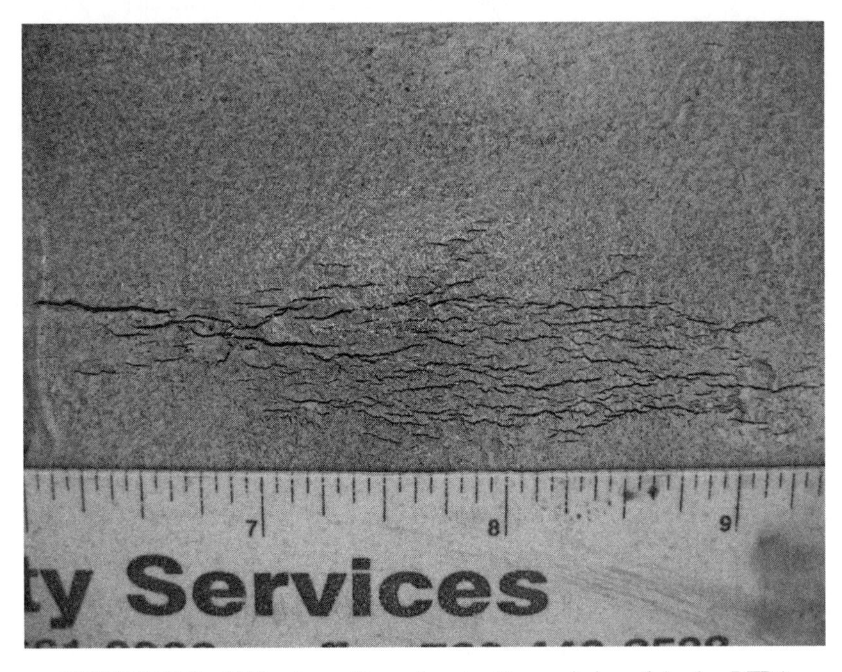


**FIGURE 38.3** SCC colony. (Reproduced with permission of Applus RTD.)

**TABLE 38.1 Overview of NDE Inspection Techniques**

| Method | Typical Feature | Advantages | Limitations |
|---|---|---|---|
| Visual and physical measurement | • External corrosion and associated deposits<br>• Gouges and scratches<br>• Buckles<br>• Relatively large surface-breaking defects<br>• Pipe texture and color<br>• Coating type and condition | • Minimal equipment requirement<br>• Reliable and fast method to detect relatively open surface-breaking defects, and potentially injurious deep corrosion pits | • Dependent on the visual acuity and experience of the technician<br>• Some cracks and crack colonies can be invisible to the eye<br>• Manual mapping of large external corrosion features is tedious and not as accurate as laser |
| Magnetic particle inspection | • Surface or near-surface cracks, voids, or inclusions | • Inexpensive and relatively simple<br>• Sensitive to surface and near-surface flaws<br>• Detects tight, closed cracks not seen by visual inspection | • Limited to ferromagnetic materials<br>• Dependent on the visual acuity and experience of the technician<br>• Surface preparation and post-survey demagnetization might be required |
| Ultrasonics:<br>• Zero degree (straight beam)<br>• Shear wave<br>• Phased array<br>• Time of flight diffraction | • Pipe wall thickness<br>• Cracks, delaminations, inclusions, and interfaces<br>• Internal diameter pipe-wall loss | • Can penetrate thick materials<br>• Suited to crack detection and characterization<br>• Can be automated | • Liquid couplant required<br>• Smooth surface finish needed<br>• Limited usage for some thin-wall applications<br>• Difficult to detect some defects in areas of complex geometry (such as welds) |
| Laser profilometry:<br>• Automated travel<br>• Handheld | • External metal loss<br>• Dents, scratches, and gouges | • Most accurate method of measuring surface profile<br>• Automated data acquisition minimizes potential for errors<br>• Digital data file is well suited for remaining strength calculations<br>• Can be quicker than a manual method on large features | • Addition cost and complexity<br>• Some techniques require special surface preparation |

NDE measurement technologies are described in the following sections.

### 38.6.1 Visual Assessment

Visual assessment is the first level of assessment and probably the most basic, yet it is perhaps the most critical. A technician:

- visually scans the exposed pipe to identify surface anomalies and assess the general pipe condition.
- identifies and maps areas of interest and anomalies according to the information provided in the excavation package.
- looks for potentially unsafe conditions (such as excessively deep corrosion pits) that are not identified in the scope of the excavation package provided by the pipeline operator.

The pipeline operator is responsible to provide inspection criteria. A typical assessment criterion requires the identification of all surface anomalies deeper than 10% of actual wall thickness and the assessment of all anomalies >20% of the actual pipe wall thickness. Tools required for this assessment include a:

- pit gauge;
- ultrasonic pen probe;
- tape measure;
- ruler;
- camera.

### 38.6.2 Manual Measurement

External metal loss can be effectively mapped using a grid and measuring the minimum pipe wall thickness in each square. To make the grid, a chicken wire of the specified grid size (usually 12.5 mm [0.5 in.] square) is placed over the area to be measured, and white paint is sprayed over the grid and onto the pipe. A manual pit gauge is used to measure the

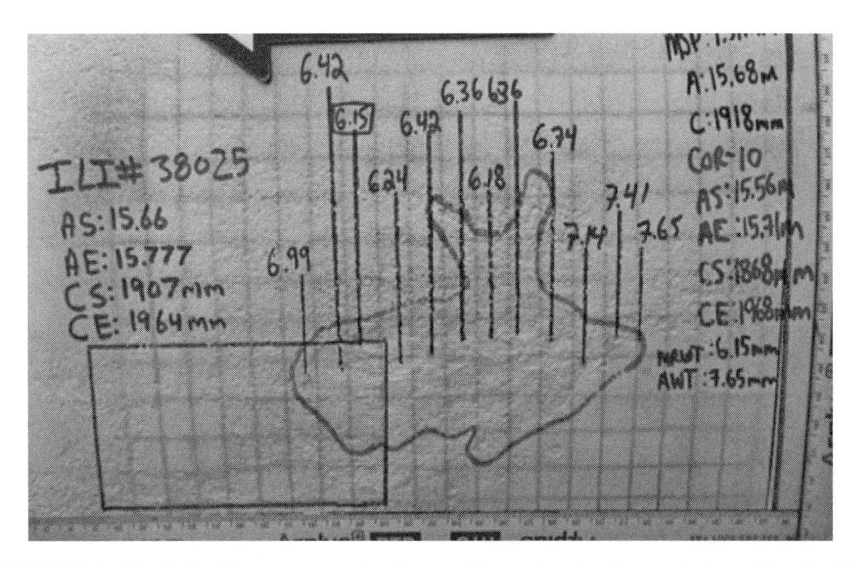


**FIGURE 38.4** Typical grid over a corrosion feature. (Reproduced with permission of Applus RTD.)

maximum pit depth in each square of the grid, or an ultrasonic pen probe is used to measure the minimum wall thickness in each grid square. These measurements are recorded and used to calculate the remaining strength of the pipeline. External metal loss features can also be mapped by surface profilometry systems, such as lasers. Figure 38.4 illustrates a typical manual grid.

### 38.6.3 Magnetic Particle Inspection

Magnetic particle inspection (MPI) is conducted by applying an aqueous mixture that contains iron particles onto a pipe that has been painted white, while an electromagnetic yoke is placed on the pipe and energized. The magnetic energy generates visible lines of magnetic flux between the legs of the yoke. Any surface discontinuity, such as a surface-breaking crack, shows as a disturbance in the visible lines of magnetic flux. This technology is very effective in detecting stress corrosion cracking. Figure 38.5 illustrates the use of a magnetic yoke on a pipe coupon that has been painted white and sprayed with iron particles in an aqueous solution.

### 38.6.4 Ultrasonic Inspection (UT)

When a transducer is placed on a pipe, ultrasonic sound energy is projected into the pipe wall. This energy reflects off

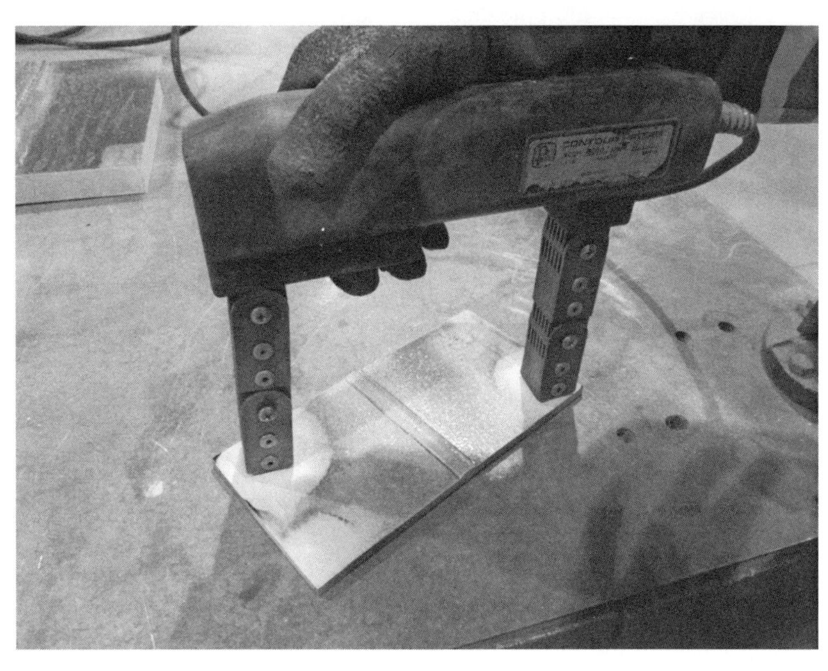

**FIGURE 38.5** Magnetic yoke in use on a pipe coupon. (Reproduced with permission of Applus RTD.)

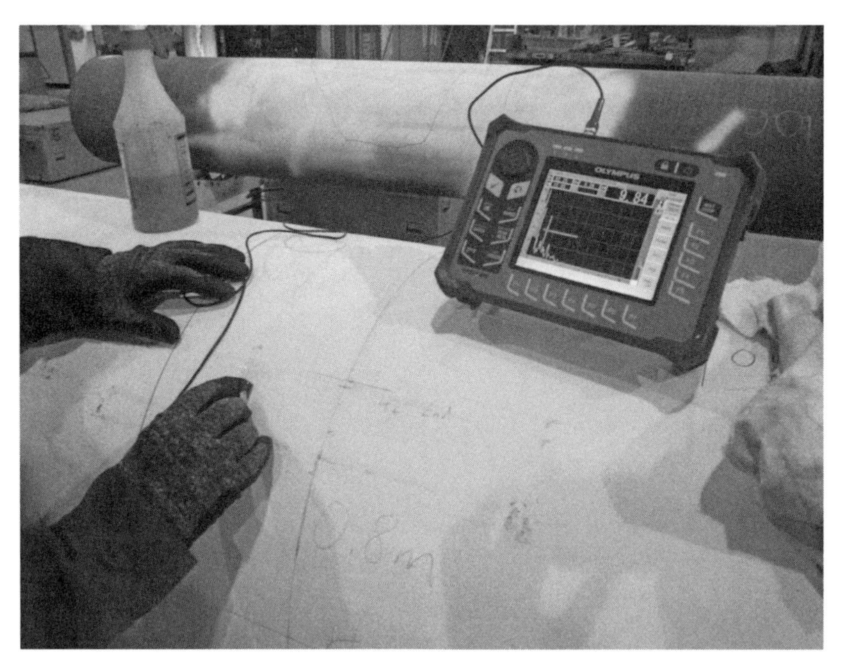


**FIGURE 38.6** Typical ultrasonic device in use on a pipe. (Reproduced with permission of Applus RTD.)

internal features and back to the transducer. A technician interprets the ultrasonic reflection that is displayed on a screen. This method is used to identify potentially dangerous internal or subsurface anomalies, such as cracks and crack-like features, delaminations and geometric features, such as the internal diameter pipe wall and welds. Some ultrasonic inspection (UT) technologies are also used to size surface-breaking cracks. Figure 38.6 illustrates a typical UT set being used for pipe assessment.

The common types of UT technologies used in pipeline integrity assessment include:

- zero degree or straight beam, which is used for wall thickness, internal metal loss, and internal delaminations.
- shear wave, which is best suited for crack identification and sizing, and internal wall loss.
- phased array, which is most commonly used for crack identification and sizing.
- time of flight diffraction (TOFD), which is ideally suited for long-seam weld defects, such as lack of fusion.

All UT methods require a clean pipe surface and a viscous couplant between the transducer and the pipe wall to conduct the sound energy into and back out of the steel. These transducers are handheld probes, or they are mounted on an automated traveler, which takes a systematic path over the pipe.

### 38.6.5 Laser Profilometry

The profile of an external pipe wall is measured by an instrument that projects laser light onto the surface of the pipe. This light is reflected back to the device, which measures the travel time of the light to measure the distance between the sensor and the pipe surface. This process occurs at a specific time, or spacing, and provides a three-dimensional surface profile of the external pipe wall and the anomaly of interest. The sensor is handheld or mounted on an automated traveler, which moves in a specific grid pattern. The data are recorded in a digital file, which results in two benefits—data entry errors are eliminated and the digital file is suitable to calculate the remaining pipe strength.

Laser profilometry is:

- more accurate than most manual methods and is used in place of the manual grid method.
- best used for extensive external volumetric features, such as corrosion or dents.

## 38.7 EXCAVATION PACKAGE

The excavation package is the first step in the in-the-ditch inspection. It is generated by the operator and provided to the NDE crew. It is a concise set of information that includes the following:

Metadata, which at a minimum, includes:

- a unique identifier for the project and individual excavation;
- the location of the excavation;
- pipe characteristics, including pipe diameter, wall thickness, and grade of steel;

- coating type;
- source of initial data set, usually an ILI or aboveground survey;
- the date of the inspection;
- a scope of work statement. This includes the type of assessment, project objectives and operator specific criterion for the assessment and repair of the pipe.

The location and the dimension of the anomalies on the pipe will be reported in a systematic convention of measurement. The most common convention measures the position relative to an upstream girth weld and the top, or 12 o'clock, position of the pipe. These are also the normal reference points for ILI data. The same convention is used to record the data collected by the in-the-ditch assessment.

A listing of the source data to compare to the in-the-ditch results usually consists of the pertinent portion of an ILI log that describes the location and the dimensions of the anomalies of interest. Figure 38.7 is an example of a pipe drawing with a single anomaly of interest from an excavation package.

## 38.8 DATA COLLECTION

There are three data sets related to in-the-ditch assessment. The first set is collected from the results of an assessment of the ambient pipe condition. This includes data from soil and water sampling, microbiologically influenced corrosion or pH testing, and the measurement of pipe to soil potentials. This data collection and analysis is discussed in Chapters 20, 27, and 32.

The second data set describes the coating type and condition. A failure of the coating can lead to the formation of external pipe wall anomalies. An assessment of the coating will provide a measure of the health of the coating as well as the effectiveness of the cathodic protection system. Consequently, the type and condition of the coating is critical data that is important to the assessment, and it should be

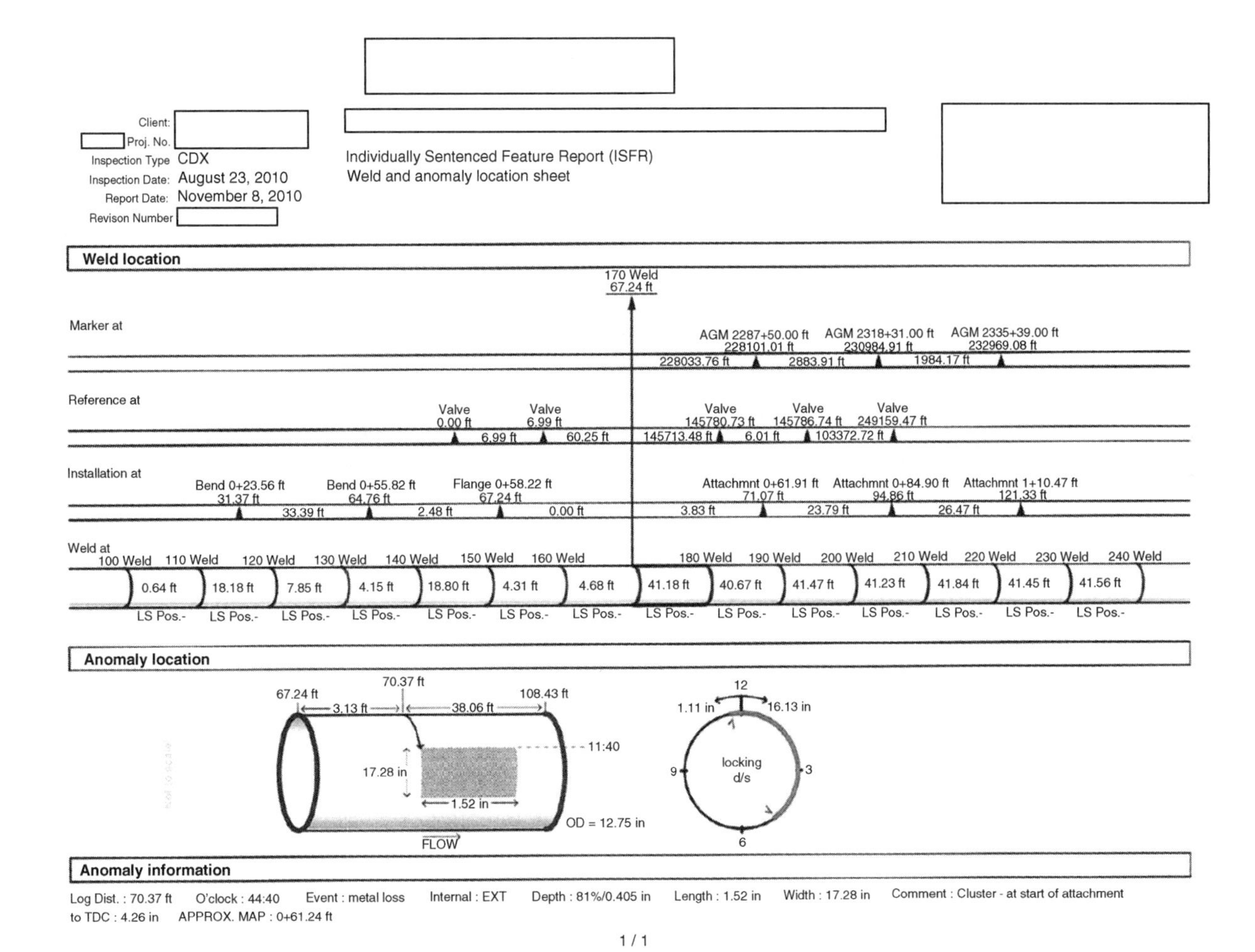


FIGURE 38.7 Drawing from a typical excavation package. (Reproduced with permission of Applus RTD.)

carefully noted. Areas of coating damage should be clearly labeled and photographed for inclusion in the final report. Coating samples or samples of deposits in the area of coating failures might also be collected as part of the assessment. Typical coating data collected during an in the ditch integrity assessment include the following:

- Coating type;
- Vintage;
- Condition;
- Description of any deposits.

The third data set, and the primary focus of this chapter, is the NDE data collected in the assessment of the condition of the pipe. These data include the size of the anomaly, which is described by the width, length, and maximum depth. Depending on the anomaly and the type of dig, more detailed sizing information might be required.

There are various ways to collect this data. Depending on the type of anomaly, different technologies, such as laser profilometry, MPI or UT are used to assess the anomalies and provide data. Laser profilometry and UT tools can be automated to speed the measurement and data acquisition and improve data resolution and quality. While these automated tools have been a boon to in-the-ditch assessment by speeding the data gathering and providing more accurate data, a good assessment will always rely on good field notes and a good photographic record. Some tablet-based data collection tools are available, but currently, most technicians rely on written field notes to ensure their reports are concise and complete. Technicians are advised to use an NDE data collection form that the operator has approved, to encourage consistent and complete data collection.

## 38.9 CONDUCTING IN-THE-DITCH ASSESSMENT

Pipeline integrity is perhaps the most stimulating and challenging of all NDE work. It requires a technician to look beyond the bounds of the scope of work to ensure there are no unexamined significant threats to a pipeline, and it requires the organization of a relatively complex data set into a concise final report.

This work requires two sets of procedures. They are a standard set of procedures for a specific type of NDE (e.g., MPI or UT); and a set of procedures to put these technologies to work in the context of pipeline integrity. This second set of procedures should include such things as specific criteria for anomaly assessment, data collection, photographic methods and reporting. The purpose of the entire group of procedures is to provide a concise list of steps for in-the-ditch inspections that result in a safe, accurate, and repeatable outcome.

The second set of pipeline integrity specific procedures is developed from the following sequence of events and tasks for conducting in-the-ditch inspection.

First is the preparation for the excavation:

1. Identify the dig location. The location is selected according to an operator-specific criterion that is applied to another inspection such as an ILI or aboveground survey.
2. Summarize this information in an excavation package. This excavation package contains the pipeline metadata and the specific anomaly data. The pipeline is then safely excavated to expose the anomaly.

   Next, the NDE crews are engaged:
3. Before the crew of NDE technicians can conduct the in-the-ditch assessment, the exact anomaly location must be confirmed using the following data:
   - GPS spatial data that were recorded by the ILI tool and the tool tracking crew or the survey technicians.
   - Unique pipe identifiers, such as the orientation of the long seam, length of the exposed pipe joint, and pipe wall thickness.
   - Distance from nearby aboveground pipeline appurtenances, such as valves or aboveground monuments.
   - The location and size of the anomalies of interest from the ILI log.

   It is best to cross-reference as many data sets as possible to ensure the excavation is in the correct location. Figure 38.8 shows a typical pipeline integrity excavation.

   Once the location has been confirmed and before pipe coating is removed by mechanical methods, sandblasting, or other methods:
4. A preliminary assessment may be conducted to ensure there are no anomalies that represent an immediate safety risk to the pipeline crew. This is also the time to gather any coating assessment data as required. If the initial assessment determines the pipe is safe to work on, the coating is removed and any ILI anomalies identified in the excavation package are marked on the pipe.

   It is important to understand the measurement convention when marking and measuring the integrity anomalies. Most conventions reference the upstream girth weld and the top dead center of the pipe looking downstream. This convention is usually provided by the operator.

   This is the baseline for the NDE crew to begin to evaluate these anomalies.
5. After the pipe coating is removed and the ILI anomalies have been marked on the pipe, a detailed anomaly assessment is conducted with the more advanced

**FIGURE 38.8** Typical pipeline excavation. (Reproduced with permission of Applus RTD.)

assessment tools. The NDE technicians will also inspect the rest of the exposed pipe to ensure that other anomalies were not missed by the ILI tool nor noted on the excavation package.

External metal loss features, such as external corrosion or gouges, are assessed with a manual grid and pit gauge, UT pen probes, laser profilometry tools, or a combination of the three.

Cracks and crack fields are assessed using a combination of techniques. If the operator is interested in quantifying the cracks, the following is the most common approach:

- Identify the cracks and mark them using MPI.
- Use an ultrasonic technique, such as shear wave or phased array, to estimate the length and depth of the cracks.
- Remove pipe wall in the area of the cracks by buffing or grinding the pipe wall in small increments to a maximum depth or until the crack has disappeared. These increments are normally no more than 10% of the original pipe wall thickness. Most operators define the maximum grind depth as 40% of the original pipe wall thickness.
- After every increment, the crack or crack field is redefined by using a combination of MPI and UT methods.

Operators who are not interested in correlating crack depths to ILI data might choose to simply install a reinforcement sleeve on the pipe over the cracks.

6. Every in-the-ditch inspection ends in a repair. Depending on the result of the assessment and the Rstreng calculations and the operator criteria, this repair may simply consist of a recoat of the pipe or it may be necessary to install a reinforcement sleeve on the outside of the pipe.

## 38.10 DATA MANAGEMENT

Good NDE data management begins long before an NDE technician enters the ditch. A pipeline integrity management plan is incomplete without a solid plan for the acquisition, organization, and analysis of the NDE data set. The requirement for the NDE data set should be reflected in the pipeline integrity excavation package, which is the primary source of direction for the technician.

### 38.10.1 Quality Control

NDE data is used in the calculation of the remaining strength of a pipeline. An error in the data will result in an incorrect calculation, which will lead to either an unsafe pipeline condition or a waste of resources. Automated measurement and data collection improve the accuracy of measurement and virtually eliminate errors in data collection. All the processes and procedures, from the initial assessment to final reporting must be built on the principles of robust quality control and assurance.

### 38.10.2 Reporting

The report is the final result of the assessment, and it is critical that the report is concise and complete. An unclear report will cause confusion or even incorrect analysis. It is also critical to ensure the report and associated data is labeled according to a coherent naming convention and stored in a secure, searchable location. A missing or lost assessment report leaves no record or evidence that the assessment ever happened.

A good and complete report begins with a clear definition of the objectives of the dig, and includes a clear set of reporting instructions. A typical reporting procedure should specify

- the convention used to describe the position of individual defects.
- the instructions for labeling and photographing individual defects.
- the location, type, size (including depth), and repair information of all features that are reported.
- engineering calculations, such as Rstreng values
- the time allowed for the completion of the draft and final report.
- the report content and delivery format (most commonly electronic in the form of an Excel spreadsheet, database entry, or a Word or PDF document).
- the quality control process that indicates who is responsible for the creation and review of the report.
- naming conventions for the individual data files and the compiled reports.
- method of transmission, such as uploading to an ftp or SharePoint site.

## 38.11 RECENT TECHNOLOGICAL DEVELOPMENTS

What is next for in-the-ditch pipeline anomaly assessment? The world of pipelines and pipeline integrity has become significantly more demanding in the last 5 years. It is more important than ever to continue to improve existing technologies and develop new ones.

Most of the recent advancements have been evolutionary rather than revolutionary. For example, the primary technologies have not changed while, in some cases, the resolution and accuracy have improved. These advances in data improvement have been made possible by the ongoing miniaturization and efficiency of digital technology. These developments have allowed the digitization and automation of more parts of the assessment process. This technology advancement has led to better analysis and a higher degree of confidence due to quicker results, higher resolution, and more accurate data.

### 38.11.1 Electromagnetic Acoustic Transducer (EMAT)

Electromagnetic Acoustic Transducer (EMAT) was developed because of the need for a reliable crack detection device that can be used in a dry gas environment. Whereas most of this development has focused on ILI technology, it remains to be seen if can be applied to external, direct measurement. The commercialization and development of the EMAT ILI into a reliable inspection tool is directly related to exhaustive verification of the tool results by in-the-ditch NDE assessment.

### 38.11.2 Structured Light

Quite literally, structured-light technology is a high-resolution digital camera that takes a photograph of a prepared pipe and uses the shades of the image to model the surface profile of the pipe. In principle, the result is similar to a laser profilometry device. The data output is suitable for use in an automated program that calculates the remaining strength of the pipe wall. This technology is most effective for describing surface metal loss and pipe deformation defects.

### 38.11.3 Ultrasonic

Faster processing and smaller transducers have allowed advancements in the analysis of the UT data. This advancement has led to the real-time, three-dimensional projection of the UT signal so that internal pipe features are more accurately displayed. This technology has the potential to make the interpretation of ultrasonic signals less dependent on the skill of the operator.

### 38.11.4 Eddy Current

Eddy current technology is not widely used in the assessment of pipeline integrity anomalies, although recent developments show some promise in the quantification of cracks. The recent state of the art in eddy current sensors shows an increase in data accuracy and repeatability.

## 38.12 SUMMARY

This chapter has presented an overview of in-the-ditch pipeline assessment, including the objectives required for the proper design of the program, the role of ILI, available NDE inspection technologies, the type of data to gather, and the steps to create a good report.

In-the-ditch pipeline inspection is a key part of any pipeline integrity program. It helps to support the safe operation of a hydrocarbon pipeline and, if properly planned and conducted, it maximizes the overall lifecycle investment by the pipeline operator.

## ACKNOWLEDGMENTS

- James Grant, Applus RTD, Edmonton, Alberta, Canada
- Rick Layh, Applus RTD, Edmonton, Alberta, Canada

- Shamus McDonnell, Hunter McDonnell Pipeline Services, Edmonton, Alberta, Canada
- Len Krissa, Enbridge Pipelines, Edmonton, Alberta, Canada
- Mike Reed, Alliance Pipeline, Calgary, Alberta, Canada

## REFERENCE

1. www.pipelinealias.com

## BIBLIOGRAPHY

Dennis, N. (2000) Pipeline coating failure—not always what you think it is. CORROSION/2000, NACE International, Houston, TX, Paper No. 00755.

Dilip, T. Wayne, F H., and Nick, F G. Field Joint Developments and Compatibility Considerations, Canusa-Canada.

# 39

# ULTRASONIC MONITORING OF PIPELINE WALL THICKNESS WITH AUTONOMOUS, WIRELESS SENSOR NETWORKS

FREDERIC CEGLA,[1] AND JON ALLIN[2]
[1]Mechanical Engineering Department, Imperial College London, South Kensington, UK
[2]Permasense Ltd., Horsham, UK

## 39.1 INTRODUCTION

Erosion and corrosion present a major threat to oil and gas industry pipeline infrastructure. Koch et al. [1] have estimated that the cost of corrosion is about 2–4% of GDP for a developed country and this is probably much higher for industries such as the oil and gas industry. For the industry as a whole, it is therefore of utmost importance to ensure the integrity and safe operation of its pipe work infrastructure.

The management of the extensive pipeline assets requires several ingredients and nondestructive evaluation (NDE) inspections play a key role in this. However, by the nature of the location of resources and production processes, most of the producing and refining assets of interest are located in inhospitable places and many inspections take place in hazardous (e.g., high temperature) and isolated environments. This limits the access to the components of interest and either requires measurements to be taken at shut downs or when access (in form of scaffolding or other maintenance work that gives access to the pipe work such as striping of insulation) is possible. Indeed, the cost of access is often a multiple of the actual cost of the NDE measurement itself. It is therefore, potentially, very desirable to permanently install NDE sensors that can be wirelessly addressed at the location where an examination has to be carried out and to interrogate when a measurement is required or at will via a wireless link.

A large proportion of wall thickness integrity checks are performed by ultrasonic thickness gauging examinations such as described by standards EN14127:2011 or ASTM E797/E797M-10. Several technologies to perform ultrasonic wall thickness monitoring are on the market (see e.g., [2–4]) and their working principle is based on the physics described earlier in the European and American standards. The authors have developed technology that is capable of functioning on most metal pipe work up to temperatures of 600 °C, automatically sending back the information via a wireless network, and most of the work presented in this chapter is focusing on this technology. The aim is to highlight a representative example of the overall capabilities of ultrasonic wall thickness monitoring sensors. This chapter describes the working principle of the developed autonomous, wireless, ultrasonic sensor and shows some example results from realistic applications on pipework in the oil and gas industry.

## 39.2 THE PHYSICS OF ULTRASONIC THICKNESS GAUGING

Ultrasonic thickness gauging is one of the most commonly used techniques to determine wall thicknesses in opaque pipe work. There are several standards such as EN14127:2011 or ASTM E797/E797M-10 that outline the working principles. The technique involves placing an ultrasonic sensor in contact with the component under test. Then the sensor is excited with a high-frequency electric voltage signal. The

*Oil and Gas Pipelines: Integrity and Safety Handbook*, First Edition. Edited by R. Winston Revie.

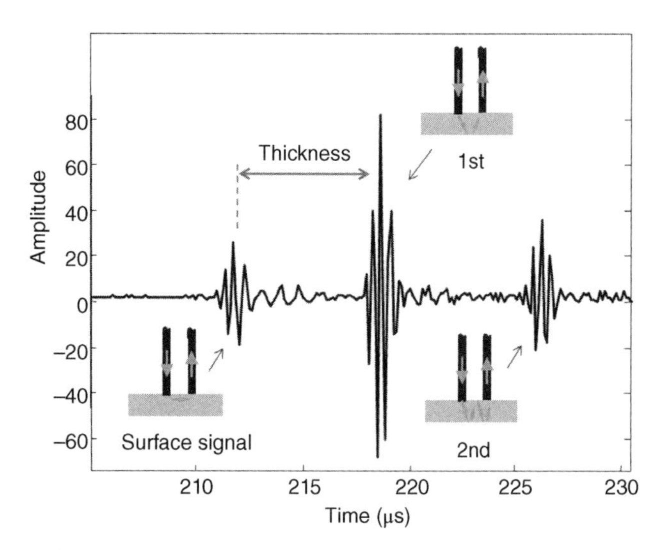


**FIGURE 39.1** Typical ultrasonic signal (voltage amplitude as a function of time) recorded with a permanently installed sensor. Insets show the different signal paths relative to the sending (left) and receiving (right) transducer and the component whose thickness is to be measured (grey box at the bottom of each inset).

sensor converts this electric signal into mechanical vibrations. The vibrations are transmitted from one surface of the test piece into the bulk of the material and travel at the ultrasonic velocity into the material until they are reflected from a feature within the material and travel back to the ultrasonic transducer. At the ultrasonic sensor these reflections are turned back into voltage signals that show up as echoes on the recorded voltage versus time plot (A-scan, see Figure 39.1). In a clean pipe of constant thickness, this results in a train of echoes separated by time.

$$T = 2d/v \tag{39.1}$$

Equation 39.1 shows that the ultrasonic travel time ($T$) or temporal separation between echoes is directly related to the wall thickness ($d$) via the ultrasonic velocity ($v$), a material property.

Figure 39.1 shows the signal path geometry and a typical signal that is recorded from a sensor that is attached to a specimen. (It is worth mentioning that this is a signal acquired with a dual element probe corresponding to a thickness measurement in Mode 2 as per EN14127 or Case 3 as per Section 8.3 of ASTM E797.) The first echo represents a reflection from the front surface (or a signal that has travelled a short distance along the surface) and the second echo represents a reflection from the internal wall of the pipe (or component under test). Subsequent echoes are due to further reverberations of the signal within the pipe/component wall. Based on the measured signal the thickness is calculated from the time difference between consecutive echoes, multiplication by the ultrasonic velocity, and division by a factor of two (to account for the double transit to the wall and back), that is, inversion of Equation 39.1.

The resulting thickness estimate, therefore, depends on the accuracy with which the arrival time of the signal can be determined and the precision to which the ultrasonic velocity is known.

It is common knowledge that the ultrasonic velocity is a function of the material and temperature [5], and therefore, readings taken at different temperatures need to be compensated for temperature changes in order to avoid introducing errors into the measurement results or the test environment must be carried out in a relatively stable temperature environment. When monitoring wall thicknesses with permanently installed sensors at regular and frequent time intervals, temperature compensation for process conditions that are stable to within ±20 °C (40 °F) becomes less important. The trend of the measured wall thicknesses will be clearly picked up albeit it with a marginal increase in scatter about the trend due to temperature variations.

One of the clear advantages of the permanent installation of ultrasonic sensors for monitoring of wall thickness data is that the uncertainties that are usually introduced due to variability in transducer position and coupling with the sample are removed. The automated, wireless data acquisition of data using permanently installed sensors also removes the need for an operator to enter a hazardous environment and eliminates potential human intervention causing errors in the measured dataset (such as wrong recording of results, location information, or recalling of setup information). Monitoring with permanently installed sensors, therefore, highlights the changes in the physical interaction of the ultrasonic wave with the tested component to a greater extent than it is the case in manual examinations. This makes it possible to make much more precise measurements and follow very small changes in thickness. With temperature compensation this has been achieved down to thickness variations of a few tenths of nanometers in very controlled lab conditions [6]. In field conditions, thickness measurements are repeatable to ±10–100 μm (0.4–4 mil).

## 39.3 AUTONOMOUS SENSOR AND NETWORK CONSIDERATIONS

Figure 39.2 shows a schematic of a wireless corrosion monitoring system (by Permasense Ltd.). The system exists in several forms, depending on remoteness of the monitoring location and the required sensor density. There is a short range dense sensor network system for facilities such as refineries and a long-range sensor network system for large area facilities such as production fields.

The short range systems can support ~100 sensors per gateway, with a wireless communication range of ~100 m from sensor node to sensor node and the gateway. The

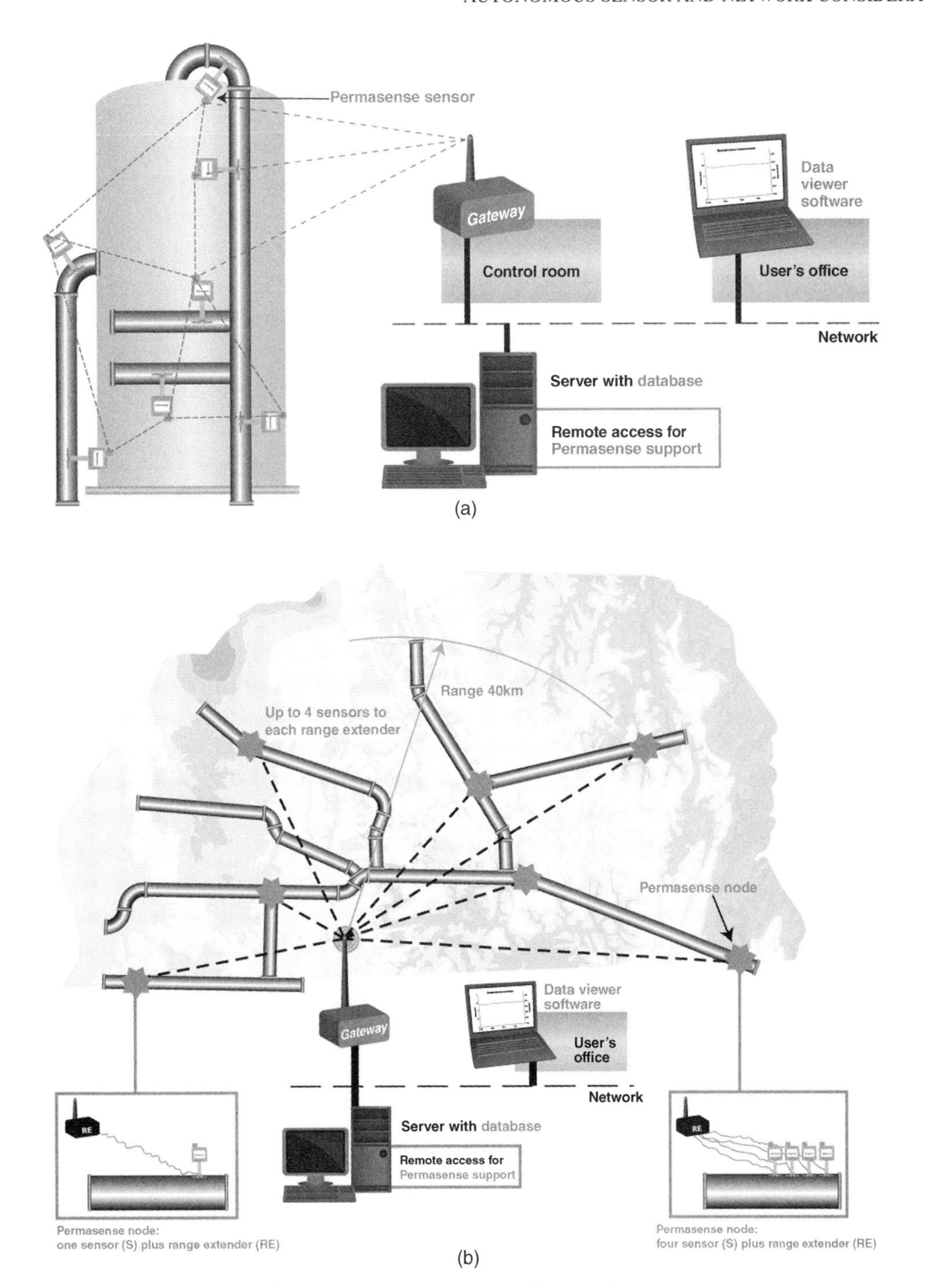


**FIGURE 39.2** Schematics of (a) short- and (b) long-range wireless ultrasonic corrosion monitoring systems (as an example the system of Permasense Ltd is shown here).

gateway is the wireless communication station that is plugged into a data network on which the database that stores all the data is housed.

The long-range system uses the same sensors; however, the sensors are connected to a range extender, which can transmit the data to a gateway on a data network that is located up to 10 s of km from the actual sensor or sensor network.

The short-range system is available in several variants compliant with one or other of the several standards for wireless communication. A key attribute is that the communications module has very low power consumption so that each sensor can be powered by a battery pack for an anticipated period of 5 years. The system can be installed in hazardous environments and has obtained intrinsic safety certification in the relevant classes and jurisdictions.

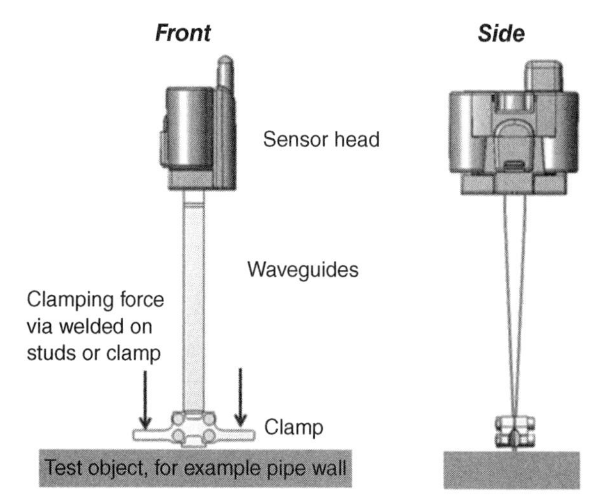


**FIGURE 39.3** Schematic of the front and side view of the permanently installable ultrasonic sensor, showing the most important features of the sensor hardware. There are two waveguides one to send the signal and the other to receive it, the total vertical dimension from the clamp to the sensor head is roughly 300 mm (~1 ft).

Figure 39.3 shows the actual ultrasonic sensor. It consists of an instrumentation (sensor) head that contains all the ultrasonic transduction; electronics and wireless communication modules; two waveguides, one of which transmits ultrasonic energy from the instrumentation section to the component under test and one of which receives returned energy; and a clamp that enables easy attachment of the whole sensor onto the component that is to be tested, for example, a pipe. The waveguides allow isolation of the transduction and electronic components that cannot withstand high temperatures and the actual measurement area, which can be at temperatures up to 600 °C.

## 39.4 TEST RESULTS

The sensor has been extensively tested in the laboratory and applied in many thousands of field locations. This section will summarize some of the key findings of laboratory tests and operating experience.

### 39.4.1 Lab Tests

The sensitivity to wall thickness and performance in tracking thickness loss in rapidly corroding environments were tested in several experiments under laboratory conditions. Figure 39.4a shows test results obtained when measuring mild steel plates of thickness 3–10 mm (0.118–0.394 in.) with the sensor and independently verified by means of a micrometer. Figure 39.4b shows test results obtained when a ~25.1 mm (0.988 in.) thick mild steel plate is milled down in thickness to about 24.6 mm (0.969 in.) in 0.1 mm (4 mil) steps with acquisition of about 50–100 thickness readings in between milling runs. The results show very good performance with respect to absolute thickness and the capability of monitoring thickness loss of <0.1 mm (4 mil).

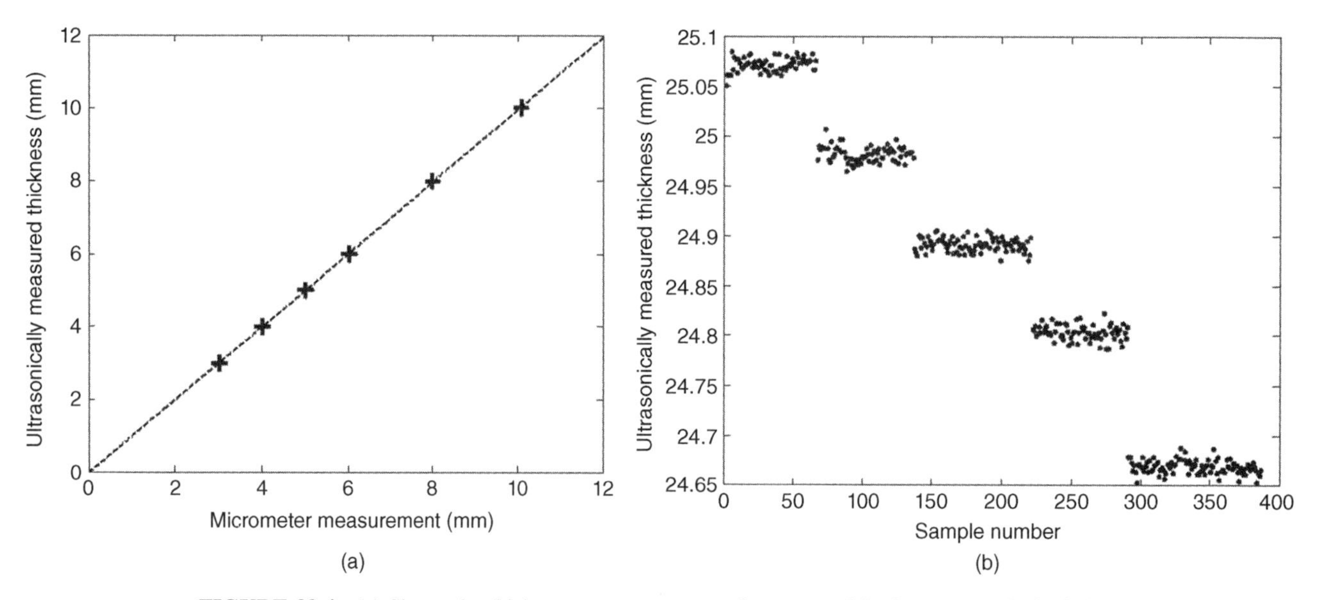


**FIGURE 39.4** (a) Shows the thickness measurement performance of the Permasense Ltd wireless ultrasonic sensor on calibration plates of 3, 4, 5, 6, 8, and 10 mm (0.118, 0.157, 0.197, 0.236, 0.315, and 0.394 in.) thickness as independently verified by a micrometer measurement; (b) shows the measurements of the thickness of a ~25.1 mm (0.988 in.) thick steel plate while it is milled down to about ~24.6 mm (0.969 in.) in 0.1 mm (4 mil) steps in a milling machine.

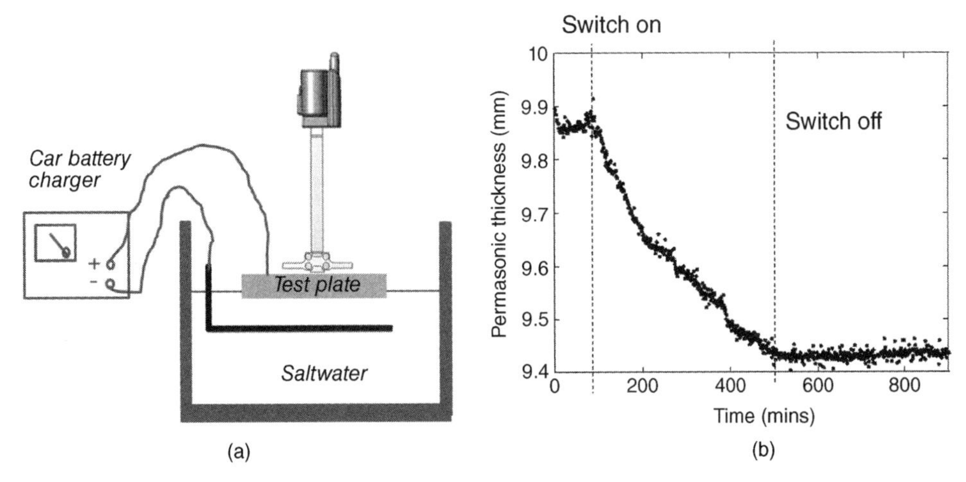


**FIGURE 39.5** (a) Schematic of the setup used to simulate accelerated corrosion, a sensor is attached to a ~10 mm (0.394 in.) thick test specimen that is partially immersed in a saltwater solution and a car battery charger is used to drive a current through the system to artificially corrode the test specimen; (b) results of the monitored thickness as a function of time after the current is switched on. A clear loss of thickness is visible while the current is applied.

Figure 39.5a shows a schematic of an experimental setup used to artificially induce corrosion in a test sample in order to illustrate the sensors capability of tracking thickness loss while corrosion is occurring. A current is driven through the test piece in order to oxidize iron and make it dissolve into a saltwater solution. The monitored thickness as a function of time is shown in Figure 39.5b. It is clearly visible that no thickness loss is recorded before and after the current is applied and very large thickness loss is recorded while the current is applied to the test piece. This further demonstrated the capability of recording sub 0.1 mm (4 mil) thickness loss with the Permasense sensor.

### 39.4.2 Operating Experience

Permanently installed ultrasonic wall thickness monitoring systems have been deployed in oil and gas facilities on every continent, by all of the oil majors, in upstream (production) and downstream (refinery) plant. Many thousands of thickness trends have been collected over the last few years. Some thickness trends that highlight the capabilities of such systems are discussed here.

Figure 39.6 shows an example of a sensor that has been installed on pipework of nominally 10 mm (0.394 in.) thickness and monitored a constant thickness for the whole period of installation (3.5 years). The trend shows that there is no active corrosion at the installation site and that there is excellent repeatability of the measured thickness over the whole 3.5-year period. This sensor was installed in an inaccessible location, saving the cost of access for routine measurements at regular intervals. Furthermore, it suggests that corrosion is not an issue at this location and that the component could tolerate more aggressive (higher total acid number (TAN)) feedstock or process conditions if the operations team should deem this desirable.

In contrast, the wall thicknesses that are shown in Figure 39.7 show evidence of a constant loss of wall thickness as a function of time over a period of about 1.5 years. Here, the determined rate of wall loss can be compared with available data on process conditions/feedstock/inhibitor dosage, and with rates of wall loss expected under those conditions, to confirm or challenge those expectations. The information can also be used to predict the retirement time/lifetime of pipework with confidence. Furthermore, in critical locations the rate measurement can be used to give confidence that pipework will remain safe until the next scheduled outage or planned maintenance event.

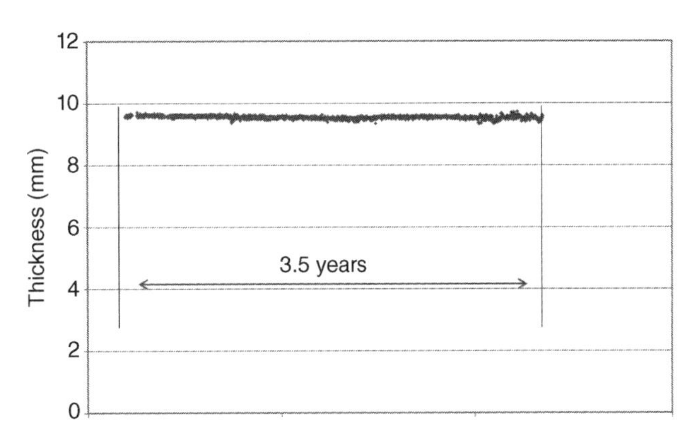


**FIGURE 39.6** A constant wall thickness monitored over a period of 3.5 years, confirming the absence of corrosion on a nominally 10 mm (0.394 in.) thick pipe in an oil and gas facility.

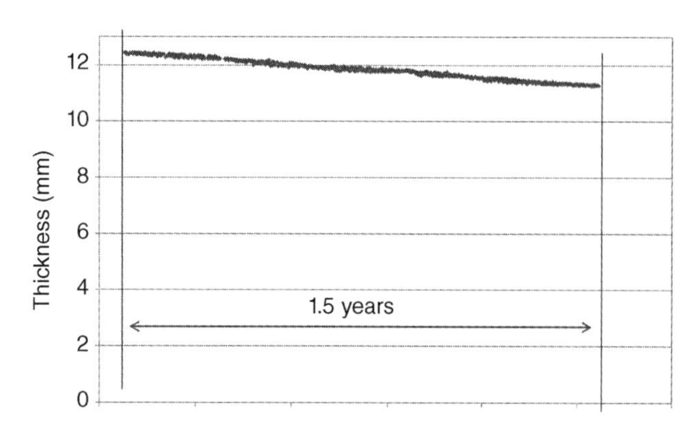


**FIGURE 39.7** Wall thickness measurements on an initially 12.5 mm (0.492 in.) thick component that shows practically linear wall loss as a function of time.

Particularly exciting is data from locations where intermittent wall loss can be determined. An example of that is shown in Figure 39.8. Here, the sensor recorded a loss of 1 mm/year (~40 mil/year) for 6 months. Following this, the wall thickness remains constant for a further 4–6 months, after which the constant wall loss of 1 mm/year (40 mil/year) resumes. This clearly highlights the opportunity to correlate wall thickness loss with plant operating parameters. The high frequency of wall thickness measurements, coupled with the excellent repeatability, gives the necessary confidence and response time (data to desk) to make it possible to study the effect of feedstock on corrosion activity, to verify and update corrosion inhibition strategies, and to optimize process conditions in order to minimize component wall thickness loss and maximize plant life.

## 39.5 APPLICATIONS

The system has been tried and tested in refineries, upstream production and processing facilities, chemical processing units, conventional power generation plants, and geothermal power generation plants. Key benefits of this type of monitoring system include understanding the causes of corrosion, optimization of prevention and mitigation strategies, improved insight in the impact of feedstock decision, and greater confidence in asset management decisions.

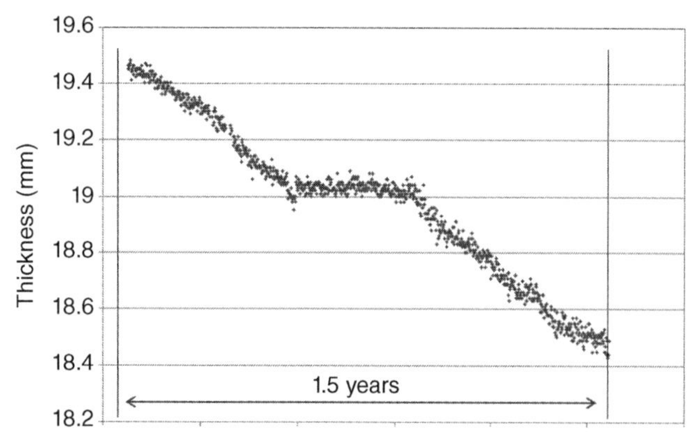


**FIGURE 39.8** Wall thickness measurements as a function of time at a location where intermittent wall loss is detected.

**TABLE 39.1 Examples of Applications**

| | |
|---|---|
| Equipment | **Refineries** |
| | Throughout the refinery, including: crude units (CDU), vacuum units (VDU), amine units, alkylation units, cokers, cracking units (FCC, cat crackers, hydrocrackers), hydrotreaters, sour water strippers—including on distillation column walls, stripper towers, inlets, outlets, and transfer lines (including reducers and elbows), air coolers—header boxes, inlets, and outlets, heat exchanger casings, inlets and outlets, pump inlets/outlets, furnaces—inlets, outlets, U-bends |
| | **Upstream production facilities** |
| | Processing facilities; platform risers |
| Materials | Full range of steels, including carbon steel, cast carbon steel, P5/5 chrome, P9/9 chrome, 1% Cr (5130), duplex, P265GH (430-161), 1.4571 (316 Ti), P295GH (17 Mn4) |
| Locations | Known thin spots, existing corrosion monitoring locations |
| | Areas of turbulence: reducers, outside of T junctions, up and downstream of valves, outside of elbows, 1–2 diameters downstream of bends |
| | Metalwork operating outside recommended operating window |

Further, major advantages of the system are that it delivers data regularly to desk, with excellent repeatability and also from inaccessible and hazardous environments (e.g., at temperatures up to 600 °C) where it would usually not be possible to obtain online data without incurring large access costs and putting operators in harm's way.

Applications examples, including the type of units and component materials, are listed in Table 39.1.

## 39.6 SUMMARY

This chapter has summarized the current capabilities of online corrosion monitoring systems that are based on autonomous wireless ultrasonic sensors. Initially, the technology was developed for locations with difficult and costly access where leaving sensors in place would present a real benefit. However, online corrosion monitoring has many more advantages, including the reduced lag time from measurement to desk, the reduced uncertainty due to the dramatic reduction of positioning and, coupling errors and increased safety since the inspector does not have to put himself in harm's way to collect data past the first installation. Furthermore, it is possible to monitor also while the plant is operating at elevated temperatures. Such monitoring makes it easier to take more informed decisions about the state of critical components along the pipe work and to verify whether the piping network performs to expectation. This chapter has shown that use of these technologies is now established practice in oil and gas facilities and numerous other process industries worldwide.

## ACKNOWLEDGMENTS

The authors would like to thank all the parties that have contributed to this article, notably, the many students and staff in the NDE group at Imperial College London, and the staff of Permasense Ltd.

## REFERENCES

1. Koch, G.H., Thompson, N.G., and Payer, J.H. (2001) Corrosion cost and preventive strategies in the United States. Technical Report FHWA-RD-01-156, CC Technologies Laboratories, Inc., 6141 Avery Road, Dublin, OH 43016-8761 NACE International, 1440 South Creek Drive, Houston, TX 77084-4906.
2. GE Measurement and Control, Available at www.ge-mcs.com, last accessed 7/11/2013.
3. Rohrback Cosasco, Available at http://www.cosasco.com/ultrasonics.html, last accessed 7/11/2013.
4. Permasense, Available at www.permasense.com, last accessed 7/11/2013.
5. Papadakis, E.P., Lynnworth, L.C., Fowler, K.A., and Carnevale, E.H. (1972) Ultrasonic attenuation and velocity in hot specimens by the momentary contact method with pressure coupling, and some results on steel to 1200 °C. *Journal of the Acoustical Society of America*, 52(3B), 850–857.
6. Gajdacsi, A. and Cegla, F. (2014) High accuracy wall thickness loss monitoring. *AIP Conference Proceedings*, 1581(1), 1687–1694.

# 40

# FLAW ASSESSMENT

TED L. ANDERSON
*Quest Integrity Group, LLC, Boulder, CO, USA*

## 40.1 OVERVIEW

Flaws in a pipeline can reduce the burst pressure and shorten the life of the line. The vast majority of releases that have occurred in pipelines can be attributed to flaws. Some flaws are introduced at the time of manufacture, while others develop in service. Table 40.1 lists the most common flaw types. Various flaw assessment procedures are available to predict the effect of flaws on burst pressure and remaining life. Such procedures are a crucial part of an effective integrity management program. For example, remaining life calculations are typically used to establish appropriate intervals for hydrostatic testing and in-line inspection(ILI).

### 40.1.1 Why Are Flaws Detrimental?

A fundamental reason why flaws can adversely affect pipeline integrity is that these imperfections act as stress concentrators. Figure 40.1 shows two finite element models that illustrate this effect. Figure 40.1a is a color stress map of a dented pipe under pressure loading. The deviation of a pipe from a cylindrical shell results in much higher stresses in the dent compared with the rest of the pipe. Figure 40.1b is a color stress map at the tip of a crack, which is the most severe type of stress concentrator. Part of the reason for the stress concentration is the local reduction in wall thickness, but the sharpness of the flaw magnifies this effect significantly. A notch-like feature, such as a lack-of-fusion flaw, has a lower stress concentration factor (SCF) than a crack, and a smooth metal loss flaw has a lower stress concentration than either a crack or a notch.

In the case of crack-like flaws and metal loss, the high local stress at the flaw reduces the burst pressure. Burst tests on dented pipes usually do not show a significant reduction in burst pressure, but dented pipes can fail over time by fatigue, which is driven by cyclic pressure. Metal loss and crack-like flaws can grow over time that gradually reduces burst pressure until it reaches the local operating pressure.

### 40.1.2 Material Properties for Flaw Assessment

The response of pipelines to flaws is influenced by material properties. The following properties affect the burst pressure:

1. Specified minimum yield strength (SMYS)
2. Specified minimum tensile strength (SMTS)
3. Toughness

When the pipe reaches its burst pressure, the failure mode may either be *ductile rupture* or *fracture*. In the former case, the burst pressure is governed by the tensile properties, while fracture is controlled by toughness. In a fracture event, the pipe essentially unzips, as a crack rapidly propagates. Toughness is defined as the ability of the material to resist rapid crack propagation. Traditionally, the toughness of pipeline steels and welds has been quantified by Charpy tests, but toughness tests based on fracture mechanics principals have seen increasing use in the pipeline industry in recent years.

Figure 40.2 schematically illustrates the effect of toughness on burst pressure of a pipe that contains a longitudinal crack-like flaw. In the fracture regime, burst pressure increases with toughness. At high toughness levels, a plateau

*Oil and Gas Pipelines: Integrity and Safety Handbook,* First Edition. Edited by R. Winston Revie.

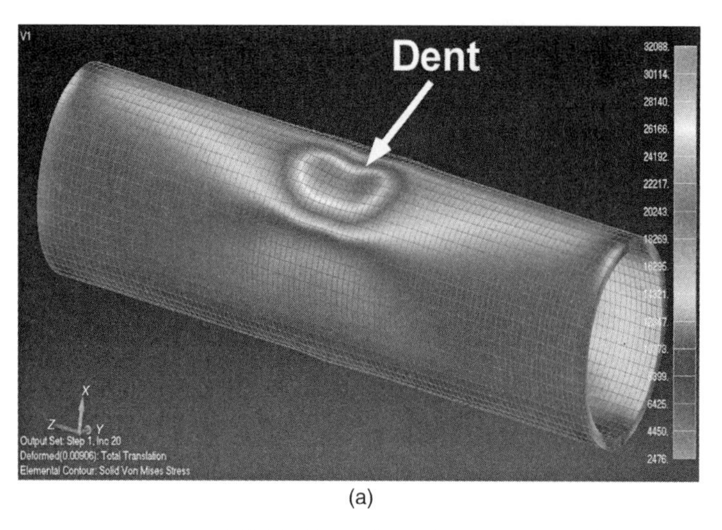


(a)

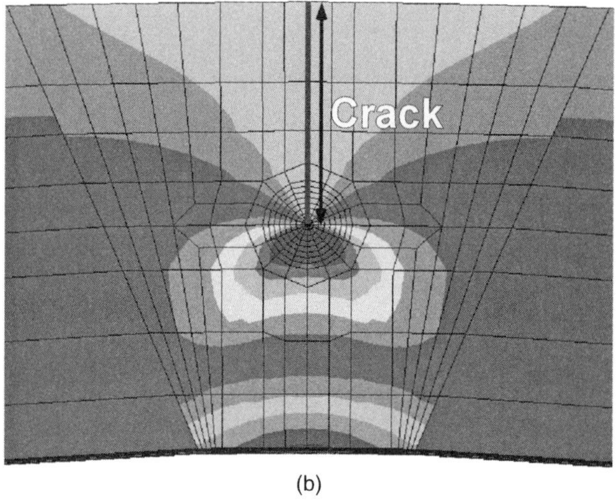


(b)

**FIGURE 40.1** Finite element models with color stress maps that show stress concentrations due to pipeline flaws. (a) Stress concentration at a dent. (b) Stress concentration at the tip of a crack.

**TABLE 40.1 Common Flaw Types**

| Flaw Classification | Examples |
|---|---|
| Metal loss | Localized corrosion and pitting |
| | General wall thinning |
| | Gouges |
| Cracks and crack-like flaws | Hook cracks in ERW seams |
| | Lack-of-fusion flaws |
| | Fatigue cracks |
| | Stress corrosion cracks |
| Dents | Constrained or unconstrained dents |
| | Sharp or smooth dents |
| | Dents with metal loss |
| | Dents at welds |

is reached where the failure mode changes to ductile rupture, which is controlled by tensile properties. In other words, increases in toughness eventually produce diminishing returns at high toughness levels.

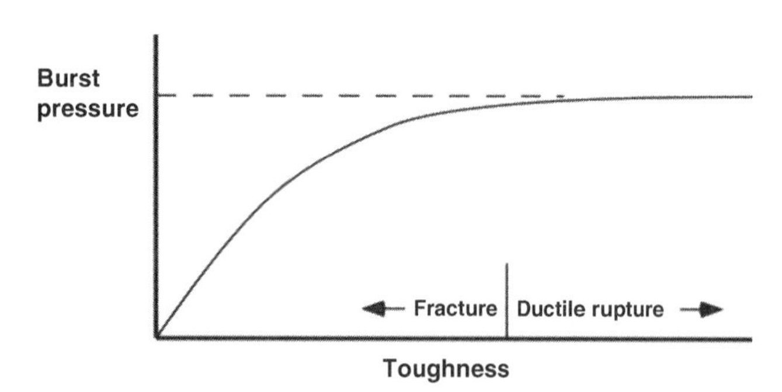


**FIGURE 40.2** Effect of toughness on burst pressure in a pipe that contains a longitudinal crack.

For time-dependent damage mechanisms, such as fatigue, corrosion, and stress corrosion cracking (SCC), material properties in addition to those listed previously may influence the damage rate. Initiation of fatigue is influenced by strength, but fatigue crack growth rates are insensitive to tensile properties. Environmental degradation mechanisms such as corrosion and SCC are driven by complex interactions between the environment and the material.

### 40.1.3 Effect of Notch Acuity

At low to moderate toughness levels, the sharpness of a flaw can have a significant effect on burst pressure. Figure 40.3 illustrates the typical trend. Sharp cracks result in the lowest burst pressure because of the stress concentration effect described earlier. As the notch radius increases, the failure mode transitions from fracture to ductile rupture. Failure at smooth metal loss flaws generally occurs by ductile rupture, where tensile properties govern the strength of the pipe.

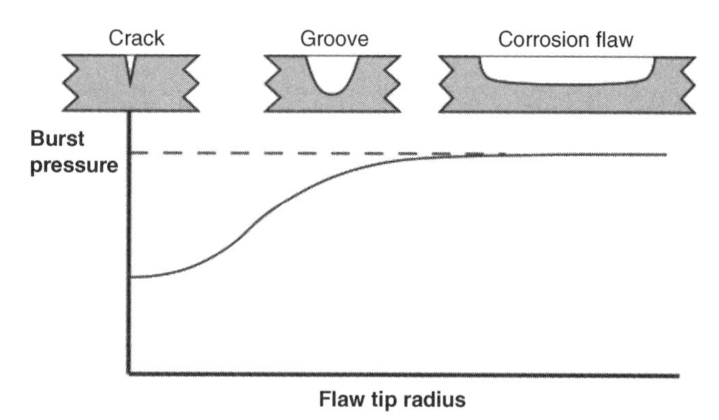


**FIGURE 40.3** Effect of notch acuity (sharpness) on burst pressure in a pipe that contains a longitudinal flaw.

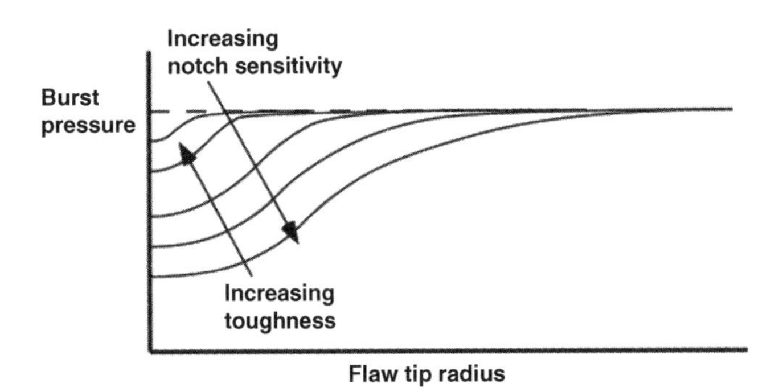


**FIGURE 40.4** Combined effect of toughness and notch acuity on burst pressure.

Figure 40.4 illustrates the combined effect of toughness and notch acuity. Low toughness materials are *notch sensitive*, in that the burst pressure is sensitive to the notch tip radius. By contrast, the burst pressure for high toughness materials is insensitive to notch acuity (sharpness). In this case, a pipe with a longitudinal crack of a given length and depth will have essentially the same burst pressure as a pipe (made from the same material) with a metal loss flaw with the same length and depth.

## 40.2 ASSESSING METAL LOSS

A number of procedures are available for assessing metal loss in pipelines, including B31G [1], API 579 [2], as well as a DNV Recommended Practice [3]. These methods differ somewhat in minor details, but they all are based on research performed at Battelle Laboratories in the late 1960s and early 1970s [4]. The modified B31G procedure, which is presented ahead, is the most commonly used method in the pipeline industry.

Consider a metal loss flaw with longitudinal extent $L$ and maximum depth $d$, as Figure 40.5 illustrates. According to the modified B31G method (level 1), the remaining strength of a corroded pipe is given by

$$\frac{S_F}{S_{flow}} = \frac{1 - 0.85\left(\frac{d}{t}\right)}{1 - \frac{0.85}{M}\left(\frac{d}{t}\right)} \tag{40.1}$$

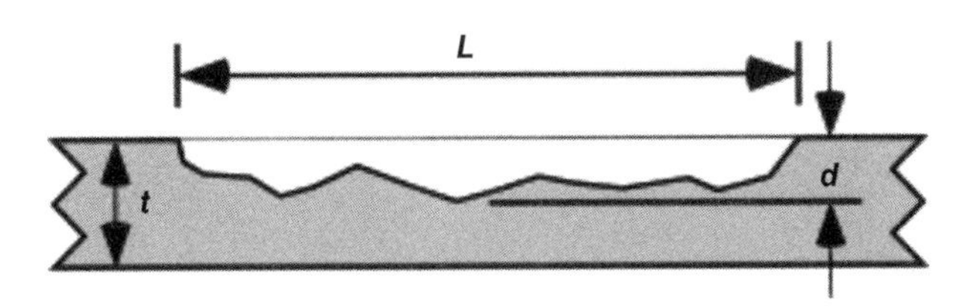


**FIGURE 40.5** Sketch of a metal loss flaw, where $L$ is the length in the longitudinal direction and $d$ is the depth.

where $S_F$ is the hoop stress at failure in the corroded pipe and $S_{flow}$ is the flow stress. The parameter $M$ is called the Folias factor or the bulging factor and is given by

$$M = \begin{cases} \left(1 + 0.6275z - 0.003375z^2\right) & z \le 50 \\ 0.32z + 3.3 & z > 50 \end{cases} \tag{40.2}$$

where $z$ is a dimensionless function of flaw length, $L$, wall thickness, $t$, and pipe outside diameter, $D$:

$$z = \frac{L^2}{Dt} \tag{40.3}$$

The underlying assumption of Equation 40.1 is that a pipe without corrosion will fail when the hoop stress reaches $S_{flow}$. The B31G method gives several alternative definitions of flow stress, including 1.1 SMYS, SMYS + 69 MPa (10 ksi), and the average of SMYS and SMTS. All of these definitions are conservative, as an unflawed pipe will typically burst when the hoop stress is approximately equal to the tensile strength.

API 579 refers to the ratio on the left-hand side of Equation 40.1 as the *remaining strength factor* (RSF). The RSF is the ratio of the strength of the corroded pipe to that of the uncorroded pipe. The reciprocal of RSF can be viewed as a SCF. In other words, the nominal hoop stress and the local hoop stress in the remaining ligament of the flaw are related through the RSF (=1/SCF):

$$S_{hoop}^{local} = \frac{S_{hoop}^{nominal}}{RSF} \tag{40.4}$$

Equation 40.1 predicts the burst pressure that must be divided by a safety factor to obtain the maximum allowable operating pressure (MAOP) for the corroded pipe. B31G recommends a safety factor of 1.39 for pipelines with an allowable hoop stress of 72% SMYS. Since 1.39 = 1/0.72, this implies that the corroded pipe can have a *local* hoop stress, $S_{hoop}^{local}$, equal to 0.72 $S_{flow}$ rather than 0.72 SMYS. Note that if the corrosion flaw is small or shallow, the B31G acceptance criterion can result in an allowable pressure that is greater than the original design pressure, in which case the pressure reverts to the original MAOP.

When wall thickness has been mapped with multiple readings, usually on a grid, a more advanced method called *effective area* can be used to assess metal loss. Both B31G and API 579 include the effective area method as a level 2 assessment. Given a thickness map over the area of interest, a "river-bottom" profile is taken, which corresponds to the longitudinal path of minimum thickness through the flaw. The thickness readings along the river bottom profile are then projected onto a 2D plane. The resulting thickness profile is subdivided into sections and subjected to a

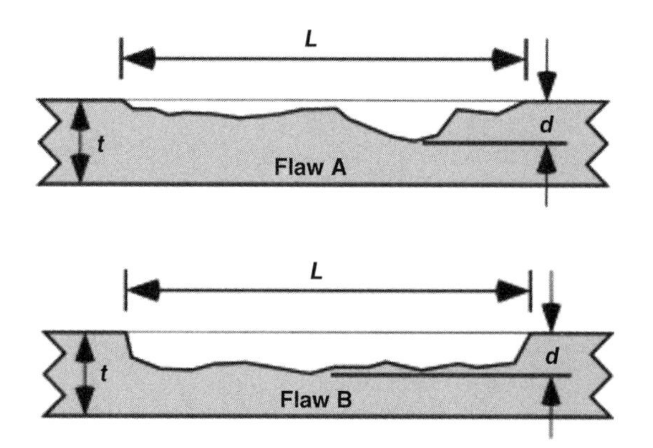


**FIGURE 40.6** Comparison of two metal loss flaws with the same length and maximum depth. The B31G level 1 method (Equation 40.1) predicts that both flaws will have the same burst pressure, but the level 2 effective area method accounts for the actual flaw profile.

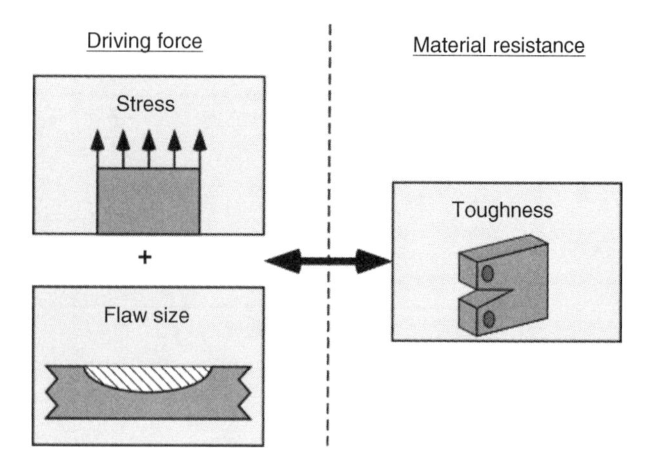


**FIGURE 40.7** The three key variables in a fracture mechanics analysis. The stress and crack size influence the driving force for fracture (i.e., the forces that are trying to propagate the crack) while the toughness is a measure of the material resistance to fracture.

series of RSF calculations using a modified version of Equation 40.1.

Note that in the level 1 procedure already discussed, only the length, $L$, and maximum depth, $d$, are used in the calculation, while the effective area method accounts for the shape of the thickness profile. Figure 40.6 depicts two corrosion flaws with the same $L$ and $d$, but with different thickness profiles. The level 1 method would predict the same burst pressure for the two flaws, but the effective area method would account for the additional reinforcement in flaw A, and thus predict a higher burst pressure for this case.

## 40.3 CRACK ASSESSMENT

Cracks can pose a greater threat to pipeline integrity than metal loss flaws. A sharp crack is a much more severe stress concentrator than a metal loss flaw of the same length and depth. Cracks and other planar flaws result in a significant decrease in burst pressure for materials that are notch sensitive (Figure 40.4). Moreover, cracks can grow in service by fatigue or through environmentally assisted corrosion mechanisms such as SCC.

Fracture mechanics is a relatively new engineering discipline, which predicts the effect of cracks on structural integrity [5]. There are three key variables in a fracture assessment, as Figure 40.7 illustrates. A typical fracture mechanics model consists of an equation that relates stress, flaw size, and toughness. Given one equation and three unknowns, it is necessary to specify two quantities in order to calculate the third. For example, if stress and toughness are known, it is possible to calculate the critical crack size for fracture.

Note that the variables in Figure 40.7 are grouped into two categories: *driving force* and *material resistance*. The driving force, which is a function of stress and crack size, is promoting crack propagation, while the toughness is a measure of the ability of the material to resist crack propagation. Fracture occurs when the driving force exceeds the material resistance. Fracture is more likely as either stress or crack size increase, since both quantities affect the driving force. As an analogy, consider plastic deformation (yielding) in metals. In this case, the applied stress can be viewed as the driving force for yielding, while the yield strength is a measure of the material resistance to yielding.

A suitable fracture mechanics model can be used to compute the burst pressure in a pipe, given the crack dimensions and material toughness. Figure 40.8 is sketch of a crack stability diagram for longitudinal cracks in a pipeline. For a fixed material toughness, burst pressure is plotted against crack length for part-through surface cracks of various depths. The diagram also shows the burst pressure versus crack length curve for a through-wall crack. Note that the predicted failure pressure for short, deep cracks is less than the failure pressure for through-wall cracks of the same length. The physical meaning of this result is that the remaining ligament of the short, deep crack can fail, but the resulting through-wall crack is stable, in which case a leak occurs rather than rupture of the pipe. If the failure pressure for a part-through crack is above that of the corresponding through-wall crack, the analysis predicts that the pipe will rupture when the critical pressure is reached.

### 40.3.1 The Log-Secant Model for Longitudinal Cracks

During the Battelle research that resulted in a model for volumetric metal loss flaws, the authors [4] observed that the

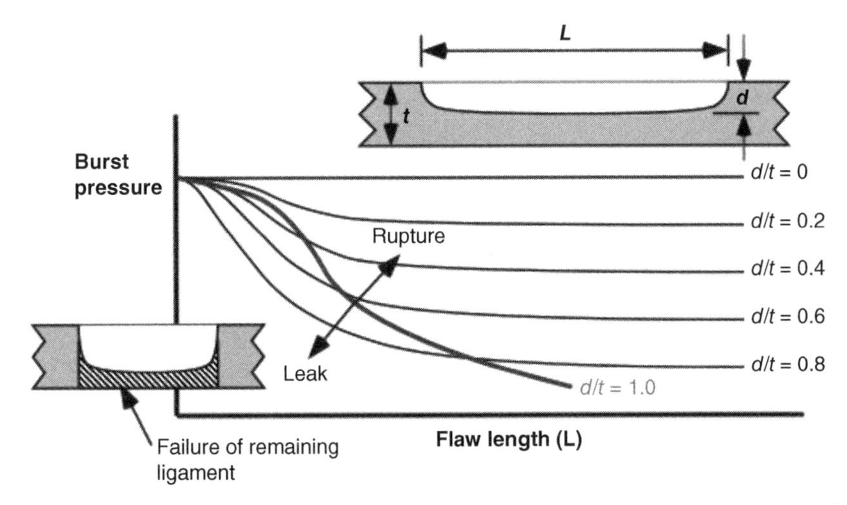


**FIGURE 40.8** Crack stability diagram that is a plot of burst pressure versus flaw dimensions.

metal loss equation overpredicted the burst pressure of pipes with sharp notches, especially for pipes with low Charpy toughness. They concluded that a burst pressure equation for crack-like flaws needed to account for material toughness.

At the time of the Battelle project (circa 1970), the field of fracture mechanics was in its infancy, and the fundamental understanding of failure of structures containing cracks was limited. A rigorous model for axial cracks in pipelines was simply beyond the state of knowledge at the time. The Battelle research team developed a semiempirical burst pressure equation for longitudinal cracks that was based on a simplistic model that was published in 1966 [6]. The original model, as well as the Battelle equation, contained a secant function as the argument of a natural logarithm. Consequently, the Battelle equation became known as the *log-secant model*. This model predicts the burst pressure versus toughness trend depicted in Figure 40.2, and it reduces to the B31G metal loss equation when toughness is large.

The log-secant model has been used extensively by the pipeline industry since it was first published. Over time, weaknesses in the model became apparent. For example, the model is excessively conservative for long, shallow flaws. In fact, this model predicts that a superficial longitudinal scratch on the pipe ($d/t \rightarrow 0$) will result in a large drop in burst pressure relative to a pristine pipe.

In 2008, Kiefner [7] published a modified version of the log-secant model that was intended to address shortcomings in the original equation. While the modification helped to mitigate the problem with long, shallow cracks, it introduced another problem. Namely, the modified log-secant equation predicts that burst pressure is insensitive to toughness, which is an obviously incorrect result. Figure 40.9 is a plot that compares the burst pressure predictions for the original and modified log-secant equations.

The field of fracture mechanics has advanced considerably since 1970, and better models are available to predict the relationship between burst pressure, crack size, and toughness in pipelines. The modified log-secant model does not incorporate recent advances in fracture mechanics. Rather, it merely applies a multiplying factor to the original equation. For reasons described ahead, the failure assessment diagram (FAD) approach is vastly superior to both the original and modified log-secant models for assessing crack-like flaws in pipelines.

### 40.3.2 The Failure Assessment Diagram (FAD)

The FAD method is based on sound fracture mechanics principals. This methodology appears in a number of Standards, including API 579 [2] and the British Standard BS 7910 [8]. A recent study [9] demonstrated that the API 579 FAD method was significantly better at predicting burst test results than were the original and modified log-secant methods.

In addition to being based on modern fracture mechanics principals, the FAD method is more versatile than the log-secant methods. The FAD method applies to both longitudinal flaws and circumferential flaws, but the log-secant methods pertain only to longitudinal flaws. Also, the FAD method accounts for the differences between cracks on the OD versus ID of the pipe, and it incorporates weld residual stresses when appropriate. The log-secant methods do not distinguish between ID and OD cracks, nor do these methods account for weld residual stress.

Figure 40.10 schematically illustrates the FAD. Given a particular pressure and crack size, the pipe is assessed by calculating the toughness ratio (the $y$ coordinate) and the load ratio (the $x$ coordinate), and then plotting the point on the FAD. In simplified terms, the toughness ratio is a check on

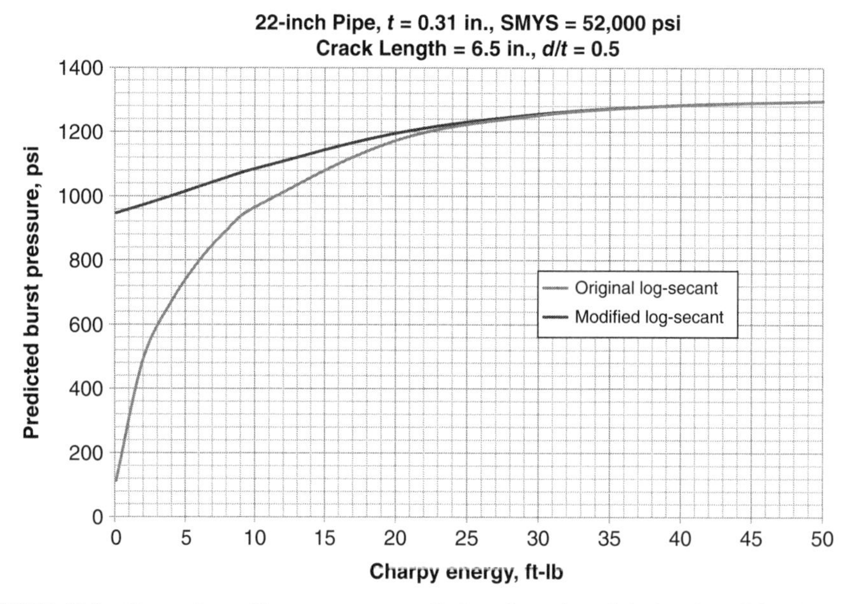


**FIGURE 40.9** Comparison of burst pressure predictions from the original and modified log-secant models.

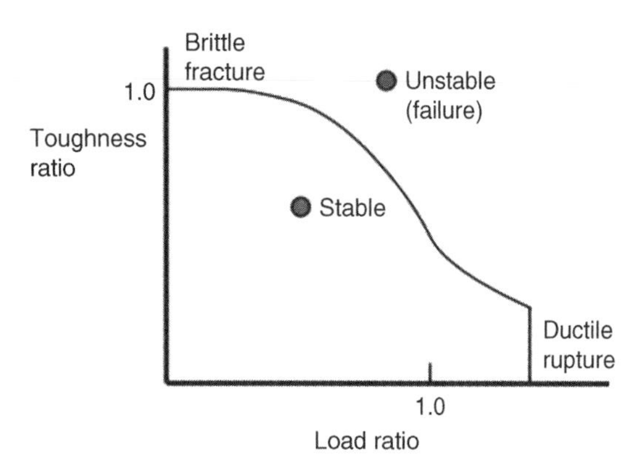


**FIGURE 40.10** The FAD. If the assessment point falls within the FAD curve, the flaw is stable. Failure is predicted when the assessment point falls outside of the FAD curve.

brittle fracture,[1] while the load ratio is a check on yielding and ductile rupture. The FAD curve interpolates between the extremes of fully brittle and fully ductile failure. If the assessment point falls inside of the curve, the crack is predicted to be stable. Failure is predicted if the point falls outside of the curve.

The assessment point is computed for a given pressure and crack size. An iterative procedure is required to compute the critical crack size. First, the assessment point is computed for an initial guess of the critical crack size. If the point falls inside the FAD curve, the crack size is gradually increased until the point falls on the curve. If the initial guess falls outside of the FAD curve, the crack size is decreased until the assessment point falls on the curve.

Figure 40.11 illustrates the procedure for determining burst pressure with the FAD method. Given an arbitrary pressure, $p$ that results in an assessment point $A$, the burst pressure is inferred from the distance from the origin to $A$ relative to the distance from the origin to $B$, which lies on the FAD curve:

$$p_{burst} = p\frac{OB}{OA} \tag{40.5}$$

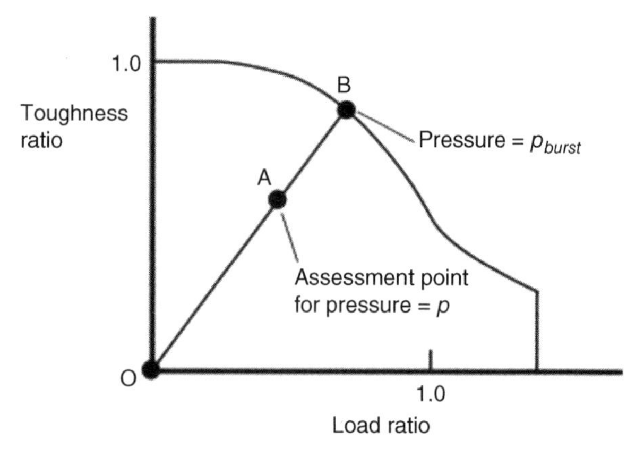


**FIGURE 40.11** Calculation of burst pressure with the FAD method. Given an arbitrary pressure, $p$, the ratio of burst pressure to $p$ is equal to the ratio of the line segments OB and OA.

[1] In the present context, "brittle" means fracture without significant plastic deformation. This macroscopic definition differs from the microscopic view, where "brittle" corresponds to cleavage fracture. A macroscopically brittle fracture ($L_r \rightarrow 0$ on the FAD) may or may not occur by cleavage. By the same token, cleavage fracture may be preceded by macroscopic plastic deformation.

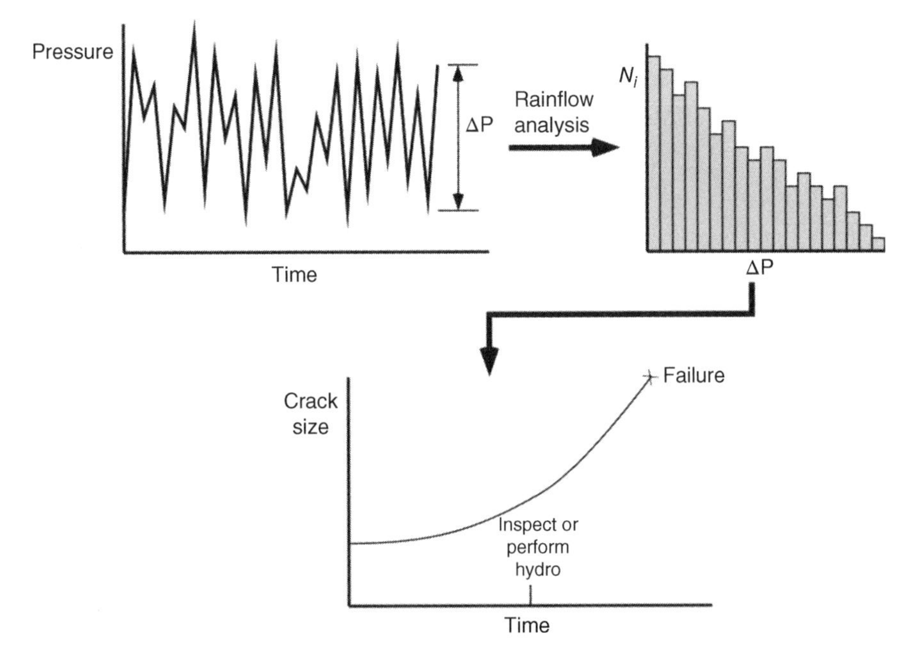


**FIGURE 40.12** The steps involved in a pressure cycle fatigue analysis. Pressure versus time data are processed through a rainflow cycle counting algorithm, resulting in a cyclic pressure histogram. The histogram is then input into a fatigue crack growth analysis.

A variation of this procedure is used when weld residual stresses are included in the analysis.

### 40.3.3 Pressure Cycle Fatigue Analysis

Preexisting flaws in a pipeline can grow by fatigue when the pressure oscillates frequently. Fatigue crack growth is predominately a problem in liquid lines but can also occur in gas lines.

In addition to predicting the critical conditions for burst in a pipe that contains a crack-like flaw, fracture mechanics concepts can be used to predict the rate at which a crack grows due to cyclic pressure. Figure 40.12 illustrates the steps involved in pressure cycle fatigue analysis. First, the pressure versus time data are processed through an algorithm called rainflow counting [10]. This results in a histogram of cyclic pressure ($\Delta p$) versus number of binned cycles at each $\Delta p$. The cyclic pressure histogram is then input into a fracture mechanics-based fatigue crack growth model. An initial crack size must also be input into this model, along with relevant material properties. The initial flaw size can be based on ILI data or it can be inferred from a hydrostatic test. That is, if a section of a pipeline passes a hydrotest, then it is possible to infer the size of the largest cracks that could have survived the test. The fatigue crack growth analysis predicts how long it will take for the flaw to grow from the initial size to a critical size at which point failure is predicted.

The calculated remaining life is typically used to set the interval for the next ILI run or hydrostatic test. In most cases, this interval is set to half the estimated remaining life.

## 40.4 DENTS

Dents are the most common source of failure in pipelines, but they are the least understood flaw type. Although there are established models and methods to assess metal loss and cracks, dents are usually handled with rules-based approaches. Existing regulations for dents are based primarily on experience and general rules of thumb.

Leaks and ruptures at dents are usually caused by fatigue.[2] The dent results in a stress concentration,that magnifies cyclic stresses. A 3D finite element analysis of the dented pipe, such as shown in Figure 40.1(a), is usually required to estimate remaining life. Simple formulas based on dent dimensions are insufficient for estimating accurate stresses for life assessment.

The primary dent characterizing parameter in rules-based dent acceptance criteria is the ratio of the dent depth to the pipe diameter. This parameter does not tell the complete story of a dent, however. The sketches in Figure 40.13 depict two dents of the same depth/diameter ratio. In one case, the dent is

[2] Mechanical damage that occurs when the dent is created can compromise the coating and lead to SCC in some cases.

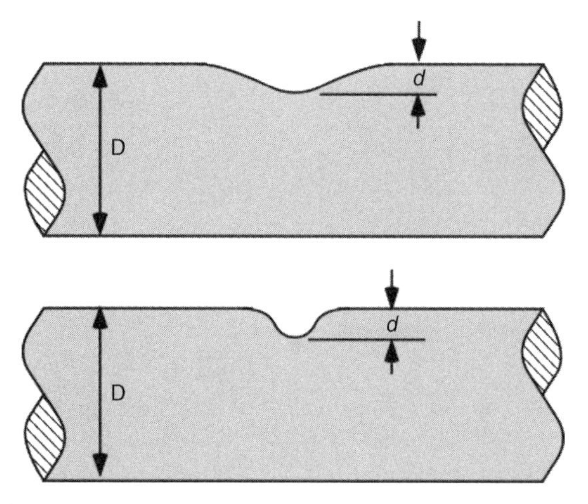


**FIGURE 40.13** Sketches of two dents with the same depth/diameter ratio but with different shapes. The sharper and more abrupt dent will have a shorter fatigue life than the smooth and gradual dent.

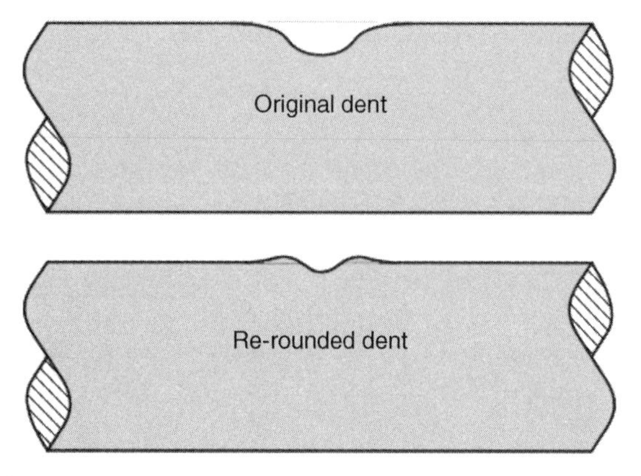


**FIGURE 40.14** Illustration of an unconstrained dent. The shape and depth of the dent changes over time, as it re-rounds with pressure cycling.

smooth and gradual, while the dent is sharper and more abrupt in the other case. The latter dent has a greater SCF than the former, and will have a shorter fatigue life.

The constraint on the pipe can also affect fatigue life, along with the size and shape of the dent. Figures 40.14 and 40.15 illustrate unconstrained and constrained dents, respectively. An unconstrained dent tends to re-round with pressure cycling, as Figure 40.14 illustrates. Consequently, by the time a dent is detected by ILI, it may look considerably different than when it first formed. In the case of a constrained dent, such as a rock dent at the 6 o'clock position (Figure 40.15), the dent typically retains its original shape. Although an unconstrained dent that has re-rounded often

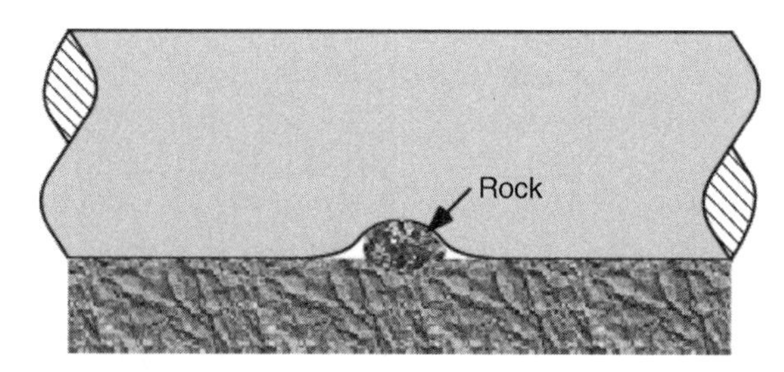


**FIGURE 40.15** Illustration of a dent that is constrained by a rock at the bottom of the trench in which the pipe was buried.

looks less severe than a constrained rock dent, the latter may actually have a longer fatigue life. The constraint on the dent that prevents it from re-rounding with pressure cycling also reduces the stresses in the dent.

Dent constraint tends to vary with position around a buried pipe. A dent at the 12 o'clock position usually experiences less constraint from the overburden relative to a dent at 6 o'clock that is resisted by a solid foundation.

## REFERENCES

1. American Society of Mechanical Engineers (2012) ASME B31G-2012: Manual for determining the remaining strength of corroded pipelines, October 2012.
2. American Petroleum Institute and the American Society for Mechanical Engineers (2007) API 579-1/ASME FFS-1: Fitness-for-service, June 2007.
3. Det Norske Veritas Recommended Practice DNV-RP-F101: Corroded pipelines, October, 2010.
4. Kiefner, J.F., Maxey, W.A., Eiber, R.J., and Duffy, A.R. (1973) ASTM STP 536: Failure stress levels of flaws in pressurized cylinders, American Society for Testing and Materials.
5. Anderson, T.L. (2005) *Fracture Mechanics: Fundamentals and Applications*, 3rd ed., Taylor & Francis, Boca Raton, FL.
6. Burdekin, F.M. and Stone, D.E.W. (1966) The crack opening displacement approach to fracture mechanics in yielding materials. *Journal of Strain Analysis*, 1, 145–153.
7. Kiefner, J.F. (2008) Modified equation helps integrity management. *Oil and Gas Journal*, 106(37), 76–82.
8. British Standards Institution (2013) BS 7910: 2013: Guide to methods of assessing the acceptability of flaws in metallic structures, December 2013.
9. Anderson, T.L. (2013) Pipeline crack assessment: new technology versus old-school methods. Proceedings of the 25th Pipeline Pigging & Integrity Management Course, February, 2013, Clarion Technical.
10. American Society for Testing and Materials (2005) ASTM E1049-85: Standard practices for cycle counting in fatigue analysis, ASTM International, 2005.

# 41

# INTEGRITY MANAGEMENT OF PIPELINE FACILITIES

Keith G. Leewis
*Dynamic Risk Assessment Systems, Inc., Calgary, Alberta, Canada*

## 41.1 OVERVIEW

Pipeline facilities in the design codes are generally considered to be the high-pressure pipe above- and belowground, inside the fence and not the rotating pumps and compressor equipment. Design standards cover pipe and associated headers, main line valves with sensors, controls and sometimes valve operators, meter station connections with pressure reductions, product heating and conditioning systems that operate at full line pressure inside the fence at multifunctional compressor and pump stations. ASME B31.4 and ASME B31.8 provide guidance for the design, testing, operations, and mitigation of pipelines. The rotating equipment, such as compressors, engine controls with the associated fuel feeds, any product treatment or chemical processing equipment, and other none pipeline facilities, are covered by other codes and regulations and not discussed in this chapter.

Public pressure to reduce pipeline risk increases with each incident. The oil and gas operating companies that are members of American Petroleum Institute (API), Canadian Energy Pipeline Association (CEPA), and Interstate Natural Gas Association of North America (INGAA) have stated that their integrity management goal is to reach zero incidents. The public and the regulators in Canada and the United States would also like to see zero leaks and rupture incidents, an enviable target. Safety management cultures are being supported through standards development and regulatory workshop. The liquid operators expect to release their API 1173 Pipeline Safety Management System Requirements soon, while Canadian Standards Association (CSA) has Z662 Appendix A for safety and loss management systems for pipelines, including facilities and ISO 55,000 Asset Management Standard and PAS-55 Asset Management provide alternate suggestions for implementing safety and loss management systems. These are overarching standards for management systems of which the integrity management facilities is a component.

Integrity management is a worthwhile investment of energies. The process minimizes business risks associated with accidents and loss of production. It maintains shareholder values and it maximizes the return on the investments. Many of the operators of petroleum facilities use robust risk-based inspection (RBI) standards such as API 581 and ASME PCC-3 to ensure safety and reliability.

In Canada, the pipeline design, operations, maintenance, abandonment, and integrity management expectations are all included in CSA Z662, and this code is incorporated by reference into the Canadian regulations. In the United States, the regulations 49 CFR 192 and 195 set out all the pipeline requirements and include the integrity management regulations. The federal language calls up selected sections of the ASME codes, B31.4, and B31.8, for the design, construction, operations, maintenance, and abandonment of pipelines, and then uses language in the integrity management standards, API 1160 (liquids), and ASME B31.8 S (gas), to support their periodic integrity assessment interval (beyond inspection) and improve the overall safety requirement. Readers can also find helpful guidance beyond these regulations, codes and standards, in the referenced American Society of Mechanical Engineers (ASME), Canadian Energy Pipeline Association (CEPA), and Pipeline Research Council International (PRCI) publications.

*Oil and Gas Pipelines: Integrity and Safety Handbook,* First Edition. Edited by R. Winston Revie.

## 41.2 OUTLINE

An integrity management program for gas pipelines and facilities or liquids pipelines and facilities, including storage tanks, is required to have the five plans as specified in ASME B31.8S:

1. Integrity management plan;
2. Performance plan;
3. Communications plan;
4. Management of change plan;
5. Quality control plan.

The key components of a hazardous liquid facility integrity management program are described differently in the liquids standards API 580, API 581, and ASME PCC-3, all for risk based design, and the key components are the same but detailed differently:

1. Program scope;
2. Corporate policies, goals, objectives;
3. Description of facilities;
4. Records and data;
5. Change management;
6. Competency and training;
7. Hazard identification and control;
8. Risk assessment;
9. Facilities Integrity Management Program (FIMP) program planning;
10. Evaluation of inspection, testing, patrols, and monitoring;
11. Mitigation and repair;
12. Continual improvement;
13. Incident investigations.

Even though the two may look vastly different as described, they require managing the same subcomponents to ensure that root cause of leaks and incidents are investigated and there is a continuous improvement system in place to ensure that these high-pressure facilities are well managed.

In Table 41.1, a listing of codes, standards, and practices used for managing the integrity of facilities is presented.

## 41.3 SCOPING A FACILITIES INTEGRITY PLAN

Many operators find that ASME B31.8S best helps scope the threats in the risk assessment process. Threats have always managed by operators the prevention and mitigation options found in ASME B31.8S Table 4. Consequences tend to be fixed by physics and for pipelines encroachment and these have different prevention and mitigation options. The Threat section discussion of B31.8S and Appendix provides helpful guidance for addressing each of the nine prescriptive threats and helps with understanding and scoping the consequences of a leak or rupture. B31.8S guidance helps determine the major threats to be addressed and verify that these have been a concern as supported by historical records. Operators should check that the failure analyses involved a root cause analysis to ensure the failures are properly classified in the historical records. An excellent listing of expected damage mechanisms that lead to failure can be found in PCC-3.

### 41.3.1 Generic Threats

An integrity assessment must address all nine prescriptive threats as outlined in ASME B31.8S and other national regulations, codes, and standards:

1. External corrosion;
2. Internal corrosion;
3. Stress corrosion cracking;
4. Manufacturing (pipe seam and pipe);
5. Construction (girth and fabrication welds, bends buckles, threads, broken pipe couplings);
6. Equipment (gaskets, O-rings, control, relief seal, and pump packing);
7. Third-party damage (immediate 1st, 2nd, and 3rd party, vandalism, previously damaged pipe);
8. Incorrect operations;
9. Weather and outside force (WOF), earth movement, (landslides, subsidence, faulting, earthquake, or volcanism) hydrotechnical (heavy rains or floods) extreme hot or cold weather, or lightening.

These threats may all be present in facilities, although we generally think of pipelines as being the primary source of problems. However, these nasty occurrences are not all happening out in the right-of-way. Other chapters in this publication provide helpful information on determining the magnitude each of these threats and assistance for prevention and mitigation strategies.

### 41.3.2 Interactive Threats

Failures can be caused from these nine threats directly or they can be interacting threats such as having some coating damage coincident with a resident threat, such as a manufacturing or construction or equipment imperfection. Minor imperfections are generally caused by poor quality control in the pipe mill, at the fabricator, in multiple handling transfers, and/or during the construction process particularly during the girth welding. Since these are left behind after construction, inspection and testing, they are collectively labeled resident

**TABLE 41.1 Codes, Standards, and Practices Used to Manage Facilities Integrity**

| Useful Codes Standards and Practices | Current Version—Focus | Overview |
|---|---|---|
| ASME B31.4 Pipeline Transportation Systems for Liquids and Slurries | 2013 Design operations and maintenance | ASME B31.4 covers piping transporting liquids between production facilities, tank farms, natural gas processing plants, refineries, pump stations, ammonia plants, terminals (marine, rail, and truck), and other delivery and receiving points.<br>Piping here consists of pipe, flanges, bolting, gaskets, valves, relief devices, fittings, and the pressure-containing parts of other piping components. It also includes hangers and supports, and other equipment items necessary to prevent overstressing the pressure-containing parts. |
| ASME B31.8 Gas Transmission and Distribution Piping Systems | 2014 Design operations and maintenance | ASME B31.8 covers gas transmission and distribution piping systems, including gas pipelines, gas compressor stations, gas metering and regulation stations, gas mains, and service lines up to the outlet of the customer's meter set assembly. It includes gas transmission and gathering pipelines, including appurtenances that are installed offshore for the purpose of transporting gas from production facilities to onshore locations; gas storage equipment of the closed pipe type that is fabricated or forged from pipe or fabricated from pipe and fittings; and gas storage lines. |
| ASME B31.8 S Managing System Integrity of Gas Pipelines | 2014 Risk management | This companion standard to B31.8 is generally for pipelines outside the fence and has the best overall guidance for assessment and criteria. ASME B31.8S is specifically designed to provide the operator with the information necessary to develop and implement an effective integrity management program utilizing proven industry practices and processes. Effective system management can decrease repair and replacement costs, prevent malfunctions, and minimize system downtime. |
| ASME PCC 2 Repair of Pressure Equipment and Piping | 2011 Repair systems | This standard provides methods for repair of equipment and piping within the scope of ASME Pressure Technology Codes and Standards after it has been placed in service. These repair methods include relevant design, fabrication, examination, and testing practices and may be temporary or permanent, depending on the circumstances. The methods provided in this standard address the repair of components when repair is deemed necessary based on appropriate inspection and flaw assessment. These inspection and flaw evaluation methods are not covered in this document, but are covered in other post-construction codes and standards. |
| ASME PCC 3 Inspection Planning Using Risk-Based Methods | 2012 Risk-based inspection | This is probably provides the most general guidance and there are selections for different facilities. The document provides guidance to owners, operators, and designers of pressure-containing equipment for developing and implementing an inspection program. These guidelines include means for assessing a risk-based inspection program and its plan. The approach emphasizes safe and reliable operation through cost-effective inspection. A spectrum of complimentary risk analysis approaches (qualitative through fully quantitative) should be considered as part of the inspection planning process. |
| API 579/ASME FFS-1Fitness for Service | 2009 Repair criteria | Fitness-for-service assessment is a multidisciplinary engineering approach that is used to determine if equipment is fit to continue operation for some desired future period. The equipment may contain flaws, have sustained damage, or have aged so that it cannot be evaluated by use of the original construction codes. API 579-1/ASME FFS-1 is a comprehensive consensus industry recommended practice that can be used to analyze, evaluate, and monitor equipment for continued operation. The main types of equipment covered by this standard are pressure vessels, piping, and tanks. |

*(continued)*

**TABLE 41.1** ***(Continued)***

| Useful Codes Standards and Practices | Current Version—Focus | Overview |
|---|---|---|
| API 580 Recommended Practice for Risk-Based Inspection | 2009 Risk-based inspection | API Recommended Practice 580 provides users with the basic elements for developing, implementing and maintaining a RBI program. It provides guidance to owners, operators, and designers of pressure-containing equipment for developing and implementing an inspection program. These guidelines include means for assessing an inspection program and its plan. The approach emphasizes safe and reliable operation through risk-prioritized inspection. A spectrum of complementary risk analysis approaches (qualitative through fully-quantitative) can be considered as part of the inspection planning process. RBI guideline issues covered include an introduction to the concepts and principles of risk-based inspection for risk management; and individual sections that describe the steps in applying these principles within the framework of the RBI process |
| API 581 Base resource Document on RBI | 2008 Risk-based inspection | This document is typical for refining, and petrochemical plants. It is very detailed and descriptive. It details the procedures and methodology in RBI. RBI is an integrated methodology that uses risk as a basis for prioritizing and managing an in-service equipment inspection program by combining both the likelihood of failure and the consequence of failure. Utilizing the output of the RBI, the user can design an inspection program that manages or maintains the risk of equipment failures. The following are three major goals of the RBI program:<br>Provide the capability to define and quantify the risk of process equipment failure, creating an effective tool for managing many of the important elements of a process plant;<br>Allow management to review safety, environmental, and business-interruption risks in an integrated, cost-effective manner;<br>Systematically reduce the likelihood and consequence of failure by allocating inspection resources to high-risk equipment.<br>The RBI methodology provides the basis for managing risk, by making informed decisions on the inspection method, coverage required and frequency of inspections. In most plants, a large percent of the total unit risk will be concentrated in a relatively small percent of the equipment items. These potential high-risk components may require greater attention, perhaps through a revised inspection plan. With an RBI program in place, inspections will continue to be conducted as defined in existing working documents, but priorities and frequencies will be guided by the RBI procedure. The RBI analysis looks not only at inspection, equipment design, and maintenance records, but also at numerous process safety management issues and all other significant issues that can affect the overall mechanical integrity and safety of a process unit. |
| API 650 Welded Tanks for Oil Storage | 2013 Design operations and maintenance | This standard establishes minimum requirements for material, design, fabrication, erection, and inspection for vertical, cylindrical, aboveground, closed- and open-top, welded storage tanks in various sizes and capacities for internal pressures approximating atmospheric pressure (internal pressures not exceeding the weight of the roof plates), but a higher internal pressure is permitted when additional requirements are met. This standard applies only to tanks whose entire bottom is uniformly supported and to tanks in non-refrigerated service, which have a maximum design temperature of 93 °C (200 °F) or less.<br>This standard is designed to provide industry with tanks of adequate safety and reasonable economy for use in the storage of petroleum, petroleum products, and other liquid products. This standard does not present or establish a fixed series of allowable tank sizes; instead, it is intended to permit the purchaser to select whatever size tank may best meet their needs. This standard is intended to help purchasers and manufacturers in ordering, fabricating, and erecting tanks; it is not intended to |

| | | |
|---|---|---|
| | | prohibit purchasers and manufacturers from purchasing or fabricating tanks that meet specifications other than those contained in this standard. |
| API 653 Tank Inspection Repair, Alteration, and Reconstruction | 2013<br>Repair | This standard covers steel storage tanks built to API 650 and its predecessor API 12C. It provides minimum requirements for maintaining the integrity of such tanks after they have been placed in service and addresses inspection, repair, alteration, relocation, and reconstruction.<br>The scope is limited to the tank foundation, bottom, shell, structure, roof, attached appurtenances, and nozzles to the face of the first flange, first threaded joint, or first welding-end connection. Many of the design, welding, examination, and material requirements of API 650 can be applied in the maintenance inspection, rating, repair, and alteration of in-service tanks. In the case of apparent conflicts between the requirements of this standard and API 650 or its predecessor API 12C, for tanks that have been placed in service.<br>This standard employs the principles of API 650; however, storage tank owner/operators, based on consideration of specific construction and operating details, may apply this standard to any steel tank constructed in accordance with a tank specification. |
| API 1160 Managing System Integrity for Hazardous Liquid Pipelines | 2013<br>Risk management | The document includes a description of an integrity management program that forms the basis of the Recommended Practice (RP). The RP also supports the development of integrity management programs required under 49 CFR 195.452 of the U.S. federal pipeline safety regulations. The guidance is largely targeted to the pipeline along the right-of-way, but the process and approach can be applied to pipeline facilities, including pipeline stations, terminals, and delivery facilities associated with pipeline systems. |
| CSA Z662 Oil and Gas Pipeline Systems | 2011<br>Design operations and maintenance<br>Plus risk management | This standard covers the design, construction, operation, and maintenance of oil and gas industry pipeline systems that convey liquid hydrocarbons, including crude oil, multiphase fluids, condensate, liquid petroleum products, natural gas liquids, and liquefied petroleum gas; oilfield water; oilfield steam; carbon dioxide used in oilfield enhanced recovery schemes; or gas. Note: Designers are cautioned that the requirements in this Standard might not be appropriate for gases other than natural gas, manufactured gas, or synthetic natural gas. |
| NEB Onshore Pipeline Regulations (OPR) | Current<br>Supplemental to Z662 | The National Energy Board Onshore Pipeline Regulations (OPR) was made under the National Energy Board Act. Companies are responsible for meeting the requirements of the OPR to manage safety, security, and environmental protection throughout the entire lifecycle of their facilities, from design, through to construction, operation and abandonment.<br>(1) When a company designs, constructs, operates or abandons a pipeline, or contracts for the provision of those services, the company shall ensure that the pipeline is designed, constructed, operated, or abandoned in accordance with the applicable provisions of<br>(a) these regulations;<br>(b) CSA Z276, if the pipeline transports liquefied natural gas;<br>(c) CSA Z341 for underground storage of hydrocarbons; and<br>(d) CSA Z662, if the pipeline transports liquid or gaseous hydrocarbons.<br>(2) Without limiting the generality of subsection (1), the company shall ensure that the pipeline is designed, constructed, operated, or abandoned in accordance with the design, specifications, programs, manuals, procedures, measures and plans developed and implemented by the company in accordance with these regulations. |

(*continued*)

**TABLE 41.1 (*Continued*)**

| Useful Codes Standards and Practices | Current Version—Focus | Overview |
|---|---|---|
| 49 CFR 192 Transportation of natural and other gas by pipeline: minimum federal safety standards | 2014 | These are the U.S. regulations that prescribe the minimum safety requirements for pipeline facilities and the transportation of gas, including pipeline facilities and the transportation of gas within the limits of the outer continental shelf as that term is defined in the Outer Continental Shelf Lands Act (43U.S.C. 1331). |
| 49 CFR 195 Transportation of hazardous liquids by pipeline | 2014 | These are the U.S. regulations that provide the expected safety standards and reporting requirements for pipeline facilities used in the transportation of hazardous liquids or carbon dioxide. |
| CEPA Facilities Integrity Management Program Recommended Practice | 2013 | This report outlines a comprehensive methodology to address the integrity management of oil and gas facilities. It covers more than the Integrity Management Program (IMP) and it requires the establishment and demonstration that there is a strong safety culture across the company. It requires the operator to set out and then ensure the contents are well documented, followed, and performance measures are being continually improved. |

threats. These minor resident imperfections or defects were all subjected to a post construction hydrostatic test. After 1970, every pipeline facility has met regulatory requirements; however older fabrications may not have been tested to the same rigorous high pressures. Therefore all the resident defects are qualified, but a lower pressure. A lower test pressure means these resident imperfections are larger in size and represent an easier deterioration initiation site.

If the performance of the cathodic protection system (coating and applied potential) is insufficient then corrosion may occur allowing the wall thickness to diminish over time. Historical mill and construction records[1] can indicate the possibility of systemic resident defects. Correlate mill and construction dates with the facility's original construction records. Humans have yet to build a perfect structure and there are always resident imperfections in the final product. All the other six threats can initiate damage at a resident imperfection. Land movement can act on imperfections such as porosity and inclusions in the girth weld, which make the weld strain sensitive, and eventually the strain load can cause the line to part at that girth weld. In other cases, the anticorrosion coating can deteriorate allowing corrosion to initiate at an imperfection site eventually leading to a leak.

Not all inspection techniques can find all threats. In most cases, the congestion inside a facility precludes the use of pigs for internal inspection, complicates the interpretation of aboveground external corrosion direct assessment (ECDA) electrical inspections and therefore requires other techniques such as guided wave in an excavation to locate wall thinning. A combination of ECDA principles, internal corrosion direct assessment (ICDA) principles, bell hole inspections with ultrasonic, guided wave, and X-ray inspections have been used to provide the inspections and help the operator complete the integrity and risk assessment of their congested facility.

### 41.3.3 Root Cause

Note that some of the inspection activities can address more than one type of threat and some threats are interactive. The root causes are not addressed; however, the possible root causes can be inferred in the Pipeline and Hazardous Materials Safety Administration (PHMSA) incident reports by scanning and interpreting a number of column entries for each separate incident. It would help if there was a movement to record the root cause rather than the leading suspect at the time. Sometimes it seems like a punishment to go back and update each record but this is necessary to discover systemic or other root causes. Root cause analysis can be used to design future facilities and in many cases make older equipment perform safer and better.

[1] Interstate Natural Gas Association of America, "Integrity Characteristics of Vintage Pipelines," October, 2004, www.INGAA.org.
ASME CRTD #43 History of Line Pipe Manufacturing in North America, 1996, www.ASME.org.

### 41.3.4 Failure Frequency

More than half the liquid releases reported are inside the fence of facilities (see PRCI L52317 in the essential reading). This is expected because these facilities are all very diverse and equipment is concentrated. There are more bolted joints and other mechanical fittings that can leak or rupture. Oil lines tend to have many more valves than gas systems and these extra valves assist in batching product for shipment. Releases are reported in to PHMSA and are generally small and compared to the rupture releases along the right-of-way, stay within the fence line on the operator property. As the number of pipeline ruptures trends down, operators expect that facility releases will become a much greater focus in performance management. Many incidents can be related to operator error and these threats require procedures and training to ensure these standard operating procedures (SOPs) are consistently followed in the assembly, operations, and repair of the whole range of components that make up the facility. These were some of the concerns that drove the move to incorporate safety and loss management systems to help get to zero failures.

More incidents occur along the right-of-way than within gas transmission facilities; however injuries, fires and explosions are much more common in gas facilities. This is to be expected because there are more mechanical joints that can leak or rupture, and it is much more likely that people are present when these incidents occur or are caused by activities. Like liquids, many of the incidents can be related to control and relief as equipment failures and incorrect operations. These threats require SOP for the purchase, installation, and maintenance. Periodic training helps ensure these SOPs are consistently followed in the materials supply, construction, operations, and repair of the wide range of components that make up the facility. Corrosion is also shown to be a small proportion of the incidents. While liquid lines report all leaks, the gas lines have a threshold below which consequences are small. Pipes seem to be associated with many of the incidents; however, the PHMSA records have insufficiently detail to develop a performance record for pipe or the wide variety of brands, fittings, and equipment used in oil and gas pipelines. In the past, operators have gathered to share incident data in an attempt to locate systemic problems.

## 41.4 SPECIFIC FACILITY THREATS

### 41.4.1 Fatigue

Vibration is typical at pump and compressor stations and the resulting larger physical displacement of a small diameter pipe concentrates these strains where the small pipe is welded to the larger diameter pipe. If the smaller diameter pipe is not properly restrained, vibration-induced displacement can lead to fatigue at the joint between the two eventually generating a leak or worse. In facilities there are many instances of small diameter pipes or appurtenances being welded to a much larger pipe. These connections are used as controls and bypasses to equalize pressures across valves or with parallel systems, to mount sensors for pressure and other controls, and inject methyl mercaptan, or to provide blow down protection.

### 41.4.2 Temperature

Product compression and expansion generates temperature differences. In stations, the discharge end of the compressor exhausts hot gas and this gas is cooled by using a large heat exchanger pressure vessel. Excess heat if not removed will damage the anticorrosion coatings on the pipeline downstream of the compressor station. Natural gas transfer connections also include a pressure reduction and this expansion necessitates auxiliary heating equipment to heat the cold gas.

### 41.4.3 Complexity

Pump and compressor station facilities contain redundant, complex, and congested interconnects especially if there are a number of parallel pipelines in the same right-of-way. Similarly, custody transfer and metering interconnects, pressure reduction equipment, and product conditioning facilities are located inside a facility with a perimeter fence. The yard plumbing consist of multiple runs of buried and above ground heavier wall pipe plus cast or forged fittings, complex ILI inspection tool launchers and receivers, valve complexes such as the suction and discharge headers feeding multiple compressors and pumps.

Main line valves are generally colocated with the corresponding smaller interconnection valves in tie-overs to looped adjacent parallel transmission lines. Today, more and more of these manually operated main line valves and crossover valves are being retrofitted with remote or automatic operators. In addition to the remote operators, pressure sensors, flow sensors, and valve monitoring electronics are retrofitted to speed control room decision making and reduce reaction times to line breaks. Early valve closure helps reduce the total discharge of product and if the discharged was ignited, help shorten the delay interval before the fire dies so that emergency responders can approach the rupture origin, and begin fire mitigation.

Hazardous liquid pipeline facilities have increased complexity especially if products are shipped in batches. Additional valving with associated control systems are required to introduce the different products in sequence while maintaining both flow and pressure. Similarly additional system complexity is required to separate the batches and deal with composition diffusional changes at the beginning and end of each product delivery.

## 41.5 FACILITY SAFETY CONSEQUENCES

Safety is paramount. Risk reduction is achieved by avoiding harming the people or environment near the probable incident location. The industry has a target of zero releases.

Encroachment increases as the number of structures outside the fence grow with time. These consequence factors are generally not in control of the operator. Sometimes these nearby structures also contain dangerous materials that can exacerbate the damages an incident. In Canada, both oil and liquids companies need to consider class location and this infers that periodic aerial and other surveys are needed to locate all the adjacent structures and ascertain the number of occupants, type of business, and if it contains hazardous contents.

Today, the majority of the facilities are unmanned. Larger regional facilities colocate the operator's regional office within the fence perimeter of the larger compressor/pump stations. Colocation means that clerks, operators, mechanics, corrosion techs, plus other employees will be present for some time throughout the day. Maintenance employees are present throughout the day, plus there is a large concentration and more variety of equipment that needs attention than out on the right-of-way. This higher employee presence means that the safety consequences of failure inside a compressor station are much higher than out along the right-of-way.

Over time encroachment has occurred, with neighbors sharing the road and living nearby facilities. These neighbors can become affected by an emergency at the facility and may require evacuation in addition to the employees. The potential impact radius (PIR) defines the damage perimeter of a circle for a given thermal radiation threshold. For natural gas fires, the PIR is defined in 49 CFR 192.903. There are two additional PHMSA reports that provide PIR calculation guidance when the burning gas has a different composition than natural gas. The major consequence of a natural gas rupture is a fire and people within the PIR are at risk for death or injury. Note fire damage to dwellings has sometimes exceeded the PIR distance.

Liquid spills are not round but form a more complicated polygon shape with the boundary controlled by the local geography and elevation changes as gravity drives the liquid perimeter. The intersection of the polygon determines if there is an impact to adjacent structures or locations of interest. Liquid leaks and ruptures may feed a pool fire and thermal radiation damage can spread as the pool drains downhill if the outflow is not naturally confined.

Simply, if a neighboring structure can be impacted by thermal damage from a burning source inside the facility then there is a risk. Similarly, if a polygon path from a leak interacts with a structure or crosses a potable water source, then those sites need to be part of the operator's emergency plan for the facility.

### 41.5.1 Environmental Consequences

Liquid petroleum crude oil and other refined products are more safely shipped by pipelines. If these pipelines leak or rupture then their contents represent a hazard to the potable water supplies, agricultural, scenic, wetland ecosystems, and so on. Emergency plans need to anticipate how to confine the damage and then remediate the damaged regions to help them return to some semblance of normalcy.

### 41.5.2 Business Consequences

Some lines, if shut in, will remove customers from their energy supply. Businesses depend on reliable supplies of energy and there are contractual penalties for interruptions. Identified sites as defined in ASME B31.8S and 49 CFR 192 Part O are locations where we expect to find individuals with impaired mobility such as daycares, retirement homes, hospitals, schools, prisons, and so on. Alternate energy sources would be required for extended durations especially to hospitals and prisons were evacuation is so much more complicated.

System reliability is affected if incidents occur at compressor stations. The effect of losing one or more stations due to lighting or other WOF cause will reduce the throughput all along the line. Although it might not look to be much, a 5% reduction in mass flow might have a very significant impact on the bottom line.

Sometimes the emergency site impacts vehicle and other transportation routes. Line breaks have been known to shut down, a high-tension electrical transmission lines, a rail line, or a major interstate highway. This related interactive damage is at a minimum an inconvenience for the general public and if extended can generate deteriorating general business consequences. Fire can spread and extend in area while the emergency responders are waiting for the fuel source to be shut in before they can safely begin to take control the fires and search for victims. Thermal radiation has been known to cause exposed pipelines or other containers of combustible material to overheat and also rupture greatly adding to the burning volumes of fuel and causing further delay and hindrance for emergency response and mitigation.

### 41.5.3 Reputation Consequences

News worthiness is a simple concept. The incident reputational damage factor increases from local newsworthiness, to regional news casts, to state wide TV footage, to national TV coverage, to worldwide TV coverage. The number of days that the incident remains in the news increases the adverse consequences. Although this difficult to quantify it is a very significant consideration in developing a complete description of the consequences to be used in a risk assessment of a probable event.

## 41.6 BUILDING A FACILITY INTEGRITY PLAN

The failure analysis helps to scope the integrity management plan for priority locations that will address 80% of the possible failure sites. Use the following series of steps to construct the plan:

- Study the facility drawings, including the design of the facility. These design parameters are essential in predicting where problems may be expected.
- Review the mass flow as material moves in and out of the system and other parameters for weak points.
- Produce a comprehensive inspection program.
- Visit the site to validate details in the inspection program.
- Supply an inspection plan overview, with essential inspection protocols, and procedures.
- Inspect to measure and develop a baseline survey of the facility.

API 581 and ASME PCC-3 describe RBI, and it is used in many facilities to inspect and measure, evaluate, and then help decide how to monitor or repair any large damage before it becomes a safety problem. A good system-based approach will attempt to include in the inspection plan the principles found in the simple PDCA circle principle:

- Plan—data gathering and process system analysis, select the failure modes, define areas with similar degradation modes, perform damage and corrosion (degradation) mechanism analyses, formulate the inspection test plan.
- Do—perform inspections and tests according to the already mentioned plan.
- Check—analyses the results, check compliance, perform fitness for service analyses, select mitigation, monitoring, or repair, plan start of next cycle.
- Act—implement actions decided in the previous phase.

Think of these four forming a continuous circle and each trip around the loop drives the risk ever down toward the goal of no leaks and ruptures. This system-based PDCA approach underlies most plans to provide continuous improvement.

PCC-3 addresses many generic varieties of facilities such as rotating equipment, pumps and compressors. It suggests inspection approaches for RBI programs. Rotating equipment generally includes the monitoring and detection of excess vibration and includes processes to detect accelerated wear by monitoring/analysis of suspended particulates in the lubrication systems. Similarly, not all connections are welded. In many cases flanges are used to join sections. Inspection includes the bolting sequence and bolt torque, for the range of compressed gasket materials. If these steps were done poorly the flanges become sources of leaks. Valves, pressure controls, pressure reliefs, and other mechanical systems provide safety but can also be a source of leaks if not inspected and maintained.

The incoming product sometimes contains gas composition impurities and condensed water that need to be removed before compression and these internal corrosion threats are discussed in detail in the pipeline threat segment of this book. However, treatment requires chemical conditioning technologies to remove water, $CO_2$, $H_2S$, and other corrosive impurities. These pressure vessel facilities are tall structures that consist of counter flow towers, absorption chambers, or other chemical processes to remove water, sour gas, and other harmful impurities. These are fed by pressurized headers to receive the incoming product and then conduct the cleaner product streams back into the pipeline system to be pumped downstream. Any chemical processes used for product conditioning such as the removal of water and $H_2S$ are not pipe, but follow pressure vessel designs using similar codes and standards. When using other pressure containment design standards, including those for bolted connections, the same risk-based principles found in B31.8S or PCC-3 can be consistently applied.

## 41.7 INTEGRITY ASSURANCE

There are four accepted integrity assessment methodologies:

- Pressure testing: Pressurizing the line by filling it with water, removing the air, and using a pump to raise the internal pressure to values above the maximum allowable operating pressure (MAOP). This internal pressure test ensures the pipeline is leak tight and as the pressure increases increasingly smaller imperfections are burst. With a successful test (now generally to 125% MAOP or even to specified minimum yield stress (SMYS)), the operator knows the maximum size of any surviving imperfections that still reside in the tested segment.
- In-line inspection: ILI or pigging can comprise several body structures to house magnets, sensors, recording, and processing information. Magnetic flux sensors read the increase in field lines due to local wall loss, ultrasonic sensors measure time of flight to determine wall thickness and look for cracks, piezoelectric require a couplant while electromagnetic acoustic transducer (EMATs) are coupled through a magnetic field. Geometry is measured mechanically and location in three-dimensional space is tracked by inertial guidance and above ground markers. API 1163 outlines the requirements for the ILI methodology.
- Direct assessment: Uses electrical resistance, voltage and/or current to determine the performance of the coating and applied voltage corrosion protection systems that mitigate to prevent external corrosion. Internal corrosion requires water to be introduced and possible

downstream traps or hold up regions are inspected for internal corrosion. If no corrosion is found in the last two susceptible locations then the line is considered free of internal corrosion until the next water introduction site. NACE standards outline the methodologies for the Direct Assessments; SP-0502 for ECDA, SP-0206 for DG-ICDA, and SP 0204for SCCDA.
- New technology: No technology has yet to be approved as a standalone integrity assessment system; however, there are two useful guided wave standards NACE SP0313 and ASTM E2775 for an external estimate of wall loss and magnetic tomography looks very promising.

## BIBLIOGRAPHY: ESSENTIAL READING

**API Standards** The following two API publications outline a methodology for Facility risk reduction.

**API 580** ***Recommended Practice for Risk-based Inspection (RBI)***

Increasing failure rates, combined with pressures in meeting regulatory and corporate compliance (safety, environmental, business), are causing the refining and petrochemical sectors, to place an increased emphasis on enhancing RBI.

This API standard contains the basic concepts of risk assessment typically targeted within the fence perimeter. There is a threat assessment process, a consequence assessment and a plan to manage risk through inspection and mitigation. There are three main approaches:

1. Simplified adjustment factor, which is less data intensive.
2. Reliability relies on inspection, design, materials, and operating data.
3. Appendix C provides a workbook for quantitative risk based inspection analysis.

This document is a supplement to API 510 Pressure Vessel Inspection, API RP570 Piping Inspection, and API RP653 Tank Inspection Repair Alteration. Qualitative risk analysis for facilities is analogous to system-wide qualitative risk assessment of pipeline assets, while Quantitative Risk Analysis supports optimization of risk mitigation through Risk Sensitivity Studies and Cost Benefit Analysis.

**API 581** ***Base Resource Document on RBI Levels of Analysis*** (builds on 580)

The API RBI procedure has three levels of analysis:

- Level I—screening tool that quickly highlights the high-risk equipment, which users may wish to assess in greater detail.
- Level II—a step closer to being a quantitative analysis than Level I, and it is a scaled down approach of Level III. Provides most of the benefit of Level III analysis, but it requires less input than Level III.
- Level III—quantitative approach to RBI providing the most detailed analysis of the three levels.

How does a 581 assessment help operators?

- Evaluate current inspection plans to determine priorities for inspections.
- Evaluate future plans for decision making.
- Evaluate changes to basic operations as they affect equipment integrity.
- Identify critical contributors to risk that may otherwise be overlooked.
- Establish economic optimum levels of inspection as weighed against risk reduction.
- Incorporate "acceptable risk" levels.

**ASME** PCC-3 *Inspection Planning Using Risk-Based Methods*

This is probably provides the most general guidance and there are selections for different facilities. The document provides guidance to owners, operators, and designers of pressure-containing equipment for developing and implementing an inspection program. These guidelines include means for assessing a RBI program and its plan. The approach emphasizes safe and reliable operation through cost-effective inspection. A spectrum of complimentary risk analysis approaches (qualitative through fully quantitative) should be considered as part of the inspection planning process.

**CEPA**: Canadian Energy Pipeline Association *Facilities Integrity Management Program Recommended Practice*, 1st Edition, May 2013 (available as a free publication downloadable from the CEPA web site).

This report outlines a comprehensive methodology to address the integrity management of oil and gas facilities. It covers more than IMP. It requires the establishment and demonstration that there is a strong safety culture across the company. It requires the operator to set out and then ensure the following chapter contents are well documented, followed, and performance measures are being continually improved:

- Scope;
- Corporate policies goals objectives and organization;
- Description of facilities;
- Records;
- Change management;
- Competency and training;
- Hazard identification and control;
- Risk assessment;

- Options for reducing uncertainty in the frequency or consequences of incident;
- Planning and execution;
- Repair and repair execution;
- Continual improvement.

The recommended practice is well illustrated and has multiple helpful tables and other suggestions.

**PRCI** Report L52317—Carolyn Kolovich and Harvey Haines (Kiefner & Associates, Inc.) Cheryl Trench (Allegro Energy Consulting) *Pipeline Facility Incident Data Review and Statistical Analysis*, December 2008.

This report provides an overview of the failure statistics to 2006 for different kinds of facilities and can be a help in determining the risk assessment.

# PART VI

# MAINTENANCE, REPAIR, REPLACEMENT, AND ABANDONMENT

# 42

# PIPELINE CLEANING

RANDY L. ROBERTS
*N-SPEC Pipeline Services, Business Unit of Coastal Chemical Co., L.L.C./A Brenntag Company, Broussard, LA, USA*

## 42.1 INTRODUCTION

Internal pipeline cleaning requires knowledge of any given segment(s) and their operating parameters, for example, pipe length, pipe diameter, bend radiuses, single or multi-diameter, types of valves, taps (in/out) and their o'clock position on the pipe, gas/liquid volume (velocity), operating pressure and maximum allowable operating pleasure (MAOP), and of course pig traps. The decision then is to determine if cleaning should be on-line or off-line or if mechanical cleaning or liquid (chemical) cleaning should be used. What type of pigs and cleaners should be used, and what is to be removed?

The list can go on and on requiring multiple decisions and a great deal of time. Thankfully, there are pipeline cleaning vendors available to guide one through this maze of obstacles and requirements. Additional factors in the cleaning decision should be based on the overall thought of what intelligent tool(s) are to be used and what data are required. The question to be answered is "how clean do I need to be for that particular in-line inspection (ILI) tool to secure a good run and receive maximum data?" A frequently asked question is what kind of debris is in the pipeline and how much? This question is unanswerable; clean is subjective, and the pipeline is dirty. There are no required industry standards to date, only actual cleaning results known throughout the industry, which have generally been found to be acceptable.

This chapter contains information on seven general topics on internally cleaning pipelines to prepare for running ILI tools and also to increase pipeline efficiency. Topics included in this chapter are contaminates, progressive pigging, pig types, mechanical versus liquid (chemical) cleaning, suggested typical pigging procedures, pipeline cleaners and diluents, and "how clean do I need to be?"

The pipeline operator is advised to include the local or known pipeline cleaner vendor for assistance and guidance on practices that the vendors encounter on a daily basis.

## 42.2 CONTAMINATES

In liquid lines (crude, NPL, product, or other) and natural gas (production, gathering, or transmission spec gas), the one common contaminates are iron compounds, commonly known as black powder. Black powder is by far the number one contaminate experienced in 99% of all pipelines [1]. Black powder is a name given in the pipeline industry in the early 1990s [2]. Black powder can consist of the following compounds in any combination and percentages by weight and/or volume (Cormier, Danny, Coastal Chemical Co., Broussard, Louisiana, Multiple analytical lab results, personal communication):

- iron sulfides;
- iron carbonate;
- iron oxide;
- iron chlorides;
- salts: calcium chloride, sodium chloride;
- free water ($H_2O$) mixed with carbon dioxide ($CO_2$) to form carbonic acid ($H_2CO_3$);
- glycols and ethanolamines, (~1–4 lb TEG/amine per mmscf treated);

*Oil and Gas Pipelines: Integrity and Safety Handbook*, First Edition. Edited by R. Winston Revie.

- corrosion inhibitors, flow promoters, methanol, biocides, construction debris, and so on;
- compressor and turbine oils and lubricants; (~1 gallon/1000 horsepower/day;
- FeS formed from $H_2S$.

Black powder is brittle and abrasive. The more the pipelines are mechanically (dry) pigged, the finer the black powder becomes, causing not only downstream nuisance problems, but also the compressing of these fine iron compounds into any pits and/or anomalies that may exist in the pipelines. The results of iron compound packing could deflect actual magnetic flux leakage (MFL) data indicating potential pipeline problems, or at a minimum, have the potential to mask the severity of possible anomalies.* Other contaminates not listed in the previous paragraph are welding rods, wooden skids, hard hats, grinders, animal bones, tools, pry-bars, bolts and studs, workman's gloves, sand, rocks, stuck pigs, parts of pigs, and others too numerous to list. Any and all these objects and contaminates must be removed prior to running any ILI tools, to protect the tool and to eliminate or minimize ILI sensor liftoff.

## 42.3 PROGRESSIVE PIGGING

Progressive pigging, simply stated, is running less aggressive pigs to more aggressive pigs in a sequential order based on the amount of debris found in the pipeline. The question then is to decide what is the least aggressive pig to start with? The general rule is, if the pipeline has never been mechanically pigged, then start with a 2-pound (0.91 kg) poly criss-cross pig. Pig types will be discussed in the Section 42.4. The 2-pound poly pig is one of the least aggressive types to run in the pigging world. They, as well as other poly pigs, can traverse 1.0D bends, require minimum pressure to launch, run on 5–10 $\Delta P$ pressure drop, and can be destroyed if lodged in the line with approximately 50–60 $\Delta P$ pressure exerted. Poly pigs are made using open cell urethane technology and are characterized in pounds of urethane per cubic foot ($lb/ft^3$). The higher the pounds, the tighter or harder the urethane.

The physical condition of the 2-pound poly should be assessed, looking particularly for deep folds, cuts, amount of debris, gashes, and wear of the pig. Mechanical pigging in gas and liquid lines is generally good for 100 km (~60 miles) to 130 km (~80 miles). When liquid (chemical) cleaning (synergistically using mechanical pigs and cleaning liquid products) is used, approximately 25% additional usage over mechanical is the rule-of-thumb. Once the pig condition evaluation is done, the same type of pig can be run again or the next level of aggressiveness type can be used.

An example would be using next a 5-pound (~2.27 kg) poly criss-cross wire brush pig then a 10-pound (~4.54 kg) type.

After running various types of poly pigs, then the next step would be to utilize steel body mandrel discs/cup type pigs. Careful attention should be noted that before running steel body pigs, the minimum bend radius must be at least 1.5D bends. Most pig manufacturers design their steel body pigs to have a pig length of 1.5 times the diameter, thus, the 1.5D minimum bends. In using the steel body types, the characteristics of discs/cup polyurethanes and durometer measurements [3] need to be understood. Polyurethanes and durometer tests will be discussed further in the Section 42.4. Within the steel body pig family of polyurethane type and durometer types, progress is normally from least aggressive to more aggressive and from sealing type to scraping type. Basically, the pipeline operator should work with the pig supplier to assist in these decisions.

## 42.4 PIG TYPES

There are many pig types and functions. As a rule, most pigs, of any type, are a standard designed length-to-diameter ratio of 1.5 times the OD of the pipe, that is, the 24 in. (~61 cm) pig is 36 in. (~91 cm) in length. This is why the lowest bend of 1.5D is important. If the pipeline has <1.5D bends, then consideration may be required to replace such bends with greater radius bends if the objective is to make the line piggable for ILI tools, or specially designed tandem pigs may be used. Pig types are of three basic designs: poly foam, unibody urethane, and steel mandrel discs/cups.

### 42.4.1 Poly Foam

Poly open cell polyurethane foam types are usually made the full OD of the pipeline, with little to no concerns for wall thickness. Poly pigs have the ability to negotiate short radius L-shaped bends, miter bends, tees, multidimensional piping, and reduced port valves. Foam pigs come in various densities determined in pounds of urethane per cubic foot, but the most common ranges are 2 $lb/ft^3$, 5–8 $lb/ft^3$, and 9–10 $lb/ft^3$ (1 $lb/ft^3$ equals 16 $kg/m^3$). These densities are usually color coded: yellow for 2 lb, red for 5 lb, and scarlet or blue for 10 lb depending on the manufacturer, but most follow these rules. The poly open cell is the least aggressive of the pig design family. They are great for sealing and light abrasion removal and can reduce in diameter up to approximately 35%. Length can be increased to allow maneuverability through large tees, some older designed ORBIT® valves, and other types of gate valves.

Wire strip brushes, nose pull rope, transmitter cavity, and jetting ports can be incorporated in each density and type of

* Actual field data noted after running MFL tool, then liquid cleaning, then running second MFL run and comparing data. Second MFL run showed greater anomaly pipe wall loss than first run. Dennis Breaux, N-SPEC Pipeline Services, Broussard, Louisiana.

poly foam pigs. Because of the many available configurations (poly criss-cross, poly criss-cross wire brush, bidirectional, bullet shape, and bare swab) of each density, the pipeline operator should check with the manufacturer's representative and the pipeline cleaning service company for help in developing a design to meet the requirements.

### 42.4.2 Unibody

The single-body cast-polyurethane pigs are designed to be more aggressive than polys but more forgiving than the steel body mandrel type. These pigs are effective in removing liquids from wet gas systems and in removing denser liquids from liquid pipelines (e.g., water from crude oil lines) and/or for displacement, and they also can be used to help control paraffin buildup in crude oil lines, as well as for separation of refined products, pipeline commissioning, and product evacuation. The unibody design can also maneuver in <1.5D radius L-shaped bends and is usually, but not limited to, a multi-disc cup configuration. The multidisc design in a bullet concave nose type or bidirectional type can have wire brushes attached along with other configurations and add-ons. The unibody cast polyurethane with hollow shaft can handle up to a 20% reduction in pipe ID [4]. These pigs can be cast in material of various durometer strengths, as will be discussed in Section 42.5.

### 42.4.3 Steel Mandrel

Steel body mandrel type pigs are the most aggressive type available made by any manufacturer. The configuration of the steel body allows for multiple designs for multiple usages.

Steel body mandrel pigs are built around a steel constructed mandrel. Three basic designs, cleaning pigs, batch and gauging, and conical cup, are usually available. In this chapter, only the cleaning pig type is discussed.

Cleaning pigs can be configured with all discs, disc with scraping cups, disc with conical cups, with any combination of all, and all types with various kinds of wire brushes and urethane scraper blades. Any of the cast polyurethane products can also be made from material of various strength levels as measured using a durometer.

Polyurethane discs are cast and molded to the desired diameter of the pipeline. There are basically three types of discs: sealing disc, scraping disc, and slotted disc. The sealing disc is usually thinner, $\leq$1 in. (25 mm), and is designed for low to medium scraping characteristics, but high on liquid sealing. The scraping disc is usually >1 in. (25 mm) in thickness and compared with the description of the sealing disc functions, just the opposite. Sometimes, a combination of both types is required. A slotted disc (or feathered type disc) is generally used on multi-diameter pipelines. A special design may be required for each pipeline condition. Considerations of pipeline length and pipe wall roughness to be pigged will also determine the type required. When all multi-type discs are used, the pig can also be used bi-directionally.

Just like the disc, cups come in two basic types: scraper cups and conical cups. Scraper cups are as the name implies but allow for greater surface forces to be exerted on the pipe walls especially in out-of-round pipe while maintaining the ability to seal. These cups can reduce on average 15–20% of the design diameter. Conical cups allow for maximum sealing with minimum scraping to remove solids. This type is normally seen on gauging plate pigs and multi-diameter and out-of-round pipelines. This type of cup can reduce up to approximately 30–35% and maintain adequate seal. Again, conical and scraper cups can be made in materials of various durometer hardness levels.

### 42.4.4 Polyurethanes

There are many types of polyurethanes, but only cast elastomers are discussed in this chapter. For more detailed information on various types of polyurethanes; see Ref. [3].

Mixing and pouring together two liquids, a prepolymer and a curative, make castable urethanes. There are basically two chemical structure types of polyurethane prepolymers:

1. Methylenebisdiphenyl diisocyanate (MDI)
2. Tolylenediisocyanate (TDI)

Both types use a curative and a prepolymer that, when mixed together, cause a chemical reaction forming the castable urethane. Each manufacturer has their own guarded ratio mixture, other additives, dyes, and process that differentiate them from each other in the market.

Following are some advantages of the polyurethane [3]:

1. Non-brittle (if kept out of ultraviolet (UV) rays for storage)
2. Elastomeric memory
3. Abrasion resistance

Some disadvantages are as follows:

1. Breakdown at high temperature, 220–225 °F (104–107 °C);
2. Moist hot environment (hydrolysis in the presence of moisture and elevated temperatures);
3. Certain chemical environments dissolve urethane (very strong acids and bases, aromatic solvents, for example, toluene, ketones, methanol, and esters);
4. UV exposure >6 months as a rule (cover and indoor storage prolong life).

A few differences between MDI and TDI are chemical makeup, but in general, MDI urethane is a little more expensive but is more durable, for example, on longer cleaning runs, >75 miles (>121 km), than TDI. However, TDI has a better compression set than MDI and functions at higher temperatures better than MDI do. Which type is best to use depends on the application.

## 42.5 DUROMETER [5]

Polyurethanes are characterized mostly by the Shore (durometer) test or Rockwell hardness test. The Rockwell test is usually used for "harder" elastomers, such as nylons, polycarbonate, polystyrene and acetyl, and so on. Shore hardness uses the Shore A or Shore D scale as the preferred method of testing for rubbers/elastomers (polyurethanes). The durometer Shore test indicates only the indentation made by the indenter foot upon the urethane. Other properties, such as strength or resistance to scratches, abrasion, and/or wear, are not indicated. The durometer measurement is expressed by a number system. The higher the durometer number the harder the urethane. TDI urethane is good in the range from 50 to 90A [3], MDI type in the 70–85A range. Combinations of materials with different durometer readings can be incorporated in a pig design to maximize desired conditions and/or results. The rule-of-thumb is the harder the durometer measurement, the better scraping capability; the softer the durometer measurement, the better the sealing characteristics.

## 42.6 MECHANICAL AND LIQUID (CHEMICAL) CLEANING

ILI tool companies require that data on all bends, diameters, wall thicknesses, ovality, and pipeline cleanliness be known before running their tool. Generally, either the ILI companies or other caliper companies will offer a caliper pig to be run first to retrieve this information. The multi-channel tool identifies multiple data points, welds, taps, valves, types of nineties, bends, direction of bends, wall thicknesses, and other data—all in the o'clock position with pipeline linear footage location.

The ILI tool companies have different tolerances for different tools, and pipeline operators need to discuss the required data for each. Once tolerances are known and approved by an ILI company, a date is scheduled to run their dummy tool, then the ILI tool. Most pipeline companies discover their pipeline is contaminated with solids and debris and needs cleaning during the installation of launcher/receiver and/or block valve replacement.

## 42.7 ON-LINE OR OFF-LINE

Once the decision to clean a pipeline is made, a decision is made on whether the line is to be cleaned on-line or off-line. ***On-line*** is defined as operating the pipeline under normal conditions while cleaning and ***off-line*** with the pipeline out of service and depressurized using other propellants to run the pigs.

As a rule, off-line cleaning can be twice as expensive as on-line cleaning due to additional required equipment and lost revenues from the out-of-service line. In general, the extra costs are caused by several factors: slower pig runs—generating more man hours, more cleaning runs, continuous nitrogen and air to propel the cleaning trains, and the fuel cost to generate that propellant over the duration of cleaning. An exception would be the use of natural gas at low pressure to propel the pig cleaning trains instead of nitrogen and compressed air. In either option, a cleaning program of a pipeline section <100 miles (~161 km) long can be expected to take 4–6 days of actual cleaning runs. Of course, this depends on the cleanliness condition of the pipeline that is required.

On-line cleaning allows the pipeline company to continue to operate and service their customers with uninterrupted service while cleaning their pipeline. This cleaning procedure is usually quicker, safer, and less costly than off-line. The general rule-of-thumb—velocity, for any size diameter pipeline, is >4 ft/s (1.2 m/s) but <15 ft/s (4.6 m/s). Although velocities >15 ft/s. (4.6 m/s) can be used, experience and pig manufacturers' studies have shown that, at those elevated speeds, hydroplaning of the pigs will occur in the presence of liquids, which causes greater blow by leaving greater volumes of liquid and entrained solids in the pipeline. Of course, the objective is to remove the solids and minimize free liquids in the pipeline, and so special cleaning procedures must be designed with the cleaning service company to address this concern.

## 42.8 CLEANING A PIPELINE

There are two types of pipeline cleaning programs available. One type is referred to as ***mechanical cleaning*** (the running of mechanical pigs dry progressively) and is the most common answer when asked, "Do you clean your pipelines?" The other cleaning program is ***liquid (chemical) cleaning***, discussed later. Most debris in a pipeline is in the 4–7 o'clock position due to gravity. In most cases, mechanical cleaning will only displace the debris from the 6 o'clock position to 360° around the pipe walls. Even if more mechanical pigs are run, the solids (i.e., black powder), iron compounds, and other organic and inorganic compounds can be broken down to sub-microparticles causing downstream problems, such as plugged meters, fouled turbine/compressor inlet filter elements, and the customer's treating process equipment. Again, the more one mechanically dry pigs, the smaller the particles can become. If the solids are iron compounds (iron sulfides, iron carbonate, iron oxides, etc.), these particles and sub-microparticles will be pressed by the pig's disc and cups, at pipeline pressure, into any pipe wall anomalies and/or pits, possibly interfering with the magna-flux readability of the actual pipe wall metal loss areas, especially if those iron compounds are magnetite.

### 42.8.1 Typical Pigging Procedures

Achieving greater solids removal from pipelines with fewer pig runs requires ***liquid (chemical) cleaning***. This type of cleaning is becoming more popular in the industry. Liquid cleaning in tandem with mechanical pigs will remove a greater volume of debris with fewer cleaning runs. Liquid cleaning by definition means the use of liquid cleaners mixed in a diluent (water, diesel, methanol, isopropyl alcohol (IPA), methyl ethyl ketone (MEK), etc.), to form a cleaning solution pushed through a pipeline using mechanical pigs. Most cleaning companies will calculate the volume of liquid solution to coat the interior walls based on a given diameter and length of pipeline with a coating thickness of 2–3 mils (0.051—0.076 mm). A volume of 10–20% of that calculated would be used as pipeline cleaner. A minimum of 10% by volume of pipeline cleaner is suggested not only to allow the cleaner to penetrate and permeate the solids, but also to have enough percentage volume to be able to carry the solids out of the pipeline. Usually, a minimum of 10% by volume is suggested. Diluents alone have no to little solids-carrying capabilities. The cleaner mixed with a diluent should be enough in volume to form a froth to keep the solids in suspension so as not to allow the solids to settle before being expelled from the pipeline. The other criterion for calculating the amount of cleaning solution is based on residence time to allow for enough contact time to penetrate the debris. A 10 s or greater resident time is the suggested required time, with 10 s being the minimum. This liquid column length is based on percent liquid cleaner in the solution, gas velocity, and/or liquid barrels per hour rate. For liquid product and crude oil lines, the minimum cleaner percent should be 20% by volume of the calculated 2–3 mil (0.051–0.076 mm) film based on the diameter and length of the pipeline, taking into consideration also the flow rates to achieve the minimum 10-s residence time.

### 42.8.2 Pipeline Cleaners and Diluents

There are various manufacturers of cleaners. However, a careful choice of designed pipeline cleaners should be based upon the following suggested characteristics:

1. Neutral pH;
2. Permeating and penetrating capabilities;
3. Original design parameters of the cleaner;
4. Case histories;
5. Seal(s) compatibility.

Pipeline acid cleaners that completely dissolve solids can be used, but may release harmful gases and could also attack pipe wall metals. Other liquid cleaners that dissolve iron sulfide **MUST** have water in them in order to facilitate the reaction. One popular liquid cleaner contains an active ingredient called tetrakis(hydroxymethyl)phosphonium sulfate (THPS), and can be formulated with an ammonium ion or organic phosphonate to speed up the dissolution of iron compounds [6]. THPS is well known as a highly effective biocide for a variety of water treatment applications, including oilfield downhole treating. However, in those applications, THPS was used in parts per million by volume (ppmv) quantities and only lately is being used in percentages with ethylene glycol to mask the water dew point effects in natural gas for the dissolution of some iron sulfide compounds. THPS dissolves certain iron sulfides by the mechanism of chelation, thereby avoiding the production of any insoluble by-products and any significant amounts of hydrogen sulfide.

Tests have shown that 1000 ml of 20% concentrate of formulated THPS solution will dissolve approximately 120 g of iron sulfides (two types tested: trolite [FeS] and pyrite [$FeS_2$]) [6]. In other words, 1 gallon of 20% THPS solution dissolves approximately 1 pound of iron sulfide. Although, THPS is environmentally benign and is completely and rapidly deactivated in the presence of free oxygen and/or contact with high-pH products (including alkaline corrosion inhibitors), there may be at least a few hazards associated with its use as a pipeline cleaner. Customers have reported that the use of THPS has coincided with increased volumes of natural gas odorants being added downstream of the cleaning process. THPS is suspected of neutralizing ethyl or methylmercaptans and tert-butylmercaptan odorants. A second issue has been that the THPS may, in the presence of arsenic, form arsine gas. "THPS under elevated water pressure, inside the pipeline, without question will cause the sulfates to easily ionize. Bacteria, if present, will consume the sulfate ion and convert it to a sulfide ion, which will cause any arsenic present to precipitate. Because phosphorus and arsenic have the same group characteristics, ion exchange will occur if arsenic is present. Phosphorus is unique since it can support five bonds while arsenic can support only four bonds. This is why there is a positive charge on the ion. In the presence of arsenic, the methylene hydroxide molecules will likely polymerize and form glycols or methyl alcohol. Also, if welding on a pipeline with arsenic and THPS has been used, and arsenic has precipitated, arsine gas will easily be formed" (Charles W. Williams Ph.D., Gibsons Environmental Services, USA Personal Communication, 2005 & 2013).

Another liquid cleaner is the gel type. Gels are very good carriers, due to their viscosity, but rely heavily on mechanical pigs to disassociate solids from the pipe walls, and then the gels carry the debris out of the line. Thousands of feet, even miles of gel used with mechanical pigs are not unusual while cleaning. Temperature, concentration, and pH can affect the stability of the gel formation.

Liquid cleaning can be carried out with two types of cleaning products, liquid and gel, with distinct physical property advantages/disadvantages of both. The term ***liquid cleaning*** is used because the term ***chemical cleaning*** implies the use of acids and caustics. Even though inhibited acids may be required under special conditions, most liquid cleaners are based on a surfactant and on the relationship between the viscosity and shear rate of the gel [7]. Both types, liquid and gel, require different cleaning procedures, number of personnel, amount of equipment for the project, and the disposal/recycle cost concerns of spent cleaners and solids. Both disposal and recycling, considered off-pipeline costs, are issues that must also be discussed as parts of an overall total package. Off-pipeline costs consist of cleaning of contractor's separation equipment, cleaning of frac tanks, disposal of urethane pig parts, and any third party confined space service charges to complete these tasks.

## 42.9 HOW CLEAN DO I NEED TO BE?

This brings us to another most asked question in our industry. "What does 'clean' mean?" A more realistic question would be, "What degree of cleanliness is required?" Today, there is no known required level of cleanliness; only a suggested level depending on pipeline diameter and type of ILI tool to be run, which at best is a suggestive opinion. Also, pipeline companies are realizing, as a result of cleaning prior to ILI tool runs, that an additional benefit of cleaning is the cost saving realized due to increased gas and liquid flow with less horsepower required. Service pipelines can be cleaned to a level better than mill-grade new pipe steel and internally coated pipelines [8]. Typical surface roughness data are shown in Table 42.1.

Pipeline companies are performing pre-cleaning efficiency tests and post-cleaning efficiency tests, and the results have been very revealing.

**TABLE 42.1 Typical Surface Roughness of Internal Pipeline Surfaces**

| | Surface Roughness | |
|---|---|---|
| Pipeline Condition | μ-in. | μm |
| Internally coated (new) | <250 | 6.3 |
| Standard non-coated (new) | ≤1800 | 46 |
| Non-coated after 10 years (average) | ±4200 | 110 |
| Potential corroded pipe (internal) | ~18,000 | 460 |

*Source:* Data in μ-inches from Ref. [9].
*Note:* 1000 μ-in. = 1.0 mil (1/1000 of an in.) [0.0254 mm]

### 42.9.1 Single Diameter Pipelines

The degree of cleanliness on single diameter lines is the least stringent for most ILI tool venders. Field experience, of this author's pipeline cleaning company, has been based on the appearance of cleaning pig(s), both for mechanical cleaning only and synergetic liquid (chemical) cleaning using both cleaning solutions and mechanical cleaning pigs. The appearance for mechanical cleaning is little to no solids in the brushes, little to no folds or major creases on the poly pigs, wear and tear on the steel body discs/cups urethane, and no large metal objects on the magnets. Also, the amount of or lack of solids and/or liquids (appearance) on the back of the pig is also an indicator along with zero to little objects brought in by the pig(s).

The appearance of the cleaning pigs used while liquid (chemical) cleaning is the same as mechanical mentioned earlier, but with an obviously cleaner appearance. These pigs may look as good as they were before launching. The added detection indicator in liquid cleaning is the spent cleaning solution itself. Generally, any solids level below 10% by volume, with the degree of clean pig appearance mentioned earlier, is considered acceptably clean for MFL tool runs. Again, the pig magnets should also be void of any large metal objects that could cause sensor liftoff.

Cleanliness for ultrasonic type and some circumferential tools is the most stringent. All the aforementioned conditions for liquid cleaning apply, but with the exception of two areas. First, the percent of solids in the spent cleaning solution should be 2% by volume or less versus 10% or less. Second, the magnets on cleaning pigs, in conjunction with the 2% or less solids in the cleaning solution, must be visibly very clean with only US quarter size (~25 mm) or less spots of metal fines on the magnets. The metal fines will look like buttons on the surfaces of the magnets.

### 42.9.2 Multi-Diameter Pipelines

In multi-diameter pipeline segments, cleanliness is equal to that in single diameter lines in mechanical cleaning, with the exception that the transitional areas will be hard to clean but not impossible. Standard designed poly foam pigs have an open cell urethane memory, and the transition zone from smaller pipe diameter to larger diameter means that the cleanliness area will depend on the urethane density and pig speed velocity. Coming from the smaller size diameter, the poly pig requires time to expand to its original size. The distance not cleaned by the pig is based on the gas velocity (pig speed). It is not uncommon for the pig to travel 60–80 ft (18–24 m) before fully expanding. These results have been proven by caliper tool data on these segments after cleaning. One recommendation when using poly foam pigs is to utilize multi-density pigs with the heavier density in the core and the lighter urethane density surrounding the core. Less dense open cell urethane expands quicker. Down side, multi-density polies are not as aggressive. More mechanical cleaning runs may be required.

The concerns expressed earlier while using poly pigs also relate to steel body discs/cup cleaning pigs for these multi-diameter lines. Designed flexibility in the discs and cups means less aggressiveness, but the upside is the shorter transition zones due to the quick ability of the pig to expand. On average, caliper data have shown the uncleaned transitional areas to be in single digit feet (<2 m). For mechanical cleaning, the cleanliness of the segment is still based on pig appearance and caliper data.

Again, all things being equal, the appearance of metal fines on magnets is the same as previously mentioned for the single diameter parameters, both for mechanical and liquid cleaning procedures.

The big difference between the two types of cleaning programs (mechanical only versus liquid synergetic cleaning) is that on liquid cleaning the appearance of the cleaning pigs remains the same, but the solids loadings of the cleaning solution is 4% or less with again the magnets having only quarter-size metal deposits attached. For ultrasonic and some circumferential tools, the acceptance is the same as for single diameter segments.

## 42.10 SUMMARY

In summary, the pigging process consists of the following steps:

1. Cleaning the pipeline;
2. Running a gauge plate pig;
3. Running a caliper tool;
4. Making any pipeline corrections if required;
5. Running the ILI dummy tool (optional with approval of the ILI Company);
6. Running the ILI tool.

The overall objective in pipeline integrity is to retrieve accurate data on the internal and external condition(s) of the pipeline. For maximum data results, the internal condition(s) of the pipeline can be improved by liquid cleaning to greatly increase the chances of a more reliable and accurate ILI tool performance. Cleaner pipelines yield better data from successful ILI tool runs.

Given the aforementioned field cleanliness parameters for both single and/or multi-diameter pipelines, the end results are still based on subjective visual appearance(s) of cleaning pigs along with field tests of solids removed in cleaning solutions. These are the best practices today, based on confirmation from caliper tool data. It is also true that the cleaner the requirement, the higher the cleaning cost. The higher the degree of cleanliness achieved, the higher the pipeline efficiency. The higher the efficiency, the lower the operating cost will be. The cleaner the pipeline, the better the chances for greater ILI tool data results, with better pipeline efficiency operating conditions as an added benefit.

## REFERENCES


1. Roberts, R.L. (2009) Pipeline pigging and cleaning: what do you really know about it? *Pipeline and Gas Journal*, 236(10), 30, 57 and 236(10), 30, August and October.
2. Baldwin, R.M. (1998) Technical assessment: black powder in the gas industry—sources, characteristics and treatment. Report Prepared for the Gas Machinery Research Council, Report No. TA 97-4, Southwest Research Institute, San Antonio, TX, May 1998, pp. 6–4.
3. Fuest, R.W. (1983) What Polyurethane? Where? Selecting the Right Polyurethane for Various Applications, Polyurethane Manufacturers Association, Milwaukee, Wisconsin, Volume 105, pp. 6–9, 11 (1983).
4. Girard Industries (2000–2012) Product Literature, TURBO Series. Houston, TX.
5. Material Property Data (2014) Available at http://www.matweb.com.
6. Gilbert, P.D., et al. (2002) Tetrakishydroxymethylphosphonium sulfate (THPS) for dissolving iron sulfides downhole and topside—a study of the chemistry influencing dissolution. Proceedings of CORROSION/2002, NACE International, Houston, TX, Paper No. 02030.
7. Union Carbide Corporation (1999) CELLOSIZE® Hydroxyethyl Cellulose. Union Carbide Corporation, Danbury, CT 06817-0001, USA, Brochure, 1999, pp. 16–17.
8. Michini, A., Mori, M., and Roberts, R.L. (2008) *On the Way to Implementing Our Integrity Program, We Discovered Pipeline Efficiencies beyond Our Expectations*, Pipeliners Association of Houston, Houston, TX.
9. Smart, J. and Winters, R.H. (2011) Black powder assessment. Proceedings of Pipeline Pigging & Integrity Management Conference, February 14–17, 2011, Houston, TX.

# 43

# MANAGING AN AGING PIPELINE INFRASTRUCTURE

Brian N. Leis
*B N Leis Consultant, Inc., Worthington, OH, USA*

## 43.1 INTRODUCTION

This chapter considers the circumstances and processes involved in managing an aging pipeline infrastructure. Background to this topic is presented in terms of the high-level factors that drive the pipeline industry, the products it transports, and some of the challenges it faces as supply basins become increasingly remote to the markets. Thereafter, the circumstances and evolution of technology that determine the construction processes and define the initial condition of the infrastructure are reviewed. These aspects are then discussed and evaluated in regard to the threats to pipeline integrity. The development and role of Codes and Standards is outlined as is the role of the regulatory framework in regard to actions that affect system integrity. Finally, the processes involved in managing an aging pipeline infrastructure are reviewed as the basis to identify process limitations and related technology gaps. The chapter closes with the finding that the current process that prioritizes on defect size could make a step improvement through use of local properties, but this is constrained by a technology gap. A near-term stopgap is suggested, as elaborated in the chapter.

## 43.2 BACKGROUND

The earliest pipelines transporting oil and natural gas in the United States first moved natural gas from a local source to and between nearby houses through hollow logs in Fredonia, NY circa 1825 [1], while oil was transported first near Oil City, PA through a short 2-in. diameter cast-iron system in the early 1860s [2]. Pipelines came into use in Canada in 1862, moving oil from the Petrolia field near the city of Sarnia, ON into the city, with natural gas being transported first by pipeline a year later moving the gas through a cast-iron system into Trois Rivières, Québec [3]. Pipelines have since been increasingly involved in raw energy transportation. However, when lower volumes were required, petroleum products such as oil also were moved from the wellhead to nearby railheads via horse-drawn carts. In those modes of transportation the product was contained in large wooden barrels, whose significance remains apparent today as the unit measure for the quantity of liquid being transported, or marketed. However, use of these transport modes diminished as the volume transported grew with demand, and the distance to markets increased.

Demand grew as petroleum-based products were found increasingly merchantable for lighting and other purposes, and as the population expanded, which as time passed became focused in larger urban centers. As the demand grew, the locally available supply from nearby shallow wells dwindled, which forced deeper wells and eventually led to the discovery and development of major supply basins. With demand growth, and the easily accessible product exhausted, the supply basins have become increasingly remote to the markets. More critically, in many such cases the desired resources (i.e., oil and natural gas) often lie in reservoirs associated with other aggressive fluids and gases, with higher temperatures involved for the deeper wells, whereas very low temperatures and ice or permafrost complicate Arctic exploration and production (E&P).

While many factors can be involved, where large volumes of oil or gas had to be transported over long distances, pipelines became recognized as the most economical

*Oil and Gas Pipelines: Integrity and Safety Handbook,* First Edition. Edited by R. Winston Revie.

mode of transportation from the wellhead to the market. Such considerations paved the way for pipelines to replace the early modes of shipping, although the barrel remains the metric for the quantity of liquid delivered. Pipelines remain the best method to transport such products when feasible. However, where the geography or other factors make this mode impractical, transport by ship was adopted. As onshore resources near to market became inadequate, E&P moved offshore and into sometimes very remote pristine areas, such as Alaska or further out into the Gulf of Mexico—with safety in those settings being more in jeopardy. This is evident from Figure 43.1 (Stiff, J. (February 12, 2014) private communication, ABSG Consulting), which presents the risk involved in oil and gas E&P and that of other industries in terms of incident likelihood on the *y*-axis and their dollar consequences on the *x*-axis, wherein offshore E&P lies along the upper bound of these metrics.

Onshore in the lower 48 US states, pipelines are used to move the imported as well as domestic supplies to market, which is also true onshore in other countries around the world. As the supply basins became increasingly remote, a system of high-capacity pipelines was developed to transport crude and refined products, and natural gas, from the supply points to the market centers, hubs, or terminals. Thus, a cross-country infrastructure of pipelines developed that traverses throughout the United States and adjacent countries, with eventual cross-border interconnects. Such development continues today in the United States, particularly in regard to shale E&P and related processing, and also worldwide, wherever demand must be tied into supply.

At present, the natural gas transmission system comprises more than 210 natural gas pipeline systems and in excess of 300,000 miles of interstate and intrastate pipelines [4]. The system that is committed to moving crude and other hazardous liquids also involves a large number of operators but is somewhat shorter, being in excess of 170,000 miles [5]. As contrasted elsewhere in detail [6], differences exist in the design and components of gas versus liquids transmission systems because of key differences in the products being transported, and the circumstances involved in maintaining efficient operation. Critical in this context is compressibility and the major differences in the phase boundaries of the transported products in pressure–temperature–volume space. Equally important are the terms and conditions of the tariffs and product quality specifications, and the need to keep liquid water out of the systems to manage internal corrosion (IC).

Transmission pipeline systems exist "mid-stream" in the oil and gas value chain that begins "upstream" with E&P and ends "downstream" with a distribution network to the end user. As such, a transmission system comprises end-point and distributed storage in various forms, as well as prime-movers, pressure controls, and other related support facilities that include a supervisory control and data acquisition (SCADA) system, and measurement and regulator stations. These system elements can be discriminated in regard to their location and function. These elements are either (1) located at discrete sites

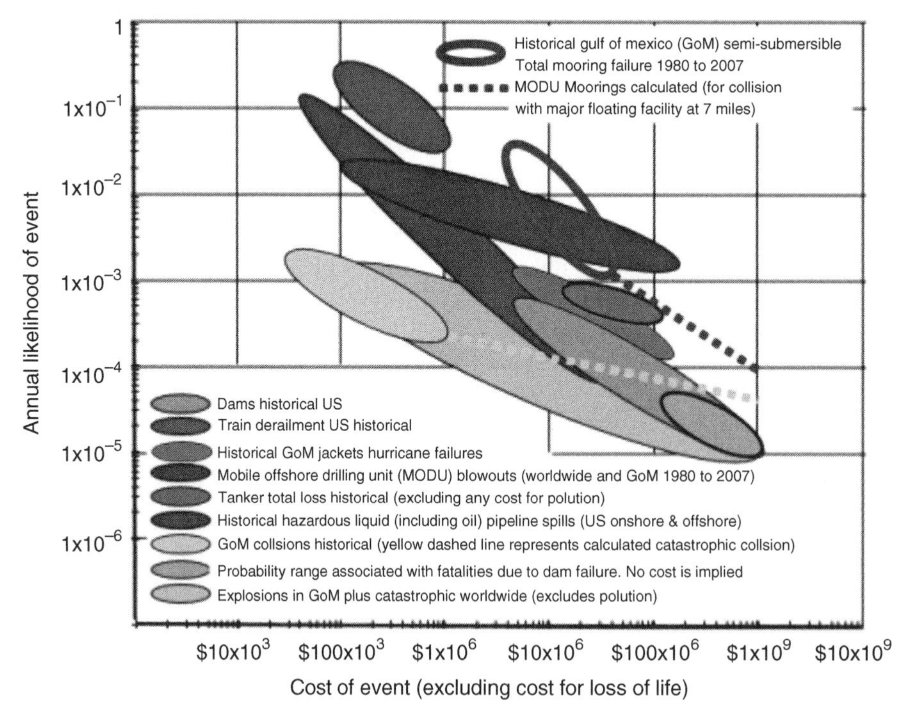


**FIGURE 43.1** Risk in energy development versus other scenarios. (Adapted from Ref. (Stiff, J. (February 12, 2014) private communication, ABSG Consulting).)

(e.g., facilities, compressor/pump stations, storage) or (2) are part of the often remote lineal aspects (e.g., pipeline, main-line valves). In regard to function, and its implications for integrity, the elements either do not involve moving parts (i.e., they are static, such as the line pipe) or they involve moving parts (like prime movers and other equipment). This chapter is specific to the network of transmission line pipe, and how that line pipe was built into a pipeline.

Logically, supply basins were sought as close to the markets as possible. Pipelines constructed between these points tended to develop first along routes that were most easily traversed, which also targeted the largest markets—tracking the so-called path of least resistance. Early pipelines often traversed open country, such as farm fields. Construction made use of ditching machines and hand shovels, with mechanized equipment used as it became available. The pipeline was constructed beside the ditch as a string and then lowered onto the ditch bottom. A-frames were used to lower-in the pipe string, which gave way to side-booms beginning in the 1930s. Where the ditch bottom was smooth, it also served as the foundation for the pipeline, with native soil returned to the ditch as bedding and padding if needed, as well as cover for the pipeline.

Given that the history of pipelines in the United States traces back more than a century, it is instructive to contrast the changes over time, as these affect all aspects of pipelines, from routing and design to construction and operations and maintenance (O&M). Figures 43.2–43.4, adapted from Ref. [7], provide perspective for such changes and indicate the timeline over which these practices were prevalent, specifically in regard to the United States. These figures, respectively, indicate the evolution in the type of steel used to make the body of the line pipe, how that steel was manufactured into line pipe, and how that line pipe has been joined, or otherwise built into pipeline systems. To facilitate later discussion of the evolution over time, each figure has been divided into three timeframes—up through ~1940, which is labeled therein "early," between 1940 and 1970, which is labeled "vintage," and beyond 1970, which is labeled "modern." The last in this sequence of dates, 1970, has been chosen in part because that is when the current regulatory framework was introduced in the United States, and in part because that date can be associated with a transition in steel and pipe making. The first of these dates is somewhat more arbitrary, being chosen to reflect a transition in many aspects of construction, the beginning of a period of change in regard to steel and pipe making, and the onset of a period that opened to the construction of a significant portion of the current pipeline network. Following discussion in regard to Figures 43.2–43.4, this chapter focuses on the pipeline network prior to what has been termed the "modern" era.

## 43.3 EVOLUTION OF LINE PIPE STEEL, PIPE MAKING, AND PIPELINE CONSTRUCTION

Figure 43.2, adapted from Ref. [7], illustrates aspects of the evolution of the steel-making processes used to make line pipe, over a timeline from circa 1900 through modern technology. The steel-making process is central to pipeline design and its in-service integrity, as it affects the chemistry of the steel, which coupled with processing affects the microstructure

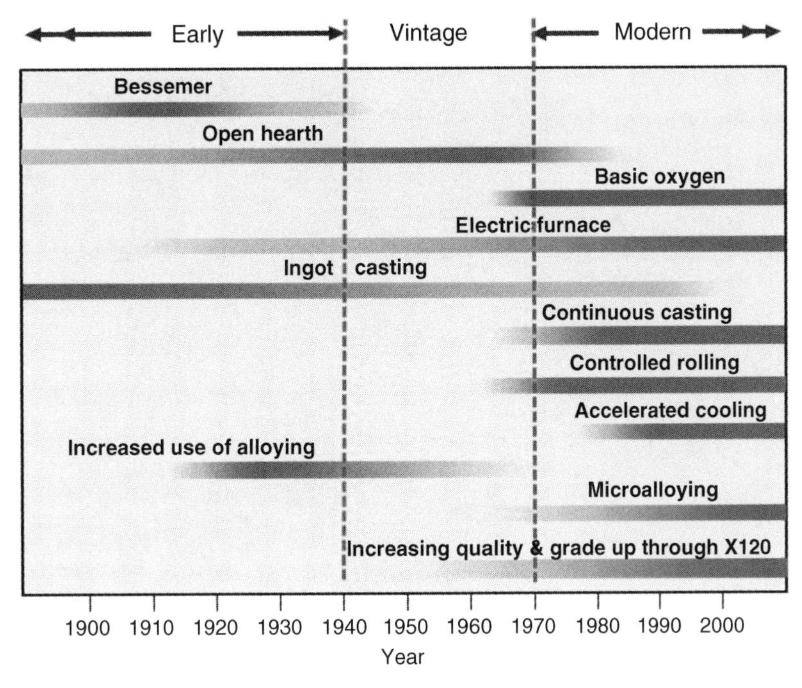


**FIGURE 43.2** Evolution of the steel-making processes. (Adapted from Ref. [7].)

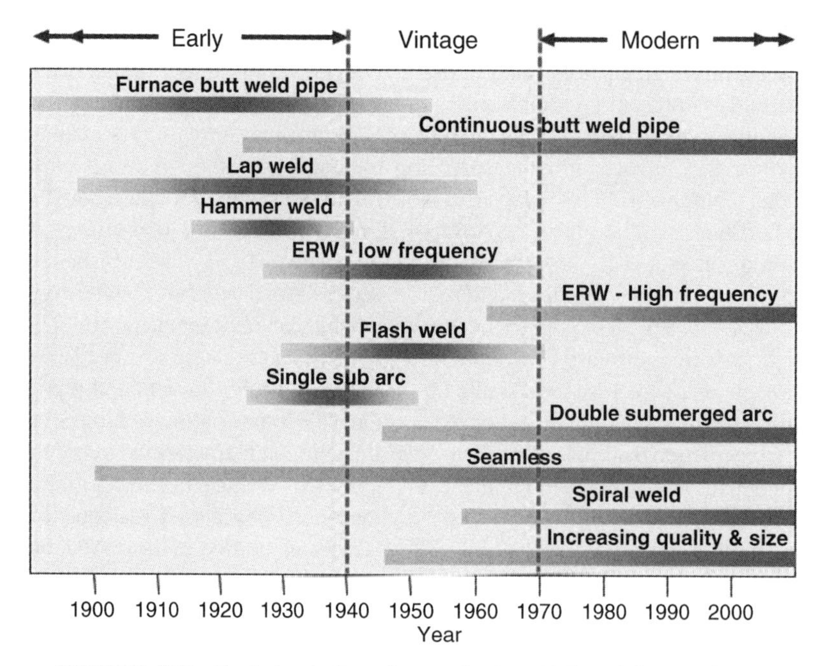


**FIGURE 43.3** Evolution in line-pipe production. (Adapted from Ref. [7].)

and the presence and location of impurities or undesirable constituents (cleanliness). All such aspects couple to control the steel's flow and fracture response, details for which can be found in Ref. [8].

Significantly, the processing of the steel in regard to rolling and other thermal mechanical (TM) controlled processing can have a beneficial effect on the flow and fracture response, but also can have a negative effect—depending on the TM conditions and history. What follows is a thumbnail sketch of such aspects, which as mentioned earlier have been addressed elsewhere in regard to pipeline applications in great detail [8].

Steel manufacturing in the United States began with the introduction of the Bessemer process in the mid-1800s, which found use for the production of butt welded, lap welded, and seamless (SMLS) pipe. The open hearth process

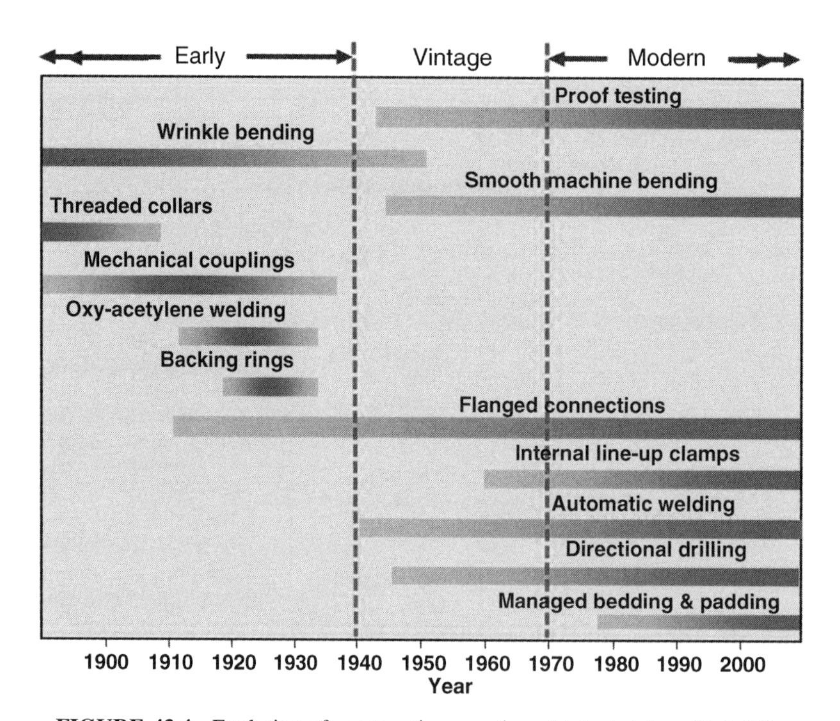


**FIGURE 43.4** Evolution of construction practices. (Adapted from Ref. [7]).

quickly followed (1870s–1880s) and evolved into the primary method for steel production in the world by the early 1900s. These developments were followed by the electric furnace (1899), which began production between 1900 and 1910. Subsequently, basic oxygen steel-making was implemented by the mid-1950s, and by 1969 accounted for nearly half of the annual steel production. Of these processes, steel produced for line pipe through the 1960s included the open hearth, basic oxygen, and Bessemer processes, with electric furnace steel used only in limited quantities.

Prior to the 1960s, ingot casting and hot rolling were typically used to produce plate and skelp for line pipe production. Both partially deoxidized (semi-killed) and fully deoxidized (killed) steels were used, and depending on the deoxidation practice, ingot structural soundness varied. During this same period, higher yield strength materials were introduced, wherein the primary method used to increase strength was increased alloy content, which tended to increase the hardness and reduce the weldability. Pre-1960 steels have higher residual impurity levels and more frequent internal anomalies as compared to post-1970s production. In many cases, the impurities are strung-out parallel to the pipe surfaces, which while they have limited influence on strength, they can act as initiation sites for some forms of corrosion and/or cracking.

From the late 1960s and into the 1970s, several major improvements were implemented in steel manufacturing and in plate/skelp rolling. Microalloyed steels with additions of niobium, vanadium, and other elements, coupled with improved steel rolling practices (controlled rolling) and improved impurity controls (desulfurization, inclusion shape control, vacuum degassing), resulted in "cleaner" steels with higher yield strengths, increased toughness levels, and improved weldability. Continuous casting (concast) began to be used in the same time period, further improving steel quality, while providing more efficient production. Nevertheless, centerline segregation in the concast strand can remain a concern in demanding applications (e.g., [8]).

Further developments in steel production and rolling practices came in the 1970s and 1980s through control of steel microstructures, additional rolling method improvements (accelerated cooling), and chemical additions. These gave rise to higher yield strengths, quantified by the "Grade" of steel, which means stronger pipe for the same wall thickness. For line, pipe Grade is denoted by the prefix X followed by the specified minimum yield stress (SMYS) in ksi units. These changes in the steel also led to fewer impurities, and improved weldability. Sophisticated steel manufacturing controls and improved nondestructive inspection (NDI) systems have also resulted in a reduction of material-related anomalies found in modern line pipe, with Grade X100 now considered a commercial product. Key in this context is that steel has evolved to address the needs of the pipeline industry with significant quality and related developments being triggered by the need to manage concerns such as brittle fracture and running ductile fracture.

Development was also spurred by the desire to manage capital expenditures (CAPEX), which is evident in the increase in grade as mentioned earlier, which allows reduced wall thickness and so increased productivity on the spread during construction. It is also evident in changes in the rolling mills to provide wider plate and skelp, which permit producing larger diameter pipelines, which diminishes the time on the spread, as well as the scope of related O&M work, to help manage operating expenditures (OPEX).

Figure 43.3 illustrates the evolution in line-pipe production. This involves transforming flat stock as plate or skelp into either a cylindrical shape that is closed by a longitudinal seam, which can be straight or helical, or a SMLS cylinder. Depending on how the line pipe is produced, and the need for dimensional control, pipe can be cold expanded up to 1.5% following completion of the pipe-making process. Thereafter, it is tested and accepted, then beveled and otherwise prepared, in accordance with contract and other specifications and regulations. Early pipe making processes noted in Figure 43.3 include furnace butt-welded, lap and hammer welded, low-frequency (LF) electric resistance welded (ERW), flash weld, and continuous butt-welded, with the latter still in use.

Subsequently used processes included single-submerged-arc welded (SAW), and some early variations on SMLS pipe, with high-frequency ERW (HF-ERW), and double submerged arc welding (DSAW) coming thereafter. The DSAW process has been used to produce both straight seam as well as spiral (helical) seam pipe. Of these, modern construction relies on SMLS pipe and the HF-ERW and DSAW processes. In contrast, the "early" category involves several long-since-abandoned processes, as does the "vintage" category. The vintage category can be viewed as a period of transition away from steel and seam making practices that were prone to process variation and production upsets, which lead to seam-related defects. Thus, as noted in regard to the evolution of steel, changes have come in pipe making to improve quality. Changes also have been made in regard to capacity of the mills to produce larger diameter, and heavier wall pipe, as needed to manage CAPEX and OPEX.

Finally, in analogy to Figures 43.2 and 43.3, Figure 43.4 illustrates the evolution of construction practices that have been used to join the line pipe, and to create the directional changes needed to track a routing via bending. It is the evolution in practices to join line pipe, and to create the needed directional changes that account for the improvements in pipeline construction—much of which has been driven by the desire to control CAPEX upfront and the OPEX over the life cycle of the pipeline network. As early practices were found to be problematic in regard to managing CAPEX or OPEX, they were quickly discontinued, with the related issues promptly resolved. In this context, it is not a surprise that much of the vintage construction remains in service today without issue. For this reason, a later section illustrates novel construction practices, wherein their utility and continued serviceability are evident.

Early joining practices relied on threaded collars, which were used on the earliest crude oil pipelines, with mechanical couplings coming into use as well. The earliest welds were made via the oxy-acetylene process, which as time passed incorporated a backing ring below the girth weld. The use of flanged connections, which were approved by the ASME circa 1894 and require welding to attach the fitting to the line pipe, followed as the early welding technology was more broadly adopted. Of these, only flanged connections continued in use beyond the period denoted "early" in the figure. Figure 43.4 also indicates the role of wrinklebends, which were used with other historical field-bending practices, such as welded segmented (miter) bends, through the 1940s. As evident in this figure, the period denoted "vintage" also was a time of transition from the early practices onto those in use today—such as submerged arc and other more advanced methods that in modern applications are used under automated conditions and work in conjunction with line-up clamps. The transition into and through the vintage era on into the modern era also gave rise to the use of machine-made field bends, with such machines emerging in the 1940s and continuing to develop since, for straight as well as helical seamed pipe. Directional drilling also traces back to a similar timeline, and has recently become viable for long-distance drills involving large diameter construction, providing a viable eco-friendly approach to deal with routings sensitive to such issues. And, where difficult routings move through rocky and hilly terrain, bedding and padding and other trench-bottom technology have evolved largely in the modern era to manage such construction.

The transition through the vintage into the modern era opened to the use of cathodic protection, (CP) beginning in the 1940s, the use of which expanded until it was eventually mandated in the transition into the modern era. As the threat of corrosion was recognized, over-the-ditch coating methods also went into use in the 1940s. Step improvements in coating technology began with the early versions of today's resin-based formulations that appeared in the 1970s. Continuous improvement in these technologies, coupled with the realization that appropriate preparation was as critical as the nature of the coating system, and its application, have now effectively controlled external corrosion (EC). Likewise, tools and technologies have evolved to better manage IC, which range from designed inhibitors through internal coatings. Field applied coatings likewise have improved to deal with surfaces left uncoated in the mill to facilitate girth-welding.

Finally, the transition through the vintage in to the modern era gave rise to significant changes in pipeline system monitoring and control. The first monitoring via in-line inspection (ILI) came late in the transition from the vintage to modern era, in 1965, which since has gone through several generations of development, and expanded its initial role in regard to metal loss to deal selectively with cracking, mechanical damage, alignment, and deformation. SCADA introduced initially to better track product delivery was subsequently coupled with sensors, remotely controlled valves, and system-specific software and found other uses, such as tracking pressure and other operational parameters, which facilitate better control, and also leak detection and outflow control.

## 43.4 PIPELINE SYSTEM EXPANSION AND THE IMPLICATIONS FOR "OLDER" PIPELINES

The evolution in the aspects of pipeline systems that was apparent in Figures 43.2–43.4 has been "fueled" by the competitive need to manage both CAPEX and OPEX as pipeline companies expanded their capacity in the competition to meet the ever growing demand for energy. In this context, that competition to get off the construction spread and into production led to improved steels that could be produced in stronger grades that also were more weldable. Such steels also were tougher and had lower transition temperatures, whose use opened to mechanized more controllable welding processes. The push to build more efficient systems over the life cycle led to improved operational controls as well as better coating processes, and improved control of CP. As such, overall system safety has the potential to improve as a consequence of the evolution evident in Figures 43.2–43.4.

### 43.4.1 System Expansion and Construction Era

The extent of the expansion to meet the market demands is evident in Figure 43.5, wherein the incremental mileage constructed is shown on the left-$y$-axis as a function of the period of construction over the interval from before 1920 into the new millennium, which is shown on the $x$-axis.

The right-$y$-axis presents the cumulative mileage constructed over this same interval. Figure 43.5a shows these data for interstate natural gas transmission pipelines, whereas Figure 43.5b shows the corresponding data for interstate hazardous liquid transmission systems. The timeline in these figures denoted >2000 includes results up through 2010, while the timeline used in Figures 43.2–43.4 ends at the start of the new millennia. Little is lost in this context as that decade saw little technology growth—except for developments in inspection capabilities. Differences also exist between Figures 43.2–43.5 in regard to the start of the trending. But, again little is lost because the accumulated mileage prior to the 1920s represents a trivial fraction of the total natural gas and hazardous liquid mileage thereafter. Inspection of Figure 43.5a and b indicates that the "early" era comprises a nominal fraction of both the networks.

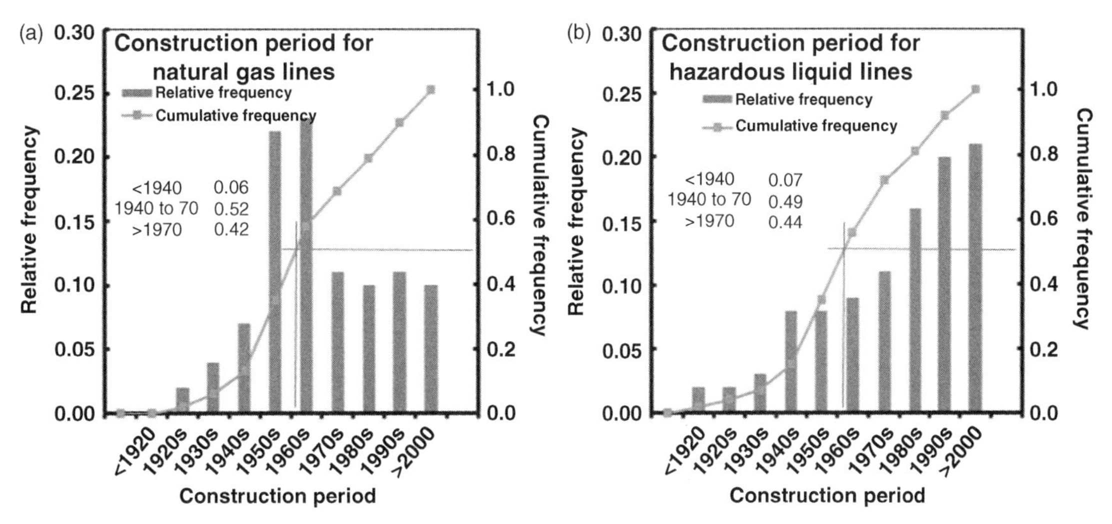


**FIGURE 43.5** Analysis of construction mileage relative to year built. (Derived from raw data in Ref. [9].)

It is apparent from Figure 43.5a that the natural gas system experienced strong growth beginning in the 1920s, which culminated in the fastest growth in the 1950s and 1960s at a relative rate that has since not been matched. The trends in Figure 43.5b indicate that the hazardous liquid system also experienced steady incremental growth since the 1920s, which led to the fastest growth in the 1950s and 1960s, that like the natural gas system its growth has been leveled and steady since the 1960s. As such, it is not surprising that the amount of mileage split between the periods designated in prior discussion as early, vintage, and modern, is comparable. This is evident by comparing the split mileage built within these three timeframes, which is apparent in the corresponding sets of numbers added into Figure 43.5a and b. For the natural gas system this split is 6% prior to 1940, with 52% from 1940 to 1970, and 42% since 1970, while for the hazardous liquid system construction prior to 1940 involved 7%, with 49% in the interval from 1940 to 1970 and 44% thereafter.

The horizontal lines shown in Figure 43.5a and b corresponds to one-half of the current total system mileage. The vertical lines that drop from its intersection with the cumulative mileage trend to the timeline shown on the $x$-axis indicate that for the hazardous liquid and natural gas transmission systems roughly one-half of those networks were constructed prior to ~1961, with about two-thirds of these networks being constructed prior to 1970. Recognizing that the amount of the network built in the "early" era is but a nominal fraction of the total network, it is evident that roughly one-half of the US transmission network was built in the era designated as "vintage" in the context of Figures 43.1–43.3.

Figure 43.5 indicates that the step increase in mileage for the natural gas system began in the 1950s, which is paralleled by the growth in the liquids mileage. While length provides an indication of transmission pipeline system expansion, the volume transported provides a better measure of its growth. Volume or capacity is controlled in part by line-pipe diameter, which is dictated by mill capacity to produce wider skelp and plate. It is also controlled in part by pressure, which under working stress design (WSD) (e.g., [10]) is controlled by SMYS, or equally the pipe grade. Suffice it here to note that by the start of the 1960s, the mills were broadly capable of producing pipe in Grade X52 for which SMYS is roughly double that for Grade A, with diameters in the order of 30 in. (or more)—much larger than several decades earlier. In this regard the capacity per unit length increases with the square of diameter and linearly with pressure. Thus, even though the incremental length per decade remains flat for these transportation systems since the 1970s, the network capacity has increased. Considering that diameters up the order of 48 in. have entered service in the United States, these systems in grades up to X80 since then, an almost four fold capacity increase has occurred relative to a 30-in. X52 pipeline that is otherwise transparent when capacity is assessed in terms of mileage.

Pipe diameter affects the wetted area on the inside diameter (ID) surface and so increases the threat of ID corrosion. It also affects the outside diameter (OD) surface area that must be coated and protected to control the threat of OD corrosion. Accounting for this difference in surface area over time should be considered as improvements in the Pipeline and Hazardous Materials Safety Administration (PHMSA) database lead to a more comprehensive framework and better quality results free of nulls. Until then, mileage remains the only viable metric to assess expansion and normalize trends over time.

It follows from Figure 43.5 that roughly one-half of the total transportation network has capitalized on the evolution in steel and line pipe production, and the technology and schemes used to build them in what has been called the modern system. What does this imply for the remainder of the transmission systems that was constructed prior to 1970? There are two basic approaches to address this question. The first is by analogy to the circumstances evident for parallel applications, such as other transportation infrastructure, systems, and components. The second is by direct analysis of the oil and gas transportation infrastructure and systems. Both approaches will be considered in turn.

### 43.4.2 Qualitative Assessment of Construction Era and Incident Frequency

Consider next parallel transport systems whose infrastructure involves components whose age spans many decades, as occurs for some commercial aircraft and ground vehicles and some pipelines, and portions of other transportation systems, such as bridges and rail transport, whose age approaches that of early pipelines. Some parts of the transportation infrastructure, such as bridges, share the same design basis that is used for pipelines. Bridges, pipelines, and many other structures, initially relied on WSD, wherein a factor of safety (FoS) was applied to the allowable stress to achieve a safe margin against uncertainty. As time passed, WSD gave way to plastic design or ultimate strength design (USD) for many new bridges and other structures (e.g., [11]). The virtue of USD was that it capitalized on the significant reserve strength evident in the post-yield response of typical early-vintage construction steels. It was accepted into practice, even though it offered less reserve capacity as compared to WSD, and is accepted also as a design basis for pipelines in a few countries. Reliability (risk)-based design (RBD (e.g., [12]), which relies on a probabilistic view of structural loading and resistance, also has found favor for structural design in some countries. Canada and a few other countries also allow RBD for pipelines. However, in the United States, pipeline design remains tied to the WSD concept, as does general boiler and pressure vessel design.

Other transportation systems (such as the rail and air transport systems) operating in most countries worldwide, also parallel key aspects of pipeline systems in many ways. Like pipelines, such systems provide for long-distance transport of commodities, but unlike pipelines also provide for public transportation. Similar to a pipeline, the paired ribbons of rail that are the backbone of railroad systems are not structurally redundant—if one of the pair of rails fails it is quite possible that the train will derail. And as for a pipeline, such failures can be as catastrophic depending on where the failure occurs and what commodity was being transported. In contrast to rail and pipeline systems, designs underlying air transport seek structural and subsystem redundancy. Yet even though redundant in concept, related failures occur in air transport causing casualties. Human factors likewise contribute to such failures, as they have in some pipeline incidents.

As for pipeline systems, both the rail and air transport systems rely on periodic inspection to detect system faults prior to their becoming significant. Both rail and air transport systems are "damage tolerant" (e.g., [13]) in that they continue to operate even though cracks have formed and are growing in structurally critical elements. Both rail and air transport systems are subject to periodic re-inspection and condition assessment. Finally, both rail and air transport systems continue to operate equipment and facilities provided, they are maintained and so "fit for service" (e.g., [14]) and also "fit for their intended purpose" (e.g., [15]) in that system. But, both rail and air transport systems also consider functionality and efficiency and in some cases will "retire assets for cause" (e.g., [16])—where that action can be motivated by brand concerns, inefficiency, and other considerations, including structural reasons. Competitive drivers also can lead to the sale of assets and to modernization in the rail and air transport industries, just as they have for pipeline systems—or for any other business.

It follows that pipelines built under conditions prevalent in the eras defined as "early" or "vintage" can be as viable as those built in the "modern" era as long as they are fit for service and for purpose. This is clear in the observation that some cars and big-rig trucks continue to operate for many hundreds of thousands of miles, whereas absent adequate maintenance they could fall short of their expected design life. Because a vehicle is old does not mean it is unsafe, but it can mean that it could require more care, or might not be as fast as its modern counterpart. This also applies in the context of vintage aircraft, many of which are still in service, but for brand reasons, economics, and other considerations they do not operate in the fleet of a major US carrier. This same brand-motivated decision is made when a two-year-old car is sold solely because the owner does not want to be seen in an "older" vehicle.

### 43.4.3 Quantitative Assessment of Construction Era and Incident Frequency

While early or vintage pipelines are analogous in most ways to other early transportation infrastructure that remains in daily use today, whose serviceability depends on adequate maintenance, the best metric of serviceability over time is a quantitative analysis of incident statistics. For pipelines, the most comprehensive basis for statistical analysis and incident trending is the pipeline incident database that has been developing since 1971 under the management of the PHMSA. The PHMSA functions under the US Department of Transportation [17] (DoT), which is responsible for all modes of transportation, such as motor carriers (under PHMSA) and air transport (under the Federal Aviation Administration). PHMSA's pipeline incident database is publicly available and can be downloaded directly from the PHMSA website [9].

Fundamental to any statistical analysis is the quality and consistency of the data, and which data should be pooled or parsed for the purposes of a given analysis. A check of the

PHMSA website indicates that their database has been packaged over time intervals during which the definition of a reportable incident (e.g., see [9]) and their input form and scope (e.g., see [18]) remained constant. As pipelines became increasingly encroached and other factors such as the growth of social media came into play, the sensitivity to the consequences of an incident has increased. Such factors have prompted changes over time in (1) the threat categories used, (2) the criteria used to define a reportable incident, and (3) the scope of data that must be reported. It must be noted in this context that as reporting criteria become more stringent, an event that might not have been reportable in past becomes reportable. More stringent criteria open to a bias toward increased incident frequency as time passes. Because the scope of the data required to offset this bias was not always a required input, and so is often not available, either this bias remains or the database must be culled to exclude it—which precludes trending over time. Over the period since 1971, three major changes can be identified in reporting structure and requirements, which can complicate trending, and attempts to offset any bias due to increasingly stringent requirements. Data quality issues also exist, as do many nulls (or empty reporting cells), which also can affect the utility of the trending. As nothing can be done to resolve the data quality concerns, the results are used as is without regard to the possible effects of the nulls.

Because the focus is a possible effect of the era of pipeline construction on the incident rate for the onshore pipeline system, the construction or manufacturing data listed in the PHMSA's database for onshore interstate pipelines has been pooled and then sorted by construction era as "early" (<1940), "vintage" (1940–1970), and "modern" (>1970), specifically for the liquid pipeline system. The raw outcomes were then normalized relative to the amount of mileage constructed in each of those eras, as reported in Figure 43.5b. Such analysis has been done in regard to the incident data for the modern era up through 2010, as well as up through 2002, because at this time there was a marked shift in the reporting requirements for the liquid system. Such analysis was done in regard to the elements of a pipeline system that differ in regard to location and function. That breakdown in regard to location parsed discrete elements (facilities such as stations, tanks, etc.) and the lineal element (including the pipeline and its mainline valves (MLVs)). The discrete elements were further sorted by function in regard to static, such as tanks, versus dynamic, which covered equipment and related components.

Because there were no clear trends across these parsed elements, only the results for aggregate system are presented in Figure 43.6. The $y$-axis in this figure is the mileage-normalized incident frequency for each of the construction eras, which are parsed on the $x$-axis. Each era in this bar graph includes two results—the first represents the data evaluated through 2010 while the second represents the data evaluated

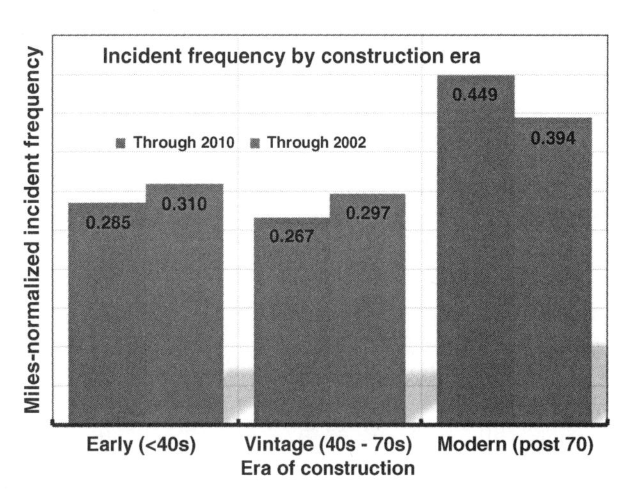


**FIGURE 43.6** Incident frequency parsed by construction era for interstate liquid transmission.

through 2002, when much more stringent reporting criteria for leaks in hazardous liquid lines were introduced.

The results of the quantitative analysis of the hydrocarbon transportation infrastructure shown in Figure 43.6 do not indicate that the portions of the pipeline system constructed in the "vintage" or the "early" era are more prone to incidents. It is evident in Figure 43.6 that the difference in the relative fractions of pipelines constructed in the vintage and early eras is small, with the incident fraction for the vintage system being ~5% less than for the early construction. While there is a modest decrease in the relative fraction of reportable incidents moving forward in construction era from early to vintage, Figure 43.6 indicates that the relative fraction of reportable incidents increases sharply moving into the modern era of pipeline construction. Contrasting the results for the database before and after the transition to more stringent reporting requirements circa 2002 accounts for roughly one-third of this increase. This outcome opens to the question: what accounts for the remaining upswing in relative incident frequency for modern construction? The results in Figure 43.5b do not indicate that this can be traced to a surge in construction, which could open to the need for contractors and so introduce a less experienced workforce. Further trending of the modern construction era indicates that while other changes leading to more stringent reporting occurred within this era, they played a minor role if any in the upswing in relative incident frequency for the modern construction. Rather, the results of trending for the modern era in 10-year increments suggest there was a period of learning and/or adaptation as the US industry incorporated the many improvements in steel making that were phased in circa 1970 (see Figure 43.2). For example, the transition from LF-ERW to HF-ERW longitudinal seam production that began in the early 1960s and was completed in the United States in the early 1970s (see Figure 43.3) experienced some developmental issues into the 1970s. Likewise, as the use of emerging construction practices broadened, there is evidence in trending the data that this aspect also experienced some developmental issues in the 1970s. If the trend evident for the 1970s was more like that of the 1980s or the 1990s, then the upswing evident in relative incident frequency for the modern construction era is diminished—with little difference is evident over time.

It follows that parallel transportation systems share comparable traits with no clear evidence that era of construction or age affects the mileage-normalized incident frequency for the hydrocarbon pipeline system assessed in regard to Figure 43.6. In contrast, a major failure within the pipeline industry and the parallel transportations systems considered earlier often opens to the discussion of a system's or component's age. Although the risk to the public or the environment associated with the trends in Figure 43.6 is not quantitatively out of line with other day to day risk exposure, it is often involuntary exposure. While such events are infrequent, such exposure can lead to environmental issues and loss of life, whose cost to the pipeline operator is not trivial. Recognizing this, since the 1930s the industry has worked to develop consensus standards and company specifications and protocols for O&M. And, as the public and environment can be adversely affected, a regulatory framework was introduced circa 1970. These aspects and their implications in regard to older pipelines (i.e., pipelines constructed in the vintage and early eras) are considered next.

## 43.5 THE EVOLUTION OF PIPELINE CODES AND STANDARDS, AND REGULATIONS

It is evident from Figures 43.2–43.4 that today's modern pipeline systems have evolved greatly, from routing and design, on through construction, and into O&M. The design Codes, related Specifications, and Regulations have likewise evolved. What began in the 1800s based on rudimentary schemes has evolved and now is directed at system safety over the life cycle of the system. Steel has replaced wood and cast iron, and couplings have been replaced by automatic welding, with inspection and controls now in use with a view to ensure the product remains in the pipeline, and moves in a safe, serviceable, modernized pipeline network. Yet, as just discussed, incidents occur on new as well as the vintage or early aspects of this transmission network—so factors other than age are involved.

### 43.5.1 Pipeline Codes and Standards

Using the aforementioned history and evolution of the pipeline network as a backdrop, this chapter focuses now on what have been termed earlier "older" pipelines constructed during both the "early" and "vintage" eras. The evolution of consensus codes and standards directed at increased safety and a reduction in incident occurrence to achieve the industry's goal of zero incidents (e.g., [19]) are outlined. Regulations promulgated to ensure safety and conformity across these systems—whether small or large are considered in parallel. Thereafter, this chapter closes with a range of illustrations of construction features that are unique to early and vintage construction eras yet remain serviceable and safe by virtue of the operator's diligence and concern for safety and serviceability. Serviceability and the safety that follow are essential regardless of the era of construction. Quite simply, a breached pipeline does not move any product and more importantly opens to significant risks for both the public and the environment—with consequent negative impacts on both OPEX and CAPEX, as well as corporate brand. The motivation to maintain and improve in that context is apparent. It is the impetus for the evolution evident in Figures 43.2 to 43.4 and equally that for the development and evolution of industry consensus codes and related standards, and also the regulations and related legislation.

The current Code for hazardous liquid and natural gas transmission pipelines, which respectively are ASME B31.4 and ASME B31.8, share a common origin as outlined in the foreword to each of these documents [20,21]. The Foreword for the current editions of these codes parallel to each other, opening by noting that "the need for a national code for pressure piping became increasingly evident from 1915 to 1925" and going on to indicate that this need was met in 1935 when the first edition of B31 titled "American Tentative Standard Code for Pressure Piping" was published. It is noted that following reorganization in 1948, an intensive review was made of the 1942 Code, with that revised edition published as ASA B31.1 in early 1951. That edition was augmented, with a new edition of B31.1.8 published in 1955. Later that year, it is noted that the decision was taken to separate the Code into parts dealing with gas transmission and distribution piping systems and liquid systems, which appeared in the 1959 release of the precursor to the current standards. Further details can be found in Refs [20,21]. Suffice it to state here that by the mid-1950s industry consensus standards had emerged to address the design basis for pipeline systems, which addressed maximum operating pressure and other key parameters.

Other industry consensus groups have worked to develop minimum standards for the supply of the line pipe, and the steel it was made from in regard to American Petroleum Institute (API) 5L [22]. API 5L first appeared as a tentative specification in 1927, and became a standard in 1928. In 1929, the second edition of this specification comprised several sections and a few tables that dealt with the rudimentary aspects of welded pipe made from Bessemer and open hearth steels, along with SMLS steel pipe, and two forms of iron pipe. After 20 years, the document grew to comprise 33 pages made up of 14 sections and three appendices, but aside from the inclusion of electric furnace steel, the scope of the processes covered had not changed appreciably.

However, a parallel specification had emerged as a tentative standard just the year before. It was denoted API 5LX [23] and dealt with the then-approved "high strength" grades up through X46. In March 1983, the 5L and 5LX specifications were merged in the 33rd edition of API 5L, with grades up through X120 now accepted under the 44th edition. Such specifications include recommended practices that are specific to line pipe, and also incorporate-by-reference many other commonly used standards for tensile, fracture, chemistry, hardness, and other characterization tests.

Over the years, as the need and technology developments dictate, other specifications, standards, and guidelines have evolved primarily under the auspices of API, National Association of Corrosion Engineers (NACE), and American Society of Mechanical Engineering (ASME). Presently, such exist for CP, coatings, ILI, direct assessment, leak detection, and a host of other pipeline design, and O&M applications. A scan of the website stores of such organizations opens to the full scope of what is available.

### 43.5.2 Pipeline Regulations

Pipeline regulations were introduced in the US circa 1970, with a comprehensive listing of the notices of proposed rule makings (NOPRs) and the publication of the final rules in the Federal Register available online from the PHMSA [24,25]. It is evident from the documents tabulated there that the then current (1968) edition of the ASME B31 Code played an important role in formulating the regulations. For example, Notice 70-2; Docket No. OPS-3B [26], concerning 49 Code of Federal Regulations (CFR) Part 192 and Minimum Federal Safety Standards for Gas Pipelines, states that "these proposed regulations closely parallel to the presently effective interim standards that are set forth in the B31 Code" and that while "a number of differences will be noted . . . for the most part these are nonsubstantive" in nature.

The period from August 1968 when the Gas Pipeline Safety Act of 1968 became law through late in 1979 when the Hazardous Liquid Pipeline Safety Act of 1979 became law set in place the early regulatory basis for transmission pipelines [27]. Since then, other legislation has been enacted that has made those requirements much more stringent, often in response to occasional high-profile incidents. In many ways the industry has been proactive in developing technology that continues to be the basis of the evolving regulatory requirements. For example, prior to the legislation known as the Pipeline Safety Act of 2002 [28], the natural gas industry undertook a broad research program to define viable approaches to pipeline condition monitoring, re-inspection intervals, and the definition of high-consequence areas, to name but a few aspects. In addition, the industry undertook developing and codifying a series of direct assessment technologies to deal with pipeline systems and segments for which their condition could not be reasonably determined via ILI. Much of this appeared in NACE recommended practices and in a supplement to B31.8 (designated B31.8S) [29], with parallel developments by the liquid pipeline industry that culminated in API 1160 [30]. Almost concurrently, major changes occurred in the regulations, as for example the inclusion of Subpart O in 49 CFR 192.

Regulatory requirements continue to evolve as issues are identified and recommendations are presented following analysis of the regulatory framework and its effectiveness by the National Transportation Safety Board (NTSB). Advisory Bulletins are issued from time to time to broadly outline potential concerns. Where incidents occur that point to unique threats, the PHMSA holds workshops and forums, and undertakes research and technology development to assist the industry and also support their audit process. Details of their efforts can be found along with listings of their reporting [31], which can be downloaded as desired. Suffice it to note in closure that the regulatory requirements embodied in 49 CFR Part 192 [32] (gas pipelines) and Part 195 [33] (hazardous liquid pipelines) are amended as time passes to reflect the legislative and other actions enacted to ensure public and environmental safety.

## 43.6 SOME UNIQUE ASPECTS OF EARLY AND VINTAGE PIPELINES

The background and related discussion of the evolutionary timeline for steel, line pipe, and construction methods identified eras in the timeline for pipeline construction. The term "vintage" covered the period between 1940 and 1970, while "early" applied <1940 and modern applied to after 1970 to coincide with the introduction of the Federal Regulations. As discussed in regard to Figure 43.6, this date also reflects a time when many major changes occurred in steel and pipe making, as well as in construction practices. The transition date from "early" to "vintage" was similarly coupled to the start of a period of change regarding to steel, pipe making, and in construction practices, and the onset of the period when a significant portion of the current pipeline was constructed. This section presents illustrations for a range of construction features that are unique to early and vintage construction eras yet remain serviceable and safe by virtue of the operator's O&M protocols and management plans, and diligence in their implementation. While the focus here is on "older" pipelines, serviceability and the level of safety it ensures is essential regardless of the era of construction.

Hints regarding design and construction aspects that are unique to "older" pipelines—the focus of this chapter—can be identified in regard to Figures 43.2 to 43.4, as practices that have been displaced over time. Not surprisingly, some of these aspects were identified in the NOPR for parts of the then pending Federal Regulations (e.g., [26]) circa 1970. Notable concerns identified then included branch connections, bends, elbows, wrinklebends, miter bends, and dents. It is noteworthy that except for dents all of these concerns involve a change in the direction of the pipeline. Subsequently, the construction features covered in the early notices were augmented in reference to different long-seam types, and couplings. The regulatory structure dealing with such aspects developed via a "grandfather clause" that as presented circa 1970 permitted the continued operation of the pipeline under its recent operating conditions without a pressure (proof) test. Where uncertainty existed, some types of line pipe long-seams were de-rated, while others were qualified based on the results of full-scale testing (e.g., [34]). Differences between liquids transportation practices and those for natural gas, which have prompted concern for fatigue, led to some key differences between these vintage networks. It seems likely that many of these aspects will be a focus of regulatory action in the wake of the public workshop on integrity verification process (IVP) (e.g., [35]), which currently considers the gas transmission system.

In general, a process or technology is displaced over time because it is inefficient or ineffective, where such metrics (inefficient or ineffective) are generic in regard to time, cost, fitness for service, fitness for purpose, and so on. What is possible in terms of these metrics is conditioned by the materials, the manufacturing and production capabilities, and the technology available. The routing and related terrain (topography, soil, moisture/water, ice, permafrost, etc.) further condition what is feasible/plausible. Terrain and in particular the type of soil (hardpan, sands, silts, etc.), the presence of water (river, stream, creak, swamp), and the surface topography are major factors coupled with the routing, as they influence the need for vertical and horizontal bends. Topography and local conditions also determine the angle that must be achieved. Against this background Figures 43.7–43.21 introduce views of the construction process as the basis to better understand its evolution over time—and plausible integrity concerns that need to be considered. Of those, Figures 43.7–43.15 characterize the "early" process, while Figures 43.16–43.20 characterize changes as the construction process transitioned into the "vintage" era, with just one committed to illustrate the challenges addressed in the modern era. These figures and the related discussion lay the foundation to outline integrity management (IM) practices for such pipelines.

### 43.6.1 "Early" Construction Practices

Figures 43.7–43.9 present very early views of the construction process circa 1913. It is apparent in those figures that the line pipe is being transported to the spread by wagons pulled by a hitch of mules. While comparable images along the spread are not available, recognizing that the automobile appeared in the US circa 1903, it is likely that this same mode was used to string the pipe where feasible.

It is apparent from the scale of Figure 43.7 that these pipes are about 20 ft long, and so are single-random joints about 12 in. in diameter. Loading and handling that size and weight posed a challenge in 1913. Although just a single hitch of mules was used, it is clear from this photo that a large number of joints could be moved along a dry road over flat terrain. However, where hills had to be negotiated or the soil was soft along the right-of-way (RoW), the number of joints transported had to be decreased. Alternatively, it is also plausible that self-propelled equipment comparable to the ditching machine used on this 1913 construction project could have been used. As Figure 43.8 shows, the design of that machine incorporates much wider wheels and includes bladed chain-driven wheels, which could facilitate moving heavier loads under adverse conditions.

While electric arc welding had emerged in the United States prior circa 1907, oxyacetylene welding was available earlier (circa 1903 in the United States), and was better developed as compared to arc welding at that time [39,40]. However, the limited ability to weld uphill with the oxyacetylene process meant that adjacent joints of pipe had to be rolled in sync to complete a girth weld, often termed a "rolled weld." Because longer sections of jointed pipe opened the door to handling issues and the stresses induced by handling, the use of welding was limited to double jointing. Thus, completing a pipeline using larger diameter (plain-ended) pipe required couplings to join the double-jointed segments. To minimize handing issues, doubled jointing and coupling were done over the length of a string of joints adjacent to the ditch. (Construction in that era

**FIGURE 43.7** Line pipe transportation circa 1913.

**FIGURE 43.8** Mechanized trenching machine.

**FIGURE 43.9** Over-ditch A-frame fulcrum.

also made use of single joints with couplings used between each joint.) Ditching was done by hand, and facilitated by machines such as that in Figure 43.8 which pulled a ripper tooth/plow that broke the surface cover and opened the upper part of the ditch. Thereafter, the trenching was finished by powered shovels or by hand, and timbers were laid across the ditch. The string of coupled pipe segments was then rolled over the ditch onto the timbers, after which the string was lowered-in manually, with the assistance of leverage and an A-frame, as is evident in Figure 43.9.

As is evident in Figures 43.9 and 43.10, lowering-in was accomplished with the help of mechanical advantage gained through a long lever supported by a stout timber fulcrum that was set on A-frames either side of the trench. This lever-fulcrum scheme provided lift for the pipe string. The string was lifted joint by joint, such that the local cross-ditch timber could be removed with the pipe sting incrementally settling into the ditch as this process was repeated stepwise along the trench. This process was continued sequentially along the ditch with subsequent strings coupled onto the prior string that remained at ground level on the timbers—such as that evident in the foreground of Figure 43.10.

The overview of the lowering-in process in Figure 43.10 clearly shows the cross-ditch support timbers, the A-frame supported fulcrum, and the long-lever used to lift the pipe string. The pipe string evident in the foreground can be seen fully settled into the ditch at the left margin of this figure, with the transition into the ditch evident behind the crew working the leverage. Two of the couplings used to connect the single or double jointed pipe segments can be seen in the pipe string in the foreground of this figure. In particular, 1913-vintage Style 38 Dresser couplings were used for this construction. Figure 43.11 illustrates this style and its function in reference to the current form of this coupling. While the passage of time has brought minor changes in its design and production, its function and overall appearance remain the same today as they did a century ago.

As evident in Figure 43.11, installation of Style 38 couplings began by running the slip-fit collar onto one of the pipes, after which the second pipe was brought into position and the collar then shifted back centrally over the ends of the joints being connected. A seal between the pipes was created through elastomeric gaskets that were located toward the ends of the coupling, as can be seen in Figure 43.11a. The schematic in Figure 43.11b indicates that tightening the bolts across the flanges of the coupling created a compressive wedge load on the elastomer, with the radial component of that load compressing the gasket between the pipe and the coupling, which locked the coupling into position and sealed the joint. Depending on the length of the "middle ring" (which is dimension B in Figure 43.11b), company construction specifications and rehabilitation protocols typically limit the total angle formed across the coupling to the order of 4 ° or less.

It is apparent from Figure 43.10 that such couplings could support the shifting and resultant angular distortion that formed in association with this lowering-in process. As such, they were also capable of accompanying the shifting and distortion as the coupled pipe joints that lay strung along

**FIGURE 43.10** Lowering-in by hand with leveraged mechanical advantage.

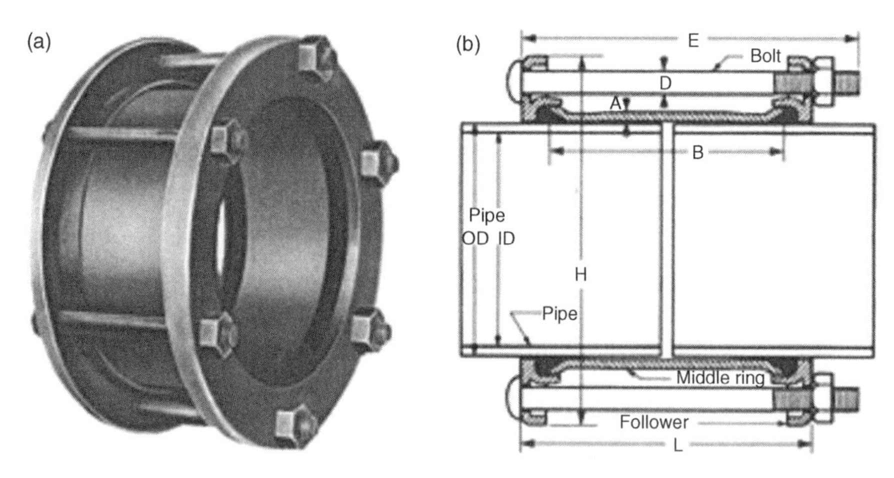


**FIGURE 43.11** Views of a recent-production Style 38 Dresser coupling (see Ref. [36]).

the ditch were rolled out over the ditch onto the cross-trench timbers. Couplings also were developed to provide for larger-angled bends, as shown in Figure 43.12. These special-order Style 56 angled Dresser couplings were based their Style 38 coupling with a mitered (sometimes spelled mitered) fabrication (Figure 43.12a) or a smooth wrinklebend in place of their usually straight middle ring. Based on Ref. [37], such mitered and smooth wrinklebend fabrications became possible in the 1930s—or toward the end of the early era.

Because the rubber/elastomeric gaskets of a half-century or more ago were prone to degrade (dry out) as time passed, or they were exposed to certain natural gas liquids or tramp species in the product stream, some joints tended to leak. Soil stresses due to shifting soils on slumping slopes likewise could cause the pipe to move relative to the coupling, and/or pullout in part or completely, which also led to leaks. Tabs welded across the coupled joint to electrically "bond" the upstream and downstream pipe for purposes of CP continuity, as illustrated in Figure 43.13, also could help to resist pullout. In contrast, leaks due to aging of the early rubber gaskets could not be easily managed given the seal materials then available. Although timely maintenance could offset such concerns, as time passed some operators found it cost effective to abandon their very early coupled high-pressure lines. Other operators dealing with lines constructed toward the end of the early era determined that they could be effectively reconditioned by a process that replaced the couplings with girth welds, and periodically inserting pups to make up lost length (e.g., [41]). Thereafter, the line could be recoated as needed.

While shop fabricated miters and smooth wrinklebends could be reliably produced toward the end of the early era, the result of their use as field construction methods could vary significantly even then, depending on the experience of the crew involved, the tools available, and the techniques used. This is illustrated in Figure 43.14, which contrasts two wrinklebends that were reportedly made in the 1930s within

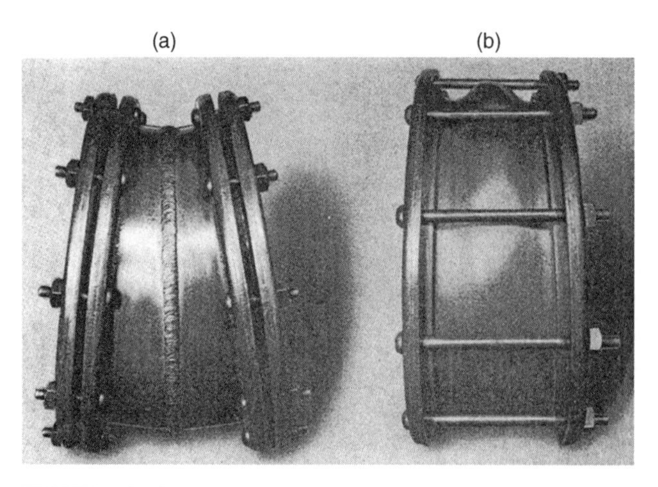


**FIGURE 43.12** 18 in. angled couplings circa 1936 (see Ref. [37]).

**FIGURE 43.13** Tab-tied coupling (at top).

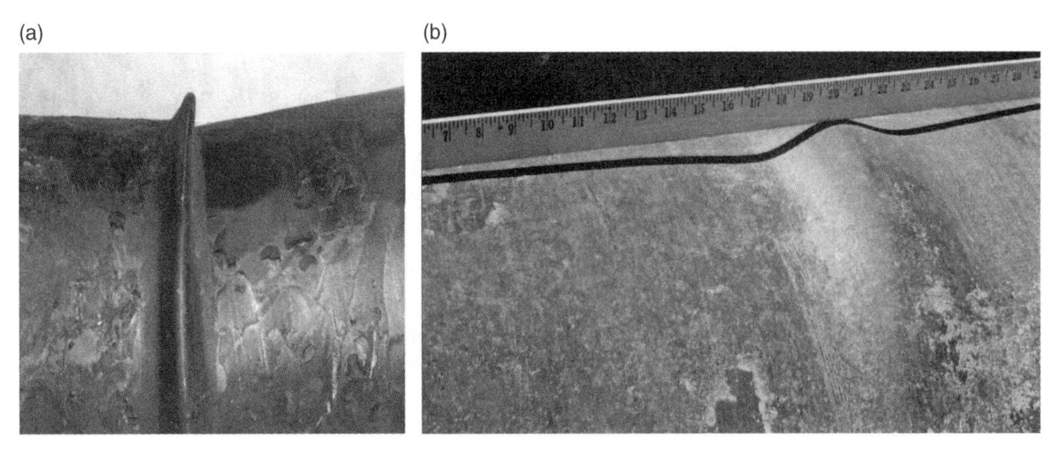


**FIGURE 43.14** Views of 1930s wrinkle bends reportedly made within years of each other.

a few years of each other. The wrinkle in Figure 43.14a (reportedly to have formed free of field-induced loads) appears to have collapsed into itself, forming a tight kinked axially asymmetric feature that, while not captured in this image, runs to varying depths the full circumference of the pipe. The aspect ratio for this bend—which is quantified by its height to its length and is one metric of wrinkle severity (e.g., [42])—is about four. In contrast to this high-aspect-ratio bend, the "smooth" wrinklebend shown in Figure 43.14b has an aspect ratio of ~0.2. Comparing these results indicates that wrinklebends that differ in terms of aspect ratio as a metric of severity by a factor of 20 could be encountered in bends produced within a few years of each another. If pressure cycling typical of some pipeline operations controlled failure for these bends, and both were cold bends made on the same line pipe steel and were free IC and/or EC, then their service lives are estimated according to Ref. [42] to differ by about five orders of magnitude. In absolute terms, the severe bend has near zero fatigue life, whereas for typical historic gas pipeline operation the life of the smooth bend is many times the system's book life.

It is known from historical archives (e.g., Chambers, A. A., private communications to G. A. Reinhardt reporting on the characteristics of construction using ERW line pipe produced by Youngstown Sheet & Tube versus the FW pipe made by A.O. Smith, Youngstown Sheet & Tube Company, 1931–Report 29, dated April 1931, and Reports 34 and 36, dated August 1931) that wrinklebends could be produced under well controlled conditions that used a stiff strong back in combination with winches and leverage about a fulcrum to produce such bends. That and other operator documentation indicates that this and similar techniques were used in 1931, and possibly earlier, to produce hot bends as well as cold bends, in construction in Ohio and elsewhere in the mid-West. It is also known that much lower stiffness systems that stored significant energy in an A-frame setup were used to make wrinkle bends in much earlier construction. As the wrinkle began to form in these more flexible schemes, that stored energy was unloaded leading to a much less controlled process and the likelihood of creating collapsed bends, like that in Figure 43.14a. Figure 43.15a and b illustrate each of these practices.

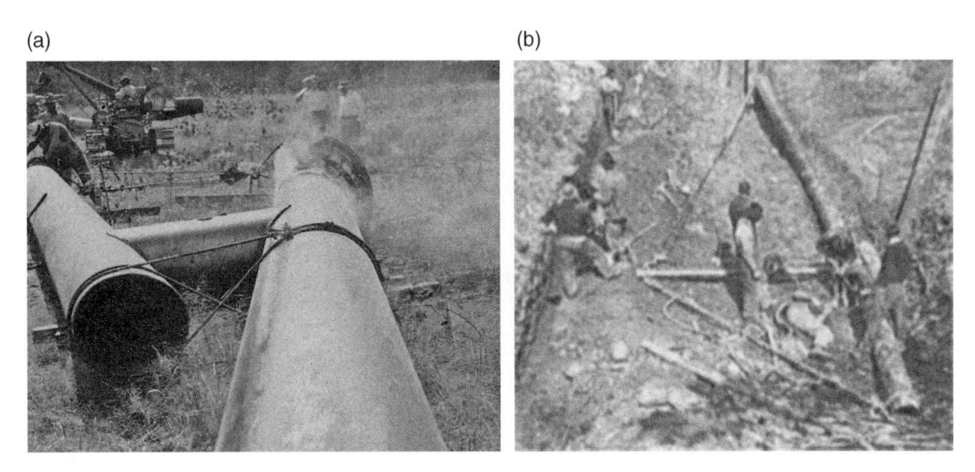


**FIGURE 43.15** Methods used to form wrinkle bends.

Where higher angle bends were needed beyond what could be achieved with one smooth wrinklebend, a sequence of such bends was introduced in the pipe, typically separated axially by a diameter or more. Shop fabricated miters are known to the author that were made during the 1930s as special-order items. It is also known that miter bends were broadly used in some field construction during the 1930s in pipelines that function at pressures up to 400 psi.

As experience along with the tools and techniques developed as time passed, some of the early construction practices used to bend pipe or fabricate fittings evolved greatly within the early era, while others were discarded. Still others, such as the design of the Dresser coupling, have remained largely as devised initially, as is evident in comparing the images and descriptions between Refs [36,37]. While the concept is largely unchanged, the early rubber gasket has been replaced by a modern elastomeric material, which underlies its continued use today.

### 43.6.2 "Vintage" Construction Practices—An Era of Change

Documentation in Dresser literature [37] circa 1936 indicates that the Dresser Company was founded in 1880, with reference made therein to the then just-patented Dresser coupling. It goes on to note that prior to 1890 all joints between plain-end pipe relied on a bell and spigot connection, which remains in common use today for water supply pipelines and drainage, or they used a threaded connection. Finally it stated that the first coupled pipeline was built in Ohio in 1891. It is not surprising then that construction circa 1913 illustrated in Figure 43.7 involved the use of the Dresser coupling.

Dresser couplings could have continued as a primary jointing method for hydrocarbon transmission pipelines well beyond the "early" construction era were it not for the evolution of welding technology, which as detailed later also involved the development and use of inspection since the 1960s, and more recently automation. It was limitations in welding technology during the "early" construction era that made a place for couplings, to both connect the double-jointed "rolled-weld" pipes, and to manage the bends necessitated by routing or topography. As shown in Figure 43.12, by the 1930s Dresser had capitalized on the ability to shop fabricate a mitered joint or make a controlled bend. As shop development and the deployment of those developments is a precursor to its field adaptation if feasible, the hand writing was on the wall for the eventual development and use of welding and more controlled bending methods under field conditions.

As electric welding became increasingly practical under field conditions for the grades of steel and the wall thicknesses then in use, the practice of girth welding displaced coupling—because in that era the gasket materials were still prone to dry out, whereas a weld seam did not. As the tools to prepare the bevel and coated electrodes emerged along with multipass welding, the door opened to making girth welds on pipe segments or pipe joints that had been cut at an angle to create a miter bend. Fittings produced using the so called orange-peel process also became more prominent. In parallel, more controlled self-propelled field bending machines had begun to emerge, with many such machines available by the mid-1950s. The improvements in welding also opened to the capability to produce a reliable DSAW long seam by the 1930s, with related improvements continuing since. Not surprisingly, some of the bending machines were mechanized, hydraulically powered adaptations of the approaches used in the early era, such as the vertical stiff-leg illustrated in the early history of field bending presented in Appendix A of Ref. [42]. Another adaptation of the early cold wrinkle-bending methods is shown in Figure 43.16a, wherein jacks are combined with reaction loads developed through cylindrical collars to create the bend. The view in Figure 43.16b makes use of the same concepts that are deployed much like many of the bending machines in use today.

The use of miter bends and the advent of self-propelled mechanized cold field bending machines beginning in the 1940s with the use of a segmented bending shoe brought an end to the use of wrinklebends for US construction by the mid-1959s. However, company specifications for some US

**FIGURE 43.16** Mechanized approaches to cold form smooth field bends circa 1955 [38].

operators indicate that its use ended selectively by the mid-1940s. In contrast, miter bends continued in use on large-diameter systems into the mid-1950s on some systems. Ref. [43] presents further details on the evolution of bending methods based on the views of one operator. Other citations of historical interest can be found in Appendix A of Ref. [42].

The combination of girth welding and field bending opened to improvements in construction practices. As time passed, some of the early machines began to look quite similar to those of the modern era. Perhaps the most striking difference evident in the photographs is the move to mechanization, which is apparent in comparing Figures 43.17 and 43.10.

Contrasting the construction techniques in Figure 43.17a with those in Figure 43.10 illustrates the shift toward mechanization, whose benefits included less time on the spread and therefore earlier return on the investment, along with much more controlled repeatable practices. In reference to Figure 43.17b, because the use of CP was only beginning in the 1940s and was not employed by all operators, the effective cleaning of the pipeline and application of a coating was an important factor in limiting EC. The benefit of replacing manual labor in this context was clear, as the mechanized approach produced consistent cleaning and coating that was continuous over the surface of the pipeline's length. More critically, competition between suppliers led to continuing improvement and therefore a better process as time passed.

Although the use of welding continued to expand as procedures and equipment improved, it is likely that the release of the 1955 edition of the ASME B31.4 and B31.8 Codes diminished the use of both miter bends and wrinklebends. This assertion reflects Code language in B31.4 that then precluded the use of wrinklebends in liquid hydrocarbon transmission pipelines and limited the use of high-angle miters to pressures below 20% of SMYS, and required their sequential spacing at a diameter or more. The Code language in B31.8 for natural gas transmission pipelines was much less restrictive as compare to that in B31.4, stating that cold wrinklebends were "not preferred" on systems operating at 40% or more of SMYS, but it did include other restrictive provisions. Thus, the Code language in 1955 effectively precluded the use of either type of bend on any new construction involving long-distance transmission pipelines.

While its effects were felt long after 1955, relative to the requirements in the B31 Codes, both 49 CFR 192 [32] and 195 [33] have been more restrictive. This remains the case today, with waivers/special permits involving legacy miter bends being denied even where full-scale test results on bends removed from service failed at stresses approaching the usable transplants (UTS) [44] and X-ray documentation existed in support of the quality of the welds (e.g., [45]).

Prior to 1955 both miter bends and wrinklebends had been built into construction in the United States where the topography and other factors led the need for bends. It is known anecdotally that thousands of miter bends remain in gas distribution service for some operators, whereas wrinklebends exist also by the thousands for some operators in gas transmission service—whose condition and safety are reestablished frequently, as discussed later. Figure 43.18a shows a view typical of an early use of a high-angle miter bend, which has remained fit-for-service in gas transmission service at 400 psi or less since its fabrication. Figure 43.18b shows a view of a more complex fabricated bend that involves a miter made on a chill or backing ring, along with a plain-end bell and spigot at either end that connects into the run pipe, which although exposed for inspection in this image remains in service today. Similar views exist for a range of wrinkle-sag-bends that also continue fit-for-service in gas transmission pipelines since their installation decades ago. However, these are less illustrative because they are just slightly exposed for purposes of inspection and so are not shown here.

While the language of the 1955 Codes and the much later promulgated Regulations limited the potential role of many early and vintage features in hydrocarbon transmission pipelines, this not deter pipeline construction in the vintage era, as the need for bends could be met by machine-made cold field bends or by shop produced fittings—depending on the

(a)

(b)

**FIGURE 43.17** Mechanized aspects in the early days of the vintage era.

**FIGURE 43.18** Views of vintage bends that through inspection and timely maintenance remain fit-for-service since their installation.

application. Where other types of fabricated fittings were required, such as reducers, the need was met by suppliers that developed the tools and capabilities necessary to address the demands for such items. The need for larger items such as MLVs was likewise met by suppliers that developed the capabilities to address the demands. Technology essential to develop this range of products benefited from parallel developments in other industries, which coupled with selected full-scale testing and the related development of Codes (e.g., [46]) and Standards (e.g., [47]) led to the fittings supply industry as it is known today.

Although the extent to which welding was used was limited somewhat by the effects of the 1955 Codes, this did not hamper the development of welding technology, the equipment, and the welding rods and coatings, and the expansion of the system capacity it facilitated during the vintage era, as discussed in regard to Figure 43.5. Given that variable weld quality would result from unqualified procedures and/or welders, the need for quality control (QC) and quality assurance (QA) was addressed along with the development of specifications and training programs and, eventually, also automation. The necessary developments in welding (e.g., see [39,40]) were paralleled by the development of NDI technologies and practices (e.g., [48]), which were supported by related specifications (e.g., [49,50]). As the principles of structural integrity (e.g., [13,14]) that were emerging in the 1960s were better understood in a welding context, the transition from workmanship standards gave way to engineering critical assessment (ECA) methods (e.g., [51,52]) to quantify the fitness of a weldment for its intended service.

The basis for making reliable welds and ensure their quality and adequacy from a design perspective paved the way to the widespread use of welding and the creative fabricated solutions that welding facilitated. The capability to predict structural response through advancements in technology that was spun out from rocket and other military technology made it possible for the pipeline industry to engineer those creative solutions. Challenging problems such as river and valley crossings could now be addressed in a safe and practical manner using pipe bridges, as opposed to open cuts that could lead to environmental and other issues, but were the method of choice in the prior era. In this context, the initial years of the "vintage" era were a time of change that opened to new horizons. The results of such synergy between analysis of structural response and the ability to reliably fabricate pipeline bridges are presented in Figures 43.19 and 43.20 as illustrations of the changes this synergy facilitated.

**FIGURE 43.19** Views of a vintage era twin stayed arch pipeline bridge fabricated with miters.

**FIGURE 43.20** Views of a pipeline bridge built as a cable-stayed catenary in the vintage era.

The first of these, which is shown in Figure 43.19 is twin-arch design. As the view in Figure 43.19a shows, the adjacent arches were laterally coupled, which provides lateral stability. The bends evident in this figure were formed as smooth wrinkles, with included angles up to ~12°. This bridge, referred to as a stayed arch in this industry, was built in 1948 to cross a stream as is evident in Figure 43.19b, which required a span of about 200 ft. In this application, both pipes were pressurized and flowed gas, whereas most stayed arches made use of one non-pressurized structural pipe adjacent to the pressurized pipe. As is the case for any arch, the pipe components that form the arch operate nominally in axial compression, while the dead and live loads and distortion of the bends due to the pipeline's pressure or thermal exposure are reacted by its thrust block foundations. Although still fit-for-service, this bridge was removed as part of a replacement program, with the crossing now made using a horizontal directional drilled (HDD) pipeline. This HDD approach avoids environmental issues that can occur with cut crossings at streams, and is free of concern for the effects of flooding and/or scouring around the bridge's abutments.

Longer spans needed to bypass geographical barriers such as ravines and/or rivers in valleys can be achieved by a cable suspension design in contrast to the shorter spans suited to the arch concept just discussed. Longer spans also can be managed by a bridge formed by as free-standing catenary if the line tension and other aspects unique to this concept are managed by the design and construction practices. As the scale of such crossing increases, so do the loads, and stresses and displacements induced. Thus, the approach used to react the loads and accommodate the displacements must reflect the local conditions, which include the effects of the operating environment and other natural forces. Such challenges are illustrated next in regard to the cable-stayed catenary span shown in Figure 43.20.

The concept that underlies a catenary pipeline bridge is today referred to as a "highly tensioned suspended pipeline" (HTSP). Six such spans exist in the United States, all constructed by one operator, with that shown above being the second longest at 430 ft. As implied by the name of this design concept, the span nominally operates in axial tension that for the catenary in Figure 43.20 is carried by a 16 in. diameter line pipe with a 0.344 in. thick wall, with this thickness dictated in large part by the line tension. The line has an operating pressure of ~1000 psig, which creates a hoop stress of ~23.3 ksi.

The approaches to the catenary span are transitions from the usual buried construction to aboveground, which expose the pipeline near the tops of the hills on opposite sides of the canyon. No special anchor supports were provided at the ends of the span, with the anchorage and support for the span developing entirely by the soil. While several of the canyons with such spans experience landslides during heavy rains, this has not adversely affected the approaches. Within this long span the line pipe has been wrinkle-bent at about 20 ft intervals, which involves smooth wrinkles that extend over the top of the pipe section that run from about the 9:00 o'clock position around to about 3:00 o'clock. Records in the archives indicate that strain gage data was gathered that indicates the wrinkles relieve the axial bending strains, resulting in more cable-like action as the span develops its catenary profile under the action of the vertical loads. It is reported by the operator that the pipeline shown in regard to Figure 43.20 has operated since 1944 with no integrity problems specific to the HTSP concept. Such catenary spans appear to be effective in traversing ravines even where heavy rainfall has caused slips and larger landslides. And given success with the concept as discussed earlier, it provides an alternative to pass environmentally-sensitive areas where a small "footprint" is desired. It should be noted in closing this illustration that the HTSP as deployed earlier makes use of cable-stays to manage lateral movement, and also includes dampers as part of its functional control package.

### 43.6.3 Summary and a Brief Look Forward at the "Modern" Construction Era

The past several sections have considered the evolution of the pipeline industry relative to Figures 43.2 to 43.4, and then illustrated that evolution during the early and vintage phases of pipeline construction in the United States. This brief subsection is a bridge between that history and the discussion of integrity and the safety and serviceability it ensures that follows in the next section. Thus, it is appropriate to develop this summary around the aspects considered to date that broadly impact integrity. Integrity is affected by (1) the regulatory requirements and statutes and (2) the diligence of the operator in adhering to those directives, and in implementing programs and best practices that carry beyond those directives. At the highest level this requires that the operator (1) identify the potential threats to asset integrity, (2) assess asset condition periodically over the scope of the relevant threats, and (3) respond on a timely basis to remediate or replace the asset as appropriate, while the operator works to improve its processes and how it implements them. Key in that context include: a consistent process for threat assessment, the ability to measure condition and quantify its severity, and an effective means to return the condition and integrity to a level adequate to ensure safe operation until the next reassessment is completed—and the technology necessary to make this happen.

The following circumstances rise to the surface in view of the aforementioned criteria:

- prior to 1920:
  - open hearth and Bessemer steels, furnace and wrought long seams, oxyacetylene welds, field-fabricated fittings, no NDI, couplings, no ILI, no pressure test, no aboveground surveys, no CP, no coatings;
- 1920s–1940:
  - LF ERW and FW seams emerge, alloying emerging, bare electrode arc welds emerge for girth and long seams, no NDI, mix of field fabricated and purchased fittings, coupled joints, wrinkle and miter bends, no pressure tests, no ILI, no aboveground surveys, no CP, rudimentary coatings emerge;
- 1940s–1970:
  - more modern welding practices and coatings emerge and existing practices improve, some NDI, some pressure testing, less likelihood of couplings and wrinkle/miter bends, purchased fittings, controlling codes and standards emerge, ILI emerges very late on, pressure tests emerge, CP emerges, aboveground surveys emerging, bending machines emerge;
- post 1970:
  - welding and NDI improve, pressure testing and bending practices modernize, as does CP and aboveground surveys, spike test is developed, advanced coatings emerge and their application is perfected, ILI grows into its third generation, arctic and offshore challenges met along with related midstream challenges, more mechanization, QC, QA, and automation, Regulations emerge while Codes and Standards broaden and improve, in the ditch inspection emerges and continues to develop, IM technology and practices formalized and continue to improve.

Many midstream companies operate mainline transmission pipelines with some combination of construction that covers from one to all of these eras. Depending on the operator and its mix of mileage and their O&M strategy, some operators will rehabilitate or replace aspects of their early and vintage mileage to harmonize its management. This could include removing oxyacetylene welds and couplings and recoating over the ditch, and include changes to make the line piggable. For pipelines reworked prior to the Regulations, the rework could involve a gas test at 1.1 times the service pressure or higher pressure hydrotest, whereas if it was done after, it would include a Code-based test or better. The laterals tend not to see the same level of investment to rehabilitate or replace fit-for-service segments—with many of the smaller diameter segments still the same as they were constructed. Depending on where laterals lie in the system, this type of line typically will have since been pressure tested.

Perhaps the most significant discriminator between the modern era and those that precede it lies in the challenges posed by construction that is moving to increasingly remote sites that involve major geographical barriers, or exposure to pristine environments. As such, a geographical barrier like that shown for the vintage era is shown in Figure 43.21 to illustrate this ever increasing challenge relative to that discussed earlier in regard to Figure 43.20.

Figure 43.21 provides one illustration of the challenges that must be dealt with as geographical barriers are encountered between the site of the E&P and the point of delivery. This figure shows an arch bridge that in concept is comparable to that in Figure 43.19. This bridge has a span on the order of 230 ft, just slightly more than the twin arch span shown from the vintage era. This bridge sits at near sea level, moving crude in a transmission system from a maritime terminal to a refinery. It spans a river that supplies fresh water to a major fraction of the population of one of the larger cities in the world. By Code this bridge operates at a maximum pressure corresponding to 50% of SMYS.

Although comparable in concept, there are several critical differences that make crossing this barrier more difficult—aside from environmental and public impacts. First, while this modern construction involves a span only slightly longer than the vintage bridge, it crosses a river that traverses locally flat land in a tropical climate, which during the rainy season makes it prone to heavy flooding and scouring. Second, this modern construction is a single arch, and therefore lacks the lateral support afforded by a twin arch, a factor that must be dealt with by the foundation design. Third, and very significant to the loadings that must be resisted, this modern construction transports largely incompressible crude through

(a)

(b)

**FIGURE 43.21** Views of a 40 in. diameter 230 span arch bridge transporting crude oil.

a 40-in. diameter line, whereas the vintage bridge transports gas and is much smaller in diameter and of lighter wall thickness. Taken together, these differences make for a much more difficult and complicated design, which would not have been attempted a few decades earlier.

Traversing hills and valleys up through the scale of mountains presents another set of challenges, including the need to protect the pipeline and its coating from rock damage during construction, as well as in service. Specialized coatings have been developed to protect the pipe in such construction, and machines have been developed that can return the acceptably sized spoil to the ditch as bedding or padding (e.g., [53]). Hills and valleys open to the effects of gravity and pressure head for liquid systems, or for gas systems during hydrotesting. Liquid systems often deal with this aspect through use of telescoped wall thicknesses, with the transitions managed by technology today much better than even a few decades ago. This is evident in comparing the guidance circa the 1950s [54] with recent developments [55]. Much more discussion and other illustrations are possible, particularly in reference to cases involving environmental exposure, but not within the current page constraints. Suffice it to note that as new challenges emerge and new threats are identified, a response will develop to ensure the system remains fit for service. In this context, the industry remains committed to the goal of zero incidents [19]).

## 43.7 MANAGEMENT APPROACH AND CHALLENGES

As noted in the last subsection, the IM approach depends on (1) diligent response to the regulatory requirements and (2) the operator's involvement in programs and best practices that carry beyond those directives. Within the current regulatory framework and the technology that has been developed to implement it and using operator-specific protocols and practices, the operator develops an integrity/change management and continuous improvement plan. Implementing that IM plan involves three steps each of which in balance support IM and the safety and serviceability it ensures. The three steps in IM are (1) identify the potential threats to asset integrity; (2) assess asset condition periodically over the scope of the relevant threats; and (3) respond on a timely basis to remediate or replace the asset as appropriate. The regulatory framework requires the operator to identify and put into play ways to improve their processes and how they apply them as the operator deploys this plan. It also requires that operators implement this plan as part of a cyclical process. The requirements include a maximum interval for this revaluation that is specified as 5 years for hazardous liquid systems and 7 years for natural gas systems, with this process initially focused on high-consequence areas (HCAs) relative to environmental and public impacts. Some perspective on reevaluation intervals and the related aspects during the time that ASME B31.8S was developing can be found in Refs [56] and [57].

Because pipeline systems often incorporate aspects from the early, vintage, and modern eras, which through timely inspection and maintenance are equally fit-for-service, each pipeline system is unique. Through mergers and acquisitions a larger operator might be running five or more systems each of which is unique. Accordingly, the regulatory framework must be generic; its requirements can be prescriptive at a higher level, but it equally must be adaptable on a case specific level in the details of their use.

Because of this uniqueness and the case-specific nature of the day-to-day aspects of IM, this section outlines the elements of the steps involved in IM as opposed to providing case-specific illustrations of the implementation of an IM plan, as a book dedicated to that topic would otherwise be needed. Key high-level documents that more broadly define these steps from a regulatory perspective are Parts 192 [32] and 195 [33] of Title 49 of the CFR, while ASME B31.8S [29] and API 1160 [30] present the Codes and Standards equivalent to the regulatory framework. The CFR, ASME B31.8S, and API 1160 all incorporate by reference a host of other codes, standards, and recommended practices and procedures, some of which are included in the list of references that closes this chapter.

### 43.7.1 Threat Identification and Assessment

Threats identification and assessment are the primary elements of the first step in the IM process. While not all information gleaned by a webcrawl is viable, there is a clear consensus evident by sorting through a few key "high-relevancy" hits in response to the search terms "pipeline threat assessment" that opens to more than 1.8 million documents. The best insight and clear guidance in regard to the scope and processes to be followed exists in the references cited in the prior section. As those citations provide clear guidance and detail the process, there is little merit in repeating that content here.

Suffice it here to note that the process relies on "adequate" documentation, where adequacy requires that the records be considered reliable, traceable, verifiable, and complete. Those records for existing systems must reflect the service experience for the system of interest, and as relevant also information that would support like-similar analysis of comparable systems. For new systems, the records develop as the system is permitted, routed, designed, specified, and then expanded as it is constructed and inspected, and pre-service tested and characterized, and then expand further as it is operated and maintained. Thus, the process is a little different but the scope and intent are comparable.

Regardless of the details the records are evaluated and trended in conjunction with like-similar analysis to identify credible threats from among those cited in the guidance. Those threats found credible are then categorized as "stable over time," "time dependent," or "time independent" and then tabulated and managed accordingly. Credibility is determined relative to factors such as (1) incident history, (2) maintenance experience of the pipeline, (3) data gathered through aboveground surveys that quantify coating condition and can indicate if and where EC is active, (4) monitoring online and in-line, (5) analysis considering the transported product, changes in elevation, flow behavior and so on to quantify the likely occurrence of IC, and other indicators that are known to affect a given threat. Threat categorization based on concepts such as "stable," "time dependent," and "time independent" has been the practice for more than a decade, and will continue in use based on presentations at the recent PHMSA workshop on IVP [35]. However, based on those presentations it also appears that the basis to qualify as "stable" will be more stringently defined, with other changes also plausible. As such, readers should continue to track these developments until the new rule is adopted.

Once credibility is determined, the significance of the potential threat is considered along with whether it is "stable," "time dependent," or "time independent," and the outcomes are prioritized as input to subsequent planning as information becomes available from inspection and condition monitoring activities, and other next steps. Table 43.1 illustrates one example outcome of the threat identification and assessment process for a hazardous liquid pipeline.

**TABLE 43.1 Example of a Threat Matrix for a Hazardous Liquid Pipeline**

| Time Dependent | Stable | Time Independent |
|---|---|---|
| External corrosion | Materials (manufacturing) | Excavation (third-party/ mechanical damage) |
| Internal corrosion | Construction (welding and fabrication) | Hydraulic event (incorrect operations) |
| Stress corrosion cracking | Equipment | Natural hazards (heavy rains/floods, earth movement) |

### 43.7.2 Inspection and Condition Monitoring

Inspection and condition monitoring are the primary elements of the second step in IM. A webcrawl on this subject also opens to about 1.8 million documents, but as noted earlier the most relevant insight and guidance exists in the references cited in the introductory subsection. Again, because those citations provide clear guidance, and detail the process, there is no need to repeat it here.[1]

In contrast to the technology that underlies the process of threat identification and assessment, which continues much as it was a decade ago, the capability to quantify the circumstances that determine threat severity has improved significantly, and the technology and the tools it supports have similarly expanded. The ability to detect as well as the ability to size anomalies has improved in regard to the pipe body, with parallel improvements in the context of girth welds and longitudinal seams. Sensors and algorithms to detect and size metal loss are moving onto a fourth generation. Some crack detection tools are now into a third generation of development, and selectively are capable of detecting and sizing cracks—including stress corrosion cracking (SCC). In-line tools now exist to quantify pipe shape and local deformations, which now selectively permit sizing expansions as well as wrinkles and buckles. The extent of this development and the viability of the tools for both the body and the seam can be assessed in the context of ERW pipelines in Refs [58–60].

Technology and tools to inspect and monitor pipeline condition for in-the-ditch applications and aboveground use have grown as rapidly as the in-line aspects. In-the-ditch tools provide ever-improving schemes and hardware to detect and size anomalies in the pipe body, as well as in girth and long seam welds. But in spite of these developments, just as for the current in-line capability, work remains

[1] This topic is developing rapidly, such that citations to specific technologies and their applications become quickly out of date. Accordingly, this section does not cite specific tools, technologies, or applications, with the added benefit that it avoids any commercial implications. Vendor websites are comprehensive, wherein as is usually the case caveat emptor applies. Many papers are also openly available online, but in some cases care must be taken to sort through the commercial aspects versus the technological facts and supporting science.

before such methods can be relied on as the only basis to quantify condition. At present the best results develop when multiple types of technology are deployed, such as time of flight diffraction (TOFD) and phased array ultrasonic testing (PAUT). Rather good outcomes have resulted using inverse wave field extrapolation, which couples the PAUT and TOFD technologies, although this work is much less mature than the tools in couples. Recent discussion of the development and the viability of such tools for both the body and the seam also can be assessed in the context of ERW pipelines in Refs [58–60].

In closure, while outside the present scope much other work involving inspection and monitoring is underway or already deployed in the context of pipeline integrity and allied aspects. SCADA has been deployed in various stages of development now for decades. What began as rudimentary tools for leak detection continue to develop, with technology also developing to detect pipeline contact or encroachment. The use of satellite images and airborne sensors is expanding with developments in this area covering ground movement, encroachment, and leak detection. It follows that much of practical value can be anticipated in reference to work that presently is still in the realm of research.

### 43.7.3 Life-Cycle Management

Life-cycle management is the focus of the third step in the IM process. Although this element is germane to a host of industries, a webcrawl on this key phrase surprisingly again developed only ~1.8 million hits—which is similar to the aforementioned two elements. While many of the documents identified had general utility, the best insight and guidance as for the aforementioned topics exist in the references cited in the introductory subsection. And, as discussed earlier, because those citations provide clear guidance and detail the process, there is no need to repeat it here.

One hierarchical approach to life-cycle management divides this element into seven major tasks, notably (1) integrate and align data and manage key data gaps; (2) segment and identify HCAs and establish units of analysis; (3) quantify severity and assess risk; (4) prioritize the response timeline for anomalies qualified as defects into immediate, scheduled, or deferred categories; (5) identify the actions and schedule repair and/or replacement; (6) identify and manage change and establish the reevaluation cycle; and (7) document all facets of the process and the actions and update the database across the full IM process. It is apparent from this or any other task breakdown that this element is where "the rubber hits the road."

The technology that underlies each of the aforementioned tasks is evolving slowly to accommodate gaps in the efficient processing and qualification of data, and its alignment, whereas the technology supporting the other tasks is less in a state of flux. As the IM process has been refined by use, as many hazardous liquid operators are now approaching its third cycle, whereas the gas operators are into its second cycle. Vendors also support several of these tasks—some with near turnkey packages for individual tasks, while others have grouped offerings. Finally, some of the inspection companies offer a full turnkey service. As noted in footnote [1] in the context of inspection services, all such offerings are caveat emptor.

The initial focus of the life-cycle management element is data integration and alignment along the pipeline assets and over time, which begins with data gathering. Data such as the design basis, the specifications, construction practices, and welding procedures, are needed as are the properties of the line pipe, its seam type and cross-weld resistance. As-built data, including field coating practices and any routing issues, are needed as are data covering any pre-service inspections and pressure tests. The O&M and incident history is needed as are CP readings and all aboveground survey records, ILI records, and in-the-ditch inspection records. This list in total, with its sub-layers, goes well beyond what can be noted here. Suffice it to say that data gaps need to be addressed, which in a stopgap setting can make use of schemes such as like-similar trending and analysis.

Once the pipeline is characterized from start to finish, HCAs are identified and the line is segmented into units of analysis, which along with condition assessment data feeds severity analysis and risk assessment. In turn, those results lay the foundation to prioritize the response timeline for anomalies qualified as defects into immediate, scheduled, or deferred categories, as guided by the Regulations and Codes, with company best practices or industry leading practices being used as appropriate. Actions and methods best suited to each of the range of features of concern are then identified and a repair/replace schedule developed and implemented. Retrospective review of the database, procedures, new technologies, and observations in the field during the work provides inputs to identify and manage change, and insight to reevaluate the frequency of this set of actions to ensure system integrity. Finally, all facets of the process and the actions of this IM cycle are documented and the databases are updated across the full IM process.

## 43.8 CLOSURE

In view of the evolution indicated in the bar graphs of Figures 43.2–43.4 and the photographic illustrations of the transformation of the industry to meet the challenges posed from the early days evident in Figures 43.7–43.14 to that shown in Figure 43.21, it is clear that the industry has come a long way. But as was apparent in regard to the discussion of the inspection technologies required to quantify the pipeline system's condition, the industry still has a way to go in order to achieve its stated goal of zero incidents.

While the capability to quantify threat severity has improved significantly through improvements in the inspection arena, this capability has evolved more in response to how a sensor can be improved or an interpretive algorithm can be streamlined than it has been driven by the data needs of the analyst who requires inputs on size and shape (and location if in a seam) to quantify anomaly severity. Critical in this context is the realization that severity assessed in terms of defect length and depth or defect area often suffices where plastic collapse controls failure, whereas shape becomes more important where fracture controls failure. Another key in this context is the observation that metrics of fracture resistance are often uncertain, and in the worst case are unknown. This uncertainty, when combined with the fact that defect shape is ill characterized, means that potential failures controlled by fracture are in double jeopardy. Fracture controlled failure occurs at a pressure less than for collapse control, so this double jeopardy develops on the lower-bound for failure, and thus is an issue for pipelines built of steels with limited fracture resistance. It follows that those developing the inspection tools need to collaborate with those that use their data to more effectively quantify pipeline condition. This collaboration in regard to (1) anomaly size, shape, and location if in a seam and (2) properties local to the anomaly is essential if we are to achieve the stated industry goal of zero incidents [19].

The second related concern involves defect severity assessment and prioritization, as follows. For the last several decades we initially sought to detect anomalies, and thereafter sought to size them so their severity could be assessed. There are two fundamental assumptions in this context. First, this assumes that schemes that would detect a given type of defect would be equally useful and effective in sizing it. Second, this assumes that defect size correlated directly with severity throughout the length of the pipeline. There is no question that size (including the effects of shape) correlate directly with severity and priority if the loads (used generically) are the same and the properties are the same. Reality is that a pipeline has a distribution of properties that also influence failure, which is overlaid on a distribution of loads, which is overlaid on a population of defects that are represented by populations of lengths, depths, and shapes, where the last three also share distributions of correlation coefficients. Therefore, length and depth do not uniquely determine severity or priority.

If the aforementioned distributions were all known in balance for a given pipeline, then the analytical tools of RBD could couple and interpret them, to assess how they combine to control failure. However, that technology cannot identify where failure will occur along the pipeline, much less when: Therefore, it cannot determine if the potential failure will occur in or near an HCA, and/or in which HCA.

Because the line pipe's properties along the pipeline and sites of unique loadings overlay the defect sizes and shapes found by ILI, the "worst" of the defects quantified by the largest feature is not necessarily the first to fail. Those that have done full-scale tests with a range of defect sizes in the pipe only to see it fail in the pipe body remote to the defect have seen graphic evidence of this reality. It follows that we need to manage integrity in a framework that addresses likely failure sites based on defect sizes, local properties, and locally unique loads. This could be accomplished through an adaptation of the approach and processes used today, wherein the "where" and "when" become the focal point in lieu of the largest defect. Until we can identify the most likely locations of the combination of the worst of the loads, the properties, and the defects, you cannot effectively avoid the next major occurrence in an area that impacts the public or the environment.

At present, defect severity and management response tend to be based solely on defect length and depth—because that is what we can measure today. At present a tool to practically quantify or even infer the fracture properties, as they are distributed in a piece of line pipe steel, or other such structural material, does not exist. While concepts exist among some thought leaders regarding the basis for such measurements at a laboratory scale—the need to quantify fracture resistance persists as a critical technology gap. This technology gap affects integrity for many aboveground structural applications: it is made more complex for buried structures such as pipelines. Thus, many would benefit by bridging this gap.

The situation is not so bleak in regard to cases where failure is controlled by plastic collapse, as tools exist to quantify the UTS in-the-ditch—so for such pipelines it is a matter of practical deployment more than a technology gap. The situation in regard to IM for pipelines subjected to unique loads is also well developed, as today many in the industry can identify sites where the threat of unique loading exist. Indeed, this aspect is already being dealt with by some operators as a layer in their risk models, with adequate numerical modeling capable of assessing integrity.

Until "where" failure is most likely better known in terms of local properties, quantifying "when" can be quite uncertain. As long as these aspects are poorly addressed, the goal of zero incidents in HCAs remains elusive. Realizing that today's "smart" pigs have evolved over almost 50 years since their first use, the first field run for a true "mensa" pig, which will report likely failure hotspots determined by use of local measurements and analysis is well down the road. But there is an upside: a near-term technology bridge can be built that quantifies the necessary resistance metrics through correlations based on (1) the fundamental "structure–property" relationships of materials science [61] and (2) data that characterize the pipe steel in terms of steel chemistry and processing harvested from "Moody reports" in the operator's archives. Clearly, this approach cannot quantify "where" to the same extent that local properties could. But, it does open to a near-term stopgap that could identify hotspots and rank severity relative to HCAs based on parameters that can be harvested from the archives held by many operators. As such, this approach has much potential to move the bar closer to the industry's goal of zero incidents.

## ACKNOWLEDGMENTS

Many individuals and operators have contributed a host of photographs and in some cases also documentation essential to build the brief history presented and to document the evolution of the tools and technology that underlie that evolution. Others have contributed thoughtful dialog. With missing someone is always a concern in citing names, a few individuals critical to this process and supporting discussions include Bill Amend, Ted Clark, Rick Gailing, and Jerry Rau. Their operating companies past and present, including Southern California Gas Company, Panhandle (Energy Transfer), and Columbia (NiSource) likewise provided some of the archival information that underlie this work, as did Petrobras. These contributions are gratefully acknowledged.

## REFERENCES

1. Anon. History of the World Petroleum Industry, Available at http://www.geohelp.net/world.html, accessed 2/11/2013.
2. Pees, S.T. Oil History: Pipelines, Failure and Success, 1861–1864. Available at http://www.petroleumhistory.org/OilHistory/pages/Pipelines/failure.html, last accessed 2/11/2013.
3. Anon. History of Pipelines, CEPA. Available at http://www.cepa.com/about-pipelines/history-of-pipelines, last accessed 2/10/2013
4. Anon. (2007) About U.S. Natural Gas Pipelines, Energy Information Administration, June 2007.
5. Anon. (2013) Annual report mileage for hazardous liquid or carbon dioxide systems. PHMSA, January 31, 2013.
6. Anon. Energy Information Administration, Available at http://www.eia.gov/. Search site resources on liquid versus natural gas and contrast the networks and their traits.
7. Clark, E.B., Leis, B.N., and Eiber, R.J. (2004) Integrity characteristics of vintage pipelines. Battelle Report to INGAA Foundation, October, 2004. Available at http://primis.phmsa.dot.gov/gasimp/docs/IntegrityCharacteristicsOfVintagePipelinesLBCover.pdf.
8. Gray, J.M. (2002) An independent view of linepipe and linepipe steel for high strength pipelines. API X-80 Pipeline Cost Workshop, Hobart, AU, October 30, 2002.
9. Anon. Available at http://primis.phmsa.dot.gov/comm/reports/safety/psi.html, last accessed 24/1/2014.
10. Beedle, L.S. (1964) *Steel Design*, Ronald Press Co.
11. Beedle, L.S. (1960) *On the Application of Plastic Design*, Fritz Engineering Laboratory (Lehigh University).
12. Turkstra, C. (1962, 1970) *Theory of Structural Design Decisions*, University of Waterloo Press. See also Hasofer, A.M. and Lind, N.C. (1973) *An Exact and Invariant First-Order Reliability Format*, University of Waterloo Press (Solid Mechanics Division).
13. Gallagher, J.P., Giessler, F.J., Berens, A.P., and Engle, J.M., Jr. (1984) USAF Damage Tolerant Design Handbook: Guidelines for the Analysis and Design of Damage Tolerant Aircraft Structures: Revision B. UDRI Final Report, AD-A 153 161, May 1984.
14. Marshall, P.W. (1979) Strategy for monitoring, inspection and repair for fixed offshore platforms. *Structural Integrity Technology*, ASME, pp. 97–110. See also Pelleni, W.S. (1976) *Principles of Structural Integrity*, Office of Naval Research.
15. Hopkins, P. (2003) The structural integrity of oil and gas transmission pipelines. In: Milne, I., Ritchie, R.O., and Karihaloo, B. (editors), *Comprehensive Structural Integrity*, Vol. 1, Elsevier.
16. Harris, J.A., Jr. (1987) Engine component retirement for cause. United Technologies Report to AFWAL, AFWAL-TR-87-4069, AD-A 192 730, August, 1987.
17. Anon. Available at http://www.dot.gov/, opens to United States Department of Transportation website that opens to websites for the many entities it administers.
18. Anon. Available at http://ops.dot.gov/forms.htm#7000-1/2.
19. Anon. Goal is zero incidents. Available at http://www.ingaa.org/6211/11460.aspx, accessed August 2011.
20. ASME (1995) ASME B31.8: Gas Transmission and Distribution Piping Systems, American Society of Mechanical Engineers.
21. ASME (1995) ASME B31.4: Pipeline Transportation Systems for Liquid Hydrocarbons and Other Liquids, American Society of Mechanical Engineers.
22. API API 5L: API Specification for Line Pipe.
23. API. API 5LX: API Specification for High-Test Line Pipe.
24. Anon. Available at http://www.phmsa.dot.gov/pipeline/historical-rulemaking, last accessed 2/11/2013.
25. Anon. Available at http://www.phmsa.dot.gov/hazmat/regs/rulemaking/archive, last accessed 2/11/2013.
26. Anon. (1970) Minimum Federal Safety Standards for Gas Pipelines, 35 FR 3237, Notice 70-2; Docket No. OPS-3B, 49 CFR Part 192.
27. Anon. (2004) A Brief History of Federal Pipeline Safety Laws, The Pipeline Safety Trust, undated but circa 2004. Available at http://pstrust.org/about-pipelines1/regulators-regulations/a-brief-history-of-federal-pipeline-safety-laws. For details see Gas Pipeline Safety Act of 1968, Public Law 90–481, August 1968 and Hazardous Liquid Pipeline Safety Act of 1979, 49 U.S.C. § 60101, et seq, 1979.
28. Anon. (2002) Pipeline Safety Act of 2002, Public Law 107–355, December 17, 2002.

29. Anon. (2002) ASME Code Supplement on Integrity Management for Pressure Piping, B31.8S, Revision 1, ASME.
30. API (2001) API Standard 1160: Managing System Integrity for Hazardous Liquid Pipelines, American Petroleum Institute, November 2001.
31. Anon. PHMSA R&D Website. Available at http://primis.phmsa.dot.gov/matrix/.
32. Anon. Transportation of Natural and Other Gas by Pipeline, 49 CRF Part 192. Available at http://www.ecfr.gov/cgi-bin/text-idx?SID=661b3e5c845115babe0ece993cb1ffeb&tpl=/ecfrbrowse/Title49/49cfr192_main_02.tpl.
33. Anon. Transportation of Hazardous Liquids by Pipeline, 49 CFR Part 195 Available at http://www.ecfr.gov/cgi-bin/text-idx?SID=661b3e5c845115babe0ece993cb1ffeb&tpl=/ecfrbrowse/Title49/49cfr195_main_02.tpl.
34. Anon. (1964) Panhandle Lines, Panhandle monthly employee publication, June 1964, Vol. 21, No. 11, Kansas City, MO.
35. Anon. (2013) Public Workshop on Integrity Verification Process: Pipeline and Hazardous Materials Safety Administration, Docket No. PHMSA–2013–0119, August 7, 2013. Available at https://primis.phmsa.dot.gov/meetings/MtgHome.mtg?mtg=91.
36. Anon. Image available at http://www.keywordpictures.com.
37. S.R. Dresser Manufacturing Company (1936) Archives that include similar images in Dresser pipe couplings general catalog No. 36, S.R. Dresser Manufacturing Company.
38. Anon. (1955) Images scanned (with permission) from articles in the *Oil and Gas Journal*.
39. Anon. Available at http://www.weldinghistory.org/whistoryfolder/welding/wh_1900-1950.html, last accessed 13/6/2012.
40. Amend, B. (2014) Early Pipeline Construction, Maintenance, and Condition Monitoring: The Evolution of Pipeline Integrity Technologies in the Early to Mid-Twentieth Century, DNV GL White Paper, February 18, 2014.
41. Anon. (1964) Keeping Line 100 Fit, Panhandle monthly employee publication, June.
42. Leis, B.N., Zhu, X.K., and Clark, E.B. (2004) Criteria to assess wrinkle bend severity for use in pipeline integrity management. Battelle Final Report on Project PR3-03113 to the Pipeline Research Committee International, August 2004.
43. Binckley, M.J. (1952) Which bending method. *Gas*, May, 122–125.
44. Feier, I.I., Leis, B.N., and Zhu, X.-K. (2010) National grid miter joint testing. Battelle Report to National Grid, June, 2009: Appendix A, Waiver/SP Petition to PHMSA, May 19, 2010.
45. Weise, J.D. (2010) Correspondence Concerning Case 09-G-0552 (National Grid), September 14, 2010. Available at http://phmsa.dot.gov/staticfiles/PHMSA/DownloadableFiles/Files/Pipeline/SP/Brooklyn%20Union%20Gas%20Company%20National%20Grid%20Complete.pdf, last accessed 13/1/2014.
46. Anon. (2012) ASME B16.9: Factory-Made Wrought Buttwelding Fittings, American Society of Mechanical Engineers.
47. MSS MSS SP-75: Specification for High-Test Wrought Butt Welding Fittings, Manufacturers Standardization Society (MSS) of the Valve and Fittings Industry, Inc.; see also API Spec6D: Specification for Pipeline Valves.
48. Berry, F.C. and ASNT ASNT recommended practices for nondestructive testing personal qualification and certification (SNT-TC-1A) and its uses. In: Berger, B. (editor), *Nondestructive Testing Standards: A Review, ASTM STP 624*, American Society of Nondestructive Testing.
49. Anon. (2003) ASTM E164: Standard Practice for Ultrasonic Contact Examination of Weldments, American Society for Testing and Materials.
50. Anon. (2010) ASTM E273: Standard Practice for Ultrasonic Examination of Longitudinal Welded Pipe and Tubing.
51. Anon. (2013) API 1104: Standard for Welding Pipelines and Related Facilities.
52. Burdekin, F.M. and Dawes, M.G. (1971) Practical use of linear elastic and yielding fracture mechanics with particular reference to pressure vessels. Proceedings of IME Conference, May 1971, London, pp. 28–37.
53. Leis, B.N., Forte, T.P., Flamberg, S., and Clark, E.B. (2005) Emerging bedding, padding and related pipeline construction practices. Battelle Report to RSPA, Contract DTRS56-03-T-0005, and Interstate Natural Gas Association of America, January 2005.
54. George, H.H. and Rodabaugh, E.C. (1959) Tests of Pups Support, Bridging Effect, Pipe Line Industry.
55. Zhu, X.K. and Leis, B.N. (2004) Plastic collapse assessment of unequal wall joints in pipeline transitions. Proceedings of the Storage Tank Integrity and Materials Evaluation, ASME PVP-Vol. 490, pp. 253–260.
56. Leis, B.N. and Bubenik, T.A. (2001) Periodic re-verification intervals for high-consequence areas. GRI Report-00/0230.
57. Leis, B.N. and Bubenik, T.A. (2000) Implementation plan for periodic re-verification intervals for high-consequence areas. GRI Report-00/0246.
58. Kiefner, J.F., Kolovich, K.M., Macrory-Dalton, C.J., Johnston, D.C., Maier, C.J., and Beavers, J.A. (2012) Track record of in-line inspection as a means of ERW seam integrity assessment. Subtask 1.3 Report, US DoT Contract No. DTPH56-11-T-000003, August 22, 2012.
59. Leis, B.N. (2013) Compare-Contrast Analysis of Inspection Data and Failure Predictions Versus Burst-Test Outcomes for ERW-Seam Defects. Subtask 4.1 Report, US DoT Contract No. DTPH56-11-T-000003, June 23, 2013.
60. Leis, B.N. (2013) Time-Trending and Like-Similar Analysis, for ERW-Seam Failures Final Interim Report. Subtask 4.2, US DoT Contract No. DTPH56-11-T-000003, June 30, 2013.
61. Moffatt, W.G., Pearsall, G.W., and Wulff, J. (1964–1966) *Structure and Properties of Materials: Structure*, John Wiley & Sons, Inc., New York, NY, Vol. I. Also see Hayden, H.W., Moffatt, L.G., and Wulff, J. (1964–1966) *Structure and Properties of Materials: Mechanical Behavior*, John Wiley & Sons, Inc., New York, NY, Vol. III.

# 44

# PIPELINE REPAIR USING FULL-ENCIRCLEMENT REPAIR SLEEVES

William A. Bruce[1] and John Kiefner[2]
[1]*Det Norske Veritas (U.S.A.), Inc., Dublin, OH, USA*
[2]*Kiefner and Associates, Inc., Worthington, OH, USA*

## 44.1 INTRODUCTION

Damage to operating pipelines is a major concern for pipeline operators. Common types of damage include corrosion, either external or internal, mechanical damage, dents, stress corrosion cracking (SCC), and so on. External corrosion is the most common need for repair of an operating pipeline. As pipelines become older, the need for repair often increases as the result of deterioration of protective coating materials and subsequent corrosion over time. There are significant economic and environmental incentives for performing pipeline repair and maintenance without removing the pipeline from service. From an economic viewpoint, a shutdown involves revenue loss from the loss of pipeline throughput, and for gas transmission pipelines, also from the gas lost to the atmosphere. Since methane is a so called greenhouse gas, there are also environmental incentives for avoiding the venting of large quantities of gas into the atmosphere. Once the need for repair has been established, the predominant method of reinforcing or encapsulating corrosion and other types of damage in cross-country pipelines is to install a welded full-encirclement steel repair sleeve. This chapter covers the repair of pipelines using full-encirclement repair sleeves, which is also covered elsewhere [1,2]. Advantages and disadvantages compared with other repair methods are discussed and the concerns for welding onto in-service pipelines are also reviewed, as are the approaches used to address them.

## 44.2 BACKGROUND

To prevent an area of corrosion or other damage from causing a pipeline to rupture, the area containing the damage must be reinforced to prevent the pipeline from bulging. Alternatively, the area containing the damage can be encapsulated to contain any leak that may occur. The predominant method of reinforcing or encapsulating damage in cross-country pipelines is to install a full-encirclement repair sleeve. There are two basic types of tight-fitting full-encirclement steel sleeves: Type A and Type B. The primary function of both types is structural reinforcement of a defective area. Type B sleeves can also contain pressure. There are also a variety of encapsulation type sleeves, the primary function of which is to contain pressure without providing structural reinforcement.

While Type A sleeves can be installed without welding to an in-service pipeline, the ends of Type B sleeves must be welded to the pipeline to contain the pressure. There are often significant incentives for avoiding interruption of pipeline service; therefore, this welding is often performed "in-service." There are two primary concerns with welding onto in-service pipelines, whether for installing repair sleeves or installing a branch connection prior to "hot tapping." The first is for "burnthrough," where the welding arc causes the pipe wall to be penetrated. The second concern is for hydrogen cracking, since welds made in-service cool at an accelerated rate as the result of the flowing contents' ability to remove heat from the pipe wall. Because the concerns are the

*Oil and Gas Pipelines: Integrity and Safety Handbook*, First Edition. Edited by R. Winston Revie.

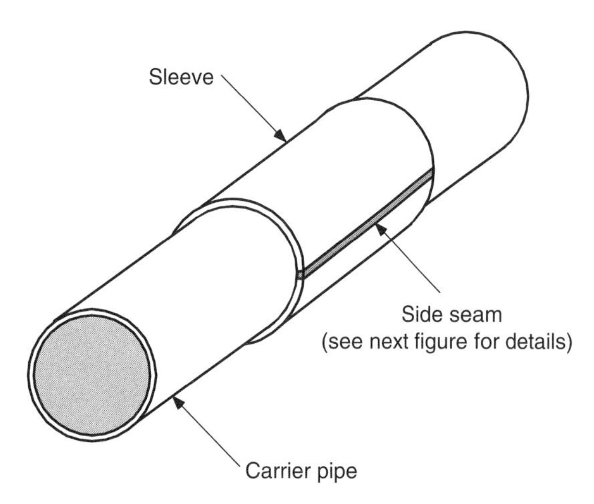


**FIGURE 44.1** Illustration of Type A (reinforcing) sleeve. (Material provided courtesy of Pipeline Research Council International. Excerpted from PRCI report PR-218-9307, "Pipeline Repair Manual", available at www.prci.org.)

same for welding repair sleeves and hot taps, this type of welding is often referred to as hot tap welding.

## 44.3 FULL-ENCIRCLEMENT STEEL SLEEVES[1]

The results of work that began in the 1970s [3–6] show that a properly fabricated steel sleeve restores the strength of a defective pipeline segment to at least 100% of the specified minimum yield strength (SMYS). Damaged pipe sections repaired using properly installed repair sleeves failed in the pipe body away from the repair.

### 44.3.1 Type A Sleeves (Reinforcing)

*Type A* sleeves are particularly attractive because they can be installed on a pipeline without welding to the carrier pipe. Such sleeves provide reinforcement for the defective area. They cannot contain pressure and are used only for non-leaking defects. To protect the repair crew from a defect that may be on the verge of failure, they should be installed at a pressure level below that at which the defect might be expected to fail (e.g., 20% less than that the pressure that was present at the time the defect was discovered).

A typical configuration and weld details for a Type A sleeve are shown in Figures 44.1 and 44.2, respectively. The sleeve consists of two halves of a cylinder of pipe or two appropriately curved pieces of plate that are placed around the carrier pipe at the damaged area and after positioning, are joined by welding the longitudinal side seams. As shown in Figure 44.2, the seams may be single-V butt welds or overlapping steel strips fillet welded to both halves. If the side seams are to be butt welded and the sleeve halves are to be made from the same diameter pipe as the carrier pipe, then each *half* should actually be more than half of the circumference of the piece of pipe (Figure 44.3). Otherwise, the gap to be filled by butt welding will be too large. With the overlapping strip concept, it is not essential that each *half* actually be more than half of the circumference because the gaps can be easily bridged by the side strips.

One major advantage of Type A sleeves over other types of repairs is that for relatively short flaws they can function effectively without necessarily being a high-integrity structural member. Relatively short flaws are those for which the length ($L$) is $\leq\sqrt{20Dt}$, where $D$ is pipe diameter, $t$ is pipe wall thickness, and all dimensions are in consistent units. This is the length beyond which hoop stress can no longer distribute itself around the flaw. For such flaws, the role of the sleeve is limited to restraining bulging of the defective area. As a result, they can be fabricated simply and require no rigorous nondestructive inspection to ensure their effectiveness. Also, because Type A sleeves do not carry much hoop stress in the case of short flaws, they can fulfill their restraining role without necessarily being as thick as the carrier pipe. As a rule of thumb, the thickness of Type A sleeves should not be less than two-thirds of the thickness of the carrier pipe when used to repair short flaws (as already discussed), assuming that the sleeve is at least as strong as the carrier pipe. When designing a Type A sleeve, it is common to simply match the grade and wall thickness of the carrier pipe.

With flaws longer than $\sqrt{20Dt}$, the sleeve thickness should be at least as great as that of the carrier pipe, again assuming that the sleeve is at least as strong as the carrier pipe. Since hoop stress cannot redistribute itself around these longer areas, they may plastically yield transferring some of the hoop stress to the sleeve.

It is acceptable to use sleeves thinner than the carrier pipe if their strength is increased by an amount sufficient to compensate for their thickness being less than that of the carrier pipe. In a similar fashion, the sleeve could be of an acceptable material with lower strength than that of the carrier pipe if its thickness is increased to compensate for the difference in strength. To assure that adequate reinforcement exists and that the sleeve will indeed prevent a rupture, the sleeve should be extended beyond the ends of the defect (i.e., onto sound, full-thickness pipe material) at least 50 mm (2 in.).

Type A sleeves have some minor disadvantages. First, they are not useful for circumferentially oriented defects because they have no effect on the longitudinal stress in the carrier pipe. Second, they cannot be used to repair leaking defects. Third, a potential crevice is created in the form of an annular space between the sleeve and the carrier pipe that

[1] Material in Section 44.3, "Full-Encirclement Steel Sleeves," through 44.3.2.2. (up to and including Sleeve Configurations for Curved (Field-Bent) Pipe) is provided courtesy of Pipeline Research Council International. Excerpted from PRCI report PR-218-9307, Pipeline Repair Manual, available at www.prci.org.

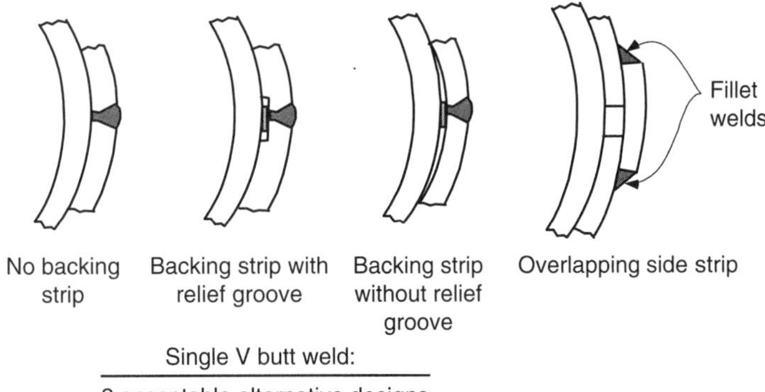


**FIGURE 44.2** Weld details for Type A sleeve. (Material provided courtesy of Pipeline Research Council International. Excerpted from PRCI report PR-218-9307, "Pipeline Repair Manual", available at www.prci.org.)

**FIGURE 44.3** Sleeve fabrication from pipe material. (© ATP Welding, Inc.)

may be susceptible to corrosion if not sealed using an effective coating. However, there have been no reported service failures caused by this latter potential problem. Because of this potential problem, however, some companies use full-encirclement sleeves as Type A sleeves, but weld the ends to the pipe to prevent further corrosion, thus making them essentially Type B sleeves. This practice should be avoided because of the potential concerns for welding onto an in-service pipeline. The best way to avoid welding problems is to avoid welding where it is not necessary.

***44.3.1.1 Assuring Effective Reinforcement*** Since the primary function of both Type A and Type B sleeves is to prevent bulging, this section pertains to assuring effective reinforcement. To be effective, Type A sleeves should reinforce the defective area, restraining it from bulging radially as much as possible. This applies to tight-fitting Type B sleeves as well. First and foremost, the sleeve should be installed as tightly as possible—that is, with a minimal gap between the sleeve and the carrier pipe in the area of the anomaly. Forming and/or positioning the sleeve so that it firmly contacts the carrier pipe, especially at the area of the defect, can assure that the gap is minimized. One or more of the following actions (discussed separately in this subsection) can further enhance the effectiveness of a Type A sleeve:

- Reduce pressure in the carrier pipe during sleeve installation.
- Externally load the sleeve to force it to fit tightly against the carrier pipe.
- Use a hardenable material that will fill in any gaps in the annular space between the sleeve and the carrier pipe.

*Pressure Reduction* Pressure reduction is essential if the defect being repaired is at or near its predicted failure pressure at the start of the repair operation. Because the extent or nature of a given defect may not be apparent upon initial excavation, the pressure should be reduced to protect the repair crew. When the pressure is reduced, the radial bulging at the defect location is reduced and can be prevented from recurring during re-pressurization by using a tightly fitting sleeve. Research results [3,4] and industry practice indicate that a pressure reduction of 20% from the predicted failure pressure is adequate for the application of Type A sleeves.

Typically, a pressure reduction is not necessary if it can be shown that the defect is not at or near its predicted failure pressure. For example, if the results of a defect assessment show that the predicted failure pressure is already 20% or more above the current pressure, a Type A sleeve can effectively repair the defect without a further reduction in pressure.

Reinforcing sleeves usually do not share much of the hoop stress that is acting on the carrier pipe unless special

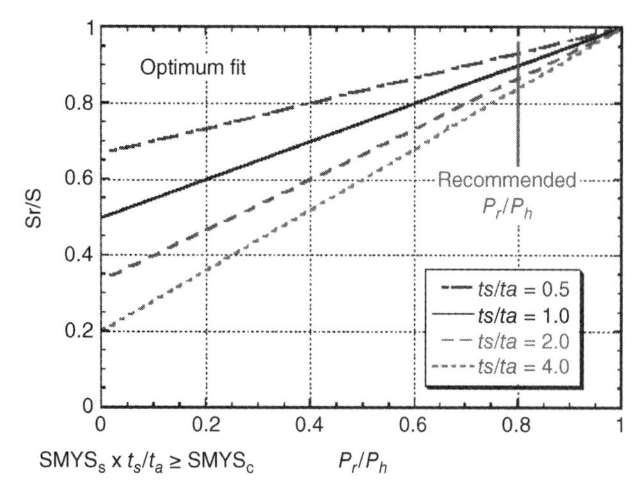


**FIGURE 44.4** Theoretical relationships between carrier pipe stress, repair pressure, and wall thickness. (Material provided courtesy of Pipeline Research Council International. Excerpted from PRCI report PR-218-9307, Pipeline Repair Manual, available at www.prci.org.)

application techniques are used. Even if the sleeve fits perfectly and has 100% efficient side seams, it will at most carry one-half of the hoop stress recovered after a pressure reduction if its wall thickness is the same as that of the carrier pipe. The optimum amounts of stress sharing produced by a snugly fitting sleeve for various amounts of pressure reduction are illustrated in Figure 44.4. The notation used in Figure 44.4 is as follows:

$t_a$ = actual wall thickness of carrier pipe
$t_s$ = wall thickness of steel sleeve
$S_o$ = initial hoop stress in carrier pipe
$S_r$ = reduced hoop stress in carrier pipe after installation of steel sleeve
$P_r$ = reduced pressure at time sleeve is applied
$P_h$ = highest pressure previously experienced by the carrier pipe after defect was present
$SMYS_c$ = specified minimum yield strength of carrier pipe
$SMYS_S$ = specified minimum yield strength of sleeve

The actual amount of hoop stress supported by the sleeves is usually much less than indicated on the graph due to variations in fit and efficiency of side seams. In spite of this, a properly fabricated sleeve can restore the strength of a defective section of pipe to at least 100% of SMYS. As noted earlier, for relatively short defects, stress sharing is not necessary for effective reinforcement. The primary function of Type A sleeves is structural reinforcement (i.e., restraint of bulging). To do this, however, they should fit snugly to the pipe.

Sleeves used to repair long defects (i.e., $L \geq \sqrt{20Dt}$) should be capable of sustaining a significant amount of hoop stress and are expected to absorb an increased amount of hoop stress compared with sleeves used to repair short defects. The reason is that a long defect-weakened region will not be able to distribute hoop stress to the areas of the carrier pipe beyond the ends of the defect. Instead, the region will tend to yield plastically and transfer circumferential stress to the sleeve.

*Mechanical Loading* The two halves of a sleeve can be forced to conform to the carrier pipe and their sides can be drawn together appropriately for welding by mechanical means such as those shown in Figures 44.5–44.7. These can consist of chains and jacks (Figures 44.5b and 44.6) or special preloading devices (Figures 44.5c and 44.7). Lugs can be preinstalled on each half (Figure 44.5a). At the option of the installer, the lugs can be cut off after installation or left in place. Cutting them off facilitates coating the sleeve, an important consideration. A third option is the special chain–clamp device shown in Figure 44.5c and Figure 44.7. The hydraulic actuator that accompanies this latter device can be used to produce a significant preload in the sleeve. A significant preload can enhance the effectiveness of a Type A and tight-fitting Type B sleeves in the same manner as a pressure reduction in the carrier pipe. However, preload should not be substituted for pressure reduction in cases where a reduction of pressure is necessary for safety prior to the start of repair operations.

*Hardenable Fillers* Hardenable fillers, such as epoxy compounds, are frequently used to ensure that no gaps exist between the sleeve and the carrier pipe, particularly in the area of the defect being repaired. These compounds are typically mixed and trowelled into depressions in the carrier pipe, such as corroded areas and dents. After the mixture hardens, the filler is shaped using files or other similar tools until the profile of the outside diameter is restored. For Type A sleeves, another alternative is discussed as follows. Before the mixture hardens (Figure 44.8), the sleeve halves are placed around the pipe, and mechanical means, such as those already discussed, are used to squeeze the excess filler material. By the time the side seams of the sleeve have been welded, the filler mixture has usually hardened and load transfer between the sleeve and the carrier pipe is assured at all defect locations. This latter technique is not appropriate for Type B sleeves as any filler that that is squeezed to the sleeve ends would cause welding difficulties. Tests performed on pipe sections with filled gouges and dents [3,4] showed that such fillers are very effective.

*Fit-Up on Submerged-Arc-Welded and Flash-Welded Line Pipe* One concern with respect to applying Type A and

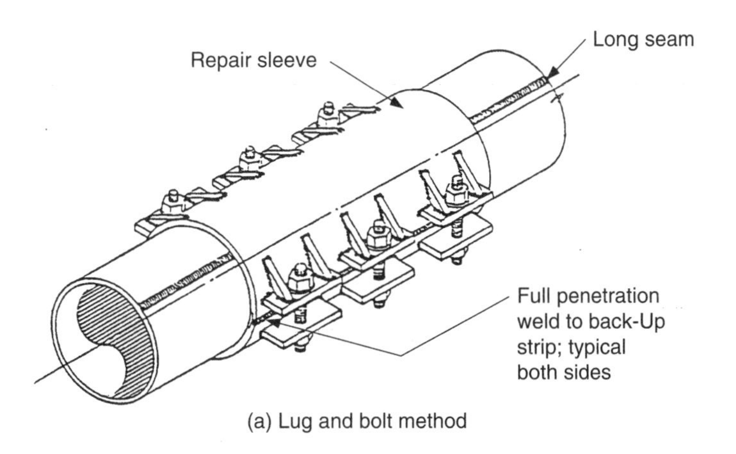


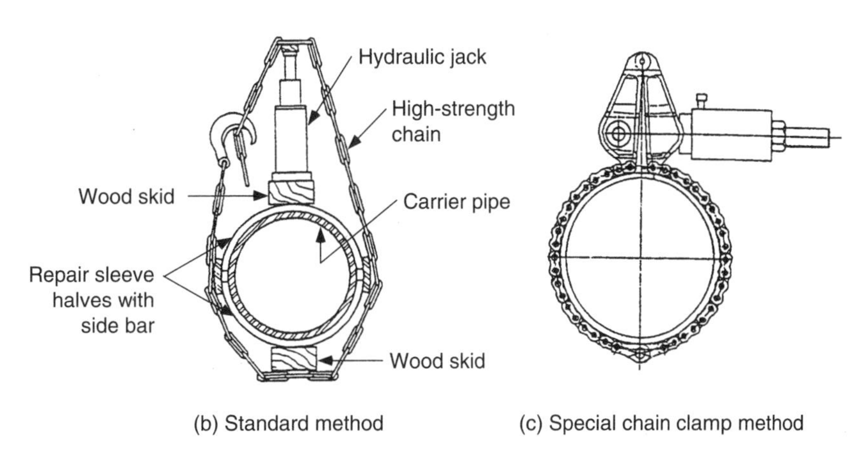


**FIGURE 44.5** Mechanical methods for assuring tight-fitting Type A sleeves. (Material provided courtesy of Pipeline Research Council International. Excerpted from PRCI report PR-218-9307, Pipeline Repair Manual, available at www.prci.org.)

**FIGURE 44.6** Photograph of jack-and-chain method shown in Figure 44.5b. (Photo provided courtesy of Det Norske Veritas (U.S.A.), Inc.)

**FIGURE 44.7** Photograph of chain–clamp device shown in Figure 44.5c. (Photo provided courtesy of PLP Pipeline Products and Services Inc.)

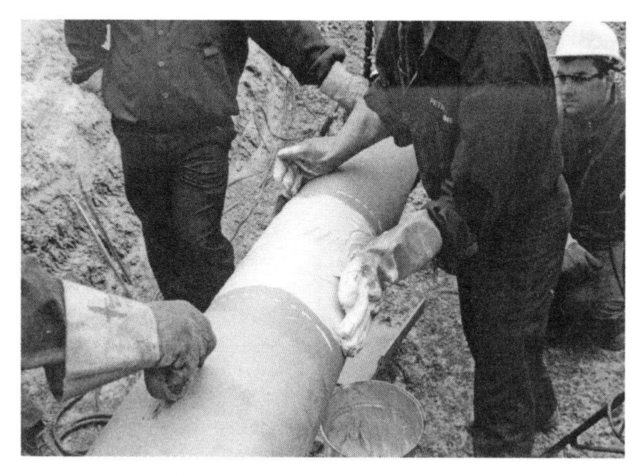

**FIGURE 44.8** Hardenable filler prior to being squeezed by sleeve halves. (Photograph courtesy of Petro-Line, Inc.)

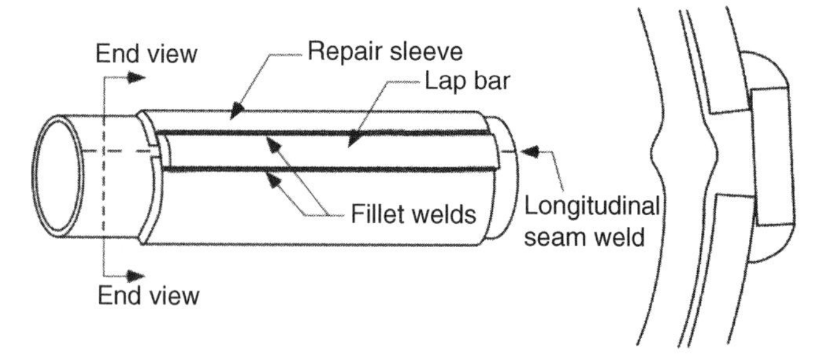


**FIGURE 44.9** Seam weld reinforcement positioned in gap created by side strip. (Drawing provided courtesy of Det Norske Veritas (U.S.A.), Inc.)

tight-fitting Type B sleeves is the presence of a crown or reinforcement on the seam weld of submerged-arc-welded (SAW) carrier pipe or the flash on flash-welded carrier pipe. To assure a tight-fitting sleeve, three options are available. The first option is to remove the weld crown or flash by grinding it flush to the surface of the carrier pipe. This option is acceptable if the pressure has been reduced as suggested in Section 3.1.1.1, or the seam weld has been inspected and found to be free from significant defects. The second option is to grind a compensating groove in one of the sleeve halves. If this second option is selected, it may be desirable or necessary (in the case of long defects or repair using a Type B sleeve) to use a sleeve that is thicker than the carrier pipe by an amount that compensates for the thickness of material removed, including any compensation needed for differences in material strength. Another option for Type A sleeves with fillet-welded overlapping side strips is to position the seam weld reinforcement in the gap created by one of the side strips (Figure 44.9).

#### *44.3.1.2 Special Type A Sleeve Configurations*

*Epoxy-Filled Shells* British Gas developed a variation of the sleeve repair concept in the form of their *epoxy-filled shell* repair method [7]. In this case, the *shell* is a sleeve with a standoff distance of several millimeters from the carrier pipe. The shell is placed on the defective pipe, and bolts are used to center it. The side seams are then welded, and the gaps at the ends are sealed with epoxy putty. After the he putty has hardened, a high-stiffness epoxy grout is pumped into the annular space until it flows out of an overflow hole at the top of the sleeve. The grout is heavily filled with graded sands, where the small particles fit into the spaces created between the large particles, so that it has a high compressive strength.

Once the epoxy filler has hardened, the radial bulging tendency of the defect is restrained by the epoxy in the same manner as a conventional Type A sleeve would have if it were directly in contact with the sleeve.

*Steel Compression Sleeves* Steel compression sleeves are a special class of Type A sleeves. They are designed, fabricated, and applied so that the repaired section of the carrier pipe is maintained under compressive hoop stress during subsequent operation. This approach is attractive for repairing longitudinally oriented crack-like defects, such as SCC, because without a tensile hoop stress there is no driving force for crack growth. This type of sleeve is not suitable for the repair of circumferential cracks or for defects in field bends.

Steel compression sleeves involve installing two sleeve halves over the defect area and drawing them together using clamps, jacks, and chains, or lugs and bolts. The sleeve halves are then welded together using conventional welding techniques. Pressure reduction during installation is normally used to induce compression in the carrier pipe. Thermal contraction of the longitudinal side seam welds also promotes compression in the carrier pipe. Epoxy filler is used between the carrier pipe and sleeve to achieve the transfer of stresses. Pressure reduction alone will only transfer a portion of the hoop stress from the carrier pipe to the sleeve.

PetroSleeve™ is a commercial product that was developed to combine pressure reduction with thermal shrinkage of the sleeve for achieving full compression in the carrier pipe. Figure 44.10 illustrates the installation process for a PetroSleeve. Two steel sleeve halves with sidebars are installed over the defect, the sleeve halves are heated, and are initially held in place with chain clamps or hydraulic jacks. The halves are then welded together using two longitudinal sidebars. During installation, an epoxy layer is applied between the sleeve and the carrier pipe. The epoxy is used as a lubricant when the halves are placed on the carrier pipe and later acts as a filler to evenly transfer the load

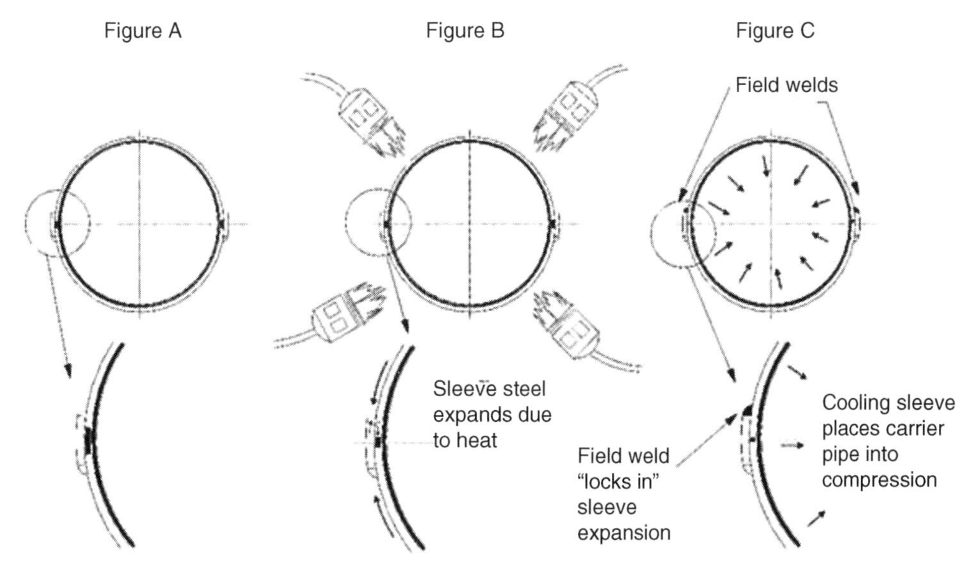


**FIGURE 44.10** Installation steps for the steel compression sleeve: (a) place half-sleeves on carrier pipe, (b) heat sleeve to expand sleeve, and (c) place field welds and cool assembly to achieve compression. (Drawing courtesy of Petro-Line, Inc. Material provided courtesy of Pipeline Research Council International. Excerpted from PRCI report PR-218-9307, Pipeline Repair Manual, available at www.prci.org.)

between the sleeve and the pipe. As with other versions of Type A sleeves, no welds are made to the carrier pipe. Thermal shrinkage of the sleeve upon cooling helps induce a compressive stress into the carrier pipe [8]. A completed PetroSleeve installation is shown in Figure 44.11.

Several factors influence the degree of stress reduction in the carrier pipe. These include the fit of the sleeve, the pipe wall thickness and diameter, the sleeve wall thickness, the internal pressure during installation, and the installation temperature. Specially developed software can be used to determine the target sleeve installation temperature and to help confirm that the desired amount of sleeve compression has been achieved [9].

**FIGURE 44.11** Example of installed and abrasively blasted steel compression sleeve. (Photograph courtesy of Petro-Line, Inc. Material provided courtesy of Pipeline Research Council International. Excerpted from PRCI report PR-218-9307, Pipeline Repair Manual, available at www.prci.org.)

Quality control procedures for PetroSleeve involve monitoring sleeve temperature during the heating process and verification of the achieved carrier pipe compression by measuring how far the two sleeve halves advance toward each other using caliper measurements. Three sets of measurements on each side of the sleeve are typically made. Nondestructive inspection of the completed welds is conducted after cooling.

PetroSleeve has been commercially available since 1994 and has been installed primarily as a means to repair SCC, corrosion, and dents. It also has been used as a permanent field repair method on long-seam indications and arc burns [10]. PetroSleeve uses a fillet welded overlapping side strip for the longitudinal seams. Since the installation of a PetroSleeve induces a compressive stress into the carrier pipe sleeve, the PetroSleeve itself must be in a high state of tension. Fillet welds are not ideal when loaded in tension because of the bending moment that results. Operators may want to consider whether fillet welded overlapping side strips are appropriate, particularly for pipelines subjected to significant cyclic pressurization. Other operators have developed their own design for steel compression sleeves that use full penetration butt welds, which have a higher joint efficiency factor than fillet welds.

### 44.3.2 Type B Sleeves (Pressure Containing)

The other type of steel sleeve used to make pipeline repairs is known as a *Type B* sleeve. The ends of Type B sleeves are

FIGURE 44.12 Installation of a Type B repair sleeve. (Material provided courtesy of Pipeline Research Council International. Excerpted from PRCI report PR-218-9307, Pipeline Repair Manual, available at www.prci.org.)

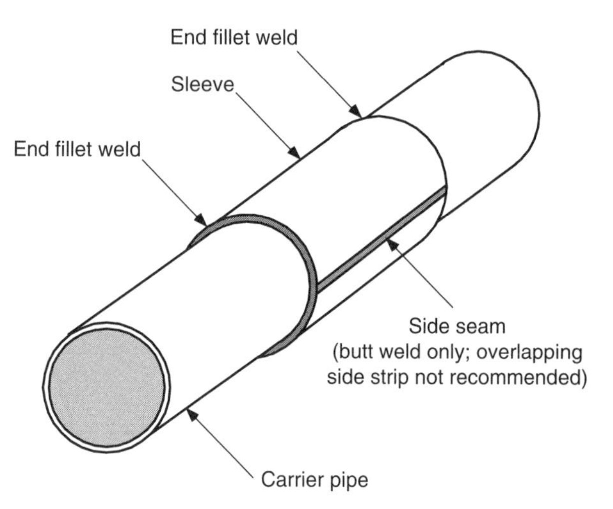


FIGURE 44.13 Illustration of a Type B sleeve. (Material provided courtesy of Pipeline Research Council International. Excerpted from PRCI report PR-218-9307, Pipeline Repair Manual, available at www.prci.org.)

fillet welded to the carrier pipe. The installation of a Type B sleeve is shown in Figure 44.12. Detailed discussions of the issues related to welding on an in-service pipeline are presented toward the end of this chapter. As with Type A sleeves, the primary purpose of tight-fitting Type B sleeves is structural reinforcement (i.e., to prevent a defect from bulging). Since their ends are attached to the carrier pipe, Type B sleeves can be used to repair leaks and to strengthen circumferentially oriented defects. In fact, Type B sleeves have been used in place of girth welds to make tie-ins on a pipeline. Because Type B sleeves may contain pressure and/or carry a substantial longitudinal stress imposed on the pipeline by lateral loads, they should be designed to carry the full pressure of the carrier pipe. Additionally, they should be carefully fabricated and inspected to ensure integrity.

***44.3.2.1 Design*** The typical configuration of a Type B sleeve is illustrated in Figure 44.13. It consists of two halves of a cylinder or pipe or two appropriately curved plates fabricated and positioned in the same manner as those of a Type A sleeve. Since the Type B sleeve is designed to contain full operating pressure, the ends are welded to the carrier pipe, and butt-welded seams are required. For sleeves that will be or may become pressurized, the overlapping strip concept of Figure 44.2d is not acceptable because fillet welds are inherently weaker than sound full-thickness butt welds. Type B sleeves should be designed to the same standard as the carrier pipe. This usually means that the wall thickness of the sleeve will be equal to that of the carrier pipe and that the grade of the sleeve material also will be equal to that of the carrier pipe. It is acceptable to use a sleeve that is thicker or thinner than the carrier pipe and is of lesser or greater yield strength than the carrier pipe as long as the pressure-carrying capacity of the sleeve is at least equal to that of the carrier pipe. Many companies simply match the wall thickness and grade of the pipe material.

For tight-fitting sleeves, the outside diameter of the sleeve is slightly greater than the outside diameter of the carrier pipe. Usually, this point is ignored in the sleeve design even though it causes the sleeve to be slightly under-designed when it's made from the same material as the carrier pipe. If material is removed from the sleeve for a groove to accommodate the carrier pipe seam weld or a backing strip for the sleeve side seam welds, the thickness of the sleeve should be greater than that of carrier pipe by an amount that compensates for the material that is to be removed.

*Sleeve Length* While it should be long enough to extend beyond both ends of the defect by at least 50 mm (2 in.), there is no inherent upper limit to the length of a Type B sleeve. However, practical considerations are likely to impose some limits on sleeve length. If the sleeve length is limited, two requirements should be satisfied. First, as mentioned previously, the sleeve should extend at least 50 mm (2 in.) beyond both ends of the defect. Second, the fillet-welded end of one sleeve should not be any closer than one-half of the carrier pipe diameter to the corresponding end of another sleeve. This latter requirement is needed to avoid a notch-like condition between the two sleeves. If two sleeves should be placed closer than one-half pipe diameter apart, the inboard ends of the sleeves should not be welded to the carrier pipe. Instead, a bridging sleeve-on-sleeve should be used (see Section 3.2.2.3) or the sleeve ends should be joined using a full penetration butt weld (Figure 44.14).

Another important factor that should be considered when installing long sleeves is the weight that is being added to the pipeline in conjunction with how it is being supported during the sleeve installation process. When the sleeve length

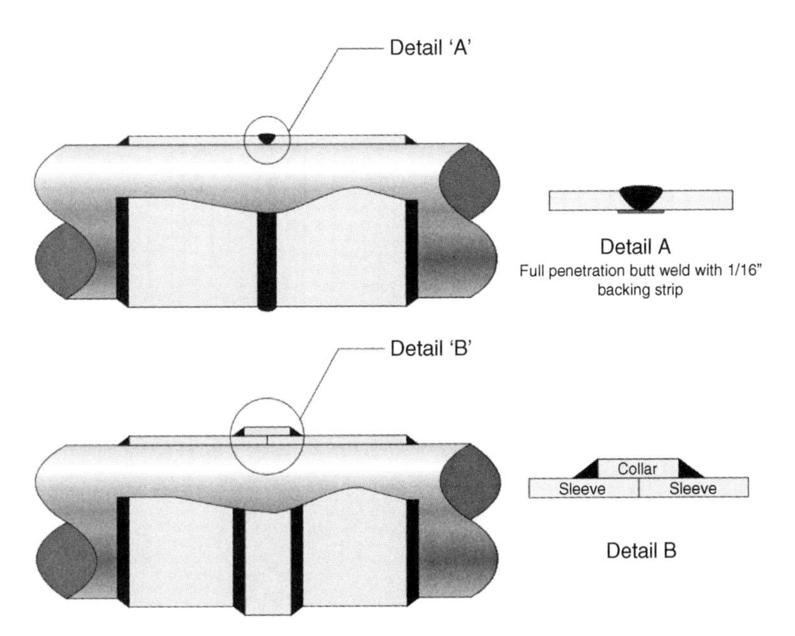


**FIGURE 44.14** Methods for joining the ends of adjacent sleeves. (Figure provided courtesy of Det Norske Veritas (U.S.A.), Inc.)

exceeds four pipe diameters or when two or more sleeves whose total length exceeds four pipe diameters are to be installed within a single excavation, the pipeline operator's written procedures should contain guidelines for support spacing, methods of temporary support (e.g., air bags, sand bags, skids), and soil conditions under the pipeline upon backfilling.

*Leaking Defects* One use for Type B sleeves is to repair leaking defects. Type B sleeves installed over leaking defects become a pressure-carrying component and should meet the same integrity requirements as any other pressure-carrying component in the system. These include the appropriateness of the design (i.e., wall thickness, material grade) and the integrity of the side seams and end fillet welds.

Type B sleeves are installed over leaks in a variety of ways. One common method is to place an elastomer seal between the carrier pipe and sleeve. Chains and hydraulic jacks are then used to force the sleeve halves against the carrier pipe. Once the leak is stopped, welding can be performed safely. Another method involves placing a small branch pipe with a valve over a hole in one of the sleeve halves. The hole and branch are located over the leak and a doughnut-shaped elastomer seal is placed between the carrier pipe and sleeve. Chains and hydraulic jacks are then used to compress the seal that forces the fluid to enter the branch. The fluid then can be released at a safe location and welding of the sleeve can be completed safely. Upon completion of sleeve fabrication, the branch valve is closed and capped. A variation of the same technique uses a plug to seal the branch through the valve, which allows the valve to be recovered.

*Nonleaking Defects* Type B sleeves are sometimes used to repair nonleaking defects. In the past, some pipeline operators used Type B sleeves exclusively because they preferred to have the ends sealed by fillet welds even when no leak existed. Other operators have installed Type B sleeves over nonleaking defects and then hot-tapped through the sleeve and pipe to pressurize the sleeve and relieve hoop stress from the defective area. This was previously common for repair of cracks and crack-like defects but is no longer considered to be necessary.

Even though a Type B sleeve may not be pressurized, any sleeve with ends welded to the carrier pipe should be designed and fabricated to be capable of sustaining the pressure in the pipeline, since there is a chance that it could later become pressurized. For example, if a Type B sleeve is used to repair internal corrosion and the internal corrosion continues, a leak could develop in the carrier pipe and pressurize the sleeve.

***44.3.2.2 Special Type B Sleeve Configurations*** The special-purpose sleeve configurations described in the following paragraphs may be useful for certain applications.

*Sleeves to Repair Girth Welds* The typical configuration of a sleeve used to repair a defective or leaking girth weld is shown in Figure 44.15. The *hump* in the sleeve is designed to accommodate the crown of the girth weld. The ends are welded to the carrier pipe so that the sleeve can share the longitudinal stress. This type of sleeve is also capable of containing pressure in the event that a leak develops. An alternative to the use of a humped sleeve is the use of a sleeve-on-sleeve repair similar to that described in Section 44.3.2.2.3

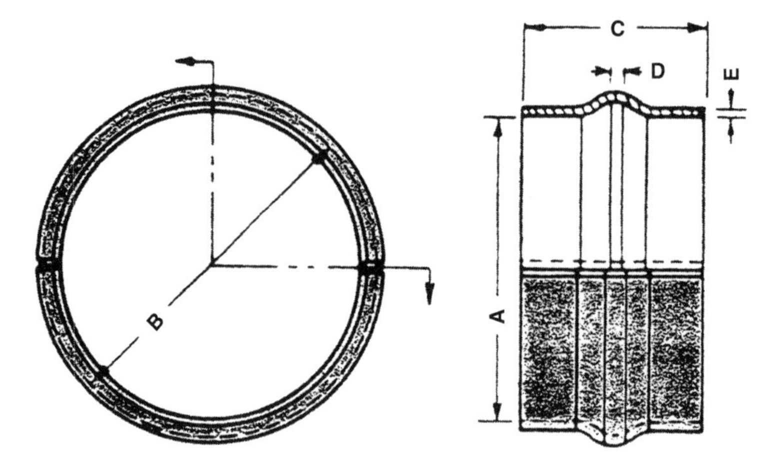


**FIGURE 44.15** Typical sleeve configuration for repair of defective girth welds. (Drawing courtesy of Allan Edwards Inc. Material provided courtesy of Pipeline Research Council International. Excerpted from PRCI report PR-218-9307, "Pipeline Repair Manual", available at www.prci.org.)

or a bridging sleeve. This technique involves two short tight-fitting sleeves that are installed on either side of the defective weld. A third sleeve is then installed over these two tight-fitting sleeves that encapsulates the defective girth weld (Figure 44.16). The third sleeve must be fillet welded to the two tight-fitting sleeves, although only one end of each tight-fitting sleeve needs to be fillet welded to the pipeline. Some operators choose to weld the in-board ends of these so that a weld on the outboard end, which can act as a stress concentration, is avoided. As with a type A sleeve, the outboard ends must be sealed using an effective coating material.

*Sleeves to Repair Couplings* Many older pipelines have joints that were constructed using mechanical couplings. For small diameter pipes, these may be threaded couplings. For large-diameter pipelines, as well as for some small-diameter ones, mechanical compression-type couplings were used. Typically, these couplings rely on longitudinally oriented bolts and collars that are used to compress packing or gaskets to seal against the pipe. These types of couplings provide negligible longitudinal stress transfer along the pipeline. As a result, they are prone to *pullout* incidents when the pipeline is subject to unusual longitudinal loads. To overcome both the pullout problem and the perennial leakage problem with this type of coupling, many pipeline operators have resorted to the repair sleeve configuration shown in Figure 44.17. This type of sleeve, often called a *pumpkin* or *balloon sleeve*, is typically welded to the pipes on both ends. The side seams are also welded so that the sleeve can contain pressure. Because the diameter of the center portion is much greater than the nominal diameter of the pipeline, this additional diameter must be taken into account when designing this type of sleeve from the standpoint of pressure containment. Because the mechanical couplings tend to transfer little or no longitudinal stress along the pipeline, the fillet welds at the ends of this type of sleeve become the primary means of longitudinal stress transfer. Therefore, the quality of the fillet welds for such a sleeve is even more critical than that of the welds at the ends of a conventional Type B sleeve.

A pumpkin sleeve may be used to repair buckles, ovalities, and wrinkle bends because of its ability to fit over such anomalies.

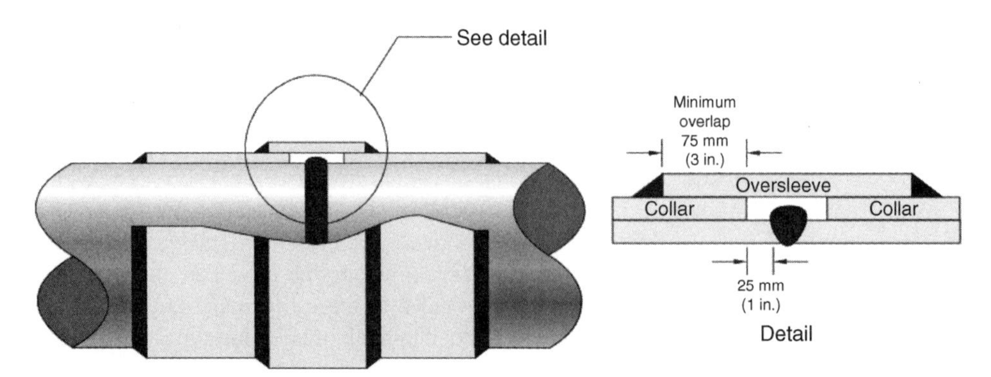


**FIGURE 44.16** Sleeve-on-sleeve repair for defective girth welds. (Figure provided courtesy of Det Norske Veritas (U.S.A.), Inc.)

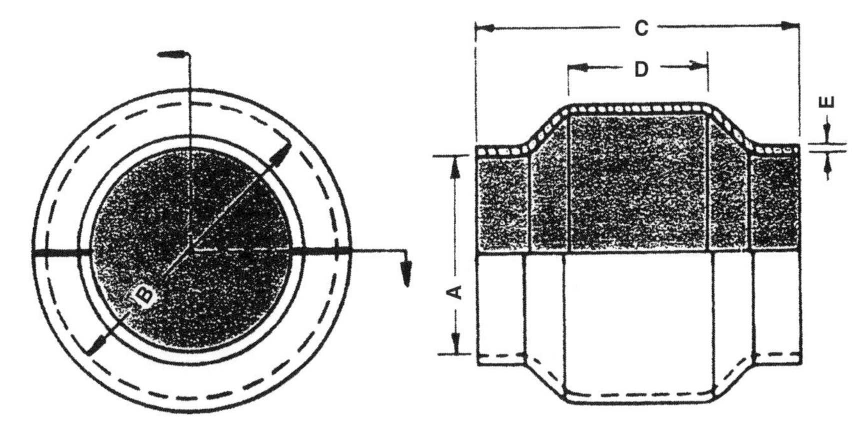


**FIGURE 44.17** Typical sleeve configuration for repair of couplings. (Drawing courtesy of Allan Edwards Inc.). (Material provided courtesy of Pipeline Research Council International. Excerpted from PRCI report PR-218-9307, Pipeline Repair Manual, available at www.prci.org.)

*Sleeve-On-Sleeve Repair* Experience with cracking at the toes of fillet welds around the ends of conventional Type B sleeves led to the development of the sleeve-on-sleeve configuration shown in Figure 44.18. This configuration consists of two rings installed outboard to the ends of the defective sleeve. Each ring is fillet welded to the carrier pipe on the end facing the end of the defective sleeve (i.e., the inboard end). If a toe crack forms at one or both of the rings, it will be contained within the space between the ring and the sleeve. The final step consists of installing two outer sleeves to bridge the gaps between the rings and the defective sleeve. These outer sleeves are fillet welded to both the rings and the defective sleeve to make a leak-tight repair in case the toe crack grows through the wall of the carrier pipe. The outboard ends must be sealed using an effective coating material. A test program [11] showed that this configuration is expected to adequately protect an existing toe crack from causing the pipeline to fail.

*Sleeve Configurations for Curved (Field-Bent) Pipe* It is possible to install a conventional Type A or Type B sleeve on a slightly curved piece of pipe. The shorter the sleeve, the better the fit will be on a curved section of pipe. For a Type A sleeve, the annular space created by the curvature could be filled with a hardenable material to provide contact with the carrier pipe. A relatively short Type B sleeve could be installed effectively, but beyond some length that depends on the pipe size and amount of curvature, a straight Type B sleeve will not fit well enough to a curved pipe to permit a satisfactory installation.

One method of installing a relatively long sleeve on a field bend is the so-called *armadillo* sleeve. The name comes from its appearance as shown in Figures 44.19 and 44.20. This sleeve is comprised of several short segments connected by bridging sleeves. A long corroded field bend could be repaired in this manner. An armadillo sleeve can be Type A if the final two ends are left un-welded to the carrier pipe as

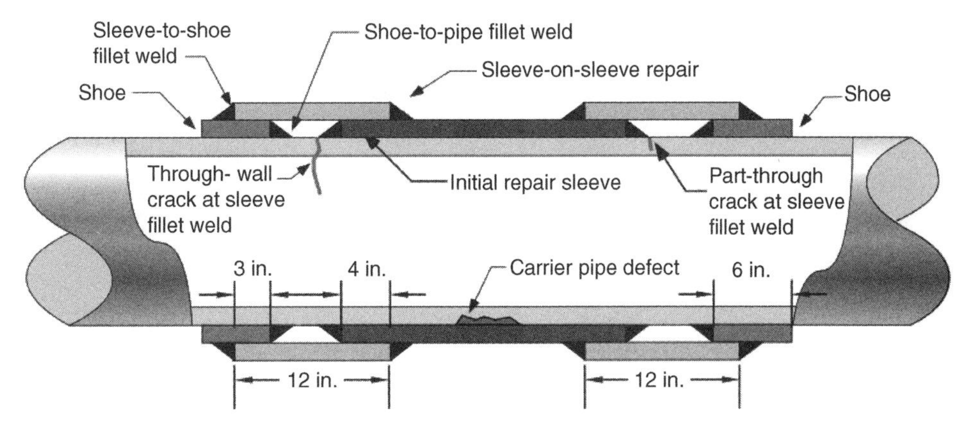


**FIGURE 44.18** Sleeve-on-sleeve repair. (Material provided courtesy of Pipeline Research Council International. Excerpted from PRCI report PR-218-9307, Pipeline Repair Manual, available at www.prci.org.)

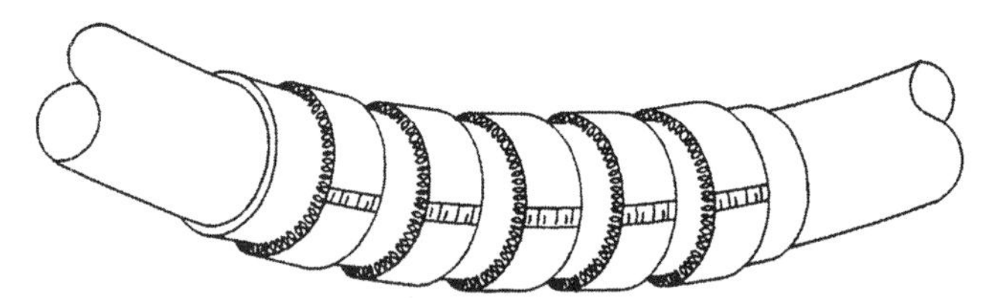

**FIGURE 44.19** Illustration of armadillo sleeve repair. (Material provided courtesy of Pipeline Research Council International. Excerpted from PRCI report PR-218-9307, Pipeline Repair Manual, available at www.prci.org.)

**FIGURE 44.20** Photographs of armadillo sleeve being installed.

shown in Figure 44.19 or Type B if they are fillet welded to the carrier pipe. The biggest disadvantage of the armadillo configuration is the large amount of welding required to fabricate it. Also, it adds considerable weight and stiffness to the repaired section of the pipeline.

*Encapsulation or Leak Box Repairs* Various pipeline components can be used to fabricate pressure containing sleeves that encapsulate a leaking defect in a non-straight section of pipe. For example, a leak in a tee can be encapsulated using a one-size-bigger tee or a leak in an elbow can be encapsulated using a one-size-bigger elbow (Figure 44.21). In both cases, the ends can be made to fit the smaller diameter pipeline using trimmed reducers or elliptical test heads with holes cut in the center. To install such an encapsulation, the components must be split longitudinally and rewelded around the leak.

**FIGURE 44.21** Encapsulation repair using a one-size-bigger elbow (after removal from service). (Photo provided courtesy of Det Norske Veritas (U.S.A.), Inc.)

### 44.3.3 Installation and Inspection of Full-Encirclement Sleeves

***44.3.3.1 Installation*** During installation, the gap between the pipe and tight-fitting sleeves should be minimized using the methods described in Section 44.3.1.1, as this improves the performance of the sleeve. Also, for Type B sleeves, tight fit-up minimizes the stress concentration that can occur at the root of the fillet weld caused by a gap (Figure 44.22).

For sleeves with longitudinal seams that consist of single-V butt welds, the use of backing strips is highly recommended (Figure 44.2b and c) as this prevents the longitudinal seam weld from penetrating into the carrier pipe. Penetration into the carrier pipe is undesirable since any crack that might develop is exposed to the hoop stress in the carrier pipe. Prior to installation, the backing strips can be tack welded to one half of the sleeve.

During installation of Type B sleeves, the longitudinal seams should be completed before beginning the circumferential fillet welds. Additionally, the circumferential weld at one end of the fitting should be completed before beginning the circumferential weld at the other end. This sequence allows the shrinkage of the longitudinal seam welds to improve the fit-up of the fitting and minimizes the stresses on the circumferential welds.

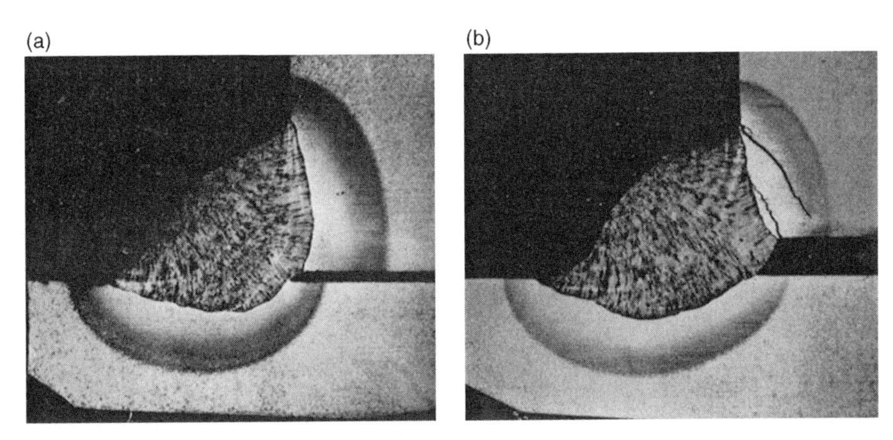


**FIGURE 44.22** Stress concentration caused by gap at fillet weld root (right). (Reproduced courtesy of TWI Ltd.)

When the thickness of a full-encirclement repair sleeve is equal to the carrier pipe wall thickness, the leg length of the fillet weld should be equal to the pipe wall thickness (i.e., a full size fillet weld). If the sleeve thickness exceeds that of the carrier pipe by more than a factor of 1.4, it is recommended that the ends be tapered to 1.4 times the carrier pipe thickness on an angle not to exceed 45°. It is recommended that the leg lengths of the fillet welds for these thicker sleeves be 1.4 times the carrier pipe wall thickness. If there is a gap between the pipe and the sleeve (or fitting), the fillet weld leg length should be increased by an amount equal to the gap (Figure 44.23).

*44.3.3.2 Inspection Requirements* The installation or fabrication of any repair requiring welding on an in-service carrier pipe should be preceded by ultrasonic inspection to determine the remaining wall thickness of the carrier pipe in the regions where welding is to be performed. For the case of fillet welds around the ends of a Type B sleeve, it is reasonable to measure the wall thickness at 50-mm (2-in.) intervals along the circumferential path where the weld is to be located.

Type B repair sleeve welds (sleeve-half butt welds and sleeve fillet welds) should be inspected after welding to assure weld integrity. Since welds that contact the in-service

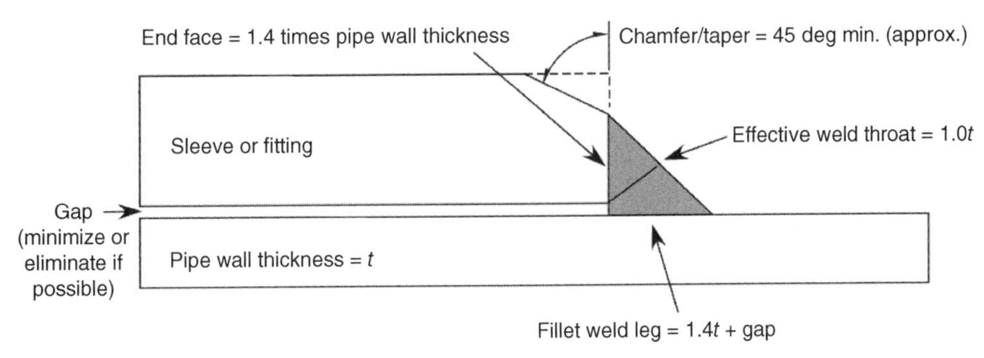


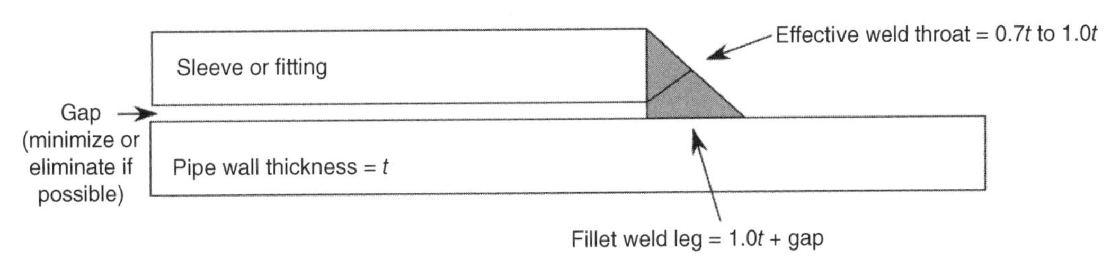


**FIGURE 44.23** Fillet weld leg length recommendations. (Figure provided courtesy of Det Norske Veritas (U.S.A.), Inc.)

carrier pipe can be particularly susceptible to hydrogen cracking, an inspection method that is capable of detecting these cracks, particularly at the carrier pipe weld toe, should be used. Weld joints are usually inspected by means of magnetic particle inspection (MPI), liquid penetrant inspection (LPI), or ultrasonic shear wave inspection. Automated and advanced ultrasonic inspection techniques are sometimes applied to assure the integrity of critical welds. Whatever method or combination of methods is employed, operator skill, training, and experience are critical to achieving a successful inspection. MPI or LPI is expected to reveal surface-connected indications, with MPI typically being more sensitive than LPI. Grinding the toe smooth facilitates the inspection of fillet welds. The best assurance of a quality repair is the use of a proven qualified procedure and a highly trained and qualified repair specialist.

### 44.3.4 Defect Repair Using Composite Materials

The use of fiber-reinforced composite materials for pipeline repairs was developed during work at Southwest Research Institute and Battelle in the late 1980s [12,13]. There are two basic types of composite repair systems; preformed (composite sleeves) and wet lay-up (composite wraps). The first commercially available system was Clock Spring®, which consists of an E-glass/polyester resin-based composite material preformed into a multilayer coil that is installed using an adhesive (Figure 44.24).

The use of wet lay-up systems for pipeline repair began in the early 1990s. These wet lay-up systems use either a resin-impregnated cloth that is activated by water in the field or a cloth that is saturated with epoxy resin in the field (Figure 44.25). The early composite systems all use E-glass as the fiber material. The use of carbon fiber composite material as a substitute for E-glass for pipeline repair was introduced in the late 1990s [14].

FIGURE 44.24 Installation of preformed composite sleeve (Clock Spring). (Reproduced courtesy of Clock Spring Company, L.P.)

FIGURE 44.25 Installation of wet lay-up composite wrap (Armor Plate® Pipe Wrap). (Reproduced courtesy of Armor Plate, Inc.)

For both types of composite repair, a high-compressive strength filler material is used to fill defect areas so that load is effectively transferred to the repair material.

The early work by Battelle showed that steel sleeves are effective because they restrain bulging, or accumulation of strain, in the defective area. Steel sleeves accomplish this while absorbing only 15–20% of the hoop stress in the carrier pipe. Steel sleeves are effective because the stiffness (elastic modulus) of the sleeve material is equivalent to that of the line pipe steel.

The mechanism by which composite materials reinforce areas of damage on operating pipelines is much more complex than that for steel sleeves because the mechanical properties of composite materials are much different than those of steel. While the strength of composite materials is similar to that of line pipe steel, the elastic modulus is significantly lower. Steel has an elastic modulus of approximately $30 \times 10^6$ psi whereas E-glass-based composite materials typically have an elastic modulus of $2.5–6 \times 10^6$ psi. Carbon fiber-based composite materials typically have an elastic modulus that is approximately twice that of E-glass based composites, but still significantly less than that of steel. Therefore, composite materials must experience a significantly greater amount of strain before a load equivalent to that of steel can be carried. For a composite material to prevent a defect from rupturing, the defect must typically plastically deform in the process of the load being transferred to the composite. Brittle pipe material or seams may only be able to tolerate a very small amount of plastic strain before a defect in the pipe or seam grows and fails.

## 44.4 COMPARISON OF STEEL SLEEVES AND FIBER REINFORCED COMPOSITE REPAIRS

When faced with the need to repair an operating pipeline, there are often a variety of repair strategies available to pipeline operators for a given repair situation. While the use of nonmetallic composite materials to repair damage has increased in recent years, the most predominant method of reinforcing damage in cross-country pipelines is to install a full-encirclement steel sleeve. The use of steel sleeves has some advantages over the use of composite materials for many applications, in terms of both performance and cost.

### 44.4.1 Applicability to Various Defect Types

Table 44.1 is an excerpt from the Pipeline Repair Manual (PRCI) [2], which provides an overview of pipeline repair methods that can typically be applied to various types of defects. The first column in Table 44.1 lists 12 types of pipeline defects to which repairs are typically applied. The remaining columns provide information regarding the applicability of steel sleeves (Type A and Type B) and composite repairs.

The notation used in Table 44.1 is as follows:

- "Yes" indicates a repair method that typically can be used to permanently repair the type of defect. In some cases, a footnote(s) is added to indicate qualifying conditions for the method to be considered permanent.
- "No" indicates a repair method that typically is not recommended for repair of the type of defect.

Table 44.1 indicates that, with very few exceptions, the applicability of Type A sleeves and composite repairs is

**TABLE 44.1 Comparison of Capabilities**

| | Repair Options for Various Types of Defects | | |
|---|---|---|---|
| Type of Defect | Type A Sleeves | Composite Repairs | Type B Sleeves |
| 1. Leak (from any cause) or defect >0.8t | No | No | Yes |
| 2. External corrosion | | | |
| 2a. Shallow to moderate pitting <0.8t | Yes | Yes | Yes |
| 2b. Deep pitting ≥0.8t | No | No | Yes |
| 2c. Selective seam attack | No | No | Yes (a) |
| 3. Internal defect corrosion | Yes (b) | Yes (b) | Yes |
| 4. Gouge of other metal loss on pipe body | Yes (c) | Yes (c) | Yes (d) |
| 5. Arc burn, inclusion, or lamination | Yes | Yes (c) | Yes |
| 6. Hard spot | Yes | No | Yes |
| 7. Dent | | | |
| 7a. Smooth dent | Yes (e) | Yes (e) | Yes |
| 7b. Dent with stress concentrator on seam weld or pipe body | Yes (c) (e) (f) | Yes (c) (e) (f) | Yes |
| 7c. Dent with stress concentrator on girth weld | No | No | Yes |
| 8. Crack or cracking | | | |
| 8a. Shallow cracking <0.4t | Yes (c) | Yes (c) | Yes (a) |
| 8b. Deep crack >0.4t and not >0.8t | Yes (c) | Yes (c) | Yes (a) |
| 9. Seam weld defect | | | |
| 9a. Volumetric defect | Yes (c) | Yes (c) | Yes |
| 9b. Linear defect | Yes (c) | Yes (c) | Yes (a) |
| 9c. Defect in or near an ERW seam | No | No | Yes (a) |
| 10. Girth weld defect | No | No | Yes |
| 11. Wrinkle bend, buckle, or coupling | No | No | Yes (g) |
| 12. Blisters, HIC | Yes | No | Yes |

*Source:* Material provided courtesy of Pipeline Research Council International. Excerpted from PRCI report PR-218-9307, Pipeline Repair Manual, available at www.prci.org.

Notes for Table 44.1:

(a) Make sure defect length is subcritical or pressurize sleeve.
(b) Make sure that internal defect or corrosion does not continue to grow beyond acceptable limits.
(c) Repair may be used for defects <0.8t deep, provided that damaged material has been removed by grinding and removal has been verified by inspection.
(d) It is recommended that the damaged material be removed with removal verified by inspection or that the carrier pipe be tapped for this repair.
(e) Use of filler material in dent and engineering assessment of fatigue are recommended.
(f) Code and regulatory restrictions on maximum dent size should be followed.
(g) Sleeve should be designed and fabricated to special "pumpkin" configuration.

similar. The two exceptions are repair of arc burns, where Type A sleeves can be used without complete removal of metallurgically altered material whereas composite repairs cannot, and repair of hydrogen-induced cracking (HIC) and blistering, where Type A sleeves can be used and composite repairs cannot. Both of these exceptions are the result of composite materials having an elastic modulus that is significantly lower than that of steel. As a result of the lower elastic modulus and the susceptibility of these defects to grow upon the application of comparatively small amounts of strain, the composite repairs allow these defects to strain sufficiently to cause them to fail or grow by fatigue.

Table 44.1 also indicates that Type B sleeves can be used to repair a wide variety of defects for which composite repairs cannot. For example, a leak imposes significant limitations on the choice of repair method. In general, only a Type B sleeve is appropriate.[2]

Rows 2a, 2b, and 2c of Table 44.1 highlight methods applicable for the repair of external corrosion, including the consideration of maximum pit depth. All three repair methods listed in Table 44.1 are acceptable for relatively shallow to moderately deep external corrosion (<80% deep). Very deep external corrosion (80% deep or greater) should be repaired only using methods that are appropriate for leaks (i.e., Type B sleeves).

### 44.4.2 Advantages and Disadvantages

In the previous section, it was shown that the applicability of Type A sleeves and composite repairs is similar and that Type B sleeves can be used where Type A sleeves and composite repairs cannot.

One of the claimed advantages of composite repairs over steel sleeves is that their installation requires no welding to an in-service pipeline. It is clear from the aforementioned discussion that the installation of Type A sleeves, which can serve the same purpose as composite repairs, also requires no welding to an in-service pipeline. Welds that do not contact the carrier pipe are not considered to be "in-service" welds according to Appendix B of API 1104 [15], even though longitudinal seam welds are made while the pipeline is in-service.

Another claimed advantage of composite repairs is that their installation is simpler than that of steel sleeves. While this may be the case for Type B sleeves where welding to an in-service pipeline using specially qualified welding procedures and specially qualified welders are required, this is certainly not the case for Type A sleeves. Type A sleeves require no in-service welding, can have fillet-welded overlapping side strips (Figure 44.11), and are very simple to fabricate and install. They can be supplied with the side strips fillet welded to the bottom half of the sleeve so that the only welds required in the field are the fillet welds on the top of the side strips. These welds can be readily made by welders without special training or qualification for in-service welding. The raw materials required to make Type A sleeves are significantly less expensive than composite materials. Type A sleeves can be fabricated in the field by simply splitting a length of pipe of equal, diameter, wall thickness, and grade as the line pipe material. Unlike composite materials, steel sleeves have no finite "shelf life." Also, the stiffness and long-term performance of Type A sleeves are equivalent to that of line pipe steel.

Federal regulations in the United States require pipeline operators to repair pipelines using methods that have been shown to *permanently* restore the serviceability of the pipeline. While there is some subjectivity in determining what constitutes "permanent," this seems to imply that the expected life of the repair should be equal to the expected life of the pipeline. While the long-term performance of steel is well established, the long-term performance of composite materials on buried pipelines beyond about 20–25 years has not yet been demonstrated. The significantly reduced stiffness of composite materials compared with steel makes the use of composite repairs questionable for application to pipelines that experience cyclic pressure fluctuations. Cyclic strains in the defect area may cause the defect to grow and eventually fail in subsequent service. Work is currently being conducted to better define the long-term performance of composite repairs [16,17].

Both Type A sleeves and composite sleeves rely upon the use of an effective sealer to keep potentially corrosive fluids (e.g., ground water) from entering the crevice area between the carrier pipe and the repair (Figure 44.26). Corrosion under Type A sleeves can be prevented by using either an elastomeric (i.e., caulk-like) sealant or a hardenable sealant.

**FIGURE 44.26** Crevice under improperly installed Clock Spring repair. (Photo provided courtesy of Det Norske Veritas (U.S.A.), Inc.)

[2] When using a steel sleeve to repair a leaking defect, venting of the leak using a small branch pipe with a valve over a hole in one of the sleeve halves may be required.

Examples of hardenable sealants include epoxy splash zone compounds that are similar to those used to adhere some composite sleeves in place or to fill areas of metal loss or denting. Some of the sealing compounds cure effectively on wet surfaces. Cure time is typically a function of the pipe temperature, with curing occurring more quickly on warmer surfaces.

While Type B sleeves do have to be fillet welded to the pipeline (Figure 44.12), they can be used where composite repairs cannot, such as for repair of defects that are 80% deep or greater, circumferentially oriented defects, leaking defects or for defects that will eventually leak (e.g., internal corrosion), and cracks. The raw materials required to make Type B sleeves are significantly less expensive than composite materials and the stiffness and long-term performance of Type B sleeves are equivalent to that of line pipe steel.

When considering the cost of various repair options, both material cost and cost of installation need to be considered. Mobilization cost and the application cost of steel sleeves may be highly variable, depending upon availability and labor rates for welding personnel. While the mobilization cost for composite repair installation may be less in some cases, the materials are more expensive. Composite sleeves or wet-layup composite repair kits are typically designed for a standard, fixed length (e.g., Clock Springs are 12 in. [305 mm] in length)). If the length of the area requiring repair is longer than that which can be repaired by a single composite sleeve, the use of steel sleeves may be significantly less expensive than composite repairs.

## 44.5 WELDING ONTO AN IN-SERVICE PIPELINE

### 44.5.1 Primary Concerns

While there are concerns that need to be considered when welding a Type B sleeve onto an in-service pipeline, this can be relatively straightforward provided that well established guidance is followed. The first of these concerns is for "burnthrough," where the welding arc causes the wall of thin-wall pipe to be penetrated. The second concern is for hydrogen cracking, since welds made in-service tend to cool at an accelerated rate as the result of the flowing contents' ability to remove heat from the pipe wall.

A burnthrough, or blowout as it is sometimes referred to, will occur when welding onto a pressurized pipe if the unmelted area beneath the weld pool has insufficient strength to contain the internal pressure of the pipe. A photograph of a typical burnthrough is shown in Figure 44.27. A burnthrough typically results in a small pin-hole in the bottom of what was the weld pool. The risk of burnthrough will increase as the pipe wall thickness decreases and the weld penetration increases.

Weld made onto in-service pipelines tend to cool at an accelerated rate as the result of the ability of the flowing contents

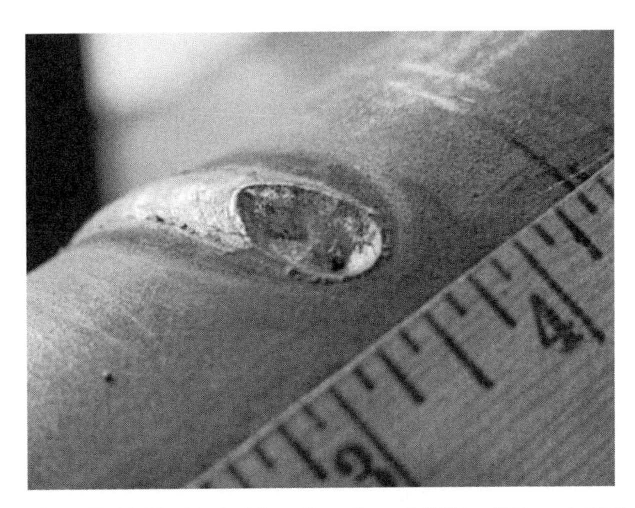

**FIGURE 44.27** Typical burnthrough on 0.125-in. (3.2-mm) thick pipe. (Photo provided courtesy of Det Norske Veritas (U.S.A.), Inc.)

to remove heat from the pipe wall. Accelerated cooling rates promote the formation of hard weld microstructures that are susceptible to hydrogen cracking. A photomicrograph of a typical hydrogen crack is shown in Figure 44.28. Hydrogen cracking requires that three primary, independent conditions be satisfied simultaneously: (1) hydrogen must be present in the weld, (2) there must be a susceptible weld microstructure, and (3) tensile stresses must be acting on the weld.

Simple guidance has been developed and is available elsewhere for preventing both burnthrough and hydrogen when welding onto an in-service pipeline [18]. This simple guidance is summarized as follows.

### 44.5.2 Preventing Burnthrough

A well-established rule of thumb indicates that penetrating the pipe wall with the welding arc (i.e., burning through) is

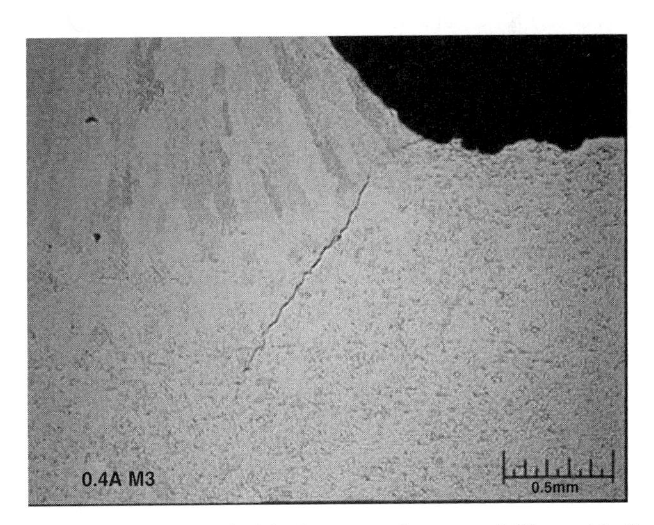


**FIGURE 44.28** Typical hydrogen crack at toe of fillet weld of full-encirclement sleeve. (Photo provided courtesy of Det Norske Veritas (U.S.A.), Inc.)

unlikely if the wall thickness is 0.250 in. (6.4 mm) or greater, provided that low-hydrogen electrodes and normal welding practices are used. This rule of thumb seems to have been lost by some companies, who have requirements for maintaining flow and/or reducing pressure even when the wall thickness is 0.250 in. (6.4 mm) or greater. If the wall thickness is 0.250 in. (6.4 mm) or greater, the primary in-service welding concern should be for hydrogen cracking and not for burnthrough.

If the wall thickness is <0.250 in. (6.4 mm), there may be a need to take special precautions to minimize the risk of burnthrough. These precautions include minimizing the penetration of the arc into the pipe wall by using small-diameter, low-hydrogen electrodes and a procedure that limits heat input.

The most useful tool for evaluating the risk of burnthrough is thermal analysis computer modeling using either the Battelle model [19,20] or the PRCI model [21]. These computer models predict inside surface temperatures, which must be kept below 1800 °F (982 °C) to minimize the risk of burnthrough, as a function of the welding parameters (current, voltage, and travel speed), geometric parameters (wall thickness, etc.) and the operating conditions (contents, pressure, flow rate, etc.). The risk of burnthrough for a given application can be evaluated or the limiting welding parameters for a given set of operating conditions can be determined.

There are several common misconceptions pertaining to operating practices required to prevent burnthrough. One is that some level of flow must always be maintained to prevent burnthrough and another is that the operating pressure must always be reduced. While maintaining flow does result in lower inside surface temperatures, it can be shown that inside surface temperatures are often <1800 °F (982 °C) due to the thermal mass of the pipe wall itself and the thermal properties of the contents, even at little or no flow. While a pressure reduction may be justified to prevent a defect from rupturing during repair on the basis of protecting the repair crew, or to further improve the tightness of the installed sleeve, previous work concluded that stress level in the pipe wall has a relatively small effect on the risk of burnthrough [22]. The reason for this is that the size of the area heated by the welding arc is small and the stress in the pipe wall can redistribute itself around the heated area, as it does around a small corrosion pit. Pressure reductions are, therefore, relatively ineffective at preventing burnthrough and are often unnecessary.

### 44.5.3 Preventing Hydrogen Cracking

To prevent hydrogen cracking, at least one of the three conditions necessary for its occurrence must be eliminated or reduced to below a threshold level. A significant amount of residual tensile stress acting on the weld cannot be avoided and must always be assumed. The first step taken by many companies toward avoiding hydrogen cracking in welds made onto in-service pipelines is to minimize the hydrogen level by using low-hydrogen electrodes or a low-hydrogen welding process. As added assurance against hydrogen cracking, many companies have developed and follow procedures that minimize the formation of crack-susceptible microstructures.

Much of the recent research and development work has been directed at predicting weld cooling rates, or more specifically, at predicting the required welding parameters to prevent the formation of crack-susceptible microstructures. Attention to this may have obscured the most important aspect of preventing hydrogen cracking, which is limiting the amount of hydrogen that enters the weld. The results of recent work show that close control of hydrogen level allows HAZ hardness in excess of 350 HV to be tolerated [23].

***44.5.3.1 Control of Weld Hydrogen Levels*** The importance of controlling hydrogen levels for welds made onto in-service pipelines is well established. Storage and handling of low-hydrogen electrodes is an inexact science at best, even though general guidelines for their use are available [24–27]. The hydrogen level of welds made using low-hydrogen electrodes can vary widely depending on a range of factors. These include the manufacturer, classification/supplemental designation, packaging, storage conditions, handling, atmospheric exposure, and drying/reconditioning practices.

Many of the potential problems associated with minimizing hydrogen levels for welds made onto in-service pipelines can be addressed at the electrode procurement stage. Supplementary designators are now available that allow a specific maximum-allowable hydrogen level to be specified. In AWS A5.1 [28], these designators are H4, H8, and H16, where "H" indicates hydrogen and "4, 8, and 16" refer to the average maximum-allowable hydrogen level in ml/100 g in the "as-received" condition. In other parts of the world, a similar system is used, although the hydrogen levels are H5, H10, and H15. In addition, AWS has introduced an "R" designator that allows a moisture resistant coating to be specified. The R designator indicates that the electrodes have passed an absorbed moisture test after exposure to an environment of 80 °F (26.7 °C) and 80% relative humidity for a period of not <9 h. Electrodes that meet this requirement have coating moisture limits that are lower than their non-moisture-resistant counterparts.

For in-service welding applications, operators should specify electrodes with the H4R designator (e.g., E7018-H4R). These are becoming more common and, while there may be a price premium, this is negligible compared with the cost for remedial action that would be required following the discovery or failure of an in-service weld with hydrogen cracks.

For in-service welding applications, the use of electrodes that are packaged in hermetically sealed cans (i.e., appropriate for use in the as-received condition) should be specified. If electrodes packaged in plastic-wrapped cardboard cartons are used, care must be taken to ensure that drying is not required by the manufacturer prior to their use in the as-received condition. If the electrodes are intended to be used in the as-received condition and they are packaged in plastic-sealed cardboard containers, care should be taken to ensure that the plastic wrap is not damaged. If drying is required, care must be taken to ensure that the drying is carried out properly.

For in-service welding applications, low-hydrogen electrodes in smaller quantities [e.g., 10-lb (4.5-kg) cans] (Figure 44.29) should be specified. This is particularly true for smaller jobs (e.g., small-diameter lines) where it would be difficult to use an entire 50-lb (22.7-kg) can. However, there may also be a price premium for this type of packaging.

*44.5.3.2 Development and Qualification of Procedures* Thermal analysis models are useful for determining welding parameters that minimize the formation of crack susceptible microstructures for specific applications, but their use is not always necessary. In fact, even if thermal analysis models are used, a welding procedure should be qualified under simulated in-service conditions to demonstrate that the predicted parameters are practical under field conditions and that sound, crack-free welds are produced.

Simple methods have been developed for qualifying welding procedures under thermal conditions that simulate those experienced when welding onto an actual in-service pipeline (i.e., conditions that result in realistic weld cooling rates and solidification characteristics). Filling a pipe section with water and letting water flow through the pipe section while the procedure qualification welds are deposited (Figure 44.30) has been shown to be more severe with respect to the resulting weld cooling rates than with most hydrocarbon liquids- and high-pressure gasses [29]. A water flow rate that is attainable from a garden hose (e.g., 10 gal/min [37.8 l/min]) is sufficient.

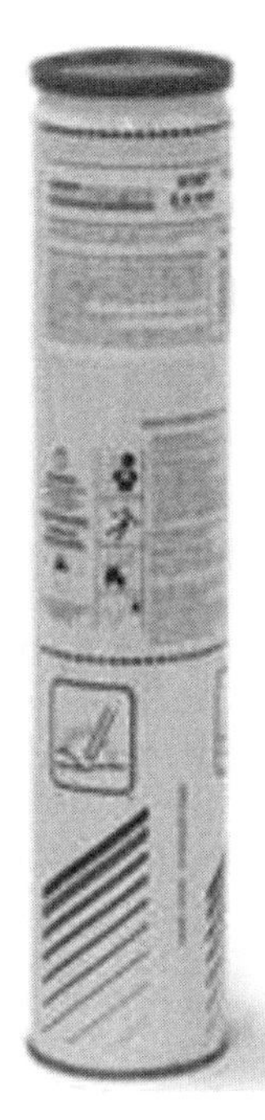

**FIGURE 44.29** E7018-H4R electrodes in 10-lb (4.5-kg) hermetically sealed can. (Photo provided courtesy of Det Norske Veritas (U.S.A.), Inc.)

**FIGURE 44.30** Setup for procedure qualification under simulated in-service conditions. (Photo provided courtesy of Det Norske Veritas (U.S.A.), Inc.)

The thermal conductivity of water more than compensates for the lack of a representative flow rate.

A standard set of procedures can be qualified that cover a range of conditions and then simple guidelines can be developed so that the simplest procedure can be selected for a specific application. The results of previous work [30] indicate that, provided that hydrogen levels are closely controlled, under worst case conditions (flowing water), a procedure that relies on a heat input of at least 25 kJ/in. (1.0 kJ/mm) is suitable for pipe materials with a carbon equivalent ($CE_{IIW}$) up to 0.35. A procedure that relies on a heat input of at least 40 kJ/in. (1.6 kJ/mm) is suitable for pipe materials with $CE_{IIW}$ up to 0.42. A procedure that relies on a properly developed temper bead deposition sequence is suitable for pipe materials with $CE_{IIW}$ up to 0.50. Temper bead procedures are also useful for welding onto thin-wall pipe where the use of a high heat input level represents a risk of burnthrough. These three procedures cover a wide range of conditions. It should be noted that there may be some conservatism associated with this approach, as not all in-service welds cool as quickly as welds made using flowing water.

## 44.6 SUMMARY AND CONCLUSIONS

The use of full-encirclement steel sleeves for pipeline repair is a relatively mature technology. As a result, relatively little is written about the use of steel sleeves compared with the use of composites where new products are continually introduced. The primary function of both Type A and Type B sleeves is structural reinforcement of a defective area. Type B sleeves are additionally useful in that they can also contain pressure. There are also a variety of encapsulation type sleeves, the primary function of which is to contain pressure without providing structural reinforcement.

The applicability of Type A sleeves is nearly identical to that for composite repairs, and their installation involves no welding to an in-service pipeline. Type B sleeves can be used where composite repairs cannot, such as for repair of defects that are 80% deep or greater, circumferentially oriented defects, leaking defects or for defects that will eventually leak (e.g., internal corrosion), and cracks. For both types of full-encirclement steel sleeve, the raw materials are relatively inexpensive and the stiffness and long-term performance are equivalent to that of line pipe steel.

The installation of Type B sleeves does involve the need to weld to an in-service pipeline. Adherence to the simple guidance summarized earlier will avoid the potential concerns associated with welding to an in-service pipeline. It is advisable to use Type A sleeves when possible, and reserve the use of Type B sleeves for only when necessary.

## REFERENCES

1. Kiefner, J.F., Bruce, W.A., and Stephens, D.R. (1994) Pipeline repair manual. Catalog Number L51716, Pipeline Research Council International, Inc., Arlington, VA, December 31, 1994.
2. Jaske, C.E., Hart, B.O., and Bruce, W.A. (2006) Updated pipeline repair manual. Final Report to Pipeline Research Council International, Inc., Contract PR-186-0324, DNV/CC Technologies, Inc. and Edison Welding Institute, August 8, 2006.
3. Kiefner, J.F. and Duffy, A.R. (1974) A study of two methods for repairing defects in line pipe. Final Report to the Pipeline Research Committee of the American Gas Association, Catalog No. L22275, October 1974.
4. Kiefner, J.F. and Whitacre, G.R. (1978) Further studies of two methods for repairing defects in line pipe. Final Report to the Pipeline Research Committee of the American Gas Association, Catalog No. L22279, March 1978.
5. Kiefner, J.F. (1977) Repair of line-pipe defects by full-encirclement sleeves. *Welding Journal*, 57(6), 26–34.
6. Kiefner, J.F. (1983) Full-encirclement sleeve repair methods given for dented or corroded pipe. *Oil and Gas Journal*, 81, 160–163.
7. Corder, I. (1991) Cost-effective on-line repair of pipeline damage using epoxy-filled shells. Proceedings of Pipeline Risk Assessment, Repair and Rehabilitation Conference, May 20–23, 1991, Houston, TX.
8. Brongers, M.P., Maier, C.J., Jaske, C.E., Vieth, P.H., Wright, M.D., and Smyth, R.J. (2001) Evaluation and use of a steel compression sleeve to repair longitudinal seam-weld defects. Proceedings of the 52nd Annual Pipeline Conference, April 17–18, 2001, San Antonio, TX.
9. PetroSleeve™ (2000) *Historical Test Results and Data*, PetroSleeve™, Inc., Nisku, Alberta.
10. Smyth, R.J. (1998) Repairing pipe defects (cracking, arc burns, corrosion, dents) without operational outages using the PetroSleeve™ Compression Sleeve Repair Technique. Proceedings of the 1998 International Pipeline Conference, Volume 1, The American Society of Mechanical Engineers, New York, 231–239.
11. Kiefner, J.F. and Maxey, W.A. (1989) Tests validate pipeline sleeve repair technology. *Oil and Gas Journal*, 87, 47–52.
12. Fawley, N.C. (1994) Development of fiberglass composite systems for natural gas pipeline service. Final Report GRI-94/0072, March 1994.
13. Stephens, D.R. and Kilinski, T. (1994) Field validation of composite repair of gas transmission pipelines. GRI Annual Report GRI-94/0139, March 1994.
14. Alexander, C. and Francini, B. (2006) State of the art assessment of composite systems used to repair transmission pipelines. Proceedings of IPC2006, 6th International Pipeline Conference (IPC 2006) September 25–29, 2006, Calgary, Alberta, Canada, Paper IPC2006-10484, Stress Engineering Services, Inc. and Kiefner & Associates, Inc.
15. API 2005 *API Standard 1104: Welding of Pipelines and Related Facilities*, 12th ed., American Petroleum Institute, Washington, DC.

16. Alexander, C. (2009) Overview and recent advances in application of composite materials for repair of transmission pipelines: evaluation of test results and ongoing research. Proceedings of Evaluation and Rehabilitation of Pipelines Conference, October 2009, Pittsburgh, PA, Stress Engineering Services, Inc.
17. Alexander, C. and Bedoya, J. (2010) Repair of dents subjected to cyclic pressure service using composite materials. Proceedings of the 8th International Pipeline Conference (IPC 2010), September 27–October 1, 2010, Calgary, Alberta, Canada, Stress Engineering Services, Inc.
18. Bruce, W.A. (2009) A simple approach to hot tap and repair sleeve welding. Proceedings of Evaluation and Rehabilitation of Pipelines, Clarion Technical Conferences and Scientific Surveys Ltd, October 19–22, 2009, Pittsburgh, PA.
19. Bubenik, T.A., Fischer, R.D., Whitacre, G.R., Jones, D.J., Kiefner, J.F., Cola, M.J., and Bruce, W.A. (1991) Investigation and prediction of cooling rates during pipeline maintenance welding. Final Report to American Petroleum Institute, December 1991.
20. Cola, M.J., Kiefner, J.F., Fischer, R.D., Jones, D.J., and Bruce, W.A. (1992) Development of simplified weld cooling rate models for in-service gas pipelines. Project Report No. J7134 to A.G.A. Pipeline Research Committee, Edison Welding Institute, Kiefner and Associates and Battelle Columbus Division, Columbus, OH, July 1992.
21. Bruce, W.A., Li, V., Citterberg, R., Wang, Y.-Y., and Chen, Y. (2001) Improved cooling rate model for welding on in-service pipelines. PRCI Contract No. PR-185-9633, EWI Project No. 42508CAP, Edison Welding Institute, Columbus, OH, July 2001.
22. Kiefner, J.F. and Fischer, R.D. (1988) Repair and hot tap welding on pressurized pipelines. Proceedings of the 11th Annual Energy Sources Technology Conference and Exhibition, January 10–13, 1988, New Orleans, LA (New York, NY: American Society of Mechanical Engineers, PD-Vol. 14, 1987). 1–10.
23. Bruce, W.A. and Etheridge, B.C. (2012) Further development of heat-affected zone hardness limits for in-service welding. Proceedings of the 9th International Pipeline Conference, September 24–28, 2012, Det Norske Veritas (U.S.A.), Inc., Dublin, OH. Paper IPC2012-90095.
24. (1999) Storing and Re-drying Electrodes. The Lincoln Electric Company, Cleveland, OH. Available at http://lincolnelectric.com/knowledge/articles/content/storing.asp.
25. ESAB AB (2009) Recommendations for the storage, re-drying and handling of ESAB consumables, Göteborg, Sweden. Available at http://esabsp.esab.net/templates/docOpen.asp?file=files/Handbooks/Welding%20Consumables/XA00097020.pdf
26. AWS (2004) *AWS D1.1:2004: Structural Welding Code—Steel*, American Welding Society, Miami, FL.
27. WTIA (1994) WTIA Technical Note 3: Care and conditioning of arc welding consumables. WTIA TN-3-94, Welding Technology Institute of Australia, Silverwater, NSW, 1994.
28. AWS (1991) *AWS A5.1-91: Specification for Carbon Steel Electrodes for Shielded Metal Arc Welding*, American Welding Society, Miami, FL.
29. Bruce, W.A. and Threadgill, P.L. (1994) Effect of procedure qualification variables for welding onto in-service pipelines. Final Report to A.G.A. Welding Supervisory Committee, Project PR-185-9329, EWI, Columbus, OH, July 21, 1994.
30. Bruce, W.A. (2001) Selecting an appropriate procedure for welding onto in-service pipelines. Proceedings of International Conference on Pipeline Repairs, March 2001, Welding Technology Institute of Australia, Wollongong, Australia.

# 45

# PIPELINE REPAIR

Robert Smyth[1] and Buddy Powers[2]
[1]Petroline Construction Group, Nisku, Alberta, Canada
[2]Clock Spring Company L.P., Houston, TX, USA

## 45.1 INTRODUCTION

This chapter introduces the various pipeline repair techniques. It is the responsibility of the operating company to review the various options that are available and select the option that provides the best result for their situation.

Common types of defects that affect the integrity of an operating pipeline include external and internal corrosion, gouges, grooves, arc burns, dents, cracking, and leaks.

Once the need for repair has been established, there are various methods for repair that can be used, depending on the type of damage. This chapter covers the repair of pipelines using pipe replacement, grinding, steel sleeves, epoxy-filled shells, steel compression sleeves, composite reinforcement sleeves, hot tapping, direct deposition welding, bolt-ons, and stopples. In addition to the information in this chapter, full-encirclement repair sleeves are further discussed in Chapter 44.

Repairs should only be undertaken with reference to the appropriate industry standards and jurisdictional regulations.

## 45.2 BACKGROUND

When undertaking a repair, the objective is to return the pipe to its original integrity. Following the repair, the excavation then can be backfilled and surface reclaimed. This repair can be accomplished by using many of the following available options:

- Pipe replacement using stoppling or frost plugs (if safe);
- Grinding;
- Type "A" reinforcing steel sleeves;
- Type "B" steel pressure containing sleeves;
- Epoxy-filled Shells;
- Steel compression sleeves;
- Composite reinforcement sleeves;
- Hot tapping;
- Direct deposition welding;
- Bolt-ons;
- Temporary leak repairs.

## 45.3 SAFETY

Defects are located by observing physical evidence (e.g., pools of product) or by utilizing internal inspection tools. Following an inspection run, the "log" is analyzed to identify the severity and location of the defects. Although, over time, the accuracy of log information has greatly improved, the actual determination of the severity of the defect is only known after the pipe has been excavated, the pipe/defect cleaned, and an engineering analysis completed.

Consequently, for safety reasons, it is advisable for the operating pressure to be reduced at the defect location during the excavation/nondestructive examination (NDE) process. As a guide, one suggestion is to reduce the operating pressure to 80% of the pressure that the site has been exposed to within the recent time frame. This criterion simulates a typical pressure test where the recognized safe operating pressure is calculated as 80% of the test pressure.

*Oil and Gas Pipelines: Integrity and Safety Handbook,* First Edition. Edited by R. Winston Revie.

## 45.4 PROTOCOLS

Pipeline operators should have properly designed repair procedures for all common defects. Then, when an integrity program is organized, these repair procedures are available.

Prior to commencing a repair program, a corporate discussion needs to be undertaken, setting forth the criteria to be used within the program. Obviously, the criteria encompass the overall integrity program. The criteria would be used to determine what defects are acceptable and only need to be recoated and what defects need to be repaired.

The defect acceptable/repair criteria would consider items such as local environmental factors, public risk, safety factor (burst pressure as a percent of SMYS), and operational concerns. Following this analysis, a protocol could be developed to convey the corporate requirement to the field.

A typical protocol for dealing with stress concentrators is presented in Figure 45.1.

Using protocols developed for various defects provides guidance to the field crews in making decisions and reducing the overall cost of the integrity repair program.

## 45.5 PIPE REPLACEMENT

Cutout and replacement is the preferred repair method for all defects. However, taking into account economics (e.g., loss of natural gas, taking line out of service), other repair techniques described later are utilized.

Pipe replacement involves the removal of the affected pipe and the replacement by welding in of a new pretested pipe section. The minimum permissible length [1] of replacement pipe is

- Around 150 mm for pipe smaller than 168.3 mm OD;
- Two times the specified pipe OD for pipe from 168.3 to 610 mm OD;
- Around 1220 mm for pipe larger than 610 mm OD.

When undertaking this work, an applicable welding procedure must be used that considers welding steel properties of the old pipe to the new pipe, as well as utilizing welders who have been tested to the procedure. In addition, usually,

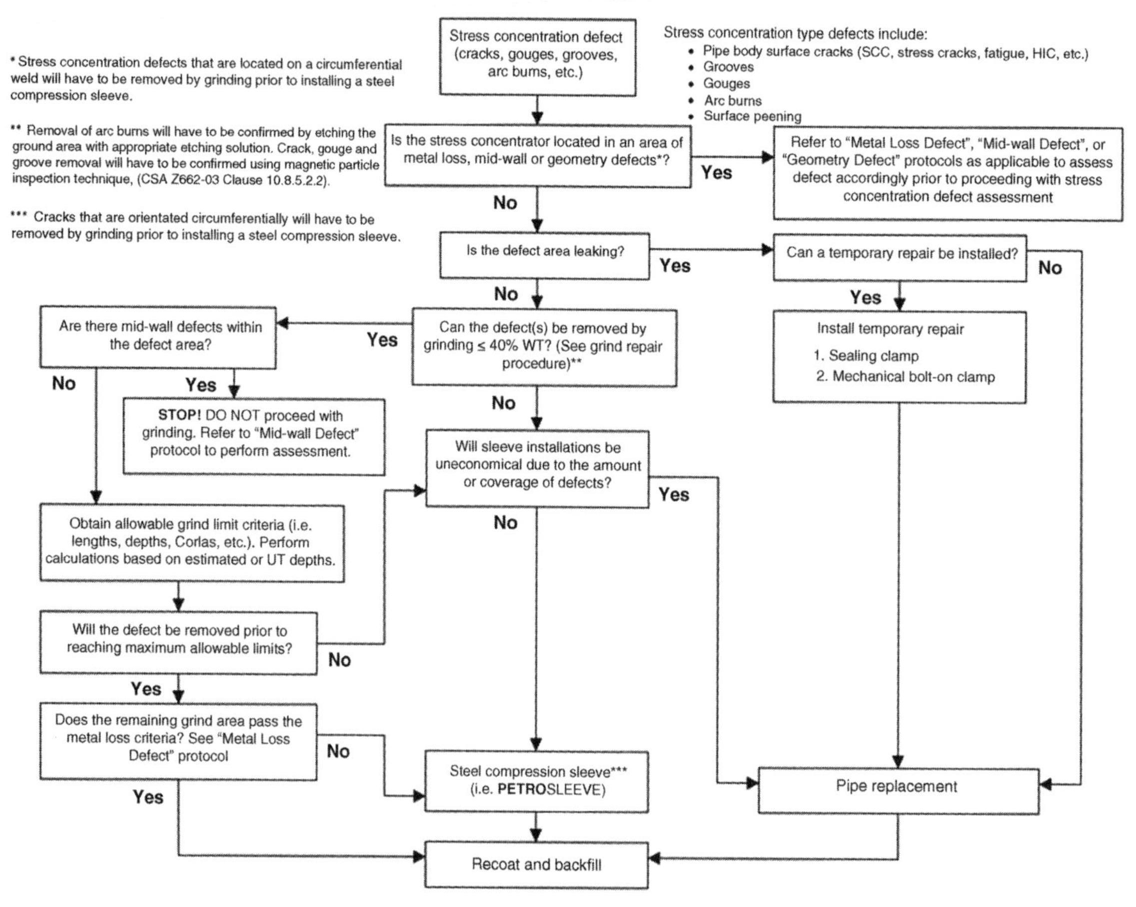


FIGURE 45.1 Engineering assessment protocol for stress concentration type defects. (Courtesy of Petro-Line, Inc.)

FIGURE 45.2 Cold cutting the pipe. (Courtesy of Petro-Line, Inc.)

FIGURE 45.3 Completing the tie-in weld on a cutout. (Courtesy of Petro-Line, Inc.)

the original pipe has become magnetic. The magnetism must be removed prior to commencing welding.

The replacement pipe needs to be pre-pressure tested to accommodate the appropriate licensed operating pressure. This pressure test is undertaken to confirm that the replacement pipe can sustain the maximum pressure of the adjacent pipe.

Prior to cold cutting the pipe (shown in Figure 45.2), the product in the pipe at that location needs to be evacuated. This product removal can be accomplished by undertaking a nitrogen purge. In this case, the product between the two shutoff valves would be swept clean by sending a pig from the upstream valve setting to the downstream setting using compressed nitrogen. When the pig arrives at the downstream valve, the nitrogen pressure then is released, leaving a safe empty pipe section of inert gas.

Another method to evacuate identified product from the defect location would be to create an isolation of the to-be-cut out section by welding to the pipe upstream and downstream of the defect "stopple" fittings. With the pipe section isolated by the installation of stopple plugs, the pipe can be evacuated of product. As well, if required, an optional by-pass line can be installed to maintain the pipeline operation independent of the isolated pipe. The welding requirements for stopple installation are explained in Chapters 18 and 44 (welding Type "B" steel sleeves).

Additionally, other isolation options have been employed. When pressure testing with water and a rupture occurs, frost plugs can be used to prevent the water on both sides of the rupture from entering the site where a new section of pipe is being welded into the line. With frost plugs, nitrogen assemblies can be placed around the pipeline isolation area and then the water under the assembly is frozen creating a frost plug.

Also, there are internal pig assemblies that can be set inside the pipe to create a "plug." These are particularly functional when undertaking a cut out of an operating pipe line. Two pigs could be launched with a distance between them. When they arrive at the cutout location, line operation would stop, the pigs would be set, and cold cutting of the pipe could be undertaken.

For all options of isolation, detailed engineering analysis should be undertaken to evaluate the isolation process, success probability, and effect on the pipeline in the long term.

Following completion of the tie-in welding of old to new pipe (shown in Figure 45.3), the welds are nondestructively inspected for defects. The replacement pipe is then coated. At the interface between the new coating and the old coating, an appropriate coating technique needs to be used. The stopple plugs are then removed and the new pipeline section is brought into service. During backfilling, the pipe needs to be supported such that no additional bending moments are applied to the pipe by the soil.

## 45.6 GRINDING/SANDING

When cracking is identified on the surface of pipe, grinding or sanding to remove the cracking can return the pipe to an acceptable integrity condition. Grinding, if successful, will remove the stress concentrating crack tip, thus converting the defect to a metal loss situation. In this case, various engineering analyses are undertaken to determine the acceptable maximum amount of steel that can be removed. This engineering analysis should consider both the current and future operation requirements of the pipe section. In Figure 45.4, a stress corrosion crack (SCC) is illustrated in Figure 45.4a, and the repair after removal by grinding is also shown in Figure 45.4b.

Prior to commencing this operation, an engineering analysis needs to be completed to determine a safe working environment, taking into account the steel properties, wall thickness, crack depth, and internal pressure. In particular, the maximum crack depth needs to be determined, as well as the maximum amount of steel removed prior to ceasing the

(a)

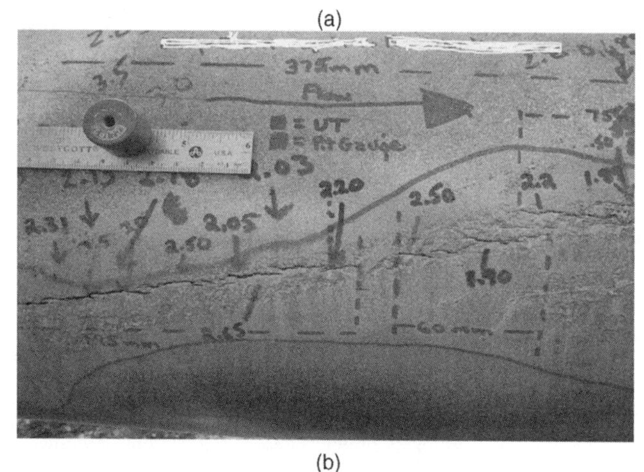


(b)

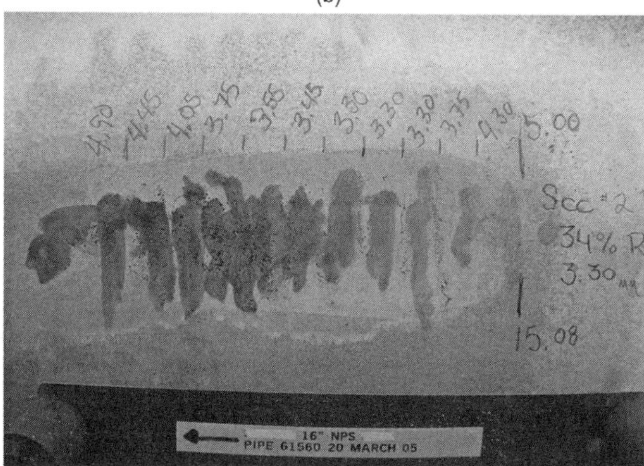


FIGURE 45.4 (a) Illustration of SCC and (b) grind repair. (Courtesy of Petro-Line, Inc.)

FIGURE 45.5 Typical Type "A" steel sleeve. (Courtesy of R. Smyth.)

FIGURE 45.6 Typical Type "B" steel sleeve. (Courtesy of R. Smyth.)

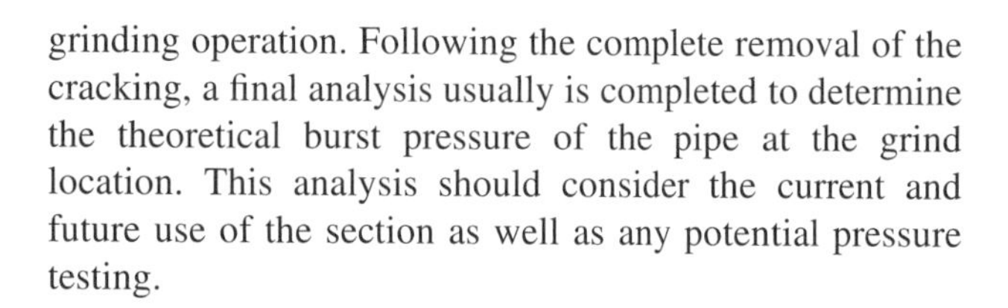

grinding operation. Following the complete removal of the cracking, a final analysis usually is completed to determine the theoretical burst pressure of the pipe at the grind location. This analysis should consider the current and future use of the section as well as any potential pressure testing.

## 45.7 FULL-ENCIRCLEMENT STEEL SLEEVES

There are two basic types of tight-fitting full-encirclement steel sleeves: Type A and Type B, shown in Figures 45.5 and 45.6, respectively. The primary function of both types is structural reinforcement of a defective area. Type B sleeves can also contain pressure. There are also various manufactured encapsulation-type sleeves (e.g., bolt-on pressure-containing fixtures), the primary function of which is to contain pressure without providing structural reinforcement.

These repair types are described in greater detail in Chapter 44.

## 45.8 EPOXY-FILLED SHELLS

British Gas developed a variation of the sleeve repair concept in the form of their epoxy-filled shell repair [2]. In this case, the shell is a steel sleeve with a significant annular gap from the carrier pipe. The system ensures structural support to the defect and has a wide tolerance for repair of complex or distorted pipe geometries.

The two sleeve halves are assembled around the pipe and then are longitudinally welded together. Following welding, the assembly is centralized around the pipe. The annular gap at each end of the sleeve is sealed with an epoxy. Following the curing of the epoxy seal, an epoxy grout (e.g., a high-stiffness epoxy/sand mixture) is pumped into the annular space from the bottom, until it flows out of an overflow hole at the top of the sleeve.

Once the epoxy filler has hardened, the pipe and defect are restrained by the epoxy/steel shell. This provides structural support to the defect, keeping it from flexing with changes in internal pressure.

## 45.9 STEEL COMPRESSION SLEEVES

Steel compression sleeves (Figure 45.7) are a special class of Type A sleeves. They are designed, fabricated, and applied so that the repaired section of the carrier pipe is maintained under compressive hoop stress during subsequent operation. This repair technique is suitable for all defects except leaking defects and cracking in the circumferential direction. It should not be used over tight radius bends. In addition, this approach is attractive for repairing longitudinally oriented crack-like defects, such as SCC, because without a tensile hoop stress there is no driving force for crack growth.

Steel compression sleeves involve assembling two sleeve halves over the defect area and drawing them together using jacking devices. The sleeve halves are then heated to a specified temperature, and then welded together using conventional welding techniques. Epoxy filler is used between the carrier pipe and sleeve to achieve the transfer of stresses and to act as a sealing agent to prevent the moisture from entering between the pipe and sleeve.

PetroSleeve™ is a commercial product that was developed to install a steel sleeve while the pipeline remained in operation, to not weld to the pipe, and to have the carrier pipe remain in a compressed state for all operational pressures. Because the carrier pipe remains in a compressed state, the sleeve/pipe/epoxy bond remains intact. Comprehensive testing [3] has demonstrated that the fillet weld design is suitable for use on pipelines with significant pressure cycling.

**FIGURE 45.7** Example of an installed 36-in. steel compression sleeve. (Courtesy of Petro-Line, Inc.)

Quality control procedures for this technique involve monitoring sleeve temperature during the heating process and verification of the achieved carrier pipe compression. Nondestructive inspection of the completed welds is conducted after cooling.

## 45.10 COMPOSITE REINFORCEMENT SLEEVES

Fiber-reinforced plastic (FRP) composites have been developed for applications in a range of industries, including transportation, chemical and both the upstream and downstream sectors of the oil and gas industry. Composites are used in these diverse applications because they can be specifically designed to provide the mechanical properties required for the application. This feature, along with a high strength-to-weight ratio, has made composites the best choice for many critical applications in advanced equipment. In addition to the discussion in this chapter, composite sleeves are further discussed in Chapter 44.

Composite repair systems work by constraining a defect as it bulges. As the internal pressure of the pipe rises toward the line maximum operating pressure (MOP) or maximum allowable operating pressure (MAOP), the pipe wall elastically stretches. Plastic deformation is possible if the wall becomes thinner because of, for example, corrosion. All properly designed and installed composite repairs work to constrain elastic/plastic distortion of the pipeline and prevent this thinner wall ligament from bulging beyond its failure limit.

There are two broad groups of composites available on the market "composite sleeves" and "wet wraps." Within both ASME B31.4 and B31.8S, composite sleeves are acceptable for the repair of wall losses of up to 80%. Standards such as ASME PCC2 Article 4.2 and ISO PDTS 24817 are typically used for design and allow the material properties of a composite to be estimated based on short-term test data, although increasing emphasis is being placed on long-term performance as the basis for design, including testing up to 10,000 h. In Equations 9 and 10 of ISO PDTS 24817, long-term test data are used to establish the long-term strength of the composite repair to ensure it will remain strong enough for the required number of years of service.

### 45.10.1 Designing an Effective Composite Repair

***45.10.1.1 Resin Selection*** FRP composites use a variety of resin formulations, including polyesters, epoxies, and vinylesters. Each of these resins has unique characteristics of cost, compatibility with reinforcing fibers, chemical resistance, strength, and moisture absorption. As an example, rigid isophthalic polyester resin has seen years of service in composite gasoline storage tanks buried in soil environments similar to that of pipeline service. It has low moisture absorption, excellent resistance to chemicals and good thermal properties, making it suitable for most pipeline applications.

#### *45.10.1.2 Fiber Type, Form, and Architecture*

Three characteristics must be considered when designing fiber reinforcement:

- fiber type (glass, carbon, Kevlar, or aramid);
- fiber form (typically roving, tow, mat, or woven fabric);
- fiber orientation or architecture.

**Fiber type:** There are a number of commercially available reinforcing fibers with a variety of chemical and mechanical properties. The most common fiber type used for the reinforcement is E-glass. E-glass was selected for the composite sleeve system studied in the Gas Research Institute program [4] because of its high strength, electrical properties, availability, and compatibility with the selected resin. E-glass is a super-cooled mixture of metallic oxides. It is brittle and transparent, but has very high tensile strength, 3400 MPa (500 ksi). These filaments are bundled together into tows. Each of the tows contains 4000 filaments.

Following processing and handling operations, the effective tensile strength is generally about 1700 MPa (246 ksi).

**Fiber form and architecture:** Reinforcement can be oriented longitudinally in a structural element, transverse to the longitudinal direction or in any direction desired by the designer. Fiber direction can be so varied that virtually isotropic material can be produced.

Chopped strand mat (CSM) is a very common form of fiberglass used in many applications. The randomly oriented, short strands of glass intertwine to form the bed of the material. The resin then binds these strands together. The strength of the mat in any given direction is unpredictable and relies largely on the strength of the epoxy or resin.

Unidirectional rovings (tows) can be used directly in composite manufacture, or they can be converted to other forms such as a glass cloth with various weaves. Cloth made from E-glass is easy to use in the manufacturing process, but is not as durable as unidirectional glass. The weave in the cloth introduces a path for moisture and reduces the modulus of elasticity.

## 45.11 HOT TAPPING

Defects can be removed using the hot tapping process. In this case a fixture (e.g., a stopple fitting) is welded to the pipe encircling the defect. The specialized flange on the stopple fitting is machined to accommodate a plug that can be secured externally and is used for a "stopple operation." Following the hot tapping process, the plug is inserted in the flange with stopple equipment, and the blind flange is installed to secure a pressure seal. Using this process, the defect is removed from the pipe and can be examined.

Welding procedures are required for this process. More information on hot tap welding is presented in Chapters 18 and 44.

## 45.12 DIRECT DEPOSITION WELDING

Direct deposition welding is a process whereby the defect is repaired or filled by welding to the operating pipeline. Significant engineering and procedure development is required to design an appropriate welding procedure for the application. In some jurisdictions, this technique is allowed only in non-sour natural gas pipelines, for a defect that is not in a dent or gouge, situated on a mill seam or circumferential weld and the remaining wall thickness at the defect location is ≥3.2 mm [5].

## 45.13 TEMPORARY REPAIR

Temporary repair can be made using commercially available "bolt-on" sleeves. The repair device usually is supplied as two halves with a bolting arrangement to mount the device onto the pipe. A typical bolt-on sleeve is shown in Figure 45.8.

**FIGURE 45.8** Typical bolt-on sleeve. (Courtesy of R. Smyth.)

**TABLE 45.1 Examples of Pipeline Defects and Applicable Repair Methods**

| Type of Defect | Repair Methods | | | | | | | | |
|---|---|---|---|---|---|---|---|---|---|
| | | | | | Sleeves | | | | |
| | Pipe Replacement | Grinding | Hot Tap | Weld Metal Deposition | Reinforcing Type A | Steel Compression Reinforcement | Pressure Containing Type B | Composite | Bolt-On |
| External corrosion ≤80% through wall | Yes | No | Note 2 | Note 3 | Note 4 | Yes | Yes | Note 4 | Yes |
| External corrosion >80% through wall | Yes | No | Note 2 | No | No | Yes | Yes | No | Yes |
| Internal corrosion ≤80% through wall | Yes | No | Note 2 | No | Note 5 | Note 5 | Yes | Note 5 | Yes |
| Internal corrosion >80% through wall | Yes | No | Note 2 | No | No | No | Yes | No | Yes |
| Gouge, groove or arc burn on pipe body | Yes | Note 1 | Note 2 | No | Note 4, 7 | Yes | Yes | Note 4, 7 | Yes |
| Crack | Yes | Note 1 | Note 2 | No | Note 1 | Note 8 | Yes | Note 1 | Yes |
| Dent on pipe body | Yes | No | Note 2 | No | Yes | Yes | Yes | Note 6 | Yes |
| Defective girth weld | Yes | No | No | Note 3 | No | No | Yes | No | Yes |
| Hard spot | Yes | No | Note 2 | No | Note 4 | | Yes | No | Yes |
| Leak | Yes | No | No | No | No | No | Yes | No | No |

*Source*: Information compiled from Ref [6]. For full details, see the current editions of these standards or of those that apply in the appropriate jurisdictions.

Note 1: Defect must be entirely removed without penetrating more than 40% of the wall thickness.

Note 2: Defect must be contained entirely within the area of the largest possible coupon of material that can be removed through the hot tap fitting.

Note 3: The welding-procedure specification shall define the minimum remaining wall thickness in the area to be repaired and the maximum internal pressure during repair. Low-hydrogen welding process must be used.

Note 4: Tight-fitting sleeve or hardenable filler must be used to fill the void between the pipe and sleeve.

Note 5: Internal corrosion must be successfully mitigated.

Note 6: Depth of dent must be ≤15% of pipe specified OD and dent must be filled before sleeve is applied [6].

Note 7: Defect must be entirely removed and removal verified.

Note 8: Any circumferential cracks shall be removed by grinding prior to application of the sleeve [6].

At each end, neoprene seals are provided so that when the bolts are tightened, a pressure seal is created circumferentially and longitudinally. After the device is installed on the pipe, at the discretion of the pipeline owner, and based on the design of the manufacturer, the device may be welded to the pipe, creating effectively a Type "B" pressure-containing sleeve.

## 45.14 TEMPORARY REPAIRS OF LEAKS

Various products are commercially available, including clamps and wrapping materials, to repair leaking defects such as a pin hole. These devices are applied around the pipe so that the leak is stopped. Following the application of these devices, a permanent repair is undertaken at a later date.

## 45.15 APPLICABILITY TO VARIOUS DEFECT TYPES

In Table 45.1, some examples of types of defects and repair methods that can be used are listed. More detailed information on defects and their repair is available in Ref [6]. In selecting a repair method to deal with a specific defect, whether the defect be a corrosion defect, a crack, a dent, a gouge, a weld defect, and so on, it is essential to consult the standard that is applicable in the jurisdiction where the pipeline is located. Once the repair method has been selected, it is important to follow the manufacturer's installation instructions, including those required for surface preparation.

## REFERENCES


1. Canadian Standards Association (2011) Canadian Standards Association Z662-11: Oil and Gas Pipeline Systems, Mississauga, Ontario, Canada, June 2011, paragraph 10.11.3, Piping replacements, p. 214.
2. Stephens, D.R., Leis, B.N., and Francini, R.B. (1998) Developments in criteria for repair or replacement decisions for pipeline corrosion defects. 3rd Congreso y Expo Internacional de Ductos, 1998, Monterrey, Mexico.
3. Brongers, M.P., Maier, C.J., Jaske, C.E., Vieth, P.H., Wright, M.D., and Smyth, R.J. (2001) Evaluation and use of a steel compression sleeve to repair longitudinal seam-weld defects. Proceedings of the 52nd Annual Pipeline Conference, April 17–18, 2001, San Antonio, TX.
4. Stephens, D.R. and Gas Research Institute. (1998) *GRI-98/0227: Summary of Validation of ClockSpring$^{TM}$ for Permanent Repair of Pipeline Corrosion Defects*, Gas Research Institute, Chicago, IL.
5. ASME (2012) Pipeline Transportation Systems for Liquids and Slurries, B31.4-2012, Table 451.6.2(b)-1, ASME, Washington, DC, 2012.
6. NACE/SSPC (2006) NACE No. 3/SSPC-SP 6: Joint Surface Preparation Standard, Commercial Blast Cleaning, NACE International, Houston, TX and SSPC, Pittsburgh, PA, 2006.

# 46

# PIPELINE OIL SPILL CLEANUP

Merv Fingas
*Spill Science, Edmonton, Alberta, Canada*

## 46.1 OIL SPILLS AND PIPELINES: AN OVERVIEW

In recent years, the media attention given to large oil spills has created a global awareness of the risks of oil spills and the damage they do to the environment. Less attention has been given to spills of lesser profile, such as spills that occur on land or those that do not cause dramatic effects.

An important part of protecting the environment is ensuring that there are as few spills as possible. While oil is a necessary risk in a complex society, recent and avoidable major oil releases have demonstrated that serious improvements in oil spill response are warranted to improve effectiveness. Industry has invoked many operating and maintenance procedures to reduce accidents that lead to spills. In fact, the rate of spillage has decreased in the past 10 years. Despite this, spill experts estimate that 30–50% of oil spills are either directly or indirectly caused by human error, with 20–40% of all spills caused by equipment failure or malfunction.

### 46.1.1 How Often Do Spills Occur?

It can sometimes be misleading to compare oil spill statistics, however, because different methods are used to collect the data. In general, statistics on oil spills are difficult to obtain and any data set should be viewed with caution. The spill volume or amount is the most difficult to determine or estimate. Sometimes the exact character or physical properties of the oil lost are not known and this leads to different estimations of the amount lost.

The number of spills reported also depends on the minimum size or volume of the spill. In both Canada and the United States, most oil spills reported are more than 4000 l (about 1000 gallons) [1]. In Canada, there are about 12 such oil spills every day from all sources, of which only about one is spilled into navigable waters. These 12 spills amount to about 40 tons of oil or petroleum product. In the United States, there are about 15 spills per day into navigable waters and an estimated 85 spills on land or into freshwater [2,3].

Despite the large number of spills, only a small percentage of oil used in the world is actually spilled. Oil spills in Canada and the United States are summarized in Figures 46.1 and 46.2 in terms of the volume of oil spilled and the actual number of spills. In terms of oil spills, it can be seen from these figures that there are differences between the two countries. The pipeline source of spillage is highlighted in these figures.

In Canada and United States, most spills take place on land and this accounts for a high volume of oil spilled. This means that the spill occurs on land, however, oil typically finds its way into a water body, be it a stream, lake, or local depression [4]. Pipeline spills account for the highest volume of oil spilled. In terms of the actual number of spills, most oil spills happen at petroleum production facilities, wells, production collection facilities, and battery sites. Pipeline spills dominate the volume spilled in both Canada and USA; one such pipeline is shown in Figure 46.3. On water, the greatest volume of oil spilled comes from marine or refinery terminals, although the largest number of spills is from the same source as in the United States—vessels other than tankers, bulk carriers, or freighters [5]. It should be noted, however, that even though many spills occur on land that the receiver of these spills may be water bodies, rivers, or creeks.

*Oil and Gas Pipelines: Integrity and Safety Handbook,* First Edition. Edited by R. Winston Revie.

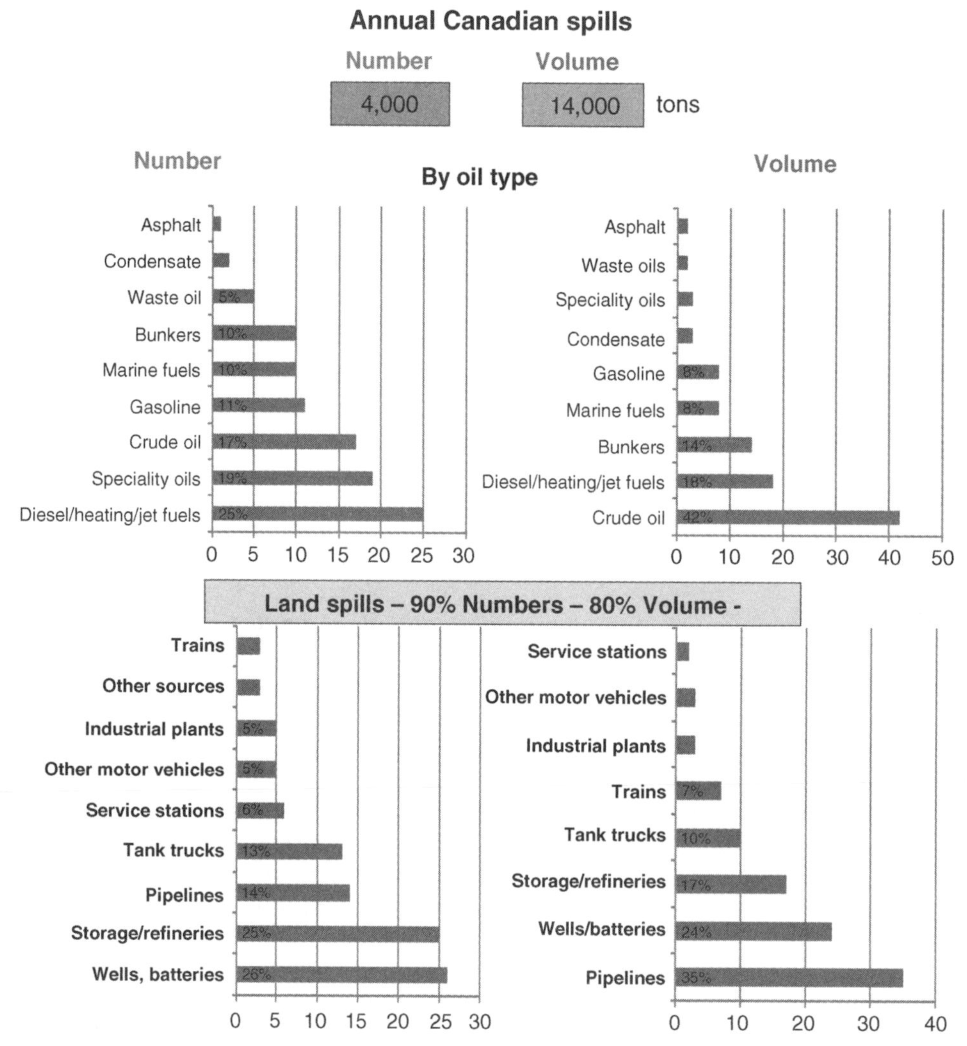


**FIGURE 46.1** Canadian total spill statistics highlighting pipeline spills. (Data compiled from Refs. [4] and [6].)

Table 46.1 shows example summary statistics of pipeline spills for Canada from the Transportation Safety Board of Canada (TSB) [4]. These statistics are for the month of November 2013, and contain year-to-date statistics as well as historical averages. "Accidents" involve some form of accidental occurrence whereas "incidents" involve other causes. As can be seen from Table 46.1, incidents exceed accidents by a factor of over 10. A larger view of Canadian accident statistics can be obtained on the TSB website [6].

### 46.1.2 Pipelines

There are many types and sizes of pipelines used for oil transport. Table 46.2 shows the categorization of these by use and size [7].

An important part of the consideration of the pipeline situation is the amount of oil transported by pipeline. Table 46.3 shows a summary of the 2013 monthly statistics on the quantity of oils transported by pipelines [8]. These data are from Statistics Canada. This is the amount of petroleum liquids transported by pipelines averaged over 1 month. Crude oil and pentanes plus includes liquids that might be described as condensates. Petroleum products such as gasoline and diesel fuel are included with liquefied petroleum gases.

These tables and Figure 46.1 allow one to estimate the overall loss of crude oils and petroleum products by pipeline in Canada. Since on a typical average 40,000 tons of material are spilled and on average 17% of this is from pipelines, in a typical year about 2400 tons of crude oil and petroleum products are estimated to be spilled. Since there are about 4000 incidents per year and about 14% of these are from pipelines, there are an estimated 560 incidents per year. The average pipeline spill is estimated to be about 4 tons. The amount spilled per transported amount is $3 \times 10^{-6}$ or about 3 in a million. Given that there are about 300,000 km of

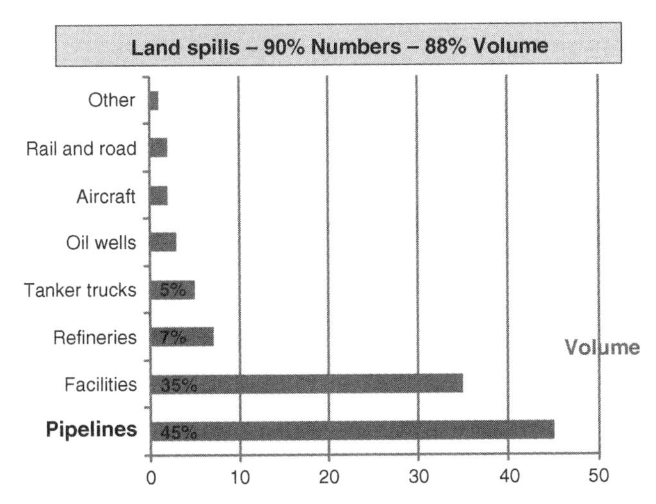


**FIGURE 46.2** USA total spill statistics highlighting pipeline spills. (Data compiled from Ref. [5].)

**FIGURE 46.3** The Trans-Alaska pipeline is unique in that approximately half of its nearly 800 mile pipeline length is aboveground. This construction protects areas of sensitive permafrost.

substantive pipelines (from Table 46.3), the estimated probability per kilometer per year is about 0.002 or about 2 in 1000.

## 46.2 RESPONSE TO OIL SPILLS

Rapid and effective response to oil spills will result in less overall damage to the environment. While it is important to focus on ways to prevent oil spills, methods for controlling them and cleaning them up must be rapidly and effectively implemented. An integrated system of contingency plans and response options can significantly reduce the environmental impact and severity of the spill.

**TABLE 46.1 Canadian Pipeline Occurrence Statistics—Example [4]**

| | January to November | | |
|---|---|---|---|
| | 2013 | 2012 | 2008–2012 Average |
| **Accidents** | **10** | **7** | **8** |
| Total, line pipe | 2 | 1 | 2 |
| Fire/ignition/explosion | 2 | 1 | 1 |
| Other damage with release | 0 | 0 | 1 |
| Total, other facilities (a) | 8 | 6 | 6 |
| Third-party damage | 1 | 1 | 1 |
| Corrosion/environmental cracking | 0 | 0 | 0 |
| Fire/ignition/explosion | 7 | 5 | 4 |
| Other damage with release | 0 | 0 | 1 |
| **Incidents** | **111** | **167** | **125** |
| Total, line pipe | 20 | 17 | 16 |
| Third-party damage/no release | 1 | 3 | 3 |
| Disturbance of supporting environment/no release | 1 | 2 | 1 |
| Uncontained release | 14 | 11 | 9 |
| Other | 4 | 1 | 4 |
| Total, other facilities[a] | 91 | 150 | 109 |
| Third-party damage/no release | 2 | 1 | 1 |
| Disturbance of supporting environment/no release | 0 | 0 | 0 |
| Uncontained release | 78 | 136 | 98 |
| Other | 11 | 13 | 11 |

[a]Includes Compressor Stations, Pump Stations, Meter Stations, Gas Processing Plants.

The purpose of contingency plans is to coordinate all aspects of the response to an oil spill. This includes stopping the flow of oil, containing the oil, and cleaning it up. Oil spills, such as forest fires and other environmental emergencies, are not predictable and can occur anytime and during any weather. Therefore, the key to effective response to an oil spill is to be prepared for the unexpected and to plan spill countermeasures that can be applied in the worst possible conditions.

### 46.2.1 Oil Spill Contingency Plans

Contingency plans can be developed for a particular facility, such as a pipeline pump station, which would include organizations and resources from the immediate area, with escalating plans for spills of greater impact. Contingency plans for larger areas usually focus more on roles, resource availability, and responsibilities and providing the basis for cooperation between the appropriate response organizations

**TABLE 46.2 Canada's Underground Oil Pipeline Network**

| Types of Pipelines | Gathering Lines | Feeder Lines | Transmission Pipelines |
|---|---|---|---|
| Description | These lines gather products from wells and move then to oil batteries | Feeder lines move products from field storage tanks to the transmission pipelines | Transmission lines transport oil over longer distances |
| Diameter | From 101.6 to 304.8 mm outside diameter (4–12 in.). | From 101.6 to 348.3 mm outside diameter (4–16 in.). | Range in size from 101.6 to 1212 mm (4–48 in.) About half are 457.2 mm (18 in.) or larger, and about one third are 254 mm. (10 in.) or smaller |
| Length | More than 250,000 km of these lines are concentrated in the producing provinces of Western Canada, primarily in Alberta. | There are more than 25,000 km of feeder pipelines in the producing areas of Western Canada. | There are more than 100,000 km of transmission lines in Canada. |
| | | | From 101.6 to 304.8 mm outside diameter (4–12 in.) |

*Source*: Table adapted from Ref. [7].

**TABLE 46.3 Typical Liquid Volume Carried by Pipeline in Canada [8]**

| Pipeline Movements | Liquid Hydrocarbons | Monthly Averages ($m^3$) |
|---|---|---|
| Total net receipts | Crude oil and pentanes plus | 23,580,000 |
| | Liquefied petroleum gases and refined petroleum products | 5,870,000 |
| Total receipts, including transfers | Crude oil and pentanes plus | 67,100,000 |
| | Liquefied petroleum gases and refined petroleum products | 9,780,000 |
| Total net deliveries | Crude oil and pentanes plus | 23,640,000 |
| | Liquefied petroleum gases and refined petroleum products | 5,880,000 |
| Transfers to interprovincial systems | Crude oil and pentanes plus | 16,510,000 |
| | Liquefied petroleum gases and refined petroleum products | 500,000 |
| Total deliveries, including transfers | Crude oil and pentanes plus | 67,170,000 |
| | Liquefied petroleum gases and refined petroleum products | 9,720,000 |
| Total disposition | Crude oil and pentanes plus | 67,070,000 |
| | Liquefied petroleum gases and refined petroleum products | 9,670,000 |
| | Total liquids | 76,740,000 |

rather than on specific response actions. Most contingency plans usually include

- a list of persons and agencies to be notified immediately after a spill occurs;
- an organization chart of response personnel and a list of their responsibilities, as well as a list of actions to be taken by them during the various phases of oil spill response;
- area-specific action plans;
- a communications network to ensure response efforts are coordinated;
- protection priorities for the affected areas;
- operational procedures for controlling and cleaning up the spill;
- reference material such as sensitivity maps and other technical data about the area;
- procedures for informing the public and keeping records;
- an inventory or database of the type and location of available equipment, supplies, and other resources; and
- scenarios for typical spills and decision trees for certain types of response actions such as using different types of countermeasures.

To remain effective, response options detailed in contingency plans must be tested frequently. This testing is conducted by responding to a practice spill as though it is real. Such exercises not only maintain and increase the skills of the response personnel, but also lead to improvements and fine tuning of the plan as weaknesses and gaps are identified in these complex plans.

### 46.2.2 Activation of Contingency Plans

The response actions defined in contingency plans, whether for spills at a single industrial facility or in an entire region, are separated into the following phases: alerting and reporting; evaluation and mobilization; containment and recovery; decontamination of equipment; disposal; and remediation or restoration. In practice, these phases often overlap rather than follow each other consecutively.

Most contingency plans also allow for a "tiered response," which means that response steps and plans escalate as the incident becomes more serious. As the seriousness of an incident is often not known in the initial phases, one of the first priorities is to determine the magnitude of the spill and its potential impact.

### 46.2.3 Training

A high-quality training program is vital for a good oil spill response program. Response personnel at all levels require training in specific operations and on using equipment for containing and cleaning up spills. To minimize injury during response, general safety training is also crucial. In many countries, response personnel are required to have 12–40 h of safety training before they can perform field work. Ongoing training and refresher courses are also essential in order to maintain and upgrade skills.

### 46.2.4 Supporting Studies and Sensitivity Mapping

A contingency plan usually includes background information on the area covered by the plan. This consists of data collected from studies and surveys and often takes the form of a sensitivity map for the area. Sensitivity maps contain information on potentially sensitive physical and biological resources that could be affected by an oil spill. This includes concentrations of wildlife such as mammals, birds, and fish; human amenities, such as recreational beaches; natural features such as types of land.

### 46.2.5 Oil Spill Cooperatives

As most oil or pipeline companies that handle oil do not have staff dedicated to cleaning up oil spills, several companies in the same area often join forces to form cooperatives. By pooling resources and expertise, these oil spill cooperatives can then develop effective and financially viable response programs. The cooperative purchases and maintains containment, cleanup, and disposal equipment and provides the training for its use.

In some spill situations, especially large spills, volunteers are an important part of the response effort. Volunteers are usually trained and their efforts coordinated with the main spill cleanup.

### 46.2.6 The Effectiveness of Cleanup

One topic of concern is the quantity of oil actually cleanup up during an oil spill. There is a public perception that most of the spills are not being cleaned up. More than 25 years ago, there was an estimate that only 10–30% of spill were actually cleaned up and removed. One must remember, however, that much of the oil may have evaporated and some lost to sedimentation or other unrecoverable processes. In the past decade, spills can be accounted for in much greater percentages, as high as 90%, once one accounts for losses such as sedimentation and evaporation. Further, there is a very large difference between small spills and very large catastrophic spills because it is much easier to remove a large portion of a small spill.

## 46.3 TYPES OF OIL AND THEIR PROPERTIES

Oil is a general term that describes a wide variety of natural substances of mineral origin, as well as a range of synthetic compounds. The many different types of oil are made up of hundreds of major compounds and thousands of minor ones. As the composition varies, each type of oil or petroleum product has certain unique properties. These properties influence how the oil behaves when it is spilled and determine the effects of the oil on living organisms in the environment. These properties also influence the efficiency of cleanup operations.

### 46.3.1 The Composition of Oil

Crude oils are mixtures of hydrocarbon compounds ranging from smaller, volatile compounds to very large compounds [5]. This mixture of compounds varies according to the geological formation of the area in which the oil is found and strongly influences the properties of the oil. For example, crude oils that consist primarily of large compounds are viscous and dense. Petroleum products such as gasoline or diesel fuel are mixtures of fewer compounds and thus their properties are more specific and less variable.

Oils also contain varying amounts of sulfur, nitrogen, oxygen, and sometimes mineral salts, as well as trace metals such as nickel, vanadium, and chromium.

The following are the oils used in this section to illustrate the fate, behavior, and cleanup of oil spills:

1. Gasoline—as used in automobiles;
2. Diesel fuel—as used in trucks, trains, and buses;
3. A light crude oil—as produced in great abundance in many countries;
4. A heavy crude oil—as produced in many counties specific examples will be given here;
5. Bunker fuel—such as Bunker C, which is a heavy residual fuel remaining after the production of gasoline and diesel fuel in refineries and often used in heating plants;
6. Dilbit—a diluted bitumen, the diluent is typically a very light oil or condensate; and
7. Bitumen—a heavy oil material typically from the oil sands operations in northern Alberta.

**TABLE 46.4 Typical Oil Properties**

| | | Light | | | Heavy | | | |
|---|---|---|---|---|---|---|---|---|
| Property | Units | Gasoline | Diesel | Crude | Crude | Bunker C | Dilbit | Bitumen |
| Viscosity | cSt or mPa.s at 15 °C | 0.5 | 2 | 5–50 | 500–50,000 | 10,000–50,000 | 150–600 | 100,000–300,000 |
| Density | g/ml at 15 °C | 0.72 | 0.84 | 0.78–0.88 | 0.88–1.00 | 0.96–1.04 | 0.9–0.95 | 98–1.015 |
| API gravity | | 65 | 35 | 30–50 | 10–23 | 5–15 | 17–25 | 8–13 |

### 46.3.2 Properties of Oil

The properties of oil discussed here are viscosity, density, and specific gravity [9,10]. These properties for the oils discussed in this chapter are listed in Table 46.4.

*Viscosity* is the resistance to flow in a liquid. The lower the viscosity, the more readily the liquid flows. For example, water has a low viscosity and flows readily, whereas peanut butter, with a high viscosity, flows poorly. The viscosity of the oil is largely determined by the amount of lighter and heavier fractions that it contains. Viscosity is affected by temperature, with a lower temperature giving a higher viscosity. Viscous oils do not spread rapidly, do not penetrate soil as readily, and affect the ability of pumps and skimmers to handle the oil.

*Density* is the weight of a given volume of oil and is typically expressed in grams per cubic centimeter ($g/cm^3$). It is the property used by the petroleum industry to define light or heavy crude oils. Density is also important as it indicates whether a particular oil will float or sink in water. As the density of water is $1.0\ g/cm^3$ at 15 °C and the density of most oils ranges from 0.7 to $0.99\ g/cm^3$, most oils will float on water. As the density of seawater is $1.03\ g/cm^3$, even heavier oils will usually float on it. Another measure of density is specific gravity, which is an oil's relative density compared to that of water at 15 °C. It is the same value as density at the same temperature. Another gravity scale is that of the American Petroleum Institute (API). The API gravity is based on the density of pure water, which has an arbitrarily assigned API gravity value of 10°. Oils with progressively lower specific gravities have higher API gravities.

The following is the formula for calculating API gravity: API gravity = [141.5 ÷ (density at 15.5 °C)] − 131.5. Oils with high densities have low API gravities and vice versa.

## 46.4 BEHAVIOR OF OIL IN THE ENVIRONMENT

When oil is spilled, whether on water or land, a number of transformation processes occur, which are referred to as the "behavior" of the oil. Two types of transformation processes are discussed in this section. The first is weathering, a series of processes whereby the physical and chemical properties of the oil change after the spill. The second is a group of processes related to the movement of oil in the environment.

### 46.4.1 An Overview of Weathering

Oil spilled on water undergoes changes in physical and chemical properties, which in combination are termed "weathering" [11,12]. Weathering processes occur at very different rates, but begin immediately after oil is spilled into the environment. Weathering rates are usually highest immediately after the spill. Both weathering processes and the rates at which they occur depend more on the type of oil than on environmental conditions. Most weathering processes are highly temperature dependent, however, and will often slow to insignificant rates as temperatures approach zero degrees.

The processes included in weathering are evaporation, emulsification, natural dispersion, dissolution, photooxidation, sedimentation, adhesion to materials, interaction with mineral fines, biodegradation, and the formation of tar balls. These processes are listed in order of importance in terms of their effect on the percentage of total mass balance, that is, the greatest loss from the slick in terms of percentage, and what is known about the process. The ones that are most important to oil spill cleanup are evaporation and emulsification.

### 46.4.2 Evaporation

Evaporation is typically the most important weathering process. Over a period of several days, a light fuel such as gasoline evaporates completely at temperatures above freezing, whereas only a small percentage of a heavier Bunker C oil evaporates. The rate at which an oil evaporates depends primarily on the oil's composition. The more volatile components an oil or fuel contains, the greater the extent and rate of its evaporation. Many components of heavier oils will not evaporate at all, even over long periods of time and at high temperatures.

Oil and petroleum products evaporate in a slightly different manner than water and the process is much less dependent on wind speed and surface area. Oil evaporation can be considerably slowed down, however, by the formation of a "crust" or "skin" on top of the oil. The skin or crust is formed when the smaller compounds in the oil are removed leaving some compounds, such as waxes and resins, at the surface. These then seal off the remainder of the oil and prevent evaporation. The rate of evaporation is very rapid immediately after a spill and then slows considerably. About 80% of evaporation occurs in the first 2 days after a spill. The properties of an oil can change significantly with the extent of evaporation. If about 40% of an oil evaporates, its viscosity could increase by as much as a thousand fold. Its density could rise by as much as 20%.

An example of how evaporation is important relates to Dilbit. Dilbit rapidly loses the more volatile diluent portion and within about a week or sooner of exposure to air, will return to the properties of the Bitumen from which it was created. This Bitumen typically has very different physical properties from the Dilbit.

### 46.4.3 Emulsification and Water Uptake

Water can enter oil through several processes [13]. Water can be present in oil in five ways. First, some oils contain about 1% water as soluble water. This water does not significantly change the physical or chemical properties of the oil. The second way is when water droplets are not held in the oil long enough to form an emulsion. These are called oils that do not form any type of water-in-oil mixtures or unstable emulsions. Unstable emulsions break down into water and oil within minutes or a few hours at most, once the sea energy diminishes. Meso-stable emulsions represent the third way water can be present in oil. Meso-stable emulsions are formed when the small droplets of water are stabilized to a certain extent by a combination of the viscosity of the oil and the interfacial action of asphaltenes and resins. The fourth way that water exists in oil is in the form of stable emulsions. These form in a way similar to meso-stable emulsions except that the oil contains sufficient asphaltenes and resins to stabilize the mixture. The viscosity of stable emulsions is 800–1000 times higher than that of the starting oil and the emulsion will remain stable for weeks and even months after formation. Stable emulsions are reddish-brown in color and appear to be nearly solid. The fifth way that oil can contain water is by viscosity entrainment. If the viscosity of the oil is such that water droplets can penetrate, but will only slowly migrate downwards, the oil can contain about 30–40% water as long as it is in an energetic body of water. Once the water calms or the oil is removed, the water slowly drains. Such water uptake is called "entrained water."

### 46.4.4 Biodegradation

A large number of microorganisms are capable of degrading petroleum hydrocarbons. Hydrocarbons metabolized by microorganisms are generally converted to an oxidized compound, which may be further degraded, may be soluble, or may accumulate in the remaining oil [14–16]. The aquatic toxicity of the biodegradation products is sometimes greater than that of the parent compounds. Biodegradation occurs slowly and is not an important removal mechanism for oil spills.

**FIGURE 46.4** A pipeline spill in Northern Alberta. The oil spreads and follows the pipeline trench and then flows into a low area below the leak.

### 46.4.5 Spreading

Oil spreads to a lesser extent and very slowly on land than on water (Figure 46.4). Oil spilled on or under ice spreads relatively rapidly but does not spread to as thin as on water. On any surface other than water, such as ice or land, a large amount of oil is retained in depressions, cracks, and other surface irregularities.

### 46.4.6 Movement of Oil Slicks on Water

In addition to their natural tendency to spread, oil slicks on water are moved along the water surface, primarily by surface currents and winds [17]. The slick generally moves at a rate that is 100% of the surface current and approximately 3% of the wind speed. If the wind is more than about 20 km/h, however, and the slick is on open water, wind predominates in determining the slick's movement. Both the wind and surface current must be considered for most situations.

### 46.4.7 Sinking and Over Washing

If oil is denser than the surface water, it may sometimes actually sink. Some rare types of heavy crudes and Bunker C can reach these densities and sink [12]. Dense materials such as Bitumen can sink in fresh water, but rarely in seawater. When oil does sink, it complicates cleanup operations as the oil can be recovered only with specialized underwater suction devices or special dredges. Sinking sometimes also occurs when sediment and oil interact. This results in a mixture that is denser than water.

### 46.4.8 Spill Modeling

Spill response personnel need to know the direction in which an oil spill is moving in order to protect sensitive resources and coastline [11,17]. To assist them with this, computerized mathematical models have been developed to predict the trajectory or pathway and fate of oil.

## 46.5 ANALYSIS, DETECTION, AND REMOTE SENSING OF OIL SPILLS

### 46.5.1 Sampling and Laboratory Analysis

Taking a sample of oil and then transporting it to a laboratory for subsequent analysis is common practice. While there are many procedures for taking oil samples, it is always important to ensure that the oil is not tainted from contact with other materials and that the sample bottles are pre-cleaned with solvents, such as hexane, that are suitable for the oil. The simplest and most common form of analysis is to measure how much oil is in water, soil, or sediment sample. Such analysis results in a value known as total petroleum hydrocarbons (TPH). A more sophisticated form of analysis is to use a gas chromatograph (GC). One type of detector used on a gas chromatogram is a mass spectrometer (MS). The method is usually called GC–MS and can be used to quantify and identify many components in oil [18].

### 46.5.2 Detection and Surveillance

***46.5.2.1 Visual Surveillance*** Oil spills are often located and surveyed from helicopters or aircraft using only human vision. There are some conditions, however, such as fog and darkness, in which oil on the surface cannot be seen. Oil can also be difficult to see in waves and among debris or weeds and it can blend into dark backgrounds, such as soil, or shorelines. In addition, many naturally occurring substances or phenomena can be mistaken for oil.

***46.5.2.2 Remote Sensing*** Remote sensing of oil involves the use of sensors other than human vision to detect or map oil spills [19]. As already noted, oil often cannot be detected in certain conditions. Remote sensing provides a timely means to map out the locations and approximate concentrations of very large spills in many conditions. Remote sensing is usually carried out with instruments onboard aircraft or from satellite. While many sensors have been developed for a variety of environmental applications, only a few are useful for oil spill work. Remote sensing of oil on land is particularly limited and only one or two sensors are useful.

***46.5.2.3 Visual Sensors*** Many devices employing the visible spectrum, including the conventional camera and video camera, are available at a low cost. As these devices are subject to the same interferences as visual surveillance, they are used primarily to document the spill or to provide a frame of reference for other sensors.

***46.5.2.4 Infrared Sensors*** Thick oil on water or land absorbs infrared radiation from the sun and appears in infrared imagery as hot on a cold ocean surface. Unfortunately, many other false targets such as weeds, biogenic oils, debris, and oceanic and riverine fronts can interfere with oil detection. The advantage of infrared sensors over visual sensors is that they give information sometimes where visual sensors do not work.

***46.5.2.5 Laser Fluorosensors*** Oils that contain aromatic compounds, as most oils do, will absorb ultraviolet light and give off visible light in response. Since very few other compounds respond in this way, this can be used as a positive method of detecting oil at sea or on land. Laser fluorosensors use a laser in the ultraviolet spectrum to trigger this fluorescing phenomenon and a sensitive light detection system to provide an oil-specific detection tool. There is also some information in the visible light return that can be used to determine whether the oil is a light or heavy oil or a lubricating oil.

***46.5.2.6 Radar*** As oil on water calms smaller waves (waves on the order of a few centimeters), radar can detect oil on water bodies as a calm area. The technique is highly prone to false targets, however, and is limited to a narrow range of wind speeds (approximately 2–6 m/s). At winds below this wind speed, there are not enough small waves to yield a difference between the oiled area and the water. At higher winds, the waves can propagate through the oil and the radar may not be able to "see" into the troughs between the waves. Radar is not useful near coastlines or between head lands because the wind shadows look like oil. There are also many natural calms on the oceans that can resemble oil. Despite its large size and expense, radar equipment is particularly well suited for searches of large areas and for work at night or in foggy or other bad weather conditions. Radar has little application to land spills or smaller inland water bodies.

## 46.6 CONTAINMENT ON WATER

Containment of an oil spill refers to the process of confining the oil, either to prevent it from spreading to a particular area, to divert it to another area where it can be recovered or treated, or to concentrate the oil so it can be recovered or burned. Containment booms are the basic and most frequently-used piece of equipment for containing an oil spill on water. Booms are generally the first equipment mobilized at a spill and are often used throughout the operation. While many pipeline spills occur on land, most of the oil ends up in water bodies and as such, booms are frequently used.

### 46.6.1 Types of Booms and Their Construction

A boom is a floating mechanical barrier designed to stop or divert the movement of oil on water. Booms resemble a vertical curtain with portions extending above and below the water line. Booms are constructed in sections, usually 15 or 30 m long, with connectors installed on each end so that sections of the boom can be attached to each other, towed, or anchored. The three basic types of booms are fence and curtain booms, which are common, and shoreline seal booms.

### 46.6.2 Uses of Booms

Booms are used to enclose floating oil and prevent it from spreading, to protect biologically sensitive areas, to divert oil to areas where it can be recovered or treated, and to concentrate oil and maintain an adequate thickness so that skimmers can be used or other cleanup techniques, such as *in-situ* burning, can be applied. Booms are used primarily to contain oil, although they are also used to deflect oil. When used for containment, booms are often arranged in a U configuration. The U-shape is created by the current pushing against the center of the boom. The critical requirement is that the current in the apex of the "U" does not exceed 0.5 m/s or 1 knot, which is referred to as the critical velocity. This is the speed of the current flowing perpendicular to the boom, above which oil will be lost from the boom.

If used in areas where the currents are likely to exceed 0.5 m/s or 1 knot, such as in rivers and estuaries, booms are often used in the deflection mode. The boom is then deployed at various angles to the current. Figure 46.5 shows the use of diversionary booms in a fast-flowing river. The oil can then be deflected to areas where it can be collected or to less sensitive areas. If strong currents prevent the best positioning of the boom in relation to the current, several booms can be deployed in a cascading pattern to progressively move oil toward one side of the watercourse. This technique is effective in wide rivers or where strong currents may cause a single boom to fail. When booms are used for deflection, the forces of the current on the boom are usually so powerful that stronger booms are required and they must be anchored along their entire length.

**FIGURE 46.5** The use of diversionary booms to deflect oil from the center of a small river. This spill was the result of a pipeline spill on a crossing of this river.

### 46.6.3 Boom Failures

A boom's performance and its ability to contain oil are affected by water currents, waves, and winds [20–22]. Either alone or in combination, these forces often lead to boom failure and loss of oil. The most critical factor is the current speed relative to the boom. Failures will occur when this exceeds 0.35 m/s (0.7 knots).

### 46.6.4 Sorbent Booms and Barriers

Sorbent booms are specialized containment and recovery devices made of porous sorbent material such as woven or fabric polypropylene, which absorbs the oil while it is being contained [23]. Sorbent booms are used when the oil slick is relatively thin, that is, for the final "polishing" of an oil spill, to remove small traces of oil or sheen, or as a backup to other booms. Sorbent booms are often placed off a shoreline that is relatively unoiled or freshly cleaned to remove traces of oil that may recontaminate the shoreline. They are not absorbent enough to be used as a primary countermeasure technique for any significant amount of oil. Oil sorbent booms must also be removed from the water carefully to ensure that oil is not forced from them and the area recontaminated. Sorbent booms are often used on land spills to contain and remove light surface oiling.

## 46.7 OIL RECOVERY ON WATER

Recovery is the next step after containment in an oil spill cleanup operation. Even though most pipeline spills occur on land, the oil most often flows to a water body from where it is recovered. As discussed in the previous section, an important objective of containment is to concentrate oil into thick layers to facilitate recovery. In fact, the containment and recovery phases of an oil spill cleanup operation are often carried out at the same time. As soon as booms are deployed at the site of a spill, equipment and personnel are mobilized to take advantage of the increased oil thickness, favorable weather, and less weathered oil. After oil spreads or becomes highly weathered, recovery becomes less viable and is sometimes impossible.

### 46.7.1 Skimmers

Skimmers are mechanical devices designed to remove floating oil from a water surface. They vary greatly in size, application, and capacity, as well as in recovery efficiency [24,25]. Skimmers are classified according to the area where they are used, for example, inshore, offshore, in shallow water, or in rivers, and by the viscosity of the oil they are intended to recover, that is heavy or light oil.

The effectiveness of a skimmer is rated according to the amount of oil that it recovers, as well as the amount of water picked up with the oil. Effectiveness depends on a variety of factors, including the type of oil spilled, the properties of the oil such as viscosity, the thickness of the slick, sea conditions, wind speed, ambient temperature, and the presence of ice or debris.

Most skimmers function best when the oil slick is relatively thick. The oil must therefore be collected in booms before skimmers can be used effectively (Figure 46.6). The skimmer is placed wherever the oil is most concentrated in order to recover as much oil as possible. Weather conditions at a spill site have a major effect on the efficiency of skimmers. Depending on the type of skimmer, most will not work effectively in waves >1 m or in currents exceeding 1 knot. Most skimmers do not operate effectively in waters with ice or debris such as branches, seaweed, and floating waste. Some skimmers have screens around the intake to prevent debris or ice from entering, conveyors or similar devices to remove or deflect debris, and cutters to deal with sea weed. Very viscous oils, tar balls, or oiled debris can clog the intake or entrance of skimmers and make it impossible to pump oil from the skimmer's recovery system.

**FIGURE 46.6** A typical boom and skimmer used together to capture and then recover oil on water.

Skimmers are also classified according to their basic operating principles: oleophilic surface skimmers; weir skimmers; suction skimmers, or vacuum devices; elevating skimmers; and submersion skimmers.

### 46.7.2 Sorbents

Sorbents are materials that recover oil through either absorption or adsorption [23,26]. They play an important role in oil spill cleanup and are used in the following ways: to clean up the final traces of oil spills on water or land; as a backup to other containment means, such as sorbent booms; as a primary recovery means for very small spills; and as a passive means of cleanup. An example of such passive cleanup is when sorbent booms are anchored off lightly oiled shorelines to absorb any remaining oil released from the shore and prevent further contamination or reoiling of the shoreline.

Sorbents can be natural or synthetic materials. Natural sorbents are divided into organic materials, such as peat moss or wood products, and inorganic materials, such as vermiculite or clay. Sorbents are available in a loose form, which includes granules, powder, chunks, and cubes, often contained in bags, nets, or socks. Sorbents are also available formed into pads, rolls, blankets, and pillows. The use of synthetic sorbents in oil spill recovery has increased in the last few years. These sorbents are often used to wipe other oil spill recovery equipment, such as skimmers and booms, after a spill cleanup operation. Sheets of sorbent are often used for this purpose.

### 46.7.3 Manual Recovery

Small oil spills or those in remote areas are sometimes recovered by hand. Heavier oils are easier to manually remove than lighter oils. Spills on water close to shorelines are sometimes cleaned up with shovels, rakes, or by cutting the oiled vegetation. Hand bailers, which resemble a small bucket on the end of a handle, are sometimes used to recover oil from the water surface. Manual recovery is tedious and may involve dangers such as physical injury from falls.

## 46.8 SEPARATION, PUMPING, DECONTAMINATION, AND DISPOSAL

After oil is recovered from a water surface or from land, it must be temporarily stored, the water and debris separated from it, and the oil recycled or disposed of. Pumps are used to move the oil from one process to another. Storage, separation, and disposal are all crucial parts of a cleanup operation. In many cleanups, recovery has actually stopped because there was no place to put the recovered oil.

### 46.8.1 Temporary Storage

When oil is recovered, sufficient storage space must be available for the recovered product. The recovered oil often contains large amounts of water and debris, which increase the amount of storage space required. Several types of specially built tanks are available to store recovered oil. Flexible portable tanks are common type of storage used for spills recovered on land and from rivers and lakes.

### 46.8.2 Pumps

Pumps play an important role in oil spill recovery. They are an integral part of most skimmers and are also used to transfer oil from skimmers to storage tanks. Pumps used for recovered oil differ from water pumps in that they must be capable of pumping very viscous oils and dealing with water, air, and debris.

### 46.8.3 Vacuum Systems

Vacuum systems consist of vacuum pumps and tanks mounted on a skid or truck. The vacuum pump creates a vacuum in the tank and the oil moves directly through a hose or pipe to the tank from the skimmer or the source of the oil. The oil does not go through the pump, but moves directly from its source into the tank. Vacuum systems can cope with debris, viscous oils, and the intake of air or water.

### 46.8.4 Recovery from the Water Subsurface

Recovery of submerged oil on the bottom has been carried out in many different ways in the past [27]. Diver-directed pumping has been used often because divers are efficient in finding and contacting the oil when the oil coverage varies in size and thickness. The biggest problem is dealing with the large amount of recovered water and sediment along with viscous oils. Remotely operated vehicle pumping systems have largely been used to set up other equipment, but recently are proposed to cleanup oil. Several types of dredges have used to recover oil from the bottom. Where the oil is solidified, environmental clamshell dredges have been used successfully. Hopper dredges have been proposed in the past, but the large volumes of water and sediment generated, compared to the amount of oil recovered, is a significant factor. Figure 46.7 shows the recovery of oil using a backhoe.

### 46.8.5 Separation

As all skimmers recover some water with the oil, a device to separate oil and water is usually required [28]. The oil must be separated from the recovery mixture for disposal, recycling, or direct reuse by a refinery. Sometimes, settling tanks or gravity separators are incorporated into skimmers, but separators are more often installed on recovery ships or barges. Portable storage tanks are often used as separators, with outlets installed on the bottom of the tanks so that water that has settled to the bottom of the tank can be drained off, leaving the oil in the tank. Vacuum trucks are also used in this way to separate oil and water.

### 46.8.6 Decontamination

Equipment and vessels used during spills often become “contaminated” or covered with oil. Before transporting

**FIGURE 46.7** Use of a backhoe on a barge to recover sunken Dilbit, now just Bitumen, from a Michigan pipeline spill.

this equipment further, it is decontaminated. This typically involves removal to a lined area, a high-pressure wash and treatment of the recovered water. Special areas are prepared for the decontamination of vessels, booms, or skimmers. Large vessels must of necessity, be decontaminated on the water and this involves circling the vessel with booms and recovering the oil released from the vessel. Often lightly-contaminated vessels are cleaned by hand using sorbent cloths. This procedure avoids the extra procedures of booming and oil recovery.

The primary tool for oil removal is high-pressure water. The water released from decontamination is treated as recovered oil would be. It is separated and the oil placed in recovered oil collection tanks.

A final note on this topic is that workers must also decontaminate their boots and clothing if covered with oil. Stations are often set up very close to embarkation points to avoid carrying contamination further.

### 46.8.7 Disposal

Disposing of the recovered oil and oiled debris is one of the most difficult aspects of an oil spill cleanup operation [29]. Any form of disposal is subject to a complex system of local, provincial or state, and federal legislation. Unfortunately, most recovered oil consists of a wide range of contents and cannot be classified as simply liquid or solid waste. The recovered oil may contain water that is difficult to separate from the oil and many types of debris, including vegetation, sand, gravel, logs, branches, garbage, and pieces of containment booms. Incineration is a frequent means of disposal for recovered material. Approval must be obtained from government regulatory authorities. Emission guidelines for incinerators may preclude simply placing material into the incinerator.

## 46.9 SPILL-TREATING AGENTS

Treating the oil with specially-formulated chemicals is another option for dealing with oil spills. An assortment of chemical spill-treating agents is available to assist in cleaning up oil. Approval must be obtained from the appropriate authorities before these chemical agents can be used. In addition, these agents are not always effective and the treated oil may be toxic to aquatic and other wildlife. Dispersants are chemical spill-treating agents that promote the formation of small droplets of oil, which disperse throughout the top layer of the water column [30,31]. The use of dispersants remains a controversial issue and special permission is required in most jurisdictions. Generally, in freshwater or land applications, their use is banned. Surface-washing agents are intended to be applied to shorelines or surfaces to release the oil from the surface [32].

## 46.10 *IN SITU* BURNING

*In situ* burning is an oil spill cleanup technique that involves controlled burning of the oil at or near the spill site [33,34]. The major advantage of this technique is its potential for removing large amounts of oil over an extensive area in less or about the same time than other techniques but with a distinct advantage of being a final solution. The technique has been used at actual spill sites for some time, especially on land and in ice-covered waters where the oil is contained by the ice. During the 2010 oil spill in the Gulf of Mexico, it was used extensively and contributed greatly to the removal of oil from the water surface.

### 46.10.1 Advantages

Burning has some advantages over other spill cleanup techniques, the most significant of which is its ability to be a final solution and its capacity to rapidly remove large amounts of oil. Burning can prevent oil from spreading to other areas and contaminating shorelines and biota.

Burning oil is a final, one-step solution. When oil is recovered mechanically, it must be transported, stored, and disposed off, which requires equipment, personnel, time, and money. Often not enough of these resources are available when large spills occur. Burning generates a small amount of burn residue that can be recovered or further reduced through repeated burns.

It can be applied in remote areas where other methods cannot be used because of distances and lack of infrastructure. In some circumstances, such as when oil is mixed with or on ice, it may be the only available option for dealing with an oil spill. Finally, while the efficiency of a burn varies with a number of physical factors, removal efficiencies are generally greater than those for other response methods such as skimming and the use of chemical dispersants. During several test and actual burns, efficiency rates as high as 98% were achieved.

### 46.10.2 Disadvantages

The most obvious disadvantage of burning oil is the large black smoke plume. The concerns revolve around toxic emissions from the large black smoke plume. These emissions have been studied and can be dealt with. The second disadvantage is that the oil will not ignite and burn quantitatively unless it is thick enough. Most oils spread rapidly on water and the slick quickly becomes too thin for burning to be feasible. Fire-resistant booms are used to concentrate the oil on water into thicker slicks so that the oil can be burned. Burning oil is sometimes not viewed as an appealing alternative to collecting the oil and processing it for reuse. Reprocessing facilities for this purpose, however, are not readily accessible in most parts of the world. Another factor that discourages reuse of oil is that

**FIGURE 46.8** An oil burn after a spill on a northern pipeline.

recovered oil often contains too many contaminants for reuse and is incinerated instead.

### 46.10.3 Ignition and What Will Burn

Early studies of *in situ* burning focused on ignition as being the key to successful burning of oil on water [34]. It has since been found that ignition can be difficult, but only under certain circumstances. Figure 46.8 shows a land burn that is generally easier to ignite than oil on water [34]. Ignition may be difficult, however, at winds >20 m/s (40 knots). An important fact of *in situ* burning is that oils can be readily ignited if they are at least 1–3 mm thick and will continue to burn down to slicks about ½–1 mm thick. Sufficient heat is required to vaporize material so the fire will continue to burn. In very thin slicks, most of the heat is lost to the water and vaporization/combustion is not sustained.

Often a primer, such as diesel fuel, may be needed to ignite heavy oils. This is also the case for oil that contains water, although oil that is completely emulsified with water is difficult to ignite. Several burns have been conducted in which some emulsion or high water content in the oil did not affect the efficiency of the burn.

### 46.10.4 Burn Efficiency and Rates

Burn efficiency is measured as the percentage of starting oil removed compared to the amount of residue left. The amount of soot produced is usually ignored as it is a small amount and difficult to measure. Burn efficiency is largely a function of oil thickness. If a 2-mm thick slick is ignited and burns down to 1 mm, the maximum burn efficiency is 50%. If a 20-mm thick pool of oil is ignited, however, and burns down to 1 mm, the burn efficiency is about 95%. Recent research has shown that these efficiency values are affected by other factors such as the type of oil and the amount of water content. Higher efficiency is usually achieved when towing a fire-resistant boom as the oil is continually driven to the rear, burned, and leaving only a small amount of residue unburned at the end.

Most of the residue from burning oil is unburned oil with some lighter or more volatile products removed. The residue is adhesive and therefore can be recovered manually. Residue from burning heavier oils and from very efficient burns may sometimes sink in water, although this rarely happens as the residue, when cooled, is only slightly denser than sea water.

Most oil pools burn at a rate of about 2–4 mm per minute, which means that the depth of oil is reduced by 2–4 mm each minute. Table 46.5 shows the burning characteristics of several oils.

An optimal burn rate for diesel fuel and light crudes is about 5000 l of oil per $m^2$ per day.

**TABLE 46.5 Burning Properties of Various Fuels**

| Fuel | Burnability | Ease of Ignition | Flame Spread | Burning Rate[a] (mm/min) | Sootiness of Flame | Efficiency Range (%) |
|---|---|---|---|---|---|---|
| Gasoline | Very high | Very easy | Very rapid—through vapors | 4 | Medium | 95–99 |
| Diesel fuel | High | Easy | Moderate | 3.5 | Very high | 90–98 |
| Light crude | High | Easy | Moderate | 3.5 | High | 85–98 |
| Medium crude | Moderate | Easy | Moderate | 3.5 | Medium | 80–95 |
| Heavy crude | Moderate | Medium | Moderate | 3 | Medium | 75–90 |
| Weathered crude | Low | Difficult, add primer | Slow | 2.8 | Low | 50–90 |
| Crude oil with ice | Low | Difficult, add primer | Slow | 2 | Medium | 50–90 |
| Light fuel oil | Low | Difficult, add primer | Slow | 2.5 | Low | 50–80 |
| Heavy fuel oil | Very low | Difficult, add primer | Slow | 2.2 | Low | 40–70 |
| Dilbit | Moderate | Easy if fresh | Moderate | 2.5 | Medium | 40–60 |
| Lube oil | Very low | Difficult, add primer | Slow | 2 | Medium | 40–60 |
| Waste Oil | Low | Difficult, add primer | Slow | 2 | Medium | 30–60 |
| Emulsified Oil | Low | Difficult, add primer | Slow | 1–2 | Low | 30–60 |

[a]typical rates.

### 46.10.5 Use of Containment

As previously discussed, oil can be burned on water without using containment booms if the slick is thick enough (2–3 mm) to ignite. For most crude oils, however, this thickness is only maintained for a few hours, at most, after the spill occurs. Most oil on the open sea rapidly spreads to an equilibrium thickness, which is about 0.01–0.1 mm for light crude oils and about 0.05–0.5 mm for heavy crudes and residual oils. Such slicks are too thin to ignite and containment is required to concentrate the oil so it is thick enough to ignite and burn efficiently.

Fire-resistant booms are also used by spill responders to isolate the oil from the source of the spill. When considering burning as a spill cleanup technique, the integrity of the source of the spill and the possibility of further spillage is always a priority. If there is any possibility that the fire could flash back to the source of the spill, such as pumping station, the oil is not ignited.

Fire-resistant booms are made of a variety of materials, including ceramic, stainless steel, and water-cooled fiber glass. Fire booms must withstand the high temperatures, high heat flux as well as the mechanical forces during an oil spill burn. In addition, it is expected that a particular fire-resistant boom should withstand a multi-hour burn and be able to be reused several times.

During the Deepwater Horizon spill in the Gulf of Mexico, more than 400 burns were carried out using fire-resistant booms [27].

Oil is sometimes contained by natural barriers such as land forms, shorelines, offshore sand bars, or ice. Several successful experiments and burns of actual spills have shown that ice acts as a natural boom so that *in situ* burning can be carried out successfully for spills in ice.

### 46.10.6 Emissions from Burning Oil

The possibility of releasing toxic emissions into the atmosphere or the water is a barrier to the widespread acceptance of burning oil as a spill countermeasure. Some atmospheric emissions of concern are particulate matter precipitating from the smoke plume, combustion gases, and unburned hydrocarbons. While soot particles consist primarily of carbon particles, they also contain a number of adsorbed chemicals. The residue left at the burn site is also a matter of concern. Possible water emissions include sinking or floating burn residue and soluble organic compounds.

Extensive studies have been conducted to measure and analyze all these components of emissions from oil spill burns [34]. The emphasis in sampling has been on air emissions at ground level as these are the primary human health concern and the regulated value. Most burns produce an abundance of particulate matter. Particulate matter at ground level is a health concern close to the fire and under the plume, although concentrations decline rapidly downwind from the fire. The greatest concern is the smaller or respirable particles that are 2.5 μm or less in size. Concentrations at ground level (1.5 m) can still be above normal health concern levels (35 $\mu m^3/m^3$) as close downwind as 500 m, such as from the amount of oil that could be contained in a typical burn.

Polyaromatic hydrocarbons (PAHs) are a primary concern in the emissions from burning oil, both in the soot particles and as a gaseous emission. All crude oils contain PAHs, varying from as much as 1% down to about 0.001%. Most of these PAHs are burned to fundamental gases except those left in the residue and the soot. In summary, PAHs are not a serious concern when assessing the impact of burning oil.

Current thinking on burning oil as an oil spill cleanup technique is that the airborne emissions are not a serious health or environmental concern, especially at distances greater than a few kilometers from the fire. Studies have shown that emissions are low compared to other sources and generally result in concentrations of air contaminants that are below health concern levels 500 m downwind from the fire [34].

## 46.11 SHORELINE CLEANUP AND RESTORATION

Oil spilled on water, be it at sea or inland, is seldom completely contained and recovered and some of it eventually reaches the shoreline or margin of a water body. It is more difficult and time-consuming to clean up shoreline areas than it is to carry out containment and recovery operations at sea. Physically removing oil from some types of shoreline can also result in more ecological and physical damage than if oil removal is left to natural processes. The decision to initiate cleanup and restoration activities on oil-contaminated shorelines is based on careful evaluation of socio-economic, aesthetic, and ecological factors. These include the behavior of oil in shoreline regions, the types of shoreline and their sensitivity to oil spills, the assessment process, shoreline protection measures, and recommended cleanup methods. Similarly, some of the shoreline types are related to land types and the same cleanup applies.

### 46.11.1 Behavior of Oil on Shorelines

The fate and behavior of oil on shorelines are influenced by many factors, some of which relate to the oil itself, some to characteristics of the margin or shoreline, and others to conditions when the oil is deposited on the shoreline [35,36]. These factors include the type and amount of oil, the degree of weathering of the oil, both before it reaches the shoreline and while on the shoreline, the temperature, the water level when the oil washes onshore, the type of beach substrate, that is, its material composition, the type and sensitivity of biota on the beach, and the steepness of the shore.

The extent that an oil penetrates and spreads, its adhesiveness, and how much the oil mixes with the type of material on the shoreline or land surface are all important factors in terms of cleanup. Cleanup is more difficult if the oil penetrates deeply into the shoreline. Penetration varies with the type of oil and the type of material on the shoreline. For example, oil does not penetrate much into fine beach material such as sand or clay, but will penetrate extensively on a shore consisting of coarse boulders. A very light oil such as diesel on a cobble beach can penetrate to about a meter under some conditions and is difficult to remove. On the other hand, a weathered crude deposited on fine sand can remain on the surface indefinitely and is removed fairly easily using mechanical equipment. The adhesiveness of the stranded oil varies with the type of oil and the degree of weathering. Most fresh oils are not highly adhesive, whereas weathered oils often are. Diesel and gasoline are relatively nonadhesive, crudes are generally moderately adhesive when fresh and more adhesive when weathered, and Bunker C is adhesive when fresh and highly adhesive when weathered. An oil that is not adhesive when it reaches the shore may get washed off, at least partially, on the next tidal cycle.

### 46.11.2 Types of Shorelines

The type of shoreline or similar land surface is crucial in determining the fate and effects of an oil spill as well as the cleanup methods to be used. In fact, the shoreline's basic structure and the size of material present are the most important factors in terms of oil spill cleanup. There are many types of shorelines, all of which are classified in terms of sensitivity to oil spills and ease of cleanup. The types of interest are: bedrock, man-made solid structures, boulder beaches, pebble–cobble, mixed sand-gravel beaches, sand beaches, sand tidal flats, mud tidal flats, marshes, and peat and low-lying tundra. These types occur on both seashore and freshwater shores.

Marshes are important ecological habitats that often serve as nurseries for water and bird life in the area. Marshes range from fringing marshes, which are narrow areas beside a main water body, to wide marsh meadows. Marshes are rich in vegetation that traps oil. Light oils can penetrate into marsh sediments through animal burrows or cracks. Heavier oils tend to remain on the surface and smother plants or animals. Oiled marshes may take years or even decades to recover. Marshes are difficult to access and entering them by foot or by vehicle can cause more damage than the oil itself.

Peat and low-lying tundra are similar types of shoreline or land forms found in the Arctic regions. Although different, they have similar sensitivity and cleanup methodologies. Peat is a spongy, fibrous material formed by the incomplete decomposition of plant materials. Peat erodes from tundra cliffs and often accumulates in sheltered areas as does oil. Oil does not penetrate wet peat, but dry peat can absorb large amounts of oil.

### 46.11.3 Shoreline Cleanup Assessment Technique (SCAT)

Priorities for shoreline cleanup are based on a sophisticated shoreline assessment procedure. A systematic evaluation of oiled shorelines can minimize damage to the most sensitive shorelines. When an oil spill occurs, site assessment surveys are usually conducted in direct support of spill response operations. These surveys rely heavily on previously-obtained data, maps, and photographs. For example, the structure of the beach is usually already mapped and recorded as part of the sensitivity mapping exercise for the area.

The following are the objectives of site assessment surveys:

- to document the oiling conditions and the physico-ecological character of the oiled shoreline, using standardized procedures;
- to identify and describe human use and effects on the shoreline's ecological and cultural resources;
- to identify constraints on cleanup operations;
- to verify existing information on environmental sensitivities or compare it to observations from the aerial survey; and
- to define the most appropriate cleanup techniques for each segment of the shoreline.

### 46.11.4 Cleanup Methods

Many methods are available for removing oil from shorelines or land forms. All of them are costly and take a long time to carry out. The selection of the appropriate cleanup technique is based on the type of substrate, the depth of oil in the sediments, the amount and type of oil and its present form/condition, the ability of the shoreline to support traffic, the environmental, human, and cultural sensitivity of the shoreline, and the prevailing ocean and weather conditions. The cleanup techniques suitable for use on the various types of land forms or shoreline are listed in Table 46.6.

The primary objective of cleanup operations is to minimize the effects of the stranded oil and accelerate the natural recovery of affected areas. Obviously, a cleanup technique should be safe and effective and not be so intrusive as to cause more damage than the oil itself. In general, cleanup techniques should not be used if they endanger human life or safety, leave toxic residue or contaminate other shorelines or lower tidal areas, or kill plants and animals on the shoreline. In addition, excessive amounts of shoreline material should not be removed and the structure of the shoreline should not

**TABLE 46.6 Cleanup Techniques and Shoreline**

| Shoreline type | Condition of the oil | Natural recovery | Flooding | Low-pressure cold water | Low-pressure warm water | Manual removal | Vacuums | Mechanical removal | Sorbents | Tilling/ aeration | Sediment reworking/ surf-washing | Cleaning agents |
|---|---|---|---|---|---|---|---|---|---|---|---|---|
| Bedrock | fluid | ✚ | ✚ | ✚ | ▲ | ▲ | ▲ | | ▲ | | | ◆ |
| | solid | | | | ◆ | ▲ | | | | | | ✚ |
| Man-made | fluid | ✚ | ✚ | ✚ | ▲ | ▲ | | | ▲ | | | ◆ |
| | solid | | | | ◆ | ▲ | | | | | | ✚ |
| Boulder | fluid | ✚ | ✚ | ✚ | ▲ | ▲ | | | ▲ | | | ◆ |
| | solid | | | | ◆ | | | | | | | ✚ |
| Pebble-cobble | fluid | ✚ | ✚ | ✚ | | ▲ | | | ▲ | ▲ | ▲ | |
| | solid | | | | | ▲ | | ▲ | | | | ◆ |
| Mixed sand-gravel | fluid | ✚ | ✚ | ✚ | | ▲ | | | ▲ | ▲ | ▲ | |
| | solid | | | | | ▲ | | ✚ | | | | ◆ |
| Sand beach | fluid | ✚ | ✚ | ✚ | | ▲ | | ✚ | ▲ | ▲ | ▲ | |
| | solid | | | | | ▲ | | ✚ | | | | |
| Sand tidal flats | fluid | ✚ | ✚ | ✚ | | ▲ | ▲ | | ▲ | | | |
| | solid | | | | | ▲ | | ▲ | | | | |
| Mud tidal flats | fluid | ✚ | ✚ | ✚ | | | ▲ | | ▲ | | | |
| | solid | | | | | | | ▲ | | | | ◆ |
| Marshes | fluid | ✚ | ✚ | ✚ | | | | | ▲ | | | |
| | solid | ✚ | ✚ | ✚ | | ▲ | | | | | | ◆ |
| Peat Shorelines or low-lying tundra | fluid | ✚ | ✚ | ✚ | | | ▲ | | ▲ | | | |
| | solid | | | | | | | ▲ | | | | |
| Mangroves | fluid | ✚ | ✚ | ✚ | ▲ | ▲ | | | ▲ | | | ◆ |
| | solid | | | | ◆ | ▲ | | | | | | ✚ |

✚ Acceptable method.
▲ Suitable method for small quantities.
◆ Conditional method, may only work under special circumstances.

be changed so as to make it unstable. Finally, any technique that generates a lot of waste material should not be used. In the past, heavy equipment used on beaches resulted in thousands of tons of contaminated beach material, which then required disposal.

The length of time required to complete the cleanup is another important criteria when selecting a cleanup technique. The longer oil is on a beach, the harder it is to clean up. A method that removes most of the mobile oil rapidly is much better, in many circumstances, than a more thorough one that takes weeks to carry out. Time often dictates the cleanup method used.

### 46.11.5 Recommended Cleanup Methods

Some recommended shoreline cleanup methods are natural recovery, manual removal, flooding or washing, use of vacuums, mechanical removal, tilling and aeration, sediment reworking or surf washing, and the use of sorbents or chemical cleaning agents.

Sometimes the best response to an oil spill on a shoreline may be to leave the oil and monitor the natural recovery of the affected area. This would be the case if more damage would be caused by cleanup than by leaving the environment to recover on its own. This option is suitable for small spills in sensitive environments and on a beach that will recover quickly on its own such as on exposed shorelines and with nonpersistent oils such as diesel fuel on impermeable beaches. This is not an appropriate response if important ecological or human resources are threatened by long-term persistence of the oil.

Manual removal is the most common method of shoreline cleanup. Teams of workers pick up oil, oiled sediments, or oily debris with gloved hands, rakes, forks, trowels, shovels, sorbent materials, hand bailers, or poles. It may also include scraping or wiping with sorbent materials or sifting sand to remove tar balls. Material is usually collected directly into plastic bags, drums, or buckets for transfer. Figure 46.9 shows the manual removal of oil weeds on a lake shore.

**FIGURE 46.9** Manual removal of oiled reeds and heavy oil after a heavy oil spill in a lake. The shoreline here is a sandy shore that also was manually cleaned.

Flooding or washing shorelines or landforms are common cleanup methods. Low-pressure washing with cool or lukewarm water causes little ecological damage and removes oil quickly. Warmer water removes more oil, but causes more damage. High pressure and temperature cause severe ecological damage and recovery may take years. It is preferable to leave some oil on the shoreline than to remove more oil but kill all the biota with high pressure or temperature water.

Low-pressure cool or warm water washing uses water at pressures less than about 200 kPa (50 psi) and temperatures less than about 30 °C. Water is applied with hoses that do not focus the water excessively, avoiding the loss of plants and animals. Flooding is a process in which a large flow or deluge of water is released on the upper portion of the beach.

Low-pressure washing and flooding are often combined to ensure that oil is carried down the beach to the water, where it can be recovered with skimmers. Washing and flooding are best done on impermeable shoreline types and are not useful for shorelines with fine sediments such as sand or mud. These techniques are not used on shorelines where sensitive plant species are growing.

Several sizes of vacuum systems are useful for removing liquid oil that has pooled or collected in depressions on beaches and shorelines. Vacuum trucks used for collecting domestic waste are often used to remove large pools of recovered oil rather than for recovering oil directly on the spill site. For safety reasons, vacuum devices are not used for oils that are volatile.

Mechanical removal involves removing the surface oil and oiled debris with tractors, front-end loaders, scrapers, or larger equipment such as road graders and excavators. There are many types of specialized beach cleaning now available. These machines can screen out tar balls and return clean sand directly to the beaches. Front-end loaders and backhoes are used on a variety of beaches to move oiled materials and to expose buried material as well as to remove materials recovered manually from the beach. While mechanical devices remove oil quickly from shorelines, they also remove large amounts of other material and generate more waste than other techniques, unless specialized cleaners are used. Sand and sand-gravel shorelines are best suited to this technique as they can support mechanical equipment and are not usually damaged by the removal of material. Mechanical equipment should not be used on sensitive shorelines, shorelines with an abundance of plant and animal life, and shorelines that would become unstable if large amounts of material are removed.

Sorbents are used in several ways in beach cleanup. In a passive role, sorbents are left in place, on or near a beach, to absorb oil that is released from the beach by natural processes and prevent it from recontaminating other beaches or contacting wildlife. Sorbent booms as well as "pom-poms" designed for heavy oil can be staked on the beach or in the water on the beach face to catch oil released naturally. Loose sorbents such as peat moss and wood chips are generally not used because they may sink and migrate into non-oiled areas and are difficult to recover.

*In situ* burning is useful if the water level is high and the burn residue is either removed or does not suppress future plant growth. Oil will not burn on a typical beach unless the oil is pooled or concentrated in sumps or trenches. In fact, burning is a useful restorative method for marshes if done in spring when the water level is high so that the heat does not affect the plant roots. Burning in late summer or early fall, however, can kill much of the plant life. An alternative is to flood the marsh using berms and pumps, which will raise up some of the oil for burning.

## 46.12 OIL SPILLS ON LAND

While the vast majority of oil spills in North America occur on land, land spills are less dramatic than spills on water and receive less attention from the media and the public. It should be noted that much of oil spilled on land ends up in some form of water body. These water bodies include creeks, rivers, sloughs, lakes, wetlands, and marshes. In this section, two types of land spills are discussed—those that occur primarily on the surface of the land and those that occur partially or totally in the subsurface [1]. The sources and the cleanup methods differ for these types of spills. Most surface spills in Canada are the result of oil production, such as spills from pipelines and battery sites, whereas most subsurface spills are from leaking underground fuel storage tanks or pipelines. Whether on the surface or subsurface, however, each spill is unique in terms of the type of material spilled, the habitat in which the oil is spilled, its location, and the weather conditions during and after the spill [37]. One concern that should be borne in mind is the movement of spills to water bodies. Such movement on water can spread contamination rapidly over a wide area.

Protecting human health and safety is still the top priority when cleaning up oil spills on land and in the subsurface, although this is primarily an issue with some fuels, such as gasoline, condensate-containing materials such as Dilbit or diesel fuel. Minimizing long-term damage to the environment and protecting agricultural land are more often the main concerns with spills on land. This is followed by protecting nonessential uses, such as recreation.

### 46.12.1 Behavior of Oil on Land

The spreading of oil across the surface and its movement downwards through soil and rock is far more complicated and unpredictable on land than the spreading of oil on water. The movement of the oil varies for different types of oil and in different habitats and is influenced by conditions at the spill site, including the specific soil types and their arrangement, moisture conditions in the soil, the slope of the land, and the level and flow rate of the groundwater. Other factors, which vary in different habitats, are the presence of vegetation and its type and growth phase, the temperature, the presence of snow and ice, and the presence of micro-features, such as rock outcrops.

The basic types of soil to consider in relation to land oil spills are sand/gravel, loam, clay, and silt. Soil is defined as the loose unconsolidated material located near the surface, while rock is the hard consolidated material, that is, bedrock, usually found beneath the soil. Most soils consist of small fragments or grains that form openings or pores when compacted together. If these pores are sufficiently large and interconnected, the soil is said to be "permeable" and oil or water can pass through it. Gravel and sand are the most permeable type of soil. Materials such as clay, silt, or shale are termed impervious as they have extremely small, poorly interconnected pores and allow only limited passage of fluids. Soils also vary in terms of long-term retentivity. Loam tends to retain the most water or oil due to its high organic content.

As most soils are an inhomogeneous mixture of these different types of soil, the degree of spreading and penetration of oil can vary considerably even in a single location. The types of soil are often arranged in layers, with loam on top and less permeable materials such as clay or bedrock underneath. If rock is fractured and contains fissures, oil can readily pass through it. The oil's ability to permeate soils and its adhesion properties also vary significantly. Viscous oils, such as bunker fuel oil, often form a tarry mass when spilled and move slowly, particularly when the ambient temperature is low. Nonviscous products, such as gasoline, move in a manner similar to water in both summer and winter. For such light products, most spreading occurs immediately after a spill.

Crude oils have intermediate adhesion properties. In an area with typical agricultural loam, spilled crude oil usually saturates the upper 10–20 cm of soil and rarely penetrates more than 60 cm. Generally, the oil only penetrates to this depth if it has formed pools in dry depressions. If the depressions contain water, the oil may not penetrate at all.

### 46.12.2 Movement of Oil on Land Surfaces

Both the properties of the oil and the nature of the soil materials affect how rapidly the oil penetrates the soil and how much the oil adheres to the soil. For example, a low-viscosity oil penetrates rapidly into a dry porous soil such as coarse sand and therefore its rate of spreading over the surface is reduced.

When oil is spilled on land, it runs off the surface in the same direction and manner as water. The oil continues to move horizontally down-gradient until either blocked by an impermeable barrier or all the oil is absorbed by the soil. The oil will also sink into any depressions and penetrate into permeable soils.

The process whereby oil penetrates through permeable soils includes several steps. The bulk of the oil moves downward through permeable material under the influence of gravity until it is stopped by either the groundwater or an impermeable layer. It then moves down-gradient along the top of the impermeable layer or groundwater until it encounters another impermeable barrier or all the product is absorbed in the soil. Once in contact with the water-soluble material, the oil dissolves into and is transported away with the groundwater. Oils and fluids can flow along the top of the groundwater and reappear much later in springs or rivers. The descending oil is often referred to as a slug, of oil. As the slug moves through the soil, it leaves material behind that adheres to the soil. This depends on the adhesion properties of the spilled product and the nature of the soil. More of the adhered oil is moved downwards by rainfall percolating through the soil. The rain water carries dissolved components with it to the water table. The movement of the oil will be greatest where the water drainage is good.

***46.12.2.1 Movement of Oil in the Subsurface*** Regardless of its source, oil released into the subsurface soil moves along the path of least resistance and downwards, under the influence of gravity. Oil often migrates toward excavated areas such as pipeline trenches, filled-in areas around building foundations, utility corridors, and roadbeds. Such areas are often filled with material that is more permeable or less compacted than the material removed during the excavation.

The oil may continue to move downwards until it reaches the groundwater or another impermeable layer. If the soil is absorptive and capillary action occurs, however, the oil can also move upwards and even reappear at the surface, sometimes as far as a kilometer away from the spill. This is what happens when pipeline spills appear at the surface of the trench in which the pipeline is laid.

### 46.12.3 Habitats/Ecosystems

As the effects of oil and its behavior vary in different habitats, cleanup techniques and priorities are tailored to the habitat in which the spill takes place. Returning the habitat as much and as quickly as possible to its original condition is always a high priority when cleaning up oil spills.

When spills occur in the urban environment, protecting human health and safety and quickly restoring the land use are top priorities. Environmental considerations are generally not important as endangered species or ecosystems are not often found in the urban habitat. The urban environment usually includes a range of ecosystems, from natural forest to paved parking lots. Thus, a spill in an urban area often affects several ecosystems, each of which is treated individually.

The roadside environment is similar to the urban one in that restoring the use and surficial appearance of the land is given top priority. Roadside habitats are varied and include all the other ecosystems. The roadside habitat is different from the urban one, however, in that it is exposed to many emissions and is not generally viewed as a threatened or sensitive environment.

On agricultural land, the priority in cleaning up oil spills is to restore land use, for example, crop production. In this habitat, oil is more likely to penetrate deeply into the subsurface as plowing the fields creates macropores through which petroleum products and crude oils can rapidly penetrate. As oil penetrates deeper into dry agricultural land, the danger of groundwater contamination is greater than in other habitats.

On mineral soils, however, oil can make the soil non-wettable, so that water runs off rather than soaking into the soil. This causes a soil water shortage, which can result in poor rehabilitation in the area. The opposite occurs in low-lying sites or poorly drained soils where water fills the macropores of the soil, but is not absorbed into the soil itself because of the presence of the oil. This excludes air from the soil and the site becomes difficult to treat or cultivate and anaerobic conditions quickly develop.

Anaerobic conditions and restricted plant growth can also develop when oil on the surface weathers and forms an impermeable crust, which again reduces the air exchange. Recovery is affected by the amount of oil spilled on a given area. Lightly oiled soil recovers much faster than a heavier oiled area as the soil is not completely saturated and both air and water can still penetrate. Residual oil in the soil can also slow recovery by inhibiting seed germination.

Dry grassland is similar to agricultural land in that the priority for cleanup is restoring the soil so that the crop, in this case grass, can continue to grow. The surface of the grassland is often less permeable than agricultural land. Once the surface is penetrated, however, the substrate may be permeable and groundwater can be affected. Dry grassland recovers quickly from spills if the oil runs off or if the excess oil is removed without too much surface damage. The presence of dead vegetation is viewed as a symptom, not a problem. When excess oil is removed, replanting and fertilization can speed recovery of an oiled grassland. As with agricultural land, oil on the surface of grassland can sometimes weather and form an impermeable crust, which reduces air exchange and causes anaerobic conditions.

Unlike most habitats, the forest has two distinct levels of vegetation: low-lying vegetation such as shrubs and grasses and trees. The low-lying vegetation is much more sensitive to oiling than trees, but is much easier to replant and recovers much faster. Most species of trees are not seriously affected by light oil spills. If enough oil is spilled to affect the tree's roots, most trees will be killed and the forest will not recover fully for decades. It is therefore very important to rapidly remove excess oil that has not yet been absorbed by the soil.

If a forest has mineral soil, the oil can make it non-wettable so that water runs off the soil rather than soaking in. In low-lying sites or forests with poorly drained soil, the opposite occurs. Water fills the macropores of the soil, but not the soil itself because of the presence of the oil. This excludes air from the soil and the site does not revegetate quickly. Oil on the surface of forest soils can weather and form an impermeable crust, which also reduces air exchange or restricts the growth of plants. Forest is far more difficult to access and treat than most other habitats.

Wetlands are the habitat most affected by oil spills because they are at the bottom of the gravity drainage scheme. Usually, oil cannot flow out of a wetland system and oil from other areas flows into the system. Although there are variety of wetlands, oil tends to collect in all of them, creating anaerobic conditions, which slow oil degradation. Wetlands are also extremely sensitive to physical disturbance as many plants in this habitat propagate through root systems. If these root systems are damaged by the oil or the cleanup process, it takes years or even decades for the plants to grow back. Wetlands are the habitat of many species of birds and fish, as well as other aquatic resources. Wetlands are difficult to access and to clean up.

Taiga, which is characterized by coniferous trees and swampy land, generally forms the transition between northern forests and the tundra farther to the north. It is either underlain by permafrost or has a high water table. Many of the plants propagate through root systems and are highly sensitive to physical disturbance. Over a period of time, heavy loadings of oil will kill the coniferous trees. Oil on the surface of the taiga can weather and form an impermeable crust, which reduces the air exchange and restricts plant growth. Degradation of remaining oil is slow in this habitat, which takes a long time to recover. The presence of trees and

**TABLE 46.7 Cleanup Methods for Surface Land Spills**

| Habitat | Removal of Excess Oil | Natural Recovery | Manual Oil Removal | Mechanical Oil/ Surface Removal | Enhanced Biodegradation | *In-situ* Burning | Hydraulic Measures |
|---|---|---|---|---|---|---|---|
| Urban | ✔ | ✚ | ✔ | ✚ | ✚ | ❍ | ✔ |
| Roadside | ✔ | ✚ | ✔ | ✚ | ✚ | ✚ | ✚ |
| Agricultural land | ✔ | ✚ | ✔ | ✚ | ✚ | ✚ | ✔ |
| Grassland | ✔ | ✚ | ✔ | ✚ | ✚ | ✚ | ✚ |
| Forest | ✔ | ✔ | ✔ | ✚ | ✚ | ❍ | ✔ |
| Wetland | ✔ | ✔ | ✔ | ✖ | ✚ | ✔ | ✚ |
| Taiga | ✔ | ✔ | ✔ | ✖ | ✚ | ✚ | ✚ |
| Tundra | ✔ | ✔ | ✔ | ✖ | ✚ | ✚ | ✚ |

✔ Acceptable or recommended.
✚ Can be used under certain circumstances.
✖ Should not be used.
❍ Only marginally applicable.

the high moisture level make the taiga more difficult to access and clean up than most other habitats.

Tundra is a far northern habitat, characterized by low plant growth and no trees. Tundra is underlain by permafrost, which is generally impermeable to oil. Vegetation on the tundra grows in tufts, which are generally grouped into polygons. Oil spilled on the surface drains into the spaces between the tufts and polygons and eventually kills the vegetation. Without the layer of vegetation, the permafrost melts and serious land damage results. Degradation of remaining oil on the tundra is very slow and could take decades.

In all the more sensitive habitats, which include the forest, the taiga, and the tundra, the priority for cleanup operations is to remove the excess oil as rapidly as possible and without causing physical damage.

### 46.12.4 Cleanup of Surface Spills

When dealing with oil spills on land, cleanup operations should begin as soon as possible. It is important to prevent the oil from spreading by containing it and to prevent further contamination by removing the source of the spill. It is also important to prevent the oil from penetrating the surface and possibly contaminating the groundwater.

Berms or dikes can be built to contain oil spills and prevent oil from spreading horizontally. Caution must be exerted, however, that the oil does not back up behind the berm and permeate the soil. Berms can be built with soil from the area, sand bags, or construction materials. Berms are removed after cleanup to restore the area's natural drainage patterns. Sorbents can also be used to recover some of the oil and to prevent further spreading. The contaminated area can sometimes be flooded with water to slow penetration and possibly float oil to the surface, although care must be taken not to increase spreading and to ensure that water-soluble components of the oil are not carried down into the soil with the water. Shallow trenches can be dug as a method of containment, which is particularly effective if the water table is high and oil will not permeate the soil. Oil can either be recovered directly from the trenches or burned in the trenches. After the cleanup, trenches are filled in to restore natural water levels and drainage patterns.

There are a variety of methods for cleaning up surface oil spills on land, with the method used depending on the habitat in which the spill occurs. The various cleanup methods that can be used in the different habitats are shown in Table 46.7.

### 46.12.5 Natural Recovery

This is the process of leaving the spill site to recover on its own. This method is sometimes chosen for extremely sensitive habitats such as wetlands, taiga, and tundra and is always done after the excess oil has been removed from the site. In these cases, the excess oil that can be recovered is removed using techniques that do not disturb the surface or physically damage the environment. This is important as it can take years for wetlands or tundra to recover from vehicular traffic. In habitats such as wetlands and taiga where the vegetation propagates through root systems, more damage can be done by the cleanup operation than by the oil.

### 46.12.6 Removal of Excess Oil

Any excess oil that can be recovered without causing physical damage to the environment is always removed from a spill site, using techniques that do not disturb the surface. If excess oil on the surface is not removed quickly, the oil can penetrate into the soil, contaminating the groundwater and destroying vegetation.

Suction hoses, pumps, vacuum trucks, and certain skimmers and sorbents are generally effective in removing excess oil from the surface, especially from ditches or low areas. The use of sorbents can complicate cleanup operations, however, as contaminated sorbents must be disposed of appropriately.

Sorbents are best used to remove the final traces of oil from a water surface. Any removal of surface soil or vegetation also entails replanting.

Manual removal of oil involves removing oil and often highly oiled soil and vegetation with shovels and other agricultural tools. This is always followed by fertilization, selective reseeding, or transplanting plugs of vegetation from nearby unaffected areas. This form of removal is labor intensive and can severely damage the surface, especially in sensitive environments.

Mechanical recovery equipment, such as bulldozers, scrapers, and front-end loaders, can cause severe and long-lasting damage to sensitive environments. It can be used in a limited capacity to clean oil from urban areas, roadsides, and possibly on agricultural land. The unselective removal of a large amount of soil leads to the problem of disposing of the contaminated material. Contaminated soil must be treated, washed, or contained before it can safely be disposed of in a landfill site. This could cost thousands of dollars per ton.

### 46.12.7 Other Cleanup Methods

Enhanced biodegradation is another possible method for cleaning up spills on land. Certain portions of oil are biodegradable and the rate of biodegradation can sometimes be accelerated as much as 10-fold by the proper application of fertilizers.

The amount of degradation varies with the type of oil. Diesel fuel may largely evaporate and degrade on the land surface, whereas Bunker C will only slightly degrade. Under ideal conditions and using fertilizers to enhance degradation, however, it can still take decades for more than half of some oils to degrade, dependent on conditions and the type of oil. During this time, some of the oil will be removed by other processes such as evaporation or simply by movement

Scientists are now exploring the use of plants and their associated microorganisms for remediation, which is called phytoremediation. This is a low cost process that is proving effective for a wide variety of contaminants, including petroleum hydrocarbons. It can be used in combination with other remediation technologies and may prove useful in the future for treating oiled soils or wetlands. It takes several years to remediate a site and cleanup is limited to the depth of the soil within reach of the plant's roots.

*In-situ* burning has been used for several years to deal with oil spills on land. This technique removes oil quickly and without disturbing the area extensively, although it does damage or kill shrubs and trees. The heat from burning can also destroy propagating root systems and change the soil's properties. In addition, it can leave a hard crust of residual material that inhibits plant growth and changes natural water levels and drainage patterns.

These disadvantages can be overcome in some habitats. Some areas can be flooded before burning to minimize the effect of the heat and to remove oil by floating it out of the ground. Crust formation can be avoided by physically removing residue after a burn. On wetlands and in areas with high water levels, sorbents can be used to remove residues left after burning to ensure that they do not coat plants or soil after water levels fall. In marshes, burning is best done in spring when the water table is high.

Hydraulic measures, such as flooding and cold or warm water sprays, can be used to deal with land spills, although they are only effective in limited circumstances. Flooding an area where the oil is not strongly retained can cause the oil to rise to the water's surface where it can subsequently be removed using skimmers or suction devices, or by burning. This is effective in areas where the water table is high or the top layer of ground is underlain by impermeable material. Flooding may not work on soils that are high in organic material, however, as they strongly retain oil. Cold or warm water sprays can be used to clean oil from hard surfaces. Catchment basins and interceptor trenches are built to capture the released oil, which is then skimmed or pumped from the trenches.

A number of other techniques have been tried for cleaning up oil spills on land, with varying degrees of success. Tilling or aeration of soil is done to break up the crust surface and re-aerate the soil. In areas where vegetation propagates by root systems, however, tilling kills all plants and destroys the potential for regrowth. Tilling oil into the soil can actually slow natural degradation because the soil can become anaerobic when it recompresses. Vegetation cutting is useful only if there is a risk that oil on vegetation could recontaminate other areas. Many plants cannot survive cutting, however, and growth is not reestablished in the area. To date, there are no effective chemical agents for cleaning up oil spills. Surfactant agents can actually increase oil penetration into the soil and could result in the more serious problem of groundwater contamination. Figure 46.10 shows that penetration of oil liquids occurs below the soil surface.

All cleanup methods include site restoration, which involves returning the site as closely as possible to the prespill conditions. The drainage pattern of the site is restored by removing dykes, dams, and berms, and filling in ditches or drains. It may be necessary to replace any soil that was removed and to revegetate the site by fertilizing, reseeding, or transplanting vegetation from nearby.

### 46.12.8 Cleanup of Subsurface Spills

Oil spills in the subsurface are much more complicated and expensive to clean up than those on the surface and the risk of groundwater contamination is greater. Spills in the subsurface can be difficult to locate and without knowledge of the geology of the area, it can be difficult to predict the horizontal and downward movement of the oil. The first step is typically to engage a hydrogeologist to map and assess

**FIGURE 46.10** Oil seeps into a hole dug for taking a sample. This spill is more than 15 years old at the time this sample hole was dug.

both the subsurface and the oil location with respect to the soil geology.

In terms of countermeasures, the oil must be contained and its horizontal and downward movement stopped or slowed. Containment methods are difficult to implement and may cause physical damage to the site. Digging an interceptor trench can be effective in reducing horizontal spread. Such trenches are filled in after the cleanup operation to restore the natural drainage patterns of the land. Another method is to place walls around the spill source to stop its spreading. These can be slurry walls consisting of clay or cement mixtures that solidify and retain the oil, or solid sheets of steel or concrete can be positioned to retain the oil.

Once the subsurface spill is contained, there are a number of cleanup methods that can be used. The most appropriate method for a particular spill depends on the type of oil spilled and the type of soil at the site, as shown in Table 46.8.

Hydraulic measures for cleaning up subsurface spills include flooding, flushing, sumps, and subsurface drains. These methods are most effective in permeable soil and with nonadhesive oils. They all leave residual material in the soil that may be acceptable, depending on the land use. Flooding is the application of water either directly to the surface or to an interceptor trench in order to float out the oil. Flooding is effective only if the spilled oil has not already been absorbed into the soil, if sufficient water can be applied to perform the function, and if the oil is not accidentally moved into another area. Flushing involves the use of water to flush oil into a sump, recovery well, or interceptor trench. Placement of a sump or a deep hole is only effective for a light fuel in permeable soil above an impermeable layer of soil. A subsurface drain is a horizontal drain placed under the contamination, from which the fuel and often water are pumped out. Although effective in permeable soils, they are expensive and difficult to install.

Interceptor trenches are ditches or trenches dug downgradient from the spill, or in the direction in which the spill is flowing, to catch the flow of oil. They are placed just below the depth of the groundwater so that oil flowing on top of the groundwater will flow into the trench. Both water and oil are removed from the trench to ensure that flow will continue. Interceptor trenches are effective if the groundwater is very close to the surface and the soil above the groundwater is permeable.

Soil venting is done to remove vapors from permeable soil above a subsurface spill. This is effective for gasoline in warm climates and for portions of very light crude oils. Other oils do not have a high enough rate of evaporation to achieve a high recovery rate. Venting can be passive, in which vapors are released as a result of their own natural vapor pressure, or

**TABLE 46.8 Cleanup Methods for Subsurface Spills**

| Product Type in Soil Type | Hydraulic Measures | Interceptor Trench | Soil Venting | Soil Excavation | Recovery Wells |
|---|---|---|---|---|---|
| Gasoline in sand or mixed till | ✔ | ✚ | ✔ | ✚ | ✚ |
| Gasoline in loam or clay | ✚ | ✚ | ❍ | ✚ | ✔ |
| Diesel fuel in sand or mixed till | ✔ | ✚ | ❍ | ✚ | ✚ |
| Diesel fuel in loam or clay | ✔ | ✚ | ❍ | ✚ | ✔ |
| Light crude in sand or mixed till | ✔ | ✚ | ✚ | ✚ | ✚ |
| Light crude in loam or clay | ✚ | ✚ | ❍ | ✚ | ✚ |
| Heavier oils in sand or mixed till | ✚ | ✚ | ❍ | ✚ | ❍ |
| Heavier oils in loam or clay | ✚ | ✚ | ❍ | ✚ | ❍ |

✔ Acceptable or recommended.
✚ Can be used under certain circumstances.
❍ Only marginally applicable.

active, in which air is blown through the soil and/or drawn out with a vacuum pump. The fuel vapors are subsequently removed from the air to prevent air pollution. Soil venting is also done to enhance biodegradation.

Excavation is a commonly used technique for cleaning up subsurface spills, especially in urban areas where human safety is an issue. Vapors from gasoline can travel through the soil and explode if ignited. These vapors can also penetrate houses and buildings, forcing evacuation of the area. To prevent these situations, contaminated soils are often quickly excavated and treated or packaged for disposal in a landfill. Excavation may not always be possible, however, depending on the size of the spill and prevailing conditions at the site.

Recovery wells are frequently used in cleaning up subsurface spills. The well is drilled or dug to the depth of the water table so that oil flowing along the top of the water table will also enter the well. The water table is sometimes lowered, by pumping, to speed the recovery of the oil and to increase the area of the collection zone. The oil is recovered from the surface of the water by a pump or a specially designed skimmer.

Other methods are constantly being proposed or tried for cleaning up subsurface spills. One such method is biodegradation *in situ,* although its effectiveness is very much restricted by the availability of oxygen in the soil and the degradability of the oil itself. An adaptation of the venting method has been used to try to solve the oxygen problem. So far, however, biodegradation methods have not been rapid enough to be an acceptable solution. Chemical agents have also been proposed for cleaning up subsurface spills, although most of them actually make the problem worse. For example, surfactants can release the oil from soil, but then render that same oil dispersible in the groundwater.

If the groundwater does become contaminated, it is pumped to the surface and treated to remove the dissolved components. Common treatment methods include reverse osmosis and carbon filtration. Groundwater treatment is expensive and generally involves a lengthy process before contamination levels are below acceptable standards.

## REFERENCES


1. Fingas, M.F. (2012) *The Basics of Oil Spill Cleanup*, Taylor and Francis, New York, NY.
2. (2012) *Response to Marine Oil Spills*, 2nd ed., ITOPF, Witherby Publishing Co., Edinburgh, UK.
3. (2011) *International Petroleum Encyclopedia*, Pennwell Publishing Company, Tulsa, OK.
4. Transportation Safety Board of Canada Available at http://www.bst-tsb.gc.ca/eng/stats/pipeline/. Accessed December, 2013.
5. Etkin, D.S. (2011) Spill occurrences: a world overview, Chapter 2. In: *Oil Spill Science and Technology*, Gulf Publishing Company, New York, NY, pp. 7–48.
6. Transportation Safety Board of Canada Available at http://www.bst-tsb.gc.ca/eng/stats/pipeline/2012/ss12.asp. Accessed December, 2013.
7. Canadian Energy Pipeline Association Available at http://www.cepa.com/about-pipelines/types-of-pipelines. Accessed December, 2013.
8. Statistics Canada Data on Pipeline Transfers Available at http://www5.statcan.gc.ca/cansim/a26?lang=eng&retrLang=eng&id=1330003&pattern=133-0001..133-0005&tabMode=dataTable&srchLan=-1&p1=-1&p2=31. Accessed December, 2013.
9. NOAA (2013) *Transporting Alberta Oil Sands Products: Defining the Issues and Assessing the Risks*, National Oceanic and Atmospheric Administration, Washington, DC.
10. Hollebone, B. (2011) Measurement of oil physical properties, Chapter 4. In: *Oil Spill Science and Technology*, Gulf Publishing Company, New York, NY, pp. 63–86.
11. (2003) *Oil in The Sea*, National Academy Press, Washington, DC.
12. Fingas, M. (2011) Introduction to oil spill modeling, Chapter 8. In: *Oil Spill Science and Technology*, Gulf Publishing Company, New York, NY, pp. 187–200.
13. Fingas, M. and Fieldhouse, B. (2009) Studies on crude oil and petroleum product emulsions: water resolution and rheology. *Colloids Surface A: Physicochemical and Engineering Aspects*, 333, 67–81.
14. Sun, J., Khelifa, A., Zheng, X., Wang, Z., So, L.L., Wong, S., Yang, C., and Fieldhouse, B. (2010) A laboratory study on the kinetics of the formation of oil-suspended particulate matter aggregates using the NIST-1941b sediment. *Marine Pollution Bulletin*, 60(10), 1701–1707.
15. Prince, R.C., McFarlin, K.M., Butler, J.D., Febbo, E.J., Wang, F.C.Y., and Nedwed, T.J. (2013) The primary biodegradation of dispersed crude oil in the sea. *Chemosphere*, 90(2), 521–526.
16. Chandra, S., Sharma, R., Singh, K., and Sharma, A. (2012) Application of bioremediation technology in the environment contaminated with petroleum hydrocarbon. *Annals of Microbiology*, 63, 1–15.
17. Huang, J.C. (2005) A review of the state-of-the-art of oil spill fate/behavior models 2005. Proceedings of International Oil Spill Conference, IOSC. 2005; 7275.
18. Wang, Z. and Fingas, M.F. (1999) Oil spill identification. *Journal of Chromatography A*, 843, 369–411.
19. Fingas, M. and Brown, C.E. (2011) Oil spill remote sensing: a review, Chapter 6. In: *Oil Spill Science and Technology*, Gulf Publishing Company, New York, NY, 111–169.
20. Castro, A., Iglesias, G., Carballo, R., and Fraguela, J.A. (2010) Floating boom performance under waves and currents. *Journal of Hazardous Materials*, 174(1–3), 226–235.
21. Yang, X. and Liu, M. (2013) Numerical modeling of oil spill containment by boom using SPH. *Science China: Physics, Mechanics and Astronomy*, 56, 1–7.
22. Chebbi, R. (2009) Profile of oil spill confined with floating boom. *Chemical Engineering Science*, 64(3), 467–473.
23. Cooper, D., Flood, K., and Brown, C.E. (2006) Multi-track sorbent boom testing with loose sorbent material. *Proceedings of the AMOP Technical Seminar*, 173–194.

24. Moxness, V.W., Gaseidnes, K., and Asheim, H. (2011) *Skimmer capacity for viscous oil SPE J.*, 16(1), 155–161.
25. DeVitis, D., Delgado, J.-E., Meyer, P., Schmidt, W., Potter, S., and Crickard, M. (2008) Development of an American Society of Testing and Materials (ASTM) stationary skimmer test protocol—phase 2 development. *Proceedings of the AMOP Technical Seminar*, 1, 321–326.
26. Cooper, D. and Gausemel, I. (2005) Oil spill sorbents: testing protocol and certification listing program, Proceedings of 2005 IOSC. p. 5771.
27. Michel, J. (2011) Submerged oil, Chapter 26. In: M. Fingas (editor), *Oil Spill Science and Technology*, Gulf Publishing Company, New York, NY, 959–982.
28. Nordvik, A.B., Simmons, J.L., Bitting, K.R., Lewis, A., and Strøm-Kristiansen, T. (1996) Oil and water separation in marine oil spill clean-up operations. *Spill Science and Technology Bulletin*, 3, 107–122.
29. McDonagh, M., Abbott, J., Swannell, R., Gundlach, E., and Nordvik, A. (2005) Handling and disposal of oily waste from oil spills at sea, Proceedings of 2005 IOSC., pp. 3428–3440.
30. Fingas, M. (2011) Oil spill dispersants: a technical summary, Chapter 15. In: *Oil Spill Science and Technology*, Gulf Publishing Company, New York, NY, 435–582.
31. Wise, J. and Wise, J.P., Sr. (2011) A review of the toxicity of chemical dispersants. *Reviews on Environmental Health*, 26, 281–300.
32. Fingas, M. and Fieldhouse, B. (2011) Surface-washing agents, Chapter 21.. In: *Oil Spill Science and Technology*, Gulf Publishing Company, New York, NY, 683–711.
33. Mabile, N. (2012) Controlled *in-situ* burning: transition from alternative technology to conventional spill response option, Proceedings of 35th AMOP Technology, pp. 584–605.
34. Fingas, M. (2011) *In-situ* burning, Chapter 23. In: Fingas, M. (editor), *Oil Spill Science and Technology*, Gulf Publishing Company, New York, NY, pp. 737–903.
35. Owens, E.H. and Sergy, G. (2009) *Field Guide for the Protection and Cleanup of Oiled Shorelines*, Environment Canada, Ottawa, ON.
36. Taylor, E. and Reimer, D. (2008) Oil persistence on beaches in Prince William Sound—a review of SCAT surveys conducted from 1989 to 2002. *Marine Pollution Bulletin*, 56, 458–474.
37. Wang, Z., Fingas, M., Blenkinsopp, S., Sergy, G., Landriault, M., Sigouin, L., and Lambert, P. (1998) Study of the 25-year-old Nipisi oil spill: persistence of oil residues and comparisons between surface and subsurface sediments. *Environmental Science and Technology*, 32(15), 2222–2232.

# 47

# PIPELINE ABANDONMENT

ALAN PENTNEY[1] AND DEAN CARNES[2]
[1]*National Energy Board of Canada, Calgary, Alberta, Canada*
[2]*Canadian Natural Resources Ltd., Calgary, Alberta, Canada*

## 47.1 WHAT IS PIPELINE ABANDONMENT?

Oil and gas pipeline abandonment occurs when the pipeline is no longer operated and its service is permanently ceased. Depending on the regulatory jurisdiction, a pipeline may no longer be in service for legal reasons and reside under temporary designations such as being deactivated or decommissioned; but it is technically not abandoned.

Oil and gas pipelines are typically abandoned because there is no further need for the pipeline or it is being replaced for capacity reasons. As a result, most abandoned pipeline lengths are only a few kilometers and are <324 mm in diameter. In some cases the pipeline is abandoned because it is no longer safe for continued operation. Currently many major transmission pipelines in North America have been in service for more than 50 years. The age of oil and gas infrastructure is a consideration for abandonment only if the cost to maintain its integrity is not financially sound.

## 47.2 ABANDONMENT PLANNING

Pipeline abandonment has three distinct phases. The first is the planning phase that can take place any time during the pipeline operation life but should take place as early as possible and may even occur as part of the design stage. Regardless when the planning is performed, it should be reviewed at the time just before abandonment to ensure its validity. Unanticipated changes may have occurred (e.g., population encroachment where development was never expected). The second phase is the actual abandonment process that can include approvals, agreements, deconstruction, and restoration. The last phase is postabandonment where any assets left in place are monitored for liability purposes and possible future restoration may have to occur.

### 47.2.1 Removal or Abandon in Place

Removal of an abandoned pipeline may cost as much as current installation costs. It can also disrupt full land productivity for as much as 5 years and in some cases result in soil admixing, which has the potential to affect soil quality. However, there are instances where removal is necessary such as; the right-of-way is required for another pipeline or utility, future land use dictates it, a proposed development requires subsurface excavation or the pipeline owner wants to remove any liability. If the pipeline is small enough it may be removed relatively easily by pulling it from the ground or from beneath water ways. Precautions must be taken to not disturb sensitive ecosystems or if necessary the pipeline is left in place at sensitive ecosystems with suitable safeguards such as pipeline cleaning, filling with inert material or inert gas, and then plugging the pipe [1].

Cathodic protection will be disconnected from any pipe abandoned in place. Since there will be no other measures to protect the structural integrity of the pipe there is a potential for abandonment in place to create issues in the future. Common landowner concerns are that

1. the presence of the pipe could interfere with future development plans;

*Oil and Gas Pipelines: Integrity and Safety Handbook*, First Edition. Edited by R. Winston Revie.

2. the pipe could become exposed or interfere with farming, including the need for One-Call locates;
3. the pipe could interfere with drainage by becoming a conduit;
4. the pipe could collapse and cause damage to land, infrastructure or farm equipment; and
5. the liability for residual contamination or pipeline removal costs in the future may not be funded.

As a result of these concerns, early engagement with potentially affected parties and persons with right-of-way agreements is encouraged to provide information and discuss how concerns will be managed during the abandonment process and in the future.

### 47.2.2 Consultation

Typically, a regulator will want an application for approval to abandon a pipeline. The application should include an abandonment plan. A plan must be tailored to the specifics of the project and comply with current regulatory requirements. It is also necessary that the plan be broad in scope and encompasses postabandonment responsibilities in the form of right-of-way monitoring and remediation of problems associated with the abandonment. The application should identify if landowners, aboriginal groups, occupants, land managers, lessees, municipal agencies, upstream and downstream users, and other persons potentially affected, were sufficiently notified and consulted. This consultation would include

- details on what areas require soil and groundwater remediation or cleanup;
- discussion about what portions of the pipeline or aboveground facilities must be removed;
- information about what land reclamation will be provided;
- whether or not the correct land use is being accommodated; and
- how potential issues will be mitigated.

The degree of engagement should be as great as possible to develop the abandonment plan as well as to finalize an application for abandonment. There may be third party interest in reusing the pipeline, access road, or associated facilities such as a pump station where a jurisdiction permits that.

An abandonment plan is the best means to reduce the risk to public safety, property, the pipeline owner, and the environment to an acceptable level. Section 47.2.3 provides an example of an outline for a plan.

### 47.2.3 Abandonment Plan Outline

The following outline is provided from the National Energy Board guide *Regulating Pipeline Abandonment* [2].

1. Background
   a. General description of the pipeline and facilities, including history and product
   b. Proposed abandonment process, including timelines
2. Location map (right-of-way, pipe, stations, valves, storage, and so on)
3. Detailed description of facilities to be abandoned
   a. Pipeline composition, diameter, thickness, coatings, and so on
   b. Adjacent pipeline facilities (corridor)
   c. Facility components on company-owned land
   d. Land use along route (e.g., agricultural, urban, parkland)
   e. Natural features (e.g., water bodies, wetlands, native prairie, rare vegetation, species at risk)
   f. Landowners and land administration agencies
4. History of ruptures, leaks, and other construction occurrences
   a. Location of incidents and former contamination sites
   b. Environmental Site Assessment results at locations where contamination could have occurred
   c. Status of contamination remediation
5. Abandonment procedure
   a. Facilities to be left in place
      i. Locations and justification
      ii. Mitigation measures
         1. Cleaning (procedure and standards)
         2. Filling or plugging
         3. Removal of unnecessary surface equipment
         4. Identification of location of facilities
         5. Estimation of risk and risk reduction plans
            a. Contamination removal or management
            b. Soil subsidence
            c. Corrosion effects
            d. Pipe collapse
            e. Soil erosion effects
            f. Water conduit
            g. Water crossings
            h. Transportation and utility crossings
      iii. Access
      iv. Records

b. Facilities to be removed
   i. Locations and justification
   ii. Cleaning and removal procedure
c. Recycling and reuse plans
d. Reclamation procedure
   i. Restoration of Facility sites and access roads
   ii. Right-of-way reclaimed to a state comparable with surrounding environment
   iii. Reestablishment of habitat to a native state for sensitive plant species and communities
6. Consultation for developing the plan
   a. Guiding principles and goals for the consultation program
   b. Design of the consultation program
   c. Reporting on the results of the consultation
7. Performance measures
   a. Maintenance period
   b. Monitoring procedure
   c. Measures of success
8. Statement of responsibility for any facilities left in place
9. Abandonment costs
   a. Initial
   b. Ongoing
   c. Sources of abandonment funding

The document created with this table of contents can be subject to ongoing revisions throughout the life of the pipeline. More information will be available on pipeline abandonment based on research and more certainty will evolve for abandonment plans as the pipeline nears the end of its useful life.

## 47.3 PROCEDURES FOR ABANDONING PIPELINES AND RELATED FACILITIES

### 47.3.1 Contamination Remediation

When a pipeline spill occurs, the resulting contamination is expected to be cleaned up immediately to the most stringent environmental standards required by the regulators to prevent harm to the public and the environment. However, there may be residual contamination in groundwater and soil that is difficult to remove in the short term. The resulting remediation process must meet the most stringent standards and is to be conducted within a schedule agreed on with the regulators.

The schedule for remediating contamination on sites owned by the pipeline company, such as at a pump station or tank farm, may be over the course of several years or at the abandonment stage if the regulator is satisfied with the application for that schedule. In the interim, contamination must be contained on site and monitored in accordance with all regulatory requirements. In order for a pipeline to be considered fully abandoned all contaminated soil and groundwater must be remediated such that it meets required criteria. The regulator may provide a document to acknowledge the remediation but future changes in land use or discovery of undetected contamination may mean the site will require further work.

### 47.3.2 Pipeline Cleaning

Cleaning a pipeline during the abandonment process is essential no matter whether the pipe will be left in place, removed for disposal, or removed for reuse. Each of these end conditions have regulatory cleanliness requirements to safeguard the environment and any persons coming in contact with the pipe. The pipe may corrode at some point and the contaminant life may exceed the pipeline so the environment may be put at risk if a high degree of cleanliness is not achieved. Further research on the degree of cleanliness required in varying environments will occur over time.

Pipelines of unknown condition present special challenges. If they have been dormant for some time, they may have compromised integrity, which would not stand up to traditional pig runs. In these cases, worker safety aspects need to be considered when using compressed gases for pushing pigs. In addition, releases to the environment can occur if the line fails during cleaning. Where the pipe condition is unknown, vacuum trucks and coiled tubing units should be used first to remove as much fluids as possible before running cleaning pigs or performing a fresh water flush. Fresh water displacement may be used until returns are clean and may be followed by a progressive pigging program. Before running pigs a safe pigging pressure using any gases needs to be established. In some cases, clean fresh water for pushing pigs may have to be utilized. The effectiveness of any cleaning procedure will vary with the pipeline based on size, product carried, and location.

The following general cleaning procedure [3] is recommended for dry natural gas pipelines and medium duty (relatively wax free with an occasional scraping operation) oil pipelines:

*47.3.2.1 Cleaning Guidelines General Considerations* The operating history of the pipeline to be abandoned should be reviewed to enable the planning of the specific cleaning procedures required for abandonment. Information such as oil/gas analysis, piping modifications, operating flow records, records of anomalies, and maintenance records may provide some insight into additional work needed to develop an effective pipeline cleaning plan.

The owner/operator should ensure that there are adequate sending and receiving traps in place. This may require the use of temporary assemblies. If the pipeline in question is part of a larger system, the section to be abandoned should be physically disconnected upon completion of the cleaning process. Safety precautions appropriate to the in-service product hazards (i.e., flammability and explosivity of hydrocarbons, toxicity of sour products) must be established throughout the activity.

For gas pipelines, any residual gas should be vented or flared once the pressure in the pipeline has been reduced to the extent possible using operating facilities or a pull down compressor. The residual gas should be monitored for signs of liquid.

For liquid pipelines, before line flow ceases, a sufficient number of scraper pigs should be run through the line to remove the bulk of any solids or waxy buildup. A batch of solvent-type hydrocarbons such as diesel fuel or condensate inserted between two scraper pigs is recommended as an effective method of reducing solids or waxy buildup. This process should be repeated until solids can no longer be detected on the pigs as they are removed from the receiving trap.

Specialized chemical cleaning may be required if the routine cleaning method described is not successful, if the pipeline is known to have an unusually high contamination level, or if unusually high cleanliness standards are to be met. There are specialty companies offering such services using proprietary chemicals, solvents, or gels for specific types of contaminants. Special precautions must be exercised when the pipeline is opened up to control vapor hazards of flammability, explosiveness, and toxicity (e.g., hazardous compounds such as benzene). Pipelines that have been carrying formation water (water disposal in the upstream production industries) for extended periods of time may, under certain operating conditions, develop scales that have high levels of Naturally Occurring Radioactive Materials (NORM). Such pipelines may require additional consideration in regards to cleaning processes, or perhaps even in regards to their acceptability for abandonment in place.

*Cleaning Methods for Natural Gas Pipelines* A stiff rubber scraping pig should be pushed through the pipeline (at a constant speed consistent with the pig manufacturer's recommendation) using nitrogen or some other inert gas to prevent explosive mixtures. Free liquids pushed ahead of the pig may be either pushed into the downstream pipeline section or collected in a containment tank designed and isolated according to prevailing local guidelines, for disposal in accordance with area legislation or local by-laws. This process should be repeated until free liquids are no longer evident by visual inspection. Low areas of the pipeline should be checked for the collection of liquids or other contaminants.

After these initial pigging runs, the pipeline should be checked for cleanliness. If contamination is evident, the pigging procedure should be repeated using a slug of solvent between two pigs. As with the free liquids, the solvent should be collected in a containment tank and disposed of in accordance with area legislation or local by-laws. Solvent fumes should be purged with nitrogen or a similar inert gas.

*Cleaning Methods for Liquid Pipelines* Following completion of the initial in-service cleaning efforts, a final cleaning step should be done in conjunction with line evacuation. The following procedure is commonly used, although many variations exist, which should be considered. Consultants specializing in the cleaning of contaminated facilities can advise and provide plans for both normal and unusual circumstances.

A slug of liquid hydrocarbons having solvent properties such as condensate or diesel fuel is pushed through the pipeline between two stiff rubber scraper pigs at a constant speed by an inert gas such as nitrogen. Other additives or treatment chemicals may be added if desired. As a rule of thumb, the volume should be calculated to maintain a minimum pipe wall contact time by the fluid ranging from 5 to 10 min (or longer), depending on the effectiveness of the initial in-service cleaning process.

For lines having encrusted or high paraffin buildup, an additional volume of solvent preceding the first pig can be considered. All contact times should be increased for excessive lengths of line as the solvent may become saturated with hydrocarbons before completion of the run. At the end point, the solvent and hydrocarbons are pushed into another section of pipeline or collected in a containment tank for disposal.

A repeat run of the pig train, which has been already described, should be conducted if there are any indications of liquids or contaminants remaining on the pipe wall in excess of the selected cleanliness criteria. The effectiveness of the cleaning process can be gauged by either obtaining samples of the solvent near the tail end of the passing batch (at approximate 25 km intervals) and analyzing the samples for hydrocarbon content or by monitoring the quality and quantity of the solvent hydrocarbons expelled from the line and comparing it with that injected. After running solvents, cleaning pigs need to be run to remove the solvents until no more solvents are found on pig returns.

Records of all cleaning procedures and test results are to be retained indefinitely along with records of all parts of the pipeline system removed and left in place.

### 47.3.3 Removal of Facilities and Apparatus

A buried pipeline that has been abandoned in place should be capped and plugged and be physically separated from any in-service piping. It also should have all related surface equipment that is not part of another pipeline removed to pipeline depth. Examples of such equipment could include pipeline risers, liner vent piping, casing vents, underground valve vaults or valve extenders, inspection bell holes, and cathodic protection equipment such as cathodic protection rectifiers, test posts, or anode wiring.

Abandoned aboveground pipe and all related surface equipment should be removed. This would include all mechanical and electrical equipment, buildings, associated piping, supports, and foundations. Aboveground storage tanks or pressure vessels must be emptied of liquids, purged, and safeguarded against trespassing until removal for disposal or reuse at a different site. If being removed for disposal then checks need to be performed that all hazardous residue has been effectively removed and meets regulatory disposal requirements. Caution needs to be exercised when moving or handling insulated equipment as asbestos may be present and proper containment or removal may be required.

It is also recommended that any underground structures such as underground vaults, closed-top pits, or storage tanks be removed. For those that are to remain, the walls and floor should be cleaned of contamination and the walls removed to an appropriate depth as determined in the abandonment plan. The insides of the structure can then be filled with clean soil. The abandonment of underground tanks should be conducted as specified in standards such as API 1604 *Closure of Underground Storage Tanks*. Checks for leaks on buried tanks should be performed as soil remediation may be required. There have been many instances of tank leaks that have occurred that were never detected while the tank was in service.

In addition to removal of infrastructure, final reclamation of a company owned or leased property should include removal of gravel, decompaction of soil and removal of unneeded access roads. Additional restoration considerations are identified in Section 47.3.6.

### 47.3.4 Water Bodies

Water bodies such as rivers, streams, lakes, and wetlands pose special problems for abandonment. Pipelines are often installed to cross a water body by

1. horizontal directional drilling below the bed of the water body (rarely done prior to 1990s);
2. a wet crossing that includes trenching in the bed of the water body;
3. a wet crossing where the pipeline is held in place on the bed of the water body with concrete saddle weights; or
4. suspended aerial crossing.

There are potential risks if a pipeline is left in place in these situations. One is that the pipeline is perforated and it becomes a conduit for contamination into the water body from external sources. The converse might also occur where a perforated pipeline siphons water from the water body accidentally. Another risk is that the pipeline becomes exposed so it poses a navigation risk or collects debris, which may affect river flow. In all cases, exposed pipes near or in a water body are likely to be viewed negatively by the public.

Cleaning the pipeline as described in Section 47.3.2 requires additional precautions at water crossings whether the pipe is removed or left in place. The collection of cleaning fluids must be conducted with great care. If the pipe is to be removed then all regulatory requirements for constructing a pipeline crossing would apply to provide maximum environmental protection to wildlife, fisheries, and riparian vegetation. If the pipe is to remain in place then consideration should be given to filling the pipe with an inert material and plugging and capping it to prevent a conduit effect. For a pipeline laid on the bed of a water body it is recommended the pipe be removed otherwise there is a risk of the weighting not being appropriate for an empty pipeline to prevent buoyancy issues and drilling holes in the pipe may introduce contaminants to the water body where the cleaning process either did not meet cleanliness criteria or where a criteria was not available [4].

Furthermore, as the pipe degrades large pieces may break away from anchors and be carried downstream with the water current. Exposure of the abandoned pipeline at the shore of the water body or the pipeline floating in the water crossing is a long-term potential issue. It should both be monitored and mitigated as needed or that section of pipeline is removed if it is susceptible to erosion.

Aerial crossings should also be removed as the suspension support structure will deteriorate over time allowing the abandoned pipeline to fall below.

### 47.3.5 Transportation and Utility Crossings

Transportation and utility crossings require consideration to maintain the structural integrity of that infrastructure. In the abandonment planning stage, the owners of these crossings must identify their requirements if not already dictated by crossing agreements. The need to prevent disruption in service and subsidence may mean that the pipeline is filled in place with concrete or if there is a utility way such as a tunnel or pipe casing then transfer of ownership of the remaining conduit once the pipe is removed from it may be agreed to. Otherwise a further abandonment strategy for the opening below the road, railway, or utility is required.

### 47.3.6 Right-of-Way Restoration

Abandonment may require the removal of portions of a pipeline and equipment from a right-of-way on private or public land. In addition, there will be excavations to clean the pipe and install plugs and caps. As a result the land surface must be restored or reclaimed at these locations to a state comparable to the surrounding environment unless otherwise directed by the regulators. This also means that agricultural soil is reclaimed so soil productivity is restored usually within 5 years. Vegetation must be planted and sustained to demonstrate productivity. In noncultivated areas, it may be necessary to restore the right-of-way to original state by planting native species. Caution needs to be excised and future checks done to make sure noxious weeds and foreign vegetation are not accidentally introduced into the right of way by equipment brought in. The degree of restoration of a right-of-way will depend on the regulator's requirements and any agreements arrived at in consultation with landowners. In addition, any pipeline apparatus must be removed to a depth that will not interfere with farming equipment.

All surface and subsurface apparatus (including signage) along the route of a pipeline that is to be abandoned through removal also be removed as part of the abandonment process. Conversely, if the pipeline is to remain in place, signage must be maintained to inform persons conducting ground disturbance on necessary contact information to obtain assistance when working in the vicinity of existing pipelines.

Where contamination remediation is ongoing either on or off site the abandonment and associated restoration must be held in abeyance until all standards for groundwater and soil are met.

## 47.4 POST-ABANDONMENT PHYSICAL ISSUES

Once abandonment has occurred to the satisfaction of regulators and landowners, any remaining infrastructure and associated right-of-way may have the potential to be selectively affected by natural causes such as weathering. The effects can lead to impacts on use of the land. The following situations are described with advice on mitigation.

### 47.4.1 Ground Subsidence

Any underground construction can lead to voids being left in the soil structure. The natural settling or subsidence process, often aided with water percolation, leads to depressions on the land surface. These depressions can cause impediments to traffic, water ponding or channeling, and possibly erosion and poor crop production. In most cases where a pipeline has been in operation and the pipe is not removed, the soil structure is stable and further subsidence will not immediately occur. If subsidence occurs, filling in the depression is necessary with the appropriate soil, which is usually topsoil. Subsidence may also occur where pipe deterioration occurs. This is addressed in Section 47.4.2.

### 47.4.2 Pipe Deterioration and Collapse

Given that the steel components of a pipeline are subject to corrosion where the coating deteriorates, it is a common assumption that the pipeline will perforate and eventually collapse. However, there are many factors affecting corrosion and there are many factors affecting soil collapse.

Corrosion will normally occur on the outside of the pipe because the interior is protected from the outside environment by plugs and caps. In the case of production pipelines, it is possible that some internal corrosion may occur as the process of purging may leave minor amounts of water or other contaminants inside the pipeline. This corrosion would not be expected to continue beyond a short period of time. The rate of external corrosion will depend on the presence of moisture and the soil properties, and the rate at which any external coatings deteriorate. As external coatings are generally polymeric in nature, they can exist for an extended time before eventually breaking down. This may occur locally at first but eventually could be expected to be wide spread. However, in the process of coatings deteriorating the resulting perforations would be random and not concentrated enough to cause complete failure of the pipeline structure over any significant length.

An eventual perforation could then introduce moisture to the interior of the pipeline where corrosion could occur primarily on the bottom of the pipe [4].

Given the slow corrosion rate and the ability of a pipe to maintain strength even with holes, it is possible that a pipeline will not have a structural failure for centuries [5]. At that time the soil in the vicinity of the pipeline may mitigate any surface depression. The soil surrounding a pipeline provides support to the pipeline for lateral movement. However, as a pipeline deteriorates it is likely that soil would gradually fill in the pipeline as it decomposed [4]. The result might be a gradual surface depression that would only be evident on large-diameter pipes. If the soil above a pipeline has more stable properties, such as those exhibited by clay, then it could also sustain larger vertical loads and resist collapse but eventually a form of surface subsidence would occur over time.

These assumptions are yet to be proven but research should occur to determine the factors that could shed more information on predicting the effects of pipeline deterioration.

### 47.4.3 Pipe Exposure

Soil removal can occur in a number of ways to expose a pipeline. Soil may be removed by mechanical means if human activity is not controlled. It may also occur due to a geotechnical or hydrotechnical event [4]. However, it most often occurs due to erosion by water or wind. The most susceptible sites to erosion are those where slopes or fine soils are present. Pipeline trenches themselves can preferentially erode in the presence of moving surface or subsurface water, as the unconsolidated materials surrounding the backfilled pipeline can be more easily eroded than adjacent undisturbed soils. It is sometimes necessary to install water breaking devices in the trench to inhibit the movement of groundwater through the looser materials, though this is usually done at the time of initial construction.

Pipelines abandoned adjacent to watercourses can be exposed through flood events or bank movement. Where an abandoned pipeline is susceptible to this then removing the pipe is recommended. Otherwise removal in the future will likely become necessary for aesthetic, access, and public safety reasons.

### 47.4.4 Water Conduit Effect

When pipeline abandonment in place occurs, eventual corrosion may allow water to enter the pipe. If the source is of significant volume such as from a water table or surface runoff then a serious problem can be created downslope where the water may exit the pipe and cause erosion, vegetation damage, sink holes, or otherwise affect property. Conversely, the artificial drainage can disrupt natural runoff conditions that can affect ecosystems and interrupt water supplies for persons relying on them.

This water conduit effect can be prevented with the installation of caps and pipe plugs at the time of abandonment. The plugs should be long enough to prevent corrosion. Consideration also needs to be given the location of such plugs as the pipe wall adjacent to the plug will eventually corrode away again allowing water to enter. The plugs should adhere to the pipe, be impermeable and non-shrinking, and able to resist deterioration. Examples of suitable materials are concrete grout or polyurethane foam. The use of impermeable earthen plugs may also be a viable option.

Plugs should be installed downstream of any water bodies or known high-water conditions. Plugs should also be installed upstream of water bodies and environmentally sensitive areas to prevent contamination or introduction of sediment. Guidelines for locating plugs are presented in Table 47.1.

**TABLE 47.1 Guide for Locating Pipeline Plugs**

| Terrain Feature | Plug Locations |
|---|---|
| Water bodies/watercourses | Above top of flood plain—both banks |
| Long inclines (>200 m), long river banks | At top and bottom and at mid-slope |
| Sensitive land uses (e.g., natural areas, parks) | At boundaries |
| Wet areas, including groundwater discharge and recharge zones, wetlands, high water table | At boundaries and should include an adequate buffer zone |
| Cultural features (population centers) | At boundaries |

*Source*: Adapted from Ref. [3].

### 47.4.5 Slope Stability

Pipelines should be left in place on unstable slopes as the pipe may provide support to the slope and remediation is difficult. In most cases, vehicle traffic on slopes can cause loss of cover so it should be discouraged where possible and mitigated if access is an ongoing need. If the pipeline is removed on a slope then remediation with fast-growing vegetation is recommended and frequent monitoring over a few years is necessary.

In anticipation of the pipeline caps or plugs failing, there will need to be mitigation against water conduit effects on a slope. If recapping the pipe is not possible or the risk to slope failure through possible pipe water transport, then the pipeline should be forced to exit at a location where energy dissipation can occur to prevent erosion and where release of sediment is acceptable.

## 47.5 POST-ABANDONMENT CARE

In Section 47.4 the potential for physical issues arising after abandonment was identified. This leads to questions on how these are identified; how they are addressed, and who is responsible. In addition, there is a question of what is reasonable oversight for an abandoned pipeline. The following is advice on approaches to these questions.

### 47.5.1 Monitoring and Maintenance

If a pipeline is left in place then signage should be maintained to inform affected persons of its location to prevent damage to excavating or farming equipment. Contact information for the owner and One-Call information is often required. In areas that maintain a One-Call center, the pipeline should remain on the database. The pipeline right-of-way should also be examined on a frequency to be determined to check for general issues that require maintenance such as subsidence, pipe exposure, erosion, and weed growth.

### 47.5.2 Land Use Changes

When a land use change is contemplated an abandoned in place pipeline may need to be removed to accommodate development such as a road or housing development. This is not always necessary and the pros and cons of doing so needs to be carefully considered between the pipeline owner and the project developer. Different jurisdictions may treat the costs of removal of the pipe in different ways. In some, costs of removal would be the responsibility of the owner of the pipeline unless the landowner had agreed to assume responsibility at some time. In others, a shared cost approach may be used depending on when the pipeline was originally constructed and on the circumstances of the proposed development. Regulatory jurisdictions may also provide processes to mediate or dictate the allocation of costs; this will need to be investigated thoroughly in each specific jurisdiction.

### 47.5.3 Liability

In some jurisdictions an abandoned pipeline remains the responsibility of the owner until it is removed from the right-of-way (e.g., Alberta Canada [6,7]). In other jurisdictions liability is not clear once abandonment is approved because the regulatory authority may apply only to the transportation of a product. Certainly most North American environmental regulatory jurisdictions have a "polluter pay" principle where any residual contamination is the responsibility of the entity that caused the pollution.

Liability may be further complicated by existing land use agreements that may have included conditions respecting liability. Further conflicts may exist if original land use agreements have been nullified and easements are taken off the land title. As a result, it is recommended that a company abandoning a pipeline assume it has ongoing liability, and therefore continues to maintain a right to access the abandoned pipeline for monitoring and maintenance.

### 47.5.4 Financial Resources

A large corporation that abandons a pipeline must have funds to not only conduct the physical abandonment but also to mitigate post abandonment issues where a responsibility is identified. A smaller entity will also be required to demonstrate sufficient funds to conduct the physical abandonment but once that occurs often it either transfers ownership, with regulatory approval, or assumes the pipeline has no further need for financial resources. As previously noted, there are ongoing monitoring and maintenance needs at a minimum. For all companies it is recommended that there be an ongoing budgetary allocation for monitoring, maintenance, and contingencies. An alternative is a security fund that is in trust with a third party for the care of the pipeline if the entity ceases to exist.

Notes:

- Opinions expressed in this chapter are those of the authors and do not necessarily reflect those of any regulator, industry company, or standards association.
- Readers should always refer to the applicable regulations and standards in place where a pipeline is being abandoned.

## REFERENCES


1. Canadian Standards Association CSA Standard Z662-11, Oil and Gas Pipeline Systems, Clause 10.16, June 2011.
2. National Energy Board (2011) Regulating Pipeline Abandonment, Appendix 1, June 2011, Calgary, Alberta, Cat. No. NE23-161/2011E-PDF, ISBN 978-1-100-18376-3.
3. Canadian Association of Petroleum Producers, Canadian Energy Pipeline Association, Alberta Energy Utilities Board, and the National Energy Board of Canada (1996) Pipeline Abandonment—A Discussion Paper on Technical and Environmental Issues. November 1996.
4. National Energy Board Det Norske Veritas, Pipeline Abandonment Scoping Study. Report No. EP028844, Reg. No. ENACA855, pg. 19–21 and 26–27, pg. 28 and pg. 31–32, November 2010.
5. Canadian Energy Pipeline Association Canadian Energy Pipeline Association, Pipeline Abandonment Assumptions. pp. 13–14, September, 2007.
6. Province of Alberta, Pipeline Act, Revised Statutes of Alberta 2000 Chapter P-5, current as of March 29, 2014.
7. Province of Alberta, Pipeline Act Pipeline Rules P.A 10. Alberta Regulation 91/2005.

# PART VII

# RISK MANAGEMENT

# 48

# RISK MANAGEMENT OF PIPELINES

LYNNE C. KALEY AND KATHLEEN O. POWERS
*Trinity Bridge, LLC, Houston, TX, USA*

## 48.1 OVERVIEW

### 48.1.1 Risk-Based Inspection for Pipelines

The use of risk-based inspection (RBI) for pipelines is increasing in demand as government regulators worldwide are requiring operators to use risk assessment in high-consequence areas. Historical statistics attribute piping failure as the leading cause of large property losses globally (Marsh and McLennan study from mid-1990s to 2013) [1]. Data from numerous corporations indicate that the most frequent failures (approximately 35–40%) are from piping systems.

There are a number of international design codes and standards covering pressurized equipment that should be referenced when considering a piping program. The Norsk Sokkels Konkuranseposisjon (NORSOK) standards, developed by the Norwegian petroleum industry, were intended to supersede individual company specifications as a reference for government regulators. ISO 31000, *Risk Management—Principals and Guidelines*, is a family of international risk management standards that provides generic guidance for any risk management implementation. Other ISO and OGP standards focus more specifically on risk and integrity management specifications and techniques.

This chapter focuses on in-service American Petroleum Institute (API) codes and standards that were developed by representatives from multinational oil and gas companies through an ANSI-accredited balloting process. API 510, 570, and 653 and associated documents define inspection requirements for pressurized equipment while recommended practices API 580 and 581 provide the necessary guidance on RBI programs and methodology. RBI methods and applications for inspection planning, execution standards, and recommended practices are also described by American Society of Mechanical Engineers (ASME). Finally, assessment of fitness-for-service and remaining life (RL) using API 579-1/ASME FFS-1 can be used to justify continued service when damage is found during inspection.

Risk-based principles allow low risk and low-consequence systems to run to failure. This risk concept implies that other systems are high in risk and that the consequence of failure (COF) is unacceptable; as such, these systems should receive attention even when the probabilities of failure are not significantly high. Risk-based principles challenge accepted design codes that are based on deterministic design and fitness-for-service codes, particularly where worst-case scenarios are used without consideration for COF. Differences in inspection requirements and remedial action will occur when deterministic methods are compared with recommendations from risk-based methodologies.

The purpose of a RBI program is to develop risk assessment in order to implement an integrity inspection program over a defined period with available data and detailed technical data. With inspection planning focused on high-risk items, inspection generates the necessary technical data required for risk assessment updating and fine tuning future priorities. RBI methodologies can be used on older facilities that have little or poor data available; on newer assets where technical data is available, but little inspection history exists; and on facilities with good quality data and inspection history.

A RBI program should be considered for the following pipeline assets:

- Trunk lines from manifold stations to process facilities;
- Oil and gas NGL pipelines;

*Oil and Gas Pipelines: Integrity and Safety Handbook,* First Edition. Edited by R. Winston Revie.

- Manifold stations;
- Process facilities for oil and gas separation.

Since facilities range widely in age, there is often limited construction, process, and inspection history available. Inspection and maintenance programs are sometimes reactionary, lacking a cohesive strategy or long-term plan based on expected in-service damage to equipment. Construction and process data and inspection histories are marginal or missing completely. The quality of available information on the assets determines whether a qualitative or quantitative RBI approach should be used. Less data generally requires a more qualitative approach, at least initially. The approach adopted will depend on the following:

- Design and construction data readily available;
- Known equipment condition based on inspection history;
- Knowledge of past, current, and future process operational conditions;
- Complexity and age of systems considered.

Regardless of the approach used, a continuous improvement process similar to one of the following should be implemented:

1. Conduct the initial RBI assessment using available data.
2. Use inspection results to develop inspection requirements and priorities.
3. Implement initial inspection and plan over defined multi-year time period.
4. Update RBI assessment after each inspection.
5. Update the inspection plan from the updated assessment.

As the program develops and matures, a more quantitative analysis can be incorporated, as desired.

### 48.1.2 Scope

This chapter is primarily intended to be used for the planning of in-service inspection for pipeline static mechanical pressure systems when considering failures by loss of containment of the pressure envelope. Failure modes, such as failure to operate on demand, leakage through gaskets, flanged connections, valve stem packing along with valve passing, and tube clogging are not addressed. The focus is on pipeline pressure systems involving the following types of components:

- Piping systems comprising straight pipe, bends, elbows, tees, fittings, reducers
- Pressure vessels and atmospheric tanks
- Pig launchers and receivers
- Heat exchangers
- Unfired reboilers
- Valves
- Pump casings
- Compressor casings

Excluded from the scope are

- structural items, including supports, skirts and saddles;
- seals, gaskets, flanged connections;
- failure of internal components and fittings;
- instrumentation.

In this chapter, "equipment" is defined as any of the aforementioned while a "system" refers to a combination or group in the aforementioned list.

### 48.1.3 Risk Analysis

Traditional risk analysis and risk management is based on applying a systematic approach to use available knowledge to assess, mitigate (to an acceptable level), and monitor risk. The goal is to identify all reasonably foreseeable detrimental impact on asset integrity and reliability. Risk is assessed by identifying possible failures, estimating the probability of failures (POF), and assessing the COFs. The results of the risk assessment are then used to reduce the impact of failure to as low as reasonably practicable (ALARP). The risk management process is a continuous activity that reflects changes in the operating environment and the need to monitor and maintain the performance of the resources.

The major concepts of a traditional risk analysis process comprise five tasks:

1. System definition
2. Hazard identification
3. Probability assessment
4. Consequence analysis
5. Risk results

A typical risk analysis scenario includes the following set of events: loss of containment; detection, isolation, and mitigation. Depending on the nature of the process and the detail of the study, a risk analysis can include thousands of different scenarios with evaluation of both the probability and the consequence of the set of events in each scenario.

- ***System Definition***
  In the system definition phase of a risk analysis, the ground rules are established and all pertinent

information is collected. The ground rules of the analysis typically include the following:

- Goals and objectives—stating the motivation for conducting the risk analysis. Possible objectives include satisfying regulatory requirements, conducting a cost/benefit analysis, or evaluating risks of a proposed expansion project.
- Required risk measures—defining the final results required to meet the objectives.
- System boundaries—defining the physical and operating limits of the system. Physical boundaries are the equipment included in the study while operating boundaries include the function or operating mode of the system.
- Level of detail—defining how units within the system will be analyzed. Questions, such as "will each section of piping be modeled?" or "will piping be combined into groups for easier analysis?" need to be resolved early in the program.
- Data collection—defining what data must be captured and maintained. Design drawings, operating practices, and up-to-date equipment information are gathered for review. Other pertinent data, such as weather or population, may also be gathered, depending on the objectives of the study.
- Component design details—design conditions and all special requirements such as coatings, linings, operating temperature, process stability, and material of construction properties.

• ***Hazard Identification***

A critical step is to identify hazards that can cause injury to humans, damage to the environment, damage to property, or loss in production. The hazards are identified along with how the hazard can develop into an accident, the consequences of a potential accident and the safeguards in place to prevent an accident. The major focus is to evaluate a variety of scenarios that may lead to undesirable outcomes. Both the probability and the magnitude of these outcomes are estimated and displayed as results.

A hazard and operability study (HAZOP) is a structured brainstorming exercise that uses a list of guidewords to stimulate team discussion. The guidewords initially focus on process parameters (such as flow, level, temperature, and pressure) and then branch out to include other concerns such as human factors and operating outside normal parameters. The majority of identified potential deviations are typically operability issues, but potential safety concerns and environmental considerations are also identified. For maximum effectiveness, the HAZOP is typically performed by a team familiar with the process rather than an individual. This team normally needs participation from a variety of disciplines, including mechanical design, instrumentation, machinery, and metallurgy.

The task of hazard identification has received much public attention and, as a result, is one of the most mature of the various disciplines that comprise a risk analysis.

• ***Probability Assessment***

A probability assessment is conducted to estimate the probability of occurrence for the scenarios identified in the previous phase of the risk analysis. If a scenario occurs fairly frequently, it is best to use historical data to estimate the event's probability. However, if the events are rare and sufficient data does not exist to estimate their probability based on historical data alone, a building-block approach is used. Probability estimates for all elements of the scenario are obtained and combined to predict the overall scenario probability.

The most common measure of probability for a scenario is frequency. A single fixed value is assigned for the frequency/probability term and is considered to be constant over time. Frequency can be used for a single event or a series of events. A year is most often used as the standard time interval for a frequency analysis. Frequencies may be very small numbers, such as one in a million years for infrequent events, or they may be relatively high values, such as once a month or four times a day. If, for example, a pipe is known to leak about every 5 years, it would have a leak frequency of one in five years, or 0.2 per year. The term "recurrence period" is sometimes used to refer to the reciprocal of the frequency. In our example, the recurrence period for the leak would be 5 years.

To obtain the frequency of the scenario, multiply the frequency of the leak by the probability of all events that follow. The resulting likelihood is the scenario's frequency. The mathematical representation of the likelihood of the sequence, in terms of frequency, is as follows:

$$\text{Scenario frequency} = \text{leak frequency} \times \text{outcome probability}$$

For example, consider a leak that does not ignite. If the estimated leak frequency is $10^{-5}$ and the probability of that leak not igniting was 0.5, the overall frequency for the no ignition scenario is $10^{-5} \times 0.5 = 5 \times 10^{-6}$.

• ***Consequence Analysis***

The consequence of a release from equipment varies depending on factors such as the physical properties of the material being handled, toxicity or flammability characteristics, weather conditions, release duration, and mitigation actions. The effects may impact plant personnel or equipment, nearby population, and the environment. Hazardous consequences are estimated in five phases:

- Discharge
- Dispersion
- Flammable or toxic effects
- Environmental effects
- Business interruption costs

Depending on the material released, only one of the flammable, toxic or, environmental effects is calculated, although all of them may be possible with the release of certain mixtures.

- ***Risk Results***

Historically, a number of measures have been used to express risk in the context of a risk analysis. There is no single way to measure or present an estimate of the risk of an operating process. Risks to people are normally presented in one of the three ways: risk indices, individual risk measures, and societal risks.

*Risk indices* are a single number measure of risk. Some of the more common risk indices are as follows:

1. Fatal accident rate: The estimated number of fatalities per 108 exposure hours (roughly 1000 employee working lifetimes).
2. Average individual risk: A similar concept to the fatal accident rate, but with a time period of 1 year.
3. Average rate of death: The average number of fatalities from all incidents that might be expected per unit time.

*Individual risk measures* consider the risk to an individual who might be located at any point in the effect zones of incident. Following are some of the more common individual risk measures:

- Individual risk contours: The geographical distribution of individual risk. These contours show the expected frequency of an event capable of causing a fatality at a specific location regardless of whether anyone is present at that location.
- Maximum individual risk: The individual risk to the person exposed to the highest risk in an exposed population. This can be found by calculating the individual risk at every geographical location where people are present and searching for the highest value.

*Societal risk* is a measure of risk to a group of people in the effect zones of incidents. It is most often expressed in terms of the frequency distribution of multiple fatality events.

### 48.1.4 The RBI Approach

RBI is a methodology that uses risk as a basis for prioritizing and managing the efforts of an inspection program, including recommendations for monitoring and testing inspection plans. It provides focus for inspection activity to specifically address threats to the integrity of the asset and the equipment's capacity to operate as intended. The POF and COF are assessed separately and then combined to calculate risk of failure. Risk is compared and equipment risk prioritized for inspection planning and risk mitigation.

A RBI program includes the following:

- High-risk systems or processes within an operation prioritized by risk.
- Calculated risk value or category associated with an equipment item within a system or process based on a consistent methodology.
- Prioritized equipment rank based on risk.
- Development of an appropriate inspection program to address key drivers of equipment risk.
- A method to systematically manage risk of equipment failures.

A RBI program is not a fully quantitative risk analysis, but a technique that combines the disciplines of risk analysis and mechanical integrity. However, the techniques used in RBI are similar to those seen in traditional risk analysis. RBI methodology follows the five tasks of a traditional risk analysis described in the previous section, but it uses a modified and in some cases simplified approach:

- System definition—the system definition phase of an RBI program establishes ground rules and collects all pertinent information, as outlined in the system definition section.
- Hazard identification—while hazard identification is a critical step in a traditional risk analysis, the RBI approach focuses on the pressure boundary of equipment and assumes that failures are due to identifiable mechanisms of degradation in that boundary. Secondary causes of a leak, such as instrument failure or human error, are included implicitly in the RBI program's treatment of management systems.
- Probability assessment—probability and consequence are limited to a carefully defined and limited number of scenarios. The RBI study expends a considerable effort in the assessment of the failure frequency and is generally considered to vary with time. The POF increases as degradation progresses, and can be reduced when inspection is performed to better define the actual condition of the equipment.
- Consequence analysis—RBI consequence analysis is a relatively coarse determination compared with a traditional risk analysis. The effects considered may impact plant personnel or equipment, population in the nearby residences and the environment. Depending on the ground rules established and material released, all of

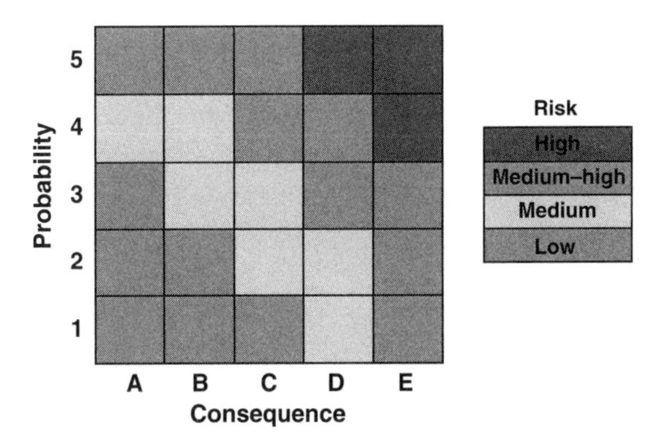


FIGURE 48.1 Risk matrix example.

the consequence effects are usually calculated and the largest used for final risk.

- Risk results—risk is presented in a risk matrix (Figure 48.1), risk categories, iso-risk (Figure 48.2), or risk values for the purpose of prioritization.

### 48.1.5 Risk Reduction and Inspection Planning

RBI focuses on the development of inspection programs to address damage types that inspection should detect; the appropriate inspection technique to detect the damage; and the application of RBI tools to reduce risk and optimize the inspection program.

Inspection influences risk primarily by reducing the POF. Many conditions (design errors, fabrication flaws, malfunction of control devices) can lead to equipment failure, but in-service inspection is primarily concerned with the detection of progressive damage. RBI considers that the POF is due to damage as a function of four factors:

- Damage mechanisms and resulting type of damage (cracking, thinning, etc.);
- Rate of damage progression;
- Probability of detecting damage and predicting future damage states with inspection techniques;
- Tolerance of the equipment to the type of damage.

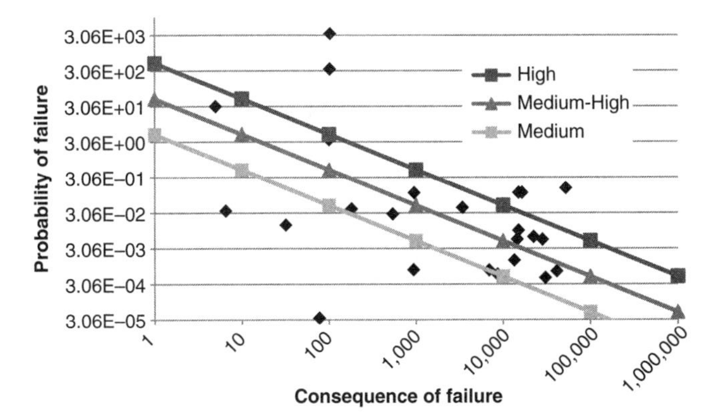


FIGURE 48.2 Iso-risk example.

RBI differs from conventional inspection management because it provides the concepts and methods to support decision making even when data is missing or uncertain. Rational decisions can be made using both quantitative and qualitative RBI methods.

The benefits of using RBI for inspection planning are as follows:

- Systematic overview of a system or process with an explicit and documented breakdown of the risks while clearly showing the risk drivers and recommending appropriate actions.
- Focused inspection efforts on items where the safety, economic, or environmental risks are identified as being high while the efforts applied to low-risk systems are reduced.
- Use of probabilistic methods to account for uncertainties in process parameters, corrosively, and degradation rates to calculate the extent of degradation.
- Consideration of consequences of failure to focus activities where they will have the most significant impact.
- Proactive and focused effort to ensure that overall system or process risk does not exceed the risk acceptance limit set by the authorities and/or operator.
- Optimized inspection or monitoring method opportunities according to the identified degradation mechanisms and the agreed inspection strategy.

The inspection planning process directs resources to accurately determine the condition of the equipment through the most effective and efficient inspection methods, balancing the necessary down time and cost of inspection against the effectiveness and benefits of inspection.

## 48.2 QUALITATIVE AND QUANTITATIVE RBI APPROACHES

### 48.2.1 API Industry Standards for RBI

Operators considering a piping inspection program should be aware of several important industry design codes and standards that cover pressurized equipment. The API in-service codes and standards, including API 510, API 570, and API 653, and their associated documents should be used to define inspection requirements for pressurized equipment and tanks. Recommended practices API 580 and API 581 provide the necessary guidance on RBI programs and methodologies. RBI methods and applications for inspection planning, execution standards, and recommended practices are also described by ASME PCC-3-2007 *Inspection Planning Using Risk-Based Methods.* Finally, assessment of fitness-for-service and RL using API 579-1/ASME FFS-1 can be used to justify continued service when damage is found during inspection.

ASME PCC-3-2007 provides guidance on the preparation and implementation of a RBI plan based on API RP 580 RBI, the earliest industry standard on the practice. ASME PCC-3 and API RP 580 are closely aligned as changes, additions, and improvements are made to each document to enhance applicability for a broader spectrum of industries. ASME-PCC-3-2007 provides recognized and generally accepted good practices that may be used in conjunction with post-construction codes, such as API 510 and API 570.

The purpose of API RP 580 is to provide users with the basic elements required for developing, implementing, and maintaining an RBI program. The risk analysis principles, guidance, and implementation strategies presented in this document are broadly applicable, but were specifically developed for applications involving fixed pressure-containing equipment and components. API RP 580 provides guidance to owners, operators, and designers of pressure-containing equipment for developing, implementing, and assessing inspection programs. API RP 580 emphasizes safe and reliable operation through cost-effective inspection using a spectrum of complimentary risk analysis approaches (qualitative through fully quantitative) as part of the inspection planning process.

API RP 580 includes an introduction to the concepts and principles of RBI for risk management as well as the steps that should be followed in applying risk management principles within the framework of the RBI process. These steps include the following:

1. Understanding the design premise;
2. Planning the RBI assessment;
3. Data and information collection;
4. Identifying damage mechanisms and failure modes;
5. Assessing POF;
6. Assessing COF;
7. Risk determination, assessment, and management;
8. Risk management with inspection activities and process control;
9. Other risk mitigation activities as needed;
10. Reassessment and updating;
11. Roles, responsibilities, training, and qualifications;
12. Documentation and recordkeeping.

The expected outcome from an RBI process should be the association of risks with appropriate inspection, process control, or other risk mitigation activities to manage the risks. The RBI process is capable of generating

- A ranking of all equipment evaluated by relative risk.
- A detailed description of the inspection plan to be employed for each equipment item, including the following:
  - Inspection method(s) that should be used (e.g., visual inspection, ultrasonic, radiography, smart pigging operations).
  - Extent of application of the inspection method (e.g., percent of total area examined or specific locations).
  - Timing of inspections/examinations (inspection intervals/due dates).
  - Risk management achieved through implementation of the inspection plan.
- Description of any other risk mitigation activities, such as repairs, replacements or safety equipment upgrades, equipment redesign or maintenance.
- Expected risk levels of all equipment after the inspection plan and other risk mitigation activities have been implemented.
- Identification of risk drivers.

API RP 581 [2] was developed to provide users with a specific approach for developing, implementing, and maintaining a RBI program. The methodology outlined in API RP 581 [2] was developed through a joint industry project initiated in May 1993 and conducted by API to develop practical methods for RBI. The original sponsor group was organized and administered by API and included the following original members: Amoco, ARCO, Ashland, BP, Chevron, CITGO, Conoco, Dow Chemical, DNO Heather, DSM Chemicals, Exxon, Fina, Koch, Marathon, Mobil, Petro-Canada, Phillips, Saudi Aramco, Shell, Sun, Texaco, and UNOCAL. The results of this project were published in a RBI base resource document (BRD) to the sponsor group based on development of quantitative and qualitative RBI methodologies.

The methodologies developed in the project continue to be improved and maintained in API RP 581 [2]. The target audience in the proposal for the API sponsor group project stated that the methods in it were "to be aimed at an inspection and engineering function audience." The methodology was specifically not intended to "become a comprehensive reference on the technology of quantitative risk assessment (QRA)."

For failure rate estimations, the proposal promised "methodologies to modify generic equipment item failure rates" via "modification factors." For consequence calculations, safety, monetary loss, and environmental impact were to be included. Existing algorithms in AIChE chemical process quantitative risk analysis (CPQRA) guidelines are "complex and are best suited for use in a computerized form" for safety evaluations. It was proposed that "for ease of use, that safety consequences be limited to the evaluation of burning pools of liquids, ignited high-velocity gas and liquid releases, explosions of vapor clouds, and toxic impacts."

The result of the sponsor group project and subsequent activities has been the development of simplified methods for estimating failure rates and consequences of pressure boundary failures. The methods are aimed at persons who are not expert in QRA. Subsequent computer programs were developed to further ease the application of the BRD and now API RP 581 [2] methodologies.

The approach outlined in API RP 581 [2] has been successfully implemented in major refineries, petrochemical/chemical plants, and upstream gas plants. In recent years, an increasing emphasis on pipeline process safety management and mechanical integrity using RBI has been considered for all upstream and pipeline applications following established and emerging OSHA, DOT, and API regulations and recommended practices.

### 48.2.2 Basic Risk Categories

Equipment or system failure and the subsequent release of hazardous materials can lead to many undesirable effects, including the unavailability of equipment. A RBI program considers these effects based on four basic risk categories:

- ***Flammable events*** can cause damage in two ways: thermal radiation and blast overpressure. Most of the damage from thermal effects tends to occur at close range, but blast effects can cause damage over a larger distance from the blast center.
- ***Toxic releases*** are addressed when they affect personnel. Acute, as opposed to chronic, exposure is considered. These releases can cause effects at greater distances than flammable events. Unlike flammable releases, toxic releases do not require an additional event (e.g., ignition, as in the case of flammables) to cause an undesirable event.
- ***Environmental risks*** are an important component to any consideration of overall risk in a processing plant. An RBI program focuses on acute environmental risks rather than chronic risks from low-level emissions. Environmental damage can occur with the release of many materials; however, the predominant environmental risk comes from the release of large amounts of liquid hydrocarbons.
- ***Business interruption*** can often exceed the costs of equipment and environmental damage and, therefore, should be accounted for in an RBI analysis. Equipment replacement costs (accounted for in flammable damage estimates) can be small compared with the business loss of critical equipment for an extended period of time.

The risk calculation is performed for each scenario, for all four risk categories, if desired. The risk for each equipment item is then found by summing the individual risk components from each scenario (hole size) calculation.

### 48.2.3 Alternative RBI Approaches

RBI begins with the identification of high-risk assets followed by assessment of equipment condition, evaluation of maintenance and inspection programs, study of operating protocols, and estimation of life consumption of these priority assets. This process takes into account the probability and consequences of mechanical failure. The information is used to modify and optimize inspection and maintenance programs, audit procedures, operating limits, and safety information.

The POF due to loss of containment should consider each of potential degradation mechanism that could lead to loss of containment. POF (loss of containment) can be estimated quantitatively, if any degradation model exists, or qualitatively by means of engineering judgment.

Given that loss of containment has occurred, the potential consequences depend on the size of the hole, which can vary from a pinhole to a full-bore rupture. The methodology's handling of hole size probability distribution by degradation mechanism should be understood when considering the potential consequences of any loss of containment, since the approach used can significantly affect the outcome. Finally, the probability that loss of containment can lead to ignition or pollution is an important issue.

RBI can be carried out using methods that are qualitative or more quantitative. In practice, most RBI efforts are carried out using a blend of both the methods, referred to as semi-quantitative, and is often represented as a continuum as shown in Figure 48.3. The qualitative assessments can be conducted at a system or equipment level while quantitative assessments are conducted on the equipment level, as follows:

1. RBI of systems assessments is qualitative. The system risk ranking allows for prioritization of groups of equipment for inspection.
2. RBI at the equipment level can be conducted qualitatively:
   a) Team work sessions are used to collect experience-based data.

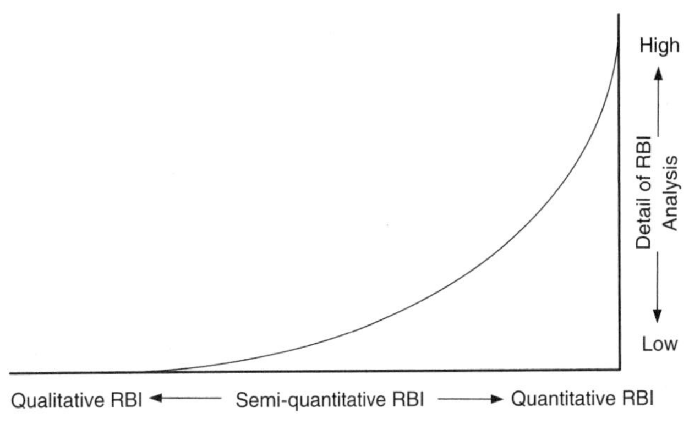


**FIGURE 48.3** Range of RBI analysis details.

b) Team work sessions are used to collect experience-based data degradation mechanisms assessment, COF assessment, POF assessments, and risk and inspection planning.
   - Each team work session follows a standard format using guidance forms and question sets to assure consistency between studies.
   - Degradation mechanisms considered in each assessment, at a minimum, consider internal, external degradation mechanisms, and mechanical damage.
   - Team work sessions include qualified subject matter experts (>10 years' experience), including an RBI expert.
   - POF and COF assessments should use only qualified specialists to assign numerical values/ranges based on engineering judgment.

c) Risk matrices with decision procedures are used for inspection planning.

3. RBI at the equipment level can be conducted quantitatively:

a) Analysis is supported and driven by equipment-specific data collected for the assessment.

b) POF determinations are model based, where values vary widely from $<10^{-5}$ to 1 with results often presented on a logarithmic scale. POF values $<10^{-5}$ are difficult to model, observe, and represent an insignificant risk.
   - Degradation mechanisms considered in each assessment, at a minimum, consider internal, external degradation mechanisms, and mechanical damage.
   - RBI team personnel should include qualified personnel knowledgeable in process operations, materials/corrosion, and inspection as well as an RBI expert.

c) COF are model based and are expressed in terms of damage to equipment and potential personnel injury.

d) Environmental COF are model based and are expressed in terms of mass or volume of a pollutant released to the environment or in financial terms as the cost of cleaning up the spill, including consideration of fines and other compensation.

e) Financial COF is expressed in financial terms and is often a combination of other consequence effects for equipment, personnel, environment, and business interruption.

Regardless of the level of assessment conducted, the team should be comprised of specialists with considerable experience with the specific installation(s) being assessed and include relevant expertise in the RBI methodology being used. Since qualitative assessments are based on rankings and numerical assignment during an experienced-based team sessions, a higher level of experience and specialist involvement is required for a credible assessment. In addition, documentation of data and any assumptions that were the basis for the assessment should be documented sufficiently for the study to be reproduced at a later date, even if the original team members are not available to participate. Conversely, RBI assessments based on actual design, operating, and inspection data require less documentation and assumptions, making future re-assessments more reproducible.

### 48.2.4 Qualitative Approaches to RBI

Qualitative approaches are similar to quantitative analysis, but require less equipment-specific data and can take less time to complete. While the results yielded are not as precise as those of the quantitative analysis, this approach provides a basis for prioritizing a RBI inspection program. However, as previously mentioned, a higher level of experience and specialist involvement is required for a credible qualitative assessment. The basis for all numerical assignments or rankings should be thoroughly documented as well as any assumptions that were the basis for the assessment in the event that the original team members are not available to participate.

Qualitative models can be interpreted as an expert judgment-based approach in which a numerical value is not calculated or assigned. A descriptive ranking is given, such as low, medium or high, or a numerical ranking such as 1, 2, or 3. Qualitative RBI approaches cover a wide spectrum of methodologies. The simplest approach would be to compile the data/information required to do a quality damage review for the POF and categorize the hazards for COF to generate a risk matrix location to recommend a relevant inspection frequency.

The advantage of using a qualitative approach is that the assessment can be completed quickly and at a lower initial cost; there is little need for detailed information and the results are easily presented and understood. A disadvantage is that the results are more subjective than quantitative analysis because they are based on the opinions and experience of the RBI team and are thus harder to update following inspection. Qualitative models provide a risk ranking of items since there is no change in risk over time and setting an acceptable risk target is not possible. As a result, an inspection interval is assigned based on position in a risk matrix that only changes if an input value, and therefore, matrix location changes.

Methodologies that would be considered qualitative generally have the following characteristics:

- Parts of the RBI assessment use category assignments and ranking methodologies. For example

1. POF assessment is quantitative and COF assessment is qualitative (and vice versa).
2. COF and POF assessments are quantitative, but risk ranking and time to inspection assessment are qualitative (i.e., do not change over time).
   - Assignment of POF and/or COF categories is done by simple algorithms based on a selected set of most relevant parameters.
   - Assignments of POF and COF are made by engineering judgment.

A qualitative analysis of a system can be conducted to quickly prioritize groups of equipment for more detailed risk analysis. The result of any qualitative analysis positions equipment within a risk matrix and provides a method for risk rank prioritization. This level of qualitative analysis is performed at any of the following levels:

- Operating plant—for example, oil/gas separation plant processing unit;
- Major area or functional section within a plant—for example, amine section of a separation unit;
- System—a major piece of equipment and associated equipment—for example, an amine absorber column, including the feed preheat, overhead, and reflux and reboiler equipment;
- Equipment or component—a major piece of equipment and associated equipment—for example, an amine absorber column or column top.

Note that the aforementioned four levels are most representative of an oil/gas separation plant. Pipeline applications would generally only apply to the system and equipment levels.

Qualitative approaches conducted at the system level are generally influenced by the number of equipment items in the system or equipment being studied because the quantity and type of inventory is considered to assign COF. As a result, assessments conducted at the system level should be based on similar equipment counts when used to compare systems for prioritization purposes. Equipment level qualitative assessments do not require any special considerations for risk prioritization purposes.

Qualitative RBI procedures have three functions:

1. Screening the system/equipment to select the level of analysis needed and to ascertain the benefit of further analyses (quantitative RBI or another technique).
2. Rating the degree of risk for system/equipment and assigning them a position on a risk matrix.
3. Identifying areas of potential concern for the system/equipment that requires enhanced inspection programs.

The qualitative analysis determines a risk ranking for systems or equipment by categorizing the two elements of risk: POF and COF. The process involved and the physical boundaries of the study area must be defined before the qualitative analysis is conducted. The following sections provide an overview of the factors that should be considered in any qualitative analysis.

***48.2.4.1 Probability of Failure Category*** POF Category is determined by estimating the susceptibility of the system/equipment to one or more degradation mechanisms that affect the probability of a leak. An adjustment is made to the POF category based on the known condition of the system/equipment, inspection program history, and process stability and design considerations. Key factors that may be considered to determine the POF based on equipment type are as follows:

- Potential damage mechanisms or damage factor (DF);
- Appropriateness of inspection or inspection factor;
- Current equipment condition factor;
- Process operation or process factor;
- Equipment design factor.

These components are combined to determine the final POF and POF category. POF is plotted on the vertical axis of the risk matrix (see Figure 48.1).

The DF is a measure of the risk associated with known or potential degradation mechanisms. The potential degradation mechanisms include levels of internal (corrosion or cracking), external (corrosion or cracking), and mechanical damage (fatigue, brittle fracture).

The *inspection factor* provides a measure of the effectiveness of the current inspection program and the ability of the program to detect and quantify active or anticipated degradation. It is based on inspection methods, coverage, and inspection program management. The inspection factor is weighted with negative numbers because the quality of the inspection program will partially offset the POF inherent in the degradation mechanisms from the DF.

The *current equipment condition factor* is a measure of the physical condition of the equipment from a maintenance and housekeeping perspective. Simple evaluations are performed on the condition and up-keep of the equipment from a visual examination.

The *process factor* is a measure of the process stability and potential for abnormal operations or upset conditions that may initiate a sequence of events leading to a loss of containment. It is a function of the process interruptions, stability of the process and potential for failure of protective devices because of plugging or other causes.

The *equipment design factor* evaluates the factor of safety equipment design, whether it is designed to current standards and the complexity of design.

***48.2.4.2 Consequence of Failure Category*** There are two major potential hazards associated with pipeline operations: (a) fire and explosion risks, flammability and (b) toxic risk. In determining the toxic consequence category, RBI considers acute and does not consider chronic effects.

***48.2.4.3 Flammable COF Category*** The key variables that affect flammable consequences are process composition, available inventory and state in the process (liquid, gas, or mixed phase). These three variables can be combined to determine flammability COF. Operating temperature and pressure can be used to further refine the COF.

The COF analysis determines a damage consequence factor and is usually determined for each process. Many processes, however, exhibit a predominate risk (either fire/explosion or toxicity); thus if the predominant risk for a given process is known, the factor for risk should be based on the known predominant risk. If there are several processes present in relatively large percentages in the area, the risk factor for each of the processes present should be considered. In general, a good practice is to determine risk for any processes with high health consequence in addition to any that comprise at least 90–95% of the total mass of process in the area.

The COF category is derived from a combination of seven elements that determine the magnitude of a fire and/or explosion hazard:

- Inherent tendency to ignite or process factor;
- Inventory available for release or inventory factor;
- Ability to flash to a vapor or state factor;
- Possibility of auto-ignition or auto-ignition factor;
- Effects of higher pressure operations or pressure factor;
- Engineered safeguards, safety factor;
- Escalation of equipment damage or damage potential factor.

The *process factor*, a process stream's inherent tendency to ignite, is derived as a combination of the material's flash factor and its reactivity factor. Flash factors correspond to the material's NFPA 1 Class rating, while the reactivity factor is a function of how readily the material can explode when exposed to an ignition source.

The *inventory factor* represents the largest amount of material that could reasonably be expected to be released from a unit in a single event. This factor is based on the largest mass of flammable inventory in the unit.

The *state factor* is a measure of how readily a material will flash to a vapor when it is released to the atmosphere. It is determined from a ratio of the average process temperature to the boiling temperature at atmospheric pressure (using absolute temperatures in the ratio).

The *auto-ignition factor* accounts for the increased probability of ignition for a fluid released at a temperature above its auto-ignition temperature.

The *pressure factor* is a measure of how quickly the fluid can escape. In general, liquids or gases processed at high pressure (>150 psig) are more likely to be released quickly and result in an instantaneous-type release, with more severe consequences than a continuous-type release.

The *safety factor* is determined to account for the safety features engineered into the unit. These safety features can play a significant role in reducing the consequences of a potentially catastrophic release. Several aspects of unit design and operation are included in this factor:

- Gas detection capabilities
- Fire-fighting systems
- Isolation capabilities
- Blast protection

The *damage potential factor* determines the potential for a fire or explosion to cause adjacent equipment damage. This is accomplished by a rough estimate of the value of equipment near large inventories of flammable or explosive materials.

The flammable COF category is determined by combining the already mentioned consequence factors and selecting the category based on ranges of these combined factors.

***48.2.4.4 Health COF Category*** The health consequence category is derived from the following elements that are combined to express the degree of a potential toxic hazard in a unit:

- Toxicity or toxic factor;
- Ability to disperse under typical conditions or dispersibility factor;
- Detection and mitigation systems, mitigation factor;
- Population in vicinity of release or population factor.

Toxic consequences depend heavily upon the percentage of the process fluid that is toxic. Highly toxic process streams containing a portion of highly toxic components use the same key variables as for flammable COF in addition to an estimate of toxic material based on the percentage of the toxic component in the process stream.

The *toxic factor* is a measure of the quantity and the toxicity of a material. The quantity portion is based on mass and the toxicity of the material is found using the NFPA toxicity factor.

The *dispersibility factor* is a measure of the ability of a material to disperse. It is determined directly from the normal boiling point of the material. The higher the boiling point, the less likely a material is to disperse.

A *mitigation factor* is determined to account for the safety features engineered into the system/equipment:

- Toxic material detection capabilities
- Isolation capabilities
- Mitigation systems

The *population factor* is a measure of the number of people that can potentially be affected by a toxic release event.

The health COF category is determined by combining the COF factors and selecting the category based on ranges of these combined factors.

***48.2.4.5 Business COF Category*** The business COF category should be assigned by a simple three-category assessment of production impact, the ability to make-up product from another source and the market value of the product.

***48.2.4.6 Final COF Category*** The COF categories (flammable, toxic, and business) are assigned letter scores. The final COF category is the highest value plotted on the horizontal axis of the risk matrix to develop a risk rating for the unit.

***48.2.4.7 Qualitative Risk Results*** The POF category rating and the highest rating from damage, toxic or business consequence categories are used to place each unit within a five-by-five risk matrix, shown as Figure 48.1. The populated matrix provides a graphic indication of risk for the systems/equipment evaluated. The system/equipment with the highest risk is a candidate for further evaluation and urgency of the evaluation.

Risk matrix results are used to locate areas of potential concern, suggest an inspection frequency and determine where additional inspection or other methods of risk reduction are necessary. The results also are used to decide whether a more quantitative study would be beneficial. It should be noted that approaching risk management and reduction by using qualitative analysis can result in higher than necessary inspection and maintenance costs, since emphasis would be placed on high-risk items only (increased inspection and maintenance activities) and not emphasize lower risk equipment (reduced inspection and maintenance activities).

The shadings provided in Figure 48.1 matrix example are guidelines for determining the degree of potential risk. These shadings are not symmetrical, as they are based on the assumption that the COF will be more heavily weighted than POF in determining total risk. It is recommended that companies develop and apply a criteria to determine when further analysis, adjustments their inspection practices or mitigation is required.

### 48.2.5 Quantitative RBI Analysis

Quantitative models are model-based approaches where numerical values are calculated. The advantage of a quantitative approach is that the results can be used to calculate with some precision when the risk acceptance limit is reached. Quantitative methods are systematic, consistent, documented, and are relatively easy to update with inspection results. A quantitative approach generally uses a program or software to calculate risk and develop inspection program recommendations. The models are initially data-intensive but use of models removes repetitive, detailed work from the traditional inspection planning process. The types of data required for quantitative RBI analyses are represented in the flow diagram in Figure 48.4. API RP 581 [2] outlines a thoroughly documented quantitative approach widely used and generally accepted in the oil and gas and petrochemical industries. API RP 581 [2] is equally applicable to offshore, midstream, and pipeline industries if a quantitative methodology is desired and supported by the data requirements. The following description of quantitative RBI analysis generally follows the methodology outlined in API RP 581 [2].

***48.2.5.1 Probability of Failure Analysis*** The POF analysis is based on using industry generic failure frequencies, ($G_{ff}$) by equipment type that is modified by two factors, reflecting identifiable differences from a generic value to an equipment-specific probability of the equipment/component being studied. A calculated equipment/component equipment factor ($F_{Equip}$) reflects the specific operation, potential in-service damage mechanisms and condition based on inspection. The $F_{Equip}$ is a combination of a DF modified by an inspection effectiveness factor (IF). The DF is a factor calculated based on the potential in-service damage mechanisms for the equipment. The inspection IF modifies the DF based on the history of inspection, the effectiveness of those inspections and the result/findings. The management systems factor, ($F_{MS}$) is based on an evaluation of the facility's management practices that affect the mechanical integrity of the equipment. The management systems evaluation tool was based on API 750 guidelines and is included as a workbook of audit questions in the API RP 581 [2].

The adjusted failure frequency or POF is calculated by multiplying the $G_{ff}$ by the two modification factors, as shown in the following equation:

$$\text{Frequency adjusted(POF)} = G_{ff} \times \text{DF} \times F_{MS}$$

The database of $G_{ff}$ was based on a compilation of available equipment failure histories from multiple industries. From these data, $G_{ff}$ were developed for each type of equipment and pipe diameter.

The DF evaluates the specific operating environment of the equipment/component and determines a factor unique to the equipment operating history and inspection condition. DF calculation includes a series of damage modules that assess the effect of specific failure mechanisms on the POF. The DF for thinning is determined by calculating an *ar/t factor*,

where

$a$ = age

$r$ = corrosion rate

$t$ = furnished thickness

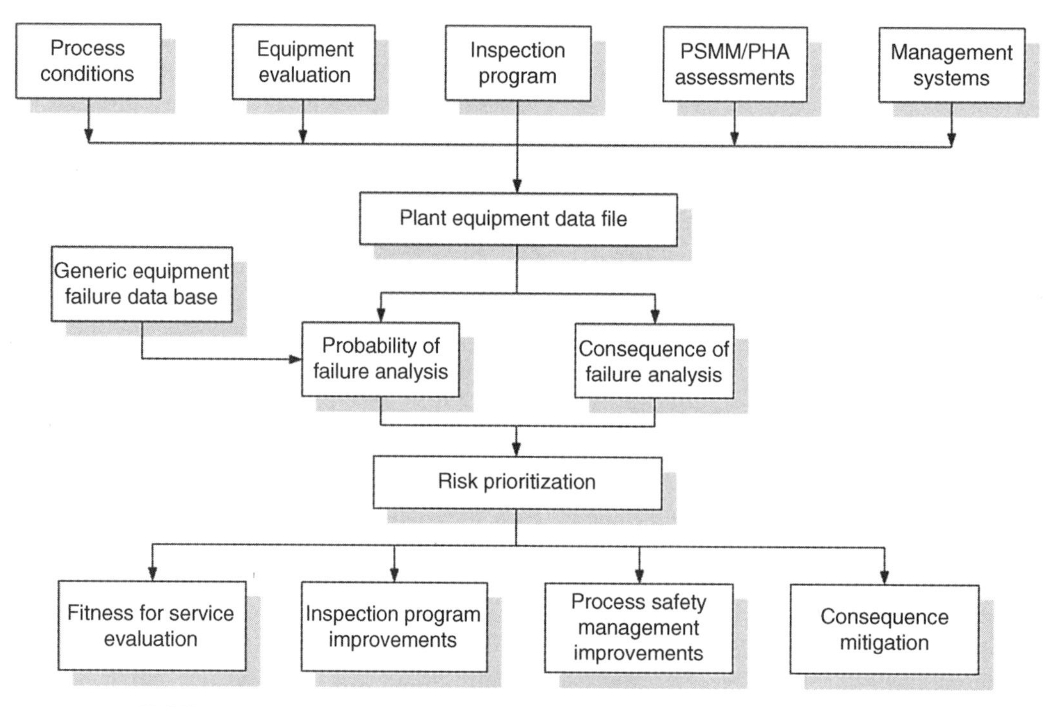


**FIGURE 48.4** Type of data required for quantitative RBI analysis flow diagram.

The *ar/t factor* is defined as the ratio of wall loss (years in service age, *a*; times damage rate, *r*) and furnished thickness, *t*. The *ar/t factor* is used with inspection effectiveness to determine a thinning DF using the table in API RP 581 [2] Part 2, Table 5.11. A similar table is provided for cracking mechanisms that can be found in API RP 581 [2] Part 2, Table 7.4. The damage modules serve the following functions:

- Screen the operation to identify the active damage mechanisms.
- Establish a damage rate in the environment and DF associated with damage, DF.
- Quantify the effectiveness of the inspection program, IF.
- Calculate the $F_{equip}$ modification factor to apply to the generic failure frequency.

The $F_{MS}$ adjusts for the influence of the facility's process safety management system on the mechanical integrity of the plant. The outline of the Management Systems Workbook from API RP 581 [2] is shown in Table 48.1. The complete audit of the workbook can be found in API RP 581 [2] Part 2, Table 4.4. The $F_{MS}$ adjustment is applied equally to all equipment/component items in the study. The $F_{MS}$ provides discrimination between risk of equipment/components under different management practices or at different locations and management systems. However, the evaluation process can be used to improve the effectiveness of any PSM program, thereby reducing overall risk.

*48.2.5.2 Consequence of Failure Analysis* COF of releasing a hazardous material are estimated in the following steps:

1. Estimate the release rate or the total mass available for release.
2. Determine if the fluid is dispersed in a rapid manner (instantaneous) or slowly (continuous).
3. Determine if the fluid disperses in the atmosphere as a liquid or a gas.
4. Estimate the impacts of any mitigation system.
5. Estimate the consequences.

**TABLE 48.1 Management Systems Workbook, $F_{MS}$**

| Title |
|---|
| Leadership and administration |
| Process safety information |
| Process hazard analysis |
| Management of change |
| Operating procedures |
| Safe work practices |
| Training |
| Mechanical integrity |
| Pre-start-up safety review |
| Emergency response |
| Incident investigation |
| Contractors |
| Audits |

Source: Adapted from Ref. [2].

Flammable, toxic, environmental, and financial impacts are calculated. Environmental consequence is calculated from the release rate or mass information. Financial consequences are calculated using the results of a combination of flammable, toxic, environmental, and business interruption.

***48.2.5.3 Estimating the Release Rate*** RBI considers releases as instantaneous or continuous release types. Instantaneous releases result in emptying the vessel contents in a relatively short period of time, for example in the case of brittle failure of a vessel. Continuous releases occur over a longer period of time at a relatively constant rate.

Note that the approach is simplified by modeling only instantaneous or continuous releases since many releases have characteristics of both.

***48.2.5.4 Predicting the Type of Outcome*** The outcome of a release refers to the physical behavior of the hazardous material. Examples of outcomes are safe dispersion, explosion, or jet fire. Outcomes should not be confused with consequences that refer to the adverse effects on people, equipment, and the environment as a result of the outcome. The actual outcome of a release depends on the nature and properties of the material released. A brief discussion of possible outcomes for various types of events is as follows.

*Flammable effects*: The following possible outcomes may result from the release of a flammable fluid:

- Safe dispersion occurs when flammable fluid is released and then disperses without ignition. The fluid disperses to concentrations below its flammable limits before it encounters a source of ignition. Although no flammable outcome occurs, it is still possible that the release of a flammable material (primarily liquids) could cause adverse environmental effects. Environmental events are addressed separately.
- Jet fires result when a high-momentum gas, liquid, or two-phase release is ignited. Radiation levels are generally high close to the jet. If a released material is not ignited immediately, a flammable plume or cloud may develop. Upon ignition, the released material will flash or burn back to form a jet fire.
- Explosions occur under certain conditions when a flame front travels very quickly. Explosions cause damage by the overpressure wave that is generated by the flame front.
- Flash fires occur when a cloud of material burns under conditions that do not generate significant overpressure. Consequences from a flash fire are only significant within the perimeter and near the burning cloud. Flash fires do not cause overpressures high enough to damage equipment.
- A fireball occurs when a large quantity of fuel ignites after it has undergone only limited mixing with the surrounding air. Thermal effects from the fireball extend well beyond the boundaries of the fireball, but they are usually short lived.
- Pool fires are caused when liquid pools of flammable materials ignite. The effects of thermal radiation are limited to a region surrounding the pool itself.

*Toxic effects*: Two outcomes are possible when a toxic material is released: safe dispersal or manifestation of toxic effects. In order for a toxic effect to occur, two conditions must be met:

- The release must reach people in a sufficient concentration.
- It must linger long enough for the effects to become harmful.

If either of the earlier conditions does not occur, the release of the toxic material results in safe dispersal. Safe dispersal is used in risk assessment to indicate that the incident falls below the pass/fail threshold. If both of the earlier conditions exist (concentration and duration) and personnel are present, toxic exposure will occur.

***48.2.5.5 Applying Effect Models to Estimate Flammable and Toxic Consequences*** RBI uses two distinct types of impact criteria to estimate consequences from a given outcome: the direct effect model and the probit (short for probability unit).

The direct effect model uses a pass/fail approach to predict the consequence from a given outcome. It assumes that no effect is observed if the outcome is below the predefined threshold and it assumes a single effect for any outcome above the threshold. This approach is fairly coarse since, in reality, spectrums of effects are observed for a range of outcomes. The probit is a statistical method of assessing the impact of a given consequence. It reflects a generalized time-dependent relationship for a variable that has a probabilistic outcome described by the normal distribution.

The initial steps in the consequence calculation predict the outcome in terms of the physical phenomena of a process release to the atmosphere. The final step converts the outcomes to consequences. Effect models, also known as impact criteria, are used to estimate consequences from an outcome. Direct effect models are used for flammable, environmental, and financial consequences, while toxic consequences are estimated using the probit.

*Environmental effects*: From an environmental standpoint, safe dispersal occurs if the released material is entirely contained within the physical boundaries of the facility. If the material cannot be contained, the release of a hazardous material will result in a spill.

*Financial effects*: Financial effects are calculated as the sum of the following consequences:

1. Equipment damage as a result of a failure and flammable effect, causing damage to adjacent equipment.
2. Injury costs as a result of the flammable or toxic effect, whichever is larger, based on affected area and population.
3. Environmental costs as a result of an uncontained spill.
4. Business interruption cost/day and outage duration for repair of holes of various sizes for each equipment type and replacement of damaged equipment adjacent to the failure.

***48.2.5.6 Quantitative Risk Results*** API RP 581 [2] outlines an industry-accepted quantitative methodology for prioritizing equipment risk at the equipment/component level. The result of the quantitative analysis positions equipment/components in a five-by-five risk matrix in addition to calculating discrete risk values for prioritization from higher to lower risk.

The POF and COF are combined to produce an estimate of risk for each equipment/component item. The equipment/components are ranked based on risk with POF, COF, and risk calculated and reported separately to aid identification of major contributors to risk or risk drivers.

Guidelines are provided to develop and modify an inspection program to manage the risks identified in the risk calculation and prioritization steps. Inspection program development will be presented in Section 48.3.

## 48.3 DEVELOPMENT OF INSPECTION PROGRAMS

### 48.3.1 Introduction

High-risk assets have the potential to cause serious disruption in operations or cause unplanned and extended production outages. The priority treatment of high-risk assets must be balanced by providing sufficient maintenance and inspection for other lower-risk equipment to avoid the unintended increased risk over time. The objective is to optimize the use of pipeline maintenance and inspection resources for all equipment/components over time to manage risk at the desired levels.

In any system or process, a relatively large percentage of risk in an operating plant is associated with a small percentage of the equipment items. RBI shifts inspection and maintenance resources to provide a higher level of coverage on the high-risk items and an appropriate effort on lower risk equipment, by systematically combining both POF and COF to establish a prioritized list of equipment/components based on total risk. From the prioritized list, an inspection program may be designed that manages (reduces or maintains) the risk of equipment failure. The potential benefit of an RBI program is to improve the operating efficiencies of systems or processes, while improving, or at least maintaining, the same level of risk.

Figures 48.5–48.7 show the impact of an RBI assessment on the inspection program. Figure 48.5 presents a comparison of the risk reduction relative to increased inspection activity associated with a typical, non-RBI program and risk using RBI. Both plans will decrease risk as the inspection activity increases to a point where the rate of risk reduction with additional activity is low. However the RBI based program achieves an overall lower level of risk with the same inspection as a result of optimization of resources based on risk. It should be noted that a certain level of residual risk is inherent in any process involving hazardous materials and additional inspection will not reduce the risk below that residual risk.

Figure 48.6 shows how refocusing inspection activities on the highest risk equipment achieve a lower risk. The lower risk is achieved in this case without additional costs to reduce

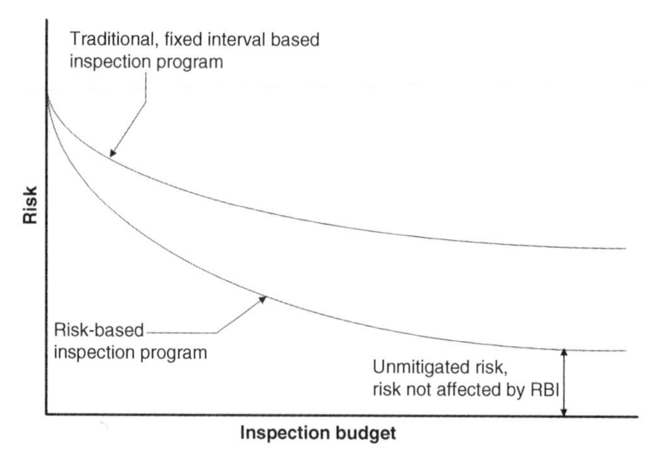


FIGURE 48.5 Impact of RBI assessment on inspection program.

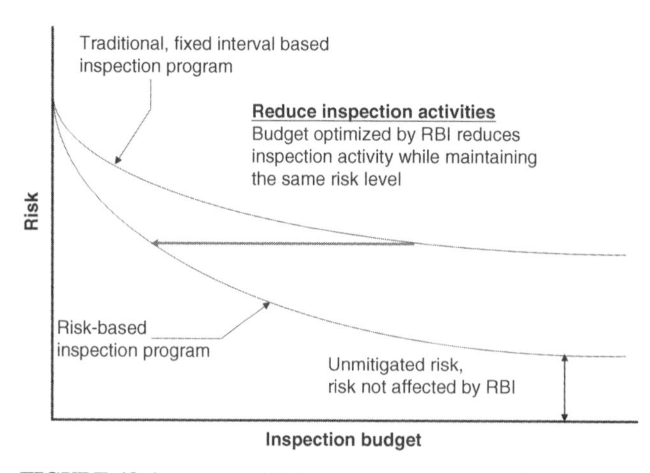


FIGURE 48.6 Impact of RBI assessment on inspection program.

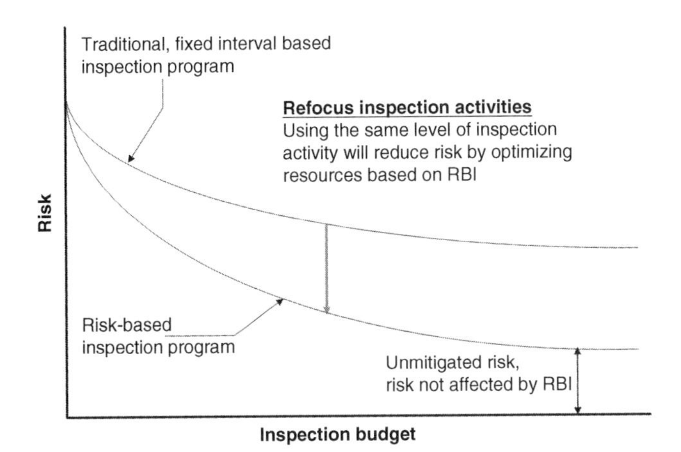


**FIGURE 48.7** Impact of RBI assessment on inspection program.

the risk. Figure 48.7 demonstrates how unnecessary or low-value inspection activities can be reduced while maintaining the same risk level when inspection priorities are based on risk.

Risk reduction or risk optimization is achieved several ways, including the following:

- Change the process, operations, and/or inventories.
- Replace equipment with other equipment that is desirable for the intended service.
- Improve management systems, training, workloads, and so on.
- Improve inspection programs.

RBI focuses on risk reduction through inspection planning by development of inspection programs that address the types of damage expected and the inspection methods appropriate to detect the expected damage.

Inspection influences risk by reducing the POF through improved knowledge of the equipment condition. Many conditions (design errors, fabrication flaws, malfunction of control devices) can lead to equipment failure, but in-service inspection is primarily concerned with the detection of progressive damage. The POF due to such damage is a function of the following factors:

- Damage mechanism and resulting type of damage (cracking, thinning, mechanical damage, etc.);
- Damage rate;
- Probability of detecting damage and predicting future damage states with inspection technique(s);
- Tolerance of the equipment to damage type.

RBI differs from conventional inspection management by providing the methods to support decision making when data is missing or uncertain. Rational decisions can be made using RBI methodologies and the application of decision making applies risk reduction through inspection.

### 48.3.2 Inspection Techniques and Effectiveness

The purpose of an inspection program is to define and perform the activities necessary to detect in-service deterioration of equipment before failures occur. An effective inspection program is developed by systematically identifying

- What types of damage are possible?
- What equipment is susceptible and what locations to look for damage?
- What inspection methods are effective to detect and measure damage?
- What frequency of inspection is required to properly detect and measure the damage?

Equipment data that must be available to properly assess the risk include equipment design and construction, the operating process conditions and the maintenance and inspection history. The following basic data are required to identify damage mechanisms of interest:

Design and construction data

- Equipment type (heat, mass, or momentum transfer) and function (shell and tube exchanger, trayed distillation column, centrifugal pump, etc.);
- Material of construction (ASTM specification, grade, and year);
- Heat treatment;
- Furnished thickness.

Operating process data

- Temperature
- Pressure
- Liquid level, %
- Corrosive components of process (such as $H_2S$, $CO_2$, amine, water, organic chlorides, etc.)
- Flow rate
- Solids contents
- Equipment history:
- Previous inspection data
- Failure analysis
- Maintenance activity
- Replacement/repair information
- Modifications

### 48.3.3 Damage Types

Damage types are the physical characteristics of damage that can be detected by an inspection technique. Damage

**TABLE 48.2 Pipeline Damage Types and Characteristics**

| Damage Type | Description |
|---|---|
| Thinning, general | Uniform loss of material thickness internally and/or externally |
| Thinning, localized and pitting | Nonuniform loss of material thickness internally and/or externally |
| Cracks connected to equipment surface | Cracking at material surface internally and/or externally |
| Cracking not connected to equipment surface | Cracking in material subsurface, not broken to surface internally and/or externally |
| Microfissuring/microvoid formation | Small fissures or voids in material subsurface |
| Metallurgical changes | Changes to the metallurgical microstructure |
| Dimensional changes | Physical dimension changes |
| Blistering | Subsurface lamellar blisters at plate inclusions, typically caused by hydrogen diffusion into material |
| Material properties changes | Changes to the material mechanical properties |

mechanisms are the corrosion or mechanical actions that produce the damage. Table 48.2 describes some familiar damage types and their characteristics. Tables 48.3–48.6 show damage mechanisms by broad categories and the types of damage associated with the mechanisms.

Damage may occur uniformly throughout a piece of equipment, or it may occur locally, depending on the mechanism at work. Uniformly occurring damage can be inspected and evaluated at any convenient location, since the results can be expected to be representative of the overall condition. Damage that occurs locally requires a more focused inspection effort. This may involve inspection of a larger area to ensure that localized damage is detected. If the damage mechanism is sufficiently well understood to allow prediction of the locations where damage will occur, the inspection effort can focus on those areas.

*48.3.3.1 Inspection Methods* Inspection methods are selected based on their ability to find the damage type; however, the mechanism that caused the damage can affect the inspection technique selection. Table 48.7 qualitatively lists the effectiveness of inspection techniques for various damage types. A range of effectiveness is given for some damage type/inspection technique combinations based on comments from various sources, including the API Subcommittee on Inspection.

**TABLE 48.3 Pipeline Corrosion Damage Mechanisms**

| Damage Mechanisms |
|---|
| Amine corrosion |
| $CO_2$ corrosion |
| Organic chloride corrosion |
| Organic sulfur corrosion |
| Microbiological corrosion |
| Sour water corrosion |
| Cooling water corrosion |
| Injection point corrosion |
| Underdeposit corrosion |
| Soil corrosion |
| Atmospheric corrosion |
| CUI |

**TABLE 48.4 Pipeline SCC Damage Mechanisms**

| Cracking Mechanism |
|---|
| Amine cracking |
| Sulfide stress cracking |
| HIC/stress-oriented hydrogen-induced cracking (SOHIC) |

**TABLE 48.5 Hydrogen Environmental Damage**

| Damage Mechanism | Damage Types |
|---|---|
| Sulfide stress cracking | Surface initiated cracking |
| Hydrogen-induced cracking | Surface and subsurface initiated cracking |
| Blistering | Subsurface lamellar blisters at plate inclusions |
| SOHIC | Surface and/or subsurface cracking, cracks can be surface connected |

**TABLE 48.6 Pipeline Mechanical Damage Mechanisms**

| Damage Mechanism | Damage Types |
|---|---|
| Erosion—solids | Thinning, localized |
| Erosion—droplets | Thinning, localized |
| Cavitation | Thinning, localized |
| Sliding wear | Thinning, general or localized |
| Fatigue | Surface or sub-surfaced initiated cracking |
| Corrosion fatigue | Surface initiated cracking |
| Overload (plastic collapse) | Thinning, localized and dimensional changes |

**TABLE 48.7 Pipeline Inspection Techniques to Detect Damage Types**

| Inspection Technique | Thinning | Surface Initiated Cracking | Subsurface Cracking | Metallurgical Changes | Dimensional Changes | Blistering |
|---|---|---|---|---|---|---|
| Visual examination | H, M, L | M, L | X | X | H, M, L | H, M, L |
| Ultrasonic pigging or manual | H, M, L | L, X | L, X | X | X | H, M |
| Ultrasonic shear wave | X | H, M | H, M | X | X | X |
| Fluorescent magnetic particle | X | H, M | L, X | X | X | X |
| Dye Penetrant | X | H, M, L | X | X | X | X |
| Acoustic emission | X | H, M, L | H, M, L | X | X | L, X |
| Eddy current | H, M | H, M | H, M | X | X | X |
| Flux leakage | H, M | X | X | X | X | X |
| Radiography | H, M, L | L, X | L, X | X | H, M | X |
| Dimensional measurements | H, M, L | X | X | X | H, M | X |
| Metallography | X | M, L | M, L | H, M | X | X |

H = high, M = medium L = low, X = ineffective.

Selection of the inspection technique will depend on not only the effectiveness of the method, but on equipment availability and access to the external or internal surfaces.

***48.3.3.2 Inspection Frequency/Interval*** Inspection frequency/interval is determined by combining the factors of RBI presented in previous sections:

1. Damage mechanism and resulting type of damage (cracking, thinning, etc.);
2. Rate of damage progression;
3. Tolerance of the equipment to the type of damage;
4. Probability of detecting damage and predicting future damage states with inspection technique(s).

Traditional inspection programs perform inspection at recommended frequencies, often selected based on some fraction of the equipment's RL defined as:

$$\text{Remaining life(years)} = \text{damage tolerance(units)} \\ (/\text{damage rate(units/year)}$$

Inspection planning using risk of loss of containment provides a tool to conduct inspection based on risk targets rather than fixed frequencies. The result is inspection intervals that vary through the life of the equipment depending on the in-service damage potential and observed rate of damage. A risk criteria defining acceptable risk based on prudent levels of safety, environmental, and financial risks must be developed for RBI decision-making. Risk acceptance may vary for different risk exposures. For example, risk tolerance for a financial risk may be higher than for a safety/health risk.

***48.3.3.3 Inspection Effectiveness*** As shown in Table 48.7, note that no inspection technique is always considered to be highly effective for all damage types and more than one inspection technique may be used for each damage type, in some cases complimentary. For example, ultrasonic thickness measurements are much more effective at locating internal corrosion if combined with an internal visual inspection.

RBI requires a quantitative evaluation of inspection effectiveness to calculate a $F_{equip}$. The following shows how the estimate of inspection effectiveness is developed.

***48.3.3.4 Qualitative Assessment of Inspection Effectiveness*** As analysis advances from observing the effectiveness of inspection techniques to quantifying the effectiveness of an inspection plan, the following factors must be evaluated:

- Damage density and variability;
- Inspection sample validity;
- Sample size;
- Detection capability of the inspection methods;
- Validity of future predictions based on past observations.

A rigorously quantitative approach would require a probabilistic description of each of these factors, permitting the inspection effectiveness to be presented as a probabilistic expression. Such an approach is costly and overly complicated for most RBI approaches.

Instead a method was developed to assess inspection effectiveness by categorizing the ability of inspection types, or common combinations of inspection types, to detect and evaluate in-service damage. An example is the combination of visual and ultrasonic inspections for the detection and measurement of general corrosion. The effectiveness categories are based on evaluating the factors identified previously. Inspection is categorized according to the ability to detect and quantify the anticipated progressive damage. The inspection effectiveness categories are described in more detail in Table 48.9:

- Highly effective (A)
- Usually effective (B)
- Fairly effective (C)
- Poorly effective (D)
- Ineffective (E)

Since a rigorously quantitative approach is not realistic, RBI relies on professional judgment and expert opinion used in conjunction with general guidance. Table 48.8 shows the levels of inspection effectiveness used in API RP 581 [2] considered in either the RBI or rigorously quantitative approach. Examples of inspection for each level of effectiveness are provided for each damage type in API RP 581 [2].

Inspection effectiveness is qualitatively evaluated by assigning the inspection methods to one of the descriptive categories listed in Table 48.9. Assignment of categories is based on professional judgment and expert opinion. Examples of each category are provided for each damage type in API RP 581 [2].

***48.3.3.5 Quantifying Inspection Effectiveness*** In order to quantitatively express the impact of inspection on the POF, it is necessary to develop a method to convert the qualitative categories in Table 48.9 into quantitative measures of inspection effectiveness. The goal is to express how effective the inspection is at correctly identifying the state of damage in the equipment examined. The handling of inspection effectiveness was constructed using Bayes' Theorem to credit inspection. Further details for the background and

**TABLE 48.8 Factors Considered in Assessing Inspection Effectiveness**

| Evaluation Factor | Rigorously Quantitative Approach | RBI Approach | Example of RBI Approach for General Corrosion |
|---|---|---|---|
| Damage density and variability | Density, mean, and extreme value distributions of damage | 1. Damage occurs over either a large or small area<br>2. Damage can occur randomly, or locations for damage can be predicted | General corrosion occurs over a substantial portion of the surface area and is relatively uniform |
| Validity of sample | Sample must be representative of the population about which a statistical inference is to be made | The inspection program is designed to concentrate on areas where damage is likely to occur | For general corrosion, most samples will be representative of the condition; however, the rate of corrosion may vary significantly within one piece of equipment |
| Sample size | Sample size must be statistically significant | The area inspected should be appropriate for damage mechanisms that are highly localized | Visual examination, combined with ultrasonics, increases the significance of the sample versus spot ultrasonic thickness readings alone |
| Detection capability | Probability of detection curves describe the capability of the method | The capability of the inspection type is evaluated qualitatively | Visual examination is rated "possibly" to "highly" effective; ultrasonic thickness measurements are rated "moderately" to "highly" effective |
| Validity of future predictions based on past observations | Damage is modeled showing rate variation with time, and so on | The past observation of damage is used to predict the future, based on increase or decrease in damage rate, or based on changes to process parameters affecting damage | General corrosion is assumed to occur at the same rate in the future as in the past, unless the process changes (feedstock, temperature, etc.) |

**TABLE 48.9 The Five Effectiveness Categories**

| Qualitative Inspection Effectiveness Category | Description of Effectiveness |
|---|---|
| Highly effective (A) | Inspection methods correctly identify the anticipated in-service damage in nearly every case (80–100%) |
| Usually effective (B) | The inspection methods will correctly identify the actual damage state most of the time (60–80%) |
| Fairly effective (C) | The inspection methods will correctly identify the true damage state about half of the time (40–60%) |
| Poorly effective (D) | The inspection methods will provide little information to correctly identify the true damage state (33–40%) |
| Ineffective (E) | The inspection method will provide no or almost no information that will correctly identify the true damage state (33%) |

construction of the thinning DF table can be found in API RP 581 [2] and other referenced publications.

### 48.3.4 Probability of Detection

Inspection techniques vary in their accuracy, depending on operator skill and test conditions. Accuracy can be measured by repeating the examinations of known flaws of various sizes and recording the results. The application of these data to in-service plant inspections is somewhat limited since they are based on examination of prepared test blocks in a comfortable laboratory environment. However, they provide two very important pieces of information:

1. They establish that there is a probability of detection (POD). Even under controlled conditions, nondestructive testing has limitations and reveals an increased probability of detection for flaws as the size of the flaws increase.
2. They establish the basis for the maximum effectiveness that can be expected from an inspection. "Real world" probabilities of detection may approach this effectiveness, but could not be expected to exceed it.

Several organizations have tried to quantify the probability of detection by performing round-robin type tests:

- Nordtest in Europe (ultrasonics [UT], MPI [MT], and radiography [RT]);
- PISC (Italy)
- EPRI (USA);
- CIPS in USA (UT-nuclear applications);
- Nippon Steel Japan (MPI);
- US Navy (UT and radiography submarines).

As discussed in the previous section on inspection effectiveness, POD data can be used in a quantitative assessment of inspection effectiveness through extreme value analysis if sufficiently detailed information on the distribution of damage states is also available.

The POD data, where available, are also useful in helping assign the inspection methods to the appropriate inspection effectiveness category.

### 48.3.5 Reducing Risk through Inspection

Presenting the RBI results in a risk matrix is a very effective way of communicating the distribution of risks throughout a system or process without showing numerical values. An example of risk matrix is shown in Figure 48.1, different sizes of matrices may be used (e.g., $5 \times 5$, $4 \times 4$, etc.). Risk categories may be assigned to the boxes on the risk matrix. The example in Figure 48.1 shows risk categorization of high, medium–high, medium, and low. A risk matrix depicts risk results at one particular point in time.

Another method of showing risk results for quantitative methodologies and when showing numeric risk values is meaningful, an iso-risk plot is used, as shown in Figure 48.2. An iso-risk plot shows POF and COF for equipment plotted on a graph with an iso-risk line (line of constant risk). If this line is the acceptable level of risk, the equipment above the line fail the established risk acceptance criteria and should be mitigated to a value below the line.

The most common presentation of the RBI results is to rank the equipment/components by risk (from high to low) in a tabular format.

Risk matrix and iso-risk plots graphically show that the equipment/components toward the upper right-hand corner of the matrix or plot have a higher priority for inspection planning because these items have the highest risk. Similarly, items residing toward the lower left-hand corner of the matrix or plot will have a lower priority. The risk matrix or plot can be used as a screening tool during the inspection prioritization process.

Established risk thresholds that divide the risk matrix, plot or prioritized risk table into acceptable and unacceptable regions of risk can be developed using corporate safety and financial policies. Inspection is conducted on or before the date that equipment reaches the risk target. As damage progresses, inspection is performed to increase confidence in the current and expected future condition of equipment. Increased confidence reduces risk, but never as low as the original risk level due to the damaged state. When inspection no longer effectively lowers the risk below the risk target, other risk mitigation activities are required to manage the risk such as FFS evaluations or plans for the equipment repair/replacement.

Regulations and laws may also specify or assist in identifying the acceptable risk thresholds. Reduction of some risks to a lower level may not be practical due to technology and cost constraints. An ALARP approach to risk management or other risk management approach may be necessary for these items.

Obviously, inspection does not mitigate damage mechanisms or in and of itself does not reduce risk. However, the knowledge of the equipment/component condition gained through effective inspection better quantifies the actual risk. Inspection does not prevent an impending failure unless the inspection initiates a risk mitigation activity that changes rate of damage and POF. Inspection serves to identify, monitor, and measure the damage rate to more accurately calculate the remaining useful life of equipment. Correct application of inspection methods improves the ability to predict the future condition of the equipment. The better the predictability, the less uncertainty there will be as to when a failure may occur. Risk mitigation activities (repair, replacement, changes, etc.) are planned and implemented prior to the predicted failure date. The reduction in uncertainty and increase in predictability through inspection translate directly into a better estimate of

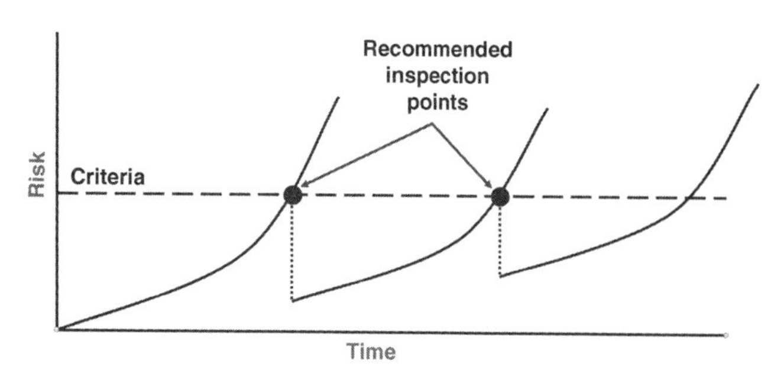


FIGURE 48.8 Equipment risk profile over time using RBI principles.

the POF and, therefore, reduce risk. Figure 48.8 shows the mitigation of risk over time that is possible by using a well-planned inspection program (Figure 48.9).

Inspection is not the only risk mitigation activity reducing risks associated. Some damage mechanisms are very difficult or impossible to monitor with inspection activities (e.g., metallurgical deterioration that may result in brittle fracture, some types of stress corrosion cracking (SCC) and even fatigue). Other damage caused by mechanical damage is event-driven and are not impacted by inspection.

Risk mitigation (by the reduction in uncertainty) achieved through inspection presumes that the organization will act on the results of the inspection in a timely manner. Risk mitigation is not achieved if gathered inspection data is not properly analyzed and acted upon. The quality of the inspection data and the analysis or interpretation significantly affects the level of risk mitigation. Proper inspection methods and data analysis tools are therefore critical.

In any system or process facility, a relatively large percentage of risk in an operating plant is associated with a small percentage of the equipment items. RBI permits the shift of inspection and maintenance resources to provide a higher level of coverage on the high-risk items and an appropriate effort on lower risk equipment, by systematically combining

FIGURE 48.9 Risk management and inspection planning.

both likelihood and COF to establish a prioritized list of pressure equipment based on total risk. From that list, users can design an inspection program that manages (reduces or maintains) the risk of equipment failure. The potential benefit of a RBI program is to increase operating times and run lengths of process facilities, while improving, or at least maintaining, the same level of risk.

## 48.4 PUTTING RBI INTO PRACTICE

### 48.4.1 A Continuum of Approach

Traditional inspection codes give prescriptive or time-based requirements for inspection recommendations, focusing on the POF over time, that is, inspection would be in a defined number of years based on estimated RL or some other criteria. Alternatively, risk-based methodology considers both the probability and COF to plan inspections, providing a basis for making informed decisions on inspection frequency and the extent of inspection.

A qualitative RBI approach uses data analysis as well as expert local knowledge and judgment to prioritize inspection, while quantitative analysis is based on a probabilistic or statistical model that calculates risk over time as the potential of in-service damage increases. Quantitative RBI requires large amounts of up-to-date and accurate data that are often not available, whereas, qualitative RBI analysis requires much less data to implement. Because it relies on less data and a simpler assessment methodology, qualitative RBI provides a more conservative analysis with greater inspection frequency. However, a qualitative program is fully compliant with API 580, the recommended practice for implementing RBI for the fixed equipment and pipeline industry, and can prepare the operator for a more in-depth quantitative analysis.

In most facilities, a large percent of the total risk is concentrated in a relatively small percent of the equipment, and these potential high-risk areas may require greater attention through a revised inspection plan. A qualitative program can help identify areas of higher risk and potentially destabilizing conditions for which a more focused quantitative approach is necessary. Thus, both qualitative and quantitative approaches can be used in a complimentary way to develop an efficient and reliable inspection program that continues over the life of the equipment. A qualitative program can be the first step toward a semi-quantitative or quantitative program as more detailed technical data become available.

### 48.4.2 Qualitative versus Quantitative Examples

These are the key elements to every RBI program, whether the approach is qualitative or quantitative:

1. Management system for maintaining documentation, history, and updated analysis;

2. Documented methodology for POF;
3. Documented methodology for COF;
4. Risk ranking for all systems/equipment;
5. Detailed inspection plan and other mitigating activities.

In the following examples, a low-pressure separation drum in sour gas service is assessed using both qualitative and qualitative methodologies to illustrate the difference in results between the two approaches.

**Background: A low-pressure separation drum is in sour gas service**

| | |
|---|---|
| Equipment type: | Drum |
| Material of construction: | Carbon steel |
| Fabrication date: | 8/2001 |
| Diameter: | 36 in. |
| Height: | 7 ft. |
| Furnished thickness: | 0.375 in. |
| $t_{min}$: | 0.285 in. |
| Corrosion allowance: | 0.090 |
| Insulation: | Yes |
| PWHT: | No |
| Process stream composition: | 12% $C_1$; 22% $C_2$; 24% $C_3$; 21% $C_4$; 13% $C_5$+; 2.8% $CO_2$; 5% $H_2O$; 2,000 ppm $H_2S$ |
| Phase: | Gas |
| Operating pressure: | 180 psi |
| Operating temperature: | 104 °F |
| Expected corrosion rate, internal thinning: | 10 mpy |
| Expected corrosion rate, corrosion under insulation (CUI) thinning: | 10 mpy |
| $H_2S$ damage susceptibility: | Medium |
| Inspection history: | B level 2006<br>B level 2010<br>B level 2013 |
| Next planned inspection: | B level 2017 |

### 48.4.3 Qualitative Example

A simple qualitative example will be used to demonstrate the use of RBI. No RBI was performed until 2005 and the first in-service inspection was scheduled for 2007.

*48.4.3.1 POF Category* When an RBI study and damage review was conducted in 2006, the expected corrosion rate of 10 mpy for internal corrosion was 10 mpy. Using Table 48.10, a corrosion rate of ≥10 mpy was assigned a POF category of 4. However once an inspection is performed the measured corrosion rate is found to actually be 5 mpy. Again, using Table 48.10 the POF category is adjusted to a category 3 in 2007.

**TABLE 48.10 Recommended Maximum Internal Thinning Interval Based on Corrosion Rate, Years**

| | | COF Category | | | |
|---|---|---|---|---|---|
| POF Category | Rate of Damage | 1 | 2 | 3 | 4 |
| 4 | CR ≥ 10 mpy | 9 | 5 | 3 | 2 |
| 3 | CR 5–10 mpy | 10 | 8 | 5 | 4 |
| 2 | CR 2 to ≤5 mpy | 10 | 10 | 8 | 6 |
| 1 | CR ≤ 2 mpy | 10 | 10 | 10 | 9 |

**TABLE 48.11 Recommended Maximum Intervals Based on Cracking Susceptibility**

| | | COF Category | | | |
|---|---|---|---|---|---|
| LOF Category | Cracking Susceptibility | 1 | 2 | 3 | 4 |
| 4 | High | 10 | 10 | 5 | 2 |
| 3 | Medium | 10 | 10 | 7 | 4 |
| 2 | Low | 15 | 15 | 10 | 6 |
| 1 | None | – | – | – | – |

While $H_2S$ was expected at fairly low concentrations and the vessel was not given any post-weld heat treatment (PWHT) during fabrication, no inspection for $H_2S$ blistering or cracking was planned. Wet $H_2S$ damage was identified as a potential damage mechanism during the RBI damage review so no inspection had been identified prior to the study. A susceptibility of "medium" was assigned for both blistering and cracking. Using Table 48.11 and the susceptibility of medium, the POF category for SCC is 3.

Note that POF category in qualitative RBI does not change with time unless a key input value is changed, such as corrosion rate.

*48.4.3.2 COF Category* The process material balance indicates handles primarily methane/ethane and propane/butane. Using Table 48.12, the flammable COF category is 4. The process also contains 0.2% $H_2S$, assigning a toxic category of 3. The final COF is the largest of the flammable and toxic or category 4.

While COF is considered to be constant with time, process stream characteristics, over the life of the system, may change significantly and should be reflected in the COF category assignments over time.

*48.4.3.3 Risk Results* The risk results for the qualitative example are summarized in the following table:

| Analysis | Basis | Results COF Category |
|---|---|---|
| COF | Flammable | 4 |
| | Health | 3 |
| | Final | 4 |
| | | POF Category |
| POF | Thinning, 10 pmy | 4 |
| | Thinning, 5 mpy | 2 |
| | Cracking | 3 |
| | | Risk Category |
| Risk | Thinning, 10 mpy | High |
| | Thinning, 5 mpy | Medium–high |
| | Cracking | High |
| | | Frequency |
| Inspection planning | Thinning, 10 mpy | 2 |
| | Thinning, 5 mpy | 6 |
| | Thinning, 5 mpy RL | 0.3 |
| | Cracking | 4 |

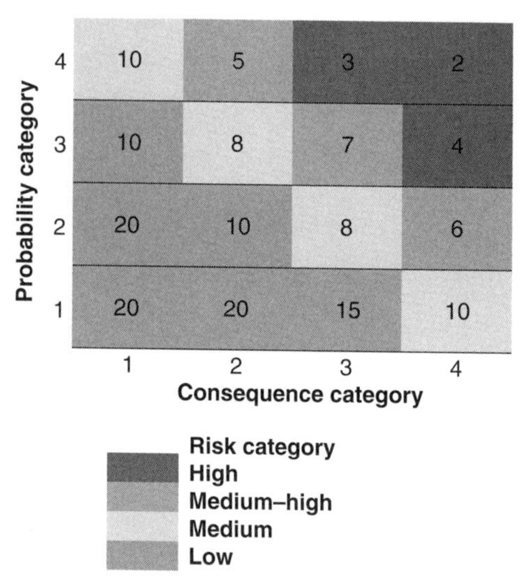


**FIGURE 48.10** Recommended risk-based inspection intervals for internal thinning.

**TABLE 48.12 Qualitative Fluid Table COF Categories**

| API Fluid Name | Applicable Materials | COF, Flammable | COF, Toxic |
|---|---|---|---|
| $C_1$–$C_2$ | Methane, ethane, ethylene, LNG, fuel gas | 4 | 0 |
| $C_3$–$C_4$ | Propane, butane, isobutane, LPG | 4 | 0 |
| $C_5$ | Pentane | 3 | 0 |
| $C_6$–$C_8$ | Gasoline, naphtha, LSR, heptane | 3 | 0 |
| $C_9$–$C_{12}$ | Jet fuel. Light kerosene | 3 | 0 |
| $C_{13}$–$C_{16}$ | Diesel, heavy kerosene, Atm gas oil | 3 | 0 |
| $C_{17}$–$C_{25}$ | Gas oil, crude | 2 | 0 |
| $C_{25+}$ | Residuum, heavy crude | 2 | 0 |
| Water | Water | 1 | 0 |
| Steam | Steam | 1 | 0 |
| Acid-LP | Low pressure acid/caustic | 3 | 0 |
| Acid-MP | Medium pressure acid/caustic | 3 | 0 |
| Acid-HP | High pressure acid/caustic | 3 | 0 |
| $H_2$ | Hydrogen only | 3 | 0 |
| $H_2S$ | Hydrogen sulfide | 3 | 4 |
| HF | Hydrofluoric acid | 0 | 4 |
| Aromatics | Benzene, toluene, xylene, cumene | 3 | 0 |
| Ammonia | Ammonia | 0 | 4 |
| HCl | Hydrochloric acid | 0 | 3 |

Notes

- $H_2S$ toxic category based on >3% in process stream. Reduce COF, toxic one category if ≤3%.
- HF toxic category based on >5% in process stream. Reduce COF, toxic one category if ≤5%.
- $C_1$–$C_2$, $C_3$–$C_4$ pressurized liquids that rapidly vaporize with atmospheric boiling temperatures below 50 °F or where the atmospheric boiling point is below the operating temperature can lead to brittle failures.
- $C_5$ and $C_6$–$C_8$ fluids should be classified as COF 4 when the operating temperature is above the atmospheric boiling point.
- Flammable services operating above their auto-ignition temperature should be assigned COF, flammable category 4.

Since the COF does not change over time in this example, the largest of the flammable and health COF category is used to set the inspection frequency for thinning and cracking mechanisms. If process conditions change, modifications can easily be made to the model to reflect those changes. Updating the expected corrosion rate of 10 mpy with an in-service measured rate reduced the POF category from 4 to 2 and results in a reduction from high to medium–high.

***48.4.3.4 Inspection Planning*** Using Figure 48.10, the change in risk category increases the inspection interval from 2 years to 6 for thinning. The recommended interval for cracking is 4 years using Figure 48.11.

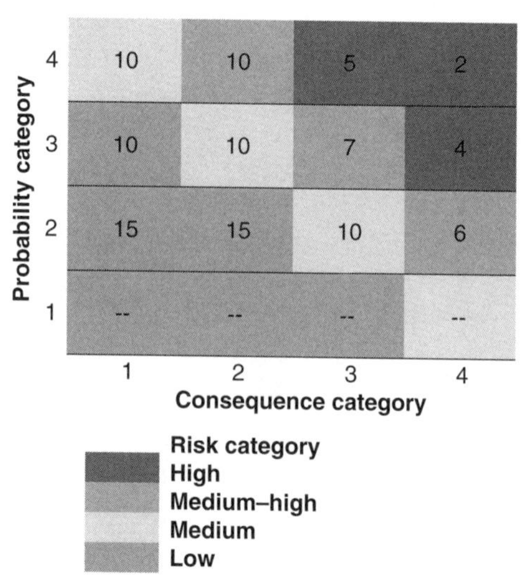


**FIGURE 48.11** Recommended risk-based inspection intervals for internal SCC.

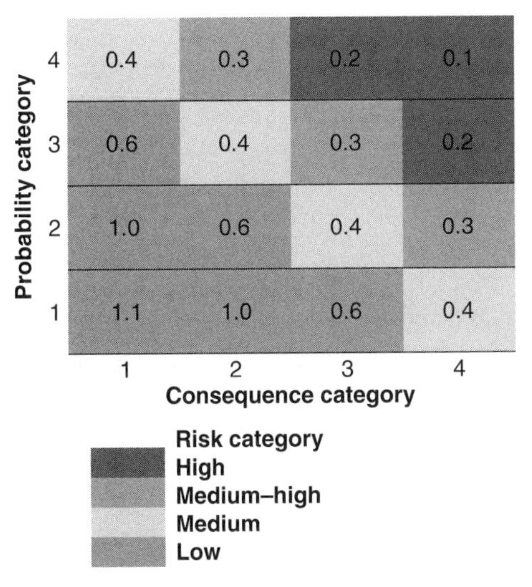


**FIGURE 48.12** Recommended risk-based inspection intervals for internal thinning as a function of remaining life fraction.

Because qualitative RBI is not quantitative enough for POF to change over time, the inspection frequency is assumed to be constant through the equipment useful life. This assumption is conservative at the start of service but becomes less conservative as the equipment reaches the end of life. As a result, the qualitative results must be based on more conservative frequencies.

Using the RL interval adjustment factor approach (Figure 48.12) adjusts the inspection interval as the equipment approaches end of life. This method requires a RL calculation based on the excess wall thickness (furnished thickness—minimum required thickness) after each inspection. The next inspection is based on the RL × *inspection factor* from Figure 48.14. In this example, the measured corrosion rate is 5 mpy with a 0.090 in. corrosion allowance at the start of service or 18 year expected life. The first inspection would be expected to occur by using the RL × 0.3 *or* 18 × 0.3 = 5.4 years. However, after 10 years of service with 5 mpy all loss, the RL is 8 years. Therefore the inspection interval at year 10 is RL × 0.3, since there is no change in

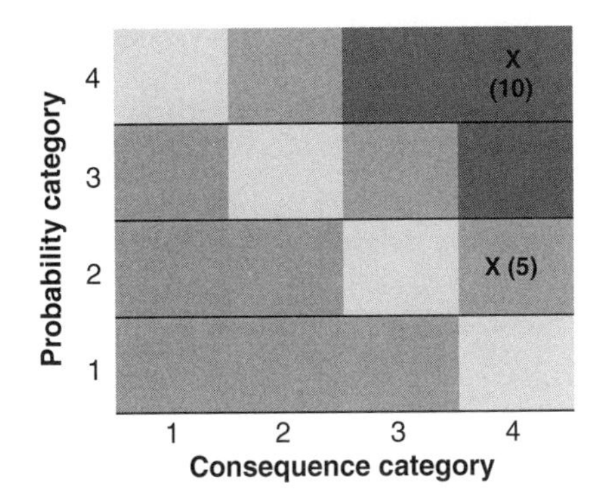


**FIGURE 48.13** Qualitative thinning inspection risk.

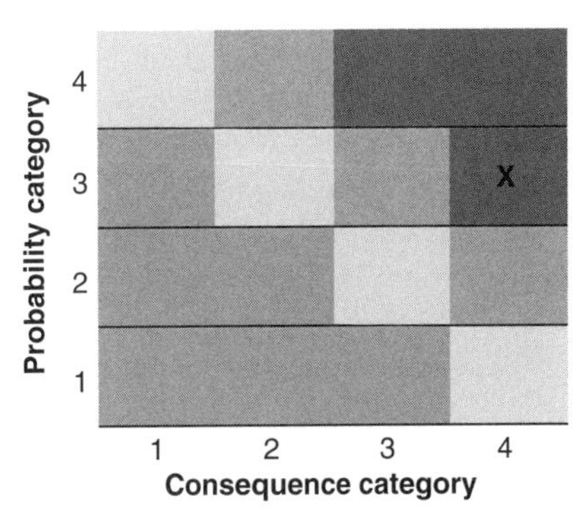


**FIGURE 48.14** Qualitative cracking inspection risk.

risk over time. But by using the RL adjustment factor approach, the new interval is 8 × 0.3 = 2.4 years. Applying the RL method for setting RBI frequency/intervals results in setting more realistic intervals over the equipment life, that is, longer intervals at the start of service and shorter intervals as the equipment approaches its end of life.

The results for thinning and cracking are shown on a risk matrix in Figures 48.13 and 48.14. Implemented correctly, a qualitative RBI study logically leads to a more conservative analysis and more inspection than a quantitative RBI analysis would suggest.

### 48.4.4 Quantitative Example

The aforementioned example will be studied using a quantitative approach to demonstrate the use of quantitative RBI. No RBI was performed until 2005 and the first in-service inspection was scheduled for 2006.

***48.4.4.1 POF Calculation*** The drum operated for 6 years with no inspection. The expected corrosion rate was 10 mpy and applied over the first 6 years of service with a starting thickness 0.375 in. The *ar/t* factor was calculated and DF determined using the inspection history. For demonstration purposes, the *ar/t* and DF's were calculated in the table before and after each B level thinning inspection assuming an inspection is performed at a 3-year frequency after the first 6 years of service.

| Age | Corrosion Rate | (*ar/t*) | DF, No Inspection | No. of B Inspections | DF | Thinning POF Category |
|---|---|---|---|---|---|---|
| 5 | 0.01 | 0.13 | 90 | 0 | 90 | 3 |
| 5 | 0.005 | 0.07 | 1 | 1 | 1 | 1 |
| 9 | 0.005 | 0.12 | 6 | 1 | 2 | 1 |
| 9 | 0.005 | 0.12 | 6 | 2 | 1 | 1 |
| 13 | 0.005 | 0.17 | 170 | 2 | 7 | 2 |
| 13 | 0.005 | 0.17 | 170 | 3 | 2 | 1 |
| 17 | 0.005 | 0.23 | 400 | 3 | 6 | 2 |
| 17 | 0.005 | 0.23 | 400 | 4 | 2 | 1 |

Similarly the cracking DF and POF must be calculated. While $H_2S$ was expected a fairly low concentrations and the vessel was not PWHT'd during fabrication, no inspection for $H_2S$ blistering or cracking was planned. Wet $H_2S$ damage was identified as a potential damage mechanism during the RBI damage review, so no inspection had been identified prior to the study. A susceptibility of medium was assigned for both blistering and cracking. A medium susceptibility has a severity index, SI, of 10. The severity index is used to determine a cracking DF with the inspection history. The DF is used to calculate the $DF_{final}$ using the following equation:

$$DF_{final} = DF \times \text{age, years since inspection}$$

| Age | Susceptibility | No. of C Inspection | DF | Cracking POF Category |
|---|---|---|---|---|
| 5 | Medium | 0 | 58.75 | 3 |
| 9 | Medium | 0 | 112.1 | 1 |
| 13 | Medium | 0 | 168.0 | 1 |
| 17 | Medium | 1 | 1 | 2 |
| 17 | Medium | 1 | 33.50 | 1 |

The final POF is the sum of the individual DFs combined with the $G_{ff}$ and $F_{MS}$ as shown in the following table:

| Year | Age | Thinning DF | Cracking DF | Combined DF | $F_{MS}$ | $G_{ff}$ | POF | POF Category |
|---|---|---|---|---|---|---|---|---|
| 2001 | 5 | 90 | 58.73 | 148.73 | 1 | 1.56E−05 | 2.32E−03 | 4 |
| 2006 | 5 | 1 | 58.73 | 59.73 | 1 | 1.56E−05 | 9.32E−04 | 3 |
| 2010 | 9 | 2 | 112.12 | 114.12 | 1 | 1.56E−05 | 1.78E−03 | 4 |
| 2010 | 9 | 1 | 112.12 | 113.12 | 1 | 1.56E−05 | 1.76E−03 | 4 |
| 2013 | 13 | 7 | 168.01 | 175.01 | 1 | 1.56E−05 | 2.73E−03 | 4 |
| 2013 | 13 | 2 | 168.01 | 170.01 | 1 | 1.56E−05 | 2.65E−03 | 4 |
| 2017 | 17 | 6 | 225.68 | 231.68 | 1 | 1.56E−05 | 3.61E−03 | 4 |
| 2017 | 17 | 2 | 1.00 | 3.00 | 1 | 1.56E−05 | 4.68E−05 | 1 |

Table 48.13 is used to determine the POF category from the combined DFs.

***48.4.4.2 COF Category*** The process material balance indicates handles primarily methane/ethane and propane/butane. Step by step calculations of COF area and category using the API 581 are beyond the scope of this discussion. The user is directed to the recommended practice for the detailed calculations. The final results of the calculations are as follows:

| Analysis | Basis | Results: COF Area | Results: COF Category |
|---|---|---|---|
| COF | Flammable, equipment | 943.66 | B |
| | Flammable, injury | 2443.84 | C |
| | Toxic | 226.84 | B |
| | Final | 2443.84 | C |

The final COF is the largest of the flammable and toxic consequences or category C as shown in Table 48.13.

While COF is considered to be constant with time, process stream characteristics can change over the life of the system may change significantly and should be reflected in the COF category assignments over time.

***48.4.4.3 Risk Results*** The risk results for the quantitative example are summarized in the following table:

| Analysis | Basis | Results: COF Area | Results: COF Category |
|---|---|---|---|
| COF | Flammable, equipment | 943.66 | B |
| | Flammable, injury | 2,443.84 | C |
| | Toxic | 226.84 | B |
| | Final | 2,443.84 | C |
| | | DF/Risk | POF Category |
| POF at 2017 | Thinning | 6 | 2 |
| | Cracking | 225.68 | 4 |
| | Total | 231.68 | 4 |
| | POF | 3.61E−03 | 4 |
| | | Risk Value, ft²/year | Risk Category |
| Risk | Total | 8.83 | 4C—high |
| Inspection Planning risk after inspection | Thinning—1B | 2 | 1 |
| | Cracking—1C | 1 | 1 |
| | Total DF | 3 | 2 |
| | POF | 4.68E−05 | 2 |
| Risk | | Risk Value, ft²/year | Risk Category |
| | | 1.14E−01 | 2C—low |

**TABLE 48.13 Example of Probability and Area-Based Consequence Categories**

| | Probability Category | | Consequence Category | |
|---|---|---|---|---|
| Category | POF Range | DF Range | Category | Range (ft²) |
| 1 | POF ≤ 1.00E − 04 | DF ≤ 1 | A | CA ≤ 100 |
| 2 | 1.00E − 04 < POF ≤ 1.00E − 03 | 1 < DF ≤ 10 | B | 100 < CA ≤ 1000 |
| 3 | 1.00E − 03 < POF ≤ 1.00E − 02 | 10 < DF ≤ 100 | C | 1000 < CA ≤ 10,000 |
| 4 | 1.00E − 02 < POF ≤ 1.00E − 01 | 100 < DF ≤ 1000 | D | 10,000 < CA ≤ 100,000 |
| 5 | POF > 1.00E − 01 | DF > 1000 | E | CA > 100,000 |

Since the COF does not change over time unless the process conditions are changed significantly, the largest of the flammable (equipment and injury) and toxic COF consequence areas are used to set the inspection frequency active damage mechanisms. If process conditions change, modifications can easily be made to the model to reflect those changes.

*48.4.4.4 Inspection Planning* As a result of implementing the recommended inspection, the risk is reduced by two POF categories as shown in Figure 48.15. The B level inspection is recommended for general thinning and a C level inspection is recommended for cracking. Reassessment projects risk into the future, predicts the future condition based on past experience and sets a next inspection date based on the target risk. Inspection effectiveness is recommended for each damage type in proportion to its contribution to risk.

Quantitative RBI analysis uses more realistic time-based degradation models, showing that POF varies with time. The result is calculated inspection intervals that vary of the life of the equipment.

Implemented correctly, a quantitative RBI study leads to less conservative results, and therefore, less inspection than qualitative RBI analysis.

### 48.4.5 Optimizing the Inspection Program

The aforementioned examples demonstrate the use of RBI to evaluate options for an inspection program and shows how inspection planning can be optimized using the following:

- Increasing activity level or frequency if insufficient reduction in risk occurs, or decreasing activity level or frequency if no gain in risk reduction results from the higher level of inspections.
- High DFs may result when an RBI program is initially implemented. Equipment showing high DF should be prioritized for inspection based on their risk and then DF/POF.
- Equipment with a history of multiple inspections and confirmed low-corrosion rates are candidates for reduced inspection activity or increased intervals.

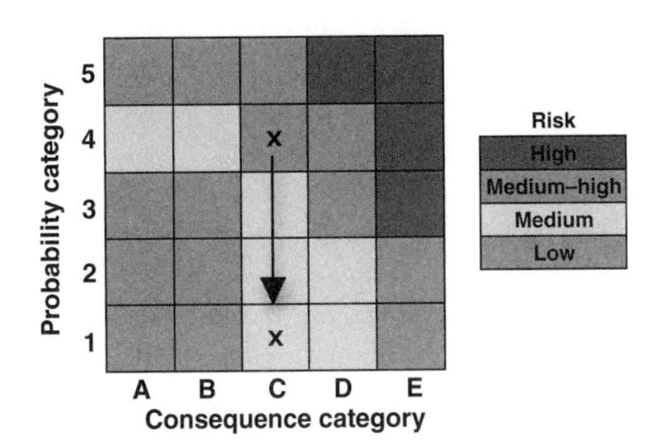


**FIGURE 48.15** Quantitative risk matrix risk reduction.

- Equipment with large uncertainty in the corrosion rate require shorter intervals and possibly more thorough inspections to maintain the target risk, at least until sufficient history has been established.
- Equipment approaching end of life due to corrosion or other deterioration requires increased inspection activity to be sure that the actual deterioration rates do not exceed expected rates. Additional inspection will not significantly reduce the DF or risk once the RL has been consumed.

### 48.4.6 Example Problem Conclusions

The purpose of this example is to show both qualitative and quantitative RBI approaches. A wide spectrum of RBI can be used, as discussed in Section 48.2.3 and shown in Figure 48.3. In addition, the example can be used to demonstrate the fact that a more qualitative approach is more conservative than a quantitative one. While qualitative approaches require less time and cost to implement, they generally recommend more equipment inspection as a result of that conservativeness. Conversely, quantitative approaches require more time and data to implement. However updating risk after implementation of an inspection plan and re-establishing the inspection requirements can require less time and cost to maintain.

*48.4.6.1 Qualitative RBI Characteristics*

- Qualitative approaches assume POF is constant over time. The disadvantage of this assumption is that the inspection recommendations generate an inspection frequency that is the same throughout the life of the equipment.
- In reality, POF increases with time and damage, which would suggest longer inspection intervals at the beginning of the equipment life that become shorter as the equipment approaches end of life.
- One way to generate changing intervals with increasing equipment life is to apply the RL factor approach to inspection interval setting. With this approach, the RL factor constant is based on the matrix location; however, as the RL decreases (as the equipment nears end of life), the inspection interval will decrease proportionately.
- When using qualitative approaches, it is difficult and not necessary to combine POF and risk for individual damage types (thinning, cracking, etc.) since frequencies are determined by location in a risk matrix, not a risk value.
- An inspection date is recommended, but no guidance is provided for methods, coverage, or location.
- While simplified in the calculation of risk, qualitative approaches require more expertise to establish a credible program. At a minimum, high-level experience is

required in corrosion/materials, inspection and process specialties, as well as RBI expertise using the approach employed.

#### 48.4.6.2 *Quantitative RBI Characteristics*

- Quantitative approaches realize and model POF as increasing over time as damage progresses. As a result, the inspection recommendations generated are an inspection interval that decreases throughout the life of the equipment. As POF increases over time, intervals are initially longer at the beginning of the equipment life and become shorter as the equipment approaches end of life.
- Quantitative approaches combine all components contributing to POF and risk (damage types such as thinning, cracking, etc.) into one risk value. A risk target is used to determine when inspection is required and the level of inspection required is proportionate to the damage types' contribution to the total risk. A risk matrix location is generally used primarily for management reporting purposes.
- Quantitative inspection planning provides an inspection date as well as guidance for effectiveness, including recommended methods, coverage, and locations based on damage mechanisms.
- Quantitative approaches are more rigorously calculated, model-based, and rely on actual equipment and process data to calculate risk. As a result, less expertise is often required to establish a credible RBI program. However high-level expertise is still required in corrosion/materials, inspection and process specialties, as well as RBI expertise using the approach employed.

Operators are encouraged to implement a quantitative approach for the aforementioned benefits, which provide the basis for more in-depth inspection planning, reliability, and financial analysis. But the qualitative approach has its place. Start there and build toward a more quantitative approach once the data and resources are in place.

## 48.5 CONCLUSION: EVALUATING RBI METHODOLOGIES

### 48.5.1 Summary

Government regulators are increasingly demanding that risk assessment techniques be applied to high-consequence areas. International standards organizations such as ISO, NORSOK, OGP, and API have each developed international design codes and standards that should be considered when implementing a risk management program for pipelines.

RBI is a structured methodology that establishes an inspection strategy over a defined period of time using detailed technical data. Based on an analysis of the probability and COF, RBI focuses inspection planning on high-risk items and generates the necessary data to determine inspection frequency and future priorities.

A RBI program should be considered for the following pipeline assets:

- Trunk lines from manifold stations to process facilities;
- Oil and gas NGL pipelines;
- Manifold stations;
- Process facilities for oil and gas separation.

The quality of available information on a facility's assets determines whether a qualitative or quantitative RBI approach should be used. The approach adopted will depend on the following:

- Design and construction data readily available.
- Known equipment condition based on inspection history.
- Knowledge of past, current, and future process operational conditions.
- Complexity and age of systems considered.

Less data generally require more qualitative approach, at least initially.

### 48.5.2 Ten Criteria for Selecting the Most Appropriate Level of RBI

RBI methodology can be used on older facilities that have little or poor data available; on newer assets that have technical data available, but little inspection history; and on facilities with good quality data and inspection history. This recommends that the following criteria be used to determine whether a qualitative–quantitative should initially be used:

1. ***Maturity of the mechanical integrity and process safety management programs***—RBI requires data for equipment design (U-1 or piping specification data), process operating conditions (temperature, pressure, material balance, including hydrocarbon, nonhydrocarbon, toxic compositions as well as liquid/gas phase information) and inspection history (inspection locations, methods, coverage) and inspection results such as measured thickness at a specific date.
2. ***Implementation of mechanical integrity program and process safety management program initiatives***—as discussed earlier, a mature mechanical integrity program may support a more quantitative approach to RBI. Conversely, in the early stages of mechanical integrity implementation, an emphasis on collected design, process, and inspection data may make a

quantitative RBI approach unrealistic. In these cases, a quantitative RBI program limits the amount of design, process, and inspection data needed for a high-quality risk analysis. But possibly more importantly, using RBI at this stage could result in a cost-effective method to risk rank and prioritize systems for implementation of the mechanical integrity program. As more data becomes available, a more quantitative assessment can be added to the RBI program, depending on the other factors considered in this section.

3. ***Maturity of inspection program***—a very mature program with all potentially active damage mechanisms properly identified can use RBI for optimizing inspection and manage/reduce risk. A program focused on optimizing risk generally results in increasing inspection for higher risk equipment and reducing inspection for lower risk equipment. The savings related to reduced inspection most often can be applied to the increased inspection areas with little negative impact on the overall cost. In the situation where little inspection has been performed, RBI may identify a much higher level of inspection required than that for a more mature program. This example highlights a situation where more inspection may be recommended when a damage mechanism (wet $H_2S$ damage potential in sour service) had not been previously identified. A quantitative approach is not required to properly identify equipment in this case.
4. ***Corrosiveness of service***—consideration of process conditions driving POF due to in-service damage needs to be considered. The key aspect of the POF is a credible corrosion/damage review performed by a knowledgeable specialist. Identify potential damage mechanisms for operation that include internal corrosion (thinning due to general or localized corrosion); external corrosion or cracking (atmospheric conditions and CUI potential as well possible cracking if austenitic materials are used); internal SCC (for sour applications with $H_2S$ or $H_2S/CO_2$, amine treating); other damage mechanisms (brittle fracture considerations for processing noncondensing hydrocarbons or cold service equipment). In non-corrosive services, RBI can justify doing little inspection to maintain acceptable risk levels.
5. ***Process operation***—considers the process conditions driving COF due to flammable and toxic consequences, as well as business interruption. The key aspect of the COF is relative consequence ranking of equipment COF does not change over time unless the process conditions change considerably. In upstream and pipeline applications, the change in process stream composition over time, and therefore COF and risk over time, is possibly as important as the COF ranking between assets at any point in time.
6. ***Complexity of the process***—reminding us that it's all about risk prioritization for inspection planning, and that absolute quantitative values are not required or even desired. Operators are looking for the best approach to properly and effectively risk rank equipment and provide the proper relative rankings. The less time and money spent on the risk analysis in order to get the desired result, the more that can be spent developing and improving the inspection program.
7. ***Personnel***—resources available to implement and maintain the risk program
8. ***Systems and training***—required over time to maintain the RBI program
9. ***Range and number of systems and assets***—what and how much equipment is being included in the RBI program. Usually piping, pressure vessels, heat exchangers, filters, externally fired heaters are included.
10. ***Cost of implementing and maintaining the RBI program versus value of program***—if the inspection program is very mature, there may be less dollar value in investing a lot of time and money to optimize it. Conversely, if the inspection program is very immature (very little high-quality inspection has been done with consideration for in-service damage), implementation of an integrity program may require more extensive inspection initially regardless of risk. Short-term benefit may not justify big investment.

Regardless of the approach used, a continuous improvement process should be implemented:

- Conduct the initial RBI assessment using available data.
- Use inspection results to develop inspection requirements and priorities.
- Implement initial inspection and plan over defined multi-year time period.
- Update RBI assessment after each inspection.
- Update the inspection plan from the updated assessment.

As the RBI program develops and matures, a more quantitative analysis can be incorporated, as desired.

### 48.5.3 Justifying Costs

Operators can justify the cost of implementing RBI because an RBI program compliments and supports other company initiatives:

- Mechanical integrity program;
- Process safely management initiatives;
- Business initiatives;
- Specialized inspection programs;

- Positive material identification;
- Corrosion under insulation.

Mechanical integrity and process safely management initiatives can use RBI to help prioritize the work over time, precluding the onerous task of implementing these programs in short periods to avoid exposures.

RBI can be used to prioritize risk and perform the activities resulting in the largest risk reduction for the risk reduction expense. Similarly, RBI can be used to support any business initiatives or specialized inspection programs (such as wet $H_2S$ inspection programs, corrosion under insulation detection/improvements, etc.) in an efficient, cost-effective manner.

## REFERENCES


1. Marsh (2014) *Report: Large Property Damage Losses in the Hydrocarbon Industry: The 100 Largest Losses 1974–2013*, 23rd ed., Available at http://usa.marsh.com.
2. API. (2008) *API RP 581 API RBI Technology*, 2nd ed., American Petroleum Institute, Washington, DC.


## BIBLIOGRAPHY


Alvarado, G.A., Kaley, L.C., and Valbuena, R.R. Det Norske Veritas (USA), Inc. (1999) Risk-based inspection demonstrating value. Proceedings of CORROSION/99 Conference, NACE International, Houston, TX, Paper No. 388.

Angelsen, S.O. and Saugerud, O.T. (1991) A Probabilistic Approach to Ammonia Pressure Vessel Integrity Analysis, ASME PVP.

Angelsen, S.O., Williams, D.J., and Damin, D.G. (1992) probabilistic remaining lifetime analysis of catalytic reformer tubes: methods and case study. In: *Fatigue, Fracture and Risk*, PVP-Vol 241, American Society of Mechanical Engineers, New York, NY, pp. 71–86.

API (2000) *API RP 581 API RBI Technology*, 1st ed., American Petroleum Institute, Washington, DC.

API (2009) *API RP 580 API Risk-Based Inspection*, 2nd ed., American Petroleum Institute, Washington, DC.

Benjamin, J.R. and Cornell, A. (1970) *Probability, Statistics and Decision for Civil Engineers*, McGraw-Hill, New York, NY.

Conley, M.J., Det Norske Veritas (USA), Inc. and Reynolds, J., Shell Oil Products Co., USA. (1997) *Using Risk-Based Inspection to Assess the Effect of Corrosive Crudes on Refining Equipment*, October 12–17, 1997, World Petroleum Congress, Beijing, China.

DNV (1996) Base Resource Document on Risk-Based Inspection for API Committee on Refinery Equipment, Det Norske Veritas (USA), Inc., October 1996 (internal committee document).

Kaley, L.C. and Henry, P.E., The Equity Engineering Group, Inc., (2008) API RP 581 risk-based inspection technology demonstrating the technology through a worked example problem for European API RBI. Proceedings of User Group Conference, Milan, Italy. October 2008.

Kaley, L.C. Trinity Bridge, LLC and Ray, B.D., Marathon Petroleum Company, LLC (2011) API 581 Ballot POF Approach Update, Presentation to API 581 Task Group, November 2011.

Kaley, L.C., Trinity Bridge, API 581 Risk-Based Inspection Methodology – Basis for Thinning Probability of Failure Calculations, September 2013. Available at http://www.trinity-bridge.com/knowledge-center.

Kaley, L.C., Trinity Bridge, API RP 581 Risk-Based Inspection Methodology – Documenting and Demonstrating the Thinning Probability of Failure, Third Edition Example, November 2014. Available at http://www.trinity-bridge.com/knowledge-center.

Madsen, H.O., Skjong, R., Tallin, A.G., and Kirkemo, F. (1987) Probabilistic fatigue crack growth analysis of offshore structures with reliability updating through inspection. Proceedings of the Marine Structural Reliability Symposium, October, 1987.

Tallin, A.G., Williams, D.J., and Gertler, S. (1993) Practical probabilistic analysis of a reformer furnace as a system. In: *Service Experience and Life Management: Nuclear, Fossil, and Petrochemical Plants*, PVP-Vol. 261, American Society of Mechanical Engineers, New York, NY.

Tallin, A. and Conley, M. (1994) DNV Industry, Inc. Assessing inspection results using Bayes' theorem. Proceedings of the 3rd International Conference & Exhibition on Improving Reliability in Petroleum Refineries and Chemical Plants, November 15–18, 1994, Gulf Publishing Company.

Tronskar, J.P. and Kaley, L.C. Det Norske Veritas (USA), Inc. (2000) Benefits of risk based inspection to the oil & gas industry, Proceedings of the Oil & Gas Conference April 27–28, 2000, Perth, Australia.

# 49

# OFFSHORE PIPELINE RISK, CORROSION, AND INTEGRITY MANAGEMENT

Binder Singh[1] and Ben Poblete[2]
[1]*Genesis-Technip USA Inc., Houston, TX, USA*
[2]*ATKINS, Houston, TX, USA*

## 49.1 INTRODUCTION

This chapter will integrate the main areas for the successful design and pragmatic application of safe and efficient offshore pipelines and subsea assets to best practices and recognized regulatory systems. The main disciplines addressed are role of risk and corrosion within integrity management (IM); major accident events and corrosion phenomena; integration of risk and corrosion engineering practice; role of corrosion and reliability modeling; practical guidelines for corrosion in the context of hazard identification (HAZID), hazard and operability (HAZOP), failure mode effects and criticality analysis (FMECA), and as low as reasonably practicable (ALARP); development of pragmatic risk allocations (qualitative and quantitative); risk-based methodologies and technologies; risk and fit for life cycle solutions; case histories, corrosion lessons learned, and context for regulations. Additionally, interpretation of risk, health, safety and the environment (HSE), corrosion and IM; recommendations; and future strategies are also covered.

Offshore (including subsea flowlines and near onshore pipelines) are a critical part of the oil and gas infrastructure. The range of design and operations extends from shallow water <200 ft (accessible by diver) through to medium depth up to 1000 ft (diver/remotely operated vehicle(ROV) access) and deepwater 1000–7000 ft (needs ROV) through to ultra-deepwater >>7000 ft (nowadays approaching and exceeding 8000 ft) and indeed above, ideally with minimum ROV campaigns. The environments are dangerous and very difficult to monitor or inspect. Worldwide surveys consistently show that the greatest threat to the integrity of upstream pipelines is corrosion; see Figure 49.1. Moreover, it is clear that the largest part of that is due to internal corrosion (typically >51%). More recent work has shown that most of this is due to localized corrosion of the main pipe body, although other surveys have shown that for topsides (marine atmospheres) significant leakage also occurs at flanges and connections.

For high-risk systems, more than 80% of the IM process is effectively corrosion management. For lower corrosive risk systems, the corrosion component may be 30 or 50%, depending on the application.

The inset in Figure 49.1 shows a graphic of the "Swiss cheese" effect, which basically postulates that final failure is usually the result of many undesirable events aligning at the same time (poor design, bad materials, risky operations, etc.), and that is certainly the case for corrosion failures as we will see later in this chapter.

It is essential to get designs right with minimal intervention over the life cycle (typically 20–30 years). This means that for the deeper more complex cases, it is preferable and often "mandated" by the operators to use inherently safer designs (ISD) with greater corrosion allowance (CA) if steel is to be used, or more likely the argument is made to use more corrosion and degradation resistant materials, such as corrosion-resistant alloys (CRAs). These may be used in solid form (very expensive) or in the clad form. For the latter, clad case, the flowline is usually made of steel, such as API 5L

*Oil and Gas Pipelines: Integrity and Safety Handbook,* First Edition. Edited by R. Winston Revie.

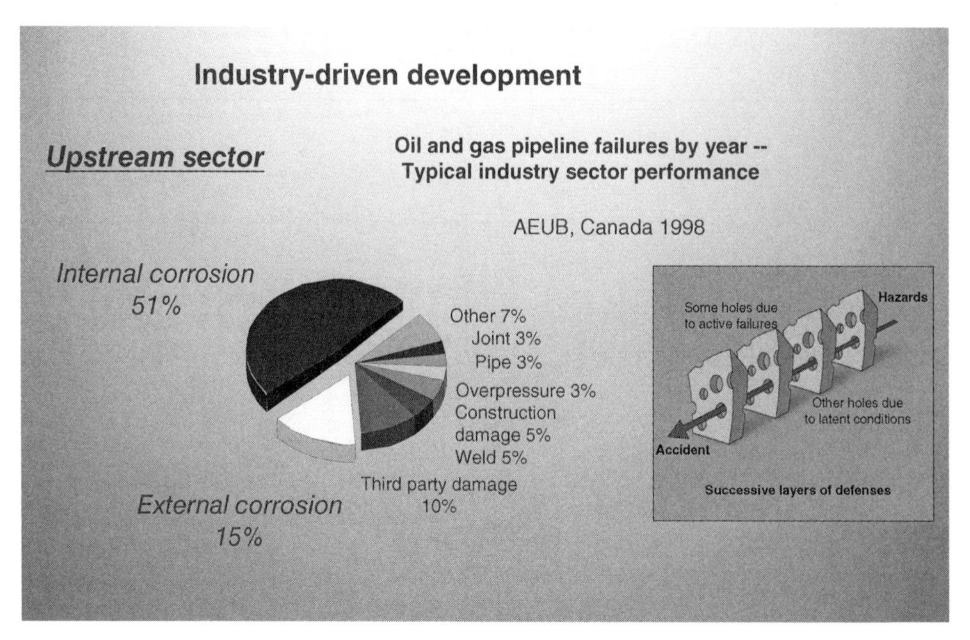


**FIGURE 49.1** Main failure modes for offshore pipelines in the upstream sector.

X65 or X70, and so on, but is internally weld clad or mechanically clad with a 3–4 mm thick internal CRA sleeve. This can be a cost-effective solution, although new problems can arise; for example, the CRA may lead to mixed material issues or galvanic corrosion at interfaces to carbon steel or individual susceptibilities to localized corrosion, such as pitting, crevice or microbially influenced corrosion (MIC).

## 49.2 CHALLENGES, LESSONS, AND SOLUTIONS

A number of offshore/subsea failures, including major accident events (MAEs), have been linked to corrosion; and in particular the role of localized corrosion has been identified as an important, often latent, step. The principles of addressing this mode of failure can be assessed for all marine/subsea/offshore assets, including pipelines and risers. As a result, corrosion assessment and control, via a corrosion management strategy (CMS) may be considered the critical part of life cycle mechanical ntegrity, which may be equated to IM.

"Fit for purpose" solutions must include an emphasis on a safety culture, application of existing knowledge, and lessons learned. Most engineering codes, standards, and recommended practices provide only partial guidance to corrosion assessment and control; it is incumbent on designers and operators to utilize best industry knowledge and experience.

In contrast, offshore/subsea seawater side or onshore soil side corrosion is well addressed by cathodic protection (CP) specifications and codes, and thus is generally very effectively "policed," and problems usually only occur when code criteria are not met. Furthermore, retrofitting is often available to "top up" CP systems with extra sacrificial anodes or perhaps redesigned with hybrid (sacrificial plus impressed current) systems, tailored for each case under code guidance, such as ISO, DNV F103, or DNV RP B401. For those reasons, CP is not generally risk based unless the applications are outside the limits of the standards given.

Nevertheless, defining greater sensitivities and margins of safety have become more important in the context of CP and coatings that are expected to perform at the high pressures, high temperatures, and high velocities (HP, HT, HV) because of the greater reservoir depths and more sophisticated subsea technology required (greater stress on materials from outside (seawater side) as well as inside turbulence (hydrocarbon production side)).

Overall, it has been recognized that internal corrosion issues can be similar for all large- and small-diameter pipes. Recent high profile corrosion-related failures have shown how incorrect or mismanaged internal corrosion can lead to failure. The main challenges are considered to be misunderstanding of the severity of damage mechanisms, incomplete development of codes and standards, and a lack of appreciation of the near misses, which should not be overlooked as a prime source of data. Risk-based approaches are required to satisfy the fast track nature of deep offshore/subsea projects.

Other challenges to deepwater CP designs are related to inappropriate deepwater cold anode chemistries (usually aluminum based with tightly prescribed indium, zinc, and iron contents), and designers should take great care to verify that the required performance can be met, both by theory and by practical testing of the anode materials; see Figure 49.2 for examples.

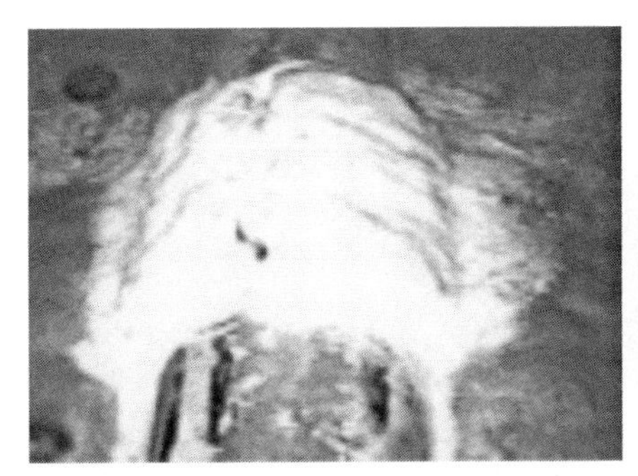
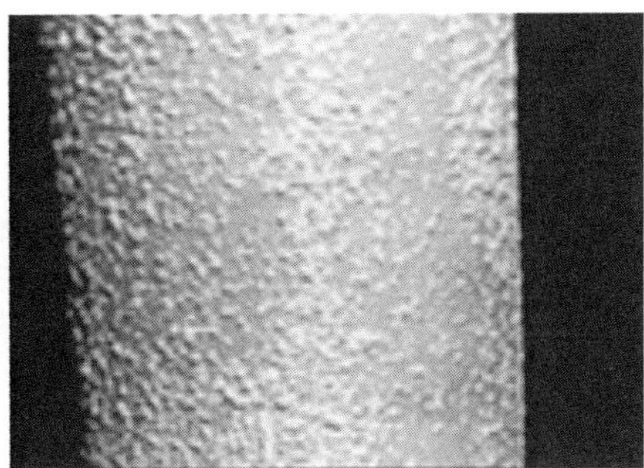

**FIGURE 49.2** Typical "over active" anode (left) and "over active" TSA coating (right), both sometimes observed in deepwater applications, due to improper anode chemistry and possible underestimation of bare steel or inordinate amount of CRA in the subsea infrastructure.

Overactive anodes can reflect rapid consumption even after 1–2 years, and the need for designers to accurately determine the surface area of seawater-wetted steel is crucial in the design calculations. The use of coatings must be supported with reliable coating breakdown factors (typically 2–10%) for most scenarios; information is usually given by most of the CP standards and recommended practices available. Other reasons for early deterioration of anodes and thermal spray aluminum (TSA) coatings may be connected to high dissolved oxygen in warmer subsea loop currents >5 ppm (rather than the expected <2 ppm at depth).

Key performance indicators (KPIs) can be used to engage and make life somewhat easier for busy technicians, inspectors, and operational field staff. The accepted methodology now seems to favor a traffic light KPI interpretation, visually color coded (Table 49.1), so that operators are alerted to significant changes and do not miss any adverse trends in the potential data being accrued.

More recent CP performance data have revealed the presence of onerous microbial corrosion activity even under ostensibly good CP; see Figure 49.3. Here, we show the stage-by-stage corrosion development of bare chain steel as metal dissolution is propagated. The same principles may be interpreted as possible degradation mechanisms for deep sea buried pipelines, in areas where microbes may thrive, perhaps without being impacted by CP. This approach may be described as corrosion risk prediction by analogy and is a very useful way forward for deep sea applications.

So to recap, even in the confident world of good coatings and excellent CP standards, the advent of deep subsea exploration and HP/HT/HV criteria, there is much work to do in refining and readjusting the codes of practice to attend to critical corrosion control issues. In practice, one new approach is the placing of anodes in clusters at various staging posts of subsea structures, PLETs, PLEMs/SCR riser bases, and so on. Here, the anode arrays are successful in propagating the CP current and protective potential along great distances, up to and beyond 25,000 ft with good coatings in place. The reader is directed to the many CP codes and standards (post 2005 issues recommended) in that regard.

Figure 49.4 summarizes this aspect, showing how modeling design data and monitoring data confirm this trend and provide confidence in reliability for subsea pipeline corrosion control.

The corrosion challenges faced by the offshore industry are considerable. For example, note the size of the spar facility under tow, compared to the coastal buildings in Figure 49.5. Observe the density of anodes dotted around the structure and the strakes designed to quell motion induced vortices.

**TABLE 49.1 Color coded KPI bands depicting critical bands of potential, for bold surfaces (excludes shielded areas and crevice formers). Noting that −800 mV is still code compliant, but −820 mV is the suggested action point**

| Interpret Code Criteria for Carbon Steel >−800 mV: Review Evolving Industry Practices | |
|---|---|
| Potential mV (Ag/AgCl Ref cell) | KPI—Interpretation |
| −450 to −550 | Freely corrosive |
| −550 to −650 | Partially corrosive |
| −650 to −750 | Inadequate/partial CP |
| −800 to −900 | Good CP (<−820 KPI threshold for action) |
| −900 to −1000 | Optimum KPI (should also combat MIC), including newly observed "Rusticles" |
| −1000 to −1100 | CP ok but watch care ~$H_2$ ↑ |
| >>−1100 mV overprotection zone | Red flag (HE/coat disbondment, etc.) |

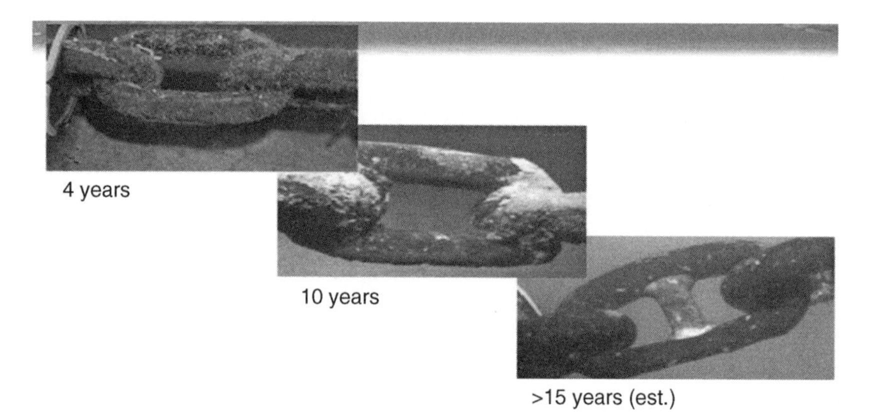


**FIGURE 49.3** Microbial corrosion (sulfate-reducing bacteria and iron oxidizing bacteria) can corrode deepwater mooring chains even under code designed CP. This type of attack will apply to deep subsea pipelines by analogy.

This is made possible by the vast amount of experience accumulated with coatings and CP monitoring as well as retrofit possibilities if indeed deemed to be required, and many subsea assets have been so retrofitted often with hybrid CP systems based on a mixture of sacrificial and impressed current systems. The reader may look into that since most of the work is now in the public domain. In addition, some useful internet links are provided in the References and Bibliography sections at the end of this chapter.

All offshore assets, such as the spar shown in Figure 49.5, are connected to horizontal subsea pipelines via vertical pipe risers; thus the corrosion control, CP, and coating systems must be compatible. Subsea pipelines corrode excessively if the structure CP system is "under" designed. Hence, pipeline corrosion engineers must ensure compatibility of the two CP systems or ensure isolation of the two systems. That isolation must be assured over the full life cycle, and that is usually verified at least biannually by taking separate survey CP potential measurements simultaneously on the structure and the pipeline or riser. The principles of corrosion/CP codes and standards can be applied to all configurations, whether structures, pipelines, pressure vessels, piping, and so on.

Regarding risk-based solutions, the important point is always to observe and not violate the principles such as the basic thermodynamics (what can happen) and the essential kinetics (what will probably happen). In contrast, if risk assessments are applied in contravention to such fundamentals then they will not succeed, an important point to remember when assigning personnel of the correct competency level in both risk and corrosion.

The remainder of this chapter will concentrate on the more difficult subject of internal corrosion and other corrosion

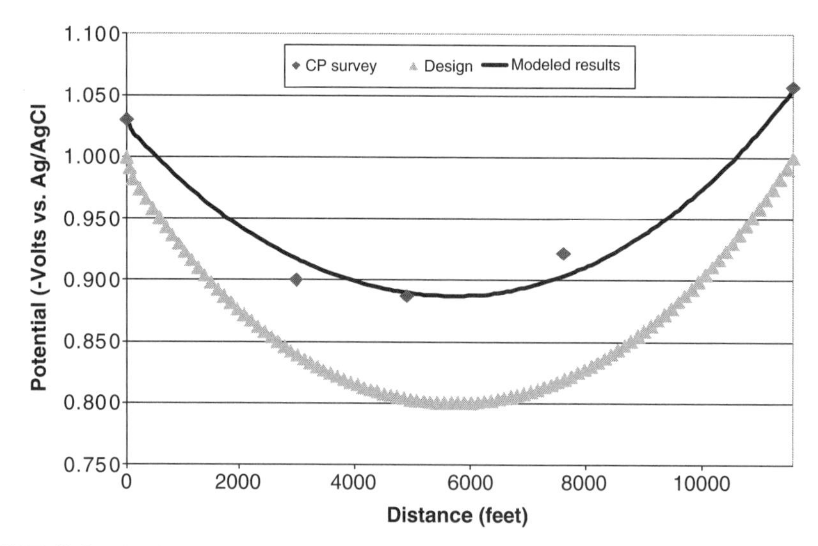


**FIGURE 49.4** Pipeline potential attenuation data depicting good agreement of annual CP pipeline survey against design and modeling trends between two anode clusters (typically 6–15 in total) usually quite tightly packed at each end.

**FIGURE 49.5** Offshore spar under tow to field location; note vast array of evenly spaced sacrificial anodes welded to the structure. Observe strakes designed to stabilize against sea state motions and vortices.

challenges not so well code driven. In that regard, and in reality, many corrosion mechanisms are understood relevant, but not always applied for realistic prediction; typically such analyses focus on uniform corrosion, which can be effectively modeled; and various software (private, JIP, public domain, etc.) exist to facilitate that quite well. In real world applications, general or uniform thinning is rarely a major threat, and corrosion-related failures are almost entirely localized and multi-mechanistic, thus the disparity between predictions, lab testing, and field observations.

Design (pipeline/riser/pressure plant, etc.) must be attended to as early as possible for optimum CAPEX and best management of change (MoC), rather than at the OPEX stage, when it is too late, too expensive, or production schedule sensitive to be meaningful. The concept of addressing all critical localized corrosion mechanisms, including mixed mechanisms as early as possible is reflected in the modified P-F diagram Figure 49.6. Here, the traditional potential failure versus time diagram is translated to read initiation–propensity–failure diagram (I–P–F), with the emphasis on early identification and recognition of corrosion and/or cracking phenomena. The "I" is not always easily measurable, but may be predicted with existing knowledge (often a design reappraisal), with discrete applied testing, forensic analyses, and by the creative use of existing but nonstandard approaches or analogies. If the deterioration mechanisms are identified and caught early enough, this can be equated to "prevention equivalence"; furthermore any mitigation actions such as inhibitor, biocide, or scale inhibitor injections, are known to work better at such incubation periods associated with localized damage. The method also allows for a better definition of CA and the much vaunted corrosion KPI approach. Thus, design reappraisal, and independent third-party (ITP) review or corrosion audit are quite important.

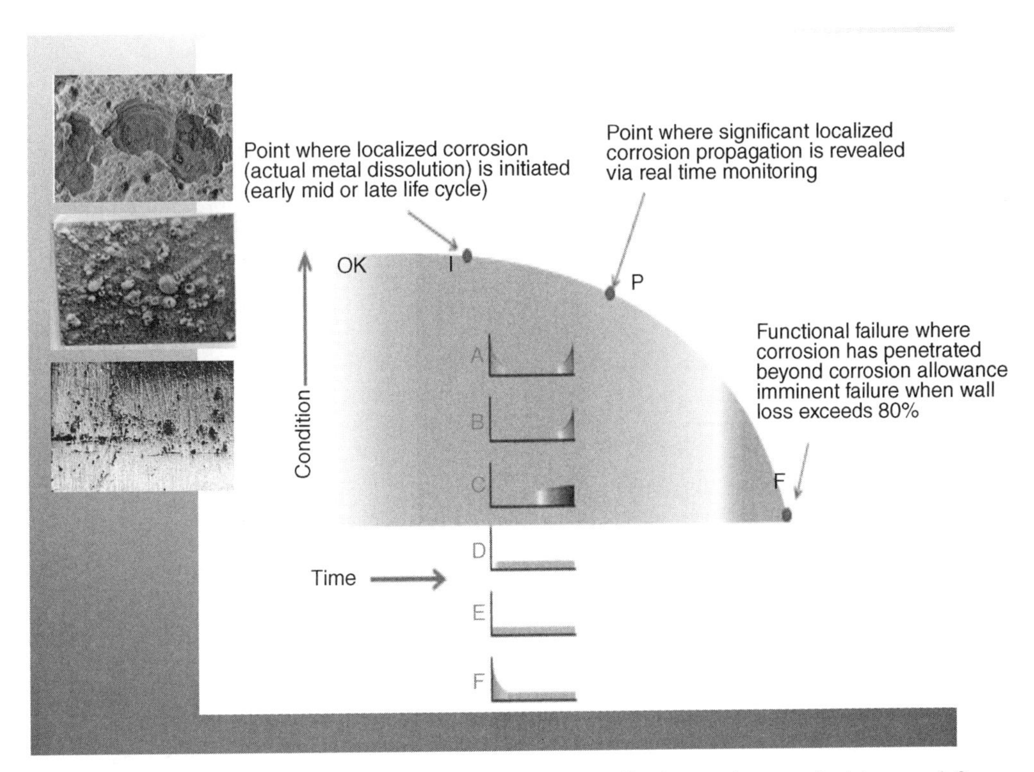


**FIGURE 49.6** Adapted I–P–F curve interpreted for localized corrosion, emphasizing need for responses (solutions) when corrosion initiated, for best measurable and thus actionable KPIs.

Most if not all failures, including corrosion failures, are invariably prompted or related to a multiple sequence of events. These may be major (macro) or minor (micro). This is sometimes referred to as the Swiss cheese effect (inset in Figure 49.1) and is a good visualization for failure analysts to bring the steps, modes, or effects into the risk discipline.

Clearly, subject matter expert (SME) effort is vital in navigating the interactions between fundamental research, applied studies, and the so-called ALARP condition; whereupon the integrity risks to the asset are stimulated to be "ALARP." Hence, it becomes realistic to attend to corrosion risk, relevant failure mechanisms, and CA within corrosion and IM on a qualitative, semiquantitative, or quantitative manner—high, medium, low (HML) risk, depending on the time/cost/schedule constraints. In this way ISD can be better instigated for the betterment of life cycle performance, akin to the safety case type approach. In that regard, ISD designs may mean those designed with better CRAs or indeed conventional carbon steels with best practice physical controls (pressure, temperature, velocity (PTV)), inhibitors (chemical controls), coatings (intervention controls), applied at the earliest point or critical onset of localized corrosion. The CA parameter often used as a safety margin can be construed as a first step in the corrosion management exercise.

Through advanced monitoring and inspection, with feedback cycles, carried out frequently, it is possible to develop a risk-based assessment of corrosion threats to mechanical integrity, in the form of relevant, measurable, and controllable KPIs. Ultimately, the decision has to be taken whether to run the asset/component/part to failure, run to degrade to the CA, or run with prevention in mind. The latter is clearly best, but often meets resistance due to immediate costs; it is usually quite easily justifiable over the life cycle; and inherently safer materials selection (CRAs) or inherently reliable designs can be very effective solutions. It is not the elimination necessarily but the control and management of failure that allows for the ALARP process to be successful.

Figure 49.6 depicts an adaptation of the common "condition versus time curve" with the initiation point (I) added to reflect the importance of identifying the corrosion starting point in addition to the monitored propagation point (P) both prior to actual failure approach or tendency (F).

The annotated sub curves ABCDEF illustrate typical renditions used in RAM (reliability, availability, and maintenance) type studies. These curves show the higher risk failure segments over the design or operational life, and these are equipment or asset specific for each environment encountered. Hence, they may reflect a bathtub shape or be relatively linear for more benign fluid contact. The inset pictures show possible initiation surface features (at point I) if the material or samples/coupons thereof were made available for close examination. The examples given show an expression of such early activity, namely, top, early microbial pitting striations; middle, early scale nucleation meteorite patterns; and bottom, early stage pitting.

With regard to Figure 49.6, it should be noted that a separate I–P–F curve can be constructed for mechanism(s) discerned, including dangerous random failures (time independent) caused by mixed or unidentified mechanisms.

For pipeline risk-based analysis, the corrective component of "fine-tuning" as new data, applied research, or field data are generated becomes a critical part of the "safety case". The HML criteria may be adjusted accordingly as *bona fide* SME judgments are made and KPI's readdressed. Subsea pipelines present a special challenge since internal coupons or probes are generally not placed subsea, for safety reasons, and topsides probes and coupons must be relied upon to determine internal corrosion condition by analogy. However, at the time of writing, many companies are working hard to offer such services in a cost-effective, safe, and ROV-retrievable manner. Hopefully that will be seen to be workable in the near future especially for SCR (risers), which are safety critical and hitherto not accessible for internal monitoring. If that happens, an array of coupons and/or probes may be placed at the touch down zone (TDZ) to great advantage for the IM process. It is believed that a significant improvement to existing predictive methods is thus possible and indeed expected over the next few years, using more practical risk-based methods. Most corrosion analyses and solutions are highly judgmental and often rely on imperfect sources of data, and so performance predictability over the life cycle tends to be inadequate. The practical risk-based analyses offer an alternative approach based on a high level of knowledge and experience. The methodology uses numerically assigned logic, prioritization, and probability criteria expressed in the form of precise risk judgment sheets.

This chapter gives a view of the relationships between corrosion, risk and reliability, and is targeted to focus on how risk analysis may be customized for an integrated corrosion and IM approach. Guiding rules are presented and applied to areas of particular interest, namely, localized corrosion, nonsteady design excursions, and differentiation between pertinent corrosion mechanisms, where most problems are noted. It is rare to have a failure due to uniform or general corrosion; invariably the mode of corrosion failure is related to a localized mechanism. In the same context, as a corollary, an important interpretation of practical loss prevention and risk theory pertinent to corrosion is also included along with case histories from the oil and gas industry. The specialized input of risk analysts and in-service corrosion/inspection practitioners is highlighted, and it is shown that such a risk-based approach offers sound, cost-effective answers and fosters the development of creative track initiatives. The role of such corrosion risk decision making within risk-based safety and verification regimes is also presented, and the methodology is argued to be a very powerful tool in life cycle corrosion management. And since the approach is relatively new, it is invariably based on qualitative reasoning, modeling and judgment rather than quantitative methods, especially for challenging deepwater assets where critical data are not so readily available. However, as the assets move into production mode and meaningful experience is accrued, the balance tilts in favor of quantitative methods. Typically this expresses itself in the continued use of existing corrosion models, but with a corrective fine-tuning based on field representative inspection and corrosion monitoring data

It is important to recognize that risk studies are not directly instinctive, and these are skills that we must learn, put into practice and continuously improve; and corrosion risk being an almost unlimited subset requires a high order of attention. In the context of integrity (effectively corrosion) management the tenets of corrosion control are predominantly knowledge based. If the processes are reasonably well understood, corrosion prediction can be "modeled" quite well and solutions can be offered with some confidence. However, in instances where root causes and corrosion mechanisms are not fully known, corrosion control essentially becomes a risk-oriented discipline. Localized corrosion mechanisms are continually being researched and so both the science and the resulting applied engineering are subject to change. Thus, in practice, all solutions to corrosion issues tend to be risk based, whether stated or not. Typically for any problem in the first instance an attempt is made to identify the main mechanisms of corrosion attack, and thereafter a reliable intervention program is considered. This may include a material change, inhibitor selection, or a design change to the component geometry or indeed the physical conditions expected. Often this is successfully done for uniform corrosion via the concept of a design CA alone, but problems remain for localized corrosion phenomena

## 49.3 LIFE CYCLE

In terms of life cycle, we follow the same pattern by assigning consequences as a time estimate of the consequential event (e.g., leak) being categorized as high risk <3 years, medium risk 3–5 years, and low risk >10 years.

### 49.3.1 Fitness for Corrosion Service

Fitness for purpose or fitness for service are commonly used phrases to define overall engineering performance. It seems logical to adjust the definition for corrosion concerns to something akin to fitness for corrosion service thus putting emphasis on the need to acquire the most suitable corrosion solution under the circumstances met, through the life cycle. Once the predictive risk judgments are made, the problem of convincing the asset holder (client) for a logical way forward is made easier and more formally documented; furthermore the project controllers will find that accountability and responsibility are better served when risk allocations are made in a consistent and repeatable way.

### 49.3.2 Conventional and Performance-Based Corrosion Management

There are currently two types of corrosion hazard management plans that are practiced, namely, conventional- and performance-based processes. The conventional entails the authorities domain (governing or certifying body) to provide the prescriptive requirements that the engineering domain (owner, designer, or operator) must comply with in the design of a facility. This depends on industry-wide accidents and experience to provide data for maintaining the regulations. Compliance in this case is the response that meets a prescriptive solution, and this is the practice of many operators today. A more recent trend is a risk- or performance-based approach to encompass the following:

- Demonstrate continuous improvement.
- Readily gain approvals for modifications, new designs, and novel procedures.
- Accurately provide a successful and critical corrosion hazard identification.
- Provide involvement and consultation by all employees.
- Provide a systematic thinking and documentation of the corrosion risk.
- Provide a method of focusing effort proportionately on critical areas.
- Promote better fit with latest formal safety assessment (FSA) or safety case.

The process places the responsibility for the corrosion risk management (CRM) plan firmly on the hands of the owners. The approval or certifying authority is responsible only for the approval and monitoring of the process. The owner or lessee must address the most serious corrosion threats to the pipeline in question. For subsea oil and gas pipelines, that threat is mainly sweet $CO_2$ and sour $H_2S$ corrosion and cracking as exemplified in Figure 49.7.

### 49.3.3 Corrosion Risk-Based Performance Goals

Conventional is gradually being phased out, and the performance basis is now preferred. More specifically, methods have been developed by the retrospective analysis of corrosion and failure reports particularly over the past several years and are better used in solo or conjunction with other risk-based tools. In the context of better corrosion management, risk-based performance goals are achievable in relatively short periods provided effective corrosion risk quantifiable tools are used. The corrosion risk techniques described are dynamic in that they are live and subject to

**FIGURE 49.7** Montage of $CO_2$ corrosion as it may develop through the various stages of initiation, development and propagation or accelerated growth.

continuous updating, thus the chances of responding to problems are far higher than for existing non-risk or prescriptive approaches, which tend to be relatively static between inspections or analyses. Thus, we anticipate better control in the following areas:

- Monitoring the unit efficiency and general condition;
- Reduction in general and localized corrosion rates;
- Better knowledge transfer, reduced downtime, and increased productivity;
- Better understanding and reduction of failures and frequencies;
- Reduction in perceived and calculated risk evaluations;
- Resolution of conflicts between marine and petrochemical practices;
- Greater confidence in plant operability, revenue, and HSE concerns;
- Better attention to regulatory, verification, and compliance issues.

Using an overall risk ranking method will provide for more satisfactory risk-based inspection (RBI). Using overall risk rank assures that the most critical components (high consequence, higher likelihood) are easily distinguishable and are assured to be prioritized accordingly. The parts ranked as medium risk may then be prioritized in a manner that places a greater emphasis on the consequence, a judgment issue.

### 49.3.4 Inherent Safe Design (ISD) and Project Phases of a Production Development

Inherent Safe Design (ISD) is an important concept that was initiated by the Flixborough chemical plant accident in the United Kingdom (1974), whereupon a critical chain of cascading events, including unverified design changes, in tandem with weak materials compatibility under an extreme excursion following a fire were identified as the root cause. The post Piper Alpha period in the North Sea (>1988) also augmented this, and prompted the "goal setting" schemes by the government [1,2]. It was concluded that better permit to work and independent verification procedures regarding both plant modifications and ongoing inspection would aid the elimination of such dangers. The main emphasis coming out of the inherently safe philosophy was to avoid the hazards rather than keeping them under control.

From a corrosion hazard perspective it is imperative that we avoid the concerns early in the design of the facility (FEED) rather than trying to control or take protective measures later during operations. This means entry should be before the start of detailed engineering and certainly before the initiation of any hazard identification studies. It has been proven that this action will result in a cheaper design from a life-cycle perspective due to less intense efforts to be added later to address the corrosion hazards. If the design of a facility follows the line:

Concept phase ⇒ Select phase ⇒ Define phase ⇒ Execute phase

The first aim is to eliminate the corrosion hazard by substituting vulnerable alloys with more suitable materials, such as CRAs, and also considering nonmetallic materials.

As the design phases progress toward commissioning, it becomes more difficult to incorporate cost-effective corrosion control solutions, largely due to loss of design flexibility, and cost and schedule constraints. Hence, there is a need to engage experienced corrosion risk specialists early in the design development. In practice, since most installations require the construction materials to be steel, the CRM process will likely be an important factor for the foreseeable future.

There are many different levels of CRM programs, the four main ones being: compliance based, experience based, data based, model based, or indeed a fifth option that is any combination of the aforementioned. Criteria for each can be defended on a case-by-case basis, and are often driven by cost and schedule.

### 49.3.5 Link between ISD and Corrosion Management

ISD and corrosion management combined reflect the ideal approach to the elimination or mitigation of corrosion hazards. The identification and understanding of the corrosion hazard mechanism is a critical aspect in the decision-making process to address this operational concern. Overall, the practical quantifiable risk-based methods are useful to promote that ideal approach.

The CRM strategies and plans are meant to be live auditable documents that are updated (feedback) during the life cycle of the facility as more experience is gained. The documents could also be used as tools to verify and communicate the corrosion hazard that may affect the facility to all affected employees. This increase in understanding by the working population usually results in a more effective and successful implementation. The CRM provides a means to step back and visualize the full multitude of corrosion concerns, per life-cycle perspective, rather than a single corrosion issue.

This approach also provides opportunities to discover benefits to overall design, rather than being singularly focused on corrosion risks, giving opportunities to significantly save on cost, weight, or schedules at start-up and during operation.

Table 49.2 shows an example of risk-based assessments pertaining to localized corrosion. Here, the SME would allocate risk levels based on theory and practice. An aggregate or average value can be enumerated to help utilize a correction factor rather than generalized corrosion rates as calculated from modeling.

**TABLE 49.2 Localized Corrosion Threat Levels—Example of pipeline risk assessment**

| Localized Corrosion Stimulus | Threat Level (High/Medium/Low) |
|---|---|
| $CO_2$ | M |
| $H_2S$ | L (possible water injection and other life cycle changes, such as reservoir scouring, bacterial presence/dependency) |
| Dissolved oxygen | L |
| Chlorides | M |
| Crevice attack | L |
| Extent of filming/scaling/ passivation breakdown | M (possible electrochemical testing akin to ASTM G01 preferably in site sampled fluids |
| Pitting (Mesa) attack | M |
| Erosion | L (potentially H) |
| Flow-assisted corrosion (FAC) | M |
| Liquid impingement | L |
| Corrosion fatigue | L |
| Stress (cold work, residual, installation, operational), including internal SCC, and so on | L |
| Surface roughness | L |
| MIC (planktonic/sessile) | M/H |
| Wettability (water/oil/ condensate) | H |
| Pre-Corrosion (via poor preservation) | L |
| Physical conditions (P, T, V) | L |
| Flow assurance phenomenon | L |
| Galvanic corrosion | L |
| Weldment corrosion (PM, WM, HAZ) | L |
| TOL | L |
| BOL | L/M |
| Any other (?) | NA |
| **Weighted Aggregate** | M |

### 49.3.6 Risk-Based Inspection and Monitoring

The goal for offshore operators is to assure that the producing asset continues to deliver its desired performance (availability) in the safest and most practical cost-effective manner. In support of this goal, an IM scheme should be devised and implemented. Strategically, the tools available to support integrity are primarily maintenance and inspection. By inspection, an understanding of condition and an assessment of continued fitness for the purpose of equipment and structure can be achieved. For most regions of the globe, the requirement to have in place an appropriate in-service inspection scheme is both mandatory and regulated.

The process of performing inspections is one of the most risky work activities that is performed on a facility (especially entry into enclosed spaces). Justification of inspection requires that a judgment be made where the risks to which the inspection resources may be exposed are balanced against the value/benefits that would be gained from the inspection results these resources are expected to deliver. For regulated schemes (such as those required by class societies or government bodies), the roots of best practice inspection programs are primarily experience driven. In this case, the in-service inspection requirements (written as company specifications, classification rules, recommended practices, codes, or standards) typically embody the decades of collective experiences gained by these bodies, although with many of these bodies this experience has been culled from across industries, disciplines, or assets.

### 49.3.7 Life Extension

The culmination of life preservation is often the development of an effective RBI plan. In order to develop and implement the RBI scheme, many factors and work activities are usually required, including

- Performing a full risk assessment for the pipe structure, identifying the consequences and likelihood of failure for major and minor parts.
- Identifying which specific connections or components provide the greatest threat to failure.
- Identifying the degradation mechanisms that are most prevalent and the impact each may bring to bear.
- Using engineering critical assessment (ECA) and fitness for service (FFS) reliability techniques to identify the most appropriate frequencies, extent of inspection and to provide a method through which the resultant findings of the inspections would positively influence the overall program.
- Identifying inadequacies or shortfalls within the existing dataset for gauged thickness values of pipe wall, vessel, or plate within the asset that is careful knowledge management.
- Predicting the rates of the deterioration mechanisms.
- Development of a site-specific corrosion model to encompass the locations and critical connections, determining how this factor affected both local and global strength of the asset.
- Using the developed corrosion model, fine tuned with historical inspection results and previous repairs, to predict areas where further thickness values would give greater understanding of condition or to identify areas that require repair prior to full implementation of the RBI program.
- Evaluating the influences or relationships of differing failure modes, such as localized grooving, that can potentially propagate by fatigue.
- Additional corrosion monitoring requirements identified and installed (corrosion coupons, probes, etc. as part of the preservation program against inspection target reference points).

In this way the representation can be used to predict corrosion for components over the life cycle. Where components fail, the new projected repairs can be rescheduled in advance of implementation of the RBI. The output is therefore central to the development of the final (evergreen) product. And delivery of best practice productivity can thus be more confidently delivered.

## 49.4 CASE HISTORIES

Engineering challenges demand that the scope for error within design, operations, and maintenance must be greatly reduced. The rush to produce and meet milestones is often the main driver, but the corrosion investigator must still adhere to the necessary scope and objectives defined. Apart from the disquiet associated with early corrosion failures, the consequent redesign or retrofit becomes more difficult and costly due to the logistics and redistribution of resources early in the operational life. This premature requirement of effort is greatly magnified for deepwater facilities, whereupon such input has been traditionally predicted as minimal.

This section looks at a number of corrosion failure investigations, including the more dangerous localized mechanisms, such as crevice, pitting, microbial, mixed alloy galvanic, and so on. Analysis using established or preferred techniques in an extended manner can provide insight into a variety of parameters and often creative solutions. The key is to maintain focus on the facts, and utilize the field, empirical, and/or test data to reason through the most likely root causes and mechanisms to derive the best solution option. And we shall see from the case examples how nominally well-engineered designs can be compromised by the inadvertent (or indeed expected) presence or outbreak of unplanned-for excursions pushing the operations beyond the original design envelope. When such failures occur within the early period of service, the client usually realizes the futility of a straight material or parts replacement, and the heavy penalties in terms of downtime, safety, and corrective mitigation or retrofit are accepted.

Judicious failure or corrosion performance analysis can provide a powerful tool for more efficient designs and for rationalizing damage control, thus helping to prevent corrosion failures, rather than just reacting to failures in an *ad hoc* fashion. The role of failure analysis is now accepted as an essential part of any corrosion management program.

### 49.4.1 Fit-for-Purpose Solutions

Fitness for purpose is usually associated with a design premise; however, the definition must also apply to the concluding solutions and recommendations made, hence the term fit-for-purpose solution. The objective is to give the best safe/economic and appropriate solution for the life cycle envisioned. The decision making is mostly judgment based, and multiple codes of practice and standards are used as reference points. Aside from certain aspects of CP, anodic protection and threshold water treatments, where definitive boundary conditions exist, there are no performance standards per se. The relevant risk principles are as follows:

ALARP: The risk of failure must be ALARP, or effectively at an agreed acceptable level. Solutions must be rationalized to an agreed semiquantifiable level. Regulation, certification, and classification have been used to endorse fit-for-purpose designs and solutions. Ultimate responsibility remains with the asset or duty holder. Typically it has been found that the solution to a corrosion problem is often driven by the core background and discipline of the investigating group; for example,

- A metallurgist may instinctively focus on materials reselection, frequently recommending an alternative substitute alloy as first option.
- The mechanical and structural engineers tend to look to the Codes of Practice for guidance alone.
- A CP or coatings specialist may focus on introducing anodes and or coatings as the most viable solution.
- The chemical engineer or applied scientist may look to process reaction changes or inhibition as a viable resolution.

All of these approaches are correct in their own way, but the objective of a fit-for-purpose resolution must be clearly and rigorously targeted. The analyst must examine the facts and let the data drive the solution direction, or in certain instances it is legitimate to generally deduce a likely mechanism and/or root cause and then generate the analytical data to fit the curve. Being knowledge based and if used judiciously, the latter can give a more cost-effective approach saving much in direct costs and effort expended.

Once solutions are defined, at minimum the asset holder should ensure that any changes or design modifications implemented are "signed off" by the relevant authority (engineering/inspection/certification/verification). Design for construction and design for performance while having overlap are quite distinct, with corrosion appraisal coming into the equation more during service. The latter point is considered important for both preventative and reactive options, and, in most if not all cases, it would be prudent to keep the original designers and design reviewers in the loop. Thus, there is a clear need for asset managers and field operators to interface closely. Ideally this should be driven by best practice requirements, but will most probably need regulatory encouragement. And we may expect similar guidance and a natural convergence to be forthcoming for pipelines in the Gulf of Mexico in due course.

### 49.4.2 Methods and Techniques of Failure Analysis

No two failures are exactly the same, and this element of the unknown makes for an exciting challenge for the investigators; however, most failures do follow a pattern with a near precedent often found in most cases.

The creative use of existing and established techniques such as Close Visual (optical stereo microscopy—longitudinal and transverse sectioning), scanning electron microscopy (SEM), energy dispersive spectroscopy (EDS), X-ray diffraction (XRD) and novel "dry" corrosion potential profiling, and signature polarization "fingerprinting" can be very useful. The SEM provides a powerful tool to examine the morphology and topography of critical as-received and ultrasonically cleaned surfaces. And we find that by comparing surfaces at bold surfaces to those at localized corrosion zones (pre/post clean) one can discover powerful clues to the dominant mechanisms at work. The data so acquired can complement EDS (semiquantative elemental analyses) and XRD (quantitative compound analyses) of scales at critical wall surfaces. Once confidence in the techniques has been established, the most important aspect is ensuring that the surface scale analyses are done at the most relevant sites with adequate controls for comparison. Physical and chemical process stream data can also be very usefully correlated with other data.

In complex case histories, cyclic polarization diagrams allowing the "fingerprinting" of alloys in the actual process fluid of interest may be used to establish pertinent electrochemical parameters, such as pitting potential, crevice/protection/re-passivation potentials, and so on. The generation of such data to support the failure analysis is a unique means of particular interest if the need to reproduce the failure is warranted. Although there is frequently a link between the root cause and the failure mechanism itself, this should not be assumed, because a relevant solution can often be provided from options outside the corrosion discipline. The best source of guidance on this link is often the field staff, who can help keep the potential solutions in perspective.

### 49.4.3 Failure Mechanisms and Excursions outside the Design Envelope

All known corrosion mechanisms occur in the Gulf of Mexico. A great preponderance of corrosion failures can be traced to mechanisms that initiated (also propagated) during out-of-service conditions or excursions outside the expected design envelope. It is important to identify general service conditions and to capture relevant preservice, temporary shut-in, nonsteady state, and excursionary data.

Typically highlighted are the poor preservation procedures (often complete lack thereof) allowing preservice corrosion during storage, before/after commissioning, during refit or plant modifications. These conditions, which are frequently overlooked by designers and operators alike, often promote severe deposit related crevice, pitting or MIC, and differentiation can often be a challenge, but careful scrutiny and multitasking the evidence can prove to be a powerful aid to the diagnosis; see Table 49.3. The differentiation between artificial, raw, and contaminated (dirty) seawaters has continued and much useful work has been done; a recent study has identified interesting characteristics of the scales/biofilms formed, relating the predominance of FeS (mackinawite) and

**TABLE 49.3 List of Main Corrosion Mechanisms in Offshore Pipelines**

| |
|---|
| General thinning (uniform corrosion)—rarely problematic—issues when localized |
| Pitting attack (pre-corrosion*/BOL/TOL) |
| Crevice corrosion (occluded cells/dead legs)* |
| Stress-induced corrosion (SCC, SSC, HE/HIC, SOHIC, etc.) |
| Corrosion fatigue* |
| MIC/fungal corrosion* |
| Galvanic corrosion (mixed materials) |
| Filiform/under film corrosion |
| Intergranular corrosion |
| Fretting/galling |
| Corrosion under insulation*, including under deposits |
| Weld corrosion (HAZ, PM, WM)* |
| Erosion/flow-induced corrosion (Cav/Imp, etc.)* |
| $CO_2/H_2S$ localized corrosion (major threat)* |

*Most serious (see Section 49.4.6).

$FeS_2$ (pyrite) for biogenic and abiotic films, respectively, and this type of applied result can prove very useful to the analyst.

Additionally, in-service erosion/cavitation/impingement and other flow-assisted phenomena are also often underestimated, but temporary changes in differential pressure, spikes in flow velocity, increased sand content, and temperature transients often provide the impetus for such activity, leading to failures of newly installed riser/piping/valve, often within months, rather than years.

Physical parameters can be particularly insidious as the forces involved can well exceed local material/filming properties and corrosion resistance. Comparing data across different systems can be difficult, but nondimensional analyses engaging Reynolds and Froude type analogies can prove invaluable in the determination of safe ceilings for such interpretation. Other more traditional mechanisms, such as dissimilar metal/galvanic corrosion, wear/fretting/vibration, intergranular/selective phase corrosion, corrosion under insulation (CUI) and uniform thinning under normal operating profiles are also documented, and tend to be well managed with diligent materials selection, inspection, inhibition, and so on. The insidious cracking failure modes, such as stress corrosion cracking (SCC), hydrogen damage (HE/HIC), and corrosion fatigue are less common, and appear to be tackled and arrested ahead of time perhaps because more attention has been devoted to these issues at the detail design stage.

### 49.4.4 Corrosion and Integrity Risk

Risk studies encompass a vast scale of work, and there are many methodologies and techniques to address risk. The authors believe a simplified 5×5 matrix logic makes the most sense, namely, having five accepted risk categories defined as very high, high, medium, low, and very low.

Risk can be described further as follows:

**Failure Risk**

Very high risk: An event having >75% probability.
High risk: An event with 30–50% probability.
Medium risk: An event with 15–30% probability.
Low risk: An event with 5–15% probability (ALARP condition).
Very low risk: An event with <5% probability.

These are semiquantitative criteria, and are generally agreed by most SMEs. The consequences side of the equation is more likely company specific, but a good basis might be construed as listed as follows, again the VH, H, M, L, VL criteria are used for consistency.

**Consequences Risk**

Very high risk: An event having >$500m business impact.
High risk: An event with $100–500m business impact.
Medium risk: An event with $ 5–100m business impact.
Low risk: An event with $1–5m business impact (ALARP condition).
Very low risk: An event with <$1m business impact.

Note: These are interpretations pertinent to offshore pipelines. Other industry tolerances, such as for onshore and power plant values, are quite different.

Bearing the aforementioned criteria in mind, overall project corrosion or mechanical integrity risk can also be illustrated by using the red, amber, and low traffic light color analogy, with notation for extreme situations at both ends of the spectrum. See Figure 49.8.

While the HML type matrix has received most attention in the risk management field, an alternative or supplementary

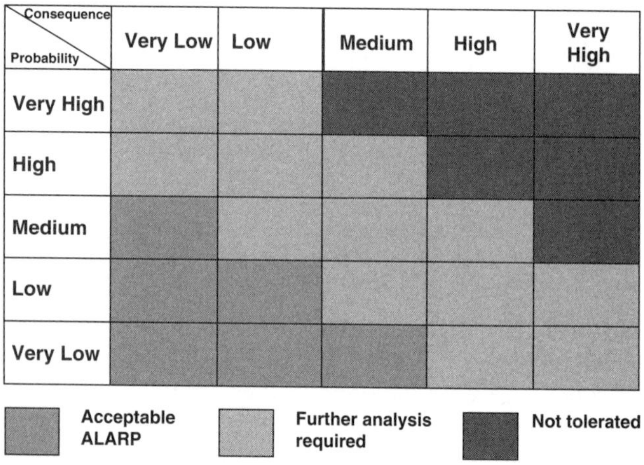

| Consequence / Probability | Very Low | Low | Medium | High | Very High |
|---|---|---|---|---|---|
| Very High | | | | | |
| High | | | | | |
| Medium | | | | | |
| Low | | | | | |
| Very Low | | | | | |

Acceptable ALARP | Further analysis required | Not tolerated

**FIGURE 49.8** HML risk and traffic light system analogy, with extreme cases very high (VH) and very low (VL) incorporated.

approach to the analogy is also very useful, namely, the concept of layers of protection analysis (LOPA). This concept can be effectively used to protect and preserve the asset and have redundancy of action so that there are always realistic barriers to prevent or control the worst case or top event scenarios.

### 49.4.5 Corrosion Failures

Many interpretations may be drawn from individual corrosion failure case histories. The following presents some conclusions and recommendations.

- A high proportion of corrosion failures are related to poor preservice preservation procedures, which must be developed and implemented, particularly for susceptible and hidden surfaces, such as internal piping/vessels/valves, and so on. The methods could utilize system flushes, and temporary inhibition.
- Designers and operators should ensure that loading and process excursions outside the originally expected design envelope must be quantified and evaluated in terms of impact on corrosion performance and failure. Nondimensional analyses and intensive on-line monitoring may be useful in this regard.
- The codes of practice pertinent to design for construction should be individually interpreted and extended to include design for corrosion performance. This may engage risk-based methods and should take into account safe access, on-line monitoring, and inspectability as protection against unexpected failure.
- Any critical design modifications or changes resulting from failure investigations specifically impacting the mechanical integrity of the component, part or member, must go through an approval or verification process as part of the asset corrosion management system.
- Stainless steels (AISI 300 series) continue to be used with untreated seawaters, although these are not compatible combinations and must be avoided, especially under low flowing or stagnating conditions. Similarly, bare carbon steel piping systems must only be used with moving seawater under suitably inhibited and biocidal treatments.
- The root cause and the corrosion failure mechanisms are not always linked, and this linkage must not be assumed. This also means that fit-for-purpose solutions may engage noncorrosion discipline options.
- All fit-for-purpose solutions developed and implemented must utilize on-line monitoring to validate life cycle performance. If this is not feasible, then parallel field simulated scaled down versions allowing retrieval and disassembly/inspection should be contemplated.

The failure investigation aspect of IM is a vital discipline and it is important to recognize where in the life cycle the failure has occurred; see Figure 49.9, which reflects the type of corrosion or mechanical degradation threats that might be in play. Here, we may surmise in general terms that storage and preservation issues are prevalent in the early stages of the design life; localized cracking and corrosion is dominant in the early to midlife periods, and general corrosion thinning tends to be concentrated in the last quarter of the life cycle. Of course any mechanism can begin at any time, especially if design operational envelopes are exceeded via upsets, or process excursions occur, which may stimulate corrosion initiation and development mechanisms.

In the context of Figure 49.9 for subsea and offshore assets, design lives are typically 25–30 years with life

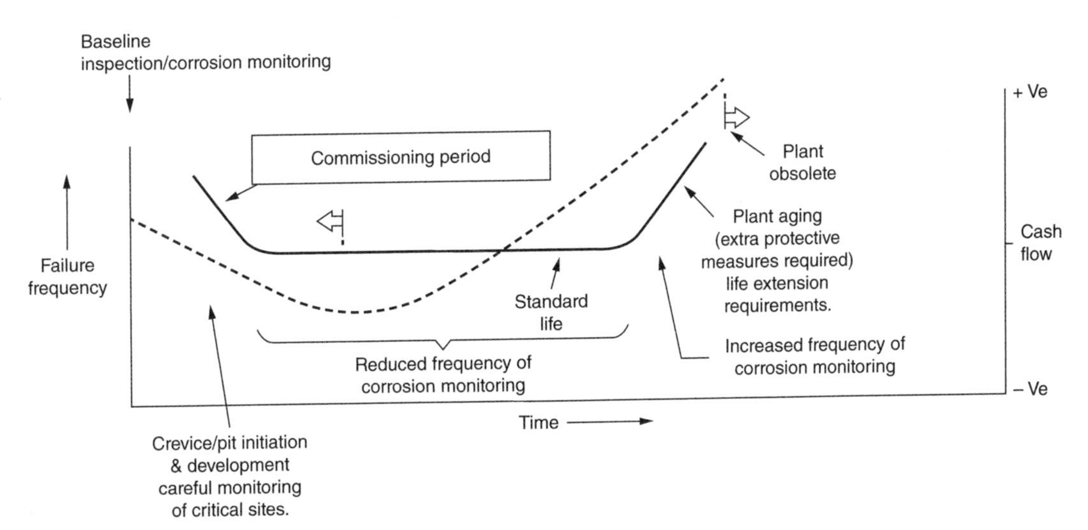


**FIGURE 49.9** Interactions between various failure modes, mechanisms, and fair wear and tear over a typical life cycle.

extension studies often pushing the limits beyond 40 years depending on productivity of the facilities. In that regard, modern plays with new alloys and new geometries have pushed the envelope of design, but have also led to some unusual failures. With the advent of modern computer programs, errors due to bad calculations are greatly minimized since software programs are exhaustively validated, and engineers utilize cross checking of critical calculations using alternative programs, and independent review and verification to ensure meaningful, pragmatic, and schedule/cost acceptable results. However, since technology is always pushing the envelope, with deeper drilling, higher pressures and temperatures, higher applied stresses, and so on, the risk of failure is always a major consideration, and typically the investigator will utilize a materials or corrosion laboratory to carry out the failure analysis.

### 49.4.6 Localized Corrosion Mechanisms in the Offshore Oil and Gas Industry

In Table 49.3, a list is presented of the main corrosion-related failure mechanisms applicable to the offshore oil and gas industry. Those that are the most serious and important in offshore pipeline and related facilities are indicated with an asterisk.

Some of the localized damage mechanisms are expanded on further in the following paragraphs, starting with some guidance notes on the more serious cracking phenomena, and closing with the more localized metal wastage mechanisms. The main difference is time, in that cracking mechanisms or modes of failure often lead to catastrophic bursting type failure; whereas metal loss tends to require more time, although the bursting analogy still applies especially at high pressures when metal will crack and burst once the allowable pressure retention is reached. Note: All examples are applicable to pipelines, flowlines or risers pertinent to offshore assets.

### 49.4.7 Pictorial Gallery of Localized Corrosion and Cracking

Illustrations of corrosion failures are very effective in communicating the message. Most of the illustrations shown are taken from actual pipeline or related failures worked on by the authors, and the intent is for educational purposes of benefit to the oil and gas community as well as new students to this subject matter. Having said that, most are available through the public domain, various papers, and presentations made by the authors, and where feasible references to the original sources are provided at the end of the chapter.

***49.4.7.1 Mechanical and Environmental Cracking*** Identifying the root cause can be a challenge, and the best guidance is to deliberate on the major and minor events that would have prevented failure. For example, the Piper Alpha disaster would have benefitted from a more serious approach to corrosion control (since it was ostensibly plagued with corrosion issues), a better permit-to-work system, and better "lockout-tagout" procedures. The root cause can therefore be argued, but one should focus on which root cause should be the attention of detailed changes to safety systems and better work industry-wide practices. Better corrosion management procedures and applications are certainly preferred actions. There are many examples in the materials and corrosion disciplines whereupon such cases have led to major commercial losses, as well as engineering "snafus" and disasters alike. In virtually all cases, having the right blend of multidisciplinary expertise would have been the correct course of action, but time, cost and/or schedules often led to cutting corners, and thus inevitable failure. Here are some common mistakes in that regard (these are all real world examples):

- Risk taken when design changes are made without a legitimate stress analysis, often by a nonprofessionally trained engineer.
- Risk taken when incompatible replacement materials used to facilitate speedy repairs.
- Risk taken when poor unverified corrosion control procedures used.
- Independent corrosion circuits require independent corrosion control.
- Risks taken when outdated or even obsolete standards, modeling, and software used—a common mistake due to professionals not staying abreast of their subject area.
- Not using ITP reviews or services of experienced classification societies, unless mandated.

Failure analysts must appreciate and recognize the main types of corrosion cracking and failure; that is, ductile, brittle, and their morphologies both in a macro sense and in the micro sense.

In Figure 49.10, localized pitting corrosion that led to branched SCC is illustrated. Such failures can be initiated at any surface where a corrosive environment meets the steel, or even CRA, under the right circumstances.

The SEM images in Figures 49.11–49.13 may be taken as representative in a very general sense. The fracture surface characteristics and morphologies observed can be interpreted and applied to pipeline and subsea structures.

Typical brittle fracture morphologies are represented in Figure 49.12, and in comparison a representation of hydrogen embrittlement fracture is shown in Figure 49.13. The use of such recognition can be a powerful tool in the failure investigation discipline.

Localized corrosion invariably leads to metal failure due to stress overload, or fatigue type mechanisms, and the structure of the alloy will be an important factor determining

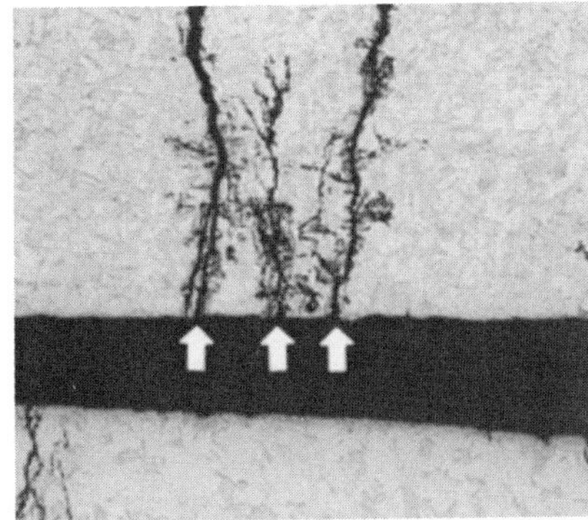

**FIGURE 49.10** Micro pitting corrosion initiation sites leading to branched SCC in steel.

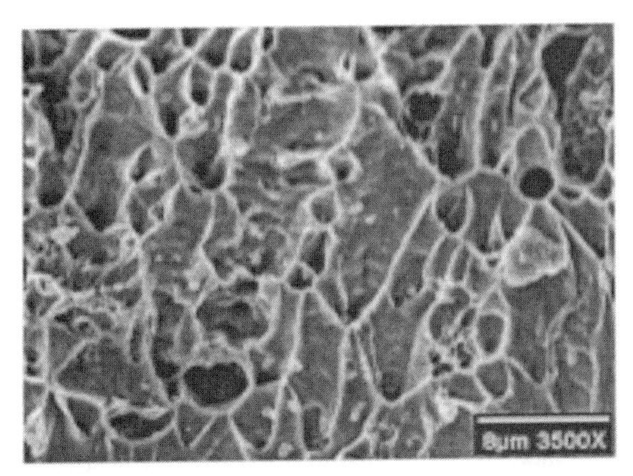


**FIGURE 49.11** Fibrous gray appearance of ductile failure, at micro level "cup and cone" (>3000 × SEM); the microvoid coalescence may also be referred to as "dimpling."

the type of fracture observed. The subjects of fracture analysis and fracture mechanics are very important for subsea flowlines, and corrosion and metallurgical engineers have resolved many mechanical integrity issues with a collaborative approach, often in cooperation with flow assurance engineers. This triangle of SME expertise is a powerful team whenever mobilized. Good exchange of critical information between operators is a good sign of the benefit to the industry as a whole.

Once the root cause(s) of a failure or problem are determined and verified, then a fit-for-purpose solution can be formulated. The best corrosion control principles and practices can be most usefully utilized by influential offshore and subsea designers.

The other area of much concern is that of corrosion fatigue, since there is still a tendency in the offshore industry

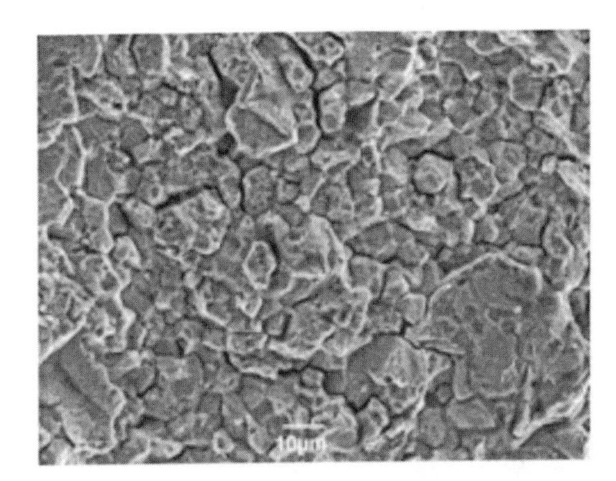

**FIGURE 49.13** Typical intergranular hydrogen embrittlement (HISC) of drilling riser pipe considered due to excessive hardness HRC>35 under high tensile stress with at least partial CP present (hydrogen evolution at wall) from local sacrificial anodes.

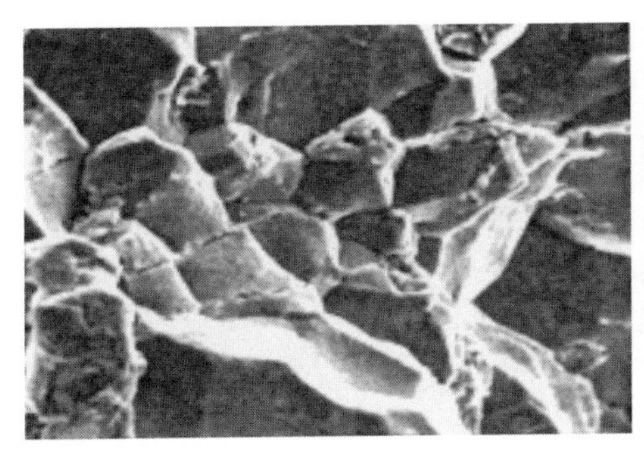
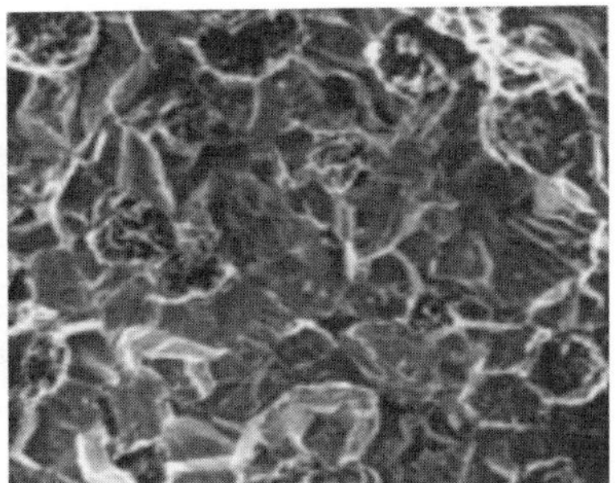

**FIGURE 49.12** Brittle fracture, granular, crack propagation, no gross plasticity, bright nonfibrous, no necking, plain strain condition—mainly intergranular (LHS) or even transgranular cleavage (RHS).

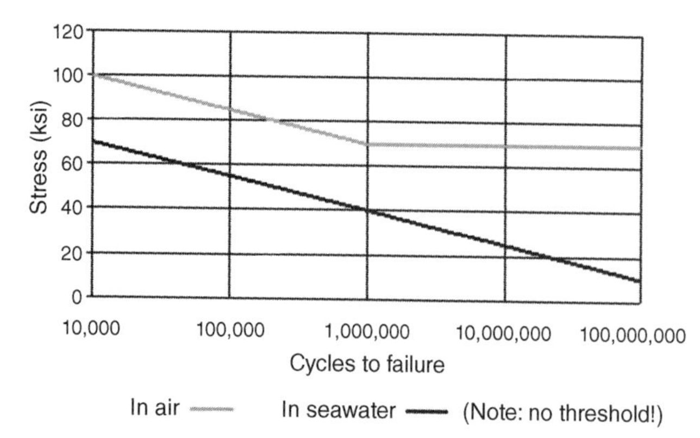


**FIGURE 49.14** Comparative S–N curves reflecting the clear distinctions between fatigue in air versus corrosion fatigue in seawater environments. Seawater or marine data must therefore always take precedence.

to assume that atmospheric fatigue data can be correctly deployed for marine air or subsea environments. That is evidently a dangerous assumption as shown in Figure 49.14 by the comparative S–N curves depicting the differences in cycles to failure for inert air and marine/seawater exposures.

Microbial activity can vary in terms of corrosion thinning from 1–7 mm/y, varying as the dissolution process typically evolves from initiation (micro) to easily measurable development and propagation, ultimately ending in perforation (macro). It is vital to take action at initiation where possible so that effective control (inhibition) can be instigated. After that, during development or propagation, in particular, such interventions tend to be less efficient and often of little real use in the IM process.

Erosion is highly dependent on the flow velocity as often reflected by the flow pattern and location of the damage. Since erosion is predominantly a physical phenomenon, chemical inhibition tends to be of little benefit unless the corrosion component is dominant.

Cavitation (Figure 49.15) can exert stresses in excess of 140 ksi, beyond the yield stress of most alloys. The role of high-velocity degradation mechanisms can be difficult to discern during the degradation event unless sensors are in place and flow regimes are being monitored and compared to the design characteristics specified. The example given is a subsea choke that was forced outside its design envelope by episodic sand bursts over a short period of time (days) with catastrophic results. The enormity of the damage is reflected in the gouging damage to the hard tungsten carbide sleeve at inset on the right hand side.

Crevices form at occluded surfaces created at crevice geometries, such as lap joints, fasteners, and so on, and lead to a localized form of corrosion, crevice corrosion. The same conditions of low pH, low oxygen, and high chloride are often found under pipeline sand deposits or dead legs and weld/parent metal interfaces with similar results, as shown in Figure 49.16.

As with all corrosion phenomena, it is vital to establish the existence and persistence of an electrolyte, such as a free water phase. Regarding multiphase production flows, the term oil or water wetting has been coined to assist in this derivation; see Figure 49.17.

The problems of flowline CUI in marine atmospheres and at, for example, topside supporting members can also be sources of failure. The problem is usually real time monitoring and the lack of effective inspection, although the latter seems to be well addressed by advanced thermal imaging; however, such risks can now be greatly reduced by using the best practice of the latest field experiences utilizing strong coatings, such as TSA coatings and thermoplastic supports such as I-Rods. These methods are gaining recognition in oil field applications.

The issues shown in Figure 49.18 are commonly witnessed in tropical or subtropical offshore installations as the pipelines transition from subsea to topside pressure plant. Such corrosion is often discovered during the annual topsides inspection "walk-round." Note: Very careful inspection and corrosion product cleaning are required to minimize the possibility of premature perforation.

### 49.4.8 Failure Analysis Check Sheet Listing

The following is check listing of the main requirements for a successful failure analysis. Here, the emphasis is on the use

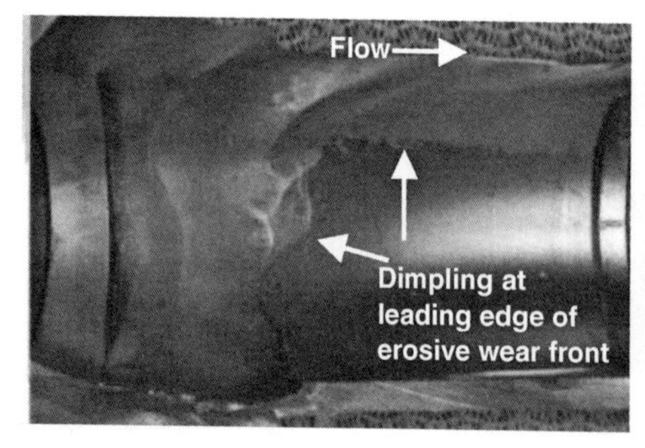


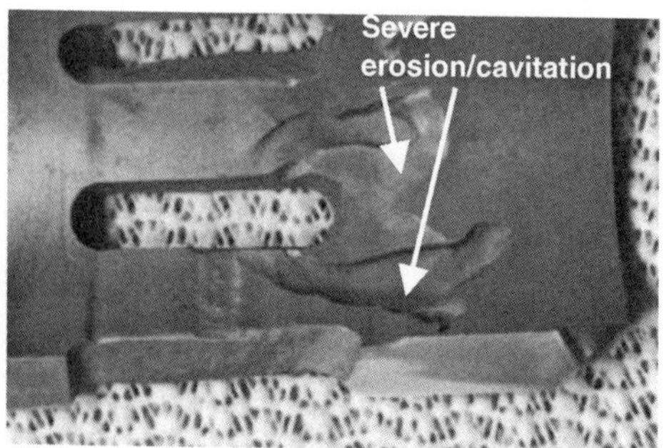


**FIGURE 49.15** Effects of ultra high velocity cavitation, erosion, and impingement on even hard alloy material surfaces as exemplified by the dimpling effect [3].

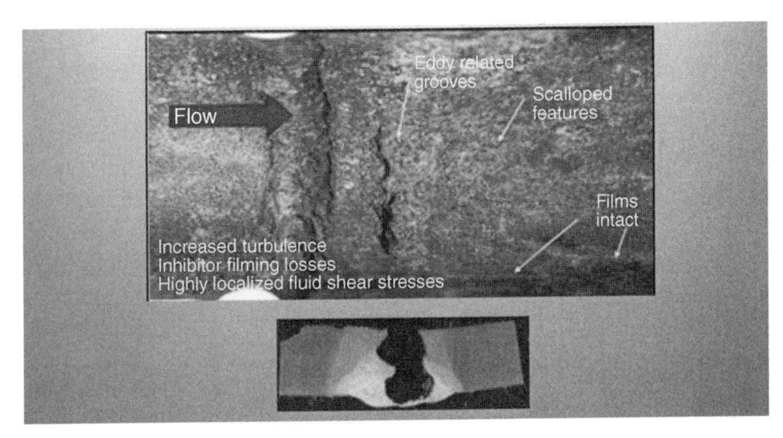


**FIGURE 49.16** Weldment corrosion, acting on parent metal and, at inset, on the weld and heat-affected zones.

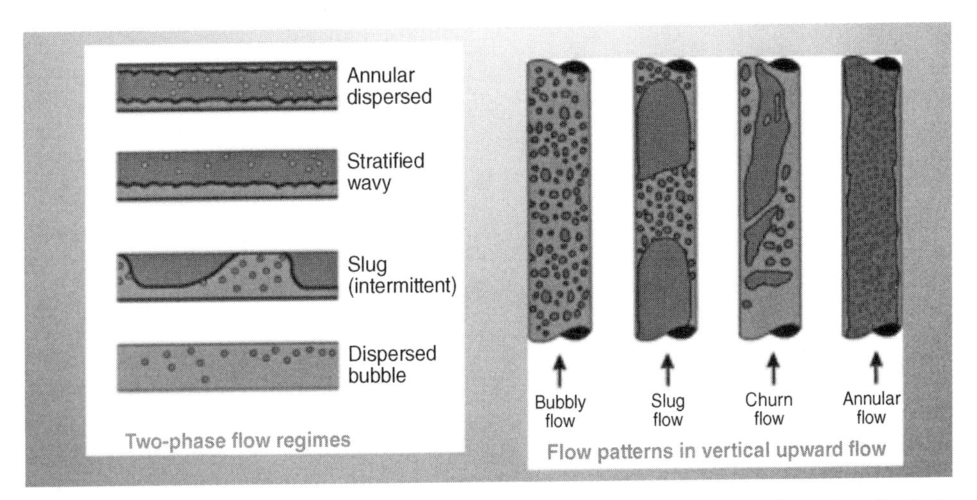


**FIGURE 49.17** Critical relationships between flow regimes and water or oil wetting for both horizontal and vertical pipelines (oil is shown in dark gray/black, and water in light gray).

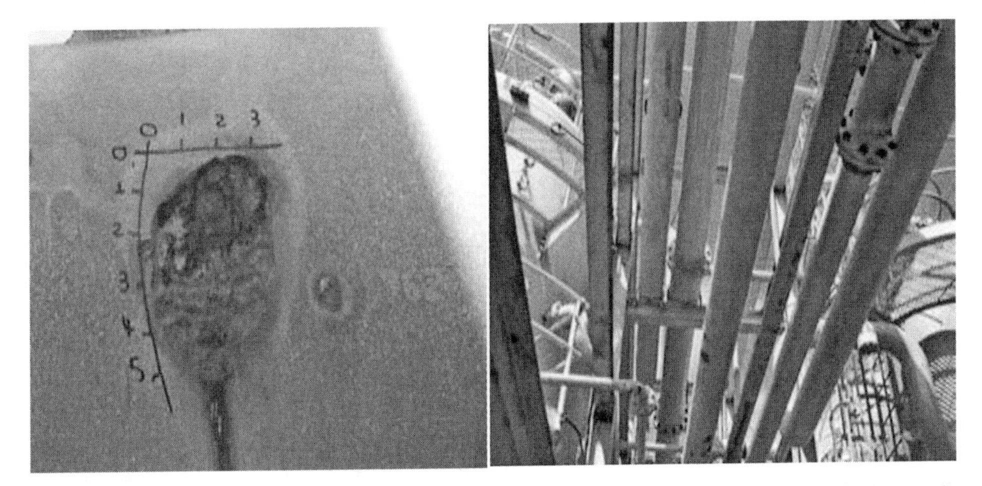

**FIGURE 49.18** Serious offshore topsides in flowline relief header; marine atmospheric (external) localized corrosion, on top of horizontal piping, reflecting the corrosive power of high humidity (>80%) year round marine atmospheres in the Gulf of Mexico.

of practical and well-established techniques rather than newer more experimental techniques.

**Operating Conditions**
Design certified/verified
Design reappraisal per corrosion
Original design life service/age at failure
History of problems; previous test data used any precedent
Material
Environments
Conditioning/inhibition pressure
Temperature
Flow velocity
Geometry
Any previous changes/retrofit, on-line monitoring
Corrosion rates determined; failure site—internal/external
Failure description

**Analyses**
Physical modeling (commercial/in-house)
Nondimensional analyses (Reynolds, Froude, type analogies) materials compatibility review
Stereo microscopic SEM–EDS
XRD–XRF
Corrosion products, metallography
Microbial sampling/tests (usually important to maintain sterile handling and analyses within 24 h of sample extraction)
Corrosion mechanism(s) determination
Root cause failure diagnosis
Fit-for-purpose resolution
Additional tests to validate
Solution implementation/on-line monitoring/field verification

## 49.5 CODES, STANDARDS, RECOMMENDED PRACTICES, AND REGULATIONS

The use of IM has been mandated for many years per the pipeline onshore regulations (DOT 193/195 series). However, the situation for offshore pipelines was based on voluntary recommended practices. That has now changed for the Gulf of Mexico region, and IM is required via the vehicle of the new Safety and Environmental Management Plan (SEMP) regulations mandated after the Macondo disaster in 2010. The following is a summary interpretation of the main features of this requirement. Arguably, mechanical integrity, read as corrosion and IM, is the most important.

Bowtie methods are an effective means to convince clients and operators through persuasive visual techniques of the need to utilize inherently safer or more reliable designs and the means of timely mitigation, both before an accident event and even after if it occurs.

As illustrated in Figure 49.19, barriers to threats and risks can be erected, for example, through materials selection, CAs, inhibitors, coatings, CP, and so on.

These are often supported by planned corrosion management strategies as shown in Table 49.4. A multidisciplinary team effort is usually required to define each step in the corrosion management sequence.

## 49.6 CORROSION RISK ANALYSIS, INSPECTION, AND MONITORING METHODOLOGIES

For a reliable and effective corrosion and IM program, it is important to generate accurate and realistic data for collation and interpretation to allow good reasoned ALARP-based solutions to the corrosion problems likely to be faced. A full

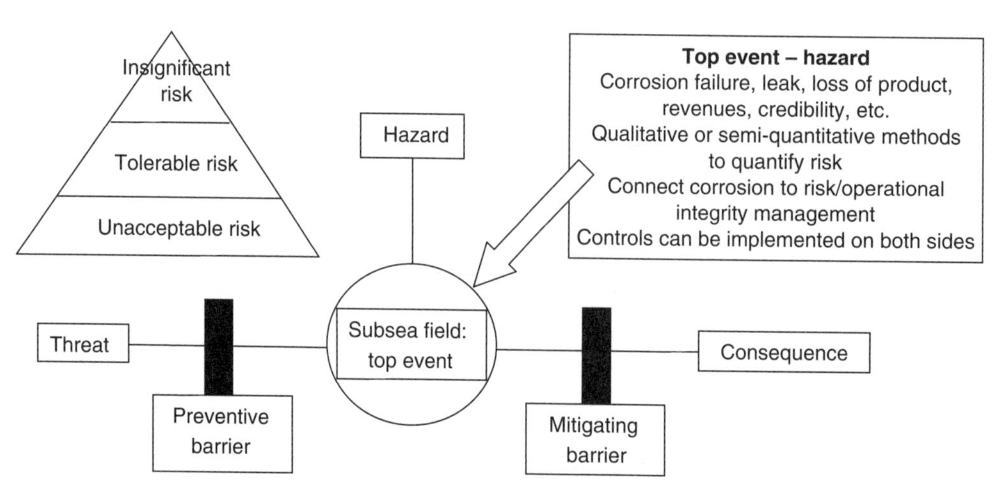


**FIGURE 49.19** Use of bowtie techniques to achieve and maintain the ALARP condition.

**TABLE 49.4 The 10-step risk-based pipeline integrity management (IM) methodology with key gate decisions as verified by independent sources, both within a company or outside via, for example, a classification body [4]**

| | |
|---|---|
| Step 1 | Define scope |
| Step 2 | Corrosion review |
| Step 3 | Develop strategy |
| Step 4 | Develop bowties |
| Step 5 | Inspection |
| Step 6 | Data management |
| Step 7 | Reporting |
| Step 8 | Corrosion and integrity statement |
| Step 9 | Fit-for-purpose solutions |
| Step 10 | Verification—feedback lessons learned; to step 1 |

understanding of the general and localized corrosion phenomena and risk-based techniques is required. Typically initial analyses will point to the next action of progress, and many methods will have been prepared to meet that need. The 10-step methodology, Table 49.4, is a useful guide to facilitate risk-based corrosion control and mechanical IM. The key learning over the past 10 years has been the need to use integrated methods and techniques along with laboratory testing and analogical approaches to transfer knowledge across assets and systems (e.g., from piping to pipelines and the reverse).

### 49.6.1 Risk and Reliability in the Corrosion Context

In the development of oil and gas projects, IM issues are minimally addressed through the design development, but, in 2010, the US Bureau of Safety and Environmental Enforcement (BSEE) promulgated the rule on Safety Environment Management Systems (SEMS), API RP 75, which now formally recognizes mechanical integrity during design life cycle.

> "CFR 249.1916 What criteria for mechanical integrity must my SEMS program meet?
>
> You must develop and implement written procedures that provide instructions to ensure the mechanical integrity and safe operation of equipment through inspection, testing, and quality assurance. The purpose of mechanical integrity is to ensure that equipment is fit for service. Your mechanical integrity program must encompass all equipment and systems used to prevent or mitigate uncontrolled releases of hydrocarbons, toxic substances, or other materials that may cause environmental or safety consequence."

Similar to occupational safety and risk assessment, corrosion risk is the set of requirements that the engineering designer must address just like any specifications. The guidance in [5] provides the template on how corrosion risk has to be included into the engineering design. What is missing is how to integrate corrosion risk into the life cycle engineering design process. To accomplish this, we have to go back to the basic principles of what we are trying to achieve with a facility design. There are three focal points when dealing with the importance of addressing corrosion risk in projects and operations: Production availability, incident realization, and life cycle perspective.

The confusion between risk/hazard analysis and assessment was addressed by the Royal Society [6]. The process sequence of risk management is:

a) Identification of undesirable events;
b) Analysis of the mechanisms of the undesirable events;
c) Consideration of the extent of the any harmful events;
d) Consideration of the likelihood of the undesirable events and specific consequences (probability/frequency);
e) Judgment about the significance of the identified hazards and estimated risks;
f) Developing and implementing decisions on the courses of action.

The terms are grouped in this manner:

- Hazard identification: Steps (a) and (b)
- Risk estimation: Steps (a)–(d)
- Risk evaluation: Step (e)
- Risk management: Step (f)
- Risk (hazard) analysis: Steps (a)–(d)
- Risk (hazard) assessment: Steps (a)–(e)

The use of "hazard" and "risk" terms are usually used as synonyms. "Risk" should be used only to describe a study or assessment where an estimate of likelihood is involved. The two most critical steps are (a) and (b); the accuracy of these two items will determine the cost effectiveness of the risk management actions.

The prescriptive or conventional hazard management process (Figure 49.20) is the basis of most of the regulations around the world. The legislative requirements prescribed by the authorities are based on many years of experience from failures in the oil and gas industry. The authorities are responsible for developing and updating the prescriptive legislations. Engineering and field development teams are required to meet the legislation to ensure the safety of their facility or pipeline design.

Goal- or performance-based legislation is based on the demonstration by the engineering and field development teams of the integrity of their design. The authorities provide only goals and objectives, and the rest of the risk management work and demonstration must be presented for the

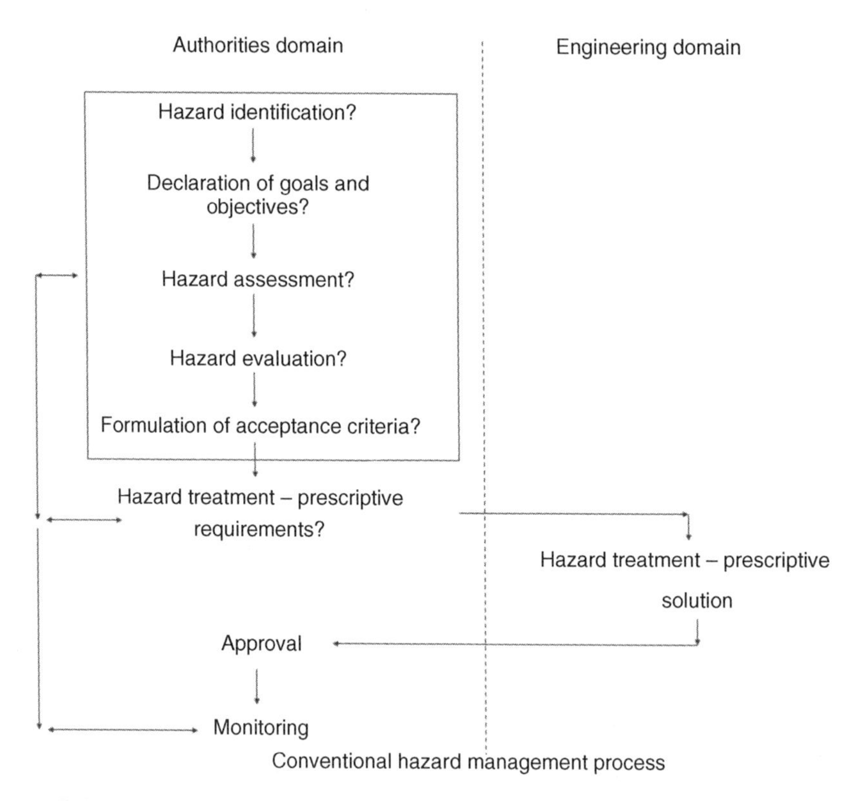


**FIGURE 49.20** Conventional hazard management process: prescriptive regulations.

approval of the engineering authority. This approach, culturally, is more amenable to the progressive and technologically changing challenges of the offshore oil and gas industry. Software (management systems) and hardware innovations can easily be incorporated as long as demonstration of ALARP can be provided. This also means that the engineering teams or operators must have a very culturally advanced HSE management system to ensure that the updates or monitoring on any of the performance-based solutions are addressed. This change, especially in the North Sea and Australian oil and gas region, gave more cultural control to the operators that provided them with opportunities and flexibility to mold their organization to meet the goals and intent of both the design and the organization. The authorities and engineering domains change as illustrated in Figure 49.21.

Prescriptive regulations will force most organizations to be complacent and work toward the minimum operational and capital cost; they have to meet the regulations and nothing more. By having to declare the organization's own goals and objectives in a performance-based management process, the creativity of the organization is opened to find means of eliminating or reducing hazards to ALARP and at the same time to create more flexibility on the operating and capital costs of a production development. The operator, now being responsible for risk treatment, will be more focused on the auditing and verification process in the management system.

It is this understanding of the role of regulation where organizations started to have control in managing their future. It must be noted that organizations have now embedded this performance-based hazard management process into their organizations with the development of their IM framework. They establish the scope and intent, and then provide the elements or business areas. In each business area, specific activities must occur for the organization to attain the desired vision. Each element will describe the intent or high level objectives and then a list of expectations or specific behaviors/actions/deliverables must be met. In the performance-based process in Figure 49.21, if the authorities domain (left column) is renamed senior management, and the engineering domain (right column) is renamed operations/design, it will reflect the current format of most IM processes existing today.

*49.6.1.1* ***Focal Point 1: Production Availability*** The intention of any new project and facility engineering design is to produce accepted revenue from the investment. The revenue and return on investment is dependent on the capacity of the facility and its planned or desired availability. The availability is dependent on both the operations reliability and maintainability of the designed equipment. Both these

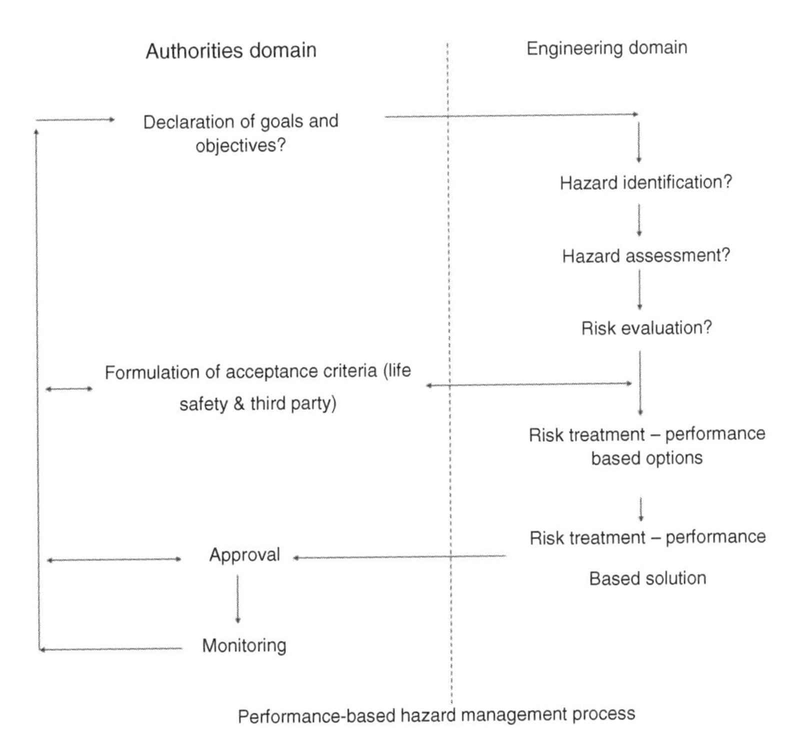


**FIGURE 49.21** Performance-based hazard management process: performance-based regulations.

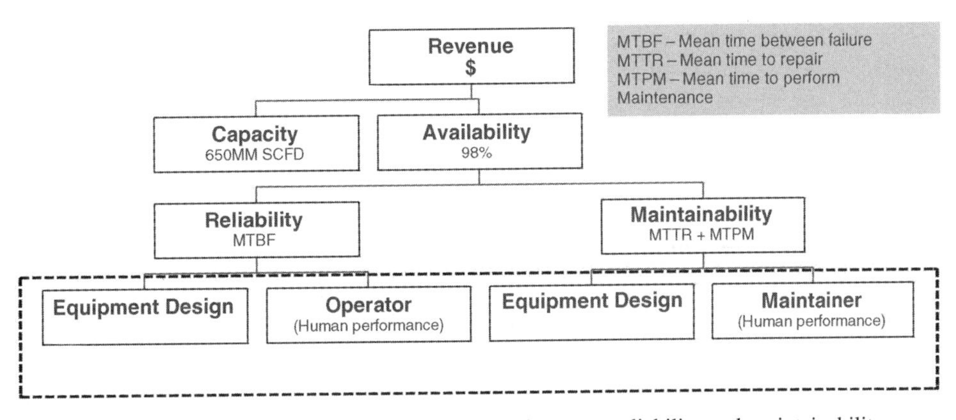


**FIGURE 49.22** Example of availability dependency on reliability and maintainability.

parameters are dependent on both the equipment (fixed or rotating equipment) and human performance (operator or maintenance staff). The equipment performance is divided into Reliability Centered Maintenance (RCM) for rotating equipment and IM for fixed equipment (vessels and piping). It must be noted that RCM and IM are still very dependent on the performance of the material chosen to be fit for service. Any unplanned degradation of the containment system of the hazard source will result in unintended facility shutdown and a potential increase in the likelihood of loss of containment accidental events. The key is reducing unplanned maintenance of rotating equipment or unplanned replacement of containment systems, such as pipelines, vessels, and piping. Understanding of the operating conditions and corrosion risk is critical in minimizing the potential for unplanned design excursions that would affect the availability and safety of the production operations. The two main drivers of increased production are reliability and maintainability. Achieving the desired performance levels will be the result of managing these two factors. Plant and process reliability contributes toward a sustainable improvement in production availability.

The production availability depends significantly on the integrity of the hazardous containment system (pipeline) providing the high reliability and maintainability of the pipeline system. To improve the efficiency of a system, especially during the design and operational phase, consideration of corrosion risk is crucial. Designing with the understanding of corrosion risk can increase the performance of the pipeline operations. Increasing the performance of the pipeline operations, by understanding the corrosion risks, improves the profitability of the business unit and reduces the likelihood of an accidental, unintended release of hydrocarbons. The availability is illustrated in Figure 49.22 Any unplanned downtime of the pipeline operations affects the revenue.

Availability = Uptime/(Downtime + Uptime)

- Availability (A) = MTBF/(MTBF + MTTR) where
- MTBF = Mean time between failures
- MTTR = Mean time to repair or replace.

From an operational perspective the same equation aforementioned can be described as follows:

- Availability (A) = MTBM/(MTBM + MDT) where
- MTBM = Mean time between maintenance
- MDT = Mean downtime.

The key in all these equations is the reduction of the mean downtime (MDT) or mean time to repair or replace (MTTR) to a point where the availability is almost 100%. With the rotating equipment, the RCM program has provided the means in the contribution to the availability of an operating facility. Normally forgotten with rotating equipment are the materials chosen for the service to be provided. Since the focus is not to lose containment, the RCM program must also include the potential loss of containment from the failure of the equipment enclosures or rotating parts. From a corrosion risk perspective, the implementation of an IM process helps to ensure that all the fixed equipment will last the whole life cycle of the facility. Because rotating or fixed equipment does not last the life of the facility, downtime to maintain the equipment will always be required.

It is critical to understand that a facility is safe when it is operating at its designed parameter. When a facility system has a change that is not managed, then we have a high likelihood for a release of containment that will put the facility at risk from an accidental fire or explosion event. The incident realization model (Section 49.6.1.2) explains how a hazardous event (equipment malfunction, human act, or external force) can cascade into a hazardous situation resulting in personnel at risk. The loss of containment also results in a delay or loss of production for the facility. A typical large offshore platform producing 100,000 barrels/day has a potential to gross $10 million/day or $116/s. This means that any delay during maintenance or operations that can affect the required throughput will be at a delay or loss of $116/s or $7000/min. Those revenue numbers can motivate a design team to look differently on a delay involving either human factors or engineering causes. For instance, providing an extra CA at the design phase is insignificant compared with the cost of an unplanned maintenance or loss of containment scenario in a pipeline; the pipeline is meant to transport hydrocarbons from location to location, and any interruption of the availability will cost the deferred revenue and will demonstrate that there is not a full understanding or control of the business risks in an operating pipeline.

***49.6.1.2 Focal Point 2: Incident Realization*** The other focus that has to be emphasized is the understanding of the mechanisms of an incident and how equipment malfunctions, such as mechanical failure due to corrosion, can be the immediate cause of a hazardous situation. A hazardous event or cause can be realized into a hazardous situation from a hazard source. The model in Figure 49.23 illustrates that the most effective prevention of an accident arises from the understanding of the mechanism that can cause a release of a hazardous source. Corrosion risk leading to equipment failure (just like human or external factors) is a direct cause of release of hazardous energy and it is these controllable causes that have to be well understood during the engineering design of a pipeline system.

This understanding of accident mechanism is an essential part of the hazard identification process required by most global pipeline HSE regulations. This model provides an approach to manage and control the corrosion risk sources of hazards, and it also provides an approach to controls that can be put in place. This is also the recommended approach to address corrosion risk in the design of systems. The approach coincides with the incident realization model in Figure 49.23 and includes the following steps, in order of preference of effectiveness:

- Design to minimize the likelihood or consequences of errors and exposure to hazards.
- Design to mitigate the consequences of error or other acts.
- Provide the user with a warning of potential error or hazard.
- Train the user to prevent error or exposure to hazards.
- Write procedures to prevent error or exposure to hazards from occurring.

There are many factors that may contribute to equipment failure due to CRM issues. Some of those factors are illustrated in the diagram in Figure 49.24, which illustrates the six categories of key issues that contribute to corrosion risk or defects leading to equipment failure: measurements, materials, personnel, environment, methods, and machines. This approach can be effective in recognizing, identifying, and managing the risks.

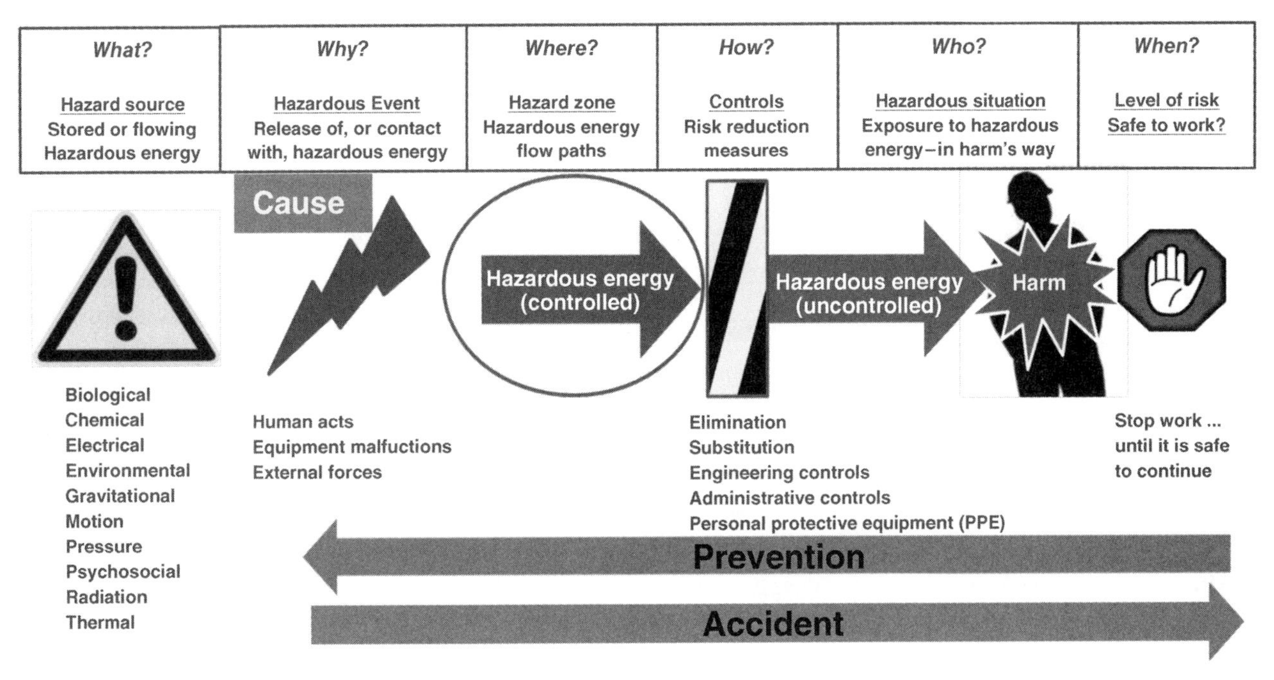


**FIGURE 49.23** Incident realization—physical model.

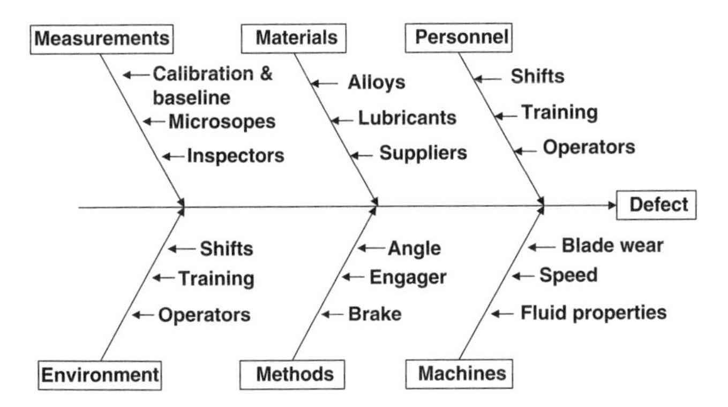


**FIGURE 49.24** Ishikawa Fishbone diagram of what can cause typical defect for equipment failure.

*49.6.1.3 Focal Point 3: Life-cycle Perspective* When dealing with a new pipeline system design, it is critical not only to understand the availability and the mechanism of an accidental event, but a project team must also appreciate the project design process. The most effective phase in affecting the design is at the earliest phases of a project. Even with the lack of detailed engineering information, the project team can maximize the attainable safety performance by utilizing an inherently safe approach to influence the design process. The influence on the magnitude of the contained hazards that can potentially be released is the first approach, and then the focus is on the mechanical or human error causes that would lead to a release of the hazard. The diagram in Figure 49.25 shows schematically the project and operational costs that are involved in attaining the maximum safety performance for an offshore facility. Influences on the design, at the concept phase, greatly improve the tendency to attain maximum safety performance. It is also apparent that any changes in design will cost more as the project progresses from concept to commissioning.

From a life-cycle perspective, it can be clearly seen how implementing corrosion risk controls at the appropriate (CAPEX) stage can save many later financial- and safety-related challenges, which affect the engineering contractor and the operator. The life-cycle perspective is a good, long-term, business approach that gives shareholders confidence in the costs and expenses of a company and its projects. The early identification and understanding of corrosion concerns and mechanisms result in the reduction of the long-term operational costs of a pipeline because of the predictability of the operating costs (inspection and maintenance) and the confidence of shareholders that the company executives have full control and predictability of expenditures. Shareholders do not like variation in expenses and income, because it demonstrates poor understanding of the business accounting. The long-term, life-cycle perspective for CRM is good for business.

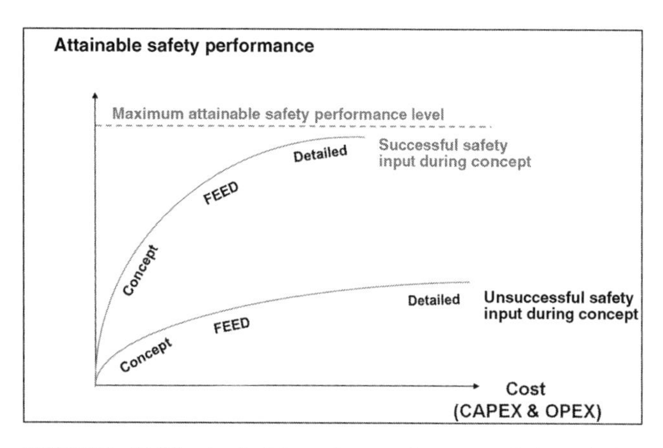


**FIGURE 49.25** Attainable safety performance during design process.

### 49.6.2 Safety Management Systems and Corrosion Risk

A SEMS requires a compilation of safety and environmental information. Some of the key elements of a SEMS where corrosion risk adds value and has a required input are detailed as follows:

- *Management of change*—changes to operating conditions and facilities during modifications or decommissioning can increase corrosion risks. For example, an undocumented change or deviation in the incoming fluids in a pipeline (more $H_2S$ or $CO_2$) or modifications in pressures and temperatures may increase the rate of corrosion in a pipeline project; operators and asset owners can benefit from corrosion expertise to identify and manage these risks.
- *Operating procedures*—corrosion inspection and management procedures are a critical portion of the operations and maintenance procedures. These procedures are the result of the design decisions (metallurgical selection) made in the detailed engineering phase of a pipeline project. It is critical that the operating procedure addresses the threat of corrosion on the process and support system.
- *Training*—the training of operators on the recognition and inspection of corrosion issues on an operating platform is very useful as another pair of eyes on the field. The same could be said about ROV operators with subsea pipeline systems; early recognition of corrosion concerns is critical to the success of a corrosion or IM plan. The skill and competency of corrosion inspectors are also invaluable since their findings and recommendations determine the forthcoming program to reduce the potential for a loss of integrity scenario in future operations.
- *Hazards analysis*—corrosion risk is one issue that is mentioned, but seldom addressed adequately during the hazard analysis process of a pipeline design and operation. Corrosion, as a cause of integrity failure, can result in significant environmental and safety effects on offshore or onshore pipelines, especially on environmentally critical landscapes and sea beds, high consequence areas. Corrosion risk inputs into hazard analysis are discussed more in the next section.
- *Incident investigation*—finding the underlying (or latent, root) causes is critical to preventing similar accidents; this is a critical assumption with a metal-based containment system dependency. Corrosion risk should be an integral question of an incident investigation process.
- *Emergency management*—this is crucial for high consequence areas where a loss of containment scenario has occurred. The physical design of the emergency systems is dependent on the understanding of the identified potential corrosion risk and predicted consequence. How an organization responds to the resultant incident is dependent on the planning for these accidental events after the prediction of the potential environmental or safety consequences.
- *Assurance of quality and mechanical integrity of equipment*—mechanical integrity addresses the activities to minimize the potential for loss of containment, whether it is the containment of fluids in rotating equipment or a fixed containment system like a pipeline. The quality assurance of a CRM program, from inspection to audits, ensures the most cost-effective means of ensuring system integrity and minimizing the potential for loss of containment.

### 49.6.3 Formal or Structured Hazard or Risk Assessment

The formal or structured hazard assessment techniques form the basis of the most effective methodology for ensuring that all the corrosion hazards and risks have been identified, assessed, and managed. In the offshore industries, the three most utilized guidelines, other than internal oil company standards, are as follows:

1 API RP 14J—Recommended Practice for Design and Hazards Analysis for Offshore Production Facilities;
2 ISO 17776:2000(E)—Petroleum and natural gas industries—Offshore production installations—Guidelines on tools and techniques for hazard identification and risk assessment;
3 NORSOK STANDARD Z-013—Risk and emergency preparedness assessment.

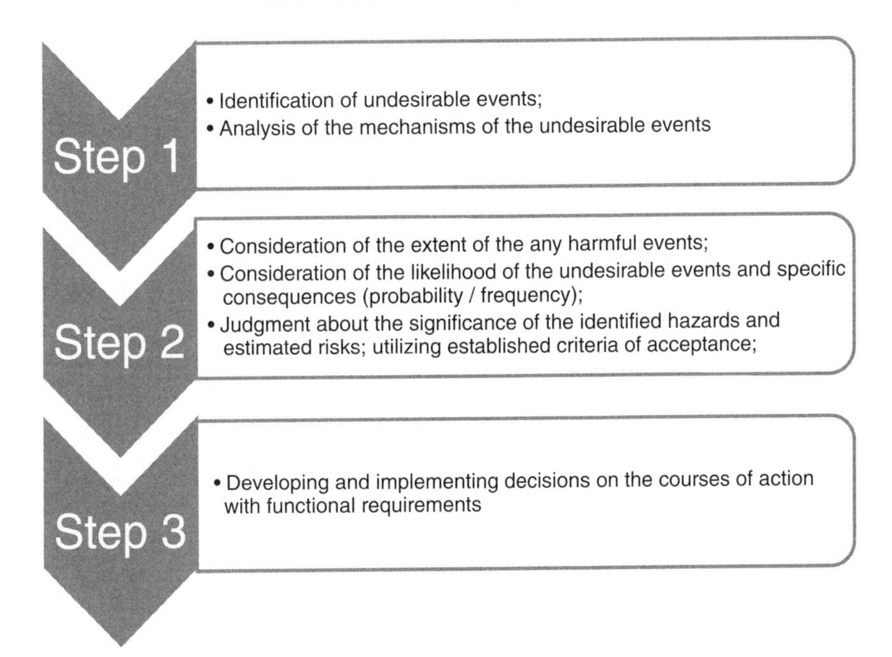


**FIGURE 49.26** Hazard analysis/management steps.

Typical hazard and risk assessments are divided into the following three steps (Figure 49.26):

1 Hazard identification;
2 Hazard analysis/assessment;
3 Hazard management.

During each step, there is a significant opportunity for corrosion risk input. Step 1—*Hazard identification*—provides the most cost-effective scenario.

***49.6.3.1 Step 1: Hazard Identification*** The main objective in hazard identification is to

a) Identify the undesirable events; and
b) Analyze the mechanism that results in an undesirable event.

An undesirable event can range from a production shutdown to a major accident; the different techniques are methods to systematically brainstorm undesirable events. The incident realization model (Figure 49.23) shows the required elements needed to describe the identified undesirable event. During a hazard identification brainstorming session, such as in a HAZID/HAZOP/What-If workshop, human factors could be found in the hazardous event or cause (Why?) or in the unintended exposure of personnel in:

- The hazardous zone (where?);
- The execution of the controls (how?); or
- In the resultant hazardous situation (who?).

The most important aspect here is first to identify the critical tasks involved in the potential release of a hazard source or unintended shutdown of critical equipment or production; these are the undesirable events. The key here is to identify, during the workshops, where corrosion management is required to maintain integrity of the process system. This means an understanding of the characteristics and internal/external environment of the contained fluids that can contribute to corrosion risks in the system. Second, it is important to establish an understanding of the potential corrosion mechanisms that will provide the guidance on how to mitigate the damaging effect. The life cycle approach should be taken during the thought process of identifying the corrosion hazard and postulating the potential mitigation solutions. This can vary from complete replacement of the containment system to modifications in operating or maintenance procedures to monitoring the decay process before any form of intervention occurs. These and other hazards should not be forgotten during hazard identification workshops.

The most important aspect during the hazard identification steps is to ensure that the facility maintains operation at its design intent. When in the operating mode, there is minimal likelihood of loss of containment because the system is maintaining and containing its operating pressure. It is only when a change in operating design intent is not properly managed that the likelihood of a loss of containment exists, e.g., due to the cooling of lines releasing all the contained hydrocarbons during a shutdown of units and production. The maintenance of the process system and design intent also includes the human operations. Any deviations from the human factors design intent, for operations and maintenance, could also result in potential undesirable events that could result in a potential production shutdown. The human factors aspects include, but are not restricted to, the inspection regime or methodology (how easy it is to inspect the area of concern), up to providing guidance on why there is a corrosion risk in the system (organization awareness). It must be reemphasized that a loss of containment scenario does result in a production shutdown of an offshore facility; it is an inclusive scenario.

The hazard identification step is the most critical step in any hazard analysis process. The typical hazard identification tools used in the oil and gas industries are:

- HAZID
  - Are corrosion risks included as the "causes" of potential loss of containment consequences (mechanical damage)?
  - Is feedback provided to personnel during a process upset or change?
  - What situational corrosion awareness do personnel have over the process?
  - What are the manning levels during the process?
  - What are the training/competency levels of personnel (inspectors) and how have these been verified? This will include corrosion risk SMEs as well.
  - Explain and define the roles and responsibilities of personnel?
  - What are the consequences of tasks omitted (inspections) or performed incorrectly?
- HAZOP
  - What type of personnel feedback is given during a process upset or change?
  - What situational corrosion awareness do personnel have over the process?
  - Are there more or no corrosion guide words in the workshop?
  - How do you ensure that personnel can inspect for corrosion easily?
  - How do you provide warnings or indications to the users?
- FMEA/FMECA
  - How do you capture the failure causes due to corrosion damage?
  - How do you capture the failure-detection methods involving corrosion damage?
  - How do you verify the decision-making times?
  - What are training/competency levels of personnel and how have these been verified? This will include corrosion risk subject matters experts?
- Bowtie assessments
  - Which are the defined barriers involving human interaction to prevent hazardous events and have they been defined to the right level?
  - What are the defined barriers involving human interaction to control consequences and have they been defined to the right level?
  - What are the training/competency levels of personnel and means of verification?
- What-if analysis
  - How do you capture occurrences of corrosion risks?
- Fault/event tree analysis (FTA/ETA)
  - How do you capture the corrosion-related tasks with the right level of detail?
  - How do you capture failure causes due to corrosion mechanical failure?
  - How do you capture failure-detection methods involving human interaction?
  - How do you verify decision-making times?
  - What are the training/competency levels of personnel and means of verification?
- Other lessons learned during the hazard identification step of major projects are
  - Inclusion of corrosion risk in the product availability leads to increased efficiency and less down time.
  - Early identification of corrosion hazards leads to less time and economical use of resources for a more appropriate design for operation and maintenance.
  - Focusing on the corrosion risks early in the conceptual and detailed design process and life cycle will increase the availability of the pipeline system and thus improve the profitability of the operations.

To conclude, the hazard identification step is the most critical step in any hazard analysis process because any mistakes in this process will propagate into the hazard analysis portion of the hazard management process.

***49.6.3.2 Step 2: Hazard Analysis/Assessment*** The main objective of hazard analysis/assessment includes

a) Hazard identification;
b) Consideration of the extent of any harmful events (consequence magnitude);
c) Consideration of the likelihood of the undesired events and specific consequences (probability/frequency);
d) Judgement about the significance of the identified hazards and estimated risks (assessment).

This step is highly dependent on the correct identification and analysis of the identified hazardous event. It is during this step where there is involvement with a screening criterion or target performance standards. This criterion is utilized for any decisions or assumptions utilized in the design process. This step evaluates or assesses the risk (to personnel and assets) utilizing the magnitude of consequences and likelihood of occurrence of the specific undesirable event. These undesirable events could be acute, such as exposure in a fire, explosion or toxic accidental event, or chronic, such as corrosion or long-term chemical/radiation exposure. All analysis or assessment methodologies have a corrosion element whether it is the cause of the hazardous event or contributes to the escalation leading to a hazardous situation (containment system having a portion of the pipeline that cannot meet the design intent). It must be emphasized that the difference between an analysis and an assessment is item d mentioned earlier—the experienced judgment about the significance of the identified hazards and estimated risks. The more relevant the experience of the individual or group is, the higher the quality of the assessment. Some of the hazard analysis work, at this step, may include the following:

- Quantitative risk analysis (QRA)
  - How do you verify the time for personnel to evacuate a facility?
  - What is the accuracy of the decisions and judgment times defined for personnel?
  - How are the alarms managed and prioritized?
- Physical effects modeling (PEM)
- Working environment studies
  - How are the levels of illumination, noise, and vibration suited to the work requirements for this system?
  - Does the access to equipment for maintenance provide sufficient work envelope?
  - How can musculoskeletal injuries be prevented during maintenance activities?
  - Explain how the manual materials handling philosophy considers user capabilities and safe limits.
  - Does the human–machine interface present the information required in a usable format?
  - Occupational exposure limits.
  - Indoor and outdoor climate limits for operations.
  - What are the vertical and horizontal clearances for the primary and secondary walkways?
  - What is the living quarter's design philosophy?
  - What are the design requirements for driller cabin's console?
  - What is the control room design philosophy?
- Human error analysis
  - Identify potential errors. (inspection errors)
  - Identify factors that may lead to the error.
  - Provide a mitigation to manage those leading factors.
  - Rank the risk for each error.
- Instrument, alarm, and valve criticality analyses
  - What are the instruments or valves needed to be used during emergencies?
  - What is the utilization frequency of the instruments or valves?
  - Do the alarms present useful information to include:
    i. Out-of-tolerance situation?
    ii. Needs to correct situation?
    iii. Which indications or warnings require action?

During this step, the hazard source has been released (controlled or uncontrolled) and the incident realization model, Figure 49.23, is progressing to the right commencing at the hazard zone and then assessing the effects of the controls in place to minimize the magnitude of the hazard situation. The QRA focuses on acute measures of personnel risk during major accidental events while the other studies involve understanding chronic risks that may potentially interfere with the operational integrity of a facility. The operational integrity not only means the equipment, but also the reliability, availability, and maintainability of personnel, which is crucial since the production capability is dependent on the training, skill, and competency of the operational staff; this is critical with corrosion inspectors in a pipeline operations environment. Incidents that could hamper the availability of critical staff or group of personnel are, for example, medical virus quarantine scenarios or Norwalk virus incidents that can shutdown accommodations due to lack of competent staff. One of the issues sometimes overlooked during the design of a facility is the unavailability of skilled personnel due to the remoteness of the facility.

Other lessons learned during the hazard identification step of major projects are

- The proposed corrosion mitigation philosophy should be established in the design phase of a pipeline project.
- Choice of materials from a life-cycle perspective is crucial; look at life cycle cost instead of CAPEX. This should include replacement and lost or deferred production costs.
- Consideration for in-pipe and outdoor operations working environment design, for facilities located not only in a cold harsh environment but in different water depths.
- Training and competence of personnel performing each job should be specified from the beginning, to reduce unnecessary special training on procedures required to operate and maintain equipment. This is critical for SMEs.

*49.6.3.3 Step 3: Hazard/Risk Management* At this step, a significant amount of iteration and sensitivity work provides value-added risk reduction measures from the analysis and assessments. The functional requirements are established based on the risk reduction measures that have been developed or progressed. At this portion of the design life cycle, any changes in the design would be cost prohibitive. Operational controls are the focus during this phase of the life cycle with the most effective solution and effort focused on meeting the original design intent. It must also be emphasized that all these assessments should be reinvestigated once more relevant data with statistical confidence from experience have been collected. Studies and decision tools utilized include, but are not restricted to,

a) Risk matrix
b) Cost benefit analysis (CBA)
   - How has redesign cost of equipment, due to human factors deficiencies, been addressed?
c) Safety integrity level (SIL) assessment
   a. Have you captured all failure causes due to human error?
d) RAM (reliability, availability, maintainability) analysis
   - How do you maintain the equipment?
   - How do you access equipment that has been maintained?
   - How do you monitor equipment upsets?
   - What are the key hazards when the equipment requires maintenance?
   - What feedback have operations personnel given for improving the safety and efficiency of the maintenance activities?
   - What working at heights is involved?
   - What manual and material handling activities are involved?
e) RCM
f) Asset integrity management (AIM)

The aspects of organization or management system design, procedural development, and training/skills/competency requirements include the human factors focus. At this stage, if the hazard identification scenario is incorrect, then the management of risks is significantly flawed. Additionally, inadequate testing for operability, after a completed design change, can seriously affect hazard management. Ineffective indications for deviations occurring in the process for operations and maintenance personnel can result in ineffective hazard management. Inadequate procedures for logging lessons learned can result in design issues that remain unresolved. These unresolved hazards may lead to near-miss incidents or major accident events depending on the extent of impact on a system. The balance of corrosion business risks in the oil and gas industry can be simply explained using Figure 49.27 describing the business flow schematic for pipeline corrosion and IM. The balance of the reactive and active corrosion monitoring and the proactive corrosion risk assessment is based on controlled and uncontrolled cost/budget. Reactive corrosion monitoring is mostly based on an unanticipated cost of a potential or a realized loss

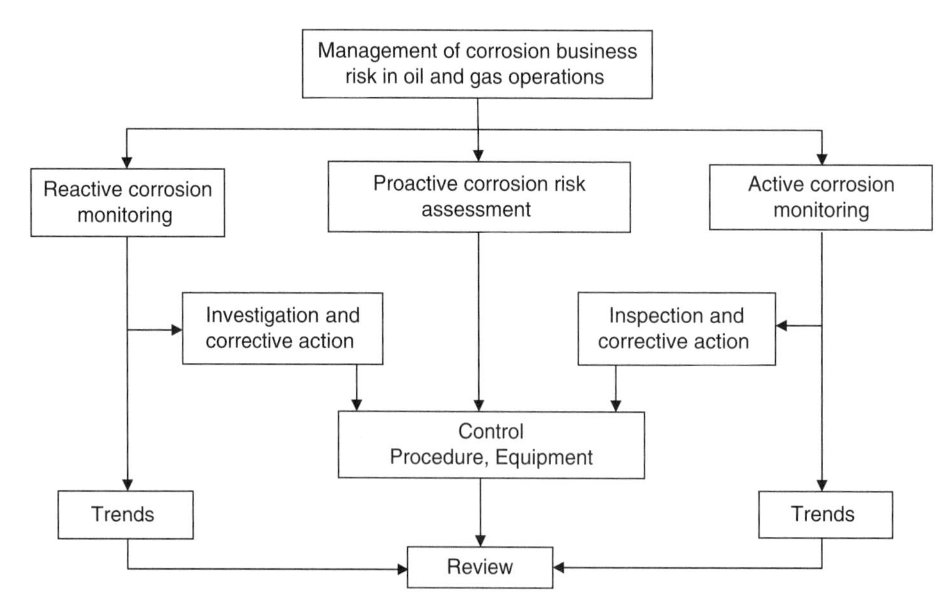


**FIGURE 49.27** Business work flow schematic for pipeline corrosion and IM.

of containment scenario. This would also include the resulting implementation of lessons learned from actual or similar field consequences and effects. This is not the ideal part of the business model, due to uncontrolled costs. The proactive portion of the business model is the active monitoring and proactive risk assessment work. These are budgeted costs or expenditures that are controlled and anticipated. The management of corrosion business risk should be focused on the controlled costs rather than the effects of the uncontrolled and unmanageable costs with a reactive method of execution. The key is to minimize the reactive corrosion monitoring process to a level that is as low as possible.

## 49.7 SUMMARY: RECOMMENDATIONS AND FUTURE STRATEGIES

The following general predictive recommendations are postulated for offshore pipelines, risers, and related topsides pressure plant:

Expect stricter offshore regulations in line with originally mandated onshore pipelines, and clearly applicable to the full life cycle.

Further develop and introduce the best combination of prescriptive and goal setting or risk-based corrosion control programs.

Anticipate the principles of the safety case to be applicable for all matters related to corrosion and IM.

Expect better use of inherently safer design and reliability and likewise better use of existing techniques, and a close observance of how new research techniques evolve, with a fair opportunity to use them when appropriate.

Encourage the use of design reappraisals for safety and environmental context; if the industry does not oblige, mandates to that effect can be expected, spurred by public opinion.

Eliminate the tendency for minimal compliance and encourage reasonable risk-based conservatism to facilitate safe operations.

Encourage third party verifications for high risk pipelines and risers.

Despite the history and guidance, it is evident that the lessons learned during the application of corrosion risk in design continue to evolve with more engineering design. This chapter has discussed relevant corrosion and IM guidance within human factors, and hazard analyses along with pragmatic lessons learned and challenges from the authors' experiences on major offshore projects. The challenge continues in integrating more corrosion risk during hazard and risk management activities in all engineering design activities. The pace of development is fast as the offshore and subsea industries tackle the more aggressive high pressures, high temperatures, and high velocities expected by deeper exploration and reservoirs. Hopefully this chapter has provided the impetus, guidance, and momentum to address these issues. But the reader should note that this is just the "tip of the spear."

## ACKNOWLEDGMENTS

The opinions and interpretations offered in this chapter rely heavily on the knowledge and experience accrued by the authors and are not connected in any way to any companies or published affiliations. The information provided is where possible generic and/or provided for educational purposes only and given in good faith. The reader should therefore validate any data or implications carefully and responsibly. The authors gratefully acknowledge the valuable support and learnings generated from many companies and institutions, as well as many researchers. In particular Atkins, Genesis-Technip, Wood Group, Deepwater Corrosion, Lloyds Register, Maritime Academy Vallejo, OIS/Oceaneering, YARD Ltd, National Nuclear Corporation (UK), Universities of Houston, Manchester, Liverpool, Loughborough (UK), UT Austin, Texas A&M, NACE International, IMECHE, IMAREST, ASM, ASME the OTC annual Conference, and most importantly to numerous industry colleagues and academics, for valuable knowledge exchange and often contentious debate.

## REFERENCES

1. Cullen, D. (1990) *The Public Inquiry into the Piper Alpha Disaster*, H.M. Stationery Office, London, UK, ISBN 0101113102.
2. Singh, B., Jukes, P., Wittkower, B., and Poblete, B. (2009) Offshore integrity management 20 years on—overview of lessons learnt Post Piper Alpha. Proceedings of Offshore Technology Conference, Houston, TX, Paper OTC 20051-PP.
3. Singh, B., Britton, J.N., and Flanery, D. (2003) Offshore corrosion failure analyses—a series of case histories. Proceedings of CORROSION/2003, NACE International, Houston, TX, Paper No. 03114.
4. Singh, B. and Britton, J.N. (2001) Offshore risk based corrosion integrity management—a new methodology. Proceedings of CORROSION/2001, NACE International, Houston, TX, Paper No. 01008.
5. International Association of Oil and Gas Producers (2008) Asset integrity—the key to managing major incident risks. Report No. 415, International Association of Oil and Gas Producers, London, December 2008. Available at: http://www.ogp.org.uk/pubs/415.pdf.

6. Royal Society Study Group (1983) Risk Assessment, Royal Society, London, January 1983, ISBN 0854032088, 9780854032082.

## BIBLIOGRAPHY

Al-Hashem, A., Carew, J., Riad, W., and Abdullah, A. (2002) Cavitation corrosion behavior of L-80 carbon steel in sea water. Proceedings of CORROSION/2002, NACE International, Houston, TX, Paper No. 02209.

American Petroleum Institute (1998) API RP 75 (R2013), Recommended Practice for Development of a Safety and Environmental Management Program for Outer Continental Shelf (OCS) Operations and Facilities. American Petroleum Institute, Washington, DC, July 1998.

American Petroleum Institute (2001) API RP 14J (R2013), Recommended Practice for Design and Hazards Analysis for Offshore Production Facilities. American Petroleum Institute, Washington, DC, April 2001.

Barton, N.A. (2003) Erosion in elbows in hydrocarbon production systems: review document. Health and Safety Executive (HSE) Research Report 115, HSE Books, Sudbury, Suffolk, U.K., 2003.

Brown, B. and Nešić, S. (2012) Aspects of localized corrosion in an $H_2S/CO_2$ environment, Proceedings of CORROSION/2012, NACE International, Houston, TX, Paper No. 0001559.

Canadian Association Petroleum Producers (CAPP) (2009) Mitigation of Internal Corrosion in Sweet Gas Gathering Systems. June 2009, CAPP, Calgary, Alberta, Canada, Paper No. 0014.

Canadian Association of Petroleum Producers (CAPP) (2009) Mitigation of Internal Corrosion in Sour Gas Gathering Systems. June 2009, CAPP, Calgary, Alberta, Canada, Paper No. 0013.

Dawson, J., Bruce, K., and John, D.G. (2001) Corrosion risk assessment and safety management for offshore processing facilities. Offshore Technology Report 1999/064, U. K. Health and Safety Executive, London, 2001.

Health and Safety Executive (HSE) (2001) *Reducing Risks, Protecting People, HSE's Decision-Making Process*, HSE Books, Sudbury, Suffolk, UK.

International Association of Oil and Gas Producers (2005) Human Factors—a means of improving HSE performance. Report No. 368, International Association of Oil and Gas Producers, London, June 2005. Available at http://www.ogp.org.uk/pubs/368.pdf.

International Association of Oil and Gas Producers (2011) Human factors engineering in projects. Report No. 454, International Association of Oil and Gas Producers, London, August 2011. Available at http://www.ogp.org.uk/pubs/454.pdf.

International Organization for Standardization (2010) ISO 17776:2000(E)—Petroleum and natural gas industries—offshore production installations—guidelines on tools and techniques for hazard identification and risk assessment. International Organization for Standardization, Geneva, Switzerland, 2010.

Jepson, P. (1999) Technical research and development presentation, $CO_2$ corrosion modeling. Proceeding of NACE International Offshore Corrosion Conference (NOOCC 99), December 7, 1999, New Orleans.

Kletz, T. (2009) *What Went Wrong? Case Histories of Process Plant Disasters and How They Could Have Been Avoided*, 5th ed., Elsevier/Gulf Professional Publishing, Burlington, MA.

Lees, F.P. (1996) *Loss Prevention in the Process Industries—Hazard Identification, Assessment and Control*, 2nd ed., Butterworth-Heinemann, Oxford, UK, Vol. 1.

Little, B.J., Wagner, P.A., and Mansfeld, F. (1997) *Microbiologically Influenced Corrosion*, NACE International, Houston, TX.

Montgomery, M.E. (1998) Proceeding of the Deepwater Conference Presentation—International Association of Drilling Contractors (IADC) August 27, 1998, Houston, TX.

National Transportation Safety Board (2003) Natural gas pipeline rupture and fire near Carlsbad, New Mexico, August 19, 2000. Pipeline Accident Report NTSB/PAR-03/01, Washington, DC, 2003.

Nyborg, R. (2010) $CO_2$ corrosion models for oil and gas production systems. Proceedings of CORROSION/2010, NACE International, Houston, TX, Paper No. 10371.

Nyborg, R. (2005) Controlling internal corrosion in oil and gas pipelines. *Oil and Gas Review*, (2), 70.

Papavinasam, S., Doiron, A., Shen, G., and Revie, R.W. (2004) Prediction of inhibitor behavior in the field from data in the laboratory. Proceedings of CORROSION/2004, NACE International, Houston, TX, Paper 04622.

Pillai, A.P. (2012) Direct assessment data categorization, integration and risk indexation—a novel approach. Proceedings of CORROSION/2012, NACE International, Houston, TX, Paper No. 0001479.

Poblete, B., Singh, B., and Dalzell, G. (2007) The inherent safe design of an offshore installation—a leap of faith. Proceeding of Chemical Plant Safety Conference, Texas A & M, Mary Kay O'Connor Process Safety Center, Oct 2007.

Singh, B. (2013) *Corrosion risk, corrosion allowance, and integrity management—a pragmatic risk based methodology*. Risk Management of Corrodible Systems, Washington DC, NACE International, Houston, TX.

Singh, B. and Krishnathasan, K. (2009) Pragmatic effects of flow on corrosion prediction. Proceedings of CORROSION/2009, NACE International, Houston, TX, Paper No. 09275.

Singh, B., Krishnathasan, K., and Ahmed, T. (2008) Predicting Pipeline Corrosion. Pipeline & Gas Technology (October 2008).

Singh, B. and Lindsay, A. (2011) From the North Sea to the Gulf of Mexico—volume 2. Research Topical Symposium: Making the Link between Corrosion Research & Best Practices, NACE International, Houston, TX.

Standards Norway (2010) NORSOK Standard Z-013: Risk and emergency preparedness assessment, 3rd edition, Standards Norway, Lysaker, Norway. October 2010.

U.S. Department of the Interior (2013) 30 CFR Part 250—U.S. Department of the Interior, Bureau of Safety and Environmental

Enforcement, 30 CFR Part 250 [Docket ID: BSEE–2012–0011], RIN 1014–AA04. Oil and Gas and Sulphur Operations in the Outer Continental Shelf—Revisions to Safety and Environmental Management Systems, April 5, 2013.

Videla, H.A., Swords, C.L., and Edyvean, R.G.J. (2002) Corrosion products and biofilm interactions in the SRB influenced corrosion of steel. Proceedings of CORROSION/2002, NACE International, Houston, TX, Paper No. 02557.

# PART VIII

# CASE HISTORIES

# 50

# BUCKLING OF PIPELINES UNDER REPAIR SLEEVES: A CASE STUDY—ANALYSIS OF THE PROBLEM AND COST-EFFECTIVE SOLUTIONS

Arnold L. Lewis II
*Research and Development Center, Saudi Aramco, Dhahran, Saudi Arabia*

## 50.1 INTRODUCTION

Saudi Aramco operates over 16,000 km of carbon steel pipeline for the transportation of hydrocarbon products. Most of these pipelines are buried as they traverse the Saudi Arabian terrain through areas of oversaturated high-salinity sand known in Saudi Arabia as sabkha. The water table depth in sabkha is affected by seasonal and tidal conditions and ranges from the surface to more than a meter below the surface. The close proximity of the water table to the surface in sabkha means that many sections of these pipelines are partially submerged in the high-salinity water at all times. It naturally follows that this hot, high-salinity environment with high oxygen concentration differentials is very corrosive to unprotected carbon steel.

Because of the corrosive environment through which the pipelines traverse, new buried pipelines were first coated externally with a coal tar enamel product. Later a tape wrap product was used followed by a characteristically higher reliability epoxy coating system in the 1980s. The epoxy coating system is still used today. For additional external corrosion mitigation, cathodic protection is also applied to the coated pipelines.

Unfortunately, the tape wrap and coal tar enamel systems combined with cathodic protection did not provide adequate long-term external corrosion protection, especially in the wetter areas. The tape wrap and coal tar enamel systems often developed areas where the coating became disbonded from the pipeline. Water would often leak and collect under the disbonded areas. Cathodic protection could not penetrate through or follow the same path as the water under the disbonded coating. This weakness of the tape wrap and enamel systems created localized areas of external corrosion on the lines that could not be detected with surface measurements of the cathodic potentials and could not be compensated for with additional cathodic protection current.

The earlier epoxy coating systems also experienced a number of initial failures primarily due to inadequate surface preparation procedures. However, failures in the epoxy systems typically resulted in a measurable cathodic potential reduction, which could be compensated for with additional cathodic protection current.

Since pipeline leaks are clearly not desirable, a solution to this problem was required. Studies concluded that the most cost-effective solution to pipeline external and internal corrosion was to carry out the following:

a) Intelligent instrument scraping is done to locate the areas with external and internal corrosion and to quantify the severity of the corrosion.
b) The locations with external or internal corrosion rated above a certain level of severity are excavated and the existing coating is removed.
c) A prefabricated repair sleeve, normally rolled from plate, is cut to a length that will cover the area of

*Oil and Gas Pipelines: Integrity and Safety Handbook,* First Edition. Edited by R. Winston Revie.

corrosion. These repair sleeves resemble a piece of pipe split horizontally with a bore slightly larger than the outside diameter of the line pipe. Repair sleeves are normally fabricated in 3 m lengths but they can be cut to cover shorter lengths of corrosion or butt welded together to cover corrosion sites longer than 3 m.

d) At the repair site, the two pieces are clamped over the exterior of the original pipeline at the location of the corrosion.
e) The repair sleeve is welded back together and then carefully welded to the exterior of the original pipeline at each end of the sleeve, creating a short length of double-walled pipeline with a sealed annular space between them.
f) Once a repair sleeve is completely welded and sealed to the pipeline, it and the adjacent pipeline area are grit-blasted and thoroughly cleaned.
g) Next, an epoxy coating is field-applied to the exterior of the pipeline and repair sleeve.
h) The pipeline and repair sleeve are again buried and fall under the influence of the cathodic protection system.

Steps a–h were done on virtually all of the company's existing scrapeable pipelines, to the point where there are over 50,000 repair sleeves on company pipelines. Obviously, the primary purpose of repair sleeves is to prevent leaks from pipelines. For this purpose, repair sleeves have proven very effective and repair sleeves are still applied today to reinforce areas with external or internal corrosion on a pipeline.

### 50.1.1 Statement of the Buckle/Collapse Problem

In early 2002, during normal scraping and cleaning operations, a scraper became stuck in a pipeline. A second scraper was sent down the line to free the first scraper, and it too became stuck at the same location. The only solution was to remove that section of pipeline. Figure 50.1 shows an example of what was found. Clearly, these scrapers rammed into a buckle/collapse in the pipeline whereupon they could travel no further. This dent, collapse, or buckle was located directly beneath a repair sleeve. Initial failure analysis concluded that the inner original pipeline failed by inward collapse, caused by high pressure in the sealed annulus between the repair sleeve and the pipeline. The repair sleeve itself suffered no apparent damage. Unfortunately, this failure mechanism became all too common with at least 30 similar failures occurring under pipeline repair sleeves within the company. Considering the more than 50,000 repair sleeves on company pipelines, an understanding of this phenomenon became an important topic.

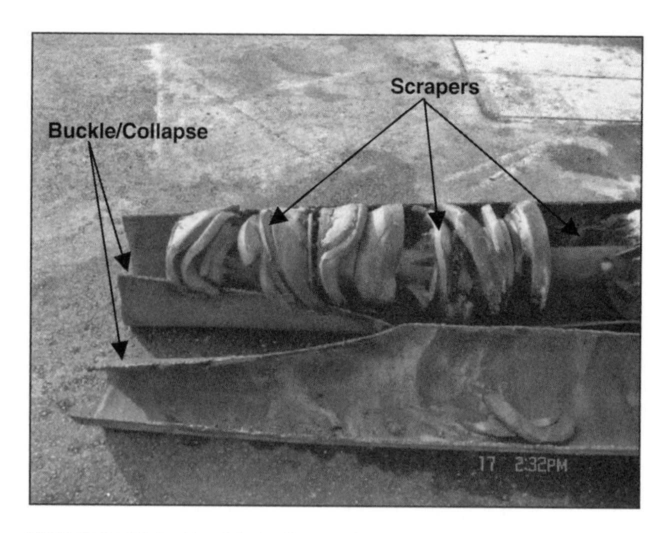


**FIGURE 50.1** Buckle/collapse along with several stuck scrapers in a 24-in. pipeline. The repair sleeve has already been removed in the photograph.

### 50.1.2 Observations

It was valuable to summarize the buckle/collapse failures with a list of general observations.

- When installed, the sealed annulus between a repair sleeve and the pipeline contains atmospheric air at ambient pressure.
- After installation of a repair sleeve, the pipeline and repair sleeve are coated externally with epoxy coating.
- The pipeline and repair sleeves are generally buried, although there are occasional sections above grade.
- The pipeline and repair sleeves are cathodically protected.
- Sometime after installation, significant elevated pressure develops in the annulus under many repair sleeves.
- The elevated pressure is due to high pressure gas sealed in the annulus.
- The high-pressure gas is 100% hydrogen.
- In at least 30 cases, the hydrogen gas pressure was sufficient to cause buckling failure of the pipeline; see Figures 50.2 and 50.3 for examples.
- The location of the buckle/collapse on the pipeline is not consistent, but has generally been at the 3 or 9 o'clock position.
- The date at which most repair sleeves or the epoxy coating was applied to the pipelines is not known with accuracy. However, a best estimate concludes that the shortest buckle/collapse failure time was approximately 5–7 years after repair sleeve installation.
- All buckle/collapse failures have occurred in areas of sabkha or high water table and where the pipeline and

**FIGURE 50.2** Buckle/collapse failure from inside a 48-in. pipeline.

**FIGURE 50.3** Buckle/collapse failure in an 18-in. pipeline after removal of the repair sleeve.

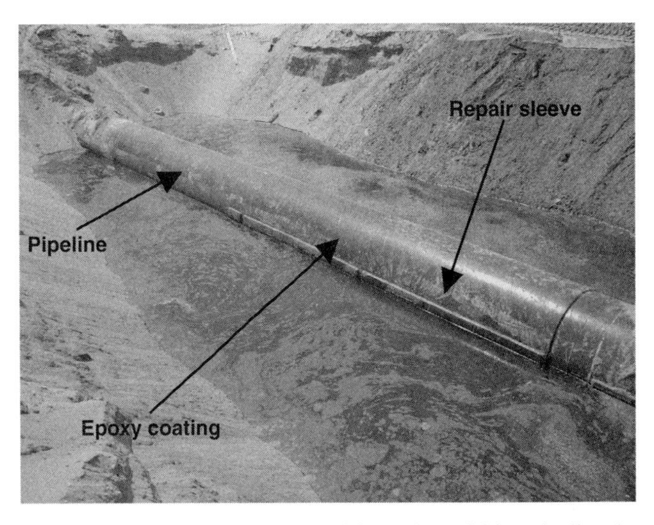


**FIGURE 50.5** Buckle/collapse failure in a 36-in. pipeline in which the epoxy coating appears visually undamaged.

repair sleeve were at least partially immersed in sabkha water; see Figure 50.4.

- Most of the buckle/collapse failures were on pipelines where the internal temperature can reach approximately 60 °C. However, one or two failures have occurred on pipelines where the internal temperature was ambient.
- Visually, the epoxy coating has been badly damaged at the site of many buckle/collapse failures, but for other cases, it appears visually undamaged; see Figures 50.5 and 50.6.
- In most failures, scale is present on the repair sleeve exterior; see Figure 50.7. A sample of scale collected from this failure site was identified as 98% calcium sulfate.

**FIGURE 50.4** Location of a buckle/collapse failure prior to pumping the sabkha water.

**FIGURE 50.6** Buckle/collapse failure in a 48-in. pipeline in which the epoxy coating is badly damaged and scaled.

**FIGURE 50.7** Buckle/collapse failure in a 48-in. pipeline showing bad coating and scale.

**FIGURE 50.8** Interior of a 36-in. pipeline at the location of a buckle/collapse failure showing no visible internal corrosion. Note that writing from the pipe manufacturer is still visible in the foreground.

**FIGURE 50.9** Photograph of an approximate 18 in. × 36 in. area in the interior of a 48-in. pipeline at the location of a buckle/collapse failure showing visible internal corrosion.

- Water was found in the annulus of at least one buckle/collapse failure.
- Visual examination of the pipeline interior of most buckle/collapse failures gives the impression that internal corrosion is not significant; see Figure 50.8. However, at least one buckle/collapse failure did have visible internal corrosion; see Figure 50.9.
- No indication of bacterial-related corrosion, biomass, iron sulfide, or odor of hydrogen sulfide gas, was observed anywhere on the exterior or interior of any buckle failure site.
- Hydrogen gas, of insufficient pressure to cause pipeline buckle/collapse, has been found in thousands of annuli under repair sleeves.
- Hydrogen gas has been found in the annuli under repair sleeves of several above-grade sections of pipelines.

It was with this short background information that a long-term experimental study for buckling under pipeline repair sleeves was initiated. The objectives for the study were

- To determine the source(s) for hydrogen gas trapped in the annuli of pipeline repair sleeves;
- To develop solutions to mitigate buckle/collapse failures under pipeline repair sleeves.

To achieve these goals, a great deal of experimental work was conducted during the course of a 4-year study of the buckle/collapse problem. However, in the interest of brevity, none of the experimental work will be discussed here. What follows are the conclusions that came from the study and answers for the two objectives of the study.

## 50.2 STUDY CONCLUSIONS

After 4 years of effort, the study team arrived at a number of conclusions related to the buckle/collapse problem. These have been arranged into seven categories and the highlights for each are shown.

### 50.2.1 Conclusions for Sources of Hydrogen in an Annulus of a Pipeline Repair Sleeve

- The quantity of hydrogen that would be released from weld metal placed during repair sleeve installation is insignificant with respect to the buckle/collapse failures [1].
- Mobile hydrogen from steel manufacturing that potentially could be released from the repair sleeve steel or the pipeline steel is insignificant with respect to the buckle/collapse failures [2–6].
- During repair sleeve installation, native water may be trapped in an annulus.
- Carbon steel corrosion within an annulus that contained water would proceed only until the oxygen trapped in the annulus was consumed.
- Due to the high pH of native waters, carbon steel corrosion within the annulus could not produce hydrogen.
- Any hydrogen that might be produced by direct or indirect microbial activity in an annulus would be insignificant with respect to the buckle/collapse failures [7,8].
- There is no source of hydrogen production within an annulus that would contribute significantly to buckle/collapse failures.
- Hydrogen in above-grade repair sleeve annuli does not come from acid rain corrosion followed by hydrogen permeation through the repair sleeve wall [9].
- Hydrogen found in annuli under pipeline repair sleeves results from pipeline internal corrosion and/or from external cathodic protection.

### 50.2.2 Factors Affecting Hydrogen Permeation from inside the Pipeline into an Annulus

- It should be assumed that most pipelines, including those considered dry by company operations, do contain sufficient condensed water for some amount of internal corrosion.
- Condensed water inside a pipeline must be <pH 7 or hydrogen will not be produced as a by-product of the corrosion process. If hydrogen is not produced as a by-product of the corrosion process, then pipeline internal corrosion does not contribute to hydrogen trapped in an annulus.
- Both hydrogen sulfide and/or carbon dioxide gases, if present in a pipeline with condensed water, will result in condensed water with a $pH < 7$.
- Hydrogen permeation through carbon steel will result from sweet, carbon dioxide only, corrosion. Hydrogen sulfide gas is not a prerequisite for hydrogen permeation due to corrosion. Previous work in Saudi Aramco's laboratories has shown the rate of hydrogen permeation from sweet, carbon dioxide, corrosion to be similar in magnitude to that from excessive cathodic protection.
- If hydrogen sulfide is present at concentrations above the low ppm level, then hydrogen permeation resulting from pipeline internal corrosion is significantly higher. Previous work in Saudi Aramco's laboratories has shown the rate of hydrogen permeation in the presence of hydrogen sulfide to be orders of magnitude greater than that from either excessive cathodic protection or sweet corrosion.
- Oxide layers on the outer surface of a pipeline can reduce hydrogen permeation into an annulus from pipeline internal corrosion. However, oxide layers are complex and easily damaged, which makes their effect on hydrogen permeation difficult to predict [10–14].
- Hydrogen permeation rate decreases with increasing thickness of the steel through which it permeates [15–18].
- Hydrogen permeation rate increases with increasing temperature [19].

### 50.2.3 Factors Affecting Hydrogen Permeation from outside the Repair Sleeve into an Annulus

- The amount of hydrogen present in an annulus, resulting from cathodic protection, is proportional to the current flowing through the repair sleeve. However, the fraction of total hydrogen, resulting from cathodic protection, that permeates into the steel, decreases with increasing cathodic protection current density.
- Hydrogen permeation rate increases with increasing temperature [19].
- Hydrogen permeation rate decreases with increasing thickness of the steel through which it permeates [15–18].
- For steels >4 mm thickness, hydrogen permeation resulting from cathodic protection is undetectable at cathodic protection current density levels less than approximately $100\ mA/m^2$.
- Oxide layers on the inner surface of a repair sleeve can reduce hydrogen permeation into an annulus resulting from cathodic protection. However, oxide layers are complex and easily damaged, which makes their effect on hydrogen permeation difficult to predict [10–14].
- Hydrogen permeation resulting from cathodic protection is lower in water that contains high levels of

dissolved calcium and magnesium. These ions, in combination with the cathodic protection reduction chemistry, produce carbonate and hydroxide scales that lower hydrogen permeation [20].

- Hydrogen permeation resulting from cathodic protection is somewhat higher when the surface of the steel is exposed to degraded epoxy coating [21–23].

### 50.2.4 Factors Affecting the Rate of Annulus Pressure Increase

- Hydrogen permeation through the pipeline wall into an annulus resulting from pipeline internal corrosion is important and this includes all of the variables discussed earlier.
- Hydrogen permeation through the repair sleeve wall into an annulus resulting from cathodic protection is important and this includes all of the variables discussed earlier [24–34].
- The rate of increase of the annulus pressure decreases as the annulus gap increases.

### 50.2.5 Factors Affecting the Time Required for a Buckle/Collapse Failure

The study concluded it is not feasible to predict the time required for an individual repair sleeve buckle/collapse failure. The degree of complexity for this failure mechanism is significant and the time to failure is a function of,

- The rate of annulus pressure increase is important and this includes all of the variables discussed earlier.
- The pressure required for a buckle/collapse failure decreases with increasing pipeline outside diameter and with increasing repair sleeve length.
- The pressure required for a buckle/collapse failure increases with increasing pipeline steel strength and with increasing pipeline wall thickness.
- The pressure required for a buckle/collapse failure will be less if the pipeline wall has defects such as general corrosion loss or pitting corrosion. This variable is valid whether the defects are on the internal or external wall of the pipeline.
- The pressure required for a buckle/collapse failure is less for a pipeline whose cross section is not circular.

### 50.2.6 Main Sources and Considerations for Hydrogen Gas Trapped in the Annulus of a Pipeline Repair Sleeve

Recall that this heading is one of the two objectives for the study. The team concluded that there are only two significant sources for hydrogen gas trapped within an annulus of a pipeline repair sleeve and these are

- Excessively high cathodic protection current density results in the production of hydrogen that permeates the repair sleeve wall into the annulus where it is trapped in sufficient quantity to be a primary source of high-pressure gas in the annulus of buckle/collapse failures.
  - The excessive cathodic protection current density occurs in low resistivity environments for poorly coated repair sleeves on otherwise well coated pipelines.
  - Under the aforementioned conditions, the excessive cathodic protection current density does not create field measurable excessive cathodic protection potentials.
- Some condensed water is found in almost all pipeline process environments. Therefore, where condensed water is present in a pipeline process environment, hydrogen produced as a by-product of pipeline internal corrosion will permeate the pipe wall into an annulus where it will be trapped in sufficient quantity to also be a primary source of high pressure gas in the annulus of buckle/collapse failures. In addition, if hydrogen sulfide is present in the pipeline process environment along with condensed water, then pipeline internal corrosion could become the most significant source of annulus hydrogen and this would produce a situation with high probability of a buckle/collapse failure.

  Excessive cathodic protection was identified as a primary source of hydrogen gas trapped in the annuli of pipeline repair sleeves. However, it is essential to understand clearly the following:
- Excessive cathodic protection on a pipeline or repair sleeve can occur even if the cathodic protection system has been designed perfectly, is operating perfectly, and is maintained perfectly.

This is because excessive cathodic protection on a pipeline or repair sleeve is the natural and predictable result for any cathodically protected pipeline that has long lengths of pipe with good or high-resistivity external coating connected directly to short lengths of pipe with bad or low-resistivity external coating. In addition, and typical for these cases, an aggravating factor is that the short lengths of pipe with bad coating are usually found in the highest conductivity soils. By itself, this combination can be entirely responsible for excessive cathodic protection because the current generated by the cathodic protection system will naturally focus to the areas of least electrical resistance, which are the areas with bad coating in high conductivity soils. In other words, cathodic protection is a primary source of hydrogen gas trapped in the annuli of pipeline repair sleeves, but bad external coating on the repair sleeve is responsible for creating the excessive cathodic protection.

### 50.2.7 Solutions to Mitigate Buckle/Collapse Failures under Pipeline Repair Sleeves

Through the course of the study, the team developed both a short-term solution for existing repair sleeves that contain high-pressure gas and two long-term actions that would help mitigate the problem.

- Short-term solution: The short-term solution is to vent trapped hydrogen gas from welded repair sleeves using a procedure based on the following steps,
  - With over 50,000 repair sleeves in the field, a list was prepared that identified those locations with higher risk of a buckle/collapse failure. Higher risk locations are sites that have suffered from previous buckle/collapse failures, sites in perennially wet sabkha, and critical pipelines that would create a problem if shut down to repair a buckle/collapse failure.
  - After the higher risk sites were identified, excavate the repair sleeves in these locations.
  - Using a purpose made remotely operated drill with nitrogen gas purge, carefully, following a specially developed procedure, drill a small hole into the repair sleeve to vent any trapped hydrogen gas therein.
  - After venting any trapped hydrogen gas, drill the hole larger, tap it, and install a standard NPT threaded plug fitting as a seal. Hydrogen gas that later permeates into the annulus of a repair sleeve equipped with a threaded plug will leak out around the threads of the plug and never build up sufficient pressure to cause a buckle/collapse. This very simple and low-cost procedure has now been successfully completed on several thousand repair sleeves in the company without incident. In addition, this is a permanent solution for the buckle/collapse problem of the particular repair sleeve.
  - Be prepared for the event in which pipeline process fluids, rather than hydrogen gas, are found when the small hole is drilled through the repair sleeve wall. In this situation, knock a short length of round steel rod of slightly larger OD than the drilled hole, but with a tapered tip, into the open hole to block the release of process fluids. This will plug the leak. Next, weld over the plug for a permanent seal. This procedure is a permanent solution for the process fluid leak of the particular repair sleeve. Note in this case that a collapse is impossible due to the open path into the pipeline.
- Long-term solutions: The team developed two long-term actions that would help to mitigate future buckle/collapse failures on company pipelines.
  - For new repair sleeve installations, place a threaded plug in the wall of the repair sleeve when it is being fabricated and prepared. Hydrogen gas that might later permeate into the annulus of the repair sleeve equipped with a threaded plug will leak out around the threads of the plug and never build up sufficient pressure to cause a buckle/collapse. This procedure is a permanent solution for the buckle/collapse problem of the particular repair sleeve.
  - Coat the pipelines and welded repair sleeves with a better coating: It was clearly pointed out that bad coating on pipeline repair sleeves is the reason for excessive cathodic protection current that ultimately leads to a buckle/collapse failure. Therefore, if the bad coating is replaced with a good coating this will solve the buckle/collapse problem under welded repair sleeves.

## 50.3 SUMMARY

Poor coating in low resistivity environments results in excess cathodic protection on pipeline repair sleeves. This, coupled with pipeline internal corrosion, caused high-pressure hydrogen gas trapped in the annulus of repair sleeves to buckle the pipeline under the repair sleeve. A simple remedial solution is to drill the repair sleeve wall, vent the trapped hydrogen, and then fill the hole with a threaded plug.

## ACKNOWLEDGMENT

The author would like to thank Saudi Aramco for the permission to publish this work and all associated figures.

## REFERENCES

1. Castillo, M., Rogers, C.E., and Payer, J.H. (2000) Hydrogen permeation and hydrogen concentration measurements to model diffusible hydrogen in weld metal. Proceedings of NACE Corrosion 2000, Paper No. 00467.
2. SAE Aerospace (2003) AMS 2759/9B: Hydrogen Embrittlement Relief (Baking) of Steel Parts, April 2003.
3. ASTM International Nisbett, E.G. (editor) (2005) Steel forgings: design production, selection, testing and application: (MNL 53), Pages 16 and 126, 2005.
4. Randerson, K., Carey, H.C., and Schomberg, K. (2003) A Model for calculating solubility of hydrogen in steel. *Steelmaking*, 30 (5), 369–378.
5. Sarychev, A.F., Gorostkin, S.V., Sarychev, B.A., and Nikolaev, O.A. (2006) Changing the hydrogen content of alloy steel in relation to the technology used to cool cast slabs. *Metallurgist*, 50(11–12), 577–580.
6. Zinchenko, S.D., Filatov, M.V., Efimov, S.V., Goshkadera, S.V., and Dub, A.V. (2004) Technological aspects of hydrogen

removal with the use of a ladle-type vacuum-degassing unit. *Metallurgist*, 48(11–12), 553–556.

7. Little, Brenda J., Ray, Richard I., and Pope, Robert K. (2000) The relationship between corrosion and the biological sulfur cycle. Proceedings of NACE Corrosion 2000, Paper No. 00394.
8. Miyamoto, K. (editor) Renewable Biological Systems for Alternative Energy Production, Food and Agricultural Organization of the United Nations, Services Bulletin—128.
9. Gordon, John D., Nilles, Mark A., and Schroder, LeRoy J. (1995) USGS Tracks Acid Rain, Fact Sheet FS-183-95, United States Geological Survey, Reston, VA, 1995.
10. Casanova, T. and Crousier, J. (1996) The influence of an oxide layer on hydrogen permeation through steel. *Corrosion Science*, 38, 1535–1544.
11. Chen, R.Y. and Chen, Y.D. (2005) Examination of oxide scales of hot rolled steel products. *ISIJ International*, 45 (1), 52–59.
12. Dey, S., Mandhyan, A.K., Sondhi, S.K., and Chattoraj, I. (2006) Hydrogen entry into pipeline steel under freely corroding conditions in two corroding media. *Corrosion Science*, 48, 2676–2688.
13. Schmitt, G., Sadlowsky, B., and Noga, J. (2000) Inhibition of Hydrogen Effusion from Steel—an Overlooked and underestimated Phenomenon, Proceedings of NACE Corrosion 2000, Paper No. 00466.
14. Song, R.H., Pyun, S.I., and Oriani, R.A. (1991) The hydrogen permeation through passivating film on iron by modulation method. *Electrochemical Acta*, 36, 825.
15. Crolet, J.L. and Bonis, M.R. (2001) Revisiting hydrogen in steel, part I: theoretical aspects of charging, stress cracking and permeation. Proceedings of NACE Corrosion 2001, Paper No. 01067.
16. Crolet, J.L. and Bonis, M.R. (2001) Revisiting hydrogen in steel, part II: experimental verifications. Proceedings of NACE Corrosion 2001, Paper No. 01072.
17. Dean, F.W.H., Carroll, M.J., and Nutty, P.A. (2001) Hydrogen Permeation flux data correlation with in-line corrosion monitoring techniques mapped over location and time. Proceedings of NACE Corrosion 2001, Paper No. 01636.
18. Tems, R.D. and Dean, F.W.H. (2000) Field application of a new, portable, non-intrusive hydrogen monitor for sour service. Proceedings of NACE Corrosion 2000, Paper No. 00471.
19. Addach, H., Bercot, P., Rezrazi, M., and Wery, M. (2005) Hydrogen permeation in iron at different temperatures. *Materials Letters*, 59, 1347–1351.
20. Ou, K.C. and Wu, J.K. (1997) Effect of calcareous deposits formation on the hydrogen absorption of steel. *Materials Chemistry and Physics*, 48, 52–55.
21. ASTM Standard Test Method for Arsenic in Paint (2002) ASTM D 2348.
22. Tendulkar, A.N.P., and Radhakrishnan, T.P. (1994) A galvanic study of the role of arsenic on the entry of hydrogen into mild steel. *Corrosion Science*, 36, 1247–1256.
23. Walker, M. (1999) Arsenic and old paint. *New Scientist*, 162 (2188), 20.
24. Banerjee, K. and Chatterjee, U.K. (2001) Hydrogen permeation and hydrogen content under cathodic charging in HSLA 80 and HSLA 100 steels. *Scripta Materialia*, 44, 213–216.
25. Benson, J., and Edyvean, R.G.J. (1998) Hydrogen permeation through protected steel in open seawater and marine mud. *CORROSION*, 54 (9), 732–739
26. Brass, A.M., and Collet-Lacoste, J.R. (1998) On the mechanism of hydrogen permeation in iron in alkaline Medium. *Acta Materialia*, 46 (3), 869–879.
27. Van Delinder, L.S. (editor) (1984) *Corrosion Basics, An Introduction*, NACE, Houston, TX, p. 120.
28. Flis, J., Zakroczymski, T., Kleshnyal, V., Kobiela, T., and Dus, R. (1999) Changes in hydrogen entry rate and in surface of iron during cathodic polarization in alkaline solutions. *Electrochimica Acta*, 44, 3989–3997.
29. Flis-Kabulska, I., Flis, J., and Zakroczymski, T. (2007) Promotion of hydrogen entry into iron from NaOH solution by iron-oxygen species. *Electrochimica Acta*, 52, 7158–7165.
30. Flis-Kabulska, I., Zakroczymski, T., and Flis, J. (2007) Accelerated entry of hydrogen into iron from NaOH solutions at low cathodic and low anodic polarizations. *Electrochimica Acta*, 52, 2966–2977.
31. Hagiwara, N. and Meyer, M. (2000) Evaluating hydrogen stress cracking of line pipe steels under cathodic protection using crack tip opening displacement tests. Proceedings of NACE Corrosion 2000, Paper No. 00380.
32. Song, F.M., Kirk, D.W., Graydon, J.W., and Cormack, D.E. (2003) A new pipeline crevice corrosion model with $O_2$ and CP. Proceedings of NACE Corrosion 2003, Paper No. 03177.
33. Tawns, A. and Oakley, R. (2000) Cathodic protection at a simulated depth of 2500 m, Proceedings of NACE Corrosion 2000, Paper No. 00134.
34. Thomason, William H. (1988) Quantitative measurement of hydrogen charging rates into steel surfaces exposed to seawater under varying cathodic protection levels. *Materials Performance*, 27, 633.

# 51

# IN-LINE INSPECTION ON AN UNPRECEDENTED SCALE

Stephan Brockhaus,[1] Hubert Lindner,[1] Tom Steinvoorte,[2] Holger Hennerkes,[3] and Ljiljana Djapic-Oosterkamp[4]
[1]*ROSEN Technology & Research Center, Lingen, Germany*
[2]*ROSEN Europe, Oldenzaal, The Netherlands*
[3]*ROSEN Headquarters, Stans, Switzerland*
[4]*Statoil, Kårstø, Norway*

## 51.1 INTRODUCTION

In 2008, Rosen inspected the world's longest subsea gas pipeline, the 1173 km long Statoil Langeled pipeline [1] connecting Nyhamna in Norway and Easington in the United Kingdom, in a single in-line inspection (ILI) run (Figure 51.1). Langeled pipeline is designed to carry more than 70 million cubic meters of dry gas per day, thus supplying the equivalent of more than 20% of the United Kingdom's gas needs. The pipeline had never before been inspected in a single ILI run.

## 51.2 CHALLENGING DESIGN AND OPERATING CONDITIONS

The planned inspection was challenging not merely because of the pipeline's unparalleled length but also on account of its challenging design and operating conditions. Langeled is a multi-diameter pipeline with high wall thickness. For the first 632 km, that is, approximately up to the Sleipner platform, the pipeline diameter is 42 in. before increasing to 44 in. for the remaining 541 km. Similarly, the pipeline's wall thickness varies from 29.1 to 62 mm in the 42 in. section and from 23.3 to 50.6 mm in the 44 in. section. Although constant internal diameters (ID) are given in the specifications, the intermediate subsea installations (valves and tees) in fact restrict tool passage at the Sleipner platform. The ID is 1016 mm in the 42 in. section and 1066 mm in the 44 in. section. The minimum bend radius is 5D. Langeled gas pipeline is coated with epoxy on the inside and asphalt and concrete on the outside. The pipeline's operating conditions make the inspection task even more complicated: Langeled is characterized by high product flow velocity and pressure. Operated at design capacities, the gas velocity increases over the pipeline length from 4 to 9.5 m/s, with associated gas pressures of 220 bar at Nyhamna and down to 75 bar at Easington. Corresponding gas densities vary approximately between 61 and 185 kg/m$^3$.

## 51.3 COMBINED TECHNOLOGIES FOR RELIABLE INSPECTION RESULTS

Given the challenging nature of the project, the most suitable technologies for the planned baseline survey were carefully selected. The two companies finally agreed to combines the two complementary corrosion measurement methods, magnetic flux leakage (MFL) and shallow internal corrosion (SIC) with high-resolution geometry inspection in one single tool [2]. MFL, which is used for measuring relative wall loss, is a versatile and reliable method for determining the geometry of metal loss in pipelines.

### 51.3.1 Magnetic Flux Leakage

The principle of MFL involves the magnetic saturation of a ferromagnetic sample, here the pipe wall, with powerful

*Oil and Gas Pipelines: Integrity and Safety Handbook*, First Edition. Edited by R. Winston Revie.

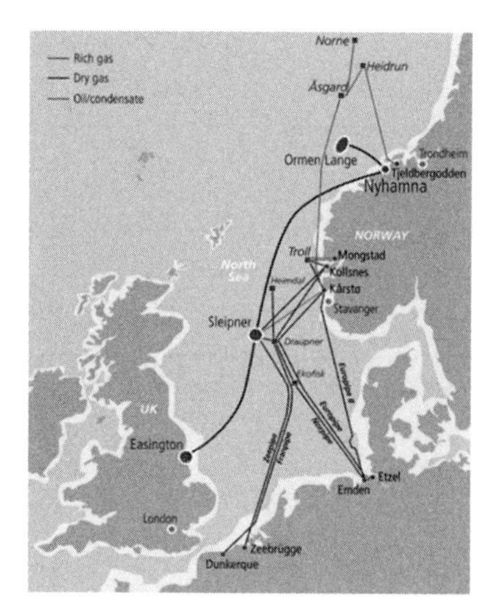


**FIGURE 51.1** Route of Langeled pipeline, the longest subsea gas pipeline in the world.

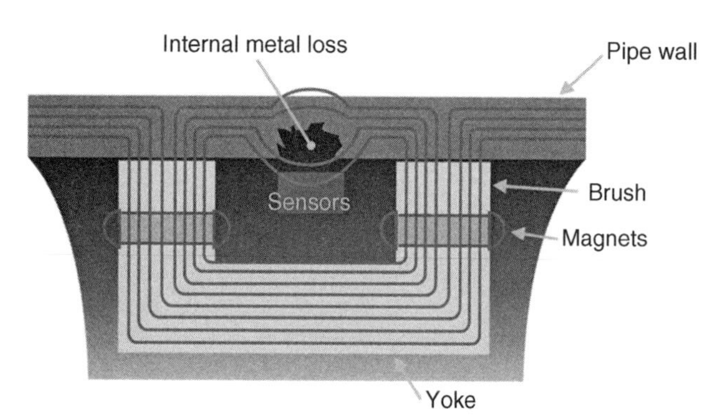


**FIGURE 51.2** MFL basic principle.

permanent magnets. Figure 51.2 illustrates the general setup. When inserting a MFL tool into a pipe, a magnetic circuit with defined magnetic flux line distribution is created (red lines). The ferrous pipe wall material guides the magnetic flux line in longitudinal direction of the pipe, also referred to as axial magnetization. Without metal loss anomaly and due to the fact that the magnetic flux density depends on the cross section of the pipe wall, that is, the wall thickness, the flux distribution remains undisturbed in the magnetically saturated pipe material. Where internal or external metal loss exists, the cross section is changed, and the flux density is altered due to the reduction in wall thickness. This forces the magnetic field to leak out of the pipe material. The magnetic configuration of the ILI tool is determined by yokes, magnets, and brushes. The latter are the contact-interfaces to the pipe wall. A measurement device with Hall effect sensors is located in between the brushes. The sensors allow a precise measurement of the leakage field. The data are used to assess feature classification, geometry, and severity.

***51.3.1.1 Magnetic Saturation of Pipe Wall*** For MFL inspections, a high magnetization level is of significant importance. The pipe wall has to be magnetically saturated for an accurate detection and sizing of metal loss features, in particular for external defects. For standard carbon steel this is typically achieved with magnetic field strength between 10 and 30 kA/m. For this range, no hysteretic effects occur and magnetic flux density and, therefore, signal amplitudes of metal loss features depend only linearly on the magnetization level. A certain duration is required until a pipe wall is magnetically saturated, since the magnetic flux evolves diffusively through the wall from the inner to the outer pipe. This duration strongly depends on the wall thickness and the permeability and conductivity of the pipe wall material. The source of the magnetic field, that is, the ILI tool, propagates with a certain tool speed through the pipeline during the inspection. Hence, not only pipe wall material properties and wall thickness determine magnetic saturation of the wall, but also the tool speed influences the magnetic flux inside and its evolution through the pipe wall. Focusing now on very heavy wall pipelines, special attention has to be attributed to the question if proper magnetization levels and, consequently, metal loss detection and sizing performance can be achieved when inspecting very heavy wall pipelines.

In response to the aforementioned situation and to address the important saturation question upfront an inspection, ideally during the design phase of the ILI tool, ROSEN developed a unique and dynamic approach on the basis of finite element method (FEM). The approach allows to take into account different material properties of the pipe material, the mechanical configuration of the pipeline, that is, diameter and wall thickness, as well as the ILI tool's speed and its magnetic configuration. Figure 51.3 illustrates the magnetization level through a pipe wall for the static case (blue), for optimum (red), and excessive (green) tool velocities, respectively. The dashed blue lines indicate the inner and outer pipe wall, respectively. The abscissa represents the distance to the centerline of the pipe, whereas the ordinate gives the magnetization values.

A closer inspection of Figure 51.3 results in the finding that magnetic saturation of the pipe wall, that is, magnetization levels above 10 kA/m, is achieved for the pure static case or speed values up to a distinct optimum tool speed. The magnetization level is well in between 10 and 30 kA/m throughout the entire pipe wall. With further increase in tool speed, a certain threshold is reached above which the magnetization level at the outer pipe wall drops below 10 kA/m, resulting in degradation in the external feature sizing capability. Proper detection of external features may still be provided. The magnetization level at the inner pipe wall slightly increases but is still below 30 kA/m.

Starting during the conceptual design phase with a first approach for the magnetic configuration, ROSEN works iteratively on the magnetic setup of MFL-based ILI tools until an optimum is achieved with the principal design. This procedure offers here an efficient and cost-effective approach

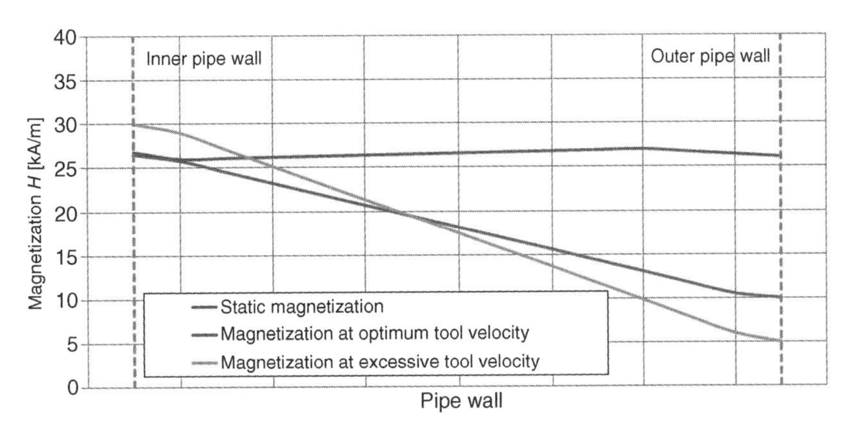


**FIGURE 51.3** Evolution of magnetization through pipe wall for different tool speeds.

for the design of high-resolution metal loss monitoring tools for the inspection of heavy walled offshore pipelines.

### 51.3.2 Eddy Current Technology

The second technology incorporated by the tool is SIC [3], an eddy current (EC)-based technology enabling high-resolution geometric measurements of internal corrosion defects, independent of wall thickness.

The EC testing method is based on the approach of generating electrical currents in conductive materials, here the pipe wall. Figure 51.4 summarizes the EC principle. An alternating current with defined amplitude and frequency is applied to a coil system. The driving current in the sending coil generates an alternating primary magnetic field, which causes ECs to flow in the surface of the nearby steel sample, here the pipe wall, by mutual inductance. The currents in the pipe wall produce a secondary magnetic field that is opposite to the primary field inducing it. Defect damages, such as corrosion, lead to a change of the ECs flow direction. This change influences the mutual inductance, that is, the resulting field of the superposed primary and induced secondary magnetic field is changed. This can be described by a variation in the electrical impedance of the coil, that is, its ohmic resistance and inductive reactance. The impedance is usually measured across a bridge circuit by which imbalance can be measured accurately. On the basis of this imbalance, metal loss can be detected and their properties determined by the evaluation of the amplitude and the phase shift between the input and output signals.

The distribution of ECs inside the steel sample depends on material properties such as conductivity, magnetic permeability and the presence of defects, and the frequency of the coil input signal. They also determine the so-called skin depth of ECs. The skin depth is a measurement of the distance where an alternating current can penetrate beneath the surface of a conductive material, that is, the pipe wall. For frequencies being suitable for ILI of pipelines, the skin depth of ECs for carbon steel lines is well below 1 mm. Thus, the EC approach can be considered to be a surface sensitive

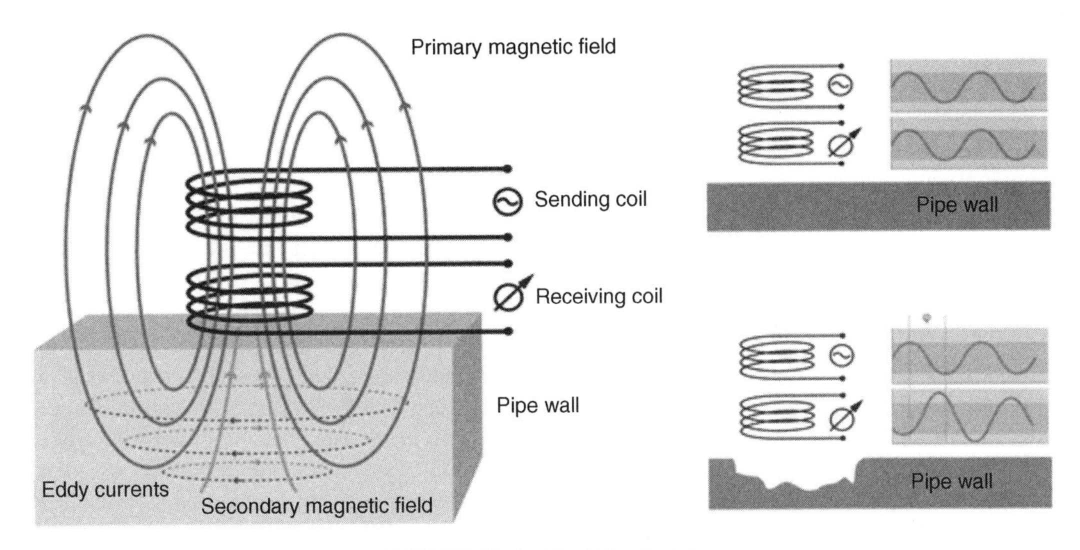


**FIGURE 51.4** The EC principle.

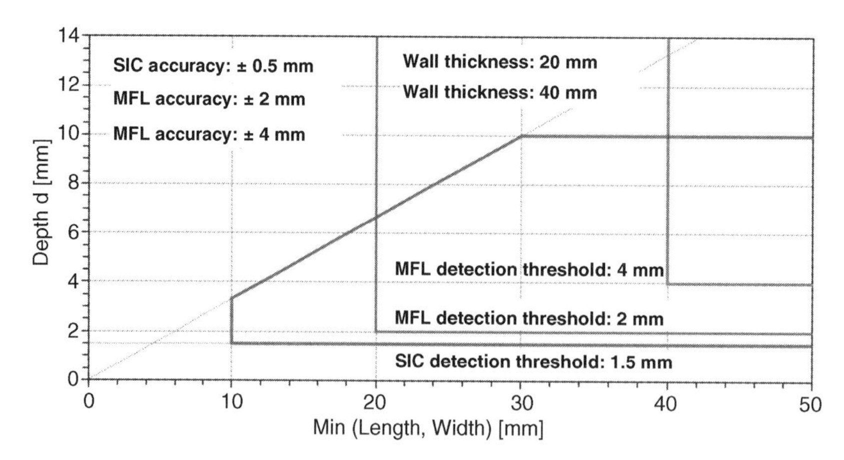


**FIGURE 51.5** Optimal sizing regimes of the MFL and SIC inspection methods.

method that is independent of the wall thickness. It allows a highly sensitive characterization of surface metal loss defects even for very heavy wall segments in a pipeline. This is in stark contrast to MFL measurements where signals are mainly affected by volume changes of metal loss and, therefore, depend strongly on the thickness of the pipe wall.

### 51.3.3 Synergetic Use of Magnetic Flux Leakage and Eddy Current

The strongest effect on the EC mutual inductance results from distance variations between sensing coils and pipe wall. This separation is also referred to as liftoff. Related signals show a saturation behavior from an increase in distance between coil and wall. For the SIC technology, saturation occurs at approximately 10 mm, the maximum liftoff value between EC sensor system and pipe wall. Vice versa, if metal loss reveals a depth comparable to the maximum allowable liftoff, the EC signal is saturated as well. Figure 51.5 illustrates the scenario and shows the optimal metal loss sizing regime of the SIC inspection technology (red). For comparison purposes, the optimal MFL sizing regime is depicted for two different wall thicknesses, 20 mm (blue) and 40 mm (green). The illustration shows the maximum sizable depth of the SIC technology and, thus, the threshold for internal metal loss defect sizing regimes between SIC and MFL. Although detection with SIC of metal loss defects with depths above the maximum depth is still possible, sizing capabilities are degraded.

In contrast to SIC, MFL is capable to detect external defects as well and is best suited for deeper and larger internal volume metal loss defects. This implies that MFL can be used at defect depths for which SIC is predestined, but with reduced performance, that is, degraded MFL detection threshold and sizing accuracy for high wall thickness pipes. The degradation is based on the dependence of the MFL technology on the wall thickness and the cross-sectional area governing the magnetic flux density. Both MFL detection threshold and sizing accuracy are in the order of 10% of the wall thickness, that is, both quantities increase for higher wall thicknesses. Since the SIC technology is based on the surface bounded EC approach, both SIC detection threshold and sizing accuracy do not depend on the thickness of heavy walled pipelines.

Results obtained with EC can likewise also assist MFL feature depth sizing algorithms for depths where MFL is favorized, through the maximum signal indications and the provision of increased resolution of feature width sizing. Both measurement approaches complement one another and show synergetic effects.

Although SIC is surface bounded and insufficient to accurately size deep features, its feature sizing can be employed to customize the MFL sizing model and improve the sizing accuracy. Figure 51.6 provides an example of top of line corrosion [4]. The figure consists of four individual panels and compares the results of SIC (1st panel) and MFL (2nd panel) with automated ultrasonic testing (AUT) data (4th panel). A profile of the MFL signal amplitude in circumferential direction through the same corrosion area is given in the 3rd panel. The horizontal and vertical axes represent longitudinal and circumferential direction, respectively. A closer inspection of the MFL panel results in the finding that MFL smears individual adjacent signals into a single feature, which makes accurate sizing difficult. The circumferential proximity of multiple features increases the MFL signal amplitudes and causes overestimation. Contrary to MFL, SIC correlates with the result of AUT data quite well, indicating the increased resolution in circumferential direction. The SIC information allows to get not only an accurate depth estimate of the metal loss, but also a precise determination of the lateral defect dimension in circumferential direction. This implies that accurate profiles of feature shapes in circumferential direction, as provided by SIC, allows an effective compensation for the signal superposition

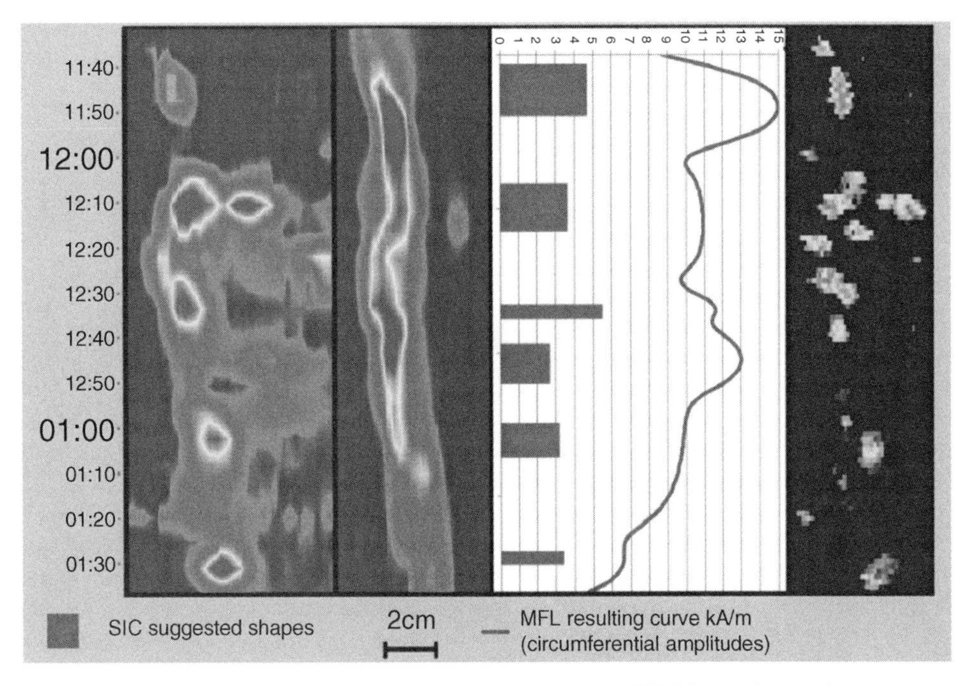


**FIGURE 51.6** Optimal sizing regimes of the MFL and SIC inspection methods.

that occurred during the MFL inspection. It converts the smeared MFL signal into several distinct features, and therefore, yields a more accurate and reliable feature sizing.

The example shown in Figure 51.6 stresses the benefits that can be achieved when combining MFL with an SIC technology for the inspection of heavy wall pipelines.

Last but not least, high-resolution geometry measurement is provided by combining EC technology-based touchless deformation measurement with systematic monitoring of the mechanical position of caliper arms. Full coverage of the internal walls of the pipeline is ensured by means of two inspection planes with a total of 108 caliper arms plus the associated SIC contour-following proximity sensors.

**FIGURE 51.7** Customized in-line inspection tool upon launching in Nyhamna, Norway.

## 51.4 OVERCOMING THE MULTI-DIAMETER CHALLENGE

To meet the specific challenges of Langeled pipeline, a customized combo tool was designed, equipped in the front segment with MFL technology for general corrosion measurement and in the rear segment with caliper arms featuring SIC sensors for SIC and high-resolution geometry measurements (Figure 51.7).

In order to minimize the wear on the internal coating, the shape of the brushes, which introduce the magnetic flux in the wall, was optimized to match the ID's of the pipeline and avoid point loads.

Before this specially designed inspection tool was launched, however, it was necessary to first ensure that it could definitely pass through all parts of Langeled pipeline.

For this purpose, a customized gauging tool was first sent through Langeled pipeline (Figure 51.8). Featuring both a gauge and a bend plate, the design was adapted to the multi-diameter application to confirm the bend sizes and ID of the pipeline. To mark the special significance of Langeled pipeline for the cooperation of the countries and companies involved, the tool featured the Norwegian and British flag as well as a "mailbox". The "mailbox" was used by the chairmen of the Norwegian administration and the collaborating companies Shell and Gassco to send letters to their

**FIGURE 51.8** The gauge tool used to verify Langeled's internal diameters (1016 and 1066 mm) and bend sizes (5D radius) also safely delivered the letters contained in its letterbox.

counterparts in England. The subsea mail safely reached its recipients.

## 51.5 OVERCOMING THE CHALLENGE OF HIGH PRODUCT FLOW VELOCITY IN THE PIPELINE

To overcome the specific challenge of the high gas velocities in Langeled pipeline and to compensate for tool accelerations due to friction changes, the combo tool was equipped with a speed control unit (SCU). The SCU enables the inspection run to be conducted at a pre-programmed target tool speed by providing controlled gas bypass to reduce tool velocity.

In controlling the bypass flow, different factors must be taken into account, since not only the tool design (e.g., bypass area, differential pressure) determines the capacity of the SCU, but also pipeline characteristics such as wall thickness, changes in diameter, coating type, and gas density. Whereas a high tool differential pressure and a large bypass area enhance the performance of the SCU, high gas densities, as present in Langeled pipeline, have the opposite effect.

To prevent the front segment of the combo tool from being pushed through the pipeline by the trailing unit, the bypass area on the rear segment must be remarkably larger. Despite this large bypass area, the rear segment must still be capable of accommodating indispensable components (electronics, batteries, storage etc.). The specially designed ILI tool meets this challenge as follows: the flow circulates around and through parts of the tender's tool body before diversion through the tool body at the front segment and final release through the SCU.

Since flow simulations indicate that the combination of high bypass velocities and high gas density leads to increased loads on the measuring unit of the rear segment (80–90 kg), the flow circulating around the tender must be systematically guided. Utilizing short caliper arms in combination with effective notches and holes strategically placed in the cups as well as specially designed lamella fins between the rear cup and the measurement unit permits partial diversion of the flow away from the measurement unit (Figure 51.9). With this measure, the load on the caliper arms was reduced by 60%, guaranteeing secure and reliable measurement, even though the space is additionally used to provide the required bypass.

If the tool is allowed to pass through the pipeline at product flow velocity, this may well mean that its speed is too high to warrant precise measurements, as in the case of Langeled. Whereas tool speed can be diminished by means of flow rate reductions, this results in costly throughput losses for the duration of the inspection. For both reasons, it was extremely important to overcome the high-flow velocity challenge posed by Langeled without compromising either measurement accuracy or pipeline capacity. The tool velocities achieved during the actual combined ILI run, which took place in August 2009, ranged from 1.5 to 3 m/s, resulting in a total travelling time of 127.3 h. The flow rate during the inspection was 40 million standard cubic meters per day.

## 51.6 OVERCOMING THE DISTANCE CHALLENGE: TOOL WEAR DURING THE LONGEST CONTINUOUS INSPECTION RUN EVER CONDUCTED

Due to the enormous length of the pipeline of 1173 km, tool wear was another major challenge posed by the Langeled inspection project. Because the tool differential pressure increases due to higher friction in the smaller section of

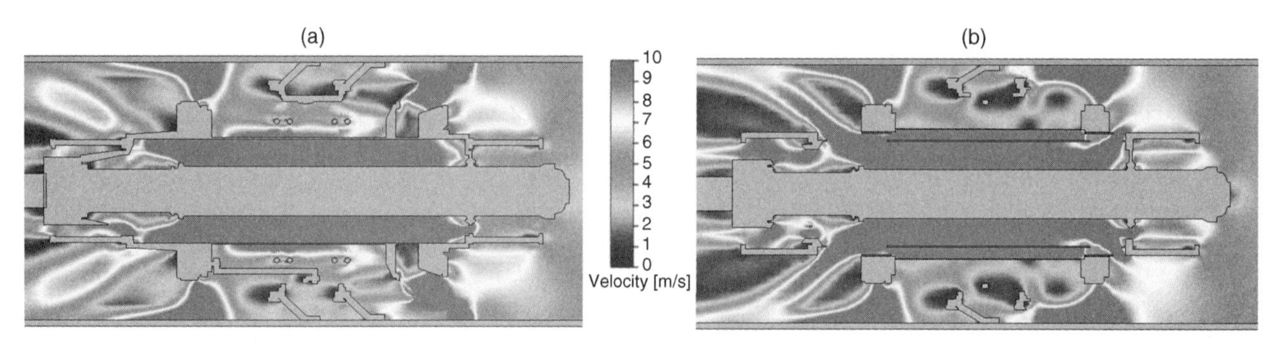


**FIGURE 51.9** Flow simulations: (a) with guided flow and (b) without guided flow.

the pipeline, this problem was exacerbated by the pipeline's dual diameter design.

To overcome this challenge, customized polyurethane cups were designed to limit tool wear during the passage through the 42 in. section while nevertheless providing sufficient sealing and carrying capabilities in the 44 in. section. Since the differential pressure of the tool is determined by the magnet forces applied at different wall thicknesses and by cup friction, the magnet strength was optimized for Langeled pipeline conditions, thereby ensuring full magnetic saturation while limiting the differential pressure and hence cup friction and wear.

The total number of polyurethane cups mounted on the front segment of the combo tool provided three sealing planes, giving a certain safety margin with regard to cup wear and unintended bypass. Additional tool centralization to reduce cup wear in the first segment was provided by the magnet forces themselves as well as by a supporting guiding disc. In the rear segment, wheels were used to support centralization and to improve run behavior (Figure 51.7).

The average differential pressures measured during the actual inspection were 2.0 bar in the 42 in. section and 0.8 bar in the 44 in. section. The tool constantly rotated and was in a good condition upon receiving.

Both the cleaning and the inspection tool were fitted with a powerful acoustic transmitter and remotely operated vehicles (ROVs) with antennas capable of detecting the acoustic signal as well as the magnetic field of the inspection tool were available to locate the tool, if needed.

## 51.7 SUMMARY

The combined inspection tool was built specifically to overcome the great challenges posed by the length, multi-diameter design, high wall thickness, high pressure and high gas velocity of Langeled Gas Pipeline. Because the SIC measurement provides accurate depth sizing of SIC while simultaneously assisting MFL in defect identification and depth sizing through its high lateral resolution of defect surface measurement, the combination of MFL and SIC is a very effective and reliable tool for monitoring corrosion growth rates. In general, the tool's combination of these two complementary corrosion measurement methods ensures high sensitivity for early feature detection, a high POI for internal/non-internal feature discrimination, accurate sizing capabilities, and notably high effectiveness in the inspection of heavy wall pipelines. Overcoming all the exceptional difficulties posed by Langeled pipeline with ease, the inspection tool successfully completed an ILI run on an unprecedented scale in August 2009.

## REFERENCES


1. Gjertveit, E., Holme R., Bruschi, R., and Zenobi, D. (2007) The Langeled project, Proceedings of the 17th International Offshore and Polar Engineering Conference (ISOPE), July 2007, Lisbon, PT.
2. Beuker, T., Brockhaus, S., Ahlbrink, R., and McGee, M. (2009) Addressing challenging environments—advanced in-line inspection solutions for gas pipelines; Proceedings of the 24th World Gas Conference, October 2009, Argentina.
3. Ahlbrink, R. and Beuker, T. (2009) Tool monitors shallow internal corrosion in pipes. *International Oil and Gas Engineer*, 47–48. Available at http://content.yudu.com/A1bo6g/IntOilGasAugust09/resources/47.htm.
4. Huyse, L., Van Roodselaar, A., Onderdonk, J., Wimolsukpirakul, B., Baker, J., Beuker, T., Palmer, J., and Jemari, N.A. (2010) Improvements in the accurate estimation of top of the line internal corrosion on subsea pipelines on the basis of in-line inspection data. Proceedings of the 8th International Pipeline Conference, September 27–October 1, 2010, Calgary, Alberta, Canada.

# 52

# DEEPWATER, HIGH-PRESSURE AND MULTIDIAMETER PIPELINES—A CHALLENGING IN-LINE INSPECTION PROJECT

HUBERT LINDNER
*ROSEN Technology & Research Center, Lingen, Germany*

## 52.1 INTRODUCTION

In-line corrosion inspections with magnetic flux leakage (MFL) or ultrasonic technology (UT) have become standard in pipeline integrity assessment worldwide. Lately, ultra-high resolution geometry inspection based on eddy current (EC) methods has been established to assess more accurately the internal geometry of pipelines.

Approximately 60% of the world's gas, oil and product pipelines can be inspected with off-the-shelf inspection tools. In the past, the remaining 40% of pipelines have often been classified as "unpiggable". A large proportion of these unpiggable pipelines are offshore, multi-diameter lines [1], with low flow conditions and often with very challenging OD ratios [2].

Today, it is possible to develop individualized inspection solutions for multi-diameter pipeline systems. The development of such solutions can already be incorporated into the FEED process for these offshore structures [3]. On the other hand, a great number of pipelines were constructed and laid in times when in-line inspection (ILI) was not available or not a requirement. Occasionally, the passage of these offshore pipelines is restricted.

This chapter presents the development of a series of 14 in./18 in. ILI tools and their successful application survey in a 101.8 mi long, high-pressure, heavy-wall offshore pipeline. In a team effort between the pipeline operator ("operator") and the inspection company (ROSEN) a solution was developed for a challenging pipeline, which would have been considered "unpiggable" in the recent past.

## 52.2 PROJECT REQUIREMENTS

The project began with a request of the operator to inspect a particular 14 in./18 in. offshore pipeline to evaluate the integrity of the pipeline via an internal inspection.

In addition to its multi-diameter design, the pipeline posed several further challenges. The essential properties are as follows:

- Length: 101.8 mi
- 14 in.: 7.4 mi (wall thickness: 0.81–0.87 in.)
- 18 in.: 94.4 mi (wall thickness: 0.87–1.26 in.)
- Bends 14 in.: 5D
- Bends 18 in.: 5D
- Maximum water depth: 6233 ft
- Maximum pressure: 4200 psi
- Known minimum ID: 11.81 in.
- Internally coated
- Medium: gas

*Oil and Gas Pipelines: Integrity and Safety Handbook,* First Edition. Edited by R. Winston Revie.

Furthermore, the pipeline has several subsea appurtenances (notably check valves, connectors, ball valves, tees, and reducers) and a jumper of the subsea connection segment and an adjacent wye-piece. The transition from 14 to 18 in. was directly incorporated in one leg of the wye-piece. This meant that a cleaning or inspection tool had to expand the driving unit and pass the cavity of the wye-piece while the rear parts were still in the 14-in. pipeline. Finally, the tools have to pass and monitor the potentially damaged sections of the 18-in. line.

## 52.3 PROJECT PLANNING

The operator's goal during project planning was to maintain its customers' product flow while ensuring the integrity and safety of its pipeline operations. With those aims in mind, the project team sought to inspect the pipeline during regular operation (no water filling or bidirectional inspection from the receiver side) through a program of geometry and MFL. Since the consequences of a possible failure of the pigging operation could have been enormous and because of the particular challenges of the pipeline and operation, an extensive test program was incorporated.

The project plan included the fields described in the following paragraphs.

### 52.3.1 Tool Development

For this project three multi-diameter tools (gauging, geometry, MFL) had to be developed, designed, and manufactured.

### 52.3.2 Testing

An extensive test program was developed that included pull and pump tests of components and parts in the existing test facilities. This included pull tests through the existing 14 in., 16 in. and 18 in. test lines as well as pump tests in the 16 in. single size pump test loop simulating heavy external pipe damage.

For a full size test of the tools, the project included a test loop containing all crucial elements and features of the original pipeline (Figure 52.1) whereby all simulations of installations would be delivered by the operator and the pipes, bends and additional elements would by supplied by the inspection vendor. The test loop was assembled and operated on the area of the Research and Technology Center (RTRC) in Lingen, Germany [4].

### 52.3.3 Simulation of Pipeline Flow Conditions

As the run conditions are very important for a reliable inspection run, the operator provided data to simulate actual

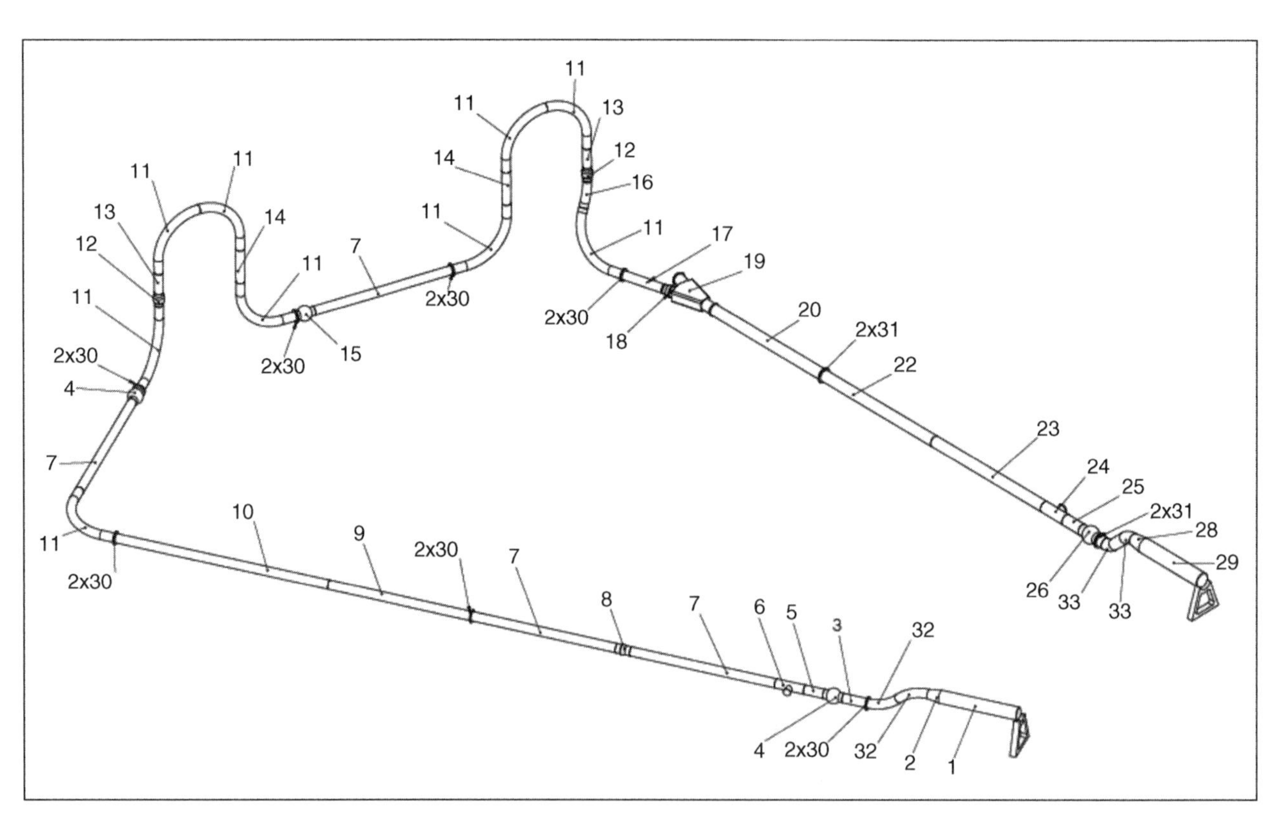


FIGURE 52.1 Final drawing of the test loop.

pipeline flow calculations. The combination of the planned flow conditions together with the effect of the pipeline topography and temperature turned out to be a critical issue with remarkable impact on the project schedule.

### 52.3.4 Preparation of On-Site Activities

During the preparation period the aspects of on-site activities were also widely considered. Based on the launcher and receiver documentation as well as on-site visit information, the launching (vertical) and receiving procedures were defined. The necessary equipment (transport frame, launching tube) was then designed and manufactured. All these procedures, parts, and actions were finalized in an on-site tool handling and launching/receiving procedure document.

Further, the project addressed a contingency plan for the unwanted case that a tool exceeded the calculated traveling time or became stationary.

### 52.3.5 Preparation of Data Evaluation

The fact that the investigation's goal was to discover the unknown condition of the pipeline required a coordinated and quick evaluation of the inspection data as well as other possible indications (gauge plate, tool condition), before progression to the next phase. Therefore, the on-site personnel were chosen to be experienced in both tool behavior and data evaluation. For a detailed analysis of the inspection data, remote capacity at the Research and Technology Center (RTRC) in Lingen, Germany was scheduled.

### 52.3.6 Project Communication

For the entire duration of the project a policy of very close communication between all parties involved was realized. This included several meetings and occasionally weekly telephone conferences. This policy allowed close cooperation and well-coordinated procedures throughout.

## 52.4 TOOL DESIGN

The main challenges for the tool designs was the wide multi-diameter working range from about 11.81–16.33 in., the high pressure (4200 psi) and the wye-piece passage. The direct consequences were small tool bodies and long sealing lengths. Since it was known that the pipeline was internally coated and that other cleaning tools had already passed the pipeline without heavy wear, it was decided to use conservative design for the gauging tool. However, it became clear after the first pump test in the 14 in./18 in. test loop that the setup had to be adjusted. The tool consisted of two bodies connected by a flexible joint. A four guiding petal disc design

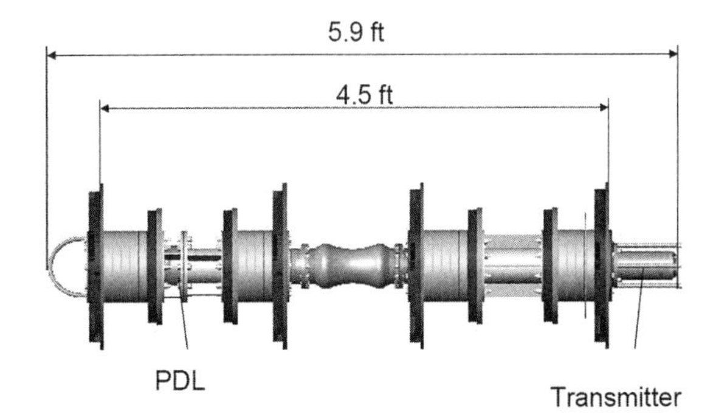


**FIGURE 52.2** 14 in./18 in. gauging tool.

was chosen (Figure 52.2). The tool was also equipped with a Pipeline Data Logger (PDL) and a transmitter.

Figure 52.2 shows the sealing length to accommodate the wye-piece passage (55.11 in.), which had to be the same for the intelligent tools. For those latter tools (essentially for the MFL tool) however, the pull unit had to provide much more stability and driving force than the petal setup could provide. Therefore, a particular segmented multi-diameter driving unit had to be developed. Although it was designed for the MFL unit, it could also be used for the geometry tool (Figure 52.3). The staggered sensor arrays of this tool provided full circumferential coverage in pipelines of both sizes (i.e., 14 in./18 in.).

In the case of the MFL tool, the driving unit has to pull two heavy-wall (up to 1.26 in. wall thickness), multi-diameter MFL units and a trailer for batteries and electronics (Figure 52.4). All components of these three tools were designed to withstand a pressure of 4350 psi.

## 52.5 TESTING

This phase was divided into three parts:

a. Pressure test of standard components and vessels (4350 psi);
b. Tool component tests (pulling force, flip over test);
c. Test for the complete tools.

Out of these, the tests for the complete tools were by far the most laborious. All tools were pulled through the 14, 16, and 18 in. pull test lines. During these runs, the intelligent tools already collected information and the pulling forces were recorded.

The most important tests were, of course, the pump tests in the special 14 in./18 in. test loop (Figure 52.5). During these runs both tool and pipeline flow and pressure data were measured to evaluate tool performance and optimize the behavior.

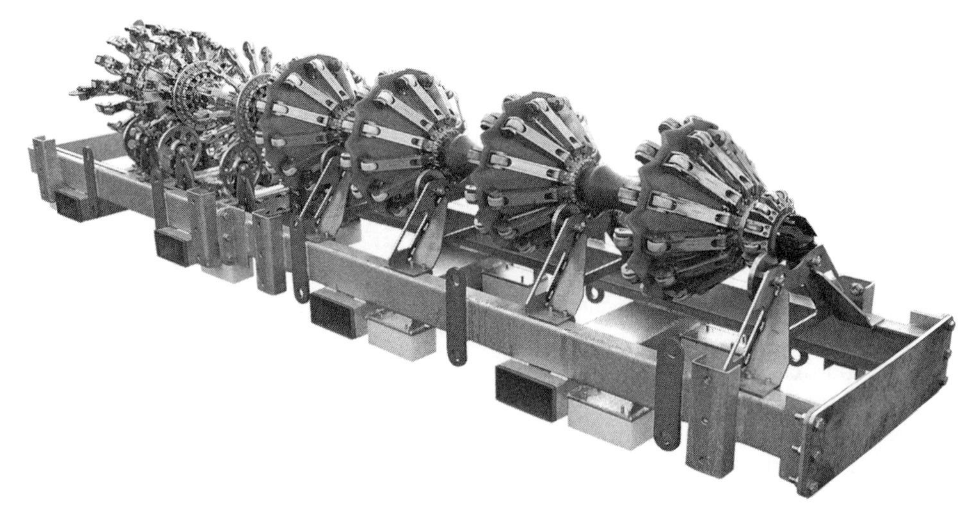

**FIGURE 52.3** The 14 in./18 in. high-resolution geometry tool.

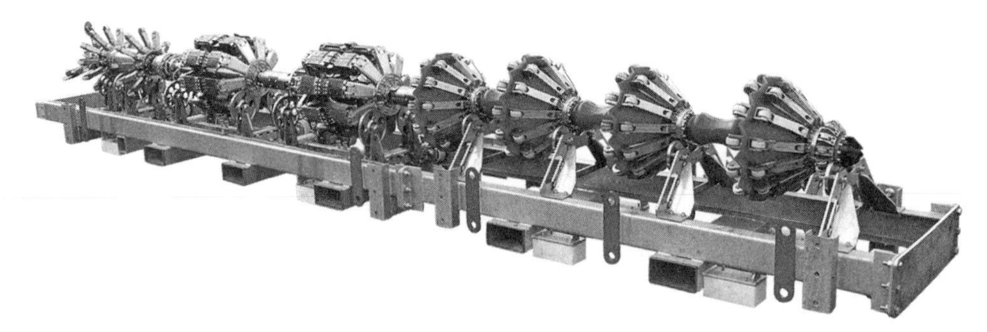

**FIGURE 52.4** The 14 in./18 in. MFL tool.

**FIGURE 52.5** The 14 in./18 in. test loop.

The test loop was put into operation at the end of October 2007. In the time until the tools were delivered in February 2008, the following runs were conducted:

| | |
|---|---|
| Gauging tool: | 4 runs (one change in discs setup) |
| Geometry tool: | 6 runs (one change to the design of the driving unit due to damage) |
| MFL tool: | 12 runs (one change to the design of the driving unit for enhanced function) |

Differential pressure and flow measurements were taken. The flow speed was the only operational parameter that could be varied during the tests. Due to the fact that the real pipeline is operated with gas and the pump test loop with water, some runs were conducted at a very low flow speed (0.82 ft/s) to approximate as closely as possible the effect of bypass in a gas line.

During the test phase all tools passed the entire flow loop at all times. Occasionally high differential pressure values were observed and analyzed and important findings were gained. These were for example an improved performance of the disc setup of the cleaning pig (CLP), and modifications of the driving unit.

Figure 52.6 illustrates the differential pressure (dp) along the high resolution geometry tool as measured by an on-board PDL during a pump test. The dp is about 29 psi in the 14 in. and 15 psi in the 18 in. section. The bottom diagram shows the inclination of the tool during the run. The inclination is a good indicator for the tool position resulting from the jumper as shown very clearly in Figure 52.6. The swan necks of launcher and receiver too are plainly visible. The pressure diagram also shows a distinct peak of nearly 145 psi.

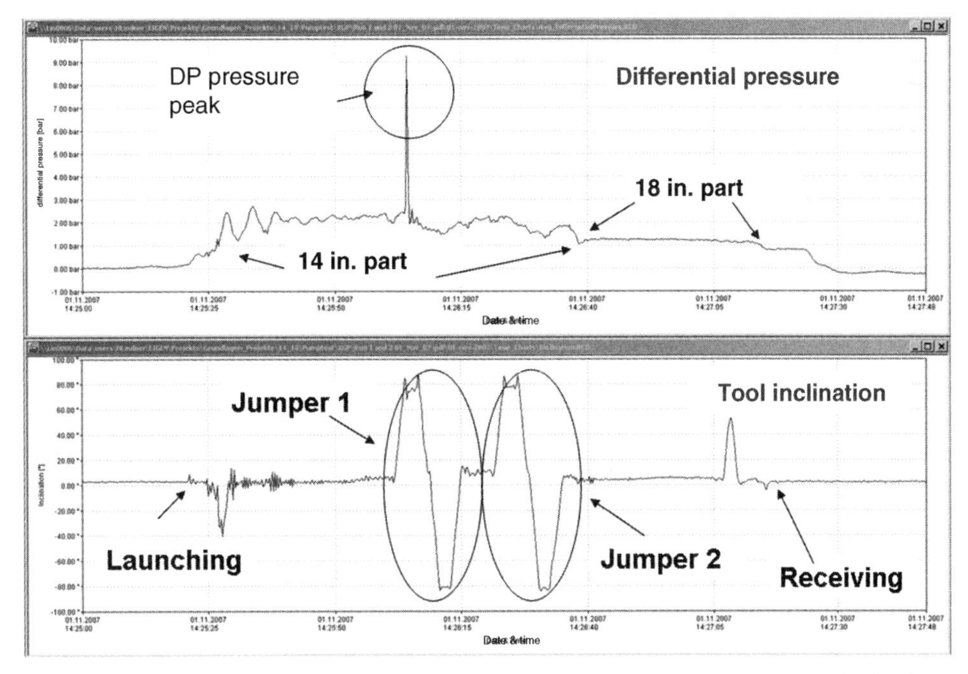


**FIGURE 52.6** Results of pump test with high-resolution geometry tool (a: dp, b: tool inclination).

This peak reflects damage to the pull unit, including missing wheel support. In combination with the synchronized display of the inclination it was possible to identify the reason for the peak and corrective design actions could be made. The subsequent tests did not show any similar effect.

Some of the measured pipeline installations are represented in Figure 52.7. The raw data furnished by the caliper arms clearly indicate the valves, connectors, tees, and of course the wye-piece.

PDL measurements of a MFL test run can be seen in Figure 52.8. Again the lower diagram shows the tool inclination in support of the information provided by the upper graph. The MFL tool needs a higher dp than expected: it is between 55 psi in an 18 in. and 80 psi in a 14 in. straight pipeline segment. For the passage of the jumper bend combination, a slightly higher differential pressure is required (up to 109 psi). The passage of the wye-piece was performed without stop and shows only little effect on the measurement.

Finally, a full-scale replica of the vertical launcher was built and the launching procedure performed as documented.

After all these tests, calculations and planning, the tools and procedures were ready for the job and mobilized. Meanwhile the operational data (1300 psi, 2,295,453 $sft^3/h$) provided good conditions for the run. As the actual pipeline was internally coated and had certain content of fluids, a lower dp was expected compared to the pump test in the non-coated test loop.

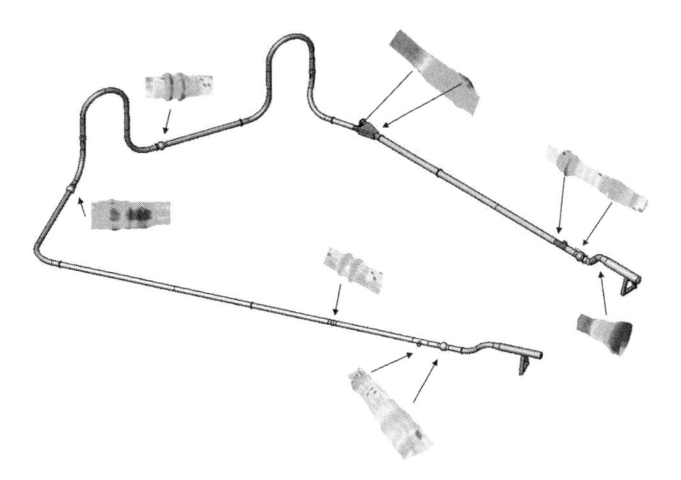

**FIGURE 52.7** Data of selected features collected by high-resolution geometry tool (raw data, not to be scaled).

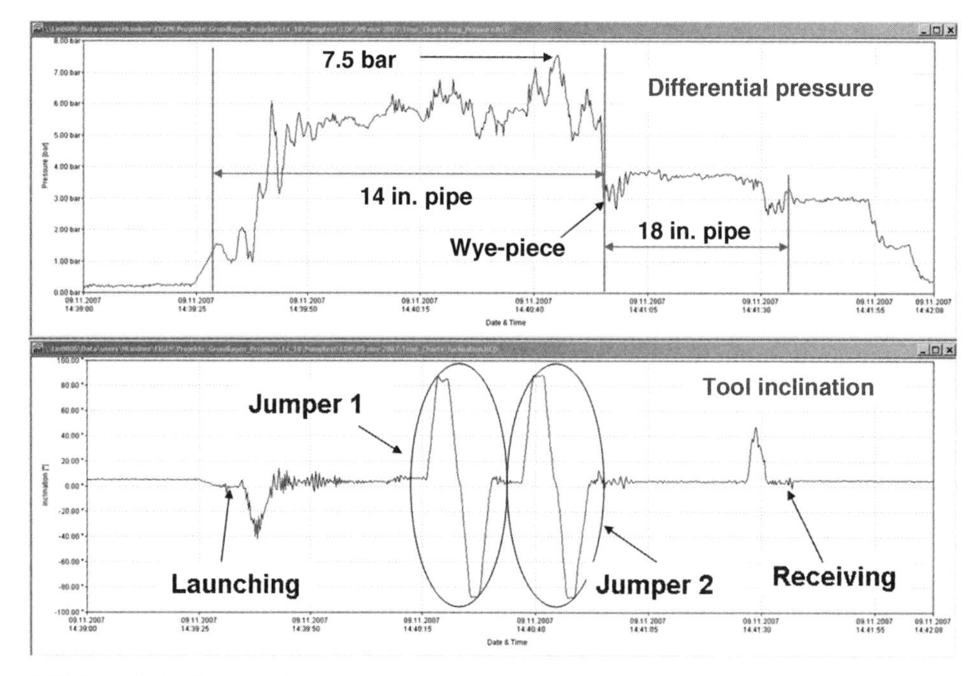


**FIGURE 52.8** Results of the pump test with the MFL tool (a: differential pressure, b: tool inclination).

## 52.6 GAUGING AND INSPECTION RUNS

For the preparation of the actual pigging project, a lot of supporting actions had to be prepared and conducted. For example, a special vessel with a remotely operated vehicle (ROV) was hired for subsea operations, subsea markers were placed and the required equipment was supplied to the platform.

Actual pigging operations started at the end of June 2008 with the gauging tool. This initial run was successful: neither the tool nor the gauge plate showed damage, and wear on the poly urethane was low. Therefore, the decision was made to run the geometry tool.

The geometry tool (Figure 52.9: left) was launched without any problem and arrived within the calculated time (Figure 52.10). The data collected during the geometry run were complete and of very high quality. The same evening initial data analysis was performed by the on-site team and data were transferred to the office-based data evaluation department. One extreme measurement was observed and isolated for detailed evaluation. Because of the known pipeline layout, it was clear that the signal was recorded in the check valve between the jumpers. The evaluation of the data revealed the internal geometry of the check valve (Figure 52.11). This information was carefully compared to the tool design and decided that it is just as expected and therefore no critical installation, though the MFL tool could be launched.

**FIGURE 52.9** Vertical launching of the geometry and MFL tool.

Because of the length of the launcher barrel, the two segments of the driving unit had to be pushed into the reduction for a certain distance. Additionally, a launching tube was used to keep the magnet units away from the wall to reduce the required pushing forces. As this was also tested on a dummy launcher with original measures, the launching procedure worked perfectly. Again, the tool was received within the allocated time and with complete data of high quality.

**FIGURE 52.10** The tool in the receiver.

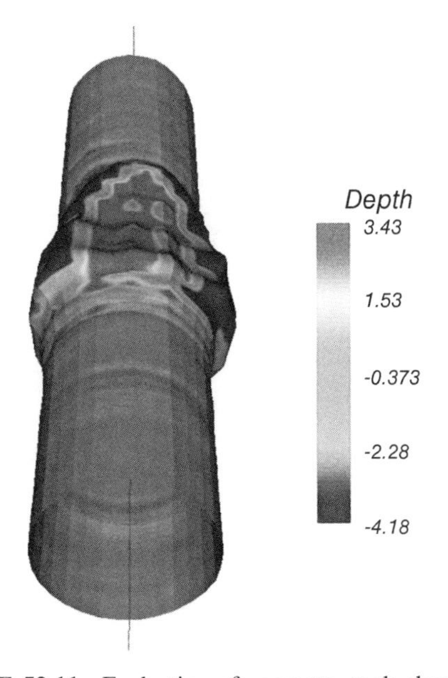


**FIGURE 52.11** Evaluation of geometry tool: check valve.

## 52.7 SUMMARY

The operator desired to perform an ILI to gather data to determine pipeline integrity. Because of the particular properties of the pipeline (high pressure, multi-diameter 14 in./18 in., high wall thicknesses and subsea installations) and the high risk of the deepwater operation, a unique development project was necessary. Therefore, two ILI tools (geometry and MFL) were developed and produced. In close cooperation with the operator, a large test program was performed at the test facility in Lingen ending with full-scale tests in a 430 ft test loop, including all crucial installation simulations.

During the test phase some deciding improvements of the design were identified and made due to pump test results. All tests, results, and changes were communicated among the project team, leading to a demonstrated solution and confidence for the inspection project.

Parallel to the tool development, manufacturing and testing, operational conditions and the on-site procedures were evaluated. Furthermore, a contingency plan was compiled regarding the definition of possible incidences and the following actions.

Due to the detailed and comprehensive preparation of the inspection and the professional and uncomplicated cooperation of the project teams all tool runs (gauging, geometry, and MFL) were launched and received without incident. The tools arrived in time with complete data of high quality.

## REFERENCES

1. Cordell, J. and Vingerhagen, A. (1998) Design and development of multidiameter pigs. Proceedings of the Pipeline Pigging Conference, 1998, Houston, TX.
2. Vingerhagen, A., Kershaw, C., and O'Donoghue, A. (1999) Pigging the Åsgard transport 42″ × 28″ pipeline—breaking new ground. Proceedings of the Pipeline Pigging Conference, 1999, Stavanger, Norway.
3. Beuker, T. and Lindner, H. (2007) Classifying pipeline complexity. World Pipelines, March 2007.
4. Lindner, H., Beuker, T., and Diekamp, M. (2011) In-line inspection of multi-diameter pipelines: standardized development and testing for a highly efficient tool fleet. Proceedings of the 6th Pipeline Technology Conference, 2011, Hannover, Germany.

# GLOSSARY

## PART 1: ABBREVIATIONS

| | |
|---|---|
| 3D | three-dimensional |
| 3RB | three-roll bending |
| AC | alternating current |
| ACFM | alternating current field measurement |
| ACPA | American Concrete Pipe Association |
| ACR | atomic concentration ratio |
| ACVG | alternating current–voltage gradient |
| AGA | American Gas Association |
| AIM | asset integrity management |
| AISI | American Iron and Steel Institute |
| ALARP | as low as reasonably practicable |
| Anon | anonymous |
| APB | acid-producing bacteria |
| API | American Petroleum Institute |
| ASME | American Society of Mechanical Engineers |
| ASNT | American Society for Nondestructive Testing |
| ASTM | American Society for Testing and Materials |
| ASWP | American Spiral Weld Pipe |
| AUT | automated ultrasonic testing |
| AWS | American Welding Society |
| AYS | actual yield strength |
| B | boron |
| BOL | bottom of line |
| BOPD | barrels of oil per day (aka bpd) |
| BRD | base resource document |
| BS | British Standard |
| BS&W | bottom solids & water (effectively electrolyte content) |
| BSEE | Bureau of Safety and Environmental Enforcement |
| BSI | British Standards Institute |
| BTCM | Battelle two-curve model (or method), a procedure developed by Maxey and coworkers at the Battelle Memorial Institute to determine critical conditions for rapid axial crack propagation in a pipe |
| C | carbon |
| CA | corrosion allowance |
| CANMET | Canada Centre for Mineral and Energy Technology |
| CAPEX | capital expenditures |
| CAPP | Canadian Association of Petroleum Producers |
| CAR | crack area ratio |
| CBA | cost benefit analysis |
| CCT | continuous cooling transformation |
| CCT | critical crevice temperature |
| CD | crack detection |
| CDF | cumulative distribution function |
| CDU | crude units |
| CEPA | Canadian Energy Pipeline Association |
| CFD | computational fluid dynamics (erosion/wetting context) |
| CFR | Code of Federal Regulations |
| CI | corrosion inhibitor |
| CIPS | close interval potential survey |
| CIS | close interval survey |

*Oil and Gas Pipelines: Integrity and Safety Handbook,* First Edition. Edited by R. Winston Revie.

| | |
|---|---|
| CLR | crack length ratio |
| CME | coronal mass ejection |
| CMFL | circumferential MFL |
| CML | corrosion monitoring location |
| CMOD | crack-mouth opening displacement |
| CMS | corrosion management strategy |
| CMT™ | cold metal transfer |
| COF | consequence of failure |
| CP | cathodic protection |
| CPQRA | chemical process quantitative risk analysis |
| CPT | critical pitting temperature |
| Cr | chromium |
| CRA | corrosion-resistant alloy |
| CRM | corrosion risk management |
| CSA | Canadian Standards Association |
| CSE | copper sulfate electrode |
| CSM | chopped strand mat |
| CSR | crack sensitivity ratio |
| CTA | central tool adjustment |
| CTOA | crack-tip-opening angle, the angle subtended by the crack faces near the tip of a propagating crack |
| CTOD | crack-tip opening displacement |
| CTR | crack thickness ratio |
| Cu | copper |
| CUD | corrosion under deposits (also under-deposit corrosion (UDC)) |
| CUI | corrosion under insulation |
| CUS | corrosion under supports |
| CWP | curved wide plate |
| *D/t* | diameter to wall thickness ratio |
| DA | direct assessment |
| DC | direct current |
| DCB | double-cantilever beam |
| DCVG | direct current–voltage gradient |
| DEG | diethylene glycol |
| DF | damage factor |
| DFDI | ductile failure damage indicator |
| DFMT | dry fluorescent magnetic testing |
| DGGE | denaturing gradient gel electrophoresis |
| DG-ICDA | dry gas ICDA |
| DMAC | Direct Arm Measurement Calipers |
| DML | dents with metal loss |
| DNV | Det Norske Veritas |
| DOE | U.S. Department of Energy |
| DOT | U.S. Department of Transportation |
| DP | design pressure |
| DSAW | double-submerged arc welding |
| dsr | dissimilatory sulfite reductase |
| DSTL | distributed source transmission line |
| DWT | drop weight tear |
| DWTT | drop weight tear test, standardized in ASTM E436 |
| E&P | exploration and production |
| EAC | environmentally assisted cracking |
| EC | external corrosion |
| EC | eddy current |
| ECA | eddy current array |
| ECA | Engineering Critical Assessment, "an analytical procedure, based upon fracture mechanics principles, that allows determination of the maximum tolerable sizes for imperfections in fusion welds" (CSA Z662) |
| ECDA | external corrosion direct assessment |
| ECN/EN | electrochemical noise |
| ECT | eddy current testing |
| EDS | energy dispersive spectroscopy |
| EDX | energy dispersive X-ray |
| EIS | electrochemical impedance spectroscopy |
| EM | electromagnetic |
| EMAT | electromagnetic acoustic transducer |
| EMF | electromotive force |
| EPRG | European Pipeline Research Group |
| ER | electrical resistance |
| ERW | electric resistance welding |
| eSCC | ethanol stress corrosion cracking |
| E-t | potential–time probes (may be derived from LPR probes) |
| EV | extreme value |
| FAC | flow-assisted corrosion |
| FAC | failure assessment curve, a curve in an FAD showing critical conditions of applied load and crack driving force for crack growth |
| FAD | failure assessment diagram, a graph of crack driving force (normalized by toughness) versus applied load (normalized by collapse load) |
| FBE | fusion-bonded epoxy |
| FCAW-G | gas-shielded flux-cored arc welding |
| FCAW-S | self-shielded flux-cored arc welding |
| FE | finite element |
| FEA | finite element analysis |
| FEED | Front End Engineering and Design |
| FEM | finite element method |
| FERC | Federal Energy Regulatory Commission |
| FFS | fitness for service |
| FFT | fast Fourier transform |
| FGE | fuel grade ethanol |
| FISH | fluorescence *in situ* hybridization |
| FMEA | failure mode and effects analysis |
| FMECA | failure mode, effects, and criticality analysis |
| FoS | factor of safety |
| FRP | fiber-reinforced plastic |

| | |
|---|---|
| FSA | formal safety assessment |
| FSM | Field Signature Method (corrosion area mapping spool) |
| FTA/ETA | fault/event tree analysis |
| FTIR | Fourier transform infrared |
| GASDECOM | a computer code for the calculation of gas decompression speed |
| GC/MS | gas chromatography/mass spectroscopy |
| GIC | geomagnetically induced current |
| GIS | geographic information system |
| GMAW-P | pulsed gas metal arc welding |
| GMAW-S | short circuit gas metal arc welding |
| GOM | Gulf of Mexico |
| GRI | Gas Research Institute |
| HAZ | heat-affected zone of a weld, in which welding heat has altered the microstructure |
| HAZID | hazard identification |
| HAZOP | hazard and operability |
| HCA | high consequence area |
| HDD | horizontal directional drilling |
| HDPE | high-density polyethylene |
| HEC | hydrogen embrittlement cracking |
| HFE | human factors and environment |
| HF-ERW | high-frequency ERW |
| HFI | high-frequency induction |
| HFW | high-frequency electric resistance welding |
| HIC | hydrogen-induced cracking |
| HISC | hydrogen-induced stress cracking |
| HML | high medium low (risk notation); also VH, VL per very high/low |
| HP/HT/HV | high pressure, high temperature, high velocity |
| HSC | hydrogen stress cracking |
| HSE | health, safety and environment |
| HT | high temperature |
| HTS | helical two-step |
| HTSP | highly tensioned suspended pipeline |
| IC | internal corrosion |
| ICDA | internal corrosion direct assessment |
| ID | internal diameter |
| IEMC | indirect electromagnetic calipers |
| IGE | Institute of Gas Engineers |
| ILI | in-line inspection |
| IM | integrity management |
| INCOTEST™ | insulated component testing |
| INGAA | Interstate Natural Gas Association of America |
| IOB | iron-oxidizing bacteria |
| IPA | isopropyl alcohol |
| I–P–F | initiation propagation failure (diagram) |
| IRO | inherently reliable operations |
| ISD | inherently safe design |
| ISO | International Organization for Standardization |
| ITP | independent third-party |
| IVP | Integrity Verification Process |
| JCO | J-forming–C-forming–O-forming |
| KP | key point (specific areas designated for inspection) |
| KPI | key performance indicator |
| LF | low frequency |
| LFET | low-frequency eddy current testing |
| LHVD | low hydrogen vertical down |
| LOPA | layers of protection analysis |
| LPI | liquid penetrant inspection |
| LP-ICDA | liquid petroleum ICDA |
| LPR | linear polarization resistance |
| LSAW | longitudinal submerged arc welded |
| MAE | major accident event |
| MAOP | maximum allowable operating pressure |
| MAWT | minimum allowable wall thickness |
| MDE | maximum distortion energy |
| MDI | methylene diphenyl diisocyanate |
| MDT | mean downtime |
| MEC™ | Magnetic Eddy Current |
| MEK | methyl ethyl ketone |
| MFL | magnetic flux leakage |
| MIC | microbiologically influenced corrosion; also called microbially induced corrosion |
| MLV | mainline valves |
| MMM | molecular microbiological method |
| MMscmd | million metric standard cubic meters per day |
| Mn | manganese |
| Mo | molybdenum |
| MoC | management of change |
| MOP | maximum operating pressure |
| MPI | magnetic particle inspection |
| MP-ICDA | multiphase flow ICDA |
| mpy | mils per year (thousandths of an inch per year) (corrosion rate); 40 mpy = 1 mm/year |
| MSC | metric standard condition |
| MSS | maximum shear stress |
| MT | magnetotellurics |
| MTTR | mean time to repair |
| MWM™ | meandering winding magnetometer |
| NACE | National Association of Corrosion Engineers |
| NAEC | narrow axial external corrosion |
| NDE | nondestructive evaluation |
| NDI | nondestructive inspection |
| NDT | nondestructive testing |
| NEB | National Energy Board of Canada |
| NGL | natural gas liquid |

| | |
|---|---|
| Ni | nickel |
| NOM | naturally occurring mercury |
| Non-PWHT | non-postweld heat treated |
| NOPR | notice of proposed rule makings |
| NORM | naturally occurring radioactive materials |
| NORSOK | Norsk Sokkels Konkuranseposisjon |
| NPS | nominal pipe size |
| NRB | nitrate-reducing bacteria |
| NTSB | National Transportation Safety Board |
| O | oxygen |
| O&M | operations and maintenance |
| OCP | open-circuit potential |
| OCTG | oil country tubular goods |
| OD | outside diameter |
| ODF | orbital die forming |
| OPEX | operational expenditure |
| OPS | Office of Pipeline Safety |
| ORP | oxidation–reduction potential |
| P&C | probes and coupons |
| P&ID | piping and instrumentation diagram |
| PA11 | polyamide 11 |
| PAH | polyaromatic hydrocarbons |
| PAUT | phased array ultrasonic testing |
| PD | published document |
| PDAM | pipeline defect assessment manual |
| PDF | probability density function |
| PDO | Petroleum Development Oman |
| PEC | pulsed eddy current |
| PEEK | polyetherether ketone |
| PFD | process flow diagram |
| PFP© | premium forged pipes |
| PHMSA | U.S. Pipeline and Hazardous Materials Safety Administration |
| PIR | potential impact radius |
| PLEM | pipe line end manifold |
| PLET | pipe line end termination |
| PM | parent metal |
| POD | probability of detection |
| POF | probability of failure |
| POF | Pipeline Operators Forum |
| POFC | probability of false call |
| POI | probability of identification |
| PPE | personal protective equipment |
| ppm | parts per million |
| ppm/ppb | parts per million/parts per billion |
| ppmv | parts per million by volume |
| PQR | procedure qualification record |
| PRCI | Pipeline Research Council International |
| PREN | pitting resistance equivalent number |
| PSP | pipe-to-soil potential |
| PTVσ | pressure, temperature, velocity, stress |
| PVDF | polyvinylidene fluoride |
| PWHT | postweld heat treatment |
| QA | quality assurance |
| QC | quality control |
| qPCR | quantitative polymeric chain reaction |
| QRA | quantitative risk assessment |
| r.d.s. | rate-determining step |
| r.i.v | rate-influencing variable (typically dominant variable under multiple variables) |
| R6 | "Assessment of the integrity of structures containing defects," an ECA code developed in the United Kingdom |
| RAO | response amplitude operator |
| RBD | reliability (risk)-based design |
| RBI | risk-based inspection |
| RCM | reliability (risk) centered maintenance |
| RFEC | remote field eddy current |
| RL | remaining life |
| RMD™ | regulated metal deposition |
| ROV | remotely operated vehicle |
| RoW | right-of-way |
| RP | recommended practice |
| RPCM | ring pair corrosion monitor (spool-based on ER principle) |
| RSF | remaining strength factor |
| RSTRENG | method for evaluating the Remaining STRENGth of corroded pipe |
| RT | radiographic testing |
| RTR | real-time radiography |
| SAW | submerged arc welding |
| SAWH | helically submerged arc welded |
| SAWL | longitudinally submerged arc welded |
| SBD | strain-based design, in which properties are specified in terms of applied strain (as opposed to stress-based design, in which applied stress is the design variable) |
| SCADA | supervisory control and data acquisition |
| SCC | stress corrosion cracking |
| SCCDA | stress corrosion cracking direct assessment |
| SCE | safety critical element |
| SCF | stress concentration factor |
| SCFD | standard cubic feet per day |
| SCR | steel catenary riser |
| SCU | speed control unit |
| SDR | standard dimension ratio |
| SE(T) (or SENT) | single-edge (notched) tension test, a fracture test to measure toughness for crack growth under low constraint (tension) |
| SEC | shallow external corrosion |
| SEM | scanning electron microscopy |
| SEMP | safety and environmental management plan |

SEMS safety and environmental management system
SFGE synthetic FGE
Si silicon
SIC shallow internal corrosion
SIF stress intensification factor
SIL safety integrity level
SLD strain limit damage
SLOFEC™ saturation low frequency eddy current
SMAW shielded metal arc welding
SME subject matter expert
SMLS seamless
SMTS specified minimum tensile strength
SMYS specified minimum yield strength
SOHIC stress-oriented hydrogen-induced cracking
SOL side of line (waterline effect, for example, inside vessels/tanks/pipes)
SOP standard operating procedure
SOW scope of work
Spec specification
SPP sulfide-producing prokaryotes
SRB sulfate-reducing bacteria
SSC sulfide (sour) stress cracking
SSCC sulfide stress corrosion cracking
STT™ surface tension transfer
T/R transformer/rectifier
TAN total acid number
TCLP toxicity characteristic leaching procedure
TDI tolylene diisocyanate
TDZ thermal diffusion zinc
TDZ touchdown zone (usually for SCR)
TEG triethylene glycol
TFI transverse field inspection
TGA thermal gas analysis
THPS tetrakis(hydroxymethyl)phosphonium sulfate
TM thermal mechanical
TMCP thermomechanically controlled processing
TOFD time-of-flight diffraction
TOFT time-of-flight diffraction
TOL top of line
TPH total petroleum hydrocarbon
TSA thermal spray aluminum
TSB Transportation Safety Board of Canada
UDC underdeposit corrosion
UKWIR United Kingdom Water Industry Research
UOE "U"ing–"O"ing–expansion process
USD ultimate strength design
UT ultrasonic testing
UTS ultimate tensile strength
UV ultraviolet
V vanadium
VDU vacuum distillation unit
WFMT wet fluorescent magnetic testing
WG-ICDA wet gas ICDA
WI work instruction
WIC Welding Institute of Canada
WM weld metal
WOF weather and outside force
WPS welding procedure specification
WSD working stress design
WSE written scheme of examination
WSIP wellhead shut-in pressure
WZ weld zone
XRD x-ray diffraction
XRF x-ray florescence
YS yield strength

# GLOSSARY

## PART 2: SELECTED TERMS

**Annulus:** The space between two cylinders, for example, the gap between the outer wall of a pipeline and the inner wall of a repair sleeve.

**Black powder:** In the gas industry, the term "black powder" is a color-descriptive term loosely used to describe a blackish material that collects in gas transmission pipelines. It consists of various forms of iron sulfide, iron oxide, and iron carbonate, mechanically mixed or chemically combined with any number of contaminants such as salts, sand, liquid hydrocarbons, and metal debris.

**Buckling:** A failure mode of a pipeline when loaded in compression. When the buckling load is reached, the initial shape is no longer stable, and a bent configuration develops.

**Charpy test:** Standardized test (ASTM E23) to measure the energy required to fracture a notched specimen in impact.

**Chemical cleaning:** Cleaning the pipeline internally through the use of chemical solutions to remove black powder and other debris.

**Code:** A set of provisions that control the design and construction in a given jurisdiction, establishing minimum structural, mechanical, and other design requirements for a pipeline and other structures.

**Consequence:** Damage of vulnerable elements subject to hazardous corrosion impact on public, work personnel, environment, property, revenue, or company image.

**Corrosion fatigue:** Cracking that results from the combined effects of corrosion and repeated or alternating stress (i.e., dynamic loading).

**Crack driving force:** Variable expressing effect of applied mechanical forces on crack propagation, for example, J-integral, K, CTOD, and CTOA.

**Crack-mouth opening displacement (CMOD):** Displacement measured at the mouth of a crack being opened in response to applied forces.

**Crack-tip opening displacement (CTOD):** Separation of crack faces near the tip of a crack being opened in response to applied forces.

**$C_V$:** Absorbed energy measured in a Charpy test.

**Design pressure:** The pressure that is used in all equations and stress calculations and represents the maximum internal pressure that the pipeline should be capable of receiving during its design life at the lowest part of the pipeline system.

**Ductility:** The ability of materials or structures to undergo large deformation without rupture. Ductile is the opposite of brittle.

**Excursion:** Deviation or alteration with respect to the design/operating envelope for specific parameters, which may lead to stimulated corrosion and/or a hazardous condition.

**Fatigue:** A form of failure that results from alternating or repeated stress.

**Hazard:** A physical or chemical condition (e.g., corrosion leak) that has the capability to damage individuals, the environment, or assets.

**Hydrogen permeation:** An expression for the entry and movement of hydrogen atoms in steel. The hydrogen atoms are created under certain conditions either by corrosion or by cathodic protection.

*Oil and Gas Pipelines: Integrity and Safety Handbook,* First Edition. Edited by R. Winston Revie.

**Inhibitor:** A chemical that reduces the rate of corrosion when added in small concentrations.

***J*-integral:** In fracture mechanics, an integral involving stresses and displacements around a crack tip, used to quantify the crack driving force.

***K*:** Stress intensity factor; in linear elastic fracture mechanics, a variable quantifying the crack driving force usually written as $K = Y\sigma\sqrt{\pi a}$, where $Y$ is a geometrical variable, $\sigma$ is the applied stress, and $a$ is the crack size (or one-half crack size for an embedded crack).

**$K_c$:** Fracture toughness, the critical value of $K$ at which crack propagation occurs.

**Live load:** Load that can be moved on or off a pipe structure, such as that caused by truck wheel loads, railway car, locomotive loads, and so on. Live loads occur during the pipeline service life due to installation, production, and maintenance operations.

**Load factor:** Part of a factor of safety applied to members that are sized in strength design where design is based on the failure strength of members.

**Material resistance:** Material parameter expressing magnitude of driving force required to propagate a crack.

**Mechanical cleaning:** Cleaning of the pipeline internally through the use of mechanical scrapers. This is normally conducted to clean the pipeline from black powder and other debris.

**Micromechanisms:** In fracture mechanics, the microscale processes by which crack propagation occurs, in pipe steels normally ductile (microvoid coalescence) or brittle (cleavage).

**Modulus of elasticity:** A measure of a material stiffness, that is, the ratio of stress divided by strain.

**Naturally occurring radioactive materials (NORM):** Naturally occurring radionuclides are abundant in the earth's crust and originate from the presence of uranium or thorium. As gas, oil, or water is produced from ground formations, some radionuclides such as radon in gas and radium in water can migrate to surface facilities and create NORM as scales or deposits.

**Performance indicators:** These are usually in the context of cracking, corrosion, or other degradation activities (KPIs: key performance indicators) or related safety, financial, or production measures. May be mechanical, metallurgical, electrochemical, or physical, singular or grouped.

**Pipeline:** A pipeline includes the first valve in and out of the launcher/receiver, the respective "kicker," bypass and pig trap control valves, line sectionalizing valves, the pipe, and generally all items of the line between the launcher and the receiver.

**Plastic collapse:** Condition for which unlimited deformation occurs without an increase in load.

**Pressure:** Force per unit area exerted by the medium in the pipeline.

**Process zone:** In fracture mechanics, the region ahead of a crack tip in which fracture micromechanisms are active.

**R curve:** In fracture mechanics, a curve showing the variation of material resistance ($J$ or CTOD) with crack growth.

**Relative roughness:** The relative roughness is a dimensionless parameter obtained by dividing the internal pipe roughness by the pipe inside diameter.

**Repair sleeve:** Repair sleeves are used to cover an area of pipeline that has suffered from either internal or external corrosion. Repair sleeves are prefabricated lengths of plate rolled into the shape of a pipe. These are sized to be slightly larger than the outside diameter of the line pipe and they are split horizontally. Repair sleeves are normally fabricated in 3 m lengths, but they can be cut to cover shorter lengths of corrosion or butt welded together to cover corrosion sites longer than 3 m. At the repair site, the two pieces are clamped over the exterior of the original pipeline at the location of the corrosion. The repair sleeve is welded back together and then carefully welded to the exterior of the original pipeline at each end of the sleeve, creating a short length of double-walled pipeline with a sealed annular space between them.

**Residual stress:** The stress that can exist in a component or structure in the absence of an applied stress.

**Sabkha:** Low-lying geographical areas of oversaturated high-salinity sand in Saudi Arabia.

**Safeguard/mitigation:** Action capable of interrupting the chain of events unleashed by damage or a corrosion-initiating event. May be prevention by inherent safety (CRA materials), physical or chemical intervention (keeping key parameters within the design PTV$\sigma$ envelopes), or via chemical inhibitors (controlling corrosion within limits), or mitigation (limiting consequences of a leak).

**Sales gas:** Natural gas that has been treated to remove excess moisture, down to a specified value, normally 7 lb/mmscf, and also hydrogen sulfide gas normally down to a few ppm.

**Strain:** The ratio of change in length divided by the original length.

**Stress:** The ratio of applied force divided by the cross-sectional area.

**Stress corrosion cracking (SCC):** Cracking that results from the combined influence of stress and corrosion.

**Torque:** The tendency of force to rotate an object about an axis or twist an object.

# INDEX




*Oil and Gas Pipelines: Integrity and Safety Handbook,* First Edition. Edited by R. Winston Revie.

## D



## E

**J**



**K**



**L**

**M**